HANDBOOK OF THE BIOLOGY OF AGING

NINTH EDITION

THE HANDBOOKS OF AGING

Consisting of Three Volumes

Critical comprehensive reviews of research knowledge, theories, concepts, and issues

Editors-in-Chief

Laura L. Carstensen and

Thomas A. Rando

Handbook of the Biology of Aging, 9th Edition
Edited by Nicolas Musi and Peter J. Hornsby

Handbook of the Psychology of Aging, 9th Edition
Edited by K. Warner Schaie and Sherry L. Willis

Handbook of Aging and the Social Sciences, 9th Edition
Edited by Kenneth F. Ferraro and Deborah Carr

HANDBOOK OF THE BIOLOGY OF AGING

NINTH EDITION

Edited by

NICOLAS MUSI

Barshop Institute for Longevity and Aging Studies, South Texas Veterans Health Care System, University of Texas Health Science Center, San Antonio, TX, United States

PETER J. HORNSBY

Barshop Institute for Longevity and Aging Studies and Department of Cellular and Integrative Physiology, University of Texas Health Science Center, San Antonio, TX, United States

Academic Press is an imprint of Elsevier
125 London Wall, London EC2Y 5AS, United Kingdom
525 B Street, Suite 1650, San Diego, CA 92101, United States
50 Hampshire Street, 5th Floor, Cambridge, MA 02139, United States
The Boulevard, Langford Lane, Kidlington, Oxford OX5 1GB, United Kingdom

British Library Cataloguing-in-Publication Data
A catalogue record for this book is available from the British Library

Library of Congress Cataloging-in-Publication Data
A catalog record for this book is available from the Library of Congress

ISBN: 978-0-12-815962-0

For Information on all Academic Press publications
visit our website at https://www.elsevier.com/books-and-journals

Publisher: Nikki Levy
Editorial Project Manager: Barbara Makinster
Production Project Manager: Swapna Srinivasan
Cover Designer: Matthew Limbert

Typeset by MPS Limited, Chennai, India

Dedication

This volume is dedicated to the memory of Dr. Edward Masoro, editor of the fifth, sixth, and seventh editions of this handbook; his name will always be associated with this important series of volumes. Dr. Masoro passed away on July 11, 2020, at the age of 95.

Dr. Masoro obtained his A.B. degree at the University of California at Berkeley in 1947 and was then awarded the PhD degree in physiology in 1950. Following early faculty positions, he was appointed as the Founding Chair of the Department of Physiology at the newly established medical school of the University of Texas system in San Antonio (see Chen, 2003, and Critser, 2010, for accounts of his early career). He persuaded other faculty members of the department to join him in a new line of research that probed the mechanisms of aging in rodents, a line of research that began to put San Antonio on the map as a major center for studies in the basic biology of aging. He established the Aging Research and Education Center of the University of Texas Health Science Center San Antonio to catalyze research on the biology of aging across basic and clinical science disciplines. The culmination of this process was the founding of the Barshop Institute for Longevity and Aging Studies, one of the first centers in the nation dedicated to basic research in the biology of aging.

Dr. Masoro's seminal work established the paradigm of extension of rodent lifespan by caloric restriction, which became a cornerstone of research into mechanisms of aging. His early research in the biology of aging focused on then-popular ideas of how calorie restriction in rodents extended lifespan. It was thought that calorie restriction affected the growth of young animals and that the smaller animals had an extended lifespan. However, he showed that calorie restriction worked in adult animals. His subsequent work tested many physiological hypotheses about the mechanisms of calorie restriction (Masoro, 2009). Investigations into this topic continue to this day, and work on the mechanisms of calorie restriction has stimulated new areas of research, such as pharmacological interventions that act as calorie restriction mimetics.

As members of the Barshop Institute, we will always be grateful to Dr. Masoro for his work on the basic research in the biology of aging at San Antonio, and we recognize that the current thriving enterprise in aging research here would never have been possible without his pioneering efforts.

References

Chen, I. (2003). Hungry for science. *Science of Aging Knowledge Environment*, *2003*(1), NF1, 8 January.

Critser, G. (2010). *Eternity soup: Inside the quest to end aging. New York:* Crown Publishing.

Masoro, E. J. (2009). Caloric restriction-induced life extension of rats and mice: A critique of proposed mechanisms. *Biochimica et Biophysica Acta*, *10*, 1040–1048.

Contents

II

Organ systems in humans and other animals, human health and longevity

List of contributors

Andrzej Bartke Southern Illinois University School of Medicine, Department of Internal Medicine, Springfield, IL, United States

Ying Ann Chiao Aging and Metabolism Research Program, Oklahoma Medical Research Foundation, Oklahoma City, OK, United States

Eileen M. Crimmins Davis School of Gerontology, University of Southern California, Los Angeles, CA, United States

Dao-Fu Dai Department of Pathology, Carver College of Medicine, University of Iowa, Iowa City, IA, United States

Justin Darcy Section on Integrative Physiology and Metabolism, Joslin Diabetes Center, Harvard Medical School, Boston, MA, United States

João Pedro de Magalhães Integrative Genomics of Ageing Group, Institute of Ageing and Chronic Disease, University of Liverpool, Liverpool, United Kingdom

Qunfeng Dong Department of Public Health Sciences, Loyola University Chicago, Chicago, IL, United States

Yimin Fang Southern Illinois University School of Medicine, Department of Neurology, Springfield, IL, United States

William Giblin Department of Pathology, University of Michigan, Ann Arbor, MI, United States

Xianlin Han Barshop Institute for Longevity and Aging Studies, University of Texas Health Science Center at San Antonio, San Antonio, Texas, United States; Division of Diabetes, Department of Medicine, University of Texas Health Science Center at San Antonio, San Antonio, Texas, United States

David E. Harrison The Jackson Laboratory, Bar Harbor, ME, United States

Erin Hascup Southern Illinois University School of Medicine, Department of Neurology, Springfield, IL, United States

Kevin Hascup Southern Illinois University School of Medicine, Department of Neurology, Springfield, IL, United States

Peter J. Hornsby University of Texas Health Science Center San Antonio, San Antonio, TX, United States

Akihiro Ikeda Department of Medical Genetics and McPherson Eye Research Institute, University of Wisconsin, Madison, WI, United States

Jamie N. Justice Sticht Center for Healthy Aging and Alzheimer's Prevention, Internal Medicine – Gerontology and Geriatric Medicine, Wake Forest School of Medicine (WFSM), Winston-Salem, NC, United States

Ajinkya S. Kawale Department of Biochemistry and Structural Biology, University of Texas Health Science Center at San Antonio, San Antonio, TX, United States

Brian K. Kennedy Department of Biochemistry and Physiology, Yong Loo School Lin School of Medicine, National University of Singapore, Singapore, Singapore; Centre for Healthy Longevity, National University Health System, Singapore, Singapore; Singapore Institute for Clinical Sciences, A*STAR, Singapore, Singapore

Jung Ki Kim Davis School of Gerontology, University of Southern California, Los Angeles, CA, United States

Ron Korstanje The Jackson Laboratory, Bar Harbor, ME, United States

Anita Krisko Department of Experimental Neurodegeneration, University Medical Center Goettingen (UMG), Goettingen, Germany

Surinder Kumar Department of Pathology, University of Michigan, Ann Arbor, MI, United States

Cyril Lagger Integrative Genomics of Ageing Group, Institute of Ageing and Chronic Disease, University of Liverpool, Liverpool, United Kingdom

Morgan E. Levine Department of Pathology, Yale University School of Medicine, New Haven, CT, United States

David B. Lombard Department of Pathology, University of Michigan, Ann Arbor, MI, United States; Institute of Gerontology, University of Michigan, Ann Arbor, MI, United States

Francesca Macchiarini Division of Aging Biology, National Institute on Aging, Bethesda, MD, United States

Samuel McFadden Southern Illinois University School of Medicine, Department of Neurology, Springfield, IL, United States

Richard A. Miller Department of Pathology and Geriatrics Center, University of Michigan, Ann Arbor, MI, United States

Ludmila Müller Department Lifespan Psychology, Max Planck Institute for Human Development, Berlin, Germany

Erin Munkácsy Barshop Institute for Longevity and Aging Studies, UT Health, San Antonio, TX, United States; Department of Molecular Medicine, UT Health, San Antonio, TX, United States

Laura J. Niedernhofer Institute on the Biology of Aging and Metabolism, Department of Biochemistry, Molecular Biology and Biophysics, University of Minnesota, Minneapolis, MN, United States

Miranda E. Orr Sticht Center for Healthy Aging and Alzheimer's Prevention, Internal Medicine – Gerontology and Geriatric Medicine, Wake Forest School of Medicine (WFSM), Winston-Salem, NC, United States; W.G. Hefner Veterans Affairs Medical Center, Salisbury, NC, United States

Juan Pablo Palavicini Barshop Institute for Longevity and Aging Studies, University of Texas Health Science Center at San Antonio, San Antonio, Texas, United States; Division of Diabetes, Department of Medicine, University of Texas Health Science Center at San Antonio, San Antonio, Texas, United States

Graham Pawelec Department of Immunology, University of Tübingen, Tübingen, Germany; Health Sciences North Research Institute, Sudbury, Ontario, Canada

Andrew M. Pickering Barshop Institute for Longevity and Aging Studies, UT Health, San Antonio, TX, United States; Department of Molecular Medicine, UT Health, San Antonio, TX, United States; Center for Neurodegeneration and Experimental Therapeutics, Department of Neurology, The University of Alabama at Birmingham, Birmingham, AL, United States; Department of Neurobiology, The University of Alabama at Birmingham, Birmingham, AL, United States

Peter S. Rabinovitch Department of Laboratory Medicine and Pathology, University of Washington, Seattle, WA, United States

Kelly R. Reveles College of Pharmacy, The University of Texas at Austin, Austin, TX, United States; Pharmacotherapy Education & Research Center, UT Health San Antonio, San Antonio, TX, United States

Nadia Rosenthal The Jackson Laboratory, Bar Harbor, ME, United States

Corinna N. Ross Population Health Program, Southwest National Primate Research Center, Texas Biomedical Research Institute, San Antonio, TX, United States; Department of Life Sciences, Texas A&M University San Antonio, San Antonio, TX, United States

Yi Sheng Department of Aging and Geriatric Research, Institute on Aging, University of Florida, Gainesville, FL, United States

Shinichi Someya Department of Aging and Geriatric Research, University of Florida, Gainsville, FL, United States

Randy Strong Department of Pharmacology, The University of Texas Health Science Center at San Antonio, and the Geriatric Research, Education and Clinical Center (GRECC) and Research Service of the South Texas Veterans Health Care System, Texas, TX, United States

Patrick Sung Department of Biochemistry and Structural Biology, University of Texas Health Science Center at San Antonio, San Antonio, TX, United States

George L. Sutphin University of Arizona, Tucson, AZ, United States

Robi Tacutu Systems Biology of Aging Group, Department of Bioinformatics and Structural Biochemistry, Institute of Biochemistry of the Romanian Academy, Bucharest, Romania

Suzette D. Tardif Population Health Program, Southwest National Primate Research Center, Texas Biomedical Research Institute, San Antonio, TX, United States

Thomas von Zglinicki Newcastle University Biosciences Institute, Campus for Ageing and Vitality, Newcastle University, Newcastle upon Tyne, United Kingdom

Robert J. Wessells Department of Physiology, Wayne State University School of Medicine, Detroit, MI, United States

Rui Xiao Department of Aging and Geriatric Research, Institute on Aging, University of Florida, Gainesville, FL, United States; Department of Pharmacology and Therapeutics, College of Medicine, University of Florida, Gainesville, FL, United States; Center for Smell and Taste, University of Florida, Gainesville, FL, United States

Guang Yang Department of Aging and Geriatric Research, Institute on Aging, University of Florida, Gainesville, FL, United States

Eric H. Young College of Pharmacy, The University of Texas at Austin, Austin, TX, United States; Pharmacotherapy Education & Research Center, UT Health San Antonio, San Antonio, TX, United States

Amina R.A.L. Zeidan College of Pharmacy, The University of Texas at Austin, Austin, TX, United States; Pharmacotherapy Education & Research Center, UT Health San Antonio, San Antonio, TX, United States

Yuan S. Zhang Carolina Population Center, The University of North Carolina at Chapel Hill, Chapel Hill, NC, United States

About the editors

Dr. Nicolas Musi is a tenured Professor of Medicine (Division of Geriatrics and Gerontology and Division of Diabetes) and Director of the Barshop Institute for Longevity and Aging Studies and the San Antonio Claude D. Pepper Older Americans Independence Center. He is also Associate Director for Research of the San Antonio Geriatric Research, Education and Clinical Center. He is an active educator and research mentor, and supervises clinical and research fellows, residents, and graduate students. In this role, he also functions as Director of a T32 Training Grant on the Biology of Aging.

Dr. Peter J. Hornsby obtained a PhD in Cell Biology at the Institute of Cancer Research of the University of London. He has held faculty positions at the University of California San Diego, the Medical College of Georgia, and Baylor College of Medicine. Currently he is Professor in the Department of Physiology and Barshop Institute for Longevity and Aging Studies, University of Texas Health Science Center, San Antonio.

Foreword

Since the inaugural publication of the *Handbooks of Aging* in 1976, the series has played a key role in promoting and guiding gerontological science. By preserving foundational knowledge and illuminating emerging areas, the series has served as a core resource for established researchers and an inspiration for students of gerontology. From its inception, gerontological science has been cross-disciplinary. The three-volume series has played a key role in maintaining cohesion in a science that spans dozens of disciplines.

The need to understand aging only increases in importance over time. The global population has now passed an important tipping point, moving from a world where children predominate to one in which there are more older people than youth. This reshaping of the age distribution in the population demands grand investments in the science of aging.

Thankfully, the science of aging is also growing faster than ever across social and biological sciences. Along with phenomenal advances in the understanding of the biology of aging as well as genetic influences on aging trajectories, and susceptibility to age-related diseases has come the awareness of the critical importance of the physical and social environments in which people age and the psychological factors that modulate and sometimes alter genetic predispositions.

The *Handbooks of Aging* series, comprised of the *Handbook of the Biology of Aging*, the *Handbook of the Psychology of Aging*, and the *Handbook of Aging and the Social Sciences*, is now in its ninth edition. The *Handbook of Aging and the Social Sciences* and the *Handbook of the Psychology of Aging* have long provided conceptual anchors and frameworks to the social and behavioral sciences while also addressing emerging topics that did not exist decades ago, such as the fluidity of race and gender, groundbreaking insights into the role of sleep in cognitive aging, and the ways that smartphones, robots, and social media can modify the experience of aging. The handbooks also provide cutting-edge updates to the understanding of genetics, built environments, and intergenerational commitments. The 9th edition of the *Handbook of the Biology of Aging* introduces geroscience, a discipline that did not exist 10 years ago and is now among the most vibrant in all of science. This edition also provides updates on the exciting advances in the genetics and integrative genomics of aging and longevity as well as the biology and therapeutic opportunities afforded by the studies of cellular senescence.

What has not changed over the editions is the superb synthesis of the field. The editors of the 9th edition extend a long tradition of giants in the field giving generously of their time and knowledge to produce consistently excellent volumes. Their thoughtful selection of topics and recruitment of deeply knowledgeable authors is reflected throughout the series. We are most grateful to Nicolas Musi and Peter J. Hornsby, editors of the *Handbook of the Biology of Aging*, Kenneth F. Ferraro and Deborah S. Carr, editors of the *Handbook of Aging and the Social Sciences*, and K. Warner Schaie and Sherry Lynn Willis, editors of the *Handbook of the Psychology of Aging*.

We also express our deep appreciation to our publishers at Elsevier, whose profound interest and dedication to the topic has facilitated the publication of the *Handbooks* through many editions. We remain eternally grateful to James Birren, for establishing the series and shepherding it through the first six editions that played a profound role in establishing the tradition of multidisciplinary science in the field of aging.

Thomas A. Rando and Laura L. Carstensen

Stanford Center on Longevity, Stanford University, Stanford, CA, United States

Preface

As this series of volumes enters its 9th edition, it may well be said that the field of the biology of aging has reached its maturity. There are several themes that are reflected in the contents of this volume.

1. In the past—now in reality the distant past—the biology of aging was a separate field of research, relatively unconnected to the mainstream of biological investigation. This has changed radically. Aging is now incorporated into a broad range of related fields, including cancer, diabetes and other metabolic diseases, immunology, and more. This has changed the way research in aging is viewed in many ways, and the benefits have been mutual on both sides. Research in aging has benefited from the input from these related areas, while those related areas have benefited from concepts developed in the biology of aging. Just one example is cellular senescence. Once considered a topic of narrow interest to a small group of investigators in technical aspects of cell culture, it has now been incorporated into multiple other areas, including cancer and metabolic diseases, as well as aging itself.
2. The second development reflected in the chapters in this volume is the degree to which the biology of aging has been incorporated into the basic biology of each organ system. For example, the basic biology of bone and the musculoskeletal system is incomplete without a consideration of the changes that take place in aging. Those include changes in the endocrine system, as well as stem cell biology. No account of the complete biology of any organ system is valid unless it incorporates the biology of aging as an integral part.
3. The third significant development is the merging of translational and clinical research with basic biology. As the field has matured, it has become possible to apply biological findings to human patients, either in clinical trials or clinical practice. Translational research has also assumed a much greater prominence. Basic biology has benefited from these clinical and translational investigations. As results have emerged from those studies, they have stimulated new avenues of investigation in basic biology. For example, studies on interventions that may increase rodent life span have already had a major impact on clinical and translational investigation, and in turn those studies have impacted new research in basic biology. mTOR inhibitors and senolytics are examples of pharmacological interventions that increase rodent life span and have already been translated to the clinic. Findings from clinical and translational studies have in turn stimulated new avenues in basic biology. We can expect this trend to continue in the future.

One of the more unfortunate consequences of these exciting developments is that it becomes impossible to produce a volume in the *Handbook of the Biology of Aging* series that comprehensively covers all aspects of the topic. The selection of topics offered in the current volume is an attempt to sample many of these exciting areas with contributions from leading scientists in their fields. But we also offer in advance our apologies to readers whose particular areas of interest we had to omit. Omission does not indicate that we felt the area of less importance, but we simply face the reality of an incredibly expanding and increasingly exciting field.

Nicolas Musi and Peter J. Hornsby

PART I

Basic mechanisms, underlying physiological changes, model organisms and interventions

CHAPTER

1

Longevity as a complex genetic trait

George L. Sutphin[1] *and Ron Korstanje*[2]

[1]University of Arizona, Tucson, AZ, United States [2]The Jackson Laboratory, Bar Harbor, ME, United States

OUTLINE

Introduction

Complex traits are phenotypic characteristics that result from the integration of many genetic loci and environmental factors. Longevity, along with the age-dependent decline in cellular and physiological processes that define aging, is quintessentially a complex genetic trait. A complete understanding of a complex trait requires both defining the range of factors that contribute to the trait and developing models for how the various factors interact. In the past several decades, hundreds of genes have been identified that are capable of influencing longevity or other age-associated phenotypes across a range of model systems. The majority of these genes can be broadly assigned to one or more of the following genetic pathways: (1) protein homeostasis, (2) insulin/IGF-1-like signaling (IIS), (3) mitochondrial metabolism, (4) sirtuins, (5) chemosensory function, or (6) dietary restriction. Pharmacologic agents targeting several of these pathways have been shown to increase lifespan and improve outcomes in age-associated disease in model systems and are either in use or in clinical trials for treatment of specific ailments. These include the mechanistic target of rapamycin (mTOR)-inhibitor rapamycin, the sirtuin activator resveratrol, and the glucose production suppressor metformin, and are discussed in greater detail in other chapters. Extragenetic but organism-intrinsic factors, such as tissue-specific gene expression, parentally inherited molecules, and epigenetics can also contribute to aging phenotypes.

Many environmental factors have been identified that impact longevity and age-associated disease. These include the abundance and composition of diet, exposure to various forms of stress, environmental temperature, social interaction, and even the presence or absence of a magnetic field. Among these, dietary restriction is by far the most widely studied. Reduction in total dietary intake or a change in the composition in the diet can have a profound impact on longevity in model systems. Short-term exposure to thermal, oxidative, endoplasmic reticulum (ER), or other forms of stress is sufficient to increase lifespan. In both worms and fruit flies, adjusting the culture temperature can dramatically influence lifespan. In each case, genes

Handbook of the Biology of Aging.
DOI: https://doi.org/10.1016/B978-0-12-815962-0.00001-9

have been identified that mediate the organism's response to the environmental stimuli.

This chapter examines aging as a complex trait. The following sections review past and ongoing efforts to define the scope of genetic, extragenetic, and environmental factors that influence aging, outline strategies for building interaction models, and discuss emerging tools that are furthering our ability to encompass the complexities of aging.

Defining the aging gene-space

A primary task in understanding the genetic complexity underlying any highly integrative phenotype is to identify the range of genes capable of impacting that phenotype. Three approaches are commonly employed to uncover novel aging factors. In models where targeted genome-scale genetic manipulation is possible and lifespan can be measured in a moderate- to high-throughput manner, screens have been carried out to identify single-gene manipulations capable of enhancing longevity. In longer-lived models and those less amenable to high-throughput targeted genetics, genetic mapping strategies are used to identify genetic loci at which natural variation is associated with differences in lifespan. A third approach is to leverage a secondary phenotype, such as stress resistance, that correlates with longevity but can be more rapidly screened to narrow the candidate gene list, and only screen genes that pass the primary threshold for longevity.

Direct screens for genetic longevity determinants

Among models commonly used in aging research, the nematode *Caenorhabditis elegans* and the budding yeast *Saccharomyces cerevisiae* possess three characteristics allowing for large-scale genetic screening for longevity: (1) genetic tools allowing for targeted genome-scale manipulation of individual genes, (2) relatively short lifespans, and (3) techniques to rapidly and inexpensively culture large populations in the laboratory. Complete genome sequences are available for both organisms (*C. elegans* Sequencing Consortium, 1998; Goffeau et al., 1996) and standardized lifespan assays can be completed in a matter of weeks (Murakami & Kaeberlein, 2009; Steffen, Kennedy, & Kaeberlein, 2009; Sutphin & Kaeberlein, 2009). Both models have been used in genome-scale screens for single-gene manipulations capable of increasing lifespan. In *Drosophila melanogaster*, while targeted gene-modification is not available at the genome-scale, random mutagenesis screens are used to identify novel longevity determinants and lifespan assays can similarly be completed in a matter of months (Linford, Bilgir, Ro, & Pletcher, 2013).

RNA interference screens in nematodes

In *C. elegans*, targeted gene knockdown by RNA interference (RNAi) can be accomplished by feeding animals bacteria expressing double-stranded RNA containing the target sequence (Timmons & Fire, 1998). Two RNAi feeding libraries targeting individual genes throughout the *C. elegans* genome have been constructed and are commercially available. The original Ahringer library contains 16,256 unique clones constructed by cloning genomic fragments targeting specific genes between two inverted T7 promoters (Fraser et al., 2000; Kamath et al., 2003). This library has recently been supplemented with an additional 3507 clones. The complete Ahringer library is commercially available through Source Bioscience (RNAi Resources | Source BioScience, n.d.). The Vidal library contains 11,511 clones produced using a full-length open reading frames (ORFs) gateway cloned into a double T7 vector (Rual et al., 2004) and is commercially available through either Source Bioscience (RNAi Resources | Source BioScience, n.d.) or Dharmacon Inc. (*C. elegans* | Dharmacon, 2019). Combined, these libraries provide single-gene clones targeting more than 20,000 unique sequences covering approximately 90% of known ORFs in *C. elegans*.

In total, more than 300 *C. elegans* genes have been identified for which reducing expression results in prolonged lifespan (Braeckman & Vanfleteren, 2007; Smith et al., 2008), the majority in longevity screens using the RNAi feeding libraries (reviewed by Yanos, Bennett, & Kaeberlein, 2012) or random mutagenesis screens (de Castro, Hegi de Castro, & Johnson, 2004; Muñoz & Riddle, 2003) (Table 1.1). These include three genome-wide screens using the Ahringer RNAi feeding library (Hamilton et al., 2005; Hansen, Hsu, Dillin, & Kenyon, 2005; Samuelson, Klimczak, Thompson, Carr, & Ruvkun, 2007), two partial screens targeting genes on specific chromosomes (Dillin et al., 2002; Lee et al., 2003), five screens of RNAi clones or mutant sets selected in a preliminary screen for a secondary longevity-associated phenotype, such as arrested development, upregulation of the mitochondrial unfolded protein response, or resistance to thermal or oxidative stress (Bennett et al., 2017; Chen, Pan, Palter, & Kapahi, 2007; Curran & Ruvkun, 2007; de Castro et al., 2004; Kim & Sun, 2007; Muñoz & Riddle, 2003), and one recent screen of *C. elegans* orthologs of human genes differentially expressed at different ages in human whole blood (Sutphin et al., 2017). Combined, these studies have identified aging factors in a range of biological processes including mitochondrial metabolism, mitochondrial unfolded protein response, cell structure, cell surface

TABLE 1.1 Invertebrate longevity screens

Study	Primary gene selection	No. genes tested	No. genes identified	Epistasis tested	Functional groups identified
Worm lifespan					
Dillin et al. (2002)	Chr. I, Ahringer RNAi Library	2445	4 + [a]	*daf-2*, *daf-16*	Metabolism, mitochondrial metabolism
Lee et al. (2003)	Chr. I & II, Ahringer RNAi Library	5690	52 + [b]	*daf-16*	Mitochondrial function, metabolism, gene expression, protein homeostasis, signal transduction, stress response
Hamilton et al. (2005)	Whole genome, Ahringer RNAi Library	16,475	89	*daf-16*, *sir-2.1*	Metabolism, signal transduction, protein homeostasis, gene expression
Hansen et al. (2005)	Whole genome, Ahringer RNAi Library	13,417	29	*daf-2*, *daf-12*, *daf-16*, *eat-2*, *glp-1*	Signal transduction, stress response, gene expression, mitochondrial metabolism
Samuelson et al. (2007)	Whole genome, Ahringer RNAi Library	16,757	115	*daf-16*	Metabolism, mitochondrial function, lysosomal functions, genomic stability, stress resistance
Chen et al. (2007)	Developmental arrest (see Kamath et al. (2003))	57	24	*daf-2*, *daf-2 daf-16*, *eat-2*	Mitochondrial function, metabolism, protein homeostasis, transcription
Curran and Ruvkun (2007)	Developmental arrest, Ahringer RNAi Library	2700	64	*daf-16*	Protein homeostasis, signal transduction, transcription, mitochondrial function, RNA processing, chromatin binding factors
Muñoz and Riddle (2003)	Thermal stress resistance, EMS mutagenesis	63	49	*daf-16*	Signal transduction, stress resistance
de Castro et al. (2004)	Oxidative stress resistance (juglone), transposon-mediated mutagenesis	6	4	*daf-16*	Stress resistance
Kim and Sun (2007)	Oxidative stress resistance (paraquat), Chr. III & IV, Ahringer RNAi Library	608	84	*daf-16*	Stress resistance, cell structure, signal transduction, metabolism, protein homeostasis, transcription, chromatin binding, mitochondrial function
Sutphin et al. (2017)	Orthologs of human genes differentially expressed with age in whole blood	82	50	*daf-16*, *eat-2*, *hif-1*, *rsks-1*, *sir-2.1*	Metabolism, calcium homeostasis, protein homeostasis, signal transduction
Yeast replicative lifespan					
Kaeberlein et al. (2005)	Random, ORF Deletion Collection	564	13		Protein homeostasis, metabolism, DNA replication
Smith et al. (2008)	Orthologs of worm pro-aging genes	264	25		Protein homeostasis, transcription, mitochondrial function
Steffen et al. (2012)	Ribosomal proteins	31	11		Protein homeostasis, metabolism
McCormick et al. (2015)	Ribosomal proteins	4698	238		Protein homeostasis, mitochondrial function, metabolism, DNA replication
Yeast chronological lifespan					
Powers et al (2006)	Whole genome, ORF Deletion Collection	~4800	90 + [c]		Protein homeostasis, stress resistance
Fabrizio et al. (2010)	Whole genome, ORF Deletion Collection	~4800	42		Protein homeostasis, metabolism, lipid production, stress tolerance, cell growth
Matecic et al. (2010)	Whole genome, ORF Deletion Collection	~4800	12		Protein homeostasis, metabolism

(*Continued*)

TABLE 1.1 (Continued)

Study	Primary gene selection	No. genes tested	No. genes identified	Epistasis tested	Functional groups identified
Burtner et al. (2011)	Random selection, ORF Deletion Collection	227	32		Mitochondrial function, stress resistance, protein homeostasis
Burtner et al. (2011)	Orthologs of worm pro-aging genes	235	18		Mitochondrial function, cell division, metabolism, protein homeostasis, exocytosis
Burtner et al. (2011)	Increased replicative life span	47	10		Cell division, transcription, DNA replication, metabolism
Burtner et al. (2011)	Increased media pH	76	20		Endocytosis, protein homeostasis, signal transduction, stress resistance
Garay et al. (2014)	Whole genome, ORF Deletion Collection	3878	262		Protein homeostasis, DNA repair, gene expression, Golgi function, metabolism
Campos et al. (2018)	Whole genome, ORF Deletion Collection	3718	254		Functional analysis not performed
Campos et al. (2018)	Whole genome, ORF Deletion Collection, nitrogen-restricted media	3718	228		Functional analysis not performed
Fruit fly lifespan					
Landis et al. (2003)	Random dox-inducible P element insertion	10,000	6		Vacuolar function, membrane transport, cell structure
Funakoshi et al. (2011)	Reduced wing and eye size; random P{GS} element insertion	716	2		Signal transduction, cell growth, protein homeostasis
Paik et al. (2012)	Random EP element insertion	27,157	15[d]		DNA replication, transcription, chromatin binding or modification, protein homeostasis, signal transduction, metabolism, immunity
Chen et al. (2014)	MicroRNA Deletion Collection	130	12[e]		MicroRNA
Ostojic et al. (2014)	Gustatory receptors	6	4		Food perception
Shaposhnikov et al. (2015)	DNA repair genes	9	8[f]		DNA damage detection and repair

[a]*The authors only pursue four genes, but do not report the total number found to significantly affect lifespan.*
[b]*The authors only report the number of significant hits on chromosome I.*
[c]*The authors pursue the 90 genes with the largest change in chronological lifespan, but do not report how many are statistically significant.*
[d]*8736 of 27,157 lines were putatively classified as long-lived; the authors selected 45 and 15 remained long-lived after validation.*
[e]*The authors tested 95* Drosophila *strains with combined deletion of 130 miRNAs.*
[f]*Genes that increased lifespan in at least one sex under at least one expression mode (constitutive vs adult only, whole body vs neuronal).*

proteins, cell signaling, protein homeostasis, RNA processing, and chromatin binding.

Notably, while a large number of genes has been identified through longevity screening in *C. elegans*, there is little overlap between screens (Yanos et al., 2012). There are several possibilities that may account for this lack of overlap. RNAi is inherently noisy, which may result in a different degree of knockdown between experiments for a given clone. Many of these screens were designed to assess maximum lifespan, scoring only the number of worms alive after all control worms had died. Between these two factors, the low overlap may reflect a high false-negative rate inherent in the methodology. Another possibility is that subtle differences in experimental design may result in a different range of factors becoming prominent. These differences include culture temperature, strain background, age at RNAi induction, and the presence or absence of floxuridine (FUdR) to prevent reproduction (Table 1.2). Recent evidence suggests that the strain of *Escherichia coli* used in RNAi experiments (HT115) may have distinct interactions with the *C. elegans* genotype in the context of lifespan from the *E. coli* strain used in most non-RNAi experiments (OP50) (Xiao et al., 2015; Yen & Curran, 2016), further complicating comparisons between RNAi screens and other categories of intervention. The

TABLE 1.2 Experimental conditions used in different *C. elegans* longevity screens

Study	Strain background	Worm stock temperature	Experiment temperature	RNAi start	FUdR?
Dillin et al. (2002)	Wild type	20°C	25°C	Egg	No
Lee et al. (2003)	Wild type	20°C	25°C	L1	Yes
Munoz et al. (2003)	*fer-15(b26ts)*	20°C	25.5°C	n/a	No
de Castro et al. (2004)	*mut-7(pk242)*	20°C	20°C	n/a	No
Hamilton et al. (2005)	Wild type	?	20°C	L1	Yes
Hansen et al. (2005)	*fer-15(b26); fem-1(hc17)*	25°C	20°C or 25°C	Egg	No
Samuelson et al. (2007)	*lin-15b(n744); eri-1(mg366)*	?	15°C to L4, 25°C thereafter	L1	Yes
Chen et al. (2007)	Wild type	20°C	20°C	L4	Yes
Curran and Ruvkun (2007)	*eri-1(mg366)*	20°C	20°C	L4	Yes
Kim and Sun (2007)	*rrf-3(pk1426)*	?	20°C	L1	No
Sutphin et al. (2017)	Wild type	20°C	15°C or 25°C	egg	Yes

Caenorhabditis Interventions Testing Program (CITP) was established in 2013 as a platform to identify novel lifespan-extending pharmacological compounds through rigorous testing across three independent laboratories, modeled after the successful Interventions Testing Program in mice (Nadon, Strong, Miller, & Harrison, 2017). Initial testing revealed significant problems in reproducibility across test sites, resulting in a multiyear project to standardize reagents, materials, and methods (Lithgow, Driscoll, & Phillips, 2017; Lucanic et al., 2017) culminating in the recent release of a set of standardized protocols for measuring worm lifespan (Lucanic et al., 2017). Wide adoption of these protocols may go some way toward limiting reproducibility and variation across labs. Petrascheck and Miller (2017) recently employed statistical models of wild-type *C. elegans* lifespan data to simulate aspects of methodological variation (e.g., scoring frequency, sample size) and population structure (e.g., hazard distribution) in longevity studies and determine their influence on reproducibility under "ideal" conditions (i.e., in the absence of methodological error and environmental variation). They conclude that statistical detection of lifespan differences is surprisingly resilient to scoring frequency and shape of the lifespan distribution within a population. Perhaps unsurprisingly, they determined that sampling size was the primary driver of poor reproducibility and that most studies are underpowered to detect small (<10%) changes in lifespan (Petrascheck & Miller, 2017). Regardless of the cause, the small degree of overlap and the fact that these screens only identified pro-aging genes—genes for which reduced expression increases lifespan—suggests that the range of genetic factors involved in *C. elegans* aging has yet to be exhaustively bounded.

Knockout screens in budding yeast

In the budding yeast *S. cerevisiae*, an analog to the *C. elegans* RNAi feeding libraries exists in the form of a genome-wide single-gene deletion strain collection. This collection contains ~4800 strains, each containing a complete open reading frame (ORF) deletion for a single nonessential gene in a common genetic background (Winzeler et al., 1999). Versions of this collection are available in both haploid mating types and in the homozygous diploid life stage. When conceptualizing longevity in a single-celled organism like *S. cerevisiae*, the first question to consider is the definition of "lifespan." Two aging paradigms are commonly studied in the budding yeast (Steinkraus, Kaeberlein, & Kennedy, 2008). Replicative lifespan refers to the number of times a cell can divide prior to undergoing senescence (Kaeberlein, 2006; Mortimer & Johnston, 1959). In contrast, chronological lifespan refers to the length of time a cell can remain in a quiescent state while retaining the ability to re-enter the cell cycle (Fabrizio & Longo, 2003; Kaeberlein, 2006; Fabrizio, Pozza, Pletcher, Gendron, & Longo, 2001).

Until recently, high-throughput techniques had only been developed for measuring chronological lifespan in yeast. Chronological lifespan is typically measured by growing yeast cells in liquid culture until they enter stationary phase, maintaining the cells in the expired media, and periodically sampling the aging culture to assess viability (Kaeberlein, 2006). Viability has traditionally been measured by plating a defined culture volume onto rich solid media and counting the number of colonies to calculate colony-forming units (CFUs). Powers, Kaeberlein, Caldwell, Kennedy, and Fields (2006) dramatically increased throughput by replacing the labor-intensive (though quantitative) process of counting CFUs with the more qualitative approach of diluting a sample from the aging culture back into rich liquid media and measuring optical density at 600 nm (OD600) after a fixed outgrowth time. They used this approach to screen the homozygous diploid deletion collection, identifying 90 chronologically long-lived mutants (Powers et al., 2006; Table 1.1). This technique was later improved to quantitatively assess outgrowth using a combined instrument that provides continuous culture agitation, temperature control, and OD600 measurement (Burtner, Murakami, & Kaeberlein, 2009; Murakami & Kaeberlein, 2009; Olsen, Murakami, & Kaeberlein, 2010) and has been used to screen selected sets of mutants from the yeast ORF deletion collection for increased chronological lifespan (Burtner, Murakami, Olsen, Kennedy, & Kaeberlein, 2011). Campos, Avelar-Rivas, Garay, Juárez-Reyes, and DeLuna (2018) used a similar method adapted for a robotic platform to measure the chronological lifespan of 3718 strains from the yeast deletion collection in media with either glutamine (a preferred nitrogen source; considered nonrestricted) or γ-aminobutyric acid (GABA; a model of nitrogen restriction) as the sole nitrogen source. They identified 573 (15%) chronologically short-lived and 254 (7%) chronologically long-lived single-gene deletion strains in the nonrestricted media, and 510 (6%) chronologically short-lived and 228 (14%) chronologically long-lived single-gene deletion strains in the nitrogen-restricted media. In a modified approach, Garay et al. (2014) monitored the chronological lifespan of 3878 fluorescently labeled deletion collection strains, each pooled into a common culture during stationary phase arrest with the wild-type strain expressing a distinct fluorescent label. Viability was assessed for each mutant strain relative to the pooled wild-type based on the relative fluorescence during outgrowth. This method identified 516 (13%) chronologically short-lived and 262 (7%) chronologically long-lived strains. Fabrizio et al. (2010) and Matecic et al. (2010) both employed an alternative competitive strategy, chronologically aging a pooled culture containing cells from each of the single-gene deletion strains in the ORF deletion collection and using microarrays to genotype the longest surviving cells. A recent improvement to further increase throughput shifts the chronologically aging yeast cultures from culture tubes to 96-deep-well plates with outgrowth analysis performed in 384-well plates (Jung, Christian, Kay, Skupin, & Linster, 2015).

Similar to the RNAi screens for increased lifespan in *C. elegans*, the screens by Powers et al. (2006), Matecic et al. (2010), and Fabrizio et al. (2010) found a remarkable lack of overlap in genes identified to affect yeast chronological lifespan (Smith, Maharrey, Carey, White, & Hartman, 2016). These yeast screens varied methodologically to a greater extent than the *C. elegans* screens. Beyond using distinct methods for measuring chronological lifespan, the three studies used different subsets of the yeast deletion collection (diploid BY4743 vs haploid BY4741) and different media types (rich vs defined media). Smith et al. (2016) recently revisited these screens, examining the methodological differences in detail, concluding that differences in media components were likely a major contributor to lack of consensus in identified genes, but that other methodological factors were also involved (aeration, method of measuring chronological lifespan, competitive vs noncompetitive environments).

In contrast to chronological lifespan, the typical method for measuring replicative lifespan in yeast involves the manual and labor-intensive removal of daughter cells from a dividing mother. To bypass this problem, a moderate-throughput iterative strategy was devised to identify long-lived mutants in the yeast deletion collection by determining replicative lifespan initially for only five cells per strain and using statistical methods to select strains for further testing (Kaeberlein et al., 2005). A preliminary report identified 13 genes for which deletion extends replicative lifespan out of the first 564 strains initially tested in the ORF deletion collection (Kaeberlein et al., 2005). Of the 13 genes, five map to the mTOR signaling pathway (*ROM2*, *RPL6B*, *RPL31A*, *TOR1*, and *URE2*). This screen was recently completed, identifying 238 long-lived deletion strains among 4698 tested, including 189 that were not previously reported to influence lifespan (McCormick et al., 2015). Pathway enrichment among the final pro-aging gene set confirmed the importance of mTOR signaling and cytosolic translation, and identified a number of other pathways and cellular processes that play an important role in replicative aging: mitochondrial translation, the tricarboxylic acid (TCA) cycle, mannosyltransferases, and the SAGA complex. Two additional replicative lifespan screens have been reported examining gene sets selected for either orthology to known worm aging genes (Smith et al., 2008) or ribosomal components (Steffen et al., 2008). Combined, longevity screens in yeast have identified approximately 250 pro-aging genes related to a range of cellular processes including protein homeostasis, metabolism, stress resistance, and mitochondrial function (Table 1.1).

Overexpression and knockout screens in fruit flies

Tools for genome-scale targeted genetic modification have yet to be used in the context of aging in *D. melanogaster*. *Drosophila* does provide a unique tool among invertebrate aging models in the form of transposable enhancer and promoter elements that can be randomly inserted into the genome allowing for unbiased identification of genes that increase lifespan when overexpressed. In an early study using this method,

Landis, Bhole, and Tower (2003) screened 10,000 lines and identified six genes for which overexpression increased longevity, including factors involved in vacuolar function, membrane transport, and cell structure (Table 1.1). Paik et al. (2012) initiated a longevity screen examining 27,157 lines and reported the first 15 long-lived transgenic strains overexpressing genes involved in transcription, translation, cell signaling, metabolism, and immunity (Table 1.1). Detailed mechanistic studies have now been published for two of these genes: (1) malic enzyme (*Men*), overexpression of which during larval development is sufficient to extend lifespan, drives a range of metabolic changes, and improves resistance to oxidative stress (Kim et al., 2015); and (2) NAD-dependent methylenetetrahydrofolate dehydrogenase-methenyltetrahydrofolate cyclohydrolase (*Nmdmc*), which increases lifespan and improves mitochondrial homeostasis when overexpressed either whole body or in fat (Yu, Jang, Paik, Lee, & Park, 2015).

A third study used growth impairment in the form of reduced wing and eye size as a surrogate marker for longevity in a screen of 716 transgenic *Drosophila* lines (Funakoshi et al., 2011). Two genes were identified with previous links to insulin/IGF-1-like and mTOR signaling (Table 1.1). Finally, Shaposhnikov, Proshkina, Shilova, Zhavoronkov, and Moskalev (2015) overexpressed nine genes involved in DNA damage detection and repair in both sexes under different temporal (constitutive vs adult-only) and tissue (whole-body vs neuronal) expression patterns and in different environmental contexts (well fed, starvation, heat stress, oxidative stress). The impact of each gene on lifespan was highly dependent on expression pattern and environmental context. Of the nine genes examined, overexpression of eight increased lifespan of well-fed, unstressed flies under at least one of the tested patterns of expression.

While genome-scale tools analogous to the *C. elegans* RNAi feeding library or the yeast deletion collection are not currently available in *Drosophila*, more targeted collections of genes have been examined in the context of aging. Chen et al. (2014) screened a collection of 95 *Drosophila* strains with deletions in 130 different microRNAs (miRNAs) (representing 99% of known *Drosophila* miRNAs) for lifespan and a variety of other phenotype. They report increased lifespan in 12 strains and decreased lifespan in 23 lines, demonstrating that miRNAs are a potentially rich source for genetic targets to modify longevity. Ostojic et al. (2014) found that knocking out four of six examined gustatory receptors resulted in increased lifespan in one or both sexes.

Genetic screening for lifespan variants using invertebrate models has been invaluable for defining the range of factors and biological processes involved in the determination of lifespan. Hundreds of genes have been identified across a range of central biological processes, the most prominent being mitochondrial metabolism, protein homeostasis, and stress resistance (Table 1.1). There is still work to be done in this area, particularly with respect to understanding how the range of factors important for lifespan is affected by different environmental conditions, such as changes in temperature or in response to dietary restriction.

Leveraging genetic diversity to identify aging loci

The previous sections describe reverse genetic approaches to identifying aging genes, in which large numbers of genes are knocked out or overexpressed individually and the effect on lifespan measured. Because of the scale, this approach has only been carried out in short-lived invertebrate models that are simple and inexpensive to maintain in the laboratory. Current genome-scale knockout efforts like the International Mouse Knockout Consortium (IMKC) and International Mouse Phenotyping Consortium (IMPC; see discussion of emerging tools later in this chapter) may lend themselves to a similar strategy in mice on a smaller scale, though the cost of maintaining statistically meaningful numbers of mice throughout life will still likely prevent full-genome mouse lifespan screens.

An alternative approach is to use forward genetics to leverage the natural phenotypic variation in genetically diverse populations to map candidate aging loci. This approach has the advantage of directly identifying genes even in long-lived mammalian systems. Genes identified in mammalian systems are more likely to be relevant to human aging than those identified in evolutionarily distant invertebrate screens. This approach also provides a complement to the screens carried out in invertebrate systems. Invertebrate screens tend to increase or decrease gene expression to levels outside of what is typically experienced from allelic variants in natural populations. Natural variants can result in large changes in gene activity, including complete inactivation of a gene, but more typically cause subtler changes in gene action or specificity. Gene mapping will therefore both identify longevity effects from less dramatic gene interventions, and point to genes that make the largest contributions to the variation in aging within the population examined.

Mapping longevity genes in human populations

The first gene-mapping studies to examine longevity employed linkage analysis in human families. Three studies of this type mapped loci associated with extreme longevity in 137 sibling pairs with one member being at least 98 years old and other members being at least 90 (males) or 95 (females) years old (Puca et al., 2001), 95 pairs of male fraternal twins with healthy aging (Reed, Dick, Uniacke, Foroud, & Nichols, 2004), and 279 families with multiple long-lived siblings (Boyden & Kunkel, 2010). All three studies identified one or more loci associated with variation in longevity, most notably a common locus on chromosome 4 (Table 1.3).

With the availability of relatively inexpensive high-density single nucleotide polymorphism (SNP) arrays and exome sequencing, most mapping efforts are now concentrating on genome-wide association studies (GWAS) with increasing population sizes. An example of this is type of study is work by Newman et al. (2010), in which more than 2 million polymorphisms were examined in a meta-analysis of four prospective cohort studies combining 1836 individuals that survived beyond 90 and a control group of 1955 individuals. Despite the large sample size, no loci reached genome-wide significance for association with longevity, though *MINPP1*, an inositol phosphatase involved in cell proliferation, approached significance. This example illustrates a common challenge in longevity mapping studies. Many loci throughout the genome have been found to be associated with variation in lifespan; however, few loci reach genome-wide statistical significance when corrected for multiple testing and little overlap is found between studies (Table 1.3). The notable exception is *APOE*, which has been identified by multiple independent genome-wide longevity mapping studies (Beekman et al., 2013; Broer et al., 2015; Deelen et al., 2014, 2011; Fortney et al., 2015; Joshi et al., 2016, 2017; McDaid et al., 2017; Nebel et al., 2011; Sebastiani et al., 2012, 2013, 2017; Timmers et al., 2019). *APOE* is located on chromosome 19 and encodes an apolipoprotein that is a major component of very low-density lipoproteins (VLDLs), which are responsible for removing excess blood cholesterol. Allelic variants in *APOE* are associated with Alzheimer's disease, atherosclerosis, and other age-associated pathologies. While not yet identified in genome-wide mapping studies, numerous targeted studies have found significant association between *FOXO3* and human longevity (Anselmi et al., 2009; Flachsbart et al., 2009; Li et al., 2009; Pawlikowska et al., 2009; Soerensen et al., 2010; Willcox et al., 2008; Zeng et al., 2010). *FOXO3* encodes a forkhead family transcription factor that has been linked to oxidative stress resistance and tumorigenesis. The *FOXO3* homologs in *C. elegans* (*daf-16*) and *D. melanogaster* (dFOXO) mediate many of the beneficial effects of reduced insulin/IGF-1-like signaling with respect to longevity and age-associated pathology. Two recent approaches to improve the power of GWAS use parental lifespan of genotyped subjects (rather than subject lifespan) to increase sample size (Joshi et al., 2016, 2017; McDaid et al., 2017; Pilling et al., 2016; Timmers et al., 2019) and/or employ methods to amplify variants previously associated with phenotypes or diseases that increase age-associated risk of mortality (Fortney et al., 2015; McDaid et al., 2017; Timmers et al., 2019). These studies have identified numerous novel longevity-associated regions and confirmed several previously identified loci (Table 1.3). Notably, while *APOE* remains the major longevity-associated gene identified through GWAS, there are now several loci that have reached genome-wide significance in at least two GWAS studies: the major histocompatibility complex (HLA) and lipoprotein(a) (*LPA*) loci on chromosome 6, the cyclin-dependent kinase inhibitor 2B (*CDKN2B*) and α-1-3-(N)-acetylgalactosaminyltransferase (*ABO*) loci on chromosome 9, and the hydroxylysine kinase (*HYKK*) locus on chromosome 15 (Table 1.3). While these loci have yet to be validated, convergence suggests that the recent improvements in GWAS power from increased sample size and new methodologies have begun to break the long-standing barrier to directly uncovering novel aging variants in human populations.

Mapping longevity genes in mouse populations

Like humans, mice have been employed as a model in studies to map genetic loci associated with variation in lifespan. While most mouse populations are maintained as inbred strains, a survey of 32 inbred strains found substantial variation in median lifespan among different inbred strains ranging from 251 days for AKR/J to 964 days for WSB/EiJ (Yuan et al., 2009). This inherent variation can be exploited by crossing strains with different lifespans and perform linkage analysis. To date, 11 such studies have been completed using crosses between seven classical inbred strains: BALB/c (BALB), C3H/HeJ (C3H), C57BL/6J (B or B6), DBA/2J (D or D2), KK/HIJ (KK), LP/J (LP), NZW/LacJ (NZW), PL/J (PL), and ST/bJ (ST); three wild-derived inbred strains: CAST/Ei (CAST), MOLD/Rk (MOLD), and POHN/DehJ (POHN); and two strains developed by crossing classical inbred strains and selecting for sleep response to ethanol: inbred long sleep (ILS/IbgTejJ or ILS) and inbred short sleep (ISS/IbgTejJ or ISS). CAST, MOLD, and POHN are particularly important, as they were inbred from wild-caught mice without domestication and provide a large fraction of the genetic diversity found among strains used to generate crosses for mapping studies. CAST and MOLD originate from the subspecies *Mus m. castaneus* and *Mus musculus molossinus*, respectively, which diverged from the subspecies of most classical inbred strains, *Mus m. musculus*, approximately half a million years ago.

TABLE 1.3 Significant and suggestive loci identified in genome-wide human mapping studies. Where available, marker names were used to standardize all genomic locations to human genome assembly GRCh37.p13. Genes listed are located within 10 kb of a marker location. Loci within 10 Mb of each other are grouped. **Bold** = significant genome-wide association; not bold = suggestive genome-wide association.

Chr	Region or marker location (Mb)	Gene(s)	References
1	3.6	TP73	Samuelson et al. (2007)
	41.2, 44.2	KCNQ4, NFYC, ST3GAL3	Kim and Sun (2007)
	55.5, 56.9, 63.4	PPAP2B	Kim and Sun (2007), Yanos et al. (2012)
	109.8, **114.2**	**CELSR2**, **PSRC1**, **MAGI3**	Curran and Ruvkun (2007), Yanos et al. (2012)
	195.9-197.3 (196.4)	**KCNT2**	Fraser et al. (2000)
	237.1, 240.9	RPL35P1	Murakami and Kaeberlein (2009), Sutphin and Kaeberlein (2009)
2	**0.6**, 8.5, 10.1	AC010969.1, RP11.254F7.1	Bennett et al. (2017), Chen et al. (2007), Curran and Ruvkun (2007)
	26.9, 28.7, **35.4**	**AC013442.1**, **KCNK3**	Curran and Ruvkun (2007), de Castro et al. (2004), RNAi Resources I Source BioScience (n.d.)
	50.5	NRXN1	Bennett et al. (2017)
	70.8	TGFA	Bennett et al. (2017)
	139.4	NXPH2	Braeckman and Vanfleteren (2007)
	162.9, 166.4	CSRNP3, DPP4	Sutphin and Kaeberlein (2009), Yanos et al. (2012)
	237.1, 238.3	**AC079135.1**, COL6A3, **GBX2**	Curran and Ruvkun (2007), Kim and Sun (2007)
3	0.2, 2.0		Linford et al. (2013), Rual et al. (2004), Samuelson et al. (2007)
	28.6–33.6 (30.2), **29.6**, **38.0**	**CTDSPL**, **RBMS3**	*C. elegans* and Dharmacon (2019), Goffeau et al. (1996), Rual et al. (2004)
	48.5, 49.9, 50.1	ATRIP, CAMKV, CCDC51, RBM6, TMA7, TRAIP	Bennett et al. (2017), Linford et al. (2013), Yanos et al. (2012)
	71.8	EIF4E3	Dillin et al. (2002), Lee et al. (2003)
	85.5	CADM2	Murakami and Kaeberlein (2009)
	96.2, 101.0	AC108697.1, IMPG2	Smith et al. (2008), Sutphin et al. (2017)
	115.1		Samuelson et al. (2007)
	162.7, 168.7		Kim and Sun (2007)
	192.5–192.6	**MB21D2**	*C. elegans* and Dharmacon (2019)
4	1.4, 2.3, **3.1**, 3.2	**HTT**, MSANTD1, RP11-478C1.7, UVSSA, ZFYVE28	Bennett et al. (2017), Curran and Ruvkun (2007), Sutphin et al. (2017), Yanos et al. (2012)
	41.6, 42.3	LIMCH1	Murakami and Kaeberlein (2009), Sutphin et al. (2017)
	60.2		Kamath et al. (2003)
	76.9	SDAD1	Bennett et al. (2017)
	89.8, 94.5	GRID2, FAM13A	Sutphin et al. (2017), Sutphin and Kaeberlein (2009)
	108.4, 110.6, 111.2, 120.7	CCDC109B, RP11-236P13.1	Chen et al. (2007), Goffeau et al. (1996), Hansen et al. (2005), Samuelson et al. (2007), Sutphin and Kaeberlein (2009)

(*Continued*)

TABLE 1.3 (Continued)

Chr	Region or marker location (Mb)	Gene(s)	References
	137.7		Bennett et al. (2017)
	160.9–162.5		*C. elegans* and Dharmacon (2019)
5	55.9	AC022431.2	Yanos et al. (2012)
	74.8–78.9 (76.2)	S100Z	Rual et al. (2004)
	96.3	LOC101929747	Linford et al. (2013)
	110.8	**CAMK4**	Smith et al., 2008
	140.2, 149.4, 149.6, **157.8**	CSF1R, SLC6A7, PCDHA1, PCDHA2, PCDHA3, PCDHA4, PCDHA5, PCDHA6, PCDHA7	Dillin et al. (2002), Kamath et al. (2003), Lee et al. (2003), Sutphin and Kaeberlein (2009)
6	8.2	MYLK4	Linford et al. (2013)
	20.0, **25.7–33.0** (**32.4**), 26.2, 29.6–29.7 (29.7), 31.9, **32.6**	**AIF1**, **BTNL2**, **C2**, **HCP5**, HIST1H2BD, HIST1H2BE, **HLA-DQA1**, **HLA-DQA2**, **HLA-DQB1**, **HLA-DRB5**, **HLADRB9**, HLA-F, HLA-F-AS1, **LY6G6F**, **MICA**, MOG, **MSH5**, **SKIV2L**, ZFP57	Chen et al. (2007), Curran and Ruvkun (2007), de Castro et al. (2004), Dillin et al. (2002), Fraser et al. (2000), Lee et al. (2003), Murakami and Kaeberlein (2009), RNAi Resources I Source BioScience (n.d.)
	48.2–62.2 (**50.2**), 50.9	**RCBTB1**	Linford et al. (2013), Yanos et al. (2012)
	102.8, 102.9, 106.8, 109.0	AIM1, FOXO3	Bennett et al. (2017), Murakami and Kaeberlein (2009)
	131.4		RNAi Resources I Source BioScience (n.d.)
	160.3–161.7 (**161.0**), **160.5**, **161.0**, 164.3, **164.4–169.6** (**164.4**, **166.2**), 166.3, 166.7	PRR18, **IGF2R**, **LPA**	Bennett et al. (2017), Curran and Ruvkun (2007), Fraser et al. (2000), Hamilton et al., (2005), Kamath et al. (2003), Linford et al., (2013), RNAi Resources I Source BioScience (n.d.), Timmons and Fire (1998), Yanos et al. (2012)
7	1.0, 1.9, **6.2**	C7orf50, MAD1L1, **USP42**	Bennett et al. (2017), Chen et al. (2007)
	22.8	**AC073072.5**, **IL6**	Sutphin et al. (2017)
	36.8	AC007349.4	Sutphin and Kaeberlein (2009)
	48.3, **49.5–75.5** (**64.7**), **75.1**, 75.2, 82.3	**AP5Z1**, **MIR4656**, **POM121C**, **RADIL**	Braeckman and Vanfleteren (2007), Curran and Ruvkun (2007), Hamilton et al. (2005), Linford et al. (2013)
	92.4, 103.9, 109.8	CDK6	Dillin et al. (2002), Kim and Sun (2007), Lee et al. (2003)
	122.8, **129.7**, 134.3	AKR1B15, SLC13A1, **ZC3HC1**	Bennett et al. (2017), Curran and Ruvkun (2007), Linford et al. (2013)
	152.6		Smith et al. (2008)
8	6.7	DEFB1	Dillin et al. (2002), Lee et al. (2003)
	19.9		Yanos et al. (2012)
	31.0	WRN	Dillin et al. (2002), Lee et al. (2003)
	41.6–67.0, 53.9, 72.3		Dillin et al. (2002), *C. elegans* Sequencing Consortium (1998), Lee et al. (2003), Sutphin and Kaeberlein (2009)
	134.5	ST3GAL1	Linford et al. (2013)
9	**5.3–5.4**	**RLN1**, **RLN2**	*C. elegans* and Dharmacon (2019)

(*Continued*)

TABLE 1.3 (Continued)

Chr	Region or marker location (Mb)	Gene(s)	References
	21.9–22.1 (**22.1**), **22.1**, 27.2, 28.1	**CDKN2B**, **CDKN2B-AS1**, **LOC729983**, **LOC100130239**, TEK	Curran and Ruvkun (2007), Dillin et al. (2002), Fraser et al. (2000), Yanos et al. (2012)
	73.8	TRPM3	Bennett et al. (2017)
	97.5, 97.6	C9orf3	Dillin et al. (2002), Lee et al. (2003)
	113.1	SVEP1	Bennett et al. (2017)
	128.3, **136.0–136.3** (**136.1**), **136.1**, 137.7, **138.3**	**ABO**, **MAPKAP1**, COL5A1	Bennett et al. (2017), Curran and Ruvkun (2007), Fraser et al. (2000), Goffeau et al. (1996), Hamilton et al. (2005)
10	**3.3–5.6** (**4.3**), 4.1, 5.1	AKR1C3, LOC101927946, R11-49907.4, U8.23	Bennett et al. (2017), Sutphin et al. (2017), Timmons and Fire (1998)
	23.4	MSRB2	Smith et al. (2008)
	51.9–52.3 (52.1)	SGMS1	Rual et al. (2004)
	89.3, 96.6	CYP2C58P	Muñoz and Riddle (2003), Sutphin and Kaeberlein (2009)
	108.6	SORCS1	Dillin et al. (2002), Lee et al. (2003)
11	**0.3–1.1**, 3.0	CARS	Bennett et al. (2017), *C. elegans* and Dharmacon (2019)
	15.9		Murakami and Kaeberlein (2009)
	50.0–51.4		*C. elegans* and Dharmacon (2019)
	90.9		Braeckman and Vanfleteren (2007)
	122.9, 124.0, 126.2	DCPS, LOC341056, VWA5A	Dillin et al. (2002), Kim and Sun (2007), Lee et al. (2003)
12	14.1	GRIN2B	Kim and Sun (2007)
	48.5, 51.7, **61.6–61.7** (**61.6**), 68.9	BIN2, **FEN1**, **FADS1**, PFKM	Fraser et al. (2000), Kim and Sun (2007), Murakami and Kaeberlein (2009), Sutphin and Kaeberlein (2009)
	83.4	**TMTC2**	Chen et al. (2007)
	110.1–113.0 (**111.9**), **112.1**, 121.4, 127.4, 131.5, 132.1	**ATXN2**, GPR133, RP11-575F12.1, **SH2B3**	Bennett et al. (2017), Curran and Ruvkun (2007), Fraser et al. (2000), Goffeau et al., (1996)
13	**19.4**	**ANKRD20A9P**, **RNU6-55P**, **RP11-38M15.11**	Sutphin et al. (2017)
	48.4, 53.8, 61.7	RN7SL618P	Bennett et al. (2017), Kim and Sun (2007), Sutphin et al. (2017)
	71.9, 73.1		Bennett et al. (2017), Smith et al. (2008)
	90.6, 93.9, 95.7, 99.1	ABCC4, GPC6, STK24	Bennett et al. (2017), Dillin et al. (2002), Lee et al. (2003)
14	**20.4–23.7**, 29.0		Bennett et al. (2017), *C. elegans* and Dharmacon (2019), *C. elegans* Sequencing Consortium (1998)
	40.7, **43.5–54.5** (**49.5**), 62.4	CTD-2277K2.1, SYT16	*C. elegans* and Dharmacon (2019), Linford et al. (2013), Sutphin and Kaeberlein (2009)
	90.8	NRDE2	Bennett et al. (2017)
	104.2–106.1		*C. elegans* and Dharmacon (2019)
15	27.0, **27.9–35.0**, 33.7	GABRB3, RYR3	Bennett et al. (2017), Dillin et al. (2002), *C. elegans* Sequencing Consortium (1998), Lee et al. (2003)

(Continued)

TABLE 1.3 (Continued)

Chr	Region or marker location (Mb)	Gene(s)	References
	53.8	WDR72	Bennett et al. (2017), Dillin et al. (2002)
	78.8–78.9 **(78.8)**, **78.8**, **78.9**	AC027228.1, **CHRNA3**, **CHRNA5**, **CHRNB4**, **HYKK**, IREB2, PSMA4, **RP11-335K5.2**	Curran and Ruvkun (2007), Hamilton et al. (2005), Kamath et al. (2003), RNAi Resources I Source BioScience (n.d.), Yanos et al. (2012)
	91.4, 94.8	**FES**, **FURIN**, MCTP2	Curran and Ruvkun (2007), Hamilton et al. (2005), Linford et al. (2013)
16	2.1	NTHL1, TSC2	Bennett et al. (2017)
	12.5, **19.0–26.6** **(24.0)**, 19.9, 28.8	ATXN2L, IQCK, **PRKCB**, RP11-1348G14.4, SNX29	Linford et al. (2013), Timmons and Fire (1998), Yanos et al. (2012)
	48.5, 50.7, 53.8, 57.9	AC023818.1, FTO, KIFC3, RN7SL54P, SIAH1, SNX20	Chen et al. (2007), Samuelson et al. (2007), Smith et al., (2008), Yanos et al. (2012)
	72.1, 78.5	**TXNL4B**, WWOX	Curran and Ruvkun (2007), Dillin et al. (2002), Lee et al. (2003)
17	8.9, **9.8–11.5** **(10.1)**	**GAS7**, NTN1	Sutphin and Kaeberlein (2009), Timmons and Fire (1998)
	31.4, **36.7–55.5**, 47.9	ASIC2, FLJ45513, RP11-40A13.1, TAC4	*C. elegans* Sequencing Consortium (1998), Rual et al. (2004), Sutphin et al. (2017)
	60.9, 61.4, 64.5, 70.0, 79.3	PRKCA, SLC38A10, TANC2	Dillin et al. (2002), Smith et al. (2008), Sutphin and Kaeberlein (2009)
18	70.2–74.1 (72.2), 71.0, 74.9	CNDP1, ZNF516	Bennett et al. (2017), Kim and Sun (2007), Rual et al. (2004)
19	**6.7–17.5**, **11.2**	**LDLR**	Curran and Ruvkun (2007), *C. elegans* Sequencing Consortium (1998)
	35.3–46.6 **(45.4)**, 38.3, **45.1–45.5** **(45.4)**, **45.4**, **49.1–53.5**	AC016582.2, **CTB-129P6.4**, **APOC1**, **APOE**, **BCAM**, **BCL3**, **CEACAM16**, **PVRL2**, **TOMM40**	Bennett et al. (2017), Chen et al. (2007), Curran and Ruvkun (2007), de Castro et al. (2004), Dillin et al., (2002), *C. elegans* Sequencing Consortium (1998), Fraser et al., (2000), Kamath et al., (2003), Lee et al., (2003), RNAi Resources I Source BioScience (n.d.), Steffen et al. (2009), Sutphin et al. (2017), Sutphin and Kaeberlein (2009)
20	**0.6–4.1** **(0.9)**	**ANGPT4**	Timmons and Fire (1998)
	52.2, 59.9, 60.2	CDH4, Y RNA	Bennett et al. (2017), Sutphin and Kaeberlein (2009)
21	14.8, 17.1, 22.9	AL050302.1, ANKRD30BP1, NCAM2, UPS24	Bennett et al. (2017), Kamath et al. (2003), Sutphin et al. (2017)
	47.6	LSS	Linford et al. (2013)
22	37.6, **44.5**, 47.5, 51.1	**PARVB**, SHANK3, SSTR3, TBC1D22A	Bennett et al. (2017), Hamilton et al. (2005)
X	141.1, 142.2		Linford et al. (2013), Samuelson et al. (2007)

One approach to gene mapping in mice is to generate recombinant inbred (RI) strain panels by crossing two or more strains and inbreeding subsequent generations to produce novel but related sets of inbred strains. Four studies have measured lifespan and mapped associated loci using either the BXD RI strain panel (de Haan, Gelman, Watson, Yunis, & Van Zant, 1998; Gelman, Watson, Bronson, & Yunis, 1988; Lang et al., 2010) or the ILSXISS RI strain panel (Rikke, Liao, McQueen, Nelson, & Johnson, 2010). A second approach is to genotype and measure lifespan of non-inbred offspring generated by crossing two or more inbred strains. Lifespan has twice been measured in the UM-HET3 four-way cross, (BALBxB6)x(C3HxD2) (Jackson, Galecki, Burke, & Miller, 2002; Miller, Chrisp, Jackson, & Burke, 1998), and once each in two four-way crosses that include a wild-derived strain, (STxB6)x(CASTxD2) and (LPxMOLD)x(NZWxBALB) (Klebanov et al., 2001). An early study by Yunis et al. (1984) and a more recent study by Yuan, Flurkey, Meng, Astle, and Harrison (2013) identified longevity-associated loci in the (B6xD2)xB6 and (POHNxB6)xPOHN backcross populations, respectively. Another study mapped loci associated with lifespan in a backcross between short-lived SAMP1 mice and long-lived B10.BR−$H2^k$/SgSnSlc (B10.BR), a congenic strain related to B57BL/10 that shares the histocompatibility 2 (H2) locus with SAMP1, removing the influence of H2 on lifespan (Guo et al., 2000). A final study mapped loci associated with longevity, circulating IGF1, and age at vaginal patency in a cross between KK/HlJ and PL/J mice (Yuan et al., 2015). These strains were selected based on a robust difference in circulating IGF1 at 6 months of age (a difference that is shown to be not dependent on differences in locus *Igf1q8*, which is known to influence IGF1 levels) and a lack of previous mapping studies using either strain background. Combined, these studies have identified more than 60 suggestive or significantly longevity-associated loci located throughout the mouse genome (Table 1.4). As with humans, few loci are identified by more than one study. Six genomic regions located on chromosomes 1, 2, 7, 8, 11, and 16 were identified by at least two studies. In nine cases—on chromosomes 1, 2, 5, 7, 10, 16, 17, and 19—markers with suggestive association with lifespan were identified by at least two studies within the same 10 Mb region. These regions represent the strongest candidates for further investigation. Two regions on chromosome 1 (120.0−136.3 and 147.0−185.5 Mb) and one region on chromosome 11 (5.2−28.9 Mb) are of particular interest, as they contain markers that reached genome-wide statistical significance in two independent studies (Table 1.4).

Mouse−human concordance

Based on the assumption that the longevity loci identified in corresponding human and mouse genomic regions result from orthologous genes, Yuan, Peters, and Paigen (2011) integrated mouse and human data by projecting the identified human loci onto the mouse genetic map. Out of 10 human GWAS regions, eight mapped within 10 Mb of a mouse QTL peak. The probability of this occurring by chance is very low ($P = .0025$) (Yuan et al., 2011), strongly suggesting that the mechanisms responsible for variation in longevity within mouse and human populations is evolutionarily conserved and validating the mouse as a model for human aging.

Age-associated gene expression studies

Instead of measuring the response of age-related processes to changes in gene activity, the impact of age on gene expression can be used to identify potential longevity factors. Comparing expression studies has the potential to provide insight into general mechanisms of aging that cross tissue and species boundaries. Microarray studies comparing young and old in flies (Landis et al., 2004; Pletcher et al., 2002; Zou, Meadows, Sharp, Jan, & Jan, 2000), mice (Weindruch, Kayo, Lee, & Prolla, 2001), and monkeys (Kayo, Allison, Weindruch, & Prolla, 2001) all found an increase in oxidative stress genes with age, which is in agreement with an observed increase in expression of oxidative stress genes in young individuals from long-lived *C. elegans* strains (McElwee, Bubb, & Thomas, 2003; Murphy et al., 2003). Age-associated gene expression changes between *C. elegans* and *D. melanogaster* were directly compared in a single study that identified a conserved expression program involving mitochondrial metabolism and DNA repair, among others (McCarroll et al., 2004). In 2009, a meta-analysis of 27 microarray datasets examining different tissues in humans, mice, and rats revealed a common age-associated gene expression signature (de Magalhães, Curado, & Church, 2009). This signature included an age-associated increase in expression of genes to immunity and inflammation, as well as lysosome-associated genes, and a decrease in expression of genes involved in mitochondrial function, cell cycle, senescence, and apoptosis. Further studies of this type will be of interest, particularly involving comparison of gene expression patterns between invertebrates and mammals. For a more detailed discussion of specific microarray aging studies and associated technical challenges, readers can refer to numerous reviews (Becker, 2002; Golden, Hubbard, & Melov, 2006; Han et al., 2004; Melov & Hubbard, 2004; Nair, Golden, & Melov, 2003; Werner, 2007).

Examining age-associated changes in gene expression is also a useful and relatively noninvasive approach to study aging in humans. The first study to examine age-dependent changes in human gene expression on a genome scale used microarrays to examine peripheral blood leukocyte transcripts and identified 295 genes with robust differential expression with age (Harries et al., 2011). Similar studies have been performed in the Framingham Heart Study and other cohorts that are part of the Cohorts for Heart and Aging Research in Genomic Epidemiology (CHARGE) consortium (Peters et al., 2015). Human gene expression studies can be used for two purposes downstream. The first is to identify specific diagnostic transcripts that are useful biomarkers of the biological age of the individual (discussed further in the "Aging biomarkers" section of

TABLE 1.4 Significant and suggestive loci identified in genome-wide mouse mapping studies. Where available, marker names were used to standardize all genomic locations to mouse genome assembly GRCm38. Genes listed are located within 10 kb of a marker location. **Bold** = significant genome-wide association; not bold = suggestive genome-wide association.

Chr	Region or marker location (Mb)	Gene(s)	References
1	13.3	Ncoa2	Murakami and Kaeberlein (2009)
	31.5–34.5 (33.8)	**Zfp451**	Linford et al. (2013)
	120.0, 121.9–133.7 (126.5)	**Gm101, Nckap5, Sctr**	Goffeau et al. (1996), Linford et al. (2013)
	147.0–185.5 (171.9), 148.8, 161.2, 167.8, **174.6**	**Cd48, Fmn2, Gm10521, Lmx1a, Prdx6**	Goffeau et al. (1996), Kamath et al. (2003), Murakami and Kaeberlein (2009)
2	**61.4–69.9 (65.4)**		Linford et al. (2013)
	102.8, 108.3	**Cd44**	Goffeau et al. (1996), Steffen et al. (2009)
	122.1, **124.6**	B2m, **Sema6d**	Goffeau et al. (1996), RNAi Resources I Source BioScience (n.d.)
4	17.4–151.1 (63.8), 22.5, **59.7–99.2 (80.5)**	Pappa, Pou3f2	Kamath et al. (2003), Rual et al. (2004), Steffen et al. (2009)
5	25.0	Prkag2	Murakami and Kaeberlein (2009)
	44.4		Murakami and Kaeberlein (2009)
	64.6, 65.4–86.6 (79.5)		Linford et al. (2013), Murakami and Kaeberlein (2009)
6	55.2	Fam188b	Murakami and Kaeberlein (2009)
	73.8-83.5 (79.6), **84.4–101.2 (93.5)**, 91.5–101.2 (99.0)	6030492E11Rik, Ctnna2, Foxp1	Linford et al. (2013)
	113.3–121.9 (113.3)	**Camk1**	Linford et al. (2013)
7	**3.5–6.9 (3.5)**	Gm7789, **Vstm1**, Zscan4c	Linford et al. (2013)
	25.9, 26.8	Cyp2a5, Cyp2b10	Goffeau et al. (1996)
	49.1–87.9 (84.6), 58.5, 65.7, 72.1, 89.3–107.3 (96.9), 103.9	Gm10624, Gm19831, Hbb-y, Hbb-bh1, Hbb-bh2, Mctp2, Tenm4, **Zfand6**	Fraser et al. (2000), Linford et al. (2013), RNAi Resources I Source BioScience (n.d.), Timmons and Fire (1998)
8	14.7–25.7 (14.7), **23.3**	**4930518F22Rik**, Dlgap2, **Golga7**	Linford et al. (2013), Sutphin and Kaeberlein (2009)
	36.0		Murakami and Kaeberlein (2009)
	102.4–115.8 (108.8)	Zfhx3	Linford et al. (2013)
9	**91.4**	**Zic1**, **Zic4**	Steffen et al. (2009)
10	48.7		Timmons and Fire (1998)
	67.0	**Reep3**	Steffen et al. (2009)
	102.9–116.1 (110.5), **119.7**	**Grip1**	Linford et al. (2013), Sutphin and Kaeberlein (2009)
11	**6.2–20.7 (8.3, 19.7)**, **16.8–24.4 (17.9)**, **17.1–26.8 (19.6)**, 35.7–39.1 (35.7)	**Etaa1**, Slit3	*C. elegans* Sequencing Consortium (1998), Linford et al. (2013)
	45.9–59.1 (55.8)		Linford et al. (2013)
12	45.2	Gm24731	RNAi Resources I Source BioScience (n.d.)
	105.1, 113.3	**Igh-7, Igh-8**	Goffeau et al. (1996), Steffen et al. (2009)
14	46.4	Bmp4, Gm15217	Murakami and Kaeberlein (2009)
16	5.7		Steffen et al. (2009), Timmons and Fire (1998)
	20.2, 29.1–37.5 (32.3)	5830438M01Rik, Rnf168, Yeats2	Linford et al. (2013), Murakami and Kaeberlein (2009)

(Continued)

TABLE 1.4 (Continued)

Chr	Region or marker location (Mb)	Gene(s)	References
	59.8–70.1 (64.6)		Linford et al. (2013)
17	13.7–39.5 (34.5), **42.7**, **50.1**	Btnl6, **Gpr111**, **Rftn1**	Murakami and Kaeberlein (2009), Rual et al. (2004)
18	53.2		Timmons and Fire (1998)
19	32.0, 39.7	Asah2, Cyp2c69	Fraser et al. (2000), Timmons and Fire (1998)
	47.2, **53.9**	Calhm1, **Pdcd4**, **Bbip1**	Murakami and Kaeberlein (2009), Timmons and Fire (1998)
X	45.4–58.6 (51.7)	2900011F02Rik, Hs6st2	Linford et al. (2013)
	122.3–155.6 (130.4)	**9530062E16Rik**, **Diap2**	Linford et al. (2013)

this chapter). Biomarkers are not necessarily causal players in the aging process, but simply consistent correlates with the health state of the individual. The second purpose is to identify candidate genes that play a causal role in the aging process. A disadvantage in human gene expression studies is the limited availability of different tissue types. The question arises as to whether gene expression in leukocytes accurately represents what happens in the whole organism. Mice and other animal models can be used to measure gene expression in different tissues in order to identify both genes associated with tissue-specific aging and genes with common age-associated expression patterns across tissues.

After GWA studies and human expression studies identify candidate genes, the next step is to validate the capability of each gene to impact aging, characterize their role in the aging process, and determine why the specific causal allele leads to variation in lifespan. For genes that are consistently identified in multiple studies, such as *APOE* and *FOXO3*, direct follow-up in a mammalian system may be appropriate. Indeed, *APOE* is under intense scrutiny, particularly in the areas of cardiovascular research, cholesterol metabolism, and neurodegenerative disease. The creation of a genome-wide knockout mouse strain collection and novel gene-editing technologies—particularly the recent and widespread adoption of the CRISPR/Cas9 system—for rapid introduction of novel variants into inbred test populations allows for rapid progression from gene identification to characterization. In the next tier, where the number of candidate genes makes direct characterization in mammalian models cost-prohibitive, the use of RNAi feeding libraries in *C. elegans* allows for rapid screening of large numbers of candidate genes for direct effects on lifespan. We recently demonstrated this approach by screening 82 genes from the CHARGE gene expression study (Peters et al., 2015) in *C. elegans*. RNAi knockdown of 50 genes (61%) significantly affected lifespan (Sutphin et al., 2017). This approach has the advantage of narrowing candidate genes to those that likely have an evolutionarily conserved role in aging at the cost of discarding genes with a human- or mammal-specific role in aging. In the same vein, this method is not suitable for all mammalian candidates, as roughly half of human and mouse genes do not have clear worm orthologs.

Nongenetic sources of complexity

Tissue-specific aging

One important aspect of aging is the differences between tissues. In human populations, mortality can result from pathology in virtually every organ system, though genetic and environmental factors can increase risk for tissue-specific diseases. This suggests that different tissues generally age at similar rates on average within the human population, but that different tissues can experience different rates of aging between individuals. In *C. elegans*, studies of tissue-specific aging have concluded that muscle cell function tends to gradually decline beginning around the transition to the post-reproductive life stage, while neurons largely retain function in old animals (Herndon et al., 2002). The decline in muscle function is accompanied by decreased pharyngeal pumping, resulting in reduced food consumption (Huang, Xiong, & Kornfeld, 2004; Kenyon, Chang, Gensch, Rudner, & Tabtiang, 1993; Smith et al., 2008) and autofluorescent age pigment accumulation throughout the body (Gerstbrein, Stamatas, Kollias, & Driscoll, 2005; Klass, 1977). Characterizing these differences, understanding the underlying molecular mechanism, and determining how these mechanisms result in age-related disease will be important in developing effective treatments.

Tissue-specific age-related DNA methylation

One source of differential aging among tissues appears to be epigenetics. Studies in rats (Thompson et al., 2010), mice (Maegawa et al., 2010), and humans (Day et al., 2013; Levine et al., 2018) have identified tissue-specific patterns of age-related DNA methylation. In mammals, DNA methylation occurs mostly within the context of CpG dinucleotides. Clusters of CpG dinucleotides—CpG islands—are often located near the 5′ end of genes. Methylation of CpG islands is associated with a closed chromatin structure and transcriptional silencing of the gene. Aging leads to epigenetic dysregulation, with different sites showing either hypermethylation or hypomethylation. This process is common across tissues for certain loci and tissue specific for others. Centenarians display delayed age-related methylation changes and can even pass the preservation of methylation states on to their offspring in a manner independent of genotype (Gentilini et al., 2013). How the observed dysregulation of methylation patterns impacts gene expression, activity in aging-relevant pathways, or progression of specific age-related diseases has yet to be fully elucidated. Which loci are controlling differences in age-related methylation and how variation in these loci can lead to differences in age-related phenotypes is also unknown. The epigenetics of aging and the use of epigenetic signatures as "clocks" (see the section on "Aging Biomarkers" later in this chapter) that measure biological age are an area of substantial research effort and the topic of recent detailed reviews (Ecker & Beck, 2019; Gabbianelli & Malavolta, 2018; Guevara & Lawler, 2018; Horvath & Raj, 2018; Wagner, 2019).

Tissue-specific responses of aging pathways

Distinct tissues experience different changes in function with age, respond differently to different forms and combinations of stress, and have differential access to endogenous and exogenous factors in circulation. Understanding how tissue-specific function and the interaction between tissues changes during aging will be critical to developing efficacious interventions to slow aging at the level of the organism. Two changes in activity through two of the central aging pathways—IIS and mTOR signaling—have dramatically different effects in different tissues. Knocking out *RAPTOR*, a component of mTOR complex 1 (mTORC1), in adipose tissue shows increased leanness and resistance to diet-induced obesity accompanied by improved glucose tolerance and insulin sensitivity (Polak et al., 2008). In contrast, knocking out *RAPTOR* in skeletal muscle leads to muscular dystrophy associated with reduced mitochondrial biogenesis and muscle oxidative capacity (Bentzinger et al., 2008). These kinds of differences are also found for ER stress and autophagy where variation leads to different outcomes depending on the tissue. Fat-specific insulin receptor knockout (FIRKO) mice live ~18% longer than wild-type mice and are protected against obesity and related glucose intolerance (Blüher et al., 2002). In contrast, pancreatic β-cell knockout (βIRKO) mice have an insulin secretion defect resembling type II diabetes (Kulkarni et al., 1999) while muscle-specific insulin receptor knockout (MIRKO) recapitulates some features of diabetes (increased fat mass, serum triglycerides, and free fatty acids) but is normal with respect to others (blood glucose, insulin, and glucose tolerance) (Brüning et al., 1998).

Many of the processes that play a primary role in aging are core regulators of cellular growth and metabolism in response to environmental queues. These examples highlight several important challenges both in developing a comprehensive understanding of the role that known aging genes play in the aging process and in translating basic aging discoveries into clinical applications. An intervention that increases lifespan may provide an overall benefit while driving disease processes in specific tissues. Conversely, a beneficial impact on aging in one tissue may be masked by a detrimental impact on another when an intervention is applied to the whole organism, resulting in a net increase in mortality. Factors with this type of outcome would not expect to be identified in the type of genetic screens or gene mapping studies described above. Clinical treatment may require altering the intervention agent or delivery method to target a subset of tissues or developing a combined therapy with a second intervention to counteract the negative impacts of the first. Interventions that target the fundamental aspects of aging—for instance, using senolytic drugs to clear accumulated senescent cells (see other chapters)—are efficacious in only a subset of tissue types and will likely need to be used in combination to achieve benefit in all tissues.

Gene–environment interaction

This chapter so far has discussed organism-intrinsic factors contributing to aging. Extrinsic factors—and the interaction between extrinsic and intrinsic factors—must also be considered when developing a complete model of organismal aging. Extrinsic factors that impact aging include any aspect of the environment that an organism interacts with: diet, weather, temperature, microflora, other members of the same species, chemical stressors such as external reactive oxygen and nitrogen species, etc. Aging research has led to the discovery of numerous genetic, pharmacological, and environmental interventions that increase lifespan in one or more model systems in the laboratory; however, interventions capable of extending lifespan in the laboratory setting may turn out to be sensitive to context. Understanding how the interaction between an organism's genetic—and epigenetic—identity interacts with its environment will be important to understanding how longevity intervention studies in the lab will translate into clinical application in human populations.

Genetic response to dietary restriction

Several methods of environmental manipulation are known to impact lifespan. Dietary restriction, typically defined as a reduction in dietary intake without malnutrition, is the most studied environmental intervention capable of increasing lifespan. Dietary restriction is commonly put forward as the most consistent means of increasing longevity; indeed reducing dietary intake has been shown to extend lifespan in most organisms where it has been attempted, including yeast, worms, fruit flies, spiders, mice, rats, hamsters, dogs, and rhesus monkeys (Colman et al., 2014; Kennedy, Steffen, & Kaeberlein, 2007; Masoro, 2005; Weindruch & Walford, 1988). The complete picture is more complicated. The outcome of dietary restriction depends on a range of factors including degree of restriction, the specific composition of both baseline and restricted diets, feeding schedule, age at onset, and the genetic identity of the subject population.

A large body of research is directed at understanding the biological processes underlying the beneficial effects of dietary restriction. A number of genetic pathways have been proposed as mediators of dietary restriction. The evidence is most consistent for mTOR signaling, a highly conserved nutrient-responsive signaling pathway that regulates many cell growth and proliferation processes. Lifespan extension by dietary restriction and reduced mTOR signaling are nonadditive in yeast, worms, and flies. Dietary restriction both reduces activity through the mTOR signaling pathway and induces mTOR-mediated changes to metabolism and protein homeostasis. Together, the accumulating invertebrate evidence places dietary restriction and mTOR signaling into at least partially overlapping pathways. In mice, reducing mTOR signaling through genetic manipulation of multiple factors increases lifespan (Lamming et al., 2012; Selman et al., 2009; Wu et al., 2013). In multiple studies using inbred and outbred populations rapamycin increases lifespan, reduces cancer incidence, and improves tissue-specific decline with age (Anisimov et al., 2011; Harrison et al., 2009; Miller et al., 2011; Neff et al., 2013; Wilkinson et al., 2012; Zhang et al., 2014). Despite similar outcomes in mice, a direct link between dietary restriction and mTOR signaling has yet to be established in a mammalian system. One goal of identifying the genes responsible for the beneficial response to dietary restriction is the development of dietary restriction mimetics, pharmacological agents capable of reproducing the beneficial outcomes without altering diet. Rapamycin and other pharmacological inhibitors of mTOR signaling (termed "rapalogs") are being pursued in basic aging research as well as clinical trials for multiple age-related diseases. Understanding the molecular impacts of mTOR signaling and the development of rapalogs are highly active areas of current investigation in aging science (Johnson et al., 2013; Kaeberlein, 2013; Lamming et al., 2013).

The second pathway that has been studied extensively with respect to the interaction with dietary restriction is IIS. In the majority of studies in *C. elegans* and *Drosophila*, epistasis analysis appears to place IIS into a pathway at least partially distinct from dietary restriction (Giannakou et al., 2008; Houthoofd et al., 2003; Kaeberlein, 2006; Lakowski & Hekimi, 1998; Lee et al., 2006); however, other studies do report interaction under specific forms of dietary restriction or genetic interventions (Clancy et al., 2002; Greer et al., 2007; Iser & Wolkow, 2007). Lifespan of the long-lived growth hormone receptor knockout (GHRKO) mice, which have reduced levels of insulin and IGF-1, is not further increased by dietary restriction (Al-Regaiey et al., 2007; Bonkowski et al., 2006). In contrast, dietary restriction is capable of increasing lifespan of mice with pituitary mutations that cause a defect secretion of several hormones including growth factor (Bartke et al., 2001). The combined evidence indicates that dietary restriction is not dependent on IIS to impact lifespan, but that the two pathways interact, perhaps by influencing overlapping sets of downstream factors or through compensatory feedback mechanisms. The genetic targets of dietary restriction are the topic of many review articles (Anderson & Weindruch, 2010; Fontana et al., 2010; Kennedy et al., 2007; Lee & Longo, 2016; Schleit et al., 2012; Swindell, 2012; Xiang & He, 2011).

A central question in dietary restriction research is whether the positive effects observed in genetically homogeneous laboratory populations will be widely realized if dietary restriction is applied to a genetically diverse population, such as humans. Several studies suggest that the genetic makeup of an individual can profoundly change the impact of dietary restriction on lifespan and healthspan. Dozens of dietary restriction experiments have been carried out in mice using a variety of strain backgrounds (reviewed in depth by Swindell, 2012). While lifespan extension is observed in most cases, there are numerous examples—most studies using DBA/2 male mice for instance—where dietary restriction appears to have no effect or even shortens lifespan. Two studies have examined the impact of dietary restriction on lifespan in ILSXISS mice, a set of recombinant inbred strains derived from eight inbred founders designed for variation in sensitivity to ethanol sedation (Williams et al., 2004). In both studies dietary-restricted mice were fed a diet with 40% fewer calories than strain-matched ad libitum-fed mice (Liao et al., 2010; Rikke et al., 2010). The results are striking: roughly half of the strains showed no significant change in lifespan, 5%–25% displayed increased lifespan, and 20%–25% displayed decreased lifespan in response to dietary restriction. In yeast, a screen of 166 single-gene deletion strains found a similar variability in response to dietary restriction ranging from a 79% decrease to a 103% increase in mean replicative lifespan (Schleit et al., 2013). A meta-analysis of reported mouse dietary restriction longevity studies found that dietary restriction may be more beneficial in outbred populations than inbred strains (Swindell, 2012), suggesting that inbreeding depression may limit the effectiveness of dietary restriction. Confounding this possibility, dietary restriction failed to increase lifespan in male wild-derived mice (grand-offspring of wild-caught mice) which were never inbred (Harper et al., 2006). Dietary restriction appeared to cause an increase in early life mortality and a decrease in late-life mortality in this population however, resulting in a significant increase in maximum lifespan despite the lack of mean lifespan extension. Taken together, these studies point to a strong genotype dependence in the individual response to dietary restriction and promote further investigation of dietary restriction in genetically heterogeneous populations.

Dietary restriction: quantity, composition, and timing

Each dietary restriction study uses a defined change in diet, usually a specific percent reduction in food, but what range of interventions does dietary restriction encompass? An earlier term—caloric restriction—reflected the idea that net caloric intake was the primary factor causing increased longevity and other health benefits. From this perspective, dietary restriction can be seen as a shift on a dietary dose–response curve and the genotype-dependent dietary response can be interpreted as some individuals being closer than others to the optimal dietary intake with respect to longevity when fed ad libitum due to genetically defined traits such as metabolic rate or appetite. In individuals for which ad libitum is near optimum, a typical dietary restriction regimen may result in malnutrition and thus be detrimental to longevity. Even this view represents a simplified model of dietary restriction that does not take into account the myriad of components that make up a diet even in the most basic organism. The current conception of dietary restriction includes a range of dietary changes from quantity to composition to temporal consumption patterns, and the range of biological processes underlying the response appears to vary according to the specific form of dietary restriction employed.

In *C. elegans*, multiple forms of dietary restriction are commonly used (Greer & Brunet, 2009; Mair et al., 2009). Most methods involve reducing the amount of bacterial food provided to the worms, but differ in the amount of bacteria provided. Other variables include whether the growth environment is liquid or solid agar-based, whether the food is alive or killed, and whether the amount of food is constant or varied (akin to feeding/fasting cycles in mammals) (Greer & Brunet, 2009). Under some conditions an extreme form of dietary restriction called bacterial deprivation, referring to the complete removal of the food during adulthood, results in optimal lifespan extension (Kaeberlein, 2006; Lee et al., 2006). Shifting a population to dietary restriction can result in early mortality in a fraction of individuals depending on whether the shift is gradual and the age at onset, and these factors can also vary from study to study. In the case of bacterial deprivation, similar median lifespan extension can be achieved for dietary restriction initiated between days 4 and 14 and similar maximal lifespan extension can be achieved for dietary restriction initiated as late as day 24 (Smith et al., 2008), while starting dietary restriction earlier than day 4 often results in early mortality in a subset of worms. A common genetic model of dietary restriction in worms is loss of function mutations in the *eat-2* gene, which results in defective pharyngeal pumping and thus reduced bacterial food consumption. An additional layer complicating the impact of dietary restriction on lifespan is food sensing. Smith et al. (2008) found that simply the ability to detect food in the local environment was sufficient to partially suppress the increased lifespan from bacterial deprivation.

In yeast, dietary restriction typically refers to reducing the media glucose concentration from 2% to 0.5% or lower (Lin et al., 2000), with optimal lifespan extension for the yeast ORF deletion collection strain background achieved at 0.05% glucose (Kaeberlein et al., 2004; Lin et al., 2000). Less common forms of dietary restriction include restricting amino acids (Jiang et al., 2000) or replacing media glucose with a nonfermentable carbon source such as glycerol, ethanol, or raffinose (Delaney et al., 2011; Kirchman & Botta, 2007). Deletion of *HXK2*, which encodes a hexokinase responsible for converting glucose into glucose-6-phosphate prior to entry into the glycolytic pathway, extends replicative lifespan (Lin et al., 2000) and has been used as a genetic model of dietary restriction (Walsh et al., 1983), though it remains unclear whether this is attributable to reduced cellular hexokinase activity (Rodríguez et al., 2001; Walsh et al., 1991).

Drosophila has been used extensively to explore nongenetic manipulations that extend lifespan and the *Drosophila* aging community has perhaps done the most to characterize the impact of dietary composition on lifespan. Flies are typically fed ad libitum on a glucose—yeast or cornmeal—yeast—sugar mixture in agar and dietary restriction accomplished by diluting one or more of the component ingredients (Bass et al., 2007; Chapman & Partridge, 1996; Good & Tatar, 2001). A standard dietary restriction experiment will examine multiple dilutions of the selected dietary component, rather than only comparing two nutrition points, thus examining the impact of the intervention under scrutiny across an entire dietary dose response. One study mapped the impact on longevity and other metabolic parameters of diets with a range of different yeast and glucose concentrations, finding that a diet with a balance of yeast and sugar results in optimal lifespan while a high-yeast diet improves fecundity (Skorupa et al., 2008). Like worms, food sensing can modulate fly longevity independent of food consumption (Libert & Pletcher, 2007).

The most common method for dietary restriction in mice is to restrict access to food based on the consumption observed in ad libitum fed control, with typical restriction levels ranging from 60% to 70% of ad libitum. An alternative method is to not limit quantity, but instead restrict the time that animals have access to food (intermittent feeding). Providing access to food only every other day has been shown to increase lifespan in both rats (Goodrick et al., 1982; Goodrick et al., 1983a, b) and mice (Goodrick et al., 1990). In mice, the extension only occurred when started in young adult mice and the degree of extension was greater in C57BL/6J and C57BL/6JxA/J F1 hybrid mice than in A/J mice (Goodrick et al., 1990). Intermittent feeding also reduced serum glucose and insulin and increase neuronal stress resistance in mice, despite a similar net caloric intake to continuously ad libitum fed mice (Anson et al., 2003). Simply, restricting the dietary methionine is sufficient to increase lifespan in both mice and rats (Miller et al., 2005). A similar outcome results from restricting tryptophan in rats (Ooka et al., 1988; Segall and Timiras, 1976; Timiras et al., 1984), although these results may simply point to an overabundance of methionine and tryptophan in the laboratory diet. Indeed, Masoro et al. (1989) found that methionine restriction did not contribute to the lifespan extension resulting from a 40% decrease in overall food intake in rats, suggesting that specific dietary components may impact phenotypes via distinct biological processes. While the baseline diet is typically standardized within an institution, and even between institutions when a common supplier and product are used, published studies use a range of diets that vary in the source and quantity of fat, protein, carbohydrate, and supplemental vitamins. A number of mouse studies highlight the interaction between genotype and dietary composition based on these differences. In 22 studies examining the impact of dietary restriction on C57BL/6 mice, the observed change in median lifespan ranges from a 32.8% decrease to a 26.8% increase (Swindell, 2012). C57BL/6 mice are susceptible to atherosclerosis and obesity and, at least in some instances, the failure of dietary restriction to extend lifespan may result from a lower fat content in the chow (Cheney et al., 1980; Harrison & Archer, 1987; Swindell, 2012; Turturro et al., 1999). Wild-derived mice are smaller but eat more as a percentage of body weight than inbred laboratory strains, suggesting a faster metabolism that may account for the less favorable response to dietary restriction (Austad & Kristan, 2003). In a study analogous to that examining *Drosophila* dietary composition by Skorupa et al. (2008), Solon-Biet et al. (2014) fed mice a range of diets that varied not only in total caloric content, but also in the balance of protein, fat, and carbohydrate, and measured lifespan and other metabolic parameters. They found that the dietary protein and carbohydrate content, and not total calories or fat content, were the primary determinants of lifespan. In particular, high-carbohydrate low-protein diets maximized lifespan. A detailed review of aging studies examining dietary composition was recently published by Le Couteur et al. (2016).

Different forms of dietary restriction interact with different genetic pathways. As mentioned in the discussion of IIS, bacterial deprivation is capable of increasing lifespan independently from the insulin receptor *daf-2* or the FOXO family transcription factor *daf-16*, placing it in a pathway parallel to IIS. sDR, a form of dietary restriction where worms are maintained on solid agar-plates in the presence of diluted bacteria, requires *daf-16* to increase lifespan, indicating that it acts, at least in part, through IIS. The precise molecular mechanisms through which dietary restriction acts to extend lifespan in yeast are not yet known; however, it is commonly thought that dietary restriction manipulates these mechanisms, at least in part, by influencing several partially redundant nutrient-responsive signaling kinases, including mTOR, cyclic AMP-dependent protein kinase (PKA), and Sch9. Mutants with reduced activity for any of these kinases have long replicative lifespans that cannot be further extended by dietary restriction (Fabrizio et al., 2004; Kaeberlein et al., 2005; Lin et al., 2000). Yeast PKA is an essential complex consisting of three catalytic subunits and regulated by two upstream sensing pathways, one involving RAS and the other a G protein-coupled receptor system. Two genes, *GPA2* and *GPR1*, encode subunits of the G protein-coupled receptor. Mutants lacking either *GPA2* or *GPR1* are replicatively long-lived relative to wild-type and are commonly used as models of reduced PKA activity (Lin et al., 2000). The third kinase, Sch9, shows sequence homology to Akt kinase, a component of IIS (Burgering & Coffer, 1995; Paradis & Ruvkun, 1998), but also functions as a ribosomal S6 kinase, a substrate of mTOR and regulator of translation in multicellular eukaryotes (Powers, 2007; Urban et al., 2007). While yeast does not possess a formal IIS pathway, Sch9 may fulfill an equivalent role in both Akt and S6 kinases in multicellular eukaryotes.

Other environmental factors that influence aging

A primary determinant of invertebrate lifespan is temperature. Culture temperature can dramatically impact lifespan in worms (Klass, 1977; Lakowski & Hekimi, 1996; Leiser et al., 2011, 2016; Mendenhall et al., 2017; Miller et al., 2017; Sutphin et al., 2012; Van Voorhies & Ward, 1999), flies (Helfand & Rogina, 2003; Loeb & Northrop, 1917; Miquel et al., 1976), and chronologically aging yeast (MacLean et al., 2001). In worms there are clear differential interactions between genotype and environmental temperature (Miller et al., 2017; Sutphin et al., 2017). Knockdown of the gene encoding the hypoxia inducible factor, *hif-1*, increases lifespan at 25°C but not at 15°C or 20°C, while stabilization of HIF-1 protein by knocking down the E3 ligase VHL-1 increases lifespan at all three temperatures, but to a much greater extent at 15°C (Leiser et al., 2011). Conversely, maintaining adult worms on plates containing up to 20 mM caffeine increases lifespan at both 15°C and 20°C, but not 25°C (Sutphin et al., 2012). Knockdown of *rsks-1*, encoding the ribosomal protein S6 kinase, has opposite effects on lifespan at 15°C and 25°C, while knockdown of *daf-2*, encoding the insulin/IGF-1-like receptor, extends lifespan to a similar degree across the temperature spectrum (Miller et al., 2017). Different genetic programs therefore have variable impact on longevity depending on environmental temperature.

Whether the temperature-responsive processes that strongly influence aging in poikilothermic invertebrate species will have clear analogs in homeothermic vertebrate species remains to be determined and the few mammalian studies that have directly examined the impact of temperature on longevity do not provide a clear answer (see the review by Keil et al., 2015). Rats maintained at 34°C lived significantly longer than rats maintained at 28°C; however, the high temperature also caused a reduction in food consumption, and simply restricting food to the same degree at 28°C resulted in an even greater increase in longevity (Kibler & Johnson, 1966). Conversely, rats maintained at 6°C (Heroux & Campbell, 1960) or 9°C (Johnson et al., 1963; Kibler & Johnson, 1961; Kibler et al., 1963) throughout adult life have substantially decreased lifespan, with the caveat that these experiments were not carried out under specific pathogen-free (SPF) conditions. Immersing rats in 23°C water to reduce body temperature for 4 hours/day, 5 days/week under SPF conditions causes a nonsignificant trend toward increased lifespan, and also results in a 44% increase in food consumption and a reduced incidence of cancer offset by a higher incidence of myocardial fibrosis (Holloszy & Smith, 1986). Maintaining mice at 10°C throughout life has no effect on lifespan relative to control mice maintained at 22°C (Vaanholt et al., 2009). Additional research will be needed to determine whether animals that actively maintain body temperature will respond similarly to environmental changes in temperature as invertebrate models. The answer will likely involve a mix of responses that are both shared (e.g., the activation of heat-shock proteins and other temperature-stress response machinery) and unshared (e.g., temperature-dependent changes in the rate of catalytic process in poikilotherms).

The molecular response of organisms to a variety of environmental stressors can be leveraged to extend lifespan. Heat-shock—the short-term exposure to high temperature—increases lifespan in worms (Cypser & Johnson, 2002; Lithgow et al., 1995; Olsen et al., 2006), flies (Hercus et al., 2003; Le Bourg et al., 2004), and chronologically aging yeast (Harris et al., 2001), and replicatively aging yeast (Shama et al., 1998) through hermetic activation of heat-shock factors. A similar outcome can be achieved by short-term exposure to cold, oxidative stress, hypergravity, or radiation (Cypser & Johnson, 2002; Le Bourg, 2007; Le Bourg & Minois, 1997; Vaiserman et al., 2003).

Flies generally have a strong inverse correlation between reproduction and longevity. Strains bred for longevity by selecting offspring from late life reproduction show reduced egg laying early in life relative to ancestral strains (Luckinbill et al., 1984; Rose, 1984). Preventing mating can also double female lifespan (Smith, 1958), though in *D. melanogaster* seminal factors have been implicated in shortening female lifespan as opposed to some intrinsic cost associated with reproduction (Ueyama & Fuyama, 2003).

Emerging tools for studying complex genetic traits

The maturing field of genomics and its derivatives—epigenomics, transcriptomics, proteomics, metabolomics—in combination with genome-scale lifespan screens and gene-mapping techniques allow us to begin to glimpse in detail the changes associated with complex phenotypes like longevity at the level of a whole organism. As these fields continue to advance, new tools are being developed to help assemble the vast amounts of information generated into a form amenable to identification of novel genes and key regulatory junctions in the network of factors that integrate to determine longevity and the processes underlying age-associated disease. Leveraging these tools will inform future studies and pinpoint targets for clinical intervention. The following sections outline emerging tools that are allowing us to encompass the range of factors that contribute to age-associated phenotypes and define their interaction.

High-throughput life span assays in yeast and worms

S. cerevisiae and *C. elegans* are the primary models used in genome-scale reverse genetic screens to directly identify genes capable of influencing lifespan, due mainly to the combination of short lifespans, ease of manipulation in the laboratory setting, and the ability to systematically target individual genes in a high-throughput manner. With the exception of yeast chronological lifespan, experimental techniques for determining lifespan are labor-intensive and studies published in the past few years have established the first proof-of-principle demonstrations for methods and technology capable of true high-throughput analysis of worm lifespan and yeast replicative lifespan.

Two studies were published in 2013 for automated, high-throughput life analysis systems in *C. elegans*. WormFarm employs an integrated microfluidic platform to maintain worms in liquid culture in 30–50 worm groups using a customized transparent eight-chamber chip (Xian et al., 2013). The microfluidics chip is designed to contain adult animals while refreshing the bacterial food supply and automatically removing larva, which can be collected for further analysis. Photographs and video taken periodically throughout the experiment allow for automated collection of phenotypic data. In addition to lifespan, the transparent WormFarm chips allow automated age-dependent characterization of body size and shape, motility, development rate, and brood size, as well as quantification of fluorescent markers when combined with fluorescence microscopy. After the initial setup, the prototype WormFarm was capable of determining lifespan for ~320 worms per chip without manual intervention. Xian et al. (2013) demonstrated the experiment-to-experiment consistency and agreement with published data when WormFarm was applied to a variety of common survival assays, including comparison of mutant strains, treatment with RNAi-expressing bacteria, stress response, and changes in media composition. In the strains and under the conditions examined, WormFarm produced similar survival data to manual measurement on solid NGM-agar plates, though the worms cultured in WormFarm tended to be physically smaller.

The second system is an automated solid NGM plate-based platform called the *C. elegans* Lifespan Machine (Stroustrup et al., 2013). In the Lifespan Machine worms are cultured on solid NGM-agar plates that are maintained on a series of modified flatbed scanners. Scanners capture periodic images of worms on each plate which are processed in a custom software package to identify worms and assess viability based on movement over time. Stroustrup et al. (2013) demonstrate the capability of the Lifespan Machine to generate reproducible survival data that are in agreement with published studies when applied to mutant strains, worms treated with RNAi-expressing bacteria, and stress response. In addition to survival, the Lifespan Machine is capable of characterizing body size, body shape, and motility. The throughput of this system is ~560 worms per scanner. The Lifespan Machine has been adopted as the lifespan platform for the CITP (Lucanic et al., 2017).

In 2017, Churgin et al. (2017) presented a third system called the WorMotel. The WorMotel features a custom-molded polydimethylsiloxane (PDMS) plate that consists of a set of 240 wells laid out on a grid with the same spacing as a standard 384-well plate that allows culturing of individual *C. elegans* on solid media. Each well effectively acts as a single-worm Petri plate surrounded by a moat filled with an aversive agent (e.g., copper sulfate) to keep the animals in the well. The plate is paired with a camera system that takes images and video of the worm at set intervals, allowing for automated monitoring of activity (to assess healthspan) and longevity determination. While the initial report did not report high-throughput data, it did include a system design that paired a plate-handling robot with the WorMotel camera system to allow the potential for high-throughput monitoring of a large number of WorMotel plates.

The WormFarm, Lifespan Machine, and WorMotel demonstrate the potential for high-throughput, automated analysis of worm lifespan, and both provide a platform for measuring lifespan, body dimensions, and motility with little manual labor after initial setup. The primary advantage to the Lifespan Machine and WorMotel over WormFarm is the ability to culture worms on NGM-agar, allowing direct comparison to the majority of published worm studies. WormFarm also necessarily excludes strains that are intolerant of liquid culture. Despite this limitation, WormFarm offers a greater range of automated phenotyping, including quantification of brood size and development, monitoring of fluorescent markers, and collection of larvae for further analysis. WormFarm also provides precise control over the entire environment experienced by the worm, including temporal control over media composition, bacterial food concentration, and temperature. The ability to wash away larvae also alleviates the need to use chemical or genetic means to prevent reproduction, thus removing a possible confounding factor. The primary advantage to the WorMotel system is the capability to monitor individual worms over the course of their lifespan and correlate longevity to individual phenotypes collected earlier in life. Each system is, in principle, scalable to any desired size. The current conception of the Lifespan Machine is more limited in this regard, as it requires the plates to be left in place throughout analysis, thus necessitating a new scanner for each set of ~560 worms to be run in parallel; however, this should be remediable in future versions by using smaller and/or mobile optics or plate handling. A common challenge among these systems is the storage and analysis of large quantities of imaging data. Aside from the development of more efficient analysis or advances in storage and processing technology, there is not a clear direction for escaping this problem. Comparing the automated survival analysis data, the Lifespan Machine and WorMotel appear to produce much smoother survival curves than WormFarm. This may result partially from more closely spaced time points, but is likely also a result of the former systems providing a more precise determination of individual time of death.

In yeast, new approaches to automated quantification of both chronological and replicative lifespan are being developed. For chronological lifespan, Belak et al. (2018) used the reduction of 3-(4,5-dimethylthiazol-2-yl)-2,5-diphenyltetrazolium bromide (MTT; yellow in color) to formazan (purple in color), which only occurs in viable *S. cerevisiae* (Hodgson et al., 1994), to quantify the viable fraction in chronologically aging yeast, allowing rapid and high-throughput quantification of chronological lifespan. This method successfully recapitulated lifespan changes resulting from glucose restriction, several short- and long-lived deletion mutants, and several pro-longevity drugs (Belak et al., 2018).

In the replicative paradigm, Lindstrom and Gottschling (2009) constructed a genetic platform called the Mother Enrichment Program (MEP) to selectively enrich yeast colonies for replicatively aged mother cells. In this system, Cre-lox recombination is used to specifically disrupt two essential genes, *UBC9* and *CDC20*, in newly formed daughter cells, thus preventing cell division in the daughter cells without altering the replicative capacity of the mother. In MEP cultures, daughter cells are produced but do not continue dividing, resulting in a decaying population growth and an enrichment for replicatively aged mother cells over time relative to non-MEP cultures. This enrichment allows for single-step affinity purification of much older cell populations for biochemical and gene-expression studies than earlier size-separation techniques. In addition to applications in young versus old comparisons, the MEP has been proposed as a high-throughput method to measure replicative lifespan. The viability of MEP cultures is determined specifically by the replicative lifespan of the mother cells. The use of a plate reader system, such as that used to measure chronological lifespan (Murakami et al., 2008) can allow rapid, high-throughput measurement of replicative lifespan for strains carrying the MEP biological machinery in liquid media. Since the MEP requires multiple genetic modifications, this technique will not likely be useful for large-scale genetic screens. There is promise for the application of MEP to replicative lifespan drug screening, where current labor-intensive methods for measuring replicative lifespan limit the number of drugs that can be easily tested, and where further genetic manipulation of the strains is not necessary.

In an approach that resembles the chemical cell labeling approached used by Belak et al. (2018) to measure chronological lifespan, Sarnoski et al. (2017) developed the high-throughput replicative lifespan measurement (High-Life) system, which uses fluorescently labeled yeast in combination with flow cytometry to rapidly measure replicative age. High-Life yeasts are labeled with three fluorescent markers: (1) fluorescein conjugated N-hydroxysuccinimide-ester (NHS-ester), which is asymmetrically segregated to daughter cells and allows selection of progenitor cells; (2) the viability dye propidium iodide to identify viable cells; and (3) the protein lectin wheat germ agglutinin conjugated to the blue fluorophore CF405M, which selectively labels bud-scars and provides a quantitative measure of the replicative age of viable cells. Sarnoski et al. (2017) used High-Life to screen 2640 compounds for increased replicative lifespan and identified terreic acid and mycophenolic acid as novel pro-longevity compounds. This initial application of High-Life was carried out in 384-well plates and used the MEP strain background to prevent the yeast from exhausting the limited volume of growth media (MEP yeast exhibit linear, rather than exponential population expansion). Because of the complicated genetic background of MEP yeast, this version of High-Life will be useful for drug screening but will be of limited use for genetic screening. In principle, the High-Life methodology can be employed with any strain background using alternative strategies to replenish exhausted media.

The most prominent trend in yeast aging is the emergence of microfluidic systems to replace the time- and labor-intensive practice of manual microdissection of daughter cells to measure yeast replicative lifespan (Crane et al., 2014; Fehrmann et al., 2013; Huberts et al., 2013a, b, 2014; Jo et al., 2015; Lee et al., 2012; Liu et al., 2015; Meitinger et al., 2014; Rafelski et al., 2012; Spivey et al., 2014; Xie et al., 2012, 2015; Zhang et al., 2012; recently reviewed by Chen et al., 2017). In general, these systems use microfluidic devices to chemically and/or mechanically trap mother cells. The geometry of the device is designed to selectively trap mother cells based on size or budding pattern. Flowing media through the chamber replenishes nutrients while flushing away newly divided daughter cells. A dissection scope with an automated stage combined with time-lapse microscopy is used to monitor cells and count divisions. Several studies demonstrate the capability of these systems to automatically measure replicative lifespan and reproduce published results while simultaneously monitoring numerous other phenotypes over the course of the replicative life of each cell, including cell morphology, number of bud scars, individual cell division times, and replicative age-dependent changes in fluorescent marker expression (Chen et al., 2017). As with the worm concepts, these microfluidics-based systems are scalable, but require a dedicated dissection microscope and some level of competency with microfluidics systems, which currently limits widespread adoption. Microfluidic solutions to high-throughput yeast replicative aging is still in its infancy and technical challenges remain. One example is the dependence of these systems on size differentiation between mother and daughter cells, which is achieved by the precise geometry of the microstructures used to capture cells. Early in replicative lifespan, optimizing the size of the structures is challenging because mother and daughter cells are similar in size. Later in replicative lifespan, the increased size of the mother cell causes the cell surface to press against the microstructures, which may have unexpected effects on cellular aging. All systems to date have focused on the use of haploid cells. Distinct microstructures may be needed for yeast cells with different baseline size resulting from ploidy (e.g., diploid cells are larger than haploid cells), genetic manipulation, pharmacology, or the use of different yeast strains. Finally, the microfluidic parameters and pretreatment conditions used in these experiments—flow speed, culture growth phase, sonication—have yet to be optimized. At present, these technical challenges appear surmountable and the coming years are likely to see many or all resolved.

These recent platforms demonstrate that automated survival analysis is possible in both yeast and worms. The systems are largely touted as moderate throughput in their current configurations, primarily because the platforms require a large amount of physical space and the experimental setup still takes a fair amount of manual labor to prepare plates and culture animals. The final survival analysis is also computationally intensive and still requires some human input. Despite current limitations, the primary challenge of automated lifespan determination has been met and the remaining technical problems should be solvable in subsequent design iterations. The application of high-throughput lifespan analysis has the potential to bring invertebrate aging studies to a new level of detail. On a first order, the expansion of genome-scale longevity screens with relatively large population sizes and full survival curves will allow a more complete picture of the range of genetic factors involved in aging and a more detailed understanding of their impact on survival. Expanding these screens to compare different genetic backgrounds and environmental conditions combined with network analyses will progress our understanding of how pathways involved in aging interact. The simultaneous measurement of healthspan markers (e.g., motility in worms) and fluorescent expression markers will provide additional detail about how age-dependent changes in gene expression are influencing lifespan and age-associated decline.

Genome-scale mouse knockout collection

Inspired by the complete sequencing of the mouse genome in 2002 and motivated by the potential power of readily available knockout mouse strains for any desired gene as a genetic tool for studying virtually any aspect of mammalian biology, the International Knockout Mouse Consortium (IKMC) was formed in 2007 and set out to generate single-gene knockout mouse lines in the C57BL/6 background for each of the ~21,000 genes in the mouse genome. The IKMC consisted of an international collaboration between four programs: The Knockout Mouse Project (KOMP; www.komp.org), European Conditional Mouse Mutagenesis Program (EUCOMM; www.eucomm.org), North American Conditional Mouse Mutagenesis Project (NorCOMM; www.norcomm.org), and Texas A&M Institute for Genomic Medicine (TIGM; www.tigm.org). The first phase of the project, generating single-gene embryonic stem (ES) cell knockout lines for every gene in the mouse genome, is now complete and both ES cell lines and germ cells or live mice for recovered strains mice are available to the broader research community for purchase. The major focus of this effort is now on the second phase, a collaborative effort under the auspices of the IMPC (www.mousephenotype.org), which aims to use the ES cell lines to generate and broadly phenotype knockout mice for each gene. The IMPC consists of 17 research institutions and five national funding programs. The goal is to extensively characterize each knockout mouse strain using a standardized set of physiological, behavioral, molecular, and biochemical assays and make the resulting dataset publicly available, thus providing a genome-scale resource for comparative genetics in the mouse.

The prospect of determining lifespan for every knockout mouse strain is both financially and logistically impractical and is not part of the IMPC phenotyping pipeline (the standard pipeline includes a battery of measurements out to approximately 16 weeks of age). Regardless, the IKMC and IMPC efforts will provide an invaluable resource to the aging research community. A broadly accessible, genome-wide knockout collection removes the first time-consuming hurdle to any knockout mouse aging study: generating a knockout line. The availability of a large array of standardized phenotype data will allow researchers to make preliminary decisions about which genes to pursue based on nonlongevity age-associated traits, either starting from a preliminary set of candidate genes or in de novo analyses generated directly from IMPC phenotype data.

Collaborative cross and diversity outbred mice

The advents of relatively inexpensive genome sequencing and high-density marker arrays now allow human GWAS to map genetic loci at a hitherto unprecedented precision. A critical characteristic of human populations that allows for this high-resolution mapping is the high degree of genetic diversity among individuals. This diversity is balanced by a lack of control in other potential confounding areas such as environment and population structure. Traditional model systems provide this control at the cost of genetic diversity. The collaborative cross (CC) and diversity outbred (DO) mouse populations are a recently developed pair of complementary tools designed to capture the advantages of both approaches—high genetic diversity to allow high-precision mapping while retaining control over environmental and population factors—within a single model system.

The CC and DO mice were bred from the same eight inbred founder strains, including five classical strains: A/J (AJ), C57BL/6J (B6), 129S1/SvImJ (129), NOD/ShiLtJ (NOD), and NZO/HlLtJ (NZO); and three wild-derived strains representing different *Mus musculus* subspecies: CAST/EiJ (CAST), PWK/PhJ (PWK), and WSB/EiJ (WSB) (Churchill et al., 2012; Svenson et al., 2012). More than 36,000,000 SNPs are present in the founder strains, with ~75% of the genetic variation resulting from the inclusion of the three wild-derived strains (Keane et al., 2011; Yalcin et al., 2011).

The CC lines are generated using a two-phase funnel breeding strategy, starting with three rounds of outbreed to combine the eight found genomes followed by repeated sibling mating to eventually produce a series of inbred lines with a uniformly distributed genetic contribution from the parental genomes (Collaborative Cross Consortium, 2012). Each CC line is developed independently, meaning that recombination events are not shared among strains, a problem that has confounded association mapping studies using other RI strain panels (Manenti et al., 2009). At present, 69 CC lines are available with heterozygosity ranging from 0.5%–15% (average 8.04%) (Srivastava et al., 2017; UNC Systems Genetics, 2019). When complete, the CC panel will include an estimated 100–200 inbred lines.

The DO population was seeded using 144 partially inbred CC lines (Chesler et al., 2008) and is maintained by a random breeding strategy designed to minimize genetic drift and allele frequency selection (Churchill et al., 2012; Rockman and Kruglyak, 2008). The DO mice are currently in the 35th generation (G35). Even at G16, each individual was estimated to have approximately 500 informative recombination events per animal (Svenson et al., 2012) providing a mapping resolution well below 1 Mb, an order-of-magnitude improvement over classical genetic crosses which typically produce resolution in the 10–50 Mb range. Mapping in the DO mice also has advantages over using inbred strain panels or classical crosses in both degree and distribution of genetic variation (Rockman and Kruglyak, 2008). The DO random breeding strategy also avoids problems with confounding linkage disequilibrium resulting from historical points of common ancestry common among strains in other RI panels (Collaborative Cross Consortium, 2012; Petkov et al., 2005). Each locus in a DO mouse has 36 possible diplotypes resulting from the eight founder haplotypes. The Mouse Universal Genotyping Array (MUGA), a high-density SNP array containing 7854 pairs of allele-specific probes with an average spacing of 325 kb was used in genotyping the initial generations of the DO mice in order to capture the small haplotype blocks produced by repeated outbreeding (Collaborative Cross Consortium, 2012; Wang et al., 2012). By G8, even MUGA was missing an estimated 20% of recombination events. To address this problem MegaMUGA (containing 77,800 SNPs) and GigaMUGA (containing 143,000 SNPs) are now being employed to capture the increasingly fragmented diplotype structure in the latest DO generations (Morgan et al., 2015). Another heterozygous stock derived from CC strains (HS-CC) is available that uses circular mating to control genetic drift at the cost of map expansion (Iancu et al., 2010).

The large number of recombination events per individual, the ability to identify the originating parental strain for each haplotype, and the wide phenotypic variation make the DO mice a powerful platform for genetic mapping. The large number of possible diplotypes does make calculating LOD scores for QTL analysis computationally challenging. While a classic intercross can be analyzed by regression analysis with three free parameters (corresponding to the three possible diplotypes), applying the same strategy to the 36 diplotype structure of the DO genome results in computational instability and results in reduced power to detect a QTL. A simplified approach has been developed that regresses on an estimated dosage from each of the eight parental haplotypes. This approach assumes a combined phenotypic effect that is intermediate between the parental strains, thus neglecting more complex allelic interactions, but has proven successful in practice (Churchill et al., 2012; Logan et al., 2013; Svenson et al., 2012). Selection of sample size to achieve a reasonable power to detect QTL in DO mice is influenced by a large number of variables including population structure, variation in allele frequency, measurement variation, effect size, and background genetic effects. While this complexity makes a generalized rule for sample size difficult, a minimum of 200 mice for QTL with allele frequency and effect size similar to most intercross studies, and up to 1000 mice to detect QTL resulting from recessive alleles originating from a single of the founder strains or to detect loci that account for <5% of variation for a phenotype of interest, are suggested (Churchill et al., 2012; Gatti et al., 2014).

Publicly available analysis tools have been developed to make gene mapping accessible. R/qtl, an R package that provides an extensive set of statistical tools for QTL mapping in experimental crosses, is publicly available (Broman et al., 2003; R/qtl, n.d) along with a published guide book (Broman et al., 2009) and J/qtl, a Java-based graphical user interface (J/qtl, 2015; Smith et al., 2009). DOQTL, a new R package providing analysis tools for linkage and association mapping specifically in the DO mice and other multiparent advanced-generation intercross (MAGIC) populations, is now available for beta-testing (DOQTL, n.d; Gatti et al., 2014).

As a first-order indication of the phenotypic variation to expect from the CC and DO populations, the eight founder strains represent the range of variation in longevity and a range of age-associated phenotypes (Fig. 1.1; Svenson et al., 2007). The median lifespan for these strains ranges from 14 months for female NZO mice to 32 months for male WSB mice, while circulating IGF-1 levels at 6 months of age, which correlates with longevity, vary more than twofold from 201 ng/mL for male WSB mice to 449 ng/mL for female NZO mice (Fig. 1.1; Yuan et al., 2009). DO mice show wide-ranging variation across all measured phenotypes. New data from DO and CC studies are being collected at the Mouse Phenome Database (MPD) (Bogue et al., 2015; Mouse Phenome Database, 2019). In addition to looking at population effects to identify novel genes contributing to a particular phenotype, individual outliers in one or multiple related phenotypes of interest provide interesting case studies to look at combinatorial effects of alleles resulting in extreme phenotypic presentation. Alleles or combinations of alleles predicted to contribute to extreme phenotypes can then be

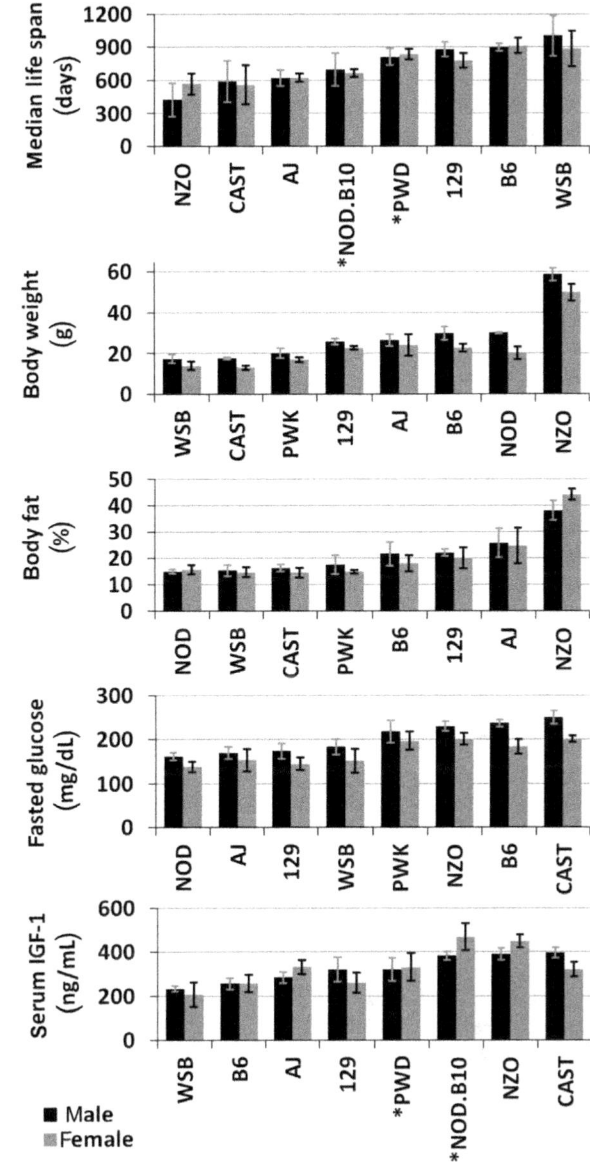

FIGURE 1.1 **Age-related phenotypes for CC and DO founder strains.** The founder strains for the CC and DO mice display a high strain-to-strain variation in lifespan, body weight (16 weeks old), percent body fat (16 weeks old), fasted circulating glucose (8 weeks old), and circulating IGF-1 (26 weeks old). 129, 129S1/SvImJ; AJ, A/J, B6 = C57BL/6J; CAST, CAST/EiJ; NOD; NOD/ShiLtJ; NOD.B10, NOD.B10Sn-$H2^b$/J; NZO, NZO/HlLtJ; PWD, PWD/PhJ; PWK, PWK/PhJ; WSB, WSB/EiJ. *PWK/PhJ was replaced with the closely related strain PWD/PhJ in this study; NOD/ShiLtJ was replaced with NOD.B10Sn-$H2^b$/J, a congenic strain in the NOD/ShiLtJ background with the $H2^b$ locus from C57BL/10SnJ, which confers resistance to diabetes. The data in this figure were compiled from the following studies: Leduc et al. (2010), Lenarcic, Svenson, Churchill, and Valdar (2012), and Yuan et al. (2009).

validated using selected CC lines or knockout strains from the IKMC collection.

The DO mice have already been used to map numerous parameters with known importance in aging including plasma cholesterol, plasma triglycerides, body weight, and body composition (Churchill et al., 2012; Svenson et al., 2012). An early study looking at plasma cholesterol provides an illustrative example of the power of DO mice for genetic mapping studies. The change in plasma cholesterol from 7 to 18 weeks of age was measured for a population of 91 G4 and G5 DO mice. A significant QTL was identified on chromosome 3 with a 2 Mb 2-LOD support interval containing only 11 genes (Svenson et al., 2012). An examination of allele effects revealed that haplotypes from 129, WSB, and NZO in this region are associated with decreased plasma cholesterol with age. Of more than 32,000 SNPs within the support interval, only seven fit the allele effect pattern, five of which are located upstream of the Foxo1 gene. More recent studies have looked at specific age-associated diseases in the DO mice, including atherosclerosis (Smallwood et al., 2014), metabolic disease (Tyler et al., 2017), and cancer (Winter et al., 2017). The ability to rapidly move from a significant QTL to one or a few strong candidate genes is one of the primary advantages that DO mice provide over traditional crosses.

The Jackson Laboratory Nathan Shock Center for Excellence in the Basic Biology of Aging is currently conducting a large collaborative aging study that will examine age-dependent changes in hundreds of phenotypes across the physiological, behavioral, and molecular spectrum—including longevity—in DO mice. The study consists of both longitudinal and cross-sectional cohorts. The longitudinal cohort will be used to examine maximum survival as the primary endpoint, with blood and urine samples collected periodically throughout life. The cross-sectional study consists of cohorts to be sacrificed at 6-month intervals through 18 months of age. Each cohort is subject to a diverse phenotyping pipeline prior to being sacrificed, with more than 15 tissues collected for analysis. While the Jackson Laboratory is leading the study and hosting the mice, investigators from a number of institutions will be involved in the phenotyping and analysis effort. When complete, this study will provide one of the richest sets of phenotypic data available to date within a single aging study.

Expression QTLs

The advent of increasingly accessible and comparatively inexpensive technology for genome-scale analysis of the molecular state of an organism, tissue, or cell type of interest has opened the door to study systems-level changes in an organism as they occur with age or in response to an intervention. Classical approaches tend to focus on specific molecules or pathways and have unquestioned value in understanding key factors contributing to longevity and age-associated disease and illuminating the mechanistic detail of the molecular processes linking these factors to phenotype. Viewed as a system, aging can be studied as a time-dependent change in the different classes of molecule within an organism (epigenetic, RNA, protein, metabolites, etc.). Determining the state of each class and developing models describing the interaction within and between classes and the response to external stimuli will lead to a more comprehensive understanding of the overall dynamics of age-specific disease processes that occur within each tissue type. The two approaches are inherently complementary, with systems biology identifying new lines of inquiry and placing the mechanistic detail of each pathway in the broader context of the aging organism, while classic molecular biology will fill in the mechanistic detail of the dynamics observed through genome-scale observation.

Emerging high-throughput technologies are allowing investigators to look at each class of molecule with ever greater resolution. These technologies include high-throughput sequencing with de novo assembly for individual genotyping, microarrays or RNA-seq for mRNA, ncRNA, or miRNA expression, chromatin immunoprecipitation combined with microarrays (ChIP-chip) or high-throughput sequencing (ChIP-seq) for examination of histone modification, bisulfide sequencing for DNA methylation, and mass spectrometry (MS) for protein or metabolite quantification. A menu of techniques available to examine interactions between molecules, such as fluorescence in situ hybridization (FISH) for DNA–DNA and RNA–RNA, yeast two hybrid for protein–protein, and yeast one hybrid for DNA–protein interaction, is also rapidly expanding as recently outlined in numerous reviews (Furey, 2012; Hou et al., 2012; König et al., 2012; Stynen et al., 2012). The application of systems-level techniques to aging is covered in detail in other chapters.

The combination of genome-scale expression analysis with gene mapping allows for the identification of expression QTLs (eQTLs), loci associated with variation in expression of one or more transcripts in a genetically diverse population. Identifying eQTLs in the context of expression changes with age or in response to a longevity intervention allows not only the identification of the direct expression changes responsible for the effects on aging, but also candidate loci for interacting partners involved in controlling the expression changes. With appropriate analysis, eQTLs are also a potentially powerful tool to narrow candidate genes identified in GWA studies, particularly when expression data are available in multiple tissues (reviewed by Montgomery & Dermitzakis, 2011).

A few studies have used eQTLs to address questions in aging research. In a recent study, Yang et al. (2013) use eQTLs in yeast to identify gene expression patterns that result from an epistatic interaction between different loci. The epistatic interaction map revealed that expression of a cluster of genes involved in oxidative phosphorylation was decreased in response to glucose, while expression of ribosomal proteins was enhanced. Silencing genes were found to be coregulated with known aging-linked genes when yeast were grown in the presence of ethanol. Using an RI panel in worms, Viñuela et al. (2010) found increased variation in gene expression with age and a decreased number of significant eQTL in aging worms relative to developing worms. In a later study, they found that regulatory loci display a higher degree of heritability in developing worms than in aging worms, and suggest that genetic control of gene expression becomes more polygenic with age (Viñuela et al., 2012). Wheeler et al. (2009) identified eQTL representing 101 genes significantly associated with age-associated gene expression changes in human kidneys. They used these 101 genes as candidates for association with glomerular filtration rate as a functional measure of kidney aging, and found a single gene with significant association, the matrix metalloproteinase 20 (*MMP20*). A study looking at sex- and age-interacting eQTL in human whole blood identified a significant age-associated change in the choline transporter-like gene *SLC44A4*, which has been previously associated with type 1 diabetes, vitiligo, age at menopause, leukocyte telomere length, and multiple sclerosis (Yao et al., 2014).

The use of eQTL analysis is growing across the spectrum of biological research and we anticipate an increasing number of studies in aging science. In principle, similar techniques can be applied to other genome-scale technologies to identify QTL associated with protein levels, metabolite abundance, or patterns of epigenetic modification.

Aging biomarkers

The range of individual lifespans experienced within a given population, even those with very little genetic variation, suggests that age-related decline is highly variable. The conceptual difference between chronological age—the time an individual has been alive—and biological age—a measure of the degree of age-associated decline that an individual has experienced—has led to a wide-ranging search for aging biomarkers, nonlongevity phenotypes indicative of the biological age of an individual (Kirkwood, 1998). Biological age is a measure of individual deviation from the chronological age with respect to mortality and is therefore necessarily a function of chronological age. The two should be indistinguishable when averaged across individuals within a population. An ideal biomarker should be able to predict individual age-specific mortality and age-associated pathology more accurately than chronological age alone and be measurable without impacting either outcome (Butler et al., 2004).

Despite a large effort to identify useful biomarkers of aging in both animal models and in humans (Johnson, 2006), no single biomarker has emerged as a clear indicator of biological age (Sprott, 2010). This is likely a reflection of the complexity of aging and the number and range of processes that ultimately contribute to mortality. Instead of using a single phenotype, combining a panel of biomarkers in a statistical model to predict biological age has potential to appropriately account for the complex nature of the aging process. This approach has two aspects: selecting an appropriate panel of biomarkers and developing a prediction model. Using a range of biomarker selection strategies, several studies have used prediction strategies based on multiple linear regression (Bae et al., 2008; Hollingsworth et al., 1965; Krøll & Saxtrup, 2000; Takeda et al., 1982), principal component analysis (Hofecker et al., 1980; Nakamura & Miyao, 2007; Nakamura et al., 1988), or the use of biomarkers to calculate deviation from chronological age (Klemera & Doubal, 2006). Levine (2013) compared these methods using a common set of human biomarkers from the National Health and Nutrition Examination Survey (NHANES III), a study population consisting of 9389 individuals ranging in age from 30–75 years, 1843 of whom were deceased. The biomarkers included 21 physiological and blood measurements selected based on availability within the selected dataset, lack of interdependence, previously identified role in aging or use as aging biomarkers, and strength of correlation with chronological age. The method proposed by Klemera and Doubal (2006), which uses a set of regression algorithms to predict individual deviation from chronological age, outperformed all of the other methods and chronological age alone in predicting mortality, particularly when combined with principal component analysis to select variables.

The selection of appropriate biomarkers is a separate challenge. One problem is that the relevant biomarkers may depend on the specific genetic background and age range of the population examined. Blood cholesterol may be an informative marker when considering a population genetically susceptible to cardiac disease and consuming a Western diet but not to a population without the genetic susceptibility consuming a Mediterranean diet. Similarly, measures of frailty are likely to be strong predictors of mortality in the elderly, but carry little or no information in young to middle-aged adults. Different biomarker panels may also be of use in predicting tissue-specific disease. Such panels will also be useful in determining whether aging interventions are generally improving health and lowering predicted mortality, or whether they are improving age-associated degradation in a specific subset of tissues and even causing differential effects in different tissues.

The identification of biomarkers is a clear application for transcriptomics, proteomics, and metabolomics. Analysis of unguided surveys of the RNA, protein, and metabolite abundance across a range of ages will provide a basis for unbiased selection of strongly age-correlated biomarkers from thousands of potential candidates. Harries et al. (2011) used microarrays to look for human age-associated changes in transcript levels using peripheral blood leukocyte samples from the InCHIANTI population. They identified a six-gene set (*LRRN3*, *CD27*, *GRAP*, *CCR6*, *VAMP5*, and *CD248*) for which expression was an accurate predictor of whether an individual was either younger than 65 years old or greater than 75 years of age. They recently validated this set by selecting individuals from the Exeter-10000 study who appeared biologically at least 8.5 years younger than their chronological age on based expression of five of the six markers (Holly et al., 2013). Compared to the general population, these biologically young individuals performed better on several other parameters associated with aging, including reduced IL-6 levels, improved muscle strength, reduced blood urea nitrogen levels, and increased blood albumin levels. Another recent study established a metabolic biomarker panel in human fasting serum samples from 6055 individuals in the United Kingdom (Menni et al., 2013). The panel consisted of 22 metabolites, the majority of which were fatty or amino acids. Different metabolites also correlated with different age-associated traits, but in all cases the correlation was improved by considering a linear regression model including all 22 metabolites, and a substantial improvement in correlation was observed for DHEAS, HDL cholesterol, and total cholesterol. Hertel et al. (2016) established a 59-metabolite human urine panel that successfully predicted patient follow-up survival and was robust to different methods of normalization.

Several studies have used metabolomics to identify biomarker panels with predictive power for biological age in mice (Calvani et al., 2014; Houtkooper et al., 2011; Tomás-Loba et al., 2013). Houtkooper et al. (2011) extensively characterized the metabolic status of 3-month-old and 22-month-old C57BL/6 N mice, including metabolomics in blood, liver, and muscle. The authors identified 10 blood metabolites or 20 muscle metabolites that were able to separate the age groups with 100% accuracy. The blood metabolite panel consisted of amino acids, acylcarnitines, and erythrocyte fatty acids, while the muscle panel was dominated by metabolites involved in polyunsaturated fatty acids and linolenic and linoleic acid metabolism, though glucose metabolism was also represented. The results in liver were less clear, with a 25-metabolite panel grouping mice by age with 93% accuracy and including components involved in glucose metabolism, phospholipid metabolism, and redox homeostasis. Calvani et al. (2014) identified 13 urine metabolites and 11 feces metabolites, each capable of 100% accurate discrimination between 3- and 16-month-old BALB/c mice using principal component analysis. Both of these studies provide a proof-of-principle that metabolite panels can be used to differentiate age groups in mice. The blood, urine, and feces are particularly attractive as samples that can be easily collected in a relatively noninvasive manner. Extrapolation of these results is limited by the use of inbred strains. It is unclear whether the metabolites identified will be useful in differentiating groups of aged mice in other inbred strains or in genetically diverse populations. Another recent study sought to develop a more general mouse metabolic signature by performing blood serum metabolomics on 117 mice from a range of inbred backgrounds with ages ranging from 2 to 30 months (Tomás-Loba et al., 2013). The authors selected 48 metabolites including phospholipids, fatty acids, and organic acids, to develop a predictive model using projection to latent structure. Mice expressing different levels of telomerase were used to validate the model. The model predicted that short-lived telomerase knockout ($TERT^{-/-}$) mice were biologically older as compared to chronologically age-matched wild-type mice. In contrast, long-lived mice overexpressing TERT and several tumor suppressor genes to counteract the carcinogenic effects of extra TERT (Sp53/Sp16/SARF/TgTERT) had a biologically younger metabolic signature compared to chronologically age-matched controls. Additional validation will be necessary to know if these metabolic biomarker panels will be generally useful as biomarkers of aging in mice. Since all three studies were conducted in inbred lines, a next step in developing a truly generalized mouse metabolic biomarker panel would be to a genetically outbred population with a high degree of genetic diversity, such as the DO mice. Measuring metabolites from multiple age points will also allow selection of subpanels that are useful for prediction across lifespan or within specific age ranges.

The use of epigenetic markers as predictors of biological age (commonly referred to as epigenetic "clocks") has become a major area of interest in the realm of aging biomarker research. Tools estimating biological age based on DNA methylation patterns have now been reported for whales (Polanowski et al., 2014), dogs (Thompson, 2017), mice (Cole et al., 2017; Han et al., 2018; Meer et al., 2018; Petkovich et al., 2017; Stubbs et al., 2017; Wang et al., 2017), and humans (Horvath, 2013; Levine et al., 2018a, 2018b). Other approaches employ panels of physiological parameters (Belsky et al., 2015; Levine, 2013; Sebastiani et al., 2017), proteomics (Menni et al., 2015), or glycans (Krištić et al., 2014).

The identification of meaningful biomarkers of aging has been a goal of the aging research community for decades and the discussion continues to be lively (Belsky et al., 2018; Hoffman et al., 2017; Jylhävä et al., 2017; Putin et al., 2016; Wagner et al., 2016; Xia et al., 2017). Recent advances in high-throughput and unbiased detection of RNA, protein, and metabolites holds promise to make truly diagnostic panels of aging biomarkers a reality. While a few studies have started to build these panels, the development of this type of approach has only just begun and the field is wide open to the application of RNA-seq, proteomics, and novel modeling techniques. In the past few years, the application of metabolomics and epigenetic techniques to biomarker discovery appears to have taken the greatest strides. Several next steps are apparent in this line of inquiry. First, different MS techniques can be used to identify different classes of metabolite. A comparison of different techniques to identify the classes that are most informative for aging and age-associated disease would help focus future efforts. Generalization of current techniques and cross-validation of identified metabolite panels using genetically heterogeneous mouse populations will allow for more general application of the biomarkers for use in other applications, such as initial testing of aging interventions without having to carry out a complete longevity study.

Conclusions

Aging is a complex process that integrates input from a wide range of organism-intrinsic and organism-extrinsic sources to produce an age-dependent increase in mortality and functional decline in virtually every tissue and organ system. In this chapter we have reviewed past and ongoing efforts to encompass the range of factors involved in aging and how they interact to influence longevity and other age-associated phenotypes. We further discussed the potential for emerging tools and technologies to contribute to our understanding and develop a more comprehensive picture of the broader aging process.

Many of the emerging tools have synergistic properties, and the prospects for combining multiple techniques to pursue novel directions of inquiry are particularly exciting. The DO, CC, and IKMC/IMPC mice provide a powerful platform for cross-validation. Discoveries made in the isogenic background of the IKMC can be verified in the genetically heterogeneous DO mice to determine whether observations are strain-specific or can be expected to be more broadly applicable across genotypes. Conversely, identification of genes through DO mapping studies can be confirmed and taken toward by a more focused genetic analysis using the available knockout mouse lines. These tools also allow rapid translation from novel discoveries made in invertebrate systems. The advent of high-throughput systems for measuring lifespan in yeast and worms will allow for large, internally consistent datasets for use in building interaction networks. The combination of the high-precision mapping capability using DO mice with "-omics" technologies will allow a much more detailed picture to be built of how genetic regulation impacts age-dependent changes in the various molecular species within cells, as we are beginning to see with eQTLs.

As aging science—and biological science generally—continues to move into the realm of systems-level analysis and the generation of big data, the technological problems of comprehensive measurement of the myriad factors contributing to complex traits will give way to challenges of data storage, analysis, and interpretation. The techniques developed in the past few years and those emerging today place us on that threshold, and we look forward to the innovation of the coming years.

References

Al-Regaiey, K. A., Masternak, M. M., Bonkowski, M. S., Panici, J. A., Kopchick, J. J., & Bartke, A. (2007). Effects of caloric restriction and growth hormone resistance on insulin-related intermediates in the skeletal muscle. *The Journals of Gerontology. Series A Biological Sciences and Medical Sciences*, *62*, 18–26.

Anderson, R. M., & Weindruch, R. (2010). Metabolic reprogramming, caloric restriction and aging. *Trends in Endocrinology & Metabolism*, *21*, 134–141. Available from https://doi.org/10.1016/j.tem.2009.11.005.

Anisimov, V. N., Zabezhinski, M. A., Popovich, I. G., Piskunova, T. S., Semenchenko, A. V., Tyndyk, M. L., et al. (2011). Rapamycin increases lifespan and inhibits spontaneous tumorigenesis in inbred female mice. *Cell Cycle*, *10*, 4230–4236. Available from https://doi.org/10.4161/cc.10.24.18486.

Anselmi, C. V., Malovini, A., Roncarati, R., Novelli, V., Villa, F., Condorelli, G., et al. (2009). Association of the FOXO3A locus with extreme longevity in a southern Italian centenarian study. *Rejuvenation Research*, *12*, 95–104. Available from https://doi.org/10.1089/rej.2008.0827.

Anson, R. M., Guo, Z., de Cabo, R., Iyun, T., Rios, M., Hagepanos, A., et al. (2003). Intermittent fasting dissociates beneficial effects of dietary restriction on glucose metabolism and neuronal resistance to injury from calorie intake. *Proceedings of the National Academy of Sciences of the United States of America*, *100*, 6216–6220. Available from https://doi.org/10.1073/pnas.1035720100.

Austad, S. N., & Kristan, D. M. (2003). Are mice calorically restricted in nature? *Aging Cell*, *2*, 201–207. Available from https://doi.org/10.1046/j.1474-9728.2003.00053.x.

Bae, C.-Y., Kang, Y. G., Kim, S., Cho, C., Kang, H. C., Yu, B. Y., et al. (2008). Development of models for predicting biological age (BA) with physical, biochemical, and hormonal parameters. *Archives of Gerontology and Geriatrics*, *47*, 253–265. Available from https://doi.org/10.1016/j.archger.2007.08.009.

Bartke, A., Wright, J. C., Mattison, J. A., Ingram, D. K., Miller, R. A., & Roth, G. S. (2001). Extending the lifespan of long-lived mice. *Nature*, *414*, 412. Available from https://doi.org/10.1038/35106646.

Bass, T. M., Grandison, R. C., Wong, R., Martinez, P., Partridge, L., & Piper, M. D. W. (2007). Optimization of dietary restriction protocols in *Drosophila*. *The Journals of Gerontology. Series A Biological Sciences and Medical Sciences*, *62*, 1071–1081.

Becker K.G. (2002). Deciphering the gene expression profile of long-lived snell mice. *Science of Aging Knowledge Environment*, 2002(11), pe4. Available from https://doi.org/10.1126/sageke.2002.11.pe4

Beekman, M., Blanché, H., Perola, M., Hervonen, A., Bezrukov, V., Sikora, E., et al. (2013). Genome-wide linkage analysis for human longevity: Genetics of Healthy Aging Study. *Aging Cell*, *12*, 184–193. Available from https://doi.org/10.1111/acel.12039.

Belak, Z. R., Harkness, T., & Eskiw, C. H. (2018). A rapid, high-throughput method for determining chronological lifespan in budding yeast. *Journal of Biological Methods*, *5*, e106. Available from https://doi.org/10.14440/jbm.2018.272.

Belsky, D. W., Caspi, A., Houts, R., Cohen, H. J., Corcoran, D. L., Danese, A., et al. (2015). Quantification of biological aging in young adults. *Proceedings of the National Academy of Sciences of the United States of America*, *112*, E4104–E4110. Available from https://doi.org/10.1073/pnas.1506264112.

Belsky, D. W., Moffitt, T. E., Cohen, A. A., Corcoran, D. L., Levine, M. E., Prinz, J. A., et al. (2018). Eleven telomere, epigenetic clock, and biomarker-composite quantifications of biological aging: Do they measure the same thing? *American Journal of Epidemiology*, *187*, 1220–1230. Available from https://doi.org/10.1093/aje/kwx346.

Bennett, C. F., Kwon, J. J., Chen, C., Russell, J., Acosta, K., Burnaevskiy, N., et al. (2017). Transaldolase inhibition impairs mitochondrial respiration and induces a starvation-like longevity response in *Caenorhabditis elegans*. *PLoS Genetics*, *13*, e1006695. Available from https://doi.org/10.1371/journal.pgen.1006695.

Bentzinger, C. F., Romanino, K., Cloëtta, D., Lin, S., Mascarenhas, J. B., Oliveri, F., et al. (2008). Skeletal muscle-specific ablation of raptor, but not of rictor, causes metabolic changes and results in muscle dystrophy. *Cell Metabolism*, *8*, 411–424. Available from https://doi.org/10.1016/j.cmet.2008.10.002.

Blüher, M., Michael, M. D., Peroni, O. D., Ueki, K., Carter, N., Kahn, B. B., et al. (2002). Adipose tissue selective insulin receptor knockout protects against obesity and obesity-related glucose intolerance. *Developmental Cell*, *3*, 25–38.

Bogue, M. A., Churchill, G. A., & Chesler, E. J. (2015). Collaborative cross and diversity outbred data resources in the mouse phenome database. *Mammalian Genome: Official Journal of the International Mammalian Genome Society*, *26*, 511–520. Available from https://doi.org/10.1007/s00335-015-9595-6.

Bonkowski, M. S., Rocha, J. S., Masternak, M. M., Al Regaiey, K. A., & Bartke, A. (2006). Targeted disruption of growth hormone receptor interferes with the beneficial actions of calorie restriction. *Proceedings of the National Academy of Sciences of the United States of America*, *103*, 7901–7905. Available from https://doi.org/10.1073/pnas.0600161103.

Boyden, S. E., & Kunkel, L. M. (2010). High-density genomewide linkage analysis of exceptional human longevity identifies multiple novel loci. *PLoS One*, *5*, e12432. Available from https://doi.org/10.1371/journal.pone.0012432.

Braeckman, B. P., & Vanfleteren, J. R. (2007). Genetic control of longevity in *C. elegans*. *Experimental Gerontology*, *42*, 90–98. Available from https://doi.org/10.1016/j.exger.2006.04.010.

Broer, L., Buchman, A. S., Deelen, J., Evans, D. S., Faul, J. D., Lunetta, K. L., et al. (2015). GWAS of longevity in CHARGE consortium confirms APOE and FOXO3 candidacy. *The Journals of Gerontology. Series A, Biological Sciences and Medical Sciences*, *70*, 110–118. Available from https://doi.org/10.1093/gerona/glu166.

Broman K.W., & Sen, S. (2009). *A guide to QTL mapping with R/qtl*. Dordrecht: Springer; 2009.

Broman, K. W., Wu, H., Sen, S., & Churchill, G. A. (2003). R/qtl: QTL mapping in experimental crosses. *Bioinformatics*, *19*, 889–890.

Brüning, J. C., Michael, M. D., Winnay, J. N., Hayashi, T., Hörsch, D., Accili, D., et al. (1998). A muscle-specific insulin receptor knockout exhibits features of the metabolic syndrome of NIDDM without altering glucose tolerance. *Molecular Cell*, *2*, 559–569.

Burgering, B. M., & Coffer, P. J. (1995). Protein kinase B (c-Akt) in phosphatidylinositol-3-OH kinase signal transduction. *Nature*, *376*, 599–602. Available from https://doi.org/10.1038/376599a0.

Burtner, C. R., Murakami, C. J., & Kaeberlein, M. (2009). A genomic approach to yeast chronological aging. *Methods in Molecular Biology*, *548*, 101–114. Available from https://doi.org/10.1007/978-1-59745-540-4_6.

Burtner, C. R., Murakami, C. J., Olsen, B., Kennedy, B. K., & Kaeberlein, M. (2011). A genomic analysis of chronological longevity factors in budding yeast. *Cell Cycle*, *10*, 1385–1396. Available from https://doi.org/10.4161/cc.10.9.15464.

Butler, R. N., Sprott, R., Warner, H., Bland, J., Feuers, R., Forster, M., et al. (2004). Biomarkers of aging: From primitive organisms to humans. *The Journals of Gerontology. Series A Biological Sciences and Medical Sciences*, *59*, B560–B567.

Calvani, R., Brasili, E., Praticò, G., Capuani, G., Tomassini, A., Marini, F., et al. (2014). Fecal and urinary NMR-based metabolomics unveil an aging signature in mice. *Experimental Gerontology*, *49*, 5–11. Available from https://doi.org/10.1016/j.exger.2013.10.010.

Campos, S. E., Avelar-Rivas, J. A., Garay, E., Juárez-Reyes, A., & DeLuna, A. (2018). Genomewide mechanisms of chronological longevity by dietary restriction in budding yeast. *Aging Cell*, *17*, e12749. Available from https://doi.org/10.1111/acel.12749.

C. elegans | Dharmacon (2019). <https://dharmacon.horizondiscovery.com/cdnas-and-orfs/non-mammalian-cdnas-and-orfs/c-elegans/#all> Accessed 1.05.19.

Chapman, T., & Partridge, L. (1996). Female fitness in *Drosophila melanogaster*: An interaction between the effect of nutrition and of encounter rate with males. *Proceedings. Biological Sciences/The Royal Society*, *263*, 755–759. Available from https://doi.org/10.1098/rspb.1996.0113.

Chen, D., Pan, K. Z., Palter, J. E., & Kapahi, P. (2007). Longevity determined by developmental arrest genes in *Caenorhabditis elegans*. *Aging Cell*, *6*, 525–533. Available from https://doi.org/10.1111/j.1474-9726.2007.00305.x.

Chen, K. L., Crane, M. M., & Kaeberlein, M. (2017). Microfluidic technologies for yeast replicative lifespan studies. *Mechanisms of Ageing and Development*, *161*, 262–269. Available from https://doi.org/10.1016/j.mad.2016.03.009.

Chen, Y.-W., Song, S., Weng, R., Verma, P., Kugler, J.-M., Buescher, M., et al. (2014). Systematic study of *Drosophila* microRNA functions using a collection of targeted knockout mutations. *Developmental Cell*, *31*, 784–800. Available from https://doi.org/10.1016/j.devcel.2014.11.029.

Cheney, K. E., Liu, R. K., Smith, G. S., Leung, R. E., Mickey, M. R., & Walford, R. L. (1980). Survival and disease patterns in C57BL/6J mice subjected to undernutrition. *Experimental Gerontology*, *15*, 237–258.

Chesler, E. J., Miller, D. R., Branstetter, L. R., Galloway, L. D., Jackson, B. L., Philip, V. M., et al. (2008). The Collaborative Cross at Oak Ridge National Laboratory: Developing a powerful resource for systems genetics. *International Mammalian Genome Society*, *19*, 382–389. Available from https://doi.org/10.1007/s00335-008-9135-8.

Churchill, G. A., Gatti, D. M., Munger, S. C., & Svenson, K. L. (2012). The diversity outbred mouse population. *International Mammalian Genome Society*, *23*, 713–718. Available from https://doi.org/10.1007/s00335-012-9414-2.

Churgin, M. A., Jung, S.-K., Yu, C.-C., Chen, X., Raizen, D. M., & Fang-Yen, C. (2017). Longitudinal imaging of *Caenorhabditis elegans* in a microfabricated device reveals variation in behavioral decline during aging. *ELife*, 6, e26652. Available from https://doi.org/10.7554/eLife.26652.

Clancy, D. J., Gems, D., Hafen, E., Leevers, S. J., & Partridge, L. (2002). Dietary restriction in long-lived dwarf flies. *Science*, *296*, 319. Available from https://doi.org/10.1126/science.1069366.

Cole, J. J., Robertson, N. A., Rather, M. I., Thomson, J. P., McBryan, T., Sproul, D., et al. (2017). Diverse interventions that extend mouse lifespan suppress shared age-associated epigenetic changes at critical gene regulatory regions. *Genome Biology*, *18*, 58. Available from https://doi.org/10.1186/s13059-017-1185-3.

Collaborative Cross Consortium. (2012). The genome architecture of the Collaborative Cross mouse genetic reference population. *Genetics*, *190*, 389–401. Available from https://doi.org/10.1534/genetics.111.132639.

Colman, R. J., Beasley, T. M., Kemnitz, J. W., Johnson, S. C., Weindruch, R., & Anderson, R. M. (2014). Caloric restriction reduces age-related and all-cause mortality in rhesus monkeys. *Nature Communications*, *5*, 3557. Available from https://doi.org/10.1038/ncomms4557.

Crane, M. M., Clark, I. B. N., Bakker, E., Smith, S., & Swain, P. S. (2014). A microfluidic system for studying ageing and dynamic single-cell responses in budding yeast. *PLoS One*, *9*, e100042. Available from https://doi.org/10.1371/journal.pone.0100042.

Curran, S. P., & Ruvkun, G. (2007). Lifespan regulation by evolutionarily conserved genes essential for viability. *PLoS Genetics*, *3*, e56. Available from https://doi.org/10.1371/journal.pgen.0030056.

Cypser, J. R., & Johnson, T. E. (2002). Multiple stressors in *Caenorhabditis elegans* induce stress hormesis and extended longevity. *The Journals of Gerontology. Series A, Biological Sciences and Medical Sciences*, *57*, B109–B114.

Day, K., Waite, L. L., Thalacker-Mercer, A., West, A., Bamman, M. M., Brooks, J. D., et al. (2013). Differential DNA methylation with age displays both common and dynamic features across human tissues that are influenced by CpG landscape. *Genome Biology*, *14*, R102. Available from https://doi.org/10.1186/gb-2013-14-9-r102.

de Castro, E., Hegi de Castro, S., & Johnson, T. E. (2004). Isolation of long-lived mutants in *Caenorhabditis elegans* using selection for resistance to juglone. *Free Radical Biology & Medicine*, *37*, 139–145. Available from https://doi.org/10.1016/j.freeradbiomed.2004.04.021.

Deelen, J., Beekman, M., Uh, H.-W., Broer, L., Ayers, K. L., Tan, Q., et al. (2014). Genome-wide association meta-analysis of human longevity identifies a novel locus conferring survival beyond 90 years of age. *Human Molecular Genetics*, *23*, 4420–4432. Available from https://doi.org/10.1093/hmg/ddu139.

Deelen, J., Beekman, M., Uh, H.-W., Helmer, Q., Kuningas, M., Christiansen, L., et al. (2011). Genome-wide association study identifies a single major locus contributing to survival into old age; the APOE locus revisited. *Aging Cell*, *10*, 686–698. Available from https://doi.org/10.1111/j.1474-9726.2011.00705.x.

de Haan, G., Gelman, R., Watson, A., Yunis, E., & Van Zant, G. (1998). A putative gene causes variability in lifespan among genotypically identical mice. *Nature Genetics*, *19*, 114–116. Available from https://doi.org/10.1038/465.

Delaney, J. R., Sutphin, G. L., Dulken, B., Sim, S., Kim, J. R., Robison, B., et al. (2011). Sir2 deletion prevents lifespan extension in 32 long-lived mutants. *Aging Cell*, *10*, 1089–1091. Available from https://doi.org/10.1111/j.1474-9726.2011.00742.x.

de Magalhães, J. P., Curado, J., & Church, G. M. (2009). Meta-analysis of age-related gene expression profiles identifies common signatures of aging. *Bioinformatics*, *25*, 875–881. Available from https://doi.org/10.1093/bioinformatics/btp073.

Dillin, A., Hsu, A.-L., Arantes-Oliveira, N., Lehrer-Graiwer, J., Hsin, H., Fraser, A. G., et al. (2002). Rates of behavior and aging specified by mitochondrial function during development. *Science*, *298*, 2398–2401. Available from https://doi.org/10.1126/science.1077780.

DOQTL. (n.d.). *DOQTL: Genotyping and QTL mapping in DO mice*. <https://rdrr.io/bioc/DOQTL/>

Ecker, S., & Beck, S. (2019). The epigenetic clock: A molecular crystal ball for human aging? *Aging*, *11*, 833–835. Available from https://doi.org/10.18632/aging.101712.

Edwards, D. R. V., Gilbert, J. R., Hicks, J. E., Myers, J. L., Jiang, L., Cummings, A. C., et al. (2013). Linkage and association of successful aging to the 6q25 region in large Amish kindreds. *Age*, *35*, 1467–1477. Available from https://doi.org/10.1007/s11357-012-9447-1.

Edwards, D. R. V., Gilbert, J. R., Jiang, L., Gallins, P. J., Caywood, L., Creason, M., et al. (2011). Successful aging shows linkage to chromosomes 6, 7, and 14 in the Amish. *Annals of Human Genetics*, *75*, 516–528. Available from https://doi.org/10.1111/j.1469-1809.2011.00658.x.

C. elegans Sequencing Consortium. (1998). Genome sequence of the nematode *C. elegans*: A platform for investigating biology. *Science*, *282*, 2012–2018.

Fabrizio, P., Hoon, S., Shamalnasab, M., Galbani, A., Wei, M., Giaever, G., et al. (2010). Genome-wide screen in *Saccharomyces cerevisiae* identifies vacuolar protein sorting, autophagy, biosynthetic, and tRNA methylation genes involved in life span regulation. *PLoS Genetics*, *6*, e1001024. Available from https://doi.org/10.1371/journal.pgen.1001024.

Fabrizio, P., & Longo, V. D. (2003). The chronological life span of *Saccharomyces cerevisiae*. *Aging Cell*, *2*, 73–81. Available from https://doi.org/10.1046/j.1474-9728.2003.00033.x.

Fabrizio, P., Pletcher, S. D., Minois, N., Vaupel, J. W., & Longo, V. D. (2004). Chronological aging-independent replicative life span regulation by Msn2/Msn4 and Sod2 in *Saccharomyces cerevisiae*. *FEBS Letters*, *557*, 136–142.

Fabrizio, P., Pozza, F., Pletcher, S. D., Gendron, C. M., & Longo, V. D. (2001). Regulation of longevity and stress resistance by Sch9 in yeast. *Science*, *292*, 288–290. Available from https://doi.org/10.1126/science.1059497.

Fehrmann, S., Paoletti, C., Goulev, Y., Ungureanu, A., Aguilaniu, H., & Charvin, G. (2013). Aging yeast cells undergo a sharp entry into senescence unrelated to the loss of mitochondrial membrane potential. *Cell Reports*, *5*, 1589–1599. Available from https://doi.org/10.1016/j.celrep.2013.11.013.

Flachsbart, F., Caliebe, A., Kleindorp, R., Blanché, H., von Eller-Eberstein, H., Nikolaus, S., et al. (2009). Association of FOXO3A variation with human longevity confirmed in *German centenarians*. *Proceedings of the National Academy of Sciences of the United States of America*, *106*, 2700–2705. Available from https://doi.org/10.1073/pnas.0809594106.

Fontana, L., Partridge, L., & Longo, V. D. (2010). Extending healthy life span—from yeast to humans. *Science*, *328*, 321–326. Available from https://doi.org/10.1126/science.1172539.

Fortney, K., Dobriban, E., Garagnani, P., Pirazzini, C., Monti, D., Mari, D., et al. (2015). Genome-wide scan informed by age-related disease identifies loci for exceptional human longevity. *PLoS Genetics*, *11*, e1005728. Available from https://doi.org/10.1371/journal.pgen.1005728.

Fraser, A. G., Kamath, R. S., Zipperlen, P., Martinez-Campos, M., Sohrmann, M., & Ahringer, J. (2000). Functional genomic analysis of *C. elegans* chromosome I by systematic RNA interference. *Nature*, *408*, 325–330. Available from https://doi.org/10.1038/35042517.

Funakoshi, M., Tsuda, M., Muramatsu, K., Hatsuda, H., Morishita, S., & Aigaki, T. (2011). A gain-of-function screen identifies wdb and lkb1 as lifespan-extending genes in *Drosophila*. *Biochemical and Biophysical Research Communications*, *405*, 667–672. Available from https://doi.org/10.1016/j.bbrc.2011.01.090.

Furey, T. S. (2012). ChIP-seq and beyond: New and improved methodologies to detect and characterize protein-DNA interactions. *Nature Reviews Genetics*, *13*, 840–852. Available from https://doi.org/10.1038/nrg3306.

Gabbianelli, R., & Malavolta, M. (2018). Epigenetics in ageing and development. *Mechanisms of Ageing and Development*, *174*, 1–2. Available from https://doi.org/10.1016/j.mad.2018.05.005.

Garay, E., Campos, S. E., González de la Cruz, J., Gaspar, A. P., Jinich, A., & Deluna, A. (2014). High-resolution profiling of stationary-phase survival reveals yeast longevity factors and their genetic interactions. *PLoS Genetics*, *10*, e1004168. Available from https://doi.org/10.1371/journal.pgen.1004168.

Gatti, D. M., Svenson, K. L., Shabalin, A., Wu, L.-Y., Valdar, W., Simecek, P., et al. (2014). Quantitative trait locus mapping methods for diversity outbred mice. *G3 Bethesda*, *4*, 1623–1633. Available from https://doi.org/10.1534/g3.114.013748.

Gelman, R., Watson, A., Bronson, R., & Yunis, E. (1988). Murine chromosomal regions correlated with longevity. *Genetics*, *118*, 693–704.

Gentilini, D., Mari, D., Castaldi, D., Remondini, D., Ogliari, G., Ostan, R., et al. (2013). Role of epigenetics in human aging and longevity: Genome-wide DNA methylation profile in centenarians and centenarians' offspring. *Age*, *35*, 1961–1973. Available from https://doi.org/10.1007/s11357-012-9463-1.

Gerstbrein, B., Stamatas, G., Kollias, N., & Driscoll, M. (2005). In vivo spectrofluorimetry reveals endogenous biomarkers that report healthspan and dietary restriction in *Caenorhabditis elegans*. *Aging Cell*, *4*, 127–137. Available from https://doi.org/10.1111/j.1474-9726.2005.00153.x.

Giannakou, M. E., Goss, M., & Partridge, L. (2008). Role of dFOXO in lifespan extension by dietary restriction in *Drosophila melanogaster*: Not required, but its activity modulates the response. *Aging Cell*, *7*, 187–198. Available from https://doi.org/10.1111/j.1474-9726.2007.00362.x.

Goffeau, A., Barrell, B. G., Bussey, H., Davis, R. W., Dujon, B., Feldmann, H., et al. (1996). Life with 6000 genes. *Science*, *274*(546), 563–567.

Golden, T. R., Hubbard, A., & Melov, S. (2006). Microarray analysis of variation in individual aging *C. elegans*: Approaches and challenges. *Experimental Gerontology*, *41*, 1040–1045. Available from https://doi.org/10.1016/j.exger.2006.06.034.

Good, T. P., & Tatar, M. (2001). Age-specific mortality and reproduction respond to adult dietary restriction in *Drosophila melanogaster*. *Journal of Insect Physiology*, *47*, 1467–1473.

Goodrick, C. L., Ingram, D. K., Reynolds, M. A., Freeman, J. R., & Cider, N. (1990). Effects of intermittent feeding upon body weight and lifespan in inbred mice: Interaction of genotype and age. *Mechanisms of Ageing and Development*, *55*, 69–87.

Goodrick, C. L., Ingram, D. K., Reynolds, M. A., Freeman, J. R., & Cider, N. L. (1982). Effects of intermittent feeding upon growth and life span in rats. *Gerontology*, *28*, 233–241. Available from https://doi.org/10.1159/000212538.

Goodrick, C. L., Ingram, D. K., Reynolds, M. A., Freeman, J. R., & Cider, N. L. (1983a). Effects of intermittent feeding upon growth, activity, and lifespan in rats allowed voluntary exercise. *Experimental Aging Research*, *9*, 203–209. Available from https://doi.org/10.1080/03610738308258453.

Goodrick, C. L., Ingram, D. K., Reynolds, M. A., Freeman, J. R., & Cider, N. L. (1983b). Differential effects of intermittent feeding and voluntary exercise on body weight and lifespan in adult rats. *Journal of Gerontology*, *38*, 36–45.

Greer, E. L., & Brunet, A. (2009). Different dietary restriction regimens extend lifespan by both independent and overlapping genetic pathways in *C. elegans*. *Aging Cell*, *8*, 113–127. Available from https://doi.org/10.1111/j.1474-9726.2009.00459.x.

Greer, E. L., Dowlatshahi, D., Banko, M. R., Villen, J., Hoang, K., Blanchard, D., et al. (2007). An AMPK-FOXO pathway mediates longevity induced by a novel method of dietary restriction in *C. elegans*. *Current Biology*, *17*, 1646–1656. Available from https://doi.org/10.1016/j.cub.2007.08.047.

Guevara, E. E., & Lawler, R. R. (2018). Epigenetic clocks. *Evolutionary Anthropology*, *27*, 256–260. Available from https://doi.org/10.1002/evan.21745.

Guo, Z., Toichi, E., Hosono, M., Hosokawa, T., Hosokawa, M., Higuchi, K., et al. (2000). Genetic analysis of lifespan in hybrid progeny derived from the SAMP1 mouse strain with accelerated senescence. *Mechanisms of Ageing and Development*, *118*, 35–44.

Hamilton, B., Dong, Y., Shindo, M., Liu, W., Odell, I., Ruvkun, G., et al. (2005). A systematic RNAi screen for longevity genes in *C. elegans*. *Genes & Development*, *19*, 1544–1555. Available from https://doi.org/10.1101/gad.1308205.

Han, E.-S., Wu, Y., McCarter, R., Nelson, J. F., Richardson, A., & Hilsenbeck, S. G. (2004). Reproducibility, sources of variability, pooling, and sample size: Important considerations for the design of high-density oligonucleotide array experiments. *The Journals of Gerontology. Series A Biological Sciences and Medical Sciences*, *59*, 306–315.

Han, Y., Eipel, M., Franzen, J., Sakk, V., Dethmers-Ausema, B., Yndriago, L., et al. (2018). Epigenetic age-predictor for mice based on three CpG sites. *ELife*, *7*, e37462. Available from https://doi.org/10.7554/eLife.37462.

Hansen, M., Hsu, A.-L., Dillin, A., & Kenyon, C. (2005). New genes tied to endocrine, metabolic, and dietary regulation of lifespan from a *Caenorhabditis elegans* genomic RNAi screen. *PLoS Genetics*, *1*, 119–128. Available from https://doi.org/10.1371/journal.pgen.0010017.

Harper, J. M., Leathers, C. W., & Austad, S. N. (2006). Does caloric restriction extend life in wild mice? *Aging Cell*, *5*, 441–449. Available from https://doi.org/10.1111/j.1474-9726.2006.00236.x.

Harries, L. W., Hernandez, D., Henley, W., Wood, A. R., Holly, A. C., Bradley-Smith, R. M., et al. (2011). Human aging is characterized by focused changes in gene expression and deregulation of alternative splicing. *Aging Cell*, *10*, 868–878. Available from https://doi.org/10.1111/j.1474-9726.2011.00726.x.

Harris, N., MacLean, M., Hatzianthis, K., Panaretou, B., & Piper, P. W. (2001). Increasing *Saccharomyces cerevisiae* stress resistance, through the overactivation of the heat shock response resulting from defects in the Hsp90 chaperone, does not extend replicative life span but can be associated with slower chronological ageing of nondividing cells. *Molecular Genetics & Genomics*, *265*, 258–263.

Harrison, D. E., & Archer, J. R. (1987). Genetic differences in effects of food restriction on aging in mice. *The Journal of Nutrition*, *117*, 376–382. Available from https://doi.org/10.1093/jn/117.2.376.

Harrison, D. E., Strong, R., Sharp, Z. D., Nelson, J. F., Astle, C. M., Flurkey, K., et al. (2009). Rapamycin fed late in life extends lifespan in genetically heterogeneous mice. *Nature*, *460*, 392–395. Available from https://doi.org/10.1038/nature08221.

Helfand, S. L., & Rogina, B. (2003). Molecular genetics of aging in the fly: Is this the end of the beginning? *BioEssays: News and Reviews in Molecular, Cellular and Developmental Biology*, *25*, 134–141. Available from https://doi.org/10.1002/bies.10225.

Hercus, M. J., Loeschcke, V., & Rattan, S. I. S. (2003). Lifespan extension of *Drosophila melanogaster* through hormesis by repeated mild heat stress. *Biogerontology*, *4*, 149–156.

Herndon, L. A., Schmeissner, P. J., Dudaronek, J. M., Brown, P. A., Listner, K. M., Sakano, Y., et al. (2002). Stochastic and genetic factors influence tissue-specific decline in ageing *C. elegans*. *Nature*, *419*, 808–814. Available from https://doi.org/10.1038/nature01135.

Heroux, O., & Campbell, J. S. (1960). A study of the pathology and life span of 6 degree C- and 30 degree C.-acclimated rats. *Laboratory Investigation A Journal of Technical Methods and Pathology*, *9*, 305–315.

Hertel, J., Friedrich, N., Wittfeld, K., Pietzner, M., Budde, K., Van der Auwera, S., et al. (2016). Measuring biological age via metabonomics: The metabolic age score. *Journal of Proteome Research*, *15*, 400–410. Available from https://doi.org/10.1021/acs.jproteome.5b00561.

Hodgson, V. J., Walker, G. M., & Button, D. (1994). A rapid colorimetric assay of killer toxin activity in yeast. *FEMS Microbiology Letters*, *120*, 201–205. Available from https://doi.org/10.1111/j.1574-6968.1994.tb07031.x.

Hofecker, G., Skalicky, M., Kment, A., & Niedermüller, H. (1980). Models of the biological age of the rat. I. A factor model of age parameters. *Mechanisms of Ageing and Development*, *14*, 345–359.

Hoffman, J. M., Lyu, Y., Pletcher, S. D., & Promislow, D. E. L. (2017). Proteomics and metabolomics in ageing research: From biomarkers to systems biology. *Essays in Biochemistry*, *61*, 379–388. Available from https://doi.org/10.1042/EBC20160083.

Hollingsworth, J. W., Hashizume, A., & Jablon, S. (1965). Correlations between tests of aging in Hiroshima subjects—an attempt to define "physiologic age". *The Yale Journal of Biology and Medicine*, *38*, 11–26.

Holloszy, J. O., & Smith, E. K. (1986). Longevity of cold-exposed rats: A reevaluation of the "rate-of-living theory". *Journal of Applied Physiology*, *61*, 1656–1660. Available from https://doi.org/10.1152/jappl.1986.61.5.1656.

Holly, A. C., Melzer, D., Pilling, L. C., Henley, W., Hernandez, D. G., Singleton, A. B., et al. (2013). Towards a gene expression biomarker set for human biological age. *Aging Cell*, *12*, 324–326. Available from https://doi.org/10.1111/acel.12044.

Horvath, S. (2013). DNA methylation age of human tissues and cell types. *Genome Biology*, *14*, R115. Available from https://doi.org/10.1186/gb-2013-14-10-r115.

Horvath, S., & Raj, K. (2018). DNA methylation-based biomarkers and the epigenetic clock theory of ageing. *Nature Reviews Genetics*, *19*, 371–384. Available from https://doi.org/10.1038/s41576-018-0004-3.

Hou, L., Huang, J., Green, C. D., Boyd-Kirkup, J., Zhang, W., Yu, X., et al. (2012). Systems biology in aging: Linking the old and the young. *Current Genomics*, *13*, 558–565. Available from https://doi.org/10.2174/138920212803251418.

Houthoofd, K., Braeckman, B. P., Johnson, T. E., & Vanfleteren, J. R. (2003). Life extension via dietary restriction is independent of the Ins/IGF-1 signalling pathway in *Caenorhabditis elegans*. *Experimental Gerontology*, *38*, 947–954.

Houtkooper, R. H., Argmann, C., Houten, S. M., Cantó, C., Jeninga, E. H., Andreux, P. A., et al. (2011). The metabolic footprint of aging in mice. *Science Reports*, *1*, 134. Available from https://doi.org/10.1038/srep00134.

Huang, C., Xiong, C., & Kornfeld, K. (2004). Measurements of age-related changes of physiological processes that predict lifespan of *Caenorhabditis elegans*. *Proceedings of the National Academy of Sciences of the United States of America*, *101*, 8084–8089. Available from https://doi.org/10.1073/pnas.0400848101.

Huberts, D. H. E. W., González, J., Lee, S. S., Litsios, A., Hubmann, G., Wit, E. C., et al. (2014). Calorie restriction does not elicit a robust extension of replicative lifespan in *Saccharomyces cerevisiae*. *Proceedings of the National Academy of Sciences of the United States of America*, *111*, 11727–11731. Available from https://doi.org/10.1073/pnas.1410024111.

Huberts, D. H. E. W., Janssens, G. E., Lee, S. S., Vizcarra, I. A., & Heinemann, M. (2013a). Continuous high-resolution microscopic observation of replicative aging in budding yeast. *Journal of Visualized Experiments*, e50143. Available from https://doi.org/10.3791/50143.

Huberts, D. H. E. W., Lee, S. S., Gonzáles, J., Janssens, G. E., Vizcarra, I. A., & Heinemann, M. (2013b). Construction and use of a microfluidic dissection platform for long-term imaging of cellular processes in budding yeast. *Nature Protocols*, *8*, 1019–1027. Available from https://doi.org/10.1038/nprot.2013.060.

Iancu, O. D., Darakjian, P., Walter, N. A. R., Malmanger, B., Oberbeck, D., Belknap, J., et al. (2010). Genetic diversity and striatal gene networks: Focus on the heterogeneous stock-collaborative cross (HS-CC) mouse. *BMC Genomics*, *11*, 585. Available from https://doi.org/10.1186/1471-2164-11-585.

Iser, W. B., & Wolkow, C. A. (2007). DAF-2/insulin-like signaling in *C. elegans* modifies effects of dietary restriction and nutrient stress on aging, stress and growth. *PLoS One*, *2*, e1240. Available from https://doi.org/10.1371/journal.pone.0001240.

Jackson, A. U., Galecki, A. T., Burke, D. T., & Miller, R. A. (2002). Mouse loci associated with life span exhibit sex-specific and epistatic effects. *The Journals of Gerontology. Series A, Biological Sciences and Medical Sciences*, *57*, B9–B15.

Jiang, J. C., Jaruga, E., Repnevskaya, M. V., & Jazwinski, S. M. (2000). An intervention resembling caloric restriction prolongs life span and retards aging in yeast. *FASEB Journal*, *14*, 2135–2137. Available from https://doi.org/10.1096/fj.00-0242fje.

Jo, M. C., Liu, W., Gu, L., Dang, W., & Qin, L. (2015). High-throughput analysis of yeast replicative aging using a microfluidic system. *Proceedings of the National Academy of Sciences of the United States of America*, *112*, 9364–9369. Available from https://doi.org/10.1073/pnas.1510328112.

Johnson, H. D., Kintner, L. D., & Kibler, H. H. (1963). Effects of 48°F. (8.9°C.) and 83°F. (28.4°C.) on longevity and pathology of male rats. *Journal of Gerontology*, *18*, 29–36. Available from https://doi.org/10.1093/geronj/18.1.29.

Johnson, S. C., Rabinovitch, P. S., & Kaeberlein, M. (2013). mTOR is a key modulator of ageing and age-related disease. *Nature*, *493*, 338–345. Available from https://doi.org/10.1038/nature11861.

Johnson, T. E. (2006). Recent results: Biomarkers of aging. *Experimental Gerontology*, *41*, 1243–1246. Available from https://doi.org/10.1016/j.exger.2006.09.006.

Joshi, P. K., Fischer, K., Schraut, K. E., Campbell, H., Esko, T., & Wilson, J. F. (2016). Variants near CHRNA3/5 and APOE have age- and sex-related effects on human lifespan. *Nature Communications*, *7*, 11174. Available from https://doi.org/10.1038/ncomms11174.

Joshi, P. K., Pirastu, N., Kentistou, K. A., Fischer, K., Hofer, E., Schraut, K. E., et al. (2017). Genome-wide meta-analysis associates HLA-DQA1/DRB1 and LPA and lifestyle factors with human longevity. *Nature Communications*, *8*, 910. Available from https://doi.org/10.1038/s41467-017-00934-5.

Jung, P. P., Christian, N., Kay, D. P., Skupin, A., & Linster, C. L. (2015). Protocols and programs for high-throughput growth and aging phenotyping in yeast. *PLoS One*, *10*, e0119807. Available from https://doi.org/10.1371/journal.pone.0119807.

J/qtl. (2015). *J/qtl: a Java graphical user interface for R/qtl*. Churchill Lab. <https://www.jax.org/research-and-faculty/research-labs/the-churchill-lab>

Jylhävä, J., Pedersen, N. L., & Hägg, S. (2017). Biological age predictors. *EBioMedicine*, *21*, 29–36. Available from https://doi.org/10.1016/j.ebiom.2017.03.046.

Kaeberlein, M. (2006). Longevity and aging in the budding yeast. In P. M. Conn (Ed.), *Handbook of models for human aging* (pp. 109–120). Boston, MA: Elsevier.

Kaeberlein, M. (2013). mTOR inhibition: From aging to autism and beyond. *Scientifica*, *2013*, 849186. Available from https://doi.org/10.1155/2013/849186.

Kaeberlein, M., Kirkland, K. T., Fields, S., & Kennedy, B. K. (2004). Sir2-independent life span extension by calorie restriction in yeast. *PLoS Biology*, *2*, E296. Available from https://doi.org/10.1371/journal.pbio.0020296.

Kaeberlein, M., Powers, R. W., Steffen, K. K., Westman, E. A., Hu, D., Dang, N., et al. (2005). Regulation of yeast replicative life span by TOR and Sch9 in response to nutrients. *Science*, *310*, 1193–1196. Available from https://doi.org/10.1126/science.1115535.

Kamath, R. S., Fraser, A. G., Dong, Y., Poulin, G., Durbin, R., Gotta, M., et al. (2003). Systematic functional analysis of the *Caenorhabditis elegans* genome using RNAi. *Nature*, *421*, 231–237. Available from https://doi.org/10.1038/nature01278.

Kayo, T., Allison, D. B., Weindruch, R., & Prolla, T. A. (2001). Influences of aging and caloric restriction on the transcriptional profile of skeletal muscle from rhesus monkeys. *Proceedings of the National Academy of Sciences of the United States of America*, *98*, 5093–5098. Available from https://doi.org/10.1073/pnas.081061898.

Keane, T. M., Goodstadt, L., Danecek, P., White, M. A., Wong, K., Yalcin, B., et al. (2011). Mouse genomic variation and its effect on phenotypes and gene regulation. *Nature*, *477*, 289–294. Available from https://doi.org/10.1038/nature10413.

Keil, G., Cummings, E., & de Magalhães, J. P. (2015). Being cool: How body temperature influences ageing and longevity. *Biogerontology*, *16*, 383–397. Available from https://doi.org/10.1007/s10522-015-9571-2.

Kennedy, B. K., Steffen, K. K., & Kaeberlein, M. (2007). Ruminations on dietary restriction and aging. *Cellular and Molecular Life Sciences*, *64*, 1323–1328. Available from https://doi.org/10.1007/s00018-007-6470-y.

Kenyon, C., Chang, J., Gensch, E., Rudner, A., & Tabtiang, R. A. (1993). *C. elegans* mutant that lives twice as long as wild type. *Nature*, *366*, 461–464. Available from https://doi.org/10.1038/366461a0.

Kerber, R. A., O'Brien, E., Boucher, K. M., Smith, K. R., & Cawthon, R. M. (2012). A genome-wide study replicates linkage of 3p22-24 to extreme longevity in humans and identifies possible additional loci. *PLoS One*, *7*, e34746. Available from https://doi.org/10.1371/journal.pone.0034746.

Kibler, H. H., & Johnson, H. D. (1961). Metabolic rate and aging in rats during exposure to cold. *Journal of Gerontology*, *16*, 13–16.

Kibler, H. H., & Johnson, H. D. (1966). Temperature and longevity in male rats. *Journal of Gerontology*, *21*, 52–56.

Kibler, H. H., Silsby, H. D., & Johnson, H. D. (1963). Metabolic trends and life span of rats living at 9C and 28C. *Journal of Gerontology*, *18*, 235–239.

Kim, G.-H., Lee, Y.-E., Lee, G.-H., Cho, Y.-H., Lee, Y.-N., Jang, Y., et al. (2015). Overexpression of malic enzyme in the larval stage extends *Drosophila* lifespan. *Biochemical and Biophysical Research Communications*, *456*, 676–682. Available from https://doi.org/10.1016/j.bbrc.2014.12.020.

Kim, Y., & Sun, H. (2007). Functional genomic approach to identify novel genes involved in the regulation of oxidative stress resistance and animal lifespan. *Aging Cell*, *6*, 489–503. Available from https://doi.org/10.1111/j.1474-9726.2007.00302.x.

Kirchman, P. A., & Botta, G. (2007). Copper supplementation increases yeast life span under conditions requiring respiratory metabolism. *Mechanisms of Ageing and Development*, *128*, 187–195. Available from https://doi.org/10.1016/j.mad.2006.10.003.

Kirkwood, T. B. (1998). Alex Comfort and the measure of aging. *Experimental Gerontology*, *33*, 135–140.

Klass, M. R. (1977). Aging in the nematode *Caenorhabditis elegans*: Major biological and environmental factors influencing life span. *Mechanisms of Ageing and Development*, *6*, 413–429.

Klebanov, S., Astle, C. M., Roderick, T. H., Flurkey, K., Archer, J. R., Chen, J., et al. (2001). Maximum life spans in mice are extended by wild strain alleles. *Experimental Biology and Medicine*, *226*, 854–859.

Klemera, P., & Doubal, S. (2006). A new approach to the concept and computation of biological age. *Mechanisms of Ageing and Development*, *127*, 240–248. Available from https://doi.org/10.1016/j.mad.2005.10.004.

König, J., Zarnack, K., Luscombe, N. M., & Ule, J. (2012). Protein-RNA interactions: New genomic technologies and perspectives. *Nature Reviews Genetics*, *13*, 77–83. Available from https://doi.org/10.1038/nrg3141.

Krištić, J., Vučković, F., Menni, C., Klarić, L., Keser, T., Beceheli, I., et al. (2014). Glycans are a novel biomarker of chronological and biological ages. *The Journals of Gerontology. Series A Biological Sciences and Medical Sciences*, *69*, 779–789. Available from https://doi.org/10.1093/gerona/glt190.

Krøll, J., & Saxtrup, O. (2000). On the use of regression analysis for the estimation of human biological age. *Biogerontology*, *1*, 363–368.

Kulkarni, R. N., Brüning, J. C., Winnay, J. N., Postic, C., Magnuson, M. A., & Kahn, C. R. (1999). Tissue-specific knockout of the insulin receptor in pancreatic beta cells creates an insulin secretory defect similar to that in type 2 diabetes. *Cell*, *96*, 329–339.

Kuningas, M., Estrada, K., Hsu, Y.-H., Nandakumar, K., Uitterlinden, A. G., Lunetta, K. L., et al. (2011). Large common deletions associate with mortality at old age. *Human Molecular Genetics*, *20*, 4290–4296. Available from https://doi.org/10.1093/hmg/ddr340.

Lakowski, B., & Hekimi, S. (1996). Determination of life-span in *Caenorhabditis elegans* by four clock genes. *Science*, *272*, 1010–1013.

Lakowski, B., & Hekimi, S. (1998). The genetics of caloric restriction in *Caenorhabditis elegans*. *Proceedings of the National Academy of Sciences of the United States of America*, *95*, 13091–13096.

Lamming, D. W., Ye, L., Katajisto, P., Goncalves, M. D., Saitoh, M., Stevens, D. M., et al. (2012). Rapamycin-induced insulin resistance is mediated by mTORC2 loss and uncoupled from longevity. *Science*, *335*, 1638–1643. Available from https://doi.org/10.1126/science.1215135.

Lamming, D. W., Ye, L., Sabatini, D. M., & Baur, J. A. (2013). Rapalogs and mTOR inhibitors as anti-aging therapeutics. *The Journal of Clinical Investigation*, *123*, 980–989. Available from https://doi.org/10.1172/JCI64099.

Landis, G. N., Abdueva, D., Skvortsov, D., Yang, J., Rabin, B. E., Carrick, J., et al. (2004). Similar gene expression patterns characterize aging and oxidative stress in *Drosophila melanogaster*. *Proceedings of the National Academy of Sciences of the United States of America*, *101*, 7663–7668. Available from https://doi.org/10.1073/pnas.0307605101.

Landis, G. N., Bhole, D., & Tower, J. (2003). A search for doxycycline-dependent mutations that increase *Drosophila melanogaster* life span identifies the VhaSFD, Sugar baby, filamin, fwd and Cctl genes. *Genome Biology*, *4*, R8.

Lang, D. H., Gerhard, G. S., Griffith, J. W., Vogler, G. P., Vandenbergh, D. J., Blizard, D. A., et al. (2010). Quantitative trait loci (QTL) analysis of longevity in C57BL/6J by DBA/2J (BXD) recombinant inbred mice. *Aging Clinical and Experimental Research*, *22*, 8–19.

Le Bourg, E. (2007). Hormetic effects of repeated exposures to cold at young age on longevity, aging and resistance to heat or cold shocks in *Drosophila melanogaster*. *Biogerontology*, *8*, 431–444. Available from https://doi.org/10.1007/s10522-007-9086-6.

Le Bourg, E., & Minois, N. (1997). Increased longevity and resistance to heat shock in *Drosophila melanogaster* flies exposed to hypergravity. *Comptes Rendus de l'Academie des Sciences. Serie III, Sciences de la vie*, *320*, 215–221.

Le Bourg, E., Toffin, E., & Massé, A. (2004). Male *Drosophila melanogaster* flies exposed to hypergravity at young age are protected against a non-lethal heat shock at middle age but not against behavioral impairments due to this shock. *Biogerontology*, *5*, 431–443. Available from https://doi.org/10.1007/s10522-004-3200-9.

Le Couteur, D. G., Solon-Biet, S., Cogger, V. C., Mitchell, S. J., Senior, A., de Cabo, R., et al. (2016). The impact of low-protein high-carbohydrate diets on aging and lifespan. *Cellular and Molecular Life Sciences*, *73*, 1237–1252. Available from https://doi.org/10.1007/s00018-015-2120-y.

Leduc, M. S., Hageman, R. S., Meng, Q., Verdugo, R. A., Tsaih, S.-W., Churchill, G. A., et al. (2010). Identification of genetic determinants of IGF-1 levels and longevity among mouse inbred strains. *Aging Cell*, *9*, 823–836. Available from https://doi.org/10.1111/j.1474-9726.2010.00612.x.

Lee C., & Longo V. (2016). Dietary restriction with and without caloric restriction for healthy aging. *F1000Research*, 5:F1000 Faculty Rev-117. Avaialble From https://doi.org/10.12688/f1000research.7136.1

Lee, G. D., Wilson, M. A., Zhu, M., Wolkow, C. A., de Cabo, R., Ingram, D. K., et al. (2006). Dietary deprivation extends lifespan in *Caenorhabditis elegans*. *Aging Cell*, *5*, 515–524. Available from https://doi.org/10.1111/j.1474-9726.2006.00241.x.

Lee, S. S., Avalos Vizcarra, I., Huberts, D. H. E. W., Lee, L. P., & Heinemann, M. (2012). Whole lifespan microscopic observation of budding yeast aging through a microfluidic dissection platform. *Proceedings of the National Academy of Sciences of the United States of America*, *109*, 4916–4920. Available from https://doi.org/10.1073/pnas.1113505109.

Lee, S. S., Lee, R. Y. N., Fraser, A. G., Kamath, R. S., Ahringer, J., & Ruvkun, G. (2003). A systematic RNAi screen identifies a critical role for mitochondria in *C. elegans* longevity. *Nature Genetics*, *33*, 40–48. Available from https://doi.org/10.1038/ng1056.

Leiser, S. F., Begun, A., & Kaeberlein, M. (2011). HIF-1 modulates longevity and healthspan in a temperature-dependent manner. *Aging Cell*, *10*, 318–326. Available from https://doi.org/10.1111/j.1474-9726.2011.00672.x.

Leiser, S. F., Jafari, G., Primitivo, M., Sutphin, G. L., Dong, J., Leonard, A., et al. (2016). Age-associated vulval integrity is an important marker of nematode healthspan. *Age*, *38*, 419–431. Available from https://doi.org/10.1007/s11357-016-9936-8.

Lenarcic, A. B., Svenson, K. L., Churchill, G. A., & Valdar, W. (2012). A general Bayesian approach to analyzing diallel crosses of inbred strains. *Genetics*, *190*, 413–435. Available from https://doi.org/10.1534/genetics.111.132563.

Levine, M., Lu, A., Quach, A., Chen, B., Baccarelli, A., Whitsel, E., et al. (2018a). An epigenetic clock for aging and life expectancy. *Innovation in Aging*, *2*, 61. Available from https://doi.org/10.1093/geroni/igy023.231.

Levine, M. E. (2013). Modeling the rate of senescence: Can estimated biological age predict mortality more accurately than chronological age? *The Journals of Gerontology. Series A Biological Sciences and Medical Sciences*, *68*, 667–674. Available from https://doi.org/10.1093/gerona/gls233.

Levine, M. E., Lu, A. T., Quach, A., Chen, B. H., Assimes, T. L., Bandinelli, S., et al. (2018b). An epigenetic biomarker of aging for lifespan and healthspan. *Aging*, *10*, 573–591. Available from https://doi.org/10.18632/aging.101414.

Li, Y., Wang, W.-J., Cao, H., Lu, J., Wu, C., Hu, F.-Y., et al. (2009). Genetic association of FOXO1A and FOXO3A with longevity trait in Han Chinese populations. *Human Molecular Genetics*, *18*, 4897–4904. Available from https://doi.org/10.1093/hmg/ddp459.

Liao, C.-Y., Rikke, B. A., Johnson, T. E., Diaz, V., & Nelson, J. F. (2010). Genetic variation in the murine lifespan response to dietary restriction: From life extension to life shortening. *Aging Cell*, *9*, 92–95. Available from https://doi.org/10.1111/j.1474-9726.2009.00533.x.

Libert, S., & Pletcher, S. D. (2007). Modulation of longevity by environmental sensing. *Cell*, *131*, 1231–1234. Available from https://doi.org/10.1016/j.cell.2007.12.002.

Lin, S. J., Defossez, P. A., & Guarente, L. (2000). Requirement of NAD and SIR2 for life-span extension by calorie restriction in *Saccharomyces cerevisiae*. *Science*, *289*, 2126–2128.

Lindstrom, D. L., & Gottschling, D. E. (2009). The mother enrichment program: A genetic system for facile replicative life span analysis in *Saccharomyces cerevisiae*. *Genetics*, *183*, 413–422. Available from https://doi.org/10.1534/genetics.109.106229, 1SI-13SI.

Linford N.J., Bilgir C., Ro J., Pletcher S. D. (2013). Measurement of lifespan in *Drosophila melanogaster*. *Journal of Visualized Experiments*. 7(71):50068, Availabel from https://doi.org/10.3791/50068.

Lithgow, G. J., Driscoll, M., & Phillips, P. (2017). A long journey to reproducible results. *Nature*, *548*, 387–388. Available from https://doi.org/10.1038/548387a.

Lithgow, G. J., White, T. M., Melov, S., & Johnson, T. E. (1995). Thermotolerance and extended life-span conferred by single-gene mutations and induced by thermal stress. *Proceedings of the National Academy of Sciences of the United States of America*, *92*, 7540–7544.

Liu, P., Young, T. Z., & Acar, M. (2015). Yeast replicator: A high-throughput multiplexed microfluidics platform for automated measurements of single-cell aging. *Cell Reports*, *13*, 634–644. Available from https://doi.org/10.1016/j.celrep.2015.09.012.

Loeb, J., & Northrop, J. H. (1917). On the influence of food and temperature upon the duration of life. *The Journal of Biological Chemistry*, *32*, 103–121.

Logan, R. W., Robledo, R. F., Recla, J. M., Philip, V. M., Bubier, J. A., Jay, J. J., et al. (2013). High-precision genetic mapping of behavioral traits in the diversity outbred mouse population. *Genes, Brain, and Behavior*, *12*, 424–437. Available from https://doi.org/10.1111/gbb.12029.

Lucanic, M., Plummer, W. T., Chen, E., Harke, J., Foulger, A. C., Onken, B., et al. (2017). Impact of genetic background and experimental reproducibility on identifying chemical compounds with robust longevity effects. *Nature Communications*, *8*, 14256. Available from https://doi.org/10.1038/ncomms14256.

Lucanic, M., Plummer, W. T., Harke, J., Lucanic, M., Chen, E., Bhaumik, D., et al. (2017). Standardized Protocols from the Caenorhabditis Intervention Testing Program 2013–2016: Conditions and assays used for quantifying the development, fertility and lifespan of hermaphroditic *Caenorhabditis Strains*. *Protocol Exchange*. Available from https://doi.org/10.1038/protex.2016.086.

Luckinbill, L. S., Arking, R., Clare, M. J., Cirocco, W. C., & Buck, S. A. (1984). Selection for delayed senescence in *Drosophila melanogaster*. *Evolution: International Journal of Organic Evolution*, *38*, 996–1003. Available from https://doi.org/10.2307/2408433.

Lunetta, K. L., D'Agostino, R. B., Karasik, D., Benjamin, E. J., Guo, C.-Y., Govindaraju, R., et al. (2007). Genetic correlates of longevity and selected age-related phenotypes: A genome-wide association study in the Framingham Study. *BMC Medical Genetics*, *8*(Suppl. 1), S13. Available from https://doi.org/10.1186/1471-2350-8-S1-S13.

MacLean, M., Harris, N., & Piper, P. W. (2001). Chronological lifespan of stationary phase yeast cells: A model for investigating the factors that might influence the ageing of postmitotic tissues in higher organisms. *Yeast*, *18*, 499–509. Available from https://doi.org/10.1002/yea.701.

Maegawa, S., Hinkal, G., Kim, H. S., Shen, L., Zhang, L., Zhang, J., et al. (2010). Widespread and tissue specific age-related DNA methylation changes in mice. *Genome Research*, *20*, 332–340. Available from https://doi.org/10.1101/gr.096826.109.

Mair, W., Panowski, S. H., Shaw, R. J., & Dillin, A. (2009). Optimizing dietary restriction for genetic epistasis analysis and gene discovery in *C. elegans*. *PLoS One*, *4*, e4535. Available from https://doi.org/10.1371/journal.pone.0004535.

Malovini, A., Illario, M., Iaccarino, G., Villa, F., Ferrario, A., Roncarati, R., et al. (2011). Association study on long-living individuals from Southern Italy identifies rs10491334 in the CAMKIV gene that regulates survival proteins. *Rejuvenation Research*, *14*, 283–291. Available from https://doi.org/10.1089/rej.2010.1114.

Manenti, G., Galvan, A., Pettinicchio, A., Trincucci, G., Spada, E., Zolin, A., et al. (2009). Mouse genome-wide association mapping needs linkage analysis to avoid false-positive loci. *PLoS Genetics*, *5*, e1000331. Available from https://doi.org/10.1371/journal.pgen.1000331.

Masoro, E. J. (2005). Overview of caloric restriction and ageing. *Mechanisms of Ageing and Development*, *126*, 913–922. Available from https://doi.org/10.1016/j.mad.2005.03.012.

Masoro, E. J., Iwasaki, K., Gleiser, C. A., McMahan, C. A., Seo, E. J., & Yu, B. P. (1989). Dietary modulation of the progression of nephropathy in aging rats: An evaluation of the importance of protein. *The American Journal of Clinical Nutrition*, *49*, 1217–1227. Available from https://doi.org/10.1093/ajcn/49.6.1217.

Matecic, M., Smith, D. L., Pan, X., Maqani, N., Bekiranov, S., Boeke, J. D., et al. (2010). A microarray-based genetic screen for yeast chronological aging factors. *PLoS Genetics*, *6*, e1000921. Available from https://doi.org/10.1371/journal.pgen.1000921.

McCarroll, S. A., Murphy, C. T., Zou, S., Pletcher, S. D., Chin, C.-S., Jan, Y. N., et al. (2004). Comparing genomic expression patterns across species identifies shared transcriptional profile in aging. *Nature Genetics*, *36*, 197–204. Available from https://doi.org/10.1038/ng1291.

McCormick, M. A., Delaney, J. R., Tsuchiya, M., Tsuchiyama, S., Shemorry, A., Sim, S., et al. (2015). A comprehensive analysis of replicative lifespan in 4,698 single-gene deletion strains uncovers conserved mechanisms of aging. *Cell Metabolism*, *22*, 895–906. Available from https://doi.org/10.1016/j.cmet.2015.09.008.

McDaid, A. F., Joshi, P. K., Porcu, E., Komljenovic, A., Li, H., Sorrentino, V., et al. (2017). Bayesian association scan reveals loci associated with human lifespan and linked biomarkers. *Nature Communications*, *8*, 15842. Available from https://doi.org/10.1038/ncomms15842.

McElwee, J., Bubb, K., & Thomas, J. H. (2003). Transcriptional outputs of the *Caenorhabditis elegans* forkhead protein DAF-16. *Aging Cell*, *2*, 111–121. Available from https://doi.org/10.1046/j.1474-9728.2003.00043.x.

Meer, M. V., Podolskiy, D. I., Tyshkovskiy, A., & Gladyshev, V. N. (2018). A whole lifespan mouse multi-tissue DNA methylation clock. *ELife*, *7*, e40675. Available from https://doi.org/10.7554/eLife.40675.

Meitinger, F., Khmelinskii, A., Morlot, S., Kurtulmus, B., Palani, S., Andres-Pons, A., et al. (2014). A memory system of negative polarity cues prevents replicative aging. *Cell*, *159*, 1056–1069. Available from https://doi.org/10.1016/j.cell.2014.10.014.

Melov, S., & Hubbard, A. (2004). Microarrays as a tool to investigate the biology of aging: A retrospective and a look to the future. *Science of Aging Knowledge Environment*, re7. Available from https://doi.org/10.1126/sageke.2004.42.re7.

Mendenhall, A., Crane, M. M., Leiser, S., Sutphin, G., Tedesco, P. M., Kaeberlein, M., et al. (2017). Environmental canalization of life span and gene expression in *Caenorhabditis elegans*. *Journals of Gerontology Series A*, *72*, 1033–1037. Available from https://doi.org/10.1093/gerona/glx017.

Menni, C., Kastenmüller, G., Petersen, A. K., Bell, J. T., Psatha, M., Tsai, P.-C., et al. (2013). Metabolomic markers reveal novel pathways of ageing and early development in human populations. *International Journal of Epidemiology*, *42*, 1111–1119. Available from https://doi.org/10.1093/ije/dyt094.

Menni, C., Kiddle, S. J., Mangino, M., Viñuela, A., Psatha, M., Steves, C., et al. (2015). Circulating proteomic signatures of chronological age. *The Journals of Gerontology. Series A Biological Sciences and Medical Sciences*, *70*, 809–816. Available from https://doi.org/10.1093/gerona/glu121.

Miller, H., Fletcher, M., Primitivo, M., Leonard, A., Sutphin, G. L., Rintala, N., et al. (2017). Genetic interaction with temperature is an important determinant of nematode longevity. *Aging Cell*, *16*, 1425–1429. Available from https://doi.org/10.1111/acel.12658.

Miller, R. A., Buehner, G., Chang, Y., Harper, J. M., Sigler, R., & Smith-Wheelock, M. (2005). Methionine-deficient diet extends mouse lifespan, slows immune and lens aging, alters glucose, T4, IGF-I and insulin levels, and increases hepatocyte MIF levels and stress resistance. *Aging Cell*, *4*, 119–125. Available from https://doi.org/10.1111/j.1474-9726.2005.00152.x.

Miller, R. A., Chrisp, C., Jackson, A. U., & Burke, D. (1998). Marker loci associated with life span in genetically heterogeneous mice. *The Journals of Gerontology. Series A Biological Sciences and Medical Sciences*, *53*, M257–M263.

Miller, R. A., Harrison, D. E., Astle, C. M., Baur, J. A., Boyd, A. R., de Cabo, R., et al. (2011). Rapamycin, but not resveratrol or simvastatin, extends life span of genetically heterogeneous mice. *The Journals of Gerontology. Series A Biological Sciences and Medical Sciences*, *66*, 191–201. Available from https://doi.org/10.1093/gerona/glq178.

Miquel, J., Lundgren, P. R., Bensch, K. G., & Atlan, H. (1976). Effects of temperature on the life span, vitality and fine structure of *Drosophila melanogaster*. *Mechanisms of Ageing and Development*, *5*, 347–370.

Montgomery, S. B., & Dermitzakis, E. T. (2011). From expression QTLs to personalized transcriptomics. *Nature Reviews Genetics*, *12*, 277–282. Available from https://doi.org/10.1038/nrg2969.

Morgan, A. P., Fu, C.-P., Kao, C.-Y., Welsh, C. E., Didion, J. P., Yadgary, L., et al. (2015). The mouse universal genotyping array: From substrains to subspecies. *G3 Bethesda*, *6*, 263–279. Available from https://doi.org/10.1534/g3.115.022087.

Mortimer, R. K., & Johnston, J. R. (1959). Life span of individual yeast cells. *Nature*, *183*, 1751–1752.

Mouse Phenome Database. (2019). Mouse Phenome Database. <https://phenome.jax.org> Accessed 26.06.19.

Muñoz, M. J., & Riddle, D. L. (2003). Positive selection of *Caenorhabditis elegans* mutants with increased stress resistance and longevity. *Genetics*, *163*, 171–180.

Murakami, C., & Kaeberlein, M. (2009). Quantifying yeast chronological life span by outgrowth of aged cells. *Journal of Visualized Experiments*. Available from https://doi.org/10.3791/1156.

Murakami, C. J., Burtner, C. R., Kennedy, B. K., & Kaeberlein, M. (2008). A method for high-throughput quantitative analysis of yeast chronological life span. *The Journals of Gerontology. Series A Biological Sciences and Medical Sciences*, *63*, 113–121.

Murphy, C. T., McCarroll, S. A., Bargmann, C. I., Fraser, A., Kamath, R. S., Ahringer, J., et al. (2003). Genes that act downstream of DAF-16 to influence the lifespan of *Caenorhabditis elegans*. *Nature*, *424*, 277–283. Available from https://doi.org/10.1038/nature01789.

Nadon, N. L., Strong, R., Miller, R. A., & Harrison, D. E. (2017). NIA interventions testing program: Investigating putative aging intervention agents in a genetically heterogeneous mouse model. *EBioMedicine*, *21*, 3–4. Available from https://doi.org/10.1016/j.ebiom.2016.11.038.

Nair, P. N., Golden, T., & Melov, S. (2003). Microarray workshop on aging. *Mechanisms of Ageing and Development*, *124*, 133–138.

Nakamura, E., & Miyao, K. (2007). A method for identifying biomarkers of aging and constructing an index of biological age in humans. *The Journals of Gerontology. Series A Biological Sciences and Medical Sciences*, *62*, 1096–1105.

Nakamura, E., Miyao, K., & Ozeki, T. (1988). Assessment of biological age by principal component analysis. *Mechanisms of Ageing and Development*, *46*, 1–18.

Nebel, A., Kleindorp, R., Caliebe, A., Nothnagel, M., Blanché, H., Junge, O., et al. (2011). A genome-wide association study confirms APOE as the major gene influencing survival in long-lived individuals. *Mechanisms of Ageing and Development*, *132*, 324–330. Available from https://doi.org/10.1016/j.mad.2011.06.008.

Neff, F., Flores-Dominguez, D., Ryan, D. P., Horsch, M., Schröder, S., Adler, T., et al. (2013). Rapamycin extends murine lifespan but has limited effects on aging. *The Journal of Clinical Investigation*, *123*, 3272–3291. Available from https://doi.org/10.1172/JCI67674.

Newman, A. B., Walter, S., Lunetta, K. L., Garcia, M. E., Slagboom, P. E., Christensen, K., et al. (2010). A meta-analysis of four genome-wide association studies of survival to age 90 years or older: The Cohorts for Heart and Aging Research in Genomic Epidemiology Consortium. *The Journals of Gerontology. Series A Biological Sciences and Medical Sciences*, *65*, 478–487. Available from https://doi.org/10.1093/gerona/glq028.

Olsen, A., Vantipalli, M. C., & Lithgow, G. J. (2006). Lifespan extension of *Caenorhabditis elegans* following repeated mild hormetic heat treatments. *Biogerontology*, *7*, 221–230. Available from https://doi.org/10.1007/s10522-006-9018-x.

Olsen, B., Murakami, C. J., & Kaeberlein, M. (2010). YODA: Software to facilitate high-throughput analysis of chronological life span, growth rate, and survival in budding yeast. *BMC Bioinformatics*, *11*, 141. Available from https://doi.org/10.1186/1471-2105-11-141.

Ooka, H., Segall, P. E., & Timiras, P. S. (1988). Histology and survival in age-delayed low-tryptophan-fed rats. *Mechanisms of Ageing and Development*, *43*, 79–98.

Ostojic, I., Boll, W., Waterson, M. J., Chan, T., Chandra, R., Pletcher, S. D., et al. (2014). Positive and negative gustatory inputs affect *Drosophila* lifespan partly in parallel to dFOXO signaling. *Proceedings of the National Academy of Sciences of the United States of America*, *111*, 8143–8148. Available from https://doi.org/10.1073/pnas.1315466111.

Paik, D., Jang, Y. G., Lee, Y. E., Lee, Y. N., Yamamoto, R., Gee, H. Y., et al. (2012). Misexpression screen delineates novel genes controlling *Drosophila* lifespan. *Mechanisms of Ageing and Development*, *133*, 234–245. Available from https://doi.org/10.1016/j.mad.2012.02.001.

Paradis, S., & Ruvkun, G. (1998). *Caenorhabditis elegans* Akt/PKB transduces insulin receptor-like signals from AGE-1 PI3 kinase to the DAF-16 transcription factor. *Genes & Development*, *12*, 2488–2498.

Pawlikowska, L., Hu, D., Huntsman, S., Sung, A., Chu, C., Chen, J., et al. (2009). Association of common genetic variation in the insulin/IGF1 signaling pathway with human longevity. *Aging Cell*, *8*, 460–472. Available from https://doi.org/10.1111/j.1474-9726.2009.00493.x.

Peters, M. J., Joehanes, R., Pilling, L. C., Schurmann, C., Conneely, K. N., Powell, J., et al. (2015). The transcriptional landscape of age in human peripheral blood. *Nature Communications*, *6*, 8570. Available from https://doi.org/10.1038/ncomms9570.

Petkov, P. M., Graber, J. H., Churchill, G. A., DiPetrillo, K., King, B. L., & Paigen, K. (2005). Evidence of a large-scale functional organization of mammalian chromosomes. *PLoS Genetics*, *1*, e33. Available from https://doi.org/10.1371/journal.pgen.0010033.

Petkovich, D. A., Podolskiy, D. I., Lobanov, A. V., Lee, S.-G., Miller, R. A., Gladyshev, V. N., & Using, D. N. A. (2017). Methylation profiling to evaluate biological age and longevity interventions. *Cell Metabolism*, *25*(954-960), e6. Available from https://doi.org/10.1016/j.cmet.2017.03.016.

Petrascheck, M., & Miller, D. L. (2017). Computational analysis of lifespan experiment reproducibility. *Frontiers in Genetics*, *8*, 92. Available from https://doi.org/10.3389/fgene.2017.00092.

Pilling, L. C., Atkins, J. L., Bowman, K., Jones, S. E., Tyrrell, J., Beaumont, R. N., et al. (2016). Human longevity is influenced by many genetic variants: Evidence from 75,000 UK biobank participants. *Aging*, *8*, 547–560. Available from https://doi.org/10.18632/aging.100930.

Pletcher, S. D., Macdonald, S. J., Marguerie, R., Certa, U., Stearns, S. C., Goldstein, D. B., et al. (2002). Genome-wide transcript profiles in aging and calorically restricted *Drosophila melanogaster*. *Current Biology*, *12*, 712–723.

Polak, P., Cybulski, N., Feige, J. N., Auwerx, J., Rüegg, M. A., & Hall, M. N. (2008). Adipose-specific knockout of raptor results in lean mice with enhanced mitochondrial respiration. *Cell Metabolism*, *8*, 399–410. Available from https://doi.org/10.1016/j.cmet.2008.09.003.

Polanowski, A. M., Robbins, J., Chandler, D., & Jarman, S. N. (2014). Epigenetic estimation of age in humpback whales. *Molecular Ecology Resources*, *14*, 976–987. Available from https://doi.org/10.1111/1755-0998.12247.

Powers, R. W., Kaeberlein, M., Caldwell, S. D., Kennedy, B. K., & Fields, S. (2006). Extension of chronological life span in yeast by decreased TOR pathway signaling. *Genes & Development*, *20*, 174–184. Available from https://doi.org/10.1101/gad.1381406.

Powers, T. (2007). TOR signaling and S6 kinase 1: Yeast catches up. *Cell Metabolism*, *6*, 1–2. Available from https://doi.org/10.1016/j.cmet.2007.06.009.

Puca, A. A., Daly, M. J., Brewster, S. J., Matise, T. C., Barrett, J., Shea-Drinkwater, M., et al. (2001). A genome-wide scan for linkage to human exceptional longevity identifies a locus on chromosome 4. *Proceedings of the National Academy of Sciences of the United States of America*, *98*, 10505–10508. Available from https://doi.org/10.1073/pnas.181337598.

Putin, E., Mamoshina, P., Aliper, A., Korzinkin, M., Moskalev, A., Kolosov, A., et al. (2016). Deep biomarkers of human aging: Application of deep neural networks to biomarker development. *Aging*, *8*, 1021–1030. Available from https://doi.org/10.18632/aging.100968.

Rafelski, S. M., Viana, M. P., Zhang, Y., Chan, Y.-H. M., Thorn, K. S., Yam, P., et al. (2012). Mitochondrial network size scaling in budding yeast. *Science*, *338*, 822–824. Available from https://doi.org/10.1126/science.1225720.

Reed, T., Dick, D. M., Uniacke, S. K., Foroud, T., & Nichols, W. C. (2004). Genome-wide scan for a healthy aging phenotype provides support for a locus near D4S1564 promoting healthy aging. *The Journals of Gerontology. Series A Biological Sciences and Medical Sciences*, *59*, 227–232.

Rikke, B. A., Liao, C.-Y., McQueen, M. B., Nelson, J. F., & Johnson, T. E. (2010). Genetic dissection of dietary restriction in mice supports the metabolic efficiency model of life extension. *Experimental Gerontology*, *45*, 691–701. Available from https://doi.org/10.1016/j.exger.2010.04.008.

RNAi Resources | Source BioScience. (n.d.). <https://www.source-bioscience.com/products/life-sciences-research/clones/rnai-resources/> Accessed 1.05.19.

Rockman, M. V., & Kruglyak, L. (2008). Breeding designs for recombinant inbred advanced intercross lines. *Genetics*, *179*, 1069–1078. Available from https://doi.org/10.1534/genetics.107.083873.

Rodríguez, A., De La Cera, T., Herrero, P., & Moreno, F. (2001). The hexokinase 2 protein regulates the expression of the GLK1, HXK1 and HXK2 genes of *Saccharomyces cerevisiae*. *The Biochemical Journal*, *355*, 625–631.

Rose, M. R. (1984). Laboratory evolution of postponed senescence in *Drosophila melanogaster*. *Evolution: An International Journal of Organic Evolution*, *38*, 1004–1010. Available from https://doi.org/10.1111/j.1558-5646.1984.tb00370.x.

R/qtl. (n.d.). R/qtl software for mapping quantitative trait loci. <https://rqtl.org/>

Rual, J.-F., Ceron, J., Koreth, J., Hao, T., Nicot, A.-S., Hirozane-Kishikawa, T., et al. (2004). Toward improving *Caenorhabditis elegans* phenome mapping with an ORFeome-based RNAi library. *Genome Research, 14*, 2162–2168. Available from https://doi.org/10.1101/gr.2505604.

Samuelson, A. V., Klimczak, R. R., Thompson, D. B., Carr, C. E., & Ruvkun, G. (2007). Identification of *Caenorhabditis elegans* genes regulating longevity using enhanced RNAi-sensitive strains. *Cold Spring Harbor Symposia on Quantitative Biology, 72*, 489–497. Available from https://doi.org/10.1101/sqb.2007.72.068.

Sarnoski, E. A., Liu, P., & Acar, M. (2017). A high-throughput screen for yeast replicative lifespan identifies lifespan-extending compounds. *Cell Reports, 21*, 2639–2646. Available from https://doi.org/10.1016/j.celrep.2017.11.002.

Schleit, J., Johnson, S. C., Bennett, C. F., Simko, M., Trongtham, N., Castanza, A., et al. (2013). Molecular mechanisms underlying genotype-dependent responses to dietary restriction. *Aging Cell, 12*, 1050–1061. Available from https://doi.org/10.1111/acel.12130.

Schleit, J., Wasko, B. M., & Kaeberlein, M. (2012). Yeast as a model to understand the interaction between genotype and the response to calorie restriction. *FEBS Letters, 586*, 2868–2873. Available from https://doi.org/10.1016/j.febslet.2012.07.038.

Sebastiani, P., Bae, H., Sun, F. X., Andersen, S. L., Daw, E. W., Malovini, A., et al. (2013). Meta-analysis of genetic variants associated with human exceptional longevity. *Aging, 5*, 653–661. Available from https://doi.org/10.18632/aging.100594.

Sebastiani, P., Solovieff, N., Dewan, A. T., Walsh, K. M., Puca, A., Hartley, S. W., et al. (2012). Genetic signatures of exceptional longevity in humans. *PLoS One, 7*, e29848. Available from https://doi.org/10.1371/journal.pone.0029848.

Sebastiani, P., Thyagarajan, B., Sun, F., Schupf, N., Newman, A. B., Montano, M., et al. (2017). Biomarker signatures of aging. *Aging Cell, 16*, 329–338. Available from https://doi.org/10.1111/acel.12557.

Segall, P. E., & Timiras, P. S. (1976). Patho-physiologic findings after chronic tryptophan deficiency in rats: A model for delayed growth and aging. *Mechanisms of Ageing and Development, 5*, 109–124.

Selman, C., Tullet, J. M. A., Wieser, D., Irvine, E., Lingard, S. J., Choudhury, A. I., et al. (2009). Ribosomal protein S6 kinase 1 signaling regulates mammalian life span. *Science, 326*, 140–144. Available from https://doi.org/10.1126/science.1177221.

Shama, S., Lai, C. Y., Antoniazzi, J. M., Jiang, J. C., & Jazwinski, S. M. (1998). Heat stress-induced life span extension in yeast. *Experimental Cell Research, 245*, 379–388. Available from https://doi.org/10.1006/excr.1998.4279.

Shaposhnikov, M., Proshkina, E., Shilova, L., Zhavoronkov, A., & Moskalev, A. (2015). Lifespan and stress resistance in drosophila with overexpressed DNA repair genes. *Science Reports, 5*, 15299. Available from https://doi.org/10.1038/srep15299.

Skorupa, D. A., Dervisefendic, A., Zwiener, J., & Pletcher, S. D. (2008). Dietary composition specifies consumption, obesity, and lifespan in *Drosophila melanogaster*. *Aging Cell, 7*, 478–490. Available from https://doi.org/10.1111/j.1474-9726.2008.00400.x.

Smallwood, T. L., Gatti, D. M., Quizon, P., Weinstock, G. M., Jung, K.-C., Zhao, L., et al. (2014). High-resolution genetic mapping in the diversity outbred mouse population identifies Apobec1 as a candidate gene for atherosclerosis. *G3 Bethesda, 4*, 2353–2363. Available from https://doi.org/10.1534/g3.114.014704.

Smith, D. L., Maharrey, C. H., Carey, C. R., White, R. A., & Hartman, J. L. (2016). Gene-nutrient interaction markedly influences yeast chronological lifespan. *Experimental Gerontology, 86*, 113–123. Available from https://doi.org/10.1016/j.exger.2016.04.012.

Smith, E. D., Kaeberlein, T. L., Lydum, B. T., Sager, J., Welton, K. L., Kennedy, B. K., et al. (2008). Age- and calorie-independent life span extension from dietary restriction by bacterial deprivation in *Caenorhabditis elegans*. *BMC Developmental Biology, 8*, 49. Available from https://doi.org/10.1186/1471-213X-8-49.

Smith, E. D., Tsuchiya, M., Fox, L. A., Dang, N., Hu, D., Kerr, E. O., et al. (2008). Quantitative evidence for conserved longevity pathways between divergent eukaryotic species. *Genome Research, 18*, 564–570. Available from https://doi.org/10.1101/gr.074724.107.

Smith, J. M. (1958). The effects of temperature and of egg-laying on the longevity of *Drosophila subobscura*. *The Journal of Experimental Biology, 35*, 832.

Smith, R., Sheppard, K., DiPetrillo, K., & Churchill, G. (2009). Quantitative trait locus analysis using J/qtl. *Methods in Molecular Biology, 573*, 175–188. Available from https://doi.org/10.1007/978-1-60761-247-6_10.

Soerensen, M., Dato, S., Christensen, K., McGue, M., Stevnsner, T., Bohr, V. A., et al. (2010). Replication of an association of variation in the FOXO3A gene with human longevity using both case-control and longitudinal data. *Aging Cell, 9*, 1010–1017. Available from https://doi.org/10.1111/j.1474-9726.2010.00627.x.

Solon-Biet, S. M., McMahon, A. C., Ballard, J. W. O., Ruohonen, K., Wu, L. E., Cogger, V. C., et al. (2014). The ratio of macronutrients, not caloric intake, dictates cardiometabolic health, aging, and longevity in ad libitum-fed mice. *Cell Metabolism, 19*, 418–430. Available from https://doi.org/10.1016/j.cmet.2014.02.009.

Spivey, E. C., Xhemalce, B., Shear, J. B., & Finkelstein, I. J. (2014). 3D-printed microfluidic microdissector for high-throughput studies of cellular aging. *Analytical Chemistry, 86*, 7406–7412. Available from https://doi.org/10.1021/ac500893a.

Sprott, R. L. (2010). Biomarkers of aging and disease: Introduction and definitions. *Experimental Gerontology, 45*, 2–4. Available from https://doi.org/10.1016/j.exger.2009.07.008.

Srivastava, A., Morgan, A. P., Najarian, M. L., Sarsani, V. K., Sigmon, J. S., Shorter, J. R., et al. (2017). Genomes of the mouse collaborative cross. *Genetics, 206*, 537–556. Available from https://doi.org/10.1534/genetics.116.198838.

Steffen K.K., Kennedy B.K., Kaeberlein M. (2009) Measuring replicative life span in the budding yeast. Journal of Visualized Experiments 28, 1209. <https://doi:10.3791/1209>.

Steffen, K. K., MacKay, V. L., Kerr, E. O., Tsuchiya, M., Hu, D., Fox, L. A., et al. (2008). Yeast life span extension by depletion of 60s ribosomal subunits is mediated by Gcn4. *Cell, 133*, 292–302. Available from https://doi.org/10.1016/j.cell.2008.02.037.

Steinkraus, K. A., Kaeberlein, M., & Kennedy, B. K. (2008). Replicative aging in yeast: The means to the end. *Annual Review of Cell and Developmental Biology, 24*, 29–54. Available from https://doi.org/10.1146/annurev.cellbio.23.090506.123509.

Stroustrup, N., Ulmschneider, B. E., Nash, Z. M., López-Moyado, I. F., Apfeld, J., & Fontana, W. (2013). The *Caenorhabditis elegans* lifespan machine. *Nature Methods, 10*, 665–670. Available from https://doi.org/10.1038/nmeth.2475.

Stubbs, T. M., Bonder, M. J., Stark, A.-K., Krueger, F., BI Ageing Clock Team, von Meyenn F., et al. (2017). Multi-tissue DNA methylation age predictor in mouse. *Genome Biology, 18*, 68. Available from https://doi.org/10.1186/s13059-017-1203-5.

Stynen, B., Tournu, H., Tavernier, J., & Van Dijck, P. (2012). Diversity in genetic in vivo methods for protein-protein interaction studies: From the yeast two-hybrid system to the mammalian split-luciferase system. *Microbiology and Molecular Biology Reviews, 76*, 331–382. Available from https://doi.org/10.1128/MMBR.05021-11.

Sutphin, G. L., Backer, G., Sheehan, S., Bean, S., Corban, C., Liu, T., et al. (2017). *Caenorhabditis elegans* orthologs of human genes differentially expressed with age are enriched for determinants of longevity. *Aging Cell, 16*, 672–682. Available from https://doi.org/10.1111/acel.12595.

Sutphin, G. L., Bishop, E., Yanos, M. E., Moller, R. M., & Kaeberlein, M. (2012). Caffeine extends life span, improves healthspan, and delays age-associated pathology in *Caenorhabditis elegans*. *Longevity & Healthspan, 1*, 9. Available from https://doi.org/10.1186/2046-2395-1-9.

Sutphin G.L., & Kaeberlein M. (2019). Measuring *Caenorhabditis elegans* life span on solid media. *Journal of Visualized Experiments* 27, 1152. Available from https://doi.org/10.3791/1152.

Svenson, K. L., Gatti, D. M., Valdar, W., Welsh, C. E., Cheng, R., Chesler, E. J., et al. (2012). High-resolution genetic mapping using the mouse diversity outbred population. *Genetics, 190*, 437–447. Available from https://doi.org/10.1534/genetics.111.132597.

Svenson, K. L., Von Smith, R., Magnani, P. A., Suetin, H. R., Paigen, B., Naggert, J. K., et al. (2007). Multiple trait measurements in 43 inbred mouse strains capture the phenotypic diversity characteristic of human populations. *Journal of Applied Physiology, 102*, 2369–2378. Available from https://doi.org/10.1152/japplphysiol.01077.2006.

Swindell, W. R. (2012). Dietary restriction in rats and mice: A meta-analysis and review of the evidence for genotype-dependent effects on lifespan. *Ageing Research Reviews, 11*, 254–270. Available from https://doi.org/10.1016/j.arr.2011.12.006.

Takeda, H., Inada, H., Inoue, M., Yoshikawa, H., & Abe, H. (1982). Evaluation of biological age and physical age by multiple regression analysis. *Medical Informatics, 7*, 221–227.

Thompson, M. J., vonHoldt, B., Horvath, S., & Pellegrini, M. (2017). An epigenetic aging clock for dogs and wolves. *Aging, 9*, 1055–1068. Available from https://doi.org/10.18632/aging.101211.

Thompson, R. F., Atzmon, G., Gheorghe, C., Liang, H. Q., Lowes, C., Greally, J. M., et al. (2010). Tissue-specific dysregulation of DNA methylation in aging. *Aging Cell, 9*, 506–518. Available from https://doi.org/10.1111/j.1474-9726.2010.00577.x.

Timiras, P. S., Hudson, D. B., & Segall, P. E. (1984). Lifetime brain serotonin: Regional effects of age and precursor availability. *Neurobiology of Aging, 5*, 235–242.

Timmers, P. R., Mounier, N., Lall, K., Fischer, K., Ning, Z., Feng, X., et al. (2019). Genomics of 1 million parent lifespans implicates novel pathways and common diseases and distinguishes survival chances. *ELife*, 8, e39856. Available from https://doi.org/10.7554/eLife.39856.

Timmons, L., & Fire, A. (1998). Specific interference by ingested dsRNA. *Nature, 395*, 854. Available from https://doi.org/10.1038/27579.

Tomás-Loba, A., Bernardes de Jesus, B., Mato, J. M., & Blasco, M. A. (2013). A metabolic signature predicts biological age in mice. *Aging Cell, 12*, 93–101. Available from https://doi.org/10.1111/acel.12025.

Turturro, A., Witt, W. W., Lewis, S., Hass, B. S., Lipman, R. D., & Hart, R. W. (1999). Growth curves and survival characteristics of the animals used in the Biomarkers of Aging Program. *The Journals of Gerontology. Series A Biological Sciences and Medical Sciences, 54*, B492–B501.

Tyler, A. L., Ji, B., Gatti, D. M., Munger, S. C., Churchill, G. A., Svenson, K. L., et al. (2017). Epistatic networks jointly influence phenotypes related to metabolic disease and gene expression in diversity outbred mice. *Genetics, 206*, 621–639. Available from https://doi.org/10.1534/genetics.116.198051.

Ueyama, M., & Fuyama, Y. (2003). Enhanced cost of mating in female sterile mutants of *Drosophila melanogaster*. *Genes & Genetic Systems, 78*, 29–36.

UNC Systems Genetics. (2019). <http://csbio.unc.edu/CCstatus/index.py> Accessed 4.06.19.

Urban, J., Soulard, A., Huber, A., Lippman, S., Mukhopadhyay, D., Deloche, O., et al. (2007). Sch9 is a major target of TORC1 in *Saccharomyces cerevisiae*. *Molecular Cell, 26*, 663–674. Available from https://doi.org/10.1016/j.molcel.2007.04.020.

Vaanholt, L. M., Daan, S., Schubert, K. A., & Visser, G. H. (2009). Metabolism and aging: Effects of cold exposure on metabolic rate, body composition, and longevity in mice. *Physiological and Biochemical Zoology, 82*, 314–324. Available from https://doi.org/10.1086/589727.

Vaiserman, A. M., Koshel, N. M., Litoshenko, A. Y., Mozzhukhina, T. G., & Voitenko, V. P. (2003). Effects of X-irradiation in early ontogenesis on the longevity and amount of the S1 nuclease-sensitive DNA sites in adult *Drosophila melanogaster*. *Biogerontology, 4*, 9–14.

Van Voorhies, W. A., & Ward, S. (1999). Genetic and environmental conditions that increase longevity in *Caenorhabditis elegans* decrease metabolic rate. *Proceedings of the National Academy of Sciences of the United States of America, 96*, 11399–11403.

Viñuela, A., Snoek, L. B., Riksen, J. A. G., & Kammenga, J. E. (2010). Genome-wide gene expression regulation as a function of genotype and age in *C. elegans*. *Genome Research, 20*, 929–937. Available from https://doi.org/10.1101/gr.102160.109.

Viñuela, A., Snoek, L. B., Riksen, J. A. G., & Kammenga, J. E. (2012). Aging uncouples heritability and expression-QTL in *Caenorhabditis elegans*. *G3 Bethesda, 2*, 597–605. Available from https://doi.org/10.1534/g3.112.002212.

Wagner, K.-H., Cameron-Smith, D., Wessner, B., & Franzke, B. (2016). Biomarkers of aging: From function to molecular biology. *Nutrients*, 8, 338. Available from https://doi.org/10.3390/nu8060338.

Wagner, W. (2019). The link between epigenetic clocks for aging and senescence. *Frontiers in Genetics*, 10, 303. Available from https://doi.org/10.3389/fgene.2019.00303.

Walsh, R. B., Clifton, D., Horak, J., & Fraenkel, D. G. (1991). *Saccharomyces cerevisiae* null mutants in glucose phosphorylation: Metabolism and invertase expression. *Genetics, 128*, 521–527.

Walsh, R. B., Kawasaki, G., & Fraenkel, D. G. (1983). Cloning of genes that complement yeast hexokinase and glucokinase mutants. *Journal of Bacteriology, 154*, 1002–1004.

Walter, S., Atzmon, G., Demerath, E. W., Garcia, M. E., Kaplan, R. C., Kumari, M., et al. (2011). A genome-wide association study of aging. *Neurobiology of Aging, 32*(2109), e15–e28. Available from https://doi.org/10.1016/j.neurobiolaging.2011.05.026.

Wang, J. R., de Villena, F. P.-M., Lawson, H. A., Cheverud, J. M., Churchill, G. A., & McMillan, L. (2012). Imputation of single-nucleotide polymorphisms in inbred mice using local phylogeny. *Genetics, 190*, 449–458. Available from https://doi.org/10.1534/genetics.111.132381.

Wang, T., Tsui, B., Kreisberg, J. F., Robertson, N. A., Gross, A. M., Yu, M. K., et al. (2017). Epigenetic aging signatures in mice livers are slowed by dwarfism, calorie restriction and rapamycin treatment. *Genome Biology, 18*, 57. Available from https://doi.org/10.1186/s13059-017-1186-2.

Weindruch, R., Kayo, T., Lee, C. K., & Prolla, T. A. (2001). Microarray profiling of gene expression in aging and its alteration by caloric restriction in mice. *The Journal of Nutrition, 131*, 918S–923S. Available from https://doi.org/10.1093/jn/131.3.918S.

Weindruch, R., & Walford, R. L. (1988). *The retardation of aging and disease by dietary restriction*. Springfield, IL: Charles C Thomas Pub Ltd.

Werner, T. (2007). Regulatory networks: Linking microarray data to systems biology. *Mechanisms of Ageing and Development, 128*, 168–172. Available from https://doi.org/10.1016/j.mad.2006.11.022.

Wheeler, H. E., Metter, E. J., Tanaka, T., Absher, D., Higgins, J., Zahn, J. M., et al. (2009). Sequential use of transcriptional profiling, expression quantitative trait mapping, and gene association implicates MMP20 in human kidney aging. *PLoS Genetics, 5*, e1000685. Available from https://doi.org/10.1371/journal.pgen.1000685.

Wilkinson, J. E., Burmeister, L., Brooks, S. V., Chan, C.-C., Friedline, S., Harrison, D. E., et al. (2012). Rapamycin slows aging in mice. *Aging Cell, 11*, 675–682. Available from https://doi.org/10.1111/j.1474-9726.2012.00832.x.

Willcox, B. J., Donlon, T. A., He, Q., Chen, R., Grove, J. S., Yano, K., et al. (2008). FOXO3A genotype is strongly associated with human longevity. *Proceedings of the National Academy of Sciences of the United States of America*, *105*, 13987–13992. Available from https://doi.org/10.1073/pnas.0801030105.

Williams, R. W., Bennett, B., Lu, L., Gu, J., DeFries, J. C., Carosone-Link, P. J., et al. (2004). Genetic structure of the LXS panel of recombinant inbred mouse strains: A powerful resource for complex trait analysis. *International Mammalian Genome Society*, *15*, 637–647. Available from https://doi.org/10.1007/s00335-004-2380-6.

Winter, J. M., Gildea, D. E., Andreas, J. P., Gatti, D. M., Williams, K. A., Lee, M., et al. (2017). Mapping complex traits in a diversity outbred F1 mouse population identifies germline modifiers of metastasis in human prostate cancer. *Cell Systems*, *4*(31-45), e6. Available from https://doi.org/10.1016/j.cels.2016.10.018.

Winzeler, E. A., Shoemaker, D. D., Astromoff, A., Liang, H., Anderson, K., Andre, B., et al. (1999). Functional characterization of the *S. cerevisiae* genome by gene deletion and parallel analysis. *Science*, *285*, 901–906.

Wu, J. J., Liu, J., Chen, E. B., Wang, J. J., Cao, L., Narayan, N., et al. (2013). Increased mammalian lifespan and a segmental and tissue-specific slowing of aging after genetic reduction of mTOR expression. *Cell Reports*, *4*, 913–920. Available from https://doi.org/10.1016/j.celrep.2013.07.030.

Xia, X., Chen, W., McDermott, J., & Han, J.-D. J. (2017). Molecular and phenotypic biomarkers of aging. *F1000Research*, *6*, 860. Available from https://doi.org/10.12688/f1000research.10692.1.

Xian, B., Shen, J., Chen, W., Sun, N., Qiao, N., Jiang, D., et al. (2013). WormFarm: A quantitative control and measurement device toward automated *Caenorhabditis elegans* aging analysis. *Aging Cell*, *12*, 398–409. Available from https://doi.org/10.1111/acel.12063.

Xiang, L., & He, G. (2011). Caloric restriction and antiaging effects. *Annals of Nutrition & Metabolism*, *58*, 42–48. Available from https://doi.org/10.1159/000323748.

Xiao, R., Chun, L., Ronan, E. A., Friedman, D. I., Liu, J., & Xu, X. Z. S. (2015). RNAi interrogation of dietary modulation of development, metabolism, behavior, and aging in *C. elegans*. *Cell Reports*, *11*, 1123–1133. Available from https://doi.org/10.1016/j.celrep.2015.04.024.

Xie, Z., Jay, K. A., Smith, D. L., Zhang, Y., Liu, Z., Zheng, J., et al. (2015). Early telomerase inactivation accelerates aging independently of telomere length. *Cell*, *160*, 928–939. Available from https://doi.org/10.1016/j.cell.2015.02.002.

Xie, Z., Zhang, Y., Zou, K., Brandman, O., Luo, C., Ouyang, Q., et al. (2012). Molecular phenotyping of aging in single yeast cells using a novel microfluidic device. *Aging Cell*, *11*, 599–606. Available from https://doi.org/10.1111/j.1474-9726.2012.00821.x.

Yalcin, B., Wong, K., Agam, A., Goodson, M., Keane, T. M., Gan, X., et al. (2011). Sequence-based characterization of structural variation in the mouse genome. *Nature*, *477*, 326–329. Available from https://doi.org/10.1038/nature10432.

Yang, H., Wang, H., Shivalila, C. S., Cheng, A. W., Shi, L., & Jaenisch, R. (2013). One-step generation of mice carrying reporter and conditional alleles by CRISPR/Cas-mediated genome engineering. *Cell*, *154*, 1370–1379. Available from https://doi.org/10.1016/j.cell.2013.08.022.

Yanos, M. E., Bennett, C. F., & Kaeberlein, M. (2012). Genome-wide RNAi longevity screens in *Caenorhabditis elegans*. *Current Genomics*, *13*, 508–518. Available from https://doi.org/10.2174/138920212803251391.

Yao, C., Joehanes, R., Johnson, A. D., Huan, T., Esko, T., Ying, S., et al. (2014). Sex- and age-interacting eQTLs in human complex diseases. *Human Molecular Genetics*, *23*, 1947–1956. Available from https://doi.org/10.1093/hmg/ddt582.

Yashin, A. I., Wu, D., Arbeev, K. G., & Ukraintseva, S. V. (2010). Joint influence of small-effect genetic variants on human longevity. *Aging*, *2*, 612–620. Available from https://doi.org/10.18632/aging.100191.

Yen, C. A., & Curran, S. P. (2016). Gene-diet interactions and aging in *C. elegans*. *Experimental Gerontology*, *86*, 106–112. Available from https://doi.org/10.1016/j.exger.2016.02.012.

Yu, S., Jang, Y., Paik, D., Lee, E., & Park, J.-J. (2015). Nmdmc overexpression extends *Drosophila* lifespan and reduces levels of mitochondrial reactive oxygen species. *Biochemical and Biophysical Research Communications*, *465*, 845–850. Available from https://doi.org/10.1016/j.bbrc.2015.08.098.

Yuan, R., Flurkey, K., Meng, Q., Astle, M. C., & Harrison, D. E. (2013). Genetic regulation of life span, metabolism, and body weight in Pohn, a new wild-derived mouse strain. *The Journals of Gerontology. Series A Biological Sciences and Medical Sciences*, *68*, 27–35. Available from https://doi.org/10.1093/gerona/gls104.

Yuan, R., Gatti, D. M., Krier, R., Malay, E., Schultz, D., Peters, L. L., et al. (2015). Genetic regulation of female sexual maturation and longevity through circulating IGF1. *The Journals of Gerontology. Series A Biological Sciences and Medical Sciences*, *70*, 817–826. Available from https://doi.org/10.1093/gerona/glu114.

Yuan, R., Peters, L. L., & Paigen, B. (2011). Mice as a mammalian model for research on the genetics of aging. *ILAR Journal/National Research Council, Institute of Laboratory Animal Resources*, *52*, 4–15.

Yuan, R., Tsaih, S.-W., Petkova, S. B., Marin de Evsikova, C., Xing, S., Marion, M. A., et al. (2009). Aging in inbred strains of mice: Study design and interim report on median lifespans and circulating IGF1 levels. *Aging Cell*, *8*, 277–287. Available from https://doi.org/10.1111/j.1474-9726.2009.00478.x.

Yunis, E. J., Watson, A. L., Gelman, R. S., Sylvia, S. J., Bronson, R., & Dorf, M. E. (1984). Traits that influence longevity in mice. *Genetics*, *108*, 999–1011.

Zeng, Y., Cheng, L., Chen, H., Cao, H., Hauser, E. R., Liu, Y., et al. (2010). Effects of FOXO genotypes on longevity: A biodemographic analysis. *The Journals of Gerontology. Series A, Biological Sciences and Medical Sciences*, *65*, 1285–1299. Available from https://doi.org/10.1093/gerona/glq156.

Zeng, Y., Nie, C., Min, J., Liu, X., Li, M., Chen, H., et al. (2016). Novel loci and pathways significantly associated with longevity. *Science Reports*, *6*, 21243. Available from https://doi.org/10.1038/srep21243.

Zhang, Y., Bokov, A., Gelfond, J., Soto, V., Ikeno, Y., Hubbard, G., et al. (2014). Rapamycin extends life and health in C57BL/6 mice. *The Journals of Gerontology. Series A Biological Sciences and Medical Sciences*, *69*, 119–130. Available from https://doi.org/10.1093/gerona/glt056.

Zhang, Y., Luo, C., Zou, K., Xie, Z., Brandman, O., Ouyang, Q., et al. (2012). Single cell analysis of yeast replicative aging using a new generation of microfluidic device. *PLoS One*, *7*, e48275. Available from https://doi.org/10.1371/journal.pone.0048275.

Zou, S., Meadows, S., Sharp, L., Jan, L. Y., & Jan, Y. N. (2000). Genome-wide study of aging and oxidative stress response in *Drosophila melanogaster*. *Proceedings of the National Academy of Sciences of the United States of America*, *97*, 13726–13731. Available from https://doi.org/10.1073/pnas.260496697.

CHAPTER

2

DNA damage and repair in aging

Ajinkya S. Kawale and Patrick Sung

Department of Biochemistry and Structural Biology, University of Texas Health Science Center at San Antonio, San Antonio, TX, United States

OUTLINE

Introduction

Our genome is under constant assault from reactive by-products of cellular metabolism and from exogenous agents such as radiation and mutagenic chemicals. Such exposures lead to the formation of myriad DNA lesions. A number of evolutionarily conserved and mechanistically distinct DNA repair pathways have evolved to remove these DNA lesions. Even with these repair systems, DNA damage accumulates stochastically with age. Mutations at specific loci and chromosomal rearrangements such as deletions, translocations, and inversions also increase with age. Several progeroid syndromes stem from mutations in DNA repair genes, which implicates DNA damage as one of the major contributing factors in aging and senility. In fact, some have argued that persistent DNA damage represents the underlying cause of aging. Since DNA lacks the capacity for self-renewal, any potential damage can engender a negative consequence permanently. As such, it seems plausible that unrepaired or improperly repaired DNA damage drives the functional decline associated with aging. In this chapter, we provide an analysis on the relationship of DNA damage, its repair, and aging.

Overview of the DNA damage response

Endogenous DNA damage

The free radical theory of aging, proposed by Denman Harman in 1956, posits that reactive oxygen species (ROS) generated during cellular metabolic reactions damage key biomolecules and that this represents a primary cause of aging (Harman, 1956). Free radicals predominantly cause the oxidation of DNA bases, with 8-oxoguanine (8-oxoG) being the most common DNA lesion in this regard. It is estimated that around 3000 8-oxoG lesions occur in each cell every day (Tubbs & Nussenzweig, 2017); 8-oxoG is believed

Handbook of the Biology of Aging.
DOI: https://doi.org/10.1016/B978-0-12-815962-0.00002-0

to be a key mutagenic driver, as it can pair with cytosine, as well as adenine, causing GC to AT transversion mutations. Free radicals also induce strand breaks in DNA by abstraction of hydrogen from the phosphodiester backbone, thus fracturing the deoxyribose sugar moiety. Such single-strand breaks (SSBs) are highly abundant lesions, due to the reactivity of ROS, and also stem from other sources. As many as 55,000 SSBs are formed per cell daily (Tubbs & Nussenzweig, 2017). Occasionally, SSBs are converted into double-strand breaks (DSBs), for instance, upon encountering an oncoming DNA replication fork. These DSBs can lead to deleterious chromosome rearrangements, such as chromosome arm translocations.

The integrity of DNA can also be altered via base hydrolysis to form abasic (AP) sites. Specifically, AP sites represent one of the most frequent endogenous lesions, formed when the N-glycosidic bond connecting the base with the sugar is cleaved, leading to the loss of the DNA base. An estimated 10,000–12,000 abasic sites are formed in each cell daily (Ciccia & Elledge, 2010). Abasic sites also arise as a result of chemical modifications induced by ROS and as an intermediate of base excision repair (BER), which entails the removal of a damaged DNA base by a DNA glycosylase (Jacobs & Schar, 2012).

Endogenous alkylating agents such as S-adenosylmethionine can lead to base modification by forming adducts that have a high mutagenic potential (Fu, Calvo, & Samson, 2012). DNA is also susceptible to deamination, causing base substitutions, even though the frequency of this is much less in comparison with other reactions that damage DNA (Kow, 2002). Other metabolic products, such as reactive nitrogen and carbonyl species, aldehydes derived from lipid peroxidation or alcohol metabolism, and estrogen metabolites, also inflict DNA damage that is mutagenic in nature (De Bont & van Larebeke, 2004).

DNA damage induced by environmental exposures

Various exogenous physical and chemical agents have the potential to cause pathological changes in DNA. Ultraviolet (UV) radiation from sunlight can induce as many as 100,000 lesions per skin cell per day (Garinis, van der Horst, Vijg, & Hoeijmakers, 2008). UV primarily generates cyclobutane pyrimidine dimers (CPDs), which are bulky helix-distorting lesions that interfere with DNA synthesis and transcription. These and other UV photoproducts are removed via nucleotide excision repair (NER) or can be replicatively bypassed by specialized DNA polymerases in a process termed translesion DNA synthesis (TLS).

Ionizing radiation induces a variety of base lesions and single-stranded DNA breaks. It is estimated that around 1000 SSBs are induced in every cell after treatment with 1 Gy ionizing radiation. When these base lesions and SSBs occur in a clustered fashion, a double-stranded break can arise, owing to the proximity of SSBs or the action of BER enzymes on the clustered base lesions. Moreover, many of the current cadre of anticancer drugs, such as alkylating agents (temozolomide), interstrand cross-linking agents (platinum salts and mitomycin C), topoisomerase poisons (e.g., camptothecin and etoposide; inhibitors of topoisomerase I and topoisomerase II, respectively), inhibitors of poly(ADP-ribose) polymerase I (e.g., olaparib), and radiomimetic drugs (e.g., neocarzinostatin and calicheamicin), are potential DSB-inducing clastogens. Some of these aforementioned chemical agents can induce DSBs either directly (e.g., when the topoisomerase II–DNA cleavage intermediate is trapped by an inhibitor such as etoposide) or indirectly, for example, when the DNA replication fork encounters a SSB resulting from the combined action of DNA glycosylases and an AP endonuclease during BER.

DSBs, if improperly repaired, could result in large deletions in chromosomes, chromosome arm translocations, acentric and dicentric chromosomes, and other gross chromosome rearrangements (Deriano & Roth, 2013). These chromosome aberrations can lead to the loss of tumor suppressor genes, activation of proto-oncogenes, cell transformation, and oncogenesis (Mitelman, Johansson, & Mertens, 2007). It should be emphasized that the DNA lesion type usually dictates the ultimate phenotypic consequence. As such, DNA lesions are classified as either mutagenic or primarily cytostatic/cytotoxic. Mutagenic lesions include base modifications and substitutions, which could in turn promote tumorigenesis. DNA lesions, such as unrepaired DSBs, induce cell death that leads to degenerative changes, disruption of tissue homeostasis, enhanced cellular senescence, and aging. It is important to note that mutagenic lesions can exert a cytotoxic effect and, conversely, primarily cytotoxic lesions, if improperly handled, can cause mutagenic changes in DNA. For example, in the absence of homologous recombination (HR), one-ended DSBs, formed due to replication fork runoff from single-stranded DNA breaks, are repaired by toxic end-joining mechanisms to generate chromosomal fusions or arm translocations.

Linkage of DNA repair mechanisms to aging

Cells have evolved highly conserved DNA repair machineries to promote genome stability and disease prevention. The DNA damage response (DDR) is responsible for the detection of DNA damage, arrest of cell cycle progression, and eliciting an orchestrated downstream response, resulting in DNA lesion removal or tolerance. DDR is mediated by key DNA damage sensors and signaling kinases that regulate various DNA lesion-specific repair pathways.

Nucleotide excision repair

NER is the primary mechanism responsible for the removal of bulky helix-distorting adducts, such as UV-induced CPDs and 6-4 pyrimidine-pyrimidone photoproducts, as well as intrastrand cross-links and cyclopurines generated by free radicals. There are two subpathways of NER, namely the global-genome NER (GG-NER) that handles lesions present in nontranscribed regions of the genome and transcription-coupled NER (TC-NER), which is specific for transcriptionally active chromatin (Scharer, 2013). In TC-NER, stalling of RNA polymerase II by a helix-distorting lesion triggers the recruitment of downstream DNA repair factors, including the TC-NER specific factors Cockayne syndrome (CS) A and B (CSA and CSB) proteins, to effect lesion removal. Damage recognition in GG-NER is carried out by the XPC-RAD23B complex. Otherwise, steps downstream of lesion recognition are similar in both pathways. Following the recruitment of XPA, which has specific affinity for the DNA damage, DNA is unwound by the transcription factor IIH complex that harbors the XPB and XPD helicases. This leads to the formation of a bubble-like DNA structure that is stabilized by the single-stranded DNA-binding factor RPA. Dual incision (i.e., two endonucleolytic incision steps) of the damage-containing strand, being catalyzed by the DNA structure-specific endonucleases XPF-ERCC1 and XPG, excises the lesion, leaving behind a single-stranded DNA gap. Repair is then completed by the action of a DNA polymerase δ or ε and DNA ligase I or III.

Although GG-NER and TC-NER differ only in the damage recognition mechanism, mutations affecting these pathways are phenotypically distinct. Individuals with mutations in genes involved in GG-NER suffer from the skin cancer-prone syndrome xeroderma pigmentosum (XP), while those who harbor defects in TC-NER suffer from CS and, interestingly, display premature aging. Surprisingly, CS patients do not show any elevated cancer susceptibility (Marteijn, Lans, Vermeulen, & Hoeijmakers, 2014).

Several independent studies have suggested a correlation between NER capacity and age. It has been estimated that the efficiency of NER declines at a rate of approximately 0.6% every year in humans (Moriwaki, Ray, Tarone, Kraemer, & Grossman, 1996). Senescent human fibroblasts show a reduced ability to remove CPDs compared to normal dermal fibroblasts (Boyle, Kill, & Parris, 2005). Proteins that are involved in NER such as XPA and ERCC3/XPD show reduced expression in older humans (Goukassian et al., 2000). Hepatocytes isolated from young rats show a greater capacity to remove CPDs than those from old animals (Gorbunova, Seluanov, Mao, & Hine, 2007). In the nematode *Caenorhabditis elegans*, mutant animals which have a naturally longer life span show a greater capacity to remove UV-induced DNA adducts as compared to normal worms (Hyun et al., 2008). Depletion of *xpa-1* (the equivalent of human *XPA*) caused enhanced sensitivity to UV radiation and reduced the lifespan of *C. elegans* (Hyun et al., 2008).

Base excision repair

BER is initiated when a DNA glycosylase recognizes and excises the damaged base. The product of glycosylase action, the AP site, is processed by an AP-endonuclease or AP-lyase to generate a one-nucleotide gap that is filled by DNA polymerase β, followed by DNA ligation that is mediated by DNA ligase I or III (Leandro, Sykora, & Bohr, 2015). The above sequence of events is collectively known as short-patch BER. An alternate, albeit minor, pathway, termed long-patch BER, entails the removal of the AP site in a 2–10 nucleotide-long fragment involving the action of APE1 and FEN1 and gap filling by DNA polymerase δ or β and DNA ligation by DNA ligase I or IIIα. This decision between short-patch versus long-patch repair is dependent on poly(ADP-ribose) polymerase, which stimulates the strand replacement synthesis activity of Pol β with the help of FEN1 in the case of blocked terminal ribose residues to promote long-patch BER (Beneke & Burkle, 2007).

BER capacity decreases with age in rodents and humans. In rodents, older animals show a significant decline in activities of DNA glycosylases, AP endonucleases, DNA polymerase β, and DNA ligase (Xu, Herzig, Rotrekl, & Walter, 2008). Human leukocytes and fibroblasts from young individuals show higher activity of 3-methyladenine DNA glycosylase, an enzyme responsible for removing alkylated bases, compared to older persons (Atamna, Cheung, & Ames, 2000). PARP activity generally declines with age and it is, interestingly, higher in cells obtained from centenarians compared to controls (Beneke & Burkle, 2007). 8-oxoG is an endogenous oxidative DNA lesion that is a useful marker for BER activity in mammalian cells. Compared to young animals, increased levels of 8-oxoG have been observed in older murine animal models (Nie et al., 2013).

DNA double-strand break repair

The signaling cascade in response to DSBs and the ultimate removal of these lesions are mechanistically intricate. ATM is a serine/threonine kinase that is recruited and activated by the MRE11-RAD50-NBS1

(MRN) complex at DSB sites (Paull, 2015). Once activated, ATM phosphorylates a plethora of substrate proteins important for DNA repair, cell cycle regulation, transcription, and chromatin remodeling. The tumor suppressor protein p53, a central player in the DDR and instigator of apoptosis, is directly phosphorylated by ATM. The phosphorylated form of p53 mediates cell cycle arrest until the DNA damage is eliminated (Blackford & Jackson, 2017). Another key substrate of ATM is the histone variant H2AX. ATM phosphorylates multiple molecules of H2AX on serine 139 (referred to as γ-H2AX) over megabase regions surrounding a DSB site. This phosphorylation is a critical event that marks the physical location of the DSB for the timely recruitment of the DNA repair machinery and promotes a signaling cascade leading to the recruitment of p53 binding protein 1 (53BP1) (Panier & Boulton, 2014). Recruitment of 53BP1 helps shepherd DSBs into the Non-Homologous End Joining (NHEJ) pathway for repair. NHEJ is an efficient but error-prone process which entails the limited nucleolytic processing of the DSB ends and their direct religation by DNA ligase IV. Immediately upon DSB formation, Ku (heterodimer of Ku70 and Ku80) engages the DSB ends, a critical event in the initiation of DSB repair by the NHEJ pathway. Ku then recruits DNA-PKcs (DNA-dependent protein kinase catalytic subunit) and together they form the DNA-PK complex that possesses kinase activity (Chang, Pannunzio, Adachi, & Lieber, 2017). If the DSB ends are modified and unligatable, an arsenal of end-processing enzymes (such as Artemis, PNKP, TDP1/2, and the DNA polymerases μ and λ) are available to process the ends and render them ligatable (Kawale et al., 2018). The terminal step involves direct ligation of the broken DNA ends by DNA ligase IV in association with XRCC4 and XLF (Conlin et al., 2017). NHEJ often entails small DNA sequence deletions at the religated junction and is thus regarded as error-prone. NHEJ is active throughout the cell cycle and the default DSB repair mechanism in G1 phase cells, wherein HR is minimally active (Daley, Niu, Miller, & Sung, 2015).

As in the case of other DNA repair pathways, the efficiency of DSB repair and the level of DDR proteins decrease with age (White & Vijg, 2016). The abundance and activities of p53 and ATM significantly decline with age in mice (Feng et al., 2007). Aging mice and human cells subject to oxidative stress show a persistent increase in γ-H2AX foci suggesting an increase in residual unrepaired DSBs (Wang et al., 2009). There are more SSBs and DSBs in the cortical neurons of the cerebrum of aging rats (Rao, 2007). DSBs accumulate in aging ovarian follicles in mice and humans and expression of key DSB repair proteins declines with age in mouse and human oocytes (Oktay, Turan, Titus, Stobezki, & Liu, 2015). Since Ku is not essential for cell survival, several studies have examined its involvement in modulating the aging process. Mice deficient in Ku80 exhibit a significantly shortened lifespan and early onset of osteoporosis, liver pathologies, atrophic skin, hair loss, and cancer (Vogel, Lim, Karsenty, Finegold, & Hasty, 1999). NHEJ-deficient mice are also immunodeficient, which stems from a defect in VDJ and class switch recombination processes that are essential for the generation of immunoglobulins. Similar to Ku80 deficiency, mice deficient in DNA-PKcs also show a premature aging phenotype and early onset of age-related pathologies (Espejel et al., 2004). Mouse embryonic fibroblasts (MEFs) with deficiency in Ku70, Ku80, DNA ligase IV, or XRCC4 senesce prematurely in culture (Lombard et al., 2005). Hematopoietic stem cells (HSC) show an age-related decline in Ku70 levels (Prall et al., 2007). Interestingly, Ku70 expression was found to be significantly elevated in a South Korean community whose average lifespan is longer than people from other communities, suggesting the possibility of using Ku70 as a longevity marker (Ju et al., 2006).

DNA damage theory of aging

The mutational theory of aging was introduced by Failla and Szillard, who posited the DNA mutation load as the primary factor driving the aging process (Failla, 1958; Szilard, 1959). Several properties of DNA help bolster this hypothesis. Foremost, DNA is the only macromolecule that is prone to changes or rearrangements that are permanent in nature. In this regard, DNA is more prone to damage owing to its large size compared to other cellular biomolecules. Second, since DNA constitutes the genetic blueprint, any alteration in coding information can exert profound phenotypic changes in an organism. Third, various progeroid syndromes, namely, Werner syndrome (WS), Bloom syndrome (BS), CS, and XFE progeroid syndrome, that are characterized by premature or accelerated aging, are caused by germline mutations in proteins involved in DNA repair or other aspects of genome maintenance.

As discussed earlier, the cellular DNA repair capacity declines with age, which is expected to contribute to accelerated senility due to cell death caused by unrepaired DNA damage. There is a significant correlation between tumorigenesis and the aging process as well. Specifically, the frequency of tumorigenesis increases exponentially with age, such that the majority of tumors are observed in the last quarter of life. Long-term cancer survivors who have undergone radiotherapy and chemotherapy show premature signs of aging. Moreover, since DNA damage can impact stem cell survival and fitness, it likely restricts the regenerative capacity of different tissues as organisms age. In support of this premise, it has been found that HSCs deficient in NHEJ or NER show a reduced differentiation capacity compared to normal HSCs. Similarly, in the brain, DNA damage induces the senescence of neural stem cells deficient in NER factors or in the DNA damage checkpoint kinase ATM.

Aging disorders associated with DNA damage

Mutations in genes involved in genome maintenance accelerate the onset of aging phenotypes in humans, providing a strong linkage between an inability to maintain genome instability and aging. The known human segmental progeroid disorders are described below.

Diseases stemming from mutations in RecQ helicases

RecQ helicases, members of the Superfamily II (SFII) helicase family, play crucial roles in the maintenance of genome stability. There are five such RecQ helicases in humans, and defects in three of them are associated with genetic disorders characterized by increased cancer susceptibility and/or features of accelerated aging. Mutations in the *WRN* and *BLM* genes cause WS and BS, respectively, whereas mutations in *RECQL4* cause Rothmund–Thomson, RAPADILINO, and Baller–Gerold syndromes (Hanada & Hickson, 2007).

Werner syndrome

WS is named after Otto Werner, who described the disorder in 1904. It is an autosomal recessive disorder and is characterized by short stature, osteoporosis, premature hair graying and hair loss, bilateral cataracts, skin ulcers, diminished fertility, and an elevated frequency of malignancies. The cardinal premature aging signs of WS are seen by the second and third decades of life. WS affects around 1 in 200,000 individuals in the United States but occurs at a frequency 10 times higher among people of Japanese descent (Hoeijmakers, 2009; Sugimoto, 2014).

WS is caused by loss of function mutations in the WRN helicase, a 160-kDa protein. WRN is unique in that, in addition to an ATP-dependent 3′–5′ helicase activity, it also possesses a 3′–5′ exonuclease attribute within its N-terminal domain. It harbors a nuclear localization signal (NLS) in its C-terminal region (Sugimoto, 2014). The majority of mutations identified in WS patients cause truncations leading to the loss of the NLS (Hanada & Hickson, 2007). Mice with the WRN helicase domain deleted ($Wrn^{\Delta hel/\Delta hel}$) show increased chromosomal instability, a reduced median lifespan, and aging-associated phenotypes similar to WS patients (Lebel, Lavoie, Gaudreault, Bronsard, & Drouin, 2003). However, mice with a complete deletion of WRN ($Wrn^{-/-}$) do not show these abnormalities and have a normal lifespan (Hanada & Hickson, 2007).

WRN has been shown to be involved in both NHEJ and HR. As mentioned before, dysregulation of NHEJ frequently leads to an aging phenotype in animal models. WRN physically interacts with the core NHEJ machinery composed of Ku, DNA-PKcs, and XRCC4-DNA ligase IV. Ku and XRCC4-DNA ligase IV stimulate the exonuclease function of WRN. WRN also plays a role in DNA resection during HR-mediated DSB repair. Moreover, WRN has been shown to protect stalled replication forks from being degraded. Multiple studies have demonstrated an increase in chromosomal aberrations, such as translocations, deletions, and inversions, in cells derived from WS patients. WRN is also important for telomere homeostasis. It localizes to the telomeres and interacts with TRF2 and POT1, subunits of the telomere-associated Shelterin complex (Sugimoto, 2014). WRN plays an important role in telomere maintenance by regulating the formation of the t-loop structure that is important for telomere stability. It is believed that the telomere maintenance function of WRN is particularly germane for preventing cellular senescence (Hisama, Oshima, & Martin, 2016).

RECQL4 and premature aging syndromes

RECQL4 is a 133-kDa protein that, like WRN, possesses a 3′–5′ helicase activity. Mutations in RECQL4 lead to three autosomal recessive disorders namely, Rothmund–Thomson syndrome (RTS), RAPADALINO, and Baller–Gerold syndrome (BGS). Of these syndromes, RTS is the most common (Hanada & Hickson, 2007). Otherwise, the three disorders show considerable overlap with respect to the clinical features of accelerated aging. RTS is characterized by several segmental progeroid manifestations such as degenerative skin changes termed poikiloderma, short stature, skeletal defects, neurodegeneration, skeletal abnormalities, hair loss, impaired development of sexual organs, juvenile cataracts, and an increased predisposition to cancer, particularly osteosarcoma (Wang et al., 2001). RAPADILINO overlaps phenotypically with RTS but without poikiloderma, whereas BGS resembles RTS but also displays craniostenosis (Hanada & Hickson, 2007).

Similar to WRN, RECQL4 has a more general role in genome maintenance with evidence implicating it in DNA replication and in DNA repair via NHEJ, HR, BER, and NER. RECQL4 biochemically interacts with the MCM helicase complex that is indispensable for DNA replication. RECQL4 interacts with Ku70-Ku80 and its depletion leads to reduced NHEJ (Shamanna et al., 2014). RECQL4 has also been shown to promote HR-dependent DSB repair by interacting with the DNA end resection factor CtIP and promoting its recruitment to DNA break sites (Lu et al., 2016). RECQL4 also plays a key role in modulating BER capacity, such that primary fibroblasts from RTS patients experience more strand breaks upon treatment with hydrogen peroxide. RECQL4 colocalizes with APE1 and FEN1, key factors in BER. Biochemically, RECQL4 stimulates the activities of APE1, FEN1, and Pol β (Schurman et al., 2009; Werner, Prahalad, Yang, & Hock, 2006). RECQL4 mutant cells also exhibit a reduction in replicative lifespan and show an elevated expression of senescent markers (Lu, Jin, & Wang, 2017).

Mouse models have been developed to study the role of RECQL4 in the prevention of premature aging. Deletion of exons 5–8 of *Recql4* in mice is embryonically lethal. These exons just precede the helicase domain and encode one of the two NLSs. Mice with deletion of exons 9–13, which code for the helicase domain, are viable and show premature aging phenotypes such as poikiloderma, defects of the skeletal system, and increased cancer susceptibility consistent with the features of RTS patients (Mann et al., 2005). Deletion of exon 13 causes 95% of animals to die within the first 2 weeks after birth. However, the surviving animals exhibit classical premature aging features of RTS patients such as growth retardation, hair loss, premature graying of hair, short stature, and bone dysplasia (Hoki et al., 2003).

Progeroid and other disorders associated with defects in nucleotide excision repair

Defective NER leads to the development of four different autosomal recessive disorders associated with either cancer predisposition or premature aging phenotypes. These are XP, CS, trichothiodystrophy (TTD), and XFE progeroid syndrome. Among these, CS, TTD, and XFE are disorders that stem from defects in the TC-NER pathway and show premature aging features but no cancer predisposition. On the other hand, XP is associated with a defect in GG-NER and leads to highly elevated cancer proneness but not progeria. CS, TTD, and XFE patients develop some but not all progressive features of premature aging and, hence, these disorders are referred to as segmental progeroid syndromes.

Xeroderma pigmentosum

XP is an autosomal recessive disorder stemming from mutations in any of the XP genes (*XPA–XPG*) that are indispensable for GG-NER. Mutations in the *XPV* gene, which encodes DNA polymerase η, capable of the replicative bypass of UV lesions, also lead to XP, with cells harboring these mutations being proficient in NER.

Clinical features associated with XP include extreme photosensitivity, dry scaly skin, abnormal pigmentation on the skin, and up to a 1000-fold increase in skin cancer (Niedernhofer, Bohr, Sander, & Kraemer, 2011). About 20–30% of XP patients develop premature aging features, such as an early onset of neurodegeneration, loss of hearing, difficulty in swallowing, and cerebral atrophy.

Cockayne syndrome

CS is an extremely debilitating disease caused by mutations in *CSA*, *CSB*, *XPB*, *XPD*, and *XPE* genes that specifically affect TC-NER. CSA is a component of an E3 ubiquitin ligase complex, whereas CSB is a DNA-dependent ATPase belonging to the SWI/SNF family of chromatin remodelers (Pascucci et al., 2018). CS is pleiotropic in phenotype, with patients exhibiting skeletal abnormalities, premature growth cessation, arrested sexual development, severe progressive neurodegeneration, loss of hearing, thinning of hair, cataracts, a prematurely aged appearance, and a reduced life span (Iyama & Wilson, 2016). Surprisingly, despite being photosensitive and harboring a DNA repair defect, CS patients do not show any predisposition to UV-induced skin cancer. CS is further divided into three types: mild, moderate, and severe. The majority of mutations in *CSB* lead to the severe disease form, whereas mutations in the *CSA* gene tend to cause a moderate disease phenotype. In the severe form of CS, prenatal development is affected, whereas in childhood CS, which is the moderate form, accelerated degeneration occurs postnatally (De Boer & Hoeijmakers, 2000; Nance & Berry, 1992). CS cells also show a defect in the transcription-coupled removal of oxidative lesions and, hence, are more sensitive to IR-induced oxidative damage than normal cells (Pascucci et al., 2018).

Trichothiodystrophy

Mutations in genes encoding TC-NER factors can lead to TTD, a multisystem disorder that is caused by mutations in the *XPB*, *XPD*, or *TTDA* genes, associated with the RNA polymerase II transcription factor II H (TFIIH) complex (De Boer & Hoeijmakers, 2000). *XPB* and *XPD* encode the helicase subunits of the TFIIH, with TTDA serving to stabilize the complex (Giglia-Mari et al., 2004). Cells from TTD patients are impaired for transcription and show reduced XPD and XPB helicase activities. Interestingly, mutations in *XPB* and *XPD* also lead to XP, which, as mentioned above, is characterized by vastly different clinical features than TTD (Niedernhofer et al., 2018). TTD patients display photosensitivity, sulfur-deficient brittle hair, dry scaly skin, mental and physical retardation, and a reduced life span, but no increased cancer incidence. Mice engineered to harbor TTD patient mutations in the *XPD* gene display a partial defect in TC-NER and show a remarkable similarity to human TTD, such as early graying of hair, cachexia, hyperplasia of the oil glands, osteoporosis, muscle loss, and a reduced life span (Diderich, Alanazi, & Hoeijmakers, 2011).

XFE progeroid syndrome

The XPF-ERCC1 complex possesses DNA structure-specific endonuclease activity and plays a crucial role in removing CPDs, other bulky DNA adducts, and inter-strand DNA cross-links during NER (Niedernhofer et al., 2004). XPF is the catalytic subunit of the nuclease complex, whereas ERCC1 contributes toward DNA binding (Enzlin & Schärer, 2002). Mutations in *XPF* can cause XP or the XFE syndrome, with the latter being characterized by a severe defect in GG-NER as well as TC-NER and an exquisite sensitivity to ICL damage, loss of movement, loss of hearing, cerebral atrophy, high blood pressure, hepatobiliary dysfunction, osteoporosis, kidney dysfunction, and anemia (Niedernhofer et al., 2006). Mouse models with a knockout of the *Ercc1* gene have been engineered to recapitulate the human XFE syndrome. These mutant mice show a striking similarity to the premature aging features in the human syndrome. It has been posited that the mutations seen in XP patients only partially affect protein function and do not lead to cell death, thus allowing the accumulation of mutations over time. On the other hand, *XFE* mutations lead to a severe defect in NER, causing cell senescence and death, thereby leading to premature aging but not cancer (Gregg, Robinson, & Niedernhofer, 2011).

MDPL syndrome

The *POLD1* gene encodes an essential subunit of DNA polymerase δ, whose primary function is to catalyze DNA synthesis during chromosomal DNA replication. Pol δ also harbors an exonuclease function and is indispensable for cell proliferation. Germline mutations in *POLD1* cause an autosomal dominant, pleiotropic disorder called the Mandibular hypoplasia, Deafness, Progeroid features, and Lipodystrophy (MDPL) syndrome. MDPL patients display progeroid features such as osteoporosis, steatosis, and severe metabolic dysfunction including diabetes mellitus within the first two decades of life (Vermeij, Hoeijmakers, & Pothof, 2016). Most MDPL patients have a deletion of a serine residue from the polymerase catalytic domain in Pol δ (Hisama et al., 2016). As a consequence, the mutant protein is devoid of polymerase activity even though it possesses an intact exonuclease function (Fiorillo et al., 2018). It is probable that the MDPL mutations lead to replication stress to trigger cellular senescence and cell death, resulting in premature aging features in patients.

Ruijs–Aalfs syndrome

Ruijs–Aalfs syndrome (RAS) is an atypical Werner-like progeroid disorder characterized by short stature, cataracts, muscle degeneration, hair-graying sarcopenia, and diabetes (Lessel et al., 2014). Patients with RAS show mutations in the *SPRTN* gene which encodes a protein involved in the repair of DNA–protein cross-links. SPRTN is associated with enhancing TLS upon UV-induced DNA damage by stabilizing the TLS polymerase, Pol η (Hisama et al., 2016). Cells from RAS patients show increased genome stability, impaired cellular proliferation and an increase in stalled replication forks. Thus, similar to the MDPL syndrome, increased replication stress likely gives rise to the premature aging phenotype in RAS (Vermeij et al., 2016).

Conclusions and future perspectives

Given its irreplaceable nature, changes in DNA often negatively impact upon cellular physiology in highly significant ways. Within the context of the

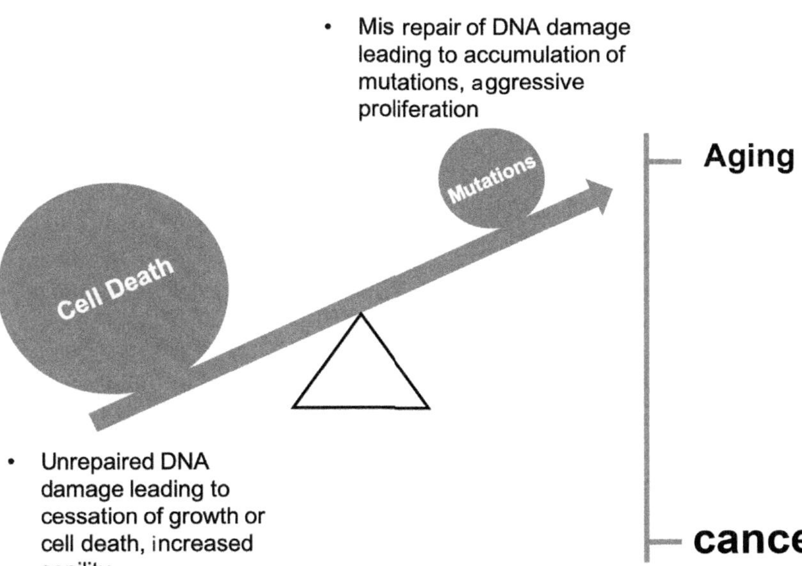

FIGURE 2.1 Aging versus cancer. Unrepaired cytotoxic DNA damage causes cell death, senility, and organismal aging, whereas a deficiency in DNA repair, which leads to the accumulation of mutations in DNA, results in carcinogenesis.

aging process, genetic mutations have been shown to affect inflammation, mitochondrial function, metabolism, autophagy, and the unfolded protein response. Dysregulation of these processes promotes cellular senescence and aging (Fig. 2.1).

As we have discussed in this chapter, a provocative hypothesis posits that unrepaired cytotoxic DNA damage causes cell death, senility, and organismal aging, whereas a deficiency in DNA repair, which leads to an accumulation of mutations in DNA, results in carcinogenesis. Ironically, aging itself represents one of the most prominent risk factors for cancer because of a general decline in DNA repair capacity and the stochastic accumulation of DNA damage. The challenge in healthy aging lies not only in finding ways to minimize the deleterious consequences of DNA damage, but also in investigating mechanisms that would improve our DNA repair capacity as we age.

Acknowledgments

Ajinkya Kawale is the recipient of a Postdoctoral Fellowship from the Cancer Prevention and Research Institute of Texas (RP170345). The research program of Patrick Sung has been supported by research grants from the US National Institutes of Health and a Team Science Grant from the Gray Foundation of New York under the Basser BRCA Initiative. Patrick Sung is a CPRIT Scholar in Cancer Research and holder of the Robert A. Welch Distinguished Chair in Chemistry.

References

Atamna, H., Cheung, I., & Ames, B. N. (2000). A method for detecting abasic sites in living cells: Age-dependent changes in base excision repair. *Proceedings of the National Academy of Sciences of the United States of America*, *97*(2), 686–691.

Beneke, S., & Burkle, A. (2007). Poly(ADP-ribosyl)ation in mammalian ageing. *Nucleic Acids Research*, *35*(22), 7456–7465. Available from https://doi.org/10.1093/nar/gkm735.

Blackford, A. N., & Jackson, S. P. (2017). ATM, ATR, and DNA-PK: The Trinity at the heart of the DNA damage response. *Molecular Cell*, *66*(6), 801–817. Available from https://doi.org/10.1016/j.molcel.2017.05.015.

Boyle, J., Kill, I. R., & Parris, C. N. (2005). Heterogeneity of dimer excision in young and senescent human dermal fibroblasts. *Aging Cell*, *4*(5), 247–255. Available from https://doi.org/10.1111/j.1474-9726.2005.00167.x.

Chang, H. H. Y., Pannunzio, N. R., Adachi, N., & Lieber, M. R. (2017). Non-homologous DNA end joining and alternative pathways to double-strand break repair. *Nature Reviews Molecular Cell Biology*, *18* (8), 495–506. Available from https://doi.org/10.1038/nrm.2017.48.

Ciccia, A., & Elledge, S. J. (2010). The DNA damage response: Making it safe to play with knives. *Molecular Cell*, *40*(2), 179–204. Available from https://doi.org/10.1016/j.molcel.2010.09.019.

Conlin, M. P., Reid, D. A., Small, G. W., Chang, H. H., Watanabe, G., Lieber, M. R., ... Rothenberg, E. (2017). DNA ligase IV guides end-processing choice during nonhomologous end joining. *Cell Reports*, *20*(12), 2810–2819. Available from https://doi.org/10.1016/j.celrep.2017.08.091.

Daley, J. M., Niu, H., Miller, A. S., & Sung, P. (2015). Biochemical mechanism of DSB end resection and its regulation. *DNA Repair*, *32*, 66–74. Available from https://doi.org/10.1016/j.dnarep.2015.04.015.

De Boer, J., & Hoeijmakers, J. H. J. (2000). Nucleotide excision repair and human syndromes. *Carcinogenesis*, *21*(3), 453–460. Available from https://doi.org/10.1093/carcin/21.3.453.

De Bont, R., & van Larebeke, N. (2004). Endogenous DNA damage in humans: A review of quantitative data. *Mutagenesis*, *19*(3), 169–185. Available from https://doi.org/10.1093/mutage/geh025.

Deriano, L., & Roth, D. B. (2013). Modernizing the nonhomologous end-joining repertoire: Alternative and classical NHEJ share the stage. *Annual Review of Genetics*, *47*(1), 433–455. Available from https://doi.org/10.1146/annurev-genet-110711-155540.

Diderich, K., Alanazi, M., & Hoeijmakers, J. H. J. (2011). Premature aging and cancer in nucleotide excision repair-disorders. *DNA Repair*, *10*(7), 772–780. Available from https://doi.org/10.1016/j.dnarep.2011.04.025.

Enzlin, J. H., & Schärer, O. D. (2002). The active site of the DNA repair endonuclease XPF-ERCC1 forms a highly conserved nuclease motif. *EMBO Journal*, *21*(8), 2045–2053. Available from https://doi.org/10.1093/emboj/21.8.2045.

Espejel, S., Martin, M., Klatt, P., Martin-Caballero, J., Flores, J. M., & Blasco, M. A. (2004). Shorter telomeres, accelerated ageing and increased lymphoma in DNA-PKcs-deficient mice. *EMBO Reports*, *5*(5), 503–509. Available from https://doi.org/10.1038/sj.embor.7400127.

Failla, G. (1958). The aging process and cancerogenesis. *Annals of the New York Academy of Sciences*, *71*(6), 1124–1140.

Feng, Z., Hu, W., Teresky, A. K., Hernando, E., Cordon-Cardo, C., & Levine, A. J. (2007). Declining p53 function in the aging process: A possible mechanism for the increased tumor incidence in older populations. *Proceedings of the National Academy of Sciences of the United States of America*, *104*(42), 16633–16638. Available from https://doi.org/10.1073/pnas.0708043104.

Fiorillo, C., D'Apice, M. R., Trucco, F., Murdocca, M., Spitalieri, P., Assereto, S., & Novelli, G. (2018). Characterization of MDPL fibroblasts carrying the recurrent p.Ser605del mutation in POLD1 gene. *DNA and Cell Biology*, *37*, 1061–1067. Available from https://doi.org/10.1089/dna.2018.4335.

Fu, D., Calvo, J. A., & Samson, L. D. (2012). Balancing repair and tolerance of DNA damage caused by alkylating agents. *Nature Reviews Cancer*, *12*(2), 104–120. Available from https://doi.org/10.1038/nrc3185.

Garinis, G. A., van der Horst, G. T., Vijg, J., & Hoeijmakers, J. H. (2008). DNA damage and ageing: New-age ideas for an age-old problem. *Nature Cell Biology*, *10*(11), 1241–1247. Available from https://doi.org/10.1038/ncb1108-1241.

Giglia-Mari, G., Coin, F., Ranish, J. A., Hoogstraten, D., Theil, A., Wijgers, N., & Vermeulen, W. (2004). A new, tenth subunit TFIIH is responsible for the DNA repair syndrome trichothiodystrophy group A. *Nature Genetics*, *36*(7), 714–719. Available from https://doi.org/10.1038/ng1387.

Gorbunova, V., Seluanov, A., Mao, Z., & Hine, C. (2007). Changes in DNA repair during aging. *Nucleic Acids Research*, *35*(22), 7466–7474. Available from https://doi.org/10.1093/nar/gkm756.

Goukassian, D., Gad, F., Yaar, M., Eller, M. S., Nehal, U. S., & Gilchrest, B. A. (2000). Mechanisms and implications of the age-associated decrease in DNA repair capacity. *FASEB Journal: Official Publication of the Federation of American Societies for Experimental Biology*, *14*(10), 1325–1334. Available from https://doi.org/10.1096/fj.14.10.1325.

Gregg, S. Q., Robinson, A. R., & Niedernhofer, L. J. (2011). Physiological consequences of defects in ERCC1-XPF DNA repair endonuclease. *DNA Repair*, *10*(7), 781–791. Available from https://doi.org/10.1016/j.dnarep.2011.04.026.

Hanada, K., & Hickson, I. D. (2007). Molecular genetics of RecQ helicase disorders. *Cellular and Molecular Life Sciences*, *64*(17), 2306–2322. Available from https://doi.org/10.1007/s00018-007-7121-z.

Harman, D. (1956). Aging: A theory based on free radical and radiation chemistry. *Journals of Gerontology*, *11*(3), 298–300.

Hisama, F. M., Oshima, J., & Martin, G. M. (2016). How research on human progeroid and antigeroid syndromes can contribute to the longevity dividend initiative. *Cold Spring Harbor Perspectives in Medicine*, *6*(4), a025882. Available from https://doi.org/10.1101/cshperspect.a025882.

Hoeijmakers, J. H. J. (2009). DNA damage, aging, and cancer. *New England Journal of Medicine*, *361*(15), 1475–1485. Available from https://doi.org/10.1056/NEJMra0804615.

Hoki, Y., Araki, R., Fujimori, A., Ohhata, T., Koseki, H., Fukumura, R., ... Abe, M. (2003). Growth retardation and skin abnormalities of the Recql4-deficient mouse. *Human Molecular Genetics*, *12* (18), 2293–2299. Available from https://doi.org/10.1093/hmg/ddg254.

Hyun, M., Lee, J., Lee, K., May, A., Bohr, V. A., & Ahn, B. (2008). Longevity and resistance to stress correlate with DNA repair capacity in *Caenorhabditis elegans*. *Nucleic Acids Research*, *36*(4), 1380–1389. Available from https://doi.org/10.1093/nar/gkm1161.

Iyama, T., & Wilson, D. M. (2016). Elements that regulate the DNA damage response of proteins defective in Cockayne syndrome. *Journal of Molecular Biology*, *428*(1), 62–78. Available from https://doi.org/10.1016/j.jmb.2015.11.020.

Jacobs, A. L., & Schar, P. (2012). DNA glycosylases: In DNA repair and beyond. *Chromosoma*, *121*(1), 1–20. Available from https://doi.org/10.1007/s00412-011-0347-4.

Ju, Y. J., Lee, K. H., Park, J. E., Yi, Y. S., Yun, M. Y., Ham, Y. H., ... Park, G. H. (2006). Decreased expression of DNA repair proteins Ku70 and Mre11 is associated with aging and may contribute to the cellular senescence. *Experimental & Molecular Medicine*, *38*(6), 686–693. Available from https://doi.org/10.1038/emm.2006.81.

Kawale, A. S., Akopiants, K., Valerie, K., Ruis, B., Hendrickson, E. A., Huang, S., ... Povirk, L. F. (2018). TDP1 suppresses misjoining of radiomimetic DNA double-strand breaks and cooperates with Artemis to promote optimal nonhomologous end joining. *Nucleic Acids Research*, *46*(17), 8926–8939. Available from https://doi.org/10.1093/nar/gky694.

Kow, Y. W. (2002). Repair of deaminated bases in DNA. *Free Radical Biology and Medicine*, *33*(7), 886–893.

Leandro, G. S., Sykora, P., & Bohr, V. A. (2015). The impact of base excision DNA repair in age-related neurodegenerative diseases. *Mutation Research*, *776*, 31–39. Available from https://doi.org/10.1016/j.mrfmmm.2014.12.011.

Lebel, M., Lavoie, J., Gaudreault, I., Bronsard, M., & Drouin, R. (2003). Genetic cooperation between the Werner syndrome protein and poly(ADP-ribose) polymerase-1 in preventing chromatid breaks, complex chromosomal rearrangements, and cancer in mice. *The American Journal of Pathology*, *162*(5), 1559–1569. Available from https://doi.org/10.1016/S0002-9440(10)64290-3.

Lessel, D., Vaz, B., Halder, S., Lockhart, P. J., Marinovic-Terzic, I., Lopez-Mosqueda, J., ... Kubisch, C. (2014). Mutations in SPRTN cause early onset hepatocellular carcinoma, genomic instability and progeroid features. *Nature Genetics*, *46*(11), 1239–1244. Available from https://doi.org/10.1038/ng.3103.

Lombard, D. B., Chua, K. F., Mostoslavsky, R., Franco, S., Gostissa, M., & Alt, F. W. (2005). DNA repair, genome stability, and aging. *Cell*, *120*(4), 497–512. Available from https://doi.org/10.1016/j.cell.2005.01.028.

Lu, H., Shamanna, R. A., Keijzers, G., Anand, R., Rasmussen, L. J., Cejka, P., ... Bohr, V. A. (2016). RECQL4 promotes DNA end resection in repair of DNA double-strand breaks. *Cell Reports*, *16*(1), 161–173. Available from https://doi.org/10.1016/j.celrep.2016.05.079.

Lu, L., Jin, W., & Wang, L. L. (2017). Aging in Rothmund-Thomson syndrome and related RECQL4 genetic disorders. *Ageing Research Reviews*, *33*, 30–35. Available from https://doi.org/10.1016/j.arr.2016.06.002.

Mann, M. B., Hodges, C. A., Barnes, E., Vogel, H., Hassold, T. J., & Luo, G. (2005). Defective sister-chromatid cohesion, aneuploidy and cancer predisposition in a mouse model of type II Rothmund-Thomson syndrome. *Human Molecular Genetics*, *14*(6), 813–825. Available from https://doi.org/10.1093/hmg/ddi075.

Marteijn, J. A., Lans, H., Vermeulen, W., & Hoeijmakers, J. H. J. (2014). Understanding nucleotide excision repair and its roles in cancer and ageing. *Nature Reviews. Molecular Cell Biology*, *15*(7), 465–481. Available from https://doi.org/10.1038/nrm3822.

Mitelman, F., Johansson, B., & Mertens, F. (2007). The impact of translocations and gene fusions on cancer causation. *Nature Reviews Cancer*, *7*(4), 233–245. Available from https://doi.org/10.1038/nrc2091.

Moriwaki, S., Ray, S., Tarone, R. E., Kraemer, K. H., & Grossman, L. (1996). The effect of donor age on the processing of UV-damaged DNA by cultured human cells: Reduced DNA repair capacity and increased DNA mutability. *Mutation Research*, *364*(2), 117–123.

Nance, M. A., & Berry, S. A. (1992). Cockayne syndrome: Review of 140 cases. *American Journal of Medical Genetics*, *42*(1), 68–84. Available from https://doi.org/10.1002/ajmg.1320420115.

Nie, B., Gan, W., Shi, F., Hu, G. X., Chen, L. G., Hayakawa, H., ... Cai, J. P. (2013). Age-dependent accumulation of 8-oxoguanine in the DNA and RNA in various rat tissues. *Oxidative Medicine and Cellular Longevity*, *2013*, 303181. Available from https://doi.org/10.1155/2013/303181.

Niedernhofer, Laura J., Bohr, V. A., Sander, M., & Kraemer, K. H. (2011). Xeroderma pigmentosum and other diseases of human premature aging and DNA repair: Molecules to patients. *Mechanisms of Ageing and Development*, *132*(6–7), 340–347. Available from https://doi.org/10.1016/j.mad.2011.06.004.

Niedernhofer, Laura J., Garinis, G. A., Raams, A., Lalai, A. S., Robinson, A. R., Appeldoorn, E., ... Hoeijmakers, J. H. J. (2006). A new progeroid syndrome reveals that genotoxic stress suppresses the somatotroph axis. *Nature*, *444*(7122), 1038–1043. Available from https://doi.org/10.1038/nature05456.

Niedernhofer, Laura J., Gurkar, A. U., Wang, Y., Vijg, J., Hoeijmakers, J. H. J., & Robbins, P. D. (2018). Nuclear genomic instability and aging. *Annual Review of Biochemistry*, *87*(1), 295–322. Available from https://doi.org/10.1146/annurev-biochem-062917-012239.

Niedernhofer, L. J., Odijk, H., Budzowska, M., van Drunen, E., Maas, A., Theil, A. F., ... Kanaar, R. (2004). The structure-specific endonuclease Ercc1-Xpf is required to resolve DNA interstrand cross-link-induced double-strand breaks. *Molecular and Cellular Biology*, *24*(13), 5776–5787. Available from https://doi.org/10.1128/mcb.24.13.5776-5787.2004.

Oktay, K., Turan, V., Titus, S., Stobezki, R., & Liu, L. (2015). BRCA mutations, DNA repair deficiency, and ovarian aging. *Biology of Reproduction*, *93*(3), 67. Available from https://doi.org/10.1095/biolreprod.115.132290.

Panier, S., & Boulton, S. J. (2014). Double-strand break repair: 53BP1 comes into focus. *Nature Reviews. Molecular Cell Biology*, *15*(1), 7–18. Available from https://doi.org/10.1038/nrm3719.

Pascucci, B., Fragale, A., Marabitti, V., Leuzzi, G., Calcagnile, A. S., Parlanti, E., ... D'Errico, M. (2018). CSA and CSB play a role in the response to DNA breaks. *Oncotarget*, *9*(14), 11581–11591. Available from https://doi.org/10.18632/oncotarget.24342.

Paull, T. T. (2015). Mechanisms of ATM activation. *Annual Review of Biochemistry*, *84*, 711–738. Available from https://doi.org/10.1146/annurev-biochem-060614-034335.

Prall, W. C., Czibere, A., Jager, M., Spentzos, D., Libermann, T. A., Gattermann, N., ... Aivado, M. (2007). Age-related transcription levels of KU70, MGST1 and BIK in CD34 + hematopoietic stem and progenitor cells. *Mechanisms of Ageing and Development*, *128*(9), 503–510. Available from https://doi.org/10.1016/j.mad.2007.06.008.

Rao, K. S. (2007). DNA repair in aging rat neurons. *Neuroscience*, *145*(4), 1330–1340. Available from https://doi.org/10.1016/j.neuroscience.2006.09.032.

Scharer, O. D. (2013). Nucleotide excision repair in eukaryotes. *Cold Spring Harbor Perspectives in Biology*, *5*(10), a012609. Available from https://doi.org/10.1101/cshperspect.a012609.

Schurman, S. H., Hedayati, M., Wang, Z., Singh, D. K., Speina, E., Zhang, Y., ... Bohr, V. A. (2009). Direct and indirect roles of RECQL4 in modulating base excision repair capacity. *Human Molecular Genetics*, *18*(18), 3470–3483. Available from https://doi.org/10.1093/hmg/ddp291.

Shamanna, R. A., Singh, D. K., Lu, H., Mirey, G., Keijzers, G., Salles, B., ... Bohr, V. A. (2014). RECQ helicase RECQL4 participates in non-homologous end joining and interacts with the Ku complex. *Carcinogenesis*, *35*(11), 2415–2424. Available from https://doi.org/10.1093/carcin/bgu137.

Sugimoto, M. (2014). A cascade leading to premature aging phenotypes including abnormal tumor profiles in Werner syndrome (Review). *International Journal of Molecular Medicine*, *33*(2), 247–253. Available from https://doi.org/10.3892/ijmm.2013.1592.

Szilard, L. (1959). On the nature of the aging process. *Proceedings of the National Academy of Sciences of the United States of America*, *45*(1), 30–45.

Tubbs, A., & Nussenzweig, A. (2017). Endogenous DNA damage as a source of genomic instability in cancer. *Cell*, *168*(4), 644–656. Available from https://doi.org/10.1016/j.cell.2017.01.002.

Vermeij, W. P., Hoeijmakers, J. H. J., & Pothof, J. (2016). Genome integrity in aging: Human syndromes, mouse models, and therapeutic options. *Annual Review of Pharmacology and Toxicology*, *56*(1), 427–445. Available from https://doi.org/10.1146/annurev-pharmtox-010814-124316.

Vogel, H., Lim, D. S., Karsenty, G., Finegold, M., & Hasty, P. (1999). Deletion of Ku86 causes early onset of senescence in mice. *Proceedings of the National Academy of Sciences of the United States of America*, *96*(19), 10770–10775.

Wang, C., Jurk, D., Maddick, M., Nelson, G., Martin-Ruiz, C., & von Zglinicki, T. (2009). DNA damage response and cellular senescence in tissues of aging mice. *Aging Cell*, *8*(3), 311–323. Available from https://doi.org/10.1111/j.1474-9726.2009.00481.x.

Wang, L. L., Levy, M. L., Lewis, R. A., Chintagumpala, M. M., Lev, D., Rogers, M., & Plon, S. E. (2001). Clinical manifestations in a cohort of 41 Rothmund-Thomson syndrome patients. *American Journal of Medical Genetics*, *102*(1), 11–17.

Werner, S. R., Prahalad, A. K., Yang, J., & Hock, J. M. (2006). RECQL4-deficient cells are hypersensitive to oxidative stress/damage: Insights for osteosarcoma prevalence and heterogeneity in Rothmund-Thomson syndrome. *Biochemical and Biophysical Research Communications*, *345*(1), 403–409. Available from https://doi.org/10.1016/j.bbrc.2006.04.093.

White, R. R., & Vijg, J. (2016). Do DNA double-strand breaks drive aging? *Molecular Cell*, *63*(5), 729–738. Available from https://doi.org/10.1016/j.molcel.2016.08.004.

Xu, G., Herzig, M., Rotrekl, V., & Walter, C. A. (2008). Base excision repair, aging and health span. *Mechanisms of Ageing an d Development*, *129*(7–8), 366–382. Available from https://doi.org/10.1016/j.mad.2008.03.001.

C H A P T E R

3

Mechanisms of cell senescence in aging

Thomas von Zglinicki

Newcastle University Biosciences Institute, Campus for Ageing and Vitality, Newcastle University, Newcastle upon Tyne, United Kingdom

O U T L I N E

What is senescence? Building blocks and feedback loops

Replicative cell senescence was first described in 1961 by Len Hayflick as reproducible, permanent loss of the replicative capacity of human primary fibroblasts during in vitro culture (Hayflick & Moorhead, 1961). The discoveries of replication-associated telomere shortening (Harley, Futcher, & Greider, 1990), rescue of senescence by telomerase-mediated telomere lengthening (Bodnar et al., 1998), and association of a canonical DNA damage response (DDR) with telomeres in senescence (d'Adda di Fagagna et al., 2003) identified cell replicative senescence (RS) as a telomere length-driven, DDR-mediated persistent activation of the cell cycle checkpoint machinery (Ben-Porath & Weinberg, 2004).

However, recent research has greatly complicated this simple picture of senescence as a permanent cell cycle arrest. First of all, senescent cells differ from their nonsenescent counterparts in many essential aspects beyond proliferation capacity. These include major reorganization of their secretome, their energy metabolism, their epigenome, nutrient signaling, autophagy, and mitophagy activities. Second, these phenotypic characteristics of senescence are interlinked and, in many cases, actually stabilize each other and the senescent DDR/growth arrest in positive feedback loops. Finally, the association between proliferation arrest and senescent phenotype can be broken in different ways: tumor cells, for instance, are not only able to enter a senescent state (e.g., following DNA-damaging stress) but can also recover from it. If and when they do so, they retain certain epigenetic characteristics which they gained as senescent cells (Milanovic, Yu, & Schmitt, 2018; Shah et al., 2013). Conversely, postmitotic somatic cells can make the transition toward a senescent phenotype in response to DNA damage at a time point long after they ceased replication (Jurk et al., 2012).

Handbook of the Biology of Aging.
DOI: https://doi.org/10.1016/B978-0-12-815962-0.00003-2

Thus, it might be helpful not to regard a persistent cell cycle arrest as the sole defining feature of senescence. Rather, senescence may be defined as a cellular stress response composed of a number of interlinked building blocks (Fig. 3.1). These building blocks give rise to phenotypic consequences (i.e., the DDR mediates the cell cycle arrest; autophagy dysfunction causes lysosome build-up and the accumulation of SA-β-galactosidase, etc.). They are interlinked by a number of connecting signaling pathways that form multiple feedback loops. These feedback loops not only maintain and stabilize the phenotype (see "The building blocks of the senescent phenotype" section for a detailed discussion), but they also ensure that very different stresses and interventions always trigger essentially a very similar phenotype. For instance, telomere uncapping, autophagy inhibition, or induction of mitochondrial dysfunction all cause cell senescence with involvement of all the six building blocks depicted in Fig. 3.1. Clearly, the degree of involvement of individual blocks and the specifics of the phenotypic consequences can differ depending on the initiating stress, the block that is its primary target, genomic and epigenomic state of the cell prior to the stress, etc. For instance, the composition of the senescence-associated secretory phenotype (SASP) may differ depending on the type of DNA damage (Rodier & Campisi, 2011) or whether a DDR or mitochondrial dysfunction was the primary inducer of senescence (Wiley et al., 2016). Importantly, we propose that senescence is defined by the involvement of all, or at least the vast majority, of these building blocks.

This definition of senescence implies that it does not prioritize certain features over others. For instance, senescence is frequently identified by reference to DDR/cell cycle arrest and SASP only, with additional features added only as an afterthought. While this may be sufficient for practical operationalization in many cases, it does not cover the full complexity of the senescent stress response and might fall short even of practical utility in more complex cases like tumor or postmitotic cell senescence mentioned above. The major advantage of defining senescence as interacting building blocks is, of course, that it easily explains how very different stressors and interventions, targeting, for instance, nuclear DNA, the proteasome, mitochondria, the inflammasome, or DNA methylation, can all cause essentially the same cellular response.

Cell senescence is a kinetic phenotype. Its induction from onset of a DDR to full-blown, "deep" senescence with participation of all the building blocks shown in Fig. 3.1 takes anything between about 2–4 weeks. Fig. 3.2 shows approximate timelines for the build-up of some marker parameters in human fibroblast senescence. While these timelines are by necessity approximate and while they are not known for all of the building blocks, or for many stressors other than DNA damage, it is clear that the different building blocks engage with different kinetics (Fig. 3.2). While kinetics of SASP induction seem to be generally similar in stress-

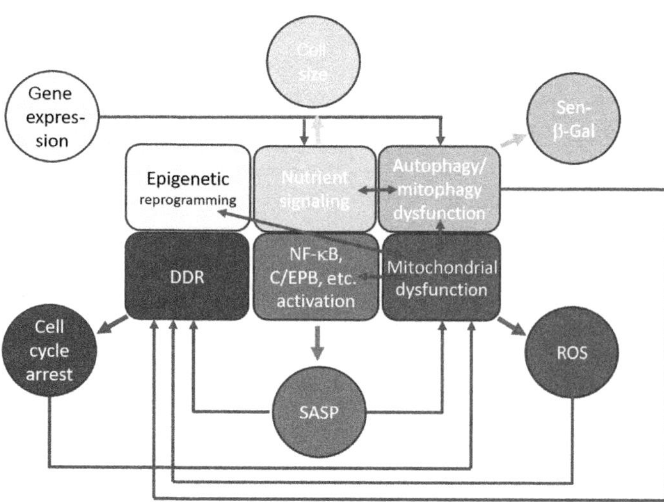

FIGURE 3.1 The senescent phenotype: building blocks, phenotypic consequences, and interlinks. The scheme shows major building blocks of the senescent phenotype (*squares*), main phenotypic consequences (*circles*), and a selection of signaling pathways interlinking them.

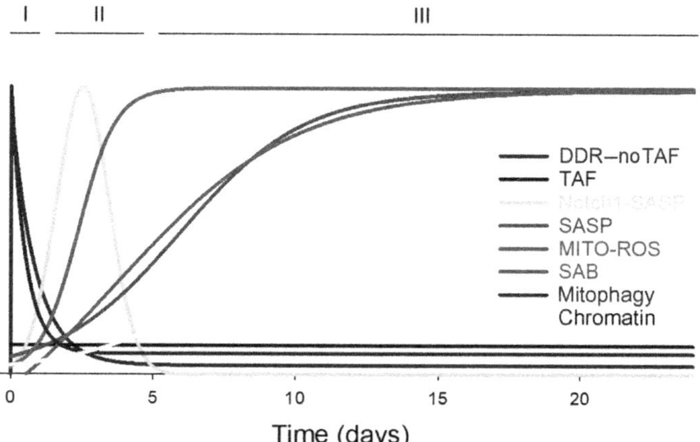

FIGURE 3.2 Time course of activation of key processes in human fibroblast senescence. Times are approximate. DDR—noTAF: frequency of γH2AX foci not colocalizing with telomeres per nucleus (Hewitt et al., 2012; Passos et al., 2010). TAF: frequency of γH2AX foci colocalizing with telomeres per nucleus (Hewitt et al., 2012). Notch1-SASP: abundance of the Notch1 target HES1 and of TGFβ1 (Hoare et al., 2016). SASP: IL-6 secretion (Rodier et al., 2009). MITO-ROS: mitochondrial mass and Mito-SOX fluorescence (Passos et al., 2010). SAB: senescence-associated β-galactosidase activity (Dalle Pezze et al., 2014). Mitophagy: mitophagic flux per hour, measured by LC3-COX4 colocalization +/− bafilomycin A for 1 h (Dalle Pezze et al., 2014). Chromatin: Chromatin reprogramming measured as nuclear granularity and HP1 foci formation (Passos et al., 2010). Senescence was triggered by induction of $HRAS^{G12V}$ or etoposide treatment (Notch1-SASP) or by ionizing radiation (10–20 Gy, all other conditions) at $t = 0$.

induced and oncogene-induced senescence (Acosta et al., 2008; Coppe et al., 2008; Hoare et al., 2016; Kuilman et al., 2008), kinetics of senescence induction following mitochondrial dysfunction, mitophagy suppression or epigenetic reprogramming might be different. The data in Fig. 3.2 suggest that DNA damage–induced senescence at least progresses through three distinct phases: (1) an early damage response phase, characterized by fast decline of the nontelomeric DDR (although not completely to prestress levels) and the mitophagic flux; (2) an intermediate senescence induction phase combining Notch1-dependent SASP and the induction of mitochondrial dysfunction; and (3) a late build-up phase showing increasing induction of a proinflammatory SASP accompanied by increasing SA-β-gal activity and even more slowly by chromatin reorganization. Importantly, growth arrest is easily reversible during the first and second phases even after high doses of radiation (Dalle Pezze et al., 2014; Passos et al., 2010), stressing the point that the cooperative engagement of multiple building blocks is necessary for senescence.

It is worth noting that we know even less about long-term kinetic changes of the senescent phenotype. Data on RS and from senescent cells in vivo, averaging over cells at very different times after senescence induction, suggest more or less stable activation of the building blocks over time. Interestingly, changes with time are being seen by anecdotal morphologic observation of long-term senescent cell cultures. Whether this indicates an aging process within senescent cells is not clear.

We do not claim that the list of building blocks and phenotypes presented in Fig. 3.1 is exhaustive. For instance, resistance to apoptosis is frequently regarded as an essential feature of senescent cells, especially in the context of antisenescence interventions (Zhu et al., 2015). A discussion in the wider community is ongoing (Gorgoulis et al., 2019) and is expected to improve the identification of the most salient features of the senescent phenotype. Awaiting the outcome of this discussion, we will here first discuss the building blocks individually as well as some of their interconnecting pathways ("The building blocks of the senescent phenotype" section). We will then try to put some frequently used terminology in the senescent field into perspective ("Terminology" section), cover accumulation and detection of senescent cells during aging in vivo ("Senescence during aging in vivo" section) and the impact of senolytic and senostatic interventions on the aging process ("Senolytics and senostatics as antiaging interventions" section). For brevity, a number of aspects of cell senescence are not covered here. The reader is directed to recent excellent reviews on senescence during development and regeneration (He & Sharpless, 2017; Hernandez-Segura, Nehme, & Demaria, 2018; Munoz-Espin & Serrano, 2014), on oncogene-induced and therapy-induced senescence (Larsson, 2011; Lee & Schmitt, 2019; Peeper, 2011; Short, Fielder, Miwa, & von Zglinicki, 2019), and on the interaction of senescent cells with the immune system (Lasry & Ben-Neriah, 2015).

The building blocks of the senescent phenotype

Telomeres and the DNA damage response

In 1990, Cal Harley demonstrated shortening of telomeres with ongoing population doublings in human fibroblasts (Harley et al., 1990) and a few years later Bodnar et al. showed that RS could be bypassed if telomere length was maintained by telomerase overexpression (Bodnar et al., 1998). Originally it was thought that the connection between replication and telomere shortening was solely due to the inability of DNA polymerases to copy the lagging strand to its very end [the so-called "end-replication problem" (Olovnikov, 1973)]. However, it was soon found that relatively mild oxidative stress could contribute significantly to replication-associated telomere shortening because of a telomere-specific single-strand break repair deficiency, indicating that telomeres might not just be passive replication counters but sentinels for cell-autonomous and nonautonomous stress (von Zglinicki, 2002). When it was found mechanistically that short uncapped telomeres in senescent cells induced a classical DDR (d'Adda di Fagagna et al., 2003; Takai, Smogorzewska, & de Lange, 2003), telomere shortening was firmly established as a driver of RS in human somatic cells.

However, shortening is not the only possible way to "uncap" a telomere such that it triggers a DDR. Depletion of shelterin components, a protein complex that binds specifically telomeric DNA and protects it from recognition by the DDR, initiated immediate senescence with long telomeres (de Lange, 2018). Physiologically more important, it was later found that apparently intact, long telomeres in the presence of good amounts of shelterin proteins could induce a DDR (Fumagalli et al., 2012; Hewitt et al., 2012). Such telomere-associated DNA damage foci (TAF) were found in both mice and human cells with aging and were associated with stress but not with telomere shortening (Anderson et al., 2019; Fumagalli et al., 2012; Hewitt et al., 2012; Jurk et al., 2014). While they have not been conclusively analyzed mechanistically, it has been suggested that they might resemble stress-induced intratelomeric double-strand breaks (Fumagalli et al., 2012). In accordance with that, they have been found in slowly proliferating and postmitotic cells, including hepatocytes (Jurk et al., 2014; Ogrodnik et al., 2017), skeletal muscle fibers (da Silva et al., 2019), neurons (Ogrodnik et al., 2019), and cardiomyocytes (Anderson et al., 2019), and might constitute a major trigger for senescence in these cells and organs.

DNA damage foci at nontelomeric sites in the genome have in general short lifespans and are typically resolved with half-lives in the order of 3–10 hours (Passos et al., 2010) following damage repair and proteasomal degradation of the foci components. In contrast, TAF are persistent and remain stable in the nucleus for at least multiple days (Anderson et al., 2019; Hewitt et al., 2012; Passos et al., 2010). This is assumed to reflect the deficiency of telomeres for multiple modes of DNA repair including nucleotide excision repair (Kruk, Rampino, & Bohr, 1995), base excision (single-strand break) repair (Richter et al., 2007; Sitte, Saretzki, & von Zglinicki, 1998) and double-strand break repair (de Lange, 2018). Given the necessity of maintaining a DDR for a couple of days to engage the different components and to secure the stability of the phenotype (see Fig. 3.2), it is obvious that TAF are more efficient triggers of senescence than nontelomeric DNA damage.

The DDR is linked intimately to all other aspects of senescence. The tumor suppressor p53 is the most ubiquitous transcription factor activated by DNA damage. It activates hundreds of genes (Hafner, Bulyk, Jambhekar, & Lahav, 2019), controls the SASP directly and indirectly (Lopes-Paciencia et al., 2019), regulates glucose metabolism, mitochondrial function and autophagy (Itahana & Itahana, 2018), and interacts with multiple chromatin modulators (Hafner et al., 2019). C/EBPβ and NF-κB activation and SASP have been associated with the DDR both as a consequence of and a contributor to it, thus stabilizing the senescent phenotype (Acosta et al., 2008; Kuilman et al., 2008; Rodier et al., 2009). However, overexpression of the DDR-driven CDKIs p21 or p16 did arrest cell proliferation without inducing an SASP (Coppe et al., 2008, 2011). However, signaling through p53 and p21 was found necessary to induce senescence-associated mitochondrial dysfunction (SAMD) through a pathway involving p38MAPK and TGFβ (Passos et al., 2010). In turn, SAMD caused hyperproduction of reactive oxygen species (ROS) in senescence, which amplified DNA damage and the DDR, constituting another example for a positive feed-forward loop-stabilizing senescence (Passos et al., 2010). Further pathways that drive SAMD in response to a DDR in senescence exist: for instance, it was recently shown that the nuclear retinoid X receptor RXRα (and possibly also related receptors) can control senescence by repressing expression of the inositol 1,4,5-triphosphate receptor type 2, an endoplasmic reticulum calcium release channel. Suppression of RXRα-induced calcium release from the ER was followed by uptake by mitochondria, enhanced ROS production, mitochondrial dysfunction, nuclear DDR, and senescence (Ma et al., 2018). Other integrative pathways are discussed below.

Senescence-associated secretory phenotype

In 2008, two back-to-back papers in *Cell* showed that, in senescence, proinflammatory master transcription factors NF-κB and C/EBPβ are activated, causing production and secretion of cytokines IL-6 and IL-8, which in turn stabilized senescence via paracrine signaling (Acosta et al., 2008; Kuilman et al., 2008). This SASP was then further characterized and shown to include, in addition to proinflammatory cytokines (including IL-6, IL-8, IL-1a, IL-1b), chemokines (CXCL1 and others), matrix modifiers (PAI-1, MMPs), and other bioactive peptides (Coppe et al., 2008; Rodier et al., 2009). While these are "typical" components of the SASP, its specific composition may vary, depending on the senescence-inducing mechanism, activity/presence of DDR components p53, p16 and p21, and cell type (Rodier & Campisi, 2011; Tchkonia, Zhu, van Deursen, Campisi, & Kirkland, 2013). More recently, the Narita group showed that senescent cells actually produce at least two mechanistically distinct and kinetically separate (see Fig. 3.2) forms of SASP. An early response is driven by the Wnt/Notch-1 pathway and contains mainly antiinflammatory peptides, specifically TGFβ (Hoare et al., 2016). The NF-κB- and C/EBPβ-dependent proinflammatory response develops only after decline of Notch-1 signaling. It is not clear yet whether the first response is necessary for development of the second in multiple variants of senescence. Both forms of SASP can attract and activate immunosurveillance of senescent cells by monocytes/macrophages (Hoare & Narita, 2018). Accumulation of cytoplasmic chromatin fragments activating the cytosolic DNA-sensing cGAS-STING pathway is a major cause of the proinflammatory SASP; deletion of either cGAS (Yang, Wang, Ren, Chen, & Chen, 2017) or STING (Dou et al., 2017) abrogated the proinflammatory SASP. Both nuclear (Dou et al., 2017) and mitochondrial (Birch & Passos, 2017) DNA are damaged in senescence and may be released into the cytoplasm. Their relative importance for activation of the cGAS-STING pathway in multiple forms of senescence is not clear.

Senescence-associated mitochondrial dysfunction

That externally triggered mitochondrial dysfunction and associated production of ROS can induce senescence, for instance by accelerating telomere shortening, has been known for a long time (Chen & Ames, 1994;

Passos et al., 2007; Saretzki, Murphy, & von Zglinicki, 2003; Toussaint, Medrano, & von Zglinicki, 2000; von Zglinicki, Saretzki, Docke, & Lotze, 1995). Genetically or pharmacologically induced mitochondrial dysfunction can also trigger senescence without involvement of ROS, although in this case the SASP appears different, that is, less proinflammatory (Wiley et al., 2016). Importantly, mitochondrial dysfunction, phenotypically characterized by increased ROS production despite reduced ETC coupling and accompanied by increased mitochondrial mass, is an inherent part of the senescent phenotype. SAMD is induced as a downstream consequence of the DDR in both oncogene-induced (Moiseeva, Bourdeau, Roux, Deschenes-Simard, & Ferbeyre, 2009), IGF-1-mediated (Bitto et al., 2010), and telomere- or stress-induced senescence (Korolchuk, Miwa, Carroll, & von Zglinicki, 2017; Passos et al., 2010). Similar to mitochondrial dysfunction during tissue aging, SAMD is primarily associated with complex I dysfunction (Korolchuk et al., 2017). A cause of complex I dysfunction in aging is decreasing efficiency of complex assembly, resulting in subcomplexes unable to participate in electron transport but still generating ROS (Miwa et al., 2014). Whether induction of senescence decreases the efficiency of complex I assembly has not been established.

The signaling pathways that cause SAMD in response to a DDR have been partially established to involve activation of p38MAPK and TGFβ (Passos et al., 2010). They also involve a shift in the net balance between mitochondrial biogenesis and mitophagy, possibly related to a readjustment of nutrient signaling through mechanistic target of rapamycin (mTOR) (Korolchuk et al., 2017; see section "Nutrient signaling"). Enhanced ROS production is one consequence of SAMD, and these ROS contribute to continuous DNA damage and DDR in senescence, forming a senescence-stabilizing forward feedback loop (Passos et al., 2010). This is one reason why mitochondria are essential for multiple facets of the senescent phenotype: when mitochondria were specifically degraded in cells after induction of senescence, the SASP, p16 and p21 activation, senescent morphology, and SA-β-gal activity were all abrogated (Correia-Melo et al., 2016).

SASP and SAMD interact closely with each other: hyperactivation of NF-κB, a master regulator of the proinflammatory SASP, by knockout of the inhibitory subunit NFKB1, induced senescence with aggravated ROS production in vitro and in vivo (Jurk et al., 2014). However, pharmacologic or genetic suppression of NF-κB activity did not reduce ROS levels in senescent cells, while, conversely, scavenging of ROS in senescence diminished NF-κB activity (Nelson, Kucheryavenko, Wordsworth, & von Zglinicki, 2018). Together with the different kinetics of SAMD and SASP induction in senescence (see Fig. 3.2) and the abrogation of the SASP by mitochondrial ablation (Correia-Melo et al., 2016) these data suggest senescence-associated ROS production as necessary and sufficient for the proinflammatory, NF-κB-dependent SASP.

Nutrient signaling

The mTOR kinase complex is the central regulator of the balance between translation, nucleotide biosynthesis, lipogenesis, glycolysis, and autophagy in response to nutrient and amino acid availability. Normally, starvation decreases mTOR activity, favoring autophagy (Laplante & Sabatini, 2012). However, in cell senescence, mTOR activity remains constitutively high, even under amino acid and nutrient starvation (Carroll et al., 2017; Zhang, Hoff, Marinucci, Cristofalo, & Sell, 2000). It has been suggested that this aberrant activation of mTOR in senescence might be driven by increased oxidative stress caused by SAMD (Nacarelli, Azar, & Sell, 2015). In fact, nutrient-independent activation of mTOR by oxidants has been shown in HEK293 cells (Sarbassov & Sabatini, 2005). Moreover, membrane depolarization and a resultant defect in cilia formation contribute to constitutive PI3K/mTORC1 signaling in senescence (Carroll et al., 2017). Furthermore, autophagic flux is enhanced in senescence (Young, Narita, & Narita, 2011), facilitating cell-autonomous replenishment of amino acid pools, which might help to maintain high levels of mTOR activity in the absence of external stimuli (Carroll et al., 2017).

High mTOR activity is an essential component of the senescent phenotype. Treatment of cells with the mTOR complex I inhibitor rapamycin delayed RS (Kolesnichenko, Hong, Liao, Vogt, & Sun, 2012; Lerner et al., 2013). mTOR activity was positively associated with senescence and PGC1α/β-driven mitochondrial biogenesis and mitochondrial dysfunction in fibroblasts (Dalle Pezze et al., 2014) and lung epithelial cells in vitro (Summer et al., 2019) and in vivo (Araya et al., 2019). Mechanistically, both mTORC1 and mTORC2 bind to p53 in competition with MDM2 and phosphorylate it at serine 15, thus enhancing its stability in the absence of DNA damage, leading to increased activation of p21 (Jung et al., 2019). If rapamycin is applied to senescent cells, it acts as a senostatic, that is, it suppresses multiple aspects of the senescent phenotype including CDKI activation, SASP, and SAMD (but not growth arrest) as efficiently as mitochondrial ablation (Aarts et al., 2017; Correia-Melo et al., 2016; Demidenko et al., 2009).

Autophagy/mitophagy

Autophagy describes several cellular pathways facilitating degradation of intracellular components by lysosomal proteases. Amongst autophagy pathways macroautophagy (hereafter for simplicity called autophagy) is the best-studied process. It begins with the engulfment of surplus or damaged proteins and entire organelles by autophagosomes that eventually fuse with lysosomes. The resulting autolysosome allows degradation of autophagy substrates and subsequent release of basic components, such as amino acids, lipids, and nucleotides, back into the cytoplasm. Autophagy activity is typically increased in conditions of nutrient deprivation sensed via the mTORC1 pathway and serves as a recycling process necessary for cell survival during starvation. Importantly, autophagy is also a quality control mechanism, selectively scavenging potentially toxic cellular components.

Autophagy is implicated in cellular senescence in a complex and often contradictory relationship (Korolchuk et al., 2017). On one hand, autophagy activation via overexpression of the autophagy gene *ULK3* promoted senescence induced by the RAS oncogene in human fibroblasts, mediated by a specialized intracellular structure called TOR-autophagy spatial coupling compartment, while knockdown of the essential autophagy genes *ATG5* and *ATG7* was found to decrease SASP and bypass senescence (Narita et al., 2011; Young et al., 2009). Likewise, overexpression of the inhibitors of cyclin-dependent kinases p16(INK4A), p19(ARF), or p21(WAF1/CIP1) induced not only senescence as expected but also autophagy (Capparelli et al., 2012). Furthermore, autophagy induction by a proteolytic cyclin E fragment (p18-CycE) was shown to facilitate DNA-damage-induced senescence (Singh, Matsuyama, Drazba, & Almasan, 2012). In contrast to these results, autophagy can also act as a mechanism to bypass oncogene-induced senescence (OIS) (Wang et al., 2012). Moreover, inhibition of autophagy in young proliferating fibroblasts led to senescence (Kang, Lee, Kim, Choi, & Park, 2011), while increased expression of autophagy genes *LC3B*, *ATG5*, and *ATG12* and enhanced mitophagy postponed senescence and enhanced replicative lifespan (Mai, Muster, Bereiter-Hahn, & Jendrach, 2012; Tai et al., 2017). Timing and intensity of changes in the autophagic flux might explain these differential impacts on the senescence process.

Mitophagy describes the selective autophagy of (dysfunctional) mitochondria. While general autophagy may become elevated or reduced depending on the model of senescence, mitophagy activity is reduced in senescent cells in vitro and in vivo (Dalle Pezze et al., 2014; Garcia-Prat et al., 2016). This might be a consequence of constitutive activation of mTORC1 in senescence (see above). Likewise, lysosomal overload, as shown by the accumulation of lipofuscin (Sitte, Merker, Grune, & von Zglinicki, 2001), occurs in senescence, despite an extensive expansion of the lysosomal compartment that gives rise to a greatly enhanced β-galactosidase activity, even at suboptimal pH (Kurz, Decary, Hong, & Erusalimsky, 2000). This will disrupt the terminal events of substrate degradation in both general autophagy and mitophagy pathways. Moreover, mitophagy may become impaired due to mechanisms independent of general autophagy. Mitophagy is governed by the PINK1 (PTEN induced putative protein kinase 1)-PRKN (parkin E3 ubiquitin protein ligase) pathway. Cytoplasmic p53, which accumulates in senescence, interacts with Parkin and prevents its translocation to dysfunctional mitochondria, thus suppressing mitophagy and stabilizing senescence. Knockout of *Prkn* or inhibition of PRKN-mediated mitophagy resulted in mitochondrial dysfunction and senescence (Araya et al., 2019; Manzella et al., 2018). Similarly, expression of PINK1 is low in aging, which suppresses the process of mitochondrial fission that is important for mitophagy by facilitating the separation of dysfunctional portions from the mitochondrial network. *Pink1* knockdown again accelerated senescence (Wang, Shen, et al., 2018). Reduced dynamics and increased fusion of the mitochondrial network in senescent cells has been proposed to result in the failure to sequester defective components of the mitochondrial network for degradation via mitophagy (Dalle Pezze et al., 2014). As a result, "old" mitochondria avoid degradation, while newly synthesized isolated mitochondria are turned over by mitophagy. The extent to which insufficient mitophagy might be a cause of SAMD remains to be established.

Epigenetic reprogramming

Cell senescence is associated with large-scale changes in chromatin organization, which overlap significantly not only with those seen during tissue aging (for review see Booth & Brunet, 2016; Sen, Shah, Nativio, & Berger, 2016) but also with those seen during transition to a more stem cell-like phenotype (Milanovic, Fan, et al., 2018). In human cell senescence, more than 30% of the chromatin is reorganized, including formation of large-scale domains (mesas) of H3K4me3 and H3K27me3 over lamin-associated domains and large-scale losses (canyons) of H3K27me3 outside these areas (Shah et al., 2013). These changes are associated with transcriptional downregulation and autophagic degradation of lamin B1 (Dou et al., 2015; Freund, Laberge, Demaria, & Campisi, 2012). Senescent cells also show enrichment of H4K16ac at promoters of active genes, overlapping with the histone chaperone HIRA (Rai et al., 2014). Globally, there is reduced DNA methylation and H3K9me3 and H3K27me3 levels (Scaffidi & Misteli, 2006), but formation of heterochromatic foci (senescence-associated heterochromatin foci, SAHF) in otherwise euchromatic regions containing H3K9me3, macroH2A, and DDR proteins like γH2AX (Narita et al., 2003). These chromatin changes are characteristic for the late-stage, fully developed senescent phenotype (Fig. 3.2).

There is strong evidence showing that epigenetic reprogramming is both driven by changes in the other building blocks of senescence and contributes to the stabilization of the phenotype. For instance, changes in mitochondrial function trigger gene expression changes in the nucleus, the retrograde response (Butow & Avadhani, 2004), which is active in cell senescence (Passos et al., 2007). At least in yeast, the retrograde response to mitochondrial stress involves large-scale changes in heterochromatin and H3K9 methylation (Tian et al., 2016). Sirtuins are histone deacetylases that affect both mitochondrial biogenesis and function (via a decline in NAD^+ levels and increased HIF-1α transcription factor activity; Gomes et al., 2013) as well as DNA damage repair and telomere maintenance (Michishita et al., 2008; Oberdoerffer & Sinclair, 2007). Availability of NAD^+ also interconnects epigenomic regulation with nutrient sensing (Berger & Sassone-Corsi, 2016). Epigenetic changes, specifically loss of H3K27me3, were strongly correlated with the upregulation of SASP genes in senescent cells (Shah et al., 2013), and inhibition of the H3K4 methyltransferase MLL1 reduced proproliferative cell cycle and DNA response genes, also resulting in reduced SASP (Capell et al., 2016).

Terminology

Replicative versus stress-induced senescence

In very broad terms, senescence can be induced by replicative exhaustion or by stress (including DNA damage, oncogenic stress, mitochondrial dysfunction, and others). To discriminate operationally between RS and stress-induced (premature) senescence (SIPS; Toussaint et al., 2000), RS is frequently equated with "telomere-dependent senescence," usually meaning "senescence induced by telomere shortening due to the end replication problem," while SIPS is equated with "telomere-independent (i.e., telomere length-independent) senescence." These are oversimplifications. First of all, telomere shortening solely due to the end-replication problem is an exception rather than the rule, occurring only under ideal conditions of very low stress and in cells with effective antistress (especially antioxidant) defenses (Martin-Ruiz et al., 2004; Serra, von Zglinicki, Lorenz, & Saretzki, 2003). More typically, telomere shortening is driven to a significant extent by telomere-specific DNA damage accumulation (von Zglinicki, 2002). Even mild stresses, just sufficient to generate base oxidations or single-strand breaks in telomeres, can accelerate telomere length-dependent senescence (Sitte et al., 1998). Conversely, SIPS induced by acute stress might not be associated with telomere shortening but typically shows DNA damage foci at telomeres as an indication of telomere uncapping (Fumagalli et al., 2012; Hewitt et al., 2012), which is again believed to be caused by the exceptional sensitivity of telomeric DNA to damage and the structural inefficiency of telomere repair (de Lange, 2018; Kruk et al., 1995; Sitte et al., 1998). It seems very probable that this type of length-independent but telomere dysfunction-mediated senescence may be a predominant form of senescence in vivo, not only in mouse tissues with very long telomeres (da Silva et al., 2019; Hewitt et al., 2012; Jurk et al., 2014; Ogrodnik et al., 2017) but possibly also in humans, especially in tissues with low cell turnover including liver (Jurk et al., 2014), lung (Birch et al., 2015), and heart (Anderson et al., 2019). Thus, RS and SIPS are not fully separate entities; rather they overlap significantly with telomere dysfunction, playing a central role throughout the whole continuum.

Senescence-associated secretory phenotype

In addition to the "classical" (proinflammatory) SASP, senescent cells generate an antiinflammatory, TGF-β-centered secretome (Hoare et al., 2016), release ROS, specifically H_2O_2 (Passos et al., 2010), oxidized proteins and lipids (Ni et al., 2016), specific endosome-encapsulated miRNAs (Terlecki-Zaniewicz et al., 2018), and other factors. While the term "SASP" is often used (and was originally coined) to describe the NF-κB- and C/EBPβ-driven secretome only, there is now a tendency to include all senescence-specific released molecules under the same name. As all these secreted factors interact with and stimulate each other, we endorse the use of "SASP" as a generic term and suggest discriminating among the different components by subterms (proinflammatory SASP, prooxidant SASP, etc.).

Senescence during aging in vivo

Since the first description of cell senescence (Hayflick & Moorhead, 1961) there has been a long-standing discussion as to whether senescence is in fact "aging under glass," as suggested by Hayflick (Hayflick, 1991; Hayflick & Moorhead, 1961), or rather a cell culture artefact with no or very limited importance for aging.

This discussion has firmly been settled in favor of Hayflick's insightful prediction by showing that targeted ablation of senescent cells can postpone and/or relieve a vast multitude of degenerative phenotypes in mice (Baker et al., 2011) (see "Senolytics and senostatics as antiaging interventions" section). This may be not surprising, given that senescent cells have been found to accumulate in practically all mammalian tissues during aging. Instead of summarizing all this evidence, this chapter will focus on senescence in postmitotic tissues/cell types and on one mechanism that contributes to senescent cell accumulation in vivo—the bystander effect. For a recent summary of methods to detect senescent cells in vivo, the reader is referred to Gorgoulis et al. (2019).

Senescence in postmitotic cells

In 2012, the author's discovered that the senescent cell response was not restricted to proliferation-competent cells. We found that DNA damage accumulation during mouse aging triggered the same senescence building blocks also in postmitotic neurons, both in the CNS and in the periphery. As in replication-competent cells (Passos et al., 2010), p21 was essential for signal transfer between an extended DDR and additional markers of the senescent phenotype (Jurk et al., 2012). More recently, a senescent phenotype was observed in neurons bearing neurofibrillary tangles from both human patients and transgenic mice, and it was shown that a senolytic intervention improved neuropathology in these mice (Musi et al., 2018).

Amongst postmitotic cells, the ability to mount a senescence stress response is not limited to neurons: senescence is induced in retinal cells in response to ischemic stress (Oubaha et al., 2016), in skeletal muscle fibers (da Silva et al., 2019), and in cardiomyocytes (Anderson et al., 2019) during aging. Similarly, significant senescence was found in slowly dividing cells like hepatocytes in vivo in both humans and mice (Aravinthan & Alexander, 2016; Jurk et al., 2014; Ogrodnik et al., 2017). Senolytic or senostatic interventions resulted in therapeutic benefits for all these tissues (Baker et al., 2011; Musi et al., 2018; Ogrodnik et al., 2017; Oubaha et al., 2016; Walaszczyk et al., 2019; Zhu et al., 2015). Thus, senescence in postmitotic cells is as much a cause of aging phenotypes and a target for "antiaging" interventions as senescence that is directly coupled to cell cycle arrest.

Senescent cell bystander effects

With the myriad of bioactive (SASP) molecules secreted by senescent cells, strong paracrine effects of senescence are no surprise. In fact, Campisi's group showed already in 2001 that senescent somatic cells stimulated growth and invasion of tumor cells (Krtolica, Parrinello, Lockett, Desprez, & Campisi, 2001), and this has been confirmed by multiple groups since (Rao & Jackson, 2016; Saleh et al., 2018). We discovered in 2012 that senescent cells also had a paracrine effect on normal, somatic cells—they induced senescence as a bystander effect (Nelson et al., 2012). This was soon confirmed by others (Acosta et al., 2013). Not dissimilar to the situation with radiation-induced bystander effects in cancer therapy (Wang, Zhou, Liu, & Zuo, 2018), the quest to identify specific, individual components of the secretome as mediators of the senescent bystander effect has been elusive. Multiple interventions targeting ROS, NF-κB activity, mTORC1 signaling, or p38MAPK activity were all capable of suppressing the bystander effect, but SAMD-derived ROS appeared to be upstream of the NF-κB-mediated proinflammatory SASP (Nelson et al., 2018). Weak bystander effects seemed dependent on cell−cell contacts via gap junctions, as they could efficiently be blocked by octanol, a disruptor of such contacts (Nelson et al., 2012).

Recently, the role of bystander-induced senescence for the accumulation of senescent cells and their functional consequences in vivo has been delineated: after transplanting very low numbers of senescent cells into either the muscle or skin of mice, enhanced frequencies of tissue-resident senescent cells were detected adjacent to the transplanted cells (da Silva et al., 2019). When slightly higher numbers of senescent cells were transplanted into fat depots, senescence spread into peripheral organs and persistent physical dysfunction was induced. With old mice as recipients, both mean lifespan and healthspan were reduced. These effects could be blocked by senolytic treatment following transplantation, indicating the causal role of the senescence-induced bystander effect (Xu et al., 2018).

Senescent cells accumulate in many tissues with age, and the rate of accumulation, at least in some tissues, predicts lifespan in mice (Jurk et al., 2014). Decreasing efficiency of immunosurveillance is one of the causes of age-associated accumulation of senescent cells (Ovadya et al., 2018), while the spread of senescence via bystander signaling is another. This was tested by comparing accumulation rates of senescent cells in wild-type and severely immunodeficient NOD scid gamma (NSG) mice under both ad libitum and dietary-restricted feeding (da Silva et al., 2019). As expected, senescent cells accumulated much faster in NSG mice fed ad libitum in agreement with the absence of a functional immune system. Suppressing the SASP (and thus the bystander effect) by dietary restriction slowed down the accumulation of senescent cells in the NSG mice, but resulted in a net loss (a senolytic effect, see section "Senolytics and senostatics as antiaging interventions") in the immunocompetent wild-type mice (da Silva et al., 2019). These data suggest that immunosurveillance and the bystander effect are the two major partners that determine the rate of accumulation of senescent cells in vivo in a dynamic equilibrium.

Senolytics and senostatics as antiaging interventions

Senolytics

Around 2008, Jim Kirkland developed the concept of using a p16 promoter-driven suicide gene for the selective ablation of p16-positive, presumably senescent, cells in mice. This led to a fruitful collaboration with Jan van Deursen showing that this approach not only reduced senescence markers in a mouse model of p16-driven accelerated aging, but, importantly, postponed multiple degenerative aging phenotypes in these animals (Baker et al., 2011). This success was confirmed in normally aging mice using two similar approaches with different suicide gene constructs (Baker et al., 2016; Chang et al., 2016). In parallel, drugs were developed that targeted antiapoptotic pathways that are specifically active in senescence, thus achieving preferential induction of apoptosis in (at least some types of) senescent cells (Baar et al., 2017; Fuhrmann-Stroissnigg et al., 2017; Yousefzadeh et al., 2018; Zhu et al., 2015, 2016, 2017). Around a dozen or so of these senolytic drugs have been described (Kirkland & Tchkonia, 2017), of which the combination of the tyrosine kinase-inhibitor dasatinib with the flavonoid quercetin (D + Q), the BCL2 family-inhibitor ABT263 (navitoclax), and the flavonoid fisetin have been tested in multiple applications. Altogether, this was a remarkable success story: by early 2019, there were at least 20 publications all reporting significant improvements of a wide range of age-associated degenerative conditions by pharmacogenetic and/or pharmacologic senolytis in mice. These conditions included muscle weakness, cataract, lipodystrophy, cardiovascular dysfunction and low stress tolerance, tumor incidence, atherosclerosis, pulmonary fibrosis, liver steatosis, osteoarthritis and osteoporosis, chemotherapy-induced multimorbidity, frailty, neurodegeneration, obesity-induced anxiety, and median lifespan (see Short et al., 2019 for a review). Even allowing for publication bias and taking major unresolved questions in the senolytics field (see below) into account, these results collectively very strongly support the notion that cellular senescence is a major and malleable driver of mammalian aging through many, if not all, tissues and organs.

Mechanistically, there are multiple pathways by which a reduction of senescent cell load may improve systemic and tissue function. Many studies have shown that senolytic interventions reduce markers of systemic inflammation in the blood. Chronic, systemic inflammation has long been recognized as a driver of aging (Franceschi, Garagnani, Parini, Giuliani, & Santoro, 2018). More evidence for systemic effects comes from experiments in which senescent cells were implanted in a single tissue (fat), leading to functional deterioration in a distant tissue (muscle weakness) (Xu et al., 2018). However, unambiguous identification of a mechanistic pathogenic role of senescence-dependent inflammation was not often successful: for instance, senolytic interventions in obese mice reduced levels of multiple proinflammatory cytokines and chemokines in blood and relieved obesity-dependent anxiety. However, manipulating systemic cytokine levels independent of senescence did not change anxiety (Ogrodnik et al., 2019). SASP factors including proinflammatory cytokines might however be pathogenetically important mediators of more local cell–cell interactions. For instance, in the obesity-driven anxiety model, senescence was observed in certain brain areas including glial cells in proximity to the lateral ventricle, an area responsible for neurogenesis, but not in lateral ventricle neurons themselves. Neurogenesis was reduced in obesity but restored following senescent cell ablation (Ogrodnik et al., 2019), pointing to the importance of local interactions between senescent and nonsenescent cells. Moreover, in a transgenic mouse model for human tau-mediated neurodegenerative disease, progressive senescence was observed in astrocytes and microglia but not in neurons. However, senolytic interventions improved the ability of neurons to handle mutated tau and neurofibrillary tangle deposition and maintained cognitive function, indicating the importance of cross-talk between neurons and senescent nonneuronal cells (Bussian et al., 2018). However, a direct role for neuron senescence in tau-mediated neurodegeneration should not be excluded: neurons from Alzheimer patients bearing neurofibrillary tangles as well as neurons from tau transgenic mice were found to be positive for senescence markers, and treatment with senolytics (D + Q) resulted in a reduction in total NFT density, neuron loss, and ventricular enlargement (Musi et al., 2018).

Senolytic interventions can also improve energy metabolism by reducing SAMD. For various types of senescent cells, including fibroblasts, hepatocytes, and brain cells (Ogrodnik et al., 2017, 2019), a reduced ability to metabolize fat is an important part of the SAMD, resulting in intracellular lipid accumulation. This way, hepatocyte senescence can cell-autonomously cause liver steatosis. Moreover, lipid accumulation in senescent cells might be sufficient to induce a SASP via GAS-STING activation, which might mediate local pathogenic cell–cell interactions (Ogrodnik et al., 2019). In general, the role of these different mechanistic pathways and their interaction during senolytic intervention is not well understood. Cell-type specific transgenic approaches are eagerly awaited for better clarification.

Another area of insufficient understanding is the question of specificity of senolytics. Most senolytic drugs described so far have only been tested in a few cell types. As testament to the complexity of the senescent phenotype and the variability in the engagement of individual senescence building blocks and pathways between cells of different origin, senolytics typically show a clear differential in their apoptotic efficiency between senescent and nonsenescent cells for one or few cell types, but not in others. Although the fact that senolytics have broad beneficial effects throughout many different organ systems is comforting, it is not sufficiently informative regarding their specificity. While the possibility of cell-autonomous effects has been shown in some studies (for instance, Ogrodnik et al., 2017), a significant impact for systemic effects has so far neither been proven nor ruled out. It still remains possible that many beneficial effects of senolytics may be mediated through reduction of senescence in a single major tissue like fat or muscle, which in turn then improves function in the target tissue(s) by reducing systemic inflammation.

Related to the issue of specificity is the question of the side effects of senolytics. Two different types of side effects need to be considered, resulting from the effects of the drugs on either nonsenescent or senescent cells. Side effects of the cancer drugs navitoclax and dasatinib on nonsenescent cells have been well documented and are dose-limiting in cancer therapy. They include anemia and thrombocytopenia and, for dasatinib, pulmonary edema and heart failure. It is assumed that these risks might be minimized by short-term treatment, which should be sufficient for long-term senescent cell ablation. However, more data are needed to establish the long-term efficiency of short senolytic treatments in a preclinical setting. While there are first data indicating that a single dose of senolytic can result in long-lasting physical improvement, these data were generated in mice transplanted with senescent cells (Xu et al., 2018) and strong evidence for a curative effect of a single course of a senolytic intervention is as yet missing.

Even less is known, so far, about the second potential type of side effect, arising from ablation of the senescent cells themselves. Senescence plays an important role for tissue remodeling in development and regeneration (He & Sharpless, 2017; Hernandez-Segura et al., 2018; Munoz-Espin & Serrano, 2014), especially for the resolution of the fibrotic response (Demaria, Desprez, Campisi, & Velarde, 2015; Krizhanovsky et al., 2008). Accordingly, it has been shown that acute ablation of senescent cells postponed wound healing (Demaria et al., 2015), but, so far, no other "negative" effects of senescent cell ablation seem to have been noted.

In summary, the impressive efficiency of senolytic interventions to relieve a wide range of age-related disabilities and degenerative conditions calls for major efforts to address the issues of specificity, mechanism of action and side effects, preclinically, and to translate the results into clinical trials. The first clinical trials are under way, and encouraging results have been reported.

Senostatics

Senostatics (sometimes also called senomorphics) are interventions that do not directly kill senescent cells but suppress the senescent phenotype by modifying one or more of the buildings blocks shown in Fig. 3.1. Ideally, they should not interfere with the DDR and leave the senescent growth arrest and thus the tumor suppressor function of senescence intact. Accordingly, typical senostatics will suppress nutrient signaling, improve auto/mitophagy and/or mitochondrial function. Thus, dietary restriction (da Silva et al., 2019; Wang et al., 2010) and DR mimetics including mTORC1 inhibitors like rapamycin (Correia-Melo et al., 2016; Wang et al., 2017) or Torin-1 (Dalle Pezze et al., 2014), mild mitochondrial uncouplers like metformin (Halicka et al, 2011; Jadhav, Dungan, & Williamson, 2013; Moiseeva et al., 2013) and 2,4-dinitrophenol (Passos et al., 2007), or monoamine oxidase-A inhibitors (Manzella et al., 2018) are potential or proven senostatics. Moreover, antioxidants or inhibitors of NF-κB can be efficient senostatics (Nelson et al., 2012, 2018), and there is evidence that multiple flavonoids, polyphenols, and other phytochemicals may have senostatic activity (Janubova & Zitnanova, 2017; Malavolta et al., 2018). Significant beneficial effects on healthspan in mice (Anisimov, Berstein, et al., 2011; Anisimov, Zabezhinski, et al., 2011; Martin-Montalvo et al., 2013; Miller et al., 2014) and humans (Campbell, Bellman, Stephenson, & Lisy, 2017; Gandini et al., 2014) have been demonstrated; however, it is not clear to what extent these were caused by their impact on senescence.

Cell-autonomous effects of senostatics on the senescent phenotype are reversible. Therefore it is generally assumed that senostatics would have to be given continuously to have lasting effects. However, this might not be true: short-term interventions with, for instance, dietary restriction (Cameron, Miwa, Walker, & von Zglinicki, 2012; Mitchell et al., 2015) or rapamycin (Bitto et al., 2016) have resulted in long-term beneficial healthspan effects. There is an important difference between the effects of senostatic interventions in vitro and in vivo. None of these interventions ablates senescent cells in in vitro assays. However, short-term (2–4 months) treatment of mice with rapamycin (Iglesias-Bartolome et al., 2012), metformin (our unpublished observations), or dietary restriction (da Silva et al., 2019; Ogrodnik et al., 2017; Wang et al., 2010) decreased frequencies of cells positive for multiple senescence markers below the levels measured before the intervention. This was dependent on intact immunosurveillance: in contrast to wild-type mice, dietary restriction of severely immunocompromised NSD mice resulted only in a slowing of accumulation of senescent hepatocytes in the liver, but not an actual decrease of their numbers (da Silva et al., 2019). Importantly, reduction of senescent cell frequencies under dietary restriction remained irreversible (at least for 3 months) when animals were returned to ad libitum feeding (Ogrodnik et al., 2017). This suggested that senostatic drugs might actually exert a net senolytic effect in immunocompetent hosts. By suppressing the senescent phenotype, they might shift the balance between senescent cell biogenesis via the bystander effect and destruction via immunosurveillance toward the latter (da Silva et al, 2019). If this is true, one might expect similar beneficial effects of relatively short senostatic interventions as with senolytics. This would be advantageous given the better safety profiles (on average) of senostatic drugs. However, essential experiments remain lacking; the persistence of reduced frequencies of senescent cells after short- or medium-term senostatic interventions other than dietary restriction has not been shown yet.

Conclusions

Senescence is a complex stress response phenotype involving a complete reprogramming of core cellular functions. Senescent cells accumulate with age in the vast majority of tissues, including slowly dividing and postmitotic cell types. They are among the causes for many age-associated diseases and disabilities. Specific ablation of senescent cells or suppression of their phenotype, and thus suppression of the signaling pathways that spread senescence intercellularly, can postpone and potentially relieve age-associated multimorbidity and frailty.

References

Aarts, M., Georgilis, A., Beniazza, M., Beolchi, P., Banito, A., Carroll, T., ... Gil, J. (2017). Coupling shRNA screens with single-cell RNA-seq identifies a dual role for mTOR in reprogramming-induced senescence. *Genes & Development, 31*, 2085–2098.

Acosta, J. C., Banito, A., Wuestefeld, T., Georgilis, A., Janich, P., Morton, J. P., ... Gil, J. (2013). A complex secretory program orchestrated by the inflammasome controls paracrine senescence. *Nature Cell Biology, 15*, 978–990.

Acosta, J. C., O'Loghlen, A., Banito, A., Guijarro, M. V., Augert, A., Raguz, S., ... Gil, J. (2008). Chemokine signaling via the CXCR2 receptor reinforces senescence. *Cell, 133*, 1006–1018.

Anderson, R., Lagnado, A., Maggiorani, D., Walaszczyk, A., Dookun, E., Chapman, J., ... Passos, J. F. (2019). Length-independent telomere damage drives post-mitotic cardiomyocyte senescence. *The EMBO Journal, 38*(5), e100492.

Anisimov, V. N., Berstein, L. M., Popovich, I. G., Zabezhinski, M. A., Egormin, P. A., Piskunova, T. S., ... Poroshina, T. E. (2011). If started early in life, metformin treatment increases life span and postpones tumors in female SHR mice. *Aging, 3*, 148–157.

Anisimov, V. N., Zabezhinski, M. A., Popovich, I. G., Piskunova, T. S., Semenchenko, A. V., Tyndyk, M. L., ... Blagosklonny, M. V. (2011). Rapamycin increases lifespan and inhibits spontaneous tumorigenesis in inbred female mice. *Cell Cycle, 10*, 4230–4236.

Aravinthan, A. D., & Alexander, G. J. M. (2016). Senescence in chronic liver disease: Is the future in aging? *Journal of Hepatology, 65*, 825–834.

Araya, J., Tsubouchi, K., Sato, N., Ito, S., Minagawa, S., Hara, H., ... Kuwano, K. (2019). PRKN-regulated mitophagy and cellular senescence during COPD pathogenesis. *Autophagy, 15*, 510–526.

Baar, M. P., Brandt, R. M., Putavet, D. A., Klein, J. D., Derks, K. W., Bourgeois, B. R., ... de Keizer, P. L. (2017). Targeted apoptosis of senescent cells restores tissue homeostasis in response to chemotoxicity and aging. *Cell, 169*, 132–147, e116.

Baker, D. J., Childs, B. G., Durik, M., Wijers, M. E., Sieben, C. J., Zhong, J., ... van Deursen, J. M. (2016). Naturally occurring p16(Ink4a)-positive cells shorten healthy lifespan. *Nature, 530*, 184–189.

Baker, D. J., Wijshake, T., Tchkonia, T., LeBrasseur, N. K., Childs, B. G., van de Sluis, B., ... van Deursen, J. M. (2011). Clearance of p16Ink4a-positive senescent cells delays ageing-associated disorders. *Nature, 479*, 232–236.

Ben-Porath, I., & Weinberg, R. A. (2004). When cells get stressed: An integrative view of cellular senescence. *The Journal of Clinical Investigation, 113*, 8–13.

Berger, S. L., & Sassone-Corsi, P. (2016). Metabolic signaling to chromatin. *Cold Spring Harbor Perspectives in Biology, 8*(11), a019463.

Birch, J., Anderson, R. K., Correia-Melo, C., Jurk, D., Hewitt, G., ... Belvisi, M. G. (2015). DNA damage response at telomeres contributes to lung aging and chronic obstructive pulmonary disease. *American Journal of Physiology-Lung Cellular and Molecular Physiology, 309*, L1124–L1137.

Birch, J., & Passos, J. F. (2017). Targeting the SASP to combat ageing: Mitochondria as possible intracellular allies? *Bioessays: News and Reviews in Molecular, Cellular and Developmental Biology, 39*(5), 1600235.

Bitto, A., Ito, T. K., Pineda, V. V., LeTexier, N. J., Huang, H. Z., Sutlief, E., ... Kaeberlein, M. (2016). Transient rapamycin treatment can increase lifespan and healthspan in middle-aged mice. *Elife, 5*, e16351.

Bitto, A., Sell, C., Crowe, E., Lorenzini, A., Malaguti, M., Hrelia, S., & Torres, C. (2010). Stress-induced senescence in human and rodent astrocytes. *Experimental Cell Research, 316*, 2961–2968.

Bodnar, A. G., Ouellette, M., Frolkis, M., Holt, S. E., Chiu, C. P., Morin, G. B., ... Wright, W. E. (1998). Extension of life-span by introduction of telomerase into normal human cells. *Science, 279*, 349–352.

Booth, L. N., & Brunet, A. (2016). The aging epigenome. *Molecular Cell, 62*, 728–744.

Bussian, T. J., Aziz, A., Meyer, C. F., Swenson, B. L., van Deursen, J. M., & Baker, D. J. (2018). Clearance of senescent glial cells prevents tau-dependent pathology and cognitive decline. *Nature, 562*, 578–582.

Butow, R. A., & Avadhani, N. G. (2004). Mitochondrial signaling: The retrograde response. *Molecular Cell*, *14*, 1–15.

Cameron, K. M., Miwa, S., Walker, C., & von Zglinicki, T. (2012). Male mice retain a metabolic memory of improved glucose tolerance induced during adult onset, short-term dietary restriction. *Longevity & Healthspan*, *1*, 3.

Campbell, J. M., Bellman, S. M., Stephenson, M. D., & Lisy, K. (2017). Metformin reduces all-cause mortality and diseases of ageing independent of its effect on diabetes control: A systematic review and meta-analysis. *Ageing Research Reviews*, *40*, 31–44.

Capell, B. C., Drake, A. M., Zhu, J., Shah, P. P., Dou, Z., Dorsey, J., ... Berger, S. L. (2016). MLL1 is essential for the senescence-associated secretory phenotype. *Genes & Development*, *30*, 321–336.

Capparelli, C., Chiavarina, B., Whitaker-Menezes, D., Pestell, T. G., Pestell, R. G., Hulit, J., ... Lisanti, M. P. (2012). CDK inhibitors (p16/p19/p21) induce senescence and autophagy in cancer-associated fibroblasts, "fueling" tumor growth via paracrine interactions, without an increase in neo-angiogenesis. *Cell Cycle*, *11*, 3599–3610.

Carroll, B., Nelson, G., Rabanal-Ruiz, Y., Kucheryavenko, O., Dunhill-Turner, N. A., Chesterman, C. C., ... Korolchuk, V. I. (2017). Persistent mTORC1 signaling in cell senescence results from defects in amino acid and growth factor sensing. *The Journal of Cell Biology*, *216*, 1949–1957.

Chang, J., Wang, Y., Shao, L., Laberge, R. M., Demaria, M., Campisi, J., ... Zhou, D. (2016). Clearance of senescent cells by ABT263 rejuvenates aged hematopoietic stem cells in mice. *Nature Medicine*, *22*, 78–83.

Chen, Q., & Ames, B. N. (1994). Senescence-like growth arrest induced by hydrogen peroxide in human diploid fibroblast F65 cells. *Proceedings of the National Academy of Sciences of the United States of America*, *91*, 4130–4134.

Coppe, J. P., Patil, C. K., Rodier, F., Sun, Y., Munoz, D. P., Goldstein, J., ... Campisi, J. (2008). Senescence-associated secretory phenotypes reveal cell-nonautonomous functions of oncogenic RAS and the p53 tumor suppressor. *PLoS Biology*, *6*, 2853–2868.

Coppe, J. P., Rodier, F., Patil, C. K., Freund, A., Desprez, P. Y., & Campisi, J. (2011). Tumor suppressor and aging biomarker p16 (INK4a) induces cellular senescence without the associated inflammatory secretory phenotype. *The Journal of Biological Chemistry*, *286*, 36396–36403.

Correia-Melo, C., Marques, F. D., Anderson, R., Hewitt, G., Hewitt, R., Cole, J., ... Passos, J. F. (2016). Mitochondria are required for pro-ageing features of the senescent phenotype. *The EMBO Journal*, *35*, 724–742.

d'Adda di Fagagna, F., Reaper, P. M., Clay-Farrace, L., Fiegler, H., Carr, P., Von Zglinicki, T., Saretzki, G., Carter, N. P., & Jackson, S. P. (2003). A DNA damage checkpoint response in telomere-initiated senescence. *Nature*, *426*, 194–198.

da Silva, P. F. L., Ogrodnik, M., Kucheryavenko, O., Glibert, J., Miwa, S., Cameron, C., ... von Zglinicki, T. (2019). The bystander effect contributes to the accumulation of senescent cells *in vivo*. *Aging Cell*, *18*(1), e12848.

Dalle Pezze, P., Nelson, G., Otten, E. G., Korolchuk, V. I., Kirkwood, T. B., von Zglinicki, T., & Shanley, D. P. (2014). Dynamic modelling of pathways to cellular senescence reveals strategies for targeted interventions. *PLoS Computational Biology*, *10*, e1003728.

de Lange, T. (2018). Shelterin-mediated telomere protection. *Annual Review of Genetics*, *52*, 223–247.

Demaria, M., Desprez, P. Y., Campisi, J., & Velarde, M. C. (2015). Cell autonomous and non-autonomous effects of senescent cells in the skin. *The Journal of Investigative Dermatology*, *135*, 1722–1726.

Demidenko, Z. N., Zubova, S. G., Bukreeva, E. I., Pospelov, V. A., Pospelova, T. V., & Blagosklonny, M. V. (2009). Rapamycin decelerates cellular senescence. *Cell Cycle*, *8*, 1888–1895.

Dou, Z., Ghosh, K., Vizioli, M. G., Zhu, J., Sen, P., Wangensteen, K. J., ... Berger, S. L. (2017). Cytoplasmic chromatin triggers inflammation in senescence and cancer. *Nature*, *550*, 402–406.

Dou, Z., Xu, C., Donahue, G., Shimi, T., Pan, J. A., Zhu, J., ... Berger, S. L. (2015). Autophagy mediates degradation of nuclear lamina. *Nature*, *527*, 105–109.

Franceschi, C., Garagnani, P., Parini, P., Giuliani, C., & Santoro, A. (2018). Inflammaging: A new immune-metabolic viewpoint for age-related diseases. *Nature Reviews Endocrinology*, *14*, 576–590.

Freund, A., Laberge, R. M., Demaria, M., & Campisi, J. (2012). Lamin B1 loss is a senescence-associated biomarker. *Molecular Biology of the Cell*, *23*, 2066–2075.

Fuhrmann-Stroissnigg, H., Ling, Y. Y., Zhao, J., McGowan, S. J., Zhu, Y., Brooks, R. W., ... Robbins, P. D. (2017). Identification of HSP90 inhibitors as a novel class of senolytics. *Nature Communications*, *8*, 422.

Fumagalli, M., Rossiello, F., Clerici, M., Barozzi, S., Cittaro, D., Kaplunov, J. M., ... d'Adda di Fagagna, F. (2012). Telomeric DNA damage is irreparable and causes persistent DNA-damage-response activation. *Nature Cell Biology*, *14*, 355–365.

Gandini, S., Puntoni, M., Heckman-Stoddard, B. M., Dunn, B. K., Ford, L., DeCensi, A., & Szabo, E. (2014). Metformin and cancer risk and mortality: A systematic review and meta-analysis taking into account biases and confounders. *Cancer Prevention Research (Philadelphia, Pa.)*, *7*, 867–885.

Garcia-Prat, L., Martinez-Vicente, M., Perdiguero, E., Ortet, L., Rodriguez-Ubreva, J., Rebollo, E., ... Munoz-Canoves, P. (2016). Autophagy maintains stemness by preventing senescence. *Nature*, *529*, 37–42.

Gomes, A. P., Price, N. L., Ling, A. J., Moslehi, J. J., Montgomery, M. K., Rajman, L., ... Sinclair, D. A. (2013). Declining NAD(+) induces a pseudohypoxic state disrupting nuclear-mitochondrial communication during aging. *Cell*, *155*, 1624–1638.

Gorgoulis, V., Adams, P. D., Alimonti, A., Bennett, D. C., Bischof, O., Bishop, C., ... Demaria, M. (2019). Cellular senescence: Defining a path forward. *Cell*, *179*, 813–827.

Hafner, A., Bulyk, M. L., Jambhekar, A., & Lahav, G. (2019). The multiple mechanisms that regulate p53 activity and cell fate. *Nature Reviews. Molecular Cell Biology*, *20*(4), 199–210.

Halicka, H. D., Zhao, H., Li, J., Traganos, F., Zhang, S., Lee, M., & Darzynkiewicz, Z. (2011). Genome protective effect of metformin as revealed by reduced level of constitutive DNA damage signaling. *Aging (Albany NY)*, *3*(10), 1028–1038.

Harley, C. B., Futcher, A. B., & Greider, C. W. (1990). Telomeres shorten during ageing of human fibroblasts. *Nature*, *345*, 458–460.

Hayflick, L. (1991). Aging under glass. *Mutation Research*, *256*, 69–80.

Hayflick, L., & Moorhead, P. S. (1961). The serial cultivation of human diploid cell strains. *Experimental Cell Research*, *25*, 585–621.

He, S., & Sharpless, N. E. (2017). Senescence in health and disease. *Cell*, *169*, 1000–1011.

Hernandez-Segura, A., Nehme, J., & Demaria, M. (2018). Hallmarks of cellular senescence. *Trends in Cell Biology*, *28*, 436–453.

Hewitt, G., Jurk, D., Marques, F. D., Correia-Melo, C., Hardy, T., Gackowska, A., ... Passos, J. F. (2012). Telomeres are favoured targets of a persistent DNA damage response in ageing and stress-induced senescence. *Nature Communications*, *3*, 708.

Hoare, M., Ito, Y., Kang, T. W., Weekes, M. P., Matheson, N. J., Patten, D. A., ... Narita, M. (2016). NOTCH1 mediates a switch between two distinct secretomes during senescence. *Nature Cell Biology*, *18*, 979–992.

Hoare, M., & Narita, M. (2018). Notch and senescence. *Advances in Experimental Medicine and Biology*, *1066*, 299–318.

Iglesias-Bartolome, R., Patel, V., Cotrim, A., Leelahavanichkul, K., Molinolo, A. A., Mitchell, J. B., & Gutkind, J. S. (2012). mTOR inhibition prevents epithelial stem cell senescence and protects from radiation-induced mucositis. *Cell Stem Cell*, *11*, 401–414.

Itahana, Y., & Itahana, K. (2018). Emerging roles of p53 family members in glucose metabolism. *International Journal of Molecular Sciences*, *19*(3), 776.

Jadhav, K. S., Dungan, C. M., & Williamson, D. L. (2013). Metformin limits ceramide-induced senescence in C2C12 myoblasts. *Mechanisms of Ageing and Development*, *134*(11–12), 548–559.

Janubova, M., & Zitnanova, I. (2017). Effects of bioactive compounds on senescence and components of senescence associated secretory phenotypes *in vitro*. *Food Function*, *8*, 2394–2418.

Jung, S. H., Hwang, H. J., Kang, D., Park, H. A., Lee, H. C., Jeong, D., ... Lee, J. S. (2019). mTOR kinase leads to PTEN-loss-induced cellular senescence by phosphorylating p53. *Oncogene*, *38*, 1639–1650.

Jurk, D., Wang, C., Miwa, S., Maddick, M., Korolchuk, V., Tsolou, A., ... von Zglinicki, T. (2012). Postmitotic neurons develop a p21-dependent senescence-like phenotype driven by a DNA damage response. *Aging Cell*, *11*, 996–1004.

Jurk, D., Wilson, C., Passos, J. F., Oakley, F., Correia-Melo, C., Greaves, L., ... von Zglinicki, T. (2014). Chronic inflammation induces telomere dysfunction and accelerates ageing in mice. *Nature Communications*, *2*, 4172.

Kang, H. T., Lee, K. B., Kim, S. Y., Choi, H. R., & Park, S. C. (2011). Autophagy impairment induces premature senescence in primary human fibroblasts. *PLoS One*, *6*, e23367.

Kirkland, J. L., & Tchkonia, T. (2017). Cellular senescence: A translational perspective. *EBioMedicine.*, *21*, 21–28.

Kolesnichenko, M., Hong, L., Liao, R., Vogt, P. K., & Sun, P. (2012). Attenuation of TORC1 signaling delays replicative and oncogenic RAS-induced senescence. *Cell Cycle*, *11*, 2391–2401.

Korolchuk, V. I., Miwa, S., Carroll, B., & von Zglinicki, T. (2017). Mitochondria in cell senescence: Is mitophagy the weakest link? *EBioMedicine.*, *21*, 7–13.

Krizhanovsky, V., Yon, M., Dickins, R. A., Hearn, S., Simon, J., Miething, C., ... Lowe, S. W. (2008). Senescence of activated stellate cells limits liver fibrosis. *Cell.*, *134*, 657–667.

Krtolica, A., Parrinello, S., Lockett, S., Desprez, P. Y., & Campisi, J. (2001). Senescent fibroblasts promote epithelial cell growth and tumorigenesis: A link between cancer and aging. *Proceedings of the National Academy of Sciences of the United States of America*, *98*, 12072–12077.

Kruk, P. A., Rampino, N. J., & Bohr, V. A. (1995). DNA damage and repair in telomeres: Relation to aging. *Proceedings of the National Academy of Sciences of the United States of America*, *92*, 258–262.

Kuilman, T., Michaloglou, C., Vredeveld, L. C., Douma, S., van Doorn, R., Desmet, C. J., ... Peeper, D. S. (2008). Oncogene-induced senescence relayed by an interleukin-dependent inflammatory network. *Cell*, *133*, 1019–1031.

Kurz, D. J., Decary, S., Hong, Y., & Erusalimsky, J. D. (2000). Senescence-associated (beta)-galactosidase reflects an increase in lysosomal mass during replicative ageing of human endothelial cells. *Journal of Cell Science*, *113*(Pt. 20), 3613–3622.

Laplante, M., & Sabatini, D. M. (2012). mTOR signaling in growth control and disease. *Cell*, *149*, 274–293.

Larsson, L. G. (2011). Oncogene- and tumor suppressor gene-mediated suppression of cellular senescence. *Seminars in Cancer Biology*, *21*, 367–376.

Lasry, A., & Ben-Neriah, Y. (2015). Senescence-associated inflammatory responses: Aging and cancer perspectives. *Trends in Immunology*, *36*, 217–228.

Lee, S., & Schmitt, C. A. (2019). The dynamic nature of senescence in cancer. *Nature Cell Biology*, *21*, 94–101.

Lerner, C., Bitto, A., Pulliam, D., Nacarelli, T., Konigsberg, M., Van Remmen, H., ... Sell, C. (2013). Reduced mammalian target of rapamycin activity facilitates mitochondrial retrograde signaling and increases life span in normal human fibroblasts. *Aging Cell*, *12*, 966–977.

Lopes-Paciencia, S., Saint-Germain, E., Rowell, M. C., Ruiz, A. F., Kalegari, P., & Ferbeyre, G. (2019). The senescence-associated secretory phenotype and its regulation. *Cytokine*, *117*, 15–22.

Ma, X., Warnier, M., Raynard, C., Ferrand, M., Kirsh, O., Defossez, P. A., ... Bernard, D. (2018). The nuclear receptor RXRA controls cellular senescence by regulating calcium signaling. *Aging Cell*, *17*, e12831.

Mai, S., Muster, B., Bereiter-Hahn, J., & Jendrach, M. (2012). Autophagy proteins LC3B, ATG5 and ATG12 participate in quality control after mitochondrial damage and influence lifespan. *Autophagy*, *8*, 47–62.

Malavolta, M., Bracci, M., Santarelli, L., Sayeed, M. A., Pierpaoli, E., Giacconi, R., ... Provinciali, M. (2018). Inducers of senescence, toxic compounds, and senolytics: The multiple faces of Nrf2-activating phytochemicals in cancer adjuvant therapy. *Mediators of Inflammation*, *2018*, 4159013.

Manzella, N., Santin, Y., Maggiorani, D., Martini, H., Douin-Echinard, V., Passos, J. F., ... Mialet-Perez, J. (2018). Monoamine oxidase-A is a novel driver of stress-induced premature senescence through inhibition of parkin-mediated mitophagy. *Aging Cell*, *17*, e12811.

Martin-Montalvo, A., Mercken, E. M., Mitchell, S. J., Palacios, H. H., Mote, P. L., Scheibye-Knudsen, M., ... de Cabo, R. (2013). Metformin improves healthspan and lifespan in mice. *Nature Communications*, *4*, 2192.

Martin-Ruiz, C., Saretzki, G., Petrie, J., Ladhoff, J., Jeyapalan, J., Wei, W., ... von Zglinicki, T. (2004). Stochastic variation in telomere shortening rate causes heterogeneity of human fibroblast replicative life span. *The Journal of Biological Chemistry*, *279*, 17826–17833.

Michishita, E., McCord, R. A., Berber, E., Kioi, M., Padilla-Nash, H., Damian, M., ... Chua, K. F. (2008). SIRT6 is a histone H3 lysine 9 deacetylase that modulates telomeric chromatin. *Nature*, *452*, 492–496.

Milanovic, M., Fan, D. N. Y., Belenki, D., Dabritz, J. H. M., Zhao, Z., Yu, Y., ... Schmitt, C. A. (2018). Senescence-associated reprogramming promotes cancer stemness. *Nature*, *553*, 96–100.

Milanovic, M., Yu, Y., & Schmitt, C. A. (2018). The senescence-stemness alliance - a cancer-hijacked regeneration principle. *Trends in Cell Biology*, *28*, 1049–1061.

Miller, R. A., Harrison, D. E., Astle, C. M., Fernandez, E., Flurkey, K., Han, M., ... Strong, R. (2014). Rapamycin-mediated lifespan increase in mice is dose and sex dependent and metabolically distinct from dietary restriction. *Aging Cell*, *13*, 468–477.

Mitchell, S. E., Tang, Z., Kerbois, C., Delville, C., Konstantopedos, P., Bruel, A., ... Speakman, J. R. (2015). The effects of graded levels of calorie restriction: I. impact of short term calorie and protein restriction on body composition in the C57BL/6 mouse. *Oncotarget*, *6*, 15902–15930.

Miwa, S., Jow, H., Baty, K., Johnson, A., Czapiewski, R., Saretzki, G., ... von Zglinicki, T. (2014). Low abundance of the matrix arm of complex I in mitochondria predicts longevity in mice. *Nature Communications*, *5*, 3837.

Moiseeva, O., Bourdeau, V., Roux, A., Deschenes-Simard, X., & Ferbeyre, G. (2009). Mitochondrial dysfunction contributes to oncogene-induced senescence. *Molecular and Cellular Biology*, *29*, 4495–4507.

Moiseeva, O., Deschênes-Simard, X., St-Germain, E., Igelmann, S., Huot, G., Cadar, A. E., ... Ferbeyre, G. (2013). Metformin inhibits the senescence-associated secretory phenotype by interfering with IKK/NF-κB activation. *Aging Cell, 12*(3), 489–498. Available from https://doi.org/10.1111/acel.12075, Epub 2013 Apr 23.

Munoz-Espin, D., & Serrano, M. (2014). Cellular senescence: From physiology to pathology. *Nature Reviews. Molecular Cell Biology, 15*, 482–496.

Musi, N., Valentine, J. M., Sickora, K. R., Baeuerle, E., Thompson, C. S., Shen, Q., & Orr, M. E. (2018). Tau protein aggregation is associated with cellular senescence in the brain. *Aging Cell, 17*, e12840.

Nacarelli, T., Azar, A., & Sell, C. (2015). Aberrant mTOR activation in senescence and aging: A mitochondrial stress response? *Experimental Gerontology, 68*, 66–70.

Narita, M., Nunez, S., Heard, E., Narita, M., Lin, A. W., Hearn, S. A., ... Lowe, S. W. (2003). Rb-mediated heterochromatin formation and silencing of E2F target genes during cellular senescence. *Cell, 113*, 703–716.

Narita, M., Young, A. R., Arakawa, S., Samarajiwa, S. A., Nakashima, T., Yoshida, S., ... Narita, M. (2011). Spatial coupling of mTOR and autophagy augments secretory phenotypes. *Science, 332*, 966–970.

Nelson, G., Kucheryavenko, O., Wordsworth, J., & von Zglinicki, T. (2018). The senescent bystander effect is caused by ROS-activated NF-kappaB signalling. *Mechanisms of Ageing and Development, 170*, 30–36.

Nelson, G., Wordsworth, J., Wang, C., Jurk, D., Lawless, C., Martin-Ruiz, C., & von Zglinicki, T. (2012). A senescent cell bystander effect: Senescence-induced senescence. *Aging Cell, 11*, 345–349.

Ni, C., Narzt, M. S., Nagelreiter, I. M., Zhang, C. F., Larue, L., Rossiter, H., ... Gruber, F. (2016). Autophagy deficient melanocytes display a senescence associated secretory phenotype that includes oxidized lipid mediators. *The International Journal of Biochemistry & Cell Biology, 81*, 375–382.

Oberdoerffer, P., & Sinclair, D. A. (2007). The role of nuclear architecture in genomic instability and ageing. *Nature Reviews: Molecular Cell Biology, 8*, 692–702.

Ogrodnik, M., Miwa, S., Tchkonia, T., Tiniakos, D., Wilson, C. L., Lahat, A., ... Jurk, D. (2017). Cellular senescence drives age-dependent hepatic steatosis. *Nature Communications, 8*, 15691.

Ogrodnik, M., Zhu, Y., Langhi, L. G. P., Tchkonia, T., Kruger, P., Fielder, E., ... Jurk, D. (2019). Obesity-induced cellular senescence drives anxiety and impairs neurogenesis. *Cell Metabolism, 29*, 1061–1077.

Olovnikov, A. M. (1973). A theory of marginotomy. The incomplete copying of template margin in enzymic synthesis of polynucleotides and biological significance of the phenomenon. *Journal of Theoretical Biology, 41*, 181–190.

Oubaha, M., Miloudi, K., Dejda, A., Guber, V., Mawambo, G., Germain, M. A., ... Sapieha, P. (2016). Senescence-associated secretory phenotype contributes to pathological angiogenesis in retinopathy. *Science Translational Medicine, 8*, 362ra144.

Ovadya, Y., Landsberger, T., Leins, H., Vadai, E., Gal, H., Biran, A., ... Krizhanovsky, V. (2018). Impaired immune surveillance accelerates accumulation of senescent cells and aging. *Nature Communications, 9*, 5435.

Passos, J. F., Nelson, G., Wang, C., Richter, T., Simillion, C., Proctor, C. J., ... von Zglinicki, T. (2010). Feedback between p21 and reactive oxygen production is necessary for cell senescence. *Molecular Systems Biology, 6*, 347.

Passos, J. F., Saretzki, G., Ahmed, S., Nelson, G., Richter, T., Peters, H., ... von Zglinicki, T. (2007). Mitochondrial dysfunction accounts for the stochastic heterogeneity in telomere-dependent senescence. *PLoS Biology, 5*, e110.

Peeper, D. S. (2011). Oncogene-induced senescence and melanoma: Where do we stand? *Pigment Cell & Melanoma Research, 24*, 1107–1111.

Rai, T. S., Cole, J. J., Nelson, D. M., Dikovskaya, D., Faller, W. J., Vizioli, M. G., ... Adams, P. D. (2014). HIRA orchestrates a dynamic chromatin landscape in senescence and is required for suppression of neoplasia. *Genes & Development, 28*, 2712–2725.

Rao, S. G., & Jackson, J. G. (2016). SASP: Tumor suppressor or promoter? Yes!. *Trends Cancer, 2*, 676–687.

Richter, T., Saretzki, G., Nelson, G., Melcher, M., Olijslagers, S., & von Zglinicki, T. (2007). TRF2 overexpression diminishes repair of telomeric single-strand breaks and accelerates telomere shortening in human fibroblasts. *Mechanisms of Ageing and Development, 128*, 340–345.

Rodier, F., & Campisi, J. (2011). Four faces of cellular senescence. *The Journal of Cell Biology, 192*, 547–556.

Rodier, F., Coppe, J. P., Patil, C. K., Hoeijmakers, W. A., Munoz, D. P., Raza, S. R., ... Campisi, J. (2009). Persistent DNA damage signalling triggers senescence-associated inflammatory cytokine secretion. *Nature Cell Biology, 11*, 973–979.

Saleh, T., Tyutynuk-Massey, L., Cudjoe, E. K., Jr., Idowu, M. O., Landry, J. W., & Gewirtz, D. A. (2018). Non-cell autonomous effects of the senescence-associated secretory phenotype in cancer therapy. *Frontiers in Oncollogy, 8*, 164.

Sarbassov, D. D., & Sabatini, D. M. (2005). Redox regulation of the nutrient-sensitive raptor-mTOR pathway and complex. *The Journal of Biological Chemistry, 280*, 39505–39509.

Saretzki, G., Murphy, M. P., & von Zglinicki, T. (2003). MitoQ counteracts telomere shortening and elongates lifespan of fibroblasts under mild oxidative stress. *Aging Cell, 2*, 141–143.

Scaffidi, P., & Misteli, T. (2006). Lamin A-dependent nuclear defects in human aging. *Science, 312*, 1059–1063.

Sen, P., Shah, P. P., Nativio, R., & Berger, S. L. (2016). Epigenetic mechanisms of longevity and aging. *Cell, 166*, 822–839.

Serra, V., von Zglinicki, T., Lorenz, M., & Saretzki, G. (2003). Extracellular superoxide dismutase is a major antioxidant in human fibroblasts and slows telomere shortening. *The Journal of Biological Chemistry, 278*, 6824–6830.

Shah, P. P., Donahue, G., Otte, G. L., Capell, B. C., Nelson, D. M., Cao, K., ... Berger, S. L. (2013). Lamin B1 depletion in senescent cells triggers large-scale changes in gene expression and the chromatin landscape. *Genes & Development, 27*, 1787–1799.

Short, S., Fielder, E., Miwa, S., & von Zglinicki, T. (2019). Senolytics and senostatics as adjuvant tumour therapy. *EBioMedicine, 41*, 683–692.

Singh, K., Matsuyama, S., Drazba, J. A., & Almasan, A. (2012). Autophagy-dependent senescence in response to DNA damage and chronic apoptotic stress. *Autophagy, 8*, 236–251.

Sitte, N., Merker, K., Grune, T., & von Zglinicki, T. (2001). Lipofuscin accumulation in proliferating fibroblasts *in vitro*: An indicator of oxidative stress. *Experimental Gerontology, 36*, 475–486.

Sitte, N., Saretzki, G., & von Zglinicki, T. (1998). Accelerated telomere shortening in fibroblasts after extended periods of confluency. *Free Radical Biology & Medicine, 24*, 885–893.

Summer, R., Shaghaghi, H., Schriner, D., Roque, W., Sales, D., Cuevas-Mora, K., ... Romero, F. (2019). Activation of mTORC1/PGC1 axis promotes mitochondrial biogenesis and induces cellular senescence in the lung epithelium. *American Journal of Physiology. Lung Cellular and Molecular Physiology, 316*, L1049–L1060.

Tai, H., Wang, Z., Gong, H., Han, X., Zhou, J., Wang, X., ... Xiao, H. (2017). Autophagy impairment with lysosomal and mitochondrial dysfunction is an important characteristic of oxidative stress-induced senescence. *Autophagy, 13*, 99–113.

Takai, H., Smogorzewska, A., & de Lange, T. (2003). DNA damage foci at dysfunctional telomeres. *Current Biology: CB*, *13*, 1549–1556.

Tchkonia, T., Zhu, Y., van Deursen, J., Campisi, J., & Kirkland, J. L. (2013). Cellular senescence and the senescent secretory phenotype: Therapeutic opportunities. *The Journal of Clinical Investigation*, *123*, 966–972.

Terlecki-Zaniewicz, L., Lammermann, I., Latreille, J., Bobbili, M. R., Pils, V., Schosserer, M., ... Grillari, J. (2018). Small extracellular vesicles and their miRNA cargo are anti-apoptotic members of the senescence-associated secretory phenotype. *Aging (Albany NY)*, *10*, 1103–1132.

Tian, Y., Garcia, G., Bian, Q., Steffen, K. K., Joe, L., Wolff, S., ... Dillin, A. (2016). Mitochondrial stress induces chromatin reorganization to promote longevity and UPR(mt). *Cell*, *165*, 1197–1208.

Toussaint, O., Medrano, E. E., & von Zglinicki, T. (2000). Cellular and molecular mechanisms of stress-induced premature senescence (SIPS) of human diploid fibroblasts and melanocytes. *Experimental Gerontology*, *35*, 927–945.

von Zglinicki, T. (2002). Oxidative stress shortens telomeres. *Trends in Biochemical Sciences*, *27*, 339–344.

von Zglinicki, T., Saretzki, G., Docke, W., & Lotze, C. (1995). Mild hyperoxia shortens telomeres and inhibits proliferation of fibroblasts: A model for senescence? *Experimental Cell Research*, *220*, 186–193.

Walaszczyk, A., Dookun, E., Redgrave, R., Tual-Chalot, S., Victorelli, S., Spyridopoulos, I., ... Richardson, G. D. (2019). Pharmacological clearance of senescent cells improves survival and recovery in aged mice following acute myocardial infarction. *Aging Cell*, *18*, e12945.

Wang, C., Maddick, M., Miwa, S., Jurk, D., Czapiewski, R., Saretzki, G., ... von Zglinicki, T. (2010). Adult-onset, short-term dietary restriction reduces cell senescence in mice. *Aging (Albany NY)*, *2*, 555–566.

Wang, R., Yu, Z., Sunchu, B., Shoaf, J., Dang, I., Zhao, S., ... Perez, V. I. (2017). Rapamycin inhibits the secretory phenotype of senescent cells by a Nrf2-independent mechanism. *Aging Cell*, *16*, 564–574.

Wang, R., Zhou, T., Liu, W., & Zuo, L. (2018). Molecular mechanism of bystander effects and related abscopal/cohort effects in cancer therapy. *Oncotarget*, *9*, 18637–18647.

Wang, Y., Shen, J., Chen, Y., Liu, H., Zhou, H., Bai, Z., ... Guo, X. (2018). PINK1 protects against oxidative stress induced senescence of human nucleus pulposus cells via regulating mitophagy. *Biochemical and Biophysical Research Communications*, *504*, 406–414.

Wang, Y., Wang, X. D., Lapi, E., Sullivan, A., Jia, W., He, Y. W., ... Lu, X. (2012). Autophagic activity dictates the cellular response to oncogenic RAS. *Proceedings of the National Academy of Sciences of the United States of America*, *109*, 13325–13330.

Wiley, C. D., Velarde, M. C., Lecot, P., Liu, S., Sarnoski, E. A., Freund, A., ... Campisi, J. (2016). Mitochondrial dysfunction induces senescence with a distinct secretory phenotype. *Cell Metabolism*, *23*, 303–314.

Xu, M., Pirtskhalava, T., Farr, J. N., Weigand, B. M., Palmer, A. K., Weivoda, M. M., ... Kirkland, J. L. (2018). Senolytics improve physical function and increase lifespan in old age. *Nature Medicine*, *24*, 1246–1256. Available from https://doi.org/10.1038/s41591-41018-40092-41599.

Yang, H., Wang, H., Ren, J., Chen, Q., & Chen, Z. J. (2017). cGAS is essential for cellular senescence. *Proceedings of the National Academy of Sciences of the United States of America*, *114*, E4612–E4620.

Young, A. R., Narita, M., Ferreira, M., Kirschner, K., Sadaie, M., Darot, J. F., ... Narita, M. (2009). Autophagy mediates the mitotic senescence transition. *Genes & Development*, *23*, 798–803.

Young, A. R., Narita, M., & Narita, M. (2011). Spatio-temporal association between mTOR and autophagy during cellular senescence. *Autophagy*, *7*, 1387–1388.

Yousefzadeh, M. J., Zhu, Y., McGowan, S. J., Angelini, L., Fuhrmann-Stroissnigg, H., Xu, M., ... Niedernhofer, L. J. (2018). Fisetin is a senotherapeutic that extends health and lifespan. *EBioMedicine*, *36*, 18–28.

Zhang, H., Hoff, H., Marinucci, T., Cristofalo, V. J., & Sell, C. (2000). Mitogen-independent phosphorylation of S6K1 and decreased ribosomal S6 phosphorylation in senescent human fibroblasts. *Experimental Cell Research*, *259*, 284–292.

Zhu, Y., Doornebal, E. J., Pirtskhalava, T., Giorgadze, N., Wentworth, M., Fuhrmann-Stroissnigg, H., ... Kirkland, J. L. (2017). New agents that target senescent cells: The flavone, fisetin, and the BCL-XL inhibitors, A1331852 and A1155463. *Aging (Albany NY)*, *9*, 955–963.

Zhu, Y., Tchkonia, T., Fuhrmann-Stroissnigg, H., Dai, H. M., Ling, Y. Y., Stout, M. B., ... Kirkland, J. L. (2016). Identification of a novel senolytic agent, navitoclax, targeting the Bcl-2 family of anti-apoptotic factors. *Aging Cell*, *15*, 428–435.

Zhu, Y., Tchkonia, T., Pirtskhalava, T., Gower, A. C., Ding, H., Giorgadze, N., ... Kirkland, J. L. (2015). The Achilles' heel of senescent cells: From transcriptome to senolytic drugs. *Aging Cell*, *14*, 644–658.

CHAPTER

4

The nature of aging and the geroscience hypothesis

Peter J. Hornsby

University of Texas Health Science Center San Antonio, San Antonio, TX, United States

OUTLINE

Aging and the geroscience hypothesis

There are basic questions that always arise in any discussion of the biology of aging. What is aging? Why does aging occur? And how does aging occur, that is, what are the molecular and cellular mechanisms? The last question is addressed in many of the chapters in this handbook. In this chapter these questions will be addressed in a general way, and the reader is referred to other chapters for much more detailed and specific treatment of the topic with respect to specific tissues and organ systems.

What is aging and why does aging occur?

We speak of aging with respect to many nonliving entities as well as living organisms. In doing so we imply that aging is a fundamental process applying generally to living and nonliving entities alike. In general, many things age, eventually deteriorate, and become nonfunctional. Vehicles, household objects, clothes, etc., all last a certain length of time but wear out and reach a point where they no longer fit the purpose for which they were designed.

Thus we use the term "aging" for the time-dependent, gradual, progressive deterioration of materials and inanimate objects as well as living organisms. When we use the term "aging" with respect to living things, are we using the word in the same manner as when we use it for inanimate objects?

The simple answer is yes. Engineers and material scientists refer to the aging of materials, buildings, machines, and so on. For example, a recent book "captures the essential current state-of-the-art on the subject of aging, reflecting various angles and viewpoints." "This work is an overview of the state of art on aging of materials and structures in the world. Aging of materials is a natural phenomenon. Each material we use will age. This aging will influence the performance of the object where the materials are used. Furthermore, the aging will be affected by the surroundings in which

Handbook of the Biology of Aging.
DOI: https://doi.org/10.1016/B978-0-12-815962-0.00004-4

the object is placed. The main focus of the book is on materials used in infrastructure, energy, buildings and industry. The book in effect establishes the definition of aging and its main research topics that are relevant for society" (van Breugel et al., 2018).

Some examples from engineering are now examined: consider components of an airplane engine or a skyscraper. Once a component, such as a gear (airplane engine) or reinforced concrete beam (skyscraper), is formed and incorporated into the engine or building we can confidently predict that at some point in the future it will lose its functionality and eventually either undergo a catastrophic failure or become unsuitable or unsafe for its function. This might be a very slow process—perhaps taking many decades—but eventual aging of the component is inevitable.

Similarly, if we consider components of the body, such as the cartilage of the knee or the lens of the eye, there will be some point of time when the biological material will become much less functional, or nonfunctional, compared to its functionality at an earlier age. These are straightforward examples—many biological examples are much more complex of course. Nevertheless, the simple rule is that both inanimate and biological entities predictably undergo time-dependent, progressive, gradual deterioration that eventually leads to a less functional or nonfunctional state.

Given these basic and essentially common-sense observations it is surprising that it was considered necessary in the past to explain the aging of biological entities by the positive action of genes that would cause a loss of functionality leading to an aged/senescent state and death. The history of the topic has been well reviewed (Austad, 2004; Kirkwood & Melov, 2011). Currently it is generally agreed that it is unnecessary to postulate that there are gene-encoded mechanisms for time-dependent gradual deterioration. Aging is the consequence of the operations of the normal laws of physics and chemistry, and therefore the time-dependent deterioration of biological entities is unsurprising and requires no extraordinary explanation. In many ways the search for a "special" explanation of aging in living organisms that differs from the explanation in inanimate objects is similar to the argument over vitalism that took place in the 19th century. The theory of vitalism was superseded by the concept that biology has no special laws, but inherits the basic laws of physics and chemistry, just like inanimate objects and materials (Mayr, 2010; Ramberg, 2000).

Here aging is defined as "time-dependent, gradual, progressive deterioration." This definition is a simplified version of definitions proposed earlier. Edward Masoro proposed that aging refers to the deteriorative changes that occur with time during postmaturational life, which underlie an increasing vulnerability to challenges that may lead to the death of the organism (Masoro, 1995). Bernard Strehler defined aging as "universal, progressive, intrinsic and deleterious" (Strehler, 1999). "Universal" was intended to distinguish it from pathology and disease. However, it seems unnecessary to exclude a process as "not aging" only because it seems not to occur in some individuals: actually, it may still be ongoing at a subclinical level. "Intrinsic" was intended to distinguish aging processes from processes that might be driven by environmental factors. However, this distinction easily gets very complex. For example, are processes driven by the gut microbiota considered intrinsic? It is easier to include any process that would be encountered by the typical individual of the species, whether that process is considered to be purely intrinsic or under the control of extrinsic factors.

We must distinguish death that results purely from aging versus death that results from extrinsic causes. For most populations of animals in the wild, death results mainly from extrinsic causes, although aging may play a role in the liability of the individual to succumb to those extrinsic causes. For example, aging may increase the likelihood that an individual will die as a result of an infection, or it may increase the likelihood that a predator will be able to kill the individual. As a consequence of this, population survival curves must be interpreted carefully in terms of their implications for aging. A difference in survival curves between two separate populations may or may not indicate a difference in aging rates between the two populations.

Immortal versus mortal systems and entities

Whereas many biological entities are subject to straightforward time-dependent, gradual, progressive deterioration, we must consider more complex systems that include both entities that age as well as mechanisms that can repair and replace those entities. Such a system, whether biological or inanimate, can appear to exist and function normally for an indefinite period. Thus it may operate as essentially immortal. Again, turning to examples in engineering and materials science, it is possible to maintain familiar objects for indefinite periods of time. While building components deteriorate over time, the building as a whole can be repeatedly repaired and kept in normal use more or less as long as we choose to make the effort to do so. Systems like the water distribution system of a city, or its road network, can likewise be constantly repaired and upgraded so that they can remain usable for an essentially indefinite period.

Here then is the fundamental difference between inanimate entities that age and eventually "die" versus those that are "immortal." The latter escape aging because they are complex systems that comprise both elements that undergo aging together with some form of repeated maintenance and upkeep so that the elements remain more or less as useable as they always were. But note that the individual components of the system do in fact "age and die." The water distribution system of a city may last indefinitely only because the individual pipes are replaced as needed, because those items do deteriorate. In essence, when we treat such a system as a single entity, we overlook the aging and replacement of the individual components, looking instead at the overall entity as not being subject to aging.

Many biological processes have at least some form of repair and replacement, and therefore in theory could be immortal, but in many cases those processes do not confer indefinite perfect functionality. In other words, a complex biological system may still deteriorate over time in a gradual progressive manner despite containing repair and replacement mechanisms. For example, many tissues contain stem cell systems that will continuously replace cells of the tissue over the lifespan. Nevertheless, such stem cell systems may show progressive changes over the lifespan such that in the old individual repair and replacement is deficient or fails. Despite the greater complexity of biological systems, we can generalize the statement that aging is a universal observation throughout the living and nonliving world.

The disposable soma concept

The "disposable soma" concept was first proposed in detail by Thomas Kirkwood (Kirkwood & Holliday, 1979), although elements of the concept had appeared in earlier writings. The philosopher Henri Bergson wrote "Life is like a current passing from germ to germ through the medium of a developed organism. It is as if the organism itself were only an excrescence, a bud caused to sprout by the former endeavoring to continue itself in a new germ" (Bergson, 1911). The concept that the immortal germ line should be distinguished from the mortal soma was first proposed by Augustus Weismann (reviewed by Kirkwood & Melov, 2011). In a letter to *Nature* in 1890 Weismann referred to the "immortality which I attribute both to the unicellular organisms and to the germinal cells of the multicellular" (Weismann, 1890).

Considering a species, for example, humans, the disposable soma concept makes the distinction between the mortal soma and the larger system that is functionally immortal. In essence, the soma means one individual of the species, minus the germ line that the individual carries. The larger system comprises an individual together with all the individual's ancestors and descendants, thereby including the germ line that is carried by that line of individuals. Therefore one complex system (a soma) is part of a larger and more complex system. As discussed above, complex systems can be either mortal or immortal. The soma comprises both a variety of elements (molecules, cells, tissues, and organ systems) that age, together with a variety of repair and replacement mechanisms. It is nevertheless mortal. On the other hand, the more complex system, comprising not only the individual but all ancestors and descendants, may be immortal. The germ line is passed from generation to generation and so one may consider the germ line as an immortal component and the soma as a mortal or "disposable" component. Fig. 4.1 illustrates the disposable soma concept.

We need to be careful when we make a statement such as "the germ line is immortal." All the germ line components are replaced over time; no individual component (cell, subcellular component, or molecule) is immortal. Even the bases of DNA turn over, so that it is likely that an individual harbors no single molecule directly inherited from either parent. It is the line that is immortal, that is, a system of components where each individual component must be replaced, repaired, or copied over time but the system can continue indefinitely. In fact, as discussed earlier, what is actually passed from generation to generation, although we refer to it as the germ line, is in reality the sequence of DNA. As Edward O. Wilson has written, "Samuel Butler's famous aphorism, that the chicken is only an egg's way of making another egg, has been modernized: the organism is only DNA's way of making more DNA" (Wilson, 1975). The disposable soma concept should be viewed together with another concept that originated at about the same time, the concept of the selfish gene developed by Richard Dawkins: "I shall argue that the fundamental unit of selection, and therefore of self-interest, is not the species, nor the group, nor even, strictly, the individual. It is the gene, the unit of heredity" (Dawkins, 1989). "Self-interest" here means maximizing the success of passing itself on to future generations. Thus, the soma is in essence a vehicle for housing and passing on genes, encoded in the DNA sequence.

Could the soma be maintained indefinitely? There is no theoretical reason why it could not. Proposals have been made to do just that, by repeated ingenious therapeutic interventions (engineered negligible senescence; Zealley & de Grey, 2013). Why does that not happen without intervention; why does the soma deteriorate and die, rather than exist indefinitely in a healthy state? We have to view the question from a gene's-eye point of view. As noted above, successful/selfish genes are those

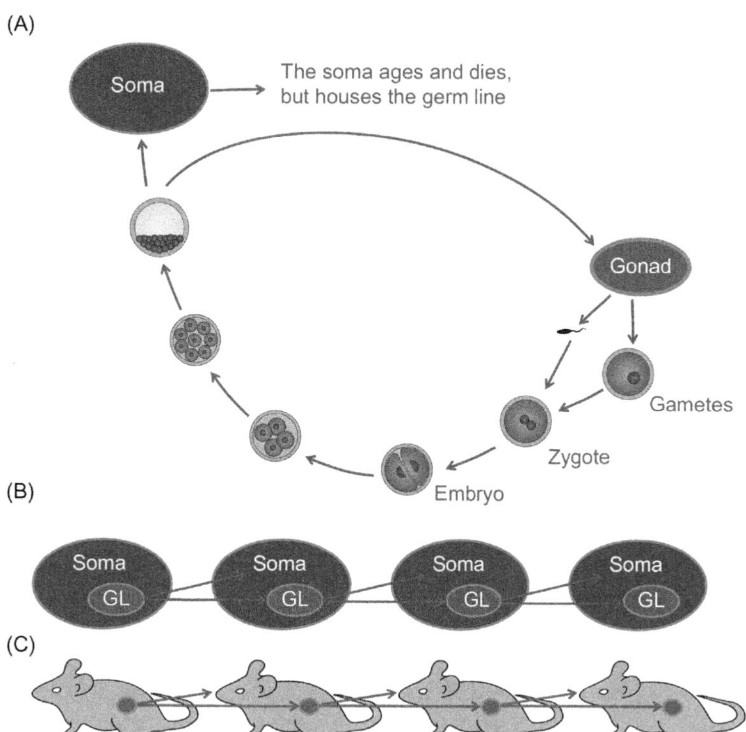

FIGURE 4.1 The disposable soma concept. (A) Male and female gametes unite to form a zygote, which develops into an embryo. As the embryo undergoes further development, a gonad differentiates; the gonad contains the gametes, that is, the germ line. The rest of the product of embryological development, the soma, ages and dies. (B) Each soma in a line of inheritance contains the germ line (GL). The germ line is passed from generation to generation, while producing a soma at each generation. (C) Illustration of the same concept with respect to the mouse. For simplicity a mouse of only one sex is shown at each generation. As in (B), the germ line (the red dot) is passed from generation to generation. In each generation it gives rise to a new soma (a mouse) while continuing as the germ line.

that maximize their success in being found in the next generation and in subsequent generations. Maintaining the soma is only in the gene's self-interest as long as the soma is useful for producing the next generation, or in other words, only up to reproductive age. But could that efficient somatic maintenance be continued beyond reproductive age? In theory it could be, but again we should look at the situation from a gene's-eye point of view. A gene cannot increase its success by maintaining a postreproductive soma. Nevertheless, it could adopt the strategy of maximizing its success by extended somatic maintenance, leading to an extended reproductive period, versus a strategy of maximizing its success by maximizing the number of copies in the next and subsequent generations. This is a tradeoff, and is a complex topic (MacArthur & Wilson, 1967), but, in any case, the degree to which it is in the gene's self-interest to ensure extended somatic maintenance is always limited. It would be impossible to imagine a situation where it is in the gene's best self-interest to maintain the soma indefinitely.

When a species has a distinct soma as part of its life history, the system comprising the individual together with all its ancestors and descendants may nevertheless be immortal. However, there are other ways that organisms may be functionally immortal. Some organisms lack a distinct soma and propagate vegetatively. For example, many vascular plants propagate by sending out runners (Hartmann et al., 2011). Another interesting example is provided by many species of fungi that comprise an extensive underground mycelium that can be functionally immortal (Petralia et al., 2014); while the fruiting bodies (the familiar mushrooms and toadstools) are disposable elements subject to aging. In such a case the organism may be, in essence, immortal, but, again, we need to be careful to note that the "immortality" applies only to the system as a whole and not to any individual cell, subcellular component, or molecule. In all these cases, after some period of time, each component of the organism would have been turned over and replaced. The same applies to entities like cell cultures. We often refer to "immortal cells" but it is the culture, not the cells, that is immortal. The HeLa cells that commonly are grown in many labs today contain no cells or molecules that actually existed in the original population of cells isolated from a cervical carcinoma.

What are the factors that determine longevity?

The disposable soma concept predicts that individual somatic maintenance mechanisms (which may also be called longevity assurance mechanisms) should operate in parallel over the same fraction of the lifespan (Fig. 4.2). When these mechanisms lose their ability to assure somatic maintenance the organism is liable to die. It is predicted that these mechanisms extend their effects over the majority of the lifespan but fail at some point late in life. One question that inevitably arises is, why do these systems

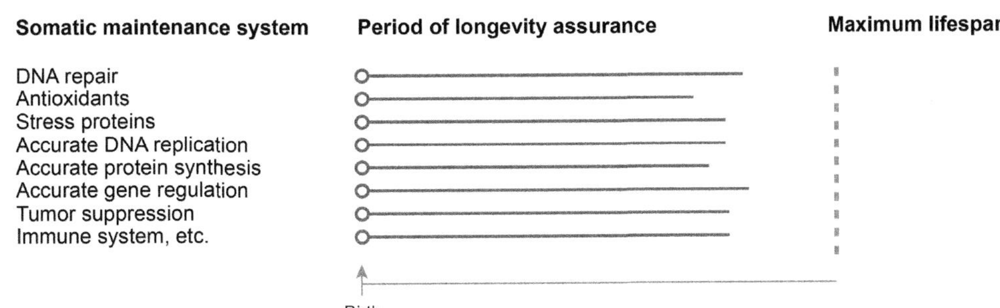

FIGURE 4.2 Somatic maintenance mechanisms (longevity assurance mechanisms). The examples of somatic maintenance mechanisms shown in this figure are those listed by Kirkwood (2010); many others exist. Each mechanism exerts its effect over the majority of the maximum lifespan, but not over the entire lifespan. For reasons discussed in the text, these mechanisms will tend to exert their effect over the same percentage of the lifespan. The red horizontal lines indicate how long the mechanism exerts an effect; the vertical gray dotted line indicates maximal lifespan. The values for the percentage of the lifespan over which these mechanisms act are hypothetical and not based on real data.

exert their effects for so long; why do they extend well beyond reproductive age? Why can an individual human female live to be more than 100 years old when her ability to reproduce has ceased more than 50 years earlier?

The reason for the extended period over which somatic maintenance is observed is the degree of robustness and reliability of the somatic maintenance mechanisms that operate in early life. They must be good enough to ensure that reproduction always takes place as efficiently as possible. To do so they must be very robust. Alex Comfort introduced the analogy of a spacecraft (Comfort, 1979). As pointed out by Hart and Turturro, "We are indebted to Alex Comfort for one of the most insightful and useful metaphors ever devised to describe aging" (Hart & Turturro, 1998). Robin Holliday in his book on the biology of aging also notes the value of this metaphor (Holliday, 1995). Comfort imagines the organism to have a trajectory like that of a spacecraft. In the case of the organism, the target for the newborn organism is reaching reproductive age successfully. In the case of the spacecraft, the target for the craft being launched from Earth is a nearby planet that it is to observe. Let us consider the spacecraft analogy with respect to the issues of robustness and reliability. To put it simply, failure is not an option. In other words, the engineers designing the spacecraft must be confident that if the mission were attempted 100 times it would be successful at least 99 times. It must be "overdesigned" to be sure that it reliably reaches its target. A second aspect of the design is that there can be no "Achilles heel" within the individual components. All components must be designed with more or less the same degree of robustness. Moreover, it does not make sense to have any component that is designed to a much higher level of robustness than the other components. Such a component may waste energy or other resources and would be refined in the design so that its robustness and reliability would be in line with that of the other components.

It would be impossible to design a craft that, more or less immediately, failed as soon as reached its target, if it is to have this degree of reliability. In fact, if appropriately designed to reach its target, it will actually continue working for some considerable time once it reaches and passes its target. It may continue working quite normally for two or three times the distance of the target before its components deteriorate and it finally becomes completely obsolete. Similarly, the design of an organism (i.e., its genome and the expression of its genes) must robustly permit reaching the target (reproduction) but if well designed (by evolution) it will pass its target by some considerable extent before starting to show deterioration. Here I use the term "design" in the sense of the design of the organism encoded in the DNA program (Dawkins, 1989). The target is reproduction, but as with the spacecraft, failure is not an option (or that line of organisms will become extinct), so there must be a robustness in the design such that typically the organism may go on living in good health well beyond reproductive age. When evolution has optimized the somatic maintenance mechanisms by strengthening any that were too weak, and toning down any that were over-robust, all maintenance mechanisms will come to be aligned in terms of timing over the lifespan (Fig. 4.2).

Although somatic maintenance mechanisms are robust and reliable over long periods of time they will not continue indefinitely, and, as discussed above, there is no mechanism whereby indefinite maintenance of the soma could evolve. On the other hand, the levels of repair and replacement required in the germ line, in order to prevent generation-to-generation deterioration, must be high and continuous, with no tendency to decline over time.

The force of natural selection over the lifespan

Ronald A. Fisher first proposed in 1930 that natural selection will have a weaker action on organisms as they become older (Fisher, 1930). Peter Medawar observed that "In the post-reproductive period of life, the direct influence of natural selection has been reduced to zero" (Medawar, 1952). Moreover, Richard Dawkins concluded that: "Another general quality that successful genes will have is a tendency to postpone the death of their survival machines at least until after reproduction. No doubt some of your cousins and great-uncles died in childhood, but not a single one of your ancestors did. Ancestors just don't die young!" (Dawkins, 1989). In other words, successful genes postpone the death of their host soma until after reproductive age, but the influence of natural selection declines to zero after that period of life. This has been phrased as "the force of natural selection declines as a function of age" and was placed on a firm mathematical footing by William D. Hamilton (Hamilton, 1966). While a consideration of this mathematical treatment is beyond the scope of this chapter, it has been stated that "Hamilton's (1966) Forces of Natural Selection are among the most significant equations in all of scientific theory" (Rose et al., 2007).

Another corollary of this concept is that maximal lifespan will tend to evolve to a value appropriate to the influence of external hazards that act on the population, such as predation. At an age where most of the population has been killed or has died due to some extrinsic cause, further somatic maintenance would be unnecessary and costly. Instead, under these circumstances, evolution will favor early reproduction and higher fecundity, strategies that may permit the population to expand over time despite the extrinsic hazards. In accordance with this analysis, Steven Austad observed that opossums (a North American marsupial) living in a predator-free island reproduced later and aged more slowly than animals of the same species on the more hazardous mainland (Austad, 1997).

Genes that exhibit antagonistic pleiotropy

Peter Medawar first suggested that evolutionary selection would favor genes that exert beneficial effects early in life even if they had harmful effects later in life (Medawar, 1952). This concept is termed "antagonistic pleiotropy" and was developed in more detail by George C. Williams (1957). Because natural selection is weaker at later ages, and is zero past reproductive age, it is possible to imagine a gene that is selected for its beneficial effect on somatic maintenance in early life, yet nevertheless exerts a negative effect late in life. For example, a possible gene in this class might be one that increases the rate of oxygen usage by mitochondria. In early life this may lead to increased vigor, possibly manifest as enhanced avoidance of predators or increased reproductive performance. However, in later life its predominant effect might be to increase oxidative damage. We should note again that this pattern of gene action can refer only to the soma, not the germ line. Any gene active in the germ line must exert a purely beneficial effect at all times or it will be subject to elimination by natural selection.

The molecular mechanisms of aging and the geroscience hypothesis

To return to the original question: what is aging? From the point of view of the genome, aging is the permitted deterioration of the soma. "Permitted" means that the deterioration does not negatively affect the passing on of the genome to the next generation. Whatever deterioration takes place is subject to neither positive nor negative evolutionary selection. Even though, in theory, the action of genes could prevent the deterioration, indefinite somatic maintenance cannot evolve because the deterioration takes place past reproductive age.

The analysis that has been laid out above does not yet address the question of what processes constitute aging in biological systems. In this chapter we are concerned only with general principles, rather than focusing on specific organs; for those the reader is referred to the specific chapters on those organs. As discussed earlier, there is no reason to suppose that aging in biological systems does not occur due to the operation of the normal laws of physics and chemistry, as it does for inanimate objects. However, aging may be much more complex than in nonliving systems, particularly because of the possible existence of repair and replacement mechanisms. In a complex system that includes repair and replacement mechanisms those systems themselves may be subject to progressive deterioration, that is, to aging. As a consequence, such complex systems, despite repair and replacement, are not necessarily immortal.

For each tissue there will be several postreproductive time-dependent processes that cause some form of deterioration of the function of the tissue (Fig. 4.3). Such processes are either unique to that tissue or are common to many or all tissues. George Martin introduced the concept of public and private markers of aging (Martin, 1988). While he introduced this concept with respect to different species (public across multiple species, private for a single species) the same general

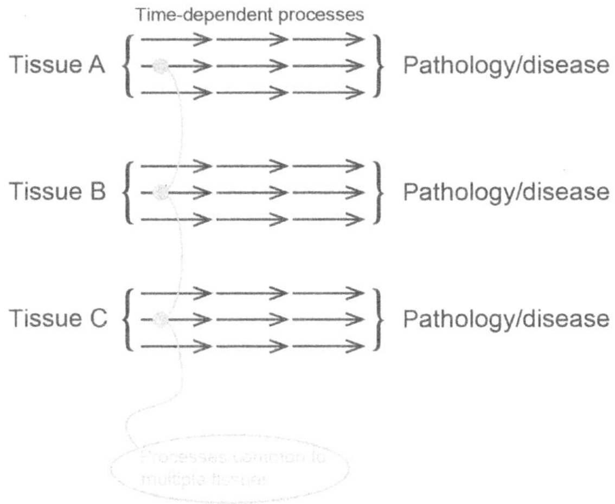

FIGURE 4.3 The basis for the geroscience hypothesis. Three tissues are illustrated as examples. In each tissue, time-dependent, gradual, progressive, deleterious processes take place. Eventually, the result of these processes is some form of identifiable pathology, leading to subclinical disease, and finally to overt disease or diseases. While some of these processes are unique to specific tissues, others, indicated by the green link, are common to multiple or all tissues. These common processes are the best targets for interventions that would maximize human health and well-being.

concept may be applied to processes in tissues. Processes unique to one tissue may be viewed as private, while those common to many may be viewed as public. The situation is similar to the origins of cancer in different tissues. Some genes are involved in cancers of a wide variety of tissues—for example, *TP53* and *RAS*. Many others are specific to one cancer type.

Similarly, some processes that occur over time in tissues and cause deterioration, pathology, and eventually contribute to diseases, are common across many tissues. Some of these processes are sterile inflammation ("inflammaging"), oxidative damage, advanced glycation end products (AGE), decreased efficiency of proteostasis, damage to mitochondria, accumulation of senescent cells, etc. (Fig. 4.3). Sometimes these common processes are termed the "pillars" or "hallmarks" of aging (Lopez-Otin et al., 2013). The recognition of such processes has already stimulated the development of promising pharmacological therapies, such as senolytics that target senescent cells (see other chapters).

The existence of these common processes across multiple tissues is the basis for the geroscience hypothesis (Sierra & Kohanski, 2017). The geroscience hypothesis posits that because these common processes likely play a major role in many or possibly all chronic diseases, applying therapies that address them directly will prevent the onset or mitigate the severity of multiple chronic diseases. Because older adults rarely suffer from a single disease, but commonly experience multimorbidity, reducing the rate of progression or the time of onset of these common processes may delay the onset of multiple diseases at once. The main goal is to develop feasible, practical, and safe interventions. Interventions that interfere with one or more of these common processes may dramatically lower healthcare costs, perhaps more than the cure of any one single disease, while significantly improving quality of life (Sierra & Kohanski, 2017).

Consider these three common mechanisms: sterile inflammation ("inflammaging"), AGE, and accumulation of senescent cells; and these four organs: arteries, bone, kidney, and brain. All three processes have been shown to occur in aging in these four organs (Al-Khazraji et al., 2018; Chaudhuri et al., 2018; Ferrucci & Fabbri, 2018; Musi et al., 2018; Palmer et al., 2019; Pignolo et al., 2019; Rowan et al., 2018; Smykiewicz et al., 2018; Ungvari et al., 2018). In arteries, these processes contribute to age-related arterial stiffness and hypertension, pathological processes including arteriosclerosis and atherosclerosis, and disease states including myocardial infarction and stroke. In bone, these processes contribute to age-related osteopenia, pathological processes such as osteoporosis, and increased liability to fractures. In kidney, these processes contribute to age-related loss of glomerular filtration rate, chronic kidney disease, and end-stage renal failure. In the brain, these processes contribute to age-related loss of cognitive function, neurodegeneration, and age-related diseases such as Alzheimer's disease. It is clear that the successful targeting of one or more of these underlying common processes would have a major impact on human health by delaying or preventing multiple age-related diseases simultaneously.

Conclusions

The geroscience hypothesis forms a useful framework for further studies on the basic mechanisms of aging. Prioritizing the discovery and characterization of those processes that act to cause the deterioration in function of multiple tissues and organs will lead to the development of new therapies that target those processes. These therapies may enable the delay or prevention of multiple diseases of late life and improve the period of the lifespan during which the individual experiences optimal health and well-being.

References

Al-Khazraji, B. K., Appleton, C. T., Beier, F., Birmingham, T. B., & Shoemaker, J. K. (2018). Osteoarthritis, cerebrovascular dysfunction and the common denominator of inflammation: A narrative review. *Osteoarthritis and Cartilage/OARS, Osteoarthritis Research Society*, *26*, 462–470.

Austad, S. N. (1997). *Why we age: What science is discovering about the body's journey through life*. New York: Wiley.

Austad, S. N. (2004). Is aging programed? *Aging Cell, 3*, 249–251.

Bergson, H. (1911). *Creative evolution* (translated by A. Mitchell). Henry Holt and Company, New York.

Chaudhuri, J., Bains, Y., Guha, S., Kahn, A., Hall, D., Bose, N., et al. (2018). The role of advanced glycation end products in aging and metabolic diseases: Bridging association and causality. *Cell Metabolism, 28*, 337–352.

Comfort, A. (1979). *The biology of senescence* (3rd ed.). New York: Elsevier.

Dawkins, R. (1989). *The selfish gene* (2nd ed.). Oxford: Oxford University Press.

Ferrucci, L., & Fabbri, E. (2018). Inflammageing: Chronic inflammation in ageing, cardiovascular disease, and frailty. *Nature Reviews Cardiology, 15*, 505–522.

Fisher, R. A. (1930). *The genetical theory of natural selection*. Oxford: Clarendon Press.

Hamilton, W. D. (1966). The moulding of senescence by natural selection. *Journal of Theoretical Biology, 12*, 12–45.

Hart, R. W., & Turturro, A. (1998). Evolution and dietary restriction. *Experimental Gerontology, 33*, 53–60.

Hartmann, H. T., Kester, D. E., Davies, F. T., Jr., & Geneve, R. L. (2011). *Hartmann & Kester's plant propagation: Principles and practices* (8th ed., Part 1). Upper Saddle River, NJ: Pearson Prentice Hall.

Holliday, R. (1995). *Understanding ageing*. Cambridge: Cambridge University Press.

Kirkwood, T. B. L. (2010). Evolution theory and the mechanisms of aging. In H. M. Fillit, K. Rockwood, K. Woodhouse, & J. C. Brocklehurst (Eds.), *Brocklehurst's textbook of geriatric medicine and gerontology* (7th ed), (pp. 18–22). New York: Elsevier.

Kirkwood, T. B. L., & Holliday, R. (1979). The evolution of ageing and longevity. *Proceedings of the Royal Society of London Series B, Containing Papers of a Biological Character. Royal Society (Great Britain), 205*, 531–546.

Kirkwood, T. B. L., & Melov, S. (2011). On the programmed/nonprogrammed nature of ageing within the life history. *Current Biology, 21*, R701–R707.

Lopez-Otin, C., Blasco, M. A., Partridge, L., Serrano, M., & Kroemer, G. (2013). The hallmarks of aging. *Cell, 153*, 1194–1217.

MacArthur, R. H., & Wilson, E. O. (1967). *The theory of island biogeography*. Princeton: Princeton University Press.

Martin, G. M. (1988). Constitutional genetic markers of aging. *Experimental Gerontology, 23*, 257–270.

Masoro, E. J. (1995). Aging: Current concepts. In: *Masoro, E.J. (Ed.), Handbook of physiology: Section 11* (pp. 3–21). New York: Oxford University Press.

Mayr, E. (2010). What is the meaning of "life"? In M. A. Bedau, & C. E. Cleland (Eds.), *The nature of life: Classical and contemporary perspectives from philosophy* (pp. 88–100). Cambridge: Cambridge University Press.

Medawar, P. (1952). *An unsolved problem of biology*. London: H.K. Lewis.

Musi, N., Valentine, J. M., Sickora, K. R., Baeuerle, E., Thompson, C. S., Shen, Q., et al. (2018). Tau protein aggregation is associated with cellular senescence in the brain. *Aging Cell, 17*, e12840.

Palmer, A. K., Xu, M., Zhu, Y., Pirtskhalava, T., Weivoda, M. M., Hachfeld, C. M., et al. (2019). Targeting senescent cells alleviates obesity-induced metabolic dysfunction. *Aging Cell, 18*, e12950.

Petralia, R. S., Mattson, M. P., & Yao, P. J. (2014). Aging and longevity in the simplest animals and the quest for immortality. *Ageing Research Reviews, 16*, 66–82.

Pignolo, R. J., Samsonraj, R. M., Law, S. F., Wang, H., & Chandra, A. (2019). Targeting cell senescence for the treatment of age-related bone loss. *Current Osteoporosis Reports, 17*, 70–85.

Ramberg, P. J. (2000). The death of vitalism and the birth of organic chemistry: Wohler's urea synthesis and the disciplinary identity of organic chemistry. *Ambix, 47*, 170–195.

Rose, M. R., Rauser, C. L., Benford, G., Matos, M., & Mueller, L. D. (2007). Hamilton's forces of natural selection after forty years. *Evolution: International Journal of Organic Evolution, 61*, 1265–1276.

Rowan, S., Bejarano, E., & Taylor, A. (2018). Mechanistic targeting of advanced glycation end-products in age-related diseases. *Biochimica et Biophysica Acta—Molecular Basis of Disease, 1864*, 3631–3643.

Sierra, F., & Kohanski, R. (2017). Geroscience and the trans-NIH Geroscience Interest Group, GSIG. *GeroScience, 39*, 1–5.

Smykiewicz, P., Segiet, A., Keag, M., & Zera, T. (2018). Proinflammatory cytokines and ageing of the cardiovascular-renal system. *Mechanisms of Ageing and Development, 175*, 35–45.

Strehler, B. (1999). *Time, cells, and aging*. Larnaca: Demetriades Brothers.

Ungvari, Z., Tarantini, S., Donato, A. J., Galvan, V., & Csiszar, A. (2018). Mechanisms of vascular aging. *Circulation Research, 123*, 849–867.

van Breugel, K., Koleva, D., & van Beek, T. (2018). *The ageing of materials and structures*. New York: Springer.

Weismann, A. (1890). Prof. Weismann's theory of heredity. *Nature, 41*, 317–323.

Williams, G. C. (1957). Pleiotropy, natural selection, and the evolution of senescence. *Evolution: International Journal of Organic Evolution, 11*, 389–411.

Wilson, E. O. (1975). *Sociobiology. The new synthesis*. Cambridge: Harvard University Press.

Zealley, B., & de Grey, A. D. (2013). Strategies for engineered negligible senescence. *Gerontology, 59*, 183–189.

C H A P T E R

5

Sirtuins, healthspan, and longevity in mammals

Surinder Kumar[1], *William Giblin*[1] *and David B. Lombard*[1,2]

[1]Department of Pathology, University of Michigan, Ann Arbor, MI, United States [2]Institute of Gerontology, University of Michigan, Ann Arbor, MI, United States

O U T L I N E

Handbook of the Biology of Aging.
DOI: https://doi.org/10.1016/B978-0-12-815962-0.00005-6

Introduction

Identifying interventions to slow the aging process and prolong lifespan has been a goal of humanity since ancient times. Tremendous progress in medicine and public health has allowed individuals in industrialized countries to live much longer than our ancestors, largely through reduced mortality from infectious diseases. However, now, as we comparatively live longer, we suffer from a distinct set of age-associated, usually chronic diseases—including cancer, type II diabetes (T2D), and cardiovascular and neurodegenerative diseases—which not only reduce quality of life but also now represent the leading causes of mortality worldwide.

A large body of evidence generated in model organisms has now conclusively shown that modulation of specific signaling pathways—for example, insulin/insulin-like growth factor-1 (IGF-1) signaling (IIS), mechanistic target of rapamycin (mTOR), and sirtuins—can extend lifespan in a manner that is conserved across distantly related organisms (Fontana, Partridge, & Longo, 2010; Kenyon, 2010; Lopez-Otin, Blasco, Partridge, Serrano, & Kroemer, 2013). In mammals, the aging process can be slowed by interventions that concomitantly reduce the onset and pace of degenerative, neoplastic, metabolic, and other age-related diseases (Kumar & Lombard, 2016; Lombard & Miller, 2014). Thus, these interventions not only extend lifespan, but more importantly, prolong healthspan (Gems, 2011). Therefore a deep understanding of these pathways and how they regulate the aging process may permit the development of therapeutics with beneficial effects against a wide spectrum of age-associated degenerative diseases.

Over the past two decades sirtuins have emerged as regulators of aging and age-associated diseases in organisms ranging from yeast to mammals, through their roles in the regulation of important physiological processes in each organism. Owing to their NAD^+ dependence, sirtuins have evolved as sensors and responders to environmental stressors to modulate diverse cellular processes such as genome maintenance, cell proliferation, and metabolism. In this chapter, we primarily discuss mammalian sirtuins, with an emphasis on their ability to influence healthspan and lifespan in mammals. We highlight current progress in the generation of potent and isoform-selective modulators of sirtuin activity, as valuable research tools, and as potential therapeutic agents.

Sirtuin-driven lifespan extension in invertebrates

Although our focus is mammalian sirtuins, we first briefly review some of the major findings on invertebrate sirtuins, to provide context. *SIR2* was first identified in yeast as a gene whose loss-of-function conferred meiotic sterility, due to loss of transcriptional repression at the silent mating-type loci (Klar, Fogel, & Macleod, 1979). Subsequent screens in yeast also identified *SIR2* and other phenotypically similar mutants, which have come to define the four known yeast *SIR* genes (Haber & George, 1979; Rine & Herskowitz, 1987; Rine, Strathern, Hicks, & Herskowitz, 1979; Shore, Squire, & Nasmyth, 1984). These enzymes comprise the SIR complex that maintains transcriptional silencing at the mating-type loci, telomeres, and the ribosomal DNA (rDNA) arrays (Kueng, Oppikofer, & Gasser, 2013). Sir2p, an NAD^+-dependent histone deacetylase (HDAC) (Imai, Armstrong, Kaeberlein, & Guarente, 2000), is the only yeast Sir protein whose activity is required for silencing at all of these three loci, while the other *SIR* genes are critical for silencing only the mating-type locus and telomeres.

Initial studies revealed that a specific mutation in one of the *S. cerevisiae SIR* genes, *Sir4–42*, increases replicative lifespan, defined as the number of times an individual yeast mother cell divides (Kennedy, Austriaco, Zhang, & Guarente, 1995). Later work showed that this mutation allows relocalization of the SIR complex (containing SIR2, SIR3, and SIR4) from telomeres to the nucleolar rDNA (Kennedy et al., 1997). This focused attention on the only strongly evolutionarily conserved member of this complex, *SIR2*. Modest *SIR2* overexpression (OE) on its own increases replicative lifespan in yeast by 30%, whereas *sir2* deletion shortens it by 50% (Kaeberlein, McVey, & Guarente, 1999). Initial investigation into the mechanism of this pro-longevity effect revealed that Sir2p protects yeast cells from a toxic accumulation of self-replicating extrachromosomal rDNA circles (ERCs) (Kaeberlein et al., 1999; Sinclair & Guarente, 1997). Deleting the gene encoding the replication fork-blocking protein, Fob1p, enhances replicative lifespan by attenuating recombination at the rDNA arrays (Defossez et al., 1999; Kaeberlein et al., 1999). This rescues the shortened lifespan of Sir2p-deficient mother yeast cells.

Data from subsequent studies suggest that additional distinct mechanisms by which Sir2p contributes to yeast replicative lifespan exist. For example, Sir2p facilitates asymmetrical partitioning of oxidized proteins, misfolded protein aggregates, and dysfunctional mitochondria to mother cells by maintaining the polarity machinery responsible for retention of damaged macromolecules (Higuchi et al., 2013; Liu et al., 2010). Sir2p may also promote yeast replicative lifespan through an epigenetic mechanism involving the heterochromatic regions adjacent to telomeres. At the telomere–euchromatin boundary, Sir2p opposes the activity of the Sas2p histone acetyltransferase to maintain silencing, by regulating histone H4 lysine 16 (H4K16) acetylation (Suka, Luo, & Grunstein, 2002). Derepression of subtelomeric silencing via a mutation-mimicking constitutive H4K16 acetylation reduces yeast lifespan (Dang et al., 2009). Indeed, Sir2p expression levels steadily decline over the yeast lifespan, concomitant with an increase in H4K16 acetylation. Altogether, these studies have cast Sir2p in multiple, distinct roles for promoting yeast replicative lifespan: first, by stabilizing the rDNA array; second, by protecting daughter cells from inheriting damaged proteins and dysfunctional mitochondria; and third, by maintaining transcriptional silencing at subtelomeric regions. These studies prompted investigations into whether *SIR2*-like genes exist and show analogous functions in higher eukaryotes.

SIR2 homologs have been identified in diverse species, and are present in both eukaryotes and prokaryotes. Collectively, these Sir2p-like proteins are termed sirtuins, and are of ancient evolutionary origin (Frye, 2000). *Caenorhabditis elegans* and *Drosophila melanogaster* have been used as experimental tools to elucidate the roles sirtuins may play in promoting longevity and to some extent disease susceptibility in multicellular organisms. The *C. elegans* genome encodes four sirtuin genes, *SIR-2.1* though *SIR-2.4*, that are homologous to mammalian *SIRT1* (*SIR-2.1*), *SIRT4* (*SIR-2.2/2.3*), and *SIRT6/SIRT7* (*SIR-2.4*) (Frye, 2000). OE of the SIRT1 homologs, SIR-2.1, modestly extends mean lifespan in worms (by 10%–15%), an effect that is dependent upon DAF-16. DAF-16 is the single worm forkhead box O (FOXO) transcription factor homolog, a major target of the IIS cascade (Berdichevsky, Viswanathan, Horvitz, & Guarente, 2006; Mouchiroud et al., 2013; Rizki et al., 2011; Tissenbaum & Guarente, 2001, 2002; Viswanathan & Guarente, 2011). Independent of IIS signaling through DAF-16, pure synthetic ascaroside administration increases worm lifespan and resistance to stress in a SIR-2.1-dependent manner. *C. elegans* ascarosides are secreted molecules that control developmental timing and various social behaviors in this organism. Inhibition of sensory neurons abolishes this phenotype, linking sensing of endogenous small molecules to sirtuin-dependent longevity and stress resistance (Ludewig et al., 2013). Also, studies of the *C. elegans* SIRT6/SIRT7 homolog SIR-2.4 have revealed that this protein promotes DAF-16 function in response to stress, and is required for resistance to cellular dysfunction induced by expression of a polyglutamine tract-containing protein (Chiang et al., 2012).

The *D. melanogaster* genome encodes five sirtuins, corresponding to mammalian SIRT1 (dSir2), SIRT2, SIRT4, SIRT6, and SIRT7 (Frye, 2000). OE of dSIR2, in the nervous system, the fat body, or whole organism increases longevity (Banerjee et al., 2012; Hoffmann, Romey, Fink, Yong, & Roeder, 2013; Rogina & Helfand, 2004). Importantly, one prominent report did not reproduce the pro-longevity effects of either *C. elegans* or *D. melanogaster* sirtuins (Burnett et al., 2011). Differing husbandry conditions and/or genetic backgrounds may explain the discrepancies between laboratories.

An increase in mobilization of transposable DNA elements (TEs), which are generally silenced due to the repressive nature of the heterochromatic regions in which they typically reside, has been associated with aging of somatic cells (De Cecco et al., 2013). Loss of constitutive heterochromatin and a concomitant increase in TE expression is observed in the fat body, which is the major metabolic regulatory organ, of aged *D. melanogaster*. Although lifespan was not directly assessed in dSIR2 overexpressors, a delay in TE expression was observed in this model, while maintenance of heterochromatic regions and enhanced TE silencing were observed upon calorie restriction (CR), implicating both CR, which extends lifespan in most organisms, and dSIR2 in maintaining genome integrity and promoting lifespan (Wood et al., 2016).

Of the five *D. melanogaster* sirtuins, only dSirt4 has been described to contain a mitochondrial targeting sequence, and localizes to mitochondria (Wood et al., 2018). Flies of either sex, genetically null for dSirt4, are short-lived compared to controls (a 28% and 17% median lifespan decrease in females and males, respectively), but experience a normal lifespan if exposed to a CR diet. OE of dSirt4, either globally or solely in the fat body, extends fly lifespan to varying degrees (13%–20% median lifespan extension in females, and 0%–20% in males). This discrepancy is likely the result of differing drivers of ubiquitous dSirt4 expression (*da-GAL4*, significant in males and females; *tub-GAL4*, significant in females only). Although mitochondria from dSirt4-null flies maintain wild-type rates of respiration and ATP levels, dSirt4-null animals are more sensitive to starvation conditions and display reduced fertility compared to their wild-type counterparts (Wood et al., 2018). Thus, mitochondrial sirtuin activity is critical for maintaining metabolic homeostasis and healthy lifespan in *D. melanogaster*.

Sirtuin enzymatic activity

Mammalian sirtuins have varied enzymatic activities, working as NAD^+-dependent deacetylases/deacylases (SIRT1, SIRT2, SIRT3, SIRT4, SIRT5, SIRT6, SIRT7), and ADP-ribosyltransferases (SIRT4, SIRT6) (Fig. 5.1). They function as cellular stress sensors to modify histones and a plethora of other proteins—including transcription factors—to modulate diverse cellular processes (Imai et al., 2000; Jiang et al., 2013; Lin et al., 2009). Although the seven mammalian sirtuins share a fairly conserved NAD^+-binding catalytic domain, the amino and carboxy regions that flank this domain are highly divergent. Sirtuins differ in expression pattern, catalytic activity, biological function, and subcellular localization (Fig. 5.1, Table 5.1), residing predominantly in the nucleus

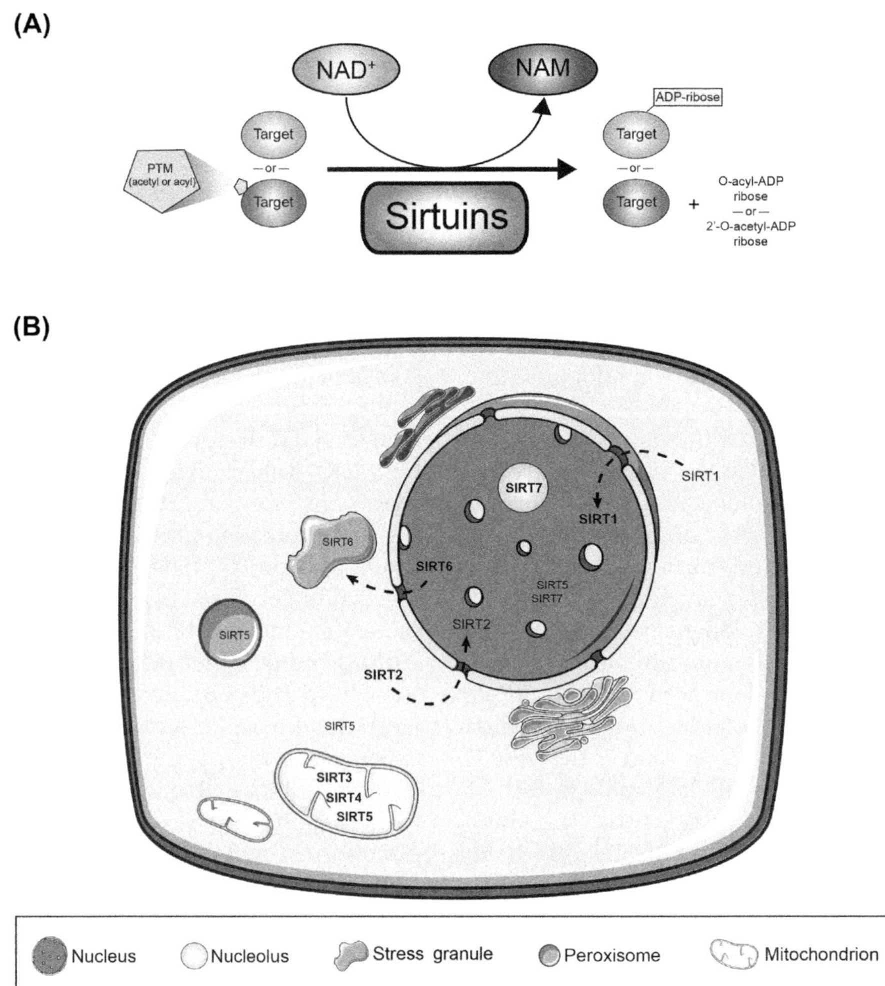

FIGURE 5.1 Sirtuin enzymatic activity and subcellular localization. (A) Sirtuin-catalyzed deacetylation/deacylation or ADP-ribosylation reactions consume nicotinamide adenine dinucleotide (NAD$^+$) as a cosubstrate and generate the sirtuin feedback inhibitor nicotinamide (NAM), and a deacetylated/deacylated or ADP-ribosylated product. Deacetylation/deacylation reactions produce 2′-O-acetyl-ADP-ribose/2′-O-acyl-ADP-ribose as a by-product. (B) SIRT1 is localized to the nucleus and can shuttle to the cytoplasm under certain conditions. SIRT2 is a cytosolic protein present in the nucleus during the G2/M transition of the cell cycle. SIRT3, SIRT4, and SIRT5 are primarily mitochondrial sirtuins. SIRT5 is also present outside the mitochondria, in peroxisomes and cytosol, where it deacylates specific targets. SIRT6 is chromatin-bound in the nucleus, but has been found associated with cytoplasmic stress granules. SIRT7 is nucleolar and associates with ribosomal DNA, but also regulates extranucleolar nuclear processes.

(SIRT1, SIRT6, and SIRT7), cytosol (SIRT2), or mitochondrial matrix (SIRT3, SIRT4, and SIRT5) (Canto, Sauve, & Bai, 2013). Among the mammalian sirtuins, only SIRT6 and SIRT7 are homologous to one another across their entire coding sequences (Frye, 2000). In this section we briefly review the salient enzymatic features and subcellular context of each sirtuin.

The best-characterized sirtuin, SIRT1, deacetylates a variety of proteins to modulate many cellular processes such as metabolism and cellular stress responses (Rahman & Islam, 2011). SIRT1 also possesses deacylase activity in vitro, though the in vivo significance of this activity is not known (Feldman, Baeza, & Denu, 2013). SIRT1 possesses two nuclear export and two nuclear localization signals, allowing it to translocate to the cytosol in transformed cell lines (Byles et al., 2010; Jin et al., 2007), adult cardiomyocytes (Tanno, Sakamoto, Miura, Shimamoto, & Horio, 2007), and neurons (Hisahara et al., 2008; Li, Xu, McBurney, & Longo, 2008); however, the biological relevance of this shuttling is not fully understood.

Cytosolic SIRT2 deacetylates α-tubulin and has been found in the nucleus during the mitotic G2/M transition, where it localizes to chromatin and deacetylates H4K16 (Vaquero et al., 2006). In this regard, SIRT2 has been identified as a regulator of mitotic exit in mammalian cells (Harting & Knoll, 2010). Also, SIRT2 deacetylase activity regulates chromatin dynamics and gene expression, by opposing autoacetylation of the histone acetyltransferase p300 in vivo (Black, Mosley,

TABLE 5.1 Sirtuin subcellular localization, enzymatic activities and known histone substrates.

Sirtuin	Subcellular localization	Enzymatic activity	Confirmed histone substrates	Examples of major nonhistone interactors and substrates
1	**Nuclear**/cytoplasmic	Deacetylase/deacylase	H1K26, H3K9, H3K14, H4K16	p300, SUV39H1, EZH2, p53, FOXOs, NF-κB, c-Fos, c-Jun, c-MYC, HIF-1α, Ku70, NBS1, PARP1, PGC-1α, many others
2	Nuclear/**cytosolic**	Deacetylase/deacylase/debenzoylase	H2B, H3, H3K56, H4K16	TUBA, PEPCK1, FOXOs, PAR3, p300, p53, p65, NF-κB, c- MYC, CDH1, CDC20, PGC-1α, PKM2
3	**Mitochondrial**	Deacetylase/decrotonylase/dehydroxybutyrylase	H3K4, H3K9, H3K18, H3K23 H3K27, H4K16	LCAD, HMGCS2, SOD2, IDH2, p53, Ku70, ACS2, ALDH2, NDUFA9, SDHA, GDH, SHMT2, PYCR1 many others
4	**Mitochondrial**	Deacetylase/ADP-ribosyltransferase/lipoamidase/deacylase	–	MCD, PDH, MCCC, GDH, IDE, ANT2, ANT3
5	**Mitochondrial**/cytosolic/nuclear/peroxisomal	Deacetylase (?)/demalonylase/desuccinylase/deglutarylase	–	CPS1, HMGCS2, SOD1, PDH, SDH, PKM2, SHMT2, ECHA many others
6	**Nuclear chromatin**/cytoplasmic	Deacetylase/deacylase/ADP-ribosyltransferase	H3K9, H3K56	c-MYC, CtIP, GCN5, c-Jun, DNA-PK, PARP, HIF-1α, TNFα PKM2
7	**Nucleolar**	Deacetylase/desuccinylase/debutyrylase/ADP-ribosyltransferase	H3K18, H3K36, H3K37, H3K122	c-MYC, p53, GABPβ1

Also, examples of major nonhistone interacting partners or substrates are indicated: *ACS2*, mitochondrial acetyl-CoA synthetase 2; *ALDH2*, mitochondrial aldehyde dehydrogenase 2; *ANT2*, adenine nucleotide translocase 2; *ANT3*, adenine nucleotide translocase 3; CDC20, cell-division cycle protein 20; *CDH1*, E-cadherin; *c-Fos*, cellular oncogene fos; *c-Jun*, jun proto-oncogene; *CPS1*, carbamoyl-phosphate synthase 1, mitochondrial; c-MYC, myc proto-oncogene protein; CtIP, C-terminal binding protein (CtBP) interacting protein; *DNA*-PK, DNA-dependent protein kinase; *ECHA*, Enoyl-CoA Hydratase Alpha; *EZH2*, enhancer of zeste homolog 2; *FOXOs*, forkhead box protein O family; *GABPβ1*, GA repeat binding protein, beta 1; *GCN5*, or K(lysine) acetyltransferase 2A (KAT2A); GDH, glutamate dehydrogenase; *HIF-1α*, hypoxia-inducible factor 1 alpha; *HMGCS2*, 3-hydroxy-3-methylglutaryl CoA synthase 2; *IDE*, insulin-degrading enzyme; IDH2, isocitrate dehydrogenase; *Ku70*, X-ray repair cross-complementing protein 6 (XRCC6); *LCAD*, long-chain specific acyl-CoA dehydrogenase; *MCCC*, methyl crotonyl- CoA carboxylase complex; *MCD*, malonyl-CoA decarboxylase; *NBS1*, Nijmegen breakage syndrome protein 1 homolog; *NDUFA9*, NADH dehydrogenase (ubiquinone) 1 alpha subcomplex, 9; *NF-κB*, nuclear factor of kappa light polypeptide gene enhancer in B cells 1; *p300*, histone acetyltransferase p300; *p53*, tumor suppressor p53; *p65*, transcription factor p65 (NF-κB p65 subunit); *PAR3*, partitioning defective 3 homolog; *PARP1*, poly [ADP-ribose] polymerase 1; *PDH*, pyruvate dehydrogenase; *PEPCK1*, phosphoenolpyruvate carboxykinase; *PGC-1α*, peroxisome proliferator-activated receptor gamma coactivator 1 alpha; *PKM2*, pyruvate kinase isoform M2; *PYCR1*, pyrroline-5-carboxylate reductase 1; *SDH*, succinate dehydrogenase; *SDHA*, succinate dehydrogenase complex, subunit A, flavoprotein; *SHMT2*, serine hydroxymethyltransferase 2; *SOD1*, superoxide dismutase 1; *SOD2* (MnSOD) superoxide dismutase; *SUV39H1*, suppressor of variegation 3–9 homolog 1; TNFα, tumor necrosis factor alpha; *TUBA*, tubulin alpha-3 chain.

Kitada, Washburn, & Carey, 2008). SIRT2 also activates the NADPH-generating enzyme, glucose-6-phosphate dehydrogenase (G6PD) by deacetylating lysine 403, which lies within its $NADP^+$ binding domain (Wang et al., 2014). The SIRT2–G6PD interaction is further enhanced upon oxidative stress, and results in activation of the pentose phosphate pathway, which supplies the cell with NADPH to maintain redox homeostasis. Recently, Huang et al. revealed the existence of a novel noncanonical histone mark, lysine benzoylation (Kbz), and identified 22 Kbz sites on histones (Huang, Zhang et al., 2018). Kbz is associated with gene expression and has physiological relevance distinct from histone acetylation. Huang and colleagues further show that SIRT2 catalyzes the removal of histone Kbz both in vitro and in vivo.

SIRT3 is the major mitochondrial deacetylase that plays essential roles in mitochondrial function, ATP production, reactive oxygen species (ROS) management, β-oxidation, ketogenesis, and cell death among other processes (Carrico, Meyer, He, Gibson, & Verdin, 2018; Lombard et al., 2007; Lombard & Zwaans, 2014). SIRT3 is also reported to act as a histone deacylase. In this context, SIRT3 functions as a histone lysine decrotonylase to regulate transcription (Bao et al., 2014). Lysine crotonylation is a novel histone posttranslational modification (PTM) enriched at active gene promoters and potential enhancers in mammalian cell genomes (Tan et al., 2011). Most recently, Zhang et al. reported that SIRT3 also functions as a histone de-β-hydroxybutyrylase (Zhang, Cao et al., 2019). The levels of histone lysine-β-hydroxybutyrylation, a mark associated with active gene expression, are elevated under the conditions of starvation or streptozotocin-induced diabetic ketosis (Xie et al., 2016).

Initially, SIRT4 was reported to be an ADP-ribosyltransferase (Haigis et al., 2006); however Laurent et al. subsequently found that SIRT4 deacetylates and inhibits malonyl-CoA decarboxylase (MCD), an enzyme that generates acetyl-CoA from malonyl-CoA to regulate fatty acid synthesis (Laurent, German et al., 2013). Later studies revealed additional activities for SIRT4 (Kumar & Lombard, 2017). Mathias et al. showed that SIRT4 acts as a lipoamidase to remove lipoyl groups from the E2 component of dihydrolipoyllysine acetyltransferase of pyruvate dehydrogenase (PDH), resulting in reduced PDH activity (Mathias et al., 2014). Recently, Anderson et al. revealed that SIRT4 catalyzes the removal of 3-hydroxy-3-methylglutaryl (HMG) and related modifications, 3-methylglutaryl (MG), and 3-methylglutaconyl (MGc), from lysine residues on its target proteins (Anderson et al., 2017). Furthermore, the authors showed that the methyl crotonyl-CoA carboxylase complex (MCCC), involved in leucine catabolism, is decorated with these novel modifications. SIRT4 deacylates and activates MCCC, and thus plays an important role in the leucine oxidation pathway. SIRT4 loss resulted in increased acylation and reduced stability of MCCC, reducing leucine flux. As a consequence, *Sirt4* knockout (KO) mice display elevated basal and stimulated insulin secretion, and eventually develop glucose intolerance and insulin resistance (Anderson et al., 2017).

Like SIRT4, SIRT5 also possesses only very weak deacetylase activity. Initial characterization of SIRT5 found that it deacetylates and activates carbamoyl phosphate synthetase 1 (CPS1), to promote urea cycle function. Subsequent analyses have revealed that the major biochemical function of SIRT5 is to remove newly discovered, noncanonical PTMs (i.e., succinyl, malonyl, and glutaryl moieties) from lysine residues, including those on CPS1 (Du et al., 2011; Kumar & Lombard, 2018; Nakagawa, Lomb, Haigis, & Guarente, 2009; Park et al., 2013; Peng et al., 2011; Rardin et al., 2013; Tan et al., 2014).

Nuclear SIRT6 deacetylates H3K9 and H3K56 (Michishita et al., 2009, 2008; Yang, Zwaans, Eckersdorff, & Lombard, 2009), the DNA repair factor CtIP (Kaidi, Weinert, Choudhary, & Jackson, 2010), and the acetyltransferase GCN5 (Dominy et al., 2012). The presence of long-chain fatty acids stimulates SIRT6 function (Feldman et al., 2013), hinting at a distinct means of coupling SIRT6 function to nutrient status independent of NAD^+ levels. SIRT6 also activates poly[ADP-ribose] polymerase 1 (PARP1) via mono-ADP-ribosylation (Mao, Hine et al., 2011), and promotes secretion of the proinflammatory cytokine tumor necrosis factor alpha (TNF-α) via deacylation (Jiang et al., 2013). SIRT6 and its invertebrate homolog SIR-2.4 promote formation of cytoplasmic stress granules, and associate with these structures (Jedrusik-Bode et al., 2013; Michishita, Park, Burneskis, Barrett, & Horikawa, 2005; Simeoni et al., 2013). SIRT6 regulates hepatic circadian gene expression by directly interacting with the circadian control proteins, CLOCK (circadian locomotor output cycles kaput), and BMAL (or ARNTL for aryl hydrocarbon receptor nuclear translocator-like) to modulate their recruitment to chromatin. SIRT6 also restricts sterol regulatory element binding protein 1 (SREBP-1)-mediated transcription of target genes, thereby controlling circadian-dependent metabolism, including fatty acid synthesis and β-oxidation (Masri et al., 2014).

SIRT7 is mainly localized to the nucleolus, where it associates with RNA polymerase I and nucleolar transcription activator upstream-binding factor 1, and occupies the promoters of rDNA loci (Ford et al., 2006). The c-MYC transcription factor interacts with and targets SIRT7 to the rDNA promoters. By reducing ribosomal protein expression, SIRT7 mitigates ER stress brought about by the unfolded protein response (Shin et al., 2013). By interacting with the transcription factor ELK4, SIRT7 suppresses transcription of target genes by deacetylating H3K18 at specific gene promoters (Barber et al., 2012). SIRT7 is also responsible for the reduction in genome-wide H3K18 acetylation associated with oncogenic transformation. Recently, Wang et al. identified H3K36 as another SIRT7 deacetylation target (Wang, Angulo-Ibanez et al., 2019). H3K36 acetylation usually decorates weakened nucleosome structures that might contribute to euchromatin formation and active transcription. SIRT7-deficient cells exhibit H3K36 hyperacetylation in whole-cell extracts, at rDNA sequences in nucleoli, and at specific SIRT7 target loci, suggesting the physiologic relevance of SIRT7 in regulating endogenous H3K36 acetylation levels (Wang, Angulo-Ibanez et al., 2019).

SIRT7 protects the testicular receptor 4 (TR4) transcription factor from ubiquitin-mediated degradation by binding to components of the E3 ubiquitin ligase complex (Yoshizawa et al., 2014). SIRT7 also localizes outside the nucleolus. In vivo and mass spectrometry analyses have demonstrated that SIRT7 can directly deacetylate GA repeat binding protein, beta 1 (GABPβ1) on three lysine resides, K69, K340, and K369. GABPβ1 along with its alpha subunit (GABPα) comprises a heterotetrameric complex that stimulates transcription of target mitochondrial genes. Deacetylation of these residues is required for GABPα/GABPβ complex formation and subsequent transcriptional activation. This complex governs mitochondrial function, and SIRT7, through GABPβ1 signaling, is a nuclear regulator of mitochondrial activity (Ryu et al., 2014). Similar to other sirtuin family members, SIRT7 also exhibits enzymatic activities other than deacetylation. Li et al. found that SIRT7 is a NAD^+-dependent histone desuccinylase, and desuccinylates H3K122 to regulate chromatin remodeling during DNA repair (Li, Shi et al., 2016). In another study by Tanabe et al., SIRT7 exhibited robust H3K36/K37 deacetylase and debutyrylase activity (Tanabe et al., 2018). Nucleosome binding of SIRT7 through its C-terminal basic region was critical for efficient H3K36/37 deacylation in the nucleosome (Tanabe et al., 2018). Most recently, SIRT7 was shown to possess an auto mono-ADP-ribosyl transferase activity at an alternative catalytic site, in addition to its primary deacetylation catalytic site (Simonet et al., 2020). ADP-ribosyl-SIRT7 is recognized by the ADP-ribose reader macroH2A1.1, which regulates SIRT7 binding to specific chromatin regions. The interaction between SIRT7 and macroH2A1.1 plays a key role in epigenetically regulating the expression of genes involved in differentiation, proliferation, metabolism, and aging pathways (Simonet et al., 2020).

Sirtuin enzymatic activity requires and consumes NAD^+, synthesized de novo or regenerated through the salvage pathway, to produce the noncompetitive feedback inhibitor nicotinamide (NAM), 2′-O-acetyl-ADP-ribose (or the corresponding acyl derivative), and the modified substrate (Tanner, Landry, Sternglanz, & Denu, 2000) (Fig. 5.1). Thus, the requirement of sirtuins for NAD^+ distinguishes them from other classes of mammalian deacetylases, linking their enzymatic activity to the cellular nutritional milieu and metabolic state of the organism (Bordone & Guarente, 2005).

Sirtuins and mammalian longevity

Several groups have generated sirtuin KO and overexpressor mouse models (Finkel, Deng, & Mostoslavsky, 2009). In this section, we review currently existing data showing that OE of two mammalian sirtuins, SIRT1 and SIRT6, can extend mammalian lifespan. SIRT2 OE also extends longevity in the context of impaired chromosomal segregation.

SIRT1

The closest mammalian homolog of yeast Sir2p is SIRT1. Multiple groups have investigated whether elevated SIRT1 expression can increase mammalian lifespan. Initial characterization of transgenic mice overexpressing SIRT1 from an ectopic promoter in multiple tissues revealed improvements in several metabolic parameters, somewhat resembling the benefits of CR (Bordone et al., 2007). Subsequent characterization of transgenic mouse strains modestly overexpressing SIRT1—from a bacterial artificial chromosome containing the *Sirt1* genomic locus and its putative regulatory elements—revealed that SIRT1 protects against hepatosteatosis and preserves hepatic insulin sensitivity in a diabetic mouse model and in mice fed a high-fat diet (HFD) (Banks et al., 2008; Pfluger, Herranz, Velasco-Miguel, Serrano, & Tschop, 2008). Overexpressing SIRT1 in the brain rescues the age-related decline in circadian rhythm adaptation by regulating expression of the central circadian clock genes, *Bmal1* and *Per2* (period circadian protein homolog 2) in the suprachiasmatic nucleus (SCN) (Chang & Guarente, 2013). Given that mutations in genes controlling the circadian rhythm are associated with features of accelerated aging, and SIRT1, BMAL1, and PER2 expression all decrease in the SCN with age, it will be important to determine if maintaining SIRT1 expression in this region of the brain is critical for promoting longevity in mammals (Chang & Guarente, 2013; Masri & Sassone-Corsi, 2014).

Later studies demonstrated that SIRT1 OE in most tissues does not extend mouse lifespan (Herranz et al., 2010; Jeong et al., 2012). However, Herranz et al. reported that *Sirt1* transgenic mice were healthier in some respects than littermate controls. Improved maintenance of glucose homeostasis, wound healing, and neuromuscular function, as well as delayed bone loss, and a reduced incidence of carcinomas and sarcomas, were observed in SIRT1-overexpressing mice (Herranz et al., 2010). The inability of global SIRT1 OE to extend longevity has been ascribed to its failure to protect against age-associated lymphoma, a major cause of death in many inbred mouse strains (Herranz et al., 2010).

In contrast, using a brain-specific SIRT1-overexpressing mouse line (BRASTO mice), Satoh et al. demonstrated a median lifespan extension of 11%, and a delay in the incidence of cancer-related death in both sexes (Satoh et al., 2013). Increased SIRT1 expression specifically in the dorsomedial and lateral hypothalamic nuclei (DMH/LH) extends lifespan of both male and female mice by upregulating expression of the orexin type 2 receptor, through deacetylation of the transcription factor Nk2 homeobox 1 (Fig. 5.2). DMH/LH SIRT1 OE increases physical activity and whole-body oxygen consumption, maintains healthy mitochondrial morphology in skeletal muscle of aged animals, and promotes body temperature maintenance. Aged BRASTO mice are protected against age-associated degenerative changes at the neuromuscular junction (NMJ), the interface between motor nerves and muscle, and exhibit increased numbers of NMJ components; nonmyelinating terminal Schwann cells and preserved acetylcholine receptors (Snyder-Warwick, Satoh, Santosa, Imai, & Jablonka-Shariff, 2018). Sympathetic NMJ innervation, which controls muscle metabolism, maintenance, and function of nerve–muscle contact, is also enhanced in BRASTO mice. Increasing SIRT1 expression in motor neurons protects NMJ from some of the deleterious effects of aging, and delays motor neuron degeneration in a mouse model of amyotrophic lateral sclerosis (ALS) (Herskovits et al., 2018). Based on the characterization of a second independently derived BRASTO transgenic line in which no lifespan extension was observed, Satoh et al. hypothesize that relative differences in SIRT1 OE levels in specific regions of the brain may be required for increased longevity. In this regard, dose-dependent effects of sirtuin expression have been

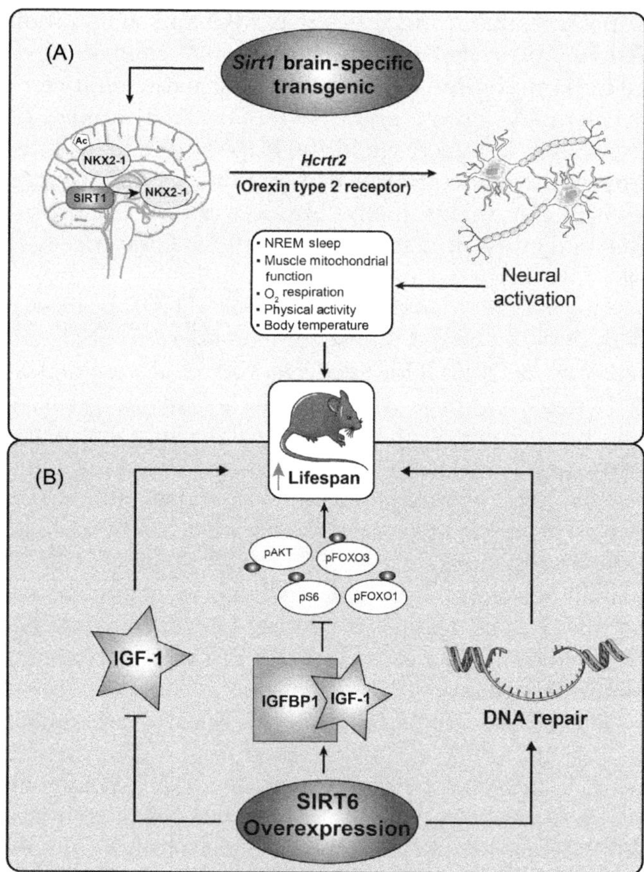

FIGURE 5.2 Lifespan extension in brain-specific SIRT1 and global SIRT6 overexpressing mice. (A) SIRT1 overexpression in the dorsomedial and lateral hypothalamic nuclei (DMH/LH) extends lifespan of both male and female mice by upregulating expression of *Hcrtr2*, the gene encoding the orexin type 2 receptor (OX2R), through deacetylation and activation of the Nk2 homeobox 1 transcription factor (NKX2-1). Increased neural activation results in better quality of sleep, and preservation of youthful mitochondrial morphology and function. (B) Global SIRT6 overexpression increases lifespan in male mice by inhibition of the insulin-like growth factor 1 (IGF-1) signaling cascade. Phosphorylation of downstream IGF-1 signaling effectors are affected as indicated: phosphorylated AKT (pAKT), phosphorylated forkhead box protein O1 and O3A (pFOXO1 and pFOXO3A), and phosphorylated S6 kinase (pS6). IGFBP1 (IGF-1 binding protein 1) binds to IGF-1 to limit the amount available to bind to receptor molecules. SIRT6 also promotes DNA DSB repair, which is a major factor in organismal lifespan. Overexpressing beaver SIRT6 in mice increases lifespan and prevents premature aging in mice.

previously documented, both in invertebrates and in mammals. For example, in the mouse heart, a 7.5-fold increase in SIRT1 expression protects against cardiac dysfunction, apoptosis, and oxidative stress, but an approximately 12.5-fold increase in SIRT1 levels causes hypertrophy and impairs cardiac function (Alcendor et al., 2007).

Global deletion of *Sirt1* in inbred 129/Sv mice causes late prenatal or early postnatal lethality. Homozygous null embryos are developmentally delayed, with a subset displaying exencephaly (Cheng et al., 2003; McBurney et al., 2003). However, in a genetically outbred strain, or an FVB background, *Sirt1* nullizygosity is compatible with adult viability (McBurney et al., 2003; Satoh et al., 2010). *Sirt1* KO mice are runted, sterile, and show a significantly reduced lifespan (Boily, He, Pearce, Jardine, & McBurney, 2009; Li et al., 2008; Mercken, Hu et al., 2014).

Via deacetylation of the rate-limiting NAD^+ biosynthetic enzyme, NAM phosphoribosyltransferase (NAMPT), SIRT1 promotes its extracellular secretion from adipocytes, which in turn enhances NAD^+ levels and promotes SIRT1 activity in the hypothalamus (Yoon et al., 2015). An age-associated decline in circulating extracellular NAMPT (eNAMPT) levels contributes to the reduction in hypothalamic NAD^+ levels and SIRT1 activity (Yoshida et al., 2019). Notably, Zhang et al. have reported a progressive age-dependent hypothalamic activation of nuclear factor kappa-light-chain-enhancer of activated B cells (NF-κB), associated with an increase in the expression of proinflammatory markers in the aging brain (Zhang et al., 2013). Suppression of hypothalamic NF-κB activity extends mouse lifespan. Given that SIRT1 is an inhibitor of NF-κB transcriptional output (Yeung et al., 2004), it is possible that the pro-longevity function of hypothalamic SIRT1 may occur in part via suppression of NF-κB signaling. Likewise, intestinal specific deletion of *Sirt1* results in abnormal activation of Paneth cells, beginning at the age of 5–8 months, with increased activation of NF-κB, stress pathways, and spontaneous inflammation by 22–24 months of age (Wellman et al., 2017). The tissue decline due to aging is associated with deterioration of adult stem cell function. In this context, Igarashi et al. showed that the number and proliferative activity of intestinal stem cells (ISCs) decline during aging, as do levels of SIRT1 (Igarashi et al., 2019). Treatment with NAM riboside (NR), an NAD^+ precursor, rejuvenates ISCs from aged mice and improves repair of gut damage. The effect of NR is blocked by the SIRT1 inhibitor, EX527, suggesting that SIRT1 activity is involved in maintenance of ISC function during aging (Igarashi et al., 2019). In this regard, CR associated upregulation of SIRT1 results in an increase in ISC number. SIRT1 deacetylates a mTOR complex 1 (mTORC1) substrate S6K1 (Hong, Zhao, Lombard, Fingar, & Inoki, 2014), enhancing its phosphorylation by mTORC1, which leads to an increase in protein synthesis and an increase in ISC number (Igarashi & Guarente, 2016). Furthermore, SIRT1 OE in mesenchymal stem cells (MSCs) increases osteoblastic bone formation, improves skeletal growth, inhibits osteoblastic bone resorption, and results in longer lifespan in *Bmi1* KO mice (Sun, Qiao et al., 2018), where BMI1 deficiency leads to an osteoporotic phenotype. Consistently, SIRT1 pharmacological activation or CR-mediated upregulation results in a significant increase in bone mass in ovariectomized female mice and aged male mice, models for post-menopausal and age-related osteoporosis, respectively (Zainabadi, Liu, Caldwell, & Guarente, 2017). In contrast *Sirt1* depletion results in a low bone mass phenotype, implicating SIRT1 as a positive regulator of bone mass. Together these results highlight the importance of SIRT1 activity for normal mammalian development and physiologic homeostasis.

SIRT2

BUBR1 is a protein kinase involved in the spindle assembly checkpoint, and ensures proper chromosomal segregation by inhibiting anaphase until correct alignment of sister chromosomes occurs (Bolanos-Garcia & Blundell, 2011; Elowe, 2011). BUBR1 protein levels decline as a function of age, and mice carrying hypomorphic alleles (designated $BubR1^{H/H}$) are predisposed to aneuploidy and display premature aging phenotypes; postnatal developmental delay, impaired would healing, cataracts, and shortened lifespan (Baker et al., 2004; Hartman, Wengenack, Poduslo, & van Deursen, 2007; Matsumoto et al., 2007). Transgenic OE of BUBR1 rescues these progeroid phenotypes, maintains genomic integrity, and extends lifespan in wild-type mice (Baker et al., 2013).

Using biochemical and mass spectrometry approaches, North et al. demonstrated that lysine 668 (K668) in BUBR1 is a SIRT2 target. By acetylating K668, the acetyltransferase cyclic-AMP response element binding protein (CREB) binding protein (CBP) promotes BUBR1 ubiquitylation and subsequent degradation. SIRT2 stabilizes BUBR1 protein by counteracting CBP-mediated K668 acetylation; conversely inhibiting SIRT2 activity or reducing its levels diminishes BUBR1 protein levels.

Crossing *Sirt2* transgenic mice (*Sirt2tg*) to $BubR1^{H/H}$ animals on the C57BL/6J background resulted in a 58% increase in median lifespan and a 21% increase in maximal lifespan compared to $BubR1^{H/H}$ mice in analyses of combined sexes (North et al., 2014). When sexes were analyzed independently, SIRT2 increased male median lifespan by 123% but showed no effect in females. In *Sirt2tg*-$BubR1^{H/H}$ tissues, BUBR1 protein levels were restored, demonstrating that SIRT2 promotes BUBR1 stability in vivo. $BubR1^{H/H}$ mice die of cardiac dysfunction, while SIRT2 OE ameliorates their cardiac phenotypes. Also, in vivo supplementation with the NAD^+ precursor NAM mononucleotide (NMN) rescues the age-dependent decline of NAD^+ in heart tissue and elevates BUBR1 protein levels in testes from aged wild-type mice to levels observed in young mice.

The phenotypes associated with $BubR1^{H/H}$ mice are not reported to be sex-specific (Baker et al, 2004). However, male mice in general may be more susceptible to cardiac pathology than females (Du, 2004), which may partly account for the observed sex bias for SIRT2-dependent lifespan extension. These results are clinically relevant, since mutations in *BUB1B*, the human ortholog of *BubR1*, are associated with mosaic variegated aneuploidy (MVA), a syndrome characterized by progeria-like phenotypes, short stature, and shortened lifespan (Callier et al., 2005; Garcia-Castillo, Vasquez-Velasquez, Rivera, & Barros-Nunez, 2008; Wijshake et al., 2012). Given that BUBR1 protein levels decrease with age, and SIRT2 OE is able to mitigate this defect, SIRT2 OE might be predicted to extend the lifespan in wild-type mice, and merits closer examination as a potential therapeutic for human disease.

SIRT6

In two independent *Sirt6* transgenic mouse strains, whole-body SIRT6 OE extends median lifespan of males by 14.5% and 9.9%, respectively (Fig. 5.2) (Kanfi et al., 2012). In contrast, no significant increase in lifespan is observed in female transgenics. Sexually dimorphic effects of pro-longevity interventions have commonly been observed in studies of lifespan extension by small molecules in mice (e.g., Harrison et al., 2009; Strong et al., 2008). In the case of SIRT6, these differences may result from discrepant effects of SIRT6 OE on IIS. In male *Sirt6* transgenic mice, lower serum IGF-1 levels and reduced downstream signaling are present, an effect associated with longevity in many other model organisms (Fontana et al., 2010). Further analyses by Kanfi et al. revealed that male *Sirt6* transgenics are relatively protected against lung tumors and show a trend toward maintenance of glucose tolerance with age. Protection against other phenotypic characteristics of aging, such as osteopenia and adrenal cortical hyperplasia, was not observed upon detailed pathological analysis (Kanfi et al., 2012). Most recently, Tian et al. reported a role for SIRT6-dependent DNA double-strand break (DSB) repair as a major factor in determining organismal lifespan (Tian et al., 2019). By performing a comparative analysis of 18 rodent species with diverse lifespans, they found that DSB repair coevolves with maximum lifespan in rodents, and the capacity of the SIRT6 protein to promote DSB repair largely accounts for the variation in DSB repair efficacy between short- and long-lived species. Furthermore, focusing on the short-lived mouse and the long-lived beaver, the authors identified five amino acid (aa) residues that are fully responsible for differential activities of their respective SIRT6 proteins. Interestingly, transgenic *Drosophila* lines expressing either beaver SIRT6 or mutant mouse SIRT6 with beaver aa residues exhibit extended lifespan compared to those expressing mouse SIRT6 or mutant beaver SIRT6 with mouse aa residues (Tian et al., 2019). Using foreskin fibroblasts isolated from 19 donors between 20 and 64 years of age, Xu et al. observed a very strong negative correlation between age and SIRT6 expression levels (Xu, Zhang et al., 2015). Sahin and colleagues also reported an age-dependent increase in methylation of the human *SIRT6* promoter, suggesting that SIRT6 expression may decrease with age (Sahin, Yilmaz, & Gozukirmizi, 2014). A significant decline in the base excision repair (BER) efficiency is observed in aged foreskin fibroblasts, which can be rescued by the OE of SIRT6 (Xu, Zhang et al., 2015), further implicating SIRT6 as a major link between DNA repair and lifespan extension.

Germline deletion of *Sirt6* in inbred 129/Sv-strain mice causes severe growth retardation, along with progressive, lethal hypoglycemia, resulting in accelerated aging and premature death within a month (Mostoslavsky et al., 2006). Similarly, *SIRT6*-null cynomolgus monkeys, developed using a CRISPR–Cas9-based approach, die hours after birth and exhibit severe prenatal developmental retardation (Zhang et al., 2018). Importantly, a homozygous inactivating mutation (Asp63 to His63) in *SIRT6* results in severe congenital anomalies and perinatal lethality in humans (Ferrer et al., 2018). *SIRT6*^{D63H} homozygous fetus-derived human-induced pluripotent stem cells fail to differentiate into embryoid bodies, functional cardiomyocytes, and neural progenitor cells due to a failure to repress pluripotent genes. Similar effects were also observed in *Sirt6*^{D63H} mouse embryonic stem cells (Ferrer et al., 2018). SIRT6-deficient mice display a degenerative phenotype, lymphopenia, and genomic instability; although a subset of germline KO mice on an outbred 129/Sv-C57BL/6 genetic background can survive up to 1 year if supplemented with glucose early in life (Xiao et al., 2010). Recently, Peshti et al. have shown that *Sirt6* deletion in mixed 129/SvJ/BALB/c background reaches adulthood, however only 10% of male mice survived at 200 days of age, whereas more than 80% of the female mice were alive at this age (Peshti et al., 2017). In comparison to wild-type controls, SIRT6-deficient mice exhibit reduced body weight, increased glucose uptake, and an age-dependent deterioration of retinal function and severe corneal inflammation (Peshti et al., 2017). SIRT6 inhibits expression of GLUT1, which is responsible for transporting glucose across the plasma membrane in mammalian cells. In *Sirt6* KO mouse retinas, GLUT1 is upregulated, several glutamate receptor genes are downregulated, and an increase in apoptosis occurs. *Sirt6* KO retinas are less responsive to photostimulation, indicating that SIRT6 is required for normal retinal function (Silberman et al., 2014). Furthermore, heterozygosity of *Trp53* in *Sirt6* KO mice rescues several age-related phenotypes and strikingly extends the lifespan of both female and male *Sirt6* KO mice, suggesting elevated p53 activity contributes significantly to accelerated aging in SIRT6-deficient mice (Ghosh et al., 2018). In this context, SIRT6 deacetylates p53 and reduces the stability and activity of p53. Alternatively, or in addition, this finding may reflect the deleterious consequences of chronically activated cellular responses to the elevated levels of DNA damage occurring in the absence of SIRT6.

These initial results reveal that specific mammalian sirtuins can exert pro-longevity effects, similar to their invertebrate homologs. One important caveat is that disease-specific interventions can extend lifespan, if those diseases are a prevalent cause of death in the population (Lombard & Miller, 2014). In this regard, the generality of the positive effects of SIRT1 and SIRT6 on lifespan needs to be assessed in other mouse genetic backgrounds. SIRT1 and SIRT6 expression may indeed slow aging in mammals, however in order to conclusively demonstrate that these sirtuins are bona fide antiaging factors, many more studies are required to comprehensively assess the impact of increased levels of these and other sirtuins on longevity and diverse age-sensitive traits.

Genetic variants of human sirtuin genes

SIRT1

Several studies have examined the association between polymorphisms in sirtuin genes, lifespan, and disease in humans (Table 5.2). For example, an association between two single-nucleotide polymorphisms (SNPs) that flank the *SIRT1* gene and longevity has been reported in a population of almost 500 individuals of Han Chinese descent (Zhang, Bai, & Chen, 2010) and one SNP associates with longevity in a population of almost 1400 Dutch subjects (Figarska, Vonk, & Boezen, 2013). Kilic et al. analyzed *SIRT1* gene polymorphisms, and their relationship with SIRT1 levels and longevity in a Turkish population. The oldest individuals carrying AG genotypes at SNP rs7895833 in the *SIRT1* promoter region showed the highest SIRT1 levels, hinting at an association between SNP rs7895833 and lifespan (Kilic et al., 2015). However, in separate Dutch, German, and Chinese populations, no significant association between *SIRT1* variants and longevity was detected (Flachsbart et al., 2006; Kuningas, Putters, Westendorp, Slagboom, & van Heemst, 2007; Lin, Yan et al., 2016), highlighting the need for comprehensive follow-up studies in larger, genetically diverse populations.

One study found that SIRT1 in the mouse brain modulates mood and behavior by positively regulating expression of the gene encoding monoamine oxidase A, which reduces serotonin levels. This study also

TABLE 5.2 Single-nucleotide polymorphisms (SNPs) in *SIRT1*, *SIRT2*, *SIRT3*, *SIRT5*, and *SIRT6*. SNPs are enumerated, along with their associated phenotype and populations in which they have been described. Journal references describing these associations are listed in the final column

Sirtuin	SNP	Population	Phenotype	References
1	rs3758391	Han Chinese	Longevity	Zhang et al. (2010)
	rs3758391	Dutch (Leiden 85 + Study)	Cognitive functioning	Kuningas et al. (2007)
	rs3758391	Mexico City	Type II diabetes	Cruz et al. (2010)
	rs3758391	Han Chinese	Acute coronary syndrome	Hu et al. (2015)
	rs3758391	Japanese	Autoimmune thyroid diseases	Sarumaru et al. (2016)
	rs3758391	Han Chinese	Diffuse large B-cell lymphoma	Kan et al. (2018)
	rs4746720	Han Chinese	Longevity	Zhang et al. (2010)
	rs4746720	Han Chinese, Japanese	Major depressive disorder	Hirata et al. (2019), Rao et al. (2020)
	rs7895833	Turkish	Longevity	Kilic et al. (2015)
	rs12778366	Dutch (Vlagtwedde/Vlaardingen Study)	Longevity/glucose tolerance	Figarska et al. (2013)
	rs10997870	Swiss	Anxiety disorder	Libert et al. (2011)
	rs10997870	Swiss and US	Panic disorder	Libert et al. (2011)
	rs12778366	Swiss	Social phobia/panic disorder	Libert et al. (2011)
	rs10997875	Japanese	Major depressive disorder	Kishi et al. (2010)
	rs12415800	Han Chinese, Japanese	Major depressive disorder	Consortium (2015), Hirata et al. (2019)
2	rs2015	Han Chinese	Colorectal cancer	Yang, Ding et al. (2017)
	rs2015	Han Chinese	Parkinson disease	Chen, Mai et al. (2019)
	rs2241703	Japanese	Height	Haketa et al. (2013)
	rs2241703	Han Chinese	Parkinson disease	Wang, Cai et al. (2018)
	rs10410544	Italian and Swiss	Alzheimer's disease	Polito et al. (2013); Wei et al. (2014)
	rs10410544	Han Chinese	Alzheimer's disease	Wei et al. (2014); Xia et al. (2014)
	rs11879029	Japanese	Height	Haketa et al. (2013)
	rs45592833	Italian	Longevity	Crocco et al. (2016)
3	rs511744	US	Longevity	TenNapel et al. (2014)
	rs536715	Multiethnic (Northern Manhattan Study)	Carotid plaques	Dong et al. (2011)
	rs3825075	Multiethnic (Northern Manhattan Study)	Carotid intima-media thickness	Della-Morte et al. (2012)
	rs4147918	Cohorts (Framingham Heart Study)	Cancer patients' survival	Doherty et al. (2017)
	rs4758633	US	Longevity	TenNapel et al. (2014)
	rs4980329	Italian	Amyotrophic lateral sclerosis	Albani et al. (2017)
	rs4980329	Multiethnic (Northern Manhattan Study)	Carotid plaques	Dong et al. (2015)
	rs11246020	US and Finnish	Metabolic syndrome	Hirschey et al. (2011)
	rs11246020	Not reported	Pulmonary arterial hypertension	Paulin et al. (2014)
	rs11555236	Italian	Longevity	Bellizzi et al. (2005), Rose et al. (2003)
	rs11555236	Canadian	Longevity	Halaschek-Wiener et al. (2009)
	rs11555236	Italian	Longevity	Albani et al. (2014)
	rs12363280	Multiethnic (Northern Manhattan Study)	Carotid plaques	Dong et al. (2015)
	rs12363280	Multiethnic (Northern Manhattan Study)	Carotid intima-media thickness	Della-Morte et al. (2012)

(*Continued*)

TABLE 5.2 (Continued)

Sirtuin	SNP	Population	Phenotype	References
5	rs2253217	Japanese-American Men	Longevity	Donlon et al. (2017)
	rs2804924	Chinese (ethnicity unreported)	Acute myocardial infarction	Chen, Wang et al. (2018)
	rs2841505	US	Longevity	TenNapel et al. (2014)
	rs4712032	Multiethnic (Northern Manhattan Study)	Carotid plaques	Dong et al. (2011)
	rs4712047	US	Longevity	TenNapel et al. (2014)
	rs9382222	Multi-ethnic	Brain aging	Glorioso et al. (2011)
	rs12216101	Multiethnic (Northern Manhattan Study)	Carotid plaques	Dong et al. (2011)
	rs112443954	Chinese (ethnicity unreported)	Acute myocardial infarction	Chen, Wang et al. (2018)
	rs573515169	Chinese (ethnicity unreported)	Acute myocardial infarction	Chen, Wang et al. (2018)
6	rs107251	Multiethnic (Northern Manhattan Study)	Carotid plaques	Dong et al. (2011)
	rs107251	US	Longevity	TenNapel et al. (2014)
	rs352493	Han Chinese	Coronary artery disease	Tang et al. (2016)
	rs3760908	Han Chinese	Coronary artery disease	Tang et al. (2016)
	rs117385980	Finnish	Longevity	Hirvonen et al. (2017)

analyzed *SIRT1* SNPs in humans with specific anxiety disorders (Libert et al., 2011). In a random sample of roughly 3400 Swiss participants, *SIRT1* SNPs were significantly associated with anxiety disorder (rs10997870), panic disorder (rs12778366 and rs10997870), and social phobia (rs12778366). Association of one SNP, rs10997870, with increased risk of panic disorder was further validated in a study of 9000 twins of Caucasian descent. This study confirmed earlier analyses of a smaller Japanese cohort that revealed an association between the *SIRT1* SNP rs10997875, which is in strong linkage disequilibrium with rs10997870, and major depressive disorder (MDD) (Kishi et al., 2010). Whole-genome sequencing of 5303 Chinese women with recurrent MDD along with 5337 controls identified a SNP in the 5′ region of the *SIRT1* gene, rs12415800, which is linked to MDD (Consortium, 2015). However, rs12415800 is much rarer in European populations, with a frequency of only 3% (Major Depressive Disorder Working Group of the Psychiatric et al., 2013) compared to 45% in this study (Consortium, 2015). A marginal association between rs4746720 and schizophrenia has been reported in a small Japanese cohort, though this association does not withstand the stringent Bonferroni's correction for multiple testing (Kishi et al., 2011).

Recently, significant association between rs4746720 and gray matter density was observed in two brain regions: the orbital parts of the right and left inferior frontal gyri (Rao, Luo, Sui, Xu, & Zhang, 2020). Post hoc comparisons between different genotypes of rs4746720 revealed that the MDD carriers of the risk allele (G) predicted higher mean gray matter density than the carriers of the nonrisk allele. However, rs12415800, the top-risk SNP from the CONVERGE Consortium, was not successfully genotyped in this study and a strong linkage disequilibrium between rs4746720 and rs12415800 was observed in East Asian populations from the 1000 Genomes Project pilot study (Rao et al., 2020). Moreover, a large study of individuals of European descent was unable to replicate the strong association for SNP rs12415800 with MDD found in the Chinese population of CONVERGE Consortium (Hyde et al., 2016). In contrast, a recent study reported that 778 suicide cases and 760 controls in a Japanese population were in strong linkage disequilibrium for *SIRT1* SNPs rs12415800 and rs4746720 (Hirata et al., 2019). The authors found significant associations between both SNPs and completed suicide among women aged ≥50 years. Other recent studies report additional *SIRT1* SNPs, which confer susceptibility to MDD in Han Chinese populations (Tang et al., 2018) and Russian populations of European descent (Aftanas et al., 2018). Overall, these studies support the hypothesis that genetic variations in *SIRT1* represent a risk factor for anxiety, depression disorders, and other mental health disorders (Libert et al., 2011); however, genetic association of *SIRT1* with mental health disorders is largely dependent upon study structure and the ethnicity of the study population.

Recent studies have revealed that another *SIRT1* SNP, rs3758391, is associated with diverse human disease conditions. Hu et al. reported that rs3758391 affects *SIRT1* mRNA expression in healthy individuals; individuals with the TT genotype display significantly higher expression levels of *SIRT1* mRNA than those with TC and CC genotypes (Hu et al., 2015). In the same study, *SIRT1* mRNA expression was found to be reduced in the PBMCs from patients with acute coronary syndrome (ACS), suggesting that the *SIRT1* SNP rs3758391 TT genotype plays a protective role against ACS. Moreover, in patients with autoimmune thyroid diseases (AITD), including Graves and Hashimoto diseases, the C carriers (TC + CC genotypes) of SNP rs3758391 show significantly higher titers of antithyroid microsomal antibody than those with a TT genotype (Sarumaru et al., 2016), further implying a protective role of the TT genotype at SNP rs3758391. However, no association was observed between SNP rs3758391 genotypes and the levels of *SIRT1* mRNA in AITD patients. Recently, Kan et al. reported an association between *SIRT1* SNP rs3758391 and the risk of diffuse large B-cell lymphoma (DLBCL) in a Chinese population (Kan, Ge, Wang, Xiao, & Zhao, 2018). DLBCL patients exhibited a higher frequency of TT and TC genotypes at rs3758391. In comparison to controls, DLBCL patients showed significantly higher expression levels of *SIRT1* mRNA, and *SIRT1* mRNA expression was higher in the TT subgroup than that of TC/CC subgroup. Multivariate Cox regression analysis revealed that the TT genotype of rs3758391 is an independent poor prognostic factor for DLBCL patients; patients with TC/CC genotype displayed a higher overall survival rate than those with genotype TT (Kan et al., 2018). Thus, the *SIRT1* rs3758391 polymorphism may serve as a potential biomarker of DLBCL patients' survival in Chinese Han populations.

Several studies have also demonstrated that genetic variation in *SIRT1* is correlated with increased glucose tolerance (Figarska et al., 2013), T2D susceptibility (Cruz et al., 2010; Rai et al., 2012), obesity (Clark et al., 2012; van den Berg et al., 2009), and fatty liver disease (Hou et al., 2018) in diverse populations. An in-depth, careful investigation of the effect of these polymorphisms on *SIRT1* gene expression and activity is necessary.

SIRT2

A number of studies have investigated the association of a *SIRT2* SNP, rs10410544, with Alzheimer's disease (AD). Polito et al. proposed the rs10410544 T-allele as a genetic risk factor for late-onset AD in the two *APOE*-ε4-negative Caucasian populations (Polito et al., 2013). Subsequently, using a Northern Han Chinese cohort, Xia and colleagues confirmed that the rs10410544 T-allele is indeed associated with late-onset AD in *APOE*-ε4 noncarriers (Xia et al., 2014). In contrast, Porcelli et al. found no association between rs10410544 and AD in two independent European populations. Moreover, rs10410544 TT carriers are protected from depressive symptoms (Porcelli et al., 2013). However, a meta-analysis, including all of the above studies, demonstrates that the *SIRT2* polymorphism, rs10410544, is significantly associated with AD risk in individuals negative for *APOE*-ε4 (Wei et al., 2014). Nevertheless, further large-scale studies are required to validate these findings.

A recent study found that the rs2241703 polymorphism in the *SIRT2* gene is associated with Parkinson disease (PD) in the Chinese Han population (Wang, Cai et al., 2018). The microRNA miR-486-3p binds to the 3′ UTR of *SIRT2* and influences its translation, reducing the α-synuclein-induced aggregation and toxicity. The SNP rs2241703 A-allele disrupts miR-486-3p binding sites in the 3′ UTR of *SIRT2*, which may subsequently promote SIRT2 translation and contribute to PD (Wang, Cai et al., 2018). More recently, Chen et al. showed that SNP rs2015 in the 3′ UTR of *SIRT2* is significantly associated with the risk of PD in the Han Chinese population (Chen, Mai et al., 2019). The A-allele at rs2015 may increase *SIRT2* expression by inhibiting the binding of miR-8061 to the 3′ UTR of *SIRT2*, ultimately contributing to the risk of PD. Consistently, mRNA levels of *SIRT2* are upregulated in the peripheral blood of PD patients compared to healthy controls (Chen, Mai et al., 2019). SNP rs2015 is also associated with susceptibility to colorectal cancer (CRC). Yang et al. showed that compared to the rs2015 C-allele, the A-allele correlates with lower CRC risk in males and older patients with age greater than 60. In silico analysis predicts that the interaction between hsa-miR-376a-5p and the rs2015 C-allele results in lower expression of SIRT2 (Yang, Ding, Gao, & Wang, 2017). Haketa et al. reported an association between *SIRT2* gene polymorphisms and height in healthy, elderly Japanese subjects. Individuals with the AA genotype at SNPs, rs2241703 and rs11879029, were on average 7.4 and 3.6 cm taller than those with the GG genotype (Haketa et al., 2013). Mean height is significantly lower in young persons with PD when compared with healthy controls (Ragonese et al., 2007), suggesting that individuals with the AA genotype at SNPs rs2241703 and rs11879029 may be protected against PD. Another *SIRT2* SNP, rs45592833, which lies within a binding site recognized by three different miRNAs (miR-3170, miR-92a-1−5p, and miR-615-5p), showed a significant association with human longevity (Crocco, Montesanto, Passarino, & Rose, 2016). The T-allele of rs45592833 is predicted to be bound with high affinity to these miRNAs, likely suppressing *SIRT2* expression, and associates with reduced survival at very old ages in an allele dose-dependent manner (Crocco et al., 2016).

Yang et al. performed an analysis of the *SIRT2* gene promoter in large cohorts of acute myocardial infarction (AMI) patients and ethnic background-matched controls, and identified three heterozygous DNA sequence variants (DSVs) (g.38900888_91delTAAA, g.38900270A > G and g.38899853C > T) selectively in three AMI patients (Yang, Gao et al., 2017). These DSVs alter the transcriptional activity of the *SIRT2* gene promoter, indicating that changes in SIRT2 levels may contribute to the risk for AMI. In a subsequent study, the same group identified four heterozygous DSVs (g.38900912G > T, g.38900561C > T, g.38900359C > T, and g.38900237G > A) in four T2D patients, three of which (g.38900912G > T, g.38900359C > T, and g.38900237G > A) significantly increased the transcriptional activity of the *SIRT2* gene promoter (Liu, Yang, Pang, Yu, & Yan, 2018), which may contribute to T2D development.

SIRT3

Genetic studies in human populations testing links between *SIRT3* and human longevity have yielded conflicting results. In an Italian cohort, Rose et al. identified a significant association between a synonymous SNP in *SIRT3* and increased male survival (Rose et al., 2003). Genetic linkage analysis in a second study also found this silent-marker SNP, rs11555236, to be associated with longevity and to be in linkage disequilibrium with a variable number tandem repeat found in a putative *SIRT3* enhancer (Bellizzi et al., 2005). These findings were not replicated, however, when central and southern Italian populations were pooled with a sample from a German cohort (Lescai et al., 2009).

Another study by Albani et al. analyzed *SIRT3* genetic variants previously reported to be present in healthy elderly individuals of European ancestry (Halaschek-Wiener et al., 2009), and confirmed findings of the earlier analyses that rs11555236 is associated with longevity, though only in females when the data were stratified by sex. Furthermore, this study also confirmed the earlier result that this SNP is in linkage disequilibrium with a putative *SIRT3* enhancer, suggesting that increased SIRT3 expression may correlate with a longer lifespan in some human populations. Indeed, SIRT3 expression was found to be higher in peripheral blood mononuclear cells extracted from individuals homozygous for this variant (Albani et al., 2014). A subsequent study by TenNapel et al., which analyzed the relationship of *SIRT3* SNPs with human lifespan, reported that individuals with the TT genotype at *SIRT3* SNP rs511744 display slightly increased lifespan compared to those with the CT or the CC genotype (TenNapel et al., 2014). Another *SIRT3* SNP, rs4758633, exhibits a gender-specific impact on the lifespan; females with the AA genotype displayed an increased lifespan compared to those with a GA or GG genotype, whereas males with the AA genotype displayed decreased lifespan compared to males with the other genotypes (TenNapel et al., 2014). Moreover, a recent study examined the effect of genetic variation on survival as a function of age in cancer patients, and revealed that survival significantly decreases in patients with heterozygote or variant homozygote genotypes compared to those with a common homozygote genotype for rs4147918, a SNP commonly assigned to *SIRT3* and *GPX4* (Doherty, Kernogitski, Kulminski, & Pedro de Magalhaes, 2017). While current data are consistent with the notion that *SIRT3* genetic variation affects human lifespan, and rs11555236 is a marker for longevity in some populations, further validation of the relationship between these SNPs and longevity in other human populations is necessary.

SIRT3 may play roles in protecting humans from forms of metabolic dysfunction, including obesity, insulin resistance, and liver steatosis. A nonsynonymous SNP in the catalytic domain of *SIRT3* was associated with metabolic syndrome in a cohort of human patients with nonalcoholic fatty liver disease (NAFLD). This correlation was further validated in a human population of over 7000 Finnish males from the Metabolic Syndrome in Men cohort. In vitro analysis demonstrated that this polymorphism reduces the catalytic activity of recombinant SIRT3 by one-third. Thus, reduced SIRT3 activity may result in enhanced susceptibility to metabolic disease in humans (Hirschey et al., 2011). A second study tested this SNP, rs11246020, for association with idiopathic pulmonary arterial hypertension (PAH) in a small cohort of patients, who presented with either idiopathic PAH (iPAH) or PAH associated with collagen vascular disease, and found an increase in the frequency of homozygosity or heterozygosity at this SNP only in the iPAH patients. Examination of tissue samples from a subset of iPAH patients revealed decreased SIRT3 expression and increased mitochondrial acetylation compared to tissue from healthy donors. Ex vivo analysis of human and mouse pulmonary artery smooth muscle cells revealed decreased mitochondrial function in the absence of SIRT3, as well as a transcriptional profile associated with inhibition of apoptosis and vascular remodeling that is important for the development of PAH. Mouse and rat models of PAH demonstrated that SIRT3 expression can rescue PAH disease phenotypes in vivo, highlighting the potential for SIRT3 as a biomarker for and therapeutic target of iPAH pathology (Paulin et al., 2014). Furthermore, Albani et al. observed a potential protective role of the T-allele in *SIRT3* SNP rs4980329 in ALS; the rs4980329 T-allele exists at a lower frequency in ALS patients (Albani et al., 2017).

Furthermore, a number of studies have revealed the association of various *SIRT3* SNPs with cardiovascular diseases (CVD). Dong et al. observed that *SIRT3* SNP rs536715 is associated with carotid plaque numbers in patients with hypertension (Dong et al., 2011). Subsequently, Dong and colleagues reported a gender-specific effect of *SIRT3* variants on total carotid plaque area. The minor allele carriers of *SIRT3* SNPs rs4980329 (T-allele) or rs12363280 (G-allele) display higher total carotid plaque area in men but not in women (Dong et al., 2015). Another study reported that carriers with the CC genotype at SNP rs12363280 in *SIRT3* exhibited lower carotid intima–media thickness, whereas only women homozygous for the minor T-allele *SIRT3* SNP rs3825075 had this phenotype (Della-Morte et al., 2012). Yin et al. genetically and functionally analyzed the *SIRT3* gene promoter in large cohorts of myocardial infarction (MI) patients and ethnic-matched controls, and reported the identification of DSVs and SNPs, which are either selectively or frequently present in MI patients. Most of these DSVs and SNPs significantly decreased the transcriptional activity of the SIRT3 gene promoter, implying their impact on SIRT3 expression as a risk factor to MI development (Yin, Pang, Huang, Cui, & Yan, 2016).

SIRT5

There are few studies suggesting that *SIRT5* polymorphisms may impact human healthspan or lifespan. SNP rs9382222 (C > T), present in a conserved region of the *SIRT5* promoter, correlates with reduced levels of *SIRT5* mRNA expression in the anterior cingulate cortex (ACC) region of the brain, and associates with an "aged" ACC phenotype, as assessed by expression of age-regulated transcripts (Glorioso, Oh, Douillard, & Sibille, 2011). Two SNPs, rs4712032 and rs12216101, in the *SIRT5* gene are associated with a greater number of carotid plaques (Dong et al., 2011). Chen et al. identified two heterozygous DSVs (g.13574131C > A and g.13574287G > C) and three heterozygous SNPs [g.13573450A > G (rs573515169), g.13574110G > A (rs2804924), and g.13574259G > C (rs112443954)] selectively in AMI patients. These DSVs and SNPs significantly reduced the transcriptional activity of the *SIRT5* gene promoter, potentially contributing to AMI development (Chen, Wang et al., 2018). TenNapel et al. reported that SNP rs2841505 is associated with slightly reduced lifespan in cohort members with a GG genotype compared to those with other genotypes (TenNapel et al., 2014). Another *SIRT5* SNP, rs4712047, shows a gender-specific impact on lifespan; females with a GG genotype exhibit an increased lifespan, whereas males with the same genotype display reduced lifespan compared to individuals with the other genotypes (TenNapel et al., 2014). Recently, Donlon et al. reported that *SIRT5* SNP rs2253217 correlates with longer lifespan in cohort members with a TT genotype than those with TC or CC genotypes (Donlon et al., 2017). Taken together, these studies suggest a correlation of human lifespan with *SIRT5* SNPs, however, as with other sirtuin SNPs, further studies are needed to validate these findings in larger groups and other genetically distinct populations.

SIRT6

Similar to *SIRT5*, there are only a handful of cohort studies analyzing the correlation of *SIRT6* SNPs with human healthspan or lifespan. Dong et al. observed that T-carriers of the *SIRT6* SNP rs107251 display an increased risk for the occurrence and number of carotid plaques (Dong et al., 2011). Two other *SIRT6* SNPs, rs352493 and rs3760908, with the C- and A-alleles, respectively, were associated with the severity of coronary artery disease in Chinese Han populations (Tang et al., 2016). However, Lin et al. found that five SNPs, rs350852, rs350844, rs352493, rs4807546, and rs3760905, across the *SIRT6* gene and its 5 kb up-/downstream region, are not associated with human longevity (Lin, Zhang et al., 2016). Likewise, Li et al. reported that *SIRT6* SNP rs350846 is not associated with human longevity (Li, Qin et al., 2016). In contrast, a study by TenNapel et al. reported that individuals with either the CC or CT genotype at *SIRT6* SNP rs107251 displayed a >5-year mean survival advantage compared to those with TT genotype (TenNapel et al., 2014). A subsequent study by Hirvonen and colleagues identified a SNP, rs117385980, in the *SIRT6* gene in a cohort of Finnish men. Data from this study suggested that individuals with a T-allele at SNP rs117385980 exhibit shorter lifespan than those with the C-allele (Hirvonen et al., 2017). Since several mouse studies strongly suggest that SIRT6 not only extends lifespan but also improves healthy aging, more studies are warranted to establish a firm correlation between *SIRT6* SNPs and human healthspan and lifespan. Moreover, the functional significance of these polymorphisms on *SIRT6* expression needs to be investigated.

Sirtuins as modulators of responses to caloric restriction

Dietary or caloric restriction (CR) is the only known environmental intervention that robustly and reliably extends lifespan across phyla, including mammals. CR can extend rodent lifespan up to 50%, while suppressing diverse age-associated conditions such as cancer, autoimmune disease, T2D, neurodegeneration, and many others (Speakman & Mitchell, 2011). Mechanisms underlying this effect are still unclear. Since most individuals find that a long-term reduction in caloric intake is difficult to maintain, a great deal of current aging research focuses on elucidating mechanisms of the pro-health effects of CR, using yeast, flies, worms, and rodent models. Several physiological pathways have been proposed to regulate CR-dependent lifespan extension and healthspan maintenance, including IIS, mTOR signaling, AMPK signaling, and sirtuins (Guarente, 2013). Given that sirtuin catalytic activity is NAD^+-dependent, and NAD^+ levels rise upon nutrient stress in certain tissues, upregulation of specific sirtuin functions has been proposed to represent a means to reap some of the beneficial effects of CR (Guarente, 2013).

Conflicting results have been obtained regarding roles for sirtuins in beneficial responses to CR. Reduction of glucose in growth medium extends the replicative lifespan of *S. cerevisiae* in a strain-specific manner, though there is controversy as to whether this phenotype is Sir2p-dependent. The use of inconsistent glucose concentrations by different laboratories has further complicated the interpretation of discrepant results in this area (Kaeberlein & Powers, 2007; Lamming et al., 2005; Lin, Defossez, & Guarente, 2000). Longo and Kennedy have detailed the controversies surrounding the involvement of yeast sirtuins in the CR response (Longo & Kennedy, 2006).

In *C. elegans*, one group has reported that the SIRT1 homolog SIR-2.1 is required for increased longevity in response to CR (Wang & Tissenbaum, 2006), whereas several other labs have found that SIR-2.1 is dispensable for this effect (Greer & Brunet, 2009; Hansen et al., 2007; Kaeberlein et al., 2006; Lee et al., 2006; Mair, Panowski, Shaw, & Dillin, 2009). In *D. melanogaster*, an initial report found that dSIR2 is required for CR-induced longevity (Rogina & Helfand, 2004), a result not replicated by a subsequent study (Burnett et al., 2011). However, subsequent work using RNAi-mediated dSIR2 knockdown (KD) in specific tissues or the whole organism supported the initial finding (Banerjee et al., 2012; Bauer et al., 2009; Pallos et al., 2008). It is likely that CR and dSIR2 act at least in part through distinct pathways in flies, since gene expression pattern changes associated with dSIR2 OE differ from those induced during CR (Hoffmann et al., 2013).

More recent studies in the fly have revealed that lifespan in this species responds to the relative amounts of protein and carbohydrate in the diet, rather than to an overall reduction in calories per se. A complete lack of specific amino acids (aas) such as arginine, methionine, or isoleucine is detrimental to lifespan, while an intermediate amount of dietary methionine results in a significant extension (Tatar, Post, & Yu, 2014). This mirrors findings in rodents, where methionine restriction can also extend lifespan (Miller et al., 2005; Perrone, Malloy, Orentreich, & Orentreich, 2013; Richie et al., 1994). Little is known about the potential role of sirtuins in responding to specific dietary components in the context of lifespan.

SIRT1

Several reports support the hypothesis that SIRT1 is a mediator of aspects of the CR response in mammals, in a tissue-specific manner. For example, SIRT1 protein levels increase in white adipose tissue (WAT), skeletal muscle, and hypothalamus in response to CR in mice; however SIRT1 protein expression and NAD^+ levels decrease in the liver (Chen et al., 2008; Cohen et al., 2004; Satoh et al., 2010). Similarly, tissue-specific changes in SIRT1 activity occur during CR, likely due to changes in NAD^+ levels (Chen et al., 2008). Importantly, CR-mediated lifespan extension is lost in mice genetically null for *Sirt1* compared to *Sirt1* wild-type or heterozygous controls (Mercken, Hu et al., 2014). Transgenic mice overexpressing SIRT1 in WAT, brown adipose tissue (BAT), and the brain show metabolic phenotypes somewhat reminiscent of those resulting from a CR diet (Bordone et al., 2007). Although SIRT1 expression is not increased in the liver or skeletal muscle of these transgenics, they are more glucose tolerant, and show reduced blood cholesterol, glucose, and insulin levels, and less WAT accumulation. Consistent with these findings, deletion of *Sirt1* in skeletal muscle abrogates increased insulin sensitivity associated with CR; in contrast no apparent defect in the response to CR is observed when *Sirt1* is deleted specifically in the liver (Chen et al., 2008; Schenk et al., 2011). *Sirt1* deletion in the CNS and peripheral neurons also impairs insulin sensitivity in response to CR (Cohen, Supinski, Bonkowski, Donmez, & Guarente, 2009). Surprisingly however, *Sirt1* deletion in CNS neurons alone enhances insulin sensitivity and glucose tolerance compared to controls (Lu et al., 2013). Findings from global *Sirt1* KO and brain-specific SIRT1 transgenic mouse models demonstrate that SIRT1 supports maintenance of body temperature and neuronal activity in hypothalamic nuclei during CR (Satoh et al., 2010). Hippocampal SIRT1 protein expression was significantly elevated in female, but not male, mice, after a year of CR, highlighting important sex differences in response to CR (Wahl et al., 2018). Enhanced SIRT1 expression in the brain was observed in rats after a 4-week CR diet, and was maintained after cerebral ischemia/reperfusion (I/R) injury compared to the control cohort. Decreased infarct volume and improved neurobehavioral outcomes after injury were observed in CR rats. These effects were ablated upon siRNA-mediated SIRT1 downregulation in vivo, suggesting an important role in CR-induced neuroprotection for SIRT1 (Ran et al., 2015). SIRT1 also protects against age-related decline in renal function in mice fed a CR diet (Kume et al., 2010). Overall, SIRT1 is required for both increased physical activity and certain metabolic responses observed in response to CR (Boily et al., 2008; Chen et al., 2008; Chen, Steele, Lindquist, & Guarente, 2005; Imai, 2009; Satoh et al., 2013).

A related, but distinct question concerns how SIRT1 may affect the lifespan of specific cell populations. Ectopic SIRT1 OE or siRNA *SIRT1* KD delayed or accelerated senescence of normal human diploid fibroblasts, respectively. Cell cultures supplemented with serum from calorie-restricted rats displayed an increased replicative lifespan and increased SIRT1 protein content when compared to cultures containing serum from AL rats, suggesting a role for SIRT1 in mediating CR-dependent replicative lifespan and delayed senescence of normal human diploid fibroblasts (de Cabo et al., 2015). Generally, upon nutrient depletion, mTOR activity decreases while sirtuin activity increases (Madeo, Carmona-Gutierrez, Hofer, & Kroemer, 2019). However a recent report finds that both mTORC1 signaling and SIRT1 activity are enhanced in ISC during CR (Igarashi & Guarente, 2016). This study indicated that Paneth cells secrete cyclic ADP ribose to activate SIRT1, and cooperate with mTORC1 activation in ISCs, to stimulate ISC proliferation in response to CR. Contrary to expectations, increased activity of the mTORC/S6K1 axis results in an increased ISC population, thus supporting the notion that SIRT1 regulates the gut stem cell niche, partly contributing to the CR-dependent longevity phenotype in mammals (Igarashi & Guarente, 2016).

SIRT2

Little is known about the potential role of SIRT2 in response to a CR diet. One study found that SIRT2 expression is induced in the kidney and WAT of mice fed a CR diet. The authors demonstrate that SIRT2 interacts with and deacetylates FOXO3A in cell culture, resulting in the increased expression of several downstream targets such as mitochondrial superoxide dismutase 2 (SOD2) and the pro-apoptotic factor, BIM. Through this mechanism, SIRT2 reduced cellular ROS levels through SOD2 activation, and promoted apoptosis via BIM activity, when oxidative damage is too extensive to repair (Wang, Nguyen, Qin, & Tong, 2007). The interpretation that SIRT2 responds to CR and oxidative stress via the FOXO signaling cascade requires testing in *Sirt2* KO animals, to determine if SIRT2 regulates physiologic aspects of the CR response.

SIRT3

SIRT3 expression increases in liver, skeletal muscle, and adipose tissue during CR, suggesting that SIRT3 may play a role in regulating mitochondrial functions under these conditions (Hallows et al., 2011; Hirschey et al., 2010; Nakagawa et al., 2009; Palacios et al., 2009; Schwer et al., 2009; Shi, Wang, Stieren, & Tong, 2005; Someya et al., 2010). Although overall hepatic mitochondrial protein acetylation increases during CR (Schwer et al., 2009), deacetylation of specific mitochondrial target proteins occurs in this setting (Hebert et al., 2013), resulting from increased SIRT3 expression. As noted previously, SIRT3 plays a major role in suppressing ROS levels. Indeed, SIRT3 is required for reducing cellular ROS levels during CR (Qiu, Brown, Hirschey, Verdin, & Chen, 2010; Someya et al., 2010). As a consequence of this activity, SIRT3 is required for the protection against age-associated hearing loss conferred by CR, via preservation of cochlear cells against age-associated attrition (Someya et al., 2010). CR decreases hepatic acetyl-CoA, serum insulin, and triglyceride levels in a SIRT3-dependent manner (Hebert et al., 2013; Someya et al., 2010). SIRT3 deacetylates and activates the urea cycle enzyme ornithine transcarbamoylase (OTC) in murine liver mitochondria in response to CR (Hallows et al., 2011). OTC deficiency results in an accumulation of orotic acid due to urea cycle dysfunction. *Sirt3*-null mice fed either a normal or CR diet exhibit increased OTC acetylation and higher levels of urinary orotic acid compared to wild-type mice fed a CR diet. Also, mass spectrometry analysis of blood from fasted *Sirt3*-null mice revealed a decrease in the urea cycle metabolite, citrulline, which is a direct product of OTC activity. Bioinformatic analysis of publicly available gene expression data sets, in which induction of SIRT3 by either CR or fasting has been observed, has implicated nuclear respiratory factor 2 (NRF2) as a transcriptional regulator of SIRT3. NRF2 is responsible for driving expression of many mitochondrially expressed genes (Scarpulla, 2002). When the DNA-binding subunit of NRF2 was transiently overexpressed or KD in cell culture, *SIRT3* expression increased or decreased, respectively (Satterstrom et al., 2015). Clearly SIRT3 is necessary for multiple metabolic responses to CR. Given SIRT3's function as a tumor suppressor, it will be of interest to determine whether SIRT3 is required for improved cancer suppression induced by CR, or indeed if SIRT3 is required for longevity induced by this intervention.

SIRT4

SIRT4 protein levels decline in liver (Haigis et al., 2006; Schwer et al., 2009) but not pancreatic β cells during CR (Haigis et al., 2006). SIRT4 suppresses β-cell insulin secretion in response to glucose or aas (Ahuja et al., 2007; Haigis et al., 2006). During CR, reduced α-cell SIRT4 activity—occurring through mechanisms that are as yet unclear—may permit increased coupling of insulin secretion to aa metabolism (Haigis et al., 2006). This effect makes physiologic sense, in that aas are used as a metabolic fuel to a greater degree during CR than under normal feeding conditions.

SIRT5

Little is currently known regarding roles for SIRT5 during CR. Overall levels of lysine succinylation, a SIRT5 target modification, increase in liver during fasting (Park et al., 2013); the impact of chronic CR on succinylation, malonylation, or glutarylation has not been reported. Hepatic SIRT5 protein levels are unchanged during CR (Schwer et al., 2009), though nothing is known regarding SIRT5 biochemical activity in this context. SIRT5 has been shown to promote ketogenesis under fasting conditions (Rardin et al., 2013), suggesting that SIRT5 could potentially play a role in use of this alternative fuel during CR. Alternatively, SIRT5 suppresses mitochondrial respiration through specific mitochondrial complexes (Park et al., 2013), suggesting that a decrease in SIRT5 function during CR might contribute to increased mitochondrial metabolism in response to this intervention. Characterization of *Sirt5* KO mice under CR conditions will be required to elucidate the roles, if any, for SIRT5 during CR.

SIRT6

A role for SIRT6 in response to CR has not yet been assessed, though increased SIRT6 expression (Kim, Xiao et al., 2010) and protein stability (Kanfi et al., 2008) in response to this intervention have been described in several tissues in rodents. Fasting regimens induce metabolic reprogramming by initially upregulating gluconeogenesis to maintain blood glucose levels. Several reports indicate that SIRT6 is a negative regulator of hepatic gluconeogenesis, inhibits expression of genes involved in glycolysis (Dominy et al., 2012; Kim, Xiao et al., 2010; Xiong, Tao, DePinho, & Dong, 2013), and negatively affects lipogenesis by repressing the SREBPs, transcription factors important for lipogenesis and cholesterol biogenesis (Elhanati et al., 2013; Tao, Xiong, DePinho, Deng, & Dong, 2013b). Increased SIRT6 expression in response to nutrient deprivation in cell culture models was found to be SIRT1-dependent (Kim, Xiao et al., 2010), and expression of *Sirt1* mRNA itself increases in response to CR in rat brain, WAT, kidney, and liver (Cohen et al., 2004). Given that both of these sirtuins interact with several factors and pathways relevant to the CR response (reviewed in Guarente, 2013), it is tempting to speculate that metabolic reprogramming during CR might involve a complex relationship between SIRT1 and SIRT6. A more recent study reported attenuated age-associated kidney degeneration, enhanced SIRT6 protein expression, and reduced inflammatory cytokine production (i.e., TNF-α and IL1β) in mice after a 6-month CR regimen. Lung fibroblasts cultured in low glucose were resistant to cell senescence and exhibited increased SIRT6 expression compared to normal glucose conditions. Ectopic SIRT6 OE in these cells led to delayed replicative senescence by reducing signaling though NF-κB, a SIRT6 target pathway. Conversely, SIRT6 KD results in accelerated cell senescence and enhanced NF-κB signaling in a culture model, suggesting a role for SIRT6 in managing NF-κB signaling in response to CR (Zhang, Li, Mu et al., 2016). However, currently no evidence directly links SIRT6 activity to the induction of lifespan or healthspan benefits of a CR diet.

Roles for sirtuins in diverse disease states

Substantial evidence now exists demonstrating that activation of sirtuin activity in different contexts can confer health benefits in mammals, and in some cases can extend lifespan, as discussed previously. While the evidence directly linking sirtuins to increased longevity in mammals has emerged only recently, a large body of work exists demonstrating that sirtuins ameliorate numerous age-associated pathological conditions. In this section, we discuss recent findings relevant to sirtuin functions as promoters of mammalian healthspan, with a focus on age-associated disease (Fig. 5.3).

Cancer

A large body of research has elucidated complex, and often seemingly conflicting roles for sirtuins in cancer. The following section highlights recent evidence, largely from cell culture and mouse models, that demonstrates that sirtuins modulate tumorigenesis in a cell- and context-specific manner.

SIRT1

SIRT1 acts as both a tumor suppressor and oncoprotein, via interaction with and modification of dozens of distinct substrates relevant to cancer proliferation and survival, including the c-MYC oncoprotein and the tumor suppressor p53 (Costa-Machado & Fernandez-Marcos, 2019; Yuan, Su, & Chen, 2013). Elevated SIRT1 expression has been detected in many human malignancies, including breast, prostate, lung, colon, liver,

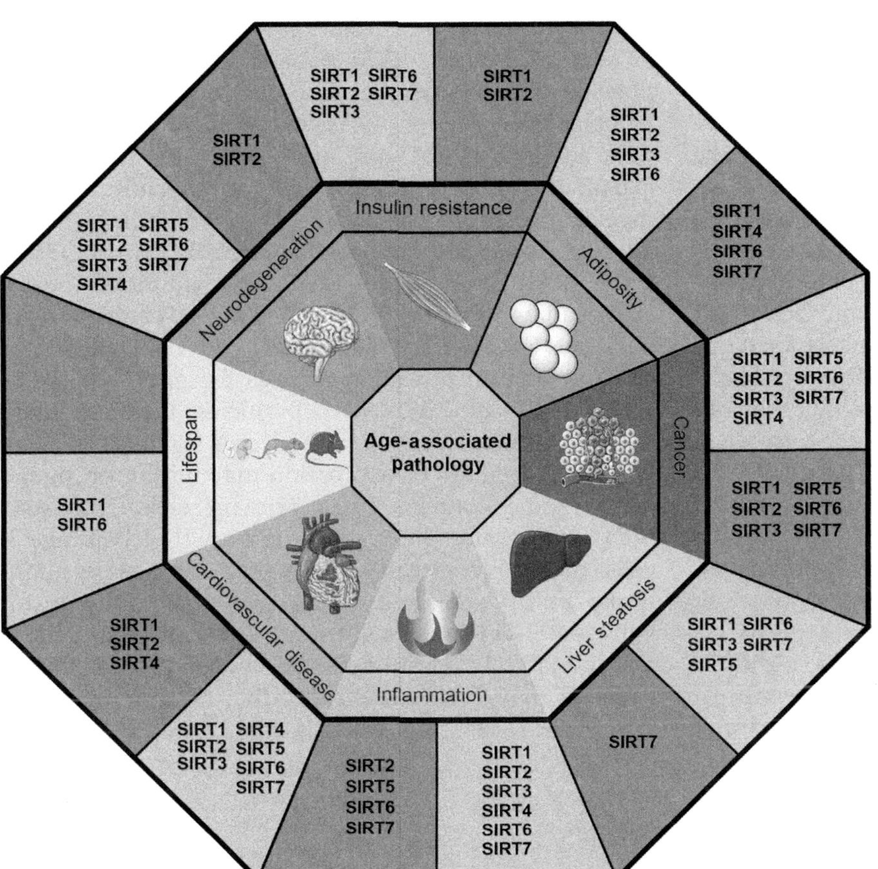

FIGURE 5.3 Summary of sirtuin involvement in inhibiting age-related pathology. Outline of sirtuin effects on the indicated age-associated diseases. Sirtuins listed in green boxes ameliorate corresponding pathologies, while those listed in red boxes aggravate them. Brain-specific SIRT1 or whole-body SIRT6 overexpression extends lifespan in mice.

pancreatic, lymphoma, leukemia, and some ovarian and cervical cancers; higher levels of SIRT1 are generally associated with poor prognosis and overall survival (Costa-Machado & Fernandez-Marcos, 2019; Yuan et al., 2013). SIRT1 deacetylates the tumor suppressor p53 and attenuates its transcriptional activity (Luo et al., 2001; Vaziri et al., 2001). In addition, SIRT1 deacetylates p300 and reduces its binding with p53, inhibiting the p300-mediated acetylation and activation of p53 (Bouras et al., 2005). SIRT1-induced deacetylation represses p53-mediated apoptosis, leading to uncontrolled cell cycle progression (Luo et al., 2001; Vaziri et al., 2001). In this context, in nonsmall-cell lung cancer, tumors with high SIRT1 levels exhibit hypoacetylated p53, correlating with decreased survival overall (Tseng, Lee, Hsu, Tzao, & Wang, 2009).

Conversely, tumor suppressor functions of SIRT1 have been observed in mouse models, wherein SIRT1 promotes the maintenance of genome stability (Yuan, Minter-Dykhouse, & Lou, 2009), and inhibition of cell growth via survivin (an inhibitor of apoptosis), β-catenin, and other pathways (Lim et al., 2010; Srisuttee et al., 2012). Global SIRT1 OE at moderate levels reduces the incidence of spontaneous carcinomas and sarcomas (Herranz et al., 2010), and mice with moderate levels of SIRT1 OE are protected from mutant *Kras*-driven lung adenocarcinomas (Costa-Machado et al., 2018). Similarly, increased expression of SIRT1 in the $Apc^{+/min}$ mouse model of intestinal polyposis decreases tumorigenesis (Firestein et al., 2008). Paradoxically, however, global or enterocyte-specific deletion of *Sirt1* reduces the size and number of intestinal polyps formed in this model (Boily et al., 2009; Leko, Park, Lao, Simon, & Bedalov, 2013). *Sirt1* deletion results in increased apoptosis in polyps arising in $Apc^{+/min}$ mice, implying that SIRT1 can play a pro-survival role during tumor progression (Leko et al., 2013).

SIRT1 positively regulates the epithelial-to-mesenchymal transition (EMT) in various cancer types (Palmirotta et al., 2016). SIRT1 promotes the invasiveness and metastasis of hepatocellular carcinoma (HCC), by activating EMT markers, such as SNAIL, TWIST, and VIMENTIN, and inhibiting E-cadherin (Hao et al., 2014; Serrano-Gomez, Maziveyi, & Alahari, 2016). In prostate cancer cells, SIRT1 interacts with EMT transcription factor ZEB1, which recruits SIRT1 to the *E-cadherin* promoter, where it deacetylates histone H3, leading to suppression of *E-cadherin* gene expression (Byles et al., 2012). In melanoma, SIRT1 facilitates metastasis by decreasing E-cadherin expression and accelerating the autophagic degradation of E-cadherin via deacetylation of Beclin-1 (Sun, Jiao, Wang, Yu, & Ming, 2018). In contrast to these reports, SIRT1 OE or activation by resveratrol inhibits migration, invasion, and metastasis of breast cancer and oral squamous cell carcinoma (OSCC) cells (Chen et al., 2014; Simic et al., 2013). Mechanistically, SIRT1-mediated deacetylation of SMAD4 leads to inhibition of TGF-β signaling and reduction of MMP7 expression, thereby preventing the degradation of E-cadherin (Chen et al., 2014; Simic et al., 2013). SIRT1 also suppresses hypoxia-induced EMT in nasal polypogenesis (Lee et al., 2016).

SIRT1 directly binds to and deacetylates hypoxia-inducible factor 1α (HIF-1α), a transcription factor that regulates gene expression in response to reduced oxygen tension or increased ROS levels, to mediate metabolic reprogramming in cancer cells or other rapidly proliferating cell types (Lim et al., 2010). Whether SIRT1-dependent deacetylation promotes HIF-1α activity or attenuates its ability to promote expression of its target genes is a matter of debate. One study demonstrated that *SIRT1* KD in vivo impairs transcriptional output of HIF-1α target genes in HCC cells (Laemmle et al., 2012). Consistently, Joo et al. reported that SIRT1 stabilizes HIF-1α during hypoxia, via direct deacetylation. SIRT1 depletion or inhibition leads to increased HIF-1α acetylation and reduced hypoxic HIF-1α accumulation, while SIRT1 OE enhances HIF-1α accumulation, resulting in increased expression of HIF-1α target genes, and promotion of cancer cell invasion (Joo et al., 2015). In contrast, others have shown that SIRT1-dependent deacetylation of HIF-1α inhibits transcriptional coactivator binding to HIF-1α target gene promoters, thereby negatively regulating tumor growth and angiogenesis in tumor xenograft models (Lim et al., 2010). Recently, HIF-1α has been shown to promote cancer stem cell-like features in human ovarian cancer cells by upregulating SIRT1 expression (Qin, Liu et al., 2017), suggesting the existence of a positive feedback loop to promote malignancy.

SIRT1 also interacts with the oncoprotein c-MYC, though the functional outcome of this interaction is controversial (Mao, Zhao et al., 2011; Menssen et al., 2012; Yuan et al., 2009). OE of c-MYC and SIRT1 is present in several cancer types, and is often associated with higher tumor grade (Yuan et al., 2013). Although SIRT1 expression is stimulated by c-MYC, deacetylated c-MYC is more susceptible to degradation, and therefore SIRT1 indirectly inhibits c-MYC function (Mao, Zhao et al., 2011; Yuan et al., 2009). In this context, SIRT1 deficiency in preadipocytes induces hyperacetylation of c-MYC, leading to an abnormal proliferation rate and differentiation into small, inflamed, and dysfunctional adipocytes (Abdesselem et al., 2016). However, others have found that deacetylation stabilizes c-MYC, and enhances SIRT1 activity by promoting NAD^+ synthesis, establishing a positive feedback loop that may reinforce cellular malignant transformation (Menssen et al., 2012). Disruption of this positive feedback loop by inhibiting NAMPT results in downregulation of c-MYC and transient derepression of p53, resulting in apoptosis induction and reduced clonogenic growth in human *BRAF*-mutant CRC cell lines and patient-derived tumoroids (Brandl et al., 2019). Moreover, p53 blocks binding of c-MYC to the *SIRT1* promoter, and thus inhibits c-MYC-mediated upregulation of SIRT1 (Yuan, Liu, Lei, & Tang, 2017). Additionally, in acute myeloid leukemia (AML) stem cells, featuring internal tandem duplication in Fms-like tyrosine kinase (FLT3-ITD), c-MYC upregulates USP22 deubiquitinase expression, leading to reduced ubiquitination and enhanced stability of SIRT1 (Li et al., 2014). SIRT1 depletion or pharmacological inhibition reduces the growth of FLT3-ITD AML stem cells and promotes tyrosine kinase inhibitor-mediated killing of these cells (Li et al., 2014).

Thus, while SIRT1 expression may inhibit oncogenesis in some contexts (e.g., primary cells), increased SIRT1 expression may confer a growth advantage in cells that have already undergone oncogenic alterations. Future research focusing on the relationship between SIRT1 and its substrates, in particular HIF-1α and c-MYC, in various tissue- and cell-type specific contexts may further delineate the multiple means by which SIRT1 affects neoplasia.

SIRT2

SIRT2 regulates chromatin assembly and promotes chromosomal stability during mitosis, thereby functioning as a tumor suppressor (Inoue et al., 2007, 2009; Vaquero et al., 2006). SIRT2 deacetylates α-tubulin, controls the early metaphase checkpoint, and affects cell cycle progression by mediating monomethylation of H4K20 through its interaction with the histone methyltransferase PR-Set7 (Dryden, Nahhas, Nowak, Goustin, & Tainsky, 2003; North, Marshall, Borra, Denu, & Verdin, 2003; North & Verdin, 2007; Pandithage et al., 2008; Serrano et al., 2013).

Evidence from *Sirt2* KO mice supports a tumor suppressor role for SIRT2 in vivo. *Sirt2*-null male mice develop liver and intestinal cancer, while females develop mammary tumors, associated with centrosome amplification, mitotic catastrophe, and eventual cellular transformation (Kim et al., 2011). Consistently, SIRT2 expression is reduced in human breast cancer, HCC, colorectal carcinoma, nonsmall-cell lung cancer (NSCLC), and prostate cancer (Damodaran et al., 2017; Kim et al., 2011; Li, Huang et al., 2016; Li, Xie, Chen, Lu, & Xia, 2013; Xu, Jiang, Shen, Qian, & Peng, 2015; Zhang, Zhan et al., 2017). As a tumor suppressor in breast cancer, SIRT2 deacetylates and activates pyruvate kinase isoform M2 (PKM2) (Park et al., 2016), which catalyzes the last step of glycolysis, resulting in formation of pyruvate and ATP. Inhibition of PKM2 activity leads to the accumulation of glycolytic intermediates to facilitate tumor growth. Ectopic expression of a deacetylation mimetic PKM2 mutant in *Sirt2*-deficient mammary tumor cells alters their glycolytic reprogramming and inhibits tumor growth (Park et al., 2016). Moreover, ectopic expression of SIRT2 sensitizes breast cancer cells to ROS-induced DNA damage and cell death, by deacetylating and inhibiting the peroxidase activity of the antioxidant protein peroxiredoxin (Psdx-1) (Fiskus et al., 2016).

In lung cancer, SIRT2 binds to the promoter region of Jumonji domain-containing 2A (JMJD2A), a histone demethylase, and negatively regulates its expression, resulting in inhibition of lung tumor growth (Xu, Jiang et al., 2015). In this context, reduced SIRT2 levels via increased expression of miR150, promote JMJD2A expression and enhance proliferation of NSCLC cells (Jiang, Shen, Chen, & Xu, 2018). Li et al. have shown that SIRT2 deacetylates SKP2 and induces its degradation, resulting in increased levels of the cell cycle inhibitor, p27 and suppression of NSCLC cell growth (Li, Huang et al., 2016). Recently, SIRT2 has been shown to remove fatty acyl groups from RalB, a member of the Ras GTPase family, and suppresses its activity, resulting in inhibition of RalB-induced migration in lung cancer cells (Spiegelman et al., 2019). Furthermore, SIRT2 is protective against KRAS-driven tumorigenesis by deacetylating and inactivating KRAS (Song, Biancucci et al., 2016; Yang et al., 2013). SIRT2 loss exacerbates pancreatic tumor formation in *Kras* mutant mice (Song, Biancucci et al., 2016). Consistently, loss of SIRT2 is associated with increased resistance to MEK inhibitors in KRAS-driven colon cancer cell lines (Bajpe et al., 2015). Two independent studies have also demonstrated a tumor suppressor role for SIRT2 in mouse skin (Ming, Qiang, Zhao, & He, 2014; Serrano et al., 2013).

However, like SIRT1, SIRT2 also has seemingly contradictory roles in cancer. In contrast to reports suggesting a tumor-suppressive role of SIRT2 in breast cancer, Zhou et al. showed that SIRT2 levels are significantly higher in human triple-negative breast cancer (TNBC). SIRT2-mediated deacetylation and stabilization of SLUG enhance the basal differentiation, stemness, and invasion of a basal-like breast cancer cell line, a model for TNBC (Zhou, Ni et al., 2016). In addition, pharmacological inhibition of SIRT2 inhibits TNBC xenograft tumor growth by promoting the ubiquitination and degradation of c-MYC (Jing et al., 2016). Most recently, SIRT2 has been shown to deacetylate the RRM2 subunit of ribonucleotide reductase (RNR) at K95, and enhance RNR activity, leading to increased dNTP pool size and cancer cell proliferation (Chen, Luo et al., 2019). Substitution of endogenous RRM2 with an acetylated mimetic RRM2 mutant (RRM2-K95Q) results in significantly reduced dNTP pool size, stalling of DNA replication fork progression, and suppression of NSCLC cell growth in vitro and in a xenograft model (Chen, Luo et al., 2019). Another recent study showed that SIRT2 inhibition or deficiency induces autophagy in NSCLC cells via acetylation of HSPA5 and subsequent activation of ATF4 and DDIT4 to inhibit the mTOR signaling pathway (Mu, Lei et al., 2019). SIRT2 is also upregulated in HCC (Chen, Chan et al., 2013; Huang et al., 2017), and promotes EMT, by modulating the glycogen synthase kinase-3β/β-catenin signaling pathway via deacetylation and activation of protein kinase B (Chen, Chan et al., 2013). AML samples exhibit upregulation of SIRT2 expression, which is associated with poor overall survival (Deng, Ning, Zhou, & Liang, 2016). SIRT2 deacetylates and activates G6PD at lysine 403, promoting NADPH production and supporting biosynthesis of macromolecules to sustain AML cell proliferation (Xu, Wang, Li, & Wang, 2016). As a consequence, inhibition of SIRT2 induces apoptosis and inhibits proliferation in both AML cell lines and primary cells from AML and chronic lymphocytic leukemia (CLL) patients (Bhalla & Gordon, 2016; Dan et al., 2012; Moniot et al., 2017). SIRT2 has also been demonstrated to deacetylate p73, inactivating its transcriptional activity, promoting glioblastoma tumorigenicity (Funato et al., 2018).

A tumor-promoting role of SIRT2 is further supported by several other studies demonstrating that *SIRT2* KD actually induces apoptosis in glioma and HeLa cervical carcinoma cell lines, and SIRT2 expression is increased in neuroblastoma and pancreatic cancer cells (Costa-Machado & Fernandez-Marcos, 2019; Yuan et al., 2013). Tenovins, a class of small molecules that increase p53 acetylation and activation in tumor cells (McCarthy et al., 2012), inhibit SIRT1 and SIRT2 activity, induce apoptosis in transformed cell lines, and impede tumor growth in vivo (Lain et al., 2008; McCarthy, Hollick, & Westwood, 2010). Given that SIRT1 and SIRT2 have been shown to deacetylate p53 to inhibit its transcriptional activity and promote its ubiquitin-mediated degradation (Langley et al., 2002; Luo et al., 2004, 2001; van Leeuwen et al., 2013; Vaziri et al., 2001), an effect lost upon tenovin treatment (van Leeuwen et al., 2013), tenovins may be therapeutically useful in treating cancers with reduced p53 activity due to elevated SIRT1 and/or SIRT2 expression.

SIRT3

SIRT3 has also been implicated as both a tumor suppressor and oncogene in various contexts. The *SIRT3* gene is deleted in up to 20% of all human cancers, and in 40% of breast and ovarian cancers specifically (Finley et al., 2011). Consistent with this finding, *SIRT3* mRNA expression is reduced in diverse human cancers, including testicular, prostate, and hepatocellular tumors, compared to nonmalignant tissue (Kim, Patel et al., 2010). An increased incidence of mammary cancer occurs in mice lacking

SIRT3, and SIRT3 protein levels are lower in human breast tumors when compared to healthy breast tissue (Finley et al., 2011; Kim, Patel et al., 2010). SIRT3 protein expression is greatly reduced in a large fraction of human breast cancers, and low SIRT3 expression is associated with poor prognosis and decreased patient survival (Desouki, Doubinskaia, Gius, & Abdulkadir, 2014).

Mechanistically, one major function of SIRT3 in the context of tumor suppression is to suppress ROS-mediated damage. Elevated ROS levels promote tumorigenesis by inducing DNA mutations and genomic instability, and activating pathways regulating metabolism, cell survival, and proliferation (Liou & Storz, 2010). SIRT3 deacetylates and activates mitochondrial SOD2 and isocitrate dehydrogenase 2 (IDH2); deacetylation of IDH2 in turn allows regeneration of the antioxidant glutathione (Qiu et al., 2010; Someya et al., 2010; Tao et al., 2010). In this context, Zou et al. have shown that SIRT3 deacetylates IDH2 at lysine 413, which promotes IDH2 dimerization, resulting in its increased enzymatic activity (Zou et al., 2017). Ectopic expression of an acetylation mimetic IDH2 mutant (IDH2-K413Q) reduces IDH2 dimerization and enzymatic activity, increases cellular ROS levels and glycolysis, and promotes cellular transformation and tumorigenesis in nude mice (Zou et al., 2017). Furthermore, *Sirt3* deletion increases ROS and stabilizes HIF-1α, resulting in pro-tumorigenic metabolic reprogramming (Bell, Emerling, Ricoult, & Guarente, 2011; Finley et al., 2011; Kim, Patel et al., 2010). In addition, deletion of *Sirt3* in mouse embryonic fibroblasts increases intracellular superoxide levels and aneuploidy in response to genotoxic stress, suggesting that SIRT3 guards against chromosomal instability and tumorigenesis by maintaining genomic integrity (Kim, Patel et al., 2010). Recently, Lee et al. showed that SIRT3-mediated downregulation of cellular ROS represses the oxidation of Src kinase and attenuates focal adhesion kinase (FAK) activation (Lee, van de Ven et al., 2018), Src/FAK signaling plays a critical role in metastasis by promoting cell migration. Consequently, SIRT3 OE inhibits migration and metastasis of breast cancer cells (Lee, van de Ven et al., 2018). Overall, suppression of ROS levels is likely a principal means by which SIRT3 inhibits tumorigenesis.

In addition to its role in regulating ROS, SIRT3 has tumor-suppressive functions by modulating other processes in cancer cells. SIRT3 has been shown to activate mitochondrial respiration via the PDH complex (PDC) (Fan et al., 2014; Jing et al., 2013), which is frequently inactivated during cancer cell metabolic reprogramming (e.g., Kaplon et al., 2013). Quan et al. reported downregulation of SIRT3 in prostate carcinomas, and SIRT3 OE inhibited prostate cancer growth by inhibiting the PI3K/Akt pathway, leading to ubiquitination and degradation of the c-MYC oncoprotein (Quan et al., 2015). Furthermore, Jeong et al. have shown that SIRT3 serves as a tumor suppressor by modulating cellular iron metabolism (Jeong et al., 2015). Many cancer cells exhibit dysregulated iron metabolism. ROS upregulation, due to SIRT3 loss, promotes the iron response protein (IRP1) binding to iron response elements (IREs), resulting in increased expression of transferrin receptor (TfR1) and hence, enhanced cellular iron uptake and content. Human pancreatic cancers exhibit reduced expression of *SIRT3*, while *TFR1* expression is upregulated. Consequently, SIRT3 OE decreases TfR1 expression by inhibiting IRP1 and represses proliferation in pancreatic cancer cells (Jeong et al., 2015). Recently, Gonzalez Herrera et al. have shown that SIRT3 suppresses breast cancer xenograft formation by inhibiting glutamine metabolism and de novo nucleotide synthesis via suppressing mTORC1 signaling (Gonzalez Herrera et al., 2018). Another recent study reported that SIRT3 suppresses the EMT and metastasis of prostatic cancer cells by promoting FOXO3A expression, via attenuating the Wnt/β-catenin signaling pathway (Li, Quan, & Xia, 2018).

However, data also exist supporting an oncogenic role for SIRT3. Li–Fraumeni syndrome (LFS) is characterized by a predisposition to a wide spectrum of cancers, and frequently results from mutations in the *TP53* gene. In a study of a family with LFS and no apparent *TP53* mutation, a genomic duplication of the entire genomic region containing the *SIRT3* locus was identified, leading to increased *SIRT3* mRNA expression (Aury-Landas et al., 2013). The authors directly evaluated the potential contribution of SIRT3 to tumorigenesis in these patients, and demonstrated that SIRT3 OE in a glioma cell line inhibits apoptosis, deregulates cell cycle progression, and results in CpG island hypermethylation, features typically associated with cancer (Hanahan & Weinberg, 2011). Furthermore, SIRT3 has been reported to inhibit p53-induced growth arrest in human bladder cancer cells (Li et al., 2010). Elevated SIRT3 expression has been detected in node-positive breast cancer (Ashraf et al., 2006) and OSCC. In OSCC, *SIRT3* KD inhibits proliferation and sensitizes cells to radiation and chemotherapeutic treatments in vitro (Alhazzazi et al., 2011). However, a conflicting report claims that SIRT3 enzymatic activity is reduced in OSCCs, and that upregulation of SIRT3 expression inhibits proliferation of OSCC cell lines (Chen, Chiang, Liu, Chen, & Chiang, 2013).

In a recent study, Wei et al. uncovered a role for SIRT3 in deacetylating and activating the mitochondrial serine hydroxymethyltransferase 2 (SHMT2) to promote CRC cell growth (Wei et al., 2018). SHMT2 catalyzes the reversible, concurrent conversion of serine to glycine and tetrahydrofolate (THF) to 5,10-methylenetetrahydrofolate (5,10-CH_2-THF) (Ducker & Rabinowitz, 2017), a pathway which is significantly upregulated in various cancers to support cell proliferation. Acetylation of SHMT2 at K95 disrupts its functional tetramer structure, inhibits its enzymatic activity, and promotes its degradation via the K63-ubiquitin-lysosome pathway, resulting in decreased CRC cell proliferation and tumor growth. In human CRC patients, increased levels of SIRT3 expression correlate with reduced levels of SHMT2-K95 acetylation and worsened overall survival (Wei et al., 2018). Another recent study by Chen et al. demonstrated that SIRT3 plays a tumor-promoting role by deacetylating and activating pyrroline-5-carboxylate reductase 1 (PYCR1) (Chen, Yang et al., 2019), which is highly expressed in various cancers, and promotes proliferation of tumor cells. PYCR1 participates in the proline synthesis process by catalyzing the reduction of proline-5C to proline with concomitant generation of NAD^+ and $NADP^+$. SIRT3 interacts with and deacetylates PYCR1 at K228, resulting in decamer formation and activation of PYCR1. Ectopic expression of an acetylation mimetic PYCR1 mutant (PYCR1-K228Q) reduces PYCR1 oligomerization and enzymatic activity, resulting in reduced proliferation (Chen, Yang et al., 2019). Most recently, Li et al. demonstrated that SIRT3 is highly expressed in DLBCLs, and high SIRT3 expression is linked to poor survival (Li, Chiang et al., 2019). SIRT3 deficiency reduced the proliferation, survival, and self-renewal of DLBCL cells in vitro and *Sirt3* KO attenuated B-cell lymphomagenesis in VavP-Bcl2 mice. Mechanistically, SIRT3 deacetylates glutamate dehydrogenase (GDH) and drives anaplerotic glutaminolysis to replenish the TCA cycle in DLBCL cells. SIRT3 depletion impaired glutamine flux to the TCA cycle and reduced the acetyl-CoA pools, which in turn induce autophagy and cell death. The authors further described the development of a mitochondrial-targeted class I sirtuin inhibitor, YC8-02, which phenocopied the effects of SIRT3 depletion and killed DLBCL cells (Li, Chiang et al., 2019). Interestingly, in a recent study, Park et al. showed that SIRT3 and the mitochondrial chaperone TRAP1 are overexpressed in glioma stem cells (GSCs), and that a cooperative interplay between these two proteins increased the mitochondrial respiratory capacity and reduced ROS production in GSCs (Park et al., 2019). SIRT3 deacetylates TRAP1 and increases its chaperone activity, which in turn stabilizes SIRT3 protein. Inhibition of TRAP1 or SIRT3 in GSCs compromises this interplay and attenuates their independent functions, leading to metabolic dysregulation, overproduction of ROS, loss of stemness, and suppression of tumor formation by GSCs in vivo (Park et al., 2019). Thus, as with SIRT1 and SIRT2, the oncogenic and tumor suppressor activities of SIRT3 are highly context specific.

SIRT4

Several recent studies have revealed that SIRT4 expression is downregulated in a variety of human cancers, and lower SIRT4 levels are often correlated with worse prognosis and poor survival in cancer patients (Betsinger & Cristea, 2019). In addition, *Sirt4* KO mice spontaneously develop tumors (Jeong et al., 2013). Taken together these observations suggest that SIRT4 typically serves as a tumor suppressor. SIRT4 functions in this manner by suppressing glutamine metabolism through distinct mechanisms (Csibi et al., 2013; Jeong, Lee, Lee, & Haigis, 2014; Jeong et al., 2013). SIRT4 inhibits cells from metabolizing glutamine to replenish tricarboxylic acid (TCA) cycle intermediates (anaplerosis). Jeong et al. showed that SIRT4 represses glutamine anaplerosis in response to DNA damage, and that consequently *Sirt4* KO MEFs do not reduce their glutamine uptake after UV exposure (Jeong et al., 2013). In response to DNA damage, SIRT4 levels increase, and repress mitochondrial glutamine metabolism. Since glutamine is critical for G1 to S cell cycle progression (Colombo et al., 2011), SIRT4 indirectly regulates the cell cycle, allowing for proper DNA repair and maintenance of genomic stability. In vivo, SIRT4 deficiency results in growth of larger tumors in mice. Correspondingly, two independently derived strains of *Sirt4*-null mice show an increased incidence of spontaneous lung tumors (Jeong et al., 2013). SIRT4 deletion hastens lymphomagenesis and death in a mouse model of Burkitt lymphoma (Jeong et al., 2014), whereas SIRT4 OE reduces glutamine consumption and glutamine-dependent growth in MYC-driven human Burkitt lymphoma cells, and sensitizes these cells to glucose depletion.

In breast cancer, the transcriptional regulator CtBP represses SIRT4 expression to promote glutaminolysis and cancer progression (Wang, Zhou, Wang, Cui, & Di, 2015). Depletion of CtBP resulted in decreased glutamine consumption, and disruption of metabolic homeostasis in cancer cells. An independent study found that the mTORC1 negatively regulates SIRT4 expression by promoting the degradation of the transcriptional regulator, CREB2, resulting in GDH activation and increased glutamine metabolism in cancer cells (Csibi et al., 2013). SIRT4 ADP-ribosylates and inactivates GDH to inhibit glutamine metabolism (Haigis et al., 2006). SIRT4 OE in MEFs lacking the negative regulator of mTORC1, TSC2, shows attenuated transformation, proliferation, and tumor development in a xenograft assay (Csibi et al., 2013). In human CRC, SIRT4 OE inhibits glutamine metabolism (Huang et al., 2016; Miyo et al., 2015), and upregulates E-cadherin expression, resulting in reduced mobility and proliferation of CRC cell lines (Miyo et al., 2015). Huang et al. showed that SIRT4 OE increases the sensitivity of CRC cells to chemotherapeutic drug 5-fluorouracil by inhibiting the cell cycle (Huang et al., 2016).

In addition to its roles in regulating glutamine metabolism, SIRT4 regulates other processes in cancer cells. In NSCLC, SIRT4 has been shown to inhibit mitochondrial fission (Fu et al., 2017), which play critical roles in malignancy. Ectopic expression of SIRT4 reduced mitogen-activated protein kinase (MAPK)/extracellular signal-regulated kinase (MEK/ERK) activity in lung cancer cells, which inhibits phosphorylation of mitochondrial fission factor Drp1 and impaired Drp1 recruitment to the mitochondrial membrane via an interaction with Fis1. As a consequence of reduced mitochondrial fission, SIRT4 overexpressing lung cancer cells display cell cycle arrest, impaired proliferation, and attenuated cell invasion and migration (Fu et al., 2017). Likewise, recent studies have shown that OE of SIRT4 suppresses the activity of ERK and induces cell cycle arrest, resulting in reduced malignant properties of gastric cancer (Hu, Lin et al., 2019) and thyroid cancer cells (Chen, Lin et al., 2019). Another recent study by Chen et al. revealed that SIRT4-overexpressing HCC cells exhibited decreased glycolytic capacity, with a concomitant decrease in the expression of key glycolytic enzymes (Chen, Ding et al., 2019).

In HCC, the histone methyltransferase SET8 promotes the Warburg effect by suppressing the expression of SIRT4 by inactivating the transcription factor Krüppel-like factor 4 (KLF4) (Chen, Ding et al., 2019). Similarly, ectopic expression of SIRT4 in pancreatic cancer cells induces a remarkable reduction in their glycolytic capacity, and level of HIF-1α protein declined in the presence of ectopically expressed SIRT4 (Hu, Qin et al., 2019). In pancreatic cancer, expression of the epigenetic modifier, UHRF1 (ubiquitin like with plant homeodomain and ring finger domains 1), is negatively correlated with SIRT4 expression. In this context, Hu et al. showed that UHRF1 suppresses SIRT4 expression in pancreatic cancer cells to promote aerobic glycolysis (Hu, Qin et al., 2019). Consequently, SIRT4-overexpressing pancreatic cells exhibited increased apoptosis, reduced proliferation and colony-formatting abilities (Hu, Qin et al., 2019). Overall, SIRT4 represses tumorigenesis by regulating glutamine metabolism, and other processes involved in cancer, including Warburg metabolism, cell cycle progression, mitochondrial dynamics, metastasis, and apoptosis.

SIRT5

A number of recent studies investigating the role of SIRT5 in cancer show that like other family members, SIRT5 also plays dual roles in tumorigenesis and can serve as both a tumor promoter and tumor suppressor in a context-specific manner. In this regard, *SIRT5* expression is elevated in some cancer types, while others display reduced levels of *SIRT5* expression (Bringman-Rodenbarger, Guo, Lyssiotis, & Lombard, 2018).

SIRT5 is overexpressed in advanced NSCLC, and high levels of SIRT5 expression correlate with poor prognosis (Lu, Zuo, Feng, & Zhang, 2014). SIRT5 promotes the expression of NRF2 and facilitates lung cancer growth and drug resistance by upregulating the expression of NRF2 targets involved in oxidative and xenobiotic defense (Lu et al., 2014). Consistently, SIRT5 promotes cisplatin resistance in ovarian cancer cells by upregulating the expression of NRF2 and its target HO-1 (Sun et al., 2019). Another study supporting the tumor promoter role of SIRT5 in lung cancer showed that SIRT5 reduces the expression of SUN2, a key component of the LINC (linker of nucleoskeleton and cytoskeleton) complex, which functions as a tumor suppressor by downregulating expression of GLUT1 and LDHA to inhibit Warburg metabolism (Lv et al., 2015). In contrast, one study suggested that SIRT5 suppresses lung tumor cell growth by reducing ROS levels, via desuccinylating and activating SOD1 (Lin, Xu et al., 2013). In this context lung tumor cells expressing a desuccinylation mimetic SOD1 mutant exhibit reduced proliferation (Lin, Xu et al., 2013).

Recently, Wang et al. demonstrated upregulation of *SIRT5* expression in CRC, where increased expression of *SIRT5* was associated with a poor prognosis (Wang, Wang et al., 2018). Mechanistically, SIRT5 promotes the malignant phenotypes of CRC cells by sustaining the anaplerotic entry of glutamine into the TCA cycle via deglutarylation and activation of glutamate dehydrogenase 1 (GLUD1) (Wang, Wang et al., 2018), an enzyme that catalyzes conversion of glutamate to α-ketoglutarate. Another recent study by Du et al. revealed that SIRT5 promotes therapeutic resistance in WT *KRAS* CRC cells by reducing oxidative damage (Du, Liu et al., 2018). SIRT5 demalonylates and inactivates SDH subunit A (SDHA), leading to succinate accumulation, which in turn activates a ROS scavenger protein, thioredoxin reductase 2 (TrxR2) (Du, Liu et al., 2018). Additionally, inactivation of SDHA increases the succinate/α-KG ratio, resulting in inhibition of α-KG-dependent dioxygenases and epigenetic dysregulation, and promoting tumorigenesis and resistance to targeted therapy (Du, Liu et al., 2018; Xu et al., 2011). Consistently, SIRT5-mediated desuccinylation and inactivation of SDHA promotes clear cell renal cell carcinoma tumorigenesis (Ma et al., 2019). Most recently, Gao et al. have shown that SIRT5 promotes chemotherapeutic resistance in mutant *KRAS* CRC cancer stem cells by demalonylating and inhibiting the glycolytic enzyme triosephosphate isomerase (TPI), routing the glycolytic intermediates to the oxidative pentose phosphate pathway, in turn upregulating NADPH production to suppress levels of chemotherapy-induced ROS (Gao et al., 2019).

As in other cancer types, SIRT5 is reported to be overexpressed in HCC, where high SIRT5 expression correlates with overall poor survival (Chang et al., 2018). SIRT5 promotes HCC growth by inhibiting the expression of the transcription factor E2F1 (Chang et al., 2018), which exhibits a tumor suppressor role in HCC (Zhan et al., 2014).

Recently, Zhang et al. have shown that SIRT5 inhibits HCC cell apoptosis by deacetylating cytochrome c, causing reduced release of cytochrome c from mitochondria (Zhang, Wang et al., 2019). In contrast, two recent studies reported reduced levels of SIRT5 expression in HCC samples, and decreased SIRT5 expression is associated with poor overall survival (Chen, Tian et al., 2018; Guo et al., 2018). Chen and colleagues revealed the presence of SIRT5 in peroxisomes, where it desuccinylates and inhibits acyl-CoA oxidase 1 (ACOX1), the rate-limiting enzyme in peroxisomal β-oxidation, and is a major contributor of H_2O_2 production and oxidative DNA damage in HCC (Chen, Tian et al., 2018). Guo et al. have shown that SIRT5 deacetylates vimentin and suppresses EMT. In this context, cell migration was enhanced by OE of an acetylation mimetic vimentin mutant, whereas cell migration was inhibited by the OE of a nonacetylation mimetic vimentin allele (Guo et al., 2018).

Similarly, Yang et al. uncovered a role of SIRT5 in folate metabolism: SIRT5 desuccinylates and activates SHMT2 to promote tumor cell growth (Yang, Wang et al., 2017). Hypersuccinylation attenuates SHMT2 activity, and is associated with strikingly reduced U2OS and HCT116 cell proliferation and xenograft growth in vitro and in vivo (Yang, Wang et al., 2018). Another study revealed that SIRT5 desuccinylates PKM2 and inhibits its pyruvate kinase activity, resulting in accumulation of glycolytic intermediates to facilitate tumor growth (Xiangyun et al., 2018). Hypersuccinylation of PKM2 decreases cellular NADPH production, sensitizes A549 cells to oxidative stress, and represses their proliferation and tumor-forming ability (Xiangyun et al., 2017). In contrast, another group showed that SIRT5-mediated desuccinylation increases PKM2 pyruvate kinase activity (Wang, Wang et al., 2017).

Taken together, these studies strongly implicate SIRT5 as a tumor promoter in a variety of cancer types. In contrast some reports highlight tumor suppressor functions of SIRT5. In this regard, Guan and colleagues reported that SIRT5, along with SIRT1, mediates H_2O_2-induced deacetylation of the tumor suppressor promyelocytic leukemia protein (PML), a role which is essential for PML's nuclear localization to exert its tumor suppressor function. SIRT1 KD HeLa cells exhibit fewer H_2O_2-induced PML-nuclear bodies and resist H_2O_2-induced apoptosis, which is restored by the ectopic expression of SIRT5 (Guan et al., 2014). Furthermore, SIRT5 desuccinylates and inhibits glutaminase, an enzyme-catalyzing hydrolysis of glutamine to glutamate and ammonia (Polletta et al., 2015). Human breast cancer cells with increased SIRT5 expression display reduced ammonia levels, resulting in decreased ammonia-induced autophagy and mitophagy (Polletta et al., 2015). Ammonia-induced autophagy can play a protective role in tumor cells, conferring resistance against chemotherapy and other environmental stresses. Consequently, breast cancer patients expressing higher levels of SIRT5 responded better to chemotherapy (Xu et al., 2018). Glioma samples and cells harboring an *IDH1*^{R132H} mutation convert α-KG to oncometabolite R-2-hydroxyglutarate (R-2HG), which competitively inhibits SDH, resulting in succinyl-CoA accumulation and mitochondrial protein hypersuccinylation (Li et al., 2015). As a consequence, dysfunctional mitochondria accumulate the antiapoptotic protein BCL-2, conferring apoptosis resistance. Strikingly, ectopic SIRT5 expression reduced mitochondrial protein succinylation levels, inhibited BCL2 accumulation, and slowed xenograft growth (Li et al., 2015). Most recently, Lu et al. have shown that SIRT5 inhibits gastric cancer cell growth and migration via desuccinylating 2-oxoglutarate dehydrogenase and inhibiting its activity, causing mitochondrial dysfunction and imbalanced redox status (Lu et al., 2019), further implying a tumor-suppressor function for SIRT5.

SIRT6

As with other sirtuins, SIRT6 plays oncogenic and tumor-suppressor roles in a context-specific manner. In human breast cancer patients, elevated SIRT6 expression is associated with reduced overall survival; in these cells, SIRT6 promotes chemoresistance (Khongkow et al., 2013). However, another report found that *SIRT6* depletion increases resistance of breast cancer cells to chemotherapy treatment ex vivo, while OE of nonphosphorylatable SIRT6 increased sensitivity. In this context, SIRT6 is phosphorylated on serine 338 (S338) by AKT1, which promotes E3 ubiquitin-protein ligase MDM2-dependent SIRT6 degradation. Proliferation of breast tumor cells is attenuated in mouse xenograft experiments and ex vivo cultures, when S338 is mutated and SIRT6 is stabilized. An inverse correlation between levels of phospho-SIRT6^{S338} and patient survival was also reported, potentially implicating SIRT6 stabilization as a target for treatment-resistant breast cancer patients (Thirumurthi et al., 2014). More recently, SIRT6 haploinsufficiency in mutant *BRAF*^{V600E} melanoma has been shown to confer resistance to MEK inhibitor treatment (Strub et al., 2018), while KD of *SIRT6* expression in prostate cancer cells sensitizes them to chemotherapy and decreases cancer cell survival (Liu, Xie et al., 2013).

Ming et al. demonstrated that SIRT6 functions as an oncogene in the epidermis by promoting the expression of COX-2 via repressing AMPK signaling (Ming, Han et al., 2014). Skin-specific deletion of SIRT6 inhibits skin tumorigenesis in the mouse. Cea et al. demonstrated that SIRT6 is highly expressed in multiple myeloma, as an adaptive response to genomic stability, and high levels of SIRT6 levels are associated with poor prognosis (Cea et al., 2016). Upregulation of SIRT6 helps multiple myeloma cells to better adapt to DNA damage, by downregulating expression of MAPK pathway genes, MAPK signaling, and proliferation. Mechanistically, SIRT6 interacts with the transcription factor ELK1 and ERK signaling-related genes, binds to their promoters, and deacetylates H3K9 at these sites. This causes the inactivation of ERK2/p90RSK signaling, increasing DNA repair via Chk1 and confers resistance to DNA damage. SIRT6 depletion enhances proliferation of multiple myeloma cells, and sensitizes them to DNA-damaging agents (Cea et al., 2016). Recently, Han et al. showed that SIRT6 levels are upregulated in HCC and are associated with poor prognosis (Han, Jia, Wu, & Huang, 2019). SIRT6 functions as an oncogene by deacetylating Beclin-1 and inducing the autophagic degradation of E-cadherin, resulting in the promotion of HCC cell invasion, migration, and EMT. Additionally, SIRT6 enhances N-cadherin and vimentin expression most likely by deacetylating FOXO3a (Han et al., 2019).

Although SIRT6 expression is elevated in some human tumors (Bauer et al., 2012; Khongkow et al., 2013; Liu. Xie et al., 2013), the *SIRT6* locus is deleted in approximately 20% of human cancers and 35% of cancer cell lines surveyed (Sebastian et al., 2012). Additionally, Kugel et al. identified rare point mutations in *SIRT6*, which are associated with either destabilization or inactivation of SIRT6 in human cancers (Kugel et al., 2015). Mechanistically, SIRT6 inhibits the transcriptional activities of HIF-1α and c-MYC, resulting in repression of glycolytic metabolic reprogramming (Sebastian et al., 2012; Zhong et al., 2010). Through these activities, SIRT6 suppresses adenoma growth and malignant transformation in $Apc^{+/min}$ mice (Sebastian et al., 2012). In this context, SIRT6 was found to be SUMOylated, which promotes the SIRT6 interaction with c-MYC (Cai et al., 2016). Reduced SUMOylation decreases SIRT6 binding to c-MYC and hence, reduced occupancy of SIRT6 at c-MYC target genes, resulting in hyperacetylation of H3K56 and upregulation of c-MYC target genes (Cai et al., 2016). SIRT6 has also been shown to deacetylate and promote nuclear export of the nuclear PKM2 (Bhardwaj & Das, 2016), a PKM2 isoform that enhances aerobic glycolysis and promotes tumor growth. As a consequence of SIRT6-dependent deacetylation, PKM2 nuclear protein kinase and transcriptional coactivator functions are abolished, resulting in the suppression of the oncogenic functions of PKM2 (Bhardwaj & Das, 2016). Moreover, in a recent study, SIRT6 was shown to downregulate the levels of PKM2 via deacetylating hnRNP A1 (Yang, Zhu et al., 2019), which is involved in the alternative splicing of the pyruvate kinase mRNA, allowing tumor cells to produce the PKM2 isoform specifically.

In pancreatic ductal adenocarcinoma (PDAC), SIRT6 serves as a tumor suppressor by downregulating the expression of oncofetal RNA-binding protein Lin28B (Kugel et al., 2016), which negatively regulates let-7 microRNA. Downstream let-7 RNA target genes, such as *HMGA2*, *IGF2BP1*, and *IGF2BP3*, promote PDAC progression and metastasis. Consequently, SIRT6 loss results in histone hyperacetylation at the Lin28b promoter, resulting in upregulation of Lin28B and let-7 target genes, leading to a highly aggressive metastatic disease in a mouse model of pancreatic cancer (Kugel et al., 2016). Han et al. reported reduced levels of SIRT6 in human NSCLC tissues and cell lines (Han, Liu, Liu, & Li, 2014). Increased expression of SIRT6 suppresses the proliferation in NSCLC cells via inhibiting the expression of Twist1 (Han et al., 2014), a member of basic helix-loop-helix transcription factor family that promotes tumor proliferation and malignant transformation. In contrast, Bai et al. reported elevated SIRT6 levels in NSCLC patient tumors and cell lines. SIRT6 depletion reduces NSCLC cell migration and invasion, whereas SIRT6 OE upregulates ERK1/2 phosphorylation, activates matrix metalloproteinase 9 (MMP9), and promotes NSCLC cell migration and invasion (Bai et al., 2016).

In human cancer cell lines, SIRT6 OE is reported to result in increased p53/p73-dependent apoptosis. Cell death requires signaling through the ataxia telangiectasia mutated (ATM) kinase, and requires SIRT6 mono-ADP-ribosyltransferase activity (Van Meter, Mao, Gorbunova, & Seluanov, 2011). Consistently, in a mouse liver cancer model, increased SIRT6 expression reduces survivin expression by deacetylating H3K9 at the survivin promoter, and induces apoptosis during cancer initiation (Min et al., 2012). In contrast, Ran et al. reported that SIRT6 inhibits HCC cell apoptosis by deacetylating H3K9 at the Bax promoter, resulting in reduced expression and mitochondrial translocation of Bax (Ran et al., 2016). Likewise, in prostate cancer cells, SIRT6 inhibits apoptosis by promoting the expression of antiapoptotic protein BCL2 (Liu. Xie et al., 2013).

Thus, the opposing actions of SIRT6 to suppress or facilitate tumorigenesis may be specific to tumor type and stage—that is, SIRT6 may inhibit transformation in normal cells, but may be recruited to serve an oncogenic function in cancer cells. Modulation of SIRT6 activity or expression in premalignant cells versus advanced tumors in the same cancer type would help further delineate these roles of SIRT6.

SIRT7

SIRT7 is overexpressed across many cancer types, and is considered to represent a predictor of poor survival, suggesting an oncogenic role for this protein (Aljada et al., 2015; Ashraf et al., 2006; Frye, 2002; Geng, Peng, Chen, Luo, & Li, 2015; Haider et al., 2017;

Han et al., 2013; Kim et al., 2013; Li, Tian et al., 2019; Malik et al., 2015; Singh et al., 2015; Wang, Lu et al., 2015; Wei, Jing, Ke, & Yi, 2017; Yu et al., 2014; Zhang et al., 2015). Barber et al. reported that acetylated H3K18 is a target of SIRT7 (Barber et al., 2012). Loss of H3K18 acetylation has been associated with oncogenic transformation and more aggressive cancers in humans (Manuyakorn et al., 2010; Seligson et al., 2009). SIRT7-mediated deacetylation of H3K18 is required for the maintenance of anchorage-independent growth and escape from contact inhibition in cell culture models. In vivo experiments using xenografts in mice further implicate *Sirt7* as a protooncogene (Barber et al., 2012). It is worth mentioning that this study only assessed a role for SIRT7 in maintaining tumorigenicity, rather than its roles during premalignant multistep transformation. Zhang et al. reported that SIRT7 is overexpressed in gastric cancer tissues, and that higher SIRT7 levels are significantly correlated with disease stage, metastasis, and poor patient survival (Zhang et al., 2015). KD of SIRT7 induced apoptosis in gastric cancer cells, resulting in their impaired growth both in vitro and in a xenograft mouse model. SIRT7 binds to the promoter of miR-34a and deacetylates H3K18, thus repressing miR-34a expression. miR-34a silencing inhibits the SIRT7-KD-induced apoptosis of gastric cancer cells, restoring their proliferation and colony formation capacity. This indicates that SIRT7 downregulates miR-34a expression to promote gastric cancer. Consistently, miR-34a levels are tightly associated with survival of patients with gastric cancer (Zhang et al., 2015). Recently, Wei et al. reported that SIRT7 is highly expressed in osteosarcoma tissues and cell lines, and that elevated SIRT7 levels are correlated with poor prognosis in osteosarcoma patients (Wei, Jing et al., 2017). The authors showed that SIRT7 binds to the promoter region of CDC4 in osteosarcoma cells and deacetylates H3K18, resulting in reduced expression of CDC4. As a consequence, SIRT7 KD resulted in increased expression of CDC4 (Wei, Jing et al., 2017). Furthermore, SIRT7 KD resulted in reduced proliferation, migration, invasion, tumor formation, and metastasis of osteosarcoma cells, which are partially rescued by silencing of CDC4, suggesting that SIRT7 promotes osteosarcoma via targeting CDC4 (Wei, Jing et al., 2017). More recently, Li et al. reported that SIRT7 expression was upregulated in papillary thyroid cancer, and depletion of SIRT7 expression resulted in reduced proliferation, colony formation, migration, and invasion of thyroid cancer cells (Li, Tian et al., 2019). Conversely, SIRT7 OE promoted the growth of thyroid tumors in a mouse xenograft model. The authors further showed that SIRT7 binds to the promoter region of deleted in breast cancer-1 (DBC1) and deacetylates H3K18, resulting in reduced expression of DBC1 (Li, Tian et al., 2019). DBC1 is an endogenous inhibitor of SIRT1 activity (Kim, Chen, & Lou, 2008). As a consequence, increased SIRT1 activity results in increased deacetylation and subsequent phosphorylation of Akt and p70S6K1, promoting thyroid tumorigenesis (Li, Tian et al., 2019).

Besides H3K18, SIRT7 has also been reported to target several nonhistone proteins for deacetylation. For instance, in HCC, SIRT7 deacetylates p53 at lysines 320 and 373 to suppress its activity, and prevents p53-dependent apoptosis in response to doxorubicin (Zhao et al., 2019). As a consequence, SIRT7 depletion significantly increased doxorubicin toxicity both in vitro and in vivo, while SIRT7 OE largely abolished doxorubicin-induced apoptosis (Zhao et al., 2019). In HCC, SIRT7 expression has been shown to be upregulated in a large cohort of HCC patients (Kim et al., 2013), and Zhao et al. demonstrated that higher levels of SIRT7 are associated with therapy resistance and poor survival (Zhao et al., 2019).

A number of other studies have reported mechanisms of action, other than deacetylation, by which SIRT7 promotes cancer cell growth. Shin et al. reported that SIRT7 is recruited by the c-MYC oncoprotein to suppress its transcriptional output, and inhibit ER stress (Shin et al., 2013). The authors hypothesize that SIRT7 depletion might represent a strategy to activate c-MYC-induced apoptosis in cancer cells that are sensitive to ER stress (Ozcan & Tabas, 2012). Yu et al. observed that SIRT7 induces ERK1/2 phosphorylation and activates the Raf–MEK–ERK signaling pathway to promote CRC cell growth (Yu et al., 2014). SIRT7 levels are upregulated in the CRC tissues, and are significantly correlated with tumor stage, lymph node metastasis, and poor patient survival (Yu et al., 2014). Kim et al. reported upregulation of SIRT7 expression in a large cohort of human HCC patients. Reducing SIRT7 expression repressed cyclin D1, and induced p21WAF1/Cip1-dependent G1 cell cycle arrest and Beclin-1 expression in HCC (Kim et al., 2013). Mu et al. found that SIRT7 expression is upregulated in human glioma tissues, and that expression levels of SIRT7 are correlated with tumor grade. Data presented in this study indicate that SIRT7 promotes malignancy of glioma cells most likely by activating the ERK and STAT3 signaling cascade (Mu, Liu et al., 2019). Wang et al. demonstrated increased levels of SIRT7 in human ovarian cancer cells compared to normal ovarian cells. Repression of SIRT7 significantly reduced ovarian cancer cell malignant properties, by inducing apoptosis via reducing Bcl-2 and NF-κB levels (Wang, Lu et al., 2015). Malik et al. identified SIRT7 as a key regulator of metastasis in human cancers. In epithelial prostate carcinomas, high SIRT7 levels are associated with aggressive cancer phenotypes, metastatic disease, and poor patient prognosis. Mechanistically, SIRT7 interacts functionally with SIRT1, and promotes prostate cancer cell migration and metastasis by repressing E-cadherin expression (Malik et al., 2015).

In contrast, only a handful of studies suggest a tumor-suppressor role for SIRT7 (McGlynn et al., 2015; Tang, Shi et al., 2017). In pancreatic cancer, low levels of nuclear SIRT7 expression are associated with aggressive tumor phenotypes and poor patient outcomes. Patients with high levels of nuclear SIRT7 had a longer lifespan than those with low levels of SIRT7. However, the precise mechanisms underlying this association are unclear (McGlynn et al., 2015). Recently, Tang et al. found that SIRT7 is significantly downregulated in breast cancer pulmonary metastases in human and mice, and reduced levels of SIRT7 are correlated with poor prognosis in breast cancer patients (Tang, Shi et al., 2017). SIRT7 deficiency promotes breast cancer cell metastasis, while ectopic expression of SIRT7 inhibits metastasis in a polyomavirus middle T antigen-driven breast cancer model. Mechanistically, SIRT7 deacetylates and promotes SMAD4 degradation mediated by β-TrCP1, and SIRT7 deficiency activates transforming growth factor-β signaling and enhances EMT. Significantly, pharmacological activation of SIRT7 deacetylase activity inhibits breast cancer lung metastases, and increases survival (Tang, Shi et al., 2017). Collectively, these studies highlight the fact that as with other sirtuins, the oncogenic and tumor suppressor activities of SIRT7 are highly context specific.

Metabolic syndrome

Metabolic syndrome is a constellation of clinical findings—elevated blood pressure, fasting glucose, serum triglycerides, and LDL cholesterol levels, along with obesity—that predispose to heart disease or failure, diabetes, and cancer (Han & Lean, 2016; Samson & Garber, 2014). The prevalence of metabolic syndrome in the adult population increases with age, along with changes in body composition, resulting in progressive loss of skeletal muscle mass and accumulation of WAT (Han & Lean, 2016; St-Onge & Gallagher, 2010). Increased visceral WAT is associated with an elevated risk of CVD, T2D, insulin resistance, hypertension, and cancer (Basen-Engquist & Chang, 2011; Morley, 2004). In the United States, it has been estimated that 4% of new cases of cancer in men and 7% in women result from obesity (Polednak, 2008). Sirtuin-regulated pathways, such as inflammatory responses, IIS, mTOR, and AMPK signaling, have all been hypothesized to mediate this obesity−cancer link (Roberts, Dive, & Renehan, 2010). A large body of evidence now supports physiological roles for sirtuins in suppressing aspects of metabolic syndrome, with the potential to affect other age-related diseases such as cancer.

SIRT1

SIRT1 is a major regulator of diverse aspects of metabolism, an area reviewed extensively by others (Boutant & Cantó, 2016; Chang & Guarente, 2014). For example, in the liver, SIRT1 regulates gluconeogenesis, fatty acid oxidation, cholesterol efflux, bile acid synthesis, and lipogenesis. Only studies linking SIRT1 with age-associated metabolic dysfunction will be emphasized here. Obese patients with NAFLD exhibit significantly lower levels of plasma SIRT1 compared to control lean patients (Mariani et al., 2015). Consistently, tissue-specific deletion of *Sirt1* in hepatocytes sensitizes mice to development of hepatosteatosis on an HFD; in humans this condition is associated with T2D and represents a common cause of liver dysfunction. Consistently, aged mice with hepatic *Sirt1* deletion manifest hepatosteatosis even under normal dietary conditions (Purushotham et al., 2009; Wang, Li, & Deng, 2010). In addition, pharmacological inhibition of SIRT1 induces lipid accumulation in human fetal hepatocytes (Tobita et al., 2016). In contrast, *Sirt1* transgenic mice are protected against genetically or diet-induced hepatosteatosis and hepatic insulin insensitivity (Banks et al., 2008; Pfluger et al., 2008).

In the liver, SIRT1 regulates lipid metabolism by deacetylating SREBPs (Ponugoti et al., 2010; Walker et al., 2010). SIRT1 destabilizes SREBP proteins. In response to fasting, when hepatic SIRT1 protein levels are increased (Rodgers et al., 2005), SREBP-dependent lipid synthesis is inhibited. SIRT1 activation by small molecules in vivo reduces SREBP-dependent gene expression and attenuates liver steatosis in diet-induced and genetically obese mice (Walker et al., 2010). Via peroxisome proliferator-activated receptor-γ (PPARγ) coactivator 1α (PGC-1α)-deacetylation, SIRT1 activates fatty acid β-oxidation genes and promotes fatty acid use (Rodgers, Lerin, Gerhart-Hines, & Puigserver, 2008). *Sirt1* deletion increases PGC-1α acetylation, resulting in reduced fatty acid β-oxidation and increased susceptibility of mice to HFD-induced hepatic steatosis (Purushotham et al., 2009). Together these results highlight the role of SIRT1 as a central player in the metabolic regulation.

Under fasting conditions, SIRT1 expression increases in skeletal muscle, where activated PGC-1α supports mitochondrial biogenesis (Gerhart-Hines et al., 2007). *Sirt1* deletion in skeletal muscle results in reduced mitochondrial gene (mtDNA) expression, while mitochondrial copy number and nuclear-encoded mitochondrial gene expression remain unchanged (Gomes et al., 2013). SIRT1 ablation in skeletal muscle reduces promoter activity of the nuclear mitochondrial transcription factor A (TFAM). In this model, an age-related reduction in NAD^+ levels and subsequent decline of SIRT1 activity generates a pseudo-hypoxic state, wherein HIF-1α is stabilized due to a reduction in activity of the von

Hippel-Lindau E3-ubiquitin ligase, which normally targets HIF-1α for proteasomal degradation. HIF-1α then leads to inhibition of TFAM expression, resulting in decreased mtDNA gene expression. Together, these results suggest that mtDNA gene expression can be regulated by a SIRT1-dependent, PGC-1α-independent pathway to maintain mitochondrial homeostasis. Via PPARγ-deacetylation, SIRT1 induces "browning" of WAT, suggesting a potential for SIRT1-dependent inhibition of adiposity (Picard et al., 2004; Qiang et al., 2012). In a mouse model of mitochondrial disease, characterized by exercise intolerance and impaired oxidative phosphorylation due to cytochrome c oxidase deficiency, NR supplementation or genetic or pharmaceutical ablation of the NAD^+-consuming enzyme, PARP1, increases the NAD^+/NADH ratio and restores motor function and mitochondrial respiration in brain and skeletal muscle tissues. Also, in vivo NR supplementation upregulates skeletal muscle TFAM expression and expression of several mtDNA-encoded genes involved in fatty acid oxidation or oxidative phosphorylation (Cerutti et al., 2014).

Adipose tissue inflammation, characterized by enhanced infiltration and altered polarization of macrophages, contributes to insulin resistance and T2D. Via deacetylating the p65 subunit of NF-κB, SIRT1 inhibits inflammatory signaling in macrophages and adipocytes, leading to improved insulin sensitivity and glucose metabolism (Schug et al., 2010; Yoshizaki et al., 2009, 2010). Myeloid-specific *Sirt1* KO mice, fed an HFD, exhibit aggravated inflammatory response in liver and adipose tissues that leads to the development of systemic insulin resistance (Schug et al., 2010). In adipocytes, SIRT1 modulates the expression and secretion of several adipokines, which regulate the recruitment and polarization of macrophages in adipose tissues (Hui et al., 2017). Adipose-specific *Sirt1* KO mice display an increased number of adipose-resident macrophages with polarization toward the proinflammatory M1 subtype, resulting in a heightened susceptibility to HFD-induced insulin resistance (Hui et al., 2017). Consistently, SIRT1 expression in human subcutaneous fat is inversely related to adipose tissue macrophage infiltration (Gillum et al., 2011). In contrast, SIRT1 OE prevents adipose tissue macrophage accumulation in response to chronic high-fat feeding (Gillum et al., 2011), and improves glucose tolerance and insulin sensitivity in mice even on a low-fat diet (Boutant et al., 2015). In addition to inflammation, impaired adipose tissue angiogenesis contributes to the pathogenesis of dysfunctional adipose tissue in obesity (Crewe, An, & Scherer, 2017). Pharmacological activation of SIRT1 induces angiogenic genes in preadipocytes, and reduces inflammation and fibrosis in WAT, thereby promoting insulin sensitivity (Nawaz et al., 2018).

Pancreatic β-cell-specific SIRT1 OE promotes insulin secretion through repression of uncoupling protein 2 (UCP2) gene expression, improving glucose tolerance in younger mice (Moynihan et al., 2005; Ramsey, Mills, Satoh, & Imai, 2008). Despite sustained SIRT1 OE, insulin secretion and glucose tolerance decline as these mice age, due to an age-associated decrease in plasma NMN levels (Ramsey et al., 2008). NMN is an NAD^+ precursor that ameliorates glucose intolerance by elevating NAD^+ levels in a mouse model of T2D (Yoshino, Mills, Yoon, & Imai, 2011). NMN supplementation restores the age-related glucose-stimulated insulin secretion defect in aged β-cell-specific *Sirt1* transgenic females (Ramsey et al., 2008). These studies suggest that maintenance of proper cellular NAD^+ levels is critical for organismal metabolic homeostasis. Indeed, an age-related decrease in skeletal muscle NAD^+ levels impairs mitochondrial homeostasis and ATP production by reducing SIRT1-dependent TFAM expression (Gomes et al., 2013). Strikingly, this defect is prevented by chronic CR, or reversed by NMN supplementation. Most recently, Yoshida et al. have shown that the circulating levels of eNAMPT significantly decline with age in mice and humans (Yoshida et al., 2019), which contributes to an age-associated decline in NAD^+ levels. SIRT1-mediated deacetylation of intracellular NAMPT (iNAMPT) promotes its extracellular secretion from adipocytes (Yoon et al., 2015). eNAMPT is carried in extracellular vesicles (EVs), which are internalized into cells to promote NAD^+ biosynthesis (Yoshida et al., 2019). Adipose tissue-specific NAMPT OE increases circulating eNAMPT levels in aged mice, resulting in increased NAD^+ levels in multiple tissues, leading to their enhanced functioning and improved median lifespan in female mice. Additionally, supplementing eNAMPT containing EVs isolated from young mice significantly improves physical activity and extends both median and maximal lifespan in aged mice (Yoshida et al., 2019). A more careful investigation of cell-type-specific effects of SIRT1 modulation will be necessary to fully elucidate the numerous mechanisms by which SIRT1 affects energy expenditure, food intake, and diet-related obesity over the lifespan of mice.

SIRT2

SIRT2 has been implicated in regulating adipose tissue development and function. OE of SIRT2 inhibits differentiation of preadipocytes in cell culture and conversely RNAi KD promotes differentiation (Jing, Gesta, & Kahn, 2007). Mechanistically, SIRT2 deacetylates FOXO1, which increases FOXO1-mediated repression of PPARγ, the master regulator of adipogenesis. This results in SIRT2-dependent inhibition of adipocyte differentiation in cell culture (Jing et al., 2007; Wang & Tong, 2009). Krishnan et al. provided evidence that reduced SIRT2 expression may promote obesity in mammals (Krishnan et al., 2012). They found reduced SIRT2 levels and HIF-1α accumulation in visceral white adipocytes of obese mice and humans. Increased HIF-1α activation in this context reduces SIRT2 protein expression, which in turn results in hyperacetylation and inactivation of PGC-1α, suppressing the expression of β-oxidation and mitochondrial genes. SIRT2 KD in adipocyte cultures results in decreased fatty acid β-oxidation, while knocking down HIF-1α has the opposite effect. Furthermore, SIRT2 suppresses fatty acid synthesis in the liver via deacetylation of ATP-citrate lyase (ACLY), leading to its ubiquitylation and degradation (Guo et al., 2019; Lin, Tao et al., 2013). ACLY generates acetyl-CoA from citrate for fatty acid and cholesterol biosynthesis, thereby linking carbohydrate to lipid metabolism (Feng, Zhang, Xu, & Shen, 2019). Under high-glucose conditions, the PCAF acetyltransferase acetylates ACLY, which increases its stability and promotes de novo lipid synthesis (Lin, Tao et al., 2013). Altogether, these findings demonstrate that SIRT2 could be a potential therapeutic target for the regulation of obesity and obesity-related disease.

Beyond its role in adipocytes, SIRT2 plays other roles in metabolic homeostasis. SIRT2 enhances hepatic gluconeogenesis during glucose deprivation by acting on multiple targets. Phosphoenolpyruvate carboxykinase (PEPCK) catalyzes the rate-limiting, irreversible step of gluconeogenesis and is a target of SIRT2 deacetylase activity; SIRT2 prevents ubiquitylation-dependent PEPCK degradation via deacetylation (Jiang et al., 2011). Interestingly, PEPCK transgenic mice develop phenotypes resembling human T2D (Valera, Pujol, Pelegrin, & Bosch, 1994), indicating that SIRT2 may promote T2D or other metabolic dysfunctions via effects on PEPCK. In this regard, pharmacological inhibition of SIRT2 with sirtinol results in hyperacetylation and degradation of PEPCK1, reducing the hepatic glucose output in vitro and in vivo (Zhang, Pan et al., 2017). In contrast, more recently, Watanabe et al. showed that hepatic SIRT2 OE increases hepatic glucose uptake in HFD-fed obese diabetic mice, mitigating their impaired glucose tolerance via deacetylating the glucokinase regulatory protein (GKRP) (Watanabe et al., 2018). GKRP binds with glucokinase and inhibits activity, and glucose dissociates GKRP from glucokinase, resulting in glucokinase activation (Raimondo, Rees, & Gloyn, 2015).

Additionally, SIRT2 has been implicated in insulin signaling. In this context, Lemos et al. demonstrated that SIRT2 is downregulated in insulin-resistant hepatocytes and liver tissues, and that SIRT2 OE improves insulin sensitivity in insulin-resistant hepatocytes (Lemos et al., 2017). Similarly, another study showed that ectopic expression of SIRT2 enhances insulin signaling via promoting AKT activity (Ramakrishnan et al., 2014). The absence of SIRT2 exacerbates insulin resistance in HFD-fed mice; *Sirt2* KO mice exhibit attenuated skeletal muscle insulin-induced glucose uptake, and this impairment is aggravated in HFD-fed *Sirt2* KO mice (Lantier et al., 2018). In contrast to these reports, Arora et al. identified an upregulation of SIRT2 expression in insulin-resistant skeletal muscle cells. Pharmacological inhibition of SIRT2 improved AKT activity and increased insulin-stimulated glucose uptake in these cells (Arora & Dey, 2014). Furthermore, SIRT2 suppresses insulin-stimulated glucose uptake by deacetylating the TUG protein, which plays a critical role in insulin-induced exocytic translocation of the glucose transporter (GLUT4) to stimulate glucose uptake in fat and muscle (Belman et al., 2015). *Sirt2* KO mice exhibit increased TUG acetylation and enhanced glucose disposal during glucose tolerance tests. Possible impacts of SIRT2 on glucose homeostasis warrant closer examination in vivo.

SIRT3

A comprehensive analysis of the mitochondrial acetyl-proteome revealed that many mitochondrial proteins are acetylated in hepatic mitochondria in the absence of SIRT3 function (Hebert et al., 2013). This study confirms that SIRT3 is a dominant mitochondrial deacetylase (Lombard et al., 2007), and demonstrates that SIRT3 deacetylates factors that regulate diverse mitochondrial pathways in response to CR, suggesting that SIRT3 allows the cell to adapt to nutrient fluctuations by regulating mitochondrial functions.

Numerous studies have shown that SIRT3 is indeed a major regulator of metabolic homeostasis during fasting or CR (Lombard & Zwaans, 2014; van de Ven, Santos, & Haigis, 2017). Under conditions of low nutrient intake, hepatic fatty acid synthesis is inhibited, and fatty acids are liberated from WAT for oxidation in the liver to produce acetyl-CoA. Acetyl-CoA is in turn converted to acetoacetic acid and β-hydroxybutyrate (termed ketone bodies) in hepatic mitochondria, to be used as an energy source in the brain, heart, and skeletal muscle when glucose is limiting (Laffel, 1999; Shimazu, Hirschey, Huang, Ho, & Verdin, 2010). The rate-limiting step in β-hydroxybutyrate production is catalyzed by the enzyme 3-hydroxy-3-methylglutaryl CoA synthase 2 (HMGCS2), a SIRT3 target. Acetylation of HMGCS2 reduces its catalytic activity and attenuates β-hydroxybutyrate production. In wild-type mice, fasting increases SIRT3 expression, and promotes HMGCS2 deacetylation. Consequently, in *Sirt3* KO mice, β-hydroxybutyrate production is impaired in response to fasting, and HMGCS2 is hyperacetylated under both basal and fasting conditions, implicating SIRT3 as an important mediator of ketogenesis in the fasting liver (Shimazu, Hirschey, Hua et al., 2010). Interestingly, a ketogenic diet reduces midlife mortality, and significantly increases median lifespan and survival in C57BL/6 male mice (Newman et al., 2017; Roberts et al., 2017). Motor function, memory, and muscle mass are preserved in aged ketogenic diet-fed mice.

SIRT3 also regulates the urea cycle via deacetylation and activation of OTC (Hallows et al., 2011), an enzyme that catalyzes the second step of the urea cycle, the key process in the detoxification of ammonia generated by amino-acid catabolism. Fasted mice lacking SIRT3 exhibit metabolite alterations in the urea cycle; CR stimulates urea cycle and upregulates OTC activity, effects lost in the absence of SIRT3. Consequently, *Sirt3* KO mice display increased levels of orotic acid (Hallows et al., 2011), a well-known outcome of human OTC deficiency (Finkelstein, Hauser, Leonard, & Brusilow, 1990).

In the context of ROS regulation, SIRT3 deacetylates and activates SOD2 (Tao et al., 2010) and IDH2 (Someya et al., 2010), two key enzymes that protect against oxidative stress. By reducing oxidative stress, CR delays the onset of age-related hearing loss, an effect that is SIRT3-dependent (Someya et al., 2010). SIRT3 is also protective against noise-induced hearing loss (Brown et al., 2014). In addition, upregulation of SIRT3 rejuvenates aged hematopoietic stem cells via suppressing ROS levels (Brown et al., 2013). Furthermore, SIRT3 plays a key role in regulating hyperlipidemia-induced oxidative stress, considered to be one of the main pathogenic factors that contribute to pancreatic β-cell dysfunction in the development of T2D. SIRT3 levels are reduced in T2D patient islets and β-cells isolated from HFD-fed mice, accompanied by abnormal elevation of ROS levels (Caton et al., 2013; Zhou et al., 2017). *Sirt3* deletion further impairs ROS detoxification via SOD2 hyperacetylation, resulting in aggravated β-cell dysfunction and impaired glucose-stimulated insulin secretion (Zhou et al., 2017). In contrast, SIRT3 OE protects against palmitate-induced β-cell dysfunction by antagonizing oxidative stress-induced cell damage (Kim et al., 2015; Zhou et al., 2017).

Another study reports that in response to a long-term HFD challenge, *Sirt3* germline KO mice develop obesity, insulin resistance, and hepatic steatosis at an accelerated rate (Hirschey et al., 2011). Even under basal conditions, *Sirt3* KO mice have increased WAT and are glucose intolerant (Jing et al., 2011). SIRT3 also maintains metabolic homeostasis by regulating fatty acid oxidation during fasting (Hirschey et al., 2010), and by promoting conversion of pyruvate to acetyl-CoA for further metabolism in the TCA cycle through activating PDC (Fan et al., 2014; Jing et al., 2013). Interestingly, *Sirt3* gene ablation in either skeletal muscle or the liver alone does not recapitulate the metabolic dysfunction phenotypes observed in the whole-body KO strains, suggesting that SIRT3 may exert metabolic control in part through effects in other tissues (e.g., the CNS) (Fernandez-Marcos et al., 2012; Lombard & Zwaans, 2014). In summary, SIRT3 has emerged as an important factor in promoting diverse mitochondrial functions, oxidative stress responses, cellular energy homeostasis, and organismal metabolic health.

SIRT4

SIRT4 is a negative regulator of fatty acid oxidation and mitochondrial respiration in hepatocytes and myocytes (Nasrin et al., 2010). AMPK plays a key role in promoting fatty acid oxidation by phosphorylating and inhibiting ACC, which catalyzes the production of malonyl-CoA from acetyl-CoA. Ho et al. have shown that SIRT4 suppresses fatty acid oxidation by inhibiting AMPK activity via modulating the activity of ANT2 (Ho et al., 2013), an inner mitochondrial membrane-associated protein that catalyzes exchange of ATP generated in the mitochondria with cytosolic ADP (Pebay-Peyroula & Brandolin, 2004). Peroxisome proliferator-activated receptor alpha (PPARα) activation promotes fatty acid oxidation during fasting. SIRT4 represses PPARα transcriptional activity, thereby inhibiting fatty acid oxidation; in *Sirt4*-null mice liver PPARα activity is enhanced, as is the transcriptional output of its targets and fatty acid oxidation (Laurent, de Boer et al., 2013). In response to fasting, SIRT4 expression decreases in liver tissue, potentially contributing to elevated PPARα activation and fatty acid oxidation in this context. Laurent et al. also report that SIRT4 regulates lipid metabolism by deacetylation and inactivation of MCD, an enzyme that converts malonyl-CoA, a precursor for lipogenesis and an inhibitor of fatty acid oxidation, to acetyl-CoA (Laurent, German et al., 2013). Thus, *Sirt4* KO mice are protected from diet-induced obesity, consistent with a role for SIRT4 in promoting lipogenesis in vivo. Moreover, in a recent study by Guo et al., SIRT4 was shown to inhibit fatty acid oxidation in the liver through deacetylation and destabilization of mitochondrial trifunctional protein α-subunit (MTPα), which catalyzes the second and third steps of fatty acid oxidation (Guo et al., 2016). Acetylation of MTPα promotes its stability, while SIRT4-mediated deacetylation induces its ubiquitylation and subsequent degradation, further implying the inhibitory role of SIRT4 in fatty acid oxidation (Guo et al., 2016). Ectopic expression of acetylation mimetic MTPα mutant or KD of SIRT4 inhibits cellular lipid accumulation in the livers of high-fat/high-sucrose diet-fed mice (Guo et al., 2016).

As previously noted, SIRT4 also negatively regulates insulin secretion via repression of GDH activity, which converts glutamate to α-ketoglutarate (Ahuja et al., 2007;

Haigis et al., 2006), and by activating the MCCC, involved in leucine catabolism (Anderson et al., 2017). Under fed conditions, SIRT4 inhibits GDH activity to repress glutamine metabolism and β-cell insulin secretion (Haigis et al., 2006). Moreover, SIRT4 deacylates and activates MCCC to promote leucine oxidation (Anderson et al., 2017). SIRT4 loss resulted in increased acylation and reduced stability of MCCC, impairing leucine flux. As a consequence, *Sirt4* KO mice display elevated basal and stimulated insulin secretion, and eventually develop glucose intolerance and insulin resistance (Anderson et al., 2017; Laurent, German et al., 2013). Since chronically elevated insulin levels can exert deleterious metabolic effects, this may help explain why *Sirt4* KO mice are relatively protected from weight gain in response to HFD, but not other negative metabolic sequelae of this intervention (Laurent, German et al., 2013).

SIRT5

Like SIRT3 and SIRT4, SIRT5 localizes primarily to the mitochondrial matrix. However, a previous study from our group identified a substantial fraction of active SIRT5 present outside the mitochondrion in both mouse liver and human cells (Park et al., 2013), a finding also reported by others (Chen, Tian et al., 2018; Geng, Li, Liu, Li, & Fu, 2011; Matsushita et al., 2011). As previously noted, SIRT5 possesses minimal lysine deacetylase function, while showing strong activity towards the noncanonical, negatively charged PTMs; succinyl, malonyl, and glutaryl moieties (Du et al., 2011; Nishida et al., 2015; Park et al., 2013; Peng et al., 2011; Rardin et al., 2013; Tan et al., 2014). Two large-scale mass spectrometry analyses of mouse liver have identified thousands of SIRT5 desuccinylation sites on hundreds of protein targets (Park et al., 2013; Rardin et al., 2013), many of which are distinct from acetylation sites. Among sites whose succinylation could be quantified, succinylation increased on >90% of these in the context of SIRT5 deficiency (Park et al., 2013). These studies have revealed that SIRT5 is a major regulator of cellular succinylation, analogous to the dominant role SIRT3 plays in regulating mitochondrial acetylation.

Surprisingly, analysis of *Sirt5*-null mice undertaken to date has generally revealed a lack of striking phenotypes (Lombard et al., 2007), though very mild resistance to HFD has been noted (Yu, Sadhukhan et al., 2013). SIRT5 has been reported to deacetylate (Nakagawa et al., 2009; Ogura et al., 2010), desuccinylate (Du et al., 2011), and deglutarylate (Tan et al., 2014) CPS1, thereby activating this enzyme to detoxify ammonia via the urea cycle for subsequent renal excretion. Consequently, *Sirt5* KO mice display elevated blood ammonia after a prolonged fast (Nakagawa et al., 2009). In nonhepatic cells, SIRT5 suppresses ammonia production by desuccinylating and inhibiting the activity of glutaminase (Polletta et al., 2015), an enzyme catalyzing the conversion of glutamine to glutamate and ammonia.

A number of studies suggest that SIRT5 plays context-specific roles in regulating the activities of various enzymes involved in glycolysis and mitochondrial respiration. In this context, Nishida et al. showed that SIRT5 promotes the activity of GAPDH and perhaps other glycolytic enzymes via demalonylation, thus accelerating glycolytic flux (Nishida et al., 2015). Wang et al. identified PKM2 as another glycolytic enzyme whose activity is regulated by SIRT5. SIRT5 desuccinylates PKM2 and increases its pyruvate kinase activity (Wang, Wang et al., 2017), further suggesting that SIRT5 promotes glycolytic flux. Consequently, hepatocytes from *Sirt5* KO mice display reduced glycolysis (Nishida et al., 2015). In contrast, *Sirt5* deletion in LPS-activated macrophages redirects cellular metabolism in favor of glycolysis (Wang, Wang et al., 2017). LPS-activated macrophages exhibit a Warburg-like metabolism, suggesting context-specific roles of SIRT5 in regulating glycolytic flux. Consistently, Xiangyun and colleagues reported that SIRT5-mediated desuccinylation inhibits PKM2 activity in tumor cells (Xiangyun et al., 2017). SIRT5 represses PDC and succinate dehydrogenase (SDH) activities; consequently SIRT5-deficient cells and mitochondria show elevated respiration (Park et al., 2013). Paradoxically, *Sirt5* KO liver homogenates display impaired enzymatic activities of Complex II and ATP synthase, and reduced Complex II-driven respiration (Zhang, Bharathi et al., 2017). Zhang et al. showed that SIRT5 electrostatically binds to cardiolipin and desuccinylates inner mitochondrial membrane proteins, including Complex II and ATP synthase, to promote respiratory chain function (Zhang, Bharathi et al., 2017). Consistently, ectopic expression of SIRT5 increased ATP synthesis and oxygen consumption in HepG2 cells (Buler, Aatsinki, Izzi, Uusimaa, & Hakkola, 2014), again highlighting the context-specific function of SIRT5.

SIRT5 promotes HMGCS2 activity and ketone body formation; *Sirt5*-null mice show a modest reduction in ketone levels during fasting (Rardin et al., 2013). An increase in medium- and long-chain acylcarnitines was reported in liver and skeletal muscle of *Sirt5*-null mice, suggesting that SIRT5 deficiency confers a defect in fatty acid oxidation (Rardin et al., 2013). Consistently, impaired fatty acid oxidation and reduced ATP production are observed in SIRT5-deficient hearts during energy-demanding conditions, such as fasting and exercise (Sadhukhan et al., 2016), further confirming the role of SIRT5 in promoting fatty acid oxidation.

SIRT5 plays a key role in attenuating cellular ROS levels. In this context, one report indicates that SIRT5 activates SOD1, a key cellular antioxidant enzyme.

Mutating the SIRT5 target site in SOD1 inhibits lung tumor cell growth, potentially implicating SIRT5 as a mediator of cancer proliferation via SOD1 activation and cellular ROS protection (Lin, Xu et al., 2013). Another report suggests that SIRT5 may directly deacetylate FOXO3A to promote its nuclear localization, which in turn promotes expression of genes involved in antioxidant defense (Wang, Zhu, Xing, Ma, & Lin, 2015). Zhou et al. have shown that SIRT5 attenuates cellular ROS levels by promoting NADPH production through desuccinylation and deglutarylation of IDH2 and G6PD, respectively, and inducing their activity (Zhou, Wang et al., 2016). SIRT5 desuccinylates and inhibits PKM2, diverting glucose flux into the pentose phosphate pathway and thus generating reducing potential for antioxidant responses (Xiangyun et al., 2017).

Recently, Ma and Fei identified an important role for SIRT5 in the pathogenesis of T2D (Ma & Fei, 2018). SIRT5 level is found to be upregulated in patients with T2D, and elevated SIRT5 levels are positively correlated with blood glucose levels. In this context, SIRT5 inhibits the expression of pancreatic and duodenal homeobox 1 (PDX1) via H4K16 deacetylation, suppressing the proliferation of pancreatic β cells and insulin secretion. Downregulation of SIRT5 expression upregulates PDX1 expression and promotes the secretion of insulin (Ma & Fei, 2018). In contrast, Wang et al. have shown that SIRT5 OE protects β cells against glucolipotoxicity-induced apoptosis and improves glucose-stimulated insulin secretion in β cells (Wang, Liu et al., 2018). Furthermore, hepatic OE of SIRT5 ameliorated the metabolic abnormalities and attenuated hepatic steatosis in an animal model of T2D (Du, Hu et al., 2018). Due to its unique thermogenic capacity, brown fat is an attractive target for treating obesity and T2D. In a recent study, Shuai and colleagues showed that SIRT5 is required for brown adipocyte differentiation and brown adipogenic gene expression (Shuai et al., 2019). *Sirt5* KO mice exhibit less browning capacity in subcutaneous WAT and show cold intolerance, suggesting that SIRT5 can modulate the browning process in vivo (Shuai et al., 2019). Most recently, Wang et al. have shown that SIRT5 plays a key role in BAT thermogenesis via promoting the stability and function of the BAT mitochondrial uncoupling protein UCP1, mediated by desuccinylation (Wang, Meyer et al., 2019).

SIRT6

SIRT6 regulates glycolytic gene expression at the chromatin level, thereby promoting glucose homeostasis (Zhong et al., 2010). In this regard, SIRT6 functions as a corepressor of HIF-1α and is able to inhibit gene expression by deacetylating H3K9 at HIF-1α target gene promoters. Together, SIRT6 and HIF-1α mediate metabolic reprogramming in favor of glycolysis in response to nutrient deprivation. SIRT6 deficiency is also associated with increased insulin signaling, through mechanisms that remain unclear (Xiao et al., 2010). In this regard, *Sirt6* global KO mice typically die from hypoglycemia by 4 weeks of age, highlighting the critical roles this enzyme plays in glucose homeostasis (Mostoslavsky et al., 2006). Adenoviral-mediated p53 OE in the liver of wild-type mice results in inefficient restoration of blood glucose levels after fasted mice were injected with pyruvate (Zhang et al., 2014). *Sirt6* expression is induced by p53, and KD of *Sirt6* when p53 is overexpressed in vivo alleviates the p53-induced defect in blood glucose restoration during the pyruvate tolerance test. This report also demonstrated that following p53 activation, FOXO1 deacetylated by SIRT6 is efficiently exported from the nucleus. A concomitant decrease in expression of FOXO1 target genes involved in gluconeogenesis is observed. Song et al. have shown that β-cell-specific *Sirt6* KO (βS6KO) mice develop glucose intolerance with severely impaired glucose-stimulated insulin secretion. Lack of SIRT6 results in FOXO1 hyperacetylation, and thus repression of glucose-sensing genes such as *Pdx1* and *Glut2*, whereas ectopic expression of SIRT6 in islets from βS6KO mice attenuates FOXO1 repression of *Pdx1* and *Glut2*, thereby maintaining the glucose-sensing ability of β cells and systemic glucose tolerance (Song, Wang, Ka, Bae, & Park, 2016).

In addition to these cell-autonomous effects on HIF-1α activity, SIRT6 regulates other metabolic processes in the liver specifically. Liver-specific *Sirt6* deletion results in liver steatosis, reduced fatty acid β-oxidation, and increased expression of genes involved in glycolysis and lipogenesis (Kim, Xiao et al., 2010). Hepatic *Sirt6* KD increases both blood glucose levels and gluconeogenic gene expression (Dominy et al., 2012; Kim, Xiao et al., 2010). SIRT6 expression is reduced in a mouse model of obesity/diabetes. In this model, gluconeogenesis is upregulated, but expression of several gluconeogenic genes is suppressed upon adenoviral-mediated SIRT6 OE (Dominy et al., 2012). SIRT6 inhibits gluconeogenesis by indirectly inactivating a key transcriptional regulator of gluconeogenic genes, PGC-1α, by promoting its acetylation (Dominy et al., 2012). This occurs through KAT2A (lysine acetyltransferase 2A)/GCN5, an acetyltransferase that represses PGC-1α transcriptional activation (Lerin et al., 2006). SIRT6 interacts with and deacetylates GCN5, increasing its acetyltransferase activity. Activated GCN5 subsequently acetylates PGC-1α to repress the gluconeogenic gene expression program. It is possible that SIRT6 targets other effectors through indirect acetylation via GCN5 (Dominy et al., 2012). Based on the roles played by SIRT6 in glucose metabolism, inhibiting SIRT6 has been proposed as an approach for treating T2D. In this context, recently, Sociali et al. demonstrated that pharmacological inhibition of SIRT6 in HFD-fed mice improves oral glucose tolerance and upregulates the expression of the glucose transporters GLUT1 and GLUT4 in the muscle and promotes glycolytic activity (Sociali et al., 2017). In addition, SIRT6 inhibition reduces the plasma levels of insulin, triglycerides, and cholesterol. Taken together, these results suggest SIRT6 inhibition as a viable approach for improving the glycemic control in T2D. Given the other crucial roles played by SIRT6 in cellular homeostasis—for example, promotion of DNA repair—such strategies should be undertaken with great caution.

SIRT6 is also a regulator of cholesterol homeostasis. SIRT6 represses expression of the gene encoding the master regulator of cholesterol biosynthesis SREPB2, by deacetylating H3K9 and H3K56 at its promoter (Tao et al., 2013b). In a genetic model of obesity, SIRT6 OE reduces hepatic and serum cholesterol levels. Elhanati et al. further demonstrate that SIRT6 represses both SREBP1 and 2 through several mechanisms: (1) inhibiting *Srebf* gene transcription, (2) inhibiting posttranslational processing that generates active SREB proteins, and (3) promoting the phosphorylated, inactive state of the SREB proteins (Elhanati et al., 2013). SIRT6 OE ameliorates aspects of age-associated metabolic decline, such as glucose intolerance in male mice (Kanfi et al., 2012). *Sirt6* transgenic mice fed an HFD are protected against WAT accumulation, elevated triglyceride and LDL cholesterol levels, and impaired glucose tolerance. Moreover, fat-specific *Sirt6* deletion results in compromised lipolysis-induced adipocyte hypertrophy and promotes obesity in HFD-fed mice (Kuang et al., 2017). SIRT6 loss results in FOXO1 activation, which indirectly suppresses the expression of adipose triglyceride lipase (ATGL), the rate-limiting enzyme in the lipolytic reaction. Indeed, obese patients exhibit decreased expression of SIRT6 along with reduction of ATGL expression (Kuang et al., 2017). Another recent study by Yao et al. also demonstrated that SIRT6 deficiency in the fat predisposes mice to obesity and metabolic syndrome (Yao et al., 2017). Cold exposure and β-adrenergic agonist administration markedly increase SIRT6 expression in brown and beige fat. Upregulated SIRT6 enhances phospho-ATF2 binding to the PGC-1α gene promoter to activate its expression, promoting the thermogenic function of brown and beige fat. Mice lacking SIRT6 in adipose tissues exhibit impaired thermogenesis, "whitening" of brown fat, increased blood glucose levels, severe insulin resistance, and hepatic steatosis (Yao et al., 2017). Most recently, Bang et al. have shown that SIRT6 confers resistance against ER stress-induced hepatic steatosis by deacetylating X-box-binding protein 1, promoting its degradation (Bang et al., 2019).

Paradoxically, a recent study by Chen et al. revealed that SIRT6 can promote adipogenesis by repressing the expression of KIF5C, a kinesin family member, which negatively regulates adipogenesis by interacting with CK2α, blocking its nuclear translocation and kinase activity (Chen, Hao et al., 2017). CK2α plays an important role in mitotic clonal expansion during adipogenesis via inducing p27 phosphorylation and degradation. Consequently, SIRT6 deficiency in preadipocytes blocks their adipogenesis (Chen, Hao et al., 2017). Overall, SIRT6 is an important, complex, regulator of glycolysis, gluconeogenesis, and lipid metabolism, and may prove relevant as a therapeutic target for acquired insulin resistance.

SIRT7

On a 129/Sv genetic background, *Sirt7* KO mice develop hepatosteatosis, along with upregulation of inflammatory markers such as TNFα and IL-6 (Shin et al., 2013). In this regard, SIRT7 interacts with c-MYC, targeting SIRT7 to rDNA promoters to reduce ribosomal protein expression, mitigating ER stress induced by the unfolded protein response. Accordingly, in a C57BL/6 genetic background SIRT7 OE in vivo mitigates the negative sequelae of a HFD, such as ER stress, fatty liver, inflammation, and lipogenesis. Consistently, Ryu et al. have shown that *Sirt7* KO C57BL/6 mice display increased levels of plasma triglycerides and free fatty acids, and develop hepatic microvascular steatosis (Ryu et al., 2014). SIRT7 deacetylates GABPb1, promoting the formation and activation of the GABP complex that induces mitochondrial gene expression to maintain mitochondrial homeostasis. As a consequence, SIRT7-ablated mice experience multisystemic mitochondrial dysfunction, as evidenced by increased blood lactate levels, reduced exercise performance, cardiac dysfunction, and age-related hearing loss (Ryu et al., 2014). However, independently derived *Sirt7* germline and liver-specific KO strains on a C57BL/6 genetic background did not show development of fatty liver. Indeed, in response to chronic high-fat feeding, these animals were actually protected against liver steatosis and diet-induced obesity (Yoshizawa et al., 2014). These *Sirt7*-null mice also manifested other beneficial metabolic effects, in particular reduced fasting glucose and insulin levels. Yoshizawa et al. showed that SIRT7 inhibits activity of the E3 ubiquitin ligase complex, decreasing protein levels of the TR4 transcription factor. TR4 is a nuclear receptor protein that plays a role in lipid homeostasis (Zhou et al., 2011). SIRT7 protects TR4 from proteasomal degradation to promote expression of genes involved in fatty acid uptake, triglyceride storage, and lipid accumulation in the liver. The reason for the striking phenotypic discrepancies between these reports is unclear; divergent strain backgrounds, different targeting constructs, and/or husbandry differences could conceivably all play a role.

A study by Cioffi et al. suggests a role for SIRT7 in promoting obesity (Cioffi et al., 2015). Cioffi and colleagues have shown that SIRT7 promotes fat mass formation by inducing the differentiation of early adipocyte precursors to mature adipocytes. Consequently, *Sirt7* KD results in reduced adipogenesis as evidenced by decreased expression of adipogenesis markers and reduced Oil Red O staining. Indeed, *Sirt7* KO mice display significantly reduced adiposity (Cioffi et al., 2015). Mechanistically, SIRT7 promotes adipogenesis by preventing SIRT1 autodeacetylation, resulting in reduced SIRT1 activity (Fang et al., 2017). SIRT1 inhibits PPARγ to block adipocyte differentiation. As a result of SIRT1 hyperactivation, *Sirt7* KO mice display diminished accumulation of white fat (Fang et al., 2017).

A handful of studies also implicate SIRT7 in regulating glucose metabolism. In this context, SIRT7 suppresses the glycolytic pathway by deacylating and inhibiting phosphoglycerate kinase 1 (PGK1) (Hu et al., 2017), a key enzyme catalyzing the first ATP-generating step of glycolysis. SIRT7 also modulates the glycolytic pathway by suppressing the expression of HIF-1 and HIF-2 (Hubbi, Hu, Kshitiz, Gilkes, & Semenza, 2013). Moreover, SIRT7 regulates gluconeogenesis by modulating glucose-6-phosphatase (G6PC) expression (Jiang et al., 2017).

Cardiovascular dysfunction

CVD, including coronary heart disease and peripheral arterial disease, accounts for roughly one-third of all deaths in the United States (Benjamin et al., 2017), and the risk of developing clinically evident CVD increases dramatically in older individuals (Lakatta & Levy, 2003). With advancing age, endothelial cells are increasingly unable to efficiently proliferate to heal vasculature injury or ischemia (Brandes, Fleming, & Busse, 2005; Ungvari, Kaley, de Cabo, Sonntag, & Csiszar, 2010). Aging is associated with atherosclerotic disease and vascular stiffening, predisposing older individuals to hypertension and ischemic injury.

SIRT1

Several studies demonstrate that SIRT1 is essential for vasorelaxation (Mattagajasingh et al., 2007), vasoprotection (Csiszar et al., 2008), endothelial ischemic recovery, and cholesterol metabolism (Potente & Dimmeler, 2008). *SIRT1* KD ex vivo and *Sirt1* genetic deletion specifically in the endothelial lineage in vivo inhibit endothelial cell migration and angiogenesis (Potente et al., 2007). In vivo studies in mouse and zebrafish models have identified SIRT1 as a key mediator of endothelial function and vascular growth. In zebrafish, SIRT1 is required for endothelial sprout formation and vessel migration. However, endothelial cell-specific *Sirt1* KO mice are developmentally unremarkable, though they are unable to revascularize tissue in response to ischemia-induced injury (Guarani et al., 2011). Conversely, cardiac-specific OE of SIRT1 is protective against ischemia-induced injury (Hsu et al., 2010). Yang et al. found that SUV39H, the mammalian histone H3K9 methyltransferase, in cooperation with heterochromatin protein 1 gamma (HP1γ), catalyzes H3K9 trimethylation on the SIRT1 promoter and represses its transcription. Consequently, inhibiting SUV39H expression or activity restored SIRT1 expression and attenuated cardiac injury following MI (Yang, Weng et al., 2017).

The pro-angiogenic and vascularization functions of SIRT1 can be rationalized mechanistically through SIRT1's interaction with the FOXO family transcription factors and the NOTCH signaling pathway (Oellerich & Potente, 2012). FOXO transcription factors, particularly FOXO1, are potent negative regulators of angiogenesis (Paik, 2006; Paik et al., 2007; Potente et al., 2005). SIRT1 deacetylates FOXO1 in endothelial cells, reducing its transcriptional activity and antiangiogenic function (Potente et al., 2007). NOTCH signaling orchestrates postnatal vascular morphogenesis in a dose-dependent manner. In endothelial stalk cells, NOTCH signaling is inversely correlated with vessel sprouting behavior (Phng & Gerhardt, 2009). SIRT1 deacetylates and destabilizes the NOTCH1 intracellular domain (NICD), promoting its proteasomal degradation and inhibiting NOTCH effector responses (Guarani et al., 2011). Consequently, endothelial cells lacking SIRT1 activity have increased NOTCH signaling, impaired vessel growth and defective sprout elongation. In vivo, SIRT1 ablation in zebrafish and mice reduces vascular branching and density due to enhanced NOTCH signaling. Therefore, activation of SIRT1 in endothelial cells may be a useful means to protect endothelial tissue from age-related functional decline, particularly after injury.

Circulating blood flow in the mammalian cardiovascular system induces mechanical stress on the vascular endothelial cells lining the vessel walls. Atheroprotective flow results from undisturbed, steady pulsatile flow in straight sections of the artery, and promotes downregulation of proinflammatory pathways (Chien, 2008). When undirected or branched flow patterns occur, for example, at bends in the arterial tree, proinflammatory and proliferative pathways are activated. Pulsatile shear (PS) stress thus promotes endothelial homeostasis and benefits vascular physiology via antioxidative and antiinflammatory effects on vascular endothelial cells. Endothelial homeostasis in response to PS stress is enhanced by SIRT1. PS stress induces expression of antiinflammatory and antioxidant genes such as *Sod1* and *Sod2*, an effect that is lost upon KD of CaMKKβ (Ca^{2+}/calmodulin-dependent protein kinase kinase). CaMKKβ is an AMPK kinase that phosphorylates and stabilizes SIRT1, and together these two proteins suppress oxidative stress and inflammation. In a sensitized genetic background and in response to an atherogenic diet, increased atherosclerotic lesions are observed in mice lacking either CaMKKβ or SIRT1, indicating that these proteins could conceivably represent clinically relevant targets for intervention to treat age-associated CVD (Wen et al., 2013). Furthermore, SIRT1 interacts directly with the RelA/p65 subunit of NF-κB and deacetylates it, resulting in loss of NF-κB-regulated gene expression (Yeung et al., 2004). Reduced levels of SIRT1 lead to increased acetylation of the p65 subunit of NF-κB, resulting in an enhanced inflammatory response in endothelial cells. Pharmacological activation of SIRT1 improves endothelial function by reducing oxidative stress, NF-κB activation, and TNF-α levels and by enhancing cyclooxygenase-2-mediated dilation and protein expression in the absence of changes in nitric oxide bioavailability (Gano et al., 2014). In apolipoprotein E KO mice, endothelial specific OE of SIRT1 resulted in upregulation of endothelial nitric oxide synthase, and suppression of aortic plaque formation in response to an atherogenic diet (Zhang et al., 2008). Additionally, SIRT1 OE in apolipoprotein E KO mice diminished the expression of lectin-like ox-LDL receptor-1 (Lox-1) via suppression of the NF-κB signaling pathway, resulting in reduced ox-LDL uptake-induced foam cell (a special form of lipid-laden macrophage) formation (Stein et al., 2010).

Pathological cardiac hypertrophy (CH) is a response to chronic hypertension or other sources of cardiac injury. Cardiomyocyte metabolism shifts from fatty acid oxidation to glycolysis during CH, likely due to repression of fatty acid oxidation and oxidative phosphorylation genes (Kolwicz & Tian, 2011). SIRT1 interacts with and activates PPARα to inhibit this metabolic switch and the development of hypertrophy via inactivation of NF-κB. Treating mice with the SIRT1 activator, resveratrol, attenuates markers of induced CH in wild-type, but not PPARα-null mice (Planavila, Iglesias, Giralt, & Villarroya, 2011). Thus, SIRT1 is able to orchestrate metabolic reprogramming and inflammatory responses to protect against the development of CH. However, OE of both PPARα and SIRT1 in mice impairs mitochondrial function and promotes heart failure during CH, by downregulating genes involved in mitochondrial respiration, oxidative stress, and cardiac contractility (Oka et al., 2011). Furthermore, agonist-mediated activation of both PPARα/γ downregulates SIRT1 levels and causes cardiac dysfunction by inhibiting the activity and expression of PGC1α, leading to reduced mitochondrial content (Kalliora et al., 2019). Resveratrol-mediated SIRT1 activation corrected myocardial mitochondrial respiration and attenuated the cardiac dysfunction in agonist-treated mice (Kalliora et al., 2019). The impact of SIRT1 on CH is highly dose-dependent. As noted previously, SIRT1 transgenic mice are protected from CH and age-dependent loss of cardiac function when SIRT1 OE is relatively modest (up to 7.5-fold above basal), but still higher levels of SIRT1 OE produce deleterious effects on myocardium (Alcendor et al., 2007).

However, another study arrived at conflicting results, and found that *Sirt1*-null mice showed resistance to the development of exercise- or agonist-induced CH (Sundaresan et al., 2011). In this regard, in response to growth factor stimulation, SIRT1 deacetylates and activates AKT and 3-phosphoinositide-dependent protein kinase-1 (PDK1), a kinase that activates AKT. AKT is a serine-threonine kinase that is a central player in a network of diverse cellular processes, such as cell proliferation, apoptosis, glucose metabolism, and angiogenesis (Manning & Cantley, 2007). In myocardium, the end result of persistent hyperactive IGF-AKT signaling is hypertrophy and eventual heart failure (Condorelli et al., 2002; Shiojima et al., 2002). SIRT1-deficient hearts are proportionally smaller than those of wild-type mice, and show hyperacetylated, hypoactive AKT (Sundaresan et al., 2011). In transgenic CD1 mice, a fourfold OE of SIRT1 specifically in the heart increases AKT activation, and results in AKT-dependent hypertrophy, indicating that SIRT1 and AKT activation can induce CH in this strain.

Endoplasmic reticulum (ER) stress is involved in the pathogenesis of CVDs including heart failure. Prola et al. reported that SIRT1 protects cardiomyocytes from ER stress-induced apoptosis by inhibiting the PERK/eIF2α pathway, via deacetylating eukaryotic translation initiation factor 2a (eIF2a) (Prola et al., 2017). Recently, SIRT1 has been shown to deacetylate and activate sarco-ER Ca^{2+}-ATPase (SERCA2a) (Gorski et al., 2019), which is a critical determinant of cardiac function. Increased acetylation and reduced activity of SERCA2a are the major features of heart failure. Pharmacological activation of SIRT1 recovered cardiac defects in failing hearts via restoration of SERCA2a activity (Gorski et al., 2019).

A recent study revealed that SIRT1, by deacetylating the voltage-gated cardiac Na^+ channel ($Na_v1.5$), plays a critical role in the generation and conduction of the cardiac electrical impulse (Vikram et al., 2017). $Na_v1.5$ conducts the inward depolarizing cardiac Na^+ current (I_{Na}) and is vital for normal cardiac electrical activity. Cardiomyocyte-specific deletion of *Sirt1* in mice causes hyperacetylation of $Na_v1.5$ and reduces its translocation to the cardiomyocyte membrane, resulting in decreased I_{Na}, leading to cardiac conduction abnormalities and premature death owing to arrhythmia (Vikram et al., 2017). Furthermore, cardiomyocyte-specific deletion of *Sirt1* leads to impaired postischemic contractile function in a Langendorff perfusion model (Wang, Quan et al., 2018). Cardiomyocyte-specific *Sirt1* KO hearts exhibited increased acetylation of LKB1, and thus impaired activation of AMPK. In this regard, an AMPK agonist rescued SIRT1-depleted hearts from ischemic insult (Wang, Quan et al., 2018), indicating that SIRT1-mediated AMPK activation via LKB1 deacetylation is protective during ischemia.

SIRT2

Only a few studies have explored the roles of SIRT2 in cardiac physiology. An early study by Lynn et al. showed that SIRT2 negatively modulated anoxia–reoxygenation tolerance in H9c2 cardiomyoblasts (Lynn, McLeod, Gordon, Bao, & Sack, 2008). SIRT2 levels were elevated in response to anoxia–reoxygenation, and SIRT2 OE increased susceptibility to anoxia–reoxygenation injury. Conversely, SIRT2 KD enhanced the anoxia–reoxygenation tolerance in H9c2 cells. SIRT2 depletion provoked the expression of an adaptor protein 14-3-3 ζ, which sequesters proapoptotic protein BAD and confers protection against anoxia–reoxygenation-induced apoptotic cell death (Lynn et al., 2008).

In contrast, in a recent study by Tang et al., SIRT2 exhibited cardioprotective effects in the development of angiotensin-II-induced and age-associated cardiac hypertrophy (Tang, Chen et al., 2017). SIRT2 protein expression levels were found to be downregulated in hypertrophic hearts. *Sirt2* deletion resulted in more severe cardiac hypertrophy and fibrosis, and decreased cardiac ejection fraction and fractional shortening in angiotensin-II-infused and aged mice. Conversely, cardiac-specific SIRT2 OE protected hearts against angiotensin-II-induced cardiac hypertrophy and fibrosis, and rescued cardiac function. Mechanistically, SIRT2 promoted the activity of AMP-activated protein kinase (AMPK) by deacetylating its upstream kinase liver kinase B1 (Tang, Chen et al., 2017), and LKB1-AMPK signal exerts antihypertrophic effects (Ikeda et al., 2009; Pillai et al., 2010). Consistently, a study by Sarikhani et al. showed that SIRT2 levels were reduced during agonist (phenylephrine or isoproterenol)-induced hypertrophy (Sarikhani et al., 2018). Additionally, *Sirt2* KO mice spontaneously developed cardiac hypertrophy, remodeling, fibrosis, and dysfunction in an age-dependent manner. SIRT2 deficiency aggravated agonist-induced hypertrophy, whereas SIRT2 OE protected against agonist-induced cardiac hypertrophy. Mechanistically, SIRT2 deacetylates and inactivates the transcription factor, nuclear factor of activated T-cells (NFATc2), which plays a key role in the expression of cardiac fetal genes. Lack of SIRT2 stabilizes NFATc2 and promotes the nuclear localization of NFATc2, resulting in its increased transcriptional activity (Sarikhani et al., 2018). The cardioprotective effects of SIRT2 were further substantiated by the fact that doxorubicin-induced cardiotoxicity involves the upregulation of microRNA miR-140-5p, which promotes myocardial oxidative stress via targeting SIRT2 (Zhao et al., 2018).

SIRT3

Studies have shown reduced levels of SIRT3 in rodent models of heart failure, following I/R injury, and human hypertension patients (Dikalova et al., 2017; Grillon, Johnson, Kotlo, & Danziger, 2012; Klishadi et al., 2015). *Sirt3* KO mice experienced a shorter lifespan than wild-type mice and exhibited accelerated cardiac damage, characterized by cardiac hypertrophy and fibrosis with age (Benigni et al., 2019). Additionally, SIRT3 loss aggravated cardiac hypertrophy, fibrosis, and increased mortality in young mice subjected to transverse aortic constriction (TAC) (Hafner et al., 2010). Conversely, overexpressing SIRT3 in cultured cardiomyocytes and transgenic mouse lines, or supplementing mice with exogenous NAD^+ precursors repressed the hallmarks of agonist-induced cardiac hypertrophy (Pillai et al., 2010; Sundaresan et al., 2009). SIRT3 protected against cardiac hypertrophy by inducing the expression of the antioxidant proteins SOD2 and catalase. The SIRT3-mediated reduction in ROS levels inhibits AKT signaling and downstream gene expression associated with induction of cardiac hypertrophy (Sundaresan et al., 2009). SIRT3 also promotes the activity of SOD2 via direct deacetylation (Tao et al., 2010). Consequently, *Sirt3* KO mice exhibited elevated mitochondrial oxidative stress that contributes to endothelial dysfunction and subsequent hypertension (Dikalova et al., 2017). Consistent with this, human hypertensive patients displayed a decrease in SIRT3 levels and an increase in SOD2 acetylation (Dikalova et al., 2017). *Sirt3* deficiency is also associated with hyperacetylation of optic atrophy 1 (OPA1), resulting in defective transmitochondrial cristae alignment and impaired mitochondrial bioenergetics in cardiomyocytes, leading to abnormalities in cardiac function (Benigni et al., 2019). Furthermore, similar to SIRT1 and SIRT2, SIRT3 can deacetylate and activate LKB1, which in turn activates AMPK (Pillai et al., 2010). SIRT3 OE decreased LKB1 acetylation and mediated antihypertrophic effects in cardiomyocytes (Pillai et al., 2010). Recently, Wei et al. showed that *Sirt3* KO mice had exacerbated angiotensin-II-induced microvascular rarefaction and mitochondrial dysfunction, which led to cardiac fibrosis as evidenced by increased collagen I and collagen III expression (Wei, Huan et al., 2017). Cardiac microvascular endothelial cell-specific *Sirt3* transgenic mice showed decreased fibrosis, as well as improved cardiac function, compared with wild-type mice in response to angiotensin-II infusion (Wei, Huan et al., 2017).

Porter et al. investigated the role of SIRT3 in I/R injury, and found that SIRT3-deficient cardiac cells exhibited reduced mitochondrial complex I activity and were more susceptible to I/R injury (Porter, Urciuoli, Brookes, & Nadtochiy, 2014). Additionally, in the Langendorff model, *Sirt3* heterozygosity resulted in reduced functional recovery and greater infarct size in adult hearts (Porter et al., 2014), hinting that an age-associated decline in SIRT3 levels may contribute to a greater level of I/R injury in the aged heart. Additionally, *Sirt3*-null mice presented a preexisting coronary microvascular dysfunction, and SIRT3 deficiency exacerbated postmyocardial ischemia, cardiac dysfunction, and impairment of cardiac recovery (He, Zeng, & Chen, 2016). In contrast, SIRT3 OE preserved cardiac function in post-myocardial ischemic mice (He et al., 2016). In this context, Zhai et al. have shown that melatonin treatment ameliorated myocardial I/R injury by upregulating SIRT3 expression (Zhai et al., 2017), which exerted cardioprotective effects by enhancing mitochondrial biogenesis and reducing oxidative stress, and myocardial apoptosis (Yu et al., 2017; Zhai et al., 2017). Moreover, OE of SIRT3 protected cardiomyocytes from doxorubicin toxicity by attenuating ROS production, and mitochondrial DNA (mtDNA) damage induced by doxorubicin (Pillai et al., 2016).

Bochaton et al. identified a role for SIRT3 in deacetylating cyclophilin D, a modulator of the mitochondrial permeability transition pore (mPTP), to suppress mPTP opening, thereby inhibiting I/R injury-induced death of cardiomyocytes (Bochaton et al., 2015).

SIRT4

The role of SIRT4 in the pathogenesis of heart disease is only beginning to be studied. Available studies have reported contradictory roles for SIRT4 in cardiac health. Using genetically modified mice, Luo et al. have shown that SIRT4 promotes the development of angiotensin-II-mediated cardiac hypertrophy (Luo et al., 2017). SIRT4 deficiency conferred resistance to angiotensin-II-induced hypertrophic growth, and the development of cardiac fibrosis. In contrast, SIRT4-overexpressing mice showed aggravated hypertrophy and exhibited reduced cardiac function in response to angiotensin-II treatment. Mechanistically, SIRT4 promotes cardiac hypertrophy by sequestering SIRT3 from its substrate, MnSOD, causing MnSOD inactivation and increased ROS levels, which are linked to enhanced risk for developing cardiac hypertrophy. Treating SIRT4-overexpressing mice with a SOD mimetic blocked the exaggerated response to angiotensin-II observed in these mice (Luo et al., 2017). Additionally, Xiao et al. found that the expression level of SIRT4 significantly increases following cardiac hypertrophy (Xiao, Zhang, Fan, Cui, & Shen, 2016). OE of microRNA miR-497 resulted in downregulation of SIRT4 expression and attenuated the adverse cardiac responses to pressure overload and angiotensin-II, suggesting that the microRNA miR-497 suppresses cardiac hypertrophy by targeting SIRT4 (Xiao et al., 2016).

In contrast to its role in promoting cardiac hypertrophy, SIRT4 has also been shown to play an ameliorative role during ischemic heart injury. In this context, Liu et al. reported that SIRT4 exerts cardio-protection against hypoxia-induced apoptosis (Liu, Che, Xue et al., 2013). SIRT4 levels were markedly downregulated in response to hypoxia treatment in H9c2 cardiomyoblasts. Ectopic expression of SIRT4 prevents the translocation of the pro-apoptotic protein Bax to the mitochondria, and inhibits apoptosis in these cells (Liu, Che, Xue et al., 2013). Since cardiomyocytes are not capable of self-renewal, apoptosis plays a critical role in the pathogenesis of ischemic heart injury by causing cardiomyocyte loss. By inhibiting hypoxia-induced apoptosis, SIRT4 increases cardiomyoblast viability and protects against the pathogenesis of ischemic heart injury (Liu, Che, Xue et al., 2013). Consistent with this, a recent study by Zeng et al. showed that SIRT4 expression is downregulated in cardiomyocytes, both in vitro and in vivo following myocardial I/R injury (Zeng, Liu, & Wang, 2018). OE of SIRT4 attenuated the effects of myocardial I/R injury via inhibiting apoptosis and preserving mitochondrial respiration (Zeng et al., 2018).

SIRT5

Recent work has revealed multiple roles for SIRT5 in maintaining cardiac homeostasis, especially under stress conditions. Intermittent hypoxia upregulates SIRT5 levels in rat cardiomyocytes (Zhu et al., 2012), and by physically interacting with the antiapoptotic protein Bcl-XL, SIRT5 protects cardiomyocytes from oxidative stress-induced apoptosis (Liu, Che, Zheng et al., 2013). A recent study revealed that succinyl-CoA, the precursor of the SIRT5 target PTM lysine succinylation, is the most abundant acyl-CoA species in cardiac tissue (Sadhukhan et al., 2016). Correspondingly, SIRT5 is expressed at higher levels in both human and murine hearts relative to other tissues (Michishita et al., 2005; Nakagawa et al., 2009). *Sirt5*-ablated hearts display prominent increase of lysine succinylated proteins, which participate in processes such as oxidative phosphorylation, fatty acid metabolism/oxidation, branched-chain aa metabolism, and the TCA cycle among others (Boylston et al., 2015; Hershberger et al., 2017; Sadhukhan et al., 2016), supporting the notion that via desuccinylation, SIRT5 plays a crucial role in regulating cardiac metabolism.

A study by Boylston et al. reported that SIRT5-deficient hearts exhibit increased susceptibility to I/R injury and impaired recovery after IR insult (Boylston et al., 2015). As reported previously, SIRT5 depletion results in SDH hypersuccinylation, which leads to increased SDH activity in *Sirt5* KO hearts (Boylston et al., 2015; Park et al., 2013). Pretreating mice with dimethyl malonate, a competitive inhibitor of SDH, reduces superoxide production and rescues *Sirt5* KO hearts from their increased susceptibility to IR injury (Boylston et al., 2015), implicating elevated SDH activity in the context of SIRT5 deficiency as causative of their increased sensitivity to IR. In this regard, recently Liu et al. showed that the administration of exogenous NAD^+ promotes SIRT5-mediated desuccinylation and inactivation of SDH, and attenuates depression of cardiac function in isolated rat hearts after IR insult (Liu, Wang, Zhao, Wu, & Wang, 2019).

Sadhukhan et al. showed that *Sirt5* KO mice exhibit both an impaired shortening fraction and ejection fraction and develop cardiac hypertrophy during normal aging (Sadhukhan et al., 2016). They further demonstrated that SIRT5 desuccinylates and activates enoyl-CoA hydratase alpha (ECHA), a subunit of mitochondrial trifunctional protein required for FAO. As a consequence, SIRT5 deficiency results in impaired myocardial fatty acid metabolism and reduced ATP production in aged hearts during energetically demanding conditions, such as fasting and exercise (Sadhukhan et al., 2016). Herschberger et al. evaluated the role of SIRT5 in cardiac stress responses, using an established model of pressure overload-induced hypertrophy induced by TAC. Whole-body *Sirt5* ablation led to the development of severe cardiac dysfunction in response to TAC, and was associated with increased mortality (Hershberger et al., 2017). Surprisingly, in contrast to global *Sirt5* KO mice, no difference was observed either in mortality, cardiac function, or hypertrophy between heart-specific *Sirt5* KO mice and their littermate controls in response to TAC (Hershberger et al., 2018), likely indicating that SIRT5 exerts some of its cardioprotective effects through functions in other cell types. Analogously, the mitochondrial sirtuin SIRT3 inhibits fibrosis in the heart and other tissues via activity in fibroblasts (Sundaresan et al., 2015). Taken together, these findings highlight important roles for SIRT5 in cardioprotection under stress conditions, potentially via regulation of metabolism, and through mechanisms that are at least in part cell nonautonomous.

SIRT6

Similar to SIRT3, SIRT6 negatively regulates cardiac hypertrophy in part by inhibiting the IGF-AKT signaling cascade (Sundaresan et al., 2012). In failing human and hypertrophic mouse hearts, SIRT6 protein expression is reduced compared to controls. *Sirt6* gene deletion, specifically in the mouse heart or globally, results in cardiac hypertrophy, whereas cardiac-specific SIRT6 OE protects mice from induction of cardiac hypertrophy. Mechanistically, SIRT6 negatively regulates cardiac hypertrophy by corepressing c-Jun-dependent transcription at the chromatin level, thereby inhibiting downstream IGF-AKT signaling. Importantly, in vivo inhibition of IGF signaling in whole-body and cardiac-specific *Sirt6* KO mice inhibits the hypertrophic response (Sundaresan et al., 2012). Impaired autophagy has been reported to contribute to the pathogenesis of cardiac hypertrophy induced by SIRT6 deficiency. A study by Lu et al. demonstrated that SIRT6 promotes autophagy in isoproterenol-treated cardiomyocytes, exerting antihypertrophic effects (Lu et al., 2016). Mechanistically, the authors proposed that attenuation of AKT signaling via SIRT6-dependent FOXO3 activation contributes to the pro-autophagic effect of SIRT6 in suppression of isoproterenol-induced cardiac hypertrophy (Lu et al., 2016). Recently, Zhang et al. showed that SIRT6 exerts antihypertrophic effects by suppressing expression of STAT3 (Zhang, Li, Shen et al., 2016), which is critical for the development of cardiac hypertrophy and heart failure. Furthermore, Yu et al. reported that SIRT6-dependent inhibition of NF-κB suppresses angiotensin-II-induced hypertrophy in neonatal rat cardiomyocytes (Yu, Cai et al., 2013).

Inhibition of NF-κB signaling by SIRT6 has also been shown to protect against angiotensin-II-induced differentiation of cardiac fibroblasts into myofibroblasts (Tian et al., 2015). As a consequence, SIRT6 depletion increases α-SMA levels and extracellular matrix (ECM) deposition, and upregulates the expression of focal adhesion-associated genes and fibrosis-related genes in angiotensin-II-stimulated cardiac fibroblasts (Tian et al., 2015). Finally, SIRT6-mediated downregulation of NF-κB also protects against hypoxia/reoxygenation-induced injury by attenuating hypoxia-induced apoptosis and mitochondrial defects (Cheng, Cheng et al., 2016; Maksin-Matveev et al., 2015). Recently, Wang et al. reported that SIRT6 protects the heart from I/R injury by activating FOXO3A in an AMP/ATP-induced AMPK-dependent manner, thus upregulating antioxidants and suppressing oxidative stress (Wang, Wang et al., 2016). Heterozygosity of *Sirt6* aggravated myocardial damage, ventricular remodeling, and oxidative stress in mice subjected to myocardial I/R, whereas restoration of SIRT6 expression reversed these deleterious effects of SIRT6 deficiency in the ischemic heart (Wang, Wang et al., 2016).

A study by Zhang et al. highlighted the importance of SIRT6 in atherosclerosis, the major cause of CVDs (Zhang, Ren et al., 2016). The authors found that SIRT6 expression levels are decreased in human atherosclerotic plaques. Additionally, *Sirt6* heterozygosity aggravated atherosclerotic plaque burden, and was associated with features of plaque instability in a mouse model of atherosclerosis. Mechanistically, SIRT6 regulates H3K9 and H3K56 acetylation marks at the regulatory region of the NKG2D ligand, a mediator of the innate immune response involved in the process of atherosclerosis. Consequently,

Sirt6 heterozygosity resulted in increased expression of NKG2D ligand, leading to NK cell activation and increased levels of inflammatory cytokines in NK cells. Blocking the NKG2D ligand–receptor interaction abolished the effect of *Sirt6* heterozygosity (Zhang, Ren et al., 2016). Moreover, SIRT6 has been shown to play a protective role against atherosclerosis by regulating the expression of PCK9 (Tao, Xiong, DePinho, Deng, & Dong, 2013a), which plays a critical role in the regulation of LDL-cholesterol levels, via control of LDL receptor degradation. SIRT6 deacetylates histone H3K9 and H3K56 in the proximal promoter region of *Pcsk9* and suppresses its expression (Tao et al., 2013a). Consequently, SIRT6-overexpressing, HFD-fed mice display lower levels of LDL-cholesterol (Tao et al., 2013a). Furthermore, SIRT6 plays a protective role against atherosclerosis by reducing ox-LDL-induced foam cell formation through an autophagy-dependent pathway (He et al., 2017).

SIRT7

SIRT7 also plays a role in protecting against the development of cardiac hypertrophy. Mice harboring a *Sirt7* germline deletion develop extensive fibrosis and succumb to cardiac hypertrophy (Vakhrusheva et al., 2008). The authors propose that SIRT7 deacetylates p53 in the myocardium to suppress p53-driven apoptosis; loss of SIRT7 leads to diminished stress responses and hypertrophy (Vakhrusheva et al., 2008). Using hypoxia/reoxygenation-treated cardiomyocytes (a cellular model of myocardial I/R injury), Sun et al. revealed that the microRNA miR-148b-3p targets SIRT7 expression, and regulates the acetylation of the p53 protein to modulate p53-mediated pro-apoptotic signaling (Sun, Zhai et al., 2018). Consequently, SIRT7 OE rescued miR-148b-3p-induced cell apoptosis in cardiomyocytes with hypoxia/reoxygenation treatment (Sun, Zhai et al., 2018). Araki et al. showed that SIRT7 modulates autophagy in cardiac fibroblasts, and contributes to myocardial tissue repair by maintaining TGF-β signaling (Araki et al., 2015). As a consequence, compared to wild-type controls, *Sirt7* KO mice exhibited susceptibility to cardiac rupture following MI (Araki et al., 2015).

SIRT7 has also been implicated in cardiac fibrosis. In a recent study by Wang et al., SIRT7 expression was found to be upregulated during angiotensin-II-mediated differentiation of cardiac fibroblasts (Wang, Liu et al., 2017). SIRT7 depletion resulted in reduced α-SMA levels and ECM deposition, and reduced the expression of focal adhesion-associated genes and fibrosis-related genes in angiotensin-II-treated cardiac fibroblasts. Conversely, OE of SIRT7 exacerbated angiotensin-II-induced cardiac fibrosis. Furthermore, the authors propose that SMAD2 signaling and ERK signaling commonly mediate the effects of SIRT7 on the process of angiotensin-II-induced cardiac fibrosis (Wang, Liu et al., 2017).

Overall, members of the sirtuin family play crucial roles in maintaining cardiovascular homeostasis, through overlapping and distinct mechanisms. Therefore it will be of great interest to assess the therapeutic potential of sirtuin activation in patients with cardiac dysfunction.

Inflammatory signaling

Chronic inflammation is both a general feature of, and a risk factor for, most common diseases of aging, including cancer, T2D, and CVD (Barbaresko, Koch, Schulze, & Nothlings, 2013). The NF-κB family of transcription factors coordinates cellular responses to diverse genotoxic, inflammatory, and oxidative stresses. NF-κB is critical in regulating cellular processes such as apoptosis, proliferation, inflammation, and immunity; a progressive age-associated increase in NF-κB activity has been directly implicated in the aging process (Adler et al., 2007; Tilstra, Clauson, Niedernhofer, & Robbins, 2011; Zhang et al., 2013). Several cellular signaling pathways involved in lifespan determination overlap considerably with sirtuin function, including IIS (especially FOXO transcription factors), mTOR, and NF-κB signaling (Tilstra et al., 2011), implying that sirtuins may affect healthspan and lifespan in part by modulating the inflammatory response through regulation of these pathways.

SIRT1 suppresses inflammation in multiple contexts by deacetylation of regulatory factors including the NF-κB p65 subunit and FOXO proteins (Haigis & Sinclair, 2010; Jung et al., 2009). Liver-specific *Sirt1* KO mice fed an HFD show increased inflammation and elevated NF-κB signaling (Purushotham et al., 2009). Adipose tissue inflammation, characterized by enhanced infiltration and altered polarization of macrophages, contributes to insulin resistance and its associated metabolic diseases. In adipocytes, SIRT1 deacetylates the transcription factor nuclear factor of activated T cells cytoplasmic 1 (NFATc1) and enhances IL-4 gene expression, which in turn regulates the recruitment and polarization of macrophages in adipose tissues (Hui et al., 2017). Adipocyte-selective deletion of *Sirt1* aggravates HFD-induced insulin resistance, associated with an increased number of adipose-resident macrophages and their polarization toward the proinflammatory M1 subtype (Hui et al., 2017).

Rheumatoid arthritis (RA), a chronic inflammatory disease, is characterized by excessive production of proinflammatory cytokines in joints, leading to progressive joint destruction. SIRT1 ameliorates the RA-associated inflammatory phenotype by promoting macrophage polarization into the M2 subtype via activating AMPKα signaling (Park et al., 2017). By contrast, osteoarthritis is an age-related inflammatory degenerative disease of cartilage, characterized by a disruption in the cartilage ECM. TNF-α and interleukin-1 beta (IL-1β) stimulate matrix metalloproteinase (MMP) activity in chondrocytes, leading to ECM breakdown. SIRT1 may play an antiinflammatory role in osteoarthritis; increased MMP expression and loss of cartilage are observed in young *Sirt1* KO mice (Gabay et al., 2013). In human chondrocyte cell models, TNF-α promotes SIRT1 inactivation by inducing its cathepsin B-mediated cleavage (Dvir-Ginzberg et al., 2011). Conversely, IL-1β, a mediator of cartilage breakdown, activates SIRT1 by upregulation of NAMPT expression. This interaction generates a positive feedback loop through ERK/p38 kinase activation to enforce SIRT1 activity (Hong et al., 2011). Thus, enhancement of SIRT1 activity may represent an attractive means of ameliorating the inflammatory response in osteoarthritic disease. Further evidence that SIRT1 plays an antiinflammatory function in humans comes from an analysis of a family with a missense mutation in the *SIRT1* coding sequence. Individuals in this pedigree suffer from the autoimmune disorders type I diabetes (T1D) and ulcerative colitis (Biason-Lauber et al., 2013). Expression of this SIRT1 mutant (SIRT1-L107P) in the β-cell line, MIN6, results in overproduction of nitric oxide, and increased expression of the cytokine TNF-α and the chemokine KC. Moreover, expression of SIRT1-L107P in myoblasts from healthy individuals results in insulin resistance similar to that observed in myoblasts obtained from the index patient (Biason-Lauber et al., 2013), providing compelling evidence that SIRT1 suppresses inflammatory diseases in humans. Indeed, administration of the SIRT1 activator, SRT2104, has yielded promising results in patients with psoriasis, an inflammatory disease of the skin (Krueger et al., 2015).

SIRT2 regulates inflammation by deacetylating the p65 subunit of NF-κB and modulating the expression of specific NF-κB target genes (Rothgiesser, Erener, Waibel, Luscher, & Hottiger, 2010). *Sirt2* KO mice exhibit NF-κB/p65 hyperacetylation and increased inflammatory responses in collagen-induced arthritis (Lin, Sun, Jiang, Hong, & Zheng, 2013), dextran sulfate sodium-induced colitis (Lo Sasso et al., 2014), and lipopolysaccharide (LPS)-induced microglia activation and neurotoxicity (Pais et al., 2013). However, subsequent studies show that reducing the expression or activity of SIRT2 suppresses LPS-induced microglial activation and neuroinflammation (Chen, Wu, Ding, & Ying, 2015; Wang, Zhang et al., 2016). In addition, SIRT2 deficiency inhibits LPS-induced activation of NF-κB in macrophages (Lee, Jung et al., 2014). Further studies are required to elucidate whether anti- or proinflammatory effects of SIRT2 may be relevant in metabolic diseases, including insulin resistance and T2D.

Emerging evidence supports that SIRT3 has an ameliorating effect in regulating nod-like receptor pyrin domain-containing 3 (NLRP3) inflammasome activation and inflammation-induced organ injury. NLRP3 triggers the release of proinflammatory cytokines IL-1β and IL-18 (Yang, Wang, Kouadir, Song, & Shi, 2019), and initiates obesity-induced inflammation and insulin resistance (Vandanmagsar et al., 2011). SIRT3 suppresses the activation of the NLRP3 inflammasome by attenuating mitochondrial ROS levels via deacetylation and activation of SOD2 (Traba et al., 2017, 2015; Zhao, Zhang, Sui, Zhu, & Zeng, 2016). In addition, Liu et al. showed that SIRT3 suppresses NLRP3 inflammasome activation by regulating autophagy in macrophages, by deacetylating ATG5 (Liu, Huang et al., 2018). In diabetic wound macrophages, upregulation of the fatty acid-binding protein FABP4 promotes inflammation by diminishing SIRT3 expression (Boniakowski et al., 2019). Indeed, *Sirt3* deletion exacerbates cecal ligation and puncture (CLP)-induced upregulation of the NLRP3 inflammasome, resulting in kidney dysfunction, renal tubular cell injury and apoptosis, mitochondrial alterations, and ROS production in *Sirt3* KO mice (Zhao et al., 2016). Consistently, *Sirt3* KO mice display mitochondrial dysfunction and elevated NLRP3 inflammasome assembly in the brain, which leads to increased microgliosis and neuroinflammation (Tyagi et al., 2018). On the other hand, SIRT3 OE ameliorates trimethylamine-N-oxide-induced endothelial NLRP3 inflammasome activation and vascular inflammation (Chen, Zhu et al., 2017). Similarly, administration of the SIRT3 activator, viniferin, reduces NLRP3 activation, and diminishes neutrophil influx and severity of endotoxin-mediated acute lung injury in mice (Kurundkar et al., 2019). Taken together, these results suggest that enhancing SIRT3 expression or activity, such as through fasting-mimetic diets, may play a pivotal role in preventing inflammation by blunting the assembly and activation of the NLRP3 inflammasome.

SIRT4 interferes with the NF-κB signaling pathway by preventing the nuclear translocation and transcriptional activity of NF-κB, thereby exerting antiinflammatory actions (Tao et al., 2015). Whether SIRT4 catalytic activity is required for its antiinflammatory function is not certain. On the other hand, SIRT5 has been shown to compete with SIRT2 to interact with p65, in a deacetylase activity-independent manner, to block the deacetylation of p65 by SIRT2, resulting in increased acetylation of p65 and the activation of the NF-κB pathway and its downstream cytokines (Qin, Han et al., 2017).

Strong evidence that SIRT6 plays an antiinflammatory role comes from studies in *Sirt6*-null mice. SIRT6 binds to the NF-κB RELA/p65 subunit and interacts with the promoters of NF-κB targets, where it deacetylates H3K9 to silence expression of these genes (Kawahara et al., 2009). *RelA/p65* heterozygosity in SIRT6-deficient mice rescues a subset of these mice from early demise, suggesting that elevated inflammation mediated through NF-κB may play an important role in the pathologies of the SIRT6-deficient mouse strain. Indeed, myeloid-specific deletion of *Sirt6* promotes the activation of NF-κB and endogenous production of IL-6, which facilitates proinflammatory M1 polarization of bone marrow macrophages and boosts their migration into adipose tissue (Lee et al., 2017). As a consequence, myeloid-specific *Sirt6* KO mice display greater accumulation of macrophages in fat, resulting in increased inflammation and insulin resistance in the context of an HFD (Lee et al., 2017). Furthermore, fat-specific *Sirt6* KO mice exhibit increased expression of various inflammatory genes including F4/80, TNF-α, IL-6, and MCP-1 in both WAT and BAT (Xiong et al., 2017), resulting in elevated adipose tissue inflammation and insulin resistance in HFD-fed fat-specific *Sirt6* KO mice (Kuang et al., 2017; Xiong et al., 2017). However, in vitro analyses have demonstrated that SIRT6 can stimulate proinflammatory TNF-α secretion by removal of long-chain fatty acyl groups from this protein (Jiang et al., 2013; Van Gool et al., 2009), whereas the presence of lysine fatty acylation on TNF-α promotes its lysosomally mediated degradation (Jiang, Zhang, & Lin, 2016). A separate study reports that efficient translation of TNF-α relies on intracellular NAD^+ levels in a SIRT6-dependent manner (Jiang et al., 2013; Van Gool et al., 2009). Additional detailed analysis in vivo is required to elucidate the role that SIRT6 may play in enhancing inflammation.

Until recently, there was no report available touching upon the possible involvement of SIRT7 in inflammatory signaling pathways. Using *Sirt7* KO mice, Miyasato et al. showed that SIRT7 is required for p65 nuclear translocation and NF-κB transcriptional activity (Miyasato et al., 2018). *Sirt7* deletion reduces the expression of TNF-α and ameliorates cisplatin-induced nephrotoxicity in mice. In contrast, Chen et al. showed that SIRT7 depletion promotes the nuclear translocation of NF-κB/p-p65 and increases secretion of inflammatory cytokines (IL-1β and IL-6) in LPS-induced dairy cow mammary epithelial cells (Chen, Li et al., 2019). More studies are warranted to delineate SIRT7 functions in inflammatory signaling.

Neurodegenerative disease

Sirtuins have been shown to mitigate age-associated neurodegenerative diseases such as AD, Huntington disease (HD), and PD (Herskovits & Guarente, 2014), suggesting that sirtuins might represent therapeutic targets in the context of these common diseases that currently lack curative treatments.

SIRT1

Accumulation of extracellular β-amyloid (Aβ) plaques and neurofibrillary tau protein tangles are hallmarks of AD. Individuals with the apolipoprotein E4 (*APOE4*) allele have an increased predisposition to developing AD. In neuronal cell culture models, it has been shown that expression of APOE4 reduces SIRT1 levels, and secretion of the nonpathogenic, soluble precursor protein alpha (sPPα). Also, a reduction in SIRT1 protein levels was reported in postmortem brain tissue from AD patients. SIRT1 OE in this model reconstitutes sPPα secretion, indicating that SIRT1 may play an important role in promoting nonpathogenic amyloid precursor protein proteolysis (Theendakara et al., 2013). Induction of Aβ peptides in a mouse neuroblastoma cell line expressing human amyloid precursor proteins results in a decrease of SIRT1 protein levels and enzymatic activity. Treatment with the phosphodiesterase inhibitor, cilostazol, decreases Aβ accumulation and upregulates *Sirt1* mRNA and protein expression in these cells (Lee et al., 2014a, 2014b). The retinoic acid receptor RARβ activates expression of the γ-secretase ADAM10. Increased expression of ADAM10 diverts amyloid precursor proteins away from processing by γ-secretase, thereby attenuating production of pathogenic Aβ available for plaque formation. Expressions of ADAM10 and RARβ are stimulated by cilostazol treatment, and are attenuated upon pharmacologic or siRNA-mediated SIRT1 inhibition. Overall, these results hint that SIRT1 may decrease Aβ production by activating RARβ and ADAM10 (Lee et al., 2014b). Similar to cilostazol treatment, treadmill exercise reduces Aβ production by increasing SIRT1 levels, which subsequently results in increased levels of ADAM10 and RARβ, ultimately facilitating the nonamyloidogenic pathway (Koo, Kang, Oh, Yang, & Cho, 2017). A separate study in AD mice overexpressing *Sirt1* shows that SIRT1 deacetylates tau, which may promote its ubiquitin-mediated degradation to suppress its pathogenic accumulation (Cohen et al., 2011; Min et al., 2010). Consequently, SIRT1 deficiency in the brain elevates acetylated-tau and leads to synapse loss and behavioral deficits, whereas SIRT1 OE reduces acetylated-tau and ameliorates propagation of tau pathology (Min et al., 2018). Neuronal specific SIRT1 OE in the dorsal CA1 region of the hippocampus strikingly reduces the presence of Aβ and phosphorylated tau in the AD model, while increasing the expression of neurotrophic factors, such as brain-derived neurotrophic factor, and enhances cognitive function and resilience against AD (Corpas et al., 2017).

HD is an autosomal dominant disease caused by polyglutamine repeat expansion in the huntingtin protein, resulting in neuronal dysfunction and death primarily in the basal ganglia. While conflicting roles have been described for SIRT1 orthologs in *C. elegans* and *D. melanogaster* HD models (Pallos et al., 2008; Parker et al., 2012), mammalian SIRT1 is clearly neuroprotective in mouse models of HD (Jeong et al., 2012; Jiang et al., 2012). SIRT1 OE in the brain of HD mice results in delayed onset and slower progression of disease, whereas *Sirt1* deletion in the brain exacerbates HD symptomatology. SIRT1-dependent deactivations of p53, mTORC1, FOXO3A, and PGC-1α pathways specifically in the brain have been proposed as the molecular mechanisms by which SIRT1 promotes cell survival and confers neuroprotection in HD (Jiang et al., 2012).

Xeroderma pigmentosum group A (XPA) is an autosomal recessive disorder that results from defects in nucleotide excision repair. XPA patients exhibit UV-induced skin cancer and, unexpectedly, neurodegeneration and sensorineural hearing loss (DiGiovanna & Kraemer, 2012), phenotypes often associated with mitochondrial disease. Recently, Fang et al. have described mitochondrial dysfunction occurring in human XPA fibroblasts, with increased ROS production, decreased membrane potential, and impaired mitophagy, leading to increased apoptosis (Fang et al., 2014). Decreased SIRT1 expression and NAD^+ were both observed, as well as increased PARP1 activity. These mitochondrial defects were rescued in mice and worms deficient for XPA upon administration of NAD^+ precursors or PARP inhibitors. Using rat neurons, human cell lines, and XPA-deficient mouse and nematode models, Fang et al. implicated NAD^+/SIRT1 in regulating mitophagy in the pathology of XPA-dependent neurodegeneration. It is clear that increased SIRT1 activity can ameliorate disease effects in animal models of neurodegeneration. Whether or not modulating SIRT1 activity in human patients might prove therapeutically useful in delaying or treating these diseases is currently unknown.

SIRT2

SIRT2 may likewise represent a target for treatment of age-associated neurodegeneration. Genetic or pharmaceutical inhibition of SIRT2, or the *D. melanogaster* homolog of mammalian SIRT1, dSIR2, improves survival of neurons expressing mutant huntingtin protein in HD *C. elegans* and *D. melanogaster* models (Luthi-Carter et al., 2010; Pallos et al., 2008). Administration of a brain-permeable SIRT2 inhibitor reduces mutant huntingtin aggregation and brain atrophy, and improves motor function and survival in two separate mouse HD models (Chopra et al., 2012). Additionally, genetic or pharmacologic SIRT2 inhibition in a striatal neuron model of HD causes reduced nuclear trafficking of SREBP-2, responsible for sterol biosynthesis, resulting in significant downregulation of its expression (Luthi-Carter et al., 2010). Mutant huntingtin fragments increase sterol levels in neuronal cells, therefore manipulation of sterol biosynthesis by inhibiting SIRT2 may prove beneficial in HD. However, *Sirt2* gene KO has no impact on the expression of the cholesterol biosynthesis enzymes, and SIRT2 ablation does not improve phenotypes in the severe R6/2 mouse model of HD (Bobrowska, Donmez, Weiss, Guarente, & Bates, 2012).

Administration of 1-methyl-4-phenyl-1,2,3,6-tetrahydropyridine (MPTP) to mice induces a PD-like condition, mitigated by genetic deletion of *Sirt2*. SIRT2 deficiency results in a decrease of neuronal apoptosis via an increase in FOXO3A acetylation and a concomitant decrease in BIM expression (Liu, Arun, Ellis, Peritore, & Donmez, 2014). In addition, administration of a selective brain-permeable SIRT2 inhibitor, AK7 prevents MPTP-induced dopamine depletion and dopaminergic neuronal loss in vivo (Chen, Wales et al., 2015). Furthermore, pharmacological inhibition or RNAi-mediated KD of SIRT2 alleviates α-synuclein toxicity in fly and human cell line models of PD (Outeiro et al., 2007). Recently, Esteves et al. reported altered NAD^+ metabolism in sporadic PD patient-derived cells, resulting in enhanced SIRT2 activation and subsequent deacetylation of α-tubulin. SIRT2 depletion or pharmacological inhibition increases α-tubulin acetylation and maintains normal autophagic flux, facilitating the trafficking and clearance of α-synuclein aggregates. Consistently, *Sirt2* KO mice resist MPTP-induced motor impairments (Esteves et al., 2018). SIRT2 also governs α-synuclein aggregation and toxicity by deacetylating it at lysine 6 and 10 (de Oliveira et al., 2017). Strikingly, deacetylation mimetic mutants (K-R) exacerbate α-synuclein toxicity in the substantia nigra of rats, while an acetylation-mimetic mutant induced less toxicity. In contrast, Singh et al. reported that SIRT2 reduces the formation of α-synuclein aggregates in a human cell line model of PD (Singh, Hanson, & Morris, 2017).

Genetic linkage analysis demonstrated that the *SIRT2* polymorphism rs10410544 is associated with a modest (1.1−1.65-fold) increased risk of developing late-onset AD in APOE4-negative carriers of Caucasian and Han Chinese descent (Polito et al., 2013; Wei et al., 2014; Xia et al., 2014). While this SNP lies within an intron of the *SIRT2* gene, and is not expected to affect protein expression or activity by obvious mechanisms, it has been hypothesized that it could be associated with a variant in proximity to the *SIRT2* genomic locus that would affect expression. SIRT2 expression analysis in these populations is required to determine if altered SIRT2 levels are associated with the risk of developing AD. However, a recent study showed that *SIRT2* mRNA expression is increased in the peripheral blood of AD patients (Wongchitrat et al., 2019). Pharmacological inhibition of SIRT2 reduces the Aβ production in amyloid precursor overexpressing H4-SW neuroglioma cells, and leads to reduced amyloidogenesis and improved cognitive performance in two AD transgenic mouse models (Biella et al., 2016). Tubulin deacetylation by SIRT2 leads to loss of stability and depolymerization of microtubules, which facilitates tau dissociation and consequent phosphorylation in AD models. Pharmacological inhibition of SIRT2 results in improved microtubule dynamics and reduced tau phosphorylation (Esteves et al., 2019). Taken together, these results suggest that the pathological roles of SIRT2 in AD are associated with Aβ and/or tau.

SIRT2 is highly expressed in the adult mammalian brain and functions during development of the central nervous system. SIRT2 protein is localized to myelinated axons and in the cytoplasm of oligodendroglial cells, where it negatively regulates differentiation of oligodendrocyte precursors by deacetylating α-tubulin (Li et al., 2007). Conversely, SIRT2 has been reported to accelerate differentiation of an oligodendroglial precursor cell line. OE of SIRT2 results in increased expression of myelin basic protein (MBP), a characteristic of mature oligodendrocytes. The homeodomain transcription factor NKX2.2 represses *Sirt2* transcription by binding to its promoter, and consequently inhibits differentiation when overexpressed (Ji, Doucette, & Nazarali, 2011). The reasons for the discrepancy between these findings, and the molecular mechanisms of how SIRT2 influences oligodendrocyte differentiation, are not clear.

A link between SIRT2 function and myelination has also been reported. Beirowski et al. reported a delay in myelination of Schwann cells in a tissue-specific *Sirt2* KO mouse model (Beirowski et al., 2011). In this context, SIRT2 deacetylates partitioning defective 3 homolog (PAR3), a regulator of cell polarity in Schwann cells, thereby inhibiting polarity control protein atypical protein kinase C (aPKC). Inhibition of aPKC activity subsequently alters myelin formation by decreasing downstream cell polarity complex signaling. Further investigation into pharmacological activation of SIRT2 to promote remyelination of injured axons could potentially allow new methods of treating human disease associated with progressive myelin loss, including multiple sclerosis.

SIRT3

In an ex vivo cell culture system, SIRT3 is neuroprotective in the context of mutant huntingtin expression. Expression of mutant huntingtin results in decreased SIRT3 activity, reduced cellular NAD^+ levels, and impaired mitochondrial biogenesis. Conversely, administration of the resveratrol dimer, viniferin, activates AMP-activated kinase (AMPK), promotes mitochondrial biogenesis, and increases SIRT3 expression. *Sirt3* KD inhibits viniferin-mediated AMPK activation and attenuates its neuroprotective effects (Fu et al., 2012). Recently, Cheng et al. reported that SIRT3 mediates adaptive responses of neurons to bioenergetic, oxidative, and excitatory stress (Cheng, Yang et al., 2016). SIRT3-deficient cortical neurons exhibit enhanced sensitivity to glutamate-induced calcium overload and excitotoxicity, and oxidative and mitochondrial stress, whereas ectopic expression of SIRT3 restores neuronal stress resistance. SIRT3 ablation results in increased degeneration of striatal and hippocampal neurons in mouse models of HD and epilepsy, respectively. Running wheel exercise increases the expression of SIRT3 in hippocampal neurons, which is mediated by excitatory glutamatergic neurotransmission, and is essential for mitochondrial homeostasis and neuroprotective effects of running (Cheng, Yang et al., 2016). A study by Shulyakova et al. showed that ectopic expression of SIRT3 protects neuronally differentiated PC12 cells, a model of sympathetic catecholaminergic neurons, from degeneration induced by oxidative stress and trophic withdrawal (Shulyakova et al., 2014). Consistently, SIRT3 attenuates nigrostriatal dopaminergic neuron damage in a mouse model of PD by improving antioxidant capacity in mitochondria (Liu, Peritore, Ginsberg, Kayhan, & Donmez, 2015).

Furthermore, *Sirt3* KO mice exhibit poor remote memory and impaired long-term potentiation. Loss of SIRT3 results in decreased neuronal number in the ACC, which seems to contribute to memory deficits in *Sirt3* KO mice (Kim, Kim, Choi et al., 2019). Recently, Lee et al. reported that SIRT3 levels are reduced in cerebral cortex of AD patients, resulting in increased acetylation of p53, which in turn leads to increased p53 occupancy of mtDNA in AD (Lee, Kim et al., 2018). Mitochondrially targeted p53 reduces mitochondria DNA-encoded ND2 and ND4 gene expression, resulting in increased ROS and reduced mitochondrial oxygen consumption. SIRT3 OE restored the expression of ND2 and ND4 and improved mitochondrial oxygen consumption by repressing mito-p53 activity (Lee, Kim et al., 2018). Most recently, a study by Liu et al. demonstrated that SIRT3 plays a role in hippocampal neurons in behavioral and GABAergic synaptic adaptations to intermittent fasting (Liu, Cheng et al., 2019). Intermittent food deprivation improves mood and cognition, and protects neurons against excitotoxic degeneration in animal models of epilepsy and AD. Germline *Sirt3* KO or hippocampal neuron-specific *Sirt3* depletion abolished fasting-induced neuronal network and behavioral adaptations. In a mouse model of AD, intermittent fasting constrained neuronal network hyperexcitability and ameliorated deficits in hippocampal synaptic plasticity in a SIRT3-dependent manner (Liu, Cheng et al., 2019), further implying a protective role of SIRT3 in AD.

SIRT4

The role of SIRT4 in alleviating neuropathology is suggested by the observation that SIRT4 loss results in reduced expression of the glutamate transporter at the cell surface of neurons, resulting in decreased glutamate uptake in *Sirt4* KO mice (Shih, Liu, Mason, Higashimori, & Donmez, 2014). Excess glutamate at synapses in the brain prevents efficient neurotransmission and results in excitotoxicity, which causes cell death and is implicated in neurodegenerative disease. Consequently, *Sirt4* KO mice display enhanced kainate (KA)-induced seizure phenotypes (Shih et al., 2014). Additionally, Komlos et al. reported that SIRT4 regulates glial development in the brain. Ectopically expressed GDH promotes the differentiation of radial glia into astrocytes, which can be inhibited by overexpressing SIRT4 (Komlos et al., 2013). Therefore reduced SIRT4 levels or activity, resulting in increased GDH activity, can cause defects in glial development in the brain. Consistently, an activating mutation in *GDH* is present in patients with congenital hyperinsulinism hyperammonemia syndrome, a form of hyperinsulinism; these patients frequently experience seizures, hypoglycemia, and developmental defects.

SIRT5

There are only few reports available in the literature exploring the effects of SIRT5 in neurodegenerative disease. A study by Liu and colleagues suggested that SIRT5 alleviates nigrostriatal dopaminergic degeneration by suppressing mitochondrial-derived ROS levels in a mouse model of PD (Liu, Peritore, Ginsberg, Shih et al., 2015). *Sirt5* KO mice brain striata exhibit decreased levels of the mitochondrial antioxidant enzyme SOD2, resulting in increased ROS levels and aggravated nigrostriatal dopaminergic degeneration in response to MPTP (Liu, Peritore, Ginsberg, Shih et al., 2015), a small molecule that induces selective loss of nigrostriatal dopaminergic neurons in mice. Furthermore, Li and Liu identified neuroprotective roles of SIRT5 in response to KA-induced epileptic seizure (Li & Liu, 2016). KA-treated *Sirt5* KO mice display strikingly increased mortality and severe response to epileptic seizures. Loss of SIRT5 is associated with worsened reactive astrogliosis in the hippocampus, and dramatically exacerbated hippocampal neuronal loss and degeneration in KA-exposed mice (Li & Liu, 2016). Notably, the protective effects of SIRT5 in KA-induced epilepsy appear to be independent of its role in managing ROS levels. Moreover, recent studies suggest that SIRT5 plays a role in protein kinase C epsilon (PKCε)-mediated tolerance against cerebral ischemia by regulating purine metabolism (Koronowski et al., 2018; Morris-Blanco et al., 2016).

SIRT6 and SIRT7

Until recently, there was no report examining the roles of SIRT6 in neurodegeneration. Recent studies reveal that SIRT6 levels are reduced in both AD patients and the 5XFAD mouse model of AD (Jung et al., 2016; Kaluski et al., 2017). Aβ42, a major component of senile plaques, reduces SIRT6 expression and induces DNA damage in AD (Jung et al., 2016). Upregulation of p53 by the small molecule Nutlin-3 prevents Aβ42-induced DNA damage by restoring the expression of SIRT6. Indeed, OE of SIRT6 in HT22 mouse hippocampal neurons is protective against Aβ42-induced injury (Jung et al., 2016). Kaluski and colleagues reported that SIRT6 regulates tau protein stability and phosphorylation through increased activation of the kinase GSK3α/β. Brain-specific deletion of *Sirt6* resulted in increased signs of DNA damage, cell death, and hyperphosphorylated tau (Kaluski et al., 2017). Taken together, these studies hint at a potential protective role of SIRT6 in AD, however additional studies are required.

In the context of neurodegenerative diseases and brain, SIRT7 is largely unexplored, however, a recent study indicated that SIRT7 has a modest protective effect on neurogenesis in a mouse model of autoimmune neuroinflammation. *Sirt7* KO mice in the experimental autoimmune encephalomyelitis (EAE) paradigm exhibit reduced survival of newly generated hippocampal neurons (Burg, Bittner, & Ellwardt, 2018).

Pharmacological modulation of sirtuin activity

As discussed throughout this chapter, sirtuins play diverse roles in modulating multiple age-associated diseases. Thus, potent and selective sirtuin modulators are highly desirable as a novel therapeutic class.

Sirtuin-activating compounds

High-throughput screens have been conducted to identify small-molecule sirtuin-activating compounds (STACs). These studies initially identified resveratrol (RSV) and other polyphenols as SIRT1 activators (Howitz et al., 2003); subsequent studies identified a large series of artificial, higher potency STACs. Concerns have been raised as to whether these polyphenols actually directly activate SIRT1 (Baur & Sinclair, 2006; Kaeberlein et al., 2005; Pacholec et al., 2010; Park et al., 2012). STACs have been reported to interact with a specific region on the SIRT1 protein (Hubbard et al., 2013). In this regard, a specific aa in SIRT1 required for STAC-mediated activation has been identified. Cells bearing a SIRT1 mutant at this site do not show the increased mitochondrial copy number and ATP content normally induced by STAC treatment (Hubbard et al., 2013).

RSV has been reported to extend longevity in yeast, worms, flies, and short-lived fish (Baur, 2010b). In mammals, RSV supplementation extends lifespan in mouse models of obesity, AD, HD, ALS, and sepsis-induced acute kidney injury (Baur et al., 2006; Gerhardt et al., 2011; Holthoff, Wang, Seely, Gokden, & Mayeux, 2012; Mancuso et al., 2014; Porquet et al., 2013; Sebai, Sani, Ghanem-Boughanmi, & Aouani, 2010; Song, Chen, Zhang, Li, & Le, 2014). RSV administration improves survival in a rat hypertension model and also attenuates catecholamine-induced mortality in obese rats (Avila et al., 2013; Rimbaud et al., 2011). However, RSV treatment does not extend longevity in mice fed a standard diet (Miller et al., 2011; Pearson et al., 2008; Strong et al., 2013). In mice, RSV induces gene expression changes similar to CR (Barger, Kayo, Pugh, Prolla, & Weindruch, 2008; Barger, Kayo, Vann et al., 2008; Pearson et al., 2008), and protects against damaging effects of high-fat/high-calorie diets (Andrade et al., 2014; Baur et al., 2006; Jimenez-Gomez et al., 2013; Mattison et al., 2014). It is also protective against multiple types of cancers, CVD, neurodegenerative diseases, both T1D and T2D, and also possesses antiinflammatory and antiviral qualities (see Dai, Sinclair, Ellis, & Steegborn, 2018; Salehi et al., 2018). Despite the apparent beneficial effects of RSV in multiple systems, in human patients with NAFLD, RSV treatment appeared to exert toxic effects on hepatocytes, and did not ameliorate liver steatosis or insulin resistance (Chachay et al., 2014). However, this study was limited in that only male participants were included, and no assessment of RSV metabolites was reported. Consistently, in another placebo-controlled, high-dose, and long-term study, RSV treatment had no beneficial effect in alleviating clinical or histological NAFLD (Heeboll et al., 2016). Another placebo-controlled clinical trial investigating the effects and safety of RSV supplementation on liver fat content and cardiometabolic risk parameters in overweight and obese and insulin-resistant subjects concluded that RSV supplementation is safe but does not considerably impact liver fat content or cardiometabolic risk parameters in humans (Kantartzis et al., 2018). Moreover, in a randomized placebo-controlled clinical trial, RSV treatment did not improve inflammatory status, glucose homeostasis, blood pressure, or hepatic lipid content in middle-aged men with metabolic syndrome (Kjaer et al., 2017). In AD patients, RSV and its major metabolites were measurable in plasma and CSF, and some alteration in AD biomarkers was observed (Turner et al., 2015). Compared with the placebo group, RSV supplementation exerted antioxidant effects in the blood and PBMCs of patients with T2D, and resulted in significant reductions in weight, BMI, and blood pressure levels (Seyyedebrahimi, Khodabandehloo, Nasli Esfahani, & Meshkani, 2018). In addition to these studies, a large number of clinical trials, investigating the effects of RSV on CVD, obesity, diabetes, AD, and cancer, have been reported in the literature (Berman, Motechin, Wiesenfeld, & Holz, 2017; Kulashekar, Stom, & Peuler, 2018). Most of the current clinical trials reveal that RSV was well tolerated.

One difficulty with the use of RSV in vivo is that it has significant off-target effects (Baur, 2010a). To elucidate the effects of SIRT1 activation more specifically, structurally unrelated synthetic STACs have been evaluated, though off-target effects, if any, of these molecules have not yet been determined. One STAC, SRT1720, extends mean mouse lifespan in response to an HFD by 18% (Minor et al., 2011) and by 8.8% in mice fed a standard diet (Mitchell et al., 2014). Lifelong SRT2104 supplementation, beginning at 6 months of age, extends mean lifespan of male C57BL/6J mice fed a standard diet by 9.7% (Mercken, Mitchell et al., 2014). Although no difference in body weight, caloric intake, or physical activity was observed, SRT2104-supplemented mice exhibited a lower percentage of fat mass, decreased fasting blood glucose and insulin levels, and increased skeletal muscle endurance. Microarray analysis revealed that SRT2104 likely has antiinflammatory properties in skeletal muscle tissue, evidenced by a decrease in expression of NF-κB target genes. SRT2104 protects against experimentally induced muscle atrophy in wild-type mice, and muscle-specific *Sirt1* KD in vivo accelerates muscle loss. Also, SIRT1-dependent stimulation of osteogenic differentiation by SRT2104 treatment was reported using myoblast cell cultures, suggesting SRT2104 activates SIRT1 to protect against age-related muscle loss and osteoporosis. There have been multiple clinical trials evaluating SRT2104 in different contexts. SRT2104 is reasonably well tolerated and holds promise in some contexts—for instance, in colitis and psoriasis—however it exhibits poor and variable pharmacokinetics (Dai et al., 2018), highlighting the need to develop newer small-molecule SIRT1 activators with an improved pharmacokinetic and tolerability profile.

Honokiol (HKL) [2-(4-hydroxy-3-prop-2-enyl-phenyl)-4-prop-2-enyl-phenol], a low-molecular-weight natural biphenolic compound is reported to be a pharmacological activator of SIRT3 (Pillai et al., 2015). HKL, derived from the bark of magnolia trees, is traditionally used in Asian medicine. Pillai et al. showed that HKL treatment upregulates SIRT3 expression and activity, and attenuates the agonist-induced hypertrophic response of cardiomyocytes in vitro, as well as pressure overload cardiac hypertrophy in vivo (Pillai et al., 2015). Subsequently, Pillai and colleagues report that HKL-mediated activation of SIRT3 also protects the heart from doxorubicin-induced cardiac damage without compromising the tumoricidal effects of this drug (Pillai et al., 2017). Hexafluoro, a fluorinated synthetic HKL analog, prevents TGF-β-induced loss of SIRT3 in normal fibroblasts; mice treated with hexafluoro display substantial attenuation of bleomycin-induced fibrosis in the lung and skin (Akamata et al., 2016). A number of studies have further shown that by upregulating the activity and expression of SIRT3, HKL ameliorates oxidative stress and mitochondrial dysfunction, and protects against many pathological conditions, including hypertension-induced kidney injury (Li, Zhang et al., 2017), Aβ toxicity (Ramesh et al., 2018) and onset of AD symptoms (Li, Jia et al., 2018), and hyperglycemic intracerebral hemorrhage-induced neuronal injury (Zheng et al., 2018), among others. Lu et al. identified another SIRT3 activator, 7-hydroxy-3-(4′-methoxyphenyl) coumarin (C12), which binds to SIRT3 with high affinity and can promote the deacetylation and activation of MnSOD (Lu et al., 2017). However, a detailed characterization of its potential SIRT3-activating effects in the context of disease conditions remains to be performed.

Until recently, there were no pharmacological activators available for SIRT6; these are potentially desirable compounds, given the reported lifespan-extending effects of SIRT6. You et al. synthesized and screened pyrrolo [1,2-a] quinoxaline derivatives, yielding the first synthetic SIRT6 activators, which bind to a SIRT6-specific acyl channel pocket, with UBCS039 being the most potent SIRT6 activator with an EC50 value of $38 \pm 13\,\mu m$ (You et al., 2017). However, UBCS039 lacks specificity; it can promote SIRT5 desuccinylase activity; moreover, there are no data available that demonstrate its activity against SIRT6 inside the cell. More recently, Huang et al. reported MDL-800 as a more selective and cellularly active SIRT6 activator with an EC50 value of $10.3 \pm 0.3\,\mu M$ (Huang, Zhao et al., 2018). MDL-800 directly activates SIRT6 deacetylase activity by increasing the binding affinities of acetylated substrates and cofactor as well as increasing the catalytic efficiency of SIRT6, resulting in decreased H3K9 and H3K56 acetylation levels. MDL-800 treatment inhibits the growth of human HCC cells via cell-cycle arrest, and suppresses HCC xenograft tumor growth in vivo (Huang, Zhao et al., 2018).

Nonallosteric methods to activate sirtuins have also received intense scrutiny as an alternative to STACs. Pharmacologically increasing cellular NAD^+ levels by supplementation with NAD^+ precursors such as NR and NMN, or inducing expression of NAMPT to convert NAM to NAD^+ more efficiently, have identified an alternative means of promoting sirtuin function (Canto et al., 2012; Yoshino et al., 2011). For example, in a mouse model of T2D, NMN supplementation mitigates negative metabolic effects—insulin insensitivity, glucose intolerance, and inflammation—of age-related or diet-induced diabetes, potentially due to the activation of SIRT1 and other sirtuins, and their downstream target pathways (Yoshino et al., 2011). In a mouse model of muscular dystrophy, NR treatment prevents loss of muscle stem cells, improves muscle function, and protects aging mice from muscle degeneration (Zhang, Ryu et al., 2016). In addition, NR treatment delays the senescence of neural and melanocyte stem cells and prolongs mouse lifespan. In a recent study, Yang et al. reported dihydronicotinamide riboside (NRH) as another NAD^+ precursor, far more potent than NR and NMN. NRH administration significantly enhances NAD^+ levels in both cells and mice with no apparent adverse effects (Yang, Mohammed, Zhang, & Sauve, 2019). Notably, NRH treatment confers resistant to cell death caused by NAD^+-depleting genotoxins such as hydrogen peroxide and methylmethane sulfonate. Also, mice lacking the NAD^+-consuming enzyme CD38 have increased cellular NAD^+ and SIRT1 activity in several metabolically active tissues (Barbosa et al., 2007). These mice are highly resistant to weight gain in response to HFD relative to controls. One caveat in interpreting any findings involving NAD^+ modulation is that increased NAD^+ levels may activate not only sirtuins, but also additional NAD^+-dependent enzymes, such as PARP1 (Li, Bonkowski et al., 2017). In the context of CD38-deficient mice, protection against weight gain was lost when animals were treated with a sirtuin inhibitor, implicating sirtuin activation in this effect.

Indeed, there have been few clinical trials of NR studying the pharmacokinetics, safety, and efficacy of this agent. In the first assessment of NR pharmacokinetics in humans, single doses of NR produced dose-dependent increases in blood NAD^+ metabolite levels (Trammell et al., 2016). Another nonrandomized pharmacokinetics study, administering oral NR in healthy volunteers for 8 days, revealed that oral NR significantly increases circulating NAD^+ and is well tolerated, with no obvious adverse effects in humans (Airhart et al., 2017). Recently, a 6-week randomized, double-blind, placebo-controlled, crossover clinical trial found that chronic supplementation of NR is well tolerated and effectively stimulates NAD^+ metabolism in healthy middle-aged and older adults (Martens et al., 2018). In addition, NR supplementation reduced blood pressure and arterial stiffness. Clinical trials examining the safety and bioavailability of NMN have been initiated (Tsubota, 2016), however the results of these trials are yet to be reported.

Sirtuin-inhibiting compounds

Compared to sirtuin-activating compounds, a large range of potent sirtuin inhibitors have been reported in the literature. In this section we selectively discuss a few of these compounds, specifically those which have been evaluated for their in vivo effects in either cells or whole animals. For details regarding other sirtuin inhibitors readers are referred elsewhere (e.g., Carafa et al., 2016; Dai et al., 2018; Kumar & Lombard, 2018; Wang, He et al., 2019).

EX-527 is a potent and SIRT1-selective inhibitor with an IC50 value of <100 nM (Napper et al., 2005). It exploits the unique sirtuin deacylation reaction mechanism to inhibit SIRT1 activity (Gertz et al., 2013), and has been widely used to investigate the pharmacological effects of SIRT1-selective inhibition in cellular and animal model systems. EX-527 has entered into clinical trials, and is considered safe and well tolerated in healthy volunteers at all tested doses (Westerberg et al., 2015). Similarly, EX-527 is well tolerated in early-stage HD patients with no adverse effects and circulating levels of soluble huntingtin were not affected by EX-527 at plasma concentrations within the anticipated therapeutic concentration range (Sussmuth et al., 2015).

AGK2 is a widely used potent SIRT2 inhibitor (IC50 = 3.5 μM), which preferentially binds in the "C-pocket" of SIRT2, mimicking the NAM-mediated sirtuin inhibition (Outeiro et al., 2007). AGK2 exhibits more than 10-fold specificity for SIRT2 over SIRT1 and is protective in models of PD (Outeiro et al., 2007) and HD (Luthi-Carter et al., 2010). SirReal2 is another potent SIRT2 inhibitor (IC50 = 140 nM) and is highly selective for SIRT2 versus SIRTs 1/3/4/5/6 (Rumpf et al., 2015). SirReal2 treatment leads to tubulin hyperacetylation in HeLa cells and induces destabilization of the checkpoint protein BubR1 in vivo. Quinti et al. reported the identification of thiazole MIND4 as a potent SIRT2 inhibitor (IC50 = 31.2 μM), with neuroprotective effects in HD models (Quinti et al., 2016). Molecular docking studies suggest selectivity of MIND4 for SIRT2 over SIRT1 and SIRT3. Jing et al. developed a thiomyristoyl lysine compound, TM, as a potent SIRT2-specific inhibitor, which exhibits broad anticancer effects in various human cancer cells and in a mouse model of breast cancer (Jing et al., 2016). Most recently, Farooqi et al. developed a series of lysine-based thioureas as mechanism-based inhibitors of SIRT2, and reported compounds AF8, AF10, and AF12 as selective SIRT2 inhibitors with IC50 values of 0.06, 0.15, and 0.08 μM, respectively (Farooqi et al., 2019). AF8 and AF10 showed cytotoxicity across a variety of cancer cell lines, with minimal toxicity toward normal cells, and potently inhibited tumor growth in a CRC xenograft mouse model.

In addition to SIRT1- or SIRT2-specific inhibitors, researchers have also identified some dual SIRT1/SIRT2 inhibitors—for instance, sirtinol is the best known such dual inhibitor, and exhibits strong antitumor activity (Grozinger, Chao, Blackwell, Moazed, & Schreiber, 2001; Ota et al., 2006). Cambinol is another dual inhibitor of SIRT1/SIRT2, the first sirtuin inhibitor capable of exerting anticancer effects in mouse xenograft models (Heltweg et al., 2006). Lain et al. identified Tenovin-6 as a SIRT1/SIRT2 dual inhibitor with IC50 values of 21 and 10 μM, respectively (Lain et al., 2008). Tenovin-6 also exhibits moderate inhibition of SIRT3 (IC50 = 67 μM). Tenovin-6 treatment increased p53 acetylation, and inhibited melanoma cell growth in vitro and in a xenograft mouse model (Lain et al., 2008), increased FOXO1 acetylation in chronic myeloid leukemia (CML) cells and impaired disease progression in mouse CML models (Yuan et al., 2012), and induced apoptosis in gastric cancer cells via upregulating death receptor 5 (Hirai et al., 2014).

In subsequent studies, some of these dual inhibitors have been modified to improve their selectivity. In this context, Lara et al. modified sirtinol to synthesize salermide, which inhibits SIRT2 more efficiently than SIRT1; 80% inhibition was achieved at 25 μM for SIRT2 as compared to 90 μM for SIRT1 (Lara et al., 2009). Salermide induces cancer-specific pro-apoptotic effects on diverse tumor cells, by attenuating the repression of pro-apoptotic genes by SIRT1-mediated H4K16 deacetylation. Mahajan et al. reported the synthesis of cambinol analogs, which display improved selectivity-specific sirtuins. Analog 17 is highly selective for SIRT1 (IC50 = 27 μM), 24 for SIRT2 (IC50 = 13 μM), and 8 for SIRT3 (IC50 = 6 μM) (Mahajan et al., 2014). Recently Alhazzazi et al. developed a selective SIRT3 inhibitor, LC-0296, which exhibits about ~20- and 10-fold greater inhibition of SIRT3 (IC50 = 3.6 μM) compared to SIRT1 (IC50 = 67 μM) and SIRT2 (IC50 = 33 μM) (Alhazzazi et al., 2016). LC-0296 reduces cell proliferation and promotes apoptosis of head and neck squamous cell carcinoma (HNSCC) via modulating cellular ROS levels, and increases the sensitivity of HNSCC cells to radiation and cisplatin treatment. Most recently, Li et al. reported JH-T4 and YC8-02 as potent SIRT1/2/3 inhibitors (Li, Chiang et al., 2019). Both JH-T4 and YC8-02 are more selective for SIRT2 (IC50 = 0.013 and 0.062 μM, respectively) than for SIRT1 (IC50 = 0.8 and 2.8 μM, respectively) or SIRT3 (IC50 = 2.5 and 0.53 μM, respectively). YC8-02 is mitochondrial permeable and phenocopies the effects of SIRT3 depletion and killed DLBCL cells (Li, Chiang et al., 2019).

Using in silico screening, Parenti and colleagues reported sirtuin inhibitors, which show significant selectivity for SIRT6 over SIRT1 and SIRT2. Treating cells with the lead compounds increases acetylation at SIRT6 target lysines on histone 3, reduces TNF-α secretion, upregulates GLUT-1 expression, and increases glucose uptake, biological effects observed in response to SIRT6 ablation or silencing (Parenti et al., 2014). Subsequently, the same group identified analogs of one of the lead compounds with quinazolinedione structure, which are more potent and highly selective SIRT6 inhibitors (Sociali et al., 2015). In addition to the biological effects exerted by the reference lead compound, these new compounds exacerbated DNA damage and cell death in response to the PARP inhibitor olaparib in BRCA2-deficient Capan-1 cells, and sensitized pancreatic cancer cells to gemcitabine cytotoxicity. Furthermore, administration of the lead SIRT6 inhibitor in a mouse model of

T2D for 10 days was well tolerated, and reduced plasma insulin levels, while improving oral glucose tolerance, increasing expression of the glucose transporters GLUT1 and -4 in muscle, and enhancing glycolysis (Sociali et al., 2017).

Derivatives of another lead compound with salicylate structure have been reported, which display improved potency and SIRT6 selectivity (Damonte et al., 2017). Similar to quinazolinedione-based derivatives, compounds with salicylate-like structure are capable of recapitulating the biological effects of SIRT6 depletion. Most recently, Kim et al. identified ID: 97491 as a potent SIRT7 inhibitor ($IC50 = 0.325 \mu M$) (Kim, Kim, Cho et al., 2019), however, its selectivity for SIRT7 over other sirtuins needs to be addressed. ID: 97491 promotes p53 acetylation, and exerts anticancer effects in uterine sarcoma cells and inhibits tumor growth in a xenograft mouse model. Overall, significant progress has been made in developing sirtuin modulators, which hold the promise of treating both common and rare diseases. However, only a handful of them have actually reached clinical trials.

Conclusions

Sirtuins comprise a group of nutrient- and stress-responsive factors that regulate diverse cellular processes to promote healthspan, and in the case of two sirtuins, SIRT1 and SIRT6, lifespan extension when overexpressed in mice. However, as this chapter has made clear, many discrepant results regarding roles for sirtuins in various disease models have been reported in the literature. These studies suggest that upregulation of mammalian sirtuins is beneficial in a highly context- and tissue-specific manner. Generically speaking, strain differences (both in invertebrates and in mice), differences in experimental protocols, and/or microbiome differences between colonies (in the case of mouse studies) may account for many of these apparent discrepancies. However, the specific biology of sirtuin proteins may offer another potential explanation for differing results between labs. Sirtuins require NAD^+ for activity, levels of which are regulated by cellular nutrient status, organismal diet, stress conditions, and even organismal age. In this regard, genetic, pathophysiological, and pharmacological studies imply that sirtuin activity can be affected by alterations in NAD^+ levels. For instance, changes in NAD^+ levels in each subcellular compartment can directly regulate deacetylase activities of SIRT1−3 (Feldman et al., 2015).

A consequence of sirtuin NAD^+-dependency is that a chronic decline in NAD^+ levels would be predicted to impair sirtuin activity, resulting in metabolic decline, progressive loss of homeostatic maintenance, ultimately facilitating the onset of disease states resembling those described in sirtuin loss-of-function models. This may be relevant during organismal aging. Notably, an age-related decrease in NAD^+ levels has been reported in *C. elegans* (Fang et al., 2016, 2014; Mouchiroud et al., 2013), in several tissues of mice and rats (Braidy et al., 2011, 2014; Mills et al., 2016; Mouchiroud et al., 2013; North et al., 2014; Ramsey et al., 2008; Zhang, Ryu et al., 2016), and in the skin and brain of elderly humans (Massudi et al., 2012; Zhu et al., 2015). Moreover, for NAD^+ availability, sirtuins must compete with other NAD^+-consuming enzymes; PARPs, and CD38. Chronic PARP activation, induced by an increase in chronic nuclear DNA damage, could deplete intracellular NAD^+, resulting in decreased sirtuin activity, contributing to age-associated pathophysiology (Bai, Canto, Oudart et al., 2011; Mouchiroud et al., 2013). Genetic deletion or pharmacologic inhibition of PARP in skeletal muscle or BAT activates SIRT1 in mice, resulting in higher energy expenditure, increased mitochondrial content and protection from metabolic dysfunction (Bai, Canto, Brunyanszki et al., 2011; Bai, Canto, Oudart et al., 2011). In contrast, nuclear PARP1 deficiency does not enhance SIRT2 or SIRT3 activity, implying that therapies directed at enhancing NAD^+ levels might have unexpected sirtuin-specific effects.

Similarly, CD38 is a cell surface receptor on lymphocytes that catalyzes the synthesis and hydrolysis of cyclic ADP-ribose using NAD^+ as the precursor. In cells lacking CD38, NAD^+ levels rise and SIRT1 activity increases, conferring protection against diet-induced obesity and glucose tolerance in response to high fat intake (Barbosa et al., 2007). Furthermore, in response to CR, fasting, and exercise, skeletal muscle exhibits increased NAD^+ levels, and thus, upregulated SIRT1 activity (Canto et al., 2010; Chen et al., 2008).

In light of all of these findings, supplementation with NAD^+ would be predicted to mitigate some age-associated phenotypes, and could promote longevity by maintaining sirtuin-mediated homeostatic control in older organisms. Restoration of cellular NAD^+ levels via supplementation with NAD^+ precursors NR or NMN has been reported in heart and testes from old mice (North et al., 2014), and can protect against diet-induced obesity (Canto et al., 2012), restore the decline in mitochondrial function in aged mice (Gomes et al., 2013; Zhang, Ryu et al., 2016), rescue the age-related depletion of neural, muscle, and melanocyte stem/progenitor cell pools (Stein & Imai, 2014; Zhang, Ryu et al., 2016), and extend lifespan of *C. elegans* and mice (Hashimoto, Horikawa, Nomura, & Sakamoto, 2010; Mouchiroud et al., 2013; Schmeisser et al., 2013; Zhang, Ryu et al., 2016). Notably, increased levels of NAD^+, achieved by NR administration, are well tolerated, with no apparent adverse effects in humans (Airhart et al., 2017; Martens et al., 2018; Trammell et al., 2016). In light of all of these findings, a new avenue of sirtuin activation via modulation of cellular NAD^+ levels has emerged.

Much attention has been given to the effects of sirtuin OE; in this regard, it is clear from work both in invertebrates (Whitaker et al., 2013) and mice (Alcendor et al., 2007) that the effects of sirtuin OE are exquisitely sensitive to expression levels, as might be predicted for pleiotropic regulators that modify a host of downstream proteins. Tissue-specific effects of sirtuins are also highly relevant in this regard. For example, in the BRASTO mouse lines, high levels of SIRT1 OE in other hypothalamic nuclei were associated with loss of the beneficial effects of DMH- and LH-specific OE. This may imply that SIRT1 exerts opposing, region-specific pro- and antilongevity effects in the hypothalamus (Satoh et al., 2013).

A great deal of interest exists in the potential for manipulating sirtuin activities as a treatment for various pathologic conditions, including cancer, neurodegeneration, metabolic dysfunction, and others. Intensive efforts by many laboratories over the past ~15 years have generated many isoform-specific small-molecule modulators of sirtuin activities (Dai et al., 2018). However, the number of selective activators identified so far is modest compared to the number of inhibitors. Moreover, nearly all of the activators are directed against SIRT1, and despite their reported beneficial effects, suffer from poor bioavailability, variable pharmacokinetics, and/or limited target specificity. In addition, the substrate sequence has been found to have a dramatic influence on RSV and other STAC-mediated modulation of SIRT1 activity (Hubbard et al., 2013; Lakshminarasimhan, Rauh, Schutkowski, & Steegborn, 2013). Compared to activators, a large variety of potent sirtuin inhibitors have been identified. Many of these inhibitors show promising results in cell culture, and in some cases in mouse models, of diverse human diseases. Assuming the remarkable effects of these compounds in rodents hold true in humans, they could provide a new way to treat many age-associated pathologies. However, to date, a very limited number of these compounds have been evaluated in clinical trials (Dai et al., 2018; Sussmuth et al., 2015; Westerberg et al., 2015). With the advent of mechanistic and structural insights gained in recent years, the development of new compounds with improved efficacy, selectivity, and pharmacological features is underway in many laboratories.

Although the known repertoire of sirtuin functions continues to expand, little research to date has focused on functional interactions between the seven sirtuins, in the context of redundancy, antagonism, or synergistic effects. Several key cellular proteins (e.g., c-MYC, p53, HIF-1α, CPS1, GDH, and many others) are targets of multiple sirtuins. In some cases, sirtuins act in opposition to one another (e.g., SIRT3 and SIRT5 on SDH, SIRT3 and SIRT4 on GDH), while in other incidences they work synergistically. For instance, a study by Lin et al. provided strong evidence that SIRT3 and SIRT5 have distinct, nonredundant functions in regulating retinal homeostasis and play synergistic roles in photoreceptor survival (Lin, Kubota et al., 2016). In a follow-up study, the authors further showed that either *Sirt3* KO or *Sirt5* KO mice with monoallelic *Nampt* deletion in rod photoreceptors exhibited no acceleration of retinal dysfunction in the context of streptozotocin-induced hyperglycemia (Lin, Lin, Chen, Chen, & Apte, 2019). However, simultaneous deficiencies of both SIRT3 and SIRT5 results in significant inner retinal dysfunction after induction of hyperglycemia compared to hyperglycemic littermate controls. Interestingly, NAD^+ deficiency causes severe SIRT3 dysfunction, while having little impact on SIRT5 enzymatic activities in photoreceptor cells (Lin, Kubota et al., 2016). How this functional regulation is achieved in vivo when all sirtuins are NAD^+-responsive remains somewhat mysterious. Other means of sirtuin regulation may be relevant in this regard, occurring through protein–protein interactions, regulation of sirtuin expression levels, PTMs, levels of metabolites such as free fatty acids, and other mechanisms. Future studies will explore these issues, for example, via analysis of compound sirtuin mutant mouse models.

Furthermore, sirtuins do not act in isolation to promote vertebrate healthspan and lifespan. mTOR signaling and IIS regulate lifespan in evolutionarily distant organisms (Lopez-Otin et al., 2013). These pathways overlap with sirtuin function in multiple contexts. For example, SIRT1 regulates insulin secretion and insulin signaling at numerous levels, and several reports point to roles for SIRT6 in suppressing IIS and mTORC1 signaling (Hong et al., 2014; Kanfi et al., 2012; Sundaresan et al., 2012; Xiao et al., 2010). To further complicate matters, some sirtuins have redundant functions that may need to be targeted simultaneously in order to elicit biological effects. Both SIRT1 and SIRT2 regulate mTORC1 signaling, for instance, at the level of S6 kinase (Hong et al., 2014).

Over the past 20 years, a large body of research has illuminated complex relationships between mammalian sirtuins, healthspan, and longevity. It is clear that these proteins can exert beneficial effects in the context of important diseases of aging, and may therefore represent therapeutic targets in this context. However, a mechanistic understanding of these effects is still incomplete; no doubt this work will provide fruitful avenues for future sirtuin research.

Acknowledgments

We thank members of the Lombard lab for helpful discussions. Work in our laboratory is supported by R01GM101171 and DoD awards CA170628 and NF170044 (to D.L.), R01HL114858 (N. Lukacs/D.L.), and by the Glenn Foundation for Medical Research. S.K. is supported in part through an award from the Pablove Foundation. W.G. was supported by NIH Training Grants T32-AG000114 and T32-HL007853. Some graphics in the figures were obtained and modified from Servier Medical Art.

References

Abdesselem, H., Madani, A., Hani, A., Al-Noubi, M., Goswami, N., Ben Hamidane, H., ... Halabi, N. (2016). SIRT1 limits adipocyte hyperplasia through c-Myc inhibition. *The Journal of Biological Chemistry*, *291*, 2119–2135.

Adler, A. S., Sinha, S., Kawahara, T. L., Zhang, J. Y., Segal, E., & Chang, H. Y. (2007). Motif module map reveals enforcement of aging by continual NF-kappaB activity. *Genes & Development*, *21*, 3244–3257.

Aftanas, L. I., Anisimenko, M. S., Berdyugina, D. A., Garanin, A. Y., Maximov, V. N., Voevoda, M. I., ... Danilenko, K. V. (2018). SIRT1 allele frequencies in depressed patients of european descent in Russia. *Frontiers in Genetics*, *9*, 686.

Ahuja, N., Schwer, B., Carobbio, S., Waltregny, D., North, B. J., Castronovo, V., ... Verdin, E. (2007). Regulation of insulin secretion by SIRT4, a mitochondrial ADP-ribosyltransferase. *The Journal of Biological Chemistry*, *282*, 33583–33592.

Airhart, S. E., Shireman, L. M., Risler, L. J., Anderson, G. D., Nagana Gowda, G. A., Raftery, D., ... O'Brien, K. D. (2017). An open-label, non-randomized study of the pharmacokinetics of the nutritional supplement nicotinamide riboside (NR) and its effects on blood NAD+ levels in healthy volunteers. *PLoS One*, *12*, e0186459.

Akamata, K., Wei, J., Bhattacharyya, M., Cheresh, P., Bonner, M. Y., Arbiser, J. L., ... Varga, J. (2016). SIRT3 is attenuated in systemic sclerosis skin and lungs, and its pharmacologic activation mitigates organ fibrosis. *Oncotarget*, *7*, 69321–69336.

Albani, D., Ateri, E., Mazzuco, S., Ghilardi, A., Rodilossi, S., Biella, G., ... Di Giorgi, E. (2014). Modulation of human longevity by SIRT3 single nucleotide polymorphisms in the prospective study "Treviso Longeva (TRELONG)". *Age*, *36*, 469–478.

Albani, D., Pupillo, E., Bianchi, E., Chierchia, A., Martines, R., Forloni, G., & Beghi, E. (2017). The role of single-nucleotide variants of the energy metabolism-linked genes SIRT3, PPARGC1A and APOE in amyotrophic lateral sclerosis risk. *Genes & Genetic Systems*, *91*, 301–309.

Alcendor, R. R., Gao, S., Zhai, P., Zablocki, D., Holle, E., Yu, X., ... Sadoshima, J. (2007). Sirt1 regulates aging and resistance to oxidative stress in the heart. *Circulation Research*, *100*, 1512–1521.

Alhazzazi, T. Y., Kamarajan, P., Joo, N., Huang, J. Y., Verdin, E., D'Silva, N. J., & Kapila, Y. L. (2011). Sirtuin-3 (SIRT3), a novel potential therapeutic target for oral cancer. *Cancer*, *117*, 1670–1678.

Alhazzazi, T. Y., Kamarajan, P., Xu, Y., Ai, T., Chen, L., Verdin, E., & Kapila, Y. L. (2016). A novel Sirtuin-3 inhibitor, LC-0296, inhibits cell survival and proliferation, and promotes apoptosis of head and neck cancer cells. *Anticancer Research*, *36*, 49–60.

Aljada, A., Saleh, A. M., Alkathiri, M., Shamsa, H. B., Al-Bawab, A., & Nasr, A. (2015). Altered Sirtuin 7 expression is associated with early stage breast cancer. *Breast Cancer*, *9*, 3–8.

Anderson, K. A., Huynh, F. K., Fisher-Wellman, K., Stuart, J. D., Peterson, B. S., Douros, J. D., ... Green, M. F. (2017). SIRT4 is a lysine deacylase that controls leucine metabolism and insulin secretion. *Cell Metabolism*, *25*, 838–855, e815.

Andrade, J. M., Paraiso, A. F., de Oliveira, M. V., Martins, A. M., Neto, J. F., Guimaraes, A. L., ... Santos, S. H. (2014). Resveratrol attenuates hepatic steatosis in high-fat fed mice by decreasing lipogenesis and inflammation. *Nutrition*, *30*, 915–919.

Araki, S., Izumiya, Y., Rokutanda, T., Ianni, A., Hanatani, S., Kimura, Y., ... Yasuda, O. (2015). Sirt7 contributes to myocardial tissue repair by maintaining transforming growth factor-beta signaling pathway. *Circulation*, *132*, 1081–1093.

Arora, A., & Dey, C. S. (2014). SIRT2 negatively regulates insulin resistance in C2C12 skeletal muscle cells. *Biochimica et Biophysica Acta*, *1842*, 1372–1378.

Ashraf, N., Zino, S., Macintyre, A., Kingsmore, D., Payne, A. P., George, W. D., & Shiels, P. G. (2006). Altered sirtuin expression is associated with node-positive breast cancer. *British Journal of Cancer*, *95*, 1056–1061.

Aury-Landas, J., Bougeard, G., Castel, H., Hernandez-Vargas, H., Drouet, A., Latouche, J. B., ... Lasset, C. (2013). Germline copy number variation of genes involved in chromatin remodelling in families suggestive of Li-Fraumeni syndrome with brain tumours. *European Journal of Human Genetics: EJHG*, *21*, 1369–1376.

Avila, P. R., Marques, S. O., Luciano, T. F., Vitto, M. F., Engelmann, J., Souza, D. R., ... De Souza, C. T. (2013). Resveratrol and fish oil reduce catecholamine-induced mortality in obese rats: Role of oxidative stress in the myocardium and aorta. *The British Journal of Nutrition*, *110*, 1580–1590.

Bai, L., Lin, G., Sun, L., Liu, Y., Huang, X., Cao, C., ... Xie, C. (2016). Upregulation of SIRT6 predicts poor prognosis and promotes metastasis of non-small cell lung cancer via the ERK1/2/MMP9 pathway. *Oncotarget*, *7*, 40377–40386.

Bai, P., Canto, C., Brunyanszki, A., Huber, A., Szanto, M., Cen, Y., ... Oudart, H. (2011). PARP-2 regulates SIRT1 expression and whole-body energy expenditure. *Cell Metabolism*, *13*, 450–460.

Bai, P., Canto, C., Oudart, H., Brunyanszki, A., Cen, Y., Thomas, C., ... Houtkooper, R. H. (2011). PARP-1 inhibition increases mitochondrial metabolism through SIRT1 activation. *Cell Metabolism*, *13*, 461–468.

Bajpe, P. K., Prahallad, A., Horlings, H., Nagtegaal, I., Beijersbergen, R., & Bernards, R. (2015). A chromatin modifier genetic screen identifies SIRT2 as a modulator of response to targeted therapies through the regulation of MEK kinase activity. *Oncogene*, *34*, 531–536.

Baker, D. J., Dawlaty, M. M., Wijshake, T., Jeganathan, K. B., Malureanu, L., van Ree, J. H., ... Shapiro, V. (2013). Increased expression of BubR1 protects against aneuploidy and cancer and extends healthy lifespan. *Nature Cell Biology*, *15*, 96–102.

Baker, D. J., Jeganathan, K. B., Cameron, J. D., Thompson, M., Juneja, S., Kopecka, A., ... Roche, P. (2004). BubR1 insufficiency causes early onset of aging-associated phenotypes and infertility in mice. *Nature Genetics*, *36*, 744–749.

Banerjee, K. K., Ayyub, C., Ali, S. Z., Mandot, V., Prasad, N. G., & Kolthur-Seetharam, U. (2012). dSir2 in the adult fat body, but not in muscles, regulates life span in a diet-dependent manner. *Cell Reports*, *2*, 1485–1491.

Bang, I. H., Kwon, O. K., Hao, L., Park, D., Chung, M. J., Oh, B. C., ... Park, B. H. (2019). Deacetylation of XBP1s by sirtuin 6 confers resistance to ER stress-induced hepatic steatosis. *Experimental & Molecular Medicine*, *51*, 107.

Banks, A. S., Kon, N., Knight, C., Matsumoto, M., Gutierrez-Juarez, R., Rossetti, L., ... Accili, D. (2008). SirT1 gain of function increases energy efficiency and prevents diabetes in mice. *Cell Metabolism*, *8*, 333–341.

Bao, X., Wang, Y., Li, X., Li, X. M., Liu, Z., Yang, T., ... Li, X. D. (2014). Identification of 'erasers' for lysine crotonylated histone marks using a chemical proteomics approach. *Elife*, *3*, e02999.

Barbaresko, J., Koch, M., Schulze, M. B., & Nothlings, U. (2013). Dietary pattern analysis and biomarkers of low-grade inflammation: A systematic literature review. *Nutrition Reviews*, *71*, 511–527.

Barber, M. F., Michishita-Kioi, E., Xi, Y., Tasselli, L., Kioi, M., Moqtaderi, Z., ... Chen, K. (2012). SIRT7 links H3K18 deacetylation to maintenance of oncogenic transformation. *Nature*, *487*, 114–118.

Barbosa, M. T., Soares, S. M., Novak, C. M., Sinclair, D., Levine, J. A., Aksoy, P., & Chini, E. N. (2007). The enzyme CD38 (a NAD glycohydrolase, EC 3.2.2.5) is necessary for the development of diet-induced obesity. *The FASEB Journal*, *21*, 3629–3639.

Barger, J. L., Kayo, T., Pugh, T. D., Prolla, T. A., & Weindruch, R. (2008). Short-term consumption of a resveratrol-containing nutraceutical mixture mimics gene expression of long-term caloric restriction in mouse heart. *Experimental Gerontology*, *43*, 859–866.

Barger, J. L., Kayo, T., Vann, J. M., Arias, E. B., Wang, J., Hacker, T. A., ... Leeuwenburgh, C. (2008). A low dose of dietary resveratrol partially mimics caloric restriction and retards aging parameters in mice. *PLoS One*, *3*, e2264.

Basen-Engquist, K., & Chang, M. (2011). Obesity and cancer risk: Recent review and evidence. *Current Oncology Reports*, *13*, 71–76.

Bauer, I., Grozio, A., Lasiglie, D., Basile, G., Sturla, L., Magnone, M., ... Poggi, A. (2012). The NAD + -dependent histone deacetylase SIRT6 promotes cytokine production and migration in pancreatic cancer cells by regulating Ca2 + responses. *The Journal of Biological Chemistry*, *287*, 40924–40937.

Bauer, J. H., Morris, S. N., Chang, C., Flatt, T., Wood, J. G., & Helfand, S. L. (2009). dSir2 and Dmp53 interact to mediate aspects of CR-dependent lifespan extension in *D. melanogaster*. *Aging*, *1*, 38–48.

Baur, J. A. (2010a). Biochemical effects of SIRT1 activators. *Biochimica et Biophysica Acta*, *1804*, 1626–1634.

Baur, J. A. (2010b). Resveratrol, sirtuins, and the promise of a DR mimetic. *Mechanisms of Ageing and Development*, *131*, 261–269.

Baur, J. A., Pearson, K. J., Price, N. L., Jamieson, H. A., Lerin, C., Kalra, A., ... Lewis, K. (2006). Resveratrol improves health and survival of mice on a high-calorie diet. *Nature*, *444*, 337–342.

Baur, J. A., & Sinclair, D. A. (2006). Therapeutic potential of resveratrol: The in vivo evidence. *Nature Reviews. Drug Discovery*, *5*, 493–506.

Beirowski, B., Gustin, J., Armour, S. M., Yamamoto, H., Viader, A., North, B. J., ... Schmidt, R. E. (2011). Sir-two-homolog 2 (Sirt2) modulates peripheral myelination through polarity protein Par-3/atypical protein kinase C (aPKC) signaling. *Proceedings of the National Academy of Sciences of the United States of America*, *108*, E952–E961.

Bell, E. L., Emerling, B. M., Ricoult, S. J., & Guarente, L. (2011). SirT3 suppresses hypoxia inducible factor 1alpha and tumor growth by inhibiting mitochondrial ROS production. *Oncogene*, *30*, 2986–2996.

Bellizzi, D., Rose, G., Cavalcante, P., Covello, G., Dato, S., De Rango, F., ... Mari, V. (2005). A novel VNTR enhancer within the SIRT3 gene, a human homologue of SIR2, is associated with survival at oldest ages. *Genomics*, *85*, 258–263.

Belman, J. P., Bian, R. R., Habtemichael, E. N., Li, D. T., Jurczak, M. J., Alcazar-Roman, A., ... Bogan, J. S. (2015). Acetylation of TUG protein promotes the accumulation of GLUT4 glucose transporters in an insulin-responsive intracellular compartment. *The Journal of Biological Chemistry*, *290*, 4447–4463.

Benigni, A., Cassis, P., Conti, S., Perico, L., Corna, D., Cerullo, D., ... Lionetti, V. (2019). Sirt3 deficiency shortens life span and impairs cardiac mitochondrial function rescued by opa1 gene transfer. *Antioxidants & Redox Signaling*, *31*, 1255–1271.

Benjamin, E. J., Blaha, M. J., Chiuve, S. E., Cushman, M., Das, S. R., Deo, R., ... Gillespie, C. (2017). Heart disease and stroke statistics-2017 update: A Report From the American Heart Association. *Circulation*, *135*, e146–e603.

Berdichevsky, A., Viswanathan, M., Horvitz, H. R., & Guarente, L. (2006). *C. elegans* SIR-2.1 interacts with 14-3-3 proteins to activate DAF-16 and extend life span. *Cell*, *125*, 1165–1177.

Berman, A. Y., Motechin, R. A., Wiesenfeld, M. Y., & Holz, M. K. (2017). The therapeutic potential of resveratrol: A review of clinical trials. *NPJ Precision Oncology*, *1*, 35.

Betsinger, C. N., & Cristea, I. M. (2019). Mitochondrial function, metabolic regulation, and human disease viewed through the prism of Sirtuin 4 (SIRT4) functions. *Journal of Proteome Research*, *18*, 1929–1938.

Bhalla, S., & Gordon, L. I. (2016). Functional characterization of NAD dependent de-acetylases SIRT1 and SIRT2 in B-cell chronic lymphocytic leukemia (CLL). *Cancer Biology & Therapy*, *17*, 300–309.

Bhardwaj, A., & Das, S. (2016). SIRT6 deacetylates PKM2 to suppress its nuclear localization and oncogenic functions. *Proceedings of the National Academy of Sciences of the United States of America*, *113*, E538–E547.

Biason-Lauber, A., Boni-Schnetzler, M., Hubbard, B. P., Bouzakri, K., Brunner, A., Cavelti-Weder, C., ... Brorsson, C. (2013). Identification of a SIRT1 mutation in a family with type 1 diabetes. *Cell Metabolism*, *17*, 448–455.

Biella, G., Fusco, F., Nardo, E., Bernocchi, O., Colombo, A., Lichtenthaler, S. F., ... Albani, D. (2016). Sirtuin 2 inhibition improves cognitive performance and acts on Amyloid-beta protein precursor processing in two Alzheimer's disease mouse models. *Journal of Alzheimer's Disease: JAD*, *53*, 1193–1207.

Black, J. C., Mosley, A., Kitada, T., Washburn, M., & Carey, M. (2008). The SIRT2 deacetylase regulates autoacetylation of p300. *Molecular Cell*, *32*, 449–455.

Bobrowska, A., Donmez, G., Weiss, A., Guarente, L., & Bates, G. (2012). SIRT2 ablation has no effect on tubulin acetylation in brain, cholesterol biosynthesis or the progression of Huntington's disease phenotypes in vivo. *PLoS One*, *7*, e34805.

Bochaton, T., Crola-Da-Silva, C., Pillot, B., Villedieu, C., Ferreras, L., Alam, M. R., ... Ovize, M. (2015). Inhibition of myocardial reperfusion injury by ischemic postconditioning requires sirtuin 3-mediated deacetylation of cyclophilin D. *Journal of Molecular and Cellular Cardiology*, *84*, 61–69.

Boily, G., He, X. H., Pearce, B., Jardine, K., & McBurney, M. W. (2009). SirT1-null mice develop tumors at normal rates but are poorly protected by resveratrol. *Oncogene*, *28*, 2882–2893.

Boily, G., Seifert, E. L., Bevilacqua, L., He, X. H., Sabourin, G., Estey, C., ... Jardine, K. (2008). SirT1 regulates energy metabolism and response to caloric restriction in mice. *PLoS One*, *3*, e1759.

Bolanos-Garcia, V. M., & Blundell, T. L. (2011). BUB1 and BUBR1: Multifaceted kinases of the cell cycle. *Trends in Biochemical Sciences*, *36*, 141–150.

Boniakowski, A. M., den Dekker, A. D., Davis, F. M., Joshi, A., Kimball, A. S., Schaller, M., ... Skinner, M. E. (2019). SIRT3 regulates macrophage-mediated inflammation in diabetic wound repair. *The Journal of Investigative Dermatology*, *139*, 2528–2537.e2.

Bordone, L., Cohen, D., Robinson, A., Motta, M. C., van Veen, E., Czopik, A., ... Luo, J. (2007). SIRT1 transgenic mice show phenotypes resembling calorie restriction. *Aging Cell*, *6*, 759–767.

Bordone, L., & Guarente, L. (2005). Calorie restriction, SIRT1 and metabolism: Understanding longevity. *Nature Reviews. Molecular Cell Biology*, *6*, 298–305.

Bouras, T., Fu, M., Sauve, A. A., Wang, F., Quong, A. A., Perkins, N. D., ... Pestell, R. G. (2005). SIRT1 deacetylation and repression of p300 involves lysine residues 1020/1024 within the cell cycle regulatory domain 1. *The Journal of Biological Chemistry*, *280*, 10264–10276.

Boutant, M., & Cantó, C. (2016). SIRT1 in metabolic health and disease. In R. H. Houtkooper (Ed.), *Sirtuins* (pp. 71–104). Dordrecht: Springer Netherlands.

Boutant, M., Joffraud, M., Kulkarni, S. S., Garcia-Casarrubios, E., Garcia-Roves, P. M., Ratajczak, J., ... Canto, C. (2015). SIRT1 enhances glucose tolerance by potentiating brown adipose tissue function. *Molecular Metabolism*, *4*, 118–131.

Boylston, J. A., Sun, J., Chen, Y., Gucek, M., Sack, M. N., & Murphy, E. (2015). Characterization of the cardiac succinylome and its role in ischemia-reperfusion injury. *Journal of Molecular and Cellular Cardiology*, *88*, 73–81.

Braidy, N., Guillemin, G. J., Mansour, H., Chan-Ling, T., Poljak, A., & Grant, R. (2011). Age related changes in NAD + metabolism oxidative stress and Sirt1 activity in wistar rats. *PLoS One*, *6*, e19194.

Braidy, N., Poljak, A., Grant, R., Jayasena, T., Mansour, H., Chan-Ling, T., ... Sachdev, P. (2014). Mapping NAD(+) metabolism in the brain of ageing Wistar rats: Potential targets for influencing brain senescence. *Biogerontology*, *15*, 177–198.

Brandes, R. P., Fleming, I., & Busse, R. (2005). Endothelial aging. *Cardiovascular Research*, *66*, 286–294.

Brandl, L., Zhang, Y., Kirstein, N., Sendelhofert, A., Boos, S. L., Jung, P., ... Menssen, A. (2019). Targeting c-MYC through interference with NAMPT and SIRT1 and their association to oncogenic drivers in murine serrated intestinal tumorigenesis. *Neoplasia*, *21*, 974–988.

Bringman-Rodenbarger, L. R., Guo, A. H., Lyssiotis, C. A., & Lombard, D. B. (2018). Emerging roles for SIRT5 in metabolism and cancer. *Antioxidants & Redox Signaling*, *28*, 677–690.

Brown, K., Xie, S., Qiu, X., Mohrin, M., Shin, J., Liu, Y., ... Chen, D. (2013). SIRT3 reverses aging-associated degeneration. *Cell Reports*, *3*, 319–327.

Brown, K. D., Maqsood, S., Huang, J. Y., Pan, Y., Harkcom, W., Li, W., ... Jaffrey, S. R. (2014). Activation of SIRT3 by the NAD(+) precursor nicotinamide riboside protects from noise-induced hearing loss. *Cell Metabolism*, *20*, 1059–1068.

Buler, M., Aatsinki, S. M., Izzi, V., Uusimaa, J., & Hakkola, J. (2014). SIRT5 is under the control of PGC-1alpha and AMPK and is involved in regulation of mitochondrial energy metabolism. *The FASEB Journal*, *28*, 3225–3237.

Burg, N., Bittner, S., & Ellwardt, E. (2018). Role of the epigenetic factor Sirt7 in neuroinflammation and neurogenesis. *Neuroscience Research*, *131*, 1–9.

Burnett, C., Valentini, S., Cabreiro, F., Goss, M., Somogyvari, M., Piper, M. D., ... McElwee, J. J. (2011). Absence of effects of Sir2 overexpression on lifespan in *C. elegans* and *Drosophila*. *Nature*, *477*, 482–485.

Byles, V., Chmilewski, L. K., Wang, J., Zhu, L., Forman, L. W., Faller, D. V., & Dai, Y. (2010). Aberrant cytoplasm localization and protein stability of SIRT1 is regulated by PI3K/IGF-1R signaling in human cancer cells. *International Journal of Biological Sciences*, *6*, 599–612.

Byles, V., Zhu, L., Lovaas, J. D., Chmilewski, L. K., Wang, J., Faller, D. V., & Dai, Y. (2012). SIRT1 induces EMT by cooperating with EMT transcription factors and enhances prostate cancer cell migration and metastasis. *Oncogene*, *31*, 4619–4629.

Cai, J., Zuo, Y., Wang, T., Cao, Y., Cai, R., Chen, F. L., ... Mu, J. (2016). A crucial role of SUMOylation in modulating Sirt6 deacetylation of H3 at lysine 56 and its tumor suppressive activity. *Oncogene*, *35*, 4949–4956.

Callier, P., Faivre, L., Cusin, V., Marle, N., Thauvin-Robinet, C., Sandre, D., ... Faber, V. (2005). Microcephaly is not mandatory for the diagnosis of mosaic variegated aneuploidy syndrome. *American Journal of Medical Genetics Part A*, *137*, 204–207.

Canto, C., Houtkooper, R. H., Pirinen, E., Youn, D. Y., Oosterveer, M. H., Cen, Y., ... Cettour-Rose, P. (2012). The NAD(+) precursor nicotinamide riboside enhances oxidative metabolism and protects against high-fat diet-induced obesity. *Cell Metabolism*, *15*, 838–847.

Canto, C., Jiang, L.Q., Deshmukh, A.S., Mataki, C., Coste, A., Lagouge, M., Zierath, J.R., Auwerx, J., (2010). Interdependence of AMPK and SIRT1 for metabolic adaptation to fasting and exercise in skeletal muscle. *Cell Metabolism 11*, 213–219

Canto, C., Sauve, A. A., & Bai, P. (2013). Crosstalk between poly (ADP-ribose) polymerase and sirtuin enzymes. *Molecular Aspects of Medicine*, *34*, 1168–1201.

Carafa, V., Rotili, D., Forgione, M., Cuomo, F., Serretiello, E., Hailu, G. S., ... Altucci, L. (2016). Sirtuin functions and modulation: From chemistry to the clinic. *Clinical Epigenetics*, *8*, 61.

Carrico, C., Meyer, J. G., He, W., Gibson, B. W., & Verdin, E. (2018). The mitochondrial acylome emerges: Proteomics, regulation by sirtuins, and metabolic and disease implications. *Cell Metabolism*, *27*, 497–512.

Caton, P. W., Richardson, S. J., Kieswich, J., Bugliani, M., Holland, M. L., Marchetti, P., ... Sugden, M. C. (2013). Sirtuin 3 regulates mouse pancreatic beta cell function and is suppressed in pancreatic islets isolated from human type 2 diabetic patients. *Diabetologia*, *56*, 1068–1077.

Cea, M., Cagnetta, A., Adamia, S., Acharya, C., Tai, Y. T., Fulciniti, M., ... Bhasin, M. K. (2016). Evidence for a role of the histone deacetylase SIRT6 in DNA damage response of multiple myeloma cells. *Blood*, *127*, 1138–1150.

Cerutti, R., Pirinen, E., Lamperti, C., Marchet, S., Sauve, A. A., Li, W., ... Auwerx, J. (2014). NAD(+)-dependent activation of Sirt1 corrects the phenotype in a mouse model of mitochondrial disease. *Cell Metabolism*, *19*, 1042–1049.

Chachay, V. S., Macdonald, G. A., Martin, J. H., Whitehead, J. P., O'Moore-Sullivan, T. M., Lee, P., ... Ferguson, M. (2014). Resveratrol does not benefit patients with nonalcoholic fatty liver disease. *Clinical Gastroenterology and Hepatology: The Official Clinical Practice Journal of the American Gastroenterological Association*, *12*, 2092–103.e1–6.

Chang, H. C., & Guarente, L. (2013). SIRT1 mediates central circadian control in the SCN by a mechanism that decays with aging. *Cell*, *153*, 1448–1460.

Chang, H. C., & Guarente, L. (2014). SIRT1 and other sirtuins in metabolism. *Trends in Endocrinology and Metabolism: TEM*, *25*, 138–145.

Chang, L., Xi, L., Liu, Y., Liu, R., Wu, Z., & Jian, Z. (2018). SIRT5 promotes cell proliferation and invasion in hepatocellular carcinoma by targeting E2F1. *Molecular Medicine Reports*, *17*, 342–349.

Chen, D., Bruno, J., Easlon, E., Lin, S. J., Cheng, H. L., Alt, F. W., & Guarente, L. (2008). Tissue-specific regulation of SIRT1 by calorie restriction. *Genes & Development*, *22*, 1753–1757.

Chen, D., Steele, A. D., Lindquist, S., & Guarente, L. (2005). Increase in activity during calorie restriction requires Sirt1. *Science*, *310*, 1641.

Chen, G., Luo, Y., Warncke, K., Sun, Y., Yu, D. S., Fu, H., ... Duong, D. M. (2019). Acetylation regulates ribonucleotide reductase activity and cancer cell growth. *Nature Communications*, *10*, 3213.

Chen, H., Wu, D., Ding, X., & Ying, W. (2015). SIRT2 is required for lipopolysaccharide-induced activation of BV2 microglia. *Neuroreport*, *26*, 88–93.

Chen, I. C., Chiang, W. F., Huang, H. H., Chen, P. F., Shen, Y. Y., & Chiang, H. C. (2014). Role of SIRT1 in regulation of epithelial-to-mesenchymal transition in oral squamous cell carcinoma metastasis. *Molecular Cancer*, *13*, 254.

Chen, I. C., Chiang, W. F., Liu, S. Y., Chen, P. F., & Chiang, H. C. (2013). Role of SIRT3 in the regulation of redox balance during oral carcinogenesis. *Molecular Cancer*, *12*, 68.

Chen, J., Chan, A. W., To, K. F., Chen, W., Zhang, Z., Ren, J., ... Cheng, S. H. (2013). SIRT2 overexpression in hepatocellular carcinoma mediates epithelial to mesenchymal transition by protein kinase B/glycogen synthase kinase-3beta/beta-catenin signaling. *Hepatology*, *57*, 2287–2298.

Chen, K. L., Li, L., Li, C. M., Wang, Y. R., Yang, F. X., Kuang, M. Q., & Wang, G. L. (2019). SIRT7 regulates lipopolysaccharide-induced inflammatory injury by suppressing the NF-kappaB signaling pathway. *Oxidative Medicine and Cellular Longevity*, *2019*, 3187972.

Chen, L., Wang, H., Gao, F., Zhang, J., Zhang, Y., Ma, R., ... Yan, B. (2018). Functional genetic variants in the SIRT5 gene promoter in acute myocardial infarction. *Gene*, *675*, 233–239.

Chen, M. L., Zhu, X. H., Ran, L., Lang, H. D., Yi, L., & Mi, M. T. (2017). Trimethylamine-N-oxide induces vascular inflammation by activating the NLRP3 inflammasome through the SIRT3-SOD2-mtROS signaling pathway. *Journal of the American Heart Association*, *6*, e006347.

Chen, Q., Hao, W., Xiao, C., Wang, R., Xu, X., Lu, H., ... Deng, C. X. (2017). SIRT6 is essential for adipocyte differentiation by regulating mitotic clonal expansion. *Cell Reports*, *18*, 3155–3166.

Chen, S., Yang, X., Yu, M., Wang, Z., Liu, B., Liu, M., ... Zou, J. (2019). SIRT3 regulates cancer cell proliferation through deacetylation of PYCR1 in proline metabolism. *Neoplasia (New York, N.Y.)*, *21*, 665–675.

Chen, X., Ding, X., Wu, Q., Qi, J., Zhu, M., & Miao, C. (2019). Monomethyltransferase SET8 facilitates hepatocellular carcinoma growth by enhancing aerobic glycolysis. *Cell Death Dis*, *10*, 312.

Chen, X., Mai, H., Chen, X., Cai, Y., Cheng, Q., Chen, X., ... Ou, M. (2019). Rs2015 polymorphism in miRNA target site of Sirtuin2 gene is associated with the risk of parkinson's disease in chinese han population. *BioMed Research International*, *2019*, 1498034.

Chen, X., Wales, P., Quinti, L., Zuo, F., Moniot, S., Herisson, F., ... Ayata, C. (2015). The sirtuin-2 inhibitor AK7 is neuroprotective in models of Parkinson's disease but not amyotrophic lateral sclerosis and cerebral ischemia. *PLoS One*, *10*, e0116919.

Chen, X. F., Tian, M. X., Sun, R. Q., Zhang, M. L., Zhou, L. S., Jin, L., ... Chen, Y. J. (2018). SIRT5 inhibits peroxisomal ACOX1 to prevent oxidative damage and is downregulated in liver cancer. *EMBO Reports*, *2019*.

Chen, Z., Lin, J., Feng, S., Chen, X., Huang, H., Wang, C., ... Zheng, L. (2019). SIRT4 inhibits the proliferation, migration, and invasion abilities of thyroid cancer cells by inhibiting glutamine metabolism. *OncoTargets and Therapy*, *12*, 2397–2408.

Cheng, A., Yang, Y., Zhou, Y., Maharana, C., Lu, D., Peng, W., ... Misiak, M. (2016). Mitochondrial SIRT3 mediates adaptive responses of neurons to exercise and metabolic and excitatory challenges. *Cell Metabolism*, *23*, 128–142.

Cheng, H. L., Mostoslavsky, R., Saito, S., Manis, J. P., Gu, Y., Patel, P., ... Chua, K. F. (2003). Developmental defects and p53 hyperacetylation in Sir2 homolog (SIRT1)-deficient mice. *Proceedings of the National Academy of Sciences of the United States of America*, *100*, 10794–10799.

Cheng, M. Y., Cheng, Y. W., Yan, J., Hu, X. Q., Zhang, H., Wang, Z. R., ... Cheng, W. (2016). SIRT6 suppresses mitochondrial defects and cell death via the NF-kappaB pathway in myocardial hypoxia/reoxygenation induced injury. *American Journal of Translational Research*, *8*, 5005–5015.

Chiang, W. C., Tishkoff, D. X., Yang, B., Wilson-Grady, J., Yu, X., Mazer, T., ... Hsu, A. L. (2012). C. elegans SIRT6/7 homolog SIR-2.4 promotes DAF-16 relocalization and function during stress. *PLoS Genetics*, *8*, e1002948.

Chien, S. (2008). Effects of disturbed flow on endothelial cells. *Annals of Biomedical Engineering*, *36*, 554–562.

Chopra, V., Quinti, L., Kim, J., Vollor, L., Narayanan, K. L., Edgerly, C., ... Silverman, R. B. (2012). The sirtuin 2 inhibitor AK-7 is neuroprotective in Huntington's disease mouse models. *Cell Reports*, *2*, 1492–1497.

Cioffi, M., Vallespinos-Serrano, M., Trabulo, S. M., Fernandez-Marcos, P. J., Firment, A. N., Vazquez, B. N., ... Cronin, U. P. (2015). MiR-93 controls adiposity via inhibition of Sirt7 and Tbx3. *Cell Reports*, *12*, 1594–1605.

Clark, S. J., Falchi, M., Olsson, B., Jacobson, P., Cauchi, S., Balkau, B., ... Jernas, M. (2012). Association of sirtuin 1 (SIRT1) gene SNPs and transcript expression levels with severe obesity. *Obesity*, *20*, 178–185.

Cohen, D. E., Supinski, A. M., Bonkowski, M. S., Donmez, G., & Guarente, L. P. (2009). Neuronal SIRT1 regulates endocrine and behavioral responses to calorie restriction. *Genes & Development*, *23*, 2812–2817.

Cohen, H. Y., Miller, C., Bitterman, K. J., Wall, N. R., Hekking, B., Kessler, B., ... Sinclair, D. A. (2004). Calorie restriction promotes mammalian cell survival by inducing the SIRT1 deacetylase. *Science*, *305*, 390–392.

Cohen, T. J., Guo, J. L., Hurtado, D. E., Kwong, L. K., Mills, I. P., Trojanowski, J. Q., & Lee, V. M. (2011). The acetylation of tau inhibits its function and promotes pathological tau aggregation. *Nature Communications*, *2*, 252.

Colombo, S. L., Palacios-Callender, M., Frakich, N., Carcamo, S., Kovacs, I., Tudzarova, S., & Moncada, S. (2011). Molecular basis for the differential use of glucose and glutamine in cell proliferation as revealed by synchronized HeLa cells. *Proceedings of the National Academy of Sciences of the United States of America*, *108*, 21069–21074.

Condorelli, G., Drusco, A., Stassi, G., Bellacosa, A., Roncarati, R., Iaccarino, G., ... Chung, C. (2002). Akt induces enhanced myocardial contractility and cell size in vivo in transgenic mice. *Proceedings of the National Academy of Sciences of the United States of America*, *99*, 12333–12338.

Consortium, C. (2015). Sparse whole-genome sequencing identifies two loci for major depressive disorder. *Nature*, *523*, 588–591.

Corpas, R., Revilla, S., Ursulet, S., Castro-Freire, M., Kaliman, P., Petegnief, V., ... Sanfeliu, C. (2017). SIRT1 overexpression in mouse hippocampus induces cognitive enhancement through proteostatic and neurotrophic mechanisms. *Molecular Neurobiology*, *54*, 5604–5619.

Costa-Machado, L. F., & Fernandez-Marcos, P. J. (2019). The sirtuin family in cancer. *Cell Cycle*, *18*, 2164–2196.

Costa-Machado, L. F., Martin-Hernandez, R., Sanchez-Luengo, M. A., Hess, K., Vales-Villamarin, C., Barradas, M., ... de Cabo, R. (2018). Sirt1 protects from K-Ras-driven lung carcinogenesis. *EMBO Reports*, *19*.

Crewe, C., An, Y. A., & Scherer, P. E. (2017). The ominous triad of adipose tissue dysfunction: Inflammation, fibrosis, and impaired angiogenesis. *The Journal of Clinical Investigation*, *127*, 74–82.

Crocco, P., Montesanto, A., Passarino, G., & Rose, G. (2016). Polymorphisms falling within putative miRNA target sites in the 3'UTR region of SIRT2 and DRD2 genes are correlated with human longevity. *The Journals of Gerontology. Series A, Biological Sciences and Medical Sciences*, *71*, 586–592.

Cruz, M., Valladares-Salgado, A., Garcia-Mena, J., Ross, K., Edwards, M., Angeles-Martinez, J., ... Wacher-Rodarte, N. (2010). Candidate gene association study conditioning on individual ancestry in patients with type 2 diabetes and metabolic syndrome from Mexico City. *Diabetes/metabolism Research and Reviews*, *26*, 261–270.

Csibi, A., Fendt, S. M., Li, C., Poulogiannis, G., Choo, A. Y., Chapski, D. J., ... Morrison, T. (2013). The mTORC1 pathway stimulates glutamine metabolism and cell proliferation by repressing SIRT4. *Cell*, *153*, 840–854.

Csiszar, A., Labinskyy, N., Podlutsky, A., Kaminski, P. M., Wolin, M. S., Zhang, C., ... de Cabo, R. (2008). Vasoprotective effects of resveratrol and SIRT1: Attenuation of cigarette smoke-induced oxidative stress and proinflammatory phenotypic alterations. *American Journal of Physiology. Heart and Circulatory Physiology*, *294*, H2721–H2735.

Dai, H., Sinclair, D. A., Ellis, J. L., & Steegborn, C. (2018). Sirtuin activators and inhibitors: Promises, achievements, and challenges. *Pharmacology & Therapeutics*, *188*, 140–154.

Damodaran, S., Damaschke, N., Gawdzik, J., Yang, B., Shi, C., Allen, G. O., ... Jarrard, D. (2017). Dysregulation of Sirtuin 2 (SIRT2) and histone H3K18 acetylation pathways associates with adverse prostate cancer outcomes. *BMC Cancer*, *17*, 874.

Damonte, P., Sociali, G., Parenti, M. D., Soncini, D., Bauer, I., Boero, S., ... Becherini, P. (2017). SIRT6 inhibitors with salicylate-like structure show immunosuppressive and chemosensitizing effects. *Bioorganic & Medicinal Chemistry*, *25*, 5849–5858.

Dan, L., Klimenkova, O., Klimiankou, M., Klusman, J. H., van den Heuvel-Eibrink, M. M., Reinhardt, D., ... Skokowa, J. (2012). The role of sirtuin 2 activation by nicotinamide phosphoribosyltransferase in the aberrant proliferation and survival of myeloid leukemia cells. *Haematologica*, *97*, 551–559.

Dang, W., Steffen, K. K., Perry, R., Dorsey, J. A., Johnson, F. B., Shilatifard, A., ... Berger, S. L. (2009). Histone H4 lysine 16 acetylation regulates cellular lifespan. *Nature*, *459*, 802–807.

de Cabo, R., Liu, L., Ali, A., Price, N., Zhang, J., Wang, M., ... Irusta, P. M. (2015). Serum from calorie-restricted animals delays senescence and extends the lifespan of normal human fibroblasts in vitro. *Aging*, *7*, 152–166.

De Cecco, M., Criscione, S. W., Peterson, A. L., Neretti, N., Sedivy, J. M., & Kreiling, J. A. (2013). Transposable elements become active and mobile in the genomes of aging mammalian somatic tissues. *Aging, 5*, 867–883.

Defossez, P. A., Prusty, R., Kaeberlein, M., Lin, S. J., Ferrigno, P., Silver, P. A., ... Guarente, L. (1999). Elimination of replication block protein Fob1 extends the life span of yeast mother cells. *Molecular Cell, 3*, 447–455.

Della-Morte, D., Dong, C., Bartels, S., Cabral, D., Blanton, S. H., Sacco, R. L., & Rundek, T. (2012). Association of the sirtuin and mitochondrial uncoupling protein genes with carotid intima-media thickness. *Translational Research: the Journal of Laboratory and Clinical Medicine, 160*, 389–390.

Deng, A., Ning, Q., Zhou, L., & Liang, Y. (2016). SIRT2 is an unfavorable prognostic biomarker in patients with acute myeloid leukemia. *Scientific Reports, 6*, 27694.

de Oliveira, R. M., Vicente Miranda, H., Francelle, L., Pinho, R., Szego, E. M., Martinho, R., ... Guerreiro, P. (2017). The mechanism of sirtuin 2-mediated exacerbation of alpha-synuclein toxicity in models of Parkinson disease. *PLoS Biology, 15*, e2000374.

Desouki, M. M., Doubinskaia, I., Gius, D., & Abdulkadir, S. A. (2014). Decreased mitochondrial SIRT3 expression is a potential molecular biomarker associated with poor outcome in breast cancer. *Human Pathology, 45*, 1071–1077.

DiGiovanna, J. J., & Kraemer, K. H. (2012). Shining a light on xeroderma pigmentosum. *The Journal of Investigative Dermatology, 132*, 785–796.

Dikalova, A. E., Itani, H. A., Nazarewicz, R. R., McMaster, W. G., Flynn, C. R., Uzhachenko, R., ... Dikalov, S. I. (2017). Sirt3 impairment and SOD2 hyperacetylation in vascular oxidative stress and hypertension. *Circulation Research, 121*, 564–574.

Doherty, A., Kernogitski, Y., Kulminski, A. M., & Pedro de Magalhaes, J. (2017). Identification of polymorphisms in cancer patients that differentially affect survival with age. *Aging, 9*, 2117–2136.

Dominy, J. E., Jr., Lee, Y., Jedrychowski, M. P., Chim, H., Jurczak, M. J., Camporez, J. P., ... Mostoslavsky, R. (2012). The deacetylase Sirt6 activates the acetyltransferase GCN5 and suppresses hepatic gluconeogenesis. *Molecular Cell, 48*, 900–913.

Dong, C., Della-Morte, D., Cabral, D., Wang, L., Blanton, S. H., Seemant, C., ... Rundek, T. (2015). Sirtuin/uncoupling protein gene variants and carotid plaque area and morphology. *International Journal of Stroke, 10*, 1247–1252.

Dong, C., Della-Morte, D., Wang, L., Cabral, D., Beecham, A., McClendon, M. S., ... Rundek, T. (2011). Association of the sirtuin and mitochondrial uncoupling protein genes with carotid plaque. *PLoS One, 6*, e27157.

Donlon, T. A., Morris, B. J., Chen, R., Masaki, K. H., Allsopp, R. C., Willcox, D. C., ... Willcox, B. J. (2017). Analysis of polymorphisms in 58 potential candidate genes for association with human longevity. *The Journals of Gerontology. Series A, Biological Sciences and Medical Sciences.*

Dryden, S. C., Nahhas, F. A., Nowak, J. E., Goustin, A. S., & Tainsky, M. A. (2003). Role for human SIRT2 NAD-dependent deacetylase activity in control of mitotic exit in the cell cycle. *Molecular and Cellular Biology, 23*, 3173–3185.

Du, J., Zhou, Y., Su, X., Yu, J. J., Khan, S., Jiang, H., ... Choi, B. H. (2011). Sirt5 is a NAD-dependent protein lysine demalonylase and desuccinylase. *Science, 334*, 806–809.

Du, X. J. (2004). Gender modulates cardiac phenotype development in genetically modified mice. *Cardiovascular Research, 63*, 510–519.

Du, Y., Hu, H., Qu, S., Wang, J., Hua, C., Zhang, J., ... Liu, P. (2018). SIRT5 deacylates metabolism-related proteins and attenuates hepatic steatosis in ob/ob mice. *EBioMedicine, 36*, 347–357.

Du, Z., Liu, X., Chen, T., Gao, W., Wu, Z., Hu, Z., ... Li, Q. (2018). Targeting a Sirt5-positive subpopulation overcomes multidrug resistance in wild-type kras colorectal carcinomas. *Cell Reports, 22*, 2677–2689.

Ducker, G. S., & Rabinowitz, J. D. (2017). One-carbon metabolism in health and disease. *Cell Metabolism, 25*, 27–42.

Dvir-Ginzberg, M., Gagarina, V., Lee, E. J., Booth, R., Gabay, O., & Hall, D. J. (2011). Tumor necrosis factor alpha-mediated cleavage and inactivation of SirT1 in human osteoarthritic chondrocytes. *Arthritis and Rheumatism, 63*, 2363–2373.

Elhanati, S., Kanfi, Y., Varvak, A., Roichman, A., Carmel-Gross, I., Barth, S., ... Cohen, H. Y. (2013). Multiple regulatory layers of SREBP1/2 by SIRT6. *Cell Reports, 4*, 905–912.

Elowe, S. (2011). Bub1 and BubR1: At the interface between chromosome attachment and the spindle checkpoint. *Molecular and Cellular Biology, 31*, 3085–3093.

Esteves, A. R., Arduino, D. M., Silva, D. F., Viana, S. D., Pereira, F. C., & Cardoso, S. M. (2018). Mitochondrial metabolism regulates microtubule acetylome and autophagy trough Sirtuin-2: Impact for Parkinson's disease. *Molecular Neurobiology, 55*, 1440–1462.

Esteves, A. R., Palma, A. M., Gomes, R., Santos, D., Silva, D. F., & Cardoso, S. M. (2019). Acetylation as a major determinant to microtubule-dependent autophagy: Relevance to Alzheimer's and Parkinson disease pathology. *Biochim Biophys Acta Mol Basis Dis, 1865*, 2008–2023.

Fan, J., Shan, C., Kang, H. B., Elf, S., Xie, J., Tucker, M., ... Chen, H. (2014). Tyr phosphorylation of PDP1 toggles recruitment between ACAT1 and SIRT3 to regulate the pyruvate dehydrogenase complex. *Molecular Cell, 53*, 534–548.

Fang, E. F., Kassahun, H., Croteau, D. L., Scheibye-Knudsen, M., Marosi, K., Lu, H., ... Wilson, M. A. (2016). NAD(+) replenishment improves lifespan and healthspan in ataxia telangiectasia models via mitophagy and DNA repair. *Cell Metabolism, 24*, 566–581.

Fang, E. F., Scheibye-Knudsen, M., Brace, L. E., Kassahun, H., SenGupta, T., Nilsen, H., ... Bohr, V. A. (2014). Defective mitophagy in XPA via PARP-1 hyperactivation and NAD(+)/SIRT1 reduction. *Cell, 157*, 882–896.

Fang, J., Ianni, A., Smolka, C., Vakhrusheva, O., Nolte, H., Kruger, M., ... Vaquero, A. (2017). Sirt7 promotes adipogenesis in the mouse by inhibiting autocatalytic activation of Sirt1. *Proceedings of the National Academy of Sciences of the United States of America, 114*, E8352–E8361.

Farooqi, A. S., Hong, J. Y., Cao, J., Lu, X., Price, I. R., Zhao, Q., ... Lin, H. (2019). Novel lysine-based thioureas as mechanism-based inhibitors of Sirtuin 2 (SIRT2) with anticancer activity in a colorectal cancer murine model. *Journal of Medicinal Chemistry, 62*, 4131–4141.

Feldman, J. L., Baeza, J., & Denu, J. M. (2013). Activation of the protein deacetylase SIRT6 by long-chain fatty acids and widespread deacylation by mammalian sirtuins. *The Journal of Biological Chemistry, 288*, 31350–31356.

Feldman, J. L., Dittenhafer-Reed, K. E., Kudo, N., Thelen, J. N., Ito, A., Yoshida, M., & Denu, J. M. (2015). Kinetic and structural basis for acyl-group selectivity and NAD(+) dependence in sirtuin-catalyzed deacylation. *Biochemistry, 54*, 3037–3050.

Feng, X., Zhang, L., Xu, S., & Shen, A. Z. (2019). ATP-citrate lyase (ACLY) in lipid metabolism and atherosclerosis: An updated review. *Progress in Lipid Research, 77*, 101006.

Fernandez-Marcos, P. J., Jeninga, E. H., Canto, C., Harach, T., de Boer, V. C., Andreux, P., ... Houten, S. M. (2012). Muscle or liver-specific Sirt3 deficiency induces hyperacetylation of mitochondrial proteins without affecting global metabolic homeostasis. *Scientific Reports, 2*, 425.

Ferrer, C. M., Alders, M., Postma, A. V., Park, S., Klein, M. A., Cetinbas, M., ... Christoffels, V. M. (2018). An inactivating mutation in the histone deacetylase SIRT6 causes human perinatal lethality. *Genes & Development*, *32*, 373–388.

Figarska, S. M., Vonk, J. M., & Boezen, H. M. (2013). SIRT1 polymorphism, long-term survival and glucose tolerance in the general population. *PLoS One*, *8*, e58636.

Finkel, T., Deng, C. X., & Mostoslavsky, R. (2009). Recent progress in the biology and physiology of sirtuins. *Nature*, *460*, 587–591.

Finkelstein, J. E., Hauser, E. R., Leonard, C. O., & Brusilow, S. W. (1990). Late-onset ornithine transcarbamylase deficiency in male patients. *The Journal of Pediatrics*, *117*, 897–902.

Finley, L. W., Carracedo, A., Lee, J., Souza, A., Egia, A., Zhang, J., ... Clish, C. B. (2011). SIRT3 opposes reprogramming of cancer cell metabolism through HIF1alpha destabilization. *Cancer Cell*, *19*, 416–428.

Firestein, R., Blander, G., Michan, S., Oberdoerffer, P., Ogino, S., Campbell, J., ... Fuchs, C. (2008). The SIRT1 deacetylase suppresses intestinal tumorigenesis and colon cancer growth. *PLoS One*, *3*, e2020.

Fiskus, W., Coothankandaswamy, V., Chen, J., Ma, H., Ha, K., Saenz, D. T., ... Huang, P. (2016). SIRT2 deacetylates and inhibits the peroxidase activity of peroxiredoxin-1 to sensitize breast cancer cells to oxidant stress-inducing agents. *Cancer Research*, *76*, 5467–5478.

Flachsbart, F., Croucher, P. J., Nikolaus, S., Hampe, J., Cordes, C., Schreiber, S., & Nebel, A. (2006). Sirtuin 1 (SIRT1) sequence variation is not associated with exceptional human longevity. *Experimental Gerontology*, *41*, 98–102.

Fontana, L., Partridge, L., & Longo, V. D. (2010). Extending healthy life span--from yeast to humans. *Science (New York, N.Y.)*, *328*, 321–326.

Ford, E., Voit, R., Liszt, G., Magin, C., Grummt, I., & Guarente, L. (2006). Mammalian Sir2 homolog SIRT7 is an activator of RNA polymerase I transcription. *Genes & Development*, *20*, 1075–1080.

Frye, R. (2002). "SIRT8" expressed in thyroid cancer is actually SIRT7. *British Journal of Cancer*, *87*, 1479.

Frye, R. A. (2000). Phylogenetic classification of prokaryotic and eukaryotic Sir2-like proteins. *Biochemical and Biophysical Research Communications*, *273*, 793–798.

Fu, J., Jin, J., Cichewicz, R. H., Hageman, S. A., Ellis, T. K., Xiang, L., ... Hotaling, K. (2012). trans-(-)-epsilon-Viniferin increases mitochondrial sirtuin 3 (SIRT3), activates AMP-activated protein kinase (AMPK), and protects cells in models of Huntington disease. *The Journal of Biological Chemistry*, *287*, 24460–24472.

Fu, L., Dong, Q., He, J., Wang, X., Xing, J., Wang, E., ... Li, Q. (2017). SIRT4 inhibits malignancy progression of NSCLCs, through mitochondrial dynamics mediated by the ERK-Drp1 pathway. *Oncogene*, *36*, 2724–2736.

Funato, K., Hayashi, T., Echizen, K., Negishi, L., Shimizu, N., Koyama-Nasu, R., ... Todo, T. (2018). SIRT2-mediated inactivation of p73 is required for glioblastoma tumorigenicity. *EMBO Reports*, *19*, e45587.

Gabay, O., Zaal, K. J., Sanchez, C., Dvir-Ginzberg, M., Gagarina, V., Song, Y., ... McBurney, M. W. (2013). Sirt1-deficient mice exhibit an altered cartilage phenotype. *Joint, Bone, Spine: Revue du Rhumatisme*, *80*, 613–620.

Gano, L. B., Donato, A. J., Pasha, H. M., Hearon, C. M., Jr., Sindler, A. L., & Seals, D. R. (2014). The SIRT1 activator SRT1720 reverses vascular endothelial dysfunction, excessive superoxide production, and inflammation with aging in mice. *American Journal of Physiology. Heart and Circulatory Physiology*, *307*, H1754–H1763.

Gao, W., Xu, Y., Chen, T., Du, Z., Liu, X., Hu, Z., ... Li, Q. (2019). Targeting oxidative pentose phosphate pathway prevents recurrence in mutant Kras colorectal carcinomas. *PLoS Biology*, *17*, e3000425.

Garcia-Castillo, H., Vasquez-Velasquez, A. I., Rivera, H., & Barros-Nunez, P. (2008). Clinical and genetic heterogeneity in patients with mosaic variegated aneuploidy: Delineation of clinical subtypes. *American Journal of Medical Genetics Part A*, *146A*, 1687–1695.

Gems, D. (2011). Tragedy and delight: The ethics of decelerated ageing. *Philosophical transactions of the Royal Society of London Series B, Biological sciences*, *366*, 108–112.

Geng, Q., Peng, H., Chen, F., Luo, R., & Li, R. (2015). High expression of Sirt7 served as a predictor of adverse outcome in breast cancer. *International Journal of Clinical and Experimental Pathology*, *8*, 1938–1945.

Geng, Y. Q., Li, T. T., Liu, X. Y., Li, Z. H., & Fu, Y. C. (2011). SIRT1 and SIRT5 activity expression and behavioral responses to calorie restriction. *Journal of Cellular Biochemistry*, *112*, 3755–3761.

Gerhardt, E., Graber, S., Szego, E. M., Moisoi, N., Martins, L. M., Outeiro, T. F., & Kermer, P. (2011). Idebenone and resveratrol extend lifespan and improve motor function of HtrA2 knockout mice. *PLoS One*, *6*, e28855.

Gerhart-Hines, Z., Rodgers, J. T., Bare, O., Lerin, C., Kim, S. H., Mostoslavsky, R., ... Puigserver, P. (2007). Metabolic control of muscle mitochondrial function and fatty acid oxidation through SIRT1/PGC-1alpha. *The EMBO Journal*, *26*, 1913–1923.

Gertz, M., Fischer, F., Nguyen, G. T., Lakshminarasimhan, M., Schutkowski, M., Weyand, M., & Steegborn, C. (2013). Ex-527 inhibits Sirtuins by exploiting their unique NAD + -dependent deacetylation mechanism. *Proceedings of the National Academy of Sciences of the United States of America*, *110*, E2772–E2781.

Ghosh, S., Wong, S. K., Jiang, Z., Liu, B., Wang, Y., Hao, Q., ... Zhou, Z. (2018). Haploinsufficiency of Trp53 dramatically extends the lifespan of Sirt6-deficient mice. *Elife*, *7*, e32127.

Gillum, M. P., Kotas, M. E., Erion, D. M., Kursawe, R., Chatterjee, P., Nead, K. T., ... Yonemitsu, S. (2011). SirT1 regulates adipose tissue inflammation. *Diabetes*, *60*, 3235–3245.

Glorioso, C., Oh, S., Douillard, G. G., & Sibille, E. (2011). Brain molecular aging, promotion of neurological disease and modulation by sirtuin 5 longevity gene polymorphism. *Neurobiology of Disease*, *41*, 279–290.

Gomes, A. P., Price, N. L., Ling, A. J., Moslehi, J. J., Montgomery, M. K., Rajman, L., ... Hubbard, B. P. (2013). Declining NAD(+) induces a pseudohypoxic state disrupting nuclear-mitochondrial communication during aging. *Cell*, *155*, 1624–1638.

Gonzalez Herrera, K. N., Zaganjor, E., Ishikawa, Y., Spinelli, J. B., Yoon, H., Lin, J. R., ... Souza, A. (2018). Small-molecule screen identifies De Novo nucleotide synthesis as a vulnerability of cells lacking SIRT3. *Cell Reports*, *22*, 1945–1955.

Gorski, P. A., Jang, S. P., Jeong, D., Lee, A., Lee, P., Oh, J. G., ... Eom, S. H. (2019). Role of SIRT1 in modulating acetylation of the sarco-endoplasmic reticulum Ca(2 +)-ATPase in heart failure. *Circulation Research*, *124*, e63–e80.

Greer, E. L., & Brunet, A. (2009). Different dietary restriction regimens extend lifespan by both independent and overlapping genetic pathways in C. elegans. *Aging Cell*, *8*, 113–127.

Grillon, J. M., Johnson, K. R., Kotlo, K., & Danziger, R. S. (2012). Non-histone lysine acetylated proteins in heart failure. *Biochimica et Biophysica Acta*, *1822*, 607–614.

Grozinger, C. M., Chao, E. D., Blackwell, H. E., Moazed, D., & Schreiber, S. L. (2001). Identification of a class of small molecule inhibitors of the sirtuin family of NAD-dependent deacetylases by phenotypic screening. *The Journal of Biological Chemistry*, *276*, 38837–38843.

Guan, D., Lim, J. H., Peng, L., Liu, Y., Lam, M., Seto, E., & Kao, H. Y. (2014). Deacetylation of the tumor suppressor protein PML regulates hydrogen peroxide-induced cell death. *Cell Death & Disease*, *5*, e1340.

Guarani, V., Deflorian, G., Franco, C. A., Kruger, M., Phng, L. K., Bentley, K., ... Schmidt, M. H. (2011). Acetylation-dependent regulation of endothelial Notch signalling by the SIRT1 deacetylase. *Nature, 473*, 234–238.

Guarente, L. (2013). Calorie restriction and sirtuins revisited. *Genes & Development, 27*, 2072–2085.

Guo, D., Song, X., Guo, T., Gu, S., Chang, X., Su, T., ... Huang, D. (2018). Vimentin acetylation is involved in SIRT5-mediated hepatocellular carcinoma migration. *American Journal of Cancer Research, 8*, 2453–2466.

Guo, L., Guo, Y. Y., Li, B. Y., Peng, W. Q., Chang, X. X., Gao, X., & Tang, Q. Q. (2019). Enhanced acetylation of ATP-citrate lyase promotes the progression of nonalcoholic fatty liver disease. *The Journal of Biological Chemistry, 294*, 11805–11816.

Guo, L., Zhou, S. R., Wei, X. B., Liu, Y., Chang, X. X., Liu, Y., ... Qian, S. W. (2016). Acetylation of mitochondrial trifunctional protein alpha-subunit enhances its stability to promote fatty acid oxidation and is decreased in nonalcoholic fatty liver disease. *Molecular and Cellular Biology, 36*, 2553–2567.

Haber, J. E., & George, J. P. (1979). A mutation that permits the expression of normally silent copies of mating-type information in *Saccharomyces cerevisiae*. *Genetics, 93*, 13–35.

Hafner, A. V., Dai, J., Gomes, A. P., Xiao, C. Y., Palmeira, C. M., Rosenzweig, A., & Sinclair, D. A. (2010). Regulation of the mPTP by SIRT3-mediated deacetylation of CypD at lysine 166 suppresses age-related cardiac hypertrophy. *Aging, 2*, 914–923.

Haider, R., Massa, F., Kaminski, L., Clavel, S., Djabari, Z., Robert, G., ... Ricci, J. E. (2017). Sirtuin 7: A new marker of aggressiveness in prostate cancer. *Oncotarget, 8*, 77309–77316.

Haigis, M. C., Mostoslavsky, R., Haigis, K. M., Fahie, K., Christodoulou, D. C., Murphy, A. J., ... Blander, G. (2006). SIRT4 inhibits glutamate dehydrogenase and opposes the effects of calorie restriction in pancreatic beta cells. *Cell, 126*, 941–954.

Haigis, M. C., & Sinclair, D. A. (2010). Mammalian sirtuins: Biological insights and disease relevance. *Annual Review of Pathology, 5*, 253–295.

Haketa, A., Soma, M., Nakayama, T., Kosuge, K., Aoi, N., Hishiki, M., ... Hinohara, S. (2013). Association between SIRT2 gene polymorphism and height in healthy, elderly Japanese subjects. *Translational Research: the Journal of Laboratory and Clinical Medicine, 161*, 57–58.

Halaschek-Wiener, J., Amirabbasi-Beik, M., Monfared, N., Pieczyk, M., Sailer, C., Kollar, A., ... Oliveira, L. (2009). Genetic variation in healthy oldest-old. *PLoS One, 4*, e6641.

Hallows, W. C., Yu, W., Smith, B. C., Devries, M. K., Ellinger, J. J., Someya, S., ... Smith, L. M. (2011). Sirt3 promotes the urea cycle and fatty acid oxidation during dietary restriction. *Molecular Cell, 41*, 139–149.

Han, L. L., Jia, L., Wu, F., & Huang, C. (2019). SIRT6 promotes the EMT of hepatocellular carcinoma by stimulating autophagic degradation of E-cadherin. *Molecular Cancer Research: MCR, 17*, 2267–2280.

Han, T. S., & Lean, M. E. (2016). A clinical perspective of obesity, metabolic syndrome and cardiovascular disease. *JRSM Cardiovascular Disease, 5*, 2048004016633371.

Han, Y., Liu, Y., Zhang, H., Wang, T., Diao, R., Jiang, Z., ... Cai, Z. (2013). Hsa-miR-125b suppresses bladder cancer development by down-regulating oncogene SIRT7 and oncogenic long non-coding RNA MALAT1. *FEBS Letters, 587*, 3875–3882.

Han, Z., Liu, L., Liu, Y., & Li, S. (2014). Sirtuin SIRT6 suppresses cell proliferation through inhibition of Twist1 expression in non-small cell lung cancer. *International Journal of Clinical and Experimental Pathology, 7*, 4774–4781.

Hanahan, D., & Weinberg, R. A. (2011). Hallmarks of cancer: The next generation. *Cell, 144*, 646–674.

Hansen, M., Taubert, S., Crawford, D., Libina, N., Lee, S. J., & Kenyon, C. (2007). Lifespan extension by conditions that inhibit translation in *Caenorhabditis elegans*. *Aging Cell, 6*, 95–110.

Hao, C., Zhu, P. X., Yang, X., Han, Z. P., Jiang, J. H., Zong, C., ... Fan, T. T. (2014). Overexpression of SIRT1 promotes metastasis through epithelial-mesenchymal transition in hepatocellular carcinoma. *BMC Cancer, 14*, 978.

Harrison, D. E., Strong, R., Sharp, Z. D., Nelson, J. F., Astle, C. M., Flurkey, K., ... Carter, C. S. (2009). Rapamycin fed late in life extends lifespan in genetically heterogeneous mice. *Nature, 460*, 392–395.

Harting, K., & Knoll, B. (2010). SIRT2-mediated protein deacetylation: An emerging key regulator in brain physiology and pathology. *European Journal of Cell Biology, 89*, 262–269.

Hartman, T. K., Wengenack, T. M., Poduslo, J. F., & van Deursen, J. M. (2007). Mutant mice with small amounts of BubR1 display accelerated age-related gliosis. *Neurobiology of Aging, 28*, 921–927.

Hashimoto, T., Horikawa, M., Nomura, T., & Sakamoto, K. (2010). Nicotinamide adenine dinucleotide extends the lifespan of *Caenorhabditis elegans* mediated by sir-2.1 and daf-16. *Biogerontology, 11*, 31–43.

He, J., Zhang, G., Pang, Q., Yu, C., Xiong, J., Zhu, J., & Chen, F. (2017). SIRT6 reduces macrophage foam cell formation by inducing autophagy and cholesterol efflux under ox-LDL condition. *The FEBS Journal, 284*, 1324–1337.

He, X., Zeng, H., & Chen, J. X. (2016). Ablation of SIRT3 causes coronary microvascular dysfunction and impairs cardiac recovery post myocardial ischemia. *International Journal of Cardiology, 215*, 349–357.

Hebert, A. S., Dittenhafer-Reed, K. E., Yu, W., Bailey, D. J., Selen, E. S., Boersma, M. D., ... Higbee, A. J. (2013). Calorie restriction and SIRT3 trigger global reprogramming of the mitochondrial protein acetylome. *Molecular Cell, 49*, 186–199.

Heeboll, S., Kreuzfeldt, M., Hamilton-Dutoit, S., Kjaer Poulsen, M., Stodkilde-Jorgensen, H., Moller, H. J., ... Bonlokke Pedersen, S. (2016). Placebo-controlled, randomised clinical trial: High-dose resveratrol treatment for non-alcoholic fatty liver disease. *Scandinavian Journal of Gastroenterology, 51*, 456–464.

Heltweg, B., Gatbonton, T., Schuler, A. D., Posakony, J., Li, H., Goehle, S., ... Simon, J. A. (2006). Antitumor activity of a small-molecule inhibitor of human silent information regulator 2 enzymes. *Cancer Research, 66*, 4368–4377.

Herranz, D., Munoz-Martin, M., Canamero, M., Mulero, F., Martinez-Pastor, B., Fernandez-Capetillo, O., & Serrano, M. (2010). Sirt1 improves healthy ageing and protects from metabolic syndrome-associated cancer. *Nature Communications, 1*, 3.

Hershberger, K. A., Abraham, D. M., Liu, J., Locasale, J. W., Grimsrud, P. A., & Hirschey, M. D. (2018). Ablation of Sirtuin5 in the postnatal mouse heart results in protein succinylation and normal survival in response to chronic pressure overload. *The Journal of Biological Chemistry, 293*, 10630–10645.

Hershberger, K. A., Abraham, D. M., Martin, A. S., Mao, L., Liu, J., Gu, H., ... Hirschey, M. D. (2017). Sirtuin 5 is required for mouse survival in response to cardiac pressure overload. *The Journal of Biological Chemistry, 292*, 19767–19781.

Herskovits, A. Z., & Guarente, L. (2014). SIRT1 in neurodevelopment and brain senescence. *Neuron, 81*, 471–483.

Herskovits, A. Z., Hunter, T. A., Maxwell, N., Pereira, K., Whittaker, C. A., Valdez, G., & Guarente, L. P. (2018). SIRT1 deacetylase in aging-induced neuromuscular degeneration and amyotrophic lateral sclerosis. *Aging Cell, 17*, e12839.

Higuchi, R., Vevea, J. D., Swayne, T. C., Chojnowski, R., Hill, V., Boldogh, I. R., & Pon, L. A. (2013). Actin dynamics affect mitochondrial quality control and aging in budding yeast. *Current Biology: CB, 23*, 2417–2422.

Hirai, S., Endo, S., Saito, R., Hirose, M., Ueno, T., Suzuki, H., ... Hyodo, I. (2014). Antitumor effects of a sirtuin inhibitor, tenovin-6, against gastric cancer cells via death receptor 5 up-regulation. *PLoS One, 9*, e102831.

Hirata, T., Otsuka, I., Okazaki, S., Mouri, K., Horai, T., Boku, S., ... Shirakawa, O. (2019). Major depressive disorder-associated SIRT1 locus affects the risk for suicide in women after middle age. *Psychiatry Research*, *278*, 141–145.

Hirschey, M. D., Shimazu, T., Goetzman, E., Jing, E., Schwer, B., Lombard, D. B., ... Ilkayeva, O. R. (2010). SIRT3 regulates mitochondrial fatty-acid oxidation by reversible enzyme deacetylation. *Nature*, *464*, 121–125.

Hirschey, M. D., Shimazu, T., Jing, E., Grueter, C. A., Collins, A. M., Aouizerat, B., ... Schwer, B. (2011). SIRT3 deficiency and mitochondrial protein hyperacetylation accelerate the development of the metabolic syndrome. *Molecular Cell*, *44*, 177–190.

Hirvonen, K., Laivuori, H., Lahti, J., Strandberg, T., Eriksson, J. G., & Hackman, P. (2017). SIRT6 polymorphism rs117385980 is associated with longevity and healthy aging in Finnish men. *BMC Medical Genetics*, *18*, 41.

Hisahara, S., Chiba, S., Matsumoto, H., Tanno, M., Yagi, H., Shimohama, S., ... Horio, Y. (2008). Histone deacetylase SIRT1 modulates neuronal differentiation by its nuclear translocation. *Proceedings of the National Academy of Sciences of the United States of America*, *105*, 15599–15604.

Ho, L., Titus, A. S., Banerjee, K. K., George, S., Lin, W., Deota, S., ... Verdin, E. (2013). SIRT4 regulates ATP homeostasis and mediates a retrograde signaling via AMPK. *Aging*, *5*, 835–849.

Hoffmann, J., Romey, R., Fink, C., Yong, L., & Roeder, T. (2013). Overexpression of Sir2 in the adult fat body is sufficient to extend lifespan of male and female Drosophila. *Aging*, *5*, 315–327.

Holthoff, J. H., Wang, Z., Seely, K. A., Gokden, N., & Mayeux, P. R. (2012). Resveratrol improves renal microcirculation, protects the tubular epithelium, and prolongs survival in a mouse model of sepsis-induced acute kidney injury. *Kidney International*, *81*, 370–378.

Hong, E. H., Yun, H. S., Kim, J., Um, H. D., Lee, K. H., Kang, C. M., ... Hwang, S. G. (2011). Nicotinamide phosphoribosyltransferase is essential for interleukin-1beta-mediated dedifferentiation of articular chondrocytes via SIRT1 and extracellular signal-regulated kinase (ERK) complex signaling. *The Journal of Biological Chemistry*, *286*, 28619–28631.

Hong, S., Zhao, B., Lombard, D. B., Fingar, D. C., & Inoki, K. (2014). Cross-talk between sirtuin and mammalian target of rapamycin complex 1 (mTORC1) signaling in the regulation of S6 kinase 1 (S6K1) phosphorylation. *The Journal of Biological Chemistry*, *289*, 13132–13141.

Hou, Y., Su, B., Chen, P., Niu, H., Zhao, S., Wang, R., & Shen, W. (2018). Association of SIRT1 gene polymorphism and its expression for the risk of alcoholic fatty liver disease in the Han population. *Hepatology International*, *12*, 56–66.

Howitz, K. T., Bitterman, K. J., Cohen, H. Y., Lamming, D. W., Lavu, S., Wood, J. G., ... Zhang, L. L. (2003). Small molecule activators of sirtuins extend *Saccharomyces cerevisiae* lifespan. *Nature*, *425*, 191–196.

Hsu, C. P., Zhai, P., Yamamoto, T., Maejima, Y., Matsushima, S., Hariharan, N., ... Sadoshima, J. (2010). Silent information regulator 1 protects the heart from ischemia/reperfusion. *Circulation*, *122*, 2170–2182.

Hu, H., Zhu, W., Qin, J., Chen, M., Gong, L., Li, L., ... Zhou, H. (2017). Acetylation of PGK1 promotes liver cancer cell proliferation and tumorigenesis. *Hepatology*, *65*, 515–528.

Hu, Q., Qin, Y., Ji, S., Xu, W., Liu, W., Sun, Q., ... Yu, X. (2019). UHRF1 promotes aerobic glycolysis and proliferation via suppression of SIRT4 in pancreatic cancer. *Cancer Letters*, *452*, 226–236.

Hu, Y., Lin, J., Lin, Y., Chen, X., Zhu, G., & Huang, G. (2019). Overexpression of SIRT4 inhibits the proliferation of gastric cancer cells through cell cycle arrest. *Oncol Letters*, *17*, 2171–2176.

Hu, Y., Wang, L., Chen, S., Liu, X., Li, H., Lu, X., ... Gu, D. (2015). Association between the SIRT1 mRNA expression and acute coronary syndrome. *Journal of Atherosclerosis and Thrombosis*, *22*, 165–182.

Huang, G., Cheng, J., Yu, F., Liu, X., Yuan, C., Liu, C., ... Peng, Z. (2016). Clinical and therapeutic significance of sirtuin-4 expression in colorectal cancer. *Oncology Reports*, *35*, 2801–2810.

Huang, H., Zhang, D., Wang, Y., Perez-Neut, M., Han, Z., Zheng, Y. G., ... Zhao, Y. (2018). Lysine benzoylation is a histone mark regulated by SIRT2. *Nature Communications*, *9*, 3374.

Huang, S., Zhao, Z., Tang, D., Zhou, Q., Li, Y., Zhou, L., ... Dorfman, R. G. (2017). Downregulation of SIRT2 inhibits invasion of hepatocellular carcinoma by inhibiting energy metabolism. *Translational Oncology*, *10*, 917–927.

Huang, Z., Zhao, J., Deng, W., Chen, Y., Shang, J., Song, K., ... Yang, X. (2018). Identification of a cellularly active SIRT6 allosteric activator. *Nature Chemical Biology*, *14*, 1118–1126.

Hubbard, B. P., Gomes, A. P., Dai, H., Li, J., Case, A. W., Considine, T., ... Lamming, D. W. (2013). Evidence for a common mechanism of SIRT1 regulation by allosteric activators. *Science*, *339*, 1216–1219.

Hubbi, M. E., Hu, H., Kshitiz., Gilkes, D. M., & Semenza, G. L. (2013). Sirtuin-7 inhibits the activity of hypoxia-inducible factors. *The Journal of Biological Chemistry*, *288*, 20768–20775.

Hui, X., Zhang, M., Gu, P., Li, K., Gao, Y., Wu, D., ... Xu, A. (2017). Adipocyte SIRT1 controls systemic insulin sensitivity by modulating macrophages in adipose tissue. *EMBO Reports*, *18*, 645–657.

Hyde, C. L., Nagle, M. W., Tian, C., Chen, X., Paciga, S. A., Wendland, J. R., ... Winslow, A. R. (2016). Identification of 15 genetic loci associated with risk of major depression in individuals of European descent. *Nature Genetics*, *48*, 1031–1036.

Igarashi, M., & Guarente, L. (2016). mTORC1 and SIRT1 cooperate to foster expansion of gut adult stem cells during calorie restriction. *Cell*, *166*, 436–450.

Igarashi, M., Miura, M., Williams, E., Jaksch, F., Kadowaki, T., Yamauchi, T., & Guarente, L. (2019). NAD(+) supplementation rejuvenates aged gut adult stem cells. *Aging Cell*, *18*, e12935.

Ikeda, Y., Sato, K., Pimentel, D. R., Sam, F., Shaw, R. J., Dyck, J. R., & Walsh, K. (2009). Cardiac-specific deletion of LKB1 leads to hypertrophy and dysfunction. *The Journal of Biological Chemistry*, *284*, 35839–35849.

Imai, S. (2009). SIRT1 and caloric restriction: An insight into possible trade-offs between robustness and frailty. *Current Opinion in Clinical Nutrition and Metabolic Care*, *12*, 350–356.

Imai, S., Armstrong, C. M., Kaeberlein, M., & Guarente, L. (2000). Transcriptional silencing and longevity protein Sir2 is an NAD-dependent histone deacetylase. *Nature*, *403*, 795–800.

Inoue, T., Hiratsuka, M., Osaki, M., Yamada, H., Kishimoto, I., Yamaguchi, S., ... Oshimura, M. (2007). SIRT2, a tubulin deacetylase, acts to block the entry to chromosome condensation in response to mitotic stress. *Oncogene*, *26*, 945–957.

Inoue, T., Nakayama, Y., Yamada, H., Li, Y. C., Yamaguchi, S., Osaki, M., ... Oshimura, M. (2009). SIRT2 downregulation confers resistance to microtubule inhibitors by prolonging chronic mitotic arrest. *Cell Cycle*, *8*, 1279–1291.

Jedrusik-Bode, M., Studencka, M., Smolka, C., Baumann, T., Schmidt, H., Kampf, J., ... Muller, K. M. (2013). The sirtuin SIRT6 regulates stress granule formation in C. elegans and mammals. *Journal of Cell Science*, *126*, 5166–5177.

Jeong, H., Cohen, D. E., Cui, L., Supinski, A., Savas, J. N., Mazzulli, J. R., ... Krainc, D. (2012). Sirt1 mediates neuroprotection from mutant huntingtin by activation of the TORC1 and CREB transcriptional pathway. *Nature Medicine*, *18*, 159–165.

Jeong, S. M., Lee, A., Lee, J., & Haigis, M. C. (2014). SIRT4 protein suppresses tumor formation in genetic models of Myc-induced B cell lymphoma. *The Journal of Biological Chemistry*, *289*, 4135–4144.

Jeong, S. M., Lee, J., Finley, L. W., Schmidt, P. J., Fleming, M. D., & Haigis, M. C. (2015). SIRT3 regulates cellular iron metabolism and cancer growth by repressing iron regulatory protein 1. *Oncogene*, *34*, 2115–2124.

Jeong, Seung M., Xiao, C., Finley, Lydia W. S., Lahusen, T., Souza, Amanda L., ... Natalie, J. (2013). SIRT4 Has tumor-suppressive activity and regulates the cellular metabolic response to dna damage by inhibiting mitochondrial glutamine metabolism. *Cancer Cell*, *23*, 450–463.

Ji, S., Doucette, J. R., & Nazarali, A. J. (2011). Sirt2 is a novel in vivo downstream target of Nkx2.2 and enhances oligodendroglial cell differentiation. *Journal of Molecular Cell Biology*, *3*, 351–359.

Jiang, H., Khan, S., Wang, Y., Charron, G., He, B., Sebastian, C., ... Mostoslavsky, R. (2013). SIRT6 regulates TNF-α secretion through hydrolysis of long-chain fatty acyl lysine. *Nature*, *496*, 110–113.

Jiang, H., Zhang, X., & Lin, H. (2016). Lysine fatty acylation promotes lysosomal targeting of TNF-alpha. *Scientific Reports*, *6*, 24371.

Jiang, K., Shen, M., Chen, Y., & Xu, W. (2018). miR150 promotes the proliferation and migration of nonsmall cell lung cancer cells by regulating the SIRT2/JMJD2A signaling pathway. *Oncology Reports*, *40*, 943–951.

Jiang, L., Xiong, J., Zhan, J., Yuan, F., Tang, M., Zhang, C., ... Li, Y. (2017). Ubiquitin-specific peptidase 7 (USP7)-mediated deubiquitination of the histone deacetylase SIRT7 regulates gluconeogenesis. *The Journal of Biological Chemistry*, *292*, 13296–13311.

Jiang, M., Wang, J., Fu, J., Du, L., Jeong, H., West, T., ... Cai, H. (2012). Neuroprotective role of Sirt1 in mammalian models of Huntington's disease through activation of multiple Sirt1 targets. *Nature Medicine*, *18*, 153–158.

Jiang, W., Wang, S., Xiao, M., Lin, Y., Zhou, L., Lei, Q., ... Zhao, S. (2011). Acetylation regulates gluconeogenesis by promoting PEPCK1 degradation via recruiting the UBR5 ubiquitin ligase. *Molecular Cell*, *43*, 33–44.

Jimenez-Gomez, Y., Mattison, J. A., Pearson, K. J., Martin-Montalvo, A., Palacios, H. H., Sossong, A. M., ... Allard, J. S. (2013). Resveratrol improves adipose insulin signaling and reduces the inflammatory response in adipose tissue of rhesus monkeys on high-fat, high-sugar diet. *Cell Metabolism*, *18*, 533–545.

Jin, Q., Yan, T., Ge, X., Sun, C., Shi, X., & Zhai, Q. (2007). Cytoplasm-localized SIRT1 enhances apoptosis. *Journal of Cellular Physiology*, *213*, 88–97.

Jing, E., Emanuelli, B., Hirschey, M. D., Boucher, J., Lee, K. Y., Lombard, D., ... Kahn, C. R. (2011). Sirtuin-3 (Sirt3) regulates skeletal muscle metabolism and insulin signaling via altered mitochondrial oxidation and reactive oxygen species production. *Proceedings of the National Academy of Sciences of the United States of America*, *108*, 14608–14613.

Jing, E., Gesta, S., & Kahn, C. R. (2007). SIRT2 regulates adipocyte differentiation through FoxO1 acetylation/deacetylation. *Cell Metabolism*, *6*, 105–114.

Jing, E., O'Neill, B. T., Rardin, M. J., Kleinridders, A., Ilkeyeva, O. R., Ussar, S., ... Newgard, C. B. (2013). Sirt3 regulates metabolic flexibility of skeletal muscle through reversible enzymatic deacetylation. *Diabetes*, *62*, 3404–3417.

Jing, H., Hu, J., He, B., Negron Abril, Y. L., Stupinski, J., Weiser, K., ... Giannakakou, P. (2016). A SIRT2-selective inhibitor promotes c-Myc oncoprotein degradation and exhibits broad anticancer activity. *Cancer Cell*, *29*, 297–310.

Joo, H. Y., Yun, M., Jeong, J., Park, E. R., Shin, H. J., Woo, S. R., ... Kim, J. (2015). SIRT1 deacetylates and stabilizes hypoxia-inducible factor-1alpha (HIF-1alpha) via direct interactions during hypoxia. *Biochemical and Biophysical Research Communications*, *462*, 294–300.

Jung, E. S., Choi, H., Song, H., Hwang, Y. J., Kim, A., Ryu, H., & Mook-Jung, I. (2016). p53-dependent SIRT6 expression protects Abeta42-induced DNA damage. *Scientific Reports*, *6*, 25628.

Jung, K. J., Lee, E. K., Kim, J. Y., Zou, Y., Sung, B., Heo, H. S., ... Yu, B. P. (2009). Effect of short term calorie restriction on pro-inflammatory NF-kB and AP-1 in aged rat kidney. *Inflammation Research: Official Journal of the European Histamine Research Society*, *58*, 143–150.

Kaeberlein, M., McDonagh, T., Heltweg, B., Hixon, J., Westman, E. A., Caldwell, S. D., ... Fields, S. (2005). Substrate-specific activation of sirtuins by resveratrol. *The Journal of Biological Chemistry*, *280*, 17038–17045.

Kaeberlein, M., McVey, M., & Guarente, L. (1999). The SIR2/3/4 complex and SIR2 alone promote longevity in *Saccharomyces cerevisiae* by two different mechanisms. *Genes & Development*, *13*, 2570–2580.

Kaeberlein, M., & Powers, R. W., 3rd (2007). Sir2 and calorie restriction in yeast: A skeptical perspective. *Ageing Research Reviews*, *6*, 128–140.

Kaeberlein, T. L., Smith, E. D., Tsuchiya, M., Welton, K. L., Thomas, J. H., Fields, S., ... Kaeberlein, M. (2006). Lifespan extension in *Caenorhabditis elegans* by complete removal of food. *Aging Cell*, *5*, 487–494.

Kaidi, A., Weinert, B. T., Choudhary, C., & Jackson, S. P. (2010). Human SIRT6 promotes DNA end resection through CtIP deacetylation. *Science*, *329*, 1348–1353.

Kalliora, C., Kyriazis, I. D., Oka, S. I., Lieu, M. J., Yue, Y., Area-Gomez, E., ... Chin, A. (2019). Dual peroxisome-proliferator-activated-receptor-alpha/gamma activation inhibits SIRT1-PGC1alpha axis and causes cardiac dysfunction. *JCI Insight*, *5*, e129556.

Kaluski, S., Portillo, M., Besnard, A., Stein, D., Einav, M., Zhong, L., ... Sahay, A. (2017). Neuroprotective functions for the histone deacetylase SIRT6. *Cell Reports*, *18*, 3052–3062.

Kan, Y., Ge, P., Wang, X., Xiao, G., & Zhao, H. (2018). SIRT1 rs3758391 polymorphism and risk of diffuse large B cell lymphoma in a Chinese population. *Cancer Cell International*, *18*, 163.

Kanfi, Y., Naiman, S., Amir, G., Peshti, V., Zinman, G., Nahum, L., ... Cohen, H. Y. (2012). The sirtuin SIRT6 regulates lifespan in male mice. *Nature*, *483*, 218–221.

Kanfi, Y., Shalman, R., Peshti, V., Pilosof, S. N., Gozlan, Y. M., Pearson, K. J., ... de Cabo, R. (2008). Regulation of SIRT6 protein levels by nutrient availability. *FEBS Letters*, *582*, 543–548.

Kantartzis, K., Fritsche, L., Bombrich, M., Machann, J., Schick, F., Staiger, H., ... Heni, M. (2018). Effects of resveratrol supplementation on liver fat content in overweight and insulin-resistant subjects: A randomized, double-blind, placebo-controlled clinical trial. *Diabetes, Obesity & Metabolism*, *20*, 1793–1797.

Kaplon, J., Zheng, L., Meissl, K., Chaneton, B., Selivanov, V. A., Mackay, G., ... Shlomi, T. (2013). A key role for mitochondrial gatekeeper pyruvate dehydrogenase in oncogene-induced senescence. *Nature*, *498*, 109–112.

Kawahara, T. L., Michishita, E., Adler, A. S., Damian, M., Berber, E., Lin, M., ... Chang, H. Y. (2009). SIRT6 links histone H3 lysine 9 deacetylation to NF-kappaB-dependent gene expression and organismal life span. *Cell*, *136*, 62–74.

Kennedy, B. K., Austriaco, N. R., Jr., Zhang, J., & Guarente, L. (1995). Mutation in the silencing gene SIR4 can delay aging in S. cerevisiae. *Cell*, *80*, 485–496.

Kennedy, B. K., Gotta, M., Sinclair, D. A., Mills, K., McNabb, D. S., Murthy, M., ... Guarente, L. (1997). Redistribution of silencing proteins from telomeres to the nucleolus is associated with extension of life span in S. cerevisiae. *Cell*, *89*, 381–391.

Kenyon, C. J. (2010). The genetics of ageing. *Nature*, *464*, 504–512.

Khongkow, M., Olmos, Y., Gong, C., Gomes, A. R., Monteiro, L. J., Yague, E., ... Laohasinnarong, S. (2013). SIRT6 modulates paclitaxel and epirubicin resistance and survival in breast cancer. *Carcinogenesis*, *34*, 1476–1486.

Kilic, U., Gok, O., Erenberk, U., Dundaroz, M. R., Torun, E., Kucukardali, Y., ... Dundar, T. (2015). A remarkable age-related increase in SIRT1 protein expression against oxidative stress in elderly: SIRT1 gene variants and longevity in human. *PLoS One*, *10*, e0117954.

Kim, H., Kim, S., Choi, J. E., Han, D., Koh, S. M., Kim, H. S., & Kaang, B. K. (2019). Decreased neuron number and synaptic plasticity in SIRT3-knockout mice with poor remote memory. *Neurochemical Research*, *44*, 676–682.

Kim, H. S., Patel, K., Muldoon-Jacobs, K., Bisht, K. S., Aykin-Burns, N., Pennington, J. D., ... Owens, K. M. (2010). SIRT3 is a mitochondria-localized tumor suppressor required for maintenance of mitochondrial integrity and metabolism during stress. *Cancer Cell*, *17*, 41–52.

Kim, H. S., Vassilopoulos, A., Wang, R. H., Lahusen, T., Xiao, Z., Xu, X., ... Yu, H. (2011). SIRT2 maintains genome integrity and suppresses tumorigenesis through regulating APC/C activity. *Cancer Cell*, *20*, 487–499.

Kim, H. S., Xiao, C., Wang, R. H., Lahusen, T., Xu, X., Vassilopoulos, A., ... Ki, S. H. (2010). Hepatic-specific disruption of SIRT6 in mice results in fatty liver formation due to enhanced glycolysis and triglyceride synthesis. *Cell Metabolism*, *12*, 224–236.

Kim, J. E., Chen, J., & Lou, Z. (2008). DBC1 is a negative regulator of SIRT1. *Nature*, *451*, 583–586.

Kim, J. H., Kim, D., Cho, S. J., Jung, K. Y., Kim, J. H., Lee, J. M., ... Kim, K. R. (2019). Identification of a novel SIRT7 inhibitor as anticancer drug candidate. *Biochemical and Biophysical Research Communications*, *508*, 451–457.

Kim, J. K., Noh, J. H., Jung, K. H., Eun, J. W., Bae, H. J., Kim, M. G., ... Lee, J. Y. (2013). Sirtuin7 oncogenic potential in human hepatocellular carcinoma and its regulation by the tumor suppressors MiR-125a-5p and MiR-125b. *Hepatology*, *57*, 1055–1067.

Kim, M., Lee, J. S., Oh, J. E., Nan, J., Lee, H., Jung, H. S., ... Park, K. S. (2015). SIRT3 overexpression attenuates palmitate-induced pancreatic beta-cell dysfunction. *PLoS One*, *10*, e0124744.

Kishi, T., Fukuo, Y., Kitajima, T., Okochi, T., Yamanouchi, Y., Kinoshita, Y., ... Kato, T. (2011). SIRT1 gene, schizophrenia and bipolar disorder in the Japanese population: An association study. *Genes, Brain, and Behavior*, *10*, 257–263.

Kishi, T., Yoshimura, R., Kitajima, T., Okochi, T., Okumura, T., Tsunoka, T., ... Fukuo, Y. (2010). SIRT1 gene is associated with major depressive disorder in the Japanese population. *Journal of Affective Disorders*, *126*, 167–173.

Kjaer, T. N., Ornstrup, M. J., Poulsen, M. M., Stodkilde-Jorgensen, H., Jessen, N., Jorgensen, J. O. L., ... Pedersen, S. B. (2017). No beneficial effects of resveratrol on the metabolic syndrome: A randomized placebo-controlled clinical trial. *The Journal of Clinical Endocrinology and Metabolism*, *102*, 1642–1651.

Klar, A. J., Fogel, S., & Macleod, K. (1979). MAR1-a regulator of the HMa and HMalpha loci in *Saccharomyces cerevisiae*. *Genetics*, *93*, 37–50.

Klishadi, M. S., Zarei, F., Hejazian, S. H., Moradi, A., Hemati, M., & Safari, F. (2015). Losartan protects the heart against ischemia reperfusion injury: Sirtuin3 involvement. *Journal of Pharmacy & Pharmaceutical Sciences: a Publication of the Canadian Society for Pharmaceutical Sciences, Societe Canadienne des Sciences Pharmaceutiques*, *18*, 112–123.

Kolwicz, S. C., Jr., & Tian, R. (2011). Glucose metabolism and cardiac hypertrophy. *Cardiovascular Research*, *90*, 194–201.

Komlos, D., Mann, K. D., Zhuo, Y., Ricupero, C. L., Hart, R. P., Liu, A. Y., & Firestein, B. L. (2013). Glutamate dehydrogenase 1 and SIRT4 regulate glial development. *Glia*, *61*, 394–408.

Koo, J. H., Kang, E. B., Oh, Y. S., Yang, D. S., & Cho, J. Y. (2017). Treadmill exercise decreases amyloid-beta burden possibly via activation of SIRT-1 signaling in a mouse model of Alzheimer's disease. *Experimental Neurology*, *288*, 142–152.

Koronowski, K. B., Khoury, N., Morris-Blanco, K. C., Stradecki-Cohan, H. M., Garrett, T. J., & Perez-Pinzon, M. A. (2018). Metabolomics based identification of SIRT5 and protein kinase C epsilon regulated pathways in brain. *Frontiers in Neuroscience*, *12*, 32.

Krishnan, J., Danzer, C., Simka, T., Ukropec, J., Walter, K. M., Kumpf, S., ... Pedrazzini, T. (2012). Dietary obesity-associated Hif1alpha activation in adipocytes restricts fatty acid oxidation and energy expenditure via suppression of the Sirt2-NAD + system. *Genes & Development*, *26*, 259–270.

Krueger, J. G., Suarez-Farinas, M., Cueto, I., Khacherian, A., Matheson, R., Parish, L. C., ... Haddad, J. (2015). A randomized, placebo-controlled study of SRT2104, a SIRT1 activator, in patients with moderate to severe psoriasis. *PLoS One*, *10*, e0142081.

Kuang, J., Zhang, Y., Liu, Q., Shen, J., Pu, S., Cheng, S., ... Li, R. (2017). Fat-specific Sirt6 ablation sensitizes mice to high-fat diet-induced obesity and insulin resistance by inhibiting lipolysis. *Diabetes*, *66*, 1159–1171.

Kueng, S., Oppikofer, M., & Gasser, S. M. (2013). SIR proteins and the assembly of silent chromatin in budding yeast. *Annual Review of Genetics*, *47*, 275–306.

Kugel, S., Feldman, J. L., Klein, M. A., Silberman, D. M., Sebastian, C., Mermel, C., ... Denu, J. M. (2015). Identification of and molecular basis for SIRT6 loss-of-function point mutations in cancer. *Cell Reports*, *13*, 479–488.

Kugel, S., Sebastian, C., Fitamant, J., Ross, K. N., Saha, S. K., Jain, E., ... Rivera, M. N. (2016). SIRT6 suppresses pancreatic cancer through control of Lin28b. *Cell*, *165*, 1401–1415.

Kulashekar, M., Stom, S. M., & Peuler, J. D. (2018). Resveratrol's potential in the adjunctive management of cardiovascular disease, obesity, diabetes, Alzheimer's disease, and cancer. *The Journal of the American Osteopathic Association*, *118*, 596–605.

Kumar, S., & Lombard, D. B. (2016). Finding Ponce de Leon's Pill: Challenges in screening for anti-aging molecules. *F1000Res*, *5*, F1000 Faculty Rev-406.

Kumar, S., & Lombard, D. B. (2017). For certain, SIRT4 activities!. *Trends in Biochemical Sciences*, *42*, 499–501.

Kumar, S., & Lombard, D. B. (2018). Functions of the sirtuin deacylase SIRT5 in normal physiology and pathobiology. *Critical Reviews in Biochemistry and Molecular Biology*, *53*, 311–334.

Kume, S., Uzu, T., Horiike, K., Chin-Kanasaki, M., Isshiki, K., Araki, S., ... Koya, D. (2010). Calorie restriction enhances cell adaptation to hypoxia through Sirt1-dependent mitochondrial autophagy in mouse aged kidney. *The Journal of Clinical Investigation*, *120*, 1043–1055.

Kuningas, M., Putters, M., Westendorp, R. G., Slagboom, P. E., & van Heemst, D. (2007). SIRT1 gene, age-related diseases, and mortality: The Leiden 85-plus study. *The Journals of Gerontology. Series A, Biological Sciences and Medical Sciences*, *62*, 960–965.

Kurundkar, D., Kurundkar, A. R., Bone, N. B., Becker, E. J., Jr., Liu, W., Chacko, B., ... Thannickal, V. J. (2019). SIRT3 diminishes inflammation and mitigates endotoxin-induced acute lung injury. *JCI Insight*, *4*, e120722.

Laemmle, A., Lechleiter, A., Roh, V., Schwarz, C., Portmann, S., Furer, C., ... Vorburger, S. A. (2012). Inhibition of SIRT1 impairs the accumulation and transcriptional activity of HIF-1alpha protein under hypoxic conditions. *PLoS One*, *7*, e33433.

Laffel, L. (1999). Ketone bodies: A review of physiology, pathophysiology and application of monitoring to diabetes. *Diabetes/Metabolism Research and Reviews*, *15*, 412–426.

Lain, S., Hollick, J. J., Campbell, J., Staples, O. D., Higgins, M., Aoubala, M., ... Baker, L. (2008). Discovery, in vivo activity, and mechanism of action of a small-molecule p53 activator. *Cancer Cell*, *13*, 454–463.

Lakatta, E. G., & Levy, D. (2003). Arterial and cardiac aging: Major shareholders in cardiovascular disease enterprises: Part II: The aging heart in health: Links to heart disease. *Circulation*, *107*, 346–354.

Lakshminarasimhan, M., Rauh, D., Schutkowski, M., & Steegborn, C. (2013). Sirt1 activation by resveratrol is substrate sequence-selective. *Aging*, *5*, 151–154.

Lamming, D. W., Latorre-Esteves, M., Medvedik, O., Wong, S. N., Tsang, F. A., Wang, C., ... Sinclair, D. A. (2005). HST2 mediates SIR2-independent life-span extension by calorie restriction. *Science, 309*, 1861–1864.

Langley, E., Pearson, M., Faretta, M., Bauer, U. M., Frye, R. A., Minucci, S., ... Kouzarides, T. (2002). Human SIR2 deacetylates p53 and antagonizes PML/p53-induced cellular senescence. *The EMBO Journal, 21*, 2383–2396.

Lantier, L., Williams, A. S., Hughey, C. C., Bracy, D. P., James, F. D., Ansari, M. A., ... Wasserman, D. H. (2018). SIRT2 knockout exacerbates insulin resistance in high fat-fed mice. *PLoS One, 13*, e0208634.

Lara, E., Mai, A., Calvanese, V., Altucci, L., Lopez-Nieva, P., Martinez-Chantar, M. L., ... Ropero, S. (2009). Salermide, a Sirtuin inhibitor with a strong cancer-specific proapoptotic effect. *Oncogene, 28*, 781–791.

Laurent, G., de Boer, V. C., Finley, L. W., Sweeney, M., Lu, H., Schug, T. T., ... Sauve, A. A. (2013). SIRT4 represses peroxisome proliferator-activated receptor alpha activity to suppress hepatic fat oxidation. *Molecular and Cellular Biology, 33*, 4552–4561.

Laurent, G., German, N. J., Saha, A. K., de Boer, V. C., Davies, M., Koves, T. R., ... Vaitheesvaran, B. (2013). SIRT4 coordinates the balance between lipid synthesis and catabolism by repressing malonyl CoA decarboxylase. *Molecular Cell, 50*, 686–698.

Lee, A. S., Jung, Y. J., Kim, D., Nguyen-Thanh, T., Kang, K. P., Lee, S., ... Kim, W. (2014). SIRT2 ameliorates lipopolysaccharide-induced inflammation in macrophages. *Biochemical and Biophysical Research Communications, 450*, 1363–1369.

Lee, G. D., Wilson, M. A., Zhu, M., Wolkow, C. A., de Cabo, R., Ingram, D. K., & Zou, S. (2006). Dietary deprivation extends life-span in *Caenorhabditis elegans*. *Aging Cell, 5*, 515–524.

Lee, H. R., Shin, H. K., Park, S. Y., Kim, H. Y., Lee, W. S., Rhim, B. Y., ... Kim, C. D. (2014a). Attenuation of beta-amyloid-induced tauopathy via activation of CK2alpha/SIRT1: Targeting for cilostazol. *Journal of Neuroscience Research, 92*, 206–217.

Lee, H. R., Shin, H. K., Park, S. Y., Kim, H. Y., Lee, W. S., Rhim, B. Y., ... Kim, C. D. (2014b). Cilostazol suppresses beta-amyloid production by activating a disintegrin and metalloproteinase 10 via the upregulation of SIRT1-coupled retinoic acid receptor-beta. *Journal of Neuroscience Research, 92*, 1581–1590.

Lee, J., Kim, Y., Liu, T., Hwang, Y. J., Hyeon, S. J., Im, H., ... Um, S. J. (2018). SIRT3 deregulation is linked to mitochondrial dysfunction in Alzheimer's disease. *Aging Cell, 17*(1), e12679.

Lee, J. J., van de Ven, R. A. H., Zaganjor, E., Ng, M. R., Barakat, A., Demmers, J., ... Harris, I. S. (2018). Inhibition of epithelial cell migration and Src/FAK signaling by SIRT3. *Proceedings of the National Academy of Sciences of the United States of America, 115*, 7057–7062.

Lee, M., Kim, D. W., Yoon, H., So, D., Khalmuratova, R., Rhee, C. S., ... Shin, H. W. (2016). Sirtuin 1 attenuates nasal polypogenesis by suppressing epithelial-to-mesenchymal transition. *The Journal of Allergy and Clinical Immunology, 137*, 87–98, e87.

Lee, Y., Ka, S. O., Cha, H. N., Chae, Y. N., Kim, M. K., Park, S. Y., ... Park, B. H. (2017). Myeloid Sirtuin 6 deficiency causes insulin resistance in high-fat diet-fed mice by eliciting macrophage polarization toward an M1 phenotype. *Diabetes, 66*, 2659–2668.

Leko, V., Park, G. J., Lao, U., Simon, J. A., & Bedalov, A. (2013). Enterocyte-specific inactivation of SIRT1 reduces tumor load in the APC(+/min) mouse model. *PLoS One, 8*, e66283.

Lemos, V., de Oliveira, R. M., Naia, L., Szego, E., Ramos, E., Pinho, S., ... Costa, V. (2017). The NAD + -dependent deacetylase SIRT2 attenuates oxidative stress and mitochondrial dysfunction and improves insulin sensitivity in hepatocytes. *Human Molecular Genetics, 26*, 4105–4117.

Lerin, C., Rodgers, J. T., Kalume, D. E., Kim, S. H., Pandey, A., & Puigserver, P. (2006). GCN5 acetyltransferase complex controls glucose metabolism through transcriptional repression of PGC-1alpha. *Cell Metabolism, 3*, 429–438.

Lescai, F., Blanche, H., Nebel, A., Beekman, M., Sahbatou, M., Flachsbart, F., ... Passarino, G. (2009). Human longevity and 11p15.5: A study in 1321 centenarians. *European Journal of Human Genetics: EJHG, 17*, 1515–1519.

Li, F., He, X., Ye, D., Lin, Y., Yu, H., Yao, C., ... Xu, S. (2015). NADP (+)-IDH mutations promote hypersuccinylation that impairs mitochondria respiration and induces apoptosis resistance. *Molecular Cell, 60*, 661–675.

Li, F., & Liu, L. (2016). SIRT5 deficiency enhances susceptibility to kainate-induced seizures and exacerbates hippocampal neurodegeneration not through mitochondrial antioxidant enzyme SOD2. *Frontiers in Cell Neuroscience, 10*, 171.

Li, H., Jia, J., Wang, W., Hou, T., Tian, Y., Wu, Q., ... Wang, X. (2018). Honokiol alleviates cognitive deficits of Alzheimer's disease (PS1V97L) transgenic mice by activating mitochondrial SIRT3. *Journal of Alzheimer's Disease: JAD, 64*, 291–302.

Li, H., Tian, Z., Qu, Y., Yang, Q., Guan, H., Shi, B., ... Hou, P. (2019). SIRT7 promotes thyroid tumorigenesis through phosphorylation and activation of Akt and p70S6K1 via DBC1/SIRT1 axis. *Oncogene, 38*, 345–359.

Li, J., Bonkowski, M. S., Moniot, S., Zhang, D., Hubbard, B. P., Ling, A. J., ... Gorbunova, V. (2017). A conserved NAD(+) binding pocket that regulates protein-protein interactions during aging. *Science, 355*, 1312–1317.

Li, L., Osdal, T., Ho, Y., Chun, S., McDonald, T., Agarwal, P., ... Li, L. (2014). SIRT1 activation by a c-MYC oncogenic network promotes the maintenance and drug resistance of human FLT3-ITD acute myeloid leukemia stem cells. *Cell Stem Cell, 15*, 431–446.

Li, L., Shi, L., Yang, S., Yan, R., Zhang, D., Yang, J., ... Sun, L. (2016). SIRT7 is a histone desuccinylase that functionally links to chromatin compaction and genome stability. *Nature Communications, 7*, 12235.

Li, M., Chiang, Y. L., Lyssiotis, C. A., Teater, M. R., Hong, J. Y., Shen, H., ... Chen, Z. (2019). Non-oncogene addiction to SIRT3 plays a critical role in lymphomagenesis. *Cancer Cell, 35*, 916–931, e919.

Li, N., Zhang, J., Yan, X., Zhang, C., Liu, H., Shan, X., ... Zhang, P. (2017). SIRT3-KLF15 signaling ameliorates kidney injury induced by hypertension. *Oncotarget, 8*, 39592–39604.

Li, R., Quan, Y., & Xia, W. (2018). SIRT3 inhibits prostate cancer metastasis through regulation of FOXO3A by suppressing Wnt/beta-catenin pathway. *Experimental Cell Research, 364*, 143–151.

Li, S., Banck, M., Mujtaba, S., Zhou, M. M., Sugrue, M. M., & Walsh, M. J. (2010). p53-induced growth arrest is regulated by the mitochondrial SirT3 deacetylase. *PLoS One, 5*, e10486.

Li, W., Zhang, B., Tang, J., Cao, Q., Wu, Y., Wu, C., ... Liang, F. (2007). Sirtuin 2, a mammalian homolog of yeast silent information regulator-2 longevity regulator, is an oligodendroglial protein that decelerates cell differentiation through deacetylating alpha-tubulin. *The Journal of Neuroscience, 27*, 2606–2616.

Li, Y., Qin, J., Wei, X., Liang, G., Shi, L., Jiang, M., ... Zhang, Z. (2016). Association of SIRT6 gene polymorphisms with human longevity. *Iranian Journal of Public Health, 45*, 1420–1426.

Li, Y., Xu, W., McBurney, M. W., & Longo, V. D. (2008). SirT1 inhibition reduces IGF-I/IRS-2/Ras/ERK1/2 signaling and protects neurons. *Cell Metabolism, 8*, 38–48.

Li, Z., Huang, J., Yuan, H., Chen, Z., Luo, Q., & Lu, S. (2016). SIRT2 inhibits non-small cell lung cancer cell growth through impairing Skp2-mediated p27 degradation. *Oncotarget, 7*, 18927–18939.

Li, Z., Xie, Q. R., Chen, Z., Lu, S., & Xia, W. (2013). Regulation of SIRT2 levels for human non-small cell lung cancer therapy. *Lung Cancer, 82*, 9–15.

Libert, S., Pointer, K., Bell, E. L., Das, A., Cohen, D. E., Asara, J. M., ... Otowa, T. (2011). SIRT1 activates MAO-A in the brain to mediate anxiety and exploratory drive. *Cell, 147*, 1459–1472.

Lim, J. H., Lee, Y. M., Chun, Y. S., Chen, J., Kim, J. E., & Park, J. W. (2010). Sirtuin 1 modulates cellular responses to hypoxia by deacetylating hypoxia-inducible factor 1alpha. *Molecular Cell, 38*, 864–878.

Lin, J., Sun, B., Jiang, C., Hong, H., & Zheng, Y. (2013). Sirt2 suppresses inflammatory responses in collagen-induced arthritis. *Biochemical and Biophysical Research Communications, 441*, 897–903.

Lin, J. B., Kubota, S., Ban, N., Yoshida, M., Santeford, A., Sene, A., ... Tsubota, K. (2016). NAMPT-mediated NAD(+) biosynthesis is essential for vision in mice. *Cell Reports, 17*, 69–85.

Lin, J. B., Lin, J. B., Chen, H. C., Chen, T., & Apte, R. S. (2019). Combined SIRT3 and SIRT5 deletion is associated with inner retinal dysfunction in a mouse model of type 1 diabetes. *Scientific Reports, 9*, 3799.

Lin, R., Tao, R., Gao, X., Li, T., Zhou, X., Guan, K. L., ... Lei, Q. Y. (2013). Acetylation stabilizes ATP-citrate lyase to promote lipid biosynthesis and tumor growth. *Molecular Cell, 51*, 506–518.

Lin, R., Yan, D., Zhang, Y., Liao, X., Gong, G., Hu, J., ... Cai, W. (2016). Common variants in SIRT1 and human longevity in a Chinese population. *BMC Medical Genetics, 17*, 31.

Lin, R., Zhang, Y., Yan, D., Liao, X., Gong, G., Hu, J., ... Cai, W. (2016). Lack of association between polymorphisms in the SIRT6 gene and longevity in a Chinese population. *Molecular and Cellular Probes, 30*, 79–82.

Lin, S. J., Defossez, P. A., & Guarente, L. (2000). Requirement of NAD and SIR2 for life-span extension by calorie restriction in *Saccharomyces cerevisiae*. *Science, 289*, 2126–2128.

Lin, Y. Y., Lu, J. Y., Zhang, J., Walter, W., Dang, W., Wan, J., ... Boeke, J. D. (2009). Protein acetylation microarray reveals that NuA4 controls key metabolic target regulating gluconeogenesis. *Cell, 136*, 1073–1084.

Lin, Z. F., Xu, H. B., Wang, J. Y., Lin, Q., Ruan, Z., Liu, F. B., ... Chen, X. (2013). SIRT5 desuccinylates and activates SOD1 to eliminate ROS. *Biochemical and Biophysical Research Communications, 441*, 191–195.

Liou, G. Y., & Storz, P. (2010). Reactive oxygen species in cancer. *Free Radical Research, 44*, 479–496.

Liu, B., Che, W., Xue, J., Zheng, C., Tang, K., Zhang, J., ... Xu, Y. (2013). SIRT4 prevents hypoxia-induced apoptosis in H9c2 cardiomyoblast cells. *Cellular Physiology and Biochemistry: International Journal of Experimental Cellular Physiology, Biochemistry, and Pharmacology, 32*, 655–662.

Liu, B., Che, W., Zheng, C., Liu, W., Wen, J., Fu, H., ... Xu, Y. (2013). SIRT5: A safeguard against oxidative stress-induced apoptosis in cardiomyocytes. *Cellular Physiology and Biochemistry: International Journal of Experimental Cellular Physiology, Biochemistry, and Pharmacology, 32*, 1050–1059.

Liu, B., Larsson, L., Caballero, A., Hao, X., Oling, D., Grantham, J., & Nystrom, T. (2010). The polarisome is required for segregation and retrograde transport of protein aggregates. *Cell, 140*, 257–267.

Liu, L., Arun, A., Ellis, L., Peritore, C., & Donmez, G. (2014). SIRT2 enhances 1-methyl-4-phenyl-1,2,3,6-tetrahydropyridine (MPTP)-induced nigrostriatal damage via apoptotic pathway. *Frontiers in Aging Neuroscience, 6*, 184.

Liu, L., Peritore, C., Ginsberg, J., Kayhan, M., & Donmez, G. (2015). SIRT3 attenuates MPTP-induced nigrostriatal degeneration via enhancing mitochondrial antioxidant capacity. *Neurochemical Research, 40*, 600–608.

Liu, L., Peritore, C., Ginsberg, J., Shih, J., Arun, S., & Donmez, G. (2015). Protective role of SIRT5 against motor deficit and dopaminergic degeneration in MPTP-induced mice model of Parkinson's disease. *Behavioural Brain Research, 281*, 215–221.

Liu, L., Wang, Q., Zhao, B., Wu, Q., & Wang, P. (2019). Exogenous nicotinamide adenine dinucleotide administration alleviates ischemia/reperfusion-induced oxidative injury in isolated rat hearts via Sirt5-SDH-succinate pathway. *European Journal of Pharmacology, 858*, 172520.

Liu, P., Huang, G., Wei, T., Gao, J., Huang, C., Sun, M., ... Shen, W. (2018). Sirtuin 3-induced macrophage autophagy in regulating NLRP3 inflammasome activation. *Biochimica et Biophysica Acta - Molecular Basis of Disease, 1864*, 764–777.

Liu, T., Yang, W., Pang, S., Yu, S., & Yan, B. (2018). Functional genetic variants within the SIRT2 gene promoter in type 2 diabetes mellitus. *Diabetes Research and Clinical Practice, 137*, 200–207.

Liu, Y., Cheng, A., Li, Y. J., Yang, Y., Kishimoto, Y., Zhang, S., ... Lu, D. (2019). SIRT3 mediates hippocampal synaptic adaptations to intermittent fasting and ameliorates deficits in APP mutant mice. *Nature Communications, 10*, 1886.

Liu, Y., Xie, Q. R., Wang, B., Shao, J., Zhang, T., Liu, T., ... Xia, W. (2013). Inhibition of SIRT6 in prostate cancer reduces cell viability and increases sensitivity to chemotherapeutics. *Protein Cell, 4*, 702–710.

Lo Sasso, G., Menzies, K. J., Mottis, A., Piersigilli, A., Perino, A., Yamamoto, H., ... Auwerx, J. (2014). SIRT2 deficiency modulates macrophage polarization and susceptibility to experimental colitis. *PLoS One, 9*, e103573.

Lombard, D. B., Alt, F. W., Cheng, H. L., Bunkenborg, J., Streeper, R. S., Mostoslavsky, R., ... Murphy, A. (2007). Mammalian Sir2 homolog SIRT3 regulates global mitochondrial lysine acetylation. *Molecular and Cellular Biology, 27*, 8807–8814.

Lombard, D. B., & Miller, R. A. (2014). Aging, disease, and longevity in mice. *Annual Review of Gerontology and Geriatrics, 34*, 93–138.

Lombard, D. B., & Zwaans, B. M. (2014). SIRT3: As simple as it seems? *Gerontology, 60*, 56–64.

Longo, V. D., & Kennedy, B. K. (2006). Sirtuins in aging and age-related disease. *Cell, 126*, 257–268.

Lopez-Otin, C., Blasco, M. A., Partridge, L., Serrano, M., & Kroemer, G. (2013). The hallmarks of aging. *Cell, 153*, 1194–1217.

Lu, J., Sun, D., Liu, Z., Li, M., Hong, H., Liu, C., ... Chen, S. (2016). SIRT6 suppresses isoproterenol-induced cardiac hypertrophy through activation of autophagy. *Translational Research: the Journal of Laboratory and Clinical Medicine, 172*, 96–112, e116.

Lu, J., Zhang, H., Chen, X., Zou, Y., Li, J., Wang, L., ... Zhuang, W. (2017). A small molecule activator of SIRT3 promotes deacetylation and activation of manganese superoxide dismutase. *Free Radical Biology & Medicine, 112*, 287–297.

Lu, M., Sarruf, D. A., Li, P., Osborn, O., Sanchez-Alavez, M., Talukdar, S., ... Morinaga, H. (2013). Neuronal Sirt1 deficiency increases insulin sensitivity in both brain and peripheral tissues. *The Journal of Biological Chemistry, 288*, 10722–10735.

Lu, W., Zuo, Y., Feng, Y., & Zhang, M. (2014). SIRT5 facilitates cancer cell growth and drug resistance in non-small cell lung cancer. *Tumour Biology: The Journal of the International Society for Oncodevelopmental Biology and Medicine, 35*, 10699–10705.

Lu, X., Yang, P., Zhao, X., Jiang, M., Hu, S., Ouyang, Y., ... Wu, J. (2019). OGDH mediates the inhibition of SIRT5 on cell proliferation and migration of gastric cancer. *Experimental Cell Research, 382*, 111483.

Ludewig, A. H., Izrayelit, Y., Park, D., Malik, R. U., Zimmermann, A., Mahanti, P., ... Riddle, D. L. (2013). Pheromone sensing regulates *Caenorhabditis elegans* lifespan and stress resistance via the deacetylase SIR-2.1. *Proceedings of the National Academy of Sciences of the United States of America, 110*, 5522–5527.

Luo, J., Li, M., Tang, Y., Laszkowska, M., Roeder, R. G., & Gu, W. (2004). Acetylation of p53 augments its site-specific DNA binding both in vitro and in vivo. *Proceedings of the National Academy of Sciences of the United States of America, 101*, 2259–2264.

Luo, J., Nikolaev, A. Y., Imai, S., Chen, D., Su, F., Shiloh, A., ... Gu, W. (2001). Negative control of p53 by Sir2alpha promotes cell survival under stress. *Cell, 107*, 137–148.

Luo, Y. X., Tang, X., An, X. Z., Xie, X. M., Chen, X. F., Zhao, X., ... Liu, D. P. (2017). SIRT4 accelerates Ang II-induced pathological cardiac hypertrophy by inhibiting manganese superoxide dismutase activity. *European Heart Journal, 38*, 1389–1398.

Luthi-Carter, R., Taylor, D. M., Pallos, J., Lambert, E., Amore, A., Parker, A., … Gokce, O. (2010). SIRT2 inhibition achieves neuroprotection by decreasing sterol biosynthesis. *Proceedings of the National Academy of Sciences of the United States of America*, *107*, 7927–7932.

Lv, X. B., Liu, L., Cheng, C., Yu, B., Xiong, L., Hu, K., … Sang, Y. (2015). SUN2 exerts tumor suppressor functions by suppressing the Warburg effect in lung cancer. *Scientific Reports*, *5*, 17940.

Lynn, E. G., McLeod, C. J., Gordon, J. P., Bao, J., & Sack, M. N. (2008). SIRT2 is a negative regulator of anoxia-reoxygenation tolerance via regulation of 14-3-3 zeta and BAD in H9c2 cells. *FEBS Letters*, *582*, 2857–2862.

Ma, Y., & Fei, X. (2018). SIRT5 regulates pancreatic beta-cell proliferation and insulin secretion in type 2 diabetes. *Experimental and Therapeutic Medicine*, *16*, 1417–1425.

Ma, Y., Qi, Y., Wang, L., Zheng, Z., Zhang, Y., & Zheng, J. (2019). SIRT5-mediated SDHA desuccinylation promotes clear cell renal cell carcinoma tumorigenesis. *Free Radical Biology & Medicine*, *134*, 458–467.

Madeo, F., Carmona-Gutierrez, D., Hofer, S. J., & Kroemer, G. (2019). Caloric restriction mimetics against age-associated disease: Targets, mechanisms, and therapeutic potential. *Cell Metabolism*, *29*, 592–610.

Mahajan, S. S., Scian, M., Sripathy, S., Posakony, J., Lao, U., Loe, T. K., … Bedalov, A. (2014). Development of pyrazolone and isoxazol-5-one cambinol analogues as sirtuin inhibitors. *Journal of Medicinal Chemistry*, *57*, 3283–3294.

Mair, W., Panowski, S. H., Shaw, R. J., & Dillin, A. (2009). Optimizing dietary restriction for genetic epistasis analysis and gene discovery in *C. elegans*. *PLoS One*, *4*, e4535.

Major Depressive Disorder Working Group of the Psychiatric, G. C., Ripke, S., Wray, N. R., Lewis, C. M., Hamilton, S. P., Weissman, M. M., … Boomsma, D. I. (2013). A mega-analysis of genome-wide association studies for major depressive disorder. *Molecular Psychiatry*, *18*, 497–511.

Maksin-Matveev, A., Kanfi, Y., Hochhauser, E., Isak, A., Cohen, H. Y., & Shainberg, A. (2015). Sirtuin 6 protects the heart from hypoxic damage. *Experimental Cell Research*, *330*, 81–90.

Malik, S., Villanova, L., Tanaka, S., Aonuma, M., Roy, N., Berber, E., … Chua, K. F. (2015). SIRT7 inactivation reverses metastatic phenotypes in epithelial and mesenchymal tumors. *Scientific Reports*, *5*, 9841.

Mancuso, R., del Valle, J., Modol, L., Martinez, A., Granado-Serrano, A. B., Ramirez-Nunez, O., … Navarro, X. (2014). Resveratrol improves motoneuron function and extends survival in SOD1 (G93A) ALS mice. *Neurotherapeutics: The Journal of the American Society for Experimental NeuroTherapeutics*, *11*, 419–432.

Manning, B. D., & Cantley, L. C. (2007). AKT/PKB signaling: Navigating downstream. *Cell*, *129*, 1261–1274.

Manuyakorn, A., Paulus, R., Farrell, J., Dawson, N. A., Tze, S., Cheung-Lau, G., … Horvath, S. (2010). Cellular histone modification patterns predict prognosis and treatment response in resectable pancreatic adenocarcinoma: Results from RTOG 9704. *Journal of Clinical Oncology: Official Journal of the American Society of Clinical Oncology*, *28*, 1358–1365.

Mao, B., Zhao, G., Lv, X., Chen, H. Z., Xue, Z., Yang, B., … Liang, C. C. (2011). Sirt1 deacetylates c-Myc and promotes c-Myc/Max association. *The International Journal of Biochemistry & Cell Biology*, *43*, 1573–1581.

Mao, Z., Hine, C., Tian, X., Van Meter, M., Au, M., Vaidya, A., … Gorbunova, V. (2011). SIRT6 promotes DNA repair under stress by activating PARP1. *Science*, *332*, 1443–1446.

Mariani, S., Fiore, D., Basciani, S., Persichetti, A., Contini, S., Lubrano, C., … Gnessi, L. (2015). Plasma levels of SIRT1 associate with non-alcoholic fatty liver disease in obese patients. *Endocrine*, *49*, 711–716.

Martens, C. R., Denman, B. A., Mazzo, M. R., Armstrong, M. L., Reisdorph, N., McQueen, M. B., … Seals, D. R. (2018). Chronic nicotinamide riboside supplementation is well-tolerated and elevates NAD(+) in healthy middle-aged and older adults. *Nature Communications*, *9*, 1286.

Masri, S., Rigor, P., Cervantes, M., Ceglia, N., Sebastian, C., Xiao, C., … Mostoslavsky, R. (2014). Partitioning circadian transcription by SIRT6 leads to segregated control of cellular metabolism. *Cell*, *158*, 659–672.

Masri, S., & Sassone-Corsi, P. (2014). Sirtuins and the circadian clock: Bridging chromatin and metabolism. *Science Signaling*, *7*, re6.

Massudi, H., Grant, R., Braidy, N., Guest, J., Farnsworth, B., & Guillemin, G. J. (2012). Age-associated changes in oxidative stress and NAD+ metabolism in human tissue. *PLoS One*, *7*, e42357.

Mathias, R. A., Greco, T. M., Oberstein, A., Budayeva, H. G., Chakrabarti, R., Rowland, E. A., … Cristea, I. M. (2014). Sirtuin 4 is a lipoamidase regulating pyruvate dehydrogenase complex activity. *Cell*, *159*, 1615–1625.

Matsumoto, T., Baker, D. J., d'Uscio, L. V., Mozammel, G., Katusic, Z. S., & van Deursen, J. M. (2007). Aging-associated vascular phenotype in mutant mice with low levels of BubR1. *Stroke; A Journal of Cerebral Circulation*, *38*, 1050–1056.

Matsushita, N., Yonashiro, R., Ogata, Y., Sugiura, A., Nagashima, S., Fukuda, T., … Yanagi, S. (2011). Distinct regulation of mitochondrial localization and stability of two human Sirt5 isoforms. *Genes to Cells: Devoted to Molecular & Cellular Mechanisms*, *16*, 190–202.

Mattagajasingh, I., Kim, C. S., Naqvi, A., Yamamori, T., Hoffman, T. A., Jung, S. B., … Irani, K. (2007). SIRT1 promotes endothelium-dependent vascular relaxation by activating endothelial nitric oxide synthase. *Proceedings of the National Academy of Sciences of the United States of America*, *104*, 14855–14860.

Mattison, J. A., Wang, M., Bernier, M., Zhang, J., Park, S. S., Maudsley, S., … Faulkner, S. (2014). Resveratrol prevents high fat/sucrose diet-induced central arterial wall inflammation and stiffening in nonhuman primates. *Cell Metabolism*, *20*, 183–190.

McBurney, M. W., Yang, X., Jardine, K., Hixon, M., Boekelheide, K., Webb, J. R., … Lemieux, M. (2003). The mammalian SIR2alpha protein has a role in embryogenesis and gametogenesis. *Molecular and Cellular Biology*, *23*, 38–54.

McCarthy, A. R., Hollick, J. J., & Westwood, N. J. (2010). The discovery of nongenotoxic activators of p53: Building on a cell-based high-throughput screen. *Seminars in Cancer Biology*, *20*, 40–45.

McCarthy, A. R., Pirrie, L., Hollick, J. J., Ronseaux, S., Campbell, J., Higgins, M., … Lain, S. (2012). Synthesis and biological characterisation of sirtuin inhibitors based on the tenovins. *Bioorganic & Medicinal Chemistry*, *20*, 1779–1793.

McGlynn, L. M., McCluney, S., Jamieson, N. B., Thomson, J., MacDonald, A. I., Oien, K., … Shiels, P. G. (2015). SIRT3 & SIRT7: Potential novel biomarkers for determining outcome in pancreatic cancer patients. *PLoS One*, *10*, e0131344.

Menssen, A., Hydbring, P., Kapelle, K., Vervoorts, J., Diebold, J., Luscher, B., … Hermeking, H. (2012). The c-MYC oncoprotein, the NAMPT enzyme, the SIRT1-inhibitor DBC1, and the SIRT1 deacetylase form a positive feedback loop. *Proceedings of the National Academy of Sciences of the United States of America*, *109*, E187–E196.

Mercken, E. M., Hu, J., Krzysik-Walker, S., Wei, M., Li, Y., McBurney, M. W., … Longo, V. D. (2014). SIRT1 but not its increased expression is essential for lifespan extension in caloric-restricted mice. *Aging Cell*, *13*, 193–196.

Mercken, E. M., Mitchell, S. J., Martin-Montalvo, A., Minor, R. K., Almeida, M., Gomes, A. P., … Zhang, Y. (2014). SRT2104 extends survival of male mice on a standard diet and preserves bone and muscle mass. *Aging Cell*, *13*, 787–796.

Michishita, E., McCord, R. A., Berber, E., Kioi, M., Padilla-Nash, H., Damian, M., … Barrett, J. C. (2008). SIRT6 is a histone H3 lysine 9 deacetylase that modulates telomeric chromatin. *Nature*, *452*, 492–496.

Michishita, E., McCord, R. A., Boxer, L. D., Barber, M. F., Hong, T., Gozani, O., & Chua, K. F. (2009). Cell cycle-dependent deacetylation of telomeric histone H3 lysine K56 by human SIRT6. *Cell Cycle*, *8*, 2664–2666.

Michishita, E., Park, J. Y., Burneskis, J. M., Barrett, J. C., & Horikawa, I. (2005). Evolutionarily conserved and nonconserved cellular localizations and functions of human SIRT proteins. *Molecular Biology of the Cell*, *16*, 4623–4635.

Miller, R. A., Buehner, G., Chang, Y., Harper, J. M., Sigler, R., & Smith-Wheelock, M. (2005). Methionine-deficient diet extends mouse lifespan, slows immune and lens aging, alters glucose, T4, IGF-I and insulin levels, and increases hepatocyte MIF levels and stress resistance. *Aging Cell*, *4*, 119–125.

Miller, R. A., Harrison, D. E., Astle, C. M., Baur, J. A., Boyd, A. R., de Cabo, R., ... Nelson, J. F. (2011). Rapamycin, but not resveratrol or simvastatin, extends life span of genetically heterogeneous mice. *The Journals of Gerontology. Series A, Biological Sciences and Medical Sciences*, *66*, 191–201.

Mills, K. F., Yoshida, S., Stein, L. R., Grozio, A., Kubota, S., Sasaki, Y., ... Uchida, K. (2016). Long-term administration of nicotinamide mononucleotide mitigates age-associated physiological decline in mice. *Cell Metabolism*, *24*, 795–806.

Min, L., Ji, Y., Bakiri, L., Qiu, Z., Cen, J., Chen, X., ... Qin, L. (2012). Liver cancer initiation is controlled by AP-1 through SIRT6-dependent inhibition of survivin. *Nature Cell Biology*, *14*, 1203–1211.

Min, S. W., Cho, S. H., Zhou, Y., Schroeder, S., Haroutunian, V., Seeley, W. W., ... Mukherjee, C. (2010). Acetylation of tau inhibits its degradation and contributes to tauopathy. *Neuron*, *67*, 953–966.

Min, S. W., Sohn, P. D., Li, Y., Devidze, N., Johnson, J. R., Krogan, N. J., ... Gan, L. (2018). SIRT1 deacetylates tau and reduces pathogenic tau spread in a mouse model of tauopathy. *The Journal of Neuroscience*, *38*, 3680–3688.

Ming, M., Han, W., Zhao, B., Sundaresan, N. R., Deng, C. X., Gupta, M. P., & He, Y. Y. (2014). SIRT6 promotes COX-2 expression and acts as an oncogene in skin cancer. *Cancer Research*, *74*, 5925–5933.

Ming, M., Qiang, L., Zhao, B., & He, Y. Y. (2014). Mammalian SIRT2 inhibits keratin 19 expression and is a tumor suppressor in skin. *Experimental Dermatology*, *23*, 207–209.

Minor, R. K., Baur, J. A., Gomes, A. P., Ward, T. M., Csiszar, A., Mercken, E. M., ... Scheibye-Knudsen, M. (2011). SRT1720 improves survival and healthspan of obese mice. *Scientific Reports*, *1*, 70.

Mitchell, S. J., Martin-Montalvo, A., Mercken, E. M., Palacios, H. H., Ward, T. M., Abulwerdi, G., ... Sinclair, D. A. (2014). The SIRT1 activator SRT1720 extends lifespan and improves health of mice fed a standard diet. *Cell Reports*, *6*, 836–843.

Miyasato, Y., Yoshizawa, T., Sato, Y., Nakagawa, T., Miyasato, Y., Kakizoe, Y., ... Braun, T. (2018). Sirtuin 7 deficiency ameliorates cisplatin-induced acute kidney injury through regulation of the inflammatory response. *Scientific Reports*, *8*, 5927.

Miyo, M., Yamamoto, H., Konno, M., Colvin, H., Nishida, N., Koseki, J., ... Uemura, M. (2015). Tumour-suppressive function of SIRT4 in human colorectal cancer. *British Journal of Cancer*, *113*, 492–499.

Moniot, S., Forgione, M., Lucidi, A., Hailu, G. S., Nebbioso, A., Carafa, V., ... Passeri, D. (2017). Development of 1,2,4-Oxadiazoles as potent and selective inhibitors of the human deacetylase Sirtuin 2: Structure-activity relationship, x-ray crystal structure, and anticancer activity. *Journal of Medicinal Chemistry*, *60*, 2344–2360.

Morley, J. E. (2004). The metabolic syndrome and aging. *The Journals of Gerontology. Series A, Biological Sciences and Medical Sciences*, *59*, 139–142.

Morris-Blanco, K. C., Dave, K. R., Saul, I., Koronowski, K. B., Stradecki, H. M., & Perez-Pinzon, M. A. (2016). Protein kinase C epsilon promotes cerebral ischemic tolerance via modulation of mitochondrial Sirt5. *Scientific Reports*, *6*, 29790.

Mostoslavsky, R., Chua, K. F., Lombard, D. B., Pang, W. W., Fischer, M. R., Gellon, L., ... Murphy, M. M. (2006). Genomic instability and aging-like phenotype in the absence of mammalian SIRT6. *Cell*, *124*, 315–329.

Mouchiroud, L., Houtkooper, R. H., Moullan, N., Katsyuba, E., Ryu, D., Canto, C., ... Schoonjans, K. (2013). The NAD(+)/Sirtuin pathway modulates longevity through activation of mitochondrial UPR and FOXO signaling. *Cell*, *154*, 430–441.

Moynihan, K. A., Grimm, A. A., Plueger, M. M., Bernal-Mizrachi, E., Ford, E., Cras-Meneur, C., ... Imai, S. (2005). Increased dosage of mammalian Sir2 in pancreatic beta cells enhances glucose-stimulated insulin secretion in mice. *Cell Metabolism*, *2*, 105–117.

Mu, N., Lei, Y., Wang, Y., Wang, Y., Duan, Q., Ma, G., ... Su, L. (2019). Inhibition of SIRT1/2 upregulates HSPA5 acetylation and induces pro-survival autophagy via ATF4-DDIT4-mTORC1 axis in human lung cancer cells. *Apoptosis: An International Journal on Programmed Cell Death*, *24*, 798–811.

Mu, P., Liu, K., Lin, Q., Yang, W., Liu, D., Lin, Z., ... Ji, T. (2019). Sirtuin 7 promotes glioma proliferation and invasion through activation of the ERK/STAT3 signaling pathway. *Oncology Letters*, *17*, 1445–1452.

Nakagawa, T., Lomb, D. J., Haigis, M. C., & Guarente, L. (2009). SIRT5 deacetylates carbamoyl phosphate synthetase 1 and regulates the urea cycle. *Cell*, *137*, 560–570.

Napper, A. D., Hixon, J., McDonagh, T., Keavey, K., Pons, J. F., Barker, J., ... Hamelin, E. (2005). Discovery of indoles as potent and selective inhibitors of the deacetylase SIRT1. *Journal of Medicinal Chemistry*, *48*, 8045–8054.

Nasrin, N., Wu, X., Fortier, E., Feng, Y., Bare, O. C., Chen, S., ... Bordone, L. (2010). SIRT4 regulates fatty acid oxidation and mitochondrial gene expression in liver and muscle cells. *The Journal of Biological Chemistry*, *285*, 31995–32002.

Nawaz, A., Mehmood, A., Kanatani, Y., Kado, T., Igarashi, Y., Takikawa, A., ... Yagi, K. (2018). Sirt1 activator induces proangiogenic genes in preadipocytes to rescue insulin resistance in diet-induced obese mice. *Scientific Reports*, *8*, 11370.

Newman, J. C., Covarrubias, A. J., Zhao, M., Yu, X., Gut, P., Ng, C. P., ... Verdin, E. (2017). Ketogenic diet reduces midlife mortality and improves memory in aging mice. *Cell Metabolism*, *26*, 547–557, e548.

Nishida, Y., Rardin, M. J., Carrico, C., He, W., Sahu, A. K., Gut, P., ... Gibson, B. W. (2015). SIRT5 regulates both cytosolic and mitochondrial protein malonylation with glycolysis as a major target. *Molecular Cell*, *59*, 321–332.

North, B. J., Marshall, B. L., Borra, M. T., Denu, J. M., & Verdin, E. (2003). The human Sir2 ortholog, SIRT2, is an NAD + -dependent tubulin deacetylase. *Molecular Cell*, *11*, 437–444.

North, B. J., Rosenberg, M. A., Jeganathan, K. B., Hafner, A. V., Michan, S., Dai, J., ... Sauve, A. A. (2014). SIRT2 induces the checkpoint kinase BubR1 to increase lifespan. *The EMBO Journal*, *33*, 1438–1453.

North, B. J., & Verdin, E. (2007). Mitotic regulation of SIRT2 by cyclin-dependent kinase 1-dependent phosphorylation. *The Journal of Biological Chemistry*, *282*, 19546–19555.

Oellerich, M. F., & Potente, M. (2012). FOXOs and sirtuins in vascular growth, maintenance, and aging. *Circulation Research*, *110*, 1238–1251.

Ogura, M., Nakamura, Y., Tanaka, D., Zhuang, X., Fujita, Y., Obara, A., ... Inagaki, N. (2010). Overexpression of SIRT5 confirms its involvement in deacetylation and activation of carbamoyl phosphate synthetase 1. *Biochemical and Biophysical Research Communications*, *393*, 73–78.

Oka, S., Alcendor, R., Zhai, P., Park, J. Y., Shao, D., Cho, J., ... Sadoshima, J. (2011). PPARalpha-Sirt1 complex mediates cardiac hypertrophy and failure through suppression of the ERR transcriptional pathway. *Cell Metabolism, 14*, 598–611.

Ota, H., Tokunaga, E., Chang, K., Hikasa, M., Iijima, K., Eto, M., ... Kaneki, M. (2006). Sirt1 inhibitor, Sirtinol, induces senescence-like growth arrest with attenuated Ras-MAPK signaling in human cancer cells. *Oncogene, 25*, 176–185.

Outeiro, T. F., Kontopoulos, E., Altmann, S. M., Kufareva, I., Strathearn, K. E., Amore, A. M., ... McLean, P. J. (2007). Sirtuin 2 inhibitors rescue alpha-synuclein-mediated toxicity in models of Parkinson's disease. *Science, 317*, 516–519.

Ozcan, L., & Tabas, I. (2012). Role of endoplasmic reticulum stress in metabolic disease and other disorders. *Annual Review of Medicine, 63*, 317–328.

Pacholec, M., Bleasdale, J. E., Chrunyk, B., Cunningham, D., Flynn, D., Garofalo, R. S., ... Pabst, B. (2010). SRT1720, SRT2183, SRT1460, and resveratrol are not direct activators of SIRT1. *The Journal of Biological Chemistry, 285*, 8340–8351.

Paik, J. H. (2006). FOXOs in the maintenance of vascular homoeostasis. *Biochemical Society Transactions, 34*, 731–734.

Paik, J. H., Kollipara, R., Chu, G., Ji, H., Xiao, Y., Ding, Z., ... Carrasco, D. R. (2007). FoxOs are lineage-restricted redundant tumor suppressors and regulate endothelial cell homeostasis. *Cell, 128*, 309–323.

Pais, T. F., Szego, E. M., Marques, O., Miller-Fleming, L., Antas, P., Guerreiro, P., ... Outeiro, T. F. (2013). The NAD-dependent deacetylase sirtuin 2 is a suppressor of microglial activation and brain inflammation. *The EMBO Journal, 32*, 2603–2616.

Palacios, O. M., Carmona, J. J., Michan, S., Chen, K. Y., Manabe, Y., Ward, J. L., 3rd, ... Tong, Q. (2009). Diet and exercise signals regulate SIRT3 and activate AMPK and PGC-1alpha in skeletal muscle. *Aging, 1*, 771–783.

Pallos, J., Bodai, L., Lukacsovich, T., Purcell, J. M., Steffan, J. S., Thompson, L. M., & Marsh, J. L. (2008). Inhibition of specific HDACs and sirtuins suppresses pathogenesis in a *Drosophila* model of Huntington's disease. *Human Molecular Genetics, 17*, 3767–3775.

Palmirotta, R., Cives, M., Della-Morte, D., Capuani, B., Lauro, D., Guadagni, F., & Silvestris, F. (2016). Sirtuins and cancer: Role in the epithelial-mesenchymal transition. *Oxidative Medicine and Cellular Longevity, 2016*, 3031459.

Pandithage, R., Lilischkis, R., Harting, K., Wolf, A., Jedamzik, B., Luscher-Firzlaff, J., ... Knoll, B. (2008). The regulation of SIRT2 function by cyclin-dependent kinases affects cell motility. *The Journal of Cell Biology, 180*, 915–929.

Parenti, M. D., Grozio, A., Bauer, I., Galeno, L., Damonte, P., Millo, E., ... Bruzzone, S. (2014). Discovery of novel and selective SIRT6 inhibitors. *Journal of Medicinal Chemistry, 57*, 4796–4804.

Park, H. K., Hong, J. H., Oh, Y. T., Kim, S. S., Yin, J., Lee, A. J., ... Park, C. K. (2019). Interplay between TRAP1 and Sirtuin-3 modulates mitochondrial respiration and oxidative stress to maintain stemness of glioma stem cells. *Cancer Research, 79*, 1369–1382.

Park, J., Chen, Y., Tishkoff, D. X., Peng, C., Tan, M., Dai, L., ... Skinner, M. E. (2013). SIRT5-mediated lysine desuccinylation impacts diverse metabolic pathways. *Molecular Cell, 50*, 919–930.

Park, S. H., Ozden, O., Liu, G., Song, H. Y., Zhu, Y., Yan, Y., ... Principe, D. R. (2016). SIRT2-mediated deacetylation and tetramerization of pyruvate kinase directs glycolysis and tumor growth. *Cancer Research, 76*, 3802–3812.

Park, S. J., Ahmad, F., Philp, A., Baar, K., Williams, T., Luo, H., ... Brown, A. L. (2012). Resveratrol ameliorates aging-related metabolic phenotypes by inhibiting cAMP phosphodiesterases. *Cell, 148*, 421–433.

Park, S. Y., Lee, S. W., Lee, S. Y., Hong, K. W., Bae, S. S., Kim, K., & Kim, C. D. (2017). SIRT1/adenosine monophosphate-activated protein kinase alpha signaling enhances macrophage polarization to an anti-inflammatory phenotype in rheumatoid arthritis. *Frontiers in Immunology, 8*, 1135.

Parker, J. A., Vazquez-Manrique, R. P., Tourette, C., Farina, F., Offner, N., Mukhopadhyay, A., ... Tissenbaum, H. A. (2012). Integration of beta-catenin, sirtuin, and FOXO signaling protects from mutant huntingtin toxicity. *The Journal of Neuroscience, 32*, 12630–12640.

Paulin, R., Dromparis, P., Sutendra, G., Gurtu, V., Zervopoulos, S., Bowers, L., ... Bonnet, S. (2014). Sirtuin 3 deficiency is associated with inhibited mitochondrial function and pulmonary arterial hypertension in rodents and humans. *Cell Metabolism., 20*, 827–837.

Pearson, K. J., Baur, J. A., Lewis, K. N., Peshkin, L., Price, N. L., Labinskyy, N., ... Perez, E. (2008). Resveratrol delays age-related deterioration and mimics transcriptional aspects of dietary restriction without extending life span. *Cell Metabolism, 8*, 157–168.

Pebay-Peyroula, E., & Brandolin, G. (2004). Nucleotide exchange in mitochondria: Insight at a molecular level. *Current Opinion in Structural Biology, 14*, 420–425.

Peng, C., Lu, Z., Xie, Z., Cheng, Z., Chen, Y., Tan, M., ... Yang, K. (2011). The first identification of lysine malonylation substrates and its regulatory enzyme. *Molecular & Cellular Proteomics: MCP, 10*, M111 012658, 1–12.

Perrone, C. E., Malloy, V. L., Orentreich, D. S., & Orentreich, N. (2013). Metabolic adaptations to methionine restriction that benefit health and lifespan in rodents. *Experimental Gerontology, 48*, 654–660.

Peshti, V., Obolensky, A., Nahum, L., Kanfi, Y., Rathaus, M., Avraham, M., ... Cohen, H. Y. (2017). Characterization of physiological defects in adult SIRT6-/- mice. *PLoS One, 12*, e0176371.

Pfluger, P. T., Herranz, D., Velasco-Miguel, S., Serrano, M., & Tschop, M. H. (2008). Sirt1 protects against high-fat diet-induced metabolic damage. *Proceedings of the National Academy of Sciences of the United States of America, 105*, 9793–9798.

Phng, L. K., & Gerhardt, H. (2009). Angiogenesis: A team effort coordinated by notch. *Developmental Cell, 16*, 196–208.

Picard, F., Kurtev, M., Chung, N., Topark-Ngarm, A., Senawong, T., Machado De Oliveira, R., ... Guarente, L. (2004). Sirt1 promotes fat mobilization in white adipocytes by repressing PPAR-gamma. *Nature, 429*, 771–776.

Pillai, V. B., Bindu, S., Sharp, W., Fang, Y. H., Kim, G., Gupta, M., ... Gupta, M. P. (2016). Sirt3 protects mitochondrial DNA damage and blocks the development of doxorubicin-induced cardiomyopathy in mice. *American Journal of Physiology. Heart and Circulatory Physiology, 310*, H962–H972.

Pillai, V. B., Kanwal, A., Fang, Y. H., Sharp, W. W., Samant, S., Arbiser, J., & Gupta, M. P. (2017). Honokiol, an activator of Sirtuin-3 (SIRT3) preserves mitochondria and protects the heart from doxorubicin-induced cardiomyopathy in mice. *Oncotarget, 8*, 34082–34098.

Pillai, V. B., Samant, S., Sundaresan, N. R., Raghuraman, H., Kim, G., Bonner, M. Y., ... Gius, D. (2015). Honokiol blocks and reverses cardiac hypertrophy in mice by activating mitochondrial Sirt3. *Nature Communications, 6*, 6656.

Pillai, V. B., Sundaresan, N. R., Kim, G., Gupta, M., Rajamohan, S. B., Pillai, J. B., ... Gupta, M. P. (2010). Exogenous NAD blocks cardiac hypertrophic response via activation of the SIRT3-LKB1-AMP-activated kinase pathway. *The Journal of Biological Chemistry, 285*, 3133–3144.

Planavila, A., Iglesias, R., Giralt, M., & Villarroya, F. (2011). Sirt1 acts in association with PPARalpha to protect the heart from hypertrophy, metabolic dysregulation, and inflammation. *Cardiovascular Research, 90*, 276–284.

Polednak, A. P. (2008). Estimating the number of U.S. incident cancers attributable to obesity and the impact on temporal trends in incidence rates for obesity-related cancers. *Cancer Detection and Prevention, 32*, 190–199.

Polito, L., Kehoe, P. G., Davin, A., Benussi, L., Ghidoni, R., Binetti, G., ... Clerici, F. (2013). The SIRT2 polymorphism rs10410544 and risk of Alzheimer's disease in two Caucasian case-control cohorts. *Alzheimer's & Dementia: The Journal of the Alzheimer's Association, 9*, 392–399.

Polletta, L., Vernucci, E., Carnevale, I., Arcangeli, T., Rotili, D., Palmerio, S., ... Pellegrini, L. (2015). SIRT5 regulation of ammonia-induced autophagy and mitophagy. *Autophagy, 11*, 253–270.

Ponugoti, B., Kim, D. H., Xiao, Z., Smith, Z., Miao, J., Zang, M., ... Kemper, J. K. (2010). SIRT1 deacetylates and inhibits SREBP-1C activity in regulation of hepatic lipid metabolism. *The Journal of Biological Chemistry, 285*, 33959–33970.

Porcelli, S., Salfi, R., Politis, A., Atti, A. R., Albani, D., Chierchia, A., ... Liappas, I. (2013). Association between Sirtuin 2 gene rs10410544 polymorphism and depression in Alzheimer's disease in two independent European samples. *Journal of Neural Transmission, 120*, 1709–1715.

Porquet, D., Casadesus, G., Bayod, S., Vicente, A., Canudas, A. M., Vilaplana, J., ... Pallas, M. (2013). Dietary resveratrol prevents Alzheimer's markers and increases life span in SAMP8. *Age, 35*, 1851–1865.

Porter, G. A., Urciuoli, W. R., Brookes, P. S., & Nadtochiy, S. M. (2014). SIRT3 deficiency exacerbates ischemia-reperfusion injury: Implication for aged hearts. *American Journal of Physiology. Heart and Circulatory Physiology, 306*, H1602–H1609.

Potente, M., & Dimmeler, S. (2008). Emerging roles of SIRT1 in vascular endothelial homeostasis. *Cell Cycle, 7*, 2117–2122.

Potente, M., Ghaeni, L., Baldessari, D., Mostoslavsky, R., Rossig, L., Dequiedt, F., ... Alt, F. W. (2007). SIRT1 controls endothelial angiogenic functions during vascular growth. *Genes & Development, 21*, 2644–2658.

Potente, M., Urbich, C., Sasaki, K., Hofmann, W. K., Heeschen, C., Aicher, A., ... Dimmeler, S. (2005). Involvement of Foxo transcription factors in angiogenesis and postnatal neovascularization. *The Journal of Clinical Investigation, 115*, 2382–2392.

Prola, A., Pires Da Silva, J., Guilbert, A., Lecru, L., Piquereau, J., Ribeiro, M., ... Boursier, C. (2017). SIRT1 protects the heart from ER stress-induced cell death through eIF2alpha deacetylation. *Cell Death and Differentiation, 24*, 343–356.

Purushotham, A., Schug, T. T., Xu, Q., Surapureddi, S., Guo, X., & Li, X. (2009). Hepatocyte-specific deletion of SIRT1 alters fatty acid metabolism and results in hepatic steatosis and inflammation. *Cell Metabolism, 9*, 327–338.

Qiang, L., Wang, L., Kon, N., Zhao, W., Lee, S., Zhang, Y., ... Farmer, S. R. (2012). Brown remodeling of white adipose tissue by SirT1-dependent deacetylation of Ppargamma. *Cell, 150*, 620–632.

Qin, J., Liu, Y., Lu, Y., Liu, M., Li, M., Li, J., & Wu, L. (2017). Hypoxia-inducible factor 1 alpha promotes cancer stem cells-like properties in human ovarian cancer cells by upregulating SIRT1 expression. *Scientific Reports, 7*, 10592.

Qin, K., Han, C., Zhang, H., Li, T., Li, N., & Cao, X. (2017). NAD(+) dependent deacetylase Sirtuin 5 rescues the innate inflammatory response of endotoxin tolerant macrophages by promoting acetylation of p65. *Journal of Autoimmunity, 81*, 120–129.

Qiu, X., Brown, K., Hirschey, M. D., Verdin, E., & Chen, D. (2010). Calorie restriction reduces oxidative stress by SIRT3-mediated SOD2 activation. *Cell Metabolism, 12*, 662–667.

Quan, Y., Wang, N., Chen, Q., Xu, J., Cheng, W., Di, M., ... Gao, W. Q. (2015). SIRT3 inhibits prostate cancer by destabilizing oncoprotein c-MYC through regulation of the PI3K/Akt pathway. *Oncotarget, 6*, 26494–26507.

Quinti, L., Casale, M., Moniot, S., Pais, T. F., Van Kanegan, M. J., Kaltenbach, L. S., ... Runne, H. (2016). SIRT2- and NRF2-targeting thiazole-containing compound with therapeutic activity in Huntington's disease models. *Cell Chemical Biology, 23*, 849–861.

Ragonese, P., D'Amelio, M., Callari, G., Aiello, F., Morgante, L., & Savettieri, G. (2007). Height as a potential indicator of early life events predicting Parkinson's disease: A case-control study. *Movement Disorders: Official Journal of the Movement Disorder Society, 22*, 2263–2267.

Rahman, S., & Islam, R. (2011). Mammalian Sirt1: Insights on its biological functions. *Cell Communication and Signaling: CCS, 9*, 11.

Rai, E., Sharma, S., Kaul, S., Jain, K., Matharoo, K., Bhanwer, A. S., & Bamezai, R. N. (2012). The interactive effect of SIRT1 promoter region polymorphism on type 2 diabetes susceptibility in the North Indian population. *PLoS One, 7*, e48621.

Raimondo, A., Rees, M. G., & Gloyn, A. L. (2015). Glucokinase regulatory protein: Complexity at the crossroads of triglyceride and glucose metabolism. *Current Opinion in Lipidology, 26*, 88–95.

Ramakrishnan, G., Davaakhuu, G., Kaplun, L., Chung, W. C., Rana, A., Atfi, A., ... Tzivion, G. (2014). Sirt2 deacetylase is a novel AKT binding partner critical for AKT activation by insulin. *The Journal of Biological Chemistry, 289*, 6054–6066.

Ramesh, S., Govindarajulu, M., Lynd, T., Briggs, G., Adamek, D., Jones, E., ... Amin, R. (2018). SIRT3 activator Honokiol attenuates beta-Amyloid by modulating amyloidogenic pathway. *PLoS One, 13*, e0190350.

Ramsey, K. M., Mills, K. F., Satoh, A., & Imai, S. (2008). Age-associated loss of Sirt1-mediated enhancement of glucose-stimulated insulin secretion in beta cell-specific Sirt1-overexpressing (BESTO) mice. *Aging Cell, 7*, 78–88.

Ran, L. K., Chen, Y., Zhang, Z. Z., Tao, N. N., Ren, J. H., Zhou, L., ... Li, W. Y. (2016). SIRT6 overexpression potentiates apoptosis evasion in hepatocellular carcinoma via BCL2-associated X protein-dependent apoptotic pathway. *Clinical Cancer Research: an Official Journal of the American Association for Cancer Research, 22*, 3372–3382.

Ran, M., Li, Z., Yang, L., Tong, L., Zhang, L., & Dong, H. (2015). Calorie restriction attenuates cerebral ischemic injury via increasing SIRT1 synthesis in the rat. *Brain Research, 1610*, 61–68.

Rao, S., Luo, N., Sui, J., Xu, Q., & Zhang, F. (2020). Effect of the SIRT1 gene on regional cortical grey matter density in the Han Chinese population. *The British Journal of Psychiatry: the Journal of Mental Science, 216*, 254–258.

Rardin, M. J., He, W., Nishida, Y., Newman, J. C., Carrico, C., Danielson, S. R., ... Li, B. (2013). SIRT5 regulates the mitochondrial lysine succinylome and metabolic networks. *Cell Metabolism, 18*, 920–933.

Richie, J. P., Jr., Leutzinger, Y., Parthasarathy, S., Malloy, V., Orentreich, N., & Zimmerman, J. A. (1994). Methionine restriction increases blood glutathione and longevity in F344 rats. *The FASEB Journal, 8*, 1302–1307.

Rimbaud, S., Ruiz, M., Piquereau, J., Mateo, P., Fortin, D., Veksler, V., ... Ventura-Clapier, R. (2011). Resveratrol improves survival, hemodynamics and energetics in a rat model of hypertension leading to heart failure. *PLoS One, 6*, e26391.

Rine, J., & Herskowitz, I. (1987). Four genes responsible for a position effect on expression from HML and HMR in *Saccharomyces cerevisiae*. *Genetics, 116*, 9–22.

Rine, J., Strathern, J. N., Hicks, J. B., & Herskowitz, I. (1979). A suppressor of mating-type locus mutations in *Saccharomyces cerevisiae*: Evidence for and identification of cryptic mating-type loci. *Genetics, 93*, 877–901.

Rizki, G., Iwata, T. N., Li, J., Riedel, C. G., Picard, C. L., Jan, M., ... Lee, S. S. (2011). The evolutionarily conserved longevity determinants HCF-1 and SIR-2.1/SIRT1 collaborate to regulate DAF-16/FOXO. *PLoS Genetics, 7*, e1002235.

Roberts, D. L., Dive, C., & Renehan, A. G. (2010). Biological mechanisms linking obesity and cancer risk: New perspectives. *Annual Review of Medicine*, *61*, 301–316.

Roberts, M. N., Wallace, M. A., Tomilov, A. A., Zhou, Z., Marcotte, G. R., Tran, D., ... Knotts, T. A. (2017). A ketogenic diet extends longevity and healthspan in adult mice. *Cell Metabolism*, *26*, 539–546, e535.

Rodgers, J. T., Lerin, C., Gerhart-Hines, Z., & Puigserver, P. (2008). Metabolic adaptations through the PGC-1 alpha and SIRT1 pathways. *FEBS Letters*, *582*, 46–53.

Rodgers, J. T., Lerin, C., Haas, W., Gygi, S. P., Spiegelman, B. M., & Puigserver, P. (2005). Nutrient control of glucose homeostasis through a complex of PGC-1alpha and SIRT1. *Nature*, *434*, 113–118.

Rogina, B., & Helfand, S. L. (2004). Sir2 mediates longevity in the fly through a pathway related to calorie restriction. *Proceedings of the National Academy of Sciences of the United States of America*, *101*, 15998–16003.

Rose, G., Dato, S., Altomare, K., Bellizzi, D., Garasto, S., Greco, V., ... Barbi, C. (2003). Variability of the SIRT3 gene, human silent information regulator Sir2 homologue, and survivorship in the elderly. *Experimental Gerontology*, *38*, 1065–1070.

Rothgiesser, K. M., Erener, S., Waibel, S., Luscher, B., & Hottiger, M. O. (2010). SIRT2 regulates NF-kappaB dependent gene expression through deacetylation of p65 Lys310. *Journal of Cell Science*, *123*, 4251–4258.

Rumpf, T., Schiedel, M., Karaman, B., Roessler, C., North, B. J., Lehotzky, A., ... Gajer, M. (2015). Selective Sirt2 inhibition by ligand-induced rearrangement of the active site. *Nat Commun*, *6*, 6263.

Ryu, D., Jo, Y. S., Lo Sasso, G., Stein, S., Zhang, H., Perino, A., ... Hottiger, M. O. (2014). A SIRT7-dependent acetylation switch of GABPbeta1 controls mitochondrial function. *Cell Metabolism*, *20*, 856–869.

Sadhukhan, S., Liu, X., Ryu, D., Nelson, O. D., Stupinski, J. A., Li, Z., ... Locasale, J. W. (2016). Metabolomics-assisted proteomics identifies succinylation and SIRT5 as important regulators of cardiac function. *Proceedings of the National Academy of Sciences of the United States of America*, *113*, 4320–4325.

Sahin, K., Yilmaz, S., & Gozukirmizi, N. (2014). Changes in human sirtuin 6 gene promoter methylation during aging. *Biomed Reports*, *2*, 574–578.

Salehi, B., Mishra, A. P., Nigam, M., Sener, B., Kilic, M., Sharifi-Rad, M., ... Sharifi-Rad, J. (2018). Resveratrol: A double-edged sword in health benefits. *Biomedicines*, *6*, 91.

Samson, S. L., & Garber, A. J. (2014). Metabolic syndrome. *Endocrinology and Metabolism Clinics of North America*, *43*, 1–23.

Sarikhani, M., Maity, S., Mishra, S., Jain, A., Tamta, A. K., Ravi, V., ... Kumar, S. (2018). SIRT2 deacetylase represses NFAT transcription factor to maintain cardiac homeostasis. *The Journal of Biological Chemistry*, *293*, 5281–5294.

Sarumaru, M., Watanabe, M., Inoue, N., Hisamoto, Y., Morita, E., Arakawa, Y., ... Iwatani, Y. (2016). Association between functional SIRT1 polymorphisms and the clinical characteristics of patients with autoimmune thyroid disease. *Autoimmunity*, *49*, 329–337.

Satoh, A., Brace, C. S., Ben-Josef, G., West, T., Wozniak, D. F., Holtzman, D. M., ... Imai, S. (2010). SIRT1 promotes the central adaptive response to diet restriction through activation of the dorsomedial and lateral nuclei of the hypothalamus. *The Journal of Neuroscience: The Official Journal of the Society for Neuroscience*, *30*, 10220–10232.

Satoh, A., Brace, C. S., Rensing, N., Cliften, P., Wozniak, D. F., Herzog, E. D., ... Imai, S. (2013). Sirt1 extends life span and delays aging in mice through the regulation of Nk2 homeobox 1 in the DMH and LH. *Cell Metabolism*, *18*, 416–430.

Satterstrom, F. K., Swindell, W. R., Laurent, G., Vyas, S., Bulyk, M. L., & Haigis, M. C. (2015). Nuclear respiratory factor 2 induces SIRT3 expression. *Aging Cell*, *14*, 818–825.

Scarpulla, R. C. (2002). Nuclear activators and coactivators in mammalian mitochondrial biogenesis. *Biochimica et Biophysica Acta*, *1576*, 1–14.

Schenk, S., McCurdy, C. E., Philp, A., Chen, M. Z., Holliday, M. J., Bandyopadhyay, G. K., ... Olefsky, J. M. (2011). Sirt1 enhances skeletal muscle insulin sensitivity in mice during caloric restriction. *The Journal of Clinical Investigation*, *121*, 4281–4288.

Schmeisser, K., Mansfeld, J., Kuhlow, D., Weimer, S., Priebe, S., Heiland, I., ... Kanfi, Y. (2013). Role of sirtuins in lifespan regulation is linked to methylation of nicotinamide. *Nature Chemical Biology*, *9*, 693–700.

Schug, T. T., Xu, Q., Gao, H., Peres-da-Silva, A., Draper, D. W., Fessler, M. B., ... Li, X. (2010). Myeloid deletion of SIRT1 induces inflammatory signaling in response to environmental stress. *Molecular and Cellular Biology*, *30*, 4712–4721.

Schwer, B., Eckersdorff, M., Li, Y., Silva, J. C., Fermin, D., Kurtev, M. V., ... Lombard, D. B. (2009). Calorie restriction alters mitochondrial protein acetylation. *Aging Cell*, *8*, 604–606.

Sebai, H., Sani, M., Ghanem-Boughanmi, N., & Aouani, E. (2010). Prevention of lipopolysaccharide-induced mouse lethality by resveratrol. *Food and Chemical Toxicology: an International Journal Published for the British Industrial Biological Research Association*, *48*, 1543–1549.

Sebastian, C., Zwaans, B. M., Silberman, D. M., Gymrek, M., Goren, A., Zhong, L., ... Toiber, D. (2012). The histone deacetylase SIRT6 is a tumor suppressor that controls cancer metabolism. *Cell*, *151*, 1185–1199.

Seligson, D. B., Horvath, S., McBrian, M. A., Mah, V., Yu, H., Tze, S., ... Kurdistani, S. K. (2009). Global levels of histone modifications predict prognosis in different cancers. *The American Journal of Pathology*, *174*, 1619–1628.

Serrano, L., Martinez-Redondo, P., Marazuela-Duque, A., Vazquez, B. N., Dooley, S. J., Voigt, P., ... Rabanal, R. M. (2013). The tumor suppressor SirT2 regulates cell cycle progression and genome stability by modulating the mitotic deposition of H4K20 methylation. *Genes & Development*, *27*, 639–653.

Serrano-Gomez, S. J., Maziveyi, M., & Alahari, S. K. (2016). Regulation of epithelial-mesenchymal transition through epigenetic and post-translational modifications. *Molecular Cancer*, *15*, 18.

Seyyedebrahimi, S., Khodabandehloo, H., Nasli Esfahani, E., & Meshkani, R. (2018). The effects of resveratrol on markers of oxidative stress in patients with type 2 diabetes: A randomized, double-blind, placebo-controlled clinical trial. *Acta Diabetologica*, *55*, 341–353.

Shi, T., Wang, F., Stieren, E., & Tong, Q. (2005). SIRT3, a mitochondrial sirtuin deacetylase, regulates mitochondrial function and thermogenesis in brown adipocytes. *The Journal of Biological Chemistry*, *280*, 13560–13567.

Shih, J., Liu, L., Mason, A., Higashimori, H., & Donmez, G. (2014). Loss of SIRT4 decreases GLT-1-dependent glutamate uptake and increases sensitivity to kainic acid. *Journal of Neurochemistry*, *131*, 573–581.

Shimazu, T., Hirschey, M. D., Hua, L., Dittenhafer-Reed, K. E., Schwer, B., Lombard, D. B., ... Denu, J. M. (2010). SIRT3 deacetylates mitochondrial 3-hydroxy-3-methylglutaryl CoA synthase 2 and regulates ketone body production. *Cell Metabolism*, *12*, 654–661.

Shimazu, T., Hirschey, M. D., Huang, J. Y., Ho, L. T., & Verdin, E. (2010). Acetate metabolism and aging: An emerging connection. *Mechanisms of Ageing and Development*, *131*, 511–516.

Shin, J., He, M., Liu, Y., Paredes, S., Villanova, L., Brown, K., ... Wojnoonski, K. (2013). SIRT7 represses Myc activity to suppress ER stress and prevent fatty liver disease. *Cell Reports*, *5*, 654–665.

Shiojima, I., Yefremashvili, M., Luo, Z., Kureishi, Y., Takahashi, A., Tao, J., ... Walsh, K. (2002). Akt signaling mediates postnatal heart growth in response to insulin and nutritional status. *The Journal of Biological Chemistry*, *277*, 37670–37677.

Shore, D., Squire, M., & Nasmyth, K. A. (1984). Characterization of two genes required for the position-effect control of yeast mating-type genes. *The EMBO Journal*, *3*, 2817–2823.

Shuai, L., Zhang, L. N., Li, B. H., Tang, C. L., Wu, L. Y., Li, J., & Li, J. Y. (2019). SIRT5 regulates brown adipocyte differentiation and browning of subcutaneous white adipose tissue. *Diabetes*, *68*, 1449–1461.

Shulyakova, N., Sidorova-Darmos, E., Fong, J., Zhang, G., Mills, L. R., & Eubanks, J. H. (2014). Over-expression of the Sirt3 sirtuin protects neuronally differentiated PC12 cells from degeneration induced by oxidative stress and trophic withdrawal. *Brain Research*, *1587*, 40–53.

Silberman, D. M., Ross, K., Sande, P. H., Kubota, S., Ramaswamy, S., Apte, R. S., & Mostoslavsky, R. (2014). SIRT6 is required for normal retinal function. *PLoS One*, *9*, e98831.

Simeoni, F., Tasselli, L., Tanaka, S., Villanova, L., Hayashi, M., Kubota, K., ... Chua, K. F. (2013). Proteomic analysis of the SIRT6 interactome: Novel links to genome maintenance and cellular stress signaling. *Scientific Reports*, *3*, 3085.

Simic, P., Williams, E. O., Bell, E. L., Gong, J. J., Bonkowski, M., & Guarente, L. (2013). SIRT1 suppresses the epithelial-to-mesenchymal transition in cancer metastasis and organ fibrosis. *Cell Reports*, *3*, 1175–1186.

Simonet, N. G., Thackray, J. K., Vazquez, B. N., Ianni, A., Espinosa-Alcantud, M., Morales-Sanfrutos, J., Tischfield, J., ... Vaquero, A. (2020). SirT7 auto-ADP-ribosylation regulates glucose starvation response through macroH2A1.1. Sci. Adv. 6, eaaz2590.

Sinclair, D. A., & Guarente, L. (1997). Extrachromosomal rDNA circles--a cause of aging in yeast. *Cell*, *91*, 1033–1042.

Singh, P., Hanson, P. S., & Morris, C. M. (2017). Sirtuin-2 protects neural cells from oxidative stress and is elevated in neurodegeneration. *Parkinsons Disease*, *2017*, 2643587.

Singh, S., Kumar, P. U., Thakur, S., Kiran, S., Sen, B., Sharma, S., ... Ramakrishna, G. (2015). Expression/localization patterns of sirtuins (SIRT1, SIRT2, and SIRT7) during progression of cervical cancer and effects of sirtuin inhibitors on growth of cervical cancer cells. *Tumour Biology: The Journal of the International Society for Oncodevelopmental Biology and Medicine*, *36*, 6159–6171.

Snyder-Warwick, A. K., Satoh, A., Santosa, K. B., Imai, S. I., & Jablonka-Shariff, A. (2018). Hypothalamic Sirt1 protects terminal Schwann cells and neuromuscular junctions from age-related morphological changes. *Aging Cell*, *17*, e12776.

Sociali, G., Galeno, L., Parenti, M. D., Grozio, A., Bauer, I., Passalacqua, M., ... Bellotti, M. (2015). Quinazolinedione SIRT6 inhibitors sensitize cancer cells to chemotherapeutics. *European Journal of Medicinal Chemistry*, *102*, 530–539.

Sociali, G., Magnone, M., Ravera, S., Damonte, P., Vigliarolo, T., Von Holtey, M., ... Cea, M. (2017). Pharmacological Sirt6 inhibition improves glucose tolerance in a type 2 diabetes mouse model. *The FASEB Journal*, *31*, 3138–3149.

Someya, S., Yu, W., Hallows, W. C., Xu, J., Vann, J. M., Leeuwenburgh, C., ... Prolla, T. A. (2010). Sirt3 mediates reduction of oxidative damage and prevention of age-related hearing loss under caloric restriction. *Cell*, *143*, 802–812.

Song, H. Y., Biancucci, M., Kang, H. J., O'Callaghan, C., Park, S. H., Principe, D. R., ... Raparia, K. (2016). SIRT2 deletion enhances KRAS-induced tumorigenesis in vivo by regulating K147 acetylation status. *Oncotarget*, *7*, 80336–80349.

Song, L., Chen, L., Zhang, X., Li, J., & Le, W. (2014). Resveratrol ameliorates motor neuron degeneration and improves survival in SOD1(G93A) mouse model of amyotrophic lateral sclerosis. *BioMed Research International*, *2014*, 483501.

Song, M. Y., Wang, J., Ka, S. O., Bae, E. J., & Park, B. H. (2016). Insulin secretion impairment in Sirt6 knockout pancreatic beta cells is mediated by suppression of the FoxO1-Pdx1-Glut2 pathway. *Scientific Reports*, *6*, 30321.

Speakman, J. R., & Mitchell, S. E. (2011). Caloric restriction. *Molecular Aspects of Medicine*, *32*, 159–221.

Spiegelman, N. A., Zhang, X., Jing, H., Cao, J., Kotliar, I. B., Aramsangtienchai, P., ... Lin, H. (2019). SIRT2 and lysine fatty acylation regulate the activity of RalB and cell migration. *ACS Chemical Biology*, *14*, 2014–2023.

Srisuttee, R., Koh, S. S., Kim, S. J., Malilas, W., Boonying, W., Cho, I. R., ... Seto, E. (2012). Hepatitis B virus X (HBX) protein upregulates beta-catenin in a human hepatic cell line by sequestering SIRT1 deacetylase. *Oncology Reports*, *28*, 276–282.

Stein, L. R., & Imai, S. (2014). Specific ablation of Nampt in adult neural stem cells recapitulates their functional defects during aging. *The EMBO Journal*, *33*, 1321–1340.

Stein, S., Lohmann, C., Schafer, N., Hofmann, J., Rohrer, L., Besler, C., ... Boren, J. (2010). SIRT1 decreases Lox-1-mediated foam cell formation in atherogenesis. *European Heart Journal*, *31*, 2301–2309.

St-Onge, M. P., & Gallagher, D. (2010). Body composition changes with aging: The cause or the result of alterations in metabolic rate and macronutrient oxidation? *Nutrition*, *26*, 152–155.

Strong, R., Miller, R. A., Astle, C. M., Baur, J. A., de Cabo, R., Fernandez, E., ... Nelson, J. F. (2013). Evaluation of resveratrol, green tea extract, curcumin, oxaloacetic acid, and medium-chain triglyceride oil on life span of genetically heterogeneous mice. *The Journals of Gerontology. Series A, Biological Sciences and Medical Sciences*, *68*, 6–16.

Strong, R., Miller, R. A., Astle, C. M., Floyd, R. A., Flurkey, K., Hensley, K. L., ... Ongini, E. (2008). Nordihydroguaiaretic acid and aspirin increase lifespan of genetically heterogeneous male mice. *Aging Cell*, *7*, 641–650.

Strub, T., Ghiraldini, F. G., Carcamo, S., Li, M., Wroblewska, A., Singh, R., ... Gallagher, S. J. (2018). SIRT6 haploinsufficiency induces BRAF(V600E) melanoma cell resistance to MAPK inhibitors via IGF signalling. *Nature Communications*, *9*, 3440.

Suka, N., Luo, K., & Grunstein, M. (2002). Sir2p and Sas2p opposingly regulate acetylation of yeast histone H4 lysine16 and spreading of heterochromatin. *Nature Genetics*, *32*, 378–383.

Sun, M., Zhai, M., Zhang, N., Wang, R., Liang, H., Han, Q., ... Jiao, L. (2018). MicroRNA-148b-3p is involved in regulating hypoxia/reoxygenation-induced injury of cardiomyocytes in vitro through modulating SIRT7/p53 signaling. *Chemico-Biological Interactions*, *296*, 211–219.

Sun, T., Jiao, L., Wang, Y., Yu, Y., & Ming, L. (2018). SIRT1 induces epithelial-mesenchymal transition by promoting autophagic degradation of E-cadherin in melanoma cells. *Cell Death Diseases*, *9*, 136.

Sun, W., Qiao, W., Zhou, B., Hu, Z., Yan, Q., Wu, J., ... Miao, D. (2018). Overexpression of Sirt1 in mesenchymal stem cells protects against bone loss in mice by FOXO3a deacetylation and oxidative stress inhibition. *Metabolism: Clinical and Experimental*, *88*, 61–71.

Sun, X., Wang, S., Gai, J., Guan, J., Li, J., Li, Y., ... Li, Q. (2019). SIRT5 promotes cisplatin resistance in ovarian cancer by suppressing DNA damage in a ROS-dependent manner via regulation of the Nrf2/HO-1 pathway. *Frontiers in Oncology*, *9*, 754.

Sundaresan, N. R., Bindu, S., Pillai, V. B., Samant, S., Pan, Y., Huang, J. Y., ... Verdin, E. (2015). SIRT3 blocks aging-associated tissue fibrosis in mice by deacetylating and activating glycogen synthase kinase 3beta. *Molecular and Cellular Biology*, *36*, 678–692.

Sundaresan, N. R., Gupta, M., Kim, G., Rajamohan, S. B., Isbatan, A., & Gupta, M. P. (2009). Sirt3 blocks the cardiac hypertrophic response by augmenting Foxo3a-dependent antioxidant defense mechanisms in mice. *The Journal of Clinical Investigation*, *119*, 2758–2771.

Sundaresan, N. R., Pillai, V. B., Wolfgeher, D., Samant, S., Vasudevan, P., Parekh, V., ... Gupta, M. P. (2011). The deacetylase SIRT1 promotes membrane localization and activation of Akt and PDK1 during tumorigenesis and cardiac hypertrophy. *Science Signaling*, *4*, ra46.

Sundaresan, N. R., Vasudevan, P., Zhong, L., Kim, G., Samant, S., Parekh, V., ... Jeevanandam, V. (2012). The sirtuin SIRT6 blocks IGF-Akt signaling and development of cardiac hypertrophy by targeting c-Jun. *Nature Medicine*, *18*, 1643–1650.

Sussmuth, S. D., Haider, S., Landwehrmeyer, G. B., Farmer, R., Frost, C., Tripepi, G., ... Diodato, E. (2015). An exploratory double-blind, randomized clinical trial with selisistat, a SirT1 inhibitor, in patients with Huntington's disease. *British Journal of Clinical Pharmacology*, *79*, 465–476.

Tan, M., Luo, H., Lee, S., Jin, F., Yang, J. S., Montellier, E., ... Rajagopal, N. (2011). Identification of 67 histone marks and histone lysine crotonylation as a new type of histone modification. *Cell*, *146*, 1016–1028.

Tan, M., Peng, C., Anderson, Kristin A., Chhoy, P., Xie, Z., ... Zhang, Y. (2014). Lysine glutarylation is a protein posttranslational modification regulated by SIRT5. *Cell Metabolism*, *19*, 605–617.

Tanabe, K., Liu, J., Kato, D., Kurumizaka, H., Yamatsugu, K., Kanai, M., & Kawashima, S. A. (2018). LC-MS/MS-based quantitative study of the acyl group- and site-selectivity of human sirtuins to acylated nucleosomes. *Scientific Reports*, *8*, 2656.

Tang, S. S., Xu, S., Cheng, J., Cai, M. Y., Chen, L., Liang, L. L., ... Xiong, X. D. (2016). Two tagSNPs rs352493 and rs3760908 within SIRT6 gene are associated with the severity of coronary artery disease in a Chinese Han population. *Disease Markers*, *2016*, 1628041.

Tang, W., Chen, Y., Fang, X., Wang, Y., Fan, W., & Zhang, C. (2018). SIRT1 rs3758391 and major depressive disorder: New data and meta-analysis. *Neuroscience Bulletin*, *34*, 863–866.

Tang, X., Chen, X. F., Wang, N. Y., Wang, X. M., Liang, S. T., Zheng, W., ... Zhang, Z. Q. (2017). SIRT2 acts as a cardioprotective deacetylase in pathological cardiac hypertrophy. *Circulation*, *136*, 2051–2067.

Tang, X., Shi, L., Xie, N., Liu, Z., Qian, M., Meng, F., ... Zhu, W. G. (2017). SIRT7 antagonizes TGF-beta signaling and inhibits breast cancer metastasis. *Nature Communications*, *8*, 318.

Tanner, K. G., Landry, J., Sternglanz, R., & Denu, J. M. (2000). Silent information regulator 2 family of NAD-dependent histone/protein deacetylases generates a unique product, 1-O-acetyl-ADP-ribose. *Proceedings of the National Academy of Sciences of the United States of America*, *97*, 14178–14182.

Tanno, M., Sakamoto, J., Miura, T., Shimamoto, K., & Horio, Y. (2007). Nucleocytoplasmic shuttling of the NAD+-dependent histone deacetylase SIRT1. *The Journal of Biological Chemistry*, *282*, 6823–6832.

Tao, R., Coleman, M. C., Pennington, J. D., Ozden, O., Park, S. H., Jiang, H., ... Hayes McDonald, W. (2010). Sirt3-mediated deacetylation of evolutionarily conserved lysine 122 regulates MnSOD activity in response to stress. *Molecular Cell*, *40*, 893–904.

Tao, R., Xiong, X., DePinho, R. A., Deng, C. X., & Dong, X. C. (2013a). FoxO3 transcription factor and Sirt6 deacetylase regulate low density lipoprotein (LDL)-cholesterol homeostasis via control of the proprotein convertase subtilisin/kexin type 9 (Pcsk9) gene expression. *The Journal of Biological Chemistry*, *288*, 29252–29259.

Tao, R., Xiong, X., DePinho, R. A., Deng, C. X., & Dong, X. C. (2013b). Hepatic SREBP-2 and cholesterol biosynthesis are regulated by FoxO3 and Sirt6. *Journal of Lipid Research*, *54*, 2745–2753.

Tao, Y., Huang, C., Huang, Y., Hong, L., Wang, H., Zhou, Z., & Qiu, Y. (2015). SIRT4 suppresses inflammatory responses in human umbilical vein endothelial cells. *Cardiovascular Toxicology*, *15*, 217–223.

Tatar, M., Post, S., & Yu, K. (2014). Nutrient control of *Drosophila* longevity. *Trends in Endocrinology and Metabolism: TEM*, *25*, 509–517.

TenNapel, M. J., Lynch, C. F., Burns, T. L., Wallace, R., Smith, B. J., Button, A., & Domann, F. E. (2014). SIRT6 minor allele genotype is associated with >5-year decrease in lifespan in an aged cohort. *PLoS One*, *9*, e115616.

Theendakara, V., Patent, A., Peters Libeu, C. A., Philpot, B., Flores, S., Descamps, O., ... Hart, M. (2013). Neuroprotective Sirtuin ratio reversed by ApoE4. *Proceedings of the National Academy of Sciences of the United States of America*, *110*, 18303–18308.

Thirumurthi, U., Shen, J., Xia, W., LaBaff, A. M., Wei, Y., Li, C. W., ... Yu, D. (2014). MDM2-mediated degradation of SIRT6 phosphorylated by AKT1 promotes tumorigenesis and trastuzumab resistance in breast cancer. *Science Signaling*, *7*, ra71.

Tian, K., Liu, Z., Wang, J., Xu, S., You, T., & Liu, P. (2015). Sirtuin-6 inhibits cardiac fibroblasts differentiation into myofibroblasts via inactivation of nuclear factor kappaB signaling. *Translational Research: the Journal of Laboratory and Clinical Medicine*, *165*, 374–386.

Tian, X., Firsanov, D., Zhang, Z., Cheng, Y., Luo, L., Tombline, G., ... Steffan, J. (2019). SIRT6 is responsible for more efficient DNA double-strand break repair in long-lived species. *Cell*, *177*, 622–638, e622.

Tilstra, J. S., Clauson, C. L., Niedernhofer, L. J., & Robbins, P. D. (2011). NF-kappaB in aging and disease. *Aging and Disease*, *2*, 449–465.

Tissenbaum, H. A., & Guarente, L. (2001). Increased dosage of a sir-2 gene extends lifespan in *Caenorhabditis elegans*. *Nature*, *410*, 227–230.

Tissenbaum, H. A., & Guarente, L. (2002). Model organisms as a guide to mammalian aging. *Developmental Cell*, *2*, 9–19.

Tobita, T., Guzman-Lepe, J., Takeishi, K., Nakao, T., Wang, Y., Meng, F., ... Soto-Gutierrez, A. (2016). SIRT1 disruption in human fetal hepatocytes leads to increased accumulation of glucose and lipids. *PLoS One*, *11*, e0149344.

Traba, J., Geiger, S. S., Kwarteng-Siaw, M., Han, K., Ra, O. H., Siegel, R. M., ... Sack, M. N. (2017). Prolonged fasting suppresses mitochondrial NLRP3 inflammasome assembly and activation via SIRT3-mediated activation of superoxide dismutase 2. *The Journal of Biological Chemistry*, *292*, 12153–12164.

Traba, J., Kwarteng-Siaw, M., Okoli, T. C., Li, J., Huffstutler, R. D., Bray, A., ... Sauve, A. A. (2015). Fasting and refeeding differentially regulate NLRP3 inflammasome activation in human subjects. *The Journal of Clinical Investigation*, *125*, 4592–4600.

Trammell, S. A., Schmidt, M. S., Weidemann, B. J., Redpath, P., Jaksch, F., Dellinger, R. W., ... Brenner, C. (2016). Nicotinamide riboside is uniquely and orally bioavailable in mice and humans. *Nature Communications*, *7*, 12948.

Tseng, R. C., Lee, C. C., Hsu, H. S., Tzao, C., & Wang, Y. C. (2009). Distinct HIC1-SIRT1-p53 loop deregulation in lung squamous carcinoma and adenocarcinoma patients. *Neoplasia (New York, N.Y.)*, *11*, 763–770.

Tsubota, K. (2016). The first human clinical study for NMN has started in Japan. *NPJ Aging and Mechanisms of Disease*, *2*, 16021.

Turner, R. S., Thomas, R. G., Craft, S., van Dyck, C. H., Mintzer, J., Reynolds, B. A., ... Aisen, P. S. (2015). A randomized, double-blind, placebo-controlled trial of resveratrol for Alzheimer's disease. *Neurology*, *85*, 1383–1391.

Tyagi, A., Nguyen, C. U., Chong, T., Michel, C. R., Fritz, K. S., Reisdorph, N., ... Pugazhenthi, S. (2018). SIRT3 deficiency-induced mitochondrial dysfunction and inflammasome formation in the brain. *Scientific Reports*, *8*, 17547.

Ungvari, Z., Kaley, G., de Cabo, R., Sonntag, W. E., & Csiszar, A. (2010). Mechanisms of vascular aging: New perspectives. *The Journals of Gerontology. Series A, Biological Sciences and Medical Sciences*, *65*, 1028–1041.

Vakhrusheva, O., Smolka, C., Gajawada, P., Kostin, S., Boettger, T., Kubin, T., ... Bober, E. (2008). Sirt7 increases stress resistance of cardiomyocytes and prevents apoptosis and inflammatory cardiomyopathy in mice. *Circulation Research*, *102*, 703–710.

Valera, A., Pujol, A., Pelegrin, M., & Bosch, F. (1994). Transgenic mice overexpressing phosphoenolpyruvate carboxykinase develop non-insulin-dependent diabetes mellitus. *Proceedings of the National Academy of Sciences of the United States of America*, *91*, 9151–9154.

van den Berg, S. W., Dolle, M. E., Imholz, S., van der, A. D., van 't Slot, R., Wijmenga, C., ... Hoebee, B. (2009). Genetic variations in regulatory pathways of fatty acid and glucose metabolism are associated with obesity phenotypes: A population-based cohort study. *International Journal of Obesity (2005)*, *33*, 1143–1152.

van de Ven, R. A. H., Santos, D., & Haigis, M. C. (2017). Mitochondrial sirtuins and molecular mechanisms of aging. *Trends in Molecular Medicine*, *23*, 320–331.

Van Gool, F., Galli, M., Gueydan, C., Kruys, V., Prevot, P. P., Bedalov, A., ... Leo, O. (2009). Intracellular NAD levels regulate tumor necrosis factor protein synthesis in a sirtuin-dependent manner. *Nature Medicine*, *15*, 206–210.

Vandanmagsar, B., Youm, Y. H., Ravussin, A., Galgani, J. E., Stadler, K., Mynatt, R. L., ... Dixit, V. D. (2011). The NLRP3 inflammasome instigates obesity-induced inflammation and insulin resistance. *Nature Medicine*, *17*, 179–188.

van Leeuwen, I. M., Higgins, M., Campbell, J., McCarthy, A. R., Sachweh, M. C., Navarro, A. M., & Lain, S. (2013). Modulation of p53 C-terminal acetylation by mdm2, p14ARF, and cytoplasmic SirT2. *Molecular Cancer Therapeutics*, *12*, 471–480.

Van Meter, M., Mao, Z., Gorbunova, V., & Seluanov, A. (2011). SIRT6 overexpression induces massive apoptosis in cancer cells but not in normal cells. *Cell Cycle*, *10*, 3153–3158.

Vaquero, A., Scher, M. B., Lee, D. H., Sutton, A., Cheng, H. L., Alt, F. W., ... Reinberg, D. (2006). SirT2 is a histone deacetylase with preference for histone H4 Lys 16 during mitosis. *Genes & Development*, *20*, 1256–1261.

Vaziri, H., Dessain, S. K., Ng Eaton, E., Imai, S. I., Frye, R. A., Pandita, T. K., ... Weinberg, R. A. (2001). hSIR2(SIRT1) functions as an NAD-dependent p53 deacetylase. *Cell*, *107*, 149–159.

Vikram, A., Lewarchik, C. M., Yoon, J. Y., Naqvi, A., Kumar, S., Morgan, G. M., ... Kassan, M. (2017). Sirtuin 1 regulates cardiac electrical activity by deacetylating the cardiac sodium channel. *Nature Medicine*, *23*, 361–367.

Viswanathan, M., & Guarente, L. (2011). Regulation of *Caenorhabditis elegans* lifespan by sir-2.1 transgenes. *Nature*, *477*, E1–E2.

Wahl, D., Solon-Biet, S. M., Wang, Q. P., Wali, J. A., Pulpitel, T., Clark, X., ... Cooney, G. J. (2018). Comparing the effects of low-protein and high-carbohydrate diets and caloric restriction on brain aging in mice. *Cell Reports*, *25*, 2234–2243, e2236.

Walker, A. K., Yang, F., Jiang, K., Ji, J. Y., Watts, J. L., Purushotham, A., ... Smith, J. J. (2010). Conserved role of SIRT1 orthologs in fasting-dependent inhibition of the lipid/cholesterol regulator SREBP. *Genes & Development*, *24*, 1403–1417.

Wang, B., Zhang, Y., Cao, W., Wei, X., Chen, J., & Ying, W. (2016). SIRT2 plays significant roles in lipopolysaccharides-induced neuroinflammation and brain injury in mice. *Neurochemical Research*, *41*, 2490–2500.

Wang, F., Nguyen, M., Qin, F. X., & Tong, Q. (2007). SIRT2 deacetylates FOXO3a in response to oxidative stress and caloric restriction. *Aging Cell*, *6*, 505–514.

Wang, F., & Tong, Q. (2009). SIRT2 suppresses adipocyte differentiation by deacetylating FOXO1 and enhancing FOXO1's repressive interaction with PPARgamma. *Molecular Biology of the Cell*, *20*, 801–808.

Wang, F., Wang, K., Xu, W., Zhao, S., Ye, D., Wang, Y., ... Zhang, C. (2017). SIRT5 desuccinylates and activates pyruvate kinase M2 to block macrophage IL-1beta Production and to prevent DSS-induced colitis in mice. *Cell Reports*, *19*, 2331–2344.

Wang, G., Meyer, J. G., Cai, W., Softic, S., Li, M. E., Verdin, E., ... Kahn, C. R. (2019). Regulation of UCP1 and mitochondrial metabolism in brown adipose tissue by reversible succinylation. *Molecular Cell*, *74*, 844–857, e847.

Wang, H., Liu, S., Liu, S., Wei, W., Zhou, X., Lin, F., ... Pang, Y. (2017). Enhanced expression and phosphorylation of Sirt7 activates smad2 and ERK signaling and promotes the cardiac fibrosis differentiation upon angiotensin-II stimulation. *PLoS One*, *12*, e0178530.

Wang, H. L., Lu, R. Q., Xie, S. H., Zheng, H., Wen, X. M., Gao, X., & Guo, L. (2015). SIRT7 exhibits oncogenic potential in human ovarian cancer cells. *Asian Pacific Journal of Cancer Prevention: APJCP*, *16*, 3573–3577.

Wang, L., Quan, N., Sun, W., Chen, X., Cates, C., Rousselle, T., ... Li, J. (2018). Cardiomyocyte-specific deletion of Sirt1 gene sensitizes myocardium to ischaemia and reperfusion injury. *Cardiovascular Research*, *114*, 805–821.

Wang, L., Zhou, H., Wang, Y., Cui, G., & Di, L. J. (2015). CtBP maintains cancer cell growth and metabolic homeostasis via regulating SIRT4. *Cell Death Disease*, *6*, e1620.

Wang, R. H., Li, C., & Deng, C. X. (2010). Liver steatosis and increased ChREBP expression in mice carrying a liver specific SIRT1 null mutation under a normal feeding condition. *International Journal of Biological Sciences*, *6*, 682–690.

Wang, W. W., Angulo-Ibanez, M., Lyu, J., Kurra, Y., Tong, Z., Wu, B., ... Lin, H. (2019). A click chemistry approach reveals the chromatin-dependent histone H3K36 deacylase nature of SIRT7. *Journal of the American Chemical Society*, *141*, 2462–2473.

Wang, X. X., Wang, X. L., Tong, M. M., Gan, L., Chen, H., Wu, S. S., ... Zhang, H. Y. (2016). SIRT6 protects cardiomyocytes against ischemia/reperfusion injury by augmenting FoxO3alpha-dependent antioxidant defense mechanisms. *Basic Research in Cardiology*, *111*, 13.

Wang, Y., Cai, Y., Huang, H., Chen, X., Chen, X., Chen, X., ... Yang, J. (2018). miR-486-3p influences the neurotoxicity of a-synuclein by targeting the SIRT2 gene and the polymorphisms at target sites contributing to Parkinson's disease. *Cellular Physiology and Biochemistry: International Journal of Experimental Cellular Physiology, Biochemistry, and Pharmacology*, *51*, 2732–2745.

Wang, Y., He, J., Liao, M., Hu, M., Li, W., Ouyang, H., ... Ouyang, L. (2019). An overview of Sirtuins as potential therapeutic target: Structure, function and modulators. *European Journal of Medicinal Chemistry*, *161*, 48–77.

Wang, Y., Liu, Q., Huan, Y., Li, R., Li, C., Sun, S., ... Shen, Z. (2018). Sirtuin 5 overexpression attenuates glucolipotoxicity-induced pancreatic beta cells apoptosis and dysfunction. *Experimental Cell Research*, *371*, 205–213.

Wang, Y., & Tissenbaum, H. A. (2006). Overlapping and distinct functions for a *Caenorhabditis elegans* SIR2 and DAF-16/FOXO. *Mechanisms of Ageing and Development*, *127*, 48–56.

Wang, Y., Zhu, Y., Xing, S., Ma, P., & Lin, D. (2015). SIRT5 prevents cigarette smoke extract-induced apoptosis in lung epithelial cells via deacetylation of FOXO3. *Cell Stress & Chaperones*, *20*, 805–810.

Wang, Y. P., Zhou, L. S., Zhao, Y. Z., Wang, S. W., Chen, L. L., Liu, L. X., ... Zhang, J. Y. (2014). Regulation of G6PD acetylation by SIRT2 and KAT9 modulates NADPH homeostasis and cell survival during oxidative stress. *The EMBO Journal*, *33*, 1304–1320.

Wang, Y. Q., Wang, H. L., Xu, J., Tan, J., Fu, L. N., Wang, J. L., ... Chen, Y. X. (2018). Sirtuin5 contributes to colorectal carcinogenesis by enhancing glutaminolysis in a deglutarylation-dependent manner. *Nature Communications*, *9*, 545.

Watanabe, H., Inaba, Y., Kimura, K., Matsumoto, M., Kaneko, S., Kasuga, M., & Inoue, H. (2018). Sirt2 facilitates hepatic glucose uptake by deacetylating glucokinase regulatory protein. *Nature Communications*, *9*, 30.

Wei, T., Huang, G., Gao, J., Huang, C., Sun, M., Wu, J., ... Shen, W. (2017). Sirtuin 3 deficiency accelerates hypertensive cardiac remodeling by impairing angiogenesis. *Journal of the American Heart Association*, *6*, e006114.

Wei, W., Jing, Z. X., Ke, Z., & Yi, P. (2017). Sirtuin 7 plays an oncogenic role in human osteosarcoma via downregulating CDC4 expression. *American Journal of Cancer Research*, *7*, 1788–1803.

Wei, W., Xu, X., Li, H., Zhang, Y., Han, D., Wang, Y., ... Liu, N. (2014). The SIRT2 polymorphism rs10410544 and risk of Alzheimer's disease: A meta-analysis. *Neuromolecular Medicine*, *16*, 448–456.

Wei, Z., Song, J., Wang, G., Cui, X., Zheng, J., Tang, Y., ... Liu, C. Y. (2018). Deacetylation of serine hydroxymethyl-transferase 2 by SIRT3 promotes colorectal carcinogenesis. *Nature Communications*, *9*, 4468.

Wellman, A. S., Metukuri, M. R., Kazgan, N., Xu, X., Xu, Q., Ren, N. S. X., ... Chen, W. (2017). Intestinal epithelial Sirtuin 1 regulates intestinal inflammation during aging in mice by altering the intestinal microbiota. *Gastroenterology*, *153*, 772–786.

Wen, L., Chen, Z., Zhang, F., Cui, X., Sun, W., Geary, G. G., ... Chien, S. (2013). Ca2 + /calmodulin-dependent protein kinase kinase beta phosphorylation of Sirtuin 1 in endothelium is atheroprotective. *Proceedings of the National Academy of Sciences of the United States of America*, *110*, E2420–E2427.

Westerberg, G., Chiesa, J. A., Andersen, C. A., Diamanti, D., Magnoni, L., Pollio, G., ... Zhou, M. (2015). Safety, pharmacokinetics, pharmacogenomics and QT concentration-effect modelling of the SirT1 inhibitor selisistat in healthy volunteers. *British Journal of Clinical Pharmacology*, *79*, 477–491.

Whitaker, R., Faulkner, S., Miyokawa, R., Burhenn, L., Henriksen, M., Wood, J. G., & Helfand, S. L. (2013). Increased expression of *Drosophila* Sir2 extends life span in a dose-dependent manner. *Aging*, *5*, 682–691.

Wijshake, T., Malureanu, L. A., Baker, D. J., Jeganathan, K. B., van de Sluis, B., & van Deursen, J. M. (2012). Reduced life- and healthspan in mice carrying a mono-allelic BubR1 MVA mutation. *PLoS Genetics*, *8*, e1003138.

Wongchitrat, P., Pakpian, N., Kitidee, K., Phopin, K., Dharmasaroja, P. A., & Govitrapong, P. (2019). Alterations in the expression of amyloid precursor protein cleaving enzymes mRNA in Alzheimer peripheral blood. *Current Alzheimer Research*, *16*, 29–38.

Wood, J. G., Jones, B. C., Jiang, N., Chang, C., Hosier, S., Wickremesinghe, P., ... Neretti, N. (2016). Chromatin-modifying genetic interventions suppress age-associated transposable element activation and extend life span in *Drosophila*. *Proceedings of the National Academy of Sciences of the United States of America*, *113*, 11277–11282.

Wood, J. G., Schwer, B., Wickremesinghe, P. C., Hartnett, D. A., Burhenn, L., Garcia, M., ... Helfand, S. L. (2018). Sirt4 is a mitochondrial regulator of metabolism and lifespan in *Drosophila melanogaster*. *Proceedings of the National Academy of Sciences of the United States of America*, *115*, 1564–1569.

Xia, M., Yu, J. T., Miao, D., Lu, R. C., Zheng, X. P., & Tan, L. (2014). SIRT2 polymorphism rs10410544 is associated with Alzheimer's disease in a Han Chinese population. *Journal of the Neurological Sciences*, *336*, 48–51.

Xiangyun, Y., Xiaomin, N., Linping, G., Yunhua, X., Ziming, L., Yongfeng, Y., ... Shun, L. (2017). Desuccinylation of pyruvate kinase M2 by SIRT5 contributes to antioxidant response and tumor growth. *Oncotarget*, *8*, 6984–6993.

Xiao, C., Kim, H. S., Lahusen, T., Wang, R. H., Xu, X., Gavrilova, O., ... Deng, C. X. (2010). SIRT6 deficiency results in severe hypoglycemia by enhancing both basal and insulin-stimulated glucose uptake in mice. *The Journal of Biological Chemistry*, *285*, 36776–36784.

Xiao, Y., Zhang, X., Fan, S., Cui, G., & Shen, Z. (2016). MicroRNA-497 inhibits cardiac hypertrophy by targeting Sirt4. *PLoS One*, *11*, e0168078.

Xie, Z., Zhang, D., Chung, D., Tang, Z., Huang, H., Dai, L., ... Chen, Y. (2016). Metabolic regulation of gene expression by histone lysine beta-hydroxybutyrylation. *Molecular Cell*, *62*, 194–206.

Xiong, X., Tao, R., DePinho, R. A., & Dong, X. C. (2013). Deletion of hepatic FoxO1/3/4 genes in mice significantly impacts on glucose metabolism through downregulation of gluconeogenesis and upregulation of glycolysis. *PLoS One*, *8*, e74340.

Xiong, X., Zhang, C., Zhang, Y., Fan, R., Qian, X., & Dong, X. C. (2017). Fabp4-Cre-mediated Sirt6 deletion impairs adipose tissue function and metabolic homeostasis in mice. *The Journal of Endocrinology*, *233*, 307–314.

Xu, L., Che, X., Wu, Y., Song, N., Shi, S., Wang, S., ... Qu, X. (2018). SIRT5 as a biomarker for response to anthracycline-taxane-based neoadjuvant chemotherapy in triple-negative breast cancer. *Oncology Reports*, *39*, 2315–2323.

Xu, S. N., Wang, T. S., Li, X., & Wang, Y. P. (2016). SIRT2 activates G6PD to enhance NADPH production and promote leukaemia cell proliferation. *Scientific Reports*, *6*, 32734.

Xu, W., Jiang, K., Shen, M., Qian, Y., & Peng, Y. (2015). SIRT2 suppresses non-small cell lung cancer growth by targeting JMJD2A. *Biological Chemistry*, *396*, 929–936.

Xu, W., Yang, H., Liu, Y., Yang, Y., Wang, P., Kim, S. H., ... Xiao, M. T. (2011). Oncometabolite 2-hydroxyglutarate is a competitive inhibitor of alpha-ketoglutarate-dependent dioxygenases. *Cancer Cell*, *19*, 17–30.

Xu, Z., Zhang, L., Zhang, W., Meng, D., Zhang, H., Jiang, Y., ... Gorbunova, V. (2015). SIRT6 rescues the age related decline in base excision repair in a PARP1-dependent manner. *Cell Cycle*, *14*, 269–276. Available from https://doi.org/10.4161/15384101.15382014.15980641.

Yang, B., Zwaans, B. M., Eckersdorff, M., & Lombard, D. B. (2009). The sirtuin SIRT6 deacetylates H3 K56Ac in vivo to promote genomic stability. *Cell Cycle*, *8*, 2662–2663.

Yang, G., Weng, X., Zhao, Y., Zhang, X., Hu, Y., Dai, X., ... Sun, X. (2017). The histone H3K9 methyltransferase SUV39H links SIRT1 repression to myocardial infarction. *Nature Communications*, *8*, 14941.

Yang, H., Zhu, R., Zhao, X., Liu, L., Zhou, Z., Zhao, L., ... Liu, J. (2019). Sirtuin-mediated deacetylation of hnRNP A1 suppresses glycolysis and growth in hepatocellular carcinoma. *Oncogene*, *38*, 4915–4931.

Yang, M. H., Laurent, G., Bause, A. S., Spang, R., German, N., Haigis, M. C., & Haigis, K. M. (2013). HDAC6 and SIRT2 regulate the acetylation state and oncogenic activity of mutant K-RAS. *Molecular Cancer Research: MCR*, *11*, 1072–1077.

Yang, W., Gao, F., Zhang, P., Pang, S., Cui, Y., Liu, L., ... Yan, B. (2017). Functional genetic variants within the SIRT2 gene promoter in acute myocardial infarction. *PLoS One*, *12*, e0176245.

Yang, X., Wang, Z., Li, X., Liu, B., Liu, M., Liu, L., ... Yu, M. (2018). SHMT2 desuccinylation by SIRT5 drives cancer cell proliferation. *Cancer Research*, *78*(2), 372–386.

Yang, Y., Ding, J., Gao, Z. G., & Wang, Z. J. (2017). A variant in SIRT2 gene 3′-UTR is associated with susceptibility to colorectal cancer. *Oncotarget*, *8*, 41021–41025.

Yang, Y., Mohammed, F. S., Zhang, N., & Sauve, A. A. (2019). Dihydronicotinamide riboside is a potent NAD(+) concentration enhancer in vitro and in vivo. *The Journal of Biological Chemistry*, *294*, 9295–9307.

Yang, Y., Wang, H., Kouadir, M., Song, H., & Shi, F. (2019). Recent advances in the mechanisms of NLRP3 inflammasome activation and its inhibitors. *Cell Death Disease*, *10*, 128.

Yao, L., Cui, X., Chen, Q., Yang, X., Fang, F., Zhang, J., ... Chang, Y. (2017). Cold-Inducible SIRT6 regulates thermogenesis of brown and beige fat. *Cell Reports*, *20*, 641–654.

Yeung, F., Hoberg, J. E., Ramsey, C. S., Keller, M. D., Jones, D. R., Frye, R. A., & Mayo, M. W. (2004). Modulation of NF-kappaB-dependent transcription and cell survival by the SIRT1 deacetylase. *The EMBO Journal*, *23*, 2369–2380.

Yin, X., Pang, S., Huang, J., Cui, Y., & Yan, B. (2016). Genetic and functional sequence variants of the SIRT3 gene promoter in myocardial infarction. *PLoS One, 11*, e0153815.

Yoon, M. J., Yoshida, M., Johnson, S., Takikawa, A., Usui, I., Tobe, K., ... Imai, S. (2015). SIRT1-mediated eNAMPT secretion from adipose tissue regulates hypothalamic NAD+ and function in mice. *Cell Metabolism, 21*, 706–717.

Yoshida, M., Satoh, A., Lin, J. B., Mills, K. F., Sasaki, Y., Rensing, N., ... Imai, S. I. (2019). Extracellular vesicle-contained eNAMPT delays aging and extends lifespan in mice. *Cell Metabolism, 30*, 329–342.e5.

Yoshino, J., Mills, K. F., Yoon, M. J., & Imai, S. (2011). Nicotinamide mononucleotide, a key NAD(+) intermediate, treats the pathophysiology of diet- and age-induced diabetes in mice. *Cell Metabolism, 14*, 528–536.

Yoshizaki, T., Milne, J. C., Imamura, T., Schenk, S., Sonoda, N., Babendure, J. L., ... Olefsky, J. M. (2009). SIRT1 exerts anti-inflammatory effects and improves insulin sensitivity in adipocytes. *Molecular and Cellular Biology, 29*, 1363–1374.

Yoshizaki, T., Schenk, S., Imamura, T., Babendure, J. L., Sonoda, N., Bae, E. J., ... Westphal, C. (2010). SIRT1 inhibits inflammatory pathways in macrophages and modulates insulin sensitivity. *American Journal of Physiology. Endocrinology and Metabolism, 298*, E419–E428.

Yoshizawa, T., Karim, M. F., Sato, Y., Senokuchi, T., Miyata, K., Fukuda, T., ... Kadomatsu, T. (2014). SIRT7 controls hepatic lipid metabolism by regulating the ubiquitin-proteasome pathway. *Cell Metabolism, 19*, 712–721.

You, W., Rotili, D., Li, T. M., Kambach, C., Meleshin, M., Schutkowski, M., ... Steegborn, C. (2017). Structural basis of Sirtuin 6 activation by synthetic small molecules. *Angewandte Chemie, 56*, 1007–1011.

Yu, H., Ye, W., Wu, J., Meng, X., Liu, R. Y., Ying, X., ... Huang, W. (2014). Overexpression of sirt7 exhibits oncogenic property and serves as a prognostic factor in colorectal cancer. *Clinical Cancer Research: An Official Journal of the American Association for Cancer Research, 20*, 3434–3445.

Yu, J., Sadhukhan, S., Noriega, L. G., Moullan, N., He, B., Weiss, R. S., ... Auwerx, J. (2013). Metabolic characterization of a Sirt5 deficient mouse model. *Scientific Reports, 3*, 2806.

Yu, L., Gong, B., Duan, W., Fan, C., Zhang, J., Li, Z., ... Li, B. (2017). Melatonin ameliorates myocardial ischemia/reperfusion injury in type 1 diabetic rats by preserving mitochondrial function: Role of AMPK-PGC-1alpha-SIRT3 signaling. *Scientific Reports, 7*, 41337.

Yu, S. S., Cai, Y., Ye, J. T., Pi, R. B., Chen, S. R., Liu, P. Q., ... Ji, Y. (2013). Sirtuin 6 protects cardiomyocytes from hypertrophy in vitro via inhibition of NF-kappaB-dependent transcriptional activity. *British Journal of Pharmacology, 168*, 117–128.

Yuan, F., Liu, L., Lei, Y., & Tang, P. (2017). p53 inhibits the upregulation of sirtuin 1 expression induced by c-Myc. *Oncology Letters, 14*, 4396–4402.

Yuan, H., Su, L., & Chen, W. Y. (2013). The emerging and diverse roles of sirtuins in cancer: A clinical perspective. *OncoTargets and Therapy, 6*, 1399–1416.

Yuan, H., Wang, Z., Li, L., Zhang, H., Modi, H., Horne, D., ... Chen, W. (2012). Activation of stress response gene SIRT1 by BCR-ABL promotes leukemogenesis. *Blood, 119*, 1904–1914.

Yuan, J., Minter-Dykhouse, K., & Lou, Z. (2009). A c-Myc-SIRT1 feedback loop regulates cell growth and transformation. *The Journal of Cell Biology, 185*, 203–211.

Zainabadi, K., Liu, C. J., Caldwell, A. L. M., & Guarente, L. (2017). SIRT1 is a positive regulator of in vivo bone mass and a therapeutic target for osteoporosis. *PLoS One, 12*, e0185236.

Zeng, G., Liu, H., & Wang, H. (2018). Amelioration of myocardial ischemia-reperfusion injury by SIRT4 involves mitochondrial protection and reduced apoptosis. *Biochemical and Biophysical Research Communications, 502*, 15–21.

Zhai, M., Li, B., Duan, W., Jing, L., Zhang, B., Zhang, M., ... Ren, K. (2017). Melatonin ameliorates myocardial ischemia reperfusion injury through SIRT3-dependent regulation of oxidative stress and apoptosis. *Journal of Pineal Research, 63*, e12419. Available from https://doi.org/10.1111/jpi.12419.

Zhan, L., Huang, C., Meng, X. M., Song, Y., Wu, X. Q., Miu, C. G., ... Li, J. (2014). Promising roles of mammalian E2Fs in hepatocellular carcinoma. *Cellular Signalling, 26*, 1075–1081.

Zhang, G., Li, J., Purkayastha, S., Tang, Y., Zhang, H., Yin, Y., ... Cai, D. (2013). Hypothalamic programming of systemic ageing involving IKK-beta, NF-kappaB and GnRH. *Nature, 497*, 211–216.

Zhang, H., Ryu, D., Wu, Y., Gariani, K., Wang, X., Luan, P., ... Aebersold, R. (2016). NAD(+) repletion improves mitochondrial and stem cell function and enhances life span in mice. *Science, 352*, 1436–1443.

Zhang, L. L., Zhan, L., Jin, Y. D., Min, Z. L., Wei, C., Wang, Q., ... Yuan, Q. (2017). SIRT2 mediated antitumor effects of shikonin on metastatic colorectal cancer. *European Journal of Pharmacology, 797*, 1–8.

Zhang, M., Pan, Y., Dorfman, R. G., Yin, Y., Zhou, Q., Huang, S., ... Zhao, S. (2017). Sirtinol promotes PEPCK1 degradation and inhibits gluconeogenesis by inhibiting deacetylase SIRT2. *Scientific Reports, 7*, 7.

Zhang, N., Li, Z., Mu, W., Li, L., Liang, Y., Lu, M., ... Wang, Z. (2016). Calorie restriction-induced SIRT6 activation delays aging by suppressing NF-kappaB signaling. *Cell Cycle, 15*, 1009–1018.

Zhang, P., Tu, B., Wang, H., Cao, Z., Tang, M., Zhang, C., ... Yang, Y. (2014). Tumor suppressor p53 cooperates with SIRT6 to regulate gluconeogenesis by promoting FoxO1 nuclear exclusion. *Proceedings of the National Academy of Sciences of the United States of America, 111*, 10684–10689.

Zhang, Q. J., Wang, Z., Chen, H. Z., Zhou, S., Zheng, W., Liu, G., ... Liang, C. C. (2008). Endothelium-specific overexpression of class III deacetylase SIRT1 decreases atherosclerosis in apolipoprotein E-deficient mice. *Cardiovascular Research, 80*, 191–199.

Zhang, R., Wang, C., Tian, Y., Yao, Y., Mao, J., Wang, H., ... Wang, L. (2019). SIRT5 promotes hepatocellular carcinoma progression by regulating mitochondrial apoptosis. *Journal of Cancer, 10*, 3871–3882.

Zhang, S., Chen, P., Huang, Z., Hu, X., Chen, M., Hu, S., ... Cai, T. (2015). Sirt7 promotes gastric cancer growth and inhibits apoptosis by epigenetically inhibiting miR-34a. *Scientific Reports, 5*, 9787.

Zhang, W., Wan, H., Feng, G., Qu, J., Wang, J., Jing, Y., ... Chen, Z. (2018). SIRT6 deficiency results in developmental retardation in cynomolgus monkeys. *Nature, 560*, 661–665.

Zhang, W. G., Bai, X. J., & Chen, X. M. (2010). SIRT1 variants are associated with aging in a healthy Han Chinese population. *Clinica Chimica Acta; International Journal of Clinical Chemistry, 411*, 1679–1683.

Zhang, X., Cao, R., Niu, J., Yang, S., Ma, H., Zhao, S., & Li, H. (2019). Molecular basis for hierarchical histone de-β-hydroxybutyrylation by SIRT3. *Cell Discovery, 5*, 35.

Zhang, X., Li, W., Shen, P., Feng, X., Yue, Z., Lu, J., ... Fang, S. (2016). STAT3 suppression is involved in the protective effect of SIRT6 against cardiomyocyte hypertrophy. *Journal of Cardiovascular Pharmacology, 68*, 204–214.

Zhang, Y., Bharathi, S. S., Rardin, M. J., Lu, J., Maringer, K. V., Sims-Lucas, S., ... Goetzman, E. S. (2017). Lysine desuccinylase SIRT5 binds to cardiolipin and regulates the electron transport chain. *The Journal of Biological Chemistry, 292*, 10239–10249.

Zhang, Z. Q., Ren, S. C., Tan, Y., Li, Z. Z., Tang, X., Wang, T. T., ... Liu, D. P. (2016). Epigenetic regulation of NKG2D ligands is involved in exacerbated atherosclerosis development in Sirt6 heterozygous mice. *Scientific Reports, 6*, 23912.

Zhao, J., Wozniak, A., Adams, A., Cox, J., Vittal, A., Voss, J., ... Li, Z. (2019). SIRT7 regulates hepatocellular carcinoma response to therapy by altering the p53-dependent cell death pathway. *Journal of Experimental & Clinical Cancer Research: CR, 38*, 252.

Zhao, L., Qi, Y., Xu, L., Tao, X., Han, X., Yin, L., & Peng, J. (2018). MicroRNA-140-5p aggravates doxorubicin-induced cardiotoxicity by promoting myocardial oxidative stress via targeting Nrf2 and Sirt2. *Redox Biology, 15*, 284–296.

Zhao, W. Y., Zhang, L., Sui, M. X., Zhu, Y. H., & Zeng, L. (2016). Protective effects of sirtuin 3 in a murine model of sepsis-induced acute kidney injury. *Scientific Reports, 6*, 33201.

Zheng, J., Shi, L., Liang, F., Xu, W., Li, T., Gao, L., … Zhang, J. (2018). Sirt3 ameliorates oxidative stress and mitochondrial dysfunction after intracerebral hemorrhage in diabetic rats. *Frontiers in Neuroscience, 12*, 414.

Zhong, L., D'Urso, A., Toiber, D., Sebastian, C., Henry, R. E., Vadysirisack, D. D., … Nir, T. (2010). The histone deacetylase Sirt6 regulates glucose homeostasis via Hif1alpha. *Cell, 140*, 280–293.

Zhou, L., Wang, F., Sun, R., Chen, X., Zhang, M., Xu, Q., … Guan, K. L. (2016). SIRT5 promotes IDH2 desuccinylation and G6PD deglutarylation to enhance cellular antioxidant defense. *EMBO Reports, 17*, 811–822.

Zhou, W., Ni, T. K., Wronski, A., Glass, B., Skibinski, A., Beck, A., & Kuperwasser, C. (2016). The SIRT2 deacetylase stabilizes slug to control malignancy of basal-like breast cancer. *Cell Reports, 17*, 1302–1317.

Zhou, X. E., Suino-Powell, K. M., Xu, Y., Chan, C. W., Tanabe, O., Kruse, S. W., … Xu, H. E. (2011). The orphan nuclear receptor TR4 is a vitamin A-activated nuclear receptor. *The Journal of Biological Chemistry, 286*, 2877–2885.

Zhou, Y., Chung, A. C. K., Fan, R., Lee, H. M., Xu, G., Tomlinson, B., … Kong, A. P. S. (2017). Sirt3 deficiency increased the vulnerability of pancreatic beta cells to oxidative stress-induced dysfunction. *Antioxidants & Redox Signaling, 27*, 962–976.

Zhu, W. Z., Wu, X. F., Zhang, Y., & Zhou, Z. N. (2012). Proteomic analysis of mitochondrial proteins in cardiomyocytes from rats subjected to intermittent hypoxia. *European Journal of Applied Physiology, 112*, 1037–1046.

Zhu, X. H., Lu, M., Lee, B. Y., Ugurbil, K., & Chen, W. (2015). In vivo NAD assay reveals the intracellular NAD contents and redox state in healthy human brain and their age dependences. *Proceedings of the National Academy of Sciences of the United States of America, 112*, 2876–2881.

Zou, X., Zhu, Y., Park, S. H., Liu, G., O'Brien, J., Jiang, H., & Gius, D. (2017). SIRT3-Mediated dimerization of IDH2 directs cancer cell metabolism and tumor growth. *Cancer Research, 77*, 3990–3999.

CHAPTER

6

Integrative genomics of aging

João Pedro de Magalhães[1], *Cyril Lagger*[1] *and Robi Tacutu*[2]

[1]Integrative Genomics of Ageing Group, Institute of Ageing and Chronic Disease, University of Liverpool, Liverpool, United Kingdom [2]Systems Biology of Aging Group, Department of Bioinformatics and Structural Biochemistry, Institute of Biochemistry of the Romanian Academy, Bucharest, Romania

OUTLINE

Introduction

The sequencing of genomes has revolutionized biological and biomedical research. Thanks to various technologies and approaches that take advantage of genome sequence knowledge, researchers can now focus on whole biological systems rather than being limited to studying isolated parts. Because most biological processes are complex in the sense that they involve the interplay of multiple genes and proteins with each other and with the environment, surveying systems as a whole is imperative to fully comprehending them, and more accurately pinpointing how to intervene in them. Recent breakthroughs in developing cheaper and quicker sequencing technologies have given further power to our capacity to survey biological systems in a holistic way with multiple applications in aging research (reviewed in de Magalhaes, Finch, & Janssens, 2010). In addition to genomics, other omics approaches like transcriptomics, proteomics, and epigenomics have allowed for a systematic profiling of biological processes and disease states.

Aging is widely acknowledged as a complex process involving changes at various biological levels, interactions between them, and feedback regulatory circuits. The underlying mechanistic causes of aging remain a subject of debate, and it is likely that multiple degenerative processes are involved, including organ-specific processes but also interacting cell- and organ-level communications (Cevenini et al., 2010; de Magalhaes, 2011; Lopez-Otin, Blasco, Partridge, Serrano, & Kroemer, 2013). While there are simple triggers to complex biological processes, such as telomere shortening triggering replicative senescence in human fibroblasts (de Magalhaes, 2004), most researchers would agree that organismal aging involves multiple processes and possibly the interplay between various causal mechanisms. Likewise, hundreds of genes have been associated with aging in model organisms (Tacutu et al., 2018) and yet the pathways involved are complex and often interact in nonlinear ways (de Magalhaes, Wuttke, Wood, Plank, & Vora, 2012). One hypothesis is that aging and longevity cannot be fully understood by studying individual components

Handbook of the Biology of Aging.
DOI: https://doi.org/10.1016/B978-0-12-815962-0.00006-8

and processes (Cevenini et al., 2010). To understand aging we must then account for the intrinsic complexity of biological systems.

Our goal in this chapter is to review potential large-scale technologies in the context of aging and longevity research and how data can be analyzed and integrated to advance our understanding of these complex processes. We first review the major technologies available for researchers to survey biological systems in a systematic fashion and their applications to advance the biology and genetics of aging, debate issues in data analysis and statistics, and discuss data integration between different sources, as this is one of the major challenges of the post-genome era, and also one of the most promising. Various sources of data and approaches are discussed in this context.

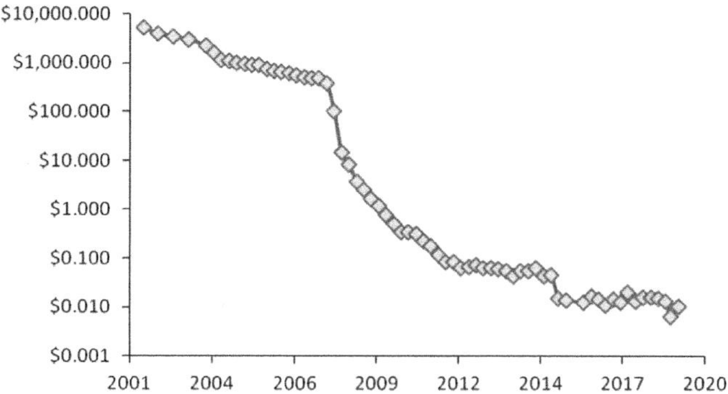

FIGURE 6.1 Exponential growth in sequencing capacity as reflected in the dropping costs of sequencing from 2001 to 2019. *Source: NHGRI (https://www.genome.gov/about-genomics/fact-sheets/Sequencing-Human-Genome-cost).*

Post-genome technologies and biogerontology

There are many open questions in biogerontology, but arguably most researchers focus on two key questions (de Magalhaes & Toussaint, 2004b): (1) what are the genetic determinants of aging, both in terms of longevity differences between individuals and species differences in aging? and (2) which changes occur across the lifetime to increase vulnerability, for example in a person from age 30 to age 70 to increase the chance of dying by roughly 30-fold? Post-genome technologies may help us answer both these questions.

Genome-wide approaches and the genetics of aging and longevity

Understanding human phenotypic variation in aging and longevity has been a long-term research goal. Studies in twins have shown that longevity in humans has a genetic component, and the heritability of longevity has been estimated at ~25% (Christensen, Johnson, & Vaupel, 2006). If we could identify genetic variants associated with exceptional human longevity, these would likely be suitable for drug discovery (de Magalhaes et al., 2012). In 1994, *APOE* was associated with longevity in a French population (Schachter et al., 1994). The sequencing of the human genome in 2001 allowed for much more powerful whole-genome genotyping platforms capable of surveying hundreds of thousands of genetic variants in a cost-effective way (de Magalhaes, 2009). In spite of these recent technological advances, the genetics of human longevity remains largely misunderstood. Several genome-wide association studies (GWAS) have been performed with hundreds of individuals, with largely disappointing results. For example, one landmark study involving several European populations with a total of over 2000 nonagenarian sibling pairs identified only *APOE* as associated with longevity (Beekman et al., 2013); and although *APOE* has been consistently associated with longevity, it only modestly explains the heritability of longevity. GWAS focused on complex diseases and processes have been on many occasions equally disappointing to date, suggesting that common genetic variants have a modest contribution to longevity and complex diseases (de Magalhaes & Wang, 2019; Manolio et al., 2009).

The dropping costs of DNA sequencing (Fig. 6.1) mean that sequencing a human genome is rapidly becoming affordable. Therefore, researchers are moving from genotyping platforms based on known genetic variants to genome sequencing of thousands of individuals. It is possible that this will reveal rare variants with strong effects on longevity, as has been predicted to be the case for complex diseases (Manolio et al., 2009). Nonetheless, considering that only *APOE* has been associated with confidence with longevity, our understanding of the genetics of longevity lags behind our understanding of the genetics of complex age-related diseases, in itself made difficult by numerous factors like multiple genes with small effects. Intrinsic difficulties in longevity studies (e.g., lack of appropriate controls) or because longevity is a more complex trait may explain why our understanding of the heritability of longevity is still poor (de Magalhaes, 2014b).

An even greater source of variation in aging and longevity than that observed between humans is observed across species. We know that mice, for example, age 25–30 times faster than humans, even under the best environmental conditions (Finch, 1990). Even when compared to chimpanzees, our closest living relative whose genome is about 95% similar to our own, aging is significantly retarded in humans (de Magalhaes, 2006).

Therefore there must be a genomic basis for species differences in aging, and again the dropping costs of sequencing have permitted much more affordable de novo sequencing of genomes (de Magalhaes et al., 2010). For example, the sequencing of long-lived species, such as the naked mole-rat and bats (Keane et al., 2014; Kim et al., 2011; Zhang et al., 2013a) can provide candidate genes for selection in long-lived species, and it is interesting to observe that genes involved in DNA damage responses and repair have emerged from such studies (de Magalhaes & Keane, 2013). For example, Keane et al. reported the sequencing and comparative analysis of the genome of the bowhead whale, the longest-lived known mammal. The analysis revealed a number of genes under positive selection and bowhead-specific mutations in genes linked to cancer and aging. The analysis also identified gene changes associated with DNA repair, cell-cycle regulation, cancer, and aging, and potentially relevant changes in genes associated with thermoregulation, sensory perception, dietary adaptations, and immune response (Keane et al., 2015).

In addition to the analysis of genomes from long-lived species, comparative analyses of genomes from species with different lifespans are also beginning to provide further candidate genes for a role in aging. We developed a method to identify candidate genes involved in species differences in aging based on detecting proteins with accelerated evolution in multiple lineages where longevity is increased (Li & de Magalhaes, 2013). Our results revealed ~100 genes and functional groups that are candidate targets of selection when longevity evolves (Li & de Magalhaes, 2013). These include DNA damage response genes and the ubiquitin pathway and thus provide evidence that at least some repair systems were selected for, and arguably optimized, in long-lived species. Other methods aimed at discovering genes associated with longevity have focused on genes showing a stronger conservation in long-lived species (Jobson, Nabholz, & Galtier, 2010), searching for protein residues that are conserved in long-lived species but not in short-lived ones (Semeiks & Grishin, 2012). One recent work employed two methods, one based on residue change that co-occurs with the evolution of longevity and the other based on changes in rates of protein evolution, to identify candidate genes and biological processes modulating longevity across primates (Muntane et al., 2018). Because all these methods are conceptually different from each other, little overlap has been observed in the results. Nonetheless, it seems that genetic alterations contributing to the evolution of longevity in mammals have common patterns (or signatures) that are detectable using cross-species genome comparisons, though much work remains in order to improve the signal-to-noise ratio of these methods. One caveat of these studies is the lack of experimental validation, and thus all of these genes must be seen as candidates. Given the declining costs of sequencing we can expect many more such studies in the near future.

Aging is a particularly difficult process to unravel because it is much harder to study in humans than most other processes and diseases. Observational studies have been conducted but are extremely time-consuming, and clinical trials for longevity itself are challenging, even though trials for aging markers are possible (de Magalhaes, Stevens, & Thornton, 2017; de Magalhaes et al., 2012). Therefore most biogerontologists rely on model systems: human cells; unicellular organisms such as the yeast *Saccharomyces cerevisiae*; the roundworm *Caenorhabditis elegans*; the fruit fly *Drosophila melanogaster*; and rodents, in particular mice (*Mus musculus*) and rats (*Rattus norvegicus*). The small size and short life cycles of these organisms—even mice do not commonly live more than 4 years—make them inexpensive subjects for aging studies, and the ability to genetically manipulate them gives researchers ample opportunities to test their theories and unravel molecular and genetic mechanisms of aging.

The aforementioned traditional biomedical model organisms are widely used in other fields and not surprisingly a variety of tools are available to study them, and recently many of these powerful tools take advantage of omics approaches. While the genetics of aging was initially unraveled using traditional genetic approaches (reviewed in Johnson, 2002), large-scale forward genetic screening approaches now allow for hundreds of genes to be tested simultaneously for phenotypes of interest, including longevity and age-related traits. Genome-wide screens for longevity have been performed (McCormick & Kennedy, 2012), in particular in worms (Hamilton et al., 2005; Hansen, Hsu, Dillin, & Kenyon, 2005; Samuelson, Carr, & Ruvkun, 2007). Hundreds of genes have been associated with life extension in this way, although the overlap between these studies has been smaller than expected. Moreover, this type of aging studies will probably become even more accessible, with the advent of specialized, automated screening devices that monitor lifespan. For example, the *Caenorhabditis elegans* Lifespan Machine was the first automated system dedicated to lifespan assays based on video tracking of worms (Stroustrup et al., 2013). Since then, a more advanced system has been created, the WorMotel, a microfabricated device for long-term cultivation and automated longitudinal imaging of large numbers of worms confined to individual wells (Churgin et al., 2017). One important observation from the worm phenotypic screening of genetic interventions is that it seems that the most important pathways that modulate lifespan when disrupted in worms (and possibly in model organisms) have been identified by now, even though there is still ample room to identify individual components.

Although usually more laborious, screens for genes affecting lifespan have also been performed in yeast, including for replicative lifespan (Smith et al., 2008), chronological lifespan (Fabrizio et al., 2010), and using pooled screen approaches (Matecic et al., 2010). Technical limitations in flies impede screens at the genome-wide level, but lifespan screens have been performed using the P-element modular-misexpression system (Paik et al., 2012) and Gene Search misexpression vector system (Funakoshi et al., 2011). Costs impede large-scale screens in mice, although large-scale knockout mouse repositories like the Knockout Mouse Project (www.komp.org) and the International Mouse Phenotype Consortium (www.mousephenotype.org) can facilitate such studies; one large-scale profiling of mouse mutants for aging-related phenotypes, called the Harwell Aging Screen, has been performed (Potter et al., 2016).

A variety of genome-wide screens have also been performed in vitro, in particular using RNAi-based technologies (Echeverri & Perrimon, 2006; Moffat & Sabatini, 2006; Mohr, Bakal, & Perrimon, 2010). These include screens focused on traits of interest for aging and longevity. For example, screens for cell lifespan have been performed in human fibroblasts revealing that senescent cells activate a self-amplifying secretory network involving CXCR2-binding chemokines (Acosta et al., 2008). A variety of readouts can be employed to assay for specific traits. For instance, screens have been performed for genes modulating resistance to oxidative stress in mammalian cells (Nagaoka-Yasuda, Matsuo, Perkins, Limbaeck-Stokin, & Mayford, 2007; Plank et al., 2013) and antioxidant responses (Liu et al., 2007). The possibilities are immense and provide another large-scale tool for deciphering biological processes.

One of the goals of biogerontology is to develop interventions that postpone degeneration, preserve health, and extend life (de Magalhaes, 2014a). Large-scale drug screening is now widespread in the pharmaceutical industry (Macarron et al., 2011). While life extension is harder and more expensive to assay than targets in high-throughput screening, systematic screens for life-extending compounds are now a distinct possibility. Petrascheck et al. assayed 88,000 chemicals for the ability to extend worm lifespan (Petrascheck, Ye, & Buck, 2007); while the success of this approach was modest (only 115 compounds significantly extended lifespan and only 13 by more than 30%), it provides proof-of-concept for large-scale screens in the context of life-extending drugs. Further investigations into drug-mediated worm longevity, using a similar protocol, even if with a smaller compounds library of known or suspected mammalian targets (many already approved for use in humans), revealed 60 promising drugs, which might provide beneficial effects on aging in mammals (Ye, Linton, Schork, Buck, & Petrascheck, 2014).

Surveying the aging phenotype on a grand scale

In addition to understanding the genetic basis for phenotypic variation in aging and longevity, it is also crucial to elucidate the changes that contribute to age-related degeneration. Several age-related changes have been described and historically this focused on broad physiological and morphological aspects and the molecular and biochemical changes for which assays existed. Thanks to genome-wide approaches we can now survey the aging phenotype with unprecedented detail (de Magalhaes, 2009; Valdes, Glass, & Spector, 2013). In particular, advances in transcriptomics have allowed researchers to survey the expression levels of all genes in the genome in a single, relatively inexpensive, experiment.

One major breakthrough in transcriptomics was the development of the microarray, which allows for the quantification of all annotated genes simultaneously. Briefly, this led in the past 20 years to a large number of gene expression profiling studies of aging (de Magalhaes, 2009; Glass et al., 2013; Lee, Klopp, Weindruch, & Prolla, 1999; Zahn et al., 2007). In a sense, however, these have been disappointing in that relatively few genes are differentially expressed with age in most tissues and few insights have emerged. As an exception, Zahn et al. observed a degree of coordination in age-related changes in gene expression. In mice, different tissues age in a coordinated fashion so that a given mouse may exhibit rapid aging while another ages slowly across multiple tissues (Zahn et al., 2007). In addition, our 2009 meta-analysis of aging gene expression studies revealed a conserved molecular signature of mammalian aging across organs and species consisting of a clear activation of inflammatory pathways accompanied by a disruption of collagen and mitochondrial genes (de Magalhaes, Curado, & Church, 2009). This molecular signature of aging maps well into established hallmarks of aging (Lopez-Otin et al., 2013). It should be noted, however, that transcriptional changes during aging may represent responses to aging rather than underlying causative mechanisms and thus their interpretation is not straightforward.

The dropping costs of sequencing have also allowed for gene expression profiling approaches that are digital in nature, as opposed to microarrays that are analog. Sequencing the transcriptome, usually referred to as RNA-seq, allows for unprecedented accuracy and power. A number of reviews have focused on the advantages of RNA-seq as compared to microarrays (de Magalhaes et al., 2010; Mortazavi, Williams, McCue, Schaeffer, & Wold, 2008; Wang, Gerstein, & Snyder, 2009), and it is very clear that RNA-seq has a superior dynamic range and provides more data than microarrays. Our lab performed one of the first RNA-seq profiling experiments in the context of aging, which revealed gene expression changes in the rat brain in various previously unknown genes, including noncoding genes and genes not yet annotated in genome databases (Wood, Craig, Li, Merry, & de Magalhaes, 2013). Although it is exciting that many changes were observed in the so-called "dark matter" transcripts, because most of these are not annotated or have little information, follow-up is complicated; this emphasizes the need to study the new genomic elements that may be phenotypically important. In this context, large-scale efforts, such as ENCODE (www.encodeproject.org), which aims to identify all functional elements in the human genome (Dunham et al., 2012), are crucial to annotate and elucidate the function of all genomic elements. Similarly, the Genotype-Tissue Expression (GTEx) Consortium (gtexportal.org) provides data from multiple human tissues and ages (GTEx Consortium, 2015), and can be used to study transcriptional changes with age in human tissues (Chatsirisupachai, Palmer, Ferreira, & de Magalhaes, 2019; Yang et al., 2015).

A number of studies have also focused on profiling gene expression changes in life-extending interventions or in long-lived strains (de Magalhaes, 2009; Lee et al., 1999; Tyshkovskiy et al., 2019; Wood et al., 2015), as well as in short-lived and/or progeroid animals. For example, a large number of studies have focused on caloric restriction (CR) to identify specific genes and processes associated both with CR and whose age-related change is ameliorated in CR. In contrast to studies of aging, CR studies have revealed substantial gene expression changes, some of which can be associated with specific pathways and processes (Lee et al., 1999; Tsuchiya et al., 2004; Wood et al., 2015). A meta-analysis of gene expression studies of CR revealed a number of conserved processes associated with CR effects like growth hormone signaling, lipid metabolism, immune response, and detoxification pathways (Plank, Wuttke, van Dam, Clarke, & de Magalhaes, 2012). In another study, midlife gene expression profiling of mice of different lifespans due to different dietary conditions revealed a possible contribution of peroxisome to aging, which was then tested experimentally in invertebrates (Zhou et al., 2012). Arguably, gene expression profiling of manipulations of aging has been more successful in providing insights than profiling of aging per se.

Another area that has recently flourished due to technological developments is comparative transcriptomics focused on species with different lifespans. In a meta-analysis, Fushan et al., for example, analyzed the RNA-seq gene expression in liver, kidney, and brain, for 33 mammalian species from different families and with varying lifespans. This uncovered parallel evolution of gene expression and lifespan, and revealed genes evolving in agreement with the gradient of life-history variation as well as several processes and pathways (Fushan et al., 2015). In a recent study, the de novo assembled transcriptome of the gray whale (considered among the top 1% longest-lived mammals) has also been compared with that of other long- and short-lived mammals (including bowhead and minke whales). The authors show that long-lived mammals share common gene expression patterns, including high expression of DNA maintenance and repair, autophagy, ubiquitination, apoptosis, and immune responses (Toren et al., 2020).

Technological and methodological advances promise to allow even more powerful surveys of the molecular state of cells. Ribosome profiling is one recent approach, also based on next-generation sequencing platforms, consisting of sequencing ribosome-protected mRNA fragments. Compared to RNA-seq using total mRNA, ribosome profiling has the advantage that it is surveying active ribosomes, known as the translatome, and thus can be used to quantify the rate of protein synthesis, which is thought to be a better predictor of protein abundance (Ingolia, Ghaemmaghami, Newman, & Weissman, 2009). Advances in sequencing technology have also allowed for quantitative surveys of changes at the DNA level, including quantifying mutation accumulation with age in the genome and at the level of the mitochondrial genome (reviewed in de Magalhaes et al., 2010). One recent study found an age-related increase in human somatic mitochondrial mutations inconsistent with oxidative damage (Kennedy, Salk, Schmitt, & Loeb, 2013). Another study in aging mice found no increase in mitochondrial DNA point mutations or deletions, questioning whether these play a role in aging (Ameur et al., 2011).

Another level of changes during the life course comes from epigenetics. These are heritable changes that are not caused by changes in the DNA sequence. Large-scale profiling of epigenetic changes with age is now becoming more common, and with the dropping costs of sequencing will no doubt become even more widespread. It is clear that epigenetic changes, like methylation, are associated with age as well as with age-related diseases (Johnson et al., 2012). For example, two recent studies found epigenetic (methylation) marks highly predictive of chronological age in humans (Hannum et al., 2013; Horvath, 2013). Recently, an additional epigenetic biomarker of aging (DNAm

PhenoAge), incorporating several clinical measures of phenotypic age assayed in whole blood, has been developed to predict a variety of aging outcomes, including all-cause mortality, cancers, healthspan, physical functioning, and Alzheimer's disease (Levine et al., 2018). While there is still debate concerning the causality of epigenetic changes, and whether they are causes or effects of age-related degeneration, this is another field that has received increased attention lately (Field et al., 2018; Schnabl, Westermeier, Li, & Klingenspor, 2018). Modern approaches that allow the epigenome to be surveyed on a genome-wide scale however, such as methyl-DNA immunoprecipitation (MeDIP) for surveying DNA methylation and ChIP-Seq for studying histone modifications, provide the tools for researchers to study the epigenetics of aging (de Magalhaes et al., 2010). As such, the epigenome is yet another layer of genomic regulation that can be studied in a high-throughput fashion across the lifespan and in manipulations of longevity.

For all the success of transcriptomics, proteins are of course the actual machines of life and the correlation between transcripts and protein levels is not perfect. Transcriptomics provides a snapshot of transcriptional responses but in the context of aging we need proteomics to truly assay what changes occur with age. Proteomics approaches are still limited, however, in that they do not allow a comprehensive survey of the proteome in a single experiment (de Magalhaes, 2009). There have been some advances, though the number of proteins surveyed is often small compared to transcriptional profiling. For example, protein profiling of aging has been performed in the mouse heart, revealing 8 and 36 protein spots whose expression was, respectively, upregulated and downregulated due to aging (Chakravarti et al., 2008); comparable results in terms of number of proteins were also found in the mouse brain (Yang et al., 2008). Insights can be gained, however, and for instance proteome profiling of aging in mice kidney revealed functional categories associated with aging related to metabolism, transport, and stress response (Chakravarti et al., 2009). More recently, plasma proteomic profiling has revealed a proteomic signature of age (Tanaka et al., 2018).

Another emerging approach to profile age-related changes involves surveying the metabolome. One study compared metabolic parameters of young and old mice, which was then integrated with gene expression and biochemical data to derive a metabolic footprint of aging (Houtkooper et al., 2011). Another study determined the sera metabolite profile of mice of different ages and different genetic backgrounds to derive a metabolic signature that predicts the biological age in mice (Tomas-Loba, Bernardes de Jesus, Mato, & Blasco, 2013). In humans, a panel of 22 metabolites was found to significantly correlate with age and with age-related clinical conditions independent of age (Menni et al., 2013), while another group found 71 and 34 metabolites significantly associated with age in, respectively, women and men (Yu et al., 2012). In another study of more than 300 unique compounds, significant changes in the relative concentration of more than 100 metabolites were associated with age (Lawton et al., 2008). Complementarily, centenarians have also been shown to be characterized by a metabolic phenotype that is distinct from that of elderly subjects, in particular regarding amino acids and lipid species (Collino et al., 2013; Montoliu et al., 2014).

The relevance of large-scale approaches is likely to increase in the near future thanks to the rapid development of single-cell sequencing. As currently the most developed class among these techniques, single-cell RNA sequencing (scRNA-seq) can typically measure the transcriptome of hundreds of thousands of individual cells simultaneously (Regev et al., 2017). Methods targeting other molecular profiles, such as the epigenome, also tend to show promising development (Clark, Lee, Smallwood, Kelsey, & Reik, 2016), with in some cases the possibility of measuring several profiles in parallel in the same cell (Angermueller et al., 2016). An important opportunity given by such technology is the capacity to better define the notion of cell type and to build a cell atlas, such as the recently created murine atlas (Tabula Muris Consortium, 2018; Han et al., 2018) and the future human cell atlas (Regev et al., 2017).

One important limitation in age-related omics profiling, for example, transcriptomics, is that changes could be due to changes in cell types within a given tissue. As such, biogerontology will benefit from single-cell sequencing and it is encouraging to notice that a few scRNA-seq studies have already been specifically designed to investigate the aging process. While results have recently been published for the human pancreas (Enge et al., 2017), the mouse lung (Angelidis et al., 2019), and the mouse brain (Ximerakis et al., 2019), two other studies provide aging data for, respectively, 3 and 18 tissues obtained from the same mice (Kimmel et al., 2019; Pisco et al., 2020). A common conclusion of these early investigations is that aging gives rise to distinct transcriptional trajectories among cell types rather than to a universal pattern. For instance, Ximerakis et al. and Kimmel et al. both report more than 3000 genes differentially expressed with age, but most of them are specific to only one cell type and some of them show opposite regulations in different cell types. Along the same lines, Kimmel et al. observe that changes in transcriptional noise depend on cell identity (however this is in contrast with Enge et al. and Angelidis et al., both reporting an overall increase of such noise in the cell types they considered). As an interesting example of multi-omics analysis, Angelidis et al. also show that the combination of scRNA-seq with proteomics reveals the cellular origin of extracellular matrix remodeling in the lungs of old mice. Although we have only considered here a few examples of the results contained in the aforementioned articles, we can already start appreciating the relevance and the potential of single-cell sequencing for aging research and large-scale analysis.

Challenges in data analysis

Although large-scale omics approaches facilitate a broad array of studies and provide an incredible amount of data, the sheer volume of data generated creates challenges in turning the data into meaningful results and novel insights. From a statistical perspective, large-scale approaches also increase the chances of false hits that need to be accounted for when analyzing and interpreting the results. The uncertainties concerning potential false results in large-scale approaches emphasize the need for further experimental validation using a different, usually low-scale, approach. In gene expression studies, qPCR validation is usually used as the gold standard (Derveaux, Vandesompele, & Hellemans, 2010). Some types of studies, like genetic association studies of longevity, are not simple to validate, and often depend on further studies in other populations, which may or may not be feasible.

In a sense, the bottleneck in research using post-genome technologies is moving away from generating data toward interpreting data. As an example, a single 3-day run from an Illumina HiSeq X platform generates up to 900 Gb of data, which must be stored, processed, quality-controlled, and analyzed. This means that the standard experiment using next-generation sequencing platforms must account for a substantial amount of time for the bioinformatics and statistical processing of the data. Although several software tools exist now for this in silico work, labs not experienced with bioinformatics might struggle to develop a suitable pipeline and have to rely on core facilities, collaborators, or commercial services. Another problem is that for many next-generation sequencing approaches, especially in the case of single-cell sequencing, there is still no gold standard for the bioinformatics and statistical analysis and multiple alternative analysis pipelines still exist. Modest alterations in statistical parameters, for which there is no established standard, can also result in significant changes in results. For example, it is important to mention that microarray platforms for gene expression profiling are at present much quicker in terms of data analysis than approaches based on next-generation sequencing; because microarrays have been used for longer, standard methods are available for them and this is not yet the case for RNA-seq. This is even more relevant for single-cell sequencing for which comparative studies and standardization have only started to be proposed very recently (Luecken & Theis, 2019; Wang, Li, Nelson, & Nabavi, 2019). Researchers planning experiments need to carefully balance the advantages of the latest next-generation sequencing platforms with the price and simpler bioinformatics and statistics of array-based platforms.

One major and long-recognized problem of large-scale approaches is multiple hypothesis testing. Even a low-density microarray platform with a few hundred genes is testing for effects a few hundred times, which by chance will generate false positives. Modern genomic approaches, for example, in GWAS studies that survey millions of genetic variants, must adequately cope with this problem to generate biologically relevant results. A standard way of dealing with multiple hypothesis testing is the Bonferroni correction, in which the P value cutoff (typically.05) is divided by the number of hypotheses being tested (e.g., for an array with 20,000 genes use .05/20,000 as cutoff). Bonferroni correction can be deemed as too stringent, and alternative methods for correcting for false positives have been developed (Storey & Tibshirani, 2003). Benjamini correction is also widely used, and is less stringent and straightforward to calculate (Benjamini & Hochberg, 1995). False discovery rate estimates based on simulations and scrambling of data have also been widely used, including by our lab (de Magalhaes et al., 2009; Plank et al., 2012, 2013), and although it requires some customization to the specific experiments, it provides an estimate of false positives based on real data captured from the experiment.

Data integration

As mentioned previously, the recent shift in biological research toward large-scale approaches has resulted in the capacity to generate huge amounts of data, much of which is publicly available. These data, however, are in most cases heterogeneous and obtained at different time scales and biological levels. Moreover, differences also often exist due to platform and methodology diversity. Still, if our aim is to obtain a global picture of complex processes, such as aging and most age-related diseases, we have to develop the computational methods and tools that allow us to integrate and analyze these diverse data. In this section, we give an overview of the online resources currently available for aging research and discuss some of the studies that aim to integrate and analyze various types of data. This can be used on its own using public databases or in combination with data from one's own experiment(s).

Data and databases

Before diving into aging-specific resources, it should be mentioned that one important prerequisite step for data integration, the existence of databases, has already seen a tremendous expansion in the past years and continues to develop at increasing speeds. Currently, there are a number of databases, for humans and model organisms, which host a plethora of information available in a standardized, computational-retrievable and-usable form (in many cases these data are even manually curated to improve quality). These databases provide access both to a wide range of -omes (including genomes, transcriptomes, proteomes, epigenomes, interactomes, reactomes, etc.) and to a multitude of functional data (including biological processes, molecular functions, appurtenance to molecular pathways, etc.). Obviously, integrating this type of information with aging-specific data leads to a more holistic perspective of the aging process and can help in a number of analyses. Although the field of biogerontology has seen a slower increase in integrative systems biology, a series of resources specific to aging have also been created in recent years (Table 6.1), in particular in the context of our Human Ageing Genomic Resources, which are arguably the benchmark in the field (Tacutu et al., 2018).

It should also be noted that although each of these databases acts as a stand-alone resource, focusing on certain facets of aging, in many cases they also show common patterns. For example, in the Human Ageing Genomic Resources, there are many genes that can be found in two or more databases (Fig. 6.2), hence also increasing the confidence of their association to aging.

Similarly, a number of databases for age-related diseases have been developed, though the quality and type of data vary greatly. For example, there are many very good databases for cancer, while the number of databases for heart diseases is still limited. A nonexhaustive list of databases for age-related diseases is provided in Table 6.2.

While some of the resources presented above integrate data related to more than one facet of aging and/or age-related diseases, the concept of multidimensional data integration, at least at the level of aging- and disease-specific databases, is still in its infancy and the task is usually left to the researchers performing integrative analyses. Some large resources, however, like NCBI and Ensembl integrate different types of data and are of course major resources for data integration.

One other aspect that should be kept in mind is that sometimes even the amount of high-throughput

TABLE 6.1 List of major online databases and resources related to aging.

Name (citation)	Web address	Short description
AnAge (Tacutu et al., 2018)	http://genomics.senescence.info/species/	Aging, longevity, and life history information in animals
CellAge (Avelar et al., 2020)	https://genomics.senescence.info/cells/	Database of genes associated with cell senescence
Comparative Cellular and Molecular Biology of Longevity Database (Stuart et al., 2013)	http://genomics.brocku.ca/ccmbl/	Database with cellular and molecular traits from vertebrate species collected to identify traits correlated with longevity
DrugAge (Barardo et al., 2017)	https://genomics.senescence.info/drugs/	Database of drugs, compounds, and supplements that extend longevity in model organisms
GenAge (Tacutu et al., 2018)	http://genomics.senescence.info/genes/	Genes associated with longevity and/or aging in model organisms and candidate aging-related human genes
GenDR (Wuttke et al., 2012)	http://genomics.senescence.info/diet/	Genes associated with dietary restriction from mutations and gene expression profiling
AgeFactDB (Huhne, Thalheim, & Suhnel, 2014)	http://agefactdb.jenage.de/	Observations on the effect of aging factors on lifespan and/or aging phenotype
Digital Ageing Atlas (Craig et al., 2014)	http://ageing-map.org/	Database of molecular, physiological, and pathological age-related changes
LongevityMap (Budovsky et al., 2013)	http://genomics.senescence.info/longevity/	Database of human genetic variants associated with longevity

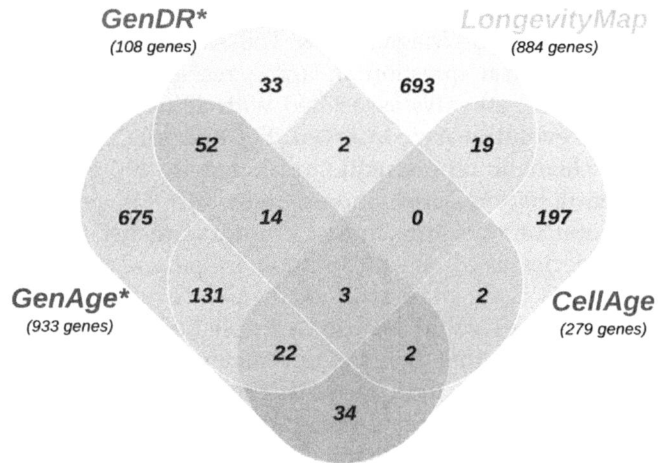

FIGURE 6.2 Venn diagram for the overlap of genomic databases in the Human Ageing Genomic Resources (Tacutu et al., 2018). Data sources used included: *GenAge, build 19*: 307 human genes and 2152 genes in model organisms; *GenDR, build 4*: 214 genes in model organisms; *LongevityMap, build 3*: 884 human genes; *CellAge, build 1*: 279 human genes. *Overlap between the four datasets was computed considering the human genes from LongevityMap and CellAge, and the nonredundant list of human genes and orthologs of genes in model organisms for GenAge and GenDR.

information for only one type of data may pose computational challenges, both in terms of handling and analyzing. Consequently, integrating and analyzing data from multiple sources will result in an even bigger challenge, the complexity increasing in most cases in a nonlinear fashion, and as such data integration comes at a cost, the haystack in which the needles have to be found increases exponentially.

Finding needles in haystacks: Network approaches and multidimensional data integration

With the expansion of large-scale approaches, and the inevitable increase in age-related data available, new hypotheses of aging trying to integrate multidimensional information have been developed. More than 20 years ago, the idea that aging was caused not simply by the failure of individual components, but rather by a network of parallel and gradual dysregulations, was proposed (Kirkwood & Kowald, 1997). While the effects of each individual event could be relatively small, the authors argued that the integrative

TABLE 6.2 Selected online databases and resources related to age-related diseases.

Name (citation)	Web address	Short description
Ageing-related Disease Genes (Fernandes et al., 2016)	https://genomics.senescence.info/diseases/	Dataset obtained from the comparison of aging-related genes with age-related disease genes
OMIM (OMIM, 2014)	http://www.omim.org/	Online Mendelian Inheritance in Man (part of NCBI)
Catalog of Published Genome-Wide Association Studies (Buniello et al., 2019)	https://www.ebi.ac.uk/gwas/	NHGRI-EBI catalog of GWAS studies
Genetic Association Database (Zhang et al., 2010)	http://geneticassociationdb.nih.gov/	Archive of human genetic association studies of complex diseases and disorders
AlzGene (Bertram, McQueen, Mullin, Blacker, & Tanzi, 2007)	http://www.alzgene.org/	Database with genetic resources for Alzheimer's disease
The COSMIC Cancer Gene Census (Sondka et al., 2018)	https://cancer.sanger.ac.uk/census/	Catalog of genes for which mutations have been causally implicated in cancer
COSMIC (Tate et al., 2019)	https://cancer.sanger.ac.uk/cosmic	Catalogue Of Somatic Mutations In Cancer
The Cancer Genome Atlas (TCGA)	http://cancergenome.nih.gov/	Portal providing access to cancer-related large-scale data from the NCI and NHGRI
The Human Metabolome Database	http://www.hmdb.ca/	Database of small-molecule metabolites found in the human body
Roadmap Epigenomics Project	http://www.roadmapepigenomics.org/	Resource of human epigenomic data
TSGene (Zhao, Kim, Mitra, Zhao, & Zhao, 2016)	https://bioinfo.uth.edu/TSGene/	Tumor suppressor gene database
Progenetix (Cai et al., 2014)	https://progenetix.org/	Copy number abnormalities in human cancer from comparative genomic hybridization experiments
MethyCancer (He et al., 2008)	http://methycancer.psych.ac.cn/	Human DNA methylation and cancer
PubMeth (Ongenaert et al., 2008)	http://www.pubmeth.org/	Cancer methylation database
LncRNADisease (Bao et al., 2019)	http://www.rnanut.net/lncrnadisease/	Long noncoding RNA and disease associations
HMDD 2.0 (Huang et al., 2019)	http://www.cuilab.cn/hmdd	Experimentally supported human microRNA and disease associations

contribution of defective mitochondria, aberrant proteins, and free radicals, taken together, could explain many of the major changes that occur during aging. Although, since it was first proposed, many other types of aging factors have been taken into consideration (and the strife to integrate more will probably continue), the "network theory of aging" might have been the onset of studying aging in a holistic way.

Network biology provides a conceptual framework to study the complex interactions between the multiple components of biological systems (Barabasi, Gulbahce, & Loscalzo, 2011). In network biology, a network is defined as a set of nodes (as a mathematical model for genes, proteins, metabolites, etc.) with some node pairs being connected through directed/asymmetric or undirected/symmetric edges (as a model for physical interactions, coexpression relationships, metabolic reactions, etc.). Depending on the type of components and the nature of the interactions that are analyzed, there is currently a large variety of network types that can be constructed; perhaps the most used being protein interaction networks, gene regulatory networks, coexpression networks, and metabolic networks.

With regards to aging research, the idea of analyzing many longevity/aging determinants at the same time has been pushed forward, mostly due to the accumulating knowledge about the genetic determinants of aging (de Magalhaes & Toussaint, 2004a; Tacutu et al., 2018), but also by the development of bioinformatics tools pertaining to network biology like Cytoscape (Saito et al., 2012). Not surprisingly, network-based approaches have been increasingly used to study aging and age-related diseases (Barabasi et al., 2011; de Magalhaes et al., 2012; Fernandes et al., 2016; Ideker & Sharan, 2008; Smita, Lange, Wolkenhauer, & Kohling, 2016; Soltow, Jones, & Promislow, 2010). Common topics addressed by these approaches include network construction for aging/longevity or various related conditions, analysis of topological features, finding functional submodules, etc.

Construction of longevity networks

The initial attempts to construct longevity networks date back more than 15 years. As a first step toward the construction of a human aging network, we used genes previously associated with aging and their interacting partners, using protein–protein interaction data, to construct networks related to DNA metabolism and the GH/IGF-1 pathway. We further suggested, based on a "guilt-by-association" methodology, that among the interacting partners of genes associated with aging there could also be other genes that are involved in aging. Additionally, functional analysis of the network revealed that many of the genes which are important during development might also regulate the rate of aging (de Magalhaes & Toussaint, 2004a).

One central question in aging research is whether genes and pathways associated with aging and longevity are evolutionary conserved. For example, in Fig. 6.3 is a schematic representation of longevity protein interaction networks across model organisms. However, the question of relevance arises: are aging-related data in one species also relevant in another species? This is an important issue since at times the data available in different species could be used complementarily. Results so far suggest that genes whose manipulation results in a lifespan effect tend to be highly evolutionary conserved across divergent eukaryotic species (Budovsky, Abramovich, Cohen, Califa-Caspi, & Fraifeld, 2007; de Magalhaes & Church, 2007; Fernandes et al., 2016; Smith et al., 2008; Yanai, Budovsky, Barzilay, Tacutu, & Fraifeld, 2017). Moreover, while not universal, some empirical data suggest that the effect on longevity of many of these genes is also conserved (Smith et al., 2008). As such, it is not completely senseless to integrate longevity-associated genes from multiple species. Using this premise, it was then shown that the human orthologs of longevity-associated genes from model organisms, together with their interacting partners, could act in a cooperative manner and form a continuous protein–protein interaction network, called the Human Longevity Network (Budovsky et al., 2007).

Topological features

One important aspect in network biology is the analysis of a network's topological characteristics (i.e., studying the way in which the nodes and edges of a network are arranged). Particular focus has been on scale-free networks, a very common type of network among social and biological networks. The scale-free topology means that the nodes in the network have a connectivity distribution p(k) given by a power-law function $k^{-\gamma}$, where p(k) is the probability that a certain node has exactly k edges, and γ is the degree exponent, a parameter value which for most of the studied networks is usually between 2 and 3 (Barabasi & Albert, 1999). The aforementioned Human Longevity Network has a scale-free topology, with a high contribution of hubs (highly connected genes) to the overall connectivity of the network. Interestingly, almost all of the hubs in the longevity network had been reported previously to be involved in at least one age-related pathology (Budovsky et al., 2007), suggesting a link between diseases and the mechanisms regulating longevity.

The scale-free design can be found in a wide range of molecular and cellular systems, largely governing their internal organization (Barabasi & Oltvai, 2004), and it appears to have been also favored by evolution (Oikonomou & Cluzel, 2006). Although a more detailed

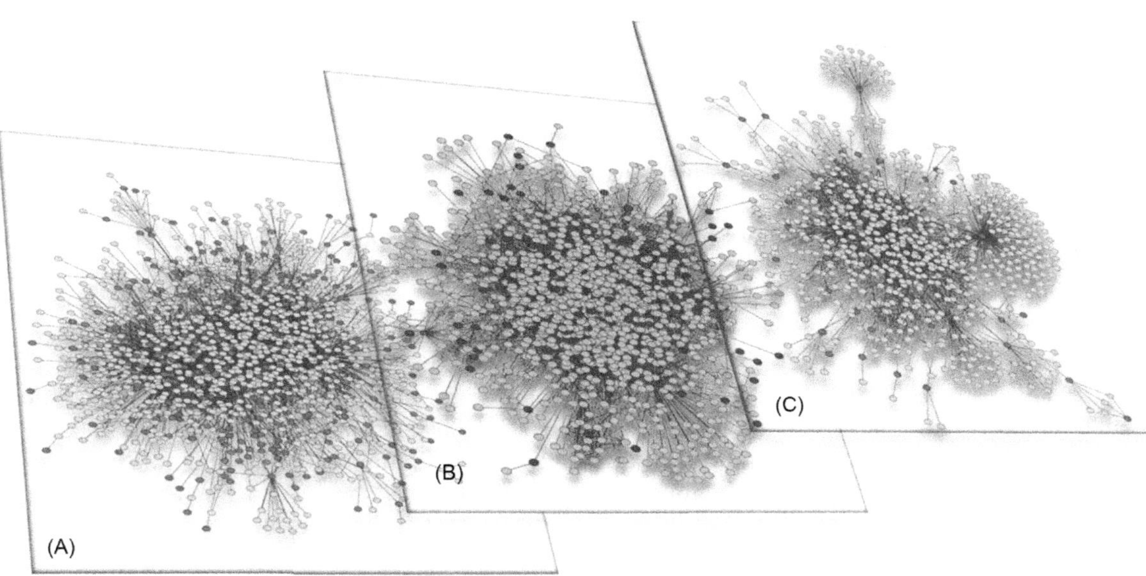

FIGURE 6.3 Schematic view on longevity networks across species. (A) Worm longevity network. (B) Fly longevity network. (C) Mouse longevity network. (A–C) Networks include longevity-associated genes (LAGs) from the GenAge database, and their protein–protein interaction from the BioGRID database (Stark et al., 2006). Dark/light colors depict LAGs and LAG-interacting partners. The number of nodes in each network (based on GenAge, build 19 and BioGRID, release 3.5.178) is summarized in the table below.

Species	LAGs in GenAge	LAGs with interactions	Longevity network	LAGs in the network
Worm	875	545	2534	525
Fly	191	162	2180	161
Mouse	136	99	834	94

discussion about the evolvability of complex networks is beyond the scope of this chapter, it should be mentioned that the properties of scale-free networks confer some net advantages in solving cellular tasks. For example, this type of architecture permits an efficient local dissipation of external perturbations, while at the same time reliably transmitting signals (and discriminating against noise) between distant elements of the network (Csermely & Soti, 2006). Additionally, the scale-free property offers an unexpected degree of robustness, maintaining the ability of nodes to communicate even under extremely high fault rates, by minimizing the effect of random failures on the entire network (Albert, Jeong, & Barabasi, 2000; Wagner, 2000).

On one hand, analyzing the aging/longevity networks can provide a framework for the conceptualization of the aging process and may reveal fundamental traits and constraints of biological systems. On the other hand, networks can help in assessing the importance of genes in a certain process. For example, differentiating between the hubs of a longevity network and all other nodes is often a very attractive way of reducing a candidate list. This approach comes as no surprise, as some components of a cellular network are more important than others with regard to aging. It was previously shown that longevity-associated genes in model organisms have a higher average connectivity, with many being network hubs (Budovsky et al., 2007; Ferrarini, Bertelli, Feala, McCulloch, & Paternostro, 2005; Promislow, 2004). Moreover, it has been established that there is a positive correlation between a protein's connectivity and its degree of pleiotropy, an elevated degree being common among proteins associated with senescence (Promislow, 2004). As such, it makes sense in choosing highly connected longevity candidates. Still, it should also be kept in mind that other topological measures besides degree also exist (e.g., closeness, eigenvector centrality, betweenness, bridging centrality, etc.) and their usage could result in a different sorting order. Ultimately, no matter what the selection criteria are, experimental validation is warranted.

Network modularity

Focusing on entire categories of genes or on network modules, and on the cross-talk between these modules, could provide valuable and unique hints regarding the system's susceptibility to failure. In relation to this, Xue et al. examined the modular structure of protein–protein interaction networks during brain aging in flies and humans. Interestingly, they found two large modules of coregulated genes, both associated with the proliferation–differentiation switch, displaying opposite age-related expression changes. A few other modules found to be associated with the oxidative–reductive metabolic switch were found, but only during fly aging. Overall, the authors found that aging is associated with a limited number of modules which are interlinked through genes more likely to affect aging/longevity (Xue et al., 2007).

Multidimensional data integration

Age-related changes can be found at many levels (expression changes, posttranslational modifications, cross-linking, or alterations in protein interactions), yet integration of multidimensional data is still in its early stages. Attempts to integrate protein–protein interaction networks with transcriptional data have already been made with relative success. As partly mentioned previously, a new analytic method permitting the integration of both transcriptome and interactome information has been employed to study network modularity in aging (Xue et al., 2007). In another study, a human protein interaction network for longevity was used in conjunction with transcriptional data from muscle aging in humans for the prediction of new longevity candidates (Bell et al., 2009).

Our meta-analysis of CR microarray studies in mammals integrated co-expression data, information on genetic mutants, and analysis of transcription factor binding sites to reveal promising candidate regulators, providing a comprehensive picture of the changes that occur during CR. In addition to the several processes previously associated with CR mentioned above, we also found novel associations, such as strong indications of the effect that caloric restriction has on circadian rhythms (Plank et al., 2012). A cross comparison of insulin signaling and CR models, using transcriptomics and metabolomics, identified some key pathways and metabolite fingerprints shared in long-lived strains, highlighting the potential role of increased amino acid metabolism and purine upregulation in longevity (Gao et al., 2018). Addressing another crucial aim in gerontology, the need to have reliable biomarkers of aging can also be done by using network-based approaches, and the integration of networks with gene expression data to create modular biomarkers of aging has been carried out (Fortney, Kotlyar, & Jurisica, 2010).

Functional classification analysis, using for example web tools like DAVID (Huang da, Sherman, & Lempicki, 2009) which analyze Gene Ontology and pathway annotations, can also generate useful information regarding the nature of longevity-associated genes. For example, several studies have already shown that inhibition of translation can be an effective modulator of lifespan extension (Curran & Ruvkun, 2007; Hansen et al., 2007; Pan et al., 2007). The integration of large-scale lists of genes with gene annotation data is therefore common in analyzing omics experiments and can provide insights concerning mechanisms, processes, and pathways (reviewed in de Magalhaes et al., 2010).

The study of aging is strongly linked to that of age-related diseases. This becomes obvious when looking at the overlap between the genes associated with major age-related diseases (including atherosclerosis, cancer, type II diabetes, and Alzheimer's disease) and the genes involved in lifespan regulation (Budovsky et al., 2007, 2009; Fernandes et al., 2016; Tacutu, Budovsky, Yanai, & Fraifeld, 2011; Wolfson, Budovsky, Tacutu, & Fraifeld, 2009), as well as when analyzing the many direct and indirect molecular interactions which exist between them (Simko, Gyurko, Veres, Nanasi, & Csermely, 2009; Tacutu et al., 2011). Networks have been extensively used for the study of diseases (Goh & Choi, 2012; Ideker & Sharan, 2008). Recently, examples of analyses of multidimensional data for age-related diseases have also started to amass. For example, based on a network of genes and diseases created by Goh et al. (2007), structural facets of proteins, such as the intrinsic disorder content, and epigenetic aspects as alternative splicing, have been studied (Midic, Oldfield, Dunker, Obradovic, & Uversky, 2009). Models of diseases–genes–drugs have also been constructed, and new insights have been found about the usage of drugs (Yildirim, Goh, Cusick, Barabasi, & Vidal, 2007). However, outside the scope of this chapter, gene–drug interaction data are thus another type of data that can be used, and indeed there are successful examples in aging research (Barardo et al., 2017; Calvert et al., 2016; Donertas, Fuentealba Valenzuela, Partridge, & Thornton, 2018; Ziehm et al., 2017). In fact, a network-based view of drug discovery and biomarkers is starting to emerge to also account for the complexity of human biology (de Magalhaes et al., 2012; Erler & Linding, 2010).

In the context of GWAS, combining GWAS with phylogenetic conservation and a complexity assessment of co-occurring transcription factor binding sites can identify *cis*-regulatory variants and elucidate their mechanistic role in disease. This has been recently carried out for type II diabetes successfully linking genetic association signals to disease-related molecular mechanisms (Claussnitzer et al., 2014). For Parkinson disease, integrative analyses of gene expression and GWAS data have also provided key insights into the genetic etiology of the disease (Edwards et al., 2011). Lastly, constructing molecular networks based on whole-genome gene expression profiling and genotyping data, together with the use of Bayesian inference, has helped to identify key causal regulators in late-onset Alzheimer's disease (Zhang et al., 2013b). Studies of proteomic and metabolomic networks, although in their infancy, may be another key step in constructing an integrative framework to study aging and age-related disease. Such integration has been suggested as needed, complementarily to the more classical genome–transcriptome–phenotype model (Hoffman, Lyu, Pletcher, & Promislow, 2017).

The above examples are only a selected few since many different types of network analyses and data integration can be performed. It is not surprising that integrative approaches are starting to be used to combine disease and aging-related data. While some studies have focused on a particular disease and its links to aging/longevity (Budovsky et al., 2009; Miller, Oldham, & Geschwind, 2008), others have attempted in a broader way to look at the common signatures of aging/longevity and all major age-related diseases (Wang, Zhang, Wang, Chen, & Zhang, 2009; Wolfson et al., 2009).

In order to better understand the gene expression and protein level changes that occur with age, other genomic and epigenetic layers should be considered. For example, age-related changes in miRNA expression profiles can have a significant impact on protein levels. In terms of data integration, combining miRNA data with protein–protein interactions has been used to analyze the molecular links between aging, longevity, and age-related diseases, and to suggest the potential role for miRNAs in targeting certain genes with features of antagonistic pleiotropy, implying thus a preferability to initiate longevity-promoting interventions in adult life (Tacutu, Budovsky, Wolfson, & Fraifeld, 2010). In another study, interpreting the methylation patterns in cancer and aging has been done using an integrative system. By developing a novel epigenome–interactome approach with differential methylation data, tissue-independent age-associated methylation hot spots targeting stem-cell differentiation pathways have been recently discovered (West, Beck, Wang, & Teschendorff, 2013).

One important aspect of data integration is that integrating multiple data sources will significantly expand our view of the aging process, and it is possible that some of the current well-accepted hypotheses will even be challenged. For example, although at the network level of protein–protein interactions it seems that hub genes are of utmost importance for the robustness of the entire network, when looking at an epigenetic level it has been suggested that the age-associated drift in DNA methylation occurs preferentially in genes that occupy peripheral network positions of exceptionally low connectivity (West, Widschwendter, & Teschendorff, 2013). Only by having a complete, multilayered picture of the aging process can we hope to fully understand its intricacies.

The emerging concept of *multilayer networks* looks therefore relevant and promising to shed new light on integrative approaches. The main idea is to go beyond the usual one-dimensional representation of networks containing only one type of nodes and one type of edges. Indeed, the above discussion makes it clear that biological systems are complex, heterogeneous, and involve various types of agents and relations between them. Interestingly, the last few years have shown increasing efforts to develop new mathematical tools to adapt the features of usual networks (such as topology and modularity) to more complex structures. As such, Kivelä et al. offer both a comprehensive review of the subject and a unifying classification of the various types of multilayer networks (Kivelä et al., 2014). Of interest, Halu et al. recently applied such techniques to create and analyze a multiplex network of human diseases (Halu, De Domenico, Arenas, & Sharma, 2019). Their model is made of two layers of similar nodes, each node representing a disease. In the first layer, diseases are connected by an edge if they share common genes, whereas in the second layer they are connected if they share common symptoms. Considering the two layers as part of a single system offers an integrative genotype–phenotype approach. The authors report, for example, that diseases sharing common genes tend to share common symptoms or also that multiplex community detection allows to confirm and find new disease associations. Inspired by such results, it is expected that similar studies applied to aging/longevity networks will appear in the future.

Predictive methods and models

Given the intrinsic costs of performing animal aging studies, particularly in mammals, developing predictive computational tools is of utmost importance. Indeed, to identify suitable drug targets with antiaging properties, methods for prioritizing them are necessary (de Magalhaes et al., 2012). Fortunately, many computational tools are already available for prioritizing candidates (Moreau & Tranchevent, 2012), and could be of great use to biogerontologists.

One of the main assumptions for many predictive methods is based on the "guilt-by-association" principle, in which new genes or drugs are considered candidates due to their relation with genes that are already known to be associated with aging or longevity. Though this premise is common to many strategies, they usually differ in the type of associations that are considered. For example, based on the finding that hubs and centrally located nodes have a higher likelihood to be associated with aging/longevity, Witten and Bonchev used a *C. elegans* network to predict new longevity-associated genes (Witten & Bonchev, 2007). Likewise, other topological measures have been employed for similar goals. Using a proximity measure in a yeast network (the shortest path to an already known gene reported to be associated with an increased lifespan), Managbanag et al. identified a set of single-gene deletions predicted to affect lifespan. Testing this experimentally, their validation showed that the predicted set was enriched for mutations conferring either increased or decreased replicative lifespan (Managbanag et al., 2008). In another example, using machine learning and classification techniques, Freitas et al. devised a predictive model to discriminate between aging-related and nonaging-related DNA repair genes. In this analysis, they found that gene connectivity together with specific gene ontology terms, having an interaction with the XRCC5 protein, and a high expression in T lymphocytes are good predictors of aging association for human DNA repair genes (Freitas, Vasieva, & de Magalhaes, 2011). Because machine learning is outside the scope of this chapter we refer readers to a recent review of machine learning in aging research (Fabris, Magalhaes, & Freitas, 2017).

In *C. elegans*, various properties of longevity genes have been analyzed and then used to verify the prediction of new longevity regulators (Li, Dong, & Guo, 2010). In one study, the authors found that longer genomic sequences, co-expression with other genes during the transition from dauer to nondauer state, enrichment in certain functions and RNAi phenotypes, higher sequence conservation, and a higher connectivity in a functional interaction network, are all predictors of an association with longevity. While the validation of the prediction was computational only, based on the precision calculated with a 10-fold cross method for a set of known positive and negative longevity-associated genes, the authors found that a few of the predicted genes had been in the meantime experimentally validated in the scientific literature (Li et al., 2010).

We have recently used a combined approach, first reasoning that the interaction partners of longevity-associated genes are more likely to modulate longevity, and second narrowing down the candidate list based on features of antagonistic pleiotropy. Although by the time of this study several genome-wide longevity assays had been performed in *C. elegans*, our prediction method, followed by experimental validation, resulted in the discovery of new longevity regulators at a frequency much higher than previously achieved (Tacutu et al., 2012).

Combining a network-based approach with transcriptional data from human aging has also been used as a method of prediction. Using a human longevity network constructed based on homologs from invertebrate species, and comparing the result with age-related transcriptional data from human muscle aging, Bell et al. determined a set of human interaction partners potentially involved in aging. Testing the homologs of these genes in *C. elegans* revealed that 33% of the candidates extended lifespan when knocked-down (Bell et al., 2009). In another, more recent study, fly longevity-associated genes from GenAge were analyzed using networks in order to predict key pathways and genes involved in lifespan regulation, which have been shown to have significant transcriptional changes in aging (Li et al., 2019).

Focusing on CR, network and systems biology approaches have also been used to predict genes necessary for the life-extending effects of CR. By looking at genes that are more connected to already-known CR-related genes, Wuttke et al. successfully predicted a set of novel genes mediating the life-extending effects of CR. Nine novel genes related to CR were validated experimentally in yeast. This revealed three novel CR mimetic genes (Wuttke et al., 2012).

In terms of metabolomics, a recent study has used network reconstruction and network analysis of metabolite relationships to associate measured plasma metabolites with sex and age, as well as to analyze the variations in the regulation of metabolic activity of amino acids, lipids, and ketone bodies (Vignoli, Tenori, Luchinat, & Saccenti, 2018). This type of methods could be used as diagnostic and predictive tools in the investigation of the human aging phenotype at a metabolic level.

While in the last few decades many studies in model organisms have successfully identified genetic factors that affect lifespan, the effect of combined interventions (epistasis), whether synergistic or antagonistic, has been evaluated to a much more limited degree. Using network features and biochemical/physicochemical features, a two-layer deletion network model has been recently developed and used for predicting the epistatic effects of double deletions on yeast longevity. Results showed that the functional features (such as mitochondrial function and chromatin silencing), the network features (such as the edge density and edge weight density of the deletion network), and the local centrality of deletion gene are important predictors for the deletion effects on longevity (Huang et al., 2012).

Candidate gene prioritization methods, such as the ones described in this section, have been instrumental in guiding various experiments that provided important insights into aging mechanisms (Lorenz, Cantor, & Collins, 2009; Wuttke et al., 2012; Xue et al., 2007). The accuracy and specificity of these in silico predictive methods is still limited, however. Similarly, while computational methods have been developed for predicting candidate drugs from gene expression data

(Iorio et al., 2010; Lamb et al., 2006; Sirota et al., 2011), these have only been partly implemented in the context of aging (Calvert et al., 2016; Donertas et al., 2018), in spite of their widespread interest.

Concluding remarks

Biological and medical research has often failed to capture the whole picture of the disease or process under study. Researchers have traditionally focused on a limited number of players that either had the greatest impact or, by chance, happened to be associated with the phenotype of interest. For some diseases (e.g., antibiotics were developed as therapies with only a modest understanding of the mechanisms involved) and processes (e.g., overcoming replicative senescence with ectopic telomerase expression), a limited mechanistic understanding may suffice to develop interventions, but for many others our understanding and models are imperfect at best and possibly even flawed. Researchers do not see the forest for the trees and for complex processes like aging, and age-related diseases like cancer and neurodegenerative diseases, this impedes the development of successful interventions. Not surprisingly, the overall rate of success of clinical trials is only about 20% (DiMasi, Feldman, Seckler, & Wilson, 2010), and to date there is no established approach to retard, even if slightly, human aging. While serendipitous discoveries like antibiotics, are always possible, it is widely acknowledged that the study of complex processes like aging stands a better chance of developing clinical interventions based on broad biological understanding (de Magalhaes, 2014a; de Magalhaes et al., 2017). In addition, the discoveries in the genetics of aging and technological advances in large-scale methodologies, like high-throughput profiling and screening, mean that it is vital now to cope with the growing amount of data in the context of drug discovery (de Magalhaes et al., 2012). As new layers of genomic regulation are uncovered (e.g., noncoding RNAs) this raises new challenges and further emphasizes the need to study biological systems in a comprehensive fashion to capture and decipher their intrinsic complexity.

Overall, our belief is that the combination of large-scale approaches to unravel both age-related changes as well as identify the causes for variability across individuals and species will drive the field forward. These require, however, adequate data and statistical analysis to avoid biases and false results. The integration of different types of data provides opportunities for synergy and discovery that we believe will result in a much deeper understanding of aging and the development of interventions to extend lifespan and preserve health.

Acknowledgments

We are thankful to Gabriela Bunu and Dmitri Toren for helping in the creation of Figs. 6.2 and 6.3. JPM is grateful to the UK Biotechnology and Biological Sciences Resource Council (BB/R014949/1), the Wellcome Trust (208375/Z/17/Z), the Royal Society (RG120543), the Leverhulme Trust (RPG-2016-015), and LongeCity for supporting his lab's development of integrative genomic resources and methods described in this chapter. RT is grateful for the funding support received through the Competitiveness Operational Programme 2014–2020, POC-A.1-A.1.1.4-E-2015. CL is grateful for the funding provided by the Human Frontier Science Program (fellowship LT000741/2019-C). We apologize to those whose relevant papers are not cited owing to lack of space.

References

Acosta, J. C., O'Loghlen, A., Banito, A., Guijarro, M. V., Augert, A., Raguz, S., ... Gil, J. (2008). Chemokine signaling via the CXCR2 receptor reinforces senescence. *Cell, 133*(6), 1006–1018. Available from https://doi.org/10.1016/j.cell.2008.03.038.

Albert, R., Jeong, H., & Barabasi, A. L. (2000). Error and attack tolerance of complex networks. *Nature, 406*(6794), 378–382. Available from https://doi.org/10.1038/35019019.

Ameur, A., Stewart, J. B., Freyer, C., Hagstrom, E., Ingman, M., Larsson, N. G., & Gyllensten, U. (2011). Ultra-deep sequencing of mouse mitochondrial DNA: Mutational patterns and their origins. *PLoS Genetics, 7*(3), e1002028. Available from https://doi.org/10.1371/journal.pgen.1002028.

Angelidis, I., Simon, L. M., Fernandez, I. E., Strunz, M., Mayr, C. H., Greiffo, F. R., ... Schiller, H. B. (2019). An atlas of the aging lung mapped by single cell transcriptomics and deep tissue proteomics. *Nature Communications, 10*(1), 963. Available from https://doi.org/10.1038/s41467-019-08831-9.

Angermueller, C., Clark, S. J., Lee, H. J., Macaulay, I. C., Teng, M. J., Hu, T. X., ... Reik, W. (2016). Parallel single-cell sequencing links transcriptional and epigenetic heterogeneity. *Nature Methods, 13*(3), 229–232. Available from https://doi.org/10.1038/nmeth.3728.

Avelar, R. A., Ortega, J. G., Tacutu, R., Tyler, E. J., Bennett, D., Binetti, P., . . . de Magalhaes, J. P. (2020). A multidimensional systems biology analysis of cellular senescence in aging and disease. *Genome Biol, 21*(1), 91. https://doi.org/10.1186/s13059-020-01990-9.

Bao, Z., Yang, Z., Huang, Z., Zhou, Y., Cui, Q., & Dong, D. (2019). LncRNADisease 2.0: An updated database of long non-coding RNA-associated diseases. *Nucleic Acids Research, 47*(D1), D1034–D1037. Available from https://doi.org/10.1093/nar/gky905.

Barabasi, A. L., & Albert, R. (1999). Emergence of scaling in random networks. *Science, 286*(5439), 509–512.

Barabasi, A. L., Gulbahce, N., & Loscalzo, J. (2011). Network medicine: A network-based approach to human disease. *Nature Reviews. Genetics, 12*(1), 56–68. Available from https://doi.org/10.1038/nrg2918.

Barabasi, A. L., & Oltvai, Z. N. (2004). Network biology: Understanding the cell's functional organization. *Nature Reviews in Genetics, 5*(2), 101–113. Available from https://doi.org/10.1038/nrg1272.

Barardo, D., Thornton, D., Thoppil, H., Walsh, M., Sharifi, S., Ferreira, S., ... de Magalhaes, J. P. (2017). The DrugAge database of aging-related drugs. *Aging Cell, 16*(3), 594–597. Available from https://doi.org/10.1111/acel.12585.

Barardo, D. G., Newby, D., Thornton, D., Ghafourian, T., de Magalhaes, J. P., & Freitas, A. A. (2017). Machine learning for predicting lifespan-extending chemical compounds. *Aging, 9*(7), 1721–1737. Available from https://doi.org/10.18632/aging.101264.

Beekman, M., Blanche, H., Perola, M., Hervonen, A., Bezrukov, V., Sikora, E., ... Franceschi, C. (2013). Genome-wide linkage analysis for human longevity: Genetics of Healthy Aging Study. *Aging Cell, 12*(2), 184–193. Available from https://doi.org/10.1111/acel.12039.

Bell, R., Hubbard, A., Chettier, R., Chen, D., Miller, J. P., Kapahi, P., ... Hughes, R. E. (2009). A human protein interaction network shows conservation of aging processes between human and invertebrate species. *PLoS Genetics, 5*(3), e1000414. Available from https://doi.org/10.1371/journal.pgen.1000414.

Benjamini, Y., & Hochberg, Y. (1995). Controlling the false discovery rate: A practical and powerful approach to multiple testing. *Journal of the Royal Statistical Society, 57*(1), 289–300.

Bertram, L., McQueen, M. B., Mullin, K., Blacker, D., & Tanzi, R. E. (2007). Systematic meta-analyses of Alzheimer's disease genetic association studies: The AlzGene database. *Nature Genetics, 39*(1), 17–23. Available from https://doi.org/10.1038/ng1934.

Budovsky, A., Abramovich, A., Cohen, R., Chalifa-Caspi, V., & Fraifeld, V. (2007). Longevity network: Construction and implications. *Mechanisms of Ageing and Development, 128*(1), 117–124. Available from https://doi.org/10.1016/j.mad.2006.11.018.

Budovsky, A., Craig, T., Wang, J., Tacutu, R., Csordas, A., Lourenco, J., ... de Magalhaes, J. P. (2013). LongevityMap: A database of human genetic variants associated with longevity. *Trends in Genetics: TIG, 29*(10), 559–560. Available from https://doi.org/10.1016/j.tig.2013.08.003.

Budovsky, A., Tacutu, R., Yanai, H., Abramovich, A., Wolfson, M., & Fraifeld, V. (2009). Common gene signature of cancer and longevity. *Mechanisms of Ageing and Development, 130*(1-2), 33–39. Available from https://doi.org/10.1016/j.mad.2008.04.002.

Buniello, A., MacArthur, J. A. L., Cerezo, M., Harris, L. W., Hayhurst, J., Malangone, C., ... Parkinson, H. (2019). The NHGRI-EBI GWAS Catalog of published genome-wide association studies, targeted arrays and summary statistics 2019. *Nucleic Acids Research, 47*(D1), D1005–D1012. Available from https://doi.org/10.1093/nar/gky1120.

Cai, H., Kumar, N., Ai, N., Gupta, S., Rath, P., & Baudis, M. (2014). Progenetix: 12 years of oncogenomic data curation. *Nucleic Acids Research, 42*(1), D1055–D1062. Available from https://doi.org/10.1093/nar/gkt1108.

Calvert, S., Tacutu, R., Sharifi, S., Teixeira, R., Ghosh, P., & de Magalhaes, J. P. (2016). A network pharmacology approach reveals new candidate caloric restriction mimetics in C. elegans. *Aging Cell, 15*(2), 256–266. Available from https://doi.org/10.1111/acel.12432.

Cevenini, E., Bellavista, E., Tieri, P., Castellani, G., Lescai, F., Francesconi, M., ... Franceschi, C. (2010). Systems biology and longevity: An emerging approach to identify innovative anti-aging targets and strategies. *Current Pharmaceutical Design, 16*(7), 802–813.

Chakravarti, B., Oseguera, M., Dalal, N., Fathy, P., Mallik, B., Raval, A., & Chakravarti, D. N. (2008). Proteomic profiling of aging in the mouse heart: Altered expression of mitochondrial proteins. *Archives of Biochemistry and Biophysics, 474*(1), 22–31. Available from https://doi.org/10.1016/j.abb.2008.02.001.

Chakravarti, B., Seshi, B., Ratanaprayul, W., Dalal, N., Lin, L., Raval, A., & Chakravarti, D. N. (2009). Proteome profiling of aging in mouse models: Differential expression of proteins involved in metabolism, transport, and stress response in kidney. *Proteomics, 9*(3), 580–597. Available from https://doi.org/10.1002/pmic.200700208.

Chatsirisupachai, K., Palmer, D., Ferreira, S., & de Magalhaes, J. P. (2019). A human tissue-specific transcriptomic analysis reveals a complex relationship between aging, cancer, and cellular senescence. *Aging Cell, 18*(6), e13041. Available from https://doi.org/10.1111/acel.13041.

Christensen, K., Johnson, T. E., & Vaupel, J. W. (2006). The quest for genetic determinants of human longevity: Challenges and insights. *Nature Reviews. Genetics, 7*(6), 436–448. Available from https://doi.org/10.1038/nrg1871.

Churgin, M. A., Jung, S. K., Yu, C. C., Chen, X., Raizen, D. M., & Fang-Yen, C. (2017). Longitudinal imaging of *Caenorhabditis elegans* in a microfabricated device reveals variation in behavioral decline during aging. *Elife*, 6, e26652. Available from https://doi.org/10.7554/eLife.26652.

Clark, S. J., Lee, H. J., Smallwood, S. A., Kelsey, G., & Reik, W. (2016). Single-cell epigenomics: Powerful new methods for understanding gene regulation and cell identity. *Genome Biology, 17*, 72. Available from https://doi.org/10.1186/s13059-016-0944-x.

Claussnitzer, M., Dankel., Simon, N., Klocke, B., Grallert, H., Glunk, V., Berulava, T., ... Laumen, H. (2014). Leveraging cross-species transcription factor binding site patterns: From diabetes risk loci to disease mechanisms. *Cell, 156*(1–2), 343–358. Available from https://doi.org/10.1016/j.cell.2013.10.058.

Collino, S., Montoliu, I., Martin, F. P., Scherer, M., Mari, D., Salvioli, S., ... Rezzi, S. (2013). Metabolic signatures of extreme longevity in northern Italian centenarians reveal a complex remodeling of lipids, amino acids, and gut microbiota metabolism. *PLoS One, 8*(3), e56564. Available from https://doi.org/10.1371/journal.pone.0056564.

GTEx Consortium. (2015). Human genomics. The Genotype-Tissue Expression (GTEx) pilot analysis: Multitissue gene regulation in humans. *Science, 348*(6235), 648–660. Available from https://doi.org/10.1126/science.1262110.

Tabula Muris Consortium. (2018). Single-cell transcriptomics of 20 mouse organs creates a *Tabula muris. Nature, 562*(7727), 367–372. Available from https://doi.org/10.1038/s41586-018-0590-4.

Craig, T., Smelick, C., Tacutu, R., Wuttke, D., Wood, S. H., Stanley, H., ... de Magalhaes, J. P. (2015). The Digital Ageing Atlas: Integrating the diversity of age-related changes into a unified resource. *Nucleic Acids Research, 43*, D873–8. Available from https://doi.org/10.1093/nar/gku843.

Csermely, P., & Soti, C. (2006). Cellular networks and the aging process. *Archives of Physiology and Biochemistry, 112*(2), 60–64. Available from https://doi.org/10.1080/13813450600711243.

Curran, S. P., & Ruvkun, G. (2007). Lifespan regulation by evolutionarily conserved genes essential for viability. *PLoS Genetics, 3*(4), e56. Available from https://doi.org/10.1371/journal.pgen.0030056.

de Magalhaes, J. P. (2004). From cells to ageing: A review of models and mechanisms of cellular senescence and their impact on human ageing. *Experimental Cell Research, 300*(1), 1–10.

de Magalhaes, J. P. (2006). Species selection in comparative studies of aging and antiaging research. In P. M. Conn (Ed.), *Handbook of models for human aging* (pp. 9–20). Burlington, MA: Elsevier Academic Press.

de Magalhaes, J. P. (2009). Aging research in the post-genome era: New technologies for an old problem. In C. H. Foyer, R. Faragher, & P. J. Thornalley (Eds.), *Redox metabolism and longevity relationships in animals and plants* (pp. 99–115). New York and Abingdon: Taylor and Francis.

de Magalhaes, J. P. (2011). The biology of ageing: A primer. In I. Stuart-Hamilton (Ed.), *An Introduction to Gerontology* (pp. 21–47). Cambridge: Cambridge University Press.

de Magalhaes, J. P. (2014a). The scientific quest for lasting youth: Prospects for curing aging. *Rejuvenation Research, 17*(5), 458–467. Available from https://doi.org/10.1089/rej.2014.1580.

de Magalhaes, J. P. (2014b). Why genes extending lifespan in model organisms have not been consistently associated with human longevity and what it means to translation research. *Cell Cycle, 13*(17), 2671–2673. Available from https://doi.org/10.4161/15384101.2014.950151.

de Magalhaes, J. P., & Church, G. M. (2007). Analyses of human-chimpanzee orthologous gene pairs to explore evolutionary hypotheses of aging. *Mechanisms of Ageing and Development*, *128*(5–6), 355–364. Available from https://doi.org/10.1016/j.mad.2007.03.004.

de Magalhaes, J. P., Curado, J., & Church, G. M. (2009). Meta-analysis of age-related gene expression profiles identifies common signatures of aging. *Bioinformatics*, *25*(7), 875–881. Available from https://doi.org/10.1093/bioinformatics/btp073.

de Magalhaes, J. P., Finch, C. E., & Janssens, G. (2010). Next-generation sequencing in aging research: Emerging applications, problems, pitfalls and possible solutions. *Ageing Research Reviews*, *9*(3), 315–323. Available from https://doi.org/10.1016/j.arr.2009.10.006.

de Magalhaes, J. P., & Keane, M. (2013). Endless paces of degeneration--applying comparative genomics to study evolution's moulding of longevity. *EMBO Reports*, *14*(8), 661–662. Available from https://doi.org/10.1038/embor.2013.96.

de Magalhaes, J. P., Stevens, M., & Thornton, D. (2017). The business of anti-aging science. *Trends in Biotechnology*, *35*(11), 1062–1073. Available from https://doi.org/10.1016/j.tibtech.2017.07.004.

de Magalhaes, J. P., & Toussaint, O. (2004a). GenAge: A genomic and proteomic network map of human ageing. *FEBS Letters*, *571*(1-3), 243–247.

de Magalhaes, J. P., & Toussaint, O. (2004b). How bioinformatics can help reverse engineer human aging. *Ageing Research Reviews*, *3*(2), 125–141.

de Magalhaes, J. P., & Wang, J. (2019). The fog of genetics: What is known, unknown and unknowable in the genetics of complex traits and diseases. *EMBO Reports*, *20*(11), e48054. Available from https://doi.org/10.15252/embr.201948054.

de Magalhaes, J. P., Wuttke, D., Wood, S. H., Plank, M., & Vora, C. (2012). Genome-environment interactions that modulate aging: Powerful targets for drug discovery. *Pharmacological Reviews*, *64*(1), 88–101. Available from https://doi.org/10.1124/pr.110.004499.

Derveaux, S., Vandesompele, J., & Hellemans, J. (2010). How to do successful gene expression analysis using real-time PCR. *Methods*, *50*(4), 227–230. Available from https://doi.org/10.1016/j.ymeth.2009.11.001.

DiMasi, J. A., Feldman, L., Seckler, A., & Wilson, A. (2010). Trends in risks associated with new drug development: Success rates for investigational drugs. *Clinical Pharmacology and Therapeutics*, *87*(3), 272–277. Available from https://doi.org/10.1038/clpt.2009.295.

Donertas, H. M., Fuentealba Valenzuela, M., Partridge, L., & Thornton, J. M. (2018). Gene expression-based drug repurposing to target aging. *Aging Cell*, *17*(5), e12819. Available from https://doi.org/10.1111/acel.12819.

Dunham, I., Kundaje, A., Aldred, S. F., Collins, P. J., Davis, C. A., Doyle, F., ... Lochovsky, L. (2012). An integrated encyclopedia of DNA elements in the human genome. *Nature*, *489*(7414), 57–74. Available from https://doi.org/10.1038/nature11247.

Echeverri, C. J., & Perrimon, N. (2006). High-throughput RNAi screening in cultured cells: A user's guide. *Nature Reviews. Genetics*, *7*(5), 373–384.

Edwards, Y. J., Beecham, G. W., Scott, W. K., Khuri, S., Bademci, G., Tekin, D., ... Vance, J. M. (2011). Identifying consensus disease pathways in Parkinson's disease using an integrative systems biology approach. *PLoS One*, *6*(2), e16917. Available from https://doi.org/10.1371/journal.pone.0016917.

Enge, M., Arda, H. E., Mignardi, M., Beausang, J., Bottino, R., Kim, S. K., & Quake, S. R. (2017). Single-cell analysis of human pancreas reveals transcriptional signatures of aging and somatic mutation patterns. *Cell*, *171*(2), 321–330, e314. Available from https://doi.org/10.1016/j.cell.2017.09.004.

Erler, J. T., & Linding, R. (2010). Network-based drugs and biomarkers. *The Journal of Pathology*, *220*(2), 290–296. Available from https://doi.org/10.1002/path.2646.

Fabris, F., Magalhaes, J. P., & Freitas, A. A. (2017). A review of supervised machine learning applied to ageing research. *Biogerontology*, *18*(2), 171–188. Available from https://doi.org/10.1007/s10522-017-9683-y.

Fabrizio, P., Hoon, S., Shamalnasab, M., Galbani, A., Wei, M., Giaever, G., ... Longo, V. D. (2010). Genome-wide screen in *Saccharomyces cerevisiae* identifies vacuolar protein sorting, autophagy, biosynthetic, and tRNA methylation genes involved in life span regulation. *PLoS Genetics*, *6*(7), e1001024. Available from https://doi.org/10.1371/journal.pgen.1001024.

Fernandes, M., Wan, C., Tacutu, R., Barardo, D., Rajput, A., Wang, J., ... de Magalhaes, J. P. (2016). Systematic analysis of the gerontome reveals links between aging and age-related diseases. *Human Molecular Genetics*, *25*(21), 4804–4818. Available from https://doi.org/10.1093/hmg/ddw307.

Ferrarini, L., Bertelli, L., Feala, J., McCulloch, A. D., & Paternostro, G. (2005). A more efficient search strategy for aging genes based on connectivity. *Bioinformatics*, *21*(3), 338–348. Available from https://doi.org/10.1093/bioinformatics/bti004.

Field, A. E., Robertson, N. A., Wang, T., Havas, A., Ideker, T., & Adams, P. D. (2018). DNA methylation clocks in aging: Categories, causes, and consequences. *Molecular Cell*, *71*(6), 882–895. Available from https://doi.org/10.1016/j.molcel.2018.08.008.

Finch, C. E. (1990). *Longevity, senescence, and the genome*. Chicago and London: The University of Chicago Press.

Fortney, K., Kotlyar, M., & Jurisica, I. (2010). Inferring the functions of longevity genes with modular subnetwork biomarkers of *Caenorhabditis elegans* aging. *Genome Biology*, *11*(2), R13. Available from https://doi.org/10.1186/gb-2010-11-2-r13.

Freitas, A. A., Vasieva, O., & de Magalhaes, J. P. (2011). A data mining approach for classifying DNA repair genes into ageing-related or non-ageing-related. *BMC Genomics*, *12*, 27. Available from https://doi.org/10.1186/1471-2164-12-27.

Funakoshi, M., Tsuda, M., Muramatsu, K., Hatsuda, H., Morishita, S., & Aigaki, T. (2011). A gain-of-function screen identifies wdb and lkb1 as lifespan-extending genes in Drosophila. *Biochemical and Biophysical Research Communications*, *405*(4), 667–672. Available from https://doi.org/10.1016/j.bbrc.2011.01.090.

Fushan, A. A., Turanov, A. A., Lee, S. G., Kim, E. B., Lobanov, A. V., Yim, S. H., ... Gladyshev, V. N. (2015). Gene expression defines natural changes in mammalian lifespan. *Aging Cell*, *14*(3), 352–365. Available from https://doi.org/10.1111/acel.12283.

Gao, A. W., Smith, R. L., van Weeghel, M., Kamble, R., Janssens, G. E., & Houtkooper, R. H. (2018). Identification of key pathways and metabolic fingerprints of longevity in *C. elegans*. *Experimental Gerontology*, *113*, 128–140. Available from https://doi.org/10.1016/j.exger.2018.10.003.

Glass, D., Vinuela, A., Davies, M. N., Ramasamy, A., Parts, L., Knowles, D., ... Spector, T. D. (2013). Gene expression changes with age in skin, adipose tissue, blood and brain. *Genome Biology*, *14*(7), R75. Available from https://doi.org/10.1186/gb-2013-14-7-r75.

Goh, K. I., & Choi, I. G. (2012). Exploring the human diseasome: The human disease network. *Brief Function in Genomics*, *11*(6), 533–542. Available from https://doi.org/10.1093/bfgp/els032.

Goh, K. I., Cusick, M. E., Valle, D., Childs, B., Vidal, M., & Barabasi, A. L. (2007). The human disease network. *Proceedings of the National Academy of Sciences of the United States of America*, *104*(21), 8685–8690. Available from https://doi.org/10.1073/pnas.0701361104.

Halu, A., De Domenico, M., Arenas, A., & Sharma, A. (2019). The multiplex network of human diseases. *NPJ Systems Biology and Applications*, *5*, 15. Available from https://doi.org/10.1038/s41540-019-0092-5.

Hamilton, B., Dong, Y., Shindo, M., Liu, W., Odell, I., Ruvkun, G., & Lee, S. S. (2005). A systematic RNAi screen for longevity genes in C. elegans. *Genes & Development*, *19*(13), 1544–1555. Available from https://doi.org/10.1101/gad.1308205.

Han, X., Wang, R., Zhou, Y., Fei, L., Sun, H., Lai, S., ... Guo, G. (2018). Mapping the mouse cell atlas by microwell-seq. *Cell, 172* (5), 1091–1107, e1017. Available from https://doi.org/10.1016/j.cell.2018.02.001.

Hannum, G., Guinney, J., Zhao, L., Zhang, L., Hughes, G., Sadda, S., ... Zhang, K. (2013). Genome-wide methylation profiles reveal quantitative views of human aging rates. *Molecular Cell, 49*(2), 359–367. Available from https://doi.org/10.1016/j.molcel.2012.10.016.

Hansen, M., Hsu, A. L., Dillin, A., & Kenyon, C. (2005). New genes tied to endocrine, metabolic, and dietary regulation of lifespan from a *Caenorhabditis elegans* genomic RNAi screen. *PLoS Genetics, 1*(1), 119–128. Available from https://doi.org/10.1371/journal.pgen.0010017.

Hansen, M., Taubert, S., Crawford, D., Libina, N., Lee, S. J., & Kenyon, C. (2007). Lifespan extension by conditions that inhibit translation in *Caenorhabditis elegans*. *Aging Cell, 6*(1), 95–110. Available from https://doi.org/10.1111/j.1474-9726.2006.00267.x.

He, X., Chang, S., Zhang, J., Zhao, Q., Xiang, H., Kusonmano, K., ... Wang, J. (2008). MethyCancer: The database of human DNA methylation and cancer. *Nucleic Acids Res, 36*(Database issue), D836–D841. Available from https://doi.org/10.1093/nar/gkm730.

Hoffman, J. M., Lyu, Y., Pletcher, S. D., & Promislow, D. E. L. (2017). Proteomics and metabolomics in ageing research: From biomarkers to systems biology. *Essays in Biochemistry, 61*(3), 379–388. Available from https://doi.org/10.1042/EBC20160083.

Horvath, S. (2013). DNA methylation age of human tissues and cell types. *Genome Biology, 14*(10), R115. Available from https://doi.org/10.1186/gb-2013-14-10-r115.

Horvath, S., & Raj, K. (2018). DNA methylation-based biomarkers and the epigenetic clock theory of ageing. *Nature Reviews. Genetics, 19*(6), 371–384. Available from https://doi.org/10.1038/s41576-018-0004-3.

Houtkooper, R. H., Argmann, C., Houten, S. M., Canto, C., Jeninga, E. H., Andreux, P. A., ... Auwerx, J. (2011). The metabolic footprint of aging in mice. *Scientific Reports, 1*, 134. Available from https://doi.org/10.1038/srep00134.

Huang da, W., Sherman, B. T., & Lempicki, R. A. (2009). Systematic and integrative analysis of large gene lists using DAVID bioinformatics resources. *Nature Protocols, 4*(1), 44–57. Available from https://doi.org/10.1038/nprot.2008.211.

Huang, T., Zhang, J., Xu, Z. P., Hu, L. L., Chen, L., Shao, J. L., ... Chou, K. C. (2012). Deciphering the effects of gene deletion on yeast longevity using network and machine learning approaches. *Biochimie, 94*(4), 1017–1025. Available from https://doi.org/10.1016/j.biochi.2011.12.024.

Huang, Z., Shi, J., Gao, Y., Cui, C., Zhang, S., Li, J., ... Cui, Q. (2019). HMDD v3.0: A database for experimentally supported human microRNA-disease associations. *Nucleic Acids Research, 47*(D1), D1013–D1017. Available from https://doi.org/10.1093/nar/gky1010.

Huhne, R., Thalheim, T., & Suhnel, J. (2014). AgeFactDB--the JenAge Ageing Factor Database--towards data integration in ageing research. *Nucleic Acids Research, 42*(1), D892–D896. Available from https://doi.org/10.1093/nar/gkt1073.

Ideker, T., & Sharan, R. (2008). Protein networks in disease. *Genome Research, 18*(4), 644–652. Available from https://doi.org/10.1101/gr.071852.107.

Ingolia, N. T., Ghaemmaghami, S., Newman, J. R., & Weissman, J. S. (2009). Genome-wide analysis in vivo of translation with nucleotide resolution using ribosome profiling. *Science, 324*(5924), 218–223. Available from https://doi.org/10.1126/science.1168978.

Iorio, F., Bosotti, R., Scacheri, E., Belcastro, V., Mithbaokar, P., Ferriero, R., ... di Bernardo, D. (2010). Discovery of drug mode of action and drug repositioning from transcriptional responses. *Proceedings of the National Academy of Sciences of the United States of America, 107*(33), 14621–14626. Available from https://doi.org/10.1073/pnas.1000138107.

Jobson, R. W., Nabholz, B., & Galtier, N. (2010). An evolutionary genome scan for longevity-related natural selection in mammals. *Molecular Biology and Evolution, 27*(4), 840–847. Available from https://doi.org/10.1093/molbev/msp293.

Johnson, A. A., Akman, K., Calimport, S. R., Wuttke, D., Stolzing, A., & de Magalhaes, J. P. (2012). The role of DNA methylation in aging, rejuvenation, and age-related disease. *Rejuvenation Research, 15*(5), 483–494. Available from https://doi.org/10.1089/rej.2012.1324.

Johnson, T. E. (2002). A personal retrospective on the genetics of aging. *Biogerontology, 3*(1–2), 7–12.

Keane, M., Craig, T., Alfoldi, J., Berlin, A. M., Johnson, J., Seluanov, A., ... de Magalhaes, J. P. (2014). The Naked Mole Rat Genome Resource: Facilitating analyses of cancer and longevity-related adaptations. *Bioinformatics, 30*, 3558–3560. Available from https://doi.org/10.1093/bioinformatics/btu579.

Keane, M., Semeiks, J., Webb, A. E., Li, Y. I., Quesada, V., Craig, T., ... de Magalhaes, J. P. (2015). Insights into the evolution of longevity from the bowhead whale genome. *Cell Reports, 10*(1), 112–122. Available from https://doi.org/10.1016/j.celrep.2014.12.008.

Kennedy, S. R., Salk, J. J., Schmitt, M. W., & Loeb, L. A. (2013). Ultra-sensitive sequencing reveals an age-related increase in somatic mitochondrial mutations that are inconsistent with oxidative damage. *PLoS Genetics, 9*(9), e1003794. Available from https://doi.org/10.1371/journal.pgen.1003794.

Kim, E. B., Fang, X., Fushan, A. A., Huang, Z., Lobanov, A. V., Han, L., ... Gladyshev, V. N. (2011). Genome sequencing reveals insights into physiology and longevity of the naked mole rat. *Nature, 479*(7372), 223–227. Available from https://doi.org/10.1038/nature10533.

Kimmel, J. C., Penland, L., Rubinstein, N. D., Hendrickson, D. G., Kelley, D. R., & Rosenthal, A. Z. (2019). Murine single-cell RNA-seq reveals cell-identity- and tissue-specific trajectories of aging. *Genome Res, 29*(12), 2088–2103. https://doi.org/10.1101/gr.253880.119.

Kirkwood, T. B., & Kowald, A. (1997). Network theory of aging. *Experimental Gerontology, 32*(4-5), 395–399.

Kivelä, M., Arenas, A., Barthelemy, M., Gleeson, J. P., Moreno, Y., & Porter, M. A. (2014). Multilayer networks. *Journal of Complex Networks, 2*(3), 203–271. Available from https://doi.org/10.1093/comnet/cnu016.

Lamb, J., Crawford, E. D., Peck, D., Modell, J. W., Blat, I. C., Wrobel, M. J., ... Golub, T. R. (2006). The Connectivity Map: Using gene-expression signatures to connect small molecules, genes, and disease. *Science, 313*(5795), 1929–1935. Available from https://doi.org/10.1126/science.1132939.

Lawton, K. A., Berger, A., Mitchell, M., Milgram, K. E., Evans, A. M., Guo, L., ... Milburn, M. V. (2008). Analysis of the adult human plasma metabolome. *Pharmacogenomics, 9*(4), 383–397. Available from https://doi.org/10.2217/14622416.9.4.383.

Lee, C. K., Klopp, R. G., Weindruch, R., & Prolla, T. A. (1999). Gene expression profile of aging and its retardation by caloric restriction. *Science, 285*(5432), 1390–1393.

Levine, M. E., Lu, A. T., Quach, A., Chen, B. H., Assimes, T. L., Bandinelli, S., ... Horvath, S. (2018). An epigenetic biomarker of aging for lifespan and healthspan. *Aging, 10*(4), 573–591. Available from https://doi.org/10.18632/aging.101414.

Li, J. Q., Duan, D. D., Zhang, J. Q., Zhou, Y. Z., Qin, X. M., Du, G. H., & Gao, L. (2019). Bioinformatic prediction of critical genes and pathways involved in longevity in *Drosophila melanogaster*. *Molecular Genetics and Genomics: MGG, 294*(6), 1463–1475. Available from https://doi.org/10.1007/s00438-019-01589-1.

Li, Y., & de Magalhaes, J. P. (2013). Accelerated protein evolution analysis reveals genes and pathways associated with the evolution of mammalian longevity. *Age*, *35*(2), 301–314. Available from https://doi.org/10.1007/s11357-011-9361-y.

Li, Y. H., Dong, M. Q., & Guo, Z. (2010). Systematic analysis and prediction of longevity genes in *Caenorhabditis elegans*. *Mechanisms of Ageing and Development*, *131*(11–12), 700–709. Available from https://doi.org/10.1016/j.mad.2010.10.001.

Liu, Y., Kern, J. T., Walker, J. R., Johnson, J. A., Schultz, P. G., & Luesch, H. (2007). A genomic screen for activators of the antioxidant response element. *Proceedings of the National Academy of Sciences of the United States of America*, *104*(12), 5205–5210. Available from https://doi.org/10.1073/pnas.0700898104.

Lopez-Otin, C., Blasco, M. A., Partridge, L., Serrano, M., & Kroemer, G. (2013). The hallmarks of aging. *Cell*, *153*(6), 1194–1217. Available from https://doi.org/10.1016/j.cell.2013.05.039.

Lorenz, D. R., Cantor, C. R., & Collins, J. J. (2009). A network biology approach to aging in yeast. *Proceedings of the National Academy of Sciences of the United States of America*, *106*(4), 1145–1150. Available from https://doi.org/10.1073/pnas.0812551106.

Luecken, M. D., & Theis, F. J. (2019). Current best practices in single-cell RNA-seq analysis: A tutorial. *Molecular Systems Biology*, *15*(6), e8746. Available from https://doi.org/10.15252/msb.20188746.

Macarron, R., Banks, M. N., Bojanic, D., Burns, D. J., Cirovic, D. A., Garyantes, T., ... Sittampalam, G. S. (2011). Impact of high-throughput screening in biomedical research. *Nature Reviews. Drug Discovery*, *10*(3), 188–195. Available from https://doi.org/10.1038/nrd3368.

Managbanag, J. R., Witten, T. M., Bonchev, D., Fox, L. A., Tsuchiya, M., Kennedy, B. K., & Kaeberlein, M. (2008). Shortest-path network analysis is a useful approach toward identifying genetic determinants of longevity. *PLoS One*, *3*(11), e3802. Available from https://doi.org/10.1371/journal.pone.0003802.

Manolio, T. A., Collins, F. S., Cox, N. J., Goldstein, D. B., Hindorff, L. A., Hunter, D. J., ... Visscher, P. M. (2009). Finding the missing heritability of complex diseases. *Nature*, *461*(7265), 747–753. Available from https://doi.org/10.1038/nature08494.

Matecic, M., Smith, D. L., Pan, X., Maqani, N., Bekiranov, S., Boeke, J. D., & Smith, J. S. (2010). A microarray-based genetic screen for yeast chronological aging factors. *PLoS Genetics*, *6*(4), e1000921. Available from https://doi.org/10.1371/journal.pgen.1000921.

McCormick, M. A., & Kennedy, B. K. (2012). Genome-scale studies of aging: Challenges and opportunities. *Current Genomics*, *13*(7), 500–507. Available from https://doi.org/10.2174/138920212803251454.

Menni, C., Kastenmuller, G., Petersen, A. K., Bell, J. T., Psatha, M., Tsai, P. C., ... Valdes, A. M. (2013). Metabolomic markers reveal novel pathways of ageing and early development in human populations. *International Journal of Epidemiology*, *42*(4), 1111–1119. Available from https://doi.org/10.1093/ije/dyt094.

Midic, U., Oldfield, C. J., Dunker, A. K., Obradovic, Z., & Uversky, V. N. (2009). Protein disorder in the human diseasome: Unfoldomics of human genetic diseases. *BMC Genomics*, *10*(Suppl. 1), S12. Available from https://doi.org/10.1186/1471-2164-10-S1-S12.

Miller, J. A., Oldham, M. C., & Geschwind, D. H. (2008). A systems level analysis of transcriptional changes in Alzheimer's disease and normal aging. *The Journal of Neuroscience*, *28*(6), 1410–1420. Available from https://doi.org/10.1523/JNEUROSCI.4098-07.2008.

Moffat, J., & Sabatini, D. M. (2006). Building mammalian signalling pathways with RNAi screens. *Nature Reviews. Molecular Cell Biology*, *7*(3), 177–187. Available from https://doi.org/10.1038/nrm1860.

Mohr, S., Bakal, C., & Perrimon, N. (2010). Genomic screening with RNAi: Results and challenges. *Annual Review of Biochemistry*, *79*, 37–64. Available from https://doi.org/10.1146/annurev-biochem-060408-092949.

Montoliu, I., Scherer, M., Beguelin, F., DaSilva, L., Mari, D., Salvioli, S., ... Collino, S. (2014). Serum profiling of healthy aging identifies phospho- and sphingolipid species as markers of human longevity. *Aging*, *6*(1), 9–25. Available from https://doi.org/10.18632/aging.100630.

Moreau, Y., & Tranchevent, L. C. (2012). Computational tools for prioritizing candidate genes: Boosting disease gene discovery. *Nature Reviews of Genetics*, *13*(8), 523–536. Available from https://doi.org/10.1038/nrg3253.

Mortazavi, A., Williams, B. A., McCue, K., Schaeffer, L., & Wold, B. (2008). Mapping and quantifying mammalian transcriptomes by RNA-Seq. *Nature Methods*, *5*(7), 621–628. Available from https://doi.org/10.1038/nmeth.1226.

Muntane, G., Farre, X., Rodriguez, J. A., Pegueroles, C., Hughes, D. A., de Magalhaes, J. P., ... Navarro, A. (2018). Biological processes modulating longevity across primates: A phylogenetic genome-phenome analysis. *Molecular Biology and Evolution*, *35*(8), 1990–2004. Available from https://doi.org/10.1093/molbev/msy105.

Nagaoka-Yasuda, R., Matsuo, N., Perkins, B., Limbaeck-Stokin, K., & Mayford, M. (2007). An RNAi-based genetic screen for oxidative stress resistance reveals retinol saturase as a mediator of stress resistance. *Free Radical Biology & Medicine*, *43*(5), 781–788. Available from https://doi.org/10.1016/j.freeradbiomed.2007.05.008.

Oikonomou, P., & Cluzel, P. (2006). Effects of topology on network evolution. *Nature Physics*, *2*, 532–536.

OMIM. (2014). Online Mendelian Inheritance in Man, OMIM®. McKusick-Nathans Institute of Genetic Medicine, Johns Hopkins University (Baltimore, MD), www.omim.org.

Ongenaert, M., Van Neste, L., De Meyer, T., Menschaert, G., Bekaert, S., & Van Criekinge, W. (2008). PubMeth: A cancer methylation database combining text-mining and expert annotation. *Nucleic Acids Research*, *36*(Database issue), D842–D846. Available from https://doi.org/10.1093/nar/gkm788.

Paik, D., Jang, Y. G., Lee, Y. E., Lee, Y. N., Yamamoto, R., Gee, H. Y., ... Park, J. J. (2012). Misexpression screen delineates novel genes controlling *Drosophila* lifespan. *Mechanisms of Ageing and Development*, *133*(5), 234–245. Available from https://doi.org/10.1016/j.mad.2012.02.001.

Pan, F., Chiu, C. H., Pulapura, S., Mehan, M. R., Nunez-Iglesias, J., Zhang, K., ... Zhou, X. J. (2007). Gene Aging Nexus: A web database and data mining platform for microarray data on aging. *Nucleic Acids Research*, *35*(Database issue), D756–D759. Available from https://doi.org/10.1093/nar/gkl798.

Petrascheck, M., Ye, X., & Buck, L. B. (2007). An antidepressant that extends lifespan in adult *Caenorhabditis elegans*. *Nature*, *450*(7169), 553–556. Available from https://doi.org/10.1038/nature05991.

Pisco, A. O., Schaum, N., McGeever, A., Karkanias, J., Neff, N. F., Darmanis, S., ... Quake, S. R. (2020). A single cell transcriptomic atlas characterizes aging tissues in the mouse. *Nature*, *583*, 590–595.

Plank, M., Hu, G., Silva, A. S., Wood, S. H., Hesketh, E. E., Janssens, G., ... Church, G. M. (2013). An analysis and validation pipeline for large-scale RNAi-based screens. *Scientific Reports*, *3*, 1076. Available from https://doi.org/10.1038/srep01076.

Plank, M., Wuttke, D., van Dam, S., Clarke, S. A., & de Magalhaes, J. P. (2012). A meta-analysis of caloric restriction gene expression profiles to infer common signatures and regulatory mechanisms. *Molecular Biosystems*, *8*(4), 1339–1349. Available from https://doi.org/10.1039/c2mb05255e.

Potter, P. K., Bowl, M. R., Jeyarajan, P., Wisby, L., Blease, A., Goldsworthy, M. E., ... Brown, S. D. (2016). Novel gene function revealed by mouse mutagenesis screens for models of age-related disease. *Nature Communications*, *7*, 12444. Available from https://doi.org/10.1038/ncomms12444.

Promislow, D. E. (2004). Protein networks, pleiotropy and the evolution of senescence. *Proceedings of the Royal Society. Biological Sciences*, *271*(1545), 1225–1234. Available from https://doi.org/10.1098/rspb.2004.2732.

Regev, A., Teichmann, S. A., Lander, E. S., Amit, I., Benoist, C., Birney, E., & Human Cell Atlas Meeting, P. (2017). The human cell atlas. *Elife*, 6, e27041. Available from https://doi.org/10.7554/eLife.27041.

Saito, R., Smoot, M. E., Ono, K., Ruscheinski, J., Wang, P. L., Lotia, S., ... Ideker, T. (2012). A travel guide to *Cytoscape plugins*. *Nature Methods*, *9*(11), 1069–1076. Available from https://doi.org/10.1038/nmeth.2212.

Samuelson, A. V., Carr, C. E., & Ruvkun, G. (2007). Gene activities that mediate increased life span of C. elegans insulin-like signaling mutants. *Genes & Development*, *21*(22), 2976–2994. Available from https://doi.org/10.1101/gad.1588907.

Schachter, F., Faure-Delanef, L., Guenot, F., Rouger, H., Froguel, P., Lesueur-Ginot, L., & Cohen, D. (1994). Genetic associations with human longevity at the APOE and ACE loci. *Nature Genetics*, *6*(1), 29–32. Available from https://doi.org/10.1038/ng0194-29.

Semeiks, J., & Grishin, N. V. (2012). A method to find longevity-selected positions in the mammalian proteome. *PLoS One*, *7*(6), e38595. Available from https://doi.org/10.1371/journal.pone.0038595.

Simko, G. I., Gyurko, D., Veres, D. V., Nanasi, T., & Csermely, P. (2009). Network strategies to understand the aging process and help age-related drug design. *Genome Medicine*, *1*(9), 90. Available from https://doi.org/10.1186/gm90.

Sirota, M., Dudley, J. T., Kim, J., Chiang, A. P., Morgan, A. A., Sweet-Cordero, A., ... Butte, A. J. (2011). Discovery and preclinical validation of drug indications using compendia of public gene expression data. *Science Translational Medicine*, *3*(96), 96ra77. Available from https://doi.org/10.1126/scitranslmed.3001318.

Smita, S., Lange, F., Wolkenhauer, O., & Kohling, R. (2016). Deciphering hallmark processes of aging from interaction networks. *Biochimica et Biophysica Acta*, *1860*(11 Pt B), 2706–2715. Available from https://doi.org/10.1016/j.bbagen.2016.07.017.

Smith, E. D., Tsuchiya, M., Fox, L. A., Dang, N., Hu, D., Kerr, E. O., ... Kennedy, B. K. (2008). Quantitative evidence for conserved longevity pathways between divergent eukaryotic species. *Genome Research*, *18*(4), 564–570. Available from https://doi.org/10.1101/gr.074724.107.

Soltow, Q. A., Jones, D. P., & Promislow, D. E. (2010). A network perspective on metabolism and aging. *Integrative and Comparative Biology*, *50*(5), 844–854. Available from https://doi.org/10.1093/icb/icq094.

Sondka, Z., Bamford, S., Cole, C. G., Ward, S. A., Dunham, I., & Forbes, S. A. (2018). The COSMIC cancer gene census: Describing genetic dysfunction across all human cancers. *Nature Reviews in Cancer*, *18*(11), 696–705. Available from https://doi.org/10.1038/s41568-018-0060-1.

Stark, C., Breitkreutz, B. J., Reguly, T., Boucher, L., Breitkreutz, A., & Tyers, M. (2006). BioGRID: A general repository for interaction datasets. *Nucleic Acids Research*, *34*(Database issue), D535–D539. Available from https://doi.org/10.1093/nar/gkj109.

Storey, J. D., & Tibshirani, R. (2003). Statistical significance for genomewide studies. *Proceedings of the National Academy of Sciences of the United States of America*, *100*(16), 9440–9445. Available from https://doi.org/10.1073/pnas.1530509100.

Stroustrup, N., Ulmschneider, B. E., Nash, Z. M., Lopez-Moyado, I. F., Apfeld, J., & Fontana, W. (2013). The *Caenorhabditis elegans* lifespan machine. *Nature Methods*, *10*(7), 665–670. Available from https://doi.org/10.1038/nmeth.2475.

Stuart, J. A., Liang, P., Luo, X., Page, M. M., Gallagher, E. J., Christoff, C. A., & Robb, E. L. (2013). A comparative cellular and molecular biology of longevity database. *Age*, *35*(5), 1937–1947. Available from https://doi.org/10.1007/s11357-012-9458-y.

Tacutu, R., Budovsky, A., Wolfson, M., & Fraifeld, V. E. (2010). MicroRNA-regulated protein-protein interaction networks: How could they help in searching for pro-longevity targets? *Rejuvenation Research*, *13*(2–3), 373–377. Available from https://doi.org/10.1089/rej.2009.0980.

Tacutu, R., Budovsky, A., Yanai, H., & Fraifeld, V. E. (2011). Molecular links between cellular senescence, longevity and age-related diseases - a systems biology perspective. *Aging*, *3*(12), 1178–1191.

Tacutu, R., Shore, D. E., Budovsky, A., de Magalhaes, J. P., Ruvkun, G., Fraifeld, V. E., & Curran, S. P. (2012). Prediction of C. elegans longevity genes by human and worm longevity networks. *PLoS One*, *7*(10), e48282. Available from https://doi.org/10.1371/journal.pone.0048282.

Tacutu, R., Thornton, D., Johnson, E., Budovsky, A., Barardo, D., Craig, T., ... de Magalhaes, J. P. (2018). Human ageing genomic resources: New and updated databases. *Nucleic Acids Research*, *46*(D1), D1083–D1090. Available from https://doi.org/10.1093/nar/gkx1042.

Tanaka, T., Biancotto, A., Moaddel, R., Moore, A. Z., Gonzalez-Freire, M., Aon, M. A., ... Ferrucci, L. (2018). Plasma proteomic signature of age in healthy humans. *Aging Cell*, *17*(5), e12799. Available from https://doi.org/10.1111/acel.12799.

Tate, J. G., Bamford, S., Jubb, H. C., Sondka, Z., Beare, D. M., Bindal, N., ... Forbes, S. A. (2019). COSMIC: The catalogue of somatic mutations in cancer. *Nucleic Acids Research*, *47*(D1), D941–D947. Available from https://doi.org/10.1093/nar/gky1015.

Tomas-Loba, A., Bernardes de Jesus, B., Mato, J. M., & Blasco, M. A. (2013). A metabolic signature predicts biological age in mice. *Aging Cell*, *12*(1), 93–101. Available from https://doi.org/10.1111/acel.12025.

Toren, D., Kulaga, A., Jethva, M., Rubin, E., Snezhkina, A. V., Kudryavtseva, A. V., ... Fraifeld, V. E. (2020). Gray whale transcriptome reveals longevity adaptations associated with DNA repair and ubiquitination. Aging Cell, 19, e13158. https://doi.org/10.1111/acel.13158.

Tsuchiya, T., Dhahbi, J. M., Cui, X., Mote, P. L., Bartke, A., & Spindler, S. R. (2004). Additive regulation of hepatic gene expression by dwarfism and caloric restriction. *Physiological Genomics*, *17*(3), 307–315. Available from https://doi.org/10.1152/physiolgenomics.00039.2004.

Tyshkovskiy, A., Bozaykut, P., Borodinova, A. A., Gerashchenko, M. V., Ables, G. P., Garratt, M., ... Gladyshev, V. N. (2019). Identification and application of gene expression signatures associated with lifespan extension. *Cell Metabolism*, *30*(3), 573–593, e578. Available from https://doi.org/10.1016/j.cmet.2019.06.018.

Valdes, A. M., Glass, D., & Spector, T. D. (2013). Omics technologies and the study of human ageing. *Nature Reviews. Genetics*, *14*(9), 601–607. Available from https://doi.org/10.1038/nrg3553.

Vignoli, A., Tenori, L., Luchinat, C., & Saccenti, E. (2018). Age and sex effects on plasma metabolite association networks in healthy subjects. *Journal of Proteome Research*, *17*(1), 97–107. Available from https://doi.org/10.1021/acs.jproteome.7b00404.

Wagner, A. (2000). Robustness against mutations in genetic networks of yeast. *Nature Genetics*, *24*(4), 355–361. Available from https://doi.org/10.1038/74174.

Wang, J., Zhang, S., Wang, Y., Chen, L., & Zhang, X. S. (2009). Disease-aging network reveals significant roles of aging genes in connecting genetic diseases. *PLoS Computational Biology*, *5*(9), e1000521. Available from https://doi.org/10.1371/journal.pcbi.1000521.

Wang, T., Li, B., Nelson, C. E., & Nabavi, S. (2019). Comparative analysis of differential gene expression analysis tools for single-cell RNA sequencing data. *BMC Bioinformatics*, *20*(1), 40. Available from https://doi.org/10.1186/s12859-019-2599-6.

Wang, Z., Gerstein, M., & Snyder, M. (2009). RNA-Seq: A revolutionary tool for transcriptomics. *Nature Reviews in Genetics*, *10*(1), 57–63. Available from https://doi.org/10.1038/nrg2484.

West, J., Beck, S., Wang, X., & Teschendorff, A. E. (2013). An integrative network algorithm identifies age-associated differential methylation interactome hotspots targeting stem-cell differentiation pathways. *Scientific Reports*, *3*, 1630. Available from https://doi.org/10.1038/srep01630.

West, J., Widschwendter, M., & Teschendorff, A. E. (2013). Distinctive topology of age-associated epigenetic drift in the human interactome. *Proceedings of the National Academy of Sciences of the United States of America*, *110*(35), 14138–14143. Available from https://doi.org/10.1073/pnas.1307242110.

Witten, T. M., & Bonchev, D. (2007). Predicting aging/longevity-related genes in the nematode *Caenorhabditis elegans*. *Chemistry & Biodiversity*, *4*(11), 2639–2655. Available from https://doi.org/10.1002/cbdv.200790216.

Wolfson, M., Budovsky, A., Tacutu, R., & Fraifeld, V. (2009). The signaling hubs at the crossroad of longevity and age-related disease networks. *The International Journal of Biochemistry & Cell Biology*, *41*(3), 516–520. Available from https://doi.org/10.1016/j.biocel.2008.08.026.

Wood, S. H., Craig, T., Li, Y., Merry, B., & de Magalhaes, J. P. (2013). Whole transcriptome sequencing of the aging rat brain reveals dynamic RNA changes in the dark matter of the genome. *Age*, *35*, 763–776. Available from https://doi.org/10.1007/s11357-012-9410-1.

Wood, S. H., van Dam, S., Craig, T., Tacutu, R., O'Toole, A., Merry, B. J., & de Magalhaes, J. P. (2015). Transcriptome analysis in calorie-restricted rats implicates epigenetic and post-translational mechanisms in neuroprotection and aging. *Genome Biology*, *16*, 285. Available from https://doi.org/10.1186/s13059-015-0847-2.

Wuttke, D., Connor, R., Vora, C., Craig, T., Li, Y., Wood, S., ... de Magalhaes, J. P. (2012). Dissecting the gene network of dietary restriction to identify evolutionarily conserved pathways and new functional genes. *PLoS Genetics*, *8*(8), e1002834. Available from https://doi.org/10.1371/journal.pgen.1002834.

Ximerakis, M., Lipnick, S. L., Innes, B. T., Simmons, S. K., Adiconis, X., Dionne, D., ... Rubin, L. L. (2019). Single-cell transcriptomic profiling of the aging mouse brain. *Nature Neuroscience*, *22*(10), 1696–1708. Available from https://doi.org/10.1038/s41593-019-0491-3.

Xue, H., Xian, B., Dong, D., Xia, K., Zhu, S., Zhang, Z., ... Han, J. D. (2007). A modular network model of aging. *Molecular Systems Biology*, *3*, 147. Available from https://doi.org/10.1038/msb4100189.

Yanai, H., Budovsky, A., Barzilay, T., Tacutu, R., & Fraifeld, V. E. (2017). Wide-scale comparative analysis of longevity genes and interventions. *Aging Cell*, *16*(6), 1267–1275. Available from https://doi.org/10.1111/acel.12659.

Yang, J., Huang, T., Petralia, F., Long, Q., Zhang, B., Argmann, C., ... Tu, Z. (2015). Synchronized age-related gene expression changes across multiple tissues in human and the link to complex diseases. *Scientific Reports*, *5*, 15145. Available from https://doi.org/10.1038/srep15145.

Yang, S., Liu, T., Li, S., Zhang, X., Ding, Q., Que, H., ... Liu, S. (2008). Comparative proteomic analysis of brains of naturally aging mice. *Neuroscience*, *154*(3), 1107–1120. Available from https://doi.org/10.1016/j.neuroscience.2008.04.012.

Ye, X., Linton, J. M., Schork, N. J., Buck, L. B., & Petrascheck, M. (2014). A pharmacological network for lifespan extension in *Caenorhabditis elegans*. *Aging Cell*, *13*(2), 206–215. Available from https://doi.org/10.1111/acel.12163.

Yildirim, M. A., Goh, K. I., Cusick, M. E., Barabasi, A. L., & Vidal, M. (2007). Drug-target network. *Nature Biotechnology*, *25*(10), 1119–1126. Available from https://doi.org/10.1038/nbt1338.

Yu, Z., Zhai, G., Singmann, P., He, Y., Xu, T., Prehn, C., ... Wang-Sattler, R. (2012). Human serum metabolic profiles are age dependent. *Aging Cell*, *11*(6), 960–967. Available from https://doi.org/10.1111/j.1474-9726.2012.00865.x.

Zahn, J. M., Poosala, S., Owen, A. B., Ingram, D. K., Lustig, A., Carter, A., ... Becker, K. G. (2007). AGEMAP: A gene expression database for aging in mice. *PLoS Genetics*, *3*(11), e201. Available from https://doi.org/10.1371/journal.pgen.0030201.

Zhang, G., Cowled, C., Shi, Z., Huang, Z., Bishop-Lilly, K. A., Fang, X., ... Wang, J. (2013a). Comparative analysis of bat genomes provides insight into the evolution of flight and immunity. *Science*, *339*(6118), 456–460. Available from https://doi.org/10.1126/science.1230835.

Zhang, B., Gaiteri, C., Bodea, L. G., Wang, Z., McElwee, J., Podtelezhnikov, A. A., ... Emilsson, V. (2013b). Integrated systems approach identifies genetic nodes and networks in late-onset Alzheimer's disease. *Cell*, *153*(3), 707–720. Available from https://doi.org/10.1016/j.cell.2013.03.030.

Zhang, Y., De, S., Garner, J. R., Smith, K., Wang, S. A., & Becker, K. G. (2010). Systematic analysis, comparison, and integration of disease based human genetic association data and mouse genetic phenotypic information. *BMC Medical Genomics*, *3*, 1. Available from https://doi.org/10.1186/1755-8794-3-1.

Zhao, M., Kim, P., Mitra, R., Zhao, J., & Zhao, Z. (2016). TSGene 2.0: An updated literature-based knowledgebase for tumor suppressor genes. *Nucleic Acids Research*, *44*(D1), D1023–D1031. Available from https://doi.org/10.1093/nar/gkv1268.

Zhou, B., Yang, L., Li, S., Huang, J., Chen, H., Hou, L., ... Han, J. D. (2012). Midlife gene expressions identify modulators of aging through dietary interventions. *Proceedings of the National Academy of Sciences of the United States of America*, *109*(19), E1201–E1209. Available from https://doi.org/10.1073/pnas.1119304109.

Ziehm, M., Kaur, S., Ivanov, D. K., Ballester, P. J., Marcus, D., Partridge, L., & Thornton, J. M. (2017). Drug repurposing for aging research using model organisms. *Aging Cell*, *16*(5), 1006–1015. Available from https://doi.org/10.1111/acel.12626.

C H A P T E R

7

Thermogenesis and aging

Justin Darcy[1], *Yimin Fang*[2], *Samuel McFadden*[2], *Kevin Hascup*[2], *Erin Hascup*[2] *and Andrzej Bartke*[3]

[1]Section on Integrative Physiology and Metabolism, Joslin Diabetes Center, Harvard Medical School, Boston, MA, United States [2]Southern Illinois University School of Medicine, Department of Neurology, Springfield, IL, United States [3]Southern Illinois University School of Medicine, Department of Internal Medicine, Springfield, IL, United States

O U T L I N E

Introduction

The biological process of aging and the regulation of longevity are deeply intertwined with energy metabolism. In ectothermic (poikilothermic or "cold-blooded") animals such as worms, insects, or fish, lowering environmental temperature (eT) leads to reduced metabolic rate, slower aging, and extended longevity (Loeb & Northrop, 1916; Miquel, Lundgren, Bensch, & Atlan, 1976; Van Voorhies & Ward, 1999). These relationships are very consistent and well documented, although, naturally, they do not apply to the extremes of eT, which can be detrimental or even lethal. In endothermic ("warm-blooded") animals, that is mammals and birds, a similar relationship of metabolic rate and aging was believed to exist, because larger species have lower metabolic rates and generally live longer than smaller species. However, this relationship, formulated as the "rate of living theory" (Pearl, 1928), is currently not considered to be an important mechanism of aging. This is due mainly to inconsistencies in a supposedly linear relationship between metabolic rate and aging. For example, humans live much longer than would be expected from their basal metabolic rate (and body size), and the same is true of bats and several species of fossorial (living underground) rodents (McNab, 1966; Wilkinson & South, 2002). Moreover, birds typically live longer than mammals of the same body size even though they have a much higher metabolic rate (Furness & Speakman, 2008). Moreover, there are various examples of increased metabolism in mammals being associated with extended, rather than shortened, longevity (Oliverio, Schmidt, & Mauer, 2016; Ortega-Molina, Efeyan, & Lopez-Guadamillas, 2012; Speakman, Talbot, & Selman, 2004; Westbrook, Bonkowski, Strader, & Bartke, 2009). Some of these findings will be discussed later in this chapter.

Research findings generated during the last decade led to a greatly increased interest in the potential benefits of increased thermogenesis on aging, longevity, and risk for chronic age-related disease (Darcy & Tseng, 2019). This was related to the demonstration that, contrary to long-held views, BAT is present in adult humans (Cypess, Lehman, & Williams, 2009; Nedergaard, Bengtsson, & Cannon, 2007; van Marken Lichtenbelt, Vanhommerig, & Smulders, 2009). Moreover, its activation can prevent

Handbook of the Biology of Aging.
DOI: https://doi.org/10.1016/B978-0-12-815962-0.00007-X

and/or reduce obesity, and improve metabolic health (Cypess, Weiner, & Roberts-Toler, 2015; Hanssen, Hoeks, & Brans, 2015). In this chapter, we will describe characteristics of BAT, as well as the so-called beige or brite adipose tissue, which are inducible thermogenic adipocyte populations within white adipose tissue (WAT). Moreover, we will discuss the control of thermogenic adipose tissue activity. Finally, we will discuss relationships between thermogenesis and aging, focusing on the potential benefits of uncoupling mitochondrial electron transport, and on the impact of longevity genes and antiaging interventions on thermogenesis.

Thermogenic adipose tissue

To a lay audience, adipose tissue is metabolically detrimental. Indeed, many WAT depots are detrimental; however, some WAT depots, such as the subcutaneous depot, can boost insulin sensitivity. Moreover, thermogenic adipose tissue, which includes brown and beige adipose tissue, is certainly metabolically beneficial. The differences between WAT, beige adipose tissue, and BAT have been thoroughly discussed previously (Lee, Mottillo, & Granneman, 2014; Lynes & Tseng, 2018; Zwick, Guerrero-Juarez, Horsley, & Plikus, 2018); however, an overly simplistic view is that WAT is unilocular, stores energy, and has few mitochondria, while BAT is multilocular, expends energy, and is rich in mitochondria. Beige adipose tissue is somewhere in between WAT and BAT in terms of its phenotype. Indeed, BAT and WAT are so different that they even arise from separate lineages, where BAT is uniquely MYF5 and PAX7 positive (Lepper & Fan, 2010; Seale, Kajimura, & Yang, 2007). Arguably, the biggest difference between thermogenic and nonthermogenic adipose tissue is the presence of uncoupling protein 1 (UCP1), which uncouples oxidative phosphorylation and ATP production in order to produce heat. Researchers have recently made major discoveries of UCP1-independent forms of nonshivering thermogenesis (Bertholet, Kazak, & Chouchani, 2017; Ikeda et al., 2017; Kazak et al., 2015; Kazak et al., 2017); however, little is known about their role in the process of aging, and thus are outside the scope of this review. Interested readers about UCP1-independent functions in BAT are directed to a relevant review (Sponton & Kajimura, 2018).

While WAT has been long known to secrete adipokines such as leptin and adiponectin, BAT secretes its own factors, which are aptly named batokines. Some of these factors, such as IL-6, have been found using BAT transplantation studies (Stanford et al., 2013). A range of other factors have been described including lipids (Leiria et al., 2019; Lynes et al., 2017; Stanford et al., 2018), proteins (Asano et al., 1999; Hondares et al., 2011; Nechad et al., 1994; Nisoli et al., 1996, 1997; Yamashita et al., 1994), and even microRNAs (Thomou et al., 2017). Through both endocrine and paracrine actions, these batokines serve to influence metabolism through a variety of mechanisms ranging from increasing lipid uptake into BAT, to improving systemic glucose homeostasis. The secretory role of BAT has even been shown to affect the female reproductive tract (Bartness & Wade, 1984; Chen et al., 2019; Yuan et al., 2016), demonstrating there is clearly much we do not understand about the crosstalk of adipose tissue with distal organ systems. Still, the secretory function of BAT remains an area of extreme interest, and readers interested in this area are directed to a relevant review (Villarroya et al., 2017).

Thermogenesis, energy metabolism, and aging

Thermogenesis, required to maintain body temperature within the very narrow physiological limits, represents a large component of energy metabolism of homeothermic animals. Energy metabolism, including mitochondrial energy generation, is involved in all life processes including development, growth, maintenance, and repair, and thus is intertwined with virtually every aspect of aging and age-related disease. As was mentioned earlier, metabolism was (and, to some extent, still is) believed to be a key driver of aging, and species-specific metabolic rate was proposed to be the determinant of lifespan (Pearl, 1928). While the somewhat simplistic original form of this hypothesis is no longer accepted, the overall concept certainly still applies to exothermic organisms, which represent a huge preponderance of animal species.

Basal metabolic rate features prominently in the studies of the pace-of-life syndrome and its role in longevity differences between species (Scholer et al., 2019). Intriguingly, some of the mutations which extend longevity of laboratory mice lead to an increase, rather than a decrease, of metabolic rate (Kumar, 2018; Leone et al., 2015; Ortega-Molina et al., 2012; Speakman et al., 2004; Vatner et al., 2018; Westbrook et al., 2009). This positive association of metabolism and longevity is particularly striking and somewhat counterintuitive. For example, hypopituitary dwarf mice lacking growth hormone (GH), as well as mice lacking GH receptors, utilize significantly more oxygen per unit of body weight than their normal ("wild-type") siblings (Westbrook et al., 2009) and exhibit remarkably extended longevity, along with an extension of healthspan and a reduced rate of aging (Aguiar-Oliveira & Bartke, 2019; Bartke et al., 2001; Brown-Borg, 2015; Junnila et al., 2013; Koopman et al., 2016; Tatar et al., 2003).

Comparing the results from different studies is complicated by the differences in methodology and analysis of data from indirect calorimetry, although some researchers have pushed for a more uniform data analysis to improve data reproducibility and transparency (Mina et al., 2018). Some studies report basal metabolic rate, while other studies report average metabolic rate including periods of wakefulness, locomotor activity, and feeding. Oxygen consumption (VO_2) is most often expressed per unit of lean body mass, thus minimizing the contribution of tissues with relatively low metabolism to the results, while other studies report VO_2 per unit of total body mass. The major impact of some life-extending mutations on adiposity and body composition (Berryman & List, 2017; Berryman et al., 2011; Darcy et al., 2017) adds to the difficulty of selecting the most appropriate method of calculating and presenting results (Westbrook et al., 2009). The fact that some life-extending mutations result in vastly altered secretory and metabolic profiles in adipose tissue depots (e.g., visceral and subcutaneous adipose tissue in long-lived vs. control littermates) adds another layer of complexity to interpreting indirect calorimetry data (Darcy et al., 2016, 2017; Masternak et al., 2012; Menon et al., 2014; Stout et al., 2015).

The complex interactions among energy metabolism, aging, and longevity are well illustrated by studies of calorie restriction (CR). Extending lifespan by reducing food intake is generally seen as the "gold standard" among antiaging interventions. While the reduction of metabolic rate in response to reduced intake of metabolic substrates appears intuitively obvious, measurements of VO_2 in calorically restricted rats indicated that after a period of decline, a new balance is reached between the available food and the body mass of an organism (which naturally declines in response to CR) such that VO_2 per gram of lean body mass does not differ from values measured in animals with ad libitum access to food (McCarter et al., 1985). However, the issue of the impact of chronic CR on metabolic rate does not appear to be fully resolved. For example, CR in nonobese adult humans was shown to reduce energy expenditure (Redman et al., 2018). Moreover, the reduced body temperature of animals subjected to CR (Weindruch & Walford, 1988) indicates that a reduced amount of energy is dedicated to thermogenesis under this condition. Yet another complication of analyzing the role of changes in energy metabolism in the beneficial effects of CR involves frequent occurrence of torpor (a period of inactivity and markedly reduced body temperature) in animals subjected to severe CR (Mitchell et al., 2015). Thus, although locomotor activity may be increased by CR, particularly around the time of feeding (Duffy et al., 1989), the average daily VO_2 and energy expenditure can be profoundly influenced by periods of torpor. Importantly, torpor is regularly employed by many species of small homeothermic animals living in their natural habitat, and serves as an important energy-conserving strategy. The relationship of greatly extended periods of torpor during hibernation to aging and longevity is outside the scope of this chapter. There is also evidence that in nonhibernating mammals living in their natural environment, body temperature can exhibit wide fluctuations (Thiel et al., 2019). Pertinent to the role of energy metabolism in aging, CR activates mitochondrial pathways in adipose tissue and liver by inducing the expression of PGC-1α, a key regulator of mitochondrial biogenesis (Guarente, 2008; Miller et al., 2019). Moreover, CR was shown to ameliorate age-related dysfunction in WAT, BAT, and beige adipose tissue (Corrales et al., 2019).

The relationships between metabolism and aging are bidirectional, and thus not limited to known or suspected effects of metabolic rate on aging. Metabolic rate, as well as the ability to increase thermogenesis in response to a cold environment and to maintain body temperature, decline with age (Berry et al., 2017; Cypess et al., 2009; Gonzales & Rikke, 2010; Reynolds et al., 1985; van Marken Lichtenbelt et al., 2009). In an interesting departure from this trend, basal metabolic rate of extremely old people tends to be elevated, rather than reduced, presumably reflecting bioenergetic consequences of chronic, low-grade inflammation (Rizzo et al., 2005).

Regulation of thermogenesis in brown adipose tissue

The discovery of functional BAT in adult humans (Cypess et al., 2009; Nedergaard et al., 2007; van Marken Lichtenbelt et al., 2009) generated great interest in using BAT as an approach to combat obesity. Much evidence has suggested that this is feasible, and that the activation of thermogenesis may also result in improved insulin sensitivity, glucose homeostasis, and other metabolic benefits (Chondronikola et al., 2014; Cypess et al., 2015; Hanssen et al., 2015). This prompted numerous studies of the regulation of BAT development and function, WAT "browning" or "beiging," and functions of beige adipose tissue, which have been thoroughly reviewed (Giralt & Villarroya, 2013; Kajimura et al., 2015; Lynes & Tseng, 2018; Townsend & Tseng, 2014; Villarroya et al., 2017). Evidence available to date indicates that in addition to the long-known role of the sympathetic nervous system (acting via β-adrenergic receptors) and thyroid hormones, the development and function of thermogenic adipose tissue can be influenced by numerous endogenous molecules, as well as by exogenous natural or pharmaceutical compounds.

Most of the studies in this area were conducted in mice using expression of UCP1 mRNA/protein to assess thermogenic activity. However, methodologies such as cold tolerance tests and analyzing maximum thermogenic capacity are now widely used to determine thermogenic activity. Hormones stimulating BAT and thermogenesis include estradiol (Clookey et al., 2019; Gonzalez-Garcia et al., 2017), leptin (Commins et al., 2001), glucagon (Townsend et al., 2019), adrenocorticotropin, ACTH (Schnabl et al., 2018), adiponectin (Qi et al., 2004), and FGF21 (Ameka et al., 2019). Orexin was reported to partially reverse age-related deterioration of BAT structure and function (Sellayah & Sikder, 2014). While hyperthyroidism increases BAT activity, hypothyroidism increases BAT volume (Begaye et al., 2018). In contrast to the stimulatory impact of ACTH on BAT (Schnabl et al., 2018), the effects of glucocorticoids on this tissue are inhibitory (Kroon et al., 2018; Schnabl et al., 2018; Thuzar et al., 2018). Inhibitory influences of GH-releasing hormone (GHRH), follicle-stimulating hormone (FSH), and RGS14 on BAT function or WAT browning, were deduced from the stimulatory effects of deleting the corresponding genes (Kumar, 2018; Leone et al., 2015; Vatner et al., 2018).

Some of the findings concerning physiological regulation of BAT, beige adipose tissue, and thermogenesis may represent mechanisms of aging. Various kinds of mutant mice with extended longevity have an increased weight and thermogenic activity of BAT (Aguiar-Oliveira & Bartke, 2019; Berryman & List, 2017; Berryman et al., 2011; Darcy et al., 2016, 2017; Leone et al., 2015), reduced somatotropic signaling, hypothyroidism, and increased levels of adiponectin and leptin.

Compounds reported to stimulate BAT and/or WAT browning include the antiinflammatory drug indometacin (Hao et al., 2018), the phosphodiesterase 5 inhibitor sildenafil (Li et al., 2018), a stimulator of energy expenditure, dihydrocapsiate (Ohyama & Suzuki, 2017), a Notch signaling inhibitor, dibenzazepine (Jiang et al., 2017), an inhibitor of Stat 3 phosphorylation, linifanib (Zhao et al., 2019), PPAR-γ agonists such as rosiglitazone, as well as natural compounds such as caffeine (Velickovic et al., 2019), β-lapachone (Kwak et al., 2019), and quercetin (Kuipers et al., 2018). Interestingly, a drug combination including quercetin leads to depletion of senescent cells (Xu et al., 2018), an effect which can promote healthy aging and extended longevity (Kirkland & Tchkonia, 2017; Kirkland et al., 2017).

Thermogenic functions of adipose tissue can be increased by exercise (Lehnig et al., 2019) and, intriguingly, also by the epigenetic impact of preconception cold exposure of the father (Sun et al., 2018). Cell signaling and molecular mechanisms of BAT stimulation include involvement of receptors for IGF-1 and insulin (Viana-Huete et al., 2018), Akt2, AS160, FK3P51 (Balsevich et al., 2017), mTORC2 (Albert et al., 2016), histone deacetylase 3 (Emmett et al., 2017; Liao et al., 2018), EERγ (Ahmadian et al., 2018), as well as Ezh2 and GSK126 (Wu et al., 2018). This list is not exhaustive and much remains to be learned about the physiological and pharmacological control of the development and function of thermogenic tissues (Castro et al., 2017; Seale, 2015; Wang & Seale, 2016).

Substrates for BAT thermogenesis are derived not only from lipids stored in brown adipocytes and from de novo lipogenesis and glycogen synthesis in this tissue (Labbe et al., 2015; Sanchez-Gurmaches et al., 2018), but also from lipolysis in WAT (Schreiber et al., 2017; Shin et al., 2017) and liver (Labbe et al., 2015; Simcox et al., 2017), and from the systemic supply of glucose, fatty acids, and acylcarnitines (Labbe et al., 2015; Simcox et al., 2017). Browning of WAT and beige adipocyte function are also influenced by other cell types in WAT and from other tissues (Wang et al., 2017) as indicated by compensatory responses to BAT removal (Piao et al., 2018).

Impact of environmental temperature on laboratory mice

Internationally accepted standards of care of laboratory mice include housing the mice at temperatures around 23°C. It appears that this standard room temperature was selected for the convenience of caretakers and investigators since humans wearing light clothing are comfortable working at this temperature, or possibly by default since most buildings devoted to biomedical research are typically maintained at this temperature. The temperature of 23°C is fairly close to the thermoneutral temperature of humans wearing light indoor clothing, but it is much lower than the thermoneutral temperature of mice, which is approximately 30°C (Gordon, 2012). It is interesting that laboratory mice, given the choice of environments with different temperatures, spend most time in areas with temperatures maintained close to their thermoneutral zone (Gordon et al., 2017; Kaikaew et al., 2017). Several investigators called attention to this issue by pointing out that mice housed at standard room temperature are chronically exposed to mild cold stress, and consequently devote a significant portion of their energy expenditure to thermogenesis needed to maintain body temperature (Gordon, 2017; Speakman & Keijer, 2012). This, of course, leads to the question, to what extent data obtained in mice under these "standard" conditions may apply to humans living and working indoors and generally being well protected from cold when they are outdoors (Speakman & Keijer, 2012). Information on the effects of different eTs on physiological functions, health, and aging of mice and other small mammals is surprisingly sparse. There are numerous studies of the responses of mice to cold, but almost all of them utilized brief (hours to days) exposure to a very low temperature (usually 4°C) rather than chronic exposure to "standard" vs. thermoneutral temperature, or to modest reductions in eT compatible with long-term survival. However, there are intriguing examples of major effects of altering eT on experimental outcome. Transgenic 3XFAD mice, which are commonly used as a model system to study Alzheimer's disease, exhibited reduced Tau phosphorylation when housed at thermoneutrality, or when exposed to repeated cold exposure over a 4-week interval (Tournissac et al., 2017, 2019). In mice exposed to CR, elevated (thermoneutral) eT antagonized the ability of this dietary intervention to protect the animals from cancer (Koizumi et al., 1996). The authors ascribed the effect to elimination of torpor by thermoneutral eT (Koizumi et al., 1996). A study from another laboratory provided evidence that housing mice at 30°C–31°C can reduce tumor formation, growth rate, and metastasis (Kokolus et al., 2013). These findings were interpreted as an indication that cold stress induced by housing at standard eT reduces antitumor immunity in these animals (Kokolus et al., 2013).

Speakman and his colleagues recently compared the impact of different eTs on VO_2 and energy expenditure in C57BL/6 mice and men (Speakman & Keijer, 2012). They concluded that housing solitary animals at 25.5°C–27.6°C, that is slightly below thermoneutrality, would be most likely to produce data comparable to results obtained in humans studied at room temperature. Importantly, solitary, as opposed to group, housing can influence the results of studying energy metabolism because small animals often huddle to reduce heat loss (Schipper et al., 2018). Moreover, Nedergaard and colleagues have published extensively arguing for the housing of mice at thermoneutrality for the best translation to humans (Fischer et al., 2018, 2019).

Our laboratory has focused on elucidating novel mechanisms of longevity in mice with mutations in the somatotropic axis. Particularly, we have become interested in the role of energy metabolism and thermogenesis in these animals. Two particular examples are the diminutive Ames dwarf and GH receptor knockout (GHRKO) mice. Ames dwarf mice not only lack GH and the downstream components of the somatotropic axis, but also thyroid-stimulating hormone and its downstream thyroid hormones, as well as prolactin (Slabaugh et al., 1981). GHRKO mice, being a murine model of Laron syndrome, have elevated GH despite greatly reduced IGF-1 levels (Zhou et al., 1997). Intriguingly, both Ames dwarf and GHRKO mice have increased levels of UCP1 in their BAT, while the short-lived bovine GH-overexpressing mice have decreased UCP1 expression (Darcy et al., 2016; Li et al., 2003). In Ames dwarf mice, genes related to lipid metabolism and mitochondrial function are also increased in BAT. We have shown that transferring these mice from standard room temperature (23°C) to a room maintained at 30°C [a temperature considered thermoneutral for mice (Gordon, 2012; Gordon et al., 2017; Speakman & Keijer, 2012)] for up to 48 hours, greatly attenuated differences between the examined parameters of energy metabolism (VO_2, heat, and RQ) (Darcy et al., 2018; Westbrook, 2012). On the basis of these findings, and other long-term studies (Darcy et al., 2016, 2018; Westbrook, 2012), we hypothesized that some of the beneficial changes in energy metabolism of the long-lived GH-related mutants are due to the increased heat loss and the consequent increase in energy expenditure for thermogenesis to maintain body temperature. We are currently testing this hypothesis by studying the effects of chronic exposure to increased or reduced eT. Preliminary results from our studies involving chronic exposure of these animals to thermoneutral or to mildly reduced eT starting at weaning suggest that the effects of eT on energy metabolism and glucose homeostasis are dependent on the age when the animals are transferred to a different eT, as well as on sex and mutations affecting somatotropic signaling. Preliminary results of an ongoing study in GHRKO mice appear to argue against a major impact of increased eT on the longevity of these mutants; however, we must await completion of this study, and others, before drawing firm conclusions of the role of thermogenesis in the longevity of GHRKO mice, or other mice with mutations in the somatotropic axis.

Concluding remarks

The simple fact that energy utilization is essential to all life processes means that there should be little doubt that altering energy metabolism will not only affect these life processes, but also longevity. After introducing thermogenesis and thermogenic adipose tissue, this review aimed to outline the bidirectional relationship between aging and energy metabolism. There is already evidence from long-lived and short-lived strains that increased thermogenesis is beneficial for both the healthspan and lifespan of an animal. However, there has been little work, to date, on the effect of thermogenesis and longevity in wild-type animals. This is an area that should be further explored. Moreover, the interaction between increased thermogenesis, decreased obesity, and decreased accumulation of age-related disease is an area that has yet to be explored. Taken together, researchers should take care to incorporate thermogenesis, housing temperatures, and related areas into their data analysis and interpretation when examining both lifespan and healthspan of any experimental model.

References

Aguiar-Oliveira, M. H., & Bartke, A. (2019). Growth hormone deficiency: Health and longevity. *Endocrine Reviews, 40*(2), 575–601.

Ahmadian, M., Liu, S., Reilly, S. M., et al. (2018). ERRgamma preserves brown fat innate thermogenic activity. *Cell Reports, 22*(11), 2849–2859.

Albert, V., Svensson, K., Shimobayashi, M., et al. (2016). mTORC2 sustains thermogenesis via Akt-induced glucose uptake and glycolysis in brown adipose tissue. *EMBO Molecular Medicine, 8*(3), 232–246.

Ameka, M., Markan, K. R., Morgan, D. A., et al. (2019). Liver derived FGF21 maintains core body temperature during acute cold exposure. *Scientific Reports, 9*(1), 630.

Asano, A., Kimura, K., & Saito, M. (1999). Cold-induced mRNA expression of angiogenic factors in rat brown adipose tissue. *The Journal of Veterinary Medical Science/The Japanese Society of Veterinary Science, 61*(4), 403–409.

Balsevich, G., Hausl, A. S., Meyer, C. W., et al. (2017). Stress-responsive FKBP51 regulates AKT2-AS160 signaling and metabolic function. *Nature Communications., 8*(1), 1725.

Bartke, A., Brown-Borg, H., Mattison, J., et al. (2001). Prolonged longevity of hypopituitary dwarf mice. *Experimental Gerontology, 36* (1), 21–28.

Bartness, T. J., & Wade, G. N. (1984). Effects of interscapular brown adipose tissue denervation on body weight and energy metabolism in ovariectomized and estradiol-treated rats. *Behavioral Neuroscience, 98*(4), 674–685.

Begaye, B., Piaggi, P., Thearle, M. S., et al. (2018). Norepinephrine and T4 are predictors of fat mass gain in humans with cold-induced brown adipose tissue activation. *The Journal of Clinical Endocrinology and Metabolism, 103*(7), 2689–2697.

Berry, D. C., Jiang, Y., Arpke, R. W., et al. (2017). Cellular aging contributes to failure of cold-induced beige adipocyte formation in old mice and humans. *Cell Metabolism, 25*(1), 166–181.

Berryman, D. E., & List, E. O. (2017). Growth hormone's effect on adipose tissue: Quality versus quantity. *Int J Mol Sci, 18*, (8).

Berryman, D. E., List, E. O., Sackmann-Sala, L., et al. (2011). Growth hormone and adipose tissue: Beyond the adipocyte. *Growth Hormone & IGF Research: Official Journal of the Growth Hormone Research Society and the International IGF Research Society, 21*(3), 113–123.

Bertholet, A. M., Kazak, L., Chouchani, E. T., et al. (2017). Mitochondrial patch clamp of beige adipocytes reveals UCP1-positive and UCP1-negative cells both exhibiting futile creatine cycling. *Cell Metabolism, 25*(4), 811–822, e4.

Brown-Borg, H. M. (2015). The somatotropic axis and longevity in mice. *American Journal of Physiology. Endocrinology and Metabolism, 309*(6), E503–E510.

Castro, E., Silva, T. E. O., & Festuccia, W. T. (2017). Critical review of beige adipocyte thermogenic activation and contribution to whole-body energy expenditure. *Hormone Molecular Biology and Clinical Investigation, 31*, 2.

Chen, L. J., Yang, Z. X., Wang, Y., et al. (2019). Single xenotransplant of rat brown adipose tissue prolonged the ovarian lifespan of aging mice by improving follicle survival. *Aging Cell*, e13024.

Chondronikola, M., Volpi, E., Borsheim, E., et al. (2014). Brown adipose tissue improves whole-body glucose homeostasis and insulin sensitivity in humans. *Diabetes, 63*(12), 4089–4099.

Clookey, S. L., Welly, R. J., Shay, D., et al. (2019). Beta 3 adrenergic receptor activation rescues metabolic dysfunction in female estrogen receptor alpha-null Mice. *Frontiers in Physiology, 10*, 9.

Commins, S. P., Watson, P. M., Frampton, I. C., & Gettys, T. W. (2001). Leptin selectively reduces white adipose tissue in mice via a UCP1-dependent mechanism in brown adipose tissue. *American Journal of Physiology. Endocrinology and Metabolism, 280* (2), E372–E377.

Corrales, P., Vivas, Y., Izquierdo-Lahuerta, A., et al. (2019). Long-term caloric restriction ameliorates deleterious effects of aging on white and brown adipose tissue plasticity. *Aging Cell, 18*(3), e12948.

Cypess, A. M., Lehman, S., Williams, G., et al. (2009). Identification and importance of brown adipose tissue in adult humans. *The New England Journal of Medicine, 360*(15), 1509–1517.

Cypess, A. M., Weiner, L. S., Roberts-Toler, C., et al. (2015). Activation of human brown adipose tissue by a beta3-adrenergic receptor agonist. *Cell Metabolism, 21*(1), 33–38.

Darcy, J., McFadden, S., & Bartke, A. (2017). Altered structure and function of adipose tissue in long-lived mice with growth hormone-related mutations. *Adipocyte., 6*(2), 69–75.

Darcy, J., McFadden, S., Fang, Y., et al. (2016). Brown adipose tissue function is enhanced in long-lived, male Ames Dwarf Mice. *Endocrinology, 157*(12), 4744–4753.

Darcy J., McFadden S., Fang Y., et al. (2018). Increased environmental temperature normalizes energy metabolism outputs between normal and Ames dwarf mice. *Aging, 10*, 2709–2722.

Darcy J., & Tseng Y. H. (2019). ComBATing aging-does increased brown adipose tissue activity confer longevity? *Geroscience, 41*, 285–296.

Duffy, P. H., Feuers, R. J., Leakey, J. A., et al. (1989). Effect of chronic caloric restriction on physiological variables related to energy metabolism in the male Fischer 344 rat. *Mechanisms of Ageing and Development, 48*, 117–133.

Emmett, M. J., Lim, H. W., Jager, J., et al. (2017). Histone deacetylase 3 prepares brown adipose tissue for acute thermogenic challenge. *Nature, 546*(7659), 544–548.

Fischer, A. W., Cannon, B., & Nedergaard, J. (2018). Optimal housing temperatures for mice to mimic the thermal environment of humans: An experimental study. *Molecular Metabolism, 7*, 161–170.

Fischer, A. W., Cannon, B., & Nedergaard, J. (2019). The answer to the question "What is the best housing temperature to translate mouse experiments to humans?" Is: Thermoneutrality. *Molecular Metabolism, 26*, 1–3.

Furness, L. J., & Speakman, J. R. (2008). Energetics and longevity in birds. *Age, 30*(2), 75.

Giralt, M., & Villarroya, F. (2013). White, brown, beige/brite: Different adipose cells for different functions? *Endocrinology, 154* (9), 2992–3000.

Gonzales, P., & Rikke, B. A. (2010). Thermoregulation in mice exhibits genetic variability early in senescence. *Age, 32*(1), 31–37.

Gonzalez-Garcia, I., Tena-Sempere, M., & Lopez, M. (2017). Estradiol regulation of brown adipose tissue thermogenesis. *Advances in Experimental Medicine and Biology, 1043*, 315–335.

Gordon, C. J. (2012). Thermal physiology of laboratory mice: Definining thermoneutrality. *Journal of Thermal Biology, 37*(8), 654–685.

Gordon, C. J. (2017). The mouse thermoregulatory system: Its impact on translating biomedical data to humans. *Physiology & Behavior, 179*, 55–66.

Gordon, C. J., Puckett, E. T., Repasky, E. S., & Johnstone, A. F. (2017). A device that allows rodents to behaviorally thermoregulate when housed in vivariums. *Journal of the American Association for Laboratory Animal Science: JAALAS, 56*(2), 173–176.

Guarente, L. (2008). Mitochondria--a nexus for aging, calorie restriction, and sirtuins? *Cell.*, *132*(2), 171–176.

Hanssen, M. J., Hoeks, J., Brans, B., et al. (2015). Short-term cold acclimation improves insulin sensitivity in patients with type 2 diabetes mellitus. *Nature Medicine*, *21*(8), 863–865.

Hao, L., Kearns, J., Scott, S., et al. (2018). Indomethacin enhances brown fat activity. *The Journal of Pharmacology and Experimental Therapeutics*, *365*(3), 467–475.

Hondares, E., Iglesias, R., Giralt, A., et al. (2011). Thermogenic activation induces FGF21 expression and release in brown adipose tissue. *The Journal of Biological Chemistry*, *286*(15), 12983–12990.

Ikeda, K., Kang, Q., Yoneshiro, T., et al. (2017). UCP1-independent signaling involving SERCA2b-mediated calcium cycling regulates beige fat thermogenesis and systemic glucose homeostasis. *Nature Medicine*, *23*(12), 1454–1465.

Jiang, C., Cano-Vega, M. A., Yue, F., et al. (2017). Dibenzazepine-loaded nanoparticles induce local browning of white adipose tissue to counteract obesity. *Molecular Therapy: The Journal of the American Society of Gene Therapy*, *25*(7), 1718–1729.

Junnila, R. K., List, E. O., Berryman, D. E., et al. (2013). The GH/IGF-1 axis in ageing and longevity. *Nature Reviews Endocrinology*, *9*(6), 366–376.

Kaikaew, K., Steenbergen, J., Themmen, A. P. N., et al. (2017). Sex difference in thermal preference of adult mice does not depend on presence of the gonads. *Biology of Sex Differences*, *8*(1), 24.

Kajimura, S., Spiegelman, B. M., & Seale, P. (2015). Brown and Beige Fat: Physiological roles beyond heat generation. *Cell Metabolism*, *22*(4), 546–559.

Kazak, L., Chouchani, E. T., Jedrychowski, M. P., et al. (2015). A creatine-driven substrate cycle enhances energy expenditure and thermogenesis in beige fat. *Cell.*, *163*(3), 643–655.

Kazak, L., Chouchani, E. T., Lu, G. Z., et al. (2017). Genetic depletion of adipocyte creatine metabolism inhibits diet-Induced thermogenesis and drives obesity. *Cell Metabolism*, *26*(4), 660–671, e3.

Kirkland, J. L., & Tchkonia, T. (2017). Cellular senescence: A translational perspective. *EBioMedicine.*, *21*, 21–28.

Kirkland, J. L., Tchkonia, T., Zhu, Y., et al. (2017). The Clinical potential of senolytic drugs. *Journal of the American Geriatrics Society*, *65*(10), 2297–2301.

Koizumi, A., Wada, Y., Tuskada, M., et al. (1996). A tumor preventive effect of dietary restriction is antagonized by a high housing temperature through deprivation of torpor. *Mechanisms of Ageing and Development*, *92*(1), 67–82.

Kokolus, K. M., Capitano, M. L., Lee, C. T., et al. (2013). Baseline tumor growth and immune control in laboratory mice are significantly influenced by subthermoneutral housing temperature. *Proceedings of the National Academy of Sciences of the United States of America*, *110*(50), 20176–20181.

Koopman, J. J., van Heemst, D., van Bodegom, D., et al. (2016). Measuring aging rates of mice subjected to caloric restriction and genetic disruption of growth hormone signaling. *Aging*, *8*(3), 539–546.

Kroon, J., Koorneef, L. L., van den Heuvel, J. K., et al. (2018). Selective glucocorticoid receptor antagonist CORT125281 activates brown adipose tissue and alters lipid distribution in male mice. *Endocrinology*, *159*(1), 535–546.

Kuipers, E. N., Dam, A. D. V., Held, N. M., et al. (2018). Quercetin lowers plasma triglycerides accompanied by white adipose tissue browning in diet-induced obese mice. *International Journal of Molecular Sciences*, *19*, (6).

Kumar, T. R. (2018). Extragonadal Actions of FSH: A critical need for novel genetic models. *Endocrinology*, *159*(1), 2–8.

Kwak, H. J., Jeong, M. Y., Um, J. Y., & Park, J. (2019). Beta-lapachone regulates obesity through modulating thermogenesis in brown adipose tissue and adipocytes: Role of AMPK Signaling Pathway. *The American Journal of Chinese Medicine*, *47*(4), 803–822.

Labbe, S. M., Caron, A., Bakan, I., et al. (2015). In vivo measurement of energy substrate contribution to cold-induced brown adipose tissue thermogenesis. *The FASEB Journal*, *29*(5), 2046–2058.

Lee, Y. H., Mottillo, E. P., & Granneman, J. G. (2014). Adipose tissue plasticity from WAT to BAT and in between. *Biochimica et Biophysica Acta*, *1842*(3), 358–369.

Lehnig, A. C., Dewal, R. S., Baer, L. A., et al. (2019). Exercise training induces depot-specific adaptations to white and brown adipose tissue. *iScience*, *11*, 425–439.

Leiria, L. O., Wang, C. H., Lynes, M. D., et al. (2019). 12-Lipoxygenase regulates cold adaptation and glucose metabolism by producing the omega-3 lipid 12-HEPE from brown fat. *Cell Metabolism*, *30*, 768–783, e7.

Leone, S., Chiavaroli, A., Shohreh, R., et al. (2015). Increased locomotor and thermogenic activity in mice with targeted ablation of the GHRH gene. *Growth Hormone & IGF Research: Official Journal of the Growth Hormone Research Society and the International IGF Research Society*, *25*(2), 80–84.

Lepper, C., & Fan, C. M. (2010). Inducible lineage tracing of Pax7-descendant cells reveals embryonic origin of adult satellite cells. *Genesis 2000*, *48*(7), 424–436.

Li, S., Li, Y., Xiang, L., et al. (2018). Sildenafil induces browning of subcutaneous white adipose tissue in overweight adults. *Metabolism: Clinical and Experimental*, *78*, 106–117.

Li, Y., Knapp, J. R., & Kopchick, J. J. (2003). Enlargement of interscapular brown adipose tissue in growth hormone antagonist transgenic and in growth hormone receptor gene-disrupted dwarf mice. *Experimental Biology and Medicine*, *228*(2), 207–215.

Liao, J., Jiang, J., Jun, H., et al. (2018). HDAC3-selective inhibition activates brown and beige fat through PRDM16. *Endocrinology*, *159*(7), 2520–2527.

Loeb, J., & Northrop, J. H. (1916). Is There a Temperature Coefficient for the Duration of Life? *Proceedings of the National Academy of Sciences of the United States of America*, *2*(8), 456–457.

Lynes, M. D., Leiria, L. O., Lundh, M., et al. (2017). The cold-induced lipokine 12,13-diHOME promotes fatty acid transport into brown adipose tissue. *Nature Medicine*, *23*(5), 631–637.

Lynes, M. D., & Tseng, Y. H. (2018). Deciphering adipose tissue heterogeneity. *Annals of the New York Academy of Sciences*, *1411*(1), 5–20.

Masternak, M. M., Bartke, A., Wang, F., et al. (2012). Metabolic effects of intra-abdominal fat in GHRKO mice. *Aging Cell*, *11*(1), 73–81.

McCarter, R., Masoro, E. J., & Yu, B. P. (1985). Does food restriction retard aging by reducing the metabolic rate? *The American Journal of Physiology*, *248*(4 Pt 1), E488–E490.

McNab, B. K. (1966). The metabolism of fossorial rodents: A study of convergence. *Ecology*, *47*(5), 712–733.

Menon, V., Zhi, X., Hossain, T., et al. (2014). The contribution of visceral fat to improved insulin signaling in Ames dwarf mice. *Aging Cell*, *13*(3), 497–506.

Miller, K. N., Clark, J. P., Martin, S. A., et al. (2019). PGC-1a integrates a metabolism and growth network linked to caloric restriction. *Aging Cell*, *18*, e12999.

Mina, A. I., LeClair, R. A., LeClair, K. B., et al. (2018). CalR: A web-based analysis tool for indirect calorimetry experiments. *Cell Metabolism*, *28*(4), 656–666, e1.

Miquel, J., Lundgren, P. R., Bensch, K. G., & Atlan, H. (1976). Effects of temperature on the life span, vitality and fine structure of Drosophila melanogaster. *Mechanisms of Ageing and Development*, *5*(5), 347–370.

Mitchell, S. E., Delville, C., Konstantopedos, P., et al. (2015). The effects of graded levels of calorie restriction: III. Impact of short term calorie and protein restriction on mean daily body temperature and torpor use in the C57BL/6 mouse. *Oncotarget.*, *6* (21), 18314–18337.

Nechad, M., Ruka, E., & Thibault, J. (1994). Production of nerve growth factor by brown fat in culture: Relation with the in vivo developmental stage of the tissue. *Comparative Biochemistry and Physiology. Comparative Physiology*, *107*(2), 381–388.

Nedergaard, J., Bengtsson, T., & Cannon, B. (2007). Unexpected evidence for active brown adipose tissue in adult humans. *American Journal of Physiology. Endocrinology and Metabolism*, *293*(2), E444–E452.

Nisoli, E., Tonello, C., Benarese, M., et al. (1996). Expression of nerve growth factor in brown adipose tissue: Implications for thermogenesis and obesity. *Endocrinology*, *137*(2), 495–503.

Nisoli, E., Tonello, C., Briscini, L., & Carruba, M. O. (1997). Inducible nitric oxide synthase in rat brown adipocytes: Implications for blood flow to brown adipose tissue. *Endocrinology*, *138*(2), 676–682.

Ohyama, K., & Suzuki, K. (2017). Dihydrocapsiate improved age-associated impairments in mice by increasing energy expenditure. *American Journal of Physiology. Endocrinology and Metabolism*, *313*(5), E586–E597.

Oliverio, M., Schmidt, E., Mauer, J., et al. (2016). Dicer1-miR-328-Bace1 signalling controls brown adiposce tissue differentiation and function. *Nature Cell Biology*, *18*(3), 328–336.

Ortega-Molina, A., Efeyan, A., Lopez-Guadamillas, E., et al. (2012). Pten positively regulates brown adipose function, energy expenditure, and longevity. *Cell Metabolism*, *15*(3), 382–394.

Pearl, R. (1928). *The rate of living*. New York: Knopf.

Piao, Z., Zhai, B., Jiang, X., et al. (2018). Reduced adiposity by compensatory WAT browning upon iBAT removal in mice. *Biochemical and Biophysical Research Communications*, *501*(3), 807–813.

Qi, Y., Takahashi, N., Hileman, S. M., et al. (2004). Adiponectin acts in the brain to decrease body weight. *Nature Medicine*, *10*(5), 524–529.

Redman, L. M., Smith, S. R., Burton, J. H., et al. (2018). Metabolic slowing and reduced oxidative damage with sustained caloric restriction support the rate of living and oxidative damage theories of aging. *Cell Metabolism*, *27*(4), 805–815, e4.

Reynolds, M. A., Ingram, D. K., & Talan, M. (1985). Relationship of body temperature stability to mortality in aging mice. *Mechanisms of Ageing and Development*, *30*(2), 143–152.

Rizzo, M. R., Mari, D., Barbieri, M., Ragno, E., Grella, R., Provenzano, R., ... Paolisso, G. (2005). Resting metabolic rate and respiratory quotient in human longevity. *The Journal of Clinical Endocrinology and Metabolism*, *90*(1), 409–413.

Sanchez-Gurmaches, J., Tang, Y., Jespersen, N. Z., et al. (2018). Brown Fat AKT2 is a cold-induced kinase that stimulates ChREBP-Mediated De Novo Lipogenesis to optimize fuel storage and thermogenesis. *Cell Metabolism*, *27*(1), 195–209, e6.

Schipper, L., Harvey, L., van der Beek, E. M., & van Dijk, G. (2018). Home alone: A systematic review and meta-analysis on the effects of individual housing on body weight, food intake and visceral fat mass in rodents. *Obesity Reviews.*, *19*(5), 614–637.

Schnabl, K., Westermeier, J., Li, Y., & Klingenspor, M. (2018). Opposing actions of adrenocorticotropic hormone and glucocorticoids on UCP1-mediated respiration in brown adipocytes. *Frontiers in Physiology*, *9*, 1931.

Scholer, M. N., Arcese, P., Puterman, M. L., et al. (2019). Survival is negatively related to basal metabolic rate in tropical Andean birds. *Functional Ecology*, *33*, 1436–1445.

Schreiber, R., Diwoky, C., Schoiswohl, G., et al. (2017). Cold-induced thermogenesis depends on ATGL-mediated lipolysis in cardiac muscle, but not brown adipose tissue. *Cell Metabolism*, *26*(5), 753–763, e7.

Seale, P. (2015). Transcriptional regulatory circuits controlling brown fat development and activation. *Diabetes*, *64*(7), 2369–2375.

Seale, P., Kajimura, S., Yang, W., et al. (2007). Transcriptional control of brown fat determination by PRDM16. *Cell Metabolism*, *6*(1), 38–54.

Sellayah, D., & Sikder, D. (2014). Orexin restores aging-related brown adipose tissue dysfunction in male mice. *Endocrinology*, *155*(2), 485–501.

Shin, H., Ma, Y., Chanturiya, T., et al. (2017). Lipolysis in brown adipocytes is not essential for cold-induced thermogenesis in mice. *Cell Metabolism*, *26*(5), 764–777, e5.

Simcox, J., Geoghegan, G., Maschek, J. A., et al. (2017). Global analysis of plasma lipids identifies liver-derived acylcarnitines as a fuel source for brown fat thermogenesis. *Cell Metabolism*, *26*(3), 509–522, e6.

Slabaugh, M. B., Lieberman, M. E., Rutledge, J. J., & Gorskin, J. (1981). Growth hormone and prolactin synthesis in normal and homozygous Snell and Ames dwarf mice. *Endocrinology*, *109*, 1040–1046.

Speakman, J. R., & Keijer, J. (2012). Not so hot: Optimal housing temperatures for mice to mimic the thermal environment of humans. *Molecular Metabolism*, *2*(1), 5–9.

Speakman, J. R., Talbot, D. A., Selman, C., et al. (2004). Uncoupled and surviving: Individual mice with high metabolism have greater mitochondrial uncoupling and live longer. *Aging Cell*, *3*(3), 87–95.

Sponton, C. H., & Kajimura, S. (2018). Multifaceted roles of beige fat in energy homeostasis beyond UCP1. *Endocrinology*, *159*(7), 2545–2553.

Stanford, K. I., Lynes, M. D., Takahashi, H., et al. (2018). 12,13-diHOME: An exercise-induced lipokine that increases skeletal muscle fatty acid uptake. *Cell Metabolism*, *27*(5), 1111–1120, e3.

Stanford, K. I., Middelbeek, R. J., Townsend, K. L., et al. (2013). Brown adipose tissue regulates glucose homeostasis and insulin sensitivity. *The Journal of Clinical Investigation*, *123*(1), 215–223.

Stout, M. B., Swindell, W. R., Zhi, X., et al. (2015). Transcriptome profiling reveals divergent expression shifts in brown and white adipose tissue from long-lived GHRKO mice. *Oncotarget.*, *6*(29), 26702–26715.

Sun, W., Dong, H., Becker, A. S., et al. (2018). Cold-induced epigenetic programming of the sperm enhances brown adipose tissue activity in the offspring. *Nature Medicine*, *24*(9), 1372–1383.

Tatar, M., Bartke, A., & Antebi, A. (2003). The endocrine regulation of aging by insulin-like signals. *Science*, *299*(5611), 1346–1351.

Thiel, A., Evans, A. L., Fuchs, B., et al. (2019). Effects of reproduction and environmental factors on body temperature and activity patterns of wolverines. *Frontiers in Zoology*, *16*, 21.

Thomou, T., Mori, M. A., Dreyfuss, J. M., et al. (2017). Adipose-derived circulating miRNAs regulate gene expression in other tissues. *Nature*, *542*(7642), 450–455.

Thuzar, M., Law, W. P., Ratnasingam, J., et al. (2018). Glucocorticoids suppress brown adipose tissue function in humans: A double-blind placebo-controlled study. *Diabetes, Obesity & Metabolism*, *20*(4), 840–848.

Tournissac, M., Bourassa, P., Martinez-Cano, R. D., et al. (2019). Repeated cold exposures protect a mouse model of Alzheimer's disease against cold-induced tau phosphorylation. *Molecular Metabolism*, *22*, 110–120.

Tournissac, M., Vandal, M., Francois, A., et al. (2017). Old age potentiates cold-induced tau phosphorylation: Linking thermoregulatory deficit with Alzheimer's disease. *Neurobiology of Aging*, *50*, 25–29.

Townsend, K. L., & Tseng, Y. H. (2014). Brown fat fuel utilization and thermogenesis. *Trends in Endocrinology and Metabolism: TEM*, *25*(4), 168–177.

Townsend, L. K., Medak, K. D., Knuth, C. M., et al. (2019). Loss of glucagon signaling alters white adipose tissue browning. *The FASEB Journal*, *33*(4), 4824–4835.

van Marken Lichtenbelt, W. D., Vanhommerig, J. W., Smulders, N. M., et al. (2009). Cold-activated brown adipose tissue in healthy men. *The New England Journal of Medicine*, *360*(15), 1500–1508.

Van Voorhies, W. A., & Ward, S. (1999). Genetic and environmental conditions that increase longevity in Caenorhabditis elegans decrease metabolic rate. *Proceedings of the National Academy of Sciences of the United States of America*, *96*(20), 11399–11403.

Vatner, D. E., Zhang, J., Oydanich, M., et al. (2018). Enhanced longevity and metabolism by brown adipose tissue with disruption of the regulator of G protein signaling 14. *Aging Cell*, *17*(4), e12751.

Velickovic, K., Wayne, D., Leija, H. A. L., et al. (2019). Caffeine exposure induces browning features in adipose tissue in vitro and in vivo. *Scientific Reports*, *9*(1), 9104.

Viana-Huete, V., Guillen, C., Garcia, G., et al. (2018). Male brown fat-specific double knockout of IGFIR/IR: Atrophy, mitochondrial fission failure, impaired thermogenesis, and obesity. *Endocrinology*, *159*(1), 323–340.

Villarroya, F., Cereijo, R., Villarroya, J., & Giralt, M. (2017). Brown adipose tissue as a secretory organ. *Nature Reviews Endocrinology*, *13*(1), 26–35.

Villarroya, F., Peyrou, M., & Giralt, M. (2017). Transcriptional regulation of the uncoupling protein-1 gene. *Biochimie*, *134*, 86–92.

Wang, W., & Seale, P. (2016). Control of brown and beige fat development. *Nature Reviews. Molecular Cell Biology*, *17*(11), 691–702.

Wang, Y., Paulo, E., Wu, D., et al. (2017). Adipocyte liver kinase b1 suppresses beige adipocyte renaissance through class IIa histone deacetylase 4. *Diabetes*, *66*(12), 2952–2963.

Weindruch, R., & Walford, R. L. (1988). *The retardation of aging and disease by dietary restriction. Springfield, IL*. Charles C. Thomas.

Westbrook, R. (2012). *The effects of altered growth hormone signaling on murine metabolism [Ph.D. Dissertation]*. Carbondale: Southern Illinois University.

Westbrook, R., Bonkowski, M. S., Strader, A. D., & Bartke, A. (2009). Alterations in oxygen consumption, respiratory quotient, and heat production in long-lived GHRKO and Ames dwarf mice, and short-lived bGH transgenic mice. *The Journals of Gerontology. Series A, Biological Sciences and Medical Sciences*, *64*(4), 443–451.

Wilkinson, G. S., & South, J. M. (2002). Life history, ecology and longevity in bats. *Aging Cell*, *1*(2), 124–131.

Wu, X., Wang, Y., Wang, Y., et al. (2018). GSK126 alleviates the obesity phenotype by promoting the differentiation of thermogenic beige adipocytes in diet-induced obese mice. *Biochemical and Biophysical Research Communications*, *501*(1), 9–15.

Xu, M., Pirtskhalava, T., Farr, J. N., et al. (2018). Senolytics improve physical function and increase lifespan in old age. *Nature Medicine*, *24*(8), 1246–1256.

Yamashita, H., Sato, Y., Kizaki, T., et al. (1994). Basic fibroblast growth factor (bFGF) contributes to the enlargement of brown adipose tissue during cold acclimation. *Pflugers Archiv: European Journal of Physiology*, *428*(3–4), 352–356.

Yuan, X., Hu, T., Zhao, H., et al. (2016). Brown adipose tissue transplantation ameliorates polycystic ovary syndrome. *Proceedings of the National Academy of Sciences of the United States of America*, *113*(10), 2708–2713.

Zhao, S., Chu, Y., Zhang, Y., et al. (2019). Linifanib exerts dual anti-obesity effect by regulating adipocyte browning and formation. *Life Sciences*, *222*, 117–124.

Zhou, Y., Xu, B. C., Maheshwari, H. G., et al. (1997). A mammalian model for Laron syndrome produced by targeted disruption of the mouse growth hormone receptor/binding protein gene (the Laron mouse). *Proceedings of the National Academy of Sciences of the United States of America*, *94*(24), 13215–13220.

Zwick, R. K., Guerrero-Juarez, C. F., Horsley, V., & Plikus, M. V. (2018). Anatomical, physiological, and functional diversity of adipose tissue. *Cell Metabolism*, *27*(1), 68–83.

CHAPTER

8

Yeast as a model organism for aging research

Anita Krisko[1] *and Brian K. Kennedy*[2,3,4]

[1]Department of Experimental Neurodegeneration, University Medical Center Goettingen (UMG), Goettingen, Germany [2]Department of Biochemistry and Physiology, Yong Loo School Lin School of Medicine, National University of Singapore, Singapore, Singapore [3]Centre for Healthy Longevity, National University Health System, Singapore, Singapore [4]Singapore Institute for Clinical Sciences, A*STAR, Singapore, Singapore

OUTLINE

Introduction

Aging is a natural phenomenon that occurs over time, despite complex pathways of maintenance and repair. As the age of an organism increases, a number of errors accumulate at different levels of biological organization, leading to altered function in most major cellular processes. These alterations, in turn, compromise tissue function and ultimately lead to organismal decline. Indeed, the events that drive aging provide the largest risk for the onset of chronic diseases in mammals.

As humans are now living longer, geroscience is a discipline receiving more and more attention. It is oriented toward understanding the pathways governing aging and uncovering possibilities for human healthspan extension (Kennedy et al., 2014). Given the impracticality of discovery-based studies in humans and even the long lifespans of mammalian model organisms, a number of nonvertebrate models have been developed and utilized to achieve scientific breakthroughs, including the fruitfly *Drosophila melanogaster*, roundworm *Caenorhabditis elegans*, and the budding yeast, *Saccharomyces cerevisiae*. Decades of

Handbook of the Biology of Aging.
DOI: https://doi.org/10.1016/B978-0-12-815962-0.00008-1

research have already unlocked many mysteries of aging; however numerous questions remain.

S. cerevisiae, budding yeast, baker's yeast, or brewer's yeast, is a unicellular eukaryote, well known for centuries for its importance in fermentation processes. To scientists, however, it is known for its simplicity in handling and manipulation, which has led to many seminal discoveries. As multicellular organisms are more complex and difficult to study, budding yeast offers an alternative to look at cellular processes in a simpler context. Its genome, consisting of 6000 genes, has been completely sequenced and mapped. Many of its genes have human orthologs, including those whose perturbation leads to disease in humans.

Yeast is characterized by relatively simple, defined growth requirements, which makes it an attractive model system to work with on a number of relevant research topics. Cells are grown in liquid medium and their growth can be divided into distinct phases after exposure to nutrients: first, they will pass through a lag phase, during which growth is limited until they adapt to the replenished nutrients, after which they will undergo an exponential growth phase, followed by a diauxic shift and stationary phase, in which they are no longer dividing (Fig. 8.1). During the exponential growth stage, as glucose is still abundant in the medium (2%), yeast, as a facultative anaerobe, relies mostly on glycolysis for energy. As the number of cells increases exponentially, the glucose is eventually consumed and yeast then undergoes a diauxic shift. In the postdiauxic growth stage, the cells switch to respiratory metabolism and their growth rate slows down until they finally reach the stationary phase. Replicative lifespan (RLS) assesses the number of times a cell divides and produces daughters when maintained in a state resembling the exponential phase, while chronological lifespan (CLS) determines the length of time a cell can remain viable in a state resembling the stationary phase (Longo, Shadel, Kaeberlein, & Kennedy, 2012).

Budding yeast can proliferate in both a diploid and haploid state. The diploid state is maintained during nutrient abundance, and these cells are able to divide by mitosis through budding, producing genetically identical daughter cells. Depletion of nutrients and starvation usually trigger meiosis and spore formation, thus giving rise to haploid spores that will continue to propagate once nutrients are reintroduced into the environment. Haploid yeast cells can appear in two different mating types: MATa and MATα. These cells are able to undergo mitosis through budding, thus producing daughter cells. The two mating types also release pheromones, thereby initiating mating that results in the formation of a stable diploid cell.

In yeast, cell-to-cell interactions are important only in the context of mating and a limited number of other contexts. In general, yeast is characterized by cell-autonomous responses, meaning that each cell controls its own growth and development. In this context, yeast is very different from multicellular organisms, which can provide clarity in interpreting research results, but is also a limitation, since intercellular communication is an important facet of mammalian aging. Moreover, the interactions with the environment are simpler in comparison with higher metazoans. Nevertheless, key pathways implicated in aging are highly conserved between yeast and mammals, which makes yeast a relevant and useful model in the studies of aging and age-related diseases (Longo et al., 2012).

Asymmetric cell division in yeast

A yeast cell begins its life by budding from a mother cell. The new daughter cell or bud is initially smaller than the mother, as well as rejuvenated, thus maintaining the potential to become a mother and engage in a full RLS.

The asymmetric cell division in yeast is thought to be the main driver of its aging. Asymmetry involves the unequal distribution of cellular contents during cell

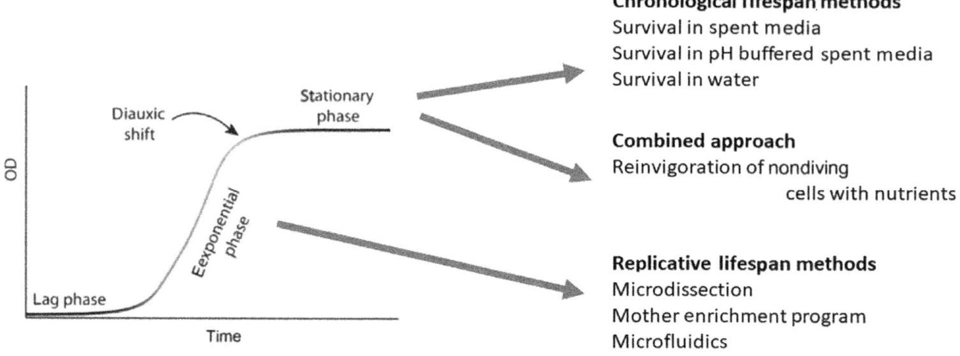

FIGURE 8.1 **Yeast growth phases and approaches to measure aging.** The yeast initially passes through a lag phase, followed by an exponential growth phase. Upon glucose depletion, it undergoes diauxic shift and enters the stationary phase, in which it is no longer dividing.

division, including even some transcription factors such as Ash1 and Ace2 (Boettcher, Marquez-Lago, Bayer, Weiss, & Barral, 2012; Dohrmann et al., 1992; Long et al., 1997; Takizawa, Sil, Swedlow, Herskowitz, & Vale, 1997). Of note, the asymmetrical distribution of Ace2 has been taken advantage of to create the mother enrichment program (MEP) (Lindstrom & Gottschling, 2009) described below. For example, during budding, transcription repressors are constrained to the bud (Gershon & Gershon, 2000), while the mother cell retains membrane proteins and protein aggregates (Baldi, Bolognesi, Meinema, & Barral, 2017). Moreover, cell compartmentalization in yeast involves the presence of a barrier that limits the diffusion of cellular components across a boundary. Typically, in yeast, such barriers consist of the plasma membrane and the membrane of the endoplasmic reticulum and are located in the bud neck. Thereby, the exchange of membrane proteins between the mother cell and the bud is limited. Furthermore, the diffusion barrier plays an important role in the segregation of cell fate determinants, such as the retention of the aging factors in the mother cell, rendering the daughter cell damage free, rejuvenated, and with potential for a full lifespan. Extensive research has shown that old nuclear pore complexes (Khmelinskii, Keller, Lorenz, Schiebel, & Knop, 2010; Winey, Yarar, Giddings, & Mastronarde, 1997), as well as the associated extrachromosomal rDNA circles (ERCs) (Sinclair & Guarente, 1997), are retained in the mother cell during division, while the new pore complexes are inserted into the nuclear envelope in the daughter cell.

The connection between rDNA recombination and replicative aging was initially centered on the formation of ERCs. ERCs are individual or multiple rDNA genes excised from the genome through homologous recombination and have the ability to self-replicate. They do not have centromeres, so they are asymmetrically segregated into mother cells during mitosis (Murray & Szostak, 1983; Sinclair & Guarente, 1997). This is enabled by a septin-dependent lateral diffusion barrier between the mother and the bud (Gehlen et al., 2011; Shcheprova, Baldi, Frei, Gonnet, & Barral, 2008). ERCs exponentially accumulate in older mother cells (Sinclair & Guarente, 1997). Together with the damaged and aggregated proteins (detailed in a separate section below), this remains one of the main aging factors recognized so far in yeast.

It is well established that aging is not a linear and immutable process: it can be hindered or promoted by both environmental and genetic factors. Several types of mild stress, including mild oxidative stress and heat shock, have shown potential to delay aging phenotypes. Which mechanisms contribute to the longevity conferred by mild stress is an area of active research. Recently, it has been described that, during mild heat shock, the disturbance of the diffusion barrier between the mother and the bud during division leads to the bud inheriting the aging factors, contributing to the longevity of the mother. As a result, yeast mother cells relax the retention of DNA circles (Baldi et al., 2017), as well as of CytoQ type protein aggregates (Miller, Mogk, & Bukau, 2015). Interestingly, the activities of certain metabolic pathways (target of rapamycin, protein kinase A pathway) have key roles in this process, therefore underscoring the importance of the crosstalk between basic cellular maintenance and metabolic activity (Baldi et al., 2017).

Emergence of yeast as a model organism in aging research

As with most microbes, budding yeast cultures and colonies can be propagated indefinitely. However, in 1959 Mortimer and Johnson proposed that yeast cells are not immortal (Mortimer & Johnston, 1959). This idea stemmed from the fact that yeast cells divide asymmetrically, with smaller daughter cell budding from the mother: if each daughter cell is immediately removed from the mother, it becomes obvious that each mother cell can divide only a limited number of times. Since then, researchers have used yeast to uncover a number of age-associated traits and genetic modifiers of lifespan (Wasko & Kaeberlein, 2014).

Aging research in yeast has come a long way since Mortimer and Johnson. The budding yeast is a very popular organism used in many branches of science as a model of the eukaryotic cell for analysis of physiological, biochemical, genetic, or molecular functions. Its usefulness in explaining the mechanisms of aging and longevity of multicellular organisms has proven essential numerous times.

The unicellularity and simplicity of yeast is at the same time a pro and con. It makes the research simple, efficient and, in the pathways that are conserved across species, also important. However, for any hallmark of aging that is above the single cell level, it has less to offer. In order to identify the aging-related pathways in yeast that may be relevant also for mammals, it is important to study these pathways in the conditions as similar as possible to those in which they have evolved.

Types of aging in yeast

As already mentioned, there are two main ways to measure the lifespan in yeast: RLS (Mortimer & Johnston, 1959), measured as the number of cell divisions that a cell goes through before senescence, and CLS (Fabrizio, Pozza, Pletcher, Gendron, & Longo, 2001b), measured as the survival of post-mitotic yeast cells in the absence of nutrients (Fig. 8.1).

Replicative lifespan

Replicative longevity is defined as the ability to produce more daughters compared to the reference strain (He, Zhou, & Kennedy, 2018). Doubts concerning the use of the number of daughter cells as a measure of age or longevity were presented very early on, suggesting that it represents a measure of fecundity rather than age or longevity (Gershon & Gershon, 2000, 2001). However, using this parameter, many genes and pathways have been implicated in the aging process. Below, they are grouped based on the hallmarks of aging to which they likely have the most impact.

The standard approach to measuring replicative aging requires continuous separation of new daughter cells from mothers using a microdissection apparatus (Steffen, Kennedy, & Kaeberlein, 2009). While this is a highly laborious process, large-scale screens have been attempted. A gene deletion library exists for yeast and a genome-wide screen has been surveyed for long-lived deletions (Kaeberlein et al., 2005b; McCormick et al., 2015a). Strikingly, 238 gene deletions were found to be long-lived, with the caveat that many false negatives existed in the screen, likely pushing the number even higher (McCormick et al., 2015a). The ease with which it is possible to extend lifespan is surprising, but data from *C. elegans* suggest that this is likely a general phenomenon, at least in nonvertebrates. An important validation of the replicative aging approach is that there is significant enrichment between aging genes identified in the yeast screen and those associated with aging in *C. elegans*, even though these are highly disparate species from an evolutionary perspective (Smith et al., 2008). Moreover, many of the most prominent pathways affecting replicative aging seem to also impact mammalian aging, including some evidence in humans (Wasko & Kaeberlein, 2014). Therefore efforts to understand replicative aging in yeast shed light on aspects of aging in more complex organisms. Genes identified in the genome-wide screen could be classified into categories based on their known functions, with ribosomal protein encoding genes or genes related to proteasomal degradation and mitochondrial function being enriched (McCormick et al., 2015b), all of which will be discussed below.

Alternative methods to measure replicative lifespan

Given the difficulty of measuring RLS using microdissection, alternative methods have been developed. Initially, biochemical approaches were used to enrich for older cells, taking advantage of the observation, alluded to earlier, that mother cells largely retain cell-surface proteins. The cell surface could then be coated with avidin and mother cells isolated with biotin attached to magnetic beads (Smeal, Claus, Kennedy, Cole, & Guarente, 1996). This approach has utility, but is not sufficient to measure the lifespan of a strain. Nevertheless, it has been used to make important discoveries, including a recent finding that aging yeast cells have lower metabolic levels and therefore switch from fermentative to respiratory metabolism (Leupold et al., 2019). This is likely consistent with an earlier finding that calorie restriction (CR) extends lifespan in part through enhancing respiratory metabolism (Lin et al., 2002). It has been speculated that aging in yeast may be a mechanism of phenotypic diversity, allowing a small population of cells to be positioned for altered environments should they occur (Knorre, Azbarova, Galkina, Feniouk, & Severin, 2018). The shift to respiratory metabolism, changes in stress resistance, and other modifications may be examples of how aging yeast cells embody this diversity.

Another device, recently developed, that allows for enrichment of old cells is the miniature chemostat aging device (MAD) (Hendrickson et al., 2018). In this device, a miniature chemostat is combined with magnetic-based streptavidin enrichment of mother cells. The principle works as follows: cells are biotinylated and attached to streptavidin beads prior to aging. Mother cells are then trapped, while the daughter cells can be released as soon as they finish budding. The device is provided with fresh media to the confined mother cells so that they can continue to divide. Mother cells could be released from the magnet at any point during the aging process, and collected for further analysis. RLS still cannot be measured using MAD, but the device provided us with a novel insight into aging in yeast and revealed unknown age-associated traits, such as showing that origins of replication become less accessible with age, and that gene expression from subtelomeric regions increases with age.

Mother enrichment program

A significant step forward in the research into aging in yeast has been made with the development of the MEP (Lindstrom & Gottschling, 2009). The MEP ensures genetic enrichment of old mother cells by preventing new daughters from budding, while allowing the mother cells to have a relatively normal RLS. MEP uses Cre-*lox* recombination to disrupt two vital genes in the daughter cells, *UBC9* and *CDC20*, thus arresting them in M phase. Specificity is ensured by expressing Cre recombinase from a daughter-specific promoter under control of *ACE2* transcription factor, which is asymmetrically inherited by the bud. Conditionality is achieved by adding an estradiol-binding domain such that Cre is confined to the cytoplasm until estradiol is added to the medium.

A variant of the MEP has recently been used to screen for drugs that extend yeast lifespan, finding two hits that extend RLS by 15%–20% (Sarnoski, Liu, & Acar, 2017). This direct approach may offer advantages over other methods to identify RLS-extending small molecules, which relied on indirect means, for instance a correlation between gene deletions that had extended RLS and also certain alterations in the G1 phase of the cell cycle, ultimately leading to the identification of ibuprofen (He et al., 2014).

Microfluidics in the research into aging in yeast

Another approach for measuring yeast lifespan has emerged with the advent of microfluidics. Recently, several microfluidics devices have been developed to measure the RLS of yeast, thus avoiding the laborious approach of microdissection (Chen, Crane, & Kaeberlein, 2017; Huberts, Janssens, Lee, Vizcarra, & Heinemann, 2013; Zou, Ren, Ou-Yang, Li, & Zheng, 2017). Microfluidics devices are objects containing submillimeter chambers and channels. This allows users to control cellular microenvironments that will allow cellular growth under specific conditions of temperature and nutrient availability. Successful construction of an aging microfluidics device in 2012 permitted measurement of yeast RLS (Lee, Vizcarra, Huberts, Lee, & Heinemann, 2012b; Xie et al., 2012; Zhang et al., 2012). Since then several other designs have become available (Chen et al., 2017, 2019). Overall, the common feature that guarantees success of such devices is the ability to trap the mother cells, enabled by the larger size of the mother compared with the daughter. Another important feature is the efficient removal of unwanted daughter cells in order to prevent overgrowth of the cells that are not being monitored. A typical RLS measurement in the microfluidics device lasts 2–5 days, a period of time for which a typical yeast cell is able to produce daughter cells. Coupled with time-lapse microscopy, microfluidics devices allow researchers to collect RLS data without the need to perform microdissection.

A fundamental advantage to the microfluidics approach is that it allows for imaging of single mother cells while they are aging, which permits a more thorough analysis of changes that occur in mother cells. For instance, aging mother cells more frequently undergo delayed nuclear segregation and can reuse budding sites (Meitinger et al., 2014). Nuclear pore complex assembly also deteriorates (Rempel et al., 2019). Aging yeast cells also lose double-strand break repair efficiency, in part due to excessive time spent transitioning through the G1 phase of the cell cycle (Young, Liu, Urbonaite, & Acar, 2019). Another study, however, found that while DNA damage foci are increased in old cells, repair efficiency was maintained (Novarina et al., 2017). A third study found that aging yeast more frequently invokes DNA damage checkpoints, which leads to disruption of histone proteins. It may be the loss of histones that limits the RLS, as restoring levels lead to increased lifespan and better genome segregation during division (Crane et al., 2019), a finding consistent with earlier results using standard microdissection (Feser et al., 2010). Genome instability at the rDNA loci is a specific instance of altered DNA integrity with aging that is discussed below.

By analyzing aging mother cells, it seems that the concept of personalized aging may extend to yeast (Crane, Chen, Blue, & Kaeberlein, 2020). An early study suggested that each cell may experience its own mode of death: via mitochondrial dysfunction, proteostasis failure, etc. (Lee, Avalos Vizcarra, Huberts, Lee, & Heinemann, 2012a). More recently, it has been reported that aging yeast cells adopt one of two different fates, which is determined at least in part stochastically (Jin et al., 2019). These differences are largely characterized morphologically and are influenced by aging genes and CR. It remains unclear what this means mechanistically, but the results are very intriguing and further studies are warranted.

Chronological lifespan

It has been suggested that, in nature, microbes spend much of their time in a stationary growth phase characterized by low metabolic activity (Werner-Washburne, Braun, Crawford, & Peck, 1996). For example, yeast in the wild exit the stationary phase only under circumstances of all essential nutrients being available, which is rare. Therefore assays to measure the maintenance of viability in nutrient-depleted states may be more in line with how a yeast cell experiences aging.

CLS is measured as the survival of a population of yeast cells that have stopped dividing (Fabrizio, Pozza, Pletcher, Gendron, & Longo, 2001a; Longo, Gralla, & Valentine, 1996). It can be monitored in either of the two stages of yeast growth: during the postdiauxic shift or in the stationary phase.

There are several methods to measure CLS, although all leave yeast suspended in a state incompatible with proliferation (Longo et al., 2012). The medium conditions also matter. For instance, cells allowed to exhaust synthetic media secrete acetic acid, which dramatically reduces pH and compromises viability (Burtner, Murakami, Kennedy, & Kaeberlein, 2009). Interventions resulting in longer lifespan often affect pH or resistance to acidic conditions. Using rich media, buffering pH, or transferring nondividing cells to water results in longer lifespan and may represent an experimental state that more closely compares to aging in other species. For CLS studies in water, yeast is grown and incubated for 3 days in synthetic medium, washed, and resuspended in sterile distilled water. Viability is monitored by measuring colony-forming units every 2 days. The cells are washed three times with water every 2 days to remove all molecules released by dead yeast. Incubation in water and the removal of nutrients released by dead organisms minimizes the chance of growth during long-term survival in the stationary phase.

What do we know about the pathways affecting CLS? Budding yeast expresses two kinds of superoxide dismutases (SODs): Sod1, a cytosolic Cu-Zn SOD, and Sod2, a mitochondrial MnSOD. In addition, it expresses catalase Ctt1. The measurements of CLS have revealed that, unlike catalase, both SODs are essential for long-term survival of yeast (Longo et al., 1996).

As mentioned above, yeast is able to grow by fermentation whereby glucose is used as the main energy source, and by respiration, where nonfermentable carbon sources (e.g., ethanol) are used. In 1999, Longo et al. published a study showing that mitochondrial damage precedes death (Longo, Liou, Valentine, & Gralla, 1999). They defined the Index of Respiratory Competence (IRC) as the percentage of live yeast cells able to respire and grow on nonfermentable carbon sources. In such a setup, it was possible to study the sequence of events that lead to the death of the Sod2-deficient mutant, as well as evaluate the role of mitochondrial damage in the death of stationary-state yeast cells. The authors revealed that a decline in IRC, as well as mitochondrial functional failure, preceded the death of both wild-type and Sod2-deficient yeast. Two mitochondrial 4Fe−4S cluster enzymes, aconitase and succinate dehydrogenase, have been identified as the primary targets of mitochondrial superoxide. In this context, the research performed in yeast has been successfully reproduced in mammals. For example, in *Sod2* knockout mice, the activity of mitochondrial aconitase was reduced in the heart and brain, and succinate dehydrogenase was reduced in the heart and skeletal muscles, thus more severely affecting postmitotic cells (Melov et al., 1998).

Moreover, using yeast as a model system, it has been found that Ras2 deletion doubles the yeast CLS, and enhances its stress resistance (Fabrizio et al., 2003). These observations were later confirmed in worms and flies, and similar observations have been made for yeast replicative aging (Lin et al., 2002), leading to a hypothesis that the longevity of organisms from yeast to humans may be regulated by a pathway modulating SOD activity through activation of stress-resistance transcription factors (Longo et al., 1999). The transcription factors in question were found to be Msn2/Msn4, essential for the longevity of the *RAS2* mutants, together with Sod2 (Fabrizio et al., 2003). Consistently, activation of the RAS pathway through deletion of glutaredoxin leads to reduced CLS (D'Aquila et al., 2019).

Deletion of the Sch9 serine-threonine kinase, one of the targets of the TORC1 kinase complex that mediates cellular nutrient and stress responses, also leads to robust CLS extension (Fabrizio et al., 2003). Deletion of the Tor1 kinase confers a similar phenotype (Powers, Kaeberlein, Caldwell, Kennedy, & Fields, 2006). Both of these deletions lead to enhanced longevity in the RLS assay (Kaeberlein et al., 2005b), and the mTORC1 kinase pathway is linked to aging in all other prominent aging model organisms (Kennedy & Lamming, 2016), as well as in humans (Mannick et al., 2014, 2018). These pathways share similarity to and/or overlap with insulin/IGF signaling, another pathway linked to aging across eukaryotes (Johnson, 2018).

The name TOR (target of rapamycin) originates from the fact that a compound called rapamycin (isolated from the bacterium *Streptomyces hygroscopicus*) is able to inhibit the kinase (Kennedy & Lamming, 2016). In yeast, the TOR pathway consists of two Tor kinases, Tor1 and Tor2. These proteins are PIK-related serine/threonine kinases with an important role in the regulation of cell metabolic activity and growth in response to nutrient availability (Gonzalez & Hall, 2017). Tor1 acts as a part of the TOR complex 1 (TORC1) and Tor2 can function in both TORC1 and TOR complex 2 (TORC2). Both complexes perform numerous functions, for example in amino acid and lipid metabolism. TORC1 complex phosphorylates, among other targets, Sch9 kinase, an ortholog of S6 kinase in mammals. More specifically, through phosphorylating Sch9, TORC1 modulates translation, carbon and nitrogen metabolism, and autophagy. The TOR pathway is discussed at several points in the context of yeast aging (below).

Replicative lifespan versus chronological lifespan

What is the relationship between the RLS and CLS? The question is far from resolved. Still, there are studies suggesting that the mechanisms underlying these aging assays might be related. Deletion of *SOD1*, encoding the cytosolic superoxide dismutase, is detrimental for both RLS and CLS (Laun et al., 2001; Longo et al., 1996). Moreover, both RLS and CLS are extended through reduced PKA and mTORC1 signaling

(Fabrizio et al., 2001a; Kaeberlein et al., 2005b; Lin, Defossez, & Guarente, 2000; Longo, Ellerby, Bredesen, Valentine, & Gralla, 1997; Powers et al., 2006). Nevertheless, quantitative analysis has yet to prove significant overlap between RLS and CLS at the genome level (Burtner, Murakami, Olsen, Kennedy, & Kaeberlein, 2011).

One intriguing line of investigation involves combining CLS and RLS in the same cell. For instance, it was shown that, during chronological aging, a cell reduces its RLS when reinvigorated with nutrients (Ashrafi, Sinclair, Gordon, & Guarente, 1999). A more recent study has confirmed this observation and investigated it further, finding that CR of cells prior to entry into the stationary phase could preserve their RLS (Delaney et al., 2013). In summary, both the RLS and CLS have something to offer in understanding mechanisms of aging and there are clearly interconnections, yet the relationship between these two types of assays is yet to be fully elucidated.

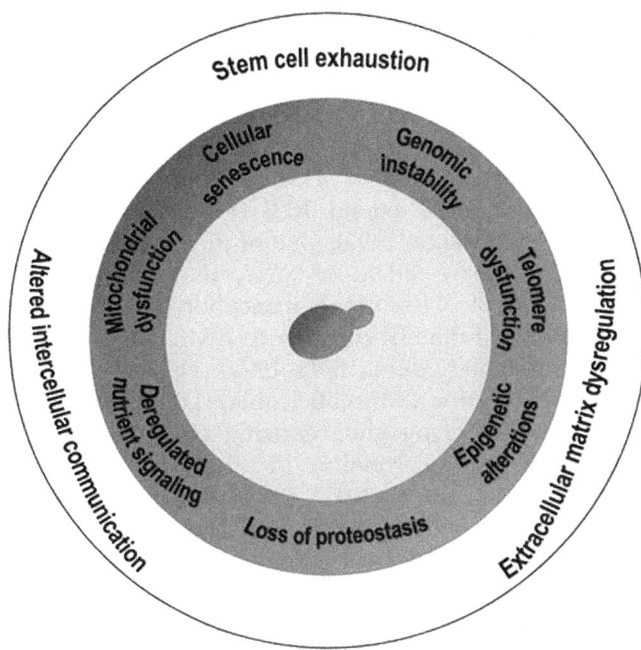

FIGURE 8.2 **Hallmarks of aging in *Saccharomyces cerevisiae*.** As a unicellular organism, yeast displays only a subset of aging hallmarks: genomic instability, telomere attrition, epigenetic alterations, loss of proteostasis, deregulated nutrient signaling, and mitochondrial dysfunction (*the inner circle*). Cellular senescence, stem cell exhaustion, and altered intercellular communication (*the outer circle*) are more related to multicellular organisms.

Hallmarks of aging and yeast

In 2013, a group of authors published an extensive analysis of phenotypes related to aging, defined as aging hallmarks (Lopez-Otin, Blasco, Partridge, Serrano, & Kroemer, 2013). These phenotypes fulfill several key criteria: they occur during normal aging, their aggravation leads to a shortened lifespan, and their amelioration extends healthspan/lifespan across species. These are genomic instability, telomere attrition, epigenetic alterations, loss of proteostasis, deregulated nutrient signaling, mitochondrial dysfunction, cellular senescence, stem cell exhaustion, and altered intercellular communication (Fig. 8.2). Even though some of the aging hallmarks are specific for multicellular organisms, yeast has proven to be an essential model organism in discovering the hallmarks that are important at the cellular level. A brief overview is given below.

Epigenetic alterations in yeast aging

Regulation of chromatin structure is essential for transcriptional activation and downstream cellular processes. Numerous studies conducted in yeast have raised interest in the sirtuin family of proteins. Sirtuins are NAD^+-dependent protein deacetylases, rather well conserved across species. In yeast aging, one particular member of this family, Sir2, is most prominent. Sir2 deacetylates histones through an NAD^+-dependent reaction in which the cleavage of a molecule of NAD^+ into nicotinamide and O-acetyl-ADP-ribose is coupled to deacetylation of a single lysine side chain (Imai, Armstrong, Kaeberlein, & Guarente, 2000; Landry, Slama, & Sternglanz, 2000; Tanny & Moazed, 2001). Overexpression of Sir2 extends yeast RLS (Kaeberlein, McVey, & Guarente, 1999), a finding later repeated in worms and fruitflies (Tissenbaum & Guarente, 2001; Viswanathan & Guarente, 2011), although these findings are likely context-dependent and have not always been replicated (Burnett et al., 2011; Viswanathan & Guarente, 2011). Increased expression of two mammalian orthologs, SIRT1 and SIRT6, is reported to extend the healthspan and/or lifespan in mice (Herranz et al., 2010; Kanfi et al., 2012). The literature on sirtuins in multicellular organisms is vast and readers are referred to recent reviews (Giblin, Skinner, & Lombard, 2014; Imai & Guarente, 2014).

Sir2 was originally identified in yeast for its role in silencing at the cryptic mating type loci, *HML* and *HMR* (Ivy, Klar, & Hicks, 1986; Rine & Herskowitz, 1987). Four additional Sir2 homologs were eventually identified from yeast (Hst1, Hst2, Hst3, and Hst4) (Brachmann et al., 1995; Derbyshire, Weinstock, & Strathern, 1996), and potential roles for these other deacetylases as compensatory factors in the absence of *SIR2* have been discussed (Giblin et al., 2014). Sir2 functions in transcriptional silencing at *HML*, *HMR*, and telomeres as part of a multiprotein silencing factor known as the SIR holocomplex, which consists of Sir2, Sir3, and Sir4 (Ghidelli, Donze, Dhillon, & Kamakaka, 2001; Moretti, Freeman, Coodly, & Shore, 1994).

The importance of Sir2 for yeast aging was initially found in the context of a screen for stress-resistant mutants that also extend RLS (Kennedy, Austriaco, Zhang, & Guarente, 1995). One of the isolated mutants was a dominant allele of *SIR4*, the *SIR4-42* allele, encoding Sir4 C-terminal truncation that prevents recruitment of the SIR complex to *HML*, *HMR*, and telomeres (Kennedy et al., 1995, 1997). This will result in Sir2 redistribution to the nucleolus/rDNA, strengthening rDNA silencing and reducing rDNA recombination, culminating finally in the RLS extension (Kennedy et al., 1995, 1997). Such redistribution of Sir proteins to the nucleolus is also a property of aged wild-type mother cells, and may be related to the sterility of old cells due to the loss of silencing at *HML* and *HMR* (Smeal et al., 1996).

As a result of *SIR2* deletion, which shortens RLS, the stability of rDNA decreases and the ERCs are increasingly formed (Kaeberlein et al., 1999). The opposite also stands: increasing *SIR2* gene dosage suppresses rDNA recombination/ERC formation and extends RLS (Kaeberlein et al., 1999). How does this relate to physiologically aged yeast? Extensive research has shown that rDNA instability in the old mother cells is sufficient to limit RLS, regardless of the absolute number of ERCs that are formed (Falcon & Aris, 2003; Ganley, Ide, Saka, & Kobayashi, 2009). Moreover, the nonfunctional DNA repair proteins could be preferentially segregated to aging mother cells, making the damage in the mother's rDNA accumulate even faster, likely also decreasing the quality of produced ribosomes (Ganley et al., 2009).

Notably, Sir2 levels decline during yeast aging (Dang et al., 2009). A recent study found that many other components of rDNA protein complexes, most notably cohesins, decline with aging, resulting in increased rDNA instability (Fine, Maqani, Li, Franck, & Smith, 2019). What is clear is that recombination and increased instability at the rDNA increase with age and negatively affect RLS. Such a key role for the rDNA in RLS regulation is underscored by two recent studies of natural lifespan variation showing the *SIR2* and rDNA loci to be major sources of RLS variation (Kwan et al., 2013; Stumpferl et al., 2012).

Further research on Sir2 has revealed roles in longevity that are independent of rDNA. One of these functions for Sir2 is its involvement in the asymmetric segregation of oxidatively damaged proteins, mitochondria, and repair machineries between mother and daughter cells during budding (Aguilaniu, Gustafsson, Rigoulet, & Nystrom, 2003; Erjavec & Nystrom, 2007; McFaline-Figueroa et al., 2011). Accumulation of protein oxidative damage is detrimental for longevity (Laun et al., 2001; Nestelbacher et al., 2000). As noted above, asymmetric cell division, whereby RLS-limiting cellular components are retained within the yeast mother cell, allows the daughter to retain full replicative potential (Shcheprova et al., 2008; Sinclair and Guarente, 1997). This phenomenon is *SIR2* dependent (Aguilaniu et al., 2003; Orlandi, Bettiga, Alberghina, Nyström, & Vai, 2010). More specifically, daughter cells inherit higher levels of the cytosolic catalase Ctt1, as well as mitochondria with higher redox potential (Aguilaniu et al., 2003; Erjavec & Nystrom, 2007; McFaline-Figueroa et al., 2011). How exactly Sir2 is involved in this asymmetric inheritance remains unclear. However, its function in promoting the actin-folding activity of the chaperonin CCT ring complex, potentially through direct deacetylation, could be one contribution (Liu et al., 2010). Polarisome-dependent retrograde transport of aggregates bearing damaged cellular components away from the daughter cell is thus supported and contributes to the restoration of complete replicative potential in the daughter cell.

Several other histone modifications have been ascribed roles in RLS. For instance, loss of Sir2 with aging leads to increased histone H4 lysine 16 acetylation at telomeres, compromising silencing at these loci (Dang et al., 2009). The histone deubiquitinase component of the SAGA/SLIK complex also affects silencing at these sites in a Sir2-dependent manner (McCormick et al., 2014). Interestingly, partial inhibition of the histone acetyltransferase component of SAGA/SLIK extends lifespan but complete deletion does not (Huang et al., 2020). How these two observations are connected remains to be determined. Loss of the chromatin-modifying enzyme, *ISW2*, also leads to RLS extension, mimicking some of the stress responses of CR (Dang et al., 2014). H3K36 methylation also promotes longevity through improved transcriptional fidelity (Sen et al., 2015). Loss of the N-α-terminal acetyltransferase Nat4 also leads to enhanced RLS and may mimic CR (Sen et al., 2015). Finally, for CLS, epigenetic mechanisms play a role in mediating mitohormesis signals to enhance telomere-proximal silencing (Schroeder, Raimundo, & Shadel, 2013). Clearly, epigenetic control mechanisms play a major role in yeast aging, and it will be important to see how they are integrated. This is analogous to mammalian aging, where recent research has emphasized the role played by epigenetic pathways.

Calorie restriction in yeast

CR is a dietary regimen that encompasses decreased calorie intake while maintaining proper nutrition. This type of strategy is one of the best-studied approaches in the longevity studies and it has shown significant effects of lifespan extension across species (Komatsu et al., 2019). In budding yeast, CR is achieved by reducing the glucose content in the growth medium below the optimal concentration of 2%. There are two most commonly used regimens: moderate CR (0.2%–0.5% glucose) and extreme CR (0.02%–0.05% glucose) (Schleit, Wasko, & Kaeberlein, 2012). In addition, the carbon source can be changed to a nonfermentable one like glycerol. Restriction of amino acids has also been reported to extend the RLS; however, the mechanism has not been studied to a similar extent (Mirzaei, Suarez, & Longo, 2014).

In contrast, the mechanisms underlying RLS extension by glucose restriction in yeast have raised more interest and remain a vibrant area of investigation. The reduction of glucose levels leads to a metabolic switch from fermentative growth, in which the cells rely on glycolysis, to mitochondrial respiration. This switch is accompanied by changes in gene expression, dominated by a repression of glycolysis and an upregulation of the electron transport chain components. Since enhanced respiratory activity is characterized by prolongevity effects, the question of whether it was required for CR-induced RLS extension was addressed with conflicting results (Kaeberlein et al., 2005a; Lin et al., 2002). Similarly, whether Sir2 is required for lifespan extension by CR is complicated. At least in the absence of *FOB1*, a condition associated with reduced rDNA recombination, *SIR2* is not required (Kaeberlein, Kirkland, Fields, & Kennedy, 2004; Kaeberlein et al., 2006; Lamming et al., 2005, 2006). Divergent findings such as these point to the complexity by which CR extends lifespan and its dependence on experimental conditions.

CR either through reduction of carbohydrates or amino acids leads to reduced signaling through the mTOR pathway, a crucial mediator of RLS and CLS (Deprez, Eskes, Winderickx, & Wilms, 2018; Gonzalez and Hall, 2017). Inhibition of TORC1 complex controls many downstream processes, including cell proliferation, protein translation, and stress responses. It also leads to the activation of autophagy, a cellular process that degrades damaged organelles and other cellular components, including protein aggregates. The origin of damage can be stress or aging, and it is known that autophagy is hampered in old cells. In yeast, the autophagy genes are not essential for the CR-induced RLS extension (Schleit et al., 2013). Nevertheless, autophagy is indispensable for complete CR-mediated extension of CLS (Aris et al., 2013), similarly to findings in the roundworm *C. elegans* (Jia & Levine, 2007).

However, several pathways were confirmed to have an essential role in yeast CR-mediated RLS extension in yeast. One such example is protein kinase A (PKA). PKA is a 3′-5′-cyclic adenosine monophosphate (cAMP)-dependent protein kinase, activated by the presence of glucose (Kim, Roy, Jouandot, & Cho, 2013). cAMP is generated following glucose uptake from adenosine monophosphate (AMP) by the activity of the enzyme adenylate cyclase (Cyr1). It then activates PKA, which plays an important role in the regulation of cell metabolism and growth, as well as in stress responses. In yeast, two activators of Cyr1 exist: G-protein-coupled receptor Gpr1, as well as Gpa2, a nucleotide-binding regulatory protein. Deletion of any of these proteins extends RLS in yeast (Lin et al., 2002); however, their double deletion does not yield an additive effect. Similarly, the deletion of hexokinase 2, the enzyme in charge of glucose phosphorylation, leads to extended longevity in yeast. In yeast, cAMP production is stimulated by two RAS GTPase proteins, Ras1 and Ras2, which are in turn activated by Cdc25, a guanine nucleotide exchange factor. Mutation in Cdc25 is also able to extend RLS in yeast, as well as the deletion of Ras1 and overexpression of Ras2. Conversely, the addition of exogenous cAMP, constitutive activation of PKA, and overexpression of Cyr1 all lead to a decrease in RLS in yeast. Clearly, these results demonstrate the importance of glucose metabolism and signaling in yeast longevity in general, however they underscore the interaction of the PKA pathway with the CR. The PKA pathway also plays a significant role in CLS extension by CR (Dolz-Edo, van der Deen, Brul, & Smits, 2019).

Another pathway implicated in CR-mediated lifespan extension that has been shown to have unparalleled importance in aging research is the TOR pathway. Deletion of either *SCH9* or *TOR1* extends the RLS of yeast (Fabrizio, Pletcher, Minois, Vaupel, & Longo, 2004; Fabrizio et al., 2001b; Kaeberlein et al., 2005b), while altering protein translation, ribosome biogenesis, and the cell cycle. However, Sch9 is not the only TOR pathway-related protein whose absence extends RLS in yeast: Ure2, Rom2, Rpl31A are only some examples (Kaeberlein et al., 2005b; McCormick et al., 2015a). In the absence of Sch9 or Tor1, the CR is not able to extend the yeast RLS further, rendering them essential for the CR-mediated RLS extension.

The mechanisms of RLS extension due to TORC1 inhibition are an area of active research. One of the most explored avenues is the role of TORC1 in translation due to the prolongevity effects of deletions of multiple ribosomal proteins and translation factors. The role of the ribosome in yeast aging has been studied in greater detail and a surprising clue is that long-lived ribosomal protein deletion strains are mostly enriched in the large subunit, with further evidence pointing to translational induction of Gcn4 as a major mechanism explaining lifespan extension (Steffen et al., 2008, 2012). The Gcn4 transcription factor activates an amino acid starvation response that likely places cells in a stress-resistant state, compatible with lifespan extension (Hinnebusch, 2005). Among the downstream pathways stimulated by Gcn4 is autophagy, and a recent report suggests that induction of autophagy underlies Gcn4-mediated lifespan extension (Hu et al., 2018; Shen, Postnikoff, & Tyler, 2019). In addition to TORC1-dependent inhibition of translation, mitochondrial respiration activation is another notable event downstream of TORC1. It has previously been shown to be important in CR-mediated CLS extension (Powers et al., 2006). In the context of RLS, TORC1-dependent enhancement of mitochondrial respiration has been found to be essential for the heat shock-mediated RLS extension (Musa et al., 2018).

Adenosine monophosphate kinase

SNF1 encodes the yeast AMP-activated protein kinase, an ortholog of mammalian AMPK (Coccetti, Nicastro, & Tripodi, 2018). It is a serine/threonine kinase that is activated in conditions of low energy, that is, a high AMP/ATP ratio. Across species, AMPK controls multiple metabolic processes via different effectors (Gonzalez, Hall, Lin, & Hardie, 2020). In yeast, Snf1 has been shown to be involved in the RLS. Deactivation of Sip2, a negative regulator of Snf1, leads to a decreased RLS as a result of increased Snf1 activity (Ashrafi, Lin, Manchester, & Gordon, 2000). In optimal conditions, acetylation of Sip2 impairs Snf1 function. During replicative aging, Sip2 is progressively deacetylated, which increasingly activates Snf1, with detrimental effects during yeast aging (Lu et al., 2011). Consistently, mutations in Sip2 that prevent acetylation of the protein shorten the RLS, and the acetylation mimetic mutants of Sip2 are characterized by increased longevity. Sip2 appears not to be essential for the CR-mediated RLS extension: CR extended the RLS of the Sip2 null and the nonacetylatable mutants. At the same time, it did not further extend the RLS of the Sip2 acetylation mimetic. Similarly, loss of Sch9 could rescue the RLS of the Sip2 null and the nonacetylatable mutants.

Loss of proteostasis

Aging and, in particular, the occurrence of age-related diseases are tightly intertwined with impaired protein maintenance, or proteostasis. The role of proteostasis machinery is to preserve the native structure and function of cellular proteins, but also to ensure timely removal of terminally damaged proteins from the cytosolic pool, either by protein degradation and/or their aggregation. Proteasome activity declines with yeast replicative aging and, expectedly, increased proteasomal activity extends RLS in yeast (Kruegel et al., 2011). Analogously, many chaperones display reduced expression levels in old cells (Moreno et al., 2019).

It is easy to assume that a decline in the proteostasis network leads to increased protein misfolding that drives age-related cellular decline. However, another theory is that there may be specific regulatory targets of these pathways that directly modulate aging, and cases have been reported. For instance, the G1 cyclin Cln3 is highly sensitive to chaperone status, and loss of Cln3 in response to a decline in chaperones promotes protein aggregation and age-associated cell cycle arrest (Moreno et al., 2019). In the case of the proteasome, increased activity results in enhanced turnover of Mig1, a transcriptional repressor of Snf1 (described above), ultimately leading to induction of respiration and a mitohormetic response (Yao et al., 2015).

More recently, evidence has been gathered suggesting that mitochondrial activity and the regulation of several metabolic pathways (glycolysis, pentose phosphate pathway, TCA cycle) can be reprogrammed depending on the quality of cell-wide protein folding (Peric et al., 2017). The metabolic response to improved proteostasis relies on the consequent deactivation of the TORC1 complex, with downstream effects on Snf1/AMPK activity, and mitochondrial respiration (Fig. 8.3). The RLS extension triggered by the manipulations of proteostasis quality is no longer present if this crosstalk is disabled, that is, if the information on the proteostasis status is not reaching the metabolic regulators. This result raises the possibility that individual hallmarks may only be fully understood if considered as parts of a whole.

As mentioned above, in addition to the soluble machineries for the protein quality control, in certain conditions the cell deposits damaged proteins into insoluble compartments termed aggregates. Again, budding yeast has continuously been playing a crucial role in understanding the role of protein aggregation in aging and diseases. The process of protein aggregation was initially described as a process of seclusion of misfolded and damaged proteins from the cytosol, presumably in order to reduce their toxicity and facilitate their clearance from the cell via a process like autophagy. Seclusion of damaged proteins into the aggregates also facilitates their retention in the mother cell during the budding, and contributes to the rejuvenation of the bud. However, recent results suggest that the formation of protein aggregates in yeast under certain stress conditions (e.g., starvation, mild heat shock) can also be an adaptive response that contributes to cell survival in the stressful conditions (Saarikangas & Barral, 2016).

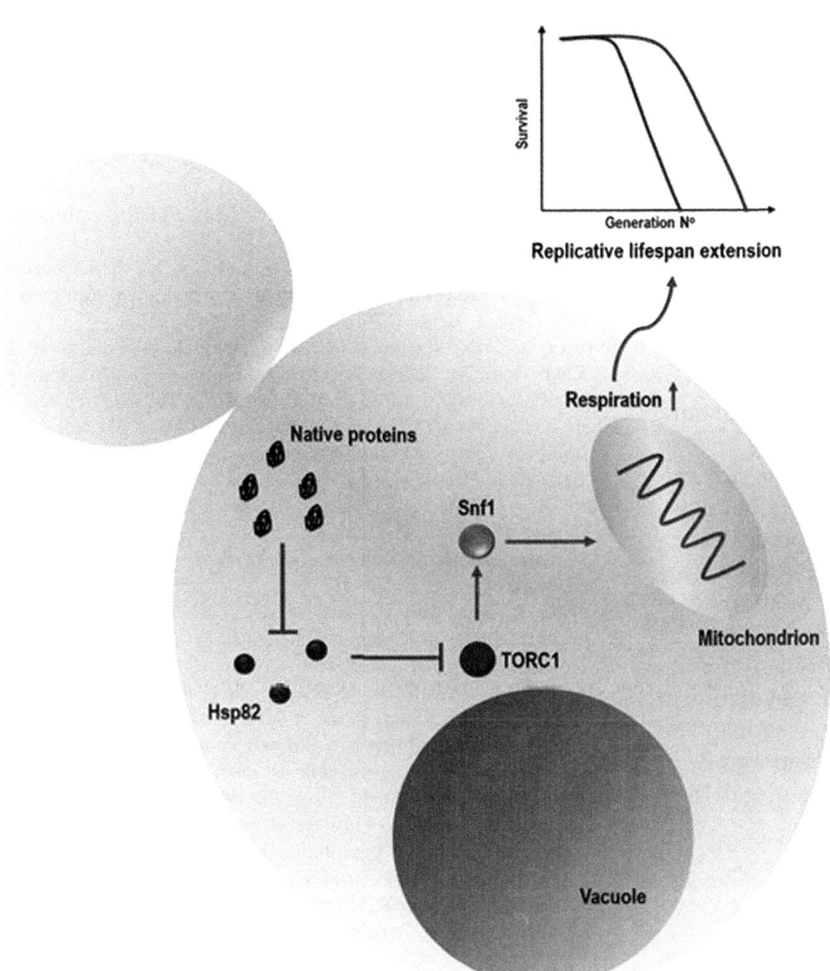

FIGURE 8.3 **Proteostasis status reprograms cellular metabolic activity.** The improved quality of protein folding reduces the levels of Hsp82 and is sensed via TORC1, leading to its deactivation. The sequential activation of Snf1 kinase results in the activation of respiration, thereby triggering replicative lifespan extension.

Systems biology and yeast aging

With the advent of RNA sequencing techniques, as well as more effective proteomics and metabolomics approaches, the understanding of the aging process in yeast has entered the era of systems biology (Crane et al., 2020). Indeed, this may be one major advantage of yeast as a model system going forward. Can a systems biology approach provide a holistic perspective of the yeast aging process, rather than the current pathway specific view?

Recently, Janssens et al. have presented a system-level model that integrates myriad information obtained by various researchers in the field over the decades of research (Janssens et al., 2015). In order to do so, they have combined computational methods with transcriptomics and proteomics approaches. The result is the first systems-level phenotype of replicatively aging yeast. Such an approach yielded a new insight: genes encoding for ribosomal proteins, tRNA synthesis and translational regulators have their transcript levels uncoupled from their protein levels. The authors propose that these changes are a root cause of aging, acting by triggering degenerative phenotypes, such as protein aggregation and increased cell size, accompanied by metabolic changes and hyperactivated stress responses. This is likely one of the first of many discoveries that will come from systems-level approaches to yeast aging.

Relevance of yeast in human aging and the future

In the early days of yeast aging research, many questioned whether replicative or CLS would have any informative value for human aging. At this point, however, it seems indisputable that yeast has introduced new pathways that are conserved in mammalian aging and provide several new insights into aging mechanisms. Yet, nonvertebrate models of aging appear to be at another crossroads, with increasing emphasis and funding targeted toward human studies. In fact, the nonvertebrates are at risk of becoming victims to their own success. Many of the interventions that we should rightly be testing in human aging have come from these models.

Moreover, with new technologies such as CRISPR/Cas9, mammalian studies are becoming more feasible. So what is the future of nonvertebrate species in aging research? Multicellular organisms like worms and flies have proven valuable recently as a means to discover noncell autonomous aging pathways, where genetic manipulation of one cell type affects aging in another location. As a single-celled organism, yeast does not avail itself to those studies. Systems biology, however, seems to be an open avenue to investigation, where yeast may again lead the way. Trying to completely understand aging in a multicellular organism may still be a step beyond, but already these approaches are being successfully applied to yeast. How do the known aging pathways weave together to control a complex process such as aging? Is it advantageous to have replicatively or chronologically old cells around? Will the final answers to the regulation of aging be beyond the ability of humans to understand? The importance of nonvertebrate models in aging research is not ending, but evolving to more complex and, perhaps even deeper, questions. One hopes that funding agencies recognize the importance of answering these new questions and do not forget their small but important friends.

Acknowledgments

A.K. is funded by the Heisenberg grant from the German Research Foundation (DFG). Yeast research in the lab of B.K.K. is supported by National Institutes of Health grants R01 GM123139 and R01 AG058742.

References

Aguilaniu, H., Gustafsson, L., Rigoulet, M., & Nystrom, T. (2003). Asymmetric inheritance of oxidatively damaged proteins during cytokinesis. *Science*, *299*, 1751–1753.

Aris, J. P., Alvers, A. L., Ferraiuolo, R. A., Fishwick, L. K., Hanvivatpong, A., Hu, D., ... Marraffini, M. (2013). Autophagy and leucine promote chronological longevity and respiration proficiency during calorie restriction in yeast. *Experimental Gerontology*, *48*, 1107–1119.

Ashrafi, K., Lin, S. S., Manchester, J. K., & Gordon, J. I. (2000). Sip2p and its partner snf1p affect aging in *S. cerevisiae*. *Genes & Development*, *14*, 1872–1885.

Ashrafi, K., Sinclair, D., Gordon, J. I., & Guarente, L. (1999). Passage through stationary phase advances replicative aging in *Saccharomyces cerevisiae*. *Proceedings of the National Academy of Sciences of the United States of America*, *96*, 9100–9105.

Baldi, S., Bolognesi, A., Meinema, A. C., & Barral, Y. (2017). Heat stress promotes longevity in budding yeast by relaxing the confinement of age-promoting factors in the mother cell. *eLife*, *6*, e28329.

Boettcher, B., Marquez-Lago, T. T., Bayer, M., Weiss, E. L., & Barral, Y. (2012). Nuclear envelope morphology constrains diffusion and promotes asymmetric protein segregation in closed mitosis. *The Journal of Cell Biology*, *197*, 921–937.

Brachmann, C. B., Sherman, J. M., Devine, S. E., Cameron, E. E., Pillus, L., & Boeke, J. D. (1995). The SIR2 gene family, conserved from bacteria to humans, functions in silencing, cell cycle progression, and chromosome stability. *Genes & Development*, *9*, 2888–2902.

Burnett, C., Valentini, S., Cabreiro, F., Goss, M., Somogyvari, M., Piper, M. D., ... McElwee, J. J. (2011). Absence of effects of Sir2 overexpression on lifespan in *C. elegans* and drosophila. *Nature*, *477*, 482–485.

Burtner, C. R., Murakami, C. J., Kennedy, B. K., & Kaeberlein, M. (2009). A molecular mechanism of chronological aging in yeast. *Cell Cycle*, *8*, 1256–1270.

Burtner, C. R., Murakami, C. J., Olsen, B., Kennedy, B. K., & Kaeberlein, M. (2011). A genomic analysis of chronological longevity factors in budding yeast. *Cell Cycle*, *10*, 1385–1396.

Chen, K. L., Crane, M. M., & Kaeberlein, M. (2017). Microfluidic technologies for yeast replicative lifespan studies. *Mechanisms of Ageing and Development*, *161*, 262–269.

Chen, K. L., Ven, T. N., Crane, M. M., Chen, D. E., Feng, Y. C., Suzuki, N., ... Kaeberlein, M. (2019). An inexpensive microscopy system for microfluidic studies in budding yeast. *Translational Medicine of Aging*, *3*, 52–56.

Coccetti, P., Nicastro, R., & Tripodi, F. (2018). Conventional and emerging roles of the energy sensor Snf1/AMPK in *Saccharomyces cerevisiae*. *Microbial Cell*, *5*, 482–494.

Crane, M. M., Russell, A. E., Schafer, B. J., Blue, B. W., Whalen, R., Almazan, J., ... Chen, K. L. (2019). DNA damage checkpoint activation impairs chromatin homeostasis and promotes mitotic catastrophe during aging. *eLife*, *8*, e50778.

Crane, M. M., Chen, K. L., Blue, B. W., & Kaeberlein, M. (2020). Trajectories of aging: How systems biology in yeast can illuminate mechanisms of personalized aging. *Proteomics*, *20*, 1800420. Available from: https://doi.org/10.1002/pmic.201800420.

Dang, W., Steffen, K. K., Perry, R., Dorsey, J. A., Johnson, F. B., Shilatifard, A., ... Berger, S. L. (2009). Histone H4 lysine 16 acetylation regulates cellular lifespan. *Nature*, *459*, 802–807.

Dang, W., Sutphin, G. L., Dorsey, J. A., Otte, G. L., Cao, K., Perry, R. M., ... Tsuchiyama, S. (2014). Inactivation of yeast Isw2 chromatin remodeling enzyme mimics longevity effect of calorie restriction via induction of genotoxic stress response. *Cell Metabolism*, *19*, 952–966.

D'Aquila, P., Montesanto, A., De Rango, F., Guarasci, F., Passarino, G., & Bellizzi, D. (2019). Epigenetic signature: Implications for mitochondrial quality control in human aging. *Aging*, *11*, 1240–1251.

Delaney, J. R., Murakami, C., Chou, A., Carr, D., Schleit, J., Sutphin, G. L., ... Goswami, S. (2013). Dietary restriction and mitochondrial function link replicative and chronological aging in *Saccharomyces cerevisiae*. *Experimental Gerontology*, *48*, 1006–1013.

Deprez, M. A., Eskes, E., Winderickx, J., & Wilms, T. (2018). The TORC1-Sch9 pathway as a crucial mediator of chronological lifespan in the yeast *Saccharomyces cerevisiae*. *FEMS Yeast Research*, *18*, foy048.

Derbyshire, M. K., Weinstock, K. G., & Strathern, J. N. (1996). HST1, a new member of the SIR2 family of genes. *Yeast*, *12*, 631–640.

Dohrmann, P. R., Butler, G., Tamai, K., Dorland, S., Greene, J. R., Thiele, D. J., & Stillman, D. J. (1992). Parallel pathways of gene regulation: Homologous regulators SWI5 and ACE2 differentially control transcription of HO and chitinase. *Genes & Development*, *6*, 93–104.

Dolz-Edo, L., van der Deen, M., Brul, S., & Smits, G. J. (2019). Caloric restriction controls stationary phase survival through Protein Kinase A (PKA) and cytosolic pH. *Aging Cell*, *18*, e12921.

Erjavec, N., & Nystrom, T. (2007). Sir2p-dependent protein segregation gives rise to a superior reactive oxygen species management in the progeny of *Saccharomyces cerevisiae*. *Proceedings of the*

National Academy of Sciences of the United States of America, *104*, 10877–10881.

Fabrizio, P., Liou, L. L., Moy, V. N., Diaspro, A., Valentine, J. S., Gralla, E. B., & Longo, V. D. (2003). SOD2 functions downstream of Sch9 to extend longevity in yeast. *Genetics*, *163*, 35–46.

Fabrizio, P., Pletcher, S. D., Minois, N., Vaupel, J. W., & Longo, V. D. (2004). Chronological aging-independent replicative life span regulation by Msn2/Msn4 and Sod2 in *Saccharomyces cerevisiae*. *FEBS Letters*, *557*, 136–142.

Fabrizio, P., Pozza, F., Pletcher, S. D., Gendron, C. M., & Longo, V. D. (2001a). Regulation of longevity and stress resistance by Sch9 in yeast. *Science*, *292*, 288–290.

Fabrizio, P., Pozza, F., Pletcher, S. D., Gendron, C. M., & Longo, V. D. (2001b). Regulation of longevity and stress resistance by Sch9 in yeast. *Science*, *292*, 288–290.

Falcon, A. A., & Aris, J. P. (2003). Plasmid accumulation reduces life span in *Saccharomyces cerevisiae*. *The Journal of Biological Chemistry*, *278*, 41607–41617.

Feser, J., Truong, D., Das, C., Carson, J. J., Kieft, J., Harkness, T., & Tyler, J. K. (2010). Elevated histone expression promotes life span extension. *Molecular Cell*, *39*, 724–735.

Fine, R. D., Maqani, N., Li, M., Franck, E., & Smith, J. S. (2019). Depletion of limiting rDNA structural complexes triggers chromosomal instability and replicative aging of *Saccharomyces cerevisiae*. *Genetics*, *212*, 75–91.

Ganley, A. R., Ide, S., Saka, K., & Kobayashi, T. (2009). The effect of replication initiation on gene amplification in the rDNA and its relationship to aging. *Molecular Cell*, *35*, 683–693.

Gehlen, L. R., Nagai, S., Shimada, K., Meister, P., Taddei, A., & Gasser, S. M. (2011). Nuclear geometry and rapid mitosis ensure asymmetric episome segregation in yeast. *Current Biology*, *21*, 25–33.

Gershon, H., & Gershon, D. (2000). The budding yeast, *Saccharomyces cerevisiae*, as a model for aging research: A critical review. *Mechanisms of Ageing and Development*, *120*, 1–22.

Gershon, H., & Gershon, D. (2001). Critical assessment of paradigms in aging research. *Experimental Gerontology*, *36*, 1035–1047.

Ghidelli, S., Donze, D., Dhillon, N., & Kamakaka, R. T. (2001). Sir2p exists in two nucleosome-binding complexes with distinct deacetylase activities. *The EMBO Journal*, *20*, 4522–4535.

Giblin, W., Skinner, M. E., & Lombard, D. B. (2014). Sirtuins: Guardians of mammalian healthspan. *Trends in Genetics*, *30*, 271–286.

Gonzalez, A., & Hall, M. N. (2017). Nutrient sensing and TOR signaling in yeast and mammals. *The EMBO Journal*, *36*, 397–408.

Gonzalez, A., Hall, M. N., Lin, S. C., & Hardie, D. G. (2020). AMPK and TOR: The Yin and Yang of cellular nutrient sensing and growth control. *Cell Metabolism*, *31*, 472–492.

He, C., Tsuchiyama, S. K., Nguyen, Q. T., Plyusnina, E. N., Terrill, S. R., Sahibzada, S., ... Tian, R. (2014). Enhanced longevity by ibuprofen, conserved in multiple species, occurs in yeast through inhibition of tryptophan import. *PLoS Genetics*, *10*, e1004860.

He, C., Zhou, C., & Kennedy, B. K. (2018). The yeast replicative aging model. *Biochimica et Biophysica Acta Molecular Basis of Disease*, *1864*, 2690–2696.

Hendrickson, D. G., Soifer, I., Wranik, B. J., Kim, G., Robles, M., Gibney, P. A., & McIsaac, R. S. (2018). A new experimental platform facilitates assessment of the transcriptional and chromatin landscapes of aging yeast. *Elife*, *7*, e39911.

Herranz, D., Munoz-Martin, M., Canamero, M., Mulero, F., Martinez-Pastor, B., Fernandez-Capetillo, O., & Serrano, M. (2010). Sirt1 improves healthy ageing and protects from metabolic syndrome-associated cancer. *Nature Communications*, *1*, 3.

Hinnebusch, A. G. (2005). Translational regulation of GCN4 and the general amino acid control of yeast. *Annual Review of Microbiology*, *59*, 407–450.

Hu, Z., Xia, B., Postnikoff, S. D., Shen, Z. J., Tomoiaga, A. S., Harkness, T. A., ... Tyler, J. K. (2018). Ssd1 and Gcn2 suppress global translation efficiency in replicatively aged yeast while their activation extends lifespan. *eLife*, *7*, e35551.

Huang, B., Zhong, D., Zhu, J., An, Y., Gao, M., Zhu, S., ... Xie, Z. (2020). Inhibition of histone acetyltransferase GCN5 extends lifespan in both yeast and human cell lines. *Aging Cell*, *19*, e13129.

Huberts, D. H., Janssens, G. E., Lee, S. S., Vizcarra, I. A., & Heinemann, M. (2013). Continuous high-resolution microscopic observation of replicative aging in budding yeast. *Journal of Visualized Experiments*, *78*, e50143.

Imai, S., Armstrong, C. M., Kaeberlein, M., & Guarente, L. (2000). Transcriptional silencing and longevity protein Sir2 is an NAD-dependent histone deacetylase. *Nature*, *403*, 795–800.

Imai, S. I., & Guarente, L. (2014). NAD and sirtuins in aging and disease. *Trends in Cell Biology*, *24*, 464–471.

Ivy, J. M., Klar, A. J., & Hicks, J. B. (1986). Cloning and characterization of four SIR genes of *Saccharomyces cerevisiae*. *Molecular and Cellular Biology*, *6*, 688–702.

Janssens, G. E., Meinema, A. C., Gonzalez, J., Wolters, J. C., Schmidt, A., Guryev, V., ... Heinemann, M. (2015). Protein biogenesis machinery is a driver of replicative aging in yeast. *eLife*, *4*, e08527.

Jia, K., & Levine, B. (2007). Autophagy is required for dietary restriction-mediated life span extension in C. elegans. *Autophagy*, *3*, 597–599.

Jin, M., Li, Y., O'Laughlin, R., Bittihn, P., Pillus, L., Tsimring, L. S., ... Hao, N. (2019). Divergent aging of isogenic yeast cells revealed through single-cell phenotypic dynamics. *Cell Systems*, *8*, 242–253, e243.

Johnson, S. C. (2018). Nutrient sensing, signaling and ageing: The role of IGF-1 and mTOR in ageing and age-related disease. *Subcellular Biochemistry*, *90*, 49–97.

Kaeberlein, M., Hu, D., Kerr, E. O., Tsuchiya, M., Westman, E. A., Dang, N., ... Kennedy, B. K. (2005a). Increased life span due to calorie restriction in respiratory deficient yeast. *PLoS Genetics*, *1*, 614–621.

Kaeberlein, M., Kirkland, K. T., Fields, S., & Kennedy, B. K. (2004). Sir2-independent life span extension by calorie restriction in yeast. *PLoS Biology*, *2*, 1381–1387.

Kaeberlein, M., McVey, M., & Guarente, L. (1999). The SIR2/3/4 complex and SIR2 alone promote longevity in *Saccharomyces cerevisiae* by two different mechanisms. *Genes & Development*, *13*, 2570–2580.

Kaeberlein, M., Powers, R. W., 3rd, Steffen, K. K., Westman, E. A., Hu, D., Dang, N., ... Kennedy, B. K. (2005b). Regulation of yeast replicative life span by TOR and Sch9 in response to nutrients. *Science*, *310*, 1193–1196.

Kaeberlein, M., Steffen, K. K., Hu, D., Dang, N., Kerr, E. O., Tsuchiya, M., ... Kennedy, B. K. (2006). Comment on "*HST2* mediates *SIR2*-independent life-span extension by calorie restriction". *Science*, *312*, 1312b.

Kanfi, Y., Naiman, S., Amir, G., Peshti, V., Zinman, G., Nahum, L., ... Cohen, H. Y. (2012). The sirtuin SIRT6 regulates lifespan in male mice. *Nature*, *483*, 218–221.

Kennedy, B. K., Austriaco, N. R., Zhang, J., & Guarente, L. (1995). Mutation in the silencing gene SIR4 can delay aging in S. cerevisiae. *Cell*, *80*, 485–496.

Kennedy, B. K., Berger, S. L., Brunet, A., Campisi, J., Cuervo, A. M., Epel, E. S., ... Pessin, J. E. (2014). Geroscience: Linking aging to chronic disease. *Cell*, *159*, 709–713.

Kennedy, B. K., Gotta, M., Sinclair, D. A., Mills, K., McNabb, D. S., Murthy, M., ... Guarente, L. (1997). Redistribution of silencing proteins from telomeres to the nucleolus is associated with extension of life span in *S. cerevisiae*. *Cell*, *89*, 381–391.

Kennedy, B. K., & Lamming, D. W. (2016). The mechanistic target of rapamycin: The grand ConducTOR of metabolism and aging. *Cell Metabolism*, *23*, 990–1003.

Khmelinskii, A., Keller, P. J., Lorenz, H., Schiebel, E., & Knop, M. (2010). Segregation of yeast nuclear pores. *Nature*, *466*, E1.

Kim, J. H., Roy, A., Jouandot, D., 2nd, & Cho, K. H. (2013). The glucose signaling network in yeast. *Biochimica et Biophysica Acta*, *1830*, 5204–5210.

Knorre, D. A., Azbarova, A. V., Galkina, K. V., Feniouk, B. A., & Severin, F. F. (2018). Replicative aging as a source of cell heterogeneity in budding yeast. *Mechanisms of Ageing and Development*, *176*, 24–31.

Komatsu, T., Park, S., Hayashi, H., Mori, R., Yamaza, H., & Shimokawa, I. (2019). Mechanisms of calorie restriction: A review of genes required for the life-extending and tumor-inhibiting effects of calorie restriction. *Nutrients*, *11*, 3068.

Kruegel, U., Robison, B., Dange, T., Kahlert, G., Delaney, J. R., Kotireddy, S., ... Schleit, J. (2011). Elevated proteasome capacity extends replicative lifespan in *Saccharomyces cerevisiae*. *PLoS Genetics*, *7*, e1002253.

Kwan, E. X., Foss, E. J., Tsuchiyama, S., Alvino, G. M., Kruglyak, L., Kaeberlein, M., ... Bedalov, A. (2013). A natural polymorphism in rDNA replication origins links origin activation with calorie restriction and lifespan. *PLoS Genetics*, *9*, e1003329.

Lamming, D. W., Latorre-Esteves, M., Medvedik, O., Wong, S. N., Tsang, F. A., Wang, C., ... Sinclair, D. A. (2005). HST2 mediates SIR2-independent life-span extension by calorie restriction. *Science*, *309*, 1861–1864.

Lamming, D. W., Latorre-Esteves, M., Medvedik, O., Wong, S. N., Tsang, F. A., Wang, C., ... Sinclair, D. A. (2006). Response to comment on "HST2 mediates SIR2-independent life-span extension by calorie restriction". *Science*, *312*, 1312c.

Landry, J., Slama, J. T., & Sternglanz, R. (2000). Role of NAD(+) in the deacetylase activity of the SIR2-like proteins. *Biochemical and Biophysical Research Communications*, *278*, 685–690.

Laun, P., Pichova, A., Madeo, F., Fuchs, J., Ellinger, A., Kohlwein, S., ... Breitenbach, M. (2001). Aged mother cells of *Saccharomyces cerevisiae* show markers of oxidative stress and apoptosis. *Molecular Microbiology*, *39*, 1166–1173.

Lee, S. S., Avalos Vizcarra, I., Huberts, D. H., Lee, L. P., & Heinemann, M. (2012a). Whole lifespan microscopic observation of budding yeast aging through a microfluidic dissection platform. *Proceedings of the National Academy of Sciences of the United States of America*, *109*, 4916–4920.

Lee, S. S., Vizcarra, I. A., Huberts, D. H., Lee, L. P., & Heinemann, M. (2012b). Whole lifespan microscopic observation of budding yeast aging through a microfluidic dissection platform. *Proceedings of the National Academy of Sciences of the United States of America*, *109*, 4916–4920.

Leupold, S., Hubmann, G., Litsios, A., Meinema, A. C., Takhaveev, V., Papagiannakis, A., ... Heinemann, M. (2019). Saccharomyces cerevisiae goes through distinct metabolic phases during its replicative lifespan. *eLife*, *8*, e41046.

Lin, S. J., Defossez, P. A., & Guarente, L. (2000). Requirement of NAD and SIR2 for life-span extension by calorie restriction in *Saccharomyces cerevisiae*. *Science*, *289*, 2126–2128.

Lin, S. J., Kaeberlein, M., Andalis, A. A., Sturtz, L. A., Defossez, P. A., Culotta, V. C., ... Guarente, L. (2002). Calorie restriction extends *Saccharomyces cerevisiae* lifespan by increasing respiration. *Nature*, *418*, 344–348.

Lindstrom, D. L., & Gottschling, D. E. (2009). The mother enrichment program: A genetic system for facile replicative life span analysis in *Saccharomyces cerevisiae*. *Genetics*, *183*, 413–422, 411SI-413SI.

Liu, B., Larsson, L., Caballero, A., Hao, X., Oling, D., Grantham, J., & Nystrom, T. (2010). The polarisome is required for segregation and retrograde transport of protein aggregates. *Cell*, *140*, 257–267.

Long, R. M., Singer, R. H., Meng, X., Gonzalez, I., Nasmyth, K., & Jansen, R. P. (1997). Mating type switching in yeast controlled by asymmetric localization of ASH1 mRNA. *Science*, *277*, 383–387.

Longo, V. D., Ellerby, L. M., Bredesen, D. E., Valentine, J. S., & Gralla, E. B. (1997). Human Bcl-2 reverses survival defects in yeast lacking superoxide dismutase and delays death of wild-type yeast. *Journal of Cell Biology*, *137*, 1581–1588.

Longo, V. D., Gralla, E. B., & Valentine, J. S. (1996). Superoxide dismutase activity is essential for stationary phase survival in *Saccharomyces cerevisiae*. Mitochondrial production of toxic oxygen species in vivo. *Journal of Biological Chemistry*, *271*, 12275–12280.

Longo, V. D., Liou, L. L., Valentine, J. S., & Gralla, E. B. (1999). Mitochondrial superoxide decreases yeast survival in stationary phase. *Archives of Biochemistry and Biophysics*, *365*, 131–142.

Longo, V. D., Shadel, G. S., Kaeberlein, M., & Kennedy, B. (2012). Replicative and chronological aging in *Saccharomyces cerevisiae*. *Cell Metabolism*, *16*, 18–31.

Lopez-Otin, C., Blasco, M. A., Partridge, L., Serrano, M., & Kroemer, G. (2013). The hallmarks of aging. *Cell*, *153*, 1194–1217.

Lu, J. Y., Lin, Y. Y., Sheu, J. C., Wu, J. T., Lee, F. J., Chen, Y., ... Berger, S. L. (2011). Acetylation of yeast AMPK controls intrinsic aging independently of caloric restriction. *Cell*, *146*, 969–979.

Mannick, J. B., Del Giudice, G., Lattanzi, M., Valiante, N. M., Praestgaard, J., Huang, B., ... Carson, S. (2014). mTOR inhibition improves immune function in the elderly. *Science Translational Medicine*, *6*, 268ra179.

Mannick, J. B., Morris, M., Hockey, H. P., Roma, G., Beibel, M., Kulmatycki, K., ... Quinn, D. (2018). TORC1 inhibition enhances immune function and reduces infections in the elderly. *Science Translational Medicine*, *10*, eaaq1564.

McCormick, M. A., Delaney, J. R., Tsuchiya, M., Tsuchiyama, S., Shemorry, A., Sim, S., ... Murakami, C. J. (2015a). A comprehensive analysis of replicative lifespan in 4,698 single-gene deletion strains uncovers conserved mechanisms of aging. *Cell Metabolism*, *22*, 895–906.

McCormick, M. A., Delaney, J. R., Tsuchiya, M., Tsuchiyama, S., Shemorry, A., Sim, S., ... Murakami, C. J. (2015b). A comprehensive analysis of replicative lifespan in 4,698 single-gene deletion strains uncovers conserved mechanisms of aging. *Cell Metabolism*, *22*, 895–906.

McCormick, M. A., Mason, A. G., Guyenet, S. J., Dang, W., Garza, R. M., Ting, M. K., ... Pillus, L. (2014). The SAGA histone deubiquitinase module controls yeast replicative lifespan via Sir2 interaction. *Cell Reports*, *8*, 477–486.

McFaline-Figueroa, J. R., Vevea, J., Swayne, T. C., Zhou, C., Liu, C., Leung, G., ... Pon, L. A. (2011). Mitochondrial quality control during inheritance is associated with lifespan and mother-daughter age asymmetry in budding yeast. *Aging Cell*, *10*, 885–895.

Meitinger, F., Khmelinskii, A., Morlot, S., Kurtulmus, B., Palani, S., Andres-Pons, A., ... Pereira, G. (2014). A memory system of negative polarity cues prevents replicative aging. *Cell*, *159*, 1056–1069.

Melov, S., Schneider, J. A., Day, B. J., Hinerfeld, D., Coskun, P., Mirra, S. S., ... Wallace, D. C. (1998). A novel neurological phenotype in mice lacking mitochondrial manganese superoxide dismutase. *Nature Genetics*, *18*, 159–163.

Miller, S. B., Mogk, A., & Bukau, B. (2015). Spatially organized aggregation of misfolded proteins as cellular stress defense strategy. *Journal of Molecular Biology*, *427*, 1564–1574.

Mirzaei, H., Suarez, J. A., & Longo, V. D. (2014). Protein and amino acid restriction, aging and disease: From yeast to humans. *Trends in Endocrinology and Metabolism: TEM*, *25*, 558–566.

Moreno, D. F., Jenkins, K., Morlot, S., Charvin, G., Csikasz-Nagy, A., & Aldea, M. (2019). Proteostasis collapse, a hallmark of aging, hinders the chaperone-Start network and arrests cells in G1. *eLife*, *8*, e48240.

Moretti, P., Freeman, K., Coodly, L., & Shore, D. (1994). Evidence that a complex of SIR proteins interacts with the silencer and telomere-binding protein RAP1. *Genes & Development*, *8*, 2257–2269.

Mortimer, R. K., & Johnston, J. R. (1959). Life span of individual yeast cells. *Nature*, *183*, 1751–1752.

Murray, A. W., & Szostak, J. W. (1983). Construction of artificial chromosomes in yeast. *Nature*, *305*, 189–193.

Musa, M., Peric, M., Bou Dib, P., Sobocanec, S., Saric, A., Lovric, A., ... Vlahovicek, K. (2018). Heat-induced longevity in budding yeast requires respiratory metabolism and glutathione recycling. *Aging*, *10*, 2407–2427.

Nestelbacher, R., Laun, P., Vondrakova, D., Pichova, A., Schuller, C., & Breitenbach, M. (2000). The influence of oxygen toxicity on yeast mother cell-specific aging. *Experimental Gerontology*, *35*, 63–70.

Novarina, D., Mavrova, S. N., Janssens, G. E., Rempel, I. L., Veenhoff, L. M., & Chang, M. (2017). Increased genome instability is not accompanied by sensitivity to DNA damaging agents in aged yeast cells. *DNA Repair*, *54*, 1–7.

Orlandi, I., Bettiga, M., Alberghina, L., Nyström, T., & Vai, M. (2010). Sir2-dependent asymmetric segregation of damaged proteins in ubp10 null mutants is independent of genomic silencing. *Biochimica et Biophysica Acta*, *1803*, 630–638.

Peric, M., Lovric, A., Saric, A., Musa, M., Bou Dib, P., Rudan, M., ... Dennerlein, S. (2017). TORC1-mediated sensing of chaperone activity alters glucose metabolism and extends lifespan. *Aging Cell*, *16*, 994–1005.

Powers, R. W., 3rd, Kaeberlein, M., Caldwell, S. D., Kennedy, B. K., & Fields, S. (2006). Extension of chronological life span in yeast by decreased TOR pathway signaling. *Genes & Development*, *20*, 174–184.

Rempel, I. L., Crane, M. M., Thaller, D. J., Mishra, A., Jansen, D. P., Janssens, G., ... van der Giessen, E. (2019). Age-dependent deterioration of nuclear pore assembly in mitotic cells decreases transport dynamics. *eLife*, *8*, e48186.

Rine, J., & Herskowitz, I. (1987). Four genes responsible for a position effect on expression from HML and HMR in *Saccharomyces cerevisiae*. *Genetics*, *116*, 9–22.

Saarikangas, J., & Barral, Y. (2016). Protein aggregation as a mechanism of adaptive cellular responses. *Curr Genet*, *62*, 711–724. Available from: https://doi.org/10.1007/s00294-016-0596-0.

Sarnoski, E. A., Liu, P., & Acar, M. (2017). A high-throughput screen for yeast replicative lifespan identifies lifespan-extending compounds. *Cell Reports*, *21*, 2639–2646.

Schleit, J., Johnson, S. C., Bennett, C. F., Simko, M., Trongtham, N., Castanza, A., ... Delaney, J. R. (2013). Molecular mechanisms underlying genotype-dependent responses to dietary restriction. *Aging Cell*, *12*, 1050–1061.

Schleit, J., Wasko, B. M., & Kaeberlein, M. (2012). Yeast as a model to understand the interaction between genotype and the response to calorie restriction. *FEBS Letters*, *586*, 2868–2873.

Schroeder, E. A., Raimundo, N., & Shadel, G. S. (2013). Epigenetic silencing mediates mitochondria stress-induced longevity. *Cell Metabolism*, *17*, 954–964.

Sen, P., Dang, W., Donahue, G., Dai, J., Dorsey, J., Cao, X., ... Lee, J. Y. (2015). H3K36 methylation promotes longevity by enhancing transcriptional fidelity. *Genes & Development*, *29*, 1362–1376.

Shcheprova, Z., Baldi, S., Frei, S. B., Gonnet, G., & Barral, Y. (2008). A mechanism for asymmetric segregation of age during yeast budding. *Nature*, *454*, 728–734.

Shen, Z. J., Postnikoff, S., & Tyler, J. K. (2019). Is Gcn4-induced autophagy the ultimate downstream mechanism by which hormesis extends yeast replicative lifespan? *Current Genetics*, *65*, 717–720.

Sinclair, D. A., & Guarente, L. (1997). Extrachromosomal rDNA circles–a cause of aging in yeast. *Cell*, *91*, 1033–1042.

Smeal, T., Claus, J., Kennedy, B., Cole, F., & Guarente, L. (1996). Loss of transcriptional silencing causes sterility in old mother cells of *S. cerevisiae*. *Cell*, *84*, 633–642.

Smith, E. D., Tsuchiya, M., Fox, L. A., Dang, N., Hu, D., Kerr, E. O., ... Welton, K. L. (2008). Quantitative evidence for conserved longevity pathways between divergent eukaryotic species. *Genome Research*, *18*, 564–570.

Steffen, K. K., Kennedy, B. K., & Kaeberlein, M. (2009). Measuring replicative lifespan in budding yeast. *Journal of Visualized Experiments*, *28*, 1209.

Steffen, K. K., MacKay, V. L., Kerr, E. O., Tsuchiya, M., Hu, D., Fox, L. A., ... Tchao, B. N. (2008). Yeast life span extension by depletion of 60s ribosomal subunits is mediated by Gcn4. *Cell*, *133*, 292–302.

Steffen, K. K., McCormick, M. A., Pham, K. M., MacKay, V. L., Delaney, J. R., Murakami, C. J., ... Kennedy, B. K. (2012). Ribosome deficiency protects against ER stress in *Saccharomyces cerevisiae*. *Genetics*, *191*, 107–118.

Stumpferl, S. W., Brand, S. E., Jiang, J. C., Korona, B., Tiwari, A., Dai, J., ... Jazwinski, S. M. (2012). Natural genetic variation in yeast longevity. *Genome Research*, *22*, 1963–1973.

Takizawa, P. A., Sil, A., Swedlow, J. R., Herskowitz, I., & Vale, R. D. (1997). Actin-dependent localization of an RNA encoding a cell-fate determinant in yeast. *Nature*, *389*, 90–93.

Tanny, J. C., & Moazed, D. (2001). Coupling of histone deacetylation to NAD breakdown by the yeast silencing protein Sir2: Evidence for acetyl transfer from substrate to an NAD breakdown product. *Proceedings of the National Academy of Sciences of the United States of America*, *98*, 415–420.

Tissenbaum, H. A., & Guarente, L. (2001). Increased dosage of a sir-2 gene extends lifespan in *Caenorhabditis elegans*. *Nature*, *410*, 227–230.

Viswanathan, M., & Guarente, L. (2011). Regulation of *Caenorhabditis elegans* lifespan by sir-2.1 transgenes. *Nature*, *477*, E1–E2.

Wasko, B. M., & Kaeberlein, M. (2014). Yeast replicative aging: A paradigm for defining conserved longevity interventions. *FEMS Yeast Research*, *14*, 148–159.

Werner-Washburne, M., Braun, E. L., Crawford, M. E., & Peck, V. M. (1996). Stationary phase in *Saccharomyces cerevisiae*. *Molecular Microbiology*, *19*, 1159–1166.

Winey, M., Yarar, D., Giddings, T. H., Jr., & Mastronarde, D. N. (1997). Nuclear pore complex number and distribution throughout the *Saccharomyces cerevisiae* cell cycle by three-dimensional reconstruction from electron micrographs of nuclear envelopes. *Molecular Biology of the Cell*, *8*, 2119–2132.

Xie, Z., Zhang, Y., Zou, K., Brandman, O., Luo, C., Ouyang, Q., & Li, H. (2012). Molecular phenotyping of aging in single yeast cells using a novel microfluidic device. *Aging Cell*, *11*, 599–606.

Yao, Y., Tsuchiyama, S., Yang, C., Bulteau, A. L., He, C., Robison, B., ... Tar, K. (2015). Proteasomes, Sir2, and Hxk2 form an interconnected aging network that impinges on the AMPK/Snf1-regulated transcriptional repressor Mig1. *PLoS Genetics*, *11*, e1004968.

Young, T. Z., Liu, P., Urbonaite, G., & Acar, M. (2019). Quantitative insights into age-associated DNA-repair inefficiency in single cells. *Cell Reports*, *28*, 2220–2230, e2227.

Zhang, Y., Luo, C., Zou, K., Xie, Z., Brandman, O., Ouyang, Q., & Li, H. (2012). Single cell analysis of yeast replicative aging using a new generation of microfluidic device. *PLoS One*, *7*, e48275.

Zou, K., Ren, D. S., Ou-Yang, Q., Li, H., & Zheng, J. (2017). Using microfluidic devices to measure lifespan and cellular phenotypes in single budding yeast cells. *J Vis Exp*, *121*, 55412.

CHAPTER

9

Model organisms (invertebrates)

Erin Munkácsy[1,2] and Andrew M. Pickering[1,2,3,4]

[1]Barshop Institute for Longevity and Aging Studies, UT Health, San Antonio, TX, United States [2]Department of Molecular Medicine, UT Health, San Antonio, TX, United States [3]Center for Neurodegeneration and Experimental Therapeutics, Department of Neurology, The University of Alabama at Birmingham, Birmingham, AL, United States [4]Department of Neurobiology, The University of Alabama at Birmingham, Birmingham, AL, United States

OUTLINE

Introduction

Invertebrate model organisms represent an invaluable tool for biology of aging research. They possess a number of advantages over vertebrate models in aging studies, including short lifespan, small size (a key advantage for large-scale lifespan investigations), ease of propagation, transparent bodies, and powerful genetics. These organisms also present advantages as simplified animal models enabling the study of individual processes. Studies using invertebrate models to investigate the biology of aging have primarily used

Handbook of the Biology of Aging.
DOI: https://doi.org/10.1016/B978-0-12-815962-0.00009-3

the nematode worm *Caenorhabditis elegans* and the fruit fly *Drosophila melanogaster*. In this chapter, we discuss the relative benefits and challenges in using these models as well as the tools these organisms contribute to aging research. Finally, a number of other invertebrate models have been developed, each with its own key advantages. These other invertebrate models are described in the third part of the chapter along with research done in these systems in the biology of aging.

Caenorhabitis elegans

Phylum: Nematoda
Class: Chromadorea
Order: Rhabditida
Family: *Rhabditidae*

History as a model organism

Nematodes, or roundworms, have been used to study basic biological processes since the 19th century. The parasitic *Ascaris* can be considered the first model organism as *Ascaris* eggs were used in early studies of development and contributed to understanding fertilization and meiosis (Nigon & Félix, 2017).

C. elegans was first described by Émile Maupas in 1900 and, along with the closely related *Caenorhabditis briggsae*, were studied as model organisms by researchers such as Victor Nigon and Ellsworth Dougherty in the 1940s and 1950s. Much early work was necessarily focused on establishing reproducible culture conditions. *Caenorhabditis* species are normally found in decaying vegetable matter where they feed on bacteria. Considerable effort went toward developing axenic culture media as had been established for *D. melanogaster*. In addition to much foundational work, these researchers studied genetic variation in heat resistance and isolated both dwarf and slow-growing mutants. Unfortunately, their work was underappreciated by the field at the time and lack of funding forced them to direct their research elsewhere (Nigon & Félix, 2017).

In the 1960s, the bacterial geneticist Sydney Brenner was interested in finding a model to study developmental neurobiology. *C. elegans* was very attractive in this regard as it is one of the simplest organisms with a nervous system. He generated and characterized several hundred mutant lines, which he published in his magnum opus in 1974 in *Genetics* (Brenner, 1974).

The potential of *C. elegans* as a model for the biology of aging was quickly seen by scientists such as Michael Klass, who described phenotypes of aging in *C. elegans* and identified a number of manipulations which altered lifespan (Klass, 1977) and subsequently isolated several long-lived mutants (Klass, 1983). These lines were found to all have mutations in the same gene, which was therefore named *age-1* (Friedman & Johnson, 1988). This gene was eventually characterized as coding a catalytic subunit of phosphatidylinositol 3-kinase, which acts downstream of the insulin-like growth factor receptor *daf-2* and upstream of the FOXO transcription factor *daf-16*, both of which have been shown to significantly impact lifespan (Kenyon et al., 1993).

Development and aging

Caenorhabditis species are small translucent nonparasitic nematodes that are ubiquitous in bacteria-rich environments in temperate regions. They are unsegmented and simple animals, lacking respiratory and circulatory systems. While many species are dioecious (having distinct female and male individuals), those most commonly used as model organisms, *C. briggsae* and *C. elegans*, are androdioecious (having male and hermaphroditic sexes). The majority of *C. elegans* in a population are hermaphrodites; they develop through four larval stages and produce sperm in the fourth larval stage before maturing and switching to producing oocytes. As with most ectotherms, development time varies greatly with temperature, but at 20°C, development from egg to mature adult takes 3 days. Cell fate is predetermined and the lineage from a single cell egg to the 959 cells that compose the eutelic (post-mitotic) adult soma of the hermaphrodite (males have 1031 somatic cells) has been mapped out. When food is scarce or population density high, worms can arrest development at the first development stage (L1) or enter an alternate third stage of larval development called the dauer stage. While dauer, the worms do not eat and take on a slender and fast-moving morphology. Dauer worms are highly stress resistant and can survive for extended time periods. When conditions become amenable to growth, the worms exit the dauer stage and resume development from L4 and proceed with normal development, reproduction, and lifespan.

Worms begin to show phenotypes characteristic of aging within a week of reaching maturity (Collins, Huang, Hughes, & Kornfeld, 2008). The worms usually shrink in size while the mid-body becomes comparatively swollen, and internally they appear increasingly disorganized. Remaining eggs are no longer well-ordered and are noticeably jumbled in the body cavity. The intestines build up autofluorescent pigments of lipofuscin and advanced glycation end-products. Electron microscopy shows that the worm's cuticle thickens and becomes increasingly wrinkled with age (Herndon et al., 2002). In addition to these changes in appearance, worms also slow down as they age, with paralysis often preceding death. Motility can be measured by a number of methods, including the rate of pharyngeal pumping and, when suspended in liquid media, the rate of thrashing.

Husbandry

WormBook is an excellent resource for protocols for maintaining and working with *C. elegans* and detailed protocols for conducting lifespan (Amrit et al., 2014) and behavior (Hart, 2006) assays have been published. Worms are most easily observed and manipulated when maintained on plates of solid agar in Petri dishes (nematode growth media, or NGM, is most commonly used) with a lawn of OP50 *E. coli* in the center of the plate for food (Wood, 1988). Although the worms can burrow in the agar, they will generally not do so as long as there is ample food. Note that if the worms are burrowing, this usually indicates that the agar is too soft or that the worms are experiencing a stress such as heat, desiccation, or starvation. In addition, some strains will be more prone to burrowing or, alternatively, climbing the sides of the dish, where they may become stuck and desiccate.

Individual worms can be transferred between plates with a worm pick. Platinum is traditionally used for these because it can be flame-sterilized and then cools quickly. If a large number, rather than individual worms, is desired, a chunk of agar can be cut from a densely populated plate and moved to a new plate. Worms will quickly crawl away from the chunked agar and into the fresh bacterial lawn. Worms can also be rinsed from a plate using osmotically balanced media such as M9 or S-Basal. This is particularly useful for removing adult worms from a plate, while leaving the eggs, or separating a large batch of adults and L1 larvae. This is a common method for maintaining large synchronous populations of worms: approximately 10,000 worms are seeded onto a 10-cm agar plate spread with OP50. The worms grow to adulthood and begin to lay eggs. Ideally, the bacteria should become depleted at this time, so that any larvae that hatch will not develop further. It may take several rounds to achieve synchronization and to determine the right number of worms for the strain used and food density. It is important that the worms do not run out of food too soon or they will withhold their eggs and burrow. Once there is a population of adults and L1 larvae, they are then collected by rinsing the plate with 3–5 mL of medium, which is transferred to a 15 mL centrifuge tube. In approximately 5 minutes, the adults will have sunk to the bottom while the L1 larvae are still suspended in the media. The L1s can then be transferred to a new collection tube and their number estimated by counting the number of worms present in a small droplet of known volume (1–10 μL). The desired number of worms can then be pipetted onto new plates to maintain the synchronous population or for experiments. This is one method to set up synchronous populations for lifespan assay.

A method to obtain a smaller synchronous population is a "limited lay," in which a number of gravid adults are transferred to a new plate and allowed to lay eggs for several hours before being removed. The number of worms and time allowed will depend on the number of offspring desired. Wild-type worms generally lay 5–9 eggs per hour when conditions are favorable. Note that when working with mutant strains, the number of parental worms required will generally be higher, as many have lower fecundity, less synchronized development, and/or a lower survival rate.

Whichever method has been used to obtain a synchronous population, worms are often moved to a new plate and counted once they have reached L4 development. This stage is easily spotted by its distinct morphology and allows further verification of a synchronous cohort. Because development is nearly complete, treatments are often begun at this stage and, if desired, worms may be moved to plates treated with 5-fluoro-2′-deoxyuridine (FUdR), which inhibits DNA synthesis and thereby production of offspring. Once lifespan plates are set up, worms are transferred and scored every few days. If FUdR is not used, it may be necessary to transfer the worms every day during the 3–4 days of intense reproduction to avoid progeny consuming all available food or being mistaken for the parental generation. Worms are scored as dead once they no longer show movement, specifically, failing to respond to touch with the worm pick.

Advantages as a model for biology of aging research

- Small genome: *C. elegans* has one of the smallest genomes (~1 million bp) of a nonparasitic metazoan and, in 1998, it was the first multicellular organism to have its entire genome sequenced (C. elegans Sequencing Consortium, 1998). The *C. elegans* genome is compact and is estimated to contain more than 20,000 protein-coding genes, approximately one-third of which have human homologs (WormBase).
- Short lifespan: Depending on temperature, the average lifespan of *C. elegans* is only 2–3 weeks, while life-extending mutations and treatments typically extend lifespan in the range of 10%–200% (Shmookler Reis et al., 2009), although one mutation has been found which extends *C. elegans* adult

lifespan 10-fold or 1000% (Ayyadevara et al., 2008). Significantly, the combination of short lifespan and suitability for forward genetic screens has enabled mutants to be identified directly by their extended lifespan, something which has not been achieved in any other animal model (Duhon, Murakami, & Johnson, 1996; Klass, 1983).

- Rapid expansion: *C. elegans* takes an average of 3 days to mature and each hermaphrodite can lay several hundred eggs over the next several days, making it easy to obtain large synchronous populations. Whole populations can be obtained from a single hermaphrodite, while males are easily generated when crosses are desired.
- Translucence: *C. elegans* are small and transparent enough to enable observation of fluorescent reporters, even when expressed in only a few cells. While the gut becomes increasingly autofluorescent with age, the level is very low in younger animals and is itself a useful biomarker of aging in the worms. Taking advantage of *C. elegans'* translucence, many lines have been created expressing fluorescent transcriptional reporters (where a fluorescent protein gene is expressed under the promoter of a gene of interest) and tagged proteins (where a fluorescent protein gene is inserted at either end of the endogenous protein, so that the protein itself has a fluorescent tag).

  Fluorescent-tagged proteins enable studies of subcellular location in vivo and have been used to view processes such as synapse repair (Nakata et al., 2005) and protein aggregation (Morley et al., 2002). Transcriptional reporters that signal activation of a cellular defense response, such as antioxidation (Link et al., 1999), the endoplasmic reticulum unfolded protein response (Calfon et al., 2002), the mitochondrial unfolded protein response (Yoneda et al., 2004) and more, have been developed and are widely used in *C. elegans* research. Fluorescent reporters combined with RNAi libraries or forward genetics screening make *C. elegans* a powerful tool for identifying genes that have roles in processes such as stress response and longevity (e.g., Shore, Carr, and Ruvkun, 2012).
- Low cost and ease of maintenance: Populations are commonly maintained on agar seeded with *E. coli* but they can also be kept in liquid culture. An agar plate can be wrapped with paraffin film and worms recovered months later, thanks to the stress-resistant dauer form that can go long periods without feeding. Cryopreservation is also routinely done with this species, with very excellent recovery. This enables nearly indefinite storage of lines, while preventing genetic drift.

Tools

Both the genome and transcriptome have been fully sequenced and annotated (WormBase). In addition, a number of other *Caenorhabditis* species have also been sequenced and studied. A systemic RNAi response greatly facilitates gene knockdown; expression of most genes can be easily knocked down by feeding worms double-stranded RNA. Finally, there are a number of databases readily available, well-characterized mutants, transgenic animals, and libraries for RNAi knockdown, all of which greatly facilitate research and large-scale screening. WormBase.org is a database for the genetic and genomic information about *C. elegans* and related nematodes. It provides a searchable and annotated genome and transcriptome, including unspliced and spliced sequences of all known isoforms, homologs in other organisms, phenotype data, and information regarding transgenic and mutant lines and RNAi constructs that have been published. Many of these lines can be obtained cheaply from the Caenorhabditis Genetics Center (cgc.umn.edu). Detailed information on biology, husbandry, and research protocols can be found at WormBook.org with additional information on morphology and cell lineage at WormAtlas.org.

In summary, *C. elegans* is a model complementary to *Drosophila* in many respects. Both have extensive toolkits enabling tissue-specific gene knockdown or overexpression. While the more complex nervous system of the fly will be a benefit to some studies, in other cases, simpler is better. It is for this reason that Sydney Brenner chose *C. elegans* as a model for studying neurobiology. The 302 neurons that compose the *C. elegans* nervous system have been mapped out and individual neurons can be observed in vivo with appropriate fluorescent markers. This is a great advantage over *Drosophila*; while many studies have successfully used fluorescent markers in fly larvae, the chitinous exoskeleton of the mature adult is, unfortunately, highly autofluorescent, as well as pigmented, and thus largely precludes in vivo observation of fluorescent markers in aging flies.

Confounding factors for aging studies

C. elegans lifespan is very malleable, making replication of experiments by different investigators or under different settings a challenge. Temperature, light, and composition of the media and food source all impact *C. elegans* lifespan and some of these effects may be transgenerational. This underscores the importance of reporting detailed methods to facilitate repeating and building on previous work. Several potential confounding factors that must be considered in using

C. elegans as a model in aging research are detailed below:

- Temperature: *C. elegans* are normally maintained and lifespans done at 20°C. Like many invertebrates, they have a shorter lifespan and faster development at higher temperatures and longer lifespan and slower development at lower temperatures (Klass, 1977). Worms are not generally kept at temperatures warmer than 25°C as they begin to experience heat stress, sterility, and increased mortality. Exposure to 30°C and above is used for heat stress assays or to increase production of males (exposure of L4 larvae to >30°C promotes nondisjunction in spermatogenesis) (Wood, 1988). Interestingly, a number of genetic manipulations have been found which alter *C. elegans* lifespan only when the worms are assayed at a specific temperature (Hartman et al., 2001; Vilchez et al., 2012).
- Nutrient status: A common method to obtain a synchronous population of worms is to ensure that eggs hatch on plates devoid of food, so that they arrest as L1s. The worms are then collected, usually by washing them off the plate with liquid medium, and seeded onto plates with food, where they will then resume normal development. This period of starvation, however, induces transcriptional changes that will not only persist throughout the worms' lifespan but be passed on for several generations. These worms and their progeny have been shown to be both more stress resistant and have extended lifespan (Jobson et al., 2015; Rechavi et al., 2014). This has also been reported in worms grown in liquid medium versus on solid agar plates (Lev et al., 2019).
- Bacterial food source: While use of axenic culture has had some success (Sutphin & Kaeberlein, 2009), worms are still most commonly and easily maintained on lawns of live *E. coli*. This introduces additional complications and variables to experiments. The species of bacteria fed to worms, as well as the exact strain, can significantly impact lifespan and interact with worm genetics in unexpected ways (Soukas et al., 2009). In addition, experiments that entail testing the effect of a drug or other compound on *C. elegans* may be confounded by that compound instead being metabolized by the *E. coli* (Mishur et al., 2016).
- Antimicrobials: Antibiotics and antifungals used to prevent contamination can have offtarget effects on animals. Tetracycline, for example, is often used to select for bacteria carrying a plasmid that can be induced to generate dsRNA against a target *C. elegans* gene. However, tetracycline also affects the worms' mitochondria and, while this effect may be negligible in wild-type worms, it is important to take into consideration when the worms will be exposed to other stressors. For example, tetracycline exacerbates the delayed development of *isp-1 (qm150)* mutants (*isp-1* encodes the iron–sulfur protein subunit of mitochondrial coenzyme Q: cytochrome c oxidoreductase) and, when it is combined with FUdR treatment or further genetic manipulations, can lead to very different experimental outcomes (personal observations).
- Offspring and FUdR: It is necessary to separate experimental animals from their offspring to prevent the effects of overcrowding and food depletion, as well as to avoid confusing the generations. This normally entails manually transferring experimental animals to new plates every day or two while they are reproducing. Alternatively, to make lifespans less labor intensive, *C. elegans* are often treated with 5-fluoro-2'-deoxyuridine (FUdR) upon maturing, which reduces or eliminates production of offspring by inhibiting DNA synthesis. While low doses have little effect on the lifespan of wild-type worms, FUdR can interact with other treatments and genetics in unexpected ways, including dramatically extending the lifespan of an otherwise short-lived mutant (Van Raamsdonk & Hekimi, 2011).

Drosophila melanogaster

Phylum: Arthropoda
Class: Insecta
Order: Diptera
Family: *Drosophilidae*
Genus: *Drosophila*

History

D. melanogaster, commonly called the fruit fly or vinegar fly, has been studied as a model organism for well over a century. There are more than 1500 species in the genus, but the majority of research has focused on *D. melanogaster*. Research on *Drosophila* has to date formed the foundation of eight Nobel prizes.

The first work on fruit flies was undertaken by Charles Woodworth and William Castle as well as Castle's students—F.W. Carpenter, A.H. Clarke, S.O. Mast, and W.M Barrows—in the early 1900s. These researchers characterized *Drosophila* as a model system and expounded the virtues of the fly as a model for studying "experimental evolution" (Castle, 1906). The use of *Drosophila* as a model system quickly expanded after this initial characterization. *Drosophila* were cheap and easy to house and their small size and rapid development made them an attractive model for studies of evolutionary biology and development. The work by Woodworth, Castle, and colleagues later inspired Thomas Hunt Morgan, who in 1933 won the Nobel Prize in Physiology or Medicine for using fruit flies to demonstrate the critical role of chromosomes in carrying genetic information (Morgan, 1911). He also discovered the white-eyed mutation that is heavily used in *Drosophila* research to this day (Bellen, Tong, & Tsuda, 2010).

Drosophila began to be utilized as an aging model in the early 1910s. The first reference to their lifespan comes from a study in 1911 by Moenkhaus which states simply "We have kept females alive for 153 days" (Abele, Brey, & Philipp, 2009; Moenkhaus, 1911). Later studies by Hyde in 1913 showed that when two inbred fly strains were crossed, the resultant hybrid strain was longer-lived than either of the two original strains (Hyde, 1913). In 1916 and 1917, Northrop and Loeb demonstrated how temperature manipulation could alter *Drosophila* lifespan (Loeb & Northrop, 1917). *Drosophila* aging studies really began in earnest with work by Raymond Pearl and colleagues in a series of papers from 1921 to 1927. This work was supposedly prompted by a catastrophic air-conditioner failure, which destroyed their mouse colony. The group were then persuaded by Morgan and Loeb to utilize *Drosophila* instead. In this work, Pearl and colleagues identified the effect on lifespan of a range of genetic mutations, inbred and cross-bred stocks, as well as environmental factors. This work was important, as it set the experimental standards for *Drosophila* experimental rigor (large animal cohorts of >1000 animals, highly standardized conditions, etc.) that have made this organism such a robust model for studying aging (e.g., Pearl, 1926; Pearl & Parker, 1922; Pearl, White, & Miner, 1929).

To this day, flies remain a highly tractable and utilized model system for aging studies. Their relatively short lifespan (~70–100 days), coupled with their ease of use and cheapness of housing, make *Drosophila* lifespan experiments a highly viable approach for rapid and high-throughput testing of interventions that can extend lifespan. *Drosophila* aging studies, along with the use of *Drosophila* as a research model in general, expanded dramatically in the late 1990s–early 2000s with the development of the upstream activation sequence (UAS)/GAL4 system in *Drosophila* (Duffy, 2002). With this approach, researchers may create and insert an overexpression, RNAi, or reporter construct into *Drosophila* connected to a UAS promoter sequence. The researcher can then cross this fly line with a GAL4 driver line to drive ubiquitous or tissue-specific transgene expression, dependent on the GAL4 promoter. Further modifications of this approach, additionally, allow temporal control over transgene expression. This genetic toolbox has enabled researchers to investigate the effect on aging of specific tissue modifications of specific genes. Other advantages of *Drosophila* include their relative level of sophistication. *Drosophila* possess a fully developed nervous system, vision, memory, motor function, muscles and digestive system as well as a heart and circulatory system. This has made the *Drosophila* model a viable and powerful system for studying a range of age-related diseases including dementias, sarcopenia, and heart disease, with a range of genetic "humanized" fly models available for studying human diseases.

Drosophila aging and development

Drosophila have a multistage lifecycle. They are ectothermic animals and thus the length of each life-stage is highly dependent on temperature. Based on temperature selections assays, 24°C–25°C is the optimal temperature for *Drosophila* and is the temperature at which they are typically housed and assays done (Dillon et al., 2009). Once mated, flies are highly reproductive and females may lay up to 120 eggs per day (Shapiro, 1932). The eggs will hatch into embryos after 12–15 hours. The larvae will go through two molts over the following 4 days (second instar and third instar larvae) and then undergo metamorphosis in the pupa stage for another 4 days before finally eclosing (emerging) as near fully developed adults. At 25°C, *Drosophila* development is expected to take 8.5 days. This rate is impacted by temperature; the fastest developmental cycle is 7 days at 28°C, while the slowest is 19 days at 18°C (Ashburner, 1989). For the purposes of recording lifespan, researchers will consider the day of eclosion as the first day of lifespan. After eclosion, flies can be expected to have a lifespan of 70–100 days, depending on genotype, food stock, care conditions, and temperature.

Drosophila lifespan husbandry

A fully outlined protocol for preparing and maintaining a *Drosophila* lifespan is outlined in Linford et al. (2013). Briefly, a mating cross should be set up between male and virgin female flies. Fly virginity is typically established by time post eclosion, as female flies will not successfully mate in the first 8 hours post eclosion at 25°C and 16 hours post eclosion at 18°C. Using virgin flies is important, as it ensures the correct progeny genotype. There are different approaches for crosses. Researchers may set up individual mating vials, each containing approximately 7 males and

10 females, in which flies are allowed to mate for 3 days. Researchers may alternatively place flies on grape-juice agar plates topped with yeast for 24 hours. The eggs may then be removed from the plate, washed, and pipetted into either vials or flasks, while the parent flies are transferred to new plates every day. With either strategy, the vials or flasks will be left for ~10 days until the offspring begin to eclose. During this time vials should be inspected daily for moisture content. If vials are too dry, cracks will appear in the food stock and may impair development and so water should be carefully added. If too many eggs are placed in a vial, abundant larvae may make the food overly soft and cause flies to become trapped in the food during collection. Once flies start to eclose, they should be collected during a tight time window. Aging studies utilize flies which have been collected over just a 48–72-hour period. Collected flies are typically allowed to mate for 2 days to ensure an equal mating status (unless the experiment seeks to investigate virgin flies). This is done by combining multiple vials of flies eclosed on a certain day into a large flask. After mating, flies will be sorted into vials containing just one sex. This ensures that mating occurs equally in all flies and does not occur after the start of the experiment. Flies should then be transferred to new food vials every 2–3 days, with the number of dead flies scored on each day of transfer.

Confounding factors for aging studies

- Mating status. Mating status appears highly impactful to *Drosophila* lifespan, particularly for female flies. Virgin female *Drosophila* live significantly longer than mated females (Kevin & Partridge, 1989). Female flies appear to incur a cost of mating. It was shown then even when spermless males were used, mated female flies still incurred a reduction in lifespan (Chapman, Hutchings, & Partridge, 1993) and is at least in part driven by signaling from seminal fluid products from the main cells of the male accessory gland (Chapman et al., 1995). Mating also appears to shift a number of other physiological measures. Flies which have been mated show increased starvation resistance but decreased oxidative stress resistance. This has been argued to potentially represent an example of the disposable soma theory of aging—that mating induces changes in resource allocation away from survival and toward reproduction (Rush et al., 2007). Because of the impact of mating on lifespan, researchers who conduct aging studies in *Drosophila* must be careful to ensure mating status is uniform amongst all flies studied. This is usually accomplished by mating flies for 2 days prior to separation of the sexes into individual vials, ensuring that all flies have mated and to a similar extent. However, some researchers may instead utilize virgin flies for their aging studies, by separating the sexes in the 8 hours post eclosion before flies are able to mate.
- Temperature. Flies possess an innate temperature preference for 24°C–25°C. When flies are placed on a temperature gradient, ~95% of the flies were found to remain in the region heated from 21°C–27°C while within this group ~80% of the population was clustered between 22.5°C and 25°C with a peak at 24°C. Binary choice temperature mazes have similarly demonstrated a strong preference in flies for this temperature range (Sayeed & Benzer, 1996). For this reason, researchers will generally culture flies between 24°C–25°C. This is particularly important for lifespan studies. As flies are ectotherms, temperature can be highly impactful. We discussed above the effects of temperature on *Drosophila* development times and temperature appears to be similarly impactful to *Drosophila* aging and lifespan. It was originally reported by Leob and Northop, and later expanded on by many others, that increasing temperature shortens fly lifespan while reducing temperature extends it (Loeb & Northrop, 1917). As one example in a paper from Atlan's group, researchers showed that their flies had a mean lifespan of 19 days at 30°C, 41 days at 27°C, 86 days at 21°C, and 130 days at 18°C. Similar results have been produced by many other researchers (Miquel et al., 1976). This was historically argued to represent an example of the rate of living theory of aging—that lowering an organism's metabolic rate will extend its lifespan. However, data on *C. elegans* argue against this; researchers have shown that cold-induced life extension represents an induced response and that ablation of cold-sensing neurons will ablate the lifespan extension (Xiao et al., 2013). The decline in lifespan at higher temperatures and increase at lower temperatures does appear to represent an actual change in the rate of aging. In addition to changes in survival, altering temperature is impactful to many measures of health and function. For these reasons, researchers should take care to ensure tight monitoring of temperature for aging studies in flies and ensure there is no temperature imbalance in *Drosophila* lifespan experiments.
- Diet. Flies are maintained on a sugar yeast solution that is boiled and solidified by agar. Dietary restriction can extend *Drosophila* lifespan and the ratio of sugar to yeast is also important to *Drosophila* survival. Flies that are maintained on 5%–10% yeast and 5%–10% sucrose have the longest lifespan, while deviation from this range in either direction is

detrimental (Skorupa et al., 2008). For this reason, many researchers utilize an SY5 or SY10 food stock for maintaining flies (Skorupa et al., 2008).

UAS/GAL4 system

A critical advantage of *Drosophila* as a research tool, including as a model for studying aging and age-related diseases, is the dual transgene UAS/GAL4 system. Researchers utilize GAL4, which is an 881-amino-acid transcriptional regulator taken from *Saccharomyces cerevisiae*, to bind to a 17-amino-acid sequence termed the UAS (Duffy, 2002). The original GAL4 sequence was subsequently mutagenized to optimize its UAS affinity (Webster et al., 1988) and, in 1988, it was demonstrated that the GAL4 gene could stimulate transcription of a reporter gene possessing a UAS sequence (Fischer et al., 1988). Importantly, expression of GAL4 does not appear innately detrimental to the fly. In 1993, Brand and Perrimon developed the UAS/GAL4 system into a genetic tool for *Drosophila* research (Brand & Perrimon, 1993) by creating a bipartite expression system. The researchers created a transgene with five tandemly arrayed optimized GAL4 binding sites upstream of their gene of interest. It was shown that in the absence of GAL4 (which is not natively expressed in flies) the transgene is not expressed. The authors then created several transgenes that expressed GAL4 under control of different promoter sequences. This approach allowed expression of GAL4 in different tissues, regions, or life stages. The two transgenes were placed in separate fly lines, such that no phenotype would be produced by either individually. However, if the two lines were crossed, the progeny would contain both a transgene that expresses GAL4 in a tissue-specific manner and a second transgene under the control of a UAS promoter sequence. In these animals, GAL4 would be expressed in a tissue- or temporal-specific manner, causing tissue- or temporal-specific expression of the UAS transgene (Brand & Perrimon, 1993). Today, vast libraries of UAS transgenes exist for overexpression, disruption, or RNAi knockdown of a high proportion of the fly genome. In addition, a myriad of UAS-linked reporter lines, humanized transgenes, and disease model transgenes have been developed. These are complemented by a similarly expansive collection of GAL4 driver lines that enable expression of different transgenes in anything from the whole animal to specific tissues or cell types or even individual clusters of cells. Through separation of the GAL4 driver and UAS promoter onto separate flies, researchers are allowed near infinite combinations of transgene expression, making this an extremely powerful research tool.

Creation of transgenic lines in flies is relatively trivial. *Drosophila* eggs may be relatively easily injected with a DNA plasmid which becomes incorporated in the genome. Typically, researchers will use a fly stock which contains a mutation causing white eyes and then insert a *mini-white* construct (causing red eye expression) into their plasmid; thus, flies that possess the construct may easily be identified by the presence of red eyes. Today it can cost as little as \$200–300 to create a fly line and lines may be created in a little as a month. This has resulted in an expansive collection of available lines. Libraries for these transgenes exist at several locations and the lines are inexpensive for academic research. Some of the major repositories are the Bloomington *Drosophila* stock center (bdsc.indiana.edu), Vienna *Drosophila* Resource Center (stockcenter.vdrc.at), FlyORF (flyorf.ch), and the Kyoto Stock Center (kyotofly.kit.jp).

GeneSwitch-GAL4

GeneSwitch-GAL4 (also called Switch or SwitchO) represents a modified form of the UAS/GAL4 system. Here, a GAL4-progesterone-receptor fusion protein was created in which the GAL4 transcription factor is able to display transcriptional activatory properties only when bound to mifepristone (RU486) (Roman et al., 2001). This allows flies possessing both a GAL4 driver and UAS transgene to be maintained without inducing transgene expression until RU486 is introduced into the flies' food. This approach enables the coupling of tissue-specific gene expression with temporal control. This is particularly important in aging studies because it enables researchers to remove developmental artifacts from effects of constructs on age by restricting RU486 feeding to time post-development. It is important to note that the progesterone agonist RU486 has been reported to mitigate the negative effects on lifespan in female flies produced by mating (Landis et al., 2015). This is a concern for aging studies, if not properly controlled for, as life extension due to RU486 could be mistaken for effects of a transgene. For this reason, it is imperative to include appropriate controls by running parallel experiments on RU486 with flies of identical genetic background but lacking the UAS transgene.

Healthspan measures

Drosophila form a powerful intermediary between *C. elegans* and mammals for the study of aging. Although the average *Drosophila* lifespan is far longer than that of *C. elegans* (80–100 days vs. 20–30 days), they possess a number of advantages over the worm. Key amongst these are more complex organs and behaviors. These allow researchers to conduct a considerably wider range of health and healthspan measures than are accomplishable in the worm. Some of the key measures are outlined below.

- Olfaction aversion training. This is a measure of associative learning and memory and is a powerful tool in the study of both dementia and age-related cognitive declines. A detailed protocol is provided in Malik and Hodge (2014). In this experiment, flies are exposed in alternation (via an air pump) to two neutral odors (3-octanol and 4-methylcyclohexanol, both diluted in mineral oil) for 5 minutes under low-level red light (flies are unable to detect red wavelengths). During this experiment, flies are held in a chamber with metal plating around the sides. When the flies are exposed to one odor they receive an electric shock, which they do not receive when exposed to the other odor. After multiple training rounds and a 1–24 hours recovery interval flies are moved to a T-maze where the two opposing odors are pumped from opposing sides. Flies are then allowed a short time to explore the maze, after which the doors of the maze are closed and the odor choice scored. In young flies there is a training-induced increase in avoidance of the "negative" odor, but this is reduced or even completely lost with age and in models of neurodegenerative disease, suggesting a decline in learning and memory (Munkacsy et al., 2019).
- Negative geotaxis (climbing capacity). Flies have an innate preference, when moved to the bottom of a container, to climb to its top (Ali et al., 2011), providing a relatively easy method of measuring muscle function in flies. The experimental design is extremely simple: Vials of flies are transferred to a transparent cylinder and allowed to acclimate for 10 minutes with the top sealed. The cylinder is then gently tapped to move all flies to the bottom of the chamber. The investigator will then pick an arbitrary point on the cylinder and score the number of flies that have crossed this line in a set amount of time. A detailed description of the approach is outlined in Ali et al. (2011). Importantly, climbing capacity declines with age (Munkacsy et al., 2019).
- Spontaneous activity. This represents a measure of muscle function that is less invasive than the negative geotaxis assay. Instead of measuring the ability of flies to climb a cylinder, the total activity of the fly is recorded over a number of hours. This is achieved by placing flies in a horizontal chamber intersected by a laser. The device is then connected to a system that will count when and how frequently the beam is broken, thus providing a measure of the flies' spontaneous activity. As with negative geotaxis, this activity declines in the fly with age (Le Bourg, 1987). A modified form of this assay may be conducted by placing a fly in a sealed Petri dish, placed on top of a grid, and scoring the number of chambers visited in a set amount of time.
- Circadian rhythmicity. Flies have a highly defined diurnal activity distribution, with high levels of activity at the start and end of the day, low activity in the middle, and almost no activity at night, although the exact activity distribution varies by sex and environment (Martin, Ernst, & Heisenberg, 1999). This can be monitored using a similar experimental design to the spontaneous activity assay. As with humans, circadian rhythmicity in the fly becomes less well-defined with age (Koh et al., 2006).
- Intestinal integrity ("Smurf" assay). Loss of intestinal integrity appears to be a strong predictor of fly mortality. In the last few days before death almost all flies will incur a loss of intestinal integrity. Furthermore, flies that have lost intestinal integrity will typically die within 2–3 days of loss. The loss of intestinal integrity increases with age and interventions which extend lifespan will typically delay loss of intestinal integrity (Rera, Clark, & Walker, 2012). Loss of intestinal integrity can be measured easily and noninvasively in flies. The animals are fed a blue dye mixed into their food. Flies with normal integrity will display a blue proboscis and gut. In contrast, the entire body of flies that have lost intestinal integrity will appear blue, and for this reason this assay is also known as a "Smurf" assay.
- Cardiac function. Although *Drosophila* possess a semiopen circulatory system, they possess a heart tube composed of several rings of contractile muscles. Cardiac function in the fly may be monitored either through dissection to expose the tube (Vogler & Ocorr, 2009) or through fluorescent reporter lines (Yu et al., 2013). Cardiac rhythmicity declines in the flies with age and on various experimental diets. This can be monitored in flies through imaging under high-speed cameras (Ocorr, Akasaka, & Bodmer, 2007).

Disease models

A wide range of transgenic *Drosophila* disease models exists, which can be used to study varied age-related diseases. The mutant lines outlined below are available from many fly repositories, including the Bloomington *Drosophila* stock center (bdsc.indiana.edu), Vienna

Drosophila Resource Center (stockcenter.vdrc.at), FlyORF (flyorf.ch), and the Kyoto Stock Center (kyotofly.kit.jp).

- Alzheimer's disease. This and other neurodegenerative diseases are particularly attractive diseases to study in *Drosophila* because of their high levels of activity and complex behavior. Alzheimer's disease may be modeled in *Drosophila* through transgenes expressing UAS-hAPP in combination with UAS-hBACE1. Lines have also been developed with constructs for UAS-hTau, UAS-hAβ, and many others, including specific mutations associated with disease risk in humans. These lines can be crossed with neuronal specific driver lines such as Elav-GAL4 and Elav-GeneSwitch-GAL4. Flies expressing these transgenes display shorter lifespan as well as deficits in learning, memory, spontaneous activity, and circadian rhythmicity, with different effects in different lines.
- Parkinson disease. This disease is a progressive neurodegenerative disease driven by loss of dopaminergic neurons in the substantia nigra, leading to tremors, coordination deficits, and loss of muscle function. *Drosophila* overexpression and mutant lines exist for α-synuclein, *PARKIN*, *PINK1*, *DJ1*, *HTRA2*, and *PLA2G6* and many of these lines can produce Parkinson-like phenotypes in the fly, including motor and coordination deficits (Feany & Bender, 2000).
- Huntington disease. This is a neurodegenerative disease producing deficits in motor coordination, loss of motor function, and neurodegeneration. It may be modeled in flies through overexpression of mutant forms of the human Huntington gene *HTT* under the control of neuronal driver lines. This will produce Huntington disease-like symptoms, including neurodegeneration, photoreceptor degeneration, and mortality (Agrawal et al., 2005)
- Amyotrophic lateral sclerosis (ALS) or Lou Gehrig disease. This disease is a neurodegenerative disease produced by degeneration of motor neurons resulting in muscle weakness and atrophy. ALS can be modeled in *Drosophila* through mutant lines available for a range of ALS risk factor genes including *SOD1*, *Als2*, *Vap33*, *TBPH*, *FIG4*, *ATX2*, and others. Expression of the mutated genes produce ALS disease-like symptoms, including motor dysfunction and shortened lifespan (Hanson et al., 2010).

Disease models also exist for a wide range of other diseases, including Angelman syndrome, cardiovascular diseases, Coffin-Lowry syndrome, Cornelia de Lange syndrome, Fragile X syndrome, galactosemia, glycerol kinase deficiency, hereditary spastic paraplegia, hexanucleotide repeat diseases, hypoparathyroidism retardation dysmorphism syndrome (HRD), L1 syndrome, Machado–Joseph disease, metabolism/diabetes/obesity, microcephaly, mitochondrial disorders, muscular dystrophies, myotonic dystrophies, neurofibromatosis, neuronal ceroid lipofuscinosis, Niemann–Pick disease, P-type ATPase disorders, peroxisome disorders, polyalanine diseases, polyglutamine diseases, primary ciliary dyskinesia, seizure disorders, spinal muscular atrophy (SMA), spinocerebellar ataxia 1, Werner syndrome, and X-linked mental retardation. All of these have publicly available lines stored in varied *Drosophila* repositories.

Other invertebrate models in aging research

In the phylum of Arthropoda alone there is a great diversity of lifespan. Of particular interest are the eusocial insects—bees, wasps, ants, and termites—because they represent models where vastly different lifespans can be observed in genetically identical animals. In many of these species, the queen can live for decades while the workers live for mere weeks or months. This trait is believed to have evolved independently at least three times, specifically in bees, ants, and termites. Several species have been sequenced, including the honeybee, *Apis mellifera*, which has also been extensively studied because of its importance to agriculture. Workers of this species can live for several months while queens live up to 8 years (Haddad, Kelbert, & Hulbert, 2007).

Despite the long evolutionary separation of nematodes and arthropods, they are still more closely related to one another than to other invertebrate phyla or to chordates. Both are currently recognized as protostomes and ecdysozomes, based on embryological development and growth through molting, respectively (Fig. 9.1). Model organisms from other clades would be expected to have retained different orthologs with chordates. Studying these organisms may grant us an insight into whether different organisms age differently and whether aging is a pathology broadly conserved across metazoan phyla or even all life forms (Table 9.1 and Fig. 9.1).

Rotifers

Phylum: Rotifera

Rotifers are transparent aquatic invertebrates. They are similar in size to *C. elegans*, ranging from 0.1 to 0.5 mm in length. Their name is a derivation of the Latin *rota* (wheel) and *fer* (bearing). This name is

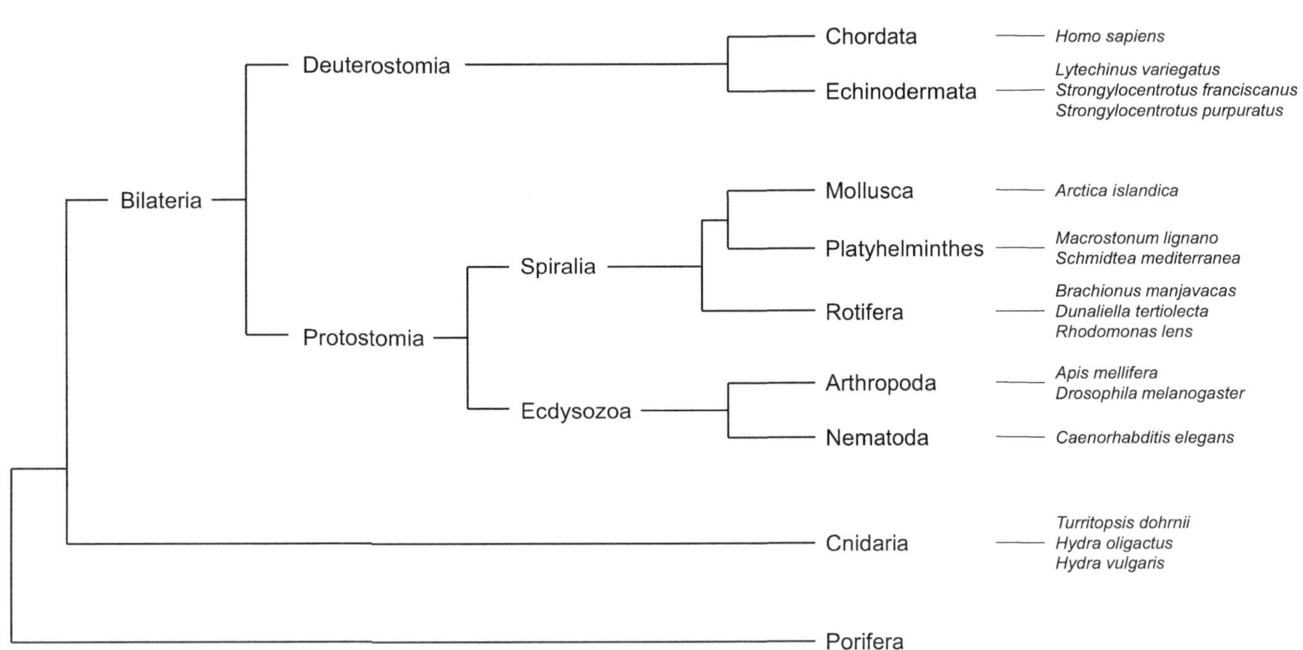

FIGURE 9.1 Metazoan phylogeny.

TABLE 9.1 List of invertebrate model organisms.

Phylum	Species	Common name	Maximum reported lifespan (years)	Genome sequenced
Arthropoda	*Drosophila melanogaster*	Fruit fly	0.3 (Finch, 1990)	Sequenced and annotated (Adams et al., 2000)
Arthropoda	*Apis mellifera*	Honeybee	8 (Haddad et al., 2007)	Genome sequenced (Honeybee Genome Sequencing Consortium, 2006)
Cnidaria	*Hydra oligactus*	Hydra	>0.6 (Yoshida et al., 2006)	No
Cnidaria	*Hydra vulgaris*	Hydra	>4 (Martinez, 1998)	No
Cnidaria	*Turritopsis dohrnii*	Immortal jellyfish	Unknown (Piraino et al., 1996)	Transcriptome only (Hasegawa et al., 2016)
Echinodermata	*Lytechinus variegatus*	Green sea urchin	4 (Beddingfield and McClintock, 2000)	Genome sequenced (Sergiev et al., 2016)
Echinodermata	*Strongylocentrotus franciscanus*	Red sea urchin	>100 (Ebert, 2008)	Genome sequenced (Sergiev et al., 2016)
Echinodermata	*Strongylocentrotus purpuratus*	Purple sea urchin	50 (Loram and Bodnar, 2012)	Genome sequenced (Sea Urchin Genome Sequencing et al., 2006)
Mollusca	*Arctica islandica*	Ocean quahog clam	>500 (Ungvari et al., 2011)	No
Nematoda	*Caenorhabditis elegans*	Roundworm	0.16 (Klass, 1977)	Sequenced and annotated (C. elegans Sequencing Consortium, 1998)
Platyhelminthes	*Macrostomum lignano*	Flatworm	>2 (Mouton et al., 2018)	Sequenced and annotated (Wasik et al., 2015)
Platyhelminthes	*Schmidtea polychroa*	Flatworm	>3 (Mouton et al., 2011)	No
Rotifera	*Brachionus manjavacas*	Rotifer	0.06 (Snell et al., 2012)	No
Rotifera	*Dunaliella tertiolecta*	Rotifer	0.08 (Snell et al., 2012)	No
Rotifera	*Rhodomonas lens*	Rotifer	0.04 (Snell et al., 2012)	No

based on the presence of a ciliated corona around their mouth, which appears similar to a wheel. They are a transparent aquatic organism. Like *C. elegans*, these organisms are eutelic (possess a fixed number of cells); they comprise approximately 1000 cells (Snell, 2014).

History

Rotifers were first described in 1696 by the Reverend John Harris, who described them as "an animal like a large maggot which could contract itself into a spherical figure and then stretch itself out again; the end of its tail appeared with a forceps like that of an earwig" (Harmer & Shipley, 1896). A more detailed description of multiple rotifer species was later produced by Antonie van Leeuwenhoek in 1702.

Role of Rotifers in research

The main use of rotifers in research has been in the context of ecology, culture conditions, and environmental toxicity. Rotifers are aerobic freshwater organisms that appear highly susceptible to environmental toxicity. For this reason, the presence or absence of different rotifer species in freshwater is used as an indicator of water toxicity. Rotifer survival in the laboratory is also used as a measure of water quality or chemical toxicity (Snell & King, 1977).

Rotifers have recently emerged as an aging model potential complementary to *C. elegans* and *Drosophila*. They have a number of key advantages in this context. A study examining the transcriptome of the rotifer *Brachionus manjavacas* across its lifespan found age-related changes that are conserved across taxa, including downregulation of many genes encoding mitochondrial proteins, but also changes in several genes that have homologs in humans but in neither *C. elegans* nor *Drosophila* (Gribble & Mark Welch, 2017). Rotifers also have a very short lifespan. Some species such as *Rhodomonas lens* live just 10–15 days, while other species such as *Dunaliella tertiolecta* can live up to 30 days (Snell, Fields, & Johnston, 2012). The rotifer has a number of distinct organs including a distinct nervous system, digestive tract, muscle, and reproductive system. Its transparent body (like *C. elegans*) enables easy study of internal processes. The genomes of several species have been sequenced and gene knockdown by exposure to dsRNA has been successful (Snell, Shearer, & Smith, 2011). Many species of rotifer have both sexual and asexual modes of reproduction, enabling experiments to be performed in genetically identical backgrounds (Snell, 2014). In addition, their fully determinant cell lineages, like *C. elegans*, are advantageous for experimental manipulations.

Rotifers in aging research

Although rotifers are not as heavily studied as *Drosophila* and *C. elegans*, a number of aging studies have been performed in these organisms. A summary of many of these studies is outlined below.

- Temperature. As with *Drosophila* and *C. elegans*, rotifer lifespan appears dependent on temperature. When researchers cultured the rotifer species *B. calyciflorus* at 16, 22, and 29°C, the animals had mean lifespans of 11.3, 6.4, and 4.1 days, respectively (Kauler & Hildegard, 2011). Animals maintained at lower temperatures also displayed reduced reproduction but an extended reproductive period (Brys, Vanfleteren, & Braeckman, 2007). The effect of temperature on rotifer lifespan appears to vary by strain; when 11 strains were examined some showed as much as a doubling in lifespan while others showed as little as a 6% increase (Gribble et al., 2018).
- Antioxidants. Exposure to specific antioxidants has been found to extend rotifer lifespan. LPBNAH1 was reported to extend rotifer mean lifespan by 150% (57 days) and increase the reproductive period (Poeggeler et al., 2005). Exposure to IPAM, which has structural similarities to melatonin, extended lifespan 270% from 24 days to 90 days (Poeggeler et al., 2010). However, in a separate study where 20 different antioxidant enzymes (including IPAM) were evaluated, all failed to extend the lifespan, though an extension of up to 20% was observed when some were used in specific combinations (Snell et al., 2012).
- Dietary restriction. As has been found in many model systems, dietary restriction appears to extend lifespan in many but not all rotifer species (Kirk, 2001). In a study of 10 rotifer species, two doubled their lifespan, three showed a modest lifespan increase (13%–29%), two experienced no change in lifespan, and three had a shortened lifespan (Kirk, 2001). Other groups have looked at dietary restriction in individual species and reported lifespan extensions from 50% to 300% (Kaneko et al., 2011; Yoshinaga, Hagiwara, & Tsukamoto, 2000, 2003). A number of gene expression differences have also been found in rotifers under dietary restriction (Oo et al., 2010; Yoshinaga et al., 2000).
- Insulin/IGF-1 signaling. Some research has looked into the role of insulin/IGF-1 signaling in rotifers. Treatment with the PI3-kinase inhibitor LY294002

has been reported to extend rotifer lifespan by 30% (Yoshinaga et al., 2005).

- Sexual dimorphism. Male rotifers are very short lived, just 4.3 days at 22°C (half that of females). The males are haploid and are produced only seasonally or under certain environmental stresses. However, as the males do not eat, it is unclear if this represents a true aging difference (Serra, Snelland, & King, 2004).
- Maternal effect. Researchers have investigated the effect of the mother's age on rotifer lifespan and found that animals born to older mothers appear to be shorter lived and less fertile. Interestingly, they also reported that dietary restriction has a greater lifespan-extending effect in animals born to older mothers than to younger mothers (Bock et al., 2019).

Rotifers represent a tractable model system that has been underutilized in aging research. Their separate evolution and retention of some genes lost in ecdysozomes would complement and strengthen continuing work in *C. elegans* and *Drosophila* systems and enable broader testing of the relevance of genes and biological processes before moving to more costly research in mammalian models.

Flatworms

Phylum: Platyhelminthes

Platyhelminthes (flatworms) are unsegmented worms that have long been studied for their incredible regenerative potential. In contrast to *C. elegans* and *D. melanogaster*, flatworms maintain a large population of totipotent stem cells. Platyhelminthes present an opportunity to study aging in an animal with high regenerative potential as well as the strategies they use to minimize the risk of cancer with age. These are processes which cannot be studied in eutelic organisms.

Flatworms in aging research

Very limited research has been done on aging in flatworms. In the few studies that have been conducted, the animals appear remarkably long-lived for a small invertebrate. It is hypothesized that the large totipotent stem cell pool, high levels of stem cell renewal, and morphological plasticity may make these organisms arguably "immortal" due to their constant regenerative capacity. Survival has been studied in both *Macrostomum lignano* and *Schmidtea polychroa*. Over 2 years, only ~30% mortality was observed in *M. lignano* (Mouton et al., 2018) and in *S. polychroa* only ~50% mortality was observed in 3 years (Mouton et al., 2011). The survival curves from both studies appear highly linear, suggesting that many of the deaths were not a result of aging. However, hazard ratio analysis has shown that there is, in fact, an increase in mortality rate with age, suggesting that aging does indeed occur but that a longer lifespan experiment would be required to observe it (Mouton et al., 2011). There are also reports of age-related gene expression changes in flatworms (Mouton et al., 2018), suggesting that although this invertebrate is very long-lived for its size, it still undergoes aging and age-related functional declines.

Bivalves

Phylum: Mollusca
Class: Bivalvia

Bivalves represent a class of mollusks with a laterally compressed body encompassed by two shells with a hinge. They exist primarily in saltwater, with a few families that live in freshwater. They are highly sedentary and have no head or central nervous system, but do possess a simple nerve network encompassing a series of paired ganglia connecting sensory organs with the foot and body. They possess rudimentary mechanoreceptors and chemoreceptors, located on tentacles, that enable them to identify touch and water. They have statocysts, which allow for control of orientation (Allen & Morgan, 1981). They have an open circulatory system, in which organs are bathed in hemolymph, and a three-chamber heart that facilitates transfer of oxygen from the gills (Allen & Morgan, 1981).

Finally, they have a strong set of muscles that open and close the shell for swimming (Barrett & Yonge, 1958). It is possible to age a bivalve by the number of growth rings or growth lines on its shell. What makes them an interesting aging model is that individuals of some species have been aged as over 500 years, making them a model of extreme longevity (Abele et al., 2009).

Bivalves in aging research

A great diversity in species lifespan has been seen in the bivalves, with some species living less than a year and others living over 500 years. The presence of the growth rings makes them an attractive model for extreme longevity studies because it is possible to determine an animal's age without maintaining it in the lab its whole life (Abele et al., 2009). A number of studies have reported these animals to have very tight control of reactive oxygen species through many mitochondria, each with quite low energetic output (Abele et al., 2007). It has also been reported that proteins derived from long-lived bivalves appear less prone to form aggregates than proteins derived from mice (Treaster et al., 2014). Metabolic rate also appears to negatively correlate with bivalve species lifespan (Philipp, Portner, & Abele, 2005). Similarly, many antioxidant enzymes appear positively correlated with bivalve species lifespan; this is most apparent in the extremely long-lived *Antarctica islandica* with a lifespan of over 200 years and high levels of the antioxidant enzymes catalase and superoxide dismutase (Zielinski and Portner, 2000). When comparing long- and short-lived bivalve species, much lower levels of hydrogen peroxide levels are released from longer-lived species. The animals also have lower levels of protein damage and appear more resistant to mortality from oxidative stress (Ungvari et al., 2011). The animals do appear to age; studies have reported an age-related decline in proteasome function and in many molecular chaperones, suggesting that the animals do still undergo an age-related loss of proteostasis (Sosnowska et al., 2014).

Honeybees

Phylum: Arthropoda
Class: Insecta
Order: Hymenoptera
Family: *Apidae*
Genus: *Apis*

Animals like the honeybee and the ant represent interesting aging models, in that not all animals are reproductive and thus the evolutionary pressures for lifespan will often differ within the species. Honeybees are a strong example of variable lifespan; the queen typically lives 8 years, with some reports of up to 13 years. In contrast, the workers will typically only live 2–6 weeks. It has been argued that these workers do indeed go through accelerated senescence; it has been reported that old worker bees become more susceptible to stressors, such as starvation and hydrogen peroxide, with potential small increases in heat susceptibility (Remolina et al., 2007), and that aging leads to a progressive increase in hazard ratio in these animals (Rueppell et al., 2007). The survival differences have also been shown not to be driven by increased foraging mortality amongst workers (Rueppell et al., 2007). In addition to variance between the queen and workers, there are also lifespan differences between other castes. It has been observed that the winter caste will live up to 10 times longer than the summer caste. These differences include both survival differences and differences in the onset of cognitive deficits. It has also been shown that alteration of environmental cues can alter the rate of onset, suggesting a plasticity to the rate of aging (Munch et al., 2013).

Sea urchins

Phylum: Echinodermata

All the invertebrate models discussed thus far and, indeed, most extant animal species, are protostomes. Chordates, however, are deuterostomes, and thus separated by some 600 million years of evolution. The other major deuterostome phyla are echinoderms (e.g., starfish, sea urchins, sea cucumbers) and hemichordates (acorn worms). The comparative closeness of these animals to humans (relative to other invertebrate models) makes them interesting models with potentially greater genetic similarity to humans than *Drosophila* and *C. elegans*.

Sea urchins have been used extensively in developmental studies, as it is relatively easy to observe developmental processes in their embryos. They have a number of distinct organs, including a defined digestive tract and a simple nervous system. The sea urchin does not have a brain, as such, but has a nerve ring encircling the mouth and is sensitive to environmental stimuli, such as touch, light, and chemicals, via numerous sensitive cells in the epithelium. Uniquely, they have a musculoskeletal system composed of a series of spines, controlled by ball and socket joints, and a series of tube feet that may be protruded through use of a water vascular system, permitting motion through hydraulic pressure. Their body cavity is filled with coelom, which forms the main fluid circulatory system in their body. They also possess a hemal circulatory system with a complex series of vessels around the gut, but the function of this system is not well understood.

Sea urchins in aging research

Sea urchins are interesting as an aging model, because of their relative closeness to vertebrates, their high regenerative capacity, and their wide range of lifespans. Similar to bivalves, sea urchin species show a dramatic range of lifespans. The red sea urchin *Strongylocentrotus franciscanus* is regarded as one of the longest-lived animals, with a maximum lifespan estimated at 200 years (Ebert, 2008; Ebert and Southon, 2003). In contrast, urchin species such as *Lytechinus variegatus* live only for about 4 years (Beddingfield and McClintock, 2000). Interestingly, sea urchins do not appear to show an increase in mortality rates or a decline in reproduction with age (Ebert, 2008; Ebert and Southon, 2003). It has been reported that sea urchins maintain regenerative capacity with age, but surprisingly show no differences in regenerative capacity between long- and short-lived species (Bodnar and Coffman, 2016). It is thus a question of whether they do indeed age in the same sense as mammals. A highly interesting area of study in these animals is how they maintain this negligible senescence. Genetic profiling has shown that these animals do indeed undergo changes in genetic profile with age, including changes in the ubiquitin–proteasome pathway, DNA metabolism, signaling pathways, and apoptosis. The changes in gene expression with age appear consistent between long- and short-lived sea urchin species. From this research, a number of pathways have been identified that may explain their negligible senescence, including changes in Notch and Wnt signaling and the lipoprotein receptor LRP4 (Bodnar, 2013; Loram and Bodnar, 2012). Other researchers have reported several genomic sequence differences that may explain the two orders of magnitude lifespan difference between sea urchin species (Sergiev et al., 2016). It has also been found that these animals do not experience age-related declines in telomerase activity (neither long-lived nor short-lived sea urchins) suggesting that these animals do not undergo repression of telomerase activity with age as an anticancer strategy, as is seen in other species (Francis et al., 2006).

Nonbilaterians

Most animals, and all those discussed thus far, are bilaterian. All bilaterians exhibit bilateral symmetry as embryos and most continue to do so as adults. Nonbilaterians represent basal animal forms; they have differentiated cell types but lack digestive organs or circulatory systems. Of the four nonbilateral phyla, Porifera (sponges) and Placozoa are the simplest metazoans, while Ctenophora (comb-jellies) and Cnidaria (e.g., jellyfish, hydra, anemones, corals) possess simple nervous systems and sensory organs. These animals are noted for having extraordinary regenerative ability. It is difficult to say whether they experience aging as we do or if some may be, in a sense, immortal. Dating of the various skeletons formed by several species of sponges indicates that some can live for hundreds, even thousands of years in the right environment (Gatti, 2002), which would make them the longest-lived animals by far and among the longest-lived multicellular organisms (Petralia et al., 2014).

Animals so long diverged evolutionarily might seem to have little relevance to understanding human biology, but surprisingly, a number of genes have been identified in Cnidarians, specifically, the coral species *Acropora millepora*, that have mammalian homologs but are not found in either *Drosophila* or *C. elegans* (Kortschak et al., 2003).

Hydra

Phylum: Cnidaria
Class: Hydrozoa
Order: Anthoathecata
Family: *Hydridae*
Genus: *Hydra*

Hydra are small freshwater organism in the Cnidaria phylum and are thus related to the jellyfish.

They are approximately 10 mm in size, with radial symmetry, and possess a simple adhesive foot. The hydra has a mouth at one end of its body surrounded by 1–12 tentacles. The tentacles contain specialized stinging cells called cnidocytes that are utilized for trapping and immobilizing prey. The hydra has a simple nervous system, composed of a nerve net, but does not have a distinct central nervous system. It is typically sessile, but under stimulation it will retract its tentacles into its body. The hydra is of interest to aging research because of its high regenerative capacity and the suggestion that it arguably does not age.

Hydra regeneration

Hydra reproduce by budding; a bud will appear midway down the animal's body which will develop into a new animal. If a hydra is cut in half, each half will regenerate to form a complete hydra. Even if the hydra is cut into many pieces, each will form a new hydra: the top ones will form a foot, the bottom ones will form a head and the middle ones will form both a head and a foot.

Hydra in aging research

In 1998, Daniel Martinez argued that hydra did not undergo senescence and could be defined as immortal. His group examined mortality rates of hydra over a period of 4 years and argued that they showed no age-related increase in mortality, at least in the 4-year window examined (Martinez, 1998). However, work by Preston W. Estep reports that over a 4-year period the hydra showed a decline in reproduction, suggesting an aging phenotype. Interestingly, the long-lived "immortal" phenotype of the hydra appears species-specific; very little mortality was detected in *Hydra vulgaris* over 4 years, while an aging curve with complete extinction was reported in *Hydra oligactis* over 250 days after its sexual differentiation (Martinez and Bridge, 2012; Yoshida et al., 2006). These animals also experience aging-like phenotypes, including reduced ability to capture shrimp, fewer contractile motions, and deterioration in actin fiber structure (Yoshida et al., 2006). The differences between these two species have been argued to represent a powerful model for understanding aging (Tomczyk et al., 2015). There has been some research into the role of FoxO signaling in hydra. It has been reported that FoxO is expressed at high levels in the hydra stem cells and that its expression is induced under conditions of stress (Bridge et al., 2010). However, while reducing FoxO expression does reduce stem cell proliferation, FoxO-deficient hydra surprisingly do not appear to lose their negligible senescence phenotype (Boehm et al., 2012).

Turritopsis dohrnii—"the immortal jellyfish"

Phylum: Cnidaria
Class: Hydrozoa
Order: Anthoathecata
Family: *Oceaniidae*
Genus: *Turritopsis*

Turritopsis dohrnii is sometimes referred to as the immortal jellyfish. It is found in the Mediterranean Sea and around Japan. It has a free-swimming adult medusa stage, in which it exists as a bell-shaped organism, about 4.5 mm in size, surrounded by up to 90 tentacles. It has a dense nerve net with a classical ring-like structure around a radial canal (Koizumi et al., 2015). Under conditions of stress or "age" it will convert back to a polyp-like state, allowing regeneration; it is this lifecycle reversal strategy that has gained this species the name "the immortal jellyfish" (Piraino et al., 2004). However, there has been limited research into whether this animal is truly immortal or if it does experience any form of age-related decline.

References

Abele, D., Brey, T., & Philipp, E. (2009). Bivalve models of aging and the determination of molluscan lifespans. *Experimental Gerontology*, *44*(5), 307–315.

Abele, E., et al. (2007). Marine invertebrate mitochondria and oxidative stress. *Frontiers in Bioscience: A Journal and Virtual Library*, *12*, 933–946.

Adams, M. D., et al. (2000). The genome sequence of Drosophila melanogaster. *Science*, *287*(5461), 2185–2195.

Agrawal, N., et al. (2005). Identification of combinatorial drug regimens for treatment of Huntington's disease using *Drosophila*. *Proceedings of the National Academy of Sciences of the United States of America*, *102*(10), 3777–3781.

Ali, Y. O., et al. (2011). Assaying locomotor, learning, and memory deficits in *Drosophila* models of neurodegeneration. *Journal of Visualized Experiments*, *49*()).

Allen, J. A., & Morgan, R. E. (1981). The functional morphology of Atlantic deep water species of the families Cuspidariidae and Poromyidae (Bivalvia): An analysis of the evolution of the septibranch condition. *Philosophical Transactions of the Royal Society B*, *294*(1073).

Amrit, F. R., et al. (2014). The *C. elegans* lifespan assay toolkit. *Methods*, *68*(3), 465–475.

Ashburner. (1989). Drosophila. *A. Laboratory Handbook*. Cold Spring Harbor Laboratory Press.

Ayyadevara, S., et al. (2008). Remarkable longevity and stress resistance of nematode PI3K-null mutants. *Aging Cell*, *7*(1), 13–22.

Barrett, J., & Yonge, C. M. (1958). *Pocket Guide to the Sea Shore*. London: William Collins Sons and Co.

Beddingfield, S. D., & McClintock, J. B. (2000). Demographic characteristics of lytechinus variegatus (echinoidea: echinodermata) from three habitats in North Florida Bay, Gulf of Mexico. *Marine Ecology*, *21*, 17–40.

Bellen, H. J., Tong, C., & Tsuda, H. (2010). 100 years of *Drosophila* research and its impact on vertebrate neuroscience: A history lesson for the future. *Nature Reviews Neuroscience*, *11*(7), 514–522.

Bock, M. J., et al. (2019). Maternal age alters offspring lifespan, fitness, and lifespan extension under caloric restriction. *Scientific Reports*, *9*(1), 3138.

Bodnar, A. G., & Coffman, J. A. (2016). Maintenance of somatic tissue regeneration with age in short- and long-lived species of sea urchins. *Aging Cell*, *15*(4), 778–787.

Bodnar, A. (2013). Proteomic profiles reveal age-related changes in coelomic fluid of sea urchin species with different life spans. *Experimental Gerontology*, *48*(5), 525–530.

Boehm, A. M., et al. (2012). FoxO is a critical regulator of stem cell maintenance in immortal Hydra. *Proceedings of the National Academy of Sciences of the United States of America*, *109*(48), 19697–19702.

Brand, A. H., & Perrimon, N. (1993). Targeted gene expression as a means of altering cell fates and generating dominant phenotypes. *Development*, *118*(2), 401–415.

Brenner, S. (1974). The genetics of *Caenorhabditis elegans*. *Genetics*, *77*(1), 71–94.

Bridge, D., et al. (2010). FoxO and stress responses in the cnidarian Hydra vulgaris. *PLoS One*, *5*(7), e11686.

Brys, K., Vanfleteren, J. R., & Braeckman, B. P. (2007). Testing the rate-of-living/oxidative damage theory of aging in the nematode model *Caenorhabditis elegans*. *Experimental Gerontology*, *42*(9), 845–851.

Calfon, M., et al. (2002). IRE1 couples endoplasmic reticulum load to secretory capacity by processing the XBP-1 mRNA. *Nature*, *415*(6867), 92–96.

Castle, W. E. (1906). Inbreeding, cross-breeding and sterility in *Drosophila*. *Science*, *23*(578), 153.

Chapman, T., et al. (1995). Cost of mating in *Drosophila melanogaster* females is mediated by male accessory gland products. *Nature*, *373*(6511), 241–244.

Chapman, T., Hutchings, J., & Partridge, L. (1993). No reduction in the cost of mating for *Drosophila melanogaster* females mating with spermless males. *Proceedings: Biological Sciences/The Royal Society*, *253*(1338), 211–217.

Collins, J. J., Huang, C., Hughes, S., & Kornfeld, K. (2008). The measurement and analysis of age-related changes in *Caenorhabditis elegans*. *WormBook: The Online Review of C. Elegans Biology*, 1–21.

C. elegans Sequencing Consortium. (1998). Genome sequence of the nematode *C. elegans*: A platform for investigating biology. *Science*, *282*(5396), 2012–2018.

Dillon, M. E., et al. (2009). Review: Thermal preference in *Drosophila*. *Journal of Thermal Biology*, *34*(3), 109119.

Duffy, J. B. (2002). GAL4 system in *Drosophila*: A fly geneticist's Swiss army knife. *Genesis*, *34*(1–2), 1–15.

Duhon, S. A., Murakami, S., & Johnson, T. E. (1996). Direct isolation of longevity mutants in the nematode *Caenorhabditis elegans*. *Developmental Genetics*, *18*(2), 144–153.

Ebert, T. A. (2008). Longevity and lack of senescence in the red sea urchin Strongylocentrotus franciscanus. *Experimental Gerontology, 43*(8), 734–738.

Ebert, T., & Southon, J. R. (2003). Red sea urchins (Strongylocentrotus franciscanus) can live over 100 years: confirmation with A-bomb 14carbon. *Fishery Bulletin, 101*(4), 915–922.

Ebert, T. A. (2008). Longevity and lack of senescence in the red sea urchin Strongylocentrotus franciscanus. *Experimental Gerontology, 43*(8), 734–738.

Feany, M. B., & Bender, W. W. (2000). A *Drosophila* model of Parkinson's disease. *Nature, 404*(6776), 394–398.

Finch, C. (1990). *Longevity, senescence, and the genome. The John D and Catherine T MacArthur Foundation series on mental health and development.* Chicago: University of Chicago Press.

Fischer, J. A., et al. (1988). GAL4 activates transcription in *Drosophila. Nature, 332*(6167), 853–856.

Francis, N., et al. (2006). Lack of age-associated telomere shortening in long- and short-lived species of sea urchins. *FEBS Letters, 580* (19), 4713–4717.

Friedman, D. B., & Johnson, T. E. (1988). A mutation in the age-1 gene in *Caenorhabditis elegans* lengthens life and reduces hermaphrodite fertility. *Genetics, 118*(1), 75–86.

Gatti, S., (2002). The Role of Sponges in High-Antarctic Carbon and Silicon Cycling - a Modelling Approach [Die Rolle der Schwamme im hochantarktischen Kohlenstoff- und Silikatkreislauf - ein Modellierungsansatz]. Berichte zur Polar- und Meeresforschung, 434. Bremerhaven Bremen: Alfred-Wegener-Institut für Polar- und Meeresforschung.

Gribble, K. E., et al. (2018). Congeneric variability in lifespan extension and onset of senescence suggest active regulation of aging in response to low temperature. *Experimental Gerontology, 114*, 99–106.

Gribble, K. E., & Mark Welch, D. B. (2017). Genome-wide transcriptomics of aging in the rotifer *Brachionus manjavacas*, an emerging model system. *BMC Genomics, 18*(1), 217.

Haddad, L. S., Kelbert, L., & Hulbert, A. J. (2007). Extended longevity of queen honey bees compared to workers is associated with peroxidation-resistant membranes. *Experimental Gerontology, 42*(7), 601–609.

Hanson, K. A., et al. (2010). Ubiquilin modifies TDP-43 toxicity in a *Drosophila* model of amyotrophic lateral sclerosis (ALS). *The Journal of Biological Chemistry, 285*(15), 11068–11072.

Harmer, S. F., & Shipley, A. E. (1896). *The Cambridge Natural History.* The Macmillan Company.

Hart, A. C. (2006). Behavior. *WormBook: The Online Review of C. elegans Biology*, 14–30.

Hartman, P. S., et al. (2001). Mitochondrial mutations differentially affect aging, mutability and anesthetic sensitivity in *Caenorhabditis elegans. Mechanisms of Ageing and Development, 122*(11), 1187–1201.

Hasegawa, Y., et al. (2016). De novo assembly of the transcriptome of turritopsis, a jellyfish that repeatedly rejuvenates. *Zoological Science, 33*(4), 366–371.

Herndon, L. A., et al. (2002). Stochastic and genetic factors influence tissue-specific decline in ageing *C. elegans. Nature, 419*(6909), 808–814.

Honeybee Genome Sequencing Consortium. (2006). Insights into social insects from the genome of the honeybee Apis mellifera. *Nature, 443*(7114), 931–949.

Hyde, Roscoe R. (1913). Inheritance of the length of life in *Drosophila ampelophila. Proceedings of the Indiana Academy of Science, 23*, 113–123.

Jobson, M. A., et al. (2015). Transgenerational effects of early life starvation on growth, reproduction, and stress resistance in *Caenorhabditis elegans. Genetics, 201*(1), 201–212.

Kaneko, G., et al. (2011). Calorie restriction-induced mother's longevity is transmitted to the daughter's generation in the rotifer *Brachionus plicatilis. Functional Ecology, 25*, 209–216.

Kauler, E., & Hildegard, E. E. (2011). The effect of temperature on life history parameters and cost of reproduction in the rotifer *Brachionus calyciflorus. Journal of Freshwater Ecology, 26*(3), 399–408.

Kenyon, C., et al. (1993). A *C. elegans* mutant that lives twice as long as wild type. *Nature, 366*(6454), 461–464.

Kevin, F., & Partridge, L. (1989). A cost of mating in female fruitflies. *Nature, 338*, 760–761.

Kirk, K. L. (2001). Dietary restriction and aging: Comparative tests of evolutionary hypotheses. *The Journals of Gerontology. Series A, Biological Sciences and Medical Sciences, 56*(3), B123–B129.

Klass, M. R. (1977). Aging in the nematode *Caenorhabditis elegans*: Major biological and environmental factors influencing life span. *Mechanisms of Ageing and Development, 6*(6), 413–429.

Klass, M. R. (1983). A method for the isolation of longevity mutants in the nematode *Caenorhabditis elegans* and initial results. *Mechanisms of Ageing and Development, 22*(3–4), 279–286.

Koh, K., et al. (2006). A *Drosophila* model for age-associated changes in sleep:wake cycles. *Proceedings of the National Academy of Sciences of the United States of America, 103*(37), 13843–13847.

Koizumi, O., et al. (2015). The nerve ring in cnidarians: its presence and structure in hydrozoan medusae. Zoology (Jena*), 118*(2), 79–88.

Kortschak, R. D., et al. (2003). EST analysis of the cnidarian Acropora millepora reveals extensive gene loss and rapid sequence divergence in the model invertebrates. *Current Biology, 13*(24), 2190–2195.

Landis, G. N., et al. (2015). The progesterone antagonist mifepristone/RU486 blocks the negative effect on life span caused by mating in female *Drosophila. Aging, 7*(1), 53–69.

Le Bourg, E. (1987). The rate of living theory: Spontaneous locomotor activity, aging and longevity in *Drosophila melanogaster. Experimental Gerontology, 22*(5), 359–369.

Lev, I., et al. (2019). Inter-generational consequences for growing *Caenorhabditis elegans* in liquid. *Philosophical Transactions of the Royal Society of London. Series B, Biological Sciences, 374*(1770), 20180125.

Linford, N. J., et al. (2013). Measurement of lifespan in *Drosophila melanogaster. Journal of Visualized Experiments* (71).

Link, C. D., et al. (1999). Direct observation of stress response in *Caenorhabditis elegans* using a reporter transgene. *Cell Stress & Chaperones, 4*(4), 235–242.

Loeb, J., & Northrop, J. H. (1917). On the influence of food and temperature upon the duration of life. *The Journal of Biological Chemistry, 32*, 103–121.

Loram, J., & Bodnar, A. (2012). Age-related changes in gene expression in tissues of the sea urchin Strongylocentrotus purpuratus. *Mechanisms of Ageing And Development, 133*(5), 338–347.

Malik, B. R., & Hodge, J. J. (2014). *Drosophila* adult olfactory shock learning. *Journal of Visualized Experiments, 90*, e50107.

Martin, J. R., Ernst, R., & Heisenberg, M. (1999). Temporal pattern of locomotor activity in *Drosophila melanogaster. Journal of Comparative Physiology A, 184*(1), 73–84.

Martinez, D. E. (1998). Mortality patterns suggest lack of senescence in hydra. *Experimental Gerontology, 33*(3), 217–225.

Martinez, D. E., & Bridge, D. (2012). Hydra, the everlasting embryo, confronts aging. *International Journal of Plant Developmental Biology, 56*(6-8), 479–487.

Miquel, J., et al. (1976). Effects of temperature on the life span, vitality and fine structure of *Drosophila melanogaster. Mechanisms of Ageing and Development, 5*(5), 347–370.

Mishur, R. J., et al. (2016). Mitochondrial metabolites extend lifespan. *Aging Cell, 15*(2), 336–348.

Moenkhaus, W. J. (1911). Inbreeding and selection on fertility and vigor. *Journal of Morphology, 22*, 126140.

Morgan, T. H. (1911). The origin of five mutations in eye color in drosophila and their modes of inheritance. *Science, 33*(849), 534–537.

Morley, J. F., et al. (2002). The threshold for polyglutamine-expansion protein aggregation and cellular toxicity is dynamic and influenced by aging in *Caenorhabditis elegans*. *Proceedings of the National Academy of Sciences of the United States of America, 99* (16), 10417–10422.

Mouton, S., et al. (2011). Lack of metabolic ageing in the long-lived flatworm *Schmidtea polychroa*. *Experimental Gerontology, 46*(9), 755–761.

Mouton, S., et al. (2018). Resilience to aging in the regeneration-capable flatworm *Macrostomum lignano*. *Aging Cell, 17*(3), e12739.

Munch, D., Kreibich, C. D., & Amdam, G. V. (2013). Aging and its modulation in a long-lived worker caste of the honey bee. *Journal of Experimental Biology, 216*(Pt 9), 1638–1649.

Munkacsy, E., et al. (2019). Neuronal-specific proteasome augmentation via Prosβ5 overexpression extends lifespan and reduces age-related cognitive decline. *Aging Cell, 18*, e13005.

Nakata, K., et al. (2005). Regulation of a DLK-1 and p38 MAP kinase pathway by the ubiquitin ligase RPM-1 is required for presynaptic development. *Cell, 120*(3), 407–420.

Nigon, V. M., & Félix, M.-A. (2017). History of research on *C. elegans* and other free-living nematodes as model organisms. WormBook: *The online review of C. elegans biology*, 1–17.

Ocorr, K., Akasaka, T., & Bodmer, R. (2007). Age-related cardiac disease model of *Drosophila*. *Mechanisms of Ageing and Development, 128*(1), 112–116.

Oo, A. K., et al. (2010). Identification of genes differentially expressed by calorie restriction in the rotifer (*Brachionus plicatilis*). *Journal of Comparative Physiology B, 180*(1), 105–116.

Pearl, R. (1926). A synthetic food medium for the cultivation of *Drosophila*: Preliminary note. *The Journal of General Physiology, 9*(4), 513–519.

Pearl, R., & Parker, S. L. (1922). On the influence of density of population upon the rate of reproduction in *Drosophila*. *Proceedings of the National Academy of Sciences of the United States of America, 8*(7), 212–219.

Pearl, R., White, F. B., & Miner, J. R. (1929). Age changes in alcohol tolerance in *Drosophila melanogaster*. *Proceedings of the National Academy of Sciences of the United States of America, 15*(5), 425–429.

Petralia, R. S., Mattson, M. P., & Yao, P. J. (2014). Aging and longevity in the simplest animals and the quest for immortality. *Ageing Research Reviews, 16*, 66–82.

Philipp, E., Portner, H. O., & Abele, D. (2005). Mitochondrial ageing of a polar and a temperate mud clam. *Mechanisms of Ageing and Development, 126*(5), 610–619.

Piraino, S., et al. (1996). Reversing the life cycle: medusae transforming into polyps and cell transdifferentiation in turritopsis nutricula (cnidaria, hydrozoa). *The Biological Bulletin, 190*(3), 302–312.

Piraino, S., et al. (2004). Reverse development in Cnidaria. *Canadian Journal of Zoology, 82*(11), 1748–1754.

Poeggeler, B., et al. (2005). Mitochondrial medicine: Neuroprotection and life extension by the new amphiphilic nitrone LPBNAH acting as a highly potent antioxidant agent. *Journal of Neurochemistry, 95*(4), 962–973.

Poeggeler, B., et al. (2010). A novel endogenous indole protects rodent mitochondria and extends rotifer lifespan. *PLoS One, 5*(4), e10206.

Rechavi, O., et al. (2014). Starvation-induced transgenerational inheritance of small RNAs in *C. elegans*. *Cell, 158*(2), 277–287.

Rera, M., Clark, R. I., & Walker, D. W. (2012). Intestinal barrier dysfunction links metabolic and inflammatory markers of aging to death in *Drosophila*. *Proceedings of the National Academy of Sciences of the United States of America, 109*(52), 21528–21533.

Remolina, S. C., et al. (2007). Senescence in the worker honey bee Apis Mellifera. *Journal of Insect Physiology, 53*(10), 1027–1033.

Roman, G., et al. (2001). P[Switch], a system for spatial and temporal control of gene expression in *Drosophila melanogaster*. *Proceedings of the National Academy of Sciences of the United States of America, 98* (22), 12602–12607.

Rueppell, O., et al. (2007). Regulation of life history determines lifespan of worker honey bees (Apis mellifera L.). *Experimental Gerontology, 42*(10), 1020–1032.

Rush, B., et al. (2007). Mating increases starvation resistance and decreases oxidative stress resistance in *Drosophila melanogaster* females. *Aging Cell, 6*(5), 723–726.

Sayeed, O., & Benzer, S. (1996). Behavioral genetics of thermosensation and hygrosensation in *Drosophila*. *Proceedings of the National Academy of Sciences of the United States of America, 93*(12), 6079–6084.

Sea Urchin Genome Sequencing, C., et al. (2006). The genome of the sea urchin Strongylocentrotus purpuratus. *Science, 314*(5801), 941–952.

Sergiev, P. V., et al. (2016). Genomes of Strongylocentrotus franciscanus and Lytechinus variegatus: are there any genomic explanations for the two order of magnitude difference in the lifespan of sea urchins? *Aging (Albany NY), 8*(2), 260–271.

Serra, M., Snelland, T. W., & King, C. E. (2004). In A. Moya, & E. Font (Eds.), *The Timing of Sex in Monogonont Rotifers, in Evolution: From Molecules to Ecosystems*. Oxford University Press.

Shapiro, H. (1932). The rate of oviposition in the fruit fly, *Drosophila*. *Biological Bulletin, 63*(3), 456–471.

Shmookler Reis, R. J., et al. (2009). Extreme-longevity mutations orchestrate silencing of multiple signaling pathways. *Biochimica et Biophysica Acta, 1790*(10), 1075–1083.

Shore, D. E., Carr, C. E., & Ruvkun, G. (2012). Induction of cytoprotective pathways is central to the extension of lifespan conferred by multiple longevity pathways. *PLoS Genetics, 8*(7), e1002792.

Skorupa, D. A., et al. (2008). Dietary composition specifies consumption, obesity, and lifespan in *Drosophila melanogaster*. *Aging Cell, 7* (4), 478–490.

Snell, T. W. (2014). Rotifers as models for the biology of aging. *International Review of Hydrobiology, 99*(1–2), 84–95.

Snell, T. W., Fields, A. M., & Johnston, R. K. (2012). Antioxidants can extend lifespan of *Brachionus manjavacas* (Rotifera), but only in a few combinations. *Biogerontology, 13*(3), 261–275.

Snell, T. W., & King, C. E. (1977). Lifespan and fecundity patterns in rotifers: The cost of reproduction. *Evolution; International Journal of Organic Evolution, 31*(4), 882–890.

Snell, T. W., Shearer, T. L., & Smith, H. A. (2011). Exposure to dsRNA elicits RNA interference in *Brachionus manjavacas* (Rotifera). *Marine Biotechnology, 13*(2), 264–274.

Sosnowska, D., et al. (2014). A heart that beats for 500 years: age-related changes in cardiac proteasome activity, oxidative protein damage and expression of heat shock proteins, inflammatory factors, and mitochondrial complexes in Arctica islandica, the longest-living noncolonial animal. *Journals of Gerontology - Series A Biological Sciences and Medical Sciences, 69*(12), 1448–1461.

Soukas, A. A., et al. (2009). Rictor/TORC2 regulates fat metabolism, feeding, growth, and life span in *Caenorhabditis elegans*. *Genes & Development, 23*(4), 496–511.

Sutphin, G. L., & Kaeberlein, M. (2009). Measuring *Caenorhabditis elegans* life span on solid media. *Journal of Visualized Experiments* (27).

Tomczyk, S., et al. (2015). Hydra, a powerful model for aging studies. *Invertebrate Reproduction & Development, 59*(sup1), 11–16.

Treaster, S. B., et al. (2014). Superior proteome stability in the longest lived animal. *Age, 36*(3), 9597.

Ungvari, Z., et al. (2011). Extreme longevity is associated with increased resistance to oxidative stress in Arctica islandica, the longest-living non-colonial animal. *Journals of Gerontology - Series A Biological Sciences and Medical Sciences, 66*(7), 741–750.

Van Raamsdonk, J. M., & Hekimi, S. (2011). FUdR causes a twofold increase in the lifespan of the mitochondrial mutant gas-1. *Mechanisms of Ageing and Development*, *132*(10), 519–521.

Vilchez, D., et al. (2012). RPN-6 determines *C. elegans* longevity under proteotoxic stress conditions. *Nature*, *489*(7415), 263–268.

Vogler, G., & Ocorr, K. (2009). Visualizing the beating heart in *Drosophila*. *Journal of Visualized Experiments* (31).

Wasik, K., et al. (2015). Genome and transcriptome of the regeneration-competent flatworm, Macrostomum lignano. *Proceedings of the National Academy of Sciences of the United States of America*, *112*(40), 12462–12467.

Webster, N., et al. (1988). The yeast UASG is a transcriptional enhancer in human HeLa cells in the presence of the GAL4 trans-activator. *Cell*, *52*(2), 169–178.

Wood, W. B. (1988). *The Nematode* Caenorhabditis elegans. Cold Spring Harbor Laboratory.

Xiao, R., et al. (2013). A genetic program promotes *C. elegans* longevity at cold temperatures via a thermosensitive TRP channel. *Cell*, *152*(4), 806–817.

Yoneda, T., et al. (2004). Compartment-specific perturbation of protein handling activates genes encoding mitochondrial chaperones. *Journal of Cell Science*, *117*(Pt 18), 4055–4066.

Yoshida, K., et al. (2006). Degeneration after sexual differentiation in hydra and its relevance to the evolution of aging. *Gene*, *385*, 64–70.

Yoshinaga, T., et al. (2005). Insulin-like growth factor signaling pathway involved in regulating longevity of rotifers. *Hydrobiologia*, *181*, 347–352.

Yoshinaga, T., Hagiwara, A., & Tsukamoto, K. (2000). Effect of periodical starvation on the life history of *Brachionus plicatilis* O.F. Muller (*Rotifera*): A possible strategy for population stability. *Journal of Experimental Marine Biology and Ecology*, *253*(2), 253–260.

Yoshinaga, T., Hagiwara, A., & Tsukamoto, K. (2003). Life history response and age-specific tolerance to starvation in *Brachionus plicatilis* O.F. Muller (Rotifera). *Journal of Experimental Marine Biology and Ecology*, *287*, 261271.

Yu, L., et al. (2013). Raf-mediated cardiac hypertrophy in adult *Drosophila*. *Disease Models & Mechanisms*, *6*(4), 964–976.

Zielinski, S., & Portner, H. O. (2000). Oxidative stress and antioxidative defense in cephalopods: a function of metabolic rate or age? *Comparative Biochemistry And Physiology B-Biochemistry & Molecular Biology*, *125*(2), 147–160.

CHAPTER

10

NIA Interventions Testing Program: A collaborative approach for investigating interventions to promote healthy aging

Francesca Macchiarini[1], *Richard A. Miller*[2], *Randy Strong*[3], *Nadia Rosenthal*[4] *and David E. Harrison*[4]

[1]Division of Aging Biology, National Institute on Aging, Bethesda, MD, United States [2]Department of Pathology and Geriatrics Center, University of Michigan, Ann Arbor, MI, United States [3]Department of Pharmacology, The University of Texas Health Science Center at San Antonio, and the Geriatric Research, Education and Clinical Center (GRECC) and Research Service of the South Texas Veterans Health Care System, Texas, TX, United States [4]The Jackson Laboratory, Bar Harbor, ME, United States

OUTLINE

Introduction

The Interventions Testing Program (ITP), funded by the National Institute on Aging (NIA), is a multisite collaborative project with the goal of identifying compounds that promote healthy aging in a genetically heterogeneous mouse model. The main goals are to provide preliminary evidence on compounds that have translational potential for human use to improve human health span with age, and to provide evidence for pathways that influence aging and are therefore targets for interventions. Funding of the ITP started in 2003, following a 1999 workshop where a panel of experts debated the pros and cons of numerous models to use as the test platform, before settling on genetically heterogeneous mice (Warner et al., 2000).

The program is funded by four cooperative agreement grants—three to support testing and one for a Data

Handbook of the Biology of Aging.
DOI: https://doi.org/10.1016/B978-0-12-815962-0.00010-X

Coordinating Center (DCC). The three testing sites, the University of Michigan (UM, PI Richard Miller), the Jackson Laboratory (JAX, PIs David Harrison and Nadia Rosenthal), and the University of Texas Health Science Center at San Antonio (UT, PI Randy Strong), work closely together with the NIA (represented by Francesca Macchiarini) to design and execute standard operating procedures (SOPs) that provide a consistent experimental protocol adhered to across the program. Each site also brings specialized expertise to the project, including statistical analysis, pharmacology, toxicology, optimal diet compounding, and nonharmful tests of age-sensitive physiological function. The DCC (JAX, PI Molly Bogue) supports the storage and analysis of primary and secondary data obtained by the ITP and provides a public-access archive of experimental outcomes and SOPs.

The ITP is an unusually collaborative program, in that it engages the broader research community in the search for agents that slow aging and extend mouse lifespan. The ITP makes an annual call for proposals to the research community, and proposals are accepted from all scientific quarters, including US and foreign universities, commercial entities, and even individuals without institutional affiliations. This open collaboration brings a diversity of outlooks and approaches to the program and taps into a wealth of experience in the greater scientific community.

Two committees provide advice to the ITP, in addition to the testing-site PIs and the sponsors of proposals. The Access Panel provides an independent review of each proposal, rating them on scientific rationale and feasibility within the ITP protocol. The ITP Steering Committee prioritizes the proposals and also provides advice on general protocol issues and specific changes under consideration by the ITP. The committees provide a breadth of expertise and help the ITP stay abreast of best practices.

Features of the Interventions Testing Program experimental design

The ITP was developed as a two-stage program. Stage I focuses on lifespan as the primary endpoint, while stage II studies follow-up on positive findings from stage I with additional lifespan studies, dose–response studies where appropriate, a wide array of health measurements, and cross-sectional pathology analysis. Although lifespan measures integrate all those processes that contribute to mortality risk, the decision to omit most physiological assessments from stage I studies could, in principle, miss discoveries of compounds that promote specific health outcomes (HOs) but do not alter mortality risk in mice. This two-stage design reflects a deliberate trade-off, in that inclusion of costly tests of health-related outcomes in stage I studies would necessarily diminish the number of compounds that can be tested each year.

In the initial year of the ITP, a SOP was developed to ensure that the same husbandry protocols were followed at the three sites with minimal variation. The SOP went through several revisions as new ideas were discussed and the importance of specific husbandry details emerged. Some aspects of laboratory mouse husbandry are fairly simple to standardize, such as light/dark cycles, humidity, HEPA air filtration, diet, documentation of pathogen control, and choice of bedding, while others are essentially impossible to standardize across sites, such as intestinal flora, light quality, barometric pressure, the levels of minerals and organic compounds in the drinking water, and the room air. These uncontrolled site-specific factors can and do lead to variations among ITP sites in lifespan and other endpoints; conversely, replication of findings across sites implies that the findings may be robust and may prove to be reproducible in other laboratories.

The mouse model used for all ITP studies is the UM-HET3 four-way cross (hereinafter termed "HET3"), produced by crossing CByB6F1 mothers (JAX stock #100009) with C3D2F1 fathers (JAX stock #100004) (Nadon et al., 2008). This genetically heterogeneous cross has been well characterized in previous aging studies (Harper et al., 2003; Miller & Chrisp, 1999), as well as in the ITP. It has the advantage that, while no two four-way cross mice are identical, each mouse shares half of its genome with every other mouse in the population and the population-level diversity can be replicated at any time by starting with the same parental strains, that is, all mice are sibs from the perspective of the nuclear genome. The genetic heterogeneity reduces the possibility that strain-specific characteristics might influence the outcome. Furthermore, the HET3 model has been found to model key aspects of human aging, such as the female survival advantage and the inverse relationship between body weight and longevity, as well as its reversal in old age (Cheng et al., 2019). Thus, the model represents the mouse species, and models the human species, far better than is possible using a single genotype.

F1 hybrid breeding stock mice are purchased from The Jackson Laboratory, and breeding is done at each of the three sites to produce each year's cohort of HET3 mice. The mice for each year's testing are bred over 6–8 months to minimize circannual effects. Each site has 44 males and 36 females in each test treatment group. Males are oversampled because of the losses expected due to fighting. The control group at each site has twice the number of mice, 88 males and 72 females. This protocol was developed after consultation with two statisticians, Andrzej Galecki and Scott Pletcher, whose power analysis indicated that groups of this size would detect a 10% change in mean lifespan with 80% power even if data from one of the sites was lost because of a systematic failure, such as a loss of pathogen-free status or a failure of temperature control (Miller et al., 2007).

The first litter from each breeding pair is used for purposes other than the lifespan studies (pilot studies, etc.) so that the mice entering into the stage I or stage II studies are all from experienced mothers. This lessens hypothetical effects of variations in early life nurturing and nutrition. Mice are weaned at 19–21 days and housed in same-sex cages, four females or three males per cage, using corn cob bedding (Bed O'Cobs, The Andersons, Maumee, Ohio) and nesting material. Mice have free access to food and acidified water at all times. At 6 weeks of age, mice are anesthetized and implanted with a microchip to provide identification, and the tail tip is collected and saved for potential genetic analysis. Each cage is then assigned to treatment or control protocols using a different random number table at each site, to provide variation from site to site in the order of assignment of mice to control and treatment groups.

In the absence of other scientific or logistical factors, the default protocol is to begin drug treatment when the mice turn 4 months old. In many cases, however, pilot studies (described below) took longer than anticipated, or issues with drug stability resulted in delays so that drug treatment was not begun until the mice were somewhat older. In other cases, drug was withheld to a later age to avoid undesirable biological effects of the compound to be tested. An example of the latter is the 17α-estradiol study, where treatment began at 10 months of age to reduce the risk of isomerization to 17β-estradiol, a conversion that appears to be limited to young animals in rodents (Hajek et al., 1997). Body weight is measured for all mice at 6, 12, 18, and 24 months of age.

Test compounds are administered in the diet, using an NIH31-based formulation. Breeding cages are fed irradiated Purina 5008 diet; weaned mice pretreatment are fed irradiated Purina LabDiet 5LG6 diet. Test compounds are formulated into ground, previously irradiated, 5LG6 by Purina TestDiet and repelleted with minimal exposure to heat, as many of the compounds are heat sensitive. Batches of food are prepared to meet the needs of all three sites for 4-month intervals so that all sites are always using the same batches of food. The use of the same diet, provided for each site, is critical, because NIH31 is an open-source formulation that specifies only the target levels of nutrients, not the source of ingredients. NIH31-based diets from different commercial vendors vary greatly in ingredients, particularly the protein source, ranging from soybean meal to fish meal to pork meal. Upon receipt of the diets, individual food pellets are tested at the ITP Analytical Pharmacology Lab at the UT site to ensure that the food contains the expected amount of the test compound. The SOP mandates culling entire cages from the longevity study if there is excessive fighting leading to open wounds. If a mouse is found with wounds covering 25% or more of its skin, all mice in the cage are euthanized and removed from the study. This policy helps to avoid oversampling of mice with atypically aggressive behavior.

Cages are checked daily for health, and mice are euthanized when an animal appears moribund, that is, unlikely to survive another 2 days. Classification of moribund status is based on the appearance of at least two of the following clinical signs: (1) severe lethargy, as indicated by reluctance to move when gently prodded with a forceps; (2) inability to eat or to drink; (3) severe balance or gait disturbance; (4) rapid weight loss over a period of 1 week or more; or (5) an ulcerated or bleeding tumor.

Triplicate testing sites are a key aspect of the ITP testing paradigm. Variation among sites in outcomes, including longevity and body weight trajectories, occurs even when testing protocols are strictly adhered to. Evidence was reported for variance in lifespan and incidence of neoplastic lesions for CD-1 mice from different facilities of one commercial mouse vendor (Engelhardt et al., 1993). Furthermore, performing the same behavioral tests in different laboratories can produce different results (Crabbe et al., 1999). And even mice in which specific clock genes have been knocked out have demonstrated significant variation in circadian rhythms in different laboratories (Van Gelder & Hogenesch, 2004).

The ITP has found site-to-site consistent, replicable differences in control male lifespan throughout at least its first 12 consecutive annual cohorts; the basis for this site-to-site variation is not understood (Fig. 10.1). Male control mice at UM are consistently longer-lived than male mice at the other two sites: based on median longevity, males at JAX live 91% as long as males at UM (786 vs. 836 days, respectively, average median survival over 12 consecutive annual cohorts), and males at UT live 88% as long (760 days), $P < .002$ in each case. In contrast, there is minimal site-to-site variation in median longevity for females (889 days at TJL; 886 at UM, and 876 at UT), with ratios of 100% (JAX) or 99% (UT) with respect to UM females.

We also see consistent and replicable site variations in body weight, but for this endpoint both males and females are equally affected. Control mice at UM, of both sexes, consistently have a lower body weight than the mice at UT or JAX (Fig. 10.2). This pattern was seen every year from 2005 to 2015 in both sexes, except for 2012 females. For reasons yet to be determined, the pattern was not seen in 2015 or 2016 in females, nor in 2016 for males. The basis for the site-to-site variation in lifespan of male but not female control mice, and

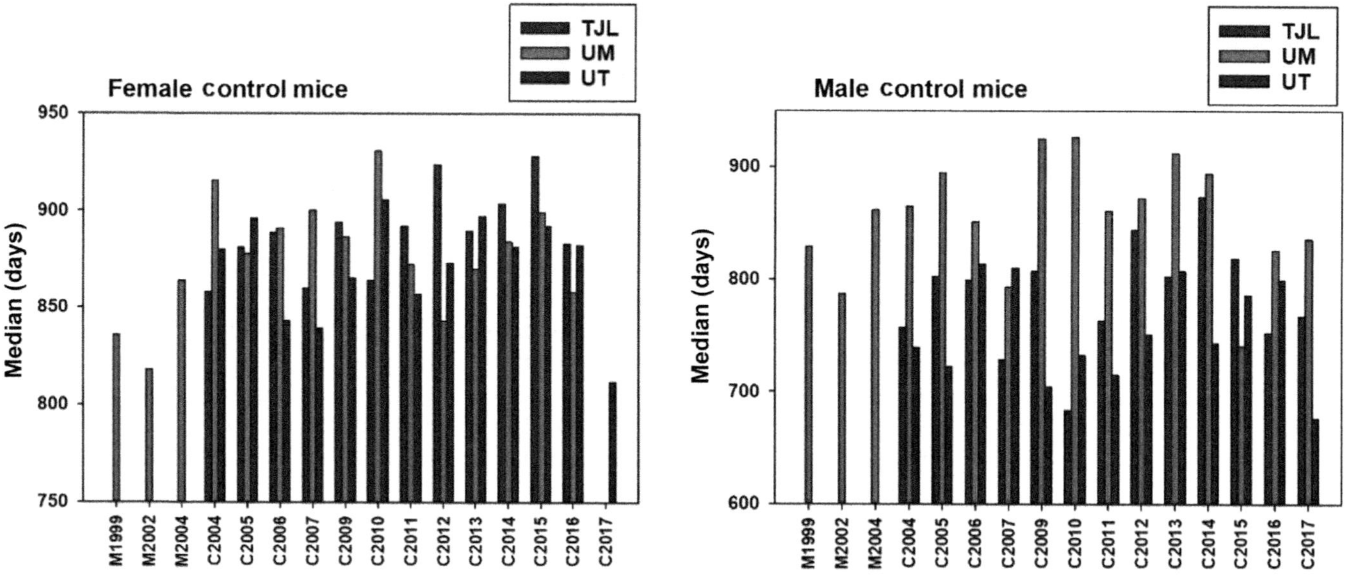

FIGURE 10.1 Lifespan of male control mice varied by site much more than did lifespan of female control mice. Sample size is approximately 100 male and 100 female mice per site.

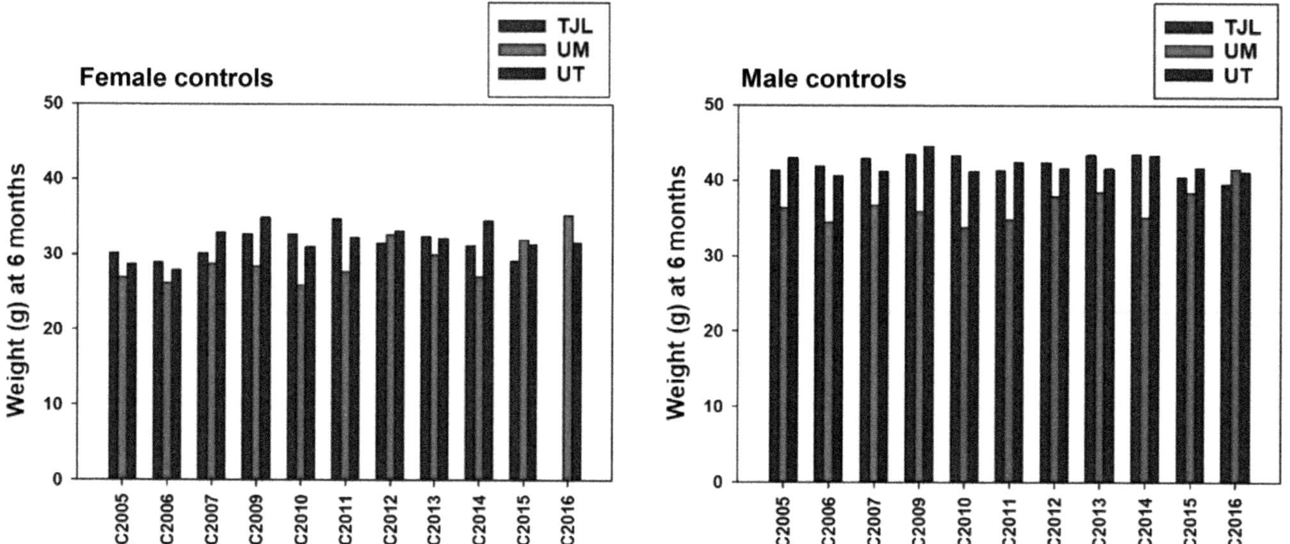

FIGURE 10.2 Body weight was measured in the entire cohort of control mice at all three sites when the mice reached 6 months of age. Sample size is approximately 100 male and 100 female mice per site.

the variation in body weights seen in both sexes, is unknown, but it illustrates the importance of triplicate testing for generating a high level of confidence in the findings of the ITP.

Types of intervention proposals sought by the Interventions Testing Program

The ITP entertains a wide range of proposals and is not focused on any one type of compound. As of the 2019 call for proposals, 157 proposals have been submitted to the ITP, with a few duplicates among them. The vast majority of proposals have been unique and have included drugs (e.g., aspirin, enalapril, angiotensin receptor subtype 1 blockers, nitroflurbiprofen, growth hormone, simvastatin, rapamycin, metformin, acarbose), common food or nutritional supplements (e.g., fish oil, vitamin B_1, curcumin, Protandim), synthetic chemical compounds that were originally discovered as components of plant extracts [e.g., nordihydroguaiaretic acid (NDGA), resveratrol], antioxidants [e.g., (4-hydroxyphenyl)-*N-tert*-butylnitrone (4-OHPBN), (3R,3′R)-zeaxanthin], and complex extracts from plants (e.g., green tea, mistletoe, and blueberry extracts). Proposals to remove components from the diet (e.g., selenium, iron, phosphate) have also been received but have not gone forward into testing because doing so would require introduction of a distinct variety of control diet, at extra cost, and because of concerns about toxicity. Proposals that require a chemically defined diet are expensive to implement, due to the cost of adding another control group. Agents that can be added to drinking water can also be considered. In principle, the ITP might also study agents that are administered by injection, but only if injections are administered for short periods of time or at infrequent intervals. Proposals that involve intermittent exposure to a test agent, or mixtures of agents, also receive consideration. With appropriate rationale, proposals can suggest alternate start ages, periodic dosing protocols, or more than a single test dose.

Applications suggesting complex plant extracts have in some cases been controversial, particularly if it is unclear what component or components of the extract are mostly likely to mediate its hypothesized effect on aging. When an extract contains a single ingredient thought to be of primary importance, this can be used to evaluate batch-to-batch consistency and to test for serum levels in the treated mice. When an extract (like Protandim®) is thought to contain several compounds acting in synergy, this strategy becomes more problematic, and when there is no consensus as to which components of a mixture are likely to be active, batch-to-batch standardization is not easily accomplished. For this reason, the ITP Access Committee gives higher priority to proposals focused on purified or semipurified components of biological extracts, although more complex mixtures can also be accepted, given adequate justification. Collaborative work involving chemists, pharmacologists, and aging researchers would be of value to learn more about which components of plant extracts are most likely to mediate their pharmacologic effects.

Some applications are rejected for stage I testing because of (1) feasibility—proposals that require daily injections, gavage, or an altered diet are incompatible with ITP protocols and funding limits; or (2) rapid metabolism—if a compound is rapidly metabolized to an inert form, it may not be possible to achieve, in mice, blood or tissue levels likely to have a pharmacological effect. To provide one example, trimethadione, an anticonvulsive medicine approved for use in humans, was shown to extend worm lifespan significantly (Evason et al., 2005), but in mice is rapidly metabolized to dimethadione (Tanaka et al., 1999), a form that has no lifespan benefit in worms.

Additional challenges that might prevent a proposal from going into stage I testing include toxicity of the compound or instability in the food preparation or storage, usually discovered during the 8-week pilot experiment conducted before the start of the full-scale longevity protocol. One way to circumvent stability issues is to encapsulate the material, as was done for rapamycin (discussed below). However, this process is expensive and may not be helpful, or necessary, for other compounds. Lastly, intellectual property issues have prevented some proposals from going into testing. The ITP is committed to publishing all findings, positive and negative, a policy that sometimes conflicts with the goals of some potential collaborators, particularly pharmaceutical companies. The ITP website lists the compounds entered into testing to date and also provides suggestions for preparing proposals (www.nia.nih.gov/research/dab/interventions-testing-program-itp).

Challenges encountered implementing testing protocols

The amount and quality of data supporting the proposal are of high importance for the development of a strong testing protocol. Deciding on which dose to test is often a substantial challenge. Drugs already in use in humans have the benefit of published therapeutic blood levels. Although the ITP does not require preliminary data in rodents or other mammalian models, such data can be very valuable, because it is difficult to extrapolate from findings in invertebrate models for determining which doses are most likely to be effective in stage I tests in mice, and because data in invertebrates do not fully predict either toxicity or effectiveness in mammals. The ITP has limited funding for preliminary dose–response studies, and therefore data that show which doses produce physiological effects in mammals, without toxicity, if provided by a sponsor, improve the likelihood that the agent will be accepted for the ITP, and improve the chances of testing an appropriate dose.

The ITP routinely conducts pilot studies to determine the stability of the compound in food, bioavailability (such as evidence for blood levels and/or biological effect after short-term feeding), and short-term toxicity. These pilot studies often require the development of new assays for detection of the test agents, an added benefit for the research community. A variety of assays have been developed for bioavailability and stability assessment, many of which make use of high-performance liquid chromatography (HPLC) coupled with UV or tandem mass spectrometry. For example, HPLC was used to measure the level, in plasma, of aspirin and its less active metabolite salicylic acid (Strong et al., 2008). An HPLC assay was developed to measure rapamycin in the blood, but an additional assay was used to demonstrate biological activity of the ingested rapamycin, measuring the phosphorylation of ribosomal protein S6 via S6 kinase 1, a downstream effector in the mechanistic target of the rapamycin (mTOR) pathway (Harrison et al., 2009). Thus, this combination of pharmacokinetic and pharmacodynamic approaches has proven very useful in guiding and justifying the stage I and stage II studies.

Stability of the compounds in food has sometimes been a major factor in determining which compounds go from pilot study to stage I testing. Some compounds proved too unstable for use, despite attempts to reformulate the compound. The most notable success at finding a workaround for instability was with rapamycin. In initial preparations, about 85% of the agent was degraded during food preparation, with correspondingly low blood values in treated mice. Through collaboration with the Southwest Research Institute, the ITP developed an encapsulation protocol that stabilized the rapamycin in the food and protected it during transit through the stomach, releasing the rapamycin in the intestine where it was efficiently absorbed (Harrison et al., 2009). The encapsulation process is expensive, so that using this method for a novel test agent would require strong justification from preliminary data. The bioanalytical assays developed for the pilot studies that demonstrate feasibility for stage I testing are also used during the execution of the stage I testing, to ensure that batches of food are consistent with regard to active compounds.

Summary of Interventions Testing Program findings

Table 10.1 lists the compounds that have entered testing as of 2020. Each set of survival data is analyzed in a standardized way. First, log-rank tests are used to compare each individual test group to the contemporaneous group of untreated controls; these tests are stratified by site and done separately for each sex. To test ideas about exceptional survival (sometimes loosely termed "maximal longevity"), we use the method of Wang and Allison (Wang et al., 2004) to evaluate the proportion of mice in the control and each test group at the age at which 90% of mice have died in the joint survival table, again with separate tests for males and females and stratification by test site. For each test we use $p = 0.05$ as our nominal significance criterion, recognizing that false positives may arise in a series of multiple comparisons, but preferring to make note of potential positive findings, which may deserve follow-up studies in ITP or other laboratories, even if some of these may later prove to represent chance effects alone (type I error). We also routinely include, as a secondary analysis, tests of each agent at each site, to identify situations in which drug effects are, or are not, reproducible across sites. Although these site-specific calculations have relatively little statistical power, they can provide reassurance when findings are seen to be robust or a cautionary note when the pooled result obscures important site-specific variations. This analytical program is conducted when the last living mouse has aged past the 90th percentile survival point for its site, because at this point the results of the Wang and Allison test are stable, and results of log-rank testing nearly so. Raw survival data sets are made publicly available on the DCC site [the Mouse Phenome Database (MPD)] so that other researchers can use them for additional statistical tests if they wish to do so.

We also routinely conduct an interim analysis at the point at which 50% of the control mice have died at each site, to give an initial indication as to possible drug efficacy. These results do not have much value as indices of effect on late-life deaths, and of course do not have the statistical power available from the completed data set, but they do provide "early warning" of agents that might merit more detailed experimentation. For example, the enhanced longevity resulting from treatment with rapamycin (Harrison et al., 2009; Miller et al., 2011) or acarbose (Harrison et al., 2014, 2019) was apparent long before the respective cohorts reached the 90th percentile, and this realization prompted early initiation of studies of the effects of these agents in the ITP laboratories, and elsewhere, a year or more before preparation of the final analysis and publication. Results of these interim analyses, at the 50th percentile, are shared with all collaborators in each year's ITP study, and can be presented at scientific meetings to stimulate discussion and new collaborative work, but are seldom published without special justification.

In special situations, we have also used additional statistical tests to address supplementary questions. For example, the survival curve for mice exposed to green tea extract (GTE) (Strong et al., 2013) suggested that this agent might benefit mice that would otherwise die at early ages, and the Gehan–Wilcoxon test, a variant of the log-rank test that gives additional weight to early deaths, was used to address this issue. We are careful to point out in our publications that such analyses, using approaches that were selected only after the data were available for inspection, carry much less weight as evidence than analysis using the protocols prespecified prior to data acquisition.

Using the approach described above, the ITP has noted increased longevity in at least one of the sexes for seven of the 43 agents tested so far: aspirin, NDGA, rapamycin, acarbose, 17α-estradiol, Protandim, and glycine. In some cases, the evidence is strong, reproducible, and internally consistent, and in others the evidence is much weaker. The original testing for some of them was followed by phase II studies, with the exception of glycine. stage II studies use approximately twice the resources (funding, cage space) of a stage I study, so the ITP must use great discretion in deciding which compounds should go into a stage II study. To date, there have been

TABLE 10.1 All compounds tested by the ITP through C2019[a] cohort.

Cohort 1: C2004—completed		
Compound	Concentration in food	Age at initiation
Aspirin[b]	20 ppm	4 months
NFP[c]	200 ppm	4 months
NDGA[d,e]	2500 ppm	9 months
4-OH-PBN[f]	315 ppm	4 months
Cohort 2: C2005—completed		
Compound	Concentration in food	Age at initiation
CAPE[g]	30 ppm	4 months
CAPE[g]	300 ppm	4 months
Enalapril maleate	120 ppm	4 months
Rapamycin[h]	14 ppm	20 month
Cohort 3: C2006—completed		
Compound	Concentration in food	Age at initiation
Rapamycin[i]	14 ppm	9 months
Simvastatin	12 ppm	10 months
Simvastatin	120 ppm	10 months
Resveratrol	300 ppm	12 months
Resveratrol	1200 ppm	12 months
Cohort 4: C2007—completed		
Compound	Concentration in food	Age at initiation
Resveratrol	300 ppm	4 months
Oxaloacetic acid	2200 ppm	4 months
Green tea extract[j]	2000 ppm	4 months
Curcumin	2000 ppm	4 months
Medium-chain triglyceride oil	60000 ppm	4 months
Cohort 5: C2009—completed		
Compound	Concentration in food	Age at initiation
17α-Estradiol[k]	4.8 ppm	10 months
Methylene blue[l]	28 ppm	4 months
Acarbose[m]	1000 ppm	4 months
Rapamycin_LoPhase II[n]	4.7 ppm	9 months
Rapamycin_MidPhase II[n]	14 ppm	9 months
Rapamycin_HiPhase II[n]	42 ppm	9 months
Cohort 6: C2010—completed		
Compound	Concentration in food	Age at initiation
Fish oil	15000 ppm	9 months
Fish oil	50000 ppm	9 months
NDGA[c] Lo_Phase II[o]	800 ppm	6 months (M only)
NDGA[c] Med_Phase II[o]	2500 ppm	6 months (M only)
NDGA[c] Hi_Phase II[o]	5000 ppm	6 months (M & F)

(Continued)

TABLE 10.1 (Continued)

Cohort 7: C2011—completed		
Compound	Concentration in food	Age at initiation
Bile Acids	5000 ppm	5 months
Metformin	1000 ppm	9 months
Metformin + rapamycin	1000 ppm M + 14 ppm R	9 months
Protandim[p]	600 ppm	10 months
17α-Estradiol	14.4 ppm	10 months
Cohort 8: C2012—completed		
Compound	Concentration in food	Age at initiation
INT-767 FXR/TG5R agonist[q]	180 ppm	10 months
Acarbose[r]	1000 ppm	16 months
HBX[s]	1 ppm	15 months
NDGA[d] cross-section study	2500 ppm (M) 5000 ppm (F)	13 months
Cohort 9: C2013—completed		
Compound	Concentration in food	Age at initiation
Ursolic acid	2000 ppm	10 months
Acarbose phase[t] II	2500 ppm	8 months
Acarbose phase[t] II	1000 ppm	8 months
Acarbose phase[t] II	400 ppm	8 months
Cohort 10: C2014—completed		
Compound	Concentration in food	Age at initiation
Supplemental glycine[u]	80,000 ppm	9 months
TM5441—inhibitor of PAI-1[v]	60 ppm	11 months
Inulin	600 ppm	11 months
Aspirin[w]	60 ppm	11 months
Aspirin[w]	200 ppm	11 months
Cohort 11: C2015—completed		
Compound	Concentration in food	Age at initiation
17-Dimethylaminoethylamino-17-demethoxygeldanamycin hydrochloride	30 ppm	6 months
MitoQ	100 ppm	7 months
Minocycline	300 ppm	6 months
β-Guanadinopropionic acid	3300 ppm	6 months
Rapamycin intermittent phase II	42 ppm	20 months
Cohort 12: C2016—almost completed		
Compound	Concentration in food	Age at initiation
MIF098	240 ppm	8 months
Nicotinamide riboside	1000 ppm	8 months
Canagliflozin—SGLT2 inhibitor	180 ppm	7 months
Candesartan cilexetil	30 ppm	8 months
Geranylgeranyl acetone	600 ppm	9 months
Hydrogen sulfide—SG1002	240 ppm	19 months
17α-Estradiol phase II at midlife	14.4 ppm	16 + 20 months

(Continued)

TABLE 10.1 (Continued)

Cohort 13: C2017—in progress		
Compound	Concentration in food	Age at initiation
R/S-1,3-butanediol	25,000 ppm	6 months
Captopril	180 ppm	5 months
L-Leucine	10,000 ppm	5 months
PB125	100 ppm	5 months
Sulindac	5 ppm	5 months
Syringaresinol	80 ppm	5 months
Rapamycin/acarbose phase II	14.7 ppm/1000 ppm	8 months
Cohort 14: C2018—in progress		
Compound	Concentration in food	Age at initiation
Fisetin on + cycling (3d on/11d off)	600 ppm	20 months
Hydrogen sulfide—SG1002 midlife	240 ppm	18 months
Cohort 15: C2019—in progress		
Compound	Concentration in food	Age at initiation
Astaxanthin	400 ppm	12 months
Dimethyl fumarate 9 + 16	120 ppm	9 months and 16 months
Meclizine	800 ppm	12 months
Mycophenolic acid	6.7 ppm	9 months
4-Phenylbutyrate	1000 ppm	9 months
Hydrogen sulfide—SG1002	240 ppm	6 months

[a]*Cohort date is the year the test mice were born.*
[b]*Increased lifespan in males but not females (Strong et al., 2008).*
[c]*Nitroflurbiprofen.*
[d]*Nordihydroguaiaretic acid.*
[e]*Increased mean lifespan in males but not females (Strong et al., 2008).*
[f]*4-OH -α-phenyl-N-tert-butyl nitrone.*
[g]*Caffeic acid phenethyl ester.*
[h]*Increased mean and maximal lifespan in both males and females (Harrison et al., 2009).*
[i]*Increased mean and maximal lifespan in males and females (Miller et al., 2011).*
[j]*Increase seen in mean lifespan in females only, that was significant only by the Gehan–Wilcoxon test, not by the log-rank test (Strong et al., 2013).*
[k]*Increased lifespan in males but not females (Harrison et al., 2014).*
[l]*Increased maximal but not mean lifespan in females; no effect in males (Harrison et al., 2014).*
[m]*Increased lifespan in both males and females, but the effects were greater in males (Harrison et al., 2014).*
[n]*Increased lifespan in both males and females (Miller et al., 2014).*
[o]*Increased lifespan in males but not females, even at doses that gave equivalent blood levels in males and females (Harrison et al., 2014).*
[p]*Protandim was increased to 1200 ppm when the mice reached 17 months of age.*
[q]*6α-ethyl-24-nor-5β-cholane-3α,7α,23-triol-23 sulfate sodium salt (dual FXR/TGR5 agonist).*
[r]*Increased lifespan, with higher maximum lifespan in both sexes and higher median lifespan in males; benefit lower than what observed with initiation of treatment at 4 months (Harrison et al., 2019).*
[s]*(2-(2-hydroxyphenyl)-benzoxazole).*
[t]*Increased lifespan in both males and females with all three doses with effects greater in males (Harrison et al., 2019).*
[u]*Increased lifespan in both males and females (Miller et al., 2019).*
[v]*Plasminogen activator inhibitor 1.*
[w]*Longevity effect observed using a 20 ppm dose was not replicated with doses of 60 and 200 ppm (Miller et al., 2019).*

three completed stage II studies, with three more initiated. The general structure of a stage II study includes a replication of the lifespan study, at multiple doses, histopathology analysis at 22–24 months, and an array of ancillary studies tailored to the compound under study. For the latter, the ITP solicits collaborations to broaden the scope of the physiological and biochemical parameters that can be measured, tapping into the wider expertise of the research community.

We summarize here the findings on the seven interventions that have shown a longevity benefit, ordered by the strength of the conclusions and depth of the supporting evidence thus far.

1. *Rapamycin* extends mean and maximal longevity, in both sexes, when started either at 9 or 20 months of age (Table 10.2; Harrison et al., 2009; Miller et al., 2011). Many age-sensitive changes in tissue structure or function were also delayed in rapamycin-treated mice (Wilkinson et al.,

TABLE 10.2 Increase in lifespan in HET3 mice treated with rapamycin initiated at 9 or 20 months of age

		% Median LS ↑	% Maximal LS ↑
Rapamycin initiation	**Cohort, dose**	**Males, females**	**Males, females**
20 months	C2005, 14 ppm	9%, 13%	9%, 14%
9 months	C2006, 14 ppm	10%, 18%	16%, 3%
9 months	C2009, 4.7 ppm	3%, 16%	6%, 5%
9 months	C2009, 14 ppm	13%, 21%	8%, 11%
9 months	C2009, 42 ppm	23%, 26%	8%, 11%

TABLE 10.3 Blood levels of rapamycin after 5 months of treatment

	4.7 ppm dose	14 ppm dose	42 ppm dose
Females	7 ng/mL	16 ng/mL	80 ng/mL
Males	6 ng/mL	9 ng/mL	23 ng.mL

2012), suggesting that this drug slows aging processes, although it is possible that effects of rapamycin on tumor cells per se also contribute to its beneficial effect on longevity (Johnson et al., 2013; Neff et al., 2013; Richardson, 2013). These initial findings have spurred a series of investigations as to the short-term and long-term effects of rapamycin on outcomes related to health and/or to postulated mechanisms of aging, including, for example: Bitto et al. (2016), Flynn et al. (2013), Lamming et al. (2013), Lesniewski et al. (2017), Sharp and Randy Strong (2010), Steinbaugh et al. (2012), and Ye et al. (2013). Follow-up studies using three different doses (14 ppm, the dose used in the stage I studies; 4.7 ppm, 3× less); and 42 ppm, 3× more), showed a dose-dependent response in both males and females, with the highest dose resulting in a 26% extension of the median lifespan in females and 23% in males (Miller et al., 2014). The similar degree of lifespan extension in males and females was in spite of the fact that blood levels of rapamycin were higher in females than in males (Table 10.3). It is noteworthy that even when each site's lifespan data are analyzed individually, the highest dose resulted in significant lifespan extension at all three sites, despite diminished statistical power, with greater variability among sites for males than for females (Miller et al., 2014).

Additional work performed by the ITP to study functional changes found that rapamycin prevented the age-associated loss of tendon elasticity (Zaseck et al., 2016) and the increase in endometriosis (Wilkinson et al., 2012). Rapamycin also produced a dose-dependent decrease in liver degeneration in male mice and a dose-independent reduction in the incidence of adrenal tumors and myocardial nuclear atypia (Wilkinson et al., 2012). Other pathologies, such as ovarian cysts and lung tumors, showed possible improvement from rapamycin treatment that did not rise to the level of statistical significance given the limited number of cases observed. It is likely that the therapeutic dose will vary for different pathologies, itself an interesting research question, that is, how does the response to inhibition of mTOR varies by cellular context. The rapamycin stage II mice at UM were also examined at 20 months of age for cataract development, with the high-dose rapamycin-treated mice of both sexes showing significantly more cataract severity than control mice (Wilkinson et al., 2012). Another detrimental effect of rapamycin is testicular degeneration, significantly elevated by all three doses of rapamycin.

These findings demonstrate that rapamycin retards the effects of aging on many organs, suggesting that it retards aging per se, as opposed to reducing one cause of death such as a specific cancer, and highlight the importance of comprehensive pathology and physiological analysis in intervention studies. The lifespan effect may be due to drug action on cancer cells per se, or to preservation of age-sensitive anticancer defenses, or to retardation of many aspects of aging, or to a combination of these factors.

2. *Acarbose* supplementation in the diet extended lifespan of both male and female mice (Harrison et al., 2014); this drug is FDA-approved for the clinical treatment of diabetes. Beneficial effects were seen at all three ITP sites (Harrison et al., 2014). Acarbose inhibits digestion of complex carbohydrates to simple sugars, blunting postprandial surges in blood glucose levels, although whether this accounts for the effect on mouse lifespan is still uncertain. The ITP is now completing a survival analysis of mice treated with canagliflozin, which also blunts postprandial glucose spikes, but through a renal mechanism; data showing a benefit from canagliflozin would strongly support protection from high glucose peaks as a modulator of aging rates in mice. Acarbose led to a much stronger effect on median lifespan in

males (22% increase $P < .0001$) than in females (5% increase, $P = .01$); increases in the age of 90% mortality were similar in both sexes (11 and 9%, both $P < .001$). Initiation of acarbose at 16 months of age gave significant, but smaller, lifespan benefits, with higher maximum lifespan in both sexes and higher median lifespan in males (Harrison et al., 2019). Follow-up work has shown male-specific retardation of hypothalamic inflammation in both sexes (Sadagurski et al., 2017), male-specific improvements in insulin sensitivity and glucose tolerance (Garratt et al., 2017) linked to changes in function of mTORC2, and reduction of spontaneous lung tumor rates (Harrison et al., 2019). The lifespan results were replicated in a separate cohort and found to be insensitive to doses between 400 ppm and 2500 ppm (Harrison et al., 2019).

3. *NDGA* increased male lifespan in two independent cohorts (Harrison et al, 2014; Strong et al., 2008). NDGA is an antiinflammatory agent with antioxidant properties available as a prescription drug in Europe. The effect was stronger at TJL and UT than at UM, possibly related to the consistently longer mean survival of control males at UM than at the other two sites. A replication confirmed the ability of this agent to increase lifespan in males but not females (Harrison et al., 2014). Males were treated with 2500 ppm NDGA (the original dose), 800 ppm (~3× less), and 5000 ppm (2× more). Females were treated with the 5000 ppm dose, resulting in blood levels of NDGA about equivalent to males receiving the 2500 ppm dose (Harrison et al., 2014). Females at 5000 ppm showed a reduction in body weight by 12 months of age of about 14%, which was not seen in the male mice. NDGA improved grip strength in males, and rotarod performance in both sexes, but did not lead to a significant effect on maximum lifespan, even in males, at any of the doses evaluated (Harrison et al., 2014; Strong et al., 2016). These sex-specific findings are provocative and provide new opportunities for investigation of how pharmacological interventions produce different effects in males and females in spite of similar blood levels. We see several possibilities: (1) NDGA might benefit males by affecting male-specific pathways related to aging and disease; (2) NDGA might be metabolized to an active agent by enzymes that are higher in males, or conversely might be converted to an inactive agent by enzymes that are higher in females; (3) NDGA might slow aging and/or cancer in both sexes, but might also exert negative effects on females that compensate for the health benefits. Follow-up studies have not been undertaken, in part because this agent is no longer commercially available from the original supplier.

4. An initial study using *17α-estradiol* (17αE2) at a dose of 4.8 ppm showed increased male lifespan, but not female lifespan (Harrison et al., 2014), using data pooled across the three ITP sites. The implications of this finding, however, were not straightforward, because the effect was far stronger at UT than at UM or TJL. A study using a higher dose (14.4 ppm) showed a strong effect ($P = .0001$) on male median lifespan, with beneficial effects at all three test sites (Strong et al., 2016). 17αE2 treatment had no significant effect on female survival, either in the pooled data or at any individual site. Follow-up studies have shown that 17αE2 retards age-related hypothalamic inflammation in male but not in female mice (Sadagurski et al., 2017), and that metabolomic changes seen in 17αE2-treated male mice were in general not seen in females or in males castrated prior to 17αE2 treatment (Garratt et al., 2018), showing that drug effects require the presence of testicular hormones even in postpubertal males. Furthermore, 17αE2-dependent improvements in insulin sensitivity and glucose tolerance were also noted only in males and prevented by castration (Garratt et al., 2017). Males treated with 17αE2 also retain more youthful levels of grip strength and rotarod performance, along with heavier muscle mass and larger individual muscle fibers. These benefits are seen even in mice in which 17αE2 is not initiated until 16 months of age (Garratt et al., 2019). Phase II studies were initiated in 2016 to assess the potential lifespan benefit from 17αE2 initiated at 16 or 20 months.

5. *Protandim*, a mixture of botanical extracts that activate Nrf2, extended median lifespan in males only (Strong et al., 2016). While a significant effect (7% increase; $P < .012$) of the drug was observed on the median survival of male mice in the pooled population from all three testing sites, there was no significant difference in the proportion of control and treated mice alive at the age at 90% mortality ($P = .1$ by the "Wang–Allison test"). A secondary analysis of the survival data from individual sites revealed a significant increase in median survival ($P = 0.03$) at the UT site but no significant effects at the TJL and UM sites; these secondary analyses have much less power than the primary test, which uses data pooled across all three sites.

6. Food supplemented with *glycine* resulted in a significant lifespan extension in both males and females (Miller et al., 2019). The effect in males was larger than in females both at the median lifespan (6.2% vs. 3.2%) and at the 90th percentile marks (6.2% vs. 2.4%). Treated females showed a 7% lower body weight than their male counterparts, but the observed lifespan advantage observed in both sexes makes it unlikely that the extension in female longevity was due to a decrease in food intake, that is, caloric restriction. The longevity effect was quite small and did not much alter median survival age. It is possible that variations in glycine supplementation, such as an earlier start or stop once mice are fully grown, might have exhibited stronger effects on lifespan.

7. *Aspirin*, a commonly used antiinflammatory agent, increased male lifespan when used at 20 ppm, with stronger effects at TJL and UT than at UM (Strong et al., 2008). Work at UT showed that females tended to have a relatively high rate of metabolism of aspirin to its less active metabolite (salicylate), potentially contributing to the sex specificity of the lifespan effect. The longevity benefit, however, was not seen in two subsequent studies, using doses of 60 and 200 ppm (Miller et al., 2019). Additional studies using these and other doses would not needed to test the robustness of these conflicting results.

Two additional treatments showed mild effects. Methylene blue increased maximum lifespan ($P = .004$) in females (Harrison et al., 2014). No effects were seen in males, and the effect on overall survival curves (by log-rank test) was not significant in either sex. While methylene blue might be of benefit at other doses, we think it possible that the effect seen was due to chance arising in a series of multiple comparisons. GTE showed a small increase (6%) in median lifespan in female mice that was not significant by the log-rank test but was significant at $P = .03$ by the Gehan–Wilcoxon test (Strong et al., 2013). Maximum lifespan was not increased in females and there was no effect on either median or maximal lifespan in male mice.

Studies nearing completion

At the time this chapter was written, several stage I studies were essentially complete and in preparation for peer review. These include: (1) longevity data for 17αE2 started at 16 or 20 months; (2) longevity data for nicotinamide riboside (NR) and canagliflozin as well as four other agents in the C2016 test group; and (3) a test of alternate dosing schemes for rapamycin, comparing continuous treatment from 20 months, 3 months exposure from 20 to 23 months, and periodic treatment in monthly cycles, 20–21, then 22–23, then 24–25, etc.

Pathology of drug-treated mice

Two distinct kinds of necropsy studies receive major emphasis in the ITP research design. Terminal ("end of life," or EOL) necropsies are conducted when the ITP leadership decides these will be useful, typically when a drug is found to increase longevity. The collaborating pathologist receives fixed carcasses from control and drug-treated mice, from each site, and slides are prepared from approximately 25–30 tissues. The pathologist, blinded to treatment group, prepares a report that includes his/her professional judgment as to likely cause of death (or cause of the morbidity that required humane euthanasia), as well as the incidence or severity of lesions seen in each tissue. The decoded reports are then evaluated to see if the drug altered the distribution of cause of death, or the incidence/severity of specific lesions. Available funding allows evaluation of 15–20 mice for each combination of sex, site, and treatment, so statistical power is only sufficient to provide useful insights into those lesions and causes of death that are quite frequent. These results provide important information as to whether a drug's effect on lifespan reflects an alteration in one or more specific lethal diseases, and also whether a drug may lead to forms of pathology not usually seen in normal mice.

In addition to the EOL necropsies, the ITP almost always includes cross-sectional ("XS") histopathology in each stage II protocol. Drug-treated and control mice are euthanized at age 22 months, fixed, and submitted for evaluation to a pathologist blinded to treatment. The pathologist scores each tissue for incidence of specific lesions, and for some lesions also includes a score for graded severity. Very few HET3 mice at this age exhibit life-threatening forms of disease, and the goal of the XS evaluation is to produce detailed information on age-dependent changes in multiple tissues, even if these are unlikely to lead to any clinical signs or symptoms. Data showing that a drug retards age-dependent changes in many tissues is critical to deciding whether its ability to extend lifespan reflects a general antiaging effect, or merely an effect on one or more common forms of lethal illness, typically neoplastic.

Materials from EOL and XS histopathology studies can and do provide a platform for many kinds of collaboration with non-ITP researchers. The dissection protocol for both kinds of studies typically generates approximately 10 slides containing 25–30 tissues for each mouse. The ITP can provide access to these slide collections and can cut new thin sections of the paraffin-embedded fixed tissues if required. From 2019 onwards, these programs have benefitted from interaction with the NIA-supported "Geropathology Network" (Ladiges et al., 2016).

In addition, carcasses of mice that die at the end of their lifespan are fixed and held at each site, to support collaborations that can make use of fixed materials from aged animals. This collection includes examples from each year of the ITP initiative. Lastly, tail-tip samples from over 8000 weanlings have been sent to a collaborating laboratory for DNA extraction and genotyping. Phenotypic data on these mice include body weights and lifespan.

Collaborative Interactions Program

The Collaborative Interactions Program (CIP) was established in 2015 to provide a source for tissues from mice treated with ITP test agents as a platform for collaboration with researchers at other institutions. Mice treated with these drugs are euthanized at age 22 months, at each of the three ITP labs, and dissected to provide multiple tissues, including liver, kidney, heart, aorta, brain cortex, skeletal muscle, three fat depots, and plasma. These samples are then frozen at −80°C for later use. The archive receives approximately eight mice of each sex at each site, and twice as many age-matched controls. Samples from young controls are also available. The process for requesting these samples is designed to be simple, and the time between request and tissue shipment is often as little as 1 month. Potential collaborators who wish access to tissues that are not on the standard dissection list are asked to contact ITP scientists to discuss their proposed collaboration. The list of drugs under test at any point is available on the NIA ITP public website, in part to facilitate such discussions.

Tests of health outcomes

Each stage II ITP study has typically included tests of age-sensitive physiological measures that can be evaluated without harming the mice, and that are relevant to human health and well-being (collectively termed health outcomes or HO tests). Each ITP site has its own specialized capabilities, ranging from DEXA scanning to echocardiography to assessments of hearing and vision. Such health measures can be important. For example, Miller et al. (2014) showed that critical endocrine and metabolic changes differed between rapamycin and diet-restricted mice—significant evidence against the idea that both use the same biological mechanisms.

Since 2016, the ITP has also incorporated some specific HO testing into each stage I protocol. The goals here are twofold: (1) to see if drugs that extend lifespan also retard age-dependent change in selected HO, and (2) to test the notion that some stage I agents might slow age-related changes in multiple HO, and yet not produce a lifespan benefit. Selection and validation of tests for the HO battery included multiple considerations, and these tests are not yet fully completed. In deciding which tests to use, we sought evidence that testing mice had no measurable effect on their lifespan, compared to untested mice in both control and drug-treated cohorts. To be included in the "standard" HO test battery, tests had to be sufficiently inexpensive to be conducted on 200 controls and approximately 400 drug-treated mice, at each site, at both 10 and 22 months of age, and had to be demonstrably age-sensitive in both sexes at all three sites. Age effects had to be sufficiently large that modest drug effects on the age trajectory could reliably be detected, in principle, in groups of about 50 mice per sex per site. Lastly, we sought test procedures that would produce data without site-specific ("batch") effects, to allow pooling of results across sites. Three tests are now in the provisionally accepted test group: grip strength (front paws and all four paws), a test of rotarod (fall latency on an accelerating rod), and rectal temperature. Other tests are also under consideration for inclusion in the HO test battery for stage I mice. Data from the 2017 cohort will, we hope, answer two important points, that is, whether HO testing alters lifespan, and whether any of the tests predict lifespan among individual mice.

Interactions with the Mouse Phenome Database

The Jackson Laboratories have a long-established database, the MPD, that provides access to a vast amount of information about the effects of mutations and interventions on behavior, health, and physiology of inbred and noninbred mice. The ITP has been collaborating with MPD scientists since 2015 to provide universal access to raw and processed data from all ITP publications. This information is most easily located at the URL phenome.jax.org/projects/ITP1. The site allows tabular and graphical display, as well as extensive real-time statistical testing, of all ITP survival statistics, as well as raw and processed data from each figure and table of ITP papers from 2015 onwards. The MPD site also has a valuable compendium of ITP SOPs and rationale for each compound tested. Goals for 2020–2021 include incorporation of ITP-related pathology data sets and tables of materials available through the CIP.

The ITP at 15 years: synopsis and future goals

As of December 2019, the ITP had initiated 79 full-lifespan longevity studies. Of these, 38 involved a single drug at a single dose; eight involved a drug tested at two doses in parallel; eight involved different doses or dosing regimens; four involved varying exposures to NDGA; five involved varying levels of 17αE2; five involved various levels of acarbose; and nine involved rapamycin at various doses, ages, and regimens. The remaining two longevity studies involved drug combinations: rapamycin plus metformin and rapamycin plus acarbose. Of these 79 experiments, 21 are incomplete (i.e., with live mice still in our colonies), and the other 58 have either been published or submitted in final form for publication.

The ITP discovery of lifespan extension by rapamycin was the first report of a pharmacological approach to modulate aging in a mammalian model that was replicated across three sites and that used a genetically heterogeneous model (Harrison et al., 2009). The surprising finding that rapamycin was equally effective when started at an age 75% of the median lifespan hinted that some aspects of aging, or at least age-dependent neoplasia, could be effectively postponed even in middle age, and, since then, data on 17αE2 and acarbose have also shown some benefits in mice started at 16 months or later (Harrison et al., 2019). As the understanding of the role of mTOR in aging and cell signaling has burgeoned in recent years, it has spurred the field searching for "rapalogs" into high gear. Molecules that modulate the activity of the mTOR pathway in precise and cell-specific ways might be able to slow aging and preserve many aspects of health with minimal side effects. A pilot study of rapamycin effects in dogs has already produced some tantalizing results (Urfer et al., 2017), with a much larger and more ambitious study just initiated (dogagingproject.org).

The survival data on acarbose, at three doses and at two starting ages, provide justification and experimental leverage for investigation of changes in blood glucose transients on aging and multiple forms of late-life diseases. An observational study of metformin in diabetic people produced the unexpected finding (Bannister et al., 2014) that mortality risks in metformin-treated patients were lower than those of untreated, nondiabetic individuals, suggesting that modification of glucose damage might also be able to slow aging and prevent diseases in humans (Chiasson et al., 2002, 2003). Although metformin by itself did not lead to significant extension of lifespan in the ITP study, the combination of metformin plus rapamycin allowed mice to live longer than mice exposed to either drug alone; the contrast with rapamycin-treated mice was at the margin of statistical significance for male mice (Strong et al., 2016). The nearly completed survival study using canagliflozin will test the idea that blunting of glucose spikes can lead to extended lifespan (at least in male mice). It is important to note that few if any of the HET3 mice die of diabetic symptoms; most die of some form of neoplasia, and investigation of the shared mechanisms downstream of acarbose and canagliflozin treatment may give key insights into the links between aging and cancer incidence.

An important goal of the ITP agenda is to give experimental biogerontologists a variety of interventions that can retard some or all aspects of aging, including multiple forms of disease, and thus help test and evaluate ideas about critical factors shared among antiaging interventions. Comparative work on acarbose and 17αE2, for example, has already documented shared male-specific prevention of hypothalamic inflammation, an observation also seen in long-lived mutant mice (Ames, Snell, GHRKO) of both sexes (Sadagurski et al., 2015). Since genetic manipulations to retard hypothalamic inflammation have been shown to increase mouse lifespan in at least one inbred strain (Zhang et al., 2013), the works on acarbose and 17αE2 provide new tools for influencing this potentially important aspect of neuroendocrine aging. Acarbose and 17αE2 also induce changes in mTORC2 function, again only in males, and again parallel to findings in long-lived mutant mice (Dominick et al., 2015; Garratt et al., 2017), providing good motivation to look at the role of mTORC2 in multisystem aging, and to search for drugs that might augment mTORC2 function specifically.

The evidence for sex-specific drug effects was unanticipated. The initial hypothesis was that an agent which slowed aging in either sex would do so in both sexes, but four of the drugs tested had effects in males that were weaker (acarbose) or absent (NDGA, 17αE2, Protandim), in female mice. Rapamycin effects in females are routinely higher than the effect of the same dose in males, but this seems likely to reflect higher blood doses in females at given doses in the food (Miller et al., 2014). Glycine, like rapamycin, has equal lifespan effects in mice of both sexes, but the glycine effects are much smaller than those of rapamycin (Miller et al., 2019). Causes of death in males differ somewhat from those that lead to death of female HET3 mice, and some of the sex specificity could reflect protection from diseases with sex-specific prevalence. It is also possible, in principle, that these sex-specific agents produce beneficial effects in both sexes, but also produce harmful effects in female mice. There is already a small amount of evidence (Garratt et al., 2017, 2018) for antiaging effects of acarbose and 17αE2 on some age-sensitive physiological tests in female mice, and a more comprehensive analysis of multiple HOs, and cellular changes, in drug-treated mice of both sexes may bring greater clarity to this set of questions. Similarly, studies on castrated, or on hormone-supplemented mice of both sexes may give additional insights, and point to strategies for improving drug efficacy in females.

The ITP has also published data showing the absence of lifespan effects of at least 17 agents that were thought to be plausible candidates for slowing the aging process (Harrison et al., 2009; Miller et al., 2011; Strong et al., 2008, 2013). These include resveratrol (two doses, two start ages), curcumin, simvastatin (two doses), fish oil (two doses), metformin, and GTE, among others less famous. Although in each case it is possible that benefits might have been observed at other doses, or in other mouse stocks, or using other base diets, and although negative results in mice do not preclude the possibility of human health benefits, negative data from a well-powered, multiinstitutional study can help to temper enthusiasm for specific classes of test agents and thus guide researchers as they decide how best to commit limited resources for their own studies.

Finding that a drug extends lifespan serves as an initial, important, step in building a case that the drug slows the aging process, or at least retards aspects of aging that contribute to mortality. The next steps are testing effects on a variety of different biological changes with age, including changes that may not be directly lethal. Published data already make a strong case that rapamycin retards age effects on multiple cells, tissues, and organ systems, including extracellular tissues, with beneficial effects on several forms of late-life pathology. Studies of the effects of 17αE2 and acarbose on multiple age-sensitive outcomes are now in progress, with promising initial results already published (Garratt et al., 2017, 2018, 2019). Once the research community develops a consensus that specific drugs are indeed slowing aging, these agents can then be used to help refine our understanding of the cells, intracellular pathways, hormones, and neurological influences that are (perhaps) shared by various classes of antiaging interventions. It will, for example, be of great interest to seek common pathways affected, in parallel, in mice exposed to low-calorie or low-methionine diets, mice bearing a mutation that slows aging, and mice treated with drugs like rapamycin, acarbose, and 17αE2. Comprehensive comparative studies of transcriptome data (Tyshkovskiy et al., 2019) have begun to demonstrate the value, but also the complexity, of such an effort, and the coming 5 years are certain to provide dramatic increases in such studies using metabolomic, proteomic, and related methods applied to multiple tissues of mice treated with antiaging drugs. Studies of epigenetic switches, including those based on DNA methylation patterns (Horvath & Raj, 2018; Petkovich et al., 2017), are also likely to be revealing, and suggest mechanistic hypotheses about critical cellular and neuroendocrine control pathways. Hypotheses about the role of mTOR, ATF4, sirtuins, glucose regulatory circuits, hormone levels, mitochondrial function, and other targets suggested as modulators of aging in mammals will be clarified using a widening range of pharmacological antiaging interventions, including those identified by the ITP laboratories.

The experience of the ITP is also helping to guide interpretation of work on interventions, including drugs, anticipated to modulate aging. We believe there is a strong case to be made for using genetically heterogeneous mice, rather than inbred or other isogenic stocks, as the test platform for initial surveys of interventions to delay aging (Miller et al., 1999a, 1999b). Although there are many alternate approaches to producing genetically heterogeneous mouse populations for experimental use, the four-way cross is both powerful and easily reproducible (Roderick, 1963). Furthermore, the accumulating data on pathology, physiology, and age-related changes in HET3 mice now provide a useful foundation on which to base future studies by ITP and other laboratory groups. Recent work using collections of related F1 hybrid stocks, also referred to as a diallel cross, have helped to illustrate the importance of using genetically heterogeneous mice instead of mice of a single, inbred genotype (O'Connell et al., 2019; Sittig et al., 2016).

The inclusion of males and females in each study allowed us to document sex-specific differences that would have been missed in single-sex designs. The use of large numbers of mice in each test cohort, though quite expensive, provides the two obvious advantages: high power to detect modest positive effects, and the ability to state negative findings with narrow confidence intervals. Inclusion of a double-sized group of control mice in each annual cohort, against which each of the treatment groups can be compared, gives increased statistical power for relatively little cost. A design in which mice are split equally among three test sites allows greater confidence when positive effects are seen at each site—even when the extent of the benefit may differ among the sites—and also leads to appropriate caution when an agent shows strong positive effects limited to a single site only.

Some findings from the ITP experience have remained perplexing, none more so than the consistent site-to-site variation in weight (males and females) and longevity (males only). Despite use of common suppliers for food, bedding, and breeder mice, and uniformity in husbandry protocols related to temperature, humidity, health surveillance, and light cycles, males from UM have been consistently longer-lived than males at the other two sites, and both males and females at UM have been lighter in weight than animals at the other two sites in each of the last seven annual cohorts. It is plausible that these effects may represent site variation in organic or inorganic impurities in the local water supply, or consistent local variation in gut microbes (Smith et al., 2019), or local sounds or odors that are undetectable by people but affect mouse health. It will be of substantial interest to see to what extent longevity effects seen in the ITP laboratories are robust when evaluated elsewhere.

The success of the ITP so far has depended to a major extent on the willingness of our colleagues to propose test agents that they felt merited lifespan analysis, and the skill of the ITP Access Committee at picking likely winners. If the success of these two groups continues into the next decade of work, we may well find ourselves with a multitude of drugs, of varying putative modes of action, which produce consistent increases in healthy mouse lifespan by delaying the aging process. Such an arsenal of drugs would provide powerful tools for learning more about the factors that modulate mammalian aging and the risk of multiple age-dependent diseases and serve as a basis for initial studies of drugs, in people, that could promote human health by delay of aging and its sequelae.

References

Bannister, C. A., Holden, S. E., Jenkins-Jones, S., Morgan, C. L., Halcox, J. P., Schernthaner, G., ... Currie, C. J. (2014). Can people with type 2 diabetes live longer than those without? A comparison of mortality in people initiated with metformin or sulphonylurea monotherapy and matched, non-diabetic controls. *Diabetes, Obesity & Metabolism, 16*, 1165–1173.

Bitto, A., Ito, T. K., Pineda, V. V., LeTexier, N. J., Huang, H. Z., Sutlief, E., ... Kaeberlein, M. (2016). Transient rapamycin treatment can increase lifespan and healthspan in middle-aged mice. *eLife*, *5*, e16351.

Cheng, C. J., Gelfond, J., Strong, R., & Nelson, J. F. (2019). Genetically heterogeneous mice exhibit a female survival advantage that is age- and site-specific: Results from a large multi-site study. *Aging Cell*, *18*, e12905.

Chiasson, J. L., Josse, R. G., Gomis, R., Hanefeld, M., Karasik, A., & Laakso, M. (2002). Acarbose for prevention of type 2 diabetes mellitus: The STOP-NIDDM randomised trial. STOP-NIDDM Trail Research Group. *Lancet*, *359*, 2072–2077.

Chiasson, J. L., Josse, R. G., Gomis, R., Hanefeld, M., Karasik, A., & Laakso, M. (2003). Acarbose treatment and the risk of cardiovascular disease and hypertension in patients with impaired glucose tolerance: The STOP-NIDDM trial. *JAMA: The Journal of the American Medical Association*, *290*, 486–494.

Crabbe, J. C., Wahlsten, D., & Dudek, B. C. (1999). Genetics of mouse behavior: Interactions with laboratory environment. *Science*, *284*, 1670–1672.

Dominick, G., Berryman, D. E., List, E. O., Kopchick, J. J., Li, X., Miller, R. A., & Garcia, G. G. (2015). Regulation of mTOR activity in Snell dwarf and GH receptor gene-disrupted mice. *Endocrinology*, *156*, 565–575.

Engelhardt, J. A., Gries, C. L., & Long, G. G. (1993). Incidence of spontaneous neoplastic and nonneoplastic lesions in Charles River CD-1 mice varies with breeding origin. *Toxicologic Pathology*, *21*, 538–541.

Evason, K., Huang, C., Yamben, I., Covey, D. F., & Kornfeld, K. (2005). Anticonvulsant medications extend worm lifespan. *Science*, *307*, 258–262.

Flynn, J. M., O'Leary, M. N., Zambataro, C. A., Academia, E. C., Presley, M. P., Garrett, B. J., ... Melov, S. (2013). Late-life rapamycin treatment reverses age-related heart dysfunction. *Aging Cell*, *12*, 851–862.

Garratt, M., Bower, B., Garcia, G. G., & Miller, R. A. (2017). Sex differences in lifespan extension with acarbose and 17-α estradiol: Gonadal hormones underlie male-specific improvements in glucose tolerance and mTORC2 signaling. *Aging Cell*, *16*, 1256–1266.

Garratt, M., Lagerborg, K. A., Tsai, Y. M., Galecki, A., Jain, M., & Miller, R. A. (2018). Male lifespan extension with 17-α estradiol is linked to a sex-specific metabolomic response modulated by gonadal hormones in mice. *Aging Cell*, *17*, e12786.

Garratt, M., Leander, D., Pifer, K., Bower, B., Herrera, J. J., Day, S. M., ... Miller, R. A. (2019). 17-α estradiol ameliorates age-associated sarcopenia and improves late-life physical function in male mice but not in females or castrated males. *Aging Cell*, *18*, e12920.

Hajek, R. A., Robertson, A. D., Johnston, D. A., Van, N. T., Tcholakian, R. K., Wagner, L. A., ... Jones, L. A. (1997). During development, 17alpha-estradiol is a potent estrogen and carcinogen. *Environmental Health Perspective*, *105*(Suppl. 3), 577–581.

Harper, J. M., Galecki, A. T., Burke, D. T., Pinkosky, S. L., & Miller, R. A. (2003). Quantitative trait loci for insulin-like growth factor-I, leptin, thyroxine, and corticosterone in genetically heterogeneous mice. *Physiological Genomics*, *15*, 44–51.

Harrison, D. E., Strong, R., Alavez, S., Astle, C. M., DiGiovanni, J., Fernandez, E., ... Miller, R. A. (2019). Acarbose improves health and lifespan in aging HET3 mice. *Aging Cell*, *18*, 12898.

Harrison, D. E., Strong, R., Allison, D. B., Ames, B. N., Astle, C. M., Atamna, H., ... Miller, R. A. (2014). Acarbose, 17-α-estradiol, and nordihydroguaiaretic acid extend mouse lifespan preferentially in males. *Aging Cell*, *13*, 273–282.

Harrison, D. E., Strong, R., Sharp, Z. D., Nelson, J. F., Astle, C. M., Flurkey, K., ... Miller, R. A. (2009). Rapamycin fed late in life extends lifespan in genetically heterogeneous mice. *Nature*, *460*, 392–395.

Horvath, S., & Raj, K. (2018). DNA methylation-based biomarkers and the epigenetic clock theory of ageing. *Nature Reviews Genetics*, *19*, 371–384.

Johnson, S. C., Martin, G. M., Rabinovitch, P. S., & Kaeberlein, M. (2013). Preserving youth: Does rapamycin deliver? *Science Translational Medicine*, *5*, 211fs40.

Ladiges, W., Ikeno, Y., Niedernhofer, L., McIndoe, R. A., Ciol, M. A., Ritchey, J., & Liggitt, D. (2016). The Geropathology Research Network: An interdisciplinary approach for integrating pathology into research on aging. *The Journals of Gerontology. Series A, Biological Sciences and Medical Sciences*, *71*, 431–434.

Lamming, D. W., Ye, L., Astle, C. M., Baur, J. A., Sabatini, D. M., & Harrison, D. E. (2013). Young and old genetically heterogeneous HET3 mice on a rapamycin diet are glucose intolerant but insulin sensitive. *Aging Cell*, *12*, 712–718.

Lesniewski, L. A., Seals, D. R., Walker, A. E., Henson, G. D., Blimline, M. W., Trott, D. W., ... Donato, A. J. (2017). Dietary rapamycin supplementation reverses age-related vascular dysfunction and oxidative stress, while modulating nutrient-sensing, cell cycle, and senescence pathways. *Aging Cell*, *16*, 17–26.

Miller, R. A., & Chrisp, C. (1999). Lifelong treatment with oral DHEA sulfate does not preserve immune function, prevent disease, or improve survival in genetically heterogeneous mice. *Journal of the American Geriatrics Society.*, *47*, 960–966.

Miller, R.A., Austad, S., Burke, D., Chrisp, C., Dysko, R., Galecki, A., Jackson, A., Monnier, V., (1999a). Exotic mice as models for aging research: polemic and prospectus. *Neurobiology of Aging*, 20, 217–231.

Miller, R.A., Burke, D., Nadon, N., (1999b). Announcement: four-way cross mouse stocks: a new, genetically heterogeneous resource for aging research. *Journal of Gerontology: Biological Sciences*, 54, B358–B360.

Miller, R. A., Harrison, D., Astle, C. M., Baur, J. A., deCabo, R., Fernandez, E., ... Strong, R. (2011). Rapamycin, but not resveratrol or simvastatin, extends lifespan of genetically heterogeneous mice. *Journals of Gerontology, Biological Sciences*, *66A*, 191–201.

Miller, R. A., Harrison, D. E., Astle, C. M., Bogue, M. A., Brind, J., Fernandez, E., ... Strong, R. (2019). Glycine supplementation extends lifespan of male and female mice. *Aging Cell*, *18*, e12953.

Miller, R. A., Harrison, D. E., Astle, C. M., Fernandez, E., Flurkey, K., Han, M., ... Strong, R. (2014). Rapamycin-mediated lifespan increase in mice is dose and sex-dependent and metabolically distinct from dietary restriction. *Aging Cell*, *13*, 468–477.

Miller, R. A., Harrison, D. E., Astle, C. M., Floyd, R. A., Flurkey, K., Hensley, K. L., ... Strong, R. (2007). NIA interventions testing program: Study design and an interim report. *Aging Cell*, *6*, 565–575.

Nadon, N. L., Strong, R., Miller, R. A., Nelson, J., Javors, M., Sharp, Z. D., ... Harrison, D. E. (2008). Design of aging intervention studies: The NIA interventions testing program. *Age*, *30*, 187–199.

Neff, F., Flores-Dominguez, D., Ryan, D. P., Horsch, M., Schröder, S., Adler, T., ... Ehninger, D. (2013). Rapamycin extends murine lifespan but has limited effects on aging. *The Journal of Clinical Investigation*, *123*, 3272–3291.

O'Connell, K., Ouellette, A. R., Neuner, S. M., Dunn, A. R., & Kaczorowski, C. C. (2019). Genetic background modifies CNS-mediated sensorimotor decline in the AD-BXD mouse model of genetic diversity in Alzheimer's disease. *Genes, Brain, and Behavior*, *18*, e12603.

Petkovich, D. A., Podolskiy, D. I., Lobanov, A. V., Lee, S. G., Miller, R. A., & Gladyshev, V. N. (2017). Using DNA methylation profiling to evaluate biological age and longevity interventions. *Cell Metabolism*, *25*, 954–960.

Richardson, A. (2013). Rapamycin, anti-aging, and avoiding the fate of Tithonus. *The Journal of Clinical Investigation*, *123*, 3204–3206.

Roderick, T. H. (1963). Selection for radiation resistance in mice. *Genetics, 48*, 205–216.

Sadagurski, M., Cady, G., & Miller, R. A. (2017). Anti-aging drugs reduce hypothalamic inflammation in a sex-specific manner. *Aging Cell, 16*, 652–660.

Sadagurski, M., Landeryou, T., Cady, G., Kopchick, J. J., List, E. O., Berryman, D. E., ... Miller, R. A. (2015). Growth hormone modulates hypothalamic inflammation in long-lived pituitary dwarf mice. *Aging Cell, 14*, 1045–1054.

Sharp, Z. D., & Randy Strong, R. (2010). The role of mTOR signaling in controlling mammalian lifespan: What a fungicide teaches us about longevity. *Journals of Gerontology. Biological Sciences and Medical Sciences, 65A*, 580–589.

Sittig, L. J., Carbonetto, P., Engel, K. A., Krauss, K. S., Barrios-Camacho, C. M., & Palmer, A. A. (2016). Genetic background limits generalizability of genotype-phenotype relationships. *Neuron, 91*, 1253–1259.

Smith, B. J., Miller, R. A., Ericsson, A. C., Harrison, D. C., Strong, R., & Schmidt, T. M. (2019). Changes in the gut microbiome and fermentation products concurrent with enhanced longevity in acarbose-treated mice. *BMC Microbiology, 19*, 130.

Steinbaugh, M. J., Sun, L. Y., Bartke, A., & Miller, R. A. (2012). Activation of genes involved in xenobiotic metabolism is a shared signature of mouse models with extended lifespan. *American Journal of Physiology. Endocrinology and Metabolism, 303*, E488–E495.

Strong, R., Miller, R. A., Antebi, A., Astle, C. M., Bogue, M., Denzel, M. S., ... Harrison, D. E. (2016). Longer lifespan in male mice treated with a weakly estrogenic agonist, an antioxidant, an α-glucosidase inhibitor or a Nrf2-inducer. *Aging Cell, 15*, 872–884.

Strong, R., Miller, R. A., Astle, C. M., Baur, J. A., de Cabo, R., Fernandez, E., ... Harrison, D. E. (2013). Evaluation of resveratrol, green tea extract, curcumin, oxaloacetic acid, and medium chain triglyceride oil on lifespan of genetically heterogeneous mice. *Journal of Gerontology, Biological Sciences, 68*, 6–16.

Strong, R., Miller, R. A., Astle, C. M., Floyd, R. A., Flurkey, K., Hensley, K. L., ... Harrison, D. E. (2008). Nordihydroguaiaretic acid and aspirin increase lifespan of genetically heterogeneous male mice. *Aging Cell, 7*, 641–650.

Tanaka, E., Ishikawa, A., & Horie, T. (1999). In vivo and in vitro trimethadione oxidation activity of the liver from various animal species including mouse, hamster, rat, rabbit, dog, monkey and human. *Human & Experimental Toxicology, 18*, 12–16.

Tyshkovskiy, A., Bozaykut, P., Borodinova, A. A., Gerashchenko, M. V., Ables, G. P., Garratt, M., ... Gladyshev, V. N. (2019). Identification and application of gene expression signatures associated with lifespan extension. *Cell Metabolism, 30*, 573–593.

Urfer, S. R., Kaeberlein, T. L., Mailheau, S., Bergman, P. J., Creevy, K. E., Promislow, D., & Kaeberlein, M. (2017). A randomized controlled trial to establish effects of short-term rapamycin treatment in 24 middle-aged companion dogs. *GeroScience, 39*, 117–127.

Van Gelder, R. N., & Hogenesch, J. B. (2004). Clean thoughts about dirty genes. *Journal of Biological Rhythms, 19*, 3–9.

Wang, C., Li, Q., Redden, D. T., Weindruch, R., & Allison, D. B. (2004). Statistical methods for testing effects on "maximum lifespan". *Mechanisms of Ageing and Development, 125*, 629–632.

Warner, H. R., Ingram, D., Miller, R. A., Nadon, N. L., & Richardson, A. G. (2000). Meeting report: Program for testing biological interventions to promote healthy aging. *Mechanisms of Ageing and Development., 115*, 199–208.

Wilkinson, J. E., Burmeister, L., Brooks, S. V., Chan, C.-C., Friedline, S., Harrison, D., ... Miller, R. A. (2012). Rapamycin slows aging in mice. *Aging Cell, 11*, 675–682.

Ye, L., Widlund, A.L., Sims, C.A., Lamming, D.W., Guan, Y., Davis, J.G., ... Baur, J.A. (2013). Rapamycin doses sufficient to extend lifespan do not compromise muscle mitochondrial content or endurance. *Aging, 5*, 539–550.

Zaseck, L. W., Miller, R. A., & Brooks, S. V. (2016). Rapamycin attenuates age-associated changes in tibialis anterior tendon viscoelastic properties. *The Journals of Gerontology. Series A, Biological Sciences and Medical Sciences, 71*, 858–865.

Zhang, G., Li, J., Purkayastha, S., Tang, Y., Zhang, H., Yin, Y., ... Cai, D. (2013). Hypothalamic programming of systemic ageing involving IKK-β, NF-κB and GnRH. *Nature, 497*, 211–216.

Further reading

Bendele, A. M., & Carlton, W. W. (1986). Incidence of obstructive uropathy in male B6C3F1 mice on a 24-month carcinogenicity study and its apparent prevention by ochratoxin A. *Laboratory Animal Science, 36*, 282–285.

Everitt, J. I., Ross, P. W., & Davis, T. W. (1988). Urologic syndrome associated with wire caging in AKR mice. *Laboratory Animal Science, 38*, 609–611.

Fast, R., Schütt, T., Toft, N., Møller, A., & Berendt, M. (2013). An observational study with long-term follow-up of canine cognitive dysfunction: Clinical characteristics, survival, and risk factors. *Journal of Veterinary Internal Medicine/American College of Veterinary Internal Medicine, 27*, 822–829.

Lee, C. K., Allison, D. B., Brand, J., Weindruch, R., & Prolla, T. A. (2002). Transcriptional profiles associated with aging and middle age-onset caloric restriction in mouse hearts. *Proceedings of the National Academy of Sciences of the United States of America, 99*, 14988–14993.

Lü, J. M., Nurko, J., Weakley, S. M., Jiang, J., Kougias, P., Lin, P. H., ... Chen, C. (2010). Molecular mechanisms and clinical applications of nordihydroguaiaretic acid (NDGA) and its derivatives: An update. *Medical Science Monitor: International Medical Journal of Experimental and Clinical Research, 16*, RA93–RA100.

Sun, L., Sadighi Akha, A. A., Miller, R. A., & Harper, J. M. (2009). Lifespan extension in mice by preweaning food restriction and by methionine restriction in middle age. *Journals of Gerontology Biological Sciences, 64*, 711–722.

Tuffery, A. A. (1966). Urogenital lesions in laboratory mice. *The Journal of Pathology and Bacteriology, 91*, 301–309.

Yamamoto, M., & Otsuki, M. (2006). Effect of inhibition of alpha-glucosidase on age-related glucose intolerance and pancreatic atrophy in rats. *Metabolism: Clinical and Experimental, 55*, 533–540.

Zhang, Y., Bokov, A., Gelfond, J., Soto, V., Ikeno, Y., Hubbard, G., ... Fischer, K. (2014). Rapamycin extends life and health in C57BL/6 mice. *Journals of Gerontology Biological Sciences, 69*, 119–130.

CHAPTER

11

Aging in nonhuman primates

Suzette D. Tardif[1] *and Corinna N. Ross*[1,2]

[1]Population Health Program, Southwest National Primate Research Center, Texas Biomedical Research Institute, San Antonio, TX, United States [2]Department of Life Sciences, Texas A&M University San Antonio, San Antonio, TX, United States

OUTLINE

Introduction

Humans have a fascination with the process of aging and it has long been noted that aging represents the single most important risk factor for the majority of chronic human diseases. As the Centers for Disease Control and Prevention (CDC) estimate that by the year 2050 the number of people over the age of 65 will double (Weir et al., 2015), studies to understand aging and develop interventions that may ameliorate its effects take on increased importance.

Studies of aging have largely focused on model organisms with extremely short lifespans (LS), such as yeast, *Caenorhabditis elegans*, and *Drosophila melanogaster*. Even in mammalian models, which are more closely related to humans, the focus has been on a species that has evolved a shorter than average LS—*Mus musculus* (i.e., the mouse). These short-lived organisms have allowed for relatively rapid answers to many basic questions regarding aging, and they are particularly well-suited for longitudinal studies. However, they have limitations in how they inform us about human aging and aging writ large. Therapeutics and interventions developed around the findings from short-lived animal models have a very high failure rate in clinical trials, creating a "translational gap" between human and animal studies (Cummings, Morstorf, & Zhong, 2014; Sabbagh, Kinney, & Cummings, 2013). One possible reason for these failures is the phylogenetic distance and evolved differences between these short-lived model organisms and the lineage in which humans evolved, that being primates. Austad and Fischer (1992) proposed that focusing virtually all studies of antiaging mechanisms in models which have demonstrably poor defenses against aging—that is, species selected for short LSs—may provide an obviously biased view of aging overall (see also Miller, 1997; Ungvari & Philipp, 2010).

There is value, then, in a thorough understanding of aging in the evolutionary lineage leading to humans—that is, primates. This chapter will provide an overview of what we know and what we speculate regarding aging in monkeys and apes.

Handbook of the Biology of Aging.
DOI: https://doi.org/10.1016/B978-0-12-815962-0.00011-1

Why do primates live so long?

Primates are often described as being long-lived—a long life associated with a very slow reproductive pace and production of a limited number of offspring. Because LS is known to be related to body size, a useful tool to understand relative LS is the longevity quotient (LQ). $LQ = LS_{actual}/LS_{expected}$, with $LS_{expected}$ determined from least-squares linear regression of *ln*(LS) versus *ln*(adult weight) for all mammals for which acceptable data are available. LQ values <1.0 indicate a shorter LS than expected, relative to body size, while values >1.0 indicate a longer than expected LS. The average LQ for primates is 1.92 (Austad & Fischer, 1992), indicating that primates live almost twice the LS that would be expected for a mammal of similar body size. However, primates are not unique among mammals in displaying an evolved long life. Both Chiroptera (bats) and Monotremata (platypus and echidna) have LQs that exceed that of primates.

What do relatively long-lived mammals have in common? In general, LS is proposed within life history theory to be the result of selection pressures that shape the amount and degree of variation in mortality in the juvenile growth period (prior to reproduction) and in the adult period. If a species has a relatively low chance of producing juveniles that will survive and reproduce, and a relatively high and stable chance of surviving in adulthood, spreading progeny out over a long LS is advantageous (Jones, 2011; Stearns, 1992). Jones (2011) proposes that the concentration on fruit as a food source may produce extreme environmental variation and population crashes that most specifically affect juvenile recruitment into reproduction. At the same time, adult primates may have reduced and stabilized mortality due to arboreality. All of the mammalian lineages with high LQs can be described as having a life style that affords significant protection from predators, with the most notable case being the evolution of flight in bats. However, simply living in trees (or arboreality) is documented to be associated with a longer LS (Shattuck & Williams, 2010).

Primates have relatively large brains and it has been proposed that large brains may drive long LS in this group through improved homeostasis and because of its specialized metabolic demands. However, a comparison of the encephalization quotient (actual vs expected brain size, based upon body size) and LQ among mammalian lineages by Austad and Fischer (1992) finds no evidence of brain size as a generalized determinant of longevity. Within primates, there is a positive relation of EQ and LQ, but that relationship is not seen in any other mammalian lineage. As Jones (2011) points out, there is not a clear rationale for choosing between large brains as a driver of long LS versus slow reproduction/long LS as a driver of large brains—that is, there are correlations but not a clear rationale for causation. He elegantly states the possibility that "the masterpiece of human higher-intelligence was, in fact, painted on the spandrel of delayed maturation that arose as a response to unpredictable food supplies."

Studies of aging in nonhuman primates are challenged by these long LSs. The primate species most commonly used in translational research (Mattison & Vaughan, 2017)—macaques (genus *Macaca*)—as well as other large-bodied Old World monkeys such as baboons (*Papio hamadryas*) and vervet monkeys (*Chlorocebus aethiops*) have maximum LSs that range from 30.8 to 40 years (genomics.senescence.info). Because primates show the predicted relationship between body size and LS, there has been a developing interest in the use of the smallest primates to study aging. The gray mouse lemur (*Microcebus murinus*), the smallest primate, is a prosimian with a maximum LS of 18.2 years (genomics.senescence.info). There are studies that suggest the species displays age-related brain pathology that may be similar to the changes seen in Alzheimer's disease (Bons, Rieger, Prudhomme, Fisher, & Krause, 2006; Mestre-Francés et al., 2018). However, this species is rarely used in translational studies, with extremely limited availability in captivity. In addition, mouse lemurs exhibit a marked seasonal metabolic pattern akin to a torpor or modified hibernation (Terrien et al., 2018). While this is a fascinating phenomenon, in and of itself, it may raise issues regarding the mouse lemur's use as a translational aging model. The common marmoset (*Callithrix jacchus*) is a New World monkey that is the smallest anthropoid primate that is commonly used in captive research. The maximum LS recorded for the common marmoset is 22.8 years (genomics.senescence.info), 57% of that of the rhesus macaque, but only 4.6 years longer than that of the less well-characterized and less available mouse lemur. In subsequent sections, each of these species will be discussed in relation to its use as an aging model.

Aging by domain

Homeostatic function

There are increasing risks of morbidity and mortality associated with age-related alterations in major homeostatic systems, most notably the cardiovascular and renal systems, in humans. Such age-related changes also occur in nonhuman primates. Increased risk of cardiovascular failure is one of the highest risk factors for mortality in aging humans (Fontana, Vinciguerra, & Longo, 2012; Mattison & Vaughan, 2017; Ungvari et al., 2008). Human aging is characterized by increased prevalence of hypertension, heart arrhythmias, atherosclerosis, and thrombosis

(Fontana et al., 2012). Chimps and rhesus monkeys have also been reported to have increased risk of deaths associated with cardiovascular disease with increasing age (Lane, 2000; Mattison & Vaughan, 2017). For marmosets, risks of cardiovascular death increase with age, although reports of this type of death are rare and are often associated with obese animals (Ross, 2019).

Macaques have served as a particularly important model in the study of the contentious issue of hormone replacement therapy and its cardiovascular benefits and risks for peri- and postmenopausal women. Initial studies in cynomolgus macaques reported a strong protective effect of estrogen replacement on coronary artery atherosclerosis (Clarkson & Appt, 2005). When the large-scale clinical trials of hormone replacement therapy suggested significant cardiovascular risks in the early 2000s, the macaque model was used to try to better define the nature of risks and benefits, eventually supporting the conclusion that estrogen-replacement benefits were critically affected by timing of replacement following estrogen withdrawal—early replacement being associated with benefits, with late replacement being associated with either no benefit or potential risks (see Clarkson & Mehaffey, 2009; Saul & Kase, 2019 for reviews on this topic).

Blood pressure is often used as a biomarker to assess heart health and risk of disease in humans and other mammals (Bartholomeusz, Hardy, Nelson, & Phillips, 1998; Mietsch et al., 2016). Otherwise healthy old marmosets have higher mean arterial pressure and diastolic pressure than young marmosets, similar to the pressure changes seen in humans (Ross et al., 2019). Increased systolic, mean arterial, and diastolic pressure may significantly predict mortality, but this may be population- and weight-dependent in humans and other nonhuman primates (Mattison et al., 2017; Ross et al., 2019).

Kidney failure is a common cause of morbidity and mortality in aging humans. The effect of age on kidney function was examined in a cross-sectional study of marmosets (Lee et al., 2018). Biomarkers of impaired kidney function (urinary albumin to creatinine ratio, urinary protein to creatinine ratio) were associated with age and were associated with marked kidney pathology in the geriatric group—specifically glomerulosclerosis, interstitial fibrosis, and arteriosclerosis. Further, renal lesions were the most common lesion and of increased severity in the aged cohort. These findings point to the marmoset as a potentially valuable model in which to study age-related changes in kidney function.

Another area of developing interest is age-related alterations in circadian functions, particularly given the ever-increasing understanding of the importance of sleep to cognitive and cardiovascular health. The applicability of findings in circadian research to humans is likely to benefit greatly from the use of diurnal animal models, such as primates, that have evolved circadian controls of neuroendocrine function likely much more similar to humans than those of nocturnal or crepuscular species, such as mice and rats. Urbanski and Sorwell (Urbanski & Sorwell, 2012) provide an overview of the effects of age on attenuation of diurnal rhythms in a number of hormones and molecular mechanisms controlling those rhythms.

Metabolic function

Changes in metabolic functioning with age in nonhuman primates are significantly associated with the animal's preexisting status of mid-life obesity and risk of type II diabetes, rather than simply a result of aging. Gray mouse lemurs have not been found to have differences in glucose tolerance associated with aging (Djelti et al., 2017). Nondiabetic aging rhesus macaques display no changes in fasting glucose concentrations with age or differential responses to an oral glucose challenge with age (Tigno, Gerzanich, & Hansen, 2004). Aging nonobese marmosets also exhibit no differences in fasting glucose or responses to oral glucose challenge (Ross et al., 2019).

However, the combined factors of obesity and aging are significantly associated with an increased risk of insulin resistance, metabolic dysfunction, metabolic syndrome, and the development of type II diabetes in humans and other nonhuman primates (Vaughan & Mattison, 2016). Obese rhesus macaques exhibit increased hyperinsulinemia, hyperglycemia, insulin resistance, and type II diabetes complications as they age (Didier et al., 2016; Vaughan & Mattison, 2016). These metabolic shifts in macaques are associated with increased neuroinflammation and decreased cognitive performance (Didier et al., 2016). The implementation of dietary restriction not only decreases obesity and risk of metabolic dysfunction, but also reduces the associated risks of brain atrophy, cancer, and cardiovascular disease in rhesus macaques (Fontana & Partridge, 2015). The risks of obesity and aging are not limited to rhesus macaques, and are described across other nonhuman primates. Cynomolgus macaques exhibit sex- and weight-dependent changes in metabolic functioning with age, with females being more susceptible to prediabetic conditions than males (Yue et al., 2016). Marmosets are also found to have increased insulin resistance associated with aging in animals weighing more than 400 g (Tardif, Mansfield, Ratnam, Ross, & Ziegler, 2011).

Metabolite pathway changes associated with aging are often complex and integrated in ways that make evaluating single pathways very difficult to interpret. Therefore it has been proposed that the study of aging in association with metabolism and metabolomics would be best achieved using a systems network approach (Soltow, Jones, & Promislow, 2010). This approach suggests that evaluating multiple metabolite pathways, metabolomics, or the influence of microbiome on host metabolomics, will reveal not only pathways of interest, but also broad targets for future intervention testing. It has been proposed that changes in gut functionality associated with an aging microbiome may be linked to broad metabolic function in the aging nonhuman primate. Aging vervet monkeys do not have differences in microbial diversity associated with age, but they do have higher microbiome loads in the gut mucosa, higher susceptibility to change in microbiome associated with westernized diet challenge, and increased innate immune responses in association with the increased gut mucosa pathogen load (Wilson et al., 2018). Geriatric marmosets have decreased gut microbiome diversity when compared to young marmosets, with increased risk of pathogenic bacterial load (Reveles, Patel, Forney, & Ross, 2019). Evaluations of marmoset metabolome have revealed that purine and pyrimidine pathways were significantly enriched with age in both sexes (Hoffman et al., 2016). Betaine metabolism and methionine metabolism are associated with aging regardless of environment for the animals, or of metabolomics assay technique (Hoffman et al., 2019). Interestingly, low levels of tryptophan metabolites were associated with risk of death in a 2-year follow-up in the marmosets, suggesting these metabolites may be used as future biomarkers of mortality (Hoffman et al., 2019). Together these studies suggest that the gut microbiome influences healthy aging and that shifts in metabolite pathways may be used to evaluate not only aging but also risk of mortality.

The changes associated with metabolic function and age in nonhuman primates suggest that targeting of metabolic pathways for interventions may significantly impact health-related changes with aging in humans. The prevention of prediabetic or type II diabetes health-related alterations should modify the increased risk for specific age- related disease and mortality.

Mobility function

The development of sarcopenia, frailty, and osteoarthritis with aging is associated with chronic disability in the elderly (Carter et al., 2012; Longo et al., 2015; Vinciguerra, Musaro, & Rosenthal, 2010). These diseases are often associated not only with chronic inflammation, but also the inability to recover subsequent to a stressful event (Collerton et al., 2012; Lipsitz, 2008; Mitnitski et al., 2015; Sanchis et al., 2015). Unfortunately, while treatments exist to relieve the symptoms of arthritis and osteoporosis, a cure or permanent remediation does not exist. Additionally, it remains difficult to accurately assess and predict which individuals may be of highest risk for the development of sarcopenia and frailty.

In humans, osteoporosis is caused by a systemic loss of bone mass and deterioration of the trabecular bone structure, due to reduced production by osteoblasts and increased absorption of bone by osteoclasts (Behrendt et al., 2016; Champ et al., 1996; Colman, Kemnitz, Lane, Abbott, & Binkley, 1999). Old World monkeys have been found to display remodeling of cortical bone that is similar to human degeneration and is associated with dysfunction of the reproductive endocrine system (Didier et al., 2016). Marmosets display age-related structural changes to bone, including degeneration of vertebrae (Bernick, Cailliet, & Levy, 1980); degeneration and softening of the bone in marmosets has also been associated with nutritional deficits and malabsorption of vitamin D (Baxter et al., 2013; Chalmers, Murgatroyd, & Wadsworth, 1983). Bone degeneration has been reversed in marmosets treated with bisphosphate therapy with increases in trabecular number and volume (Bagi et al., 2007). However, marmosets appear to be able to maintain bone mass in the absence of gonadal estrogen, as females ovariectomized for 6 months exhibited no lumbar spine bone changes; this is not what is found in macaques or humans (Colman, Beasley, Allison, & Weindruch, 2012; Colman, Kemnitz et al., 1999; Colman, Lane, Binkley, Wegner, & Kemnitz, 1999; Colman, Roecker, Ramsey, & Kemnitz, 1998; Duncan, Colman, & Kramer, 2012).

Standard measures of mobility function and ability in humans often include behavioral tasks such as grip strength, walking speed, and gait analysis. These tasks have successfully been translated for use in a variety of nonhuman primate species. Mouse lemurs have been found to display mild differences in motor strength and grip with age (Le Brazidec et al., 2017). When given a series of tasks, including open field tests and a modified rotarod task, it was found that mouse lemurs display a great deal of individual variability, with a general trend toward decreased capacities with age, such that it is possible to use performance on these tasks to predict poor agers (Languille et al., 2015). Sifakas have only subtle changes in walking speed with age, but do display slightly impaired gait (Snyder & Schmitt, 2018). Marmosets do not display age-related changes in activity patterns or the amount of time spent in movement (Ross et al., 2019; Ross, Davis, Dobek, & Tardif, 2012), but they do display modified types of movement, with decreased jumping behavior and increased time spent in a stretched hanging posture. Rhesus macaques and baboons have both been reported to display significantly altered walking speeds with aging (Huber, Gerow, & Nathanielsz, 2015; Justice, Cesari, Seals, Shively, & Carter, 2015), but reliable measures of strength, balance, and endurance do not yet exist (Justice et al., 2015).

Neurocognitive function

The study of neurocognitive decline with aging in nonhuman primates has been a particularly rich field of focus for many investigators, due to the close evolutionary relationship between nonhuman primates and humans, the similar brain architecture found in nonhuman primates, and the great need for modeling age-related diseases such as Parkinson disease, Alzheimer's disease, multiple sclerosis, and dementia.

While neurocognitive decline studies in gray mouse lemurs are nascent, these animals display significantly altered function in psychomotor tasks (hang, rotarod, balance) and anxiety tasks (open field, light/dark plus maze) with age (Languille et al., 2015). Other tasks such as spatial working memory, spatial reference memory and novel object recognition tasks do not display a clear relationship between loss of function and age, due to a widening variation in ability with age, with some old animals performing as well as young animals and others displaying significant decline (Languille et al., 2015). Finally, middle-aged female gray mouse lemurs that exhibit increased fasting glucose concentrations have impaired spatial memory performance (Djelti et al., 2017).

Marmosets have been extensively evaluated as models for Parkinson disease, Huntington disease, Alzheimer's disease, stroke, multiple sclerosis, and spinal cord injury. Marmosets display a number of degradative neural changes that are common to humans and nonhuman primates in aging. Ultrastructural examinations of the frontal cortex and hippocampus in animals more than 12 years of age showed widespread accumulation of lipofuscin in the glial cells, perivascular macrophages, and pericytes (Honavar & Lantos, 1987). Aging female marmosets have significant decreases in the suprachiasmatic nucleus cells, but the number of glial cells was significantly increased, suggesting a possible compensatory change due to neuronal loss (Engelberth et al., 2014). Induction models of dopaminergic cell death, including injection of 1-methyl-4-phenyl-1,2,3,6-tetrahydropyridine (MPTP) and 6-hydroxydopamine (6-OHDA), have been used to demonstrate the association of dopaminergic cell death with altered motor behavior and increased stress-related behaviors associated with Parkinson disease (Phillips et al., 2017; Verhave, Vanwersch, van Helden, Smit, & Philippens, 2009). Marmosets are also being assessed for natural Parkinson-like changes; for example, increases in $\alpha\alpha$-synuclein aggregations in the olfactory bulb and hippocampus have been found in aged marmosets (Kobayashi et al., 2016). Aging marmosets express increased accumulation of β-amyloid in the brain (Geula, Nagykery, & Wu, 2002; Geula et al., 1998; Marshall & Ridley, 2003; Ridley, Baker, Windle, & Cummings, 2006; Ridley, Cummings, Leow-Dyke, & Baker, 2006; Tardif et al., 2011). Conformational changes in tau increase in frequency with aging; dystrophic microglia were significantly more likely to exhibit tau hyperphosphorylation than were active microglia (Rodriguez-Callejas, Fuchs, & Perez-Cruz, 2016). Further, the number of ferritin^{+} microglia decreases with age and is significantly associated with dystrophic microglia, suggesting a strong relationship between iron presence and function in the brain (Rodríguez-Callejas et al., 2019).

Marmosets are capable of learning and performing simple cognitive tasks that can be used to assess disease progression, intervention efficacy, as well as responses to neurotoxins and pathogens (Kangas, Bergman, & Coyle, 2016; Spinelli et al., 2004; Stevens, Scott, Bowditch, Griffiths, & Pearce, 2006). Aged marmosets display reduced executive function as measured using a conveyor belt task (Ross, 2019), and detoured reach task (Ross et al., 2019). They also exhibit marked loss in spatial working memory with age (Sadoun, Rosito, Fonta, & Girard, 2019).

Recent work with aging vervet monkeys suggests they may be particularly good models for early Alzheimer's disease progression. Aged female vervets display β-amyloid plaque deposition and the presence of tau. Furthermore, the presence of cerebrospinal fluid β-amyloid was significantly associated not only with age but also with slowed gait speed. Finally, the presence of β-amyloid plaques and tau were significantly associated with brain volume and neuropathological indices from magnetic resonance imaging (Latimer et al., 2019).

Rhesus macaques are by far the best-characterized nonhuman primate model of neurocognitive decline, associated with specific age-related diseases as well as healthy aging, and the model has been well reviewed in Hainsworth et al. (2017), Hampson (2018), and Toufexis, King, and Michopoulos (2017). While cortical neurons appear to remain intact with aging in rhesus, there are significant changes in myelin and axons with age (Makris et al., 2007), with microglial phagocytosis activated in older animals (Shobin et al., 2017). Both rhesus and cynomolgus macaques exhibit β-amyloid depositions and associated cognitive deficits, but tauopathy and tangles have not been described (Cramer et al., 2018; Edler et al., 2017). Rhesus display significant alterations in spatial networks with age, resulting in deficits of spatial cognition (Engle et al., 2016), and neural network alterations are noted in animals with significantly impaired memory (Thomé, Gray, Erickson, Lipa, & Barnes, 2016). Of particular interest in the rhesus model has been the examination of the impact of ovarian hormone therapy (estrogen) on cognitive ability in female macaques. Estrogen-treated animals display better performance on the delayed response task and visuospatial attention task (Kohama et al., 2016), and have increased multisynaptic boutons (Hara et al., 2016). The timing of the estrogen treatment following postmenopausal onset may be particularly important, with spatial working memory being preserved if the estrogen is given immediately following onset, even if withdrawn after the initial treatment phases; whereas later treatment does not preserve cognitive function (Baxter, Santistevan, Bliss-Moreau, & Morrison, 2018).

As would be expected, apes also display cognitive decline associated with age; there are changes in brain structure and histology that are consistent with age-related changes found in humans (Cramer et al., 2018; Lacreuse & Herndon, 2009; Lacreuse, Parr, Chennareddi, & Herndon, 2018). There is a reduced ability to task shift and an increased risk for preservative error with age, regardless of which ape was assessed (chimps, bonobos, gorillas, orangutans) (Lacreuse et al., 2018; Manrique & Call, 2015; Toledano, Álvarez, López-Rodríguez, Toledano-Díaz, & Fernández-Verdecia, 2014).

Reproductive function

Nonhuman primates, in common with many other mammals, display lower fertility at both the beginning and end of the reproductive LS (Caro et al., 1995; Smucny et al., 2004). Anovulation, insufficient luteolysis, and impairment of gestational and lactational processes are all more common at the beginning and end of reproductive life (Atsalis & Margulis, 2008a, 2008b).

Reproductive senescence in mammals is a process through which the hypothalamic–pituitary–gonadal (HPG) axis ages, resulting ultimately in cessation of function. Walker and Herndon (2008) provide an excellent overview of what is known about reproductive senescence and menopause in nonhuman primates, with discussion of controversies stemming from differing uses of the term "menopause." Wise (2005) provides a thoughtful perspective, comparing and contrasting what is known about reproductive aging in rodents with that in women. Recent findings on nonhuman primate reproductive senescence, along with commentary, are also found in Atsalis and Margulis (2008a, 2008b).

In nonhuman primates, the loss of the follicular pool is the primary event shaping the end of reproductive life. This contrast with rodents, where there is striking variation, is seen in the size of the follicular pool remaining at the end of reproductive life as well as at maximum LS (Wise, 2005). While the loss of the follicular pool is the primary driver of primate reproductive senescence, it is clear that the entire HPG axis shows evidence of aging change. Downs and Urbanski (2006) demonstrated that rhesus macaques display age-related increases in circulating FSH prior to cessation of ovulatory cycles, a pattern also observed in perimenopausal women.

Within nonhuman primates, human females are unusual in experiencing follicular depletion relatively early in the maximum LS, resulting in an extended period of altered hormonal environments. These alterations stem from declining negative feedback signals from the ovary (reduced circulating estrogens, P_4, and inhibin), that result in elevated gonadotropic hormone concentrations for a time, followed by declining concentrations. These hormonal changes are believed to affect disease risks, with reduced circulating estrogen being the primary risk driver (Wise, 2005). The risk associated with bone loss due to decreasing estrogenic activity on osteoblasts is well described. There is both age-related and ovariectomy-induced bone loss in macaques and baboons (Colman, Lane, et al., 1999; Havill, Levine, Newman, & Mahaney, 2008; Smith, Jolette, & Turner, 2009), which has served as a model of the postmenopausal osteoporosis common in women. However, one commonly used primate aging model, the common marmoset, does not display either age-associated or ovariectomy-associated bone loss, raising the possibility that this species may "possess unique adaptations to avoid bone loss associated with estrogen depletion" (Saltzman, Abbott, Binkley, & Colman, 2019); however, cardiovascular effects continue to be hotly debated.

Monkeys and apes also experience follicular depletion and associated hormonal alterations that resemble those seen in women (Atsalis & Margulis, 2008a, 2008b; Graham, 1979; Hodgen, Goodman, O'Connor, & Johnson, 1977; Schramm, Paprocki, & Bavister, 2002; Shideler, Gee, Chen, & Lasley, 2001; Tardif & Ziegler, 1992; Tardif, 1985; Videan, Fritz, & Murphy, 2008). However, the stage of life at which this occurs is generally later than that observed in humans. Atsalis and Margulis (2008a, 2008b), reviewing the data on monkeys and apes, conclude that "potentially up to 25% of a female's life can be post-reproductive." This claim is made in reference to maximal LS; in comparison, a human female reaching the maximal LS (now around 120 years) will spend around 58% of her life in a post-reproductive state. When compared with average LS (as opposed to maximal LS), most nonhuman female primates will die at or before the point at which reproductive senescence begins. These comparisons have been controversial and will continue to be refined, given the oft-made claim that human female reproductive aging is unique and may be driven by indirect fitness advantages to postreproductive women, such as providing resources to grandchildren (the grandmother hypothesis: Hill & Hurtado, 1991; Peccei, 2001).

Manipulation of aging

Calorie restriction studies

One of the earliest interventions identified as having a robust effect on aging across a variety of short-lived species is calorie restriction (CR). While recent studies have raised questions regarding the once-touted universal nature of the effects of CR upon LS (Liao, Rikke, Johnson, Diaz, & Nelson, 2010; Vaughan et al., 2017), there remains a robust interest in understanding the effects of CR on processes of aging. There have been two studies that have attempted assessment of CR effects upon LS and healthspan in a long-lived nonhuman primate, the rhesus macaque. The limited number of primate CR studies points to the difficulties inherent in studying basic aging questions in this taxonomic group because of long life as well as the large size of many of the primate species most commonly used in research. The comparison of the two studies is a textbook example of the value of replication, given that the findings from the studies differed in important ways. The fact that two studies were initiated at a similar time with similar, but not identical, protocols has actually enriched our understanding of CR and aging.

The first study was initiated in 1987 at the National Institute on Aging (hereafter referred to as the NIA study). While preliminary reports were published on this study during its history (Mattison, Lane, Roth, & Ingram, 2003), the main results were reported in 2012. The second study was initiated in 1989 at the University of Wisconsin (hereafter referred to as the UW study), with the main results reported in 2009 after some earlier, preliminary reports (e.g., Blanc et al., 2003; Colman, Beasley, Allison, & Weindruch, 2008). The UW study population had a total of 46 males and 30 females with CR beginning in early–mid adulthood, 7–14 years of age. The NIA study involved a population of 60 males and 60 females, but ranging in age from 1 to 17 years at CR initiation. The NIA population was fed a natural-ingredient diet while the UW population received a purified diet. Of particular importance in comparing the two studies is the fact that the control populations were treated differently. The NIA control group was fed a controlled amount set at a value slightly less than ad lib consumption, while the UW control group was fed ad lib.

In 2009, Colman et al. (2009) reported that age-related mortality was decreased and LS was increased by CR in the UW study. They also reported that CR reduced the incidence of morbidities commonly associated with aging, including sarcopenia, impaired glucose homeostasis (diabetes), cardiovascular disease, and cancer. Imaging studies also suggested that CR may have reduced gray matter atrophy in selected brain regions. In contrast, Mattison et al. (2012) reported that age-related mortality and LS were not affected by CR in the NIA study. They did find that CR reduced the incidence of certain morbidities or morbidity-associated phenotypes; notably, cancer incidence and circulating triglycerides were lower in CR subjects.

The investigators for both studies then undertook a collaborative effort to better understand the similarities and differences between the studies and how those might best inform our understanding of CR effects and the potential of CR as an antiaging intervention in humans (Colman et al., 2014; Mattison et al., 2017). Numerous differences are cited as possible sources to refine our understanding of CR in a long-lived species. There were differences in the diets being fed and how those diets were presented. There were also differences in the origin of the populations, with the UW study consisting entirely of Indian-origin rhesus monkeys, while the NIA study included both Indian-origin and Chinese-origin rhesus monkeys. Disease outcomes are known to differ in Indian and Chinese rhesus (for example, in SIV/AIDS research) and therefore origin may be a factor of importance. Perhaps the most important difference in the studies was the manner in which the control population was fed. The UW control group was fed an allotment based upon the animal's daily ad lib food intake over a 3–6-month period. In contrast, the NIA control group was fed a regulated amount that was less than 100% of the expected ad lib consumption based upon published data. These different approaches to the control group reflect, to a certain degree, an underlying difference in initial study goals and point to the importance of defining such goals; for example, is CR being studied as a means to define basic mechanisms of cellular and organismal aging—in which case avoiding the potential obesogenic effects of ad lib feeding may be important—or is CR being tested as an intervention to potentially improve human health, in which case comparisons to the ad lib fed condition of many if not most Americans would be critical.

Whether due to diet composition, feeding practices, or genetic differences, the NIA control group—particularly for males—was leaner than the UW control group. The collaborative analysis concluded that "the NIA diet is associated with lower body weight, lower adiposity and improved survival and that no further advantage is gained by lowering food take or body weight" below that seen in the NIA male control group. In contrast, while the NIA control females consumed more calories than the NIA CR group, there was no difference in body weight or adiposity, suggesting a sexual dimorphism in response to CR. In the UW female population there was a significant negative effect of adiposity on survival, presumed to be due to the much greater range of adiposity in the UW versus the NIA female population.

With controls for age of CR initiation and adiposity, the two studies do report consistent effects of CR on cancer, with cancer incidence lower in the CR groups in both studies, suggesting that CR effects on cancer may be particularly robust. Effects on other age-associated conditions, such as cardiovascular disorders and metabolic disease (insulin resistance, diabetes), were less robust, primarily due to a lower incidence in the NIA control group than in the UW control group. Given that the NIA control group was basically under mild CR, these results may suggest that cardiovascular and metabolic disease risks can be ameliorated in primates with only mild CR, resulting in levels of adiposity seen in the NIA control, NIA CR, and UW CR groups.

The final conclusion of the collaborative analyses is that "age-related increase in disease vulnerability in primates is malleable and that ageing itself presents a reasonable target for intervention" (Mattison et al., 2017).

Rapamycin intervention

Rapamycin is an mTOR inhibitor that has been found to increase both mean and maximum LS in mice, even when treatment began later in life (20 months) (Harrison et al., 2009). Rapamycin has been found to delay the onset of a number of age-related diseases in mouse models, including Alzheimer's disease, cardiovascular disease, and various cancers (Cox & Mattison, 2009; Halloran et al., 2012; Harrison et al., 2009; Lamming et al., 2013; Miller et al., 2011, 2014; Reifsnyder, Doty, & Harrison, 2014; Reifsnyder, Flurkey, Te, & Harrison, 2016; Wilkinson et al., 2012; Ye et al., 2013). While its cellular actions were once thought to mimic the effects of caloric restriction, it has been determined that the suppression of the mTOR pathway acts differently from the effects of caloric restriction (Garratt, Nakagawa, & Simons, 2016; Karunadharma et al., 2015; Ross et al., 2015). Given its impact on mouse longevity and measures of healthy aging, the potential for its use as an antiaging treatment in humans is of great interest.

Rapamycin is already FDA-approved for use in humans as an immunosuppressant agent during organ transplants, but its broader use as an aging treatment may be impacted by potential side effects of the drug. Rapamycin has been reported to produce side effects, including increased insulin resistance and onset of type II diabetes, in mouse studies and in human transplant patients; however, the interpretations of the data have been contradictory (Deepa et al., 2013; Fadini, Ceolotto, Pagnin, de Kreutzenberg, & Avogaro, 2011; Miller et al., 2014; Reifsnyder et al., 2016). In order to evaluate the potential of this drug as an antiaging treatment, testing in a nonhuman primate was done. Encapsulated rapamycin has been given to marmosets to determine drug dose and efficacy (Tardif et al., 2015), to evaluate metabolic changes associated with rapamycin dosing (Ross et al., 2015; Sills, Artavia, DeRosa, Ross, & Salmon, 2019), and to examine the effects on longevity.

Encapsulated rapamycin can be delivered orally to the marmosets in a yogurt vehicle, which is readily consumed by the marmosets within their home cage, with control animals receiving empty Eudragit capsules in the yogurt vehicle. The production of the encapsulated rapamycin allows for oral dosing, while removing the potential side effect of mouth sores (Harrison et al., 2009). Marmosets given rapamycin in the morning at a dose of 1.0 mg/kg/day maintain daily serum concentrations of rapamycin near 6 ng/mL, which is similar to the level reported in human patients and in the mice tested on the drug (Tardif et al., 2015). Marmosets followed for 14 months of rapamycin dosing exhibited no signs of clinical anemia, fibrotic lung changes, mouth ulcers, or slowed wound healing, which are all potential side effects of the drug (Tardif et al., 2015). Mortality over this short-term study did not appear to be increased due to rapamycin.

Treatment with rapamycin had no overall effects on body weight, with only slight decreases in fat mass in the first few months of treatment, after which it recovered to baseline values and was maintained (Ross et al., 2015). Interestingly, at the same time that the fat mass of the animals recovered, the rapamycin animals were found to have increased food intake, perhaps as a compensation for the effects of fat mass loss. Rapamycin-treated animals also had no changes in daily activity, triglyceride concentrations, fasting glucose, or insulin response to a glucose challenge (Ross et al., 2015). Longitudinal follow-up of rapamycin-dosed animals has not revealed significant changes in metabolic or inflammatory markers (Sills et al., 2019). While there were no changes in expression of genes associated with metabolism, there was tissue-specific upregulation of some components of the proteostasis network in animals treated with rapamycin (Lelegren, Liu, Ross, Tardif, & Salmon, 2016). To date, no negative side effects of rapamycin have been described for marmosets, either in the short-term study or the continuing longevity study.

Studies of aging in nonhuman primates allow us to evaluate both the cellular and mechanistic pathways underlying the aging process, as well as the systemic changes in bodily function associated with aging in model species. As species that are adapted to longer life, they offer a comparative perspective that models many of the evolutionary pathways shared between nonhuman primates and humans. Of particular interest to the future of aging studies will be the ability to test drugs and interventions in a nonhuman primate before moving them into human trials.

References

Atsalis, S., & Margulis, S.W. (2008a). Perimenopause and menopause: Documenting life changes in aging female gorillas. In *Primate reproductive aging* (Interdisciplinary Topics in Gerontology, Vol. 36), Karger Publishers, pp. 119–146.

Atsalis, S., & Margulis, S.W. (2008b). Primate reproductive aging: From lemurs to humans. In *Primate reproductive aging* (Interdisciplinary Topics in Gerontology, Vol. 36), Karger Publishers, pp. 186–194

Austad, S. N., & Fischer, K. E. (1992). Primate longevity: Its place in the mammalian scheme. *American Journal of Primatology*, *28*(4), 251–261.

Bagi, C. M., Volberg, M., Moalli, M., Shen, V., Olson, E., Hanson, N., … Andresen, C. J. (2007). Age-related changes in marmoset trabecular and cortical bone and response to alendronate therapy resemble human bone physiology and architecture. *Anatomical Record*, *290*(8), 1005–1016. Available from https://doi.org/10.1002/ar.20561.

Bartholomeusz, B., Hardy, K. J., Nelson, A. S., & Phillips, P. A. (1998). Modulation of nitric oxide improves cyclosporin A-induced hypertension in rats and primates. *Journal of Human Hypertension*, *12*(12), 839–844.

Baxter, M. G., Santistevan, A. C., Bliss-Moreau, E., & Morrison, J. H. (2018). Timing of cyclic estradiol treatment differentially affects cognition in aged female rhesus monkeys. *Behavioral Neuroscience*, *132*(4), 213.

Baxter, V. K., Shaw, G. C., Sotuyo, N. P., Carlson, C. S., Olson, E. J., Zink, M. C., … Metcalf Pate, K. A. (2013). Serum albumin and body weight as biomarkers for the antemortem identification of bone and gastrointestinal disease in the common marmoset. *PLoS One*, *8*(12), e82747.

Behrendt, A. K., Kuhla, A., Osterberg, A., Polley, C., Herlyn, P., Fischer, D. C., … Vollmar, B. (2016). Dietary restriction-induced alterations in bone phenotype: effects of lifelong versus short-term caloric restriction on femoral and vertebral bone in C57BL/6 Mice. *Journal of Bone and Mineral Research: The Official Journal of the American Society for Bone and Mineral Research*, *31*(4), 852–863. Available from https://doi.org/10.1002/jbmr.2745.

Bernick, S., Cailliet, R., & Levy, B. M. (1980). The maturation and aging of the vertebrae of marmosets. *Spine*, *5*(6), 519–524.

Blanc, S., Schoeller, D., Kemnitz, J., Weindruch, R., Colman, R., Newton, W., … Ramsey, J. (2003). Energy expenditure of rhesus monkeys subjected to 11 years of dietary restriction. *The Journal of Clinical Endocrinology & Metabolism*, *88*(1), 16–23.

Bons, N., Rieger, F., Prudhomme, D., Fisher, A., & Krause, K. H. (2006). *Microcebus murinus*: A useful primate model for human cerebral aging and Alzheimer's disease? *Genes, Brain, and Behavior*, *5*(2), 120–130.

Caro, T., Sellen, D., Parish, A., Frank, R., Brown, D., Voland, E., & Mulder, M. B. (1995). Termination of reproduction in nonhuman and human female primates. *International Journal of Primatology*, *16*(2), 205–220.

Carter, C. S., Marzetti, E., Leeuwenburgh, C., Manini, T., Foster, T. C., Groban, L., … Morgan, D. (2012). Usefulness of preclinical models for assessing the efficacy of late-life interventions for sarcopenia. *The Journals of Gerontology. Series A, Biological Sciences and Medical Sciences*, *67*(1), 17–27. Available from https://doi.org/10.1093/gerona/glr042.

Chalmers, D. T., Murgatroyd, L. B., & Wadsworth, P. F. (1983). A survey of the pathology of marmosets (*Callithrix jacchus*) derived from a marmoset breeding unit. *Laboratory Animals*, *17*(270–279). Available from https://doi.org/10.1258/002367783781062217.

Champ, J. E., Binkley, N., Havighurst, T., Colman, R. J., Kemnitz, J. W., & Roecker, E. B. (1996). The effect of advancing age on bone mineral content of female rhesus monkeys. *Bone*, *19*(5), 485–492.

Clarkson, T. B., & Appt, S. E. (2005). Controversies about HRT—lessons from monkey models. *Maturitas*, *51*(1), 64–74.

Clarkson, T. B., & Mehaffey, M. H. (2009). Coronary heart disease of females: Lessons learned from nonhuman primates. *American Journal of Primatology: Official Journal of the American Society of Primatologists*, *71*(9), 785–793.

Collerton, J., Martin-Ruiz, C., Davies, K., Hilkens, C. M., Isaacs, J., Kolenda, C., … Kirkwood, T. B. (2012). Frailty and the role of inflammation, immunosenescence and cellular ageing in the very old: Cross-sectional findings from the Newcastle 85+ Study. *Mechanisms of Ageing and Development*, *133*(6), 456–466. Available from https://doi.org/10.1016/j.mad.2012.05.005.

Colman, R. J., Anderson, R. M., Johnson, S. C., Kastman, E. K., Kosmatka, K. J., Beasley, T. M., … Kemnitz, J. W. (2009). Caloric restriction delays disease onset and mortality in rhesus monkeys. *Science*, *325*(5937), 201–204.

Colman, R. J., Beasley, T. M., Allison, D. B., & Weindruch, R. (2008). Attenuation of sarcopenia by dietary restriction in rhesus monkeys. *The Journals of Gerontology. Series A, Biological Sciences and Medical Sciences*, *63*(6), 556–559.

Colman, R. J., Beasley, T. M., Allison, D. B., & Weindruch, R. (2012). Skeletal effects of long-term caloric restriction in rhesus monkeys. *Age*, *34*(5), 1133–1143. Available from https://doi.org/10.1007/s11357-011-9354-x.

Colman, R. J., Beasley, T. M., Kemnitz, J. W., Johnson, S. C., Weindruch, R., & Anderson, R. M. (2014). Caloric restriction reduces age-related and all-cause mortality in rhesus monkeys. *Nature Communications*, *5*, 3557. Available from https://doi.org/10.1038/ncomms4557.

Colman, R. J., Kemnitz, J. W., Lane, M. A., Abbott, D. H., & Binkley, N. (1999). Skeletal effects of aging and menopausal status in female rhesus macaques. *The Journal of Clinical Endocrinology and Metabolism*, *84*(11), 4144–4148. Available from https://doi.org/10.1210/jcem.84.11.6151.

Colman, R. J., Lane, M. A., Binkley, N., Wegner, F. H., & Kemnitz, J. W. (1999). Skeletal effects of aging in male rhesus monkeys. *Bone*, *24*(1), 17–23.

Colman, R. J., Roecker, E. B., Ramsey, J. J., & Kemnitz, J. W. (1998). The effect of dietary restriction on body composition in adult male and female rhesus macaques. *Aging*, *10*(2), 83–92.

Cox, L. S., & Mattison, J. A. (2009). Increasing longevity through caloric restriction or rapamycin feeding in mammals: Common mechanisms for common outcomes? *Aging Cell*, *8*(5), 607–613. Available from https://doi.org/10.1111/j.1474-9726.2009.00509.x.

Cramer, P. E., Gentzel, R. C., Tanis, K. Q., Vardigan, J., Wang, Y., Connolly, B., … Zerbinatti, C. (2018). Aging African green monkeys manifest transcriptional, pathological, and cognitive hallmarks of human Alzheimer's disease. *Neurobiology of Aging*, *64*, 92–106.

Cummings, J. L., Morstorf, T., & Zhong, K. (2014). Alzheimer's disease drug-development pipeline: Few candidates, frequent failures. *Alzheimer's Research & Therapy*, *6*(4), 37.

Deepa, S. S., Walsh, M. E., Hamilton, R. T., Pulliam, D., Shi, Y., Hill, S., … Van Remmen, H. (2013). Rapamycin modulates markers of mitochondrial biogenesis and fatty acid oxidation in the adipose tissue of db/db Mice. *Journal of Biochemical and Pharmacological Research*, *1*(2), 114–123.

Didier, E. S., MacLean, A. G., Mohan, M., Didier, P. J., Lackner, A. A., & Kuroda, M. J. (2016). Contributions of nonhuman primates to research on aging. *Veterinary Pathology*, *53*(2), 277–290. Available from https://doi.org/10.1177/0300985815622974.

Djelti, F., Dhenain, M., Terrien, J., Picq, J.-L., Hardy, I., Champeval, D., … Aujard, F. (2017). Impaired fasting blood glucose is associated to cognitive impairment and cerebral atrophy in middle-aged non-human primates. *Aging*, *9*(1), 173.

Downs, J. L., & Urbanski, H. F. (2006). Neuroendocrine changes in the aging reproductive axis of female rhesus macaques (*Macaca mulatta*). *Biology of Reproduction*, *75*(4), 539–546. Available from https://doi.org/10.1095/biolreprod.106.051839.

Duncan, A. E., Colman, R. J., & Kramer, P. A. (2012). Sex differences in spinal osteoarthritis in humans and rhesus monkeys (Macaca mulatta). *Spine*, *37*(11), 915–922. Available from https://doi.org/10.1097/BRS.0b013e31823ab7fc.

Edler, M. K., Sherwood, C. C., Meindl, R. S., Hopkins, W. D., Ely, J. J., Erwin, J. M., … Raghanti, M. A. (2017). Aged chimpanzees exhibit pathologic hallmarks of Alzheimer's disease. *Neurobiology of Aging*, *59*, 107–120.

Engelberth, R. C. G. J., Silva, K. D. A., Azevedo, C. V. M., Gavioli, E. C., Santos, J. R., Soares, J. G., & Cavalcante, J. S. (2014). Morphological changes in the suprachiasmatic nucleus of aging female marmosets (*Callithrix jacchus*). *BioMed Research International*, *2014*, 243825.

Engle, J. R., Machado, C. J., Permenter, M. R., Vogt, J. A., Maurer, A. P., Bulleri, A. M., & Barnes, C. A. (2016). Network patterns associated with navigation behaviors are altered in aged nonhuman primates. *Journal of Neuroscience*, *36*(48), 12217–12227.

Fadini, G. P., Ceolotto, G., Pagnin, E., de Kreutzenberg, S., & Avogaro, A. (2011). At the crossroads of longevity and metabolism: The metabolic syndrome and lifespan determinant pathways. *Aging Cell*, *10*(1), 10–17. Available from https://doi.org/10.1111/j.1474-9726.2010.00642.x.

Fontana, L., & Partridge, L. (2015). Promoting health and longevity through diet: From model organisms to humans. *Cell*, *161*(1), 106–118.

Fontana, L., Vinciguerra, M., & Longo, V. D. (2012). Growth factors, nutrient signaling, and cardiovascular aging. *Circulation Research*, *110*(8), 1139–1150. Available from https://doi.org/10.1161/CIRCRESAHA.111.246470.

Garratt, M., Nakagawa, S., & Simons, M. J. (2016). Comparative idiosyncrasies in life extension by reduced mTOR signalling and its distinctiveness from dietary restriction. *Aging Cell*, *15*(4), 737–743. Available from https://doi.org/10.1111/acel.12489.

Geula, C., Nagykery, N., & Wu, C.-K. (2002). Amyloid-β deposits in the cerebral cortex of the aged common marmoset (*Callithrix jacchus*): Incidence and chemical composition. *Acta Neuropathologica*, *103*, 48–58.

Geula, C., Wu, C.-K., Saroff, D., Lorenzo, A., Yuan, M., & Yankner, B. (1998). Aging renders the brain vulnerable to amyloid β-protein neurotoxicity. *Nature Medicine*, *4*(7), 827–831.

Graham, C. E. (1979). Reproductive function in aged female chimpanzees. *American Journal of Physical Anthropology*, *50*(3), 291–300.

Hainsworth, A. H., Allan, S. M., Boltze, J., Cunningham, C., Farris, C., Head, E., ... Oberstein, S. A. L. (2017). Translational models for vascular cognitive impairment: A review including larger species. *BMC Medicine*, *15*(1), 16.

Halloran, J., Hussong, S. A., Burbank, R., Podlutskaya, N., Fischer, K. E., Sloane, L. B., ... Galvan, V. (2012). Chronic inhibition of mammalian target of rapamycin by rapamycin modulates cognitive and non-cognitive components of behavior throughout lifespan in mice. *Neuroscience*, *223*, 102–113. Available from https://doi.org/10.1016/j.neuroscience.2012.06.054.

Hampson, E. (2018). Estrogens, aging, and working memory. *Current Psychiatry Reports*, *20*(12), 109.

Hara, Y., Yuk, F., Puri, R., Janssen, W. G., Rapp, P. R., & Morrison, J. H. (2016). Estrogen restores multisynaptic boutons in the dorsolateral prefrontal cortex while promoting working memory in aged rhesus monkeys. *Journal of Neuroscience*, *36*(3), 901–910.

Harrison, D. E., Strong, R., Sharp, Z. D., Nelson, J. F., Astle, C. M., Flurkey, K., ... Miller, R. A. (2009). Rapamycin fed late in life extends lifespan in genetically heterogeneous mice. *Nature*, *460*(7253), 392–395. Available from https://doi.org/10.1038/nature08221.

Havill, L., Levine, S., Newman, D., & Mahaney, M. (2008). Osteopenia and osteoporosis in adult baboons (*Papio hamadryas*). *Journal of Medical Primatology*, *37*(3), 146–153.

Hill, K., & Hurtado, A. M. (1991). The evolution of premature reproductive senescence and menopause in human females. *Human Nature*, *2*(4), 313–350.

Hodgen, G. D., Goodman, A. L., O'Connor, A., & Johnson, D. K. (1977). Menopause in rhesus monkeys: Model for study of disorders in the human climacteric. *American Journal of Obstetrics and Gynecology*, *127*(6), 581–584.

Hoffman, J. M., Ross, C., Tran, V., Promislow, D. E., Tardif, S., & Jones, D. P. (2019). The metabolome as a biomarker of mortality risk in the common marmoset. *American Journal of Primatology*, *81*(2), e22944.

Hoffman, J. M., Tran, V., Wachtman, L. M., Green, C. L., Jones, D. P., & Promislow, D. E. (2016). A longitudinal analysis of the effects of age on the blood plasma metabolome in the common marmoset, *Callithrix jacchus*. *Experimental Gerontology*, *76*, 17–24. Available from https://doi.org/10.1016/j.exger.2016.01.007.

Honavar, M., & Lantos, P. L. (1987). Ultrastructural changes in the frontal cortex and hippocampus in the aging marmoset. *Mechanisms of Aging and Development*, *41*, 161–175.

Huber, H. F., Gerow, K. G., & Nathanielsz, P. W. (2015). Walking speed as an aging biomarker in baboons (Papio hamadryas). *Journal of Medical Primatology*, *44*(6), 373–380.

Jones, J. H. (2011). Primates and the evolution of long, slow life histories. *Current Biology*, *21*(18), R708–R717.

Justice, J. N., Cesari, M., Seals, D. R., Shively, C. A., & Carter, C. S. (2015). Comparative approaches to understanding the relation between aging and physical function. *Journals of Gerontology Series A: Biomedical Sciences and Medical Sciences*, *71*(10), 1243–1253.

Kangas, B. D., Bergman, J., & Coyle, J. T. (2016). Touchscreen assays of learning, response inhibition, and motivation in the marmoset (*Callithrix jacchus*). *Animal Cognition*, *19*, 673–677.

Karunadharma, P. P., Basisty, N., Dai, D. F., Chiao, Y. A., Quarles, E. K., Hsieh, E. J., ... Rabinovitch, P. S. (2015). Subacute calorie restriction and rapamycin discordantly alter mouse liver proteome homeostasis and reverse aging effects. *Aging Cell*, *14*(4), 547–557. Available from https://doi.org/10.1111/acel.12317.

Kobayashi, R., Takahashi-Fujigasaki, J., Shiozawa, S., Hara-Miyauchi, C., Inoue, T., Okano, H. J., ... Okano, H. (2016). alpha-Synuclein aggregation in the olfactory bulb of middle-aged common marmoset. *Neuroscience Research*, *106*, 55–61. Available from https://doi.org/10.1016/j.neures.2015.11.006.

Kohama, S. G., Renner, L., Landauer, N., Weiss, A. R., Urbanski, H. F., Park, B., ... Neuringer, M. (2016). Effect of ovarian hormone therapy on cognition in the aged female rhesus macaque. *Journal of Neuroscience*, *36*(40), 10416–10424.

Lacreuse, A., & Herndon, J. G. (2009). Nonhuman primate models of cognitive aging. In *Animal models of human cognitive aging* Humana Press, (pp. 1–30). Available from https://doi.org/10.1007/978-1-59745-422-3_2.

Lacreuse, A., Parr, L., Chennareddi, L., & Herndon, J. G. (2018). Age-related decline in cognitive flexibility in female chimpanzees. *Neurobiology of Aging*, *72*, 83–88.

Lamming, D. W., Ye, L., Astle, C. M., Baur, J. A., Sabatini, D. M., & Harrison, D. E. (2013). Young and old genetically heterogeneous HET3 mice on a rapamycin diet are glucose intolerant but insulin sensitive. *Aging Cell*, *12*(4), 712–718. Available from https://doi.org/10.1111/acel.12097.

Lane, M. A. (2000). Nonhuman primate models in biogerontology. *Experimental Gerontology*, *35*(5), 533–541.

Languille, S., Liévin-Bazin, A., Picq, J.-L., Louis, C., Dix, S., De Barry, J., ... Schenker, E. (2015). Deficits of psychomotor and mnesic functions across aging in mouse lemur primates. *Frontiers in Behavioral Neuroscience*, *8*, 446.

Latimer, C. S., Shively, C. A., Keene, C. D., Jorgensen, M. J., Andrews, R. N., Register, T. C., ... Mintz, A. (2019). A nonhuman primate model of early Alzheimer's disease pathologic change: Implications for disease pathogenesis. *Alzheimer's & Dementia*, *15*(1), 93–105.

Le Brazidec, M., Herrel, A., Thomas, P., Grégoire, B.-A., Aujard, F., & Pouydebat, E. (2017). How aging affects grasping behavior and pull strength in captive gray *Mouse Lemurs* (*Microcebus murinus*). *International Journal of Primatology*, *38*(6), 1120–1129.

Lee, H. J., Gonzalez, O., Dick, E. J., Jr, Donati, A., Feliers, D., Choudhury, G. G., ... Kasinath, B. S. (2018). Marmoset as a model to study kidney changes associated with aging. *The Journals of Gerontology: Series A*, *74*(3), 315–324.

Lelegren, M., Liu, Y., Ross, C. N., Tardif, S. D., & Salmon, A. B. (2016). Pharmaceutical inhibiton of mTOR in the common marmoset: Effect of rapamycin on regulators of proteostasis in a nonhuman primate. *Pathobiology of Aging and Age-related Diseases*, *6*, 31793.

Liao, C. Y., Rikke, B. A., Johnson, T. E., Diaz, V., & Nelson, J. F. (2010). Genetic variation in the murine lifespan response to dietary restriction: From life extension to life shortening. *Aging Cell*, *9*(1), 92–95.

Lipsitz, L. A. (2008). Dynamic models for the study of frailty. *Mechanisms of Ageing and Development*, *129*(11), 675–676. Available from https://doi.org/10.1016/j.mad.2008.09.012.

Longo, V. D., Antebi, A., Bartke, A., Barzilai, N., Brown-Borg, H. M., Caruso, C., ... Fontana, L. (2015). Interventions to slow aging in humans: Are we ready? *Aging Cell*, *14*(4), 497–510. Available from https://doi.org/10.1111/acel.12338.

Makris, N., Papadimitriou, G. M., van der Kouwe, A., Kennedy, D. N., Hodge, S. M., Dale, A. M., ... Tuch, D. S. (2007). Frontal connections and cognitive changes in normal aging rhesus monkeys: A DTI study. *Neurobiology of Aging*, *28*(10), 1556–1567.

Manrique, H. M., & Call, J. (2015). Age-dependent cognitive inflexibility in great apes. *Animal Behaviour*, *102*, 1–6.

Marshall, J. W. B., & Ridley, R. M. (2003). Assessment of cognitive and motor deficits in a marmoset model of stroke. *Institute for Laboratory Animal Research Journal*, *44*(2), 153–160.

Mattison, J. A., Colman, R. J., Beasley, T. M., Allison, D. B., Kemnitz, J. W., Roth, G. S., ... Anderson, R. M. (2017). Caloric restriction improves health and survival of rhesus monkeys. *Nature Communications*, *8*, 14063. Available from https://doi.org/10.1038/ncomms14063.

Mattison, J. A., Lane, M. A., Roth, G. S., & Ingram, D. K. (2003). Calorie restriction in rhesus monkeys. *Experimental Gerontology*, *38* (1–2), 35–46.

Mattison, J. A., Roth, G. S., Beasley, T. M., Tilmont, E. M., Handy, A. M., Herbert, R. L., ... de Cabo, R. (2012). Impact of caloric restriction on health and survival in rhesus monkeys from the NIA study. *Nature*, *489*(7415), 318–321. Available from https://doi.org/10.1038/nature11432.

Mattison, J. A., & Vaughan, K. L. (2017). An overview of nonhuman primates in aging research. *Experimental Gerontology*, *94*, 41–45. Available from https://doi.org/10.1016/j.exger.2016.12.005.

Mestre-Francés, N., Trouche, S.G., Fontes, P., Lautier, C., Devau, G., Lasbleiz, C., ... Verdier, J.-M. (2018). Old gray mouse lemur behavior, cognition, and neuropathology. In *Conn's handbook of models for human aging* (Second Edition), Elsevier, pp. 287–300.

Mietsch, M., Baldauf, K., Reitemeier, S., Suchowski, M., Schoon, H. A., & Einspanier, A. (2016). Blood pressure as prognostic marker for body condition, cardiovascular, and metabolic diseases in the common marmoset (Callithrix jacchus). *Journal of Medical Primatology*, *45*(3), 126–138. Available from https://doi.org/10.1111/jmp.12215.

Miller, R. A. (1997). When will the biology of aging become useful? Future landmarks in biomedical gerontology. *Journal of the American Geriatrics Society*, *45*(10), 1258–1267.

Miller, R. A., Harrison, D. E., Astle, C. M., Baur, J. A., Boyd, A. R., de Cabo, R., ... Strong, R. (2011). Rapamycin, but not resveratrol or simvastatin, extends life span of genetically heterogeneous mice. *The Journals of Gerontology. Series A, Biological Sciences and Medical Sciences*, *66*(2), 191–201. Available from https://doi.org/10.1093/gerona/glq178.

Miller, R. A., Harrison, D. E., Astle, C. M., Fernandez, E., Flurkey, K., Han, M., ... Strong, R. (2014). Rapamycin-mediated lifespan increase in mice is dose and sex dependent and metabolically distinct from dietary restriction. *Aging Cell*, *13*(3), 468–477. Available from https://doi.org/10.1111/acel.12194.

Mitnitski, A., Collerton, J., Martin-Ruiz, C., Jagger, C., von Zglinicki, T., Rockwood, K., & Kirkwood, T. B. (2015). Age-related frailty and its association with biological markers of ageing. *BMC Medicine*, *13*, 161. Available from https://doi.org/10.1186/s12916-015-0400-x.

Peccei, J. S. (2001). Menopause: Adaptation or epiphenomenon. *Evolutionary Anthropology: Issues, News, and Reviews: Issues, News, and Reviews*, *10*(2), 43–57.

Phillips, K. A., Ross, C. N., Spross, J., Cheng, C. J., Izquierdo, A., Biju, K. C., ... Tardif, S. D. (2017). Behavioral phenotypes associated with MPTP induction of partial lesions in common marmosets (*Callithrix jacchus*). *Behavioural Brain Research*, *325*(Pt A), 51–62. Available from https://doi.org/10.1016/j.bbr.2017.02.010.

Reifsnyder, P. C., Doty, R., & Harrison, D. E. (2014). Rapamycin ameliorates nephropathy despite elevating hyperglycemia in a polygenic mouse model of type 2 diabetes, NONcNZO10/LtJ. *PLoS One*, *9*(12), e114324. Available from https://doi.org/10.1371/journal.pone.0114324.

Reifsnyder, P. C., Flurkey, K., Te, A., & Harrison, D. E. (2016). Rapamycin treatment benefits glucose metabolism in mouse models of type 2 diabetes. *Aging*, *8*(11), 3120–3130. Available from https://doi.org/10.18632/aging.101117.

Reveles, K. R., Patel, S., Forney, L. J., & Ross, C. N. (2019). Age-related changes in the marmoset gut micro-biome. *American Journal of Primatology, 81, e22960.*

Ridley, R. M., Baker, H. F., Windle, C. P., & Cummings, R. M. (2006). Very long term studies of the seeding of beta-amyloidosis in primates. *Journal of Neural Transmission*, *113*(9), 1243–1251. Available from https://doi.org/10.1007/s00702-005-0385-2.

Ridley, R. M., Cummings, R. M., Leow-Dyke, A., & Baker, H. F. (2006). Neglect of memory after dopaminergic lesions in monkeys. *Behavioural Brain Research*, *166*(2), 253–262. Available from https://doi.org/10.1016/j.bbr.2005.08.007.

Rodriguez-Callejas, J. D., Fuchs, E., & Perez-Cruz, C. (2016). Evidence of Tau hyperphosphorylation and dystrophic microglia in the common marmoset. *Frontiers in Aging Neuroscience*, *8*, 315.

Rodríguez-Callejas, J. d D., Cuervo-Zanatta, D., Rosas-Arellano, A., Fonta, C., Fuchs, E., & Perez-Cruz, C. (2019). Loss of ferritin-positive microglia relates to increased iron, RNA oxidation, and dystrophic microglia in the brains of aged male marmosets. *American Journal of Primatology*, *81*(2), e22956.

Ross, C.N. (2019). Marmosets in aging research. In *The Common marmoset in captivity and biomedical research*, Elsevier, pp. 355–376.

Ross, C. N., Adams, J., Gonzalez, O., Dick, E., Giavedoni, L., Hodara, V. L., ... Tardif, S. D. (2019). Cross-sectional comparison of health-span phenotypes in young versus geriatric marmosets. *American Journal of Primatology*, *81*, e22952. Available from https://doi.org/10.1002/ajp.22952.

Ross, C. N., Davis, K., Dobek, G., & Tardif, S. D. (2012). Aging phenotypes of common marmosets (Callithrix jacchus). *Journal of Aging Research*, *2012*, 567143.

Ross, C. N., Salmon, A. B., Strong, R., Fernandez, E., Javors, M., Richardson, A., & Tardif, S. D. (2015). Metabolic consequences of long-term rapamycin exposure on common marmoset monkeys (Callithrix jacchus). *Aging*, *7*(11), 1–10.

Sabbagh, J. J., Kinney, J. W., & Cummings, J. L. (2013). Animal systems in the development of treatments for Alzheimer's disease: Challenges, methods, and implications. *Neurobiology of Aging*, *34*(1), 169–183.

Sadoun, A., Rosito, M., Fonta, C., & Girard, P. (2019). Key periods of cognitive decline in a nonhuman primate model of cognitive aging, the common marmoset (Callithrix jacchus). *Neurobiology of Aging*, *74*, 1–14.

Saltzman, W., Abbott, D. H., Binkley, N., & Colman, R. J. (2019). Maintenance of bone mass despite estrogen depletion in female common marmoset monkeys (Callithrix jacchus). *American Journal of Primatology*, *81*(2), e22905.

Sanchis, J., Nunez, E., Ruiz, V., Bonanad, C., Fernandez, J., Cauli, O., ... Nunez, J. (2015). Usefulness of clinical data and biomarkers for the identification of frailty after acute coronary syndromes. *The Canadian Journal of Cardiology*, *31*(12), 1462–1468. Available from https://doi.org/10.1016/j.cjca.2015.07.737.

Saul, S. R., & Kase, N. (2019). Aging, the menopausal transition, and hormone replenishment therapy: Retrieval of confidence and compliance. *Annals of the New York Academy of Sciences, 1440*(1), 5–22.

Schramm, R. D., Paprocki, A. M., & Bavister, B. D. (2002). Features associated with reproductive ageing in female rhesus monkeys. *Human Reproduction, 17*(6), 1597–1603.

Shattuck, M. R., & Williams, S. A. (2010). Arboreality has allowed for the evolution of increased longevity in mammals. *Proceedings of the National Academy of Sciences, 107*(10), 4635–4639.

Shideler, S., Gee, N., Chen, J., & Lasley, B. (2001). Estrogen and progesterone metabolites and follicle-stimulating hormone in the aged macaque female. *Biology of Reproduction, 65*(6), 1718–1725.

Shobin, E., Bowley, M. P., Estrada, L. I., Heyworth, N. C., Orczykowski, M. E., Eldridge, S. A., … Rosene, D. L. (2017). Microglia activation and phagocytosis: Relationship with aging and cognitive impairment in the rhesus monkey. *Geroscience, 39*(2), 199–220.

Sills, A. M., Artavia, J. M., DeRosa, B. D., Ross, C. N., & Salmon, A. B. (2019). Long-term treatment with the mTOR inhibitor rapamycin has minor effect on clinical laboratory markers in middle-aged marmosets. *American Journal of Primatology, 81*(2), e22927.

Smith, S., Jolette, J., & Turner, C. (2009). Skeletal health: Primate model of postmenopausal osteoporosis. *American Journal of Primatology: Official Journal of the American Society of Primatologists, 71*(9), 752–765.

Smucny, D. A., Abbott, D. H., Mansfield, K. G., Schultz-Darken, N. J., Yamamoto, M. E., Alencar, A. I., & Tardif, S. D. (2004). Reproductive output, maternal age, and survivorship in captive common marmoset females (Callithrix jacchus). *American Journal of Primatology, 64*(1), 107–121. Available from https://doi.org/10.1002/ajp.20065.

Snyder, M. L., & Schmitt, D. (2018). Effects of aging on the biomechanics of Coquerel's sifaka (Propithecus coquereli): Evidence of robustness to senescence. *Experimental Gerontology, 111*, 235–240.

Soltow, Q. A., Jones, D. P., & Promislow, D. E. (2010). A netweork persepective on metabolism and aging. *Integrative and Comparative Biology, 50*(5), 844–854.

Spinelli, S., Pennanen, L., Dettling, A. C., Feldon, J., Higgins, G. A., & Pryce, C. R. (2004). Performance of the marmoset monkey on computerized tasks of attention and working memory. *Brain Research: Cognitive Brain Research, 19*(2), 123–137. Available from https://doi.org/10.1016/j.cogbrainres.2003.11.007.

Stearns, S.C. (1992). *The evolution of life histories*. Oxford: Oxford University Press.

Stevens, D., Scott, E. A., Bowditch, A. P., Griffiths, G. D., & Pearce, P. C. (2006). Multiple vaccine and pyridostigmine interactions: Effects on cognition, muscle function and health outcomes in marmosets. *Pharmacology, Biochemistry, and Behavior, 84*(2), 207–218. Available from https://doi.org/10.1016/j.pbb.2006.04.020.

Tardif, S., & Ziegler, T. (1992). Features of female reproductive senescence in tamarins (*Saguinus* spp.), a New World primate. *Reproduction, 94*(2), 411–421.

Tardif, S. D. (1985). Histologic evidence for age-related differences in ovarian function in tamarins (Saguinus sp.; primates). *Biology of Reproduction, 33*(4), 993–1000.

Tardif, S. D., Mansfield, K. G., Ratnam, R., Ross, C. N., & Ziegler, T. E. (2011). The marmoset as a model of aging and age-related disease. *Institute for Laboratory Animal Research Journal, 52*(1), 54–65. Available from https://doi.org/10.1093/ilar.52.1.54.

Tardif, S. D., Ross, C. N., Bergman, P., Fernandez, E., Javors, M., Salmon, A. B., … Richardson, A. (2015). Testing efficacy of administration of the antiaging drug rapamycin in a nonhuman primate, the common marmoset. *Journal of Gerontology Biological Sciences, 70*(5), 577–588.

Terrien, J., Gaudubois, M., Champeval, D., Zaninotto, V., Roger, L., Riou, J., & Aujard, F. (2018). Metabolic and genomic adaptations to winter fattening in a primate species, the grey mouse lemur (Microcebus murinus). *International Journal of Obesity, 42*(2), 221.

Thomé, A., Gray, D. T., Erickson, C. A., Lipa, P., & Barnes, C. A. (2016). Memory impairment in aged primates is associated with region-specific network dysfunction. *Molecular Psychiatry, 21*(9), 1257.

Tigno, X. T., Gerzanich, G., & Hansen, B. C. (2004). Age-related changes in metabolic parameters of nonhuman primates. *The Journals of Gerontology Series A: Biological Sciences and Medical Sciences, 59*(11), 1081–1088.

Toledano, A., Álvarez, M., López-Rodríguez, A., Toledano-Díaz, A., & Fernández-Verdecia, C. (2014). [Does Alzheimer's disease exist in all primates? Alzheimer pathology in non-human primates and its pathophysiological implications (II)] (original in Spanish). *Neurología, 29*(1), 42–55.

Toufexis, D., King, S. B., & Michopoulos, V. (2017). Socially housed female macaques: A Translational model for the interaction of chronic stress and estrogen in aging. *Current Psychiatry Reports, 19*(11), 78.

Ungvari, Z., Buffenstein, R., Austad, S. N., Podlutsky, A., Kaley, G., & Csiszar, A. (2008). Oxidative stress in vascular senescence: Lessons from successfully aging species. *Frontiers in Bioscience: A Journal and Virtual Library, 13*, 5056–5070.

Ungvari Z. Philipp E.E. (2011). Comparative gerontology—from mussels to man, *Journals of Gerontology: Series A*, 66, 295–297.

Urbanski, H. F., & Sorwell, K. G. (2012). Age-related changes in neuroendocrine rhythmic function in the rhesus macaque. *Age, 34*(5), 1111–1121.

Vaughan, K. L., Kaiser, T., Peaden, R., Anson, R. M., de Cabo, R., & Mattison, J. A. (2017). Caloric restriction study design limitations in rodent and nonhuman primate studies. *The Journals of Gerontology: Series A, 73*(1), 48–53.

Vaughan, K. L., & Mattison, J. A. (2016). Obesity and aging in humans and nonhuman primates: A mini-review. *Gerontology, 62*(6), 611–617. Available from https://doi.org/10.1159/000445800.

Verhave, P. S., Vanwersch, R. A., van Helden, H. P., Smit, A. B., & Philippens, I. H. (2009). Two new test methods to quantify motor deficits in a marmoset model for Parkinson's disease. *Behavioural Brain Research, 200*(1), 214–219.

Videan, E. N., Fritz, J., & Murphy, J. (2008). Effects of aging on hematology and serum clinical chemistry in chimpanzees (*Pan troglodytes*). *American Journal of Primatology, 70*(4), 327–338. Available from https://doi.org/10.1002/ajp.20494.

Vinciguerra, M., Musaro, A., & Rosenthal, N. (2010). Regulation of muscle atrophy in aging and disease. *Advances in Experimental Medicine and Biology, 694*, 211–233.

Walker, M. L., & Herndon, J. G. (2008). Menopause in nonhuman primates? *Biology of Reproduction, 79*(3), 398–406.

Weir, H. K., Thompson, T. D., Soman, A., Moller, B., Leadbetter, S., & White, M. C. (2015). Meeting the healthy people 2020 objectives to reduce cancer mortality. *Preventing Chronic Disease, 12*, E104. Available from https://doi.org/10.5888/pcd12.140482.

Wilkinson, J. E., Burmeister, L., Brooks, S. V., Chan, C. C., Friedline, S., Harrison, D. E., … Miller, R. A. (2012). Rapamycin slows aging in mice. *Aging Cell, 11*(4), 675–682. Available from https://doi.org/10.1111/j.1474-9726.2012.00832.x.

Wilson, Q. N., Wells, M., Davis, A. T., Sherrill, C., Tsilimigras, M. C., Jones, R. B., … Kavanagh, K. (2018). Greater microbial translocation and vulnerability to metabolic disease in healthy aged female monkeys. *Scientific Reports, 8*(1), 11373.

Wise, P.M. (2005). Aging of the female reproductive system. In *Handbook of the biology of aging*, Elsevier, pp. 570–590.

Ye, L., Widlund, A. L., Sims, C. A., Lamming, D. W., Guan, Y., Davis, J. G., … Baur, J. A. (2013). Rapamycin doses sufficient to extend lifespan do not compromise muscle mitochondrial content or endurance. *Aging, 5*(7), 539–550. Available from https://doi.org/10.18632/aging.100576.

Yue, F., Zhang, G., Tang, R., Zhang, Z., Teng, L., & Zhang, Z. (2016). Age-and sex-related changes in fasting plasma glucose and lipoprotein in cynomolgus monkeys. *Lipids in Health and Disease, 15*(1), 111.

PART II

Organ systems in humans and other animals, human health and longevity

CHAPTER

12

Senotherapeutics: Experimental therapy of cellular senescence

Jamie N. Justice[1], *Laura J. Niedernhofer*[2] *and Miranda E. Orr*[1,3]

[1]Sticht Center for Healthy Aging and Alzheimer's Prevention, Internal Medicine – Gerontology and Geriatric Medicine, Wake Forest School of Medicine (WFSM), Winston-Salem, NC, United States [2]Institute on the Biology of Aging and Metabolism, Department of Biochemistry, Molecular Biology and Biophysics, University of Minnesota, Minneapolis, MN, United States [3]W.G. Hefner Veterans Affairs Medical Center, Salisbury, NC, United States

OUTLINE

Handbook of the Biology of Aging.
DOI: https://doi.org/10.1016/B978-0-12-815962-0.00012-3

Premise for translational research and therapeutically targeting cellular senescence

Our global population is aging, and this changing demographic portends unprecedented levels of chronic diseases and healthcare burden. In response, new efforts to extend human healthspan are underway, informed by the "geroscience hypothesis," which holds that the incidence of aging-related chronic disease can be prevented, delayed, or attenuated by therapeutically targeting fundamental biology shared by the aging process and common chronic diseases (Burch et al., 2014; Kennedy et al., 2014; Sierra, 2016; Sierra & Kohanski, 2017). The promise of geroscience is that interventions will be found that target biological aging processes, thereby delaying the onset of multiple chronic diseases simultaneously.

A major geroscience accomplishment has been to frame a set of tightly interrelated cellular biologic processes that are altered with aging that could be targeted to extend lifespan and healthspan: macromolecular damage, metabolism, proteostasis, inflammation, adaptation to stress, epigenetics, cell senescence, stem cells/regeneration, and pleiotropic processes (Kennedy et al., 2014; Lopez-Otin et al., 2013). One of these biologic processes, *cellular senescence* and its related senescence-associated secretory phenotype (SASP), is an underlying contributor to age-related loss of physiologic integrity and function. Thus, interventions that mitigate senescent cell burden, termed "*senotherapeutics*," would be expected to improve function, chronic diseases, and geriatric syndromes (Niedernhofer & Robbins, 2018).

A translational approach is required to facilitate the development of senotherapeutics. Recent efforts to meet that challenge have been stimulated, in part, by investigators working in the field of basic aging biology forming highly collaborative geroscience research teams to advance the basic discovery of viable interventions to improve healthspan and possibly lifespan. "Translational research" has different connotations to different investigators. Some scientists hold the traditional perspective of a unidirectional process in which basic science laboratory discoveries are extended to humans, often culminating in a clinical trial. In contemporary views, however, this research continuum is intended to be dynamic and bidirectional, with insights gained at the translational level informing refined selection of preclinical models and new discovery at the bench (Seals, 2013; Seals & Melov, 2014).

Extending this concept to establishing the therapeutic potential of targeting cellular senescence, research activities include tandem investigations from molecular signals and in vitro studies of senescence, to evaluating interventions in model organisms, to clinical trials in patient populations that could eventually influence clinical practice guidelines and community-level health (Fig. 12.1, adapted from Seals, 2013). These investigations are intended to spur discovery at the "bench," refine interventions, improve animal

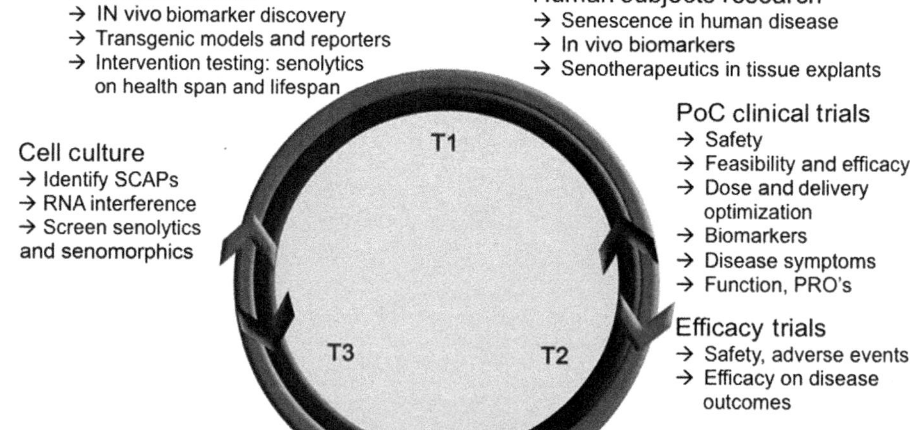

FIGURE 12.1 Senotherapeutics in translational research. Senotherapeutics was discovered through basic science research in cell culture and preclinical animal models (T1). These basic biological findings and preclinical research are now being translated into clinical human subjects research via relatively small phase I/II proof-of-concept (PoC) trials. Although results from active PoC trials registered at clinicaltrials.gov are pending (Table 12.3), phase III efficacy trials testing senotherapeutics are envisioned. Efficacy trials could lead to a new Food and Drug Administration approval, thereby providing subsequent translation for implementation into clinical practice (T2) and public health policy (T3). The process is bidirectional, in that observations made initially at higher levels of translation can be studied for underlying mechanisms or other features at the clinical and preclinical levels.

models, identify biomarkers, and develop methodologies to measure cellular senescence and drug efficacy in a clinical setting.

The potential of senotherapeutics to improve health and function (physical and mental) in older adults has resulted in an explosion in active academic translational research programs, as well as the first signs of interest from industry being interested in tackling aging as a therapeutic target. The potential of interventions targeting cellular senescence and its proinflammatory SASP are being tested in cells, preclinical models and, more recently, humans (Kirkland & Tchkonia, 2017; Kirkland et al., 2017; Niedernhofer & Robbins, 2018). In mice, senotherapeutic interventions prevent, delay the onset of, or alleviate preexisting age-related pathologies and symptoms (Tchkonia & Kirkland, 2018), including cardiovascular function, atherosclerosis, recovery from fibrosis-related illnesses and lung diseases, osteoporosis, osteoarthritis, improved adipogenesis and reduced lipotoxicity, renal function, liver cirrhosis, nonalcoholic fatty liver disease, insulin resistance, hypercholesterolemia, radioprotection, rejuvenation of aged-tissue stem cells, physical function, metabolic syndromes, obesity-induced depression, vascular function, cancer, and frailty (see Table 12.1). Moreover, senotherapeutic interventions exhibit profound effects on cognitive impairment and brain pathologies in animal models of neurodegenerative diseases and disorders including tauopathy, Alzheimer disease, Parkinson disease, and anxiety (Bussian et al., 2018; Chinta et al., 2018; Musi et al., 2018; Ogrodnik et al., 2019; Zhang et al., 2019). Though a brief overview of cellular senescence and the

TABLE 12.1 Senotherapeutic interventions prevent/delay onset or alleviate preexisting age-related pathologies and symptoms.

Pathology	Species	Evidence	References
Neurodegenerative diseases			
Amyotrophic lateral sclerosis (ALS)	Rat	Increased p16, p53, SASP in neurons, microglia, and astrocytes in lumbar spinal cord of SOD1G93A rats	Trias et al. (2019) (PMID: 30873018)
Alzheimer disease (AD)	Human	Increased expression of β-amyloid precursor protein in senescent fibroblasts in vitro	Adler et al. (1991)
		Astrocytes display a CS-like flat morphology in vivo	Nichols et al. (1993)
		Telomere shortening in microglia of AD brain with a correlation between high amyloid loads and microglial dystrophy aging-morphotype	Flanary et al. (2007)
		Association of AD with the CDKN2A locus (GWAS)	Hamshere et al. (2007), Zuchner et al. (2008)
		Increased p16INK4a and MMP-1 positive astrocytes in the frontal cortex of AD patients	Bhat et al. (2012)
		Senescence-associated transcriptome identified in neurons with tau-containing neurofibrillary tangles	Musi et al 2018
	Rat	Astrocytes display a CS-like flat morphology in vivo	Nichols et al. (1993)
		β-Amyloid induces a CS-like morphology in microglia in vitro	Korotzer et al. (1993)
	Mouse	Increased expression of p16INK4a correlates with a decline in progenitor frequency and function in the subventricular zone	Molofsky et al. (2006)
		BV2 microglia cells tend to undergo senescence after repeated inflammatory activation	Yu et al. (2012)
		Senescent oligodendrocyte precursor cells associated with A-beta plaques	Zhang et al. (2019)
Parkinson disease (PD)	Human	Reduced lamin B1 in astrocytes	Chinta et al. (2018)
	Mouse	Paraquat-induced astrocyte senescence	Chinta et al. (2018)
Tauopathy	Human	Progressive supranuclear palsy: elevated CDKN2A expression	Musi et al. (2018)
Mouse		Elevated Cdkn2a and SASP in rTg(tauP301L) mice; neurons with tau pathology display karyomegaly, DNA damage	Musi et al. (2018)

(Continued)

TABLE 12.1 (Continued)

Pathology	Species	Evidence	References
		Elevated Cdkn2a and SASP in glia derived from tauP301L mice	Bussian et al. (2018)
Traumatic brain injury	Mouse	SA-bgal and senescence markers after head injury	Tominaga et al. (2019)
Cardiovascular			
Atherosclerosis	Human	Accelerated in vitro CS of VSMCs from atherosclerotic plaques	Ross et al. (1984), Mosse et al. (1985), Bennett et al. (1995)
		Accelerated in vitro CS of VSMCs from coronary atherectomies	Bennett et al. (1998)
		Increased SA-β-Gal and low thymosin beta-10 in endothelial cells from the aorta at sites overlying atherosclerotic plaques	Vasile et al. (2001)
		Short telomeres in leukocytes from atherosclerosis patients	Samani et al. (2001), Benetos et al. (2004)
		SA-β-Gal-positive VSMCs in atherosclerotic lesions	Minamino et al. (2003)
		Prelamin-A accelerates CS in vitro, accumulates in VSMCs at atherosclerotic lesions, and colocalizes with senescent VSMCs in vivo	Ragnauth et al. (2010)
		Reduced expression and telomere binding of TRF2 and increased γH2AX-foci in VSMCs from atherosclerotic plaques	Wang et al. (2015)
	Rat	Ras-induced CS promotes vascular inflammation in the arteries	Minamino et al. (2003)
	Mouse	VSMC senescence promotes atherosclerosis	Wang et al. (2015)
		Senolytic treatment alleviates vasomotor dysfunction in atherosclerotic mice	Roos et al. (2016)
		Accumulation of foamy macrophages with CS markers at the onset of atherosclerosis; in advanced lesions, senescent cells promote plaque instability	Childs et al. (2016)
Rabbit		Intraluminal injury to the artery leads to a proliferative response followed by emergence of senescent endothelial and smooth muscle cells	Fenton et al. (2001)
Cancer		Extensive evidence (see text). Below are presented selected studies on pro- and anticancer effects of CS	Reviewed by Campisi, (2005), Campisi and d'Adda di Fagagna, (2007), Campisi (2013), Tacutu et al. (2011), Munoz-Espin and Serrano (2014), Ovadya and Krizhanovsky (2014)
Anticancer	Human	Accumulation of senescent cells in lung cancer tumors following neoadjuvant chemotherapy was associated with treatment success	Roberson et al. (2005)
		Hodgkin Reed–Sternberg cells often express a variety of CS markers. A positive correlation of Hodgkin lymphoma outcome with percentage of Hodgkin Reed–Sternberg cells expressing CS markers (p16INK4 and p21)	Caliò et al., (2015), Gopas et al. (2016)
	Mouse	Chemotherapy induces CS in lymphomas in vivo. Mice harboring tumors capable of CS had a substantially better posttherapy prognosis. Tumors with p16INK4A mutations are less responsive	Schmitt et al. (2002)
		The p53-dependent CS restricts tumorigenesis in Pten-deficient prostates	Chen et al. (2005b)
		H3K9me-mediated CS restricts aggressiveness of T-cell lymphomas	Braig et al. (2005)

(*Continued*)

TABLE 12.1 (Continued)

Pathology	Species	Evidence	References
		High-level Ras activation induces CS in vivo. Ras-induced mammary tumorigenesis requires evasion from CS p53-p16INK4A-dependent checkpoints	Sarkisian et al. (2007)
		Ablation of p53-dependent CS program in hepatic stellate enhances the transformation of adjacent epithelial cells into hepatocellular carcinoma	Lujambio et al. (2013)
Procancer	Human	Senescent fibroblasts promote proliferation of malignant but not normal epithelial cells in vitro	Krtolica et al. (2001)
		Senescent fibroblasts promote epithelial ovarian cancer cell growth in vitro and in vivo	Yang et al. (2006)
		SASP induces EMT and invasiveness of breast cancer cell lines in vitro	Coppé et al. (2008)
		Escape from CS of A549 lung cancer cells produces aggressive variants in vitro and in vivo	Yang et al. (2017)
	Human and mouse	Senescent primary fibroblasts secrete VEGF and promote angiogenesis in vitro	Coppé et al. (2006)
Fibrosis			
Pulmonary fibrosis	Human	Mutations in the telomerase components can appear as familial IPF	Armanios et al. (2007)
		Short telomeres are a risk factor for IPF	Alder et al. (2008)
		Increased SA-β-Gal and p21 in human bronchial epithelial cells from IPF patients	Minagawa et al. (2011)
		Pulmonary fibroblasts, especially myofibroblasts, from IPF patients show features of CS ex vivo	Yanai et al. (2015)
		Increased CS in IPF type II AECs	Disayabutr et al. (2016)
		Increased CS biomarkers in IPF lung	Schafer et al. (2017)
		Ex vivo cultured IPF lung fibroblasts have a senescent phenotype	Álvarez et al. (2017)
	Mouse	Bleomycin induces CS in alveolar epithelial cells with persistent SASP	Aoshiba et al. (2003), Aoshiba et al. (2013)
		Loss of Cav-1 attenuates early injury response to bleomycin by limiting stress-induced CS in epithelial cells	Shivshankar et al. (2012)
		Accumulation of senescent myofibroblasts in persistent lung fibrosis; targeting of Nox4 attenuated the CS phenotype and led to a reversal of fibrosis	Hecker et al. (2014)
		Long-term inhibition of TRF1 in type II AECs leads to increased CS and lung fibrosis	Naikawadi et al. (2016)
		Removal of CS improves pulmonary function in a bleomycin model	Schafer et al. (2017)
		Accumulation of senescent epithelial and immune cells after bleomycin treatment	Biran et al. (2017)
	Mouse and rat	Rupatadine protects against pulmonary fibrosis by attenuating PAF-mediated senescence	Lv et al. (2013)
Renal fibrosis	Mouse	INK4A (p16INK4a) KO mice exhibit exacerbated fibrosis	Wolstein et al. (2010)
		INK4A (p16INK4a) KO mice reduced interstitial fibrosis	Braun et al. (2012)
		Increased interstitial fibrosis and CS marker in aged mice post ischemia-reperfusion	Clements et al. (2013)
	Rat	Tubular cells undergo CS and express increased TGF-β1 and p21WAF1/CIP1 with advancing age	Ding et al. (2001)

(*Continued*)

TABLE 12.1 (Continued)

Pathology	Species	Evidence	References
Cardiac fibrosis	Mouse	Increased CS after myocardial infarction; p53 deficiency results in reduced CS induction and enhanced collagen deposition	Zhu et al. (2013)
Chronic pancreatitis	Rat	CS correlated with the severity of inflammation and the extension of fibrosis	Fitzner et al. (2012)
Liver fibrosis	Human	Hepatocyte telomere shortening and CS correlate with fibrosis progression	Wiemann et al. (2002)
Cirrhosis			
	Mouse	Senescent cells accumulate in acute fibrotic livers and limit acute fibrosis; inability to undergo CS leads to excessive fibrosis	Krizhanovsky et al. (2008)
		CCN1/CYR61 limits acute fibrosis and promotes CS	Borkham-Kamphorst et al. (2014)
		CCN1 induces CS of hepatic myofibroblasts and reduces acute fibrosis	Kim et al. (2013)
		IL22 induces CS of hepatic stellate cells and reduces acute fibrosis	Kong et al. (2012)
		Senolytic treatment of senescent cholangiocytes reduced liver fibrosis	Moncsek et al. (2018)
Metabolic disorders			
Diabetes	Human	High glucose promotes CS of HUVECs, ASCs, and skin fibroblasts in vitro	Blazer et al. (2002), Yokoi et al. (2006), Cramer et al. (2010)
Type II			
		Association of type II diabetes with the CDK2A/2 B locus (GWAS)	Saxena et al. (2007); Scott et al. (2007), Zeggini et al. (2007), Bao et al. (2012)
		CS-like changes in the adipose tissue of diabetic patients (increased SA-β-Gal, p16INK4a, p53, and production of proinflammatory cytokines)	Minamino et al. (2009)
		p16Ink4a-induced CS of pancreatic β-cells increases insulin secretion in vitro	Helman et al. (2016)
	Mouse	Accumulation of senescent β-cells and development of diabetes in Ptt-deficient mice; p21 deletion partially rescued these mice from diabetes development	Wang et al. (2003), Chesnokova et al. (2009)
		Accumulation of senescent β-cells in diet-induced type II diabetes accompanied by insufficient insulin release	Sone and Kagawa (2005)
		Deletion of Cdkn1b ameliorates hyperglycemia in diabetic mice	Uchida et al. (2005)
		An age-related increase in p16INK4a limits the regenerative capacity of β-cells	Krishnamurthy et al. (2006)
		Ezh2 regulates β-cell p16INK4a expression and regeneration in diabetes mellitus	Chen et al. (2009)
		Diet-induced type II diabetes-like disease promotes CS-like changes in adipose tissue (increased SA-β-Gal, p16INK4a, p53, and production of proinflammatory cytokines); inhibition of p53 ameliorates these CS-like changes and improves insulin resistance	Minamino et al. (2009)
		Rapid accumulation of senescent β-cells and development of diabetes in Lig4 − / − p53p/p mice	Tavana et al. (2010)
		Super-Ink4/Arf mice do not develop glucose intolerance with aging	Gonzalez-Navarro et al. (2013)

(Continued)

TABLE 12.1 (Continued)

Pathology	Species	Evidence	References
		PTEN regulates CS and the decline in proliferation capacity of β-cells with age	Zeng et al. (2013)
		p16Ink4a-induced CS of pancreatic β-cells increases glucose-stimulated insulin secretion in Ins2-rtTAtet-p16 transgenic mice and improves glucose tolerance in Pdx1-tTA diabetic mice	Helman et al. (2016)
		Neuronal senescence in the brain	Chow et al. (2019) (PMID: 31636448)
	Rat	Diabetic biomarker 1-deoxysphinganine triggers p21-dependent CS in Ins-1 cells in vitro	Zuellig et al. (2014)
Obesity	Human	Increased SA-β-Gal in preadipocytes from an obese subject; similarity between SASP and cytokine profile in obesity	Tchkonia et al. (2010)
		Increased CS in visceral adipose tissue endothelial cells from obese subjects	Villaret et al. (2010)
	Mouse	Diet-induced obesity promotes CS-like changes in adipose tissue leading to diabetes-like disease	Minamino et al. (2009)
		Increased vascular senescence in diet-induced obese mice	Wang et al. (2009a), Wang et al. (2009b)
		Obesity increases deoxycholic acid through the gut microbiota which provokes SASP in hepatic stellate cells	Yoshimoto et al. (2014)
		Positive correlation between growth hormone activity and white adipose tissue CS and dysfunction	Stout et al. (2014)
		Accelerated CS of T-cells in visceral adipose tissue under a high-fat diet	Shirakawa et al. (2016)
		Increased CS in a fast-food diet which was prevented by physical exercise	Schafer et al. (2016)
Hepatic steatosis	Mouse	Correlation between senescence of hepatocytes and age-related hepatic fat accumulation; elimination of senescent cells reduces hepatic steatosis	Ogrodnik et al. (2017)
Ocular degeneration			
Cataract	Mouse	Inactivation of p16ink4a or clearance of p16ink4a-positive cells delays cataract onset in BubR1 mice	Baker et al. (2008), Baker et al. (2011)
Glaucoma	Human	Increased SA-β-Gal in the outflow pathway cells of primary open-angle glaucoma patients	Liton et al. (2005)
Macular degeneration	Human	Bone morphogenetic protein-4, which is highly expressed in the retinal pigment epithelium of macular degeneration patients, can induce CS in these cells in vitro	Zhu et al. (2009)
Rhesus monkey		Increased SA-β-Gal in retinal pigment epithelial cells from older animals	Mishima et al. (1999)
Bone and joint			
Osteoarthritis (OA)	Human	Accumulation of p16INK4a-positive chondrocytes in OA in vivo; p16INK4a knockdown improved chondrocyte function in vitro	Zhou et al. (2004)
		Senescent chondrocytes accumulate with age in articular cartilage; mechanical stress to cartilage explants increases the production of CS-promoting oxidants	Martin and Buckwalter (2001); Martin et al. (2004)
		Stronger staining for SA-β-Gal in OA cartilage	Price et al. (2002)
		Oxidative stress induces telomere genomic instability, replicative senescence, and dysfunction of chondrocytes in OA cartilage	Yudoh et al. (2005)

(Continued)

TABLE 12.1 (Continued)

Pathology	Species	Evidence	References
		Telomere length associated with proximity to lesions, OA severity, and amount of senescent cells	Harbo et al. (2012)
		Correlation between SA-β-Gal in articular cartilage and OA disease severity	Gao et al. (2016)
		Accumulation of senescent cells in cartilage from OA patients	Zhao et al. (2016)
	Mouse	p16INK4a is overexpressed in OA conditions and promotes the production of the ECM proteins MMP1 and MMP13	Philipot et al. (2014)
		Transplanted senescent cartilage cells induce an OA-like condition	Xu et al. (2017)
		Clearance of senescent cells by transgenic or senolytic means attenuates the development of posttraumatic OA	Jeon et al. (2017)
	Rat	Leptin-induced CS in chondrogenic progenitor cells inhibits their chondrogenic potential and increases their osteogenic potential	Zhao et al. (2016)
Osteoporosis	Mouse	Increased CS in osteoprogenitors from old mice	Kim et al. (2017)
Intervertebral disc (IVD) degeneration	Human	Accelerated CS in IVD degeneration in vivo	Le Maitre et al. (2007)
		Accumulation of SA-β-Gal-positive cells in herniated discs	Roberts et al. (2006)
		Accumulation of senescent chondrocytes in the nucleus pulposus with aging and IVD degeneration in vivo	Kim et al. (2008), Kim et al. (2009)
	Human and rat	Accumulation of SA-β-Gal-positive cells in IVD degeneration	Gruber et al. (2007)
Muscle and frailty			
Sarcopenia	Mouse	High levels of p16ink4a and p19Arfin skeletal muscle of BubR1 mice; inactivation of p16ink4a or clearance of p16ink4a-positive cells attenuates skeletal muscle aging	Baker et al. (2008), Baker et al. (2011)
		Deficiency of skeletal muscle stem cells correlates with CS markers	Cosgrove et al. (2014)
		Senescent-like state accompanies the age-associated decline in resting satellite cell function; p16INK4a silencing restores quiescence and muscle regenerative functions	Sousa-Victor et al. (2014)
	Rat	Sarcopenia is associated with increased expression of CCN1 and induction of CS in muscle progenitor cells	Du et al. (2014)

AECs, alveolar epithelial cells; *ASCs*, adipose tissue-derived mesenchymal stem cells; *ECM*, extracellular matrix; *HUVECs*, human umbilical vein endothelial cells; *IPF*, idiopathic pulmonary fibrosis; *VSMCs*, vascular smooth muscle cells.

SASP is provided, a detailed exploration of cellular senescence is presented elsewhere in this handbook. This chapter is intended as an in-depth summary of research evaluating the clinical potential of senotherapeutics, with a focus on current testing of preclinical and clinical drugs, and practical considerations on translation of experimental therapy of cellular senescence to humans.

Cellular senescence and the senescence-associated secretory phenotype

Cellular senescence

Senescence is a cell fate in which proliferating or differentiated, nondividing cells display aberrant cell-cycle activity, resulting in a stable cell-cycle arrest and develop a profibrotic, proinflammatory, and apoptosis-inducing SASP (Campisi & d'Adda di Fagagna, 2007). Senescence is a potent tumor-suppression mechanism (Campisi, 2005; Campisi et al., 2011; Coppe et al., 2010) but the trade-off is the loss of irreplaceable, postmitotic cells. In 1961, Hayflick and Moorhead showed that human fibroblasts divide a limited number of times in vitro before entering a stable growth arrest, now called replicative senescence. Though the discovery of replicative senescence was an important milestone in understanding growth limitations, it is not considered the primary mechanism implicated in age-related loss of regenerative capacity in vivo. Many cellular stressors, oncogenic signals, and molecular damage can initiate cell senescence. Most stressors activate either or both the p53/p21 or p16^{INK4a}/retinoblastoma protein pathways (Munoz-Espin & Serrano, 2014). For example, once DNA damage is recognized by a damage sensor, the cell engages a reversible cell-cycle arrest to attempt repair. This can occur through p53 stabilization, which upregulates the cyclin-dependent kinase 2 (CDK2) inhibitor p21 to arrest the cell cycle (Herbig et al., 2003). If the DNA cannot be repaired, the cell may enter senescence by permanently exiting the cell cycle. This step requires expression of CDK inhibitors (CDKi) like the CDK4/6 inhibitor p16^{INK4a} (Herbig et al., 2003) which blocks retinoblastoma (Rb) phosphorylation to prevent S-phase entry (Takahashi et al., 2006). Cell-cycle arrest is then maintained through heterochromatinization of cell-cycle genes and the establishment of the SASP.

Senescence-associated secretory phenotype

The SASP is comprised of numerous inflammatory cytokines, chemoattractant factors, proteases, extracellular vesicles, miRNAs, metabolites, and signaling molecules, and it acts in an autocrine or paracrine fashion that is either beneficial (e.g., wound healing) or detrimental (e.g., disrupting local tissue homeostasis) depending on the context. The SASP cytokines may promote recruitment and activation of immune cells to the local environment, but if these immune cells are unable to effectively clear the senescent cells, this may result in establishing a chronically prosenescent and proinflammatory environment. Persistent senescent cells can negatively impact their immediate microenvironment through paracrine signaling to neighboring cells that hampers their regenerative capacity and disrupts the extracellular matrix. The senescent cell secretome can induce senescence in neighboring cells, referred to as the "bystander effect or secondary senescence" (da Silva et al., 2019). Senescent cell accumulation due to the bystander effect is thought to have a significant impact on the health of both immunocompromised and immunocompetent organisms. The SASP can also lead to chronic tissue inflammation, organ-level dysfunction, and even, paradoxically, carcinogenesis and tumor growth (Coppe et al., 2010; Ovadya & Krizhanovsky, 2014; Tchkonia et al., 2013).

Senescent cell accumulation

The senescent cell phenotype is characterized by failure to undergo apoptosis, due to alterations in apoptotic signaling (Childs et al., 2014). This includes altered expression of B-cell lymphoma-2 (Bcl-2) family members and prosurvival effectors, such as ephrins (EFNB1 and 3) and phosphatidylinositol-4,5-bisphosphate 3-kinase delta catalytic subunit (PI3KCD) (Wang, 1995; Zhu et al., 2015). Senescent cells accumulate and can comprise between 5%–30% of the total cells in an organ, depending on the tissue type and age of an organism. For example, the number of cells expressing markers of cellular senescence in the skin of baboons increases exponentially from negligible amounts through mid-life up to approximately 30% of cells in animals near median lifespan, but fewer are found in skeletal muscle (Herbig et al., 2006; Jeyapalan et al., 2007). Evidence suggests that even relatively few senescent cells can exert detrimental effects in vivo (Xu et al., 2017, 2018). A striking example is a model of senescent cell transplantation: Xu et al. harvested adipose progenitors or cells from ear cartilage from luciferase-expressing mice, then induced senescence by ionizing radiation. These senescent cells were transplanted into wild-type mice. Because the senescent and nonsenescent control cells express luciferase, they could be tracked in vivo using bioluminescence imaging up to 40 days after transplantation (Burd et al., 2013). These cells were highly metabolically active, secreted potent senescence secretome factors, and were positive for several biomarkers of cellular senescence. Importantly, a small number of cells were transplanted, such that only 1 in 7000–15,000 cells (or 0.01% to 0.03%) cells in the whole body, or 1 cell in 350 cells (0.28%) locally at the site of transplantation, was sufficient to elicit a host of common age-related phenotypes, including decreased grip strength, walking speed, hanging endurance, and even lifespan (Xu et al., 2018). The phenotypes were exacerbated in aged recipient mice on a high-fat diet. In 17-month-old mice transplanted with senescent cells, there was a 5.2-fold greater risk of death compared to mice injected with nonsenescent cells (Xu et al., 2018). Collectively, this supports the hypothesis that therapies targeting even a small number of these cells may be effective.

Cell senescence as a therapeutic target for age-related diseases

Notable associations exist between cellular senescence and age-related disease. If cellular senescence is a causal nexus in biological aging, then it follows that cellular senescence should be implicated in the pathogenesis of multiple, diverse age-related diseases that share few risk factors other than age (Yanai & Fraifeld, 2018), for example, nonalcoholic steatohepatitis (NASH), idiopathic pulmonary fibrosis (IPF), and cardiovascular diseases. Activating a suicide protein that clears $p16^{Ink4a+}$ cells in transgenic mice alleviates multiple aging phenotypes, including metabolic and vascular dysfunction, cataract development, lipodystrophy, and adipose tissue dysfunction and frailty; and it preserves muscle fiber diameter and prolongs time and distance achieved on a treadmill run to exhaustion (Baker et al., 2011; Roos et al., 2016; Schafer et al., 2017; Xu et al., 2015a). This expansive plurality of effects indicates that cell senescence contributes to many age-related diseases, and therefore may be a therapeutic option across a broad range of potential disease indications and/or for the comorbidities common in old age. This hypothesized link has been richly explored using animal models of disease, with transgenic/pharmacologic clearance of senescent cells, and in association studies of senescence biomarkers in human tissues from specific diseases. This topic has been the subject of numerous in-depth reviews (Childs et al., 2015, 2017; McHugh & Gil, 2018; Munoz-Espin & Serrano, 2014; Naylor et al., 2013; Tchkonia & Kirkland, 2018; Zhu et al., 2014). We have chosen not to re-review these topics, but provide an update to this impressively thorough summary of research on cell senescence and age-related diseases (Table 12.1; Yanai & Fraifeld, 2018).

Biological markers of cellular senescence

Determining the abundance and functional consequences of cellular senescence in humans is necessary to evaluate the translational potential of therapeutically targeting senescent cells (Basisty et al., 2020; Kirkland, 2013; Sharpless & Sherr, 2015). Singular biological markers pathognomonic for cellular senescence have not yet been identified. Thus, several markers are used to detect senescent cells and SASP in circulation or in tissue lysates. Senescence-associated β-galactosidase (SA-β-gal) activity is a common cellular marker; lysosomal β-galactosidase is normally active at a low pH (usually around pH 4), but activity becomes detectable at a higher pH (pH 6) in senescent cells, due to marked expansion of the lysosomal compartment (Dimri et al., 1995). CDKi $p16^{INK4a}$ expression is another established marker of cellular senescence, as are expression of p53, p21, and p38 mitogen-activated protein kinase (p38/MAPK) (Krishnamurthy et al., 2004; Sharpless, 2004, 2005). Another marker, γH2AX, reflects the activation of the DNA damage response (DDR), telomere-induced foci (TIFs), telomere-associated foci (TAFs), and DNA segments with chromatin alterations reinforcing senescence (DNA-SCARS) (d'Adda di Fagagna et al., 2003; Herbig et al., 2006; Hewitt et al., 2012; Jeyapalan et al., 2007; Rodier et al., 2011; Waaijer et al., 2018). In addition, high-mobility group A (HMGA) proteins and heterochromatin markers, including HP1 and trimethylated lysine 9 histone H3 (H3K9me3), are recognized as molecular markers of senescence-associated heterochromatin foci (SAHFs) and are considered as nonspecific markers of cellular senescence (Ivanov et al., 2013; Sadaie et al., 2013; Tsurumi & Li, 2012).

Recently, an "SASP Atlas" was constructed based on a comprehensive proteomic analysis of soluble and exosome SASP factors (Basisty et al., 2020). The SASP profile was derived from data from multiple senescence inducers and cell types, yielding hundreds of proteins. Interestingly, the secretome of senescent cells differed substantially between cell types and between methods used to induce senescence. From this analysis, several potential biomarkers of circulating SASP markers were identified, including chemokine C-X-C motif ligand 1 (CXCL1), matrix metallopeptidase 1 (MMP1), and stanniocalcin 1 (STC1), damage-associated molecular patterns (DAMPs, also known as alarmins or danger signals), HMGB1 and calreticulin (CALR); and even a noted "aging" biomarker, growth differentiating factor-15 (GDF15). These markers are in addition to canonical SASP proinflammatory, profibrotic, growth factors and matrix remodeling proteins implicated in driving negative effects on health, including interleukin-6 (IL-6), interleukin-8 (IL-8), MCP-1, platelet-derived growth factor AA (PDGF-AA), transforming growth factor-β (TGFβ), plasminogen-activating inhibitor 1 (PAI-1), other matrix metalloproteases, activin A, and osteopontin (Ovadya & Krizhanovsky, 2014; Schafer et al., 2018; Sharpless & Sherr, 2015; Tchkonia et al., 2013). Though a plethora of nonspecific markers of cell senescence and SASP exist, identification of proteins characteristic of the SASP and circulating markers of cellular senescence to assess the burden and clearance of senescent cells remains an area in critical need of development for translation of therapies targeting cellular senescence to the clinic. Biomarkers that measure on-target effects are essential for establishing drug efficacy and mechanism of action.

Extracellular vesicles, such as exosomes, represent a class of biomarkers that hold great potential to fulfill this need for translating senotherapies. Exosomes contribute to the spread of senescence by transferring senescence-driving biomolecules (proteins, lipids, and nucleic acids) that alter cell-cycle activity and confer senescence to healthy cells (Abbas et al., 2017; Davis et al., 2017; Takasugi et al., 2017; Weilner et al., 2016). They carry information regarding their native cell type and tissue location (i.e., cell type-specific membrane proteins and cargo), and are present in body fluids, including blood, saliva, urine, and cerebral spinal fluid. These qualities highlight their potential to serve as specific and accessible biomarkers for cellular senescence, even in solid tissues not amenable to biopsy, such as the brain (Akers et al., 2015; Figueroa et al., 2017; Lee et al., 2016; Thompson et al., 2016; Welton et al., 2017). As senotherapies move into human studies, exosomes and other biomarkers are becoming increasingly critical to assess efficacy and target engagement in early-phase I/II clinical studies. We anticipate that these new biomarkers will to be measured in tandem with SASP factors to provide greater specificity on senescence burden and tissue source.

Experimental therapy for cellular senescence

Given the established detrimental effects of senescent cell accumulation with advancing age, recent efforts are attempting to intervene in this process. This includes removal of p16-expressing cells through pharmacogenetic approaches in transgenic animal models for proof-of-concept and pharmacological interventions using novel senotherapeutics (Box 12.1). This new class of drugs is designed to implement strategies to interfere with detrimental effects of senescent cells, either by eliminating them (senolytics) or by shutting down their secretory machinery (senomorphics or senostatics). In addition, untargeted interventions such as exercise and diet/caloric restriction may be able to prevent or blunt the accumulation of senescent cells or modulate their SASP. To date, ~20 agents tested have shown effectiveness in clearing senescent cells, while largely sparing nonsenescent cells in vitro. A select few of these demonstrate effectiveness in vivo in mice and are being translated into early-stage clinical trials in humans (reviewed in Kirkland et al., 2017). Even taking publication bias into account, these studies collectively support the notion that cellular senescence is a significant and malleable driver of mammalian aging.

Senolytics: Foundation and discovery

Initial evidence supporting (1) a causal role of senescent cells in age-related disease, and (2) proof of principle that targeted removal of these cells could yield improvements in a wide array of age-related disease was through transgenic INK-ATTAC mice. INK-ATTAC uses a fragment of the $p16^{Ink4a}$ promoter to drive expression of a "suicide" protein, a caspase-8-FKBP fusion protein that induces apoptosis upon administration of a synthetic drug, AP20187, a molecule that dimerizes the FKBP-CASP8 fusion protein. Importantly, AP20187 activates this "suicide" gene only in $p16^{Ink4a}$-expressing cells and has no known effect in wild-type mice. Elimination of $p16^{Ink4a}$-positive cells in these transgenic mice elicits profound effects on healthspan and lifespan. The INK-ATTAC model was first used in progeroid, *Bubr1*-hypomorphic mice and revealed that chronic removal of $p16^{Ink4a}$-expressing cells in, for example, the adipose tissue, skeletal muscle, and eye, delayed the onset of age-related phenotypes, whereas late-life clearance attenuates progression of established age-related disorders (Baker et al., 2011). This was later repeated by administering AP20187 in mid-life to INK-ATTAC mice. In this study, targeted removal of $p16^{Ink4a}$-expressing cells blunted several features of aging, including glomerulosclerosis, cardiac aging and hypertrophy, cataracts, and adipose tissue dysfunction (Baker et al., 2016). The benefits of eliminating $p16^{Ink4a}$-expressing cells have been reproduced, using another suicide gene driven by

BOX 12.1

Key definitions

- *Cellular senescence*. Essentially irreversible cell-cycle arrest. Senescent cells often express antiapoptotic factors, have increased protein synthesis, ROS production, a metabolic shift, karyomegaly, and an inflammatory secretome.
- *Senescence-associated secretory phenotype (SASP, or senescence secretome)*. Proinflammatory cytokines, chemokines, growth factors, proteases, and abundant exosomes.
- *Senescent cell antiapoptotic pathways (SCAPS)*. Cellular proteins required for senescent cell viability and the target of several senolytic drugs.
- *Senotherapeutics*. Prevent the accumulation, selectively eliminate, or beneficially alter senescent cell phenotypes, selectively compared to nonsenescent cells.
- *Senolytics*. Selectively induce apoptosis of and thereby eliminate senescent cells.
- *Senomorphics (or senostatics)*. Modulate the senescent phenotype and inhibit the senescence secretome (SASP inhibitors).
- *Immune modulators*. Enhance immune cell-mediated detection and clearance of senescent cells or act via the immune–senescence axis.

the $p16^{Ink4a}$ promoter, and in a wide variety of genetic backgrounds and disease models, and often serves as the positive control for evaluating the efficacy of senolytic drugs in vivo (Farr et al., 2017; Musi et al., 2018; Ogrodnik et al., 2017; Roos et al., 2016; Schafer et al., 2017; Xu et al., 2015a; Zhu et al., 2015).

The role of senescent cells in age-associated functional decline was confirmed through the use of a separate transgenic mouse, p16-3MR (Demaria et al., 2014). This mouse line uses the *p16* promoter to drive a herpes simplex virus thymidine kinase transgene. Upon administration of ganciclovir (GCV), thymidine kinase phosphorylates the nontoxic GCV to produce phosphorylated products that result in cell apoptosis, effectively eliminating p16-expressing cells. Experiments with p16-3MR mice yielded health benefits in osteoarthritis (Jeon et al., 2017), leukemia (Abdul-Aziz et al., 2019), intervertebral disc degeneration (Patil et al., 2019), and radiation-induced immune dysfunction (Palacio et al., 2019). Importantly, the use of p16-3MR mice led to the identification of a beneficial role of senescent cells in wound healing (Demaria et al., 2014). These studies highlight the need to better understand cellular senescence in both physiological and pathological processes. Nonetheless, data derived from transgenic INK-ATTAC and p16-3MR models provide convincing evidence that cellular senescence is causally implicated in a broad range of age-related phenotypes and lends credence to the concept that removal of senescent cells can prevent or delay tissue dysfunction and extend healthspan and lifespan.

The discovery that apoptosis resistance pathways utilized by senescent cells could be targeted pharmacologically to selectively eliminate them is still relatively new. The first senescent cell antiapoptotic pathways (SCAPs) were not identified and confirmed until 2015, when Zhu et al. identified SCAPs using a bioinformatics approach based on expression profiling of senescent versus nonsenescent human cells. Key pathways identified as differentially regulated in senescent cells include tyrosine kinases/ephrins, BCL-2/BCL-X, P13K/AKT, p53/p21/PAI-1&2, and HIF-1α (Zhu et al., 2015). Later, HSP-90 was identified as a senescent cell target based on a drug screen (Fuhrmann-Stroissnigg et al., 2017). These pathways were confirmed as causal links leading to cell senescence by RNA interference studies knocking-down SCAP expression. Soon after, drugs targeting these SCAPs were repurposed from cancer therapy and tested as candidate senolytics in cells.

Dasatinib and quercetin

Dasatinib and quercetin were the first senolytic agents identified using a hypothesis-driven drug-discovery approach led by Dr. Kirkland's group at the Mayo Clinic (Zhu et al., 2015). Dasatinib is an inhibitor of multiple tyrosine kinases and is currently indicated for use as a second-line chemotherapeutic agent for treatment of chronic myeloid leukemia that is resistant to another tyrosine kinase inhibitor, imatinib (Weisberg et al., 2007). Proteomic analyses suggests that dasatinib may bind to as many as 30 different kinases (Bantscheff et al., 2007; Rix et al., 2007), indicating that its mechanism of action is linked to a broad targeting of kinases (e.g., BCRABL, SRC, ephrins, GFR, and others). This lack of specificity may underlie the utility of dasatinib as an immunomodulatory agent in other diseases beyond cancer (e.g., lymphocytosis and HIV) (Blake et al., 2008; Rivera-Torres & San Jose, 2019) and the effectiveness of dasatinib in killing multiple types of senescent cells.

Quercetin is a polyphenol, derived from plants, with a wide range of biological activities. Reported benefits include anticarcinogenic, antiinflammatory, and antiviral activities. Quercetin is a nonspecific kinase inhibitor that targets PI3K/AKT pathway modules, as well as BCL-2, insulin/IGF-1, and HIF-1α SCAP network components, and is a senolytic, possibly as a consequence of its inhibitory effects on multiple antiapoptotic genes (i.e., PI3K and other kinases) (Malavolta et al., 2016; Reyes-Farias & Carrasco-Pozo, 2019; Zhu et al., 2015). The combination of dasatinib + quercetin (D + Q) is complementary, yielding increased senolysis when coadministered (Fuhrmann-Stroissnigg et al., 2017; Zhu et al., 2017, 2016, 2015). Moreover, D + Q decreases senescent cell burden and SASP in human tissues, within 48 hours of administration, by selectively causing apoptosis of senescent versus nonsenescent cells in freshly isolated adipose tissue explants (Xu et al., 2018). D + Q alleviates a range of age- and senescence-related disorders in mice, including osteoporosis, hepatic steatosis, neurodegeneration in tau and A mouse models of Alzheimer disease, age- and high fat-diet-induced vascular calcification and hyporeactivity, pulmonary and physical function in bleomycin-induced pulmonary fibrosis, chronic kidney disease, radiation-induced skeleto-muscle dysfunction and progeroid symptoms (Farr et al., 2017; Kirkland & Tchkonia, 2017; Musi et al., 2018; Ogrodnik et al., 2017; Roos et al., 2016; Schafer et al., 2018; Xu et al., 2018; Zhu et al., 2017, 2015). Moreover, D + Q treatment can prevent the physical dysfunction, accelerated onset of age-related diseases, and early death caused by transplanting small numbers of senescent cells into young mice (Xu et al., 2018). Importantly, as a late-life intervention in aged wild-type mice, D + Q delays age-related diseases as a group, suppresses frailty, and extends median lifespan (Xu et al., 2018; Yousefzadeh et al., 2018).

Navitoclax and BCL-2/ CL-XL pathway inhibitors

Many senolytics were identified through screening target antiapoptosis factors overexpressed in senescent cells, like the BCL-2 family of proteins (Zhu et al., 2017, 2015). For example, navitoclax (ABT-263), an inhibitor of BCL-2, is an oral drug being tested in cancer, with senolytic activity (Zhu et al., 2016). Navitoclax occupies the inhibitory binding grooves on BCL-2, BCL-XL, and BCL-W, leading to activation of BAX and BCL-2 homologous antagonist/killer (BAK) proteins, triggering mitochondrial outer membrane permeabilization (MOMP) and causing cytochrome *c* release and programmed cell death (Childs et al., 2017; Kalkavan & Green, 2018; Luna-Vargas & Chipuk, 2016). Like dasatinib and quercetin, navitoclax is cell type-specific. Navitoclax induces apoptosis in human endothelial cells (HUVEC), consistent with the observation that RNA interference against BCL-XL is senolytic in HUVECs, but not in senescent adipose progenitor cells (Zhu et al., 2016). In vivo, navitoclax improves hematopoietic stem cell function in aged mice (Chang et al., 2016) and blunts atherogenesis in transgenic mice on a high-fat diet that are prone to atherosclerotic plaque formation (Childs et al., 2016). Similar to D + Q, navitoclax appears to act synergistically in combination with a second senolytic agent, piperlongumine (Wang et al., 2016). Piperlongumine is a natural product that has diverse pharmacological effects, including antitumor and senolytic activity.

Identifying novel inhibitors of the BCL-2 and BCL-XL pathway is an area of active research in the search for new and better senolytics. An early-generation analog of navitoclax, ABT-737, induces apoptosis of senescent lung epithelial cells following ionizing radiation and of senescent epidermal cells following genetic induction of senescence via p53 activation in vivo (Yosef et al., 2016). Not all BCL-2 family inhibitors are senolytic; for example, TW-37 is not (Zhu et al., 2017). However, relatively specific BCL-XL inhibitors A1331852 or A1155463 do selectively reduce the viability of senescent HUVECs and IMR90 cells (Zhu et al., 2017). Because these agents do not target BCL-2, they may clear senescent cells with fewer side effects, such as neutrophil toxicity and platelet deficiency characteristic of navitoclax. These adverse effects impede use of navitoclax as a chemotherapeutic for small-cell lung cancer or chronic lymphocytic leukemia (Leverson, 2016; Montero & Letai, 2018; Zhu et al., 2017).

UBX0101 (MDM2 inhibitor)

UBX0101 is a small-molecule inhibitor of the MDM2/p53 protein interaction (Vassilev et al., 2004), the chemical structure of which remains unpublished. MDM2, through its binding of p53, attenuates the activation and stability of the transcription factor. Small-molecule antagonists of MDM2 were developed as potential cancer therapies, envisioning them driving p53 stabilization and thereby apoptosis of tumor cells (Vassilev et al., 2004). Intraarticular injection of UBX0101 clears senescent cells induced by anterior cruciate ligament transection surgery in mice and by spontaneous osteoarthritis in aged mice (Jeon et al., 2017). In chondrocytes isolated from humans with osteoarthritis and cultured ex vivo, UBX0101 reduces markers of senescence while increasing markers of apoptosis. Finally, UBX0101 upregulates expression of type II collagen and aggrecan by cultured human chondrocytes, the two primary ECM components of healthy cartilage (Jeon et al., 2017). Based on these results, UBX0101 is being tested in clinical trials for osteoarthritis patients (see "Clinical Translation" section).

Fisetin

Based on the effectiveness of the flavanol quercetin, other natural flavonoids were tested for senotherapeutic activity (Yousefzadeh et al., 2018; Zhu et al., 2017). In vitro, fisetin was the most potent at selectively reducing senescent cell viability (Yousefzadeh et al., 2018). Fisetin significantly reduced senescence markers in human adipose tissue ex vivo. Fisetin reduced expression of senescence markers in multiple tissues of mice, restored tissue homeostasis, and extended healthspan and lifespan even when administered late in life (Yousefzadeh et al., 2018). Fisetin, like quercetin, is a polyphenol. It is found in the skins of fruits and vegetables including strawberries, apples, onions, grapes, and cucumbers (Adhami et al., 2012; Arai et al., 2000; Khan et al., 2013) and is widely available as a nutritional supplement. The relative abundance of flavonoids in the diet (> 16 mg consumed per day) supports a good safety profile. Though physiologic benefit may require larger doses than would be achieved through a common Western diet, there are no known toxicities from fisetin intake either through dietary sources or as a nutraceutical.

Fisetin is neuroprotective in numerous model systems through mechanisms that include protection against oxidative stress, reduction of reactive oxygen species, suppression of inflammation, and increased expression of the antioxidant glutathione (Ishige et al., 2001; Khan et al., 2013; Seo & Jeong, 2015; Wang et al., 2006). In a murine model of drug-induced diabetes, fisetin restored glucose homeostasis, lipid profiles, antioxidant buffering capacity, and reduced evidence of liver damage (Prasath et al., 2014; Prasath & Subramanian, 2014; Prasath et al., 2013). It is antibacterial, antiviral, and antiparasitic (Tasdemir et al., 2006), and suppresses angiogenesis (Touil et al., 2011) and tumor growth in vitro and in vivo (Adhami et al., 2012; Chen et al., 2002; Khan et al., 2013). Fisetin is senomorphic; it inhibits the activity of several proinflammatory cytokines, including TNFα, IL-6, and the transcription factor NF-κB (Gupta et al., 2014), and may act as an antioxidant by upregulating the synthesis of glutathione (an endogenous antioxidant) (Gupta et al., 2014; Khan et al., 2013). These examples underscore that flavonoids are notoriously promiscuous compounds with molecular targets associated with myriad signaling pathways, including PI3K/Akt, NF-κB, JNK, WNT, and mTOR (Khan et al., 2013). In fact, fisetin can be linked to virtually all pillars of aging. While it may be virtually impossible to define the precise, senotherapeutic target of fisetin, the promiscuity might be the source of its effectiveness in a complex process such as cellular and organismal aging.

Senotherapeutics: Broader definitions and deeper understanding

The initial bioinformatics-driven approach identified key SCAPs: tyrosine kinases/ephrins, BCL-2/BCL-X, P13K/AKT, p53/p21/PAI-1&2, HIF-1α, which could be targeted with existing drugs (Zhu et al., 2015). This was foundationally important and launched incredible activity and interest in senolytics, resulting in the first-generation agents that proved effective in rodents and human tissue explants, and which are now being translated into early-stage clinical trials (see "Clinical Translation of Senotherapeutics" section). However, this hypothesis-driven approach was not intended to conclude with a definitive list of senolytics. Rather, it served as the springboard for additional discovery to identify novel and potentially more effective chemical tools to target senescent cells and modulate their detrimental systemic effects on health and lifespan. Prodigious effort continues to improve existing senolytic agents, refine target engagement, and provide convergent evidence for efficacy across animal and disease models. In contrast to hypothesis-driven drug discovery, unbiased screenings have identified senotherapeutic drugs that either act as senolytics to selectively kill senescent cells or as senomorphics to suppress secretory products and senescence markers. It is now recognized that both classes of drugs are effective means to improve healthspan and lifespan in mice, and this broadened perspective has allowed for discovery of new agents and intervention approaches.

To obtain new insights into additional senescent cell targets beyond SCAPs, a drug screening platform was created for testing libraries of natural products, existing drug classes and unique compounds, based on measurement of SA-β-gal activity using the fluorescent substrate C_{12}FDG (5-dodecanoylaminofluorescein di-β-D-galactopyranoside) and DNA repair-deficient *Ercc1*$^{-/-}$ primary murine embryonic fibroblasts (MEFs) (Fuhrmann-Stroissnigg et al., 2017). These cells undergo premature senescence at atmospheric oxygen, possibly as a consequence of unrepaired DNA damage, which may reflect senescence induced by physiologically relevant stressors (as compared to ionizing radiation or oncogenic insult). By passage 5, ~50% of the cells are senescent, offering the advantage of screening in mixed populations of cells (senescent and proliferating) to rapidly identify senolytics that are indeed selective for senescent cells. A reduction in the ratio of senescent to nonsenescent cells can be achieved by three drug activities: (1) specifically killing senescent cells (senolytics); (2) suppressing cell senescence phenotypes (senomorphics); or (3) increasing proliferation of the nonsenescent cells (not considered senotherapeutic). The screen was validated by testing drugs previously established to be health- or life-extending in mice, including rapamycin, nordihydroguaiaretic acid, D + Q, and navitoclax. The former two proved senomorphic (reducing the number of senescent cells without affecting the total cell number), while the latter two were correctly identified as senolytic (reducing the senescent and total cell number). The screen was first applied to a library of autophagy regulators, based on the health benefits of rapamycin, including 97 drugs from 35 functional classes (Fuhrmann-Stroissnigg et al., 2017).

The screen yielded 15 hits that reduced the fraction of senescent cells by ≥50%, all of which were autophagy agonists. This was narrowed to six potential senolytics by eliminating seven senomorphic drugs and two pan-toxic drugs. Two of the six proved highly selective for killing senescent cells and both are HSP90 inhibitors (geldanamycin and 17-AAG). An analog of 17-AAG, 17-DMAG, has been tested in humans and had a 10-fold lower EC_{50} (half maximal effective concentration) compared to 17-AAG, thus was tested further in animals. When 17-DMAG was administered intermittently to a murine model of a human progeroid syndrome, the intervention reduced markers of senescence and delayed multiple age-related comorbidities, including a significant reduction in a composite score of age-related symptoms including kyphosis, dystonia, tremor, loss of forelimb grip strength, coat condition, ataxia, gait disorder, and overall body condition (Fuhrmann-Stroissnigg et al., 2017). HSPs are highly conserved proteins that stabilize unfolded or misfolded proteins, enabling tolerance of proteotoxic stress induced, for example, by heat, pH shift, UV, heavy metals and hypoxia, and regulating protein degradation (Fuhrmann-Stroissnigg et al., 2017). There are several types of HSPs, distinguished by their molecular size (HSP60, HSP70, HSP90). Many are constitutively expressed and transcriptionally upregulated in response to stress by heat shock factor 1 (HSF1). In mammals, there are multiple HSP90 isoforms with distinct subcellular localizations, encoded by multiple

HSP90 genes (Chen et al., 2005a). HSP90 is upregulated in many cancers (Taipale et al., 2010). Hence, there is strong interest in HSP90 as a drug target; therefore, multiple HSP90 inhibitors are in the drug library and in clinical studies. HSP90s are critical for stabilizing several signaling molecules and growth factor receptors required for tumor growth, including PI3K, AKT, and EGFR, many of which are SCAPs (Stebbins et al., 1997). The HSP90 inhibitor 17-DMAG causes dose-dependent, selective killing of senescent MEFs, murine mesenchymal stem cells (MSCs), and human IMR90 and WI38 fibroblasts. This highlights the utility of broad-based screens for identifying novel classes of senotherapeutics and testing drug classes targeting geroscience-guided processes. The identification of HSP90 as a senolytic unexpectedly linked three pillars of aging (proteostasis, senescence, and autophagy), in keeping with this family of proteins being intimately linked to lifespan regulation, aging, and age-related diseases (Charmpilas et al., 2017).

Other senotherapeutics

Most senolytics target known proteins and SCAPs (e.g., tyrosine kinases, HSP90, BCL-2/BCL-XL), yet not necessarily specifically. Unbiased explorations using chemical libraries or lists of drugs available for repurposing represent incredible opportunities to continue to identify new and better senotherapeutics. Numerous interventions are currently being evaluated for senolytic and senomorphic properties in various cell lines, tissues, and animal models. A brief summary of a few other notable senotherapeutics is provided below.

An unbiased transcriptomics analysis of IMR90 cells, with or without induction of senescence using ionizing radiation, revealed upregulation of proapoptotic networks, indicating transcriptional activation of apoptotic regulators, including FOXO4 (Baar et al., 2017). A Forkhead box protein O4 (FOXO4)-interacting peptide (FOXO4-D-retro-inverso, FOXO4-DRI) was designed to block the association of FOXO4 with p53 and was demonstrated to exert a senolytic effect. FOXO4-DRI induces apoptosis of senescent IMR90 and HUVECs in vitro, reduces senescence markers, and improves fitness, hair growth, and renal function in progeroid $Xpd^{TTD/TTD}$ mice and aged wild-type mice (Baar et al., 2017). Thus, FOXO4-DRI peptide is another new class of senolytics.

The first senolytics discovered were repurposed cancer drugs (Zhu et al., 2015). However, the reverse is also likely: senolytics might work in cancer. Cancer therapy frequently triggers cellular changes associated with senescence, including histone deacetylation, upregulation of BCL-2 family proteins, and greater SASP production. Chemotherapy-induced senescence has been reported to increase invasion of cancer cells in thyroid cancer (for review, see Tonnessen-Murray et al., 2019), possibly due to epithelial-to-mesenchymal transition and acquisition of stem cell-related properties in B-cell lymphoma. This suggests opportunities for discovery of new therapeutic treatment options for cancer using senolytics. Indeed, panobinostat, an FDA-approved HDAC inhibitor, has some senolytic activity in chemotherapy-induced senescence of various cancer cell lines (Samaraweera et al., 2017).

There are also immunological and genetic approaches being considered to clear senescent cells (Xue et al., 2007). Again, here, the approaches have a foundation in oncology. For example, patient-derived cytotoxic T-cells with chimeric antigen receptors (CAR T cells) can be used to target cells expressing the B-cell antigen CD19, for instance to kill lymphoma cells (Kloss et al., 2013). By analogy there is interest in developing CAR T cells that target senescent cells. Another approach is to drive natural killer cell-mediated clearance of cells expressing surface markers unique to cancer cells that are increasingly associated with senescent cells, such as decoy receptor 2 (DCR2) (Zhang et al., 2006) or programmed cell death ligand 1 (PDL1) (He et al., 2015). Oisin Biotechnology, Inc. is exploring delivery of a nonintegrating DNA plasmid to drive transient expression of an apoptotic gene. The plasmid encodes caspase 9 under a *p16* promoter to induce apoptosis only in senescent cells. This shows promising results in vitro and in old mice, with a dose-dependent reduction in $p16^{InkK4a}$ expression in kidney, lung, and inguinal fat (communications rovided through Oisin Biotechnology: see www.oisinbio.com). Though there are challenges to gene delivery in vivo, advancement to clinical trials is under consideration.

Senotherapeutics and the brain

The study of cellular senescence historically has focused on mitotically competent cells and tissues. However, recent evidence indicates that senescence occurs in postmitotic cells, tissues, and organs with aging, including in the brain (Jurk et al., 2012). Senescent cell accumulation in the brain is accelerated by the accumulation of aggregate-prone proteins common to Alzheimer disease, tauopathy, and Parkinson disease (Bussian et al., 2018; Chinta et al., 2018; Musi et al., 2018; Zhang et al., 2019). Over 20 independent preclinical studies have tested D, Q, or fisetin for disease-modifying effects in AD-relevant rodent models (Table 12.2). Some of these benefits may be attributed to nonsenolytic properties of these drugs (e.g., antioxidant, improved proteostasis, antiinflammatory,

TABLE 12.2 Preclinical in vivo studies with D, Q, and/or fisetin on brain-relevant outcome.

Drug	Neurological pathology model	Positive neurological outcomes
Q	Aβ (ICV), rats	Memory (Li et al., 2019)
	3 μg/μL of Aβ1–42 (ICV), mice	Learning and memory, neurogenesis (Karimipour et al., 2019)
	APPSWE/PS1dE9, mice	Cognition, Aβ toxicity (Hou et al., 2010) mitochondrial function, AMPK activity (Wang et al., 2014)
	3xTg-AD, mice	Reduced pathology (Aβ, tauopathy, astrogliosis, microgliosis), spatial learning, memory, and anxiety (Sabogal-Guaqueta et al., 2015) microgliosis, Aβ (Sabogal-Guáqueta et al., 2015, Vargas-Restrepo et al., 2018)
	5xFAD, mice	Stabilized apoE, reduced insoluble cortical Aβ (Zhang et al., 2016)
	SAMP8 accelerated aging, mice	Cognition, memory, astrogliosis (Moreno et al., 2017)
Fisetin	*Ercc1*$^{+/-}$ progeroid and advanced chronological age, mice	Reduced age-related histopathology as assessed by the Geropathology Grading Platform (i.e., vacuolization, inflammation, thalamic mineralization, ventricular enlargement, and tumors) (Yousefzadeh et al., 2018)
	TgAPPswe/PS1dE9, mice	Neuroprotection, synaptic function, cognition (Currais et al., 2014)
	Aβ (ICV), mice	Reduced pathology (Aβ, pTau, synaptic dysfunction, inflammation, neuronal apoptosis); cognition (Ahmad et al., 2017)
	SAMP8 accelerated aging, mice	Cognition, synaptic function, stress, inflammation (Currais et al., 2018)
	Chronological and accelerated aging, rats	Oxidative stress, inflammation, neuroprotection (Singh et al., 2018)
	Vascular dementia, rats	Neuroprotection, endothelial protection (Hemanth Kumar et al., 2017)
	Middle cerebral artery ischemia, rats	Neuroprotection (Dajas et al., 2003; Rivera et al., 2004)
	Subarachnoid hemorrhage, mice	Neuroprotection, neuroinflammation (Zhou et al., 2015)
	Intracerebral hemorrhage/brain injury, mice	Neuroprotection, neuroinflammation (Chen et al., 2018)
	Iron-induced traumatic epilepsy, rats	Seizures and cognition (Das et al., 2017)
	Traumatic brain injury, mice	Neuroprotection, reduced cerebral edema, BBB (Zhang et al., 2018)
	Aluminum chloride neurotoxicity, mice	Neuroprotection, inflammation, cognition (Prakash et al., 2013)
	Depression, mice	Behavior, neurotransmitter levels (Zhen et al., 2012)
	Huntington disease, mice	Neuroprotection, improved motor performance (Maher et al., 2011)
D	Aβ (ICV), mice	Decreased pTyr, decreased reactive microglia (Dhawan et al., 2012)
	TgAPP/PS1, mice	Decreased pTyr, reactive microglia, inflammation, improved cognitive behavior (Dhawan and Combs, 2012)
D + Q	rTg(tau$_{P301L}$), mice	Pathology (tauopathy, neurodegeneration, inflammation, brain atrophy, ventricle enlargement, white matter hyperintensities), cerebral blood flow (Musi et al., 2018)
	TgAPPswe/PS1dE9, mice	Reduced Aβ plaques, behavior, senescent OPCs (Zhang et al., 2019)
	High-fat diet-induced anxiety, mice	Reduced lipid accumulation in brain, improved anxiety (Ogrodnik et al., 2019)
	*Advanced chronological age, mice	Improved daily activity, speed and endurance (Xu et al., 2018)

D, dasatinib; *D + Q*, dasatinib + quercetin; *ICV*, intracerebroventricular; *Q*, quercetin; *brain not specifically analyzed, but suggestive of nervous system involvement.

etc.). However, recent studies provide evidence for the senolytic properties of D + Q in the brain. The cellular senescence stress response occurs as a robust and invariant response to tau accumulation as reported in five distinct tau transgenic mouse models (Bussian et al., 2018; Musi et al., 2018), and postmortem human brain tissue derived from patients with Alzheimer disease and progressive supranuclear palsy (PSP) (Musi et al., 2018). Using transcriptomic data from postmortem human brain with tau-associated senescence, D + Q was predicted to target the SCAPs expressed by senescent neurons (Musi et al., 2018). To test this mechanistically, transgenic mice aged 20 months (approximates to 70-year-old humans) with existing tau pathology, neurodegeneration, inflammation, and brain atrophy were administered D + Q or vehicle. After 3 months of intermittent treatment, the mice that received D + Q displayed significant changes in cerebral blood flow, brain volume, and ventricle volume as assessed by magnetic resonance imaging (MRI) compared to their vehicle-treated littermates. Additionally, the mice treated with D + Q had significantly fewer tau-containing neurofibrillary tangles (NFTs) and reduced senescent burden and SASP (as assessed by a composite analysis of *Cdkn2a*, *Cdkn1a*, *Cxcl1*, *Il1*, *Tlr4*, and *Tnfα*) that coincided with improved neuronal health. This study suggests that D + Q may clear senescent cells in the central nervous system and thereby stop the progression of tau-associated pathogenesis in advanced disease stages. A later study demonstrated the efficacy of D + Q as a prophylactic therapy in a mouse model of Alzheimer disease-associated Aβ plaque accumulation (Zhang et al., 2019). In this study, D + Q treatment was initiated prior to pathology or behavioral deficits. The drug combination reduced Aβ plaque accumulation, reduced the prevalence of senescent oligodendrocyte precursor cells, and prevented cognitive impairment. However, when D + Q was used as an intervention therapy (i.e., given to aged mice with elevated Aβ plaque burden), it reduced markers of senescence including SA-β-gal, *Cdkn1a*, and senescent oligodendrocyte precursor cells, but it did not clear plaques (Zhang et al., 2019). While it is unclear from this study whether senescent cells contributed to cognitive impairment or Aβ accumulation in the mice, it provides supporting evidence for the utility of D + Q to reduce senescent cells in an AD model of Aβ plaque pathology. Neuronal senescence also occurs in mouse models of metabolic imbalance [i.e., peripheral insulin resistance (Chow et al., 2019) or obesity (Ogrodnik et al., 2019)] in the absence of pathogenic protein accumulation. D + Q reduces senescent glial and neuronal cells in the brains of diet-induced or genetically predisposed obese mice. Moreover, the clearance of senescent cells corresponds with a reduction in anxiety-like behavior (Ogrodnik et al., 2019). Collectively, these studies conclusively demonstrate that various stressors can induce cellular senescence in many cell types in the brain including neurons (Chow et al., 2019; Musi et al., 2018; Ogrodnik et al., 2019), astrocytes, microglia (Bussian et al., 2018; Chinta et al., 2018; Ogrodnik et al., 2019), and oligodendrocyte precursor cells (Zhang et al., 2019) and that D + Q effectively reduces pathogenesis and improves behavior (Musi et al., 2018; Ogrodnik et al., 2019; Zhang et al., 2019). These preclinical data provide compelling evidence and rationale to begin testing senolytic therapies in older adults with incurable neurodegenerative diseases.

Senomorphics

Senomorphics refers to a wide range of agents that can modulate the phenotype, secretory products, or detrimental effects of senescent cells; these actions are independent of senescent cell clearance. Two broad categories of senomorphic agents exist: those that block SASP amplification within senescent cells, and inhibitors of extracellular SASP (Childs et al., 2017). To date only a few senomorphics have been described. However, given the relatively nonspecific criteria defining senomorphic action, a wide variety of drugs and compounds that target fundamental processes of aging (e.g., adaptation to stress) will undoubtedly prove senomorphic. This includes inhibitors of mammalian target for rapamycin (mTOR), sirtuin activators, inflammation modulators, and behavioral interventions like caloric or dietary restriction and exercise.

Nuclear factor-κB (NF-κB) inhibitors

The nuclear factor-κB (NF-κB) family of transcription factors influence inflammation, immunity, cell proliferation, differentiation, and survival; they are implicated in virtually every pillar of aging (for review, see Tilstra et al., 2011). The IKK/NF-κB signaling pathway is activated by oxidative, genotoxic, inflammatory, and other types of stress and activity is linked to numerous known regulators of lifespan, including insulin/IGF-1 and growth hormone pathways, sirtuins, FOXO, and mTOR. NF-κB transcriptionally regulates both proapoptotic (*BAX*, *BCL2L11*, *FAS* and *FAS-LG CD274* and *CASP4*) and antiapoptotic, prosurvival genes (*BCL-2*, *BCL-2L1*, *BCL2A1*, *CFLAR*, and *XIAP*) (Catz & Johnson, 2001; Guttridge et al., 1999; Stehlik et al., 1998; Tamatani et al., 1999; Yamamoto et al., 1995; Zhu et al., 2001). NF-κB activation plays an important role in maintaining and transmitting the senescent state, via regulation of SASP (Nelson et al., 2018). The role of NF-κB as the transcriptional regulator of the majority of SASP cytokines and chemokines makes it an intriguing target to modulate the proinflammatory senescent secretome. Moreover, inhibiting NF-κB can delay DNA damage-induced senescence and aging in mice (Tilstra et al., 2012).

Janus kinase/signal transducer and activator of transcription (JAK/STAT) inhibitors

The JAK/STAT pathway is a key regulator of the SASP (Meyer & Levine, 2014; Yu et al., 2009), and has potential as a therapeutic target (reviewed in Xu et al., 2016). The JAK kinase family includes JAK1, JAK2, JAK3, and tyrosine kinase 2 (TYK2), which act through STAT proteins to regulate various downstream biological effects, including inflammatory signaling and growth hormone (JAK1/2), erythropoietin signaling (JAK3), immune cell function, and host defenses (TYK2) (reviewed in Ghoreschi et al., 2009; Richard & Stephens, 2011). Pharmacologic inhibitors of JAKs have been approved for human use or are in phase I–III clinical trials for myelofibrosis (Harrison et al., 2012; Pardanani et al., 2013), acute myeloid leukemia, lymphoma, and rheumatoid arthritis (Fridman et al., 2010; Meyer & Levine, 2014). JAK inhibitors also suppress SASP, inflammation, and frailty in aged mice (Xu et al., 2015b). Treating senescent human adipose progenitor cells and HUVECs with JAK inhibitors or siRNAs targeting JAK protein expression attenuates the secretory phenotype (Xu et al., 2015a, b). Treatment of aged, but not young, mice with the JAK1/2 inhibitor ruxolitinib suppresses systemic and adipose tissue inflammation (*Il6*, *Tnfα*, *Mmp3*, *Mmp12*, and *Emr1*) and circulating SASP factors (e.g., IL-6, G-CSF, IP-10, CXCL1, and MIP-1), and enhances physical function (activity, speed, strength, and endurance) (Xu et al., 2015b). Barriers to translation include the potential side effects of JAK1/2 inhibitors. For example, anemia and thrombocytopenia can occur with ruxolitinib in patients with myelofibrosis (Verstovsek et al., 2010, 2012), although these may be avoided if intermittent administration is used to ablate senescent cells.

Senomorphic and senotherapeutic actions of geroscience-guided interventions

Given the interconnectedness of the pillars of aging, it would not be surprising if therapeutics targeting pillars other than senescence impacted senescent cell number and associated phenotypes. Numerous pharmacologic, nutraceutical, and other agents are being investigated for senomorphic properties. An example is resveratrol, a well-known nutraceutical and sirtuin activator, which suppresses SASP in vitro. Resveratrol attenuates IL-8- and IL8/CXCR2-dependent proangiogenic characteristics of the SASP from senescent human fibroblasts (Menicacci et al., 2019). Chronic treatment of killifish with resveratrol suppresses senescence, NF-κB activation, and proinflammatory SASP in the gut of the fish (Liu et al., 2018). Similarly, extracts from the root-based Asian remedy *Ophiopogonis radix* suppress IL-6 and IL-8 production by senescent human dermal fibroblasts (Kitahiro et al., 2018).

The mechanistic (formerly mammalian) target of rapamycin (mTOR) is an evolutionary conserved serine-threonine kinase that senses and integrates environmental and intracellular signals, such as growth factors and nutrients (reviewed in Saxton & Sabatini, 2017). The name TOR (target of rapamycin) is derived from its inhibitor rapamycin, which was initially isolated in the 1970s from a soil bacterium on Rapa Nui (Easter Island). mTOR signaling is vital for cellular growth and function, and directly regulates metabolism, translation, stress responses, and autophagy in cells. Both cell growth and cell senescence (cell-cycle arrest) are impacted by mTOR signaling (reviewed in Blagosklonny, 2008; Weichhart, 2018). mTOR belongs to the phosphatidylinositol-3 kinases (PI3K)-related kinase (PIKK) family and is the catalytic subunit in at least two protein complexes: mTOR complex 1 (mTORC1) and mTORC2. mTORC1 activation occurs in organelles such as peroxisomes or lysosomes and requires the presence of nutrient and energy sources, such as amino acids, glucose, lipids, oxygen, and a high ATP/AMP ratio. Aberrant mTOR activation is seen both in vitro and in vivo in a proaging environment. For example, mTORC1 becomes constitutively activated by senescence, inducing stress, replicative exhaustion, or oncogene activation, which perpetuates SASP production. Inhibition of mTOR can improve cellular function, immune responses, cardiac function, and consistently extends the lifespan of model organisms including yeast, *C. elegans*, *Drosophila*, and rodents. Rapamycin is currently the only known pharmacological treatment that increases lifespan in all model organisms studied, and in both male and female mice (Houssaini et al., 2018; Mannick et al., 2018; Nacarelli et al., 2015; Walters & Cox, 2018; Weichhart, 2018). Rapamycin blunts the proinflammatory phenotype of senescent cells through various mutually nonexclusive mechanisms. For example, mTORC1 inhibition suppresses translation of the membrane-bound cytokine IL-1α and diminishes transcription of inflammatory genes regulated by the transcription factor NF-κB (Laberge et al., 2015). Rapamycin also activates ZFP36L1, an RNA-binding protein that degrades transcripts of numerous SASP components (Herranz et al., 2015a, b).

"Lifestyle" interventions: Dietary patterns, caloric restriction, exercise

Dietary restriction (DR) is by far the most robust intervention on biological aging, with lifespan effects conserved from yeast, to *C. elegans* and *Drosophila*, mice, and nonhuman primates (Fontana et al., 2010; Liang et al., 2018; Mattison et al., 2012, 2017; Swindell, 2012). The effect of DR is nuanced and its effects on health and lifespan vary as a consequence of dose (degree of calorie reduction from ad libitum), administration (daily caloric restriction vs time-restricted feeding or intermittent fasting), onset/duration (early-life vs mid-life start), and composition (protein, methionine restriction, branch chain amino acid restriction, or diet quality) (Vaughan et al., 2017). However, it is clear that, collectively, diet modifications impact nutrient-sensing signaling cascades with a plurality of effects that impact nearly all of the major hallmarks of aging. From this vantage, it is unsurprising that DR has senotherapeutic effects. Even short-term (3-month) DR initiated in mid- to late-life in sedentary animals decreases the fraction of intestinal (14% vs 20%) and hepatic (9% vs 22%) senescent cells relative to ad libitum fed mice (Kirkland, 2010; Wang et al., 2010). DR may also reduce senescence markers in the human colonic mucosa (Fontana et al., 2018; Krishnamurthy et al., 2004; Ogrodnik et al., 2017). The senotherapeutic effects of DR may act via primary prevention by reducing oxidative stress and inflammation, and possibly by increasing autophagy (for review see Fontana et al., 2018) or secondarily by suppressing the spread of senescence via SASP.

Despite the benefits of DR, the norm continues to be nutrient excess, overfeeding, and obesity. The increased storage demands placed on adipocytes due to nutrient excess ultimately compromises their function, reflected in impaired storage of lipids and augmented release of free fatty acids and inflammatory mediators (Guilherme et al., 2008). Adipose tissue dysfunction and its sequelae are at least partially mediated through cellular senescence (Tchkonia et al., 2010). In $p16^{Ink4a}$-reporter mice, increased adipose senescent cell burden induced by aging or a "fast-food" Westernized diet is accompanied by systemic inflammation and physical dysfunction, and genetic ablation of the senescent cells improves health and physical function (Schafer et al., 2016). In sedentary mice consuming a fast-food diet, 12% of the cells in the adipose tissue were senescent compared to 2%–3% in mice fed normal chow or provided exercise while on a fast-food diet. Lower senescent cell abundance is accompanied by lower adipose cytokine expression, attenuated fat mass accumulation, robust improvements in glucose tolerance, fasting insulin, and β cell mass, and more than double the distance run to exhaustion (Schafer et al., 2016). The link between adipose tissue dysfunction and aging phenotypes is also seen in humans. In a study analyzing adipose tissue from obese older women ($n = 11$, BMI $= 31 \pm 2$ kg/m^2; 73 ± 3 years), previously enrolled in a 5-month resistance exercise ± DR lifestyle intervention, a significant negative correlation was observed between $p16^{INK4a+}$ cells and physical function, including grip strength, 400-meter walk time, gait speed, and self-reported mobility (not attenuated after adjusting for age, total/regional fat mass, and lean mass; $P \le .05$) (Justice et al., 2018a). The proportion of $p16^{INK4a+}$ cells was reduced by exercise ± DR intervention.

Clinical translation of senotherapeutics

Senolytic and senotherapeutic agents are now being tested in proof-of-concept clinical trials. This requires new paradigms for testing safety and efficacy. Senolytics and other geroscience-guided interventions that target the basic biology of aging cannot be tested in humans using long-term endpoints such as lifespan. Likewise, multimorbidity or incidence of disease, as proposed in the first aging-outcomes trial, Targeting Aging with MEtformin (TAME), are not feasible for early-stage testing of senolytics, as they require large sample sizes (thousands) and several years to monitor sufficient accumulation of disease events (Justice et al., 2018b). An alternative strategy is to examine the effects in diseases with localized accumulation of senescent cells, or accelerated aging-like conditions. The effects of a senolytic could be tested in multiple parallel trials on different disease conditions, conducted simultaneously to establish proof-of-concept that short-term administration of senolytics can affect seemingly unrelated and disparate disease conditions, analogous to the multimorbidities of old age. This section outlines current progress for the first-in-humans proof-of-concept clinical trials on senolytics and then provides a general comment on considerations for clinical translation of senolytic drugs currently in development.

Considerations for clinical translation of senotherapeutics

For clinical advancement of senotherapeutic agents, the target populations, dosing and administration, and trial endpoints must be carefully considered (for reviews see Justice et al., 2016; Kirkland et al., 2017; Newman et al., 2016).

Target population and screening

Ideally, a screening platform would identify persons with high senescent cell burden who may benefit from targeted removal of senescent cells and would be eligible for senolytic drug trials. Unfortunately, such a test does not yet exist. Therefore, current proof-of-concept trials are focused on chronic diseases or geriatric syndromes that are associated with cellular senescence and high proinflammatory SASP burden, and therapeutic options are limited to symptom management. For example, IPF, primary sclerosing cholangitis, and Alzheimer disease are age-related diseases with no effective treatment and with evidence for cellular senescence playing a contributory role in animal models (Musi et al., 2018; Schafer et al., 2017; Tabibian et al., 2014; Zhang et al., 2019). Senolytic agents hold promise as potential treatments for these conditions, as the potential benefits of treatment are highly likely to outweigh the risk in these untreatable, deadly diseases. Moreover, senotherapeutics may be effective in treating conditions with phenotypes of accelerated aging, including cancer chemotherapy, HIV infection, obesity, or genetic progeroid syndromes. However, older and more frail subjects are also more susceptible to clinically significant adverse effects, and there may be a point of diminishing benefits from senolytics. Ideally, a surrogate biomarker or biomarker panel could be used to identify persons with a high senescent cell burden in internal organs to screen for trial enrollment, though this does not yet exist.

FDA and regulatory issues

A key limitation underlying all geroscience clinical trials is that the FDA does not recognize aging, frailty, or multimorbidity of old age as drug indications. Currently, clinical trials of senolytics must pursue indications for disease-specific effects, not aging itself. While to date the small proof-of-concept trials on senolytics have not been designed to support drug registration, this will likely change as senolytic testing advances from phase I/II trials to phase III trials. The lack of an indication for aging or age-related diseases hampers progress for translation of senotherapeutics, most dramatically, by limiting the interest of the pharmaceutical industry.

Dosing and administration

First-generation senolytics have advanced to trial, and, so far, are well-tolerated. The optimal dose and administration to maximize clearance of senescent cells, improve physiologic function, and limit off-target and adverse events may require disease-specific considerations. For example, while D + Q is being evaluated for efficacy in diseases with peripheral senescent cell involvement (i.e., lung and kidney), diseases of the central nervous system, such as Alzheimer disease, may require different dosing strategies. On-going clinical studies aim to evaluate the pharmacology of D + Q in the central nervous system (NCT04063124). Results from this study will inform whether, as in mice, the dose for clearing senescent cells in the central nervous system is similar to that of peripheral senescence. While the side-effect profile of drugs with senolytic effects cannot be overlooked, these compounds may be better tolerated when used as a senotherapy, due to the possibility of using an intermittent-dosing strategy. Senolytic compounds are not required to be continuously present to exert their effect. Rather, brief disruption of prosurvival pathways may be adequate to kill senescent cells in older adults, supporting intermittent administration of senolytics. For example, D + Q have an elimination half-life of a few hours, yet a single short course alleviates physical dysfunction caused by ionizing radiation to the leg for at least 7 months (Zhu et al., 2015). Moreover, the frequency of senolytic treatment will depend on the rate of senescent cell reaccumulation, which is estimated to be on the order of 2 months. For example, conditions that cause more rapid accumulation of senescent cells (e.g., high-fat diet or exposure to genotoxic cancer therapies) may necessitate more frequent senolytic administration. While intermittent administration is anticipated to lead to fewer side effects and lower risk for off-target effects relative to continuous exposure, determining the treatment schedule is not trivial. In the next section, we discuss hurdles and potential strategies to optimize senolytic therapy.

Identifying senescent cells in a clinical population

As highlighted throughout this chapter, translating senolytic therapy through clinical trials creates an urgent need for reliable biomarkers that conclusively reflect senescent cell burden, but not other sources of inflammation. Moreover, analogous to cancer, the senescent cell phenotype is dictated by its initiating stressor, tissue type, and cell identity within the tissue. Thus, senescent cell drug sensitivity is cell-type specific. The therapeutic options to eliminate cancer cells are determined by the cellular genotype (e.g., ER-, PR-, HER2/neu-negative mammary epithelia). For solid cancers, this typically requires a tissue biopsy. Of note, collection of accessible tissue (e.g., adipose, skin, blood, synovial fluid, CSF) is already common across the early clinical studies on senotherapeutics. While collection has been justified solely to determine efficacy of the senotherapies (pre- vs posttreatment measurement of senescence biomarkers), future studies may propose to perform pregenotyping to begin understanding the heterogeneity of senescent cell burden across tissues, individuals, and diseases. A more distant goal is the reliable identification of senescent cell burden longitudinally in vivo (i.e., PET imaging techniques). While only in its infancy, the development of senescence-specific radioligands will greatly impact clinical trial design and successful implementation of senotherapies in a variety of age-associated conditions.

Treatment strategies to clear senescent cells—optimizing dose

Once senotherapeutics are discovered, the next challenge is determining how best to administer senotherapeutics (dose, route, frequency). One approach is to treat with the maximum tolerated dose to rapidly clear large numbers of senescent cells. Depending on the number of senescent cells eliminated and the response of the immune system, toxicity could occur through uncontrolled cytokine release (cytokine storm). This seems unlikely, because preclinical models with abundant senescent cells have not caused cytokine toxicity. However, this is a potentially serious side effect that cannot be overlooked. Another approach, which is currently being used in the first-in-human clinical studies, is scaling up the drug dose incrementally. While this approach may not represent the final optimal dose, it provides a safe starting point for proof-of-concept trials. Dose determination is delicately paired with dosing frequency, which currently differs between trials depending on the disease. The final doses and treatment frequency will continue to be refined throughout trials. A third approach is to administer senotherapeutics locally to determine efficacy and optimize dosing. This too can minimize the risk of systemic toxicity, although the physiology of an aged or diseased organism may compromise barrier function (e.g., joint capsule, gastrointestinal integrity, blood–brain barrier). Thus, caution must still be taken.

Treatment strategies to clear senescent cells—optimizing frequency

Senolytic therapy, in principle, does not require continuous administration of drug. Instead, a hit-and-run approach is possible; a single dose is used to remove senescent cells, analogous to a course of intravenous antibiotics. The health benefits emerge over time and the next dose need not be given until signs of disease recur. In this way, modulating dose frequency can be personalized, depending on the patient's recurrence of senescent cells and symptoms. However, this will be dependent upon better tools to measure senescent cell burden. The early clinical studies have not determined the time course of senescent cell reemergence after treatment, but future trials will begin including periodic follow-up visits to begin answering this important question.

Survey of active clinical trials using senotherapeutics

Several senolytic and senotherapeutic drugs are being advanced to clinical testing. These are indexed on clinicaltrials.gov, the web-based registry of clinical trials that is run by the United States National Library of Medicine and National Institutes of Health. A summary of active, clinicaltrials.gov registered trials currently underway is given in Table 12.3, with narrative descriptions below.

Dasatinib + quercetin

Based on the overwhelming success of D + Q in mice in a variety of disease paradigms, this drug combination is the first senolytic taken into humans. Currently, there are four clinical trials listed on clinicaltrials.gov testing D + Q for senolytic properties in different disease conditions, including IPF (NCT02874989), chronic kidney disease (NCT02848131), hematopoietic stem cell transplant survivors (NCT02652052), and Alzheimer disease (NCT04063124). Each trial is evaluating short-term, intermittent treatment with D + Q in $n = 5–26$ patients, with primary outcomes aimed at establishing safety, feasibility, tolerability, and change in biomarkers of cell senescence. Patient-reported outcomes and functional assessments are slated for the next stage trial development. At the time of writing, all trials are still active. Preliminary reports are published for the trials in IPF and diabetic kidney disease as outlined below.

Dasatinib + quercetin in idiopathic pulmonary fibrosis

The first-in-human, proof-of-concept open-label clinical trial on senolytics was conducted in patients with stable, mild to severe IPF to evaluate the safety, feasibility, and potential impact of D + Q intermittent oral dosing over 3 weeks (Justice et al., 2019). Fourteen participants were enrolled and 100% completed intermittent drug self-administration and no subjects withdrew or were otherwise lost to follow-up. Though not a primary aim of this study, statistically significant (within-subject) and clinically meaningful improvements in physical function, including 6-minute walk distance, 4-meter usual gait speed, chair-rise time, and a performance summary score, and the short physical performance battery (SPPB) were seen following D + Q treatment. These parameters were measured before and 5 days after the last D + Q dose, well beyond these drugs' elimination half-lives. A key aim of the study was to track reported symptoms and adverse events. No changes in laboratory tests suggestive of hepatic or renal toxicity were found and pulmonary function was unchanged.

Adverse event reports and thorough symptom questionnaires uncovered mostly mild-to-moderate side effects within 24 hours of the third consecutive dosing day of each week, though these were reviewed and deemed acceptable and consistent with the underlying IPF diagnosis, study procedures, or knownoff- target effects of the drugs. The most frequent reports were respiratory-related, as would be anticipated in this patient population, or as a direct result of study procedures (e.g., irritation from medical tape or bruising after blood draw or skin biopsy). Other events were recognized as anticipated side effects known to occur with dasatinib administration, including headache, nausea and gastrointestinal-related discomfort, and generally feeling unwell, but these are suspected to be transient and to self-resolve following drug discontinuation. A double-blind placebo-controlled trial is underway to confirm these findings on D + Q in IPF, including the proper control arm.

D + Q in diabetic chronic kidney disease

A phase II open-label randomized trial of 3-day administration of D + Q on biomarkers of cellular senescence, mesenchymal stem cell function, and frailty markers is being conducted in patients with diabetic chronic kidney disease, aged 40–80 years. Though exploratory, this trial does provide the first evidence that short-term dosing with D + Q is sufficient to reduce markers of senescent cells in adipose tissue, adipocyte progenitors, skin, and circulating SASP factors. In subcutaneous adipose tissue, $p16^{INK4A}$- and $p21^{CIP1}$-expressing cells and cells with increased SA-β-gal activity were reduced by 35%, 17%, and 62%, respectively (raw unadjusted data). Adipose tissue $CD68^{+}$ macrophages, which are attracted, anchored, and activated by senescent cells, were decreased by 28%, and adipose progenitors increased by senolytic treatment. Moreover, skin epidermal $p16^{INK4A+}$ and $p21^{CIP1+}$ cells were reduced by 20% and 31%, respectively, as were circulating SASP factors, including IL-1α, IL-6, and MMPs-9 and -12 (Hickson et al., 2019).

Fisetin

There are two early-stage clinical trials with information listed on clinicaltrials.gov evaluating fisetin's senolytic effects in older men and women, though neither study is complete or has published preliminary findings. Both trials employ short-term "hit and run" dosing strategies of 2 consecutive days, which is similar to that of D + Q. The phase II randomized, placebo-controlled study "Alleviation by Fisetin of Frailty, Inflammation, and Related Measures" in older women (AFFIRM; NCT03430037) will determine the effects of fisetin on measures of frailty in women aged 70–90 years. The second trial, "Inflammation and Stem Cells in Diabetic and Chronic Kidney Disease" (NCT03325322) is led by the same team of investigators as the D + Q trial in diabetic chronic kidney disease. This trial is anticipated to harmonize endpoints and key biomarkers with the D + Q trial, such as on adipose tissue-derived mesenchymal stem/stromal cell function, renal function, markers of inflammation, and physical function. These comparisons will be informative for next stage trial planning of these leading senolytic agents in chronic kidney disease.

As mentioned in a previous section, a key advantage to fisetin is its excellent anticipated safety profile. Flavonoids like fisetin are members of a broadly distributed class of plant pigments regularly consumed in the human diet; for example, strawberries, apples, grapes, onions, cucumbers, tomatoes, and persimmons contain fisetin. Milligram quantities of flavonoids are consumed daily and micromolar quantities can be detected in the human colon (Arai et al., 2000).

UBX0101

Unity Biotechnology, Inc., a major stakeholder in the senolytic biotech-space, conducted an early-stage clinical trial testing its small-molecule inhibitor of the MDM2/p53 protein interaction, UBX0101, the chemical structure of which remains unpublished. This compound is an effective senolytic in a mouse model (Jeon et al., 2017). The company recently completed a phase I, randomized, double-blinded, placebo-controlled single ascending dose study evaluating the safety, tolerability, and pharmacokinetics of UBX0101 in patients diagnosed with painful femoro-tibial osteoarthritis (NCT03513016) after intraarticular administration of the drug. The trial was subdivided into two phases, to evaluate (1) appropriate dosing of UBX0101 and clinical outcomes and (2) effects of 4 weeks treatment with UBX0101 on biomarkers of senescence and SASP in synovial fluid.

In general, the company reports that UBX0101 was well-tolerated up to the maximum dose of 4 mg and no serious or dose-dependent adverse events occurred with a single administration or up to 4 weeks local injection. Public reporting indicates some improvement in pain and function, with modulation of select inflammatory factors in synovial fluid (treatment vs placebo). Though industry-supported, the UBX0101 early-phase clinical trials add converging evidence of the safety and potential efficacy to the federally or philanthropically funded investigations using orally administered senolytics. This supports future advancement of senolytics and alternative dosing strategies for clinical trials and suggests potential clinical efficacy of senotherapeutics across a broad range of chronic diseases or conditions.

Clinical trials to extend healthy lifespan: Senotherapeutic potential

Several interventions targeting basic biological processes are being advanced to clinical trial and may demonstrate senotherapeutic properties in humans, but must be further explored. Most of the geroscience-guided clinical investigations have stored biological specimens and data that could be used to evaluate the potential off-target beneficial effects of these treatments on the burden of senescent cells or SASP. Even if these interventions do not yield strong senotherapeutic effects, evaluating biomarkers in innovative geroscience trials could validate biomarkers needed to evaluate and compare the efficacy of senotherapeutics and surrogate markers of healthspan.

mTOR inhibitors

Pharmacological TOR inhibitors, such as everolimus, are approved for human clinical use and are now

TABLE 12.3 Active clinical trials of senolytics and senotherapeutics.

Intervention	Trial, NCT#	Population	Administration and dose	Dose duration	Follow-up	Trial type	Outcome
Dasatinib + Quercetin	IPF Study, NCT02874989	Idiopathic pulmonary fibrosis ≥ 50 years $n = 26$	Oral D 100 mg/d Q 1250 mg/d	3 weeks intermittent (3d on, 4d off)	5 days post Tx	Phase II, (A) Single-arm open label (B) Double-blind RCT	Feasibility Biomarkers Physical function Pulmonary function
Dasatinib + Quercetin	HTSS Study, NCT02652052	Hematopoietic stem cell transplant survivors ≥ 18 years $n = 10$	Oral D 100 mg/d Q 1250 mg/d	3 days consecutive	--	Phase I Open label RCT	Fried Frailty Biomarkers
Dasatinib + Quercetin	Senescence in CKD, NCT02848131	Chronic kidney disease + type II diabetes 40–80 years $n = 16$	Oral D 100 mg/d Q 1000 mg/d	3 days consecutive	14 days post Tx	Phase II open-label RCT	Biomarkers Stem cell function Frailty index and kidney function (eGFR)
Dasatinib + Quercetin	SToMP-AD, NCT04063124	Alzheimer disease ≥ 65 years $n = 5$	Oral D 100 mg/d Q 1250 mg/d	12 weeks intermittent (2 d on, 14 d off)	1 day post Tx	Phase I/II open-label	Drug brain penetrance AD markers in CSF Cognitive function
Fisetin	AFFIRM, NCT03430037	Frail elderly syndrome Women ≥ 70 years $n = 40$	Oral 20 mg/kg/d	2 month intermittent (2 days on, 14 days off)	7 days post treatment and one month post treatment	Phase IIa double-blind RCT	6-min walk time Gait speed
Fisetin	Inflammation Stem Cells in Diabetic CKD, NCT03325322	Chronic kidney disease + type II diabetes 40–80 years $n = 30$	Oral 20 mg/kg/d	2 days, consecutive	2 weeks post Tx	Phase II double-blind RCT	SASP markers Stem cell function Frailty (Fried) and kidney function (eGFR)3
UBX0101	Safety and Tolerability of UBX0101, NCT03513016	Knee osteoarthritis 40–85 years $n = 78$	Intra-articular injection, 4 mg/d max	Single injection	4 weeks or 12 weeks	Phase I, double-blind RCT	Safety and tolerability Drug levels Reported pain SASP biomarkers
ABT-263	Bioavailability of a New ABT-263 Formulation NCT01053520	Healthy females 18–55 years$n = 12$	Oral 25 mg	Single dose 25 mg	13 time points over 4 days	Phase I randomized cross-over trial	Plasma drug levels

entering phase III testing based on its ability to improve vaccine response and reduced rates of flu infection in the elderly over a 9-month period (Mannick et al., 2018). mTOR inhibitors represent the first evidence that clinical trials targeting a ubiquitous aging phenotype, immunosenescence, can be treated with a drug supported by geroscience, and with preclinical evidence suggesting shared pathways with cell senescence.

Caloric restriction

Behavioral interventions are routinely implemented clinically and in well-controlled clinical trials such as the CALERIE study (Ravussin et al., 2015; Rickman et al., 2011). CALERIE is a rich potential source of specimens such as serum, peripheral blood mononuclear cells, adipose tissue, skeletal muscle, and skin, which are typically underutilized, but could be accessed for analysis to garner crucial preliminary evidence to drive future translation studies (Belsky & Harrati, 2019).

Metformin

Other clinical trials on aging using geroscience-informed interventions like metformin are in the planning stages, including the trial TAME, which is the first aging-outcomes trial designed to create a regulatory pathway for drugs targeting aging and age-related multimorbidity (Justice et al., 2018b). Additionally, a pilot and feasibility study of one-month treatment with metformin in persons aged 30–70 years with prediabetes is underway (NCT03309007); although the primary outcome is an autophagy marker, leukocyte LC3 score, the clinicaltrials.gov listing suggests that the investigators will explore the relationship of changes in LC3 to biomarkers of cell senescence, for planning of a future phase III clinical trial. Importantly, such trials provide an opportunity to discover and validate biomarkers of cellular senescence and surrogate markers of healthspan in clinical trials. Even if there are minimal effects of metformin on cell senescence, these efforts will inform our understanding of what constitutes a meaningful change in biomarkers that could be used to interpret trials using targeted clearance of senescent cells with senotherapeutics.

Conclusions

Senescent cells accumulate with age in numerous tissues and definitively contribute to chronic disease and organ dysfunction. The removal of senescent cells convincingly improves health outcomes in preclinical studies and removing senescent cells as they form extends healthy lifespan in rodent models. It is important to note that the primary impact of senotherapeutics is on median lifespan, not maximum lifespan. Thus, this class of drugs appears to compress morbidity and suppress all-cause mortality in mice, rather than extending maximum lifespan of sick mice, as proposed as a clinical goal by the geroscience hypothesis. Translating this therapeutic approach holds great promise to prevent or delay chronic diseases as a group, instead of one-at-a-time. As a step toward achieving this goal, early proof-of-concept clinical studies are first targeting chronic diseases of aging. The safety, efficacy, and tolerability of promising drug candidates are being evaluated in select age-related diseases. The diseases and conditions initially explored are those that lack effective disease-modifying treatments and are likely to benefit from senolytics, based on data from preclinical studies. A set of about five SCAPs was identified and used to identify potential drugs and interventions that yield targeted clearance of senescent cells. Additional efforts identified senotherapeutic properties of other gerocentric agents that extend healthspan and lifespan in model organisms. The collective evidence supporting the notion that it is possible to compress the period of morbidity in old age is compelling: If what can be achieved in preclinical animal models, even with late-life intervention, can be achieved in humans, it may be possibly to delay, attenuate, or even prevent multiple age-related diseases and conditions of physical decline with a single intervention. This has the potential to revolutionize geriatric medicine, where drug interventions have not been available. However, caution is warranted; translation of senotherapeutics, including natural products, is not advised until clinical trials are completed and drug administration is optimized and safety issues vetted.

References

Abbas, M., Jesel, L., Auger, C., Amoura, L., Messas, N., Manin, G., et al. (2017). Endothelial microparticles from acute coronary syndrome patients induce premature coronary artery endothelial cell aging and thrombogenicity: Role of the Ang II/AT1 receptor/NADPH oxidase-mediated activation of MAPKs and PI3-kinase pathways. *Circulation, 135*, 280–296.

Abdul-Aziz, A. M., Sun, Y., Hellmich, C., Marlein, C. R., Mistry, J., Forde, E., et al. (2019). Acute myeloid leukemia induces protumoral p16INK4a-driven senescence in the bone marrow microenvironment. *Blood, 133*, 446–456.

Adhami, V. M., Syed, D. N., Khan, N., & Mukhtar, H. (2012). Dietary flavonoid fisetin: A novel dual inhibitor of PI3K/Akt and mTOR for prostate cancer management. *Biochemical Pharmacology, 84*, 1277–1281.

Adler, M. J., Coronel, C., Shelton, E., Seegmiller, J. E., & Dewji, N. N. (1991). Increased gene expression of Alzheimer disease beta-amyloid precursor protein in senescent cultured fibroblasts. *Proceedings of the National Academy of Sciences, 88*, 16–20. Available from https://doi.org/10.1073/pnas.88.1.16.

Ahmad, A., Ali, T., Park, H. Y., Badshah, H., Rehman, S. U., & Kim, M. O. (2017). Neuroprotective effect of fisetin against amyloid-beta-induced cognitive/synaptic dysfunction, neuroinflammation, and neurodegeneration in adult mice. *Molecular Neurobiology, 54*(3), 2269–2285. Available from https://doi.org/10.1007/s12035-016-9795-4.

Akers, J. C., Ramakrishnan, V., Kim, R., Phillips, S., Kaimal, V., Mao, Y., et al. (2015). miRNA contents of cerebrospinal fluid extracellular vesicles in glioblastoma patients. *Journal of Neuro-Oncology, 123*, 205–216.

Alder, J. K., Chen, J. J., Lancaster, L., Danoff, S., Su, S. C., Cogan, J. D., et al. (2008). Short telomeres are a risk factor for idiopathic pulmonary fibrosis. *Proceedings of the National Academy of Sciences of the United States of America, 105*, 13051–13056. Available from https://doi.org/10.1073/pnas.0804280105.

Álvarez, D., Cárdenes, N., Sellarés, J., Bueno, M., Corey, C., Hanumanthu, V. S., et al. (2017). IPF lung fibroblasts have a senescent phenotype. *The American Journal of Physiology-Lung*

Cellular and Molecular Physiology, 313, L1164–L1173. Available from https://doi.org/10.1152/ajplung.00220.2017.

Aoshiba, K., Tsuji, T., & Nagai, A. (2003). Bleomycin induces cellular senescence in alveolar epithelial cells. *European Respiratory Journal, 22*, 436–443. Available from https://doi.org/10.1183/09031936.03.00011903.

Aoshiba, K., Tsuji, T., Kameyama, S., Itoh, M., Semba, S., Yamaguchi, K., & Nakamura, H. (2013). Senescence-associated secretory phenotype in a mouse model of bleomycin-induced lung injury. *Experimental and Toxicologic Pathology, 65*, 1053–1062. Available from https://doi.org/10.1016/j.etp.2013.04.001.

Arai, Y., Watanabe, S., Kimira, M., Shimoi, K., Mochizuki, R., & Kinae, N. (2000). Dietary intakes of flavonols, flavones and isoflavones by Japanese women and the inverse correlation between quercetin intake and plasma LDL cholesterol concentration. *The Journal of Nutrition, 130*, 2243–2250.

Armanios, M. Y., Chen, J. J., Cogan, J. D., Alder, J. K., Ingersoll, R. G., Markin, C., et al. (2007). Telomerase mutations in families with idiopathic pulmonary fibrosis. *The New England Journal of Medicine, 356*, 1317–1326. Available from https://doi.org/10.1056/NEJMoa066157.

Baar, M. P., Brandt, R. M. C., Putavet, D. A., Klein, J. D. D., Derks, K. W. J., Bourgeois, B. R. M., et al. (2017). Targeted apoptosis of senescent cells restores tissue homeostasis in response to chemotoxicity and aging. *Cell, 169*, 132–147, e116.

Baker, D. J., Perez-Terzic, C., Jin, F., Pitel, K. S., Niederländer, N. J., Jeganathan, K., et al. (2008). Opposing roles for p16Ink4a and p19Arf in senescence and ageing caused by BubR1 insufficiency. *Nature Cell Biology, 10*(7), 825–836. Available from https://doi.org/10.1038/ncb1744, Jul Epub 2008 May 30. Erratum in: *Nature Cell Biology* 14, 2012, 649. Pitel, Kevin [corrected to Pitel, Kevin S].

Baker, D. J., Childs, B. G., Durik, M., Wijers, M. E., Sieben, C. J., Zhong, J., et al. (2016). Naturally occurring p16(Ink4a)-positive cells shorten healthy lifespan. *Nature, 530*, 184–189.

Baker, D. J., Wijshake, T., Tchkonia, T., LeBrasseur, N. K., Childs, B. G., van de Sluis, B., et al. (2011). Clearance of p16Ink4a-positive senescent cells delays ageing-associated disorders. *Nature, 479*, 232–236.

Bantscheff, M., Eberhard, D., Abraham, Y., Bastuck, S., Boesche, M., Hobson, S., et al. (2007). Quantitative chemical proteomics reveals mechanisms of action of clinical ABL kinase inhibitors. *Nature Biotechnology, 25*, 1035–1044.

Bao, X. Y., Xie, C., & Yang, M. S. (2012). Association between type 2 diabetes and CDKN2A/B: a meta-analysis study. *Molecular Biology Reports, 39*, 1609–1616. Available from https://doi.org/10.1007/s11033-011-0900-5.

Basisty, N., Holtz, A., & Schilling, B. (2020). Accumulation of "Old Proteins" and the critical need for MS-based protein turnover measurements in aging and longevity. *Proteomics, 20*(5–6), e1800403.

Belsky, D. W., & Harrati, A. (2019). To the freezers! Stored biospecimens from human randomized trials are an important new direction for studies of biological aging. *The Journals of Gerontology. Series A, Biological Sciences and Medical Sciences, 74*, 89–90.

Bennett, M. R., Evan, G. I., & Schwartz, S. M. (1995). Apoptosis of human vascular smooth muscle cells derived from normal vessels and coronary atherosclerotic plaques. *Journal of Clinical investigation, 95*, 266–2274. Available from https://doi.org/10.1172/JCI117917.

Bennett, M. R., Macdonald, K., Chan, S. W., Boyle, J. J., & Weissberg, P. L. (1998). Cooperative interactions between RB and p53 regulate cell proliferation cell senescence, and apoptosis in human vascular smooth muscle cells from atherosclerotic plaques. *Circulation Research, 82*, 704–712. Available from https://doi.org/10.1161/01.res.82.6.704.

Benetos, A., Gardner, J. P., Zureik, M., Labat, C., Xiaobin, L., Adamopoulos, C., et al. (2004). Short telomeres are associated with increased carotid atherosclerosis in hypertensive subjects. *Hypertension, 43*, 182–185. Available from https://doi.org/10.1161/01.HYP.0000113081.42868.f4.

Bhat, R., Crowe, E. P., Bitto, A., Moh, M., Katsetos, C. D., Garcia, F. U., et al. (2012). Astrocyte senescence as a component of Alzheimer's disease. *PLoS One, 7*, e45069. Available from https://doi.org/10.1371/journal.pone.0045069.

Biran, A., Zada, L., Abou Karam, P., Vadai, E., Roitman, L., et al. (2017). Quantitative identification of senescent cells in aging and disease. *Aging Cell, 16*, 661–671. Available from https://doi.org/10.1111/acel.12592.

Blagosklonny, M. V. (2008). Aging: ROS or TOR. *Cell Cycle, 7*, 3344–3354.

Blake, S., Hughes, T. P., Mayrhofer, G., & Lyons, A. B. (2008). The Src/ABL kinase inhibitor dasatinib (BMS-354825) inhibits function of normal human T-lymphocytes in vitro. *Clinical Immunology, 127*, 330–339.

Blazer, S., Khankin, E., Segev, Y., Ofir, R., Yalon-Hacohen, M., Kra-Oz, Z., et al. (2002). High glucose-induced replicative senescence: point of no return and effect of telomerase. *Biochemical and Biophysical Research Communications, 296*, 93–101. Available from https://doi.org/10.1016/s0006-291x(02)00818-5.

Borkham-Kamphorst, E., Schaffrath, C., Van de Leur, E., Haas, U., Tihaa, L., Meurer, S. K., et al. (2014). The anti-fibrotic effects of CCN1/CYR61 in primary portal myofibroblasts are mediated through induction of reactive oxygen species resulting in cellular senescence, apoptosis and attenuated TGF-β signaling. *Biochimica et Biophysica Acta, 1843*, 902–914. Available from https://doi.org/10.1016/j.bbamcr.2014.01.023.

Braig, M., Lee, S., Loddenkemper, C., Rudolph, C., Peters, A. H., Schlegelberger, B., et al. (2005). Oncogene-induced senescence as an initial barrier in lymphoma development. *Nature, 436*, 660–665. Available from https://doi.org/10.1038/nature03841.

Braun, H., Schmidt, B. M., Raiss, M., Baisantry, A., Mircea-Constantin, D., Wang, S., & Melk, A. (2012). Cellular senescence limits regenerative capacity and allograft survival. *Journal of the American Society of Nephrology, 23*, 1467–1473. Available from https://doi.org/10.1681/ASN.2011100967.

Burch, J. B., Augustine, A. D., Frieden, L. A., Hadley, E., Howcroft, T. K., Johnson, R., et al. (2014). Advances in geroscience: Impact on healthspan and chronic disease. *The Journals of Gerontology. Series A, Biological Sciences and Medical Sciences, 69*(Suppl. 1), S1–S3.

Burd, C. E., Sorrentino, J. A., Clark, K. S., Darr, D. B., Krishnamurthy, J., Deal, A. M., et al. (2013). Monitoring tumorigenesis and senescence in vivo with a p16(INK4a)-luciferase model. *Cell, 152*, 340–351.

Bussian, T. J., Aziz, A., Meyer, C. F., Swenson, B. L., van Deursen, J. M., & Baker, D. J. (2018). Clearance of senescent glial cells prevents tau-dependent pathology and cognitive decline. *Nature, 562*, 578–582.

Caliò, A., Zamò, A., Ponzoni, M., Zanolin, M. E., Ferreri, A. J., Pedron, S., et al. (2015). Cellular senescence markers p16INK4a and p21CIP1/WAF are predictors of hodgkin lymphoma outcome. *Clinical Cancer Research, 21*, 5164–5172. Available from https://doi.org/10.1158/1078-0432.CCR-15-0508.

Campisi, J. (2005). Senescent cells, tumor suppression, and organismal aging: Good citizens, bad neighbors. *Cell, 120*, 513–522.

Campisi, J., Andersen, J. K., Kapahi, P., & Melov, S. (2011). Cellular senescence: A link between cancer and age-related degenerative disease? *Seminars in Cancer Biology, 21*, 354–359.

Campisi, J., & d'Adda di Fagagna, F. (2007). Cellular senescence: When bad things happen to good cells. *Nature Reviews. Molecular Cell Biology, 8*, 729–740.

Campisi, J. (2013). Aging, cellular senescence, and cancer. *Annual Review of Physiology, 75*, 685–705. Available from https://doi.org/10.1146/annurev-physiol-030212-183653.

Catz, S. D., & Johnson, J. L. (2001). Transcriptional regulation of bcl-2 by nuclear factor kappa B and its significance in prostate cancer. *Oncogene, 20*, 7342–7351.

Chang, J., Wang, Y., Shao, L., Laberge, R. M., Demaria, M., Campisi, J., et al. (2016). Clearance of senescent cells by ABT263 rejuvenates aged hematopoietic stem cells in mice. *Nature Medicine, 22*, 78–83.

Charmpilas, N., Kyriakakis, E., & Tavernarakis, N. (2017). Small heat shock proteins in ageing and age-related diseases. *Cell Stress & Chaperones, 22*, 481–492.

Chen, B., Piel, W. H., Gui, L., Bruford, E., & Monteiro, A. (2005a). The HSP90 family of genes in the human genome: Insights into their divergence and evolution. *Genomics, 86*, 627–637.

Chen, Z., Trotman, L. C., Shaffer, D., Lin, H. K., Dotan, Z. A., Niki, M., et al. (2005b). Crucial role of p53-dependent cellular senescence in suppression of Pten-deficient tumorigenesis. *Nature, 436*, 725–730. Available from https://doi.org/10.1038/nature03918.

Chen, C., Yao, L., Cui, J., & Liu, B. (2018). Fisetin Protects against Intracerebral Hemorrhage-Induced Neuroinflammation in Aged Mice. *Cerebrovascular Diseases, 45*(3–4), 154–161. Available from https://doi.org/10.1159/000488117.

Chen, H., Gu, X., Su, I. H., Bottino, R., Contreras, J. L., Tarakhovsky, A., & Kim, S. ,K. (2009). Polycomb protein Ezh2 regulates pancreatic beta-cell Ink4a/Arf expression and regeneration in diabetes mellitus. *Genes & Development, 23*, 975–985. Available from https://doi.org/10.1101/gad.1742509.

Chen, Y. C., Shen, S. C., Lee, W. R., Lin, H. Y., Ko, C. H., Shih, C. M., et al. (2002). Wogonin and fisetin induction of apoptosis through activation of caspase 3 cascade and alternative expression of p21 protein in hepatocellular carcinoma cells SK-HEP-1. *Archives of Toxicology, 76*, 351–359.

Chesnokova, V., Wong, C., Zonis, S., Gruszka, A., Wawrowsky, K., Ren, S. G., et al. (2009). Diminished pancreatic beta-cell mass in securin-null mice is caused by beta-cell apoptosis and senescence. *Endocrinology, 150*, 2603–2610. Available from https://doi.org/10.1210/en.2008-0972.

Childs, B. G., Baker, D. J., Kirkland, J. L., Campisi, J., & van Deursen, J. M. (2014). Senescence and apoptosis: Dueling or complementary cell fates? *EMBO Reports, 15*, 1139–1153.

Childs, B. G., Baker, D. J., Wijshake, T., Conover, C. A., Campisi, J., & van Deursen, J. M. (2016). Senescent intimal foam cells are deleterious at all stages of atherosclerosis. *Science, 354*, 472–477.

Childs, B. G., Durik, M., Baker, D. J., & van Deursen, J. M. (2015). Cellular senescence in aging and age-related disease: From mechanisms to therapy. *Nature Medicine, 21*, 1424–1435.

Childs, B. G., Gluscevic, M., Baker, D. J., Laberge, R. M., Marquess, D., Dananberg, J., et al. (2017). Senescent cells: An emerging target for diseases of ageing. *Nature Reviews Drug Discovery, 16*, 718–735.

Chinta, S. J., Woods, G., Demaria, M., Rane, A., Zou, Y., McQuade, A., et al. (2018). Cellular Senescence Is Induced by the Environmental Neurotoxin Paraquat and Contributes to Neuropathology Linked to Parkinson's Disease. *Cell Reports, 22*, 930–940.

Chow, H. M., Shi, M., Cheng, A., Gao, Y., Chen, G., Song, X., et al. (2019). Age-related hyperinsulinemia leads to insulin resistance in neurons and cell-cycle-induced senescence. *Nature Neuroscience, 22*, 1806–1819.

Clements, M. E., Chaber, C. J., Ledbetter, S. R., & Zuk, A. (2013). Increased cellular senescence and vascular rarefaction exacerbate the progression of kidney fibrosis in aged mice following transient ischemic injury. *PLoS One, 8*, e70464. Available from https://doi.org/10.1371/journal.pone.0070464.

Coppe, J. P., Desprez, P. Y., Krtolica, A., & Campisi, J. (2010). The senescence-associated secretory phenotype: The dark side of tumor suppression. *Annual Review of Pathology, 5*, 99–118.

Coppé, J. P., Kauser, K., Campisi, J., & Beauséjour, C. M. (2006). Secretion of vascular endothelial growth factor by primary human fibroblasts at senescence. *Journal of Biological Chemistry, 281*, 29568–29574. Available from https://doi.org/10.1074/jbc.M603307200.

Coppé, J. P., Patil, C. K., Rodier, F., Sun, Y., Muñoz, D. P., Goldstein, J., et al. (2008). Senescence-associated secretory phenotypes reveal cell-nonautonomous functions of oncogenic RAS and the p53 tumor suppressor. Version 2. *PLoS Biology, 6*, 2853–2868. Available from https://doi.org/10.1371/journal.pbio.0060301.

Cosgrove, B. D., Gilbert, P. M., Porpiglia, E., Mourkioti, F., Lee, S. P., Corbel, S. Y., et al. (2014). Rejuvenation of the muscle stem cell population restores strength to injured aged muscles. *Nature Medicine, 20*, 255–264. Available from https://doi.org/10.1038/nm.3464.

Cramer, C., Freisinger, E., Jones, R. K., Slakey, D. P., Dupin, C. L., Newsome, E. R., et al. (2010). Persistent high glucose concentrations alter the regenerative potential of mesenchymal stem cells. *Stem Cells and Development, 19*, 1875–1884. Available from https://doi.org/10.1089/scd.2010.0009.

Currais, A., Farrokhi, C., Dargusch, R., Armando, A., Quehenberger, O., Schubert, D., & Maher, P. (2018). Fisetin Reduces the Impact of Aging on Behavior and Physiology in the Rapidly Aging SAMP8 Mouse. *The Journals of Gerontology. Series A, Biological Sciences and Medical Sciences, 73*(3), 299–307. Available from https://doi.org/10.1093/gerona/glx104.

Currais, A., Prior, M., Dargusch, R., Armando, A., Ehren, J., Schubert, D., ... Maher, P. (2014). Modulation of p25 and inflammatory pathways by fisetin maintains cognitive function in Alzheimer's disease transgenic mice. *Aging Cell, 13*(2), 379–390. Available from https://doi.org/10.1111/acel.12185.

d'Adda di Fagagna, F., Reaper, P. M., Clay-Farrace, L., Fiegler, H., Carr, P., Von Zglinicki, T., et al. (2003). A DNA damage checkpoint response in telomere-initiated senescence. *Nature, 426*, 194–198.

Dajas, F., Rivera, F., Blasina, F., Arredondo, F., Echeverry, C., Lafon, L., ... Heinzen, H. (2003). Cell culture protection and in vivo neuroprotective capacity of flavonoids. *Neurotoxicity Research, 5*(6), 425–432.

Das, J., Singh, R., & Sharma, D. (2017). Antiepileptic effect of fisetin in iron-induced experimental model of traumatic epilepsy in rats in the light of electrophysiological, biochemical, and behavioral observations. *Nutritional Neuroscience, 20*(4), 255–264. Available from https://doi.org/10.1080/1028415X.2016.1183342.

da Silva, P. F. L., Ogrodnik, M., Kucheryavenko, O., Glibert, J., Miwa, S., Cameron, K., et al. (2019). The bystander effect contributes to the accumulation of senescent cells in vivo. *Aging Cell, 18*, e12848.

Davis, C., Dukes, A., Drewry, M., Helwa, I., Johnson, M. H., Isales, C. M., et al. (2017). MicroRNA-183-5p increases with age in bone-derived extracellular vesicles, suppresses bone marrow stromal (stem) cell proliferation, and induces stem cell senescence. *Tissue Engineering Part A, 23*, 1231–1240.

Demaria, M., Ohtani, N., Youssef, S. A., Rodier, F., Toussaint, W., Mitchell, J. R., et al. (2014). An essential role for senescent cells in optimal wound healing through secretion of PDGF-AA. *Developmental Cell, 31*, 722–733.

Dhawan, G., & Combs, C. K. (2012). Inhibition of Src kinase activity attenuates amyloid associated microgliosis in a murine model of Alzheimer's disease. *Journal of Neuroinflammation, 9*, 117. Available from https://doi.org/10.1186/1742-2094-9-117.

Dhawan, G., Floden, A. M., & Combs, C. K. (2012). Amyloid-beta oligomers stimulate microglia through a tyrosine kinase dependent mechanism. *Neurobiology of Aging, 33*(10), 2247–2261. Available from https://doi.org/10.1016/j.neurobiolaging.2011.10.027.

Dimri, G. P., Lee, X., Basile, G., Acosta, M., Scott, G., Roskelley, C., et al. (1995). A biomarker that identifies senescent human cells in culture and in aging skin in vivo. *Proceedings of the National Academy of Sciences of the United States of America*, *92*, 9363–9367.

Ding, G., Franki, N., Kapasi, A. A., Reddy, K., Gibbons, N., & Singhal, P. C. (2001). Tubular cell senescence and expression of TGF-beta1 and p21(WAF1/CIP1) in tubulointerstitial fibrosis of aging rats. *Experimental and Molecular Pathology*, *70*, 43–53. Available from https://doi.org/10.1006/exmp.2000.2346.

Disayabutr, S., Kim, E. K., Cha, S. I., Green, G., Naikawadi, R. P., Jones, K. D., et al. (2016). miR-34 miRNAs regulate cellular senescence in Type II alveolar epithelial cells of patients with idiopathic pulmonary fibrosis. *PLoS One*, *11*, e0158367. Available from https://doi.org/10.1371/journal.pone.0158367.

Du, J., Klein, J. D., Hassounah, F., Zhang, J., Zhang, C., & Wang, X. H. (2014). Aging increases CCN1 expression leading to muscle senescence. *American Journal of Physiology-Cell Physiology*, *306*, C28–C36. Available from https://doi.org/10.1152/ajpcell.00066.2013.

Farr, J. N., Xu, M., Weivoda, M. M., Monroe, D. G., Fraser, D. G., Onken, J. L., et al. (2017). Targeting cellular senescence prevents age-related bone loss in mice. *Nature Medicine*, *23*, 1072–1079.

Fenton, M., Barker, S., Kurz, D. J., & Erusalimsky, J. D. (2001). Cellular senescence after single and repeated balloon catheter denudations of rabbit carotid arteries. *Arteriosclerosis, Thrombosis, and Vascular Biology*, *21*, 220–226. Available from https://doi.org/10.1161/01.atv.21.2.220.

Figueroa, J. M., Skog, J., Akers, J., Li, H., Komotar, R., Jensen, R., et al. (2017). Detection of wild-type EGFR amplification and EGFRvIII mutation in CSF-derived extracellular vesicles of glioblastoma patients. *Neuro-Oncology*, *19*, 1494–1502.

Fitzner, B., Müller, S., Walther, M., Fischer, M., Engelmann, R., Müller-Hilke, B., et al. (2012). Senescence determines the fate of activated rat pancreatic stellate cells. *Journal of Cellular and Molecular Medicine*, *16*, 2620–2630. Available from https://doi.org/10.1111/j.1582-4934.2012.01573.x.

Flanary, B. E., Sammons, N. W., Nguyen, C., Walker, D., & Streit, W. J. (2007). Evidence that aging and amyloid promote microglial cell senescence. *Rejuvenation Research*, *10*, 61–74. Available from https://doi.org/10.1089/rej.2006.9096.

Fontana, L., Nehme, J., & Demaria, M. (2018). Caloric restriction and cellular senescence. *Mechanisms of Ageing and Development*, *176*, 19–23.

Fontana, L., Partridge, L., & Longo, V. D. (2010). Extending healthy life span--from yeast to humans. *Science*, *328*, 321–326.

Fridman, J. S., Scherle, P. A., Collins, R., Burn, T. C., Li, Y., Li, J., et al. (2010). Selective inhibition of JAK1 and JAK2 is efficacious in rodent models of arthritis: Preclinical characterization of INCB028050. *Journal of Immunology*, *184*, 5298–5307.

Fuhrmann-Stroissnigg, H., Ling, Y. Y., Zhao, J., McGowan, S. J., Zhu, Y., Brooks, R. W., et al. (2017). Identification of HSP90 inhibitors as a novel class of senolytics. *Nature Communications*, *8*, 422.

Gao, S. G., Zeng, C., Li, L. J., Luo, W., Zhang, F. J., Tian, J., et al. (2016). Correlation between senescence-associated beta-galactosidase expression in articular cartilage and disease severity of patients with knee osteoarthritis. *International Journal of Rheumatic Diseases*, *19*, 226–232. Available from https://doi.org/10.1111/1756-185X.12096.

Ghoreschi, K., Laurence, A., & O'Shea, J. J. (2009). Janus kinases in immune cell signaling. *Immunological Reviews*, *228*, 273–287.

González-Navarro, H., Vinué, Á., Sanz, M. J., Delgado, M., Pozo, M. A., & Serrano, M. (2013). et al., Increased dosage of Ink4/Arf protects against glucose intolerance and insulin resistance associated with aging. *Aging Cell*, *12*, 102–111. Available from https://doi.org/10.1111/acel.12023.

Gopas, J., Stern, E., Zurgil, U., Ozer, J., Ben-Ari, A., Shubinsky, G., et al. (2016). Reed-Sternberg cells in Hodgkin's lymphoma present features of cellular senescence. *Cell Death & Disease*, *7*, e2457. Available from https://doi.org/10.1038/cddis.2016.185.

Gruber, H. E., Ingram, J. A., Norton, H. J., & Hanley, E. N., Jr. (2007). Senescence in cells of the aging and degenerating intervertebral disc: immunolocalization of senescence-associated beta-galactosidase in human and sand rat discs. *Spine (Phila Pa 1976)*, *32*, 21–27. Available from https://doi.org/10.1097/01.brs.0000253960.57051.de.

Guilherme, A., Virbasius, J. V., Puri, V., & Czech, M. P. (2008). Adipocyte dysfunctions linking obesity to insulin resistance and type 2 diabetes. *Nature Reviews. Molecular Cell Biology*, *9*, 367–377.

Gupta, S. C., Tyagi, A. K., Deshmukh-Taskar, P., Hinojosa, M., Prasad, S., & Aggarwal, B. B. (2014). Downregulation of tumor necrosis factor and other proinflammatory biomarkers by polyphenols. *Archives of Biochemistry and Biophysics*, *559*, 91–99.

Guttridge, D. C., Albanese, C., Reuther, J. Y., Pestell, R. G., & Baldwin, A. S., Jr. (1999). NF-kappaB controls cell growth and differentiation through transcriptional regulation of cyclin D1. *Molecular and Cellular Biology*, *19*, 5785–5799.

Hamshere, M. L., Holmans, P. A., Avramopoulos, D., Bassett, S. S., Blacker, D., Bertram., et al. (2007). Genome-wide linkage analysis of 723 affected relative pairs with late-onset Alzheimer's disease. *Human Molecular Genetics*, *16*, 2703–2712. Available from https://doi.org/10.1093/hmg/ddm224.

Harbo, M., Bendix, L., Bay-Jensen, A. C., Graakjaer, J., Søe, K., Andersen, T. L., et al. (2012). The distribution pattern of critically short telomeres in human osteoarthritic knees. *Arthritis Research & Therapy*, *14*, R12. Available from https://doi.org/10.1186/ar3687.

Harrison, C., Kiladjian, J. J., Al-Ali, H. K., Gisslinger, H., Waltzman, R., Stalbovskaya, V., et al. (2012). JAK inhibition with ruxolitinib versus best available therapy for myelofibrosis. *The New England Journal of Medicine*, *366*, 787–798.

He, J., Hu, Y., Hu, M., & Li, B. (2015). Development of PD-1/PD-L1 Pathway in Tumor Immune Microenvironment and Treatment for Non-Small Cell Lung Cancer. *Scientific Reports*, *5*, 13110.

Hecker, L., Logsdon, N. J., Kurundkar, D., Kurundkar, A., Bernard, K., Hock, T., et al. (2014). Reversal of persistent fibrosis in aging by targeting Nox4-Nrf2 redox imbalance. *Sci Transl Med*, *6*, 231ra47. Available from https://doi.org/10.1126/scitranslmed.3008182.

Helman, A., Klochendler, A., Azazmeh, N., Gabai, Y., Horwitz, E., Anzi, S., et al. (2016). p16(Ink4a)-induced senescence of pancreatic beta cells enhances insulin secretion. *Nature Medicine*, *22*, 412–420. Available from https://doi.org/10.1038/nm.4054.

Hemanth Kumar, B., Arun Reddy, R., Mahesh Kumar, J., Dinesh Kumar, B., & Diwan, P. V. (2017). Effects of fisetin on hyperhomocysteinemia-induced experimental endothelial dysfunction and vascular dementia. *Canadian Journal of Physiology and Pharmacology*, *95*(1), 32–42. Available from https://doi.org/10.1139/cjpp-2016-0147.

Herbig, U., Ferreira, M., Condel, L., Carey, D., & Sedivy, J. M. (2006). Cellular senescence in aging primates. *Science*, *311*, 1257.

Herbig, U., Wei, W., Dutriaux, A., Jobling, W. A., & Sedivy, J. M. (2003). Real-time imaging of transcriptional activation in live cells reveals rapid up-regulation of the cyclin-dependent kinase inhibitor gene CDKN1A in replicative cellular senescence. *Aging Cell*, *2*, 295–304.

Herranz, N., Gallage, S., Mellone, M., Wuestefeld, T., Klotz, S., Hanley, C. J., et al. (2015a). Erratum: mTOR regulates MAPKAPK2 translation to control the senescence-associated secretory phenotype. *Nature Cell Biology*, *17*, 1370.

Herranz, N., Gallage, S., Mellone, M., Wuestefeld, T., Klotz, S., Hanley, C. J., et al. (2015b). mTOR regulates MAPKAPK2 translation to control the senescence-associated secretory phenotype. *Nature Cell Biology*, *17*, 1205–1217.

Hewitt, G., Jurk, D., Marques, F. D., Correia-Melo, C., Hardy, T., Gackowska, A., et al. (2012). Telomeres are favoured targets of a persistent DNA damage response in ageing and stress-induced senescence. *Nature Communications*, *3*, 708.

Hickson, L. J., Langhi Prata, L. G. P., Bobart, S. A., Evans, T. K., Giorgadze, N., Hashmi, S. K., et al. (2019). Senolytics decrease senescent cells in humans: Preliminary report from a clinical trial of Dasatinib plus Quercetin in individuals with diabetic kidney disease. *EBioMedicine*, *47*, 446–456.

Hou, Y., Aboukhatwa, M. A., Lei, D. L., Manaye, K., Khan, I., & Luo, Y. (2010). Anti-depressant natural flavonols modulate BDNF and beta amyloid in neurons and hippocampus of double TgAD mice. *Neuropharmacology*, *58*(6), 911–920. Available from https://doi.org/10.1016/j.neuropharm.2009.11.002.

Houssaini, A., Breau, M., Kebe, K., Abid, S., Marcos, E., Lipskaia, L., et al. (2018). mTOR pathway activation drives lung cell senescence and emphysema. *JCI Insight*, *3*, e93203.

Ishige, K., Schubert, D., & Sagara, Y. (2001). Flavonoids protect neuronal cells from oxidative stress by three distinct mechanisms. *Free Radical Biology & Medicine*, *30*, 433–446.

Ivanov, A., Pawlikowski, J., Manoharan, I., van Tuyn, J., Nelson, D. M., Rai, T. S., et al. (2013). Lysosome-mediated processing of chromatin in senescence. *The Journal of Cell Biology*, *202*, 129–143.

Jeon, O. H., Kim, C., Laberge, R. M., Demaria, M., Rathod, S., Vasserot, A. P., et al. (2017). Local clearance of senescent cells attenuates the development of post-traumatic osteoarthritis and creates a pro-regenerative environment. *Nature Medicine*, *23*, 775–781.

Jeyapalan, J. C., Ferreira, M., Sedivy, J. M., & Herbig, U. (2007). Accumulation of senescent cells in mitotic tissue of aging primates. *Mechanisms of Ageing and Development*, *128*, 36–44.

Jurk, D., Wang, C., Miwa, S., Maddick, M., Korolchuk, V., Tsolou, A., et al. (2012). Postmitotic neurons develop a p21-dependent senescence-like phenotype driven by a DNA damage response. *Aging Cell*, *11*, 996–1004.

Justice, J., Miller, J. D., Newman, J. C., Hashmi, S. K., Halter, J., Austad, S. N., et al. (2016). Frameworks for proof-of-concept clinical trials of interventions that target fundamental aging processes. *The Journals of Gerontology. Series A, Biological Sciences and Medical Sciences*, *71*, 1415–1423.

Justice, J. N., Nambiar, A. M., Tchkonia, T., LeBrasseur, N. K., Pascual, R., Hashmi, S. K., et al. (2019). Senolytics in idiopathic pulmonary fibrosis: Results from a first-in-human, open-label, pilot study. *EBioMedicine*, *40*, 554–563.

Justice, J. N., Gregory, H., Tchkonia, T., LeBrasseur, N. K., Kirkland, J. L., Kritchevsky, S. B., et al. (2018a). Cellular senescence biomarker p16INK4a + cell burden in thigh adipose is associated with poor physical function in older women. *The Journals of Gerontology. Series A, Biological Sciences and Medical Sciences*, *73*(7), 939–945.

Justice, J. N., Niedernhofer, L., Robbins, P. D., Aroda, V. R., Espeland, M. A., Kritchevsky, S. B., et al. (2018b). Development of clinical trials to extend healthy lifespan. *Cardiovascular Endocrinology & Metabolism*, *7*, 80–83.

Kalkavan, H., & Green, D. R. (2018). MOMP, cell suicide as a BCL-2 family business. *Cell Death and Differentiation*, *25*, 46–55.

Karimipour, M., Rahbarghazi, R., Tayefi, H., Shimia, M., Ghanadian, M., Mahmoudi, J., & Bagheri, H. S. (2019). Quercetin promotes learning and memory performance concomitantly with neural stem/progenitor cell proliferation and neurogenesis in the adult rat dentate gyrus. *International Journal of Developmental Neuroscience: The Official Journal of the International Society for Developmental Neuroscience*, *74*, 18–26. Available from https://doi.org/10.1016/j.ijdevneu.2019.02.005.

Krizhanovsky, V., Yon, M., Dickins, R. A., Hearn, S., Simon, J., Miething, C., et al. (2008). Senescence of activated stellate cells limits liver fibrosis. *Cell*, *134*, 657–667. Available from https://doi.org/10.1016/j.cell.2008.06.049.

Kennedy, B. K., Berger, S. L., Brunet, A., Campisi, J., Cuervo, A. M., Epel, E. S., et al. (2014). Geroscience: Linking aging to chronic disease. *Cell*, *159*, 709–713.

Khan, N., Syed, D. N., Ahmad, N., & Mukhtar, H. (2013). Fisetin: A dietary antioxidant for health promotion. *Antioxidants & Redox Signaling*, *19*, 151–162.

Kim, H. N., Chang, J., Shao, L., Han, L., Iyer, S., Manolagas, S. C., et al. (2017). DNA damage and senescence in osteoprogenitors expressing Osx1 may cause their decrease with age. *Aging Cell*, *16*, 693–703. Available from https://doi.org/10.1111/acel.12597.

Kim, K. H., Chen, C. C., Monzon, R. I., & Lau, L. F. (2013). Matricellular protein CCN1 promotes regression of liver fibrosis through induction of cellular senescence in hepatic myofibroblasts. *Molecular and Cellular Biology*, *33*, 2078–2090. Available from https://doi.org/10.1128/MCB.00049-13.

Kim, K. W., Ha, K. Y., Lee, J. S., Na, K. H., Kim, Y. Y., & Woo, Y. K. (2008). Senescence of nucleus pulposus chondrocytes in human intervertebral discs. *Asian Spine Journal*, *2*, 1–8. Available from https://doi.org/10.4184/asj.2008.2.1.1.

Kim, K. W., Chung, H. N., Ha, K. Y., Lee, J. S., & Kim, Y. Y. (2009). Senescence mechanisms of nucleus pulposus chondrocytes in human intervertebral discs. *Spine Journal*, *9*, 658–666. Available from https://doi.org/10.1016/j.spinee.2009.04.018.

Kirkland, J. L. (2010). Perspectives on cellular senescence and short term dietary restriction in adults. *Aging*, *2*, 542–544.

Kirkland, J. L. (2013). Translating advances from the basic biology of aging into clinical application. *Experimental Gerontology*, *48*, 1–5.

Kirkland, J. L., & Tchkonia, T. (2017). Cellular senescence: A translational perspective. *EBioMedicine*, *21*, 21–28.

Kirkland, J. L., Tchkonia, T., Zhu, Y., Niedernhofer, L. J., & Robbins, P. D. (2017). The clinical potential of senolytic drugs. *Journal of the American Geriatrics Society*, *65*, 2297–2301.

Kitahiro, Y., Koike, A., Sonoki, A., Muto, M., Ozaki, K., & Shibano, M. (2018). Anti-inflammatory activities of Ophiopogonis Radix on hydrogen peroxide-induced cellular senescence of normal human dermal fibroblasts. *Journal of Natural Medicines*, *72*, 905–914.

Kloss, C. C., Condomines, M., Cartellieri, M., Bachmann, M., & Sadelain, M. (2013). Combinatorial antigen recognition with balanced signaling promotes selective tumor eradication by engineered T cells. *Nature Biotechnology*, *31*, 71–75.

Kong, X., Feng, D., Wang, H., Hong, F., Bertola, A., Wang, F. S., & Gao, B. (2012). Interleukin-22 induces hepatic stellate cell senescence and restricts liver fibrosis in mice. *Hepatology*, *56*, 1150–1159. Available from https://doi.org/10.1002/hep.25744.

Korotzer, A. R., Pike, C. J., & Cotman, C. W. (1993). Beta-amyloid peptides induce degeneration of cultured rat microglia. *Brain Research*, *624*, 121–125. Available from https://doi.org/10.1016/0006-8993(93)90068-x.

Krishnamurthy, J., Torrice, C., Ramsey, M. R., Kovalev, G. I., Al-Regaiey, K., Su, L., et al. (2004). Ink4a/Arf expression is a biomarker of aging. *The Journal of Clinical Investigation*, *114*, 1299–1307.

Krishnamurthy, J., Ramsey, M. R., Ligon, K. L., Torrice, C., Koh, A., Bonner-Weir, S., & Sharpless, N. E. (2006). p16INK4a induces an age-dependent decline in islet regenerative potential. *Nature*, *443*, 453–457. Available from https://doi.org/10.1038/nature05092.

Krtolica, A., Parrinello, S., Lockett, S., Desprez, P. Y., & Campisi, J. (2001). Senescent fibroblasts promote epithelial cell growth and tumorigenesis: a link between cancer and aging. *Proceedings of the National Academy of Sciences of the United States of America*, *98*, 12072–12077. Available from https://doi.org/10.1073/pnas.211053698.

Laberge, R. M., Sun, Y., Orjalo, A. V., Patil, C. K., Freund, A., Zhou, L., et al. (2015). MTOR regulates the pro-tumorigenic senescence-associated secretory phenotype by promoting IL1A translation. *Nature Cell Biology*, *17*, 1049–1061.

Le Maitre, C. L., Freemont, A. J., & Hoyland, J. A. (2007). Accelerated cellular senescence in degenerate intervertebral discs: a possible role in the pathogenesis of intervertebral disc degeneration. *Arthritis Research & Therapy*, *9*, R45. Available from https://doi.org/10.1186/ar2198.

Lee, J., McKinney, K. Q., Pavlopoulos, A. J., Han, M. H., Kim, S. H., Kim, H. J., et al. (2016). Exosomal proteome analysis of cerebrospinal fluid detects biosignatures of neuromyelitis optica and multiple sclerosis. *Clinica Chimica Acta; International Journal of Clinical Chemistry*, *462*, 118–126.

Leverson, J. D. (2016). Chemical parsing: Dissecting cell dependencies with a toolkit of selective BCL-2 family inhibitors. *Molecular & Cellular Oncology*, *3*, e1050155.

Li, Y., Tian, Q., Li, Z., Dang, M., Lin, Y., & Hou, X. (2019). Activation of Nrf2 signaling by sitagliptin and quercetin combination against beta-amyloid induced Alzheimer's disease in rats. *Drug Development Research*, *80*(6), 837–845. Available from https://doi.org/10.1002/ddr.21567.

Liang, Y., Liu, C., Lu, M., Dong, Q., Wang, Z., Wang, Z., et al. (2018). Calorie restriction is the most reasonable anti-ageing intervention: A meta-analysis of survival curves. *Scientific Reports*, *8*, 5779.

Liton, P. B., Challa, P., Stinnett, S., Luna, C., Epstein, D. L., & Gonzalez, P. (2005). Cellular senescence in the glaucomatous outflow pathway. *Experimental Gerontology*, *40*, 745–748. Available from https://doi.org/10.1016/j.exger.2005.06.005.

Liu, S., Zheng, Z., Ji, S., Liu, T., Hou, Y., Li, S., et al. (2018). Resveratrol reduces senescence-associated secretory phenotype by SIRT1/NF-kappaB pathway in gut of the annual fish Nothobranchius guentheri. *Fish & Shellfish Immunology*, *80*, 473–479.

Lopez-Otin, C., Blasco, M. A., Partridge, L., Serrano, M., & Kroemer, G. (2013). The hallmarks of aging. *Cell*, *153*, 1194–1217.

Lujambio, A., Akkari, L., Simon, J., Grace, D., Tschaharganeh, D. F., Bolden, J. E., et al. (2013). Non-cell-autonomous tumor suppression by. *Cell*, *153*(449-60). Available from https://doi.org/10.1016/j.cell.2013.03.020.

Luna-Vargas, M. P. A., & Chipuk, J. E. (2016). Physiological and pharmacological control of BAK, BAX, and beyond. *Trends in Cell Biology*, *26*, 906–917.

Lv, X. X., Wang, X. X., Li, K., Wang, Z. Y., Li, Z., Lv, Q., et al. (2013). Rupatadine protects against pulmonary fibrosis by attenuating PAF-mediated senescence in rodents. *PLoS One*, *8*, e68631. Available from https://doi.org/10.1371/journal.pone.0068631.

Malavolta, M., Pierpaoli, E., Giacconi, R., Costarelli, L., Piacenza, F., Basso, A., et al. (2016). Pleiotropic effects of tocotrienols and quercetin on cellular senescence: Introducing the perspective of senolytic effects of phytochemicals. *Current Drug Targets*, *17*, 447–459.

Maher, P., Dargusch, R., Bodai, L., Gerard, P. E., Purcell, J. M., & Marsh, J. L. (2011). ERK activation by the polyphenols fisetin and resveratrol provides neuroprotection in multiple models of Huntington's disease. *Human Molecular Genetics*, *20*(2), 261–270. Available from https://doi.org/10.1093/hmg/ddq460.

Mannick, J. B., Morris, M., Hockey, H. P., Roma, G., Beibel, M., Kulmatycki, K., et al. (2018). TORC1 inhibition enhances immune function and reduces infections in the elderly. *Science Translational Medicine*, *10*, eaaq1564.

Mattison, J. A., Colman, R. J., Beasley, T. M., Allison, D. B., Kemnitz, J. W., Roth, G. S., et al. (2017). Caloric restriction improves health and survival of rhesus monkeys. *Nature Communications*, *8*, 14063.

Mattison, J. A., Roth, G. S., Beasley, T. M., Tilmont, E. M., Handy, A. M., Herbert, R. L., et al. (2012). Impact of caloric restriction on health and survival in rhesus monkeys from the NIA study. *Nature*, *489*, 318–321.

McHugh, D., & Gil, J. (2018). Senescence and aging: Causes, consequences, and therapeutic avenues. *The Journal of Cell Biology*, *217*, 65–77.

Martin, J. A., & Buckwalter, J. A. (2001). Telomere erosion and senescence in human articular cartilage chondrocytes. *Journals of Gerontology Series A: Biological Sciences and Medical Sciences*, *56*, B172–B179. Available from https://doi.org/10.1093/gerona/56.4.b172.

Martin, J. A., Brown, T. D., Heiner, A. D., & Buckwalter, J. A. (2004). Chondrocyte senescence, joint loading and osteoarthritis. *Clinical Orthopaedics and Related Research*, *427*, S96–S103. Available from https://doi.org/10.1097/01.blo.0000143818.74887.b1.

Menicacci, B., Margheri, F., Laurenzana, A., Chilla, A., Del Rosso, M., Giovannelli, L., et al. (2019). Chronic resveratrol treatment reduces the pro-angiogenic effect of human fibroblast "Senescent-Associated Secretory Phenotype" on endothelial colony-forming cells: The role of IL8. *The Journals of Gerontology. Series A, Biological Sciences and Medical Sciences*, *74*, 625–633.

Meyer, S. C., & Levine, R. L. (2014). Molecular pathways: Molecular basis for sensitivity and resistance to JAK kinase inhibitors. *Clinical Cancer Research: An Official Journal of the American Association for Cancer Research*, *20*, 2051–2059.

Minamino, T., Yoshida, T., Tateno, K., Miyauchi, H., Zou, Y., & Komuro, I. (2003). Ras induces vascular smooth muscle cell senescence and inflammation in human atherosclerosis. *Circulation*, *108*, 2264–2269. Available from https://doi.org/10.1161/01.CIR.0000093274.82929.22.

Minamino, T., Orimo, M., Shimizu, I., Kunieda, T., Yokoyama, M., Ito, T., et al. (2009). A crucial role for adipose tissue p53 in the regulation of insulin resistance. *Nature Medicine*, *15*, 1082–1087. Available from https://doi.org/10.1038/nm.2014.

Minagawa, S., Araya, J., Numata, T., Nojiri, S., Hara, H., Yumino, Y., et al. (2011). Accelerated epithelial cell senescence in IPF and the inhibitory role of SIRT6 in TGF-β-induced senescence of human bronchial epithelial cells. *The American Journal of Physiology-Lung Cellular and Molecular Physiology*, *300*, L391–L401. Available from https://doi.org/10.1152/ajplung.00097.2010.

Mishima, K., Handa, J. T., Aotaki-Keen, A., Lutty, G. A., Morse, L. S., & Hjelmeland, L. M. (1999). Senescence-associated beta-galactosidase histochemistry for the primate eye. *Investigative Ophthalmology & Visual Science*, *40*, 1590–1593.

Molofsky, A. V., Slutsky, S. G., Joseph, N. M., He, S., Pardal, R., Krishnamurthy, J., & Morrison, S. J. (2006). Increasing p16INK4a expression decreases forebrain progenitors and neurogenesis during ageing. *Nature*, *443*, 448–452. Available from https://doi.org/10.1038/nature05091.

Moncsek, A., Al-Suraih, M. S., Trussoni, C. E., O'Hara, S. P., Splinter, P. L., Zuber, C., et al. (2018). Targeting senescent cholangiocytes and activated fibroblasts with B-cell lymphoma-extra large inhibitors ameliorates fibrosis in multidrug resistance 2 gene knockout (Mdr2-/-) mice. *Hepatology*, *67*, 247–259. Available from https://doi.org/10.1002/hep.29464.

Montero, J., & Letai, A. (2018). Why do BCL-2 inhibitors work and where should we use them in the clinic? *Cell Death and Differentiation*, *25*, 56–64.

Moreno, L., Puerta, E., Suarez-Santiago, J. E., Santos-Magalhaes, N. S., Ramirez, M. J., & Irache, J. M. (2017). Effect of the oral administration of nanoencapsulated quercetin on a mouse model of Alzheimer's disease. *International Journal of Pharmaceutics*, *517*(1–2), 50–57. Available from https://doi.org/10.1016/j.ijpharm.2016.11.061.

Mosse, P. R., Campbell, G. R., Wang, Z. L., & Campbell, J. H. (1985). Smooth muscle phenotypic expression in human carotid arteries. I. Comparison of cells from diffuse intimal thickenings adjacent to atheromatous plaques with those of the media. *Laboratory Investigation*, *53*, 556–562.

Munoz-Espin, D., & Serrano, M. (2014). Cellular senescence: From physiology to pathology. *Nature Reviews. Molecular Cell Biology*, *15*, 482–496.

Musi, N., Valentine, J. M., Sickora, K. R., Baeuerle, E., Thompson, C. S., Shen, Q., et al. (2018). Tau protein aggregation is associated with cellular senescence in the brain. *Aging Cell*, *17*, e12840. Available from https://doi.org/10.1111/acel.12840.

Nacarelli, T., Azar, A., & Sell, C. (2015). Aberrant mTOR activation in senescence and aging: A mitochondrial stress response? *Experimental Gerontology*, *68*, 66–70.

Naikawadi, R. P., Disayabutr, S., Mallavia, B., Donne, M. L., Green, G., La, J. L., et al. (2016). Telomere dysfunction in alveolar epithelial cells causes lung remodeling and fibrosis. *JCI Insight*, *1*, e86704. Available from https://doi.org/10.1172/jci.insight.86704.

Naylor, R. M., Baker, D. J., & van Deursen, J. M. (2013). Senescent cells: A novel therapeutic target for aging and age-related diseases. *Clinical Pharmacology and Therapeutics*, *93*, 105–116.

Nelson, G., Kucheryavenko, O., Wordsworth, J., & von Zglinicki, T. (2018). The senescent bystander effect is caused by ROS-activated NF-kappaB signalling. *Mechanisms of Ageing and Development*, *170*, 30–36.

Newman, J. C., Milman, S., Hashmi, S. K., Austad, S. N., Kirkland, J. L., Halter, J. B., et al. (2016). Strategies and challenges in clinical trials targeting human aging. *The Journals of Gerontology. Series A, Biological Sciences and Medical Sciences*, *71*, 1424–1434.

Nichols, N. R., Day, J. R., Laping, N. J., Johnson, S. A., & Finch, C. E. (1993). GFAP mRNA increases with age in rat and human brain. *Neurobiology of Aging*, *14*, 421–429. Available from https://doi.org/10.1016/0197-4580(93)90100-p.

Niedernhofer, L. J., & Robbins, P. D. (2018). Senotherapeutics for healthy ageing. *Nature Reviews. Drug Discovery*, *17*, 377.

Ogrodnik, M., Miwa, S., Tchkonia, T., Tiniakos, D., Wilson, C. L., Lahat, A., et al. (2017). Cellular senescence drives age-dependent hepatic steatosis. *Nature Communications*, *8*, 15691.

Ogrodnik, M., Zhu, Y., Langhi, L. G. P., Tchkonia, T., Kruger, P., Fielder, E., ... Jurk, D. (2019). Obesity-induced cellular senescence drives anxiety and impairs neurogenesis. *Cell Metabolism*, *29*(5), 1233. Available from https://doi.org/10.1016/j.cmet.2019.01.013.

Ovadya, Y., & Krizhanovsky, V. (2014). Senescent cells: SASPected drivers of age-related pathologies. *Biogerontology*, *15*, 627–642.

Palacio, L., Goyer, M. L., Maggiorani, D., Espinosa, A., Villeneuve, N., Bourbonnais, S., et al. (2019). Restored immune cell functions upon clearance of senescence in the irradiated splenic environment. *Aging Cell*, *18*, e12971.

Pardanani, A., Laborde, R. R., Lasho, T. L., Finke, C., Begna, K., Al-Kali, A., et al. (2013). Safety and efficacy of CYT387, a JAK1 and JAK2 inhibitor, in myelofibrosis. *Leukemia: Official Journal of the Leukemia Society of America, Leukemia Research Fund, U.K*, *27*, 1322–1327.

Patil, P., Dong, Q., Wang, D., Chang, J., Wiley, C., Demaria, M., et al. (2019). Systemic clearance of p16(INK4a) - positive senescent cells mitigates age-associated intervertebral disc degeneration. *Aging Cell*, *18*, e12927.

Philipot, D., Guérit, D., Platano, D., Chuchana, P., Olivotto, E., Espinoza, F., et al. (2014). p16INK4a and its regulator miR-24 link senescence and chondrocyte terminal differentiation-associated matrix remodeling in osteoarthritis. *Arthritis Research & Therapy*, *16*, R58. Available from https://doi.org/10.1186/ar4494.

Prakash, D., Gopinath, K., & Sudhandiran, G. (2013). Fisetin enhances behavioral performances and attenuates reactive gliosis and inflammation during aluminum chloride-induced neurotoxicity. *Neuromolecular Medicine*, *15*(1), 192–208. Available from https://doi.org/10.1007/s12017-012-8210-1.

Prasath, G. S., Pillai, S. I., & Subramanian, S. P. (2014). Fisetin improves glucose homeostasis through the inhibition of gluconeogenic enzymes in hepatic tissues of streptozotocin induced diabetic rats. *European Journal of Pharmacology*, *740*, 248–254.

Prasath, G. S., & Subramanian, S. P. (2014). Antihyperlipidemic effect of fisetin, a bioflavonoid of strawberries, studied in streptozotocin-induced diabetic rats. *Journal of Biochemical and Molecular Toxicology*, *28*, 442–449.

Prasath, G. S., Sundaram, C. S., & Subramanian, S. P. (2013). Fisetin averts oxidative stress in pancreatic tissues of streptozotocin-induced diabetic rats. *Endocrine*, *44*, 359–368.

Price, J. S., Waters, J. G., Darrah, C., Pennington, C., Edwards, D. R., Donell, S. T., & Clark, I. M. (2002). The role of chondrocyte senescence in osteoarthritis. *Aging Cell*, *1*, 57–65. Available from https://doi.org/10.1046/j.1474-9728.2002.00008.x.

Ragnauth, C. D., Warren, D. T., Liu, Y., McNair, R., Tajsic, T., Figg, N., et al. (2010). Prelamin A acts to accelerate smooth muscle cell senescence and is a novel biomarker of human vascular aging. *Circulation*, *121*, 2200–2210. Available from https://doi.org/10.1161/CIRCULATIONAHA.109.902056.

Ravussin, E., Redman, L. M., Rochon, J., Das, S. K., Fontana, L., Kraus, W. E., et al. (2015). A 2-year randomized controlled trial of human caloric restriction: feasibility and effects on predictors of health span and longevity. *The Journals of Gerontology. Series A, Biological Sciences and Medical Sciences*, *70*, 1097–1104.

Reyes-Farias, M., & Carrasco-Pozo, C. (2019). The Anti-Cancer Effect of Quercetin: Molecular Implications in Cancer Metabolism. *International Journal of Molecular Sciences*, *20*(13), 3177.

Richard, A. J., & Stephens, J. M. (2011). Emerging roles of JAK-STAT signaling pathways in adipocytes. *Trends in Endocrinology and Metabolism: TEM*, *22*, 325–332.

Rickman, A. D., Williamson, D. A., Martin, C. K., Gilhooly, C. H., Stein, R. I., Bales, C. W., et al. (2011). The CALERIE Study: Design and methods of an innovative 25% caloric restriction intervention. *Contemporary Clinical Trials*, *32*, 874–881.

Rivera, F., Urbanavicius, J., Gervaz, E., Morquio, A., & Dajas, F. (2004). Some aspects of the in vivo neuroprotective capacity of flavonoids: Bioavailability and structure-activity relationship. *Neurotoxicity Research*, *6*(7–8), 543–553.

Rivera-Torres, J., & San Jose, E. (2019). Src Tyrosine Kinase Inhibitors: New Perspectives on Their Immune, Antiviral, and Senotherapeutic Potential. *Frontiers in Pharmacology*, *10*, 1011.

Rix, U., Hantschel, O., Durnberger, G., Remsing Rix, L. L., Planyavsky, M., Fernbach, N. V., et al. (2007). Chemical proteomic profiles of the BCR-ABL inhibitors imatinib, nilotinib, and dasatinib reveal novel kinase and nonkinase targets. *Blood*, *110*, 4055–4063.

Roberson, R. S., Kussick, S. J., Vallieres, E., Chen, S. Y., & Wu, D. Y. (2005). Escape from therapy-induced accelerated cellular senescence in p53-null lung cancer cells and in human lung cancers. *Cancer Research*, *65*, 2795–2803. Available from https://doi.org/10.1158/0008-5472.CAN-04-1270.

Roberts, S., Evans, E. H., Kletsas, D., Jaffray, D. C., & Eisenstein, S. M. (2006). Senescence in human intervertebral discs. *European Spine Journal*, *15*(Suppl 3), S312–S316. Available from https://doi.org/10.1007/s00586-006-0126-8.

Rodier, F., Munoz, D. P., Teachenor, R., Chu, V., Le, O., Bhaumik, D., et al. (2011). DNA-SCARS: Distinct nuclear structures that sustain damage-induced senescence growth arrest and inflammatory cytokine secretion. *Journal of Cell Science*, *124*, 68–81.

Roos, C. M., Zhang, B., Palmer, A. K., Ogrodnik, M. B., Pirtskhalava, T., Thalji, N. M., et al. (2016). Chronic senolytic treatment alleviates established vasomotor dysfunction in aged or atherosclerotic mice. *Aging Cell*, *15*, 973–977.

Ross, R., Wight, T. N., Strandness, E., & Thiele, B. (1984). Human atherosclerosis I. Cell constitution and characteristics of advanced lesions of the superficial femoral artery. *American Journal of Pathology*, *114*, 79–93.

Sabogal-Guáqueta, A. M., Munoz-Manco, J. I., Ramirez-Pineda, J. R., Lamprea-Rodriguez, M., Osorio, E., & Cardona-Gomez, G. P. (2015). The flavonoid quercetin ameliorates Alzheimer's disease pathology and protects cognitive and emotional function in aged triple transgenic Alzheimer's disease model mice. *Neuropharmacology*, *93*, 134–145. Available from https://doi.org/10.1016/j.neuropharm.2015.01.027.

Scott, L. J., Mohlke, K. L., Bonnycastle, L. L., Willer, C. J., Li, Y., Duren, W. L., et al. (2007). A genome-wide association study of type 2 diabetes in Finns detects multiple susceptibility variants. *Science*, *316*, 1341–1345. Available from https://doi.org/10.1126/science.1142382.

Sadaie, M., Salama, R., Carroll, T., Tomimatsu, K., Chandra, T., Young, A. R., et al. (2013). Redistribution of the Lamin B1 genomic binding profile affects rearrangement of heterochromatic domains and SAHF formation during senescence. *Genes & Development*, *27*, 1800–1808.

Samani, N. J., Boultby, R., Butler, R., Thompson, J. R., & Goodall, A. H. (2001). Telomere shortening in atherosclerosis. *Lancet*, *358*, 472–473. Available from https://doi.org/10.1016/S0140-6736(01)05633-1.

Samaraweera, L., Adomako, A., Rodriguez-Gabin, A., & McDaid, H. M. (2017). A novel indication for panobinostat as a senolytic drug in NSCLC and HNSCC. *Scientific Reports*, *7*, 1900.

Sarkisian, C. J., Keister, B. A., Stairs, D. B., Boxer, R. B., Moody, S. E., & Chodosh, L. A. (2007). Dose-dependent oncogene-induced senescence in vivo and its evasion during mammary tumorigenesis. *Nature Cell Biology*, *9*, 493–505. Available from https://doi.org/10.1038/ncb1567.

Saxena, R., Voight, B. F., Lyssenko, V., Burtt, N. P., de Bakker, P. I., Chen, H., et al. (2007). Genome-wide association analysis identifies loci for type 2 diabetes and triglyceride levels. *Science.*, *316*, 1331–1336. Available from https://doi.org/10.1126/science.1142358.

Saxton, R. A., & Sabatini, D. M. (2017). mTOR signaling in growth, metabolism, and disease. *Cell*, *169*, 361–371.

Schafer, M. J., Haak, A. J., Tschumperlin, D. J., & LeBrasseur, N. K. (2018). Targeting senescent cells in fibrosis: pathology, paradox, and practical considerations. *Current Rheumatology Reports*, *20*, 3.

Schafer, M. J., White, T. A., Evans, G., Tonne, J. M., Verzosa, G. C., Stout, M. B., et al. (2016). Exercise prevents diet-induced cellular senescence in adipose tissue. *Diabetes*, *65*(6), 1606–1615.

Schafer, M. J., White, T. A., Iijima, K., Haak, A. J., Ligresti, G., Atkinson, E. J., et al. (2017). Cellular senescence mediates fibrotic pulmonary disease. *Nature Communications*, *8*, 14532.

Schmitt, C. A., Fridman, J. S., Yang, M., Lee, S., Baranov, E., Hoffman, R. M., & Lowe, S. W. (2002). A senescence program controlled by p53 and p16INK4a contributes to the outcome of cancer therapy. *Cell*, *109*, 335–346. Available from https://doi.org/10.1016/s0092-8674(02)00734-1.

Seals, D. R. (2013). Translational physiology: From molecules to public health. *The Journal of Physiology*, *591*, 3457–3469.

Seals, D. R., & Melov, S. (2014). Translational geroscience: Emphasizing function to achieve optimal longevity. *Aging*, *6*, 718–730.

Seo, S. H., & Jeong, G. S. (2015). Fisetin inhibits TNF-alpha-induced inflammatory action and hydrogen peroxide-induced oxidative damage in human keratinocyte HaCaT cells through PI3K/AKT/Nrf-2-mediated heme oxygenase-1 expression. *International Immunopharmacology*, *29*, 246–253.

Sharpless, N. E. (2004). Ink4a/Arf links senescence and aging. *Experimental Gerontology*, *39*, 1751–1759.

Sharpless, N. E. (2005). INK4a/ARF: A multifunctional tumor suppressor locus. *Mutation Research*, *576*, 22–38.

Sharpless, N. E., & Sherr, C. J. (2015). Forging a signature of in vivo senescence. *Nature Reviews. Cancer*, *15*, 397–408.

Shirakawa, K., Yan, X., Shinmura, K., Endo, J., Kataoka, M., Katsumata, Y., et al. (2016). Obesity accelerates T cell senescence in murine visceral adipose tissue. *Journal of Clinical Investigation*, *126*, 4626–4639. Available from https://doi.org/10.1172/JCI88606.

Shivshankar, P., Brampton, C., Miyasato, S., Kasper, M., Thannickal, V. J., & Le Saux, C. J. (2012). Caveolin-1 deficiency protects from pulmonary fibrosis by modulating epithelial cell senescence in mice. *American Journal of Respiratory Cell and Molecular Biology*, *47*, 28–36. Available from https://doi.org/10.1165/rcmb.2011-0349OC.

Sierra, F. (2016). Moving geroscience into uncharted waters. *The Journals of Gerontology. Series A, Biological Sciences and Medical Sciences*, *71*, 1385–1387.

Sierra, F., & Kohanski, R. (2017). Geroscience and the trans-NIH Geroscience Interest Group, GSIG. *Geroscience*, *39*, 1–5.

Singh, S., Singh, A. K., Garg, G., & Rizvi, S. I. (2018). Fisetin as a caloric restriction mimetic protects rat brain against aging induced oxidative stress, apoptosis and neurodegeneration. *Life Sciences*, *193*, 171–179. Available from https://doi.org/10.1016/j.lfs.2017.11.004.

Sone, H., & Kagawa, Y. (2005). Pancreatic beta cell senescence contributes to the pathogenesis of type 2 diabetes in high-fat diet-induced diabetic mice. *Diabetologia*, *48*, 58–67. Available from https://doi.org/10.1007/s00125-004-1605-2.

Sousa-Victor, P., Gutarra, S., García-Prat, L., Rodriguez-Ubreva, J., Ortet, L., Ruiz-Bonilla, V., et al. (2014). Geriatric muscle stem cells switch reversible quiescence into senescence. *Nature*, *506*, 316–321. Available from https://doi.org/10.1038/nature13013.

Stebbins, C. E., Russo, A. A., Schneider, C., Rosen, N., Hartl, F. U., & Pavletich, N. P. (1997). Crystal structure of an Hsp90-geldanamycin complex: Targeting of a protein chaperone by an antitumor agent. *Cell*, *89*, 239–250.

Stehlik, C., de Martin, R., Kumabashiri, I., Schmid, J. A., Binder, B. R., & Lipp, J. (1998). Nuclear factor (NF)-kappaB-regulated X-chromosome-linked iap gene expression protects endothelial cells from tumor necrosis factor alpha-induced apoptosis. *The Journal of Experimental Medicine*, *188*, 211–216.

Stout, M. B., Tchkonia, T., Pirtskhalava, T., Palmer, A. K., List, E. O., Berryman, D. E., et al. (2014). Growth hormone action predicts age-related white adipose tissue dysfunction and senescent cell burden in mice. *Aging (Albany NY)*, *6*, 575–586. Available from https://doi.org/10.18632/aging.100681.

Swindell, W. R. (2012). Dietary restriction in rats and mice: A meta-analysis and review of the evidence for genotype-dependent effects on lifespan. *Ageing Research Reviews*, *11*, 254–270.

Tabibian, J. H., O'Hara, S. P., Splinter, P. L., Trussoni, C. E., & LaRusso, N. F. (2014). Cholangiocyte senescence by way of N-ras activation is a characteristic of primary sclerosing cholangitis. *Hepatology*, *59*, 2263–2275.

Tacutu, R., Budovsky, A., Yanai, H., & Fraifeld, V. E. (2011). Molecular links between cellular senescence, longevity and age-related diseases - a systems biology perspective. *Aging (Albany NY)*, *3*, 1178–1191. Available from https://doi.org/10.18632/aging.100413.

Taipale, M., Jarosz, D. F., & Lindquist, S. (2010). HSP90 at the hub of protein homeostasis: Emerging mechanistic insights. *Nature Reviews. Molecular Cell Biology*, *11*, 515–528.

Takahashi, A., Ohtani, N., Yamakoshi, K., Iida, S., Tahara, H., Nakayama, K., et al. (2006). Mitogenic signalling and the p16INK4a-Rb pathway cooperate to enforce irreversible cellular senescence. *Nature Cell Biology*, *8*, 1291–1297.

Takasugi, M., Okada, R., Takahashi, A., Virya Chen, D., Watanabe, S., & Hara, E. (2017). Small extracellular vesicles secreted from senescent cells promote cancer cell proliferation through EphA2. *Nature Communications*, *8*, 15729.

Tamatani, M., Che, Y. H., Matsuzaki, H., Ogawa, S., Okado, H., Miyake, S., et al. (1999). Tumor necrosis factor induces Bcl-2 and Bcl-x expression through NFkappaB activation in primary hippocampal neurons. *The Journal of Biological Chemistry, 274*, 8531–8538.

Tasdemir, D., Kaiser, M., Brun, R., Yardley, V., Schmidt, T. J., Tosun, F., et al. (2006). Antitrypanosomal and antileishmanial activities of flavonoids and their analogues: In vitro, in vivo, structure-activity relationship, and quantitative structure-activity relationship studies. *Antimicrobial Agents and Chemotherapy, 50*, 1352–1364.

Tavana, O., Puebla-Osorio, N., Sang, M., & Zhu, C. (2010). Absence of p53-dependent apoptosis combined with nonhomologous end-joining deficiency leads to a severe diabetic phenotype in mice. *Diabetes, 59*, 135–142. Available from https://doi.org/10.2337/db09-0792.

Tchkonia, T., & Kirkland, J. L. (2018). Aging, cell senescence, and chronic disease: Emerging therapeutic strategies. *JAMA: The Journal of the American Medical Association, 320*, 1319–1320.

Tchkonia, T., Morbeck, D. E., Von Zglinicki, T., Van Deursen, J., Lustgarten, J., Scrable, H., et al. (2010). Fat tissue, aging, and cellular senescence. *Aging Cell, 9*, 667–684.

Tchkonia, T., Zhu, Y., van Deursen, J., Campisi, J., & Kirkland, J. L. (2013). Cellular senescence and the senescent secretory phenotype: Therapeutic opportunities. *The Journal of Clinical Investigation, 123*, 966–972.

Thompson, A. G., Gray, E., Heman-Ackah, S. M., Mager, I., Talbot, K., Andaloussi, S. E., et al. (2016). Extracellular vesicles in neurodegenerative disease - pathogenesis to biomarkers. *Nature Reviews Neurology, 12*, 346–357.

Tilstra, J. S., Clauson, C. L., Niedernhofer, L. J., & Robbins, P. D. (2011). NF-kappaB in Aging and Disease. *Aging and Disease, 2*, 449–465.

Tilstra, J. S., Robinson, A. R., Wang, J., Gregg, S. Q., Clauson, C. L., Reay, D. P., et al. (2012). NF-kappaB inhibition delays DNA damage-induced senescence and aging in mice. *The Journal of Clinical Investigation, 122*, 2601–2612.

Tominaga, T., Shimada, R., Okada, Y., Kawamata, T., & Kibayashi, K. (2019). Senescence-associated-β-galactosidase staining following traumatic brain injury in the mouse cerebrum. *PLoS One, 14*, e0213673. Available from https://doi.org/10.1371/journal.pone.0213673.

Tonnessen-Murray, C. A., Frey, W. D., Rao, S. G., Shahbandi, A., Ungerleider, N. A., Olayiwola, J. O., et al. (2019). Chemotherapy-induced senescent cancer cells engulf other cells to enhance their survival. *The Journal of Cell Biology, 218*, 3827–3844.

Touil, Y. S., Auzeil, N., Boulinguez, F., Saighi, H., Regazzetti, A., Scherman, D., et al. (2011). Fisetin disposition and metabolism in mice: Identification of geraldol as an active metabolite. *Biochemical Pharmacology, 82*, 1731–1739.

Trias, E., Beilby, P. R., Kovacs, M., Ibarburu, S., Varela, V., Barreto-Núñez, R., et al. (2019). Emergence of microglia bearing senescence markers during paralysis progression in a rat model of inherited ALS. *Frontiers in Aging Neuroscience, 28*, 11:42. Available from https://doi.org/10.3389/fnagi.2019.00042.

Tsurumi, A., & Li, W. X. (2012). Global heterochromatin loss: A unifying theory of aging? *Epigenetics: Official Journal of the DNA Methylation Society, 7*, 680–688.

Uchida, T., Nakamura, T., Hashimoto, N., Matsuda, T., Kotani, K., Sakaue, H., et al. (2005). Deletion of Cdkn1b ameliorates hyperglycemia by maintaining compensatory hyperinsulinemia in diabetic mice. *Nature Medicine, 11*, 175–182. Available from https://doi.org/10.1038/nm1187.

Vargas-Restrepo, F., Sabogal-Guaqueta, A. M., & Cardona-Gomez, G. P. (2018). [Quercetin ameliorates inflammation in CA1 hippocampal region in aged triple transgenic Alzheimer s disease mice model.]. *Biomedica: Revista del Instituto Nacional de Salud, 38*(0), 69–76. Available from https://doi.org/10.7705/biomedica.v38i0.3761.

Vasile, E., Tomita, Y., Brown, L. F., Kocher, O., & Dvorak, H. F. (2001). Differential expression of thymosin beta-10 by early passage and senescent vascular endothelium is modulated by VPF/VEGF: evidence for senescent endothelial cells in vivo at sites of atherosclerosis. *FASEB Journal, 15*, 458–466. Available from https://doi.org/10.1096/fj.00-0051com.

Vassilev, L. T., Vu, B. T., Graves, B., Carvajal, D., Podlaski, F., Filipovic, Z., et al. (2004). In vivo activation of the p53 pathway by small-molecule antagonists of MDM2. *Science, 303*, 844–848.

Vaughan, K. L., Kaiser, T., Peaden, R., Anson, R. M., de Cabo, R., & Mattison, J. A. (2017). Caloric restriction study design limitations in rodent and nonhuman primate studies. *The Journals of Gerontology. Series A, Biological Sciences and Medical Sciences, 73*, 48–53.

Verstovsek, S., Kantarjian, H., Mesa, R. A., Pardanani, A. D., Cortes-Franco, J., Thomas, D. A., et al. (2010). Safety and efficacy of INCB018424, a JAK1 and JAK2 inhibitor, in myelofibrosis. *The New England Journal of Medicine, 363*, 1117–1127.

Verstovsek, S., Mesa, R. A., Gotlib, J., Levy, R. S., Gupta, V., DiPersio, J. F., et al. (2012). A double-blind, placebo-controlled trial of ruxolitinib for myelofibrosis. *The New England Journal of Medicine, 366*, 799–807.

Villaret, A., Galitzky, J., Decaunes, P., Estève, D., Marques, M. A., Sengenès, C., et al. (2010). Adipose tissue endothelial cells from obese human subjects: differences among depots in angiogenic, metabolic, and inflammatory gene expression and cellular senescence. *Diabetes, 59*, 2755–2763. Available from https://doi.org/10.2337/db10-0398.

Waaijer, M. E. C., Gunn, D. A., van Heemst, D., Slagboom, P. E., Sedivy, J. M., Dirks, R. W., et al. (2018). Do senescence markers correlate in vitro and in situ within individual human donors? *Aging, 10*, 278–289.

Walters, H. E., & Cox, L. S. (2018). mTORC inhibitors as broad-spectrum therapeutics for age-related diseases. *International Journal of Molecular Sciences, 19*, 2325.

Wang, C., Jurk, D., Maddick, M., Nelson, G., Martin-Ruiz, C., & von Zglinicki, T. (2009a). DNA damage response and cellular senescence in tissues of aging mice. *Aging Cell, 8*, 311–323. Available from https://doi.org/10.1111/j.1474-9726.2009.00481.x.

Wang, C. Y., Kim, H. H., Hiroi, Y., Sawada, N., Salomone, S., Benjamin, L. E., et al. (2009b). Obesity increases vascular senescence and susceptibility to ischemic injury through chronic activation of Akt and mTOR. *Science Signaling, 2*, ra11. Available from https://doi.org/10.1126/scisignal.2000143.

Wang, C., Maddick, M., Miwa, S., Jurk, D., Czapiewski, R., Saretzki, G., et al. (2010). Adult-onset, short-term dietary restriction reduces cell senescence in mice. *Aging, 2*, 555–566.

Wang, D. M., Li, S. Q., Wu, W. L., Zhu, X. Y., Wang, Y., & Yuan, H. Y. (2014). Effects of long-term treatment with quercetin on cognition and mitochondrial function in a mouse model of Alzheimer's disease. *Neurochemical Research, 39*(8), 1533–1543. Available from https://doi.org/10.1007/s11064-014-1343-x.

Wang, E. (1995). Senescent human fibroblasts resist programmed cell death, and failure to suppress bcl2 is involved. *Cancer Research, 55*, 2284–2292.

Wang, J., Uryga, A. K., Reinhold, J., Figg, N., Baker, L., Finigan, A., et al. (2015). Vascular smooth muscle cell senescence promotes atherosclerosis and features of plaque vulnerability. *Circulation, 132*, 1909–1919. Available from https://doi.org/10.1161/CIRCULATIONAHA.115.016457.

Wang, L., Tu, Y. C., Lian, T. W., Hung, J. T., Yen, J. H., & Wu, M. J. (2006). Distinctive antioxidant and antiinflammatory effects of flavonols. *Journal of Agricultural and Food Chemistry, 54*, 9798–9804.

Wang, Y., Chang, J., Liu, X., Zhang, X., Zhang, S., Zhang, X., et al. (2016). Discovery of piperlongumine as a potential novel lead for the development of senolytic agents. *Aging, 8*, 2915–2926.

Wang, Z., Moro, E., Kovacs, K., Yu, R., & Melmed, S. (2003). Pituitary tumor transforming gene-null male mice exhibit impaired pancreatic beta cell proliferation and diabetes. *Proceedings of the National Academy of Sciences of the United States of America*, *100*, 3428–3432. Available from https://doi.org/10.1073/pnas.0638052100.

Weichhart, T. (2018). mTOR as regulator of lifespan, aging, and cellular senescence: A mini-review. *Gerontology*, *64*, 127–134.

Weilner, S., Schraml, E., Wieser, M., Messner, P., Schneider, K., Wassermann, K., et al. (2016). Secreted microvesicular miR-31 inhibits osteogenic differentiation of mesenchymal stem cells. *Aging Cell*, *15*, 744–754.

Weisberg, E., Manley, P. W., Cowan-Jacob, S. W., Hochhaus, A., & Griffin, J. D. (2007). Second generation inhibitors of BCR-ABL for the treatment of imatinib-resistant chronic myeloid leukaemia. *Nature Reviews. Cancer*, *7*, 345–356.

Welton, J. L., Loveless, S., Stone, T., von Ruhland, C., Robertson, N. P., & Clayton, A. (2017). Cerebrospinal fluid extracellular vesicle enrichment for protein biomarker discovery in neurological disease; multiple sclerosis. *Journal of Extracellular Vesicles*, *6*, 1369805.

Wiemann, S. U., Satyanarayana, A., Tsahuridu, M., Tillmann, H. L., Zender, L., Klempnauer, J., et al. (2002). Hepatocyte telomere shortening and senescence are general markers of human liver cirrhosis. *FASEB Journal*, *16*, 935–942. Available from https://doi.org/10.1096/fj.01-0977com.

Wolstein, J. M., Lee, D. H., Michaud, J., Buot, V., Stefanchik, B., & Plotkin, M. D. (2010). INK4a knockout mice exhibit increased fibrosis under normal conditions and in response to unilateral ureteral obstruction. *American Journal of Physiology-Renal Physiology*, *299*, F1486–F1495. Available from https://doi.org/10.1152/ajprenal.00378.2010.

Xu, M., Bradley, E. W., Weivoda, M. M., Hwang, S. M., Pirtskhalava, T., Decklever, T., et al. (2017). Transplanted Senescent Cells Induce an Osteoarthritis-Like Condition in Mice. *The Journals of Gerontology. Series A, Biological Sciences and Medical Sciences*, *72*, 780–785.

Xu, M., Palmer, A. K., Ding, H., Weivoda, M. M., Pirtskhalava, T., White, T. A., et al. (2015a). Targeting senescent cells enhances adipogenesis and metabolic function in old age. *Elife*, *4*, e12997.

Xu, M., Pirtskhalava, T., Farr, J. N., Weigand, B. M., Palmer, A. K., Weivoda, M. M., ... Kirkland, J. L. (2018). Senolytics improve physical function and increase lifespan in old age. *Nature Medicine*, *24*(8), 1246–1256. Available from https://doi.org/10.1038/s41591-018-0092-9.

Xu, M., Tchkonia, T., Ding, H., Ogrodnik, M., Lubbers, E. R., Pirtskhalava, T., et al. (2015b). JAK inhibition alleviates the cellular senescence-associated secretory phenotype and frailty in old age. *Proceedings of the National Academy of Sciences of the United States of America*, *112*, E6301–E6310.

Xu, M., Tchkonia, T., & Kirkland, J. L. (2016). Perspective: Targeting the JAK/STAT pathway to fight age-related dysfunction. *Pharmacological Research: The Official Journal of the Italian Pharmacological Society*, *111*, 152–154.

Xue, W., Zender, L., Miething, C., Dickins, R. A., Hernando, E., Krizhanovsky, V., et al. (2007). Senescence and tumour clearance is triggered by p53 restoration in murine liver carcinomas. *Nature*, *445*, 656–660.

Yamamoto, K., Arakawa, T., Ueda, N., & Yamamoto, S. (1995). Transcriptional roles of nuclear factor kappa B and nuclear factor-interleukin-6 in the tumor necrosis factor alpha-dependent induction of cyclooxygenase-2 in MC3T3-E1 cells. *The Journal of Biological Chemistry*, *270*, 31315–31320.

Yanai, H., & Fraifeld, V. E. (2018). The role of cellular senescence in aging through the prism of Koch-like criteria. *Ageing Research Reviews*, *41*, 18–33.

Yanai, H., Shteinberg, A., Porat, Z., Budovsky, A., Braiman, A., Ziesche, R., & Fraifeld, V. E. (2015). Cellular senescence-like features of lung fibroblasts derived from idiopathic pulmonary fibrosis patients. *Aging (Albany NY)*, *7*, 664–672. Available from https://doi.org/10.18632/aging.100807, Erratum in: *Aging (Albany NY)* 7, 1022. Zeische, Rolf [corrected to Ziesche, Rolf].

Yang, G., Rosen, D. G., Zhang, Z., Bast, R. C., Jr, Mills, G. B., Colacino, J. A., et al. (2006). The chemokine growth-regulated oncogene 1 (Gro-1) links RAS signaling to the senescence of stromal fibroblasts and ovarian tumorigenesis. *Proceedings of the National Academy of Sciences of the United States of America*, *103*, 16472–16477. Available from https://doi.org/10.1073/pnas.0605752103.

Yang, L., Fang, J., & Chen, J. (2017). Tumor cell senescence response produces aggressive variants. *Cell Death Discovery*, *3*, 17049. Available from https://doi.org/10.1038/cddiscovery.2017.49.

Yokoi, T., Fukuo, K., Yasuda, O., Hotta, M., Miyazaki, J., Takemura, Y., et al. (2006). Apoptosis signal-regulating kinase 1 mediates cellular senescence induced by high glucose in endothelial cells. *Diabetes*, *55*, 1660–1665. Available from https://doi.org/10.2337/db05-1607.

Yosef, R., Pilpel, N., Tokarsky-Amiel, R., Biran, A., Ovadya, Y., Cohen, S., et al. (2016). Directed elimination of senescent cells by inhibition of BCL-W and BCL-XL. *Nature Communications*, *7*, 11190.

Yoshimoto, S., Loo, T. M., Atarashi, K., Kanda, H., Sato, S., Oyadomari, S., et al. (2013). Obesity-induced gut microbial metabolite promotes liver cancer through senescence secretome. *Nature*, *499*(7456), 97–101. Available from https://doi.org/10.1038/nature12347, Epub 2013 Jun 26. Erratum in: *Nature* 506, 2014, 396. Hattori, Masahisa [corrected to Hattori, Masahira].

Yousefzadeh, M. J., Zhu, Y., McGowan, S. J., Angelini, L., Fuhrmann-Stroissnigg, H., Xu, M., ... Niedernhofer, L. J. (2018). Fisetin is a senotherapeutic that extends health and lifespan. *EBioMedicine*, *36*, 18–28. Available from https://doi.org/10.1016/j.ebiom.2018.09.015.

Yu, H., Pardoll, D., & Jove, R. (2009). STATs in cancer inflammation and immunity: A leading role for STAT3. *Nature Reviews Cancer*, *9*, 798–809.

Yu, H. M., Zhao, Y. M., Luo, X. G., Feng, Y., Ren, Y., Shang, H., et al. (2012). Repeated lipopolysaccharide stimulation induces cellular senescence in BV2 cells. *Neuroimmunomodulation*, *19*, 131–136. Available from https://doi.org/10.1159/000330254.

Yudoh, K., Nguyen, V. T., Nakamura, H., Hongo-Masuko, K., Kato, T., & Nishioka, K. (2005). Potential involvement of oxidative stress in cartilage senescence and development of osteoarthritis: oxidative stress induces chondrocyte telomere instability and downregulation of chondrocyte function. *Arthritis Research & Therapy*, *7*, R380–R391. Available from https://doi.org/10.1186/ar1499.

Zhang, L., Wang, H., Zhou, Y., Zhu, Y., & Fei, M. (2018). Fisetin alleviates oxidative stress after traumatic brain injury via the Nrf2-ARE pathway. *Neurochemistry International*, *118*, 304–313. Available from https://doi.org/10.1016/j.neuint.2018.05.011.

Zhang, P., Kishimoto, Y., Grammatikakis, I., Gottimukkala, K., Cutler, R. G., Zhang, S., et al. (2019). Senolytic therapy alleviates Abeta-associated oligodendrocyte progenitor cell senescence and cognitive deficits in an Alzheimer's disease model. *Nature Neuroscience*, *22*, 719–728. Available from https://doi.org/10.1038/s41593-019-0372-9.

Zhang, X., Hu, J., Zhong, L., Wang, N., Yang, L., Liu, C. C., ... Zhuang, J. (2016). Quercetin stabilizes apolipoprotein E and reduces brain Abeta levels in amyloid model mice. *Neuropharmacology*, *108*, 179–192. Available from https://doi.org/10.1016/j.neuropharm.2016.04.032.

Zhao, X., Dong, Y., Zhang, J., Li, D., Hu, G., Yao, J., et al. (2016). Leptin changes differentiation fate and induces senescence in chondrogenic progenitor cells. *Cell Death and Disease*, *7*, e2188. Available from https://doi.org/10.1038/cddis.2016.68.

Zhang, Z., Rosen, D. G., Yao, J. L., Huang, J., & Liu, J. (2006). Expression of p14ARF, p15INK4b, p16INK4a, and DCR2 increases during prostate cancer progression. *Modern Pathology: An Official Journal of the United States and Canadian Academy of Pathology, Inc*, *19*, 1339–1343.

Zeggini, E., Weedon, M. N., Lindgren, C. M., Frayling, T. M., Elliott, K. S., Lango, H., et al. (2007). Replication of genome-wide association signals in UK samples reveals risk loci for type 2 diabetes. *Science, 316*, 1336–1341. Available from https://doi.org/10.1126/science.1142364, Erratum in: Science. 317, 2007, 1035-6.

Zhen, L., Zhu, J., Zhao, X., Huang, W., An, Y., Li, S., ... Pan, J. (2012). The antidepressant-like effect of fisetin involves the serotonergic and noradrenergic system. *Behavioural Brain Research, 228*(2), 359–366. Available from https://doi.org/10.1016/j.bbr.2011.12.017.

Zeng, N., Yang, K. T., Bayan, J. A., He, L., Aggarwal, R., Stiles, J. W., et al. (2013). PTEN controls β-cell regeneration in aged mice by regulating cell cycle inhibitor p16ink4a. *Aging Cell, 12*, 1000–1011. Available from https://doi.org/10.1111/acel.12132.

Zhou, C. H., Wang, C. X., Xie, G. B., Wu, L. Y., Wei, Y. X., Wang, Q., ... Shi, J. X. (2015). Fisetin alleviates early brain injury following experimental subarachnoid hemorrhage in rats possibly by suppressing TLR 4/NF-kappaB signaling pathway. *Brain Research, 1629*, 250–259. Available from https://doi.org/10.1016/j.brainres.2015.10.016.

Zhou, H. W., Lou, S. Q., & Zhang, K. (2004). Recovery of function in osteoarthritic chondrocytes induced by p16INK4a-specific siRNA in vitro. *Rheumatology (Oxford), 43*, 555–568. Available from https://doi.org/10.1093/rheumatology/keh127.

Zhu, D., Wu, J., Spee, C., Ryan, S. J., & Hinton, D. R. (2009). BMP4 mediates oxidative stress-induced retinal pigment epithelial cell senescence and is overexpressed in age-related macular degeneration. *Journal of Biological Chemistry, 284*, 9529–9539. Available from https://doi.org/10.1074/jbc.M809393200.

Zhu, F., Li, Y., Zhang, J., Piao, C., Liu, T., Li, H. H., & Du, J. (2013). Senescent cardiac fibroblast is critical for cardiac fibrosis after myocardial infarction. *PLoS One, 8*, e74535. Available from https://doi.org/10.1371/journal.pone.0074535.

Zhu, L., Fukuda, S., Cordis, G., Das, D. K., & Maulik, N. (2001). Anti-apoptotic protein survivin plays a significant role in tubular morphogenesis of human coronary arteriolar endothelial cells by hypoxic preconditioning. *FEBS Letters, 508*, 369–374.

Zhu, Y., Armstrong, J. L., Tchkonia, T., & Kirkland, J. L. (2014). Cellular senescence and the senescent secretory phenotype in age-related chronic diseases. *Current Opinion in Clinical Nutrition and Metabolic Care, 17*, 324–328.

Zhu, Y., Doornebal, E. J., Pirtskhalava, T., Giorgadze, N., Wentworth, M., Fuhrmann-Stroissnigg, H., et al. (2017). New agents that target senescent cells: The flavone, fisetin, and the BCL-XL inhibitors, A1331852 and A1155463. *Aging, 9*, 955–963.

Zhu, Y., Tchkonia, T., Fuhrmann-Stroissnigg, H., Dai, H. M., Ling, Y. Y., Stout, M. B., et al. (2016). Identification of a novel senolytic agent, navitoclax, targeting the Bcl-2 family of anti-apoptotic factors. *Aging Cell, 15*, 428–435.

Zhu, Y., Tchkonia, T., Pirtskhalava, T., Gower, A. C., Ding, H., Giorgadze, N., et al. (2015). The Achilles' heel of senescent cells: From transcriptome to senolytic drugs. *Aging Cell, 14*, 644–658.

Zuchner, S., Gilbert, J. R., Martin, E. R., Leon-Guerrero, C. R., Xu, P. T., Browning, C., et al. (2008). Linkage and association study of late-onset Alzheimer disease families linked to 9p21.3. *Annals of Human Genetics, 72*, 725–731. Available from https://doi.org/10.1111/j.1469-1809.2008.00474.x.

Zuellig, R. A., Hornemann, T., Othman, A., Hehl, A. B., Bode, H., Güntert, T., et al. (2014). Deoxysphingolipids, novel biomarkers for type 2 diabetes, are cytotoxic for insulin-producing cells. *Diabetes, 63*, 1326–1339. Available from https://doi.org/10.2337/db13-1042.

CHAPTER

13

The role of neurosensory systems in the modulation of aging

Guang Yang[1], *Yi Sheng*[1] *and Rui Xiao*[1,2,3]

[1]Department of Aging and Geriatric Research, Institute on Aging, University of Florida, Gainesville, FL, United States [2]Department of Pharmacology and Therapeutics, College of Medicine, University of Florida, Gainesville, FL, United States [3]Center for Smell and Taste, University of Florida, Gainesville, FL, United States

OUTLINE

Introduction

Aging, as defined by the age-related progressive decline in intrinsic physiology and survival rate, is modulated by both genetic and environmental factors (Lopez-Otin, Blasco, Partridge, Serrano, & Kroemer, 2013). In the past three decades, genetic studies in various model organisms, such as yeast, *Caenorhabditis elegans*, *Drosophila*, and mice, have revealed multiple evolutionarily conserved genetic pathways in longevity regulation (Finch & Ruvkun, 2001; Kenyon, 2010; Riera, Merkwirth, De Magalhaes Filho, & Dillin, 2016). However, less is known about how environmental factors modulate aging. Since population genetic studies suggest that only a small portion (<10%) of human lifespan is genetically heritable (Herskind et al., 1996; Ruby et al., 2018), a better understanding on how environmental factors modulate aging becomes particularly important for both scientific and practical purposes.

Most, if not all, multicellular organisms have evolved highly effective neurosensory systems to recognize, assess, and respond to distinct environmental stimuli, which give rise to the five basic senses (touch, taste, smell, sight, and hearing) as originally described by Aristotle (Sorabji, 1970). Meanwhile, thermoception, nociception, and proprioception are also widely present in many species. In addition to these acute sensations, recent studies demonstrate that neurosensory systems can transduce various environmental sensory cues into many prolonged physiological outputs, including aging. For example, pioneering studies from the Kenyon group showed that sensory perception can actively regulate the lifespan of *C. elegans* (Apfeld & Kenyon, 1999).

Handbook of the Biology of Aging.
DOI: https://doi.org/10.1016/B978-0-12-815962-0.00013-5

Specifically, suppression of genes required for the formation of certain sensory organs or for the transduction of sensory signals not only restricts the organism from responding properly to environmental stimuli, but also alters its lifespan (Alcedo & Kenyon, 2004; Apfeld & Kenyon, 1999). These important studies opened up a new avenue of research into neurosensory modulation of aging. Subsequent studies in other species such as *Drosophila* (Kapahi et al., 2004; Libert et al., 2007) and mice (Kappeler et al., 2008; Riera et al., 2014; Scherer et al., 2011) also support a critical role of neurosensory systems in aging and longevity. Very importantly, the cellular and molecular mechanisms underlying neurosensory modulation of aging often exhibit remarkable conceptual similarities across taxa (Finch & Ruvkun, 2001; Fontana, Partridge, & Longo, 2010; Kenyon, 2010; Riera et al., 2016), indicating the evolutionary conservation of neurosensory modulation of aging.

In this chapter, we integrate recent research on neurosensory modulation of aging and discuss the remaining questions in the understanding of regulatory pathways. Because many mechanistic studies of neurosensory modulation of aging were initially conducted in model organisms such as *C. elegans* and *Drosophila*, we focus our discussion on these organisms.

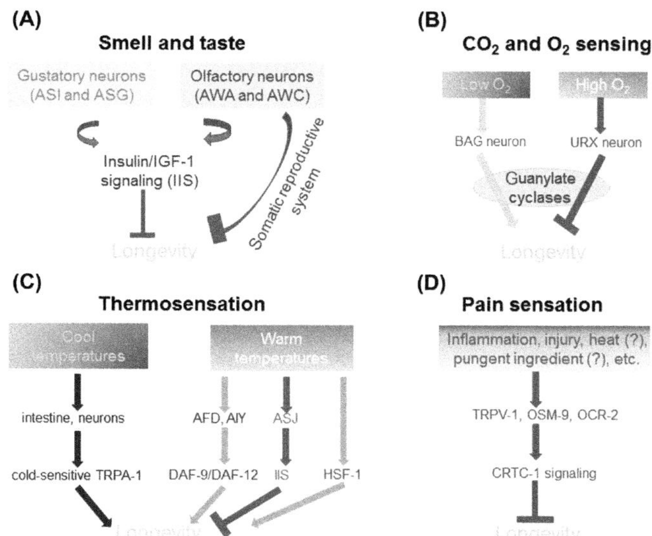

FIGURE 13.1 **Major sensations that modulate aging.** (A) Gustatory and olfactory neurons may suppress longevity with distinct mechanisms. For example, the *C. elegans* gustatory ASI and ASG neurons act through the IIS pathway while the olfactory AWA and AWC neurons can interact with the somatic reproductive system to modulate aging. (B) The BAG (low O_2-sensing) and URX neurons (high O_2-sensing) mediate O_2-modulated lifespan in *C. elegans*. (C) Multiple sensory neurons and signaling pathways are involved in temperature modulation of longevity. (D) Nociceptive TRP channels (TRPV1, OSM-9, and OCR-2) couple pain sensation with aging in mice and worms.

Sensations that regulate animal lifespan

Animals use various sensations to detect distinct environmental factors, some of which are known to modulate aging. Below we discuss various sensations that have been studied to have a great impact on animal aging (Fig. 13.1).

Smell and taste

Smell and taste are fundamentally important sensations present in nearly all animal species. On the one hand, smell and taste allow animals to sense nutrient and mating signals in the environment. On the other hand, repulsive odors and tastants are commonly associated with predators and environmental toxins (Kandel, 2013). The perception of nutrients often requires simultaneous stimulation of olfactory and gustatory organs by food-derived stimuli to form a multimodal sensory experience (Breslin, 2013). Distinct gustatory and olfactory neurons, which perceive food-derived chemical signals, can differentially regulate lifespan using different mechanisms (Fig. 13.1A) (Alcedo & Kenyon, 2004; Ostojic et al., 2014). An early study in *C. elegans* has shown that laser ablation of either gustatory ASI and ASG neurons or olfactory AWA and AWC neurons significantly extends lifespan (Alcedo & Kenyon, 2004). Interestingly, the gustatory and olfactory neurons appear to affect lifespan in parallel since the loss of olfactory AWA and AWC neurons can further extend lifespan of worms lacking the gustatory ASI neurons, suggesting that smell and taste might modulate aging via different mechanisms. Indeed, genetic studies indicate that gustatory neurons regulate lifespan through the insulin/IGF-1 signaling (IIS) pathway, while olfactory neurons may interact with the somatic reproductive system to modulate aging (Alcedo & Kenyon, 2004).

How do food-derived sensory cues modulate aging? Mechanistic studies in *C. elegans* showed that the activation of a subset of chemosensory neurons (i.e., ASI and ASJ) by food promotes the expression of insulin-like peptides (ILPs) INS-6 and DAF-28, which then inhibit the FOXO family transcription factor DAF-16 and suppress lifespan (Artan et al., 2016). By contrast, food restriction suppresses INS-6 and DAF-28 (Artan et al., 2016), which might partially explain the beneficial effect of dietary restriction in delaying aging. Although ILPs may relay longevity signals downstream of smell and taste, it remains unclear how the activation of chemosensory neurons triggers the elevated secretion of ILPs. In simple organisms like *C. elegans*, the sensory neuron itself may secret ILPs. In higher species, intermediate neuroendocrine cells might coordinate with sensory neurons during chemosensory modulation of aging.

Smell and taste also regulate lifespan in *Drosophila*. In dietary-restricted flies, which are long-lived compared to well-fed ones, odorant exposure by itself is sufficient to shorten lifespan (Libert et al., 2007). Furthermore, although flies with mutant *Or83b* odorant receptor display overall normal metabolic rates, their lifespan is extended and the triglyceride storage is significantly increased (Libert et al., 2007). This finding indicates that the sense of food, apart from its nutritional value, can directly modulate aging. Similar to *C. elegans*, exposure to odorants induces the upregulation of ILPs and the glucagon-like hormone adipokinetic hormone (AKH) in *Drosophila* (Lushchak, Carlsson, & Nässel, 2015), and this odorant-induced endocrine signaling might accelerate aging. Notably, chemosensory inputs do not always shorten lifespan. For example, the sweet taste receptor *Gr5a* seems to promote longevity, as its mutant is short-lived (Ostojic et al., 2014). Thus, different chemosensory inputs may differentially affect lifespan through multiple mechanisms.

In addition to food-derived sensory cues, sexual pheromones can modulate aging through smell and taste. Mating significantly accelerates aging in *C. elegans* hermaphrodites through multiple mechanisms (Maures et al., 2014; Shi & Murphy, 2014; Shi, Runnels, & Murphy, 2017). Intriguingly, male-secreted pheromones can shorten lifespan of hermaphrodites without mating in *C. elegans* and other species of nematode (Maures et al., 2014), implying that pheromone sensation may play a role in the trade-off between reproduction and aging. The next step to characterize the underlying mechanism of pheromone modulation of aging in *C. elegans* will require the identification of sensory neurons and chemoreceptors involved in pheromone detection. Similar to *C. elegans*, pheromone-mediated sexual perception also greatly affects *Drosophila* physiology and longevity. Specifically, female sexual pheromones suppress fat storage and longevity in male flies through a gustatory receptor, ppk23, and neuropeptide F signaling (Gendron et al., 2014).

A recent study in mice showed that conditional ablation of mature olfactory sensory neurons protects against high-fat diet-induced obesity and promotes overall metabolic health (Riera et al., 2017). Mechanistically, the reduced olfactory input elevates sympathetic nerve activity, which then activates β-adrenergic receptors in fat tissues and promotes lipolysis (Riera et al., 2017). It would be very interesting to examine whether the loss of smell promotes longevity in these mice.

Carbon dioxide and oxygen sensing

Carbon dioxide (CO_2) and oxygen (O_2) are important ambient cues for many species, especially insects. For example, mosquitoes rely on CO_2 in host-seeking (Takken & Knols, 1999), ants use CO_2 as a nest-plume component to avoid entering the wrong nest (Buehlmann, Hansson, & Knaden, 2012), and flies, more interestingly, find CO_2 attractive only during food-searching (Faucher, Forstreuter, Hilker, & Bruyne, 2006; Van Breugel, Huda, & Dickinson, 2018) while most of the time they inherently eschew high levels of CO_2 (Jones, Cayirlioglu, Grunwald Kadow, & Vosshall, 2006; Kwon, Dahanukar, Weiss, & Carlson, 2007; Suh et al., 2004). Although the sense of CO_2 and O_2 shares some similarities with smell, the signal transduction pathways for CO_2 and O_2-sensing are often different from classical chemosensation.

The low O_2 hypoxic condition significantly extends lifespan in *C. elegans* (Adachi, Fujiwara, & Ishii, 1998; Leiser, Fletcher, Begun, & Kaeberlein, 2013). Very interestingly, ablation of BAG sensory neurons that detect low levels of O_2 increases the *C. elegans* lifespan, whereas ablation of URX sensory neurons that detect high levels of O_2 decreases lifespan (Fig. 13.1B; Liu & Cai, 2013). Mechanistically, various guanylate cyclases, including GCY-31, GCY-33, and GCY-35, may act as O_2 sensors in the BAG and URX neurons to regulate lifespan (Liu & Cai, 2013). Nevertheless, the downstream signaling pathway of O_2-sensitive guanylate cyclases in longevity modulation remains unclear.

Low O_2 can also modulate aging through the transcription factor hypoxia-inducible factor 1 (HIF-1), as stabilization of HIF-1 leads to a significant increase in *C. elegans* lifespan (Mehta et al., 2009; Müller et al., 2009). However, there are also reports that *hif-1* loss-of-function mutations increase lifespan in normal laboratory culture conditions (Chen et al., 2009; Zhang et al., 2009). Thus, HIF-1 may function as both a positive and negative modulator of aging depending on the cellular context (Hwang & Lee, 2011; Hwang et al., 2015; Leiser & Kaeberlein, 2010; Mishur et al., 2016; Shamalnasab et al., 2017).

CO_2-sensitive neurons are widely present in many insect species. In *Drosophila*, the seven-transmembrane-domain chemoreceptors Gr21a and Gr63a are coexpressed in CO_2-sensitive olfactory neurons, where they act as the CO_2 sensor (Kwon et al., 2007). Intriguingly, the loss-of-function mutation of Gr63a prolongs *Drosophila* lifespan and augments its resistance toward oxidative stress (Poon, Kuo, Linford, Roman, & Pletcher, 2010). As Gr63a is expressed in a well-defined subset of sensory neurons, further mechanistic studies may help unravel the neurosensory mechanism underlying the CO_2 sensing-modulated longevity in *Drosophila*.

Thermosensation

Among various environmental factors that modulate aging, temperature has long been recognized for its significant role in affecting the lifespan of poikilotherms.

More than a century ago, an apparent inverse relationship between ambient temperatures and lifespan was observed in cold-blooded animals (Loeb & Northrop, 1916). More recently, homeothermic rodents were also reported to exhibit extended lifespan or delayed aging phenotypes upon a mild reduction in environmental temperatures (Holloszy & Smith, 1986) or core body temperatures (Conti et al., 2006; Hauck, Hunter, Danilovich, Kopchick, & Bartke, 2001; Hunter, Croson, Bartke, Gentry, & Meliska, 1999). Furthermore, a lower core body temperature is correlated with longer lifespan expectancy in humans (Roth et al., 2002). Thus, temperature appears to modulate aging in both poikilotherms and homeotherms. Traditionally, the phenomenon of temperature modulation of aging was attributed to passive thermodynamics (Lints, 1989). However, as all other aging-related paradigms are actively regulated by the interactions between genetic and environmental factors, one would argue that genetic programs should also play an active role in temperature modulation of aging.

The fundamental logic of thermosensation is conserved across species. Many temperature-sensitive neurons and ion channels (e.g., TPR channels) act as cellular and molecular thermometers with distinct temperature activation thresholds (Vriens, Nilius, & Voets, 2014). In *C. elegans*, several thermosensory neurons, including AFD, AWC, ASJ, and ASI, respond to ambient temperature changes (Aoki & Mori, 2015; Beverly, Anbil, & Sengupta, 2011; Biron, Wasserman, Thomas, Samuel, & Sengupta, 2008; Kuhara et al., 2008; Mori & Ohshima, 1995; Wang, O'Halloran, & Goodman, 2013). In a previous study, Lee and Kenyon showed that ablation of the AFD neuron or its downstream interneuron AIY results in a reduced lifespan at warm temperatures (Lee & Kenyon, 2009), suggesting that these two neurons promote longevity at warm temperatures. Mechanistically, AFD may modulate aging at warm temperatures through CMK-1/CaMKI (calcium/calmodulin-dependent kinase I)-dependent phosphorylation of CRH-1/CREB (cyclic AMP-responsive element binding protein). The activation of CRH-1/CREB then induces the secretion of a FMRFamide neuropeptide FLP-6, which targets the AIY neuron and engages DAF-9 sterol hormone signaling in lifespan regulation (Chen et al., 2016). Notably, although *C. elegans* and many other animals exhibit shortened lifespan at warm temperatures, the heat-activated AFD neuron maintains rather than shortens lifespan, suggesting that other thermosensitive neurons may function to shorten lifespan at warm temperatures. Indeed, higher temperatures activate another thermosensitive neuron, ASJ, and trigger the release of ILPs INS-6 and DAF-28, which then shorten lifespan through the IIS pathway (Zhang et al., 2018). On the other hand, overexpression of heat-shock transcription factor HSF-1 and its downstream chaperones extends lifespan in yeast (Ohtsuka, Azuma, Murakami, & Aiba, 2011) and *C. elegans* (Hsu, Murphy, & Kenyon, 2003; Morley & Morimoto, 2004). Furthermore, a p23 co-chaperone protein DAF-41 plays an important role in lifespan modulation at warm temperatures by inhibiting HSF-1 (Horikawa, Sural, Hsu, & Antebi, 2015). Thus, higher temperatures modulate aging through multiple mechanisms (Fig. 13.1C).

Opposite to the warm temperature-triggered antilongevity effect, cool temperatures extend lifespan in many species. A loss-of-function mutation in *trpa-1*, a cold-sensitive TRP channel in *C. elegans*, shortens lifespan at lower but not warmer temperatures (Xiao et al., 2013), indicating that TRPA-1 mediates the lower temperature-triggered lifespan extension. Indeed, by acting in both sensory neurons (i.e., IL1) and intestine, TRPA-1 activates the FOXO transcription factor DAF-16 through Ca^{2+}, Ca^{2+}-sensitive protein kinase C PKC-2, and serum/glucocorticoid regulated kinase SGK-1 to extend lifespan during adulthood (Fig. 13.1C) (Xiao et al., 2013; Zhang et al., 2018). Surprisingly, unlike in adulthood, cool temperatures shorten lifespan during developmental stages (Zhang et al., 2015). Again, this life stage-dependent temperature effect on longevity also involves TRPA-1 (Zhang et al., 2015), supporting the important role of thermosensitive TRP channels in temperature modulation of longevity.

Pain sensation

Noxious environmental stimuli often trigger pain sensation to alert organisms of potential impairments. In vertebrates, pain is frequently found to be accompanied with many types of injury and inflammatory diseases. Remarkably, knocking out a pain receptor TRPV1 (activated by capsaicin, noxious heat, proton, and several other noxious stimuli) promotes longevity in mice (Riera et al., 2014). Moreover, genetic disruption of *osm-9* and *ocr-2* that encode two homologs of TRPV1 in *C. elegans* also extends lifespan (Fig. 13.1D) (Riera et al., 2014), implying the evolutionary conservation of pain receptors in suppressing longevity.

TRPV1 is well established to play essential roles in nociception and pain sensation (Caterina et al., 2000; Negri et al., 2006; Špicarová & Paleček, 2008; Wang et al., 2018). Moreover, it regulates diet- and aging-induced obesity, insulin resistance, and leptin resistance (Lee et al., 2015; Motter & Ahern, 2008). Intriguingly, a recent study showed that oxytocin, a hormone that suppresses nociception of inflammatory pain, serves as an endogenous agonist for TRPV1 and desensitizes the channel (Nersesyan et al., 2017). As oxytocin has been suggested to have an antiaging effect (Elabd et al., 2014), it might modulate aging through TRPV1-mediated pain sensation. On the other hand, since TRPV1 is also a heat-activated ion channel, it would be very interesting to examine whether high temperatures shorten lifespan in mice and, if so, what the role of TRPV1 is in this process.

Light sensation

The ability to sense light is present in most organisms, including the eyeless *C. elegans* (Edwards et al., 2008; Ward, Liu, Feng, & Xu, 2008). Light perception triggers a host of responses that contribute to the regulation of photosynthesis, circadian rhythms, locomotor activity, and growth (Burgess & Granato, 2007; Legates, Fernandez, & Hattar, 2014; Peers et al., 2009; Poorter & Nagel, 2000). In *C. elegans*, ASJ, AWB, ASK, and several other sensory neurons are light sensitive and their phototransduction requires a light receptor, LITE-1, and cyclic guanosine monophosphate (cGMP)-sensitive cyclic nucleotide-gated (CNG) channels (Liu et al., 2010). Interestingly, visible light reduces *C. elegans* lifespan through light-induced oxidative stress (De Magalhaes Filho et al., 2018). Furthermore, evidence of sunlight modulation of lifespan has also been found in humans (Lowell & Davis Jr, 2008), indicating the conserved effect of light radiation on animal physiology and lifespan.

In addition to the direct impact of light radiation on aging, light can indirectly modulate aging through entraining circadian rhythms. "Circadian rhythm" refers to the endogenous clock that is synchronized to several environmental signals, including the light and dark cycle. When the effect of three different light regimes (constant light, alternating light—dark cycles, and constant darkness) on the lifespan of *Drosophila* was examined, a shortened lifespan was observed in flies under constant light (Sheeba, Sharma, Shubha, Chandrashekaran, & Joshi, 2000). By contrast, overexpression of two core circadian clock genes, *tim* and *per*, leads to enhanced fat metabolism and extended lifespan in *Drosophila* on an ad libitum (AL) diet (Katewa et al., 2016). Consistent with the effect of the circadian clock on fly aging, a profound impact of disrupted circadian cycles on the healthspan and lifespan of mice and humans was also observed (Dubrovsky, Samsa, & Kondratov, 2010; Haus & Smolensky, 2013; Leng, Musiek, Hu, Cappuccio, & Yaffe, 2019; Zelinski, Deibel, & Mcdonald, 2014; Zitting et al., 2018).

Signaling pathways involved in neurosensory modulation of aging

Insulin/IGF-1 signaling

Pioneering genetic studies demonstrated that single gene mutations of *age-1* and *daf-2*, which encode homologs of phosphatidylinositol 3-kinase and insulin receptor, respectively, can extend lifespan in *C. elegans* (Friedman & Johnson, 1988; Kenyon, Chang, Gensch, Rudner, & Tabtiang, 1993). Subsequent epistasis analysis indicated that these genes are components of the IIS pathway (Fig. 13.2A; Dorman, Albinder, Shroyer, & Kenyon, 1995; Finch & Ruvkun, 2001; Kenyon, 2010; Riera et al., 2016). As the first major signaling pathway identified in lifespan regulation, the genetic components of IIS pathway are remarkably conserved from worms to humans (An, Artan, Park, Altintas, & Lee, 2017; Barbieri, Bonafè, Franceschi, & Paolisso, 2003; Giannakou & Partridge, 2007; Van Heemst, 2010; Van Heemst et al., 2005). Although the IIS pathway does not exist in the unicellular yeast, precursors of the pathway act in the glucose/nutrient-sensing signaling cascade, which is closely connected to both replicative and chronological aging in yeast (Barbieri et al., 2003; Parrella & Longo, 2010).

The IIS pathway is highly sensitive to cellular nutrient status. Under favorable growth conditions, activation of the IIS is initiated by the binding of agonist ILPs to DAF-2, the sole insulin receptor in *C. elegans*. Thereafter, signals are transduced through a complex genetic network and consequently lead to phosphorylation, nucleus-to-cytosol translocation, and suppression of the FOXO transcription factor DAF-16, a major downstream effector of the nutrient-sensing IIS pathway (Altintas, Park, & Lee, 2016; Finch & Ruvkun, 2001; Kenyon, 2010; Riera et al., 2016). Under cellular stress conditions, reduction of IIS results in the translocation of DAF-16 from cytosol into nucleus, where the transcription factor regulates a host of target genes involved in stress resistance, pathogen

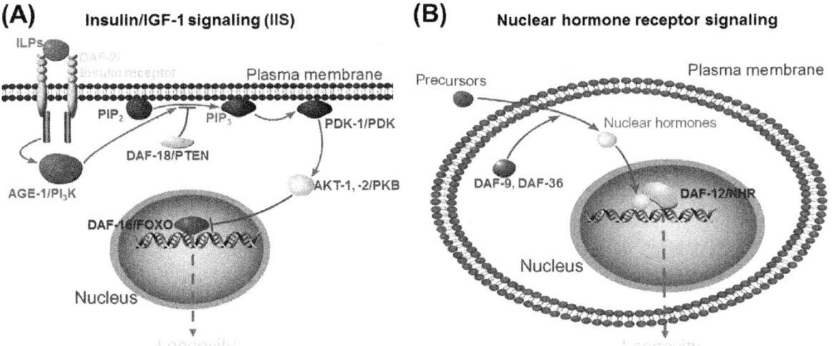

FIGURE 13.2 Signaling pathways involved in sensory modulation of aging. (A) The evolutionarily conserved insulin/IGF-1 signaling (IIS) pathway integrates multiple sensory perceptions (e.g., chemosensation, thermosensation, nociception) in neurosensory modulation of aging. Shown are the major components of IIS in *C. elegans*. (B) The DAF-12 nuclear hormone receptor signaling transduces thermosensory and chemosensory signals in aging.

defense, and longevity (Lee, Kennedy, Tolonen, & Ruvkun, 2003; Murphy et al., 2003). Besides the key downstream player DAF-16, other transcription factors also act in the IIS pathway to modulate aging. For example, HSF-1 shares some target genes with DAF-16 and promotes longevity under stress conditions, such as high temperatures (Baird et al., 2014; Chiang, Ching, Lee, Mousigian, & Hsu, 2012). Moreover, the Nrf2 transcription factor SKN-1 can act downstream of IIS to modulate aging and oxidative stress responses (Park, Tedesco, & Johnson, 2009; Tullet et al., 2008).

As a major aging-related signaling pathway, IIS integrates multiple neurosensory inputs in lifespan regulation (Apfeld & Kenyon, 1999; Artan et al., 2016; Chen et al., 2016; Delaney, Chen, Graniel, Dumas, & HU, 2017; Lin, Hsin, Libina, & Kenyon, 2001; Xiao et al., 2013). In general, after perceiving ambient stimuli, sensory neurons directly or indirectly (through interneurons) modulate the secretion of ILPs, which then act as autocrine or endocrine signals to modulate aging through the IIS pathway (Artan et al., 2016; Lin, He, Huang, & Pan, 2017). For example, sensory cilia-disrupted worms, which are defective in chemosensation, mechanosensation, and thermosensation, are generally long-lived (Apfeld & Kenyon, 1999). However, the loss-of-function mutation of *daf-16* largely suppresses the long lifespan observed in these worms, supporting the key role of IIS in neurosensory modulation of aging (Alcedo & Kenyon, 2004; Apfeld & Kenyon, 1999).

C. elegans has a compact nervous system (302 neurons in adult hermaphrodites) and many sensory neurons are polymodal neurons (Riddle, 1997). A single sensory neuron may have a big impact on *C. elegans* aging. For example, the ASJ neuron is involved in chemosensation, light sensation, and thermosensation. Interestingly, disruption of ASJ promotes the nuclear translocation of DAF-16 and this effect can be partially suppressed by food cue exposure (Artan et al., 2016). Mechanistically, food cues activate ASJ and trigger the secretion of two ILPs, ISN-6 and DAF-28, which subsequently act on DAF-16 to suppress longevity through IIS (Artan et al., 2016; Ohta, Ujisawa, Sonoda, & Kuhara, 2014). In contrast, the chemosensory ASI neuron mediates the dietary restriction-induced longevity through SKN-1, which is phosphorylated by several kinase components of IIS, including AKT-1, AKT-2, and SGK-1 (Bishop & Guarente, 2007; Tullet et al., 2008). Furthermore, the thermosensory neuron AFD prolongs lifespan at warm temperatures via FLP-6, which inhibits the ILP INS-7 and promotes DAF-16 activity (Chen et al., 2016).

Nuclear hormone receptor signaling

In *C. elegans*, a hormonal signaling pathway consisting of DAF-9/cytochrome P450, DAF-12/nuclear hormone receptor, and DAF-36/Rieske-like oxygenase regulates larval development and adult lifespan (Fig. 13.2B; Gerisch et al., 2007; Gerisch, Weitzel, Kober-Eisermann, Rottiers, & Antebi, 2001; Jia, Albert, & Riddle, 2002). In this pathway, DAF-9 integrates upstream inputs and produces or degrades a hormone that regulates DAF-12 transcriptional complexes. DAF-12, the downstream transcription factor, promotes reproductive growth in the ligand-bound state while mediating dauer arrest in the absence of ligand. The DAF-9/DAF-12 signaling pathway is used by the thermosensory AFD neuron to maintain lifespan at warm temperatures. Specifically, the heat-activated AFD neuron promotes the expression of DAF-9, which in turn affects lifespan by raising the level of sterol ligand that inhibits DAF-12 (Lee & Kenyon, 2009). In addition, DAF-36 regulates DAF-12 transcriptional activity by producing a bile acid-like steroid, 7-dehydrocholesterol, which regulates developmental timing and longevity (Wollam et al., 2011). Lastly, dietary restriction influences DAF-9/DAF-12 signaling-dependent germline plasticity and longevity by inducing a steroid hormone, Δ^7-dafachronic acid (DA) (Thondamal, Witting, Schmitt-Kopplin, & Aguilaniu, 2014).

In insects, the corpus allatum secretes a group of juvenile hormones (JHs), which modulate insect development, reproduction, diapause, and aging through both membrane and nuclear hormone receptors (Jindra, Palli, & Riddiford, 2013). Recently, two bHLH-Pas transcription factors, Met and SRC, have been proposed to act as the nuclear receptors of JHs (Zhang, Xu, Sheng, Sui, & Palli, 2011). The release of JHs is regulated by sensory inputs. For example, tactile cue-mediated social interactions accelerate female reproduction by stimulating the production of JHs in the cockroach (Uzsak & Schal, 2012; Uzsak & Schal, 2013). Notably, JHs have been suggested as an important modulator of the trade-off between reproduction and longevity, since JHs shorten lifespan in *Drosophila* (Flatt & Kawecki, 2007; Yamamoto, Bai, Dolezal, Amdam, & Tatar, 2013). Taken together, nuclear hormone receptor signaling may integrate sensory cues with reproduction and aging.

Opportunities and challenges in studying neurosensory modulation of aging

A common scheme for sensory perception is that, by using distinct membrane receptors and ion channels, specialized sensory neurons can detect various chemical and physical cues that originate from the environment. The activation of these sensory neurons then converts environmental cues into electric signals, which eventually trigger the release of various neurotransmitters, neuropeptides, and hormones. In this manner, sensory receptors and ion channels can have long-term effects on aging. Consistent with this notion, multiple receptors and ion channels involved in sensory perception have been shown to modulate aging (Sheng, Tang, Kang, & Xiao, 2017). For instance, mutations in TAX-2 and TAX-4 CNG channels (involved in chemosensation, thermosensation, and light sensation) extend lifespan in *C. elegans* (Artan et al., 2016; Lee & Kenyon, 2009). A similar effect is also found with the nociceptive TRPV1, OSM-9, and OCR-2 channels that are involved in thermosensation and pain sensation (Lee & Ashrafi, 2008; Riera et al., 2014). Because sensory receptors and ion channels largely determine the specificity of sensory perception, a systematic survey of the potential impact of these components on aging may provide novel insights into neurosensory modulation of aging.

Organismal aging is a systemic process. Distinct neurotransmitters, neuropeptides, and hormones may participate in the coordination of aging processes among different tissues and organs. A thorough understanding of how these secreted factors modulate longevity should provide key information on organismal aging. Notably, although the functions and mechanisms of several classical neurotransmitters and hormones are relatively well studied in sensory perception, much less is known about how distinct neuropeptides (over 250 distinct neuropeptides have been identified in *C. elegans*) modulate sensory perception and aging (Li & Kim, 2008; Nusbaum, Blitz, & Marder, 2017). Given that the biosynthetic processes of many neurotransmitters, neuropeptides, and hormones are evolutionarily conserved, examining the effect of distinct mutant animals with defective signaling of these molecules on aging should be very informative. As *C. elegans* features a short lifespan and rich mutant collection, we propose that these studies could be initially conducted in worms.

With the development of multiple advanced anatomical, optical, microelectronic, and optogenetic techniques, a major goal of the "Brain Initiative" is to dissect structural and functional connectomes of the nervous system (Insel, Landis, & Collins, 2013). Although *C. elegans* has the simplest nervous system among all multicellular model organisms, most neurosensory modalities are present in worms. Importantly, *C. elegans* is the only model organism in which the complete structural connectome has been established through serial electron microscopy (White, Southgate, Thomson, & Brenner, 1986). Therefore we see great potential in the functional characterization of the *C. elegans* connectome in neurosensory modulation of aging. As *C. elegans* has a well-established cellular lineage, transparent anatomy and facile genetics, tissue- and cell-specific manipulations are readily available for *C. elegans*. Optogenetic and/or chemogenetic activation and silencing of distinct components of *C. elegans* neurosensory system should help reveal the circuitry mechanism of neurosensory modulation of aging. Furthermore, tissue- and cell-specific gene disruption using CRISPR/Cas9 and RNA interference (RNAi) techniques would facilitate the verification of key signaling molecules in sensory regulation of longevity.

Due to various technical and ethical issues, mechanistic investigation of neurosensory modulation of aging is currently challenging in higher species. As many genes and signaling pathways involved in both sensory perception and aging are evolutionarily conserved across taxa, functional studies of neurosensory systems in the aging processes of model organisms such as *C. elegans* and *Drosophila* would deepen our understanding on the role of neurosensory systems in the modulation of aging.

Concluding remarks

Traditionally, sensory perception is considered to be an acute neuronal response, whereas aging is a chronic process. As a result, neurosensory modulation of aging was not well appreciated before. However, genetic studies in model organisms such as *C. elegans* and *Drosophila* have clearly demonstrated that a variety of types of sensory inputs can greatly modulate organismal aging. These pioneering findings pave the way for further mechanistic studies of neurosensory modulation of aging.

Neurosensory systems help organisms recognize distinct environmental cues and elicit appropriate physiological responses. In the past two decades, mechanistic studies have revealed several evolutionarily conserved pathways acting downstream of neurosensory modulation of aging. Notably, these downstream pathways also regulate other physiological processes besides aging. For example, although the nutrient-sensing IIS and mechanistic target of rapamycin (mTOR) pathways have a great impact on longevity, they also play essential roles in global mRNA translation, protein homeostasis, and mitochondrial function (Kaeberlein, 2013; Pitt & Kaeberlein, 2015). Therefore manipulating these downstream signaling molecules may affect different aspects of cellular physiology. To take advantage of these pathways in delaying aging, targeting the upstream sensory receptors or ion channels might achieve the best specificity.

Manipulating sensory neurons or components of sensory signaling pathways in model organisms significantly influences aging by affecting the perception of various external and internal stimuli, including food availability, temperature, oxidative stress, reproduction, and so on (Table 13.1). Nevertheless, our current understanding of sensory modulation of aging is far from complete. As the vast majority of evidence on

TABLE 13.1 Neurosensory modulation of aging in model organisms.

Sensation	Neuron/gene	Means of manipulation	Effect on lifespan	Signaling pathway	References
Caenorhabditis elegans					
Gustatory	ASI, ASG	Neuron ablation	Extend	IIS pathway	Alcedo & Kenyon, 2004
Olfactory	AWA, AWC	Neuron ablation	Extend	IIS pathway	Alcedo & Kenyon, 2004
CO_2 and O_2	BAG, URX, hif-1	Neuron ablation (BAG, URX); loss-of-function mutation (hif-1)	Extend (BAG); shorten (URX); both (hif-1)		Liu & Cai, 2013; Mehta et al., 2009; Zhang et al., 2009
Temperature	AFD, AWC, ASJ, ASI, AIY, hsf-1, trpa-1	Neuron ablation (AFD, AIY, ASJ); overexpression (hsf-1); loss-of-function mutation (trpa-1)	Shorten (AFD, AIY); extend (hsf-1); extend at higher temperatures (ASJ); shorten at lower temperatures (trpa-1)	DAF-9 hormone signaling (AFD, AIY), IIS pathway (ASJ)	Hsu et al., 2003; Xiao et al., 2013; Zhang et al., 2018
Pain	osm-9 and ocr-2	Loss-of-function mutation	Extend	CRTC/CREB signaling	Riera et al., 2014
Light	ASJ, AWB, ASK		Shorten	Oxidative stress	Liu et al., 2010 De Magalhaes Filho et al., 2018
Drosophila					
Gustatory	Gr5a, ppk23	Loss-of-function mutation	Shorten		Ostojic et al., 2014
Olfactory	Or83b	Loss-of-function mutation	Extend		Libert et al., 2007
CO2 and O2	Gr21a, Gr63a	Loss-of-function mutation	Extend		Poon et al., 2010
Mice					
Pain	TRPV1	Knockout	Extend	CRTC/CREB signaling	Riera et al., 2014

neurosensory regulation of longevity has been gathered from invertebrates, future work is required to verify whether mammalian sensory systems can also affect aging in a similar manner.

Acknowledgments

We thank members of the Xiao lab for discussion and insights. Work related to this topic was supported by grants from the US National Institutes of Health (AG063766 and AG028740), the American Cancer Society (RSG-17-171-01-DMC), the American Federation for Aging Research (AFAR), UF Older Americans Independence Center, Southeast Center for Integrated Metabolomics, and UF Center for Smell and Taste.

References

Adachi, H., Fujiwara, Y., & Ishii, N. (1998). Effects of oxygen on protein carbonyl and aging in *Caenorhabditis elegans* mutants with long (age-1) and short (mev-1) life spans. *The Journals of Gerontology Series A: Biological Sciences and Medical Sciences, 53*, B240–B244.

Alcedo, J., & Kenyon, C. (2004). Regulation of C. elegans longevity by specific gustatory and olfactory neurons. *Neuron, 41*, 45–55.

Altintas, O., Park, S., & Lee, S. J. (2016). The role of insulin/IGF-1 signaling in the longevity of model invertebrates, C. elegans and D. melanogaster. *BMB Reports, 49*, 81–92.

An, S. W. A., Artan, M., Park, S., Altintas, O., & Lee, S.-J. V. (2017). *Longevity regulation by insulin/IGF-1 signalling* In: A. Olsen, & M.D. Gill (Eds.), Ageing: Lessons from C. elegans (pp. 63–81). Springer.

Aoki, I., & Mori, I. (2015). Molecular biology of thermosensory transduction in C. elegans. *Current Opinion in Neurobiology, 34*, 117–124.

Apfeld, J., & Kenyon, C. (1999). Regulation of lifespan by sensory perception in *Caenorhabditis elegans*. *Nature, 402*, 804.

Artan, M., Jeong, D.-E., Lee, D., Kim, Y.-I., Son, H. G., Husain, Z., ... Alcedo, J. (2016). Food-derived sensory cues modulate longevity via distinct neuroendocrine insulin-like peptides. *Genes & Development, 30*, 1047–1057.

Baird, N. A., Douglas, P. M., Simic, M. S., Grant, A. R., Moresco, J. J., Wolff, S. C., ... Dillin, A. (2014). HSF-1–mediated cytoskeletal integrity determines thermotolerance and life span. *Science, 346*, 360–363.

Barbieri, M., Bonafè, M., Franceschi, C., & Paolisso, G. (2003). Insulin/IGF-I-signaling pathway: An evolutionarily conserved mechanism of longevity from yeast to humans. *American Journal of Physiology-Endocrinology and Metabolism, 285*, E1064–E1071.

Beverly, M., Anbil, S., & Sengupta, P. (2011). Degeneracy and neuromodulation among thermosensory neurons contribute to robust thermosensory behaviors in *Caenorhabditis elegans*. *Journal of Neuroscience, 31*, 11718–11727.

Biron, D., Wasserman, S., Thomas, J. H., Samuel, A. D., & Sengupta, P. (2008). An olfactory neuron responds stochastically to temperature and modulates *Caenorhabditis elegans* thermotactic behavior. *Proceedings of the National Academy of Sciences, 105*, 11002–11007.

Bishop, N. A., & Guarente, L. (2007). Two neurons mediate diet-restriction-induced longevity in *C. elegans*. *Nature, 447*, 545.

Breslin, P. A. (2013). An evolutionary perspective on food and human taste. *Current Biology*, *23*, R409–R418.

Buehlmann, C., Hansson, B. I. L. L. S., & Knaden, M. (2012). Path integration controls nest-plume following in desert ants. *Current Biology*, *22*, 645–649.

Burgess, H. A., & Granato, M. (2007). Modulation of locomotor activity in larval zebrafish during light adaptation. *Journal of Experimental Biology*, *210*, 2526–2539.

Caterina, M. J., Leffler, A., Malmberg, A., Martin, W., Trafton, J., Petersen-Zeitz, K., ... Julius, D. (2000). Impaired nociception and pain sensation in mice lacking the capsaicin receptor. *Science*, *288*, 306–313.

Chen, Y.-C., Chen, H.-J., Tseng, W.-C., Hsu, J.-M., Huang, T.-T., Chen, C.-H., & Pan, C.-L. (2016). A C. elegans thermosensory circuit regulates longevity through crh-1/CREB-dependent flp-6 neuropeptide signaling. *Developmental Cell*, *39*, 209–223.

Chen, D., Thomas, E. L., & Kapahi, P. (2009). HIF-1 modulates dietary restriction-mediated lifespan extension via IRE-1 in Caenorhabditis elegans. *PLoS Genet*, *5*(5), e1000486

Chiang, W.-C., Ching, T.-T., Lee, H. C., Mousigian, C., & Hsu, A.-L. (2012). HSF-1 regulators DDL-1/2 link insulin-like signaling to heat-shock responses and modulation of longevity. *Cell*, *148*, 322–334.

Conti, B., Sanchez-Alavez, M., Winsky-Sommerer, R., Morale, M. C., Lucero, J., Brownell, S., ... Zorrilla, E. P. (2006). Transgenic mice with a reduced core body temperature have an increased life span. *Science*, *314*, 825–828.

Delaney, C. E., Chen, A. T., Graniel, J. V., Dumas, K. J., & Hu, P. J. (2017). A histone H4 lysine 20 methyltransferase couples environmental cues to sensory neuron control of developmental plasticity. *Development*, *144*, 1273–1282.

De Magalhaes Filho, C. D., Henriquez, B., Seah, N. E., Evans, R. M., Lapierre, L. R., & Dillin, A. (2018). Visible light reduces *C. elegans* longevity. *Nature Communications*, *9*, 927.

Dorman, J. B., Albinder, B., Shroyer, T., & Kenyon, C. (1995). The age-1 and daf-2 genes function in a common pathway to control the lifespan of *Caenorhabditis elegans*. *Genetics*, *141*, 1399–1406.

Dubrovsky, Y. V., Samsa, W. E., & Kondratov, R. V. (2010). Deficiency of circadian protein CLOCK reduces lifespan and increases age-related cataract development in mice. *Aging*, *2*, 936–944.

Edwards, S. L., Charlie, N. K., Milfort, M. C., Brown, B. S., Gravlin, C. N., Knecht, J. E., & Miller, K. G. (2008). A novel molecular solution for ultraviolet light detection in *Caenorhabditis elegans*. *PLoS Biology*, *6*, e198.

Elabd, C., Cousin, W., Upadhyayula, P., Chen, R. Y., Chooljian, M. S., Li, J., ... Conboy, I. M. (2014). Oxytocin is an age-specific circulating hormone that is necessary for muscle maintenance and regeneration. *Nature Communications*, *5*, 4082.

Faucher, C., Forstreuter, M., Hilker, M., & De Bruyne, M. (2006). Behavioral responses of *Drosophila* to biogenic levels of carbon dioxide depend on life-stage, sex and olfactory context. *Journal of Experimental Biology*, *209*, 2739–2748.

Finch, C. E., & Ruvkun, G. (2001). The genetics of aging. *Annual Review of Genomics and Human Genetics*, *2*, 435–462.

Flatt, T., & Kawecki, T. J. (2007). Juvenile hormone as a regulator of the trade-off between reproduction and life span in *Drosophila melanogaster*. *Evolution; International Journal of Organic Evolution*, *61*, 1980–1991.

Fontana, L., Partridge, L., & Longo, V. D. (2010). Extending healthy life span--from yeast to humans. *Science*, *328*, 321–326.

Friedman, D. B., & Johnson, T. E. (1988). A mutation in the age-1 gene in *Caenorhabditis elegans* lengthens life and reduces hermaphrodite fertility. *Genetics*, *118*, 75–86.

Gendron, C. M., Kuo, T. H., Harvanek, Z. M., Chung, B. Y., Yew, J. Y., Dierick, H. A., & Pletcher, S. D. (2014). Drosophila life span and physiology are modulated by sexual perception and reward. *Science*, *343*, 544–548.

Gerisch, B., Rottiers, V., Li, D., Motola, D. L., Cummins, C. L., Lehrach, H., ... Antebi, A. (2007). A bile acid-like steroid modulates *Caenorhabditis elegans* lifespan through nuclear receptor signaling. *Proceedings of the National Academy of Sciences of the United States of America*, *104*, 5014–5019.

Gerisch, B., Weitzel, C., Kober-Eisermann, C., Rottiers, V., & Antebi, A. (2001). A hormonal signaling pathway influencing *C. elegans* metabolism, reproductive development, and life span. *Developmental Cell*, *1*, 841–851.

Giannakou, M. E., & Partridge, L. (2007). Role of insulin-like signalling in *Drosophila* lifespan. *Trends in Biochemical Sciences*, *32*, 180–188.

Hauck, S. J., Hunter, W. S., Danilovich, N., Kopchick, J. J., & Bartke, A. (2001). Reduced levels of thyroid hormones, insulin, and glucose, and lower body core temperature in the growth hormone receptor/binding protein knockout mouse. *Experimental Biology and Medicine*, *226*, 552–558.

Haus, E. L., & Smolensky, M. H. (2013). Shift work and cancer risk: Potential mechanistic roles of circadian disruption, light at night, and sleep deprivation. *Sleep Medicine Reviews*, *17*, 273–284.

Herskind, A. M., Mcgue, M., Holm, N. V., Sorensen, T. I., Harvald, B., & Vaupel, J. W. (1996). The heritability of human longevity: A population-based study of 2872 Danish twin pairs born 1870-1900. *Human Genetics*, *97*, 319–323.

Holloszy, J., & Smith, E. K. (1986). Longevity of cold-exposed rats: A reevaluation of the "rate-of-living theory". *Journal of Applied Physiology*, *61*, 1656–1660.

Horikawa, M., Sural, S., Hsu, A.-L., & Antebi, A. (2015). Co-chaperone p23 regulates *C. elegans* lifespan in response to temperature. *PLoS Genetics*, *11*, e1005023.

Hsu, A.-L., Murphy, C. T., & Kenyon, C. (2003). Regulation of aging and age-related disease by DAF-16 and heat-shock factor. *Science*, *300*, 1142–1145.

Hunter, W., Croson, W., Bartke, A., Gentry, M., & Meliska, C. (1999). Low body temperature in long-lived Ames dwarf mice at rest and during stress. *Physiology & Behavior*, *67*, 433–437.

Hwang, A. B., & Lee, S.-J. (2011). Regulation of life span by mitochondrial respiration: The HIF-1 and ROS connection. *Aging*, *3*, 304.

Hwang, W., Artan, M., Seo, M., Lee, D., Nam, H. G., & Lee, S. J. V. (2015). Inhibition of elongin C promotes longevity and protein homeostasis via HIF-1 in *C. elegans*. *Aging Cell*, *14*, 995–1002.

Insel, T. R., Landis, S. C., & Collins, F. S. (2013). Research priorities. The NIH BRAIN initiative. *Science*, *340*, 687–688.

Jia, K., Albert, P. S., & Riddle, D. L. (2002). DAF-9, a cytochrome P450 regulating *C. elegans* larval development and adult longevity. *Development*, *129*, 221–231.

Jindra, M., Palli, S. R., & Riddiford, L. M. (2013). The juvenile hormone signaling pathway in insect development. *Annual Review of Entomology*, *58*, 181–204.

Jones, W. D., Cayirlioglu, P., Grunwald Kadow, I., & Vosshall, L. B. (2006). Two chemosensory receptors together mediate carbon dioxide detection in *Drosophila*. *Nature*, *445*, 86.

Kaeberlein, M. (2013). Longevity and aging. *F1000prime Reports*, 5, 5.

Kandel, E. R. (2013). *Principles of neural science*. New York: McGraw-Hill.

Kapahi, P., Zid, B. M., Harper, T., Koslover, D., Sapin, V., & Benzer, S. (2004). Regulation of lifespan in *Drosophila* by modulation of genes in the TOR signaling pathway. *Current Biology*, *14*, 885–890.

Kappeler, L., De Magalhaes Filho, C., Dupont, J., Leneuve, P., Cervera, P., Périn, L., ... Epelbaum, J. (2008). Brain IGF-1 receptors control mammalian growth and lifespan through a neuroendocrine mechanism. *PLoS Biology*, *6*, e254.

Katewa, S. D., Akagi, K., Bose, N., Rakshit, K., Camarella, T., Zheng, X., ... Kapahi, P. (2016). Peripheral circadian clocks mediate dietary restriction-dependent changes in lifespan and fat metabolism in *Drosophila*. *Cell Metabolism*, *23*, 143–154.

Kenyon, C., Chang, J., Gensch, E., Rudner, A., & Tabtiang, R. (1993). A *C. elegans* mutant that lives twice as long as wild type. *Nature, 366*, 461.

Kenyon, C. J. (2010). The genetics of ageing. *Nature, 464*, 504–512.

Kuhara, A., Okumura, M., Kimata, T., Tanizawa, Y., Takano, R., Kimura, K. D., ... Mori, I. (2008). Temperature sensing by an olfactory neuron in a circuit controlling behavior of C. elegans. *Science, 320*, 803–807.

Kwon, J. Y., Dahanukar, A., Weiss, L. A., & Carlson, J. R. (2007). The molecular basis of CO2 reception in *Drosophila*. *Proceedings of the National Academy of Sciences, 104*, 3574–3578.

Lee, B. H., & Ashrafi, K. (2008). A TRPV channel modulates C. elegans neurosecretion, larval starvation survival, and adult lifespan. *PLoS Genetics, 4*, e1000213.

Lee, E., Jung, D. Y., Kim, J. H., Patel, P. R., Hu, X., Lee, Y., ... Kim, J. K. (2015). Transient receptor potential vanilloid type-1 channel regulates diet-induced obesity, insulin resistance, and leptin resistance. *FASEB Journal: Official Publication of the Federation of American Societies for Experimental Biology, 29*, 3182–3192.

Lee, S.-J., & Kenyon, C. (2009). Regulation of the longevity response to temperature by thermosensory neurons in *Caenorhabditis elegans*. *Current Biology, 19*, 715–722.

Lee, S. S., Kennedy, S., Tolonen, A. C., & Ruvkun, G. (2003). DAF-16 target genes that control *C. elegans* life-span and metabolism. *Science, 300*, 644–647.

Legates, T. A., Fernandez, D. C., & Hattar, S. (2014). Light as a central modulator of circadian rhythms, sleep and affect. *Nature Reviews Neuroscience, 15*, 443.

Leiser, S. F., Fletcher, M., Begun, A., & Kaeberlein, M. (2013). Lifespan extension from hypoxia in *Caenorhabditis elegans* requires both HIF-1 and DAF-16 and is antagonized by SKN-1. *Journals of Gerontology Series A: Biomedical Sciences and Medical Sciences, 68*, 1135–1144.

Leiser, S. F., & Kaeberlein, M. (2010). The hypoxia-inducible factor HIF-1 functions as both a positive and negative modulator of aging. *Biological Chemistry, 391*, 1131–1137.

Leng, Y., Musiek, E. S., Hu, K., Cappuccio, F. P., & Yaffe, K. (2019). Association between circadian rhythms and neurodegenerative diseases. *The Lancet Neurology, 18*, 307–318.

Li C. Kim K., Neuropeptides. In WormBook. The C. elegans Research Community, ed. http://dx.doi.org/10.1895/wormbook.1.142.1, http://www.wormbook.org.

Libert, S., Zwiener, J., Chu, X., Vanvoorhies, W., Roman, G., & Pletcher, S. D. (2007). Regulation of *Drosophila* life span by olfaction and food-derived odors. *Science, 315*, 1133–1137.

Lin, C.-T., He, C.-W., Huang, T.-T., & Pan, C.-L. (2017). Longevity control by the nervous system: Sensory perception, stress response and beyond. *Translational Medicine of Aging, 1*, 41–51.

Lin, K., Hsin, H., Libina, N., & Kenyon, C. (2001). Regulation of the *Caenorhabditis elegans* longevity protein DAF-16 by insulin/IGF-1 and germline signaling. *Nature Genetics, 28*, 139.

Lints, F. A. (1989). The rate of living theory revisited. *Gerontology, 35*, 36–57.

Liu, J., Ward, A., Gao, J., Dong, Y., Nishio, N., Inada, H., ... Xu, X. Z. S. (2010). *C. elegans* phototransduction requires a G protein–dependent cGMP pathway and a taste receptor homolog. *Nature Neuroscience, 13*, 715.

Liu, T., & Cai, D. (2013). Counterbalance between BAG and URX neurons via guanylate cyclases controls lifespan homeostasis in *C. elegans*. *The EMBO Journal, 32*, 1529–1542.

Loeb, J., & Northrop, J. H. (1916). Is there a temperature coefficient for the duration of life? *Proceedings of the National Academy of Sciences of the United States of America, 2*, 456–457.

Lopez-Otin, C., Blasco, M. A., Partridge, L., Serrano, M., & Kroemer, G. (2013). The hallmarks of aging. *Cell, 153*, 1194–1217.

Lowell, W. E., & Davis, G. E., JR (2008). The light of life: Evidence that the sun modulates human lifespan. *Medical Hypotheses, 70*, 501–507.

Lushchak, V., Carlsson, M. A., & Nässel, D. R. (2015). Food odors trigger an endocrine response that affects food ingestion and metabolism. *Cellular and Molecular Life Sciences, 72*, 3143–3155.

Maures, T. J., Booth, L. N., Benayoun, B. A., Izrayelit, Y., Schroeder, F. C., & Brunet, A. (2014). Males shorten the life span of *C. elegans* hermaphrodites via secreted compounds. *Science, 343*, 541–544.

Mehta, R., Steinkraus, K. A., Sutphin, G. L., Ramos, F. J., Shamieh, L. S., Huh, A., ... Kaeberlein, M. (2009). Proteasomal regulation of the hypoxic response modulates aging in *C. elegans*. *Science, 324*, 1196–1198.

Mishur, R. J., Khan, M., Munkácsy, E., Sharma, L., Bokov, A., Beam, H., ... Bai, Y. (2016). Mitochondrial metabolites extend lifespan. *Aging Cell, 15*, 336–348.

Mori, I., & Ohshima, Y. (1995). Neural regulation of thermotaxis in *Caenorhabditis elegans*. *Nature, 376*, 344.

Morley, J. F., & Morimoto, R. I. (2004). Regulation of longevity in *Caenorhabditis elegans* by heat shock factor and molecular chaperones. *Molecular Biology of the Cell, 15*, 657–664.

Motter, A. L., & Ahern, G. P. (2008). TRPV1-null mice are protected from diet-induced obesity. *FEBS Letters, 582*, 2257–2262.

Müller, R.-U., Fabretti, F., Zank, S., Burst, V., Benzing, T., & Schermer, B. (2009). The von Hippel Lindau tumor suppressor limits longevity. *Journal of the American Society of Nephrology, 20*, 2513–2517.

Murphy, C. T., Mccarroll, S. A., Bargmann, C. I., Fraser, A., Kamath, R. S., Ahringer, J., ... Kenyon, C. (2003). Genes that act downstream of DAF-16 to influence the lifespan of *Caenorhabditis elegans*. *Nature, 424*, 277–283.

Negri, L., Lattanzi, R., Giannini, E., Colucci, M., Margheriti, F., Melchiorri, P., ... Porreca, F. (2006). Impaired nociception and inflammatory pain sensation in mice lacking the prokineticin receptor PKR1: Focus on interaction between PKR1 and the capsaicin receptor TRPV1 in pain behavior. *The Journal of Neuroscience, 26*, 6716–6727.

Nersesyan, Y., Demirkhanyan, L., Cabezas-Bratesco, D., Oakes, V., Kusuda, R., Dawson, T., ... Chelluboina, B. (2017). Oxytocin modulates nociception as an agonist of pain-sensing TRPV1. *Cell Reports, 21*, 1681–1691.

Nusbaum, M. P., Blitz, D. M., & Marder, E. (2017). Functional consequences of neuropeptide and small-molecule co-transmission. *Nature Reviews Neuroscience, 18*, 389–403.

Ohta, A., Ujisawa, T., Sonoda, S., & Kuhara, A. (2014). Light and pheromone-sensing neurons regulates cold habituation through insulin signalling in *Caenorhabditis elegans*. *Nature Communications, 5*, 4412.

Ohtsuka, H., Azuma, K., Murakami, H., & Aiba, H. (2011). hsf1 (+) extends chronological lifespan through Ecl1 family genes in fission yeast. *Molecular Genetics and Genomics: MGG, 285*, 67–77.

Ostojic, I., Boll, W., Waterson, M. J., Chan, T., Chandra, R., Pletcher, S. D., & Alcedo, J. (2014). Positive and negative gustatory inputs affect *Drosophila* lifespan partly in parallel to dFOXO signaling. *Proceedings of the National Academy of Sciences of the United States of America, 111*, 8143–8148.

Park, S. K., Tedesco, P. M., & Johnson, T. E. (2009). Oxidative stress and longevity in *Caenorhabditis elegans* as mediated by SKN-1. *Aging Cell, 8*, 258–269.

Parrella, E., & Longo, V. D. (2010). Insulin/IGF-I and related signaling pathways regulate aging in nondividing cells: From yeast to the mammalian brain. *The Scientific World Journal, 10*, 161–177.

Peers, G., Truong, T. B., Ostendorf, E., Busch, A., Elrad, D., Grossman, A. R., ... Niyogi, K. K. (2009). An ancient light-harvesting protein is critical for the regulation of algal photosynthesis. *Nature, 462*, 518.

Pitt, J. N., & Kaeberlein, M. (2015). Why is aging conserved and what can we do about it? *PLoS Biology, 13*, e1002131.

Poon, P. C., Kuo, T.-H., Linford, N. J., Roman, G., & Pletcher, S. D. (2010). Carbon dioxide sensing modulates lifespan and physiology in *Drosophila*. *PLoS Biology, 8*, e1000356.

Poorter, H., & Nagel, O. (2000). The role of biomass allocation in the growth response of plants to different levels of light, CO2, nutrients and water: A quantitative review. *Functional Plant Biology, 27*, 1191.

Riddle, D.L. (1997). C. elegans II, In: D.L Riddle, T. Blumenthal, B.J. Meyer, & J.R. Priess (Eds), *Cold spring harbor mongraph series, Vol. 33*, 2nd edn. Cold Spring Harbor (NY): Cold Spring Harbor Laboratory Press.

Riera, C. E., Huising, M. O., Follett, P., Leblanc, M., Halloran, J., Van Andel, R., ... Dillin, A. (2014). TRPV1 pain receptors regulate longevity and metabolism by neuropeptide signaling. *Cell, 157*, 1023–1036.

Riera, C. E., Merkwirth, C., De Magalhaes Filho, C. D., & Dillin, A. (2016). Signaling networks determining life span. *Annual Review of Biochemistry, 85*, 35–64.

Riera, C. E., Tsaousidou, E., Halloran, J., Follett, P., Hahn, O., Pereira, M. M. A., ... Dillin, A. (2017). The sense of smell impacts metabolic health and obesity. *Cell Metabolism, 26*, 198–211, e5.

Roth, G. S., Lane, M. A., Ingram, D. K., Mattison, J. A., Elahi, D., Tobin, J. D., ... Metter, E. J. (2002). Biomarkers of caloric restriction may predict longevity in humans. *Science, 297*, 811, -811.

Ruby, J. G., Wright, K. M., Rand, K. A., Kermany, A., Noto, K., Curtis, D., ... Ball, C. (2018). Estimates of the heritability of human longevity are substantially inflated due to assortative mating. *Genetics, 210*, 1109–1124.

Scherer, T., O'hare, J., Diggs-Andrews, K., Schweiger, M., Cheng, B., Lindtner, C., ... Dighe, S. (2011). Brain insulin controls adipose tissue lipolysis and lipogenesis. *Cell Metabolism, 13*, 183–194.

Shamalnasab, M., Dhaoui, M., Thondamal, M., Harvald, E. B., Færgeman, N. J., Aguilaniu, H., & Fabrizio, P. (2017). HIF-1–dependent regulation of lifespan in *Caenorhabditis elegans* by the acyl-CoA–binding protein MAA-1. *Aging, 9*, 1745.

Sheeba, V., Sharma, V. K., Shubha, K., Chandrashekaran, M. K., & Joshi, A. (2000). The effect of different light regimes on adult lifespan in drosophila melanogaster is partly mediated through reproductive output. *Journal of Biological Rhythms, 15*, 380–392.

Sheng, Y., Tang, L., Kang, L., & Xiao, R. (2017). Membrane ion channels and receptors in animal lifespan modulation. *Journal of Cellular Physiology, 232*, 2946–2956.

Shi, C., & Murphy, C. T. (2014). Mating induces shrinking and death in *Caenorhabditis* mothers. *Science, 343*, 536–540.

Shi, C., Runnels, A. M., & Murphy, C. T. (2017). Mating and male pheromone kill Caenorhabditis males through distinct mechanisms. *eLife*, 6, e23493.

Sorabji, R. (1970). Aristotle on decmarcating the five senses. *The Philosophical Review, 80*, 55–79.

Špicarová, D., & Paleček, J. (2008). The role of spinal cord vanilloid (TRPV1) receptors in pain modulation. *Physiological Research/ Academia Scientiarum Bohemoslovaca, 57*, S69–S77.

Suh, G. S., Wong, A. M., Hergarden, A. C., Wang, J. W., Simon, A. F., Benzer, S., ... Anderson, D. J. (2004). A single population of olfactory sensory neurons mediates an innate avoidance behaviour in *Drosophila*. *Nature, 431*, 854.

Takken, W., & Knols, B. G. (1999). Odor-mediated behavior of Afrotropical malaria mosquitoes. *Annual Review of Entomology, 44*, 131–157.

Thondamal, M., Witting, M., Schmitt-Kopplin, P., & Aguilaniu, H. (2014). Steroid hormone signalling links reproduction to lifespan in dietary-restricted *Caenorhabditis elegans*. *Nature Communications, 5*, 4879.

Tullet, J. M. A., Hertweck, M., An, J. H., Baker, J., Hwang, J. Y., Liu, S., ... Blackwell, T. K. (2008). Direct inhibition of the longevity-promoting factor SKN-1 by insulin-like signaling in *C. elegans*. *Cell, 132*, 1025–1038.

Uzsak, A., & Schal, C. (2012). Differential physiological responses of the German cockroach to social interactions during the ovarian cycle. *The Journal of Experimental Biology, 215*, 3037–3044.

Uzsak, A., & Schal, C. (2013). Sensory cues involved in social facilitation of reproduction in *Blattella germanica* females. *PLoS One, 8*, e55678.

Van Breugel, F., Huda, A., & Dickinson, M. H. (2018). Distinct activity-gated pathways mediate attraction and aversion to CO2 in *Drosophila*. *Nature, 564*, 420–424.

Van Heemst, D. (2010). Insulin, IGF-1 and longevity. *Aging and Disease, 1*, 147.

Van Heemst, D., Beekman, M., Mooijaart, S. P., Heijmans, B. T., Brandt, B. W., Zwaan, B. J., ... Westendorp, R. G. (2005). Reduced insulin/IGF-1 signalling and human longevity. *Aging Cell, 4*, 79–85.

Vriens, J., Nilius, B., & Voets, T. (2014). Peripheral thermosensation in mammals. *Nature Reviews. Neuroscience, 15*, 573–589.

Wang, D., O'Halloran, D., & Goodman, M. B. (2013). GCY-8, PDE-2, and NCS-1 are critical elements of the cGMP-dependent thermotransduction cascade in the AFD neurons responsible for *C. elegans* thermotaxis. *The Journal of General Physiology, 142*, 437–449.

Wang, Y., Gao, Y., Tian, Q., Deng, Q., Wang, Y., Zhou, T., ... Li, Y. (2018). TRPV1 SUMOylation regulates nociceptive signaling in models of inflammatory pain. *Nature Communications, 9*, 1529.

Ward, A., Liu, J., Feng, Z., & Xu, X. Z. (2008). Light-sensitive neurons and channels mediate phototaxis in C. elegans. *Nature Neuroscience, 11*, 916–922.

White, J. G., Southgate, E., Thomson, J. N., & Brenner, S. (1986). The structure of the nervous system of the nematode *Caenorhabditis elegans*. *Philosophical Transactions of the Royal Society of London. Series B, Biological Sciences, 314*, 1–340.

Wollam, J., Magomedova, L., Magner, D. B., Shen, Y., Rottiers, V., Motola, D. L., ... Antebi, A. (2011). The Rieske oxygenase DAF-36 functions as a cholesterol 7-desaturase in steroidogenic pathways governing longevity. *Aging Cell, 10*, 879–884.

Xiao, R., Zhang, B., Dong, Y., Gong, J., Xu, T., Liu, J., & Xu, X. S. (2013). A genetic program promotes C. elegans longevity at cold temperatures via a thermosensitive TRP channel. *Cell, 152*, 806–817.

Yamamoto, R., Bai, H., Dolezal, A. G., Amdam, G., & Tatar, M. (2013). Juvenile hormone regulation of *Drosophila* aging. *BMC Biology, 11*, 85.

Zelinski, E. L., Deibel, S. H., & Mcdonald, R. J. (2014). The trouble with circadian clock dysfunction: Multiple deleterious effects on the brain and body. *Neuroscience & Biobehavioral Reviews, 40*, 80–101.

Zhang, B., Gong, J., Zhang, W., Xiao, R., Liu, J., & Xu, X. S. (2018). Brain–gut communications via distinct neuroendocrine signals bidirectionally regulate longevity in *C. elegans*. *Genes & Development, 32*, 258–270.

Zhang, Y., Shao, Z., Zhai, Z., Shen, C., & Powell-Coffman, J. A. (2009). The HIF-1 hypoxia-inducible factor modulates lifespan in C. elegans. *PloS One, 4*(7), e6348.

Zhang, B., Xiao, R., Ronan, E. A., He, Y., Hsu, A.-L., Liu, J., & Xu, X. Z. S. (2015). Environmental temperature differentially modulates *C. elegans* longevity through a thermosensitive TRP channel. *Cell Reports, 11*, 1414–1424.

Zhang, Z., Xu, J., Sheng, Z., Sui, Y., & Palli, S. R. (2011). Steroid receptor co-activator is required for juvenile hormone signal transduction through a bHLH-PAS transcription factor, methoprene tolerant. *The Journal of Biological Chemistry, 286*, 8437–8447.

Zitting, K.-M., Vujovic, N., Yuan, R. K., Isherwood, C. M., Medina, J. E., Wang, W., ... Duffy, J. F. (2018). Human resting energy expenditure varies with circadian phase. *Current Biology, 28*, 3685–3690, e3.

C H A P T E R

14

Aging of the sensory systems: hearing and vision disorders

Shinichi Someya[1] *and Akihiro Ikeda*[2]

[1]Department of Aging and Geriatric Research, University of Florida, Gainsville, FL, United States [2]Department of Medical Genetics and McPherson Eye Research Institute, University of Wisconsin, Madison, WI, United States

O U T L I N E

Introduction

Mitochondrial dysfunction, reactive oxygen species, and aging

The mammalian mitochondrial genome consists of 37 genes, encoding 13 proteins of the electron transport oxidative phosphorylation system (Kujoth, Bradshaw, Haroon, & Prolla, 2007; Wallace, 2005). Since these mitochondrial genes play an essential role in energy metabolism, mitochondrial DNA (mtDNA) mutations and associated mitochondrial dysfunction have been hypothesized to contribute to aging and age-related disorders, including age-related sensory impairments (Balaban, Nemoto, & Finkel, 2005; Finkel & Holbrook, 2000; Kujoth et al., 2007; Someya & Prolla, 2010; Wallace, 2005). Mitochondria are a major source of reactive oxygen species (ROS) and a major site of ROS-induced oxidative damage (Balaban et al., 2005; Finkel & Holbrook, 2000; Wallace, 2005). It is estimated that ~90% of intracellular ROS are generated as a by-product of mitochondrial respiration metabolism during the generation of ATP. These ROS include superoxide ($\bullet O_2^-$) and hydroxyl radical ($\bullet OH$), which are extremely unstable, and hydrogen peroxide (H_2O_2), which is freely diffusible and relatively long-lived (Beckman & Ames, 1998; Finkel & Holbrook, 2000). Of these ROS, the production of superoxide is thought to occur at three electron transport chain sites in mitochondrial respiration: complex I (NADH dehydrogenase), complex II (succinate dehydrogenase), and complex III (ubiquinone-cytochrome *c* reductase). Under

Handbook of the Biology of Aging.
DOI: https://doi.org/10.1016/B978-0-12-815962-0.00014-7

normal metabolic conditions, complex III is thought to be the main site of superoxide production. Superoxide from complex I and complex II is released into the mitochondrial matrix, whereas superoxide from complex III is released into either side of the inner mitochondrial membrane. It is also estimated that ~10% of intracellular ROS are generated within the plasma membranes and peroxisomes.

Mitochondria are protected against ROS by an interacting network of antioxidant enzymes (Balaban et al., 2005; Finkel & Holbrook, 2000; Halliwell & Gutteridge, 2007; Mari, Morales, Colell, Garcia-Ruiz, & Fernandez-Checa, 2009): mitochondrial SOD2 (superoxide dismutase 2) converts superoxide into hydrogen peroxide, which in turn is decomposed to water by two major mitochondrial antioxidant defense systems: the glutathione (GSH) and thioredoxin systems. There are three major players in the mitochondrial GSH system: GSH, GSH peroxidase 1 (GPX1), and GSH reductase (GSR). GSH acts as the major antioxidant in cells. When GSH is in a reduced state (active form), it interacts with GPX1 to decompose H_2O_2 into water. NADPH-dependent GSR then reduces oxidized GSH (GSSG) to reduced GSH. There are three major players in the mitochondrial thioredoxin system: thioredoxin 2, peroxiredoxin 3, and thioredoxin reductase 2 (TXNRD2): when thioredoxin 2 is in a reduced state, it interacts with peroxiredoxin 3 to decompose H_2O_2 into water. NADPH-dependent TXNRD2 then regenerates reduced thioredoxin 2 (active form of thioredoxin) from oxidized thioredoxin 2. These antioxidant defense systems work with each other to protect key mitochondrial components such as mtDNA, proteins, and lipids from ROS-induced damage.

A growing body of evidence suggests that ROS play a role in age-related disorders (Balaban et al., 2005; Finkel & Holbrook, 2000). This hypothesis is supported by the observations that overexpression of the mitochondrial antioxidant enzyme SOD2 significantly increases the life span of flies (Sun, Folk, Bradley, & Tower, 2002), while overexpression of a mitochondrially targeted catalase (CAT) transgene results in reduced age-related pathology in the heart and increases life span in mice (Schriner et al., 2005). Moreover, supplementation with the mitochondrial antioxidant α-lipoic acid or acetyl-L-carnitine reduces oxidative damage in the heart and brain of rats (Banaclocha, 2001; Palaniappan & Dai, 2007). It is thought that the mitochondrial antioxidant defense systems do not keep pace with the age-related increase in mitochondrial ROS production, and that during aging, the balance between the mitochondrial antioxidant defense and ROS production shifts progressively toward a more pro-oxidant state (Rebrin & Sohal, 2008). Collectively, these reports suggest that the balance between ROS production and mitochondrial antioxidant defense capacity may determine the degree of ROS-induced oxidative damage and associated mitochondrial dysfunction in the sensory receptors of cochleae and eyes, which require larger amounts of energy to process sounds or visible light. This chapter summarizes what has been learned about the roles of mitochondrial dysfunction in aging of the auditory and visual systems in rodents and humans.

Aging of the auditory system

Aging, cochlear pathology, and hearing loss

As we age, our hearing gradually declines (Someya & Prolla, 2010; Yamasoba et al., 2013). Accordingly, hearing loss is the third most prevalent chronic health condition affecting older adults; age-related hearing loss (AHL), also known as presbycusis, is the most common form of hearing impairment. The World Health Organization estimates that one-third of persons over 65 years are affected by hearing loss (WHO, 2019). Worldwide, approximately 466 million people suffer from hearing impairment and this number is expected to rise to 630 million by 2030 and over 900 million in 2050. In the United States, approximately 30 million people, or 13% of Americans 12 years and older, had bilateral hearing loss in 2001–2008, and this number rises to 48 million when individuals with unilateral hearing loss are included (Lin, Niparko, & Ferrucci, 2011). Because the prevalence of AHL is expected to rise dramatically as the world's population ages, AHL will become a major healthcare problem for which there are no established cures or treatments.

AHL is characterized by poor speech understanding, particularly in noise, impaired temporal resolution, and central auditory processing deficits (Yamasoba et al., 2013) and is also associated with dementia (Lin, Metter, et al., 2011). AHL is thought to be a multifactorial condition resulting from the interaction of numerous causes, including aging, exposure to noise and ototoxic chemicals, genetics, epigenetic variables, comorbidities, and lifestyle (Yamasoba et al., 2013). The inner ear contains two sensory systems: the auditory system (cochlea) that detects sound waves and the vestibular system (three semicircular canals, utricle, and saccule) that detects head movements and linear motion or gravity (Angelaki & Cullen, 2008; Liu & Yan, 2007). The major sites of age-related cochlear pathology include inner hair cells (IHCs), outer hair cells (OHCs), the sound receptors of the inner ear, spiral ganglion neurons (SGNs), synaptic loss, and stria vascularis cells (Fig. 14.1) (Liberman & Kujawa, 2017; Liu & Yan, 2007;

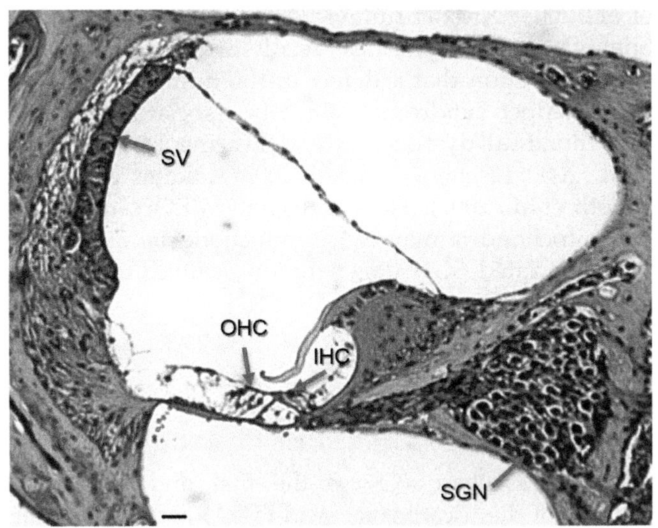

FIGURE 14.1 Cochlear cell types and tissues that are prone to age-related oxidative damage and cell death. Damage to any of these cells and tissues results in hearing impairment. *IHC*, inner hair cells; *OHC*, outer hair cells; *SGN*, spiral ganglion neuron; *SV*, stria vascularis. Scale bar: 140 μm.

Someya & Prolla, 2010; Yamasoba et al., 2013). The IHCs are the true sound receptors that relay sound wave information to the central auditory system through the SGNs (Hudspeth, 1997). Postmitotic hair cells and SGNs are particularly susceptible to injury from a combination of noise exposure, ototoxic chemicals, and oxidative damage (Yamasoba et al., 2013). The blood vessels coursing through the cochlea are also essential for transporting oxygen and nutrients such as glucose into the cochlea. Supporting this idea, AHL has been associated with significant loss of strial capillaries in the lateral wall of the gerbil cochlea (Gratton & Schulte, 1995), suggesting that an age-related decline in blood flow and oxygen transport to the cochlea plays a key role in the development of AHL. Together, age-related degeneration of the hair cells, SGN, and cochlear vasculature will disrupt auditory function and result in irreversible and permanent hearing impairment (Fig. 14.1).

Mitochondrial dysfunction and hearing loss

A central role for mitochondrial dysfunction in hearing impairment is supported by the numerous reports that a large number of genetic syndromes associated with hearing loss are due to defects in mitochondria (Chinnery et al., 2000; Fischel-Ghodsian, 2003; Kokotas, Petersen, & Willems, 2007; Someya & Prolla, 2010). This observation suggests that cochlear hair cells and SGNs are exquisitely sensitive to disturbances in energy metabolism. In support of this idea, increases in deletions, point mutations, or both, in mtDNA have been reported in human archival temporal bone samples from patients with AHL (Yamasoba et al., 2013). The 4977-bp mtDNA deletion was found to be more frequent in the archival temporal bones from patients with AHL compared to those with normal hearing (Bai, Seidman, Hinojosa, & Quirk, 1997), while specific point mutations in the mitochondrial *COX2* (mitochondrially encoded cytochrome c oxidase II) gene were found to be more frequent in the archival temporal bones from patients with AHL compared to those with normal hearing (Fischel-Ghodsian et al., 1997). In rodents, Someya and colleagues (Kim et al., 2019) have shown that mtDNA deletions accumulate with age in the inner ears of CBA/CaJ mice.

It is also well-documented that hearing loss is a common symptom in individuals harboring inherited mtDNA mutations (Chinnery et al., 2000; Fischel-Ghodsian, 2003; Kokotas et al., 2007; Xing, Chen, & Cao, 2007). The most common genetic defects observed in individuals with mitochondrial diseases are point mutations or deletions in mtDNA (Krishnan et al., 2008). Up to now, >100 different mtDNA deletions have been identified in individuals with mitochondrial diseases (MITOMAP, 2019). Among these mtDNA deletions, a 4977-bp deletion, known as the common deletion, causes several sporadic mitochondrial disorders such as Kearns–Sayre syndrome (KSS), whose symptoms include progressive sensorineural hearing loss (Chinnery et al., 2000; Fischel-Ghodsian, 2003; Kokotas et al., 2007; Taylor & Turnbull, 2005). In addition, there are numerous reports of mtDNA deletions observed in aged postmitotic tissues, such as brain, and in neurodegenerative diseases, including Parkinson disease (Bender et al., 2006; Chen et al., 2011; Copeland & Longley, 2014; Kauppila, Kauppila, & Larsson, 2017; Kraytsberg et al., 2006; Krishnan et al., 2008; Schon, DiMauro, & Hirano, 2012; Schon & Przedborski, 2011; Taylor & Turnbull, 2005). Specific point mutations in mtDNA can also cause mitochondrial disorders, including myoclonic epilepsy with ragged red fibers (MERRF) and mitochondrial encephalomyopathy, lactic acidosis, and stroke-like episodes (MELAS), the symptoms of which include progressive sensorineural hearing impairment (Chinnery et al., 2000; Fischel-Ghodsian, 2003; Kokotas et al., 2007). Collectively, these reports support the idea that mitochondrial defects or dysfunction play a central role in the pathogenesis of hearing impairment. Herein, this chapter reviews the current literature on genetic and molecular aspects of the aging of the auditory system, particularly focusing on genes involved in mitochondrial function.

Mitochondrial DNA mutations

Mitochondrial DNA polymerase γ

Previous studies have shown that mutations that impact mtDNA genomic stability, such as defects in the mtDNA polymerase γ (*POLG*), which maintains mtDNA replication fidelity (Filosto et al., 2003; Hudson & Chinnery, 2006), or the *OPA1* (optic atrophy 1) gene (Hudson et al., 2008; Liguori et al., 2008; Mancuso et al., 2004), which is involved in mitochondrial fission, lead to premature hearing loss. Mutations in *Cisd2*, which encodes a mitochondrial protein, also lead to the onset of Wolfram syndrome type 2 associated with mitochondrial dysfunction (Chen et al., 2009; Ju et al., 2008), while mutations in *ZCD2* lead to Wolfram syndrome, a recessive autosomal disorder associated with diabetes, optic atrophy, and deafness (Amr et al., 2007). Moreover, there are a number of well-characterized multisystem syndromes due to inherited mtDNA point mutations that are associated with deafness. These include maternally inherited diabetes and deafness, MELAS, MERRF, and progressive external ophthalmoplegia (PEO) (Chinnery et al., 2000; Deschauer et al., 2001; Fischel-Ghodsian, 2003; Hudson et al., 2008; Kokotas et al., 2007; Laloi-Michelin et al., 2009), which are each associated with multiple clinical phenotypes. However, as with most phenotypes observed in mtDNA genetic disorders, deafness is not an obligatory clinical feature. Possibly, nuclear modifying genes, as well as the level of heteroplasmy of the mtDNA mutation in various tissues, may determine the range of clinically relevant phenotypes in an affected individual. Importantly, these genetic observations suggest that alterations in mitochondrial function with age have the potential to play a major role in AHL.

Direct evidence for mitochondrial dysfunction in hearing loss comes from the observations that accumulation of mtDNA mutations leads to reduced life span and early onset of hearing loss in mice carrying a mutator allele of the *Polg* gene (Kujoth et al., 2005). Mitochondrial POLG is the only DNA polymerase that is active in the mitochondria and that can proofread and replicate mtDNA (Kujoth et al., 2007). Numerous mutations in *POLG* have been identified as a cause of human disorders, such as Alper syndrome, that are associated with deafness. Vermulst and co-workers (Vermulst et al., 2007) have shown that young mitochondrial mutator mice displayed a >500-fold increase in mtDNA point mutations in the brain and heart compared to age-matched wild-type (WT) mice, while middle-age mitochondrial mutator mice displayed a 7–11-fold increase in mtDNA deletions in the brain and heart compared to age-matched WT mice (Vermulst et al., 2008). A subsequent study has shown that mtDNA deletions accumulate with age in the inner ears of mtDNA mutator ($Polg^{D257A/D257A}$) mice, while young $Polg^{D257A/D257A}$ mice showed a higher burden of mtDNA point mutations in the inner ears (Kim et al., 2019). Middle-age $Polg^{D257A/D257A}$ mice also displayed a profound loss of SGNs in the cochlea and early-onset severe AHL (Kim et al., 2019; Kujoth et al., 2005; Niu, Trifunovic, Larsson, & Canlon, 2007; Someya et al., 2008). These results were consistent with the observation that a defect in the human *POLG* gene causes Alper syndrome, which is associated with both mitochondrial dysfunction and hearing loss (Chinnery et al., 2000; Fischel-Ghodsian, 2003; Kokotas et al., 2007; Kujoth et al., 2007). Thus, a decline in POLG activity in the mitochondria may cause mitochondrial dysfunction and associated SGN degeneration, leading to the onset of AHL.

Energy metabolism

Citrate synthase

Citrate synthase (CS) is the first and rate-limiting enzyme of the tricarboxylic acid (TCA) cycle in the mitochondria (Johnson, Gagnon, Longo-Guess, & Kane, 2012; Raimundo, Baysal, & Shadel, 2011). CS is encoded by a single nuclear gene and is localized in the mitochondrial matrix, where it functions as the first step of the TCA cycle, producing citric acid from acetyl-CoA and oxaloacetate. Johnson et al. (2012) mapped an *ahl4* locus on chromosome 10 that contributes to AHL in the A/J mouse strain. The authors found that a *CS* mutation underlies *ahl4*-related hearing loss in the A/J strain. These A/J mice display early onset and rapid progression of hearing loss. The authors also sequenced the entire genomes of A/J and four strains that lack an *ahl4* phenotype and found that the *rs29358506* SNP, located in exon 3 of *CS*, is the only exon or splice site variant within the 5.5 MB candidate region that is unique to the A/J strain, indicating that *CS* is the gene responsible for *ahl4*. Consistent with these results, A/J mice display poor performance in endurance exercise such as having the shortest treadmill duration time among the 10 inbred mouse strains, including the C57BL/6 strain (Lightfoot, Turner, Debate, & Kleeberger, 2001). Haramizu and co-workers (Haramizu et al., 2009) have also shown poor treadmill performance in A/J mice, estimating the endurance capacity of BALB/c mice to be 217% greater than that of A/J mice. Therefore a decline in CS activity may result in mitochondrial dysfunction and early-onset hearing loss.

Mitochondrial isocitrate dehydrogenase

Mitochondrial isocitrate dehydrogenase 2 (IDH2) participates in the TCA cycle and catalyzes the conversion of isocitrate to α-ketoglutarate and $NADP^+$ to NADPH in the mitochondrial matrix (Reitman & Yan, 2010). IDH2 also plays a role in protecting mitochondrial components from oxidative stress by supplying NADPH to both GSR and TXNRD2 within the mitochondrial matrix. Someya and co-workers (White et al., 2018) investigated the effects of *Idh2* deficiency on age-related cochlear pathology and AHL using $Idh2^{+/+}$ (WT) and $Idh2^{-/-}$ mice. The authors found that old male *Idh2*$^{-/-}$ mice displayed increased ABR thresholds, increased wave I latency, and decreased wave I amplitude compared to age-matched WT mice. This was accompanied by increased oxidative DNA damage, increased apoptotic cell death, and profound loss of SGNs and HCs in the cochlea of old *Idh2*$^{-/-}$ mice. Loss of *Idh2* also resulted in a decreased NADPH redox state and decreased activity of TXNRD2 in the mitochondria of the inner ear of young mice. Therefore IDH2 likely functions as the principal source of NADPH for the mitochondrial thioredoxin antioxidant defense and plays an essential role in protecting cochlear hair cells and neurons against oxidative stress during aging.

Mitochondrial sirtuins

Sirtuins are a family of NAD^+-dependent protein deacetylases that extend life span in worms and flies (Finkel, Deng, & Mostoslavsky, 2009). SIRT3, a member of the mammalian sirtuin family, is localized to mitochondria and regulates levels of ATP and the activity of complex I of the electron transport chain (Ahn et al., 2008). Calorie restriction (CR), or reducing food consumption by 25%–60% without malnutrition, consistently extends life span and delays the onset of age-related diseases in a variety of species, including rodents and monkeys (Fontana, Partridge, & Longo, 2010; Sohal & Weindruch, 1996; Weindruch & Sohal, 1997). Sundaresan and co-workers (Sundaresan, Samant, Pillai, Rajamohan, & Gupta, 2008) have shown that CR increases protein levels of SIRT3 in primary mouse cardiomyocytes, while overexpression of *Sirt3* protects these cells from oxidative stress-induced cell death, suggesting a role of SIRT3 in aging retardation under CR conditions. Someya and co-workers (Someya et al., 2010) investigated the effects of *Sirt3* deficiency on cochlear pathology and AHL under control diet and calorie-restricted conditions using WT and *Sirt3*$^{-/-}$ mice. Aging resulted in elevated ABR hearing thresholds in middle-age WT mice under control diet, while CR delayed the development of AHL in WT mice, but not in *Sirt3*$^{-/-}$ mice. In agreement with the ABR test results, CR reduced oxidative DNA damage and SGN degeneration in the cochlea of middle-age WT mice, but not *Sirt3*$^{-/-}$ mice. Under CR conditions, SIRT3 also activated IDH2, leading to increased NADPH levels and GSH redox state in mitochondria. Thus, SIRT3 likely acts as a key player in enhancing the mitochondrial GSH antioxidant defense system in cochlea under starvation or calorie-restricted conditions.

Mitochondrial antioxidant defense

Mitochondrial glutathione peroxidase

GPX1 plays an important role in mitochondrial antioxidant defense by decomposing H_2O_2 into water (Halliwell & Gutteridge, 2007; Mari et al., 2009). Ohlemiller and co-workers (Ohlemiller, Wright, & Dugan, 1999) have shown that cochlear ROS levels were elevated following acute noise exposure in C57BL/6 mice. Consistent with this report, mice lacking *Sod1* displayed increased hair cell loss in the cochlea compared to WT mice (McFadden, Ding, Reaume, Flood, & Salvi, 1999). Oxidative protein damage also increases with age in the cochleae of CBA/J mice (Jiang, Talaska, Schacht, & Sha, 2007), while oxidative nuclear DNA damage increases with age in the cochlea of C57BL/6 mice (Someya et al., 2009). Ohlemiller and colleagues (Ohlemiller, McFadden, Ding, Lear, & Ho, 2000) investigated whether *Gpx1* deficiency increases susceptibility to noise-induced hearing loss (NIHL) in mice. The authors found that *Gpx1*$^{-/-}$ mice showed significantly greater ABR threshold elevation after noise exposure compared to WT mice. Noise-exposed *Gpx1*$^{-/-}$ mice also showed more sensory hair cell loss compared to WT mice. Therefore a decline in GPX1 activity in the mitochondria may lead to increased levels of ROS, which in turn result in mitochondrial dysfunction, leading to cochlear cell loss and associated hearing loss.

Mitochondrial catalase

CAT converts H_2O_2 into water and oxygen, and is one of the most efficient enzymes found in cells (Halliwell & Gutteridge, 2007); each CAT molecule can decompose millions of H_2O_2 molecules every second, indicating the importance of this antioxidant enzyme in protecting the cells against ROS. CAT is usually localized in the peroxisomes, although CAT expression was also observed in the cytosol and mitochondria. Rabinovitch and colleagues (Schriner et al., 2005) generated mice overexpressing human CAT targeted to the peroxisome (PCAT), nucleus (NCAT), or mitochondria (MCAT). The authors found that overexpression of CAT in the mitochondria extended both median and maximum life span in both males and females, while overexpression of CAT in the nuclei did not show a significant extension of median nor maximum life span. PCAT animals showed a slight extension of median life span. MCAT mice also displayed increased CAT activities in the heart, skeletal muscle, and brain, reduced oxidative DNA damage, and decreased hydrogen peroxide levels in the heart. Someya and co-workers (Someya et al., 2009) have shown that young MCAT mice display normal hearing compared to age-matched WT mice; however, middle-aged MCAT mice display significantly lower ABR thresholds at low, middle, and high frequencies than those of middle-aged WT mice. Furthermore, MCAT mice display reduced oxidative DNA damage and loss of SGNs and hair cells in the cochlea. Importantly, in humans, CAT activity was significantly higher in the red blood cells of centenarians (Klapcinska et al., 2000). Thus, enhancing CAT activity in cochlear mitochondria may slow the development of age-related cochlear degeneration and hearing loss.

Antioxidant compounds

Coenzyme Q_{10} is an essential component of the mitochondrial electron transfer chain and acts as a mitochondrial antioxidant (Sohal & Forster, 2007). Guastini and co-workers (2011) investigated the effects of coenzyme Q_{10} in patients with AHL. The authors found that coenzyme Q_{10} treatment significantly improved pure tone audiometric thresholds at 1000, 2000, 4000, and 8000 Hz. Consistent with this report, supplementation with coenzyme Q_{10} delayed the onset of AHL at the high frequency in mice (Someya et al., 2009). GSH acts as the major small-molecule antioxidant in cells and the GSH system is one of the major antioxidant defense systems in the mitochondria (Anderson, 1998; Halliwell & Gutteridge, 2007; Mari et al., 2009). α-Lipoic acid and *N*-acetylcysteine are thiol compounds that have been linked to GSH production and have been shown to reduce mitochondrial ROS production (Banaclocha, 2001; Hart, Terenghi, Kellerth, & Wiberg, 2004; Palaniappan & Dai, 2007). Someya and co-workers (Someya et al., 2009) investigated the effects of α-lipoic acid or *N*-acetylcysteine on AHL in C57BL/6 mice. The authors found that supplementation with α-lipoic acid or *N*-acetylcysteine delayed the onset of AHL at the high frequency in mice. Previous studies have also shown that supplementation with α-lipoic acid slowed the development of AHL in Fisher 344 rats (Seidman, Khan, Bai, Shirwany, & Quirk, 2000) and in DBA/2J mice (Ahn, Kang, Kim, Shin, & Chung, 2008). In rats, *N*-acetyl cysteine treatment (saline injection) reduced levels of NIHL and OHC loss in the cochlea (Wu, Hsu, Cheng, & Guo, 2010). In guinea pigs, GSH treatment reduced noise-induced temporary threshold shift and protected the hair cells in the cochlea from noise exposure (Ohinata, Yamasoba, Schacht, & Miller, 2000). In humans, *N*-acetylcysteine treatment significantly reduced noise-induced temporary threshold shift in male workers with the GSH transferase μ1 (*GSTM1*) and GSH transferase θ1 (*GSTT1*) polymorphisms (Lin et al., 2010). Therefore enhancing mitochondrial antioxidant defenses through antioxidant supplementation may reduce oxidative cochlear cell damage and delay AHL.

Mitochondrial apoptosis

BCL2-antagonist/killer

An apoptosis program is thought to play a key role in aging and age-related diseases (Mattson, 2000). Neuronal death also contributes to the symptoms of many neurological disorders, including Alzheimer's disease, Parkinson disease, Huntington disease, stroke, and amyotrophic lateral sclerosis. Apoptosis can occur through two major pathways (Lindsten et al., 2000; Mattson, 2000; Youle & Strasser, 2008): the intrinsic, or mitochondrial, pathway senses intracellular damage and is initiated when the outer mitochondrial membrane loses its integrity. The extrinsic pathway is a sensor for extracellular signals, and is initiated through the ligation of tumor necrosis factor (TNF), a major mediator of extrinsic apoptosis, which is involved in immune and inflammatory responses (Chau, Chen, Wan, DeGregori, & Wang, 2004; Chen & Goeddel, 2002). Activation of TNF signaling is involved in the pathogenesis of a wide spectrum of diseases, including diabetes, cancer, osteoporosis, and autoimmune diseases, such as multiple sclerosis and inflammatory bowel disease. TNF signals through two distinct cell surface receptors, TNF-R1 and TNF-R2. Of these receptors, TNF-R1 initiates the majority of TNF's apoptotic activities.

The mitochondrial apoptosis pathway is regulated by BCL-2 family members (Lindsten et al., 2000; Youle & Strasser, 2008). Of the BCL-2 family members, the proapoptotic proteins BAX and BAK play a central role in promoting mitochondrial-mediated apoptosis. Lindsten and colleagues (Lindsten et al., 2000) have shown that $Bak^{-/-}$ mice do not exhibit any gross abnormalities or developmental defects. However, the majority of mice lacking both *Bak* and *Bax* died prenatally, with fewer than 10% of the mutants surviving into adulthood. Those $Bax^{-/-}/Bak^{-/-}$ mice displayed multiple developmental defects such as accumulation of excess cells within both the central nervous and hematopoietic systems, indicating that BAX and BAK play central roles in the regulation of apoptosis during development and tissue homeostasis. Lindsten and colleagues also found that homozygous $Bak^{-/-}$ mice are fertile and do not display any developmental abnormalities, suggesting that the role of BAK in development may be redundant with that of other proapoptotic BCL-2 family members such as BAX. Someya and co-workers (Someya et al., 2009) investigated whether BAX and BAK function in a redundant manner in the auditory system. The authors found no significant differences between middle-aged WT and $Bax^{-/-}$ mice; both had similar hearing loss indicated by higher ABR thresholds and the lower numbers of hair cells and SGNs typically seen in middle-aged C57BL/6 mice. In contrast, middle-aged $Bak^{-/-}$ mice displayed significantly lower ABR thresholds at 8, 16, and 32 kHz and fewer losses of SGNs and hair cells than those of age-matched WT. Moreover, TUNEL staining revealed that the levels of apoptotic nuclear fragmentation markers were significantly reduced in the cochlea of middle-aged $Bak^{-/-}$ mice compared to age-matched WT mice. Therefore a decline in BAK-mediated mitochondrial apoptosis may prevent mitochondrial dysfunction, which in turn prevents cochlear cell death, and the progression of hearing loss during aging.

BCL11B

The B-cell leukemia/lymphoma 11B gene (*BCL11B*) encodes a zinc-finger protein and acts as an antiapoptotic factor (Grabarczyk et al., 2007; Youle & Strasser, 2008). Grabarczyk and colleagues (Grabarczyk et al., 2007) have shown that suppression of *BCL11B* by siRNA induced apoptosis in transformed human T-cell leukemia and lymphoma cell lines, whereas normal mature T-cell lines remained unaffected, indicating that the survival of human T-cell leukemia and lymphoma cell lines is dependent on BCL11B. Okumura and colleagues (Okumura et al., 2011) examined the effects of *Bcl11b* deficiency on age-related cochlear pathology and associated hearing loss using WT and $Bcl11b^{-/-}$ mice. Immunohistological analysis revealed that BCL11B was detected in the OHCs, but not in IHCs, supporting cells, or SGNs in the cochlea of WT mice. Loss of OHCs was observed in the cochlea of young $Bcl11b^{-/-}$ mice, but not in WT mice. There were no significant differences in ABR thresholds between young WT and $Bcl11b^{-/-}$ mice. However, middle-age $Bcl11b^{-/-}$ mice displayed elevated ABR thresholds compared to age-matched WT mice. Thus, the antiapoptotic BCL11B may play a role in protecting cochlear hair cells during aging.

PTEN

Phosphatase and tensin homolog (PTEN) is a lipid phosphatase that induces cell cycle arrest and mitochondrial apoptosis through inhibition of the protein kinase B (AKT) pathway, a major survival pathway activated in cancer (Georgescu, 2010; Myers et al., 1998; Weng, Brown, & Eng, 2001; Zhu, Hoell, Ahlemeyer, & Krieglstein, 2006). Sha and colleagues (Sha, Chen, & Schacht, 2010) examined the roles of PTEN, phosphatidylinositol 3,4,5-trisphosphate (PIP3), and AKT in age-related cochlear pathology and associated hearing loss using CBA/J mice. Immunostaining analysis revealed that PIP_3 levels decreased in the IHCs, OHCs, and supporting cells with age. In contrast, levels of PTEN protein increased in the cochlea with age. Moreover, Western blotting of cochlear proteins from old mice revealed reduced levels of the phosphorylated isoforms AKT1 and AKT2 compared to young mice. Thus, increased levels of PTEN and/or decreased levels of PIP_3/AKT signaling may contribute to age-related loss of hair cells and hearing loss.

Mitochondrial thermogenesis

Uncoupling protein 2 (UCP2) is a member of the mitochondrial anion carrier protein family, which facilitates the transfer of anions from the inner to the outer mitochondrial membrane and the return transfer of protons from the outer to the inner mitochondrial membrane (Mattiasson et al., 2003). UCPs also reduce the mitochondrial membrane potential in mammalian cells. UCP2 is expressed in multiple tissues, including skeletal muscles and central nervous system and is thought to play a role in thermogenesis, obesity, and diabetes. A neuroprotective role for UCP2 was indicated by the finding that brain damage following experimental stroke was reduced in mice overexpressing human *UCP2* (Mattiasson et al., 2003). Moreover, *UCP2*, *UCP3*, and *UCP4* mRNAs were found to be expressed in the vestibular and spiral ganglions in the inner ear, while *UCP3* mRNA expression was undetectable in the brain of rats (Kitahara, Li, & Balaban, 2004). Sugiura and colleagues (Sugiura, Uchida, Nakashima, Ando, & Shimokata, 2010) investigated the association between *UCP1* and *UCP2* gene polymorphisms and hearing impairment in middle-aged and elderly males and females in Japan: detailed questionnaires, pure-tone audiometry measurements, and *UCP1 A-3826G* and *UCP2 Ala55Val* polymorphisms were examined in 1547 subjects between the ages of 40 and 79 years. Using generalized estimating equations, associations between hearing impairment and the gene polymorphisms in *UCP1* and *UCP2* with age, sex, history of occupational noise exposure, and body mass index were analyzed under dominant, recessive, and additive models. The authors found that *UCP2 Ala55Val* polymorphisms exhibited a significant association with AHL in the Japanese population. Therefore a decline in mitochondrial UCP2 activity may accelerate the progression of AHL in humans.

Aging of the visual systems

Aging, eye pathology, and visual disorders

The retina is a light-sensing organ with a well-organized layered structure that captures light, converts it to neuronal signals and transmits them to the central nervous system (Sung & Chuang, 2010; Fig. 14.2). Photoreceptor cells first capture the light and convert it to neuronal signals, which are transmitted to second-order neurons and bipolar cells through synaptic interaction. Signals from bipolar cells are transmitted to ganglion cells through another synaptic interaction. Ganglion cells reside in the inner surface of the retina, connecting dendrites with bipolar cells

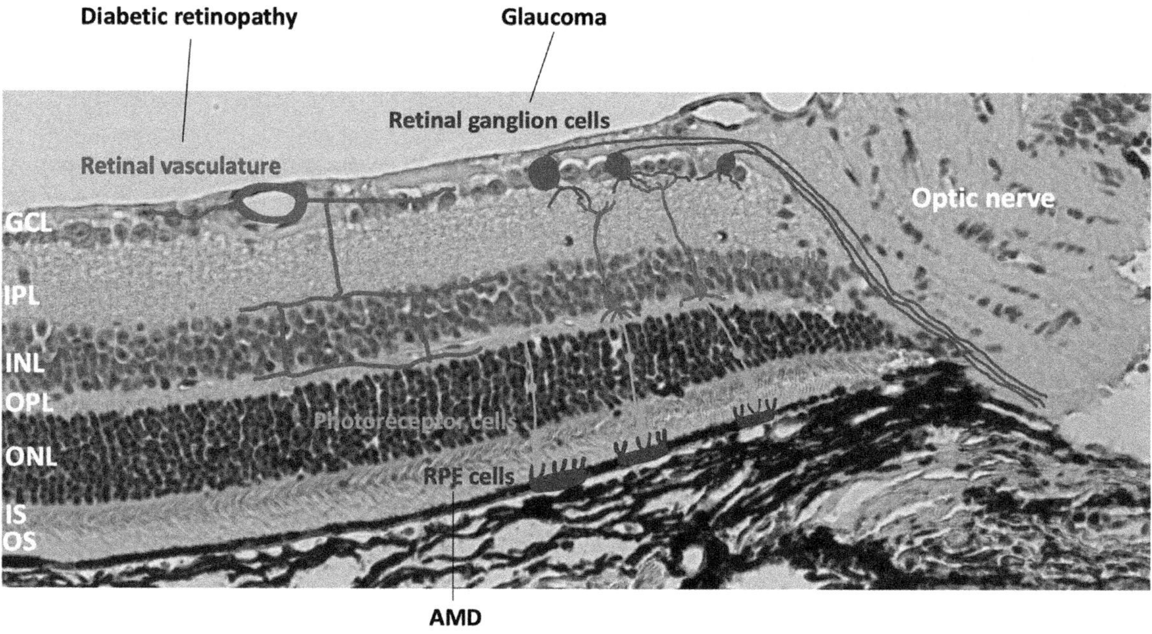

FIGURE 14.2 Major cell types and structures in the retina that are affected by age-related oxidative damage or retinal diseases. Retinal layers are labeled for orientation. *AMD*, age-related macular degeneration; *GCL*, ganglion cell layer; *INL*, inner nuclear layer; *IPL*, inner plexiform layer; IS, inner segments; *ONL*, outer nuclear layer; *OPL*, outer plexiform layer; *OS*, outer segments; *RPE*, retinal pigment epithelium.

and stretching long axons (optic nerves) into the visual cortex in the brain. The retinal pigment epithelium (RPE) cells, with their apical processes interdigitating with outer segments of photoreceptor cells, compose the outer blood–retina barrier, phagocytose debris of photoreceptor outer segments, comprise an integral part of the visual (retinoid) cycle, and supply nutrients to photoreceptor cells. Other cell types identified in the retina include horizontal cells, amacrine cells, Müller cells, and astrocytes. Although we do not discuss these cells in this chapter, they support the retina functionally and structurally.

Aging causes alterations to the retina at multiple levels (Bonnel, Mohand-Said, & Sahel, 2003; Lin, Tsubota, & Apte, 2016). Age-dependent loss of retinal neurons including photoreceptor cells (Curcio, Millican, Allen, & Kalina, 1993; Jackson, Owsley, & Curcio, 2002) and retinal ganglion cells (Curcio & Drucker, 1993; Gao & Hollyfield, 1992), as well as non-neuronal cells such as RPE cells (Gao & Hollyfield, 1992) has been observed (Fig. 14.2). A corresponding decline in retinal functions has been also observed with age (Birch & Anderson, 1992; Gerth, Garcia, Ma, Keltner, & Werner, 2002; Jackson & Owsley, 2000; Kurtenbach & Weiss, 2002). Retinal aging is also associated with the formation of intracellular and extracellular deposits. Lipofuscin, an autofluorescent lipopigment formed by lipids, metals, and misfolded proteins, accumulates within the RPE (Feeney-Burns, Hilderbrand, & Eldridge, 1984; Hohn & Grune, 2013; Terman & Brunk, 1998) and is predicted to induce ROS, resulting in cell damage and abnormalities (Wassell, Davies, Bardsley, & Boulton, 1999; Wiktor, Sarna, Wnuk, & Sarna, 2018). Extracellularly, cholesterol-rich deposits known as drusen accumulate between the RPE and Bruch's membrane (Curcio, Millican, Bailey, & Kruth, 2001). It is important to note that many of these age-dependent retinal abnormalities are also observed in age-related retinal diseases with earlier onset and/or increased severity (Fig. 14.2). Drusen formation, photoreceptor cell degeneration, and RPE degeneration are observed in age-related macular degeneration (AMD) patients (Curcio, Medeiros, & Millican, 1996; Datta, Cano, Ebrahimi, Wang, & Handa, 2017; Hageman et al., 2001), while retinal ganglion loss occurs in glaucoma (Almasieh, Wilson, Morquette, Cueva Vargas, & Di Polo, 2012; Weinreb, Aung, & Medeiros, 2014). Moreover, advanced age is the risk factor for age-related retinal diseases, including AMD and glaucoma (Klein & Klein, 2013). These observations suggest that cellular changes that occur in retinal aging may be closely associated with molecular/cellular mechanisms underlying age-related retinal diseases.

AMD is the most common cause of visual impairment in the aging population and is estimated to affect at least 11 million individuals in the United States (Pennington & DeAngelis, 2016). AMD affects the macular (central) region of the retina, causing progressive loss of central vision (Mitchell, Liew, Gopinath, & Wong, 2018). Clinically, extracellular deposits of lipid and proteins that accumulate under the retina, called "drusen," are the major characteristic lesions in the earlier stage of AMD (Mitchell et al.,

2018). The late stage of AMD can be neovascular (known as wet) or nonvascular (known as dry or atrophic). Neovascular AMD is characterized by the invasion of choroidal vasculatures into the retina. Geographic atrophy is known as end-stage dry or atrophic AMD, characterized by the loss of RPE cells, photoreceptor cells, and choriocapillaris (Fig. 14.2; Mitchell et al., 2018).

Glaucoma is the leading cause of blindness affecting more than 60 million people worldwide (Tham et al., 2014) and is characterized by degeneration of retinal ganglion cells (Fig. 14.2). Primary open-angle glaucoma, the most common type of glaucoma, affected 2.71 million individuals in the United States in 2011 and is estimated to affect more than 7 million in the United States by 2050 (Vajaranant, Wu, Torres, & Varma, 2012). Major risk factors for glaucoma are age and elevated intraocular pressure (IOP). Lowering IOP has been shown to reduce the rate of ganglion cell loss and is currently the only therapeutic approach that has been shown to slow the progression of glaucoma (Liebmann & Lee, 2017; McLaren & Moroi, 2003; Weinreb et al., 2014). However, this approach is not successful in a significant proportion of the population. Some studies have shown that the increase in the prevalence of glaucoma with age is not accounted for by the increase in IOP alone (Klein, Klein, & Linton, 1992; Kong, Van Bergen, Trounce, & Crowston, 2009; Varma, Ying-Lai, Klein, Azen, & Los Angeles Latino Eye Study, 2004), suggesting that age-dependent changes in ganglion cells predispose them to degeneration.

Diabetic retinopathy is a prevalent microvascular complication in diabetes (Beckman & Creager, 2016; Wong, Cheung, Larsen, Sharma, & Simo, 2016; Zhang et al., 2010). Chronic hyperglycemia is considered as a major risk factor, along with other diabetes-related factors, such as hypertension and dyslipidemia. Major symptoms of diabetic retinopathy include microvascular damage, neurodegeneration, and retinal dysfunction (Fig. 14.2; Wong et al., 2016). Observed effects of age on the prevalence and severity of diabetic retinopathy vary among the population studied (Tan, Gay, & Ngo, 2010). While there are studies reporting older age as the more significant risk factor for the prevalence (Namperumalsamy et al., 2009) or progression (Stratton et al., 2001) of diabetic retinopathy, other studies found younger age as the more significant risk factor (Klein, Klein, Moss, & Cruickshanks, 1994; Klein, Klein, Moss, Davis, & DeMets, 1984; Lim et al., 2008). This discrepancy in the effect of aging on diabetic retinopathy may be due to confounding variables, such as differences in genetic and/or environmental factors. Nevertheless, accumulating evidence suggests that mitochondrial dysfunction, which is also observed in the aging retina, is associated with the pathogenesis of diabetic retinopathy (Kowluru & Abbas, 2003; Kowluru & Chan, 2007), which will be also discussed in this chapter.

Mitochondrial dysfunction and visual disorders

The retina is considered to be one of the highest energy-consuming tissues in the body (Niven & Laughlin, 2008). Therefore its function is largely dependent on the capacity of mitochondrial ATP production. Within the retina, photoreceptor cells are the highest energy-producing and energy-consuming cells (Niven & Laughlin, 2008). Animal studies have shown that about 55%–65% of mitochondria are located in the photoreceptor inner segments, and oxygen consumption of photoreceptor cells is twofold higher compared with the rest of the retinal region (Alder, Ben-Nun, & Cringle, 1990; Buono & Sheffield, 1991; Chen, Soderberg, & Lindstrom, 1989). Retinal ganglion cells have a unique structure, with dendrites connected with bipolar cells in the inner plexiform layer and extremely long axons guided into the center of the retina, turning around toward the back of the eye and stretching to the visual cortex of the brain. In humans, long axons of retinal ganglion cells have two morphologically distinct areas: unmyelinated nerve fibers until axons exit the eye through the lamina cribrosa, and myelinated nerve fibers after that (Bristow, Griffiths, Andrews, Johnson, & Turnbull, 2002). Mitochondria in ganglion cells are particularly abundant along the unmyelinated axons of retinal ganglion cells, indicating the importance of mitochondrial energy production in that area (Bristow et al., 2002). This portion of retinal ganglion neurons is considered the target of axonal degeneration in glaucoma (Bristow et al., 2002). In addition to sensory neurons, such as photoreceptor cells and retinal ganglion cells, the significance of RPE cells in the pathogenesis of age-related diseases such as AMD has become more widely recognized. RPE cells are pigmented cells that facilitate multiple roles such as light absorption, selective transport of substance to the retina, processing of retinoids in the visual cycle, phagocytosis of photoreceptor outer segments, secretion, and immune modulation (Strauss, 2005). With age, a significant decrease in the number and area of mitochondria, as well as loss of cristae and matrix density, were observed in RPE cells, which were significantly greater in the RPE of AMD patients compared to normal controls (Feher et al., 2006). In addition to morphological changes in mitochondria, it has been shown that RPE cells from AMD patients exhibit reduced mitochondrial function (Ferrington et al., 2017). These observations indicate that abnormal morphology and functions of mitochondria in the RPE are tightly associated with normal aging and age-related diseases such as AMD.

In summary, mitochondrial function is critical for retinal cells, as described above, and its decline with age may cause pathological symptoms that are associated with age-related disease phenotypes (also reviewed in Eells, 2019). Herein, this chapter reviews the current literature on age-dependent mitochondrial changes in physiological and pathological aging of the retina, as well as genetic mutations causing those age-dependent retinal disease phenotypes in humans and animal models.

Mitochondrial DNA mutations

The mitochondrial theory of aging proposed that accumulation of somatic mtDNA mutations causes a decline in mitochondrial function, contributes to aging of each organ, and affects/determines life span (Kong, Trabucco, & Zhang, 2014; Ziegler, Wiley, & Velarde, 2015). It has been reported that terminally differentiated tissues with active oxidative metabolism, including the brain, tend to accumulate relatively higher levels of mtDNA mutations during aging (Lee & Wei, 2012). As a neural tissue with high metabolism, the retina belongs to this category. In support of this notion, a greater number of rearrangements and deletions were found in retinal mtDNA, compared to blood mtDNA, from normal individuals (Kenney et al., 2010). In addition, AMD retinas showed higher levels of large mtDNA deletions/rearrangements, amino-acid-changing SNPs in the coding genome, and a greater number of SNPs in the noncoding region critical for mtDNA replication and transcription (Kenney et al., 2010). These findings suggested that observed mtDNA variants may diminish energy production efficiency, alter the mtDNA copy number and impact transcription of mtDNA in AMD retinas (Kenney et al., 2010).

POLG is a mtDNA polymerase with proofreading ability (Kaguni, 2004). Mutations in the *POLG* gene cause secondary effects on mtDNA including multiple deletions in mtDNA, point mutations, and mtDNA depletion, leading to pathological symptoms (Hudson & Chinnery, 2006). In mice, the *Polg* mutation caused accumulation of mtDNA mutations and led to reduced life span and several aging-related symptoms (Kujoth et al., 2005), supporting the mitochondrial theory of aging. In humans, *POLG* mutations cause PEO and Alpers–Huttenlocher syndrome characterized by severe hepatoencephalopathy with intractable seizures and visual failure (Copeland, Ponamarev, Nguyen, Kunkel, & Longley, 2003; Hudson & Chinnery, 2006). PEO is a rare disease causing progressive loss of extraocular muscle mobility, often accompanied by other neurological or multisystem abnormalities such as limb, facial, and bulbar muscle weakness, short stature, diabetes mellitus, deafness, cardiac conduction defect, respiratory dysfunction, and endocrine organ abnormalities (Biousse & Newman, 2001). Mutations in *POLG* are responsible for autosomal dominant and autosomal recessive PEO, depending on the nature of mutations (Hudson & Chinnery, 2006).

Although mtDNA mutations have been postulated to cause age-related retinal diseases, it is difficult to directly connect somatic mutations (including secondary mtDNA defects) with retinal disease phenotypes due to their mosaic nature. However, inherited mtDNA mutations known to cause retinal pathologies provide an important indication that mtDNA mutations play critical roles in the pathogenesis of retinal diseases. Most of the inherited mtDNA mutations cause multisystem syndromic disorders, some of which show retinal abnormalities (Gorman & Taylor, 2011; Schrier & Falk, 2011). Leber hereditary optic neuropathy (LHON) is an inherited mitochondrial disorder characterized by acute and painless central vision loss (Gorman & Taylor, 2011; Schrier & Falk, 2011). Pathological characteristics of LHON include initial thickening of retinal nerve fiber layers, with disc pseudoedema and retinal ganglion cell loss within the optic nerve. Ninety-five per cent of LHON cases are accounted for by three mtDNA point mutations within mitochondrial respiratory chain complex I subunit genes (G11778A in *MTND4*, G3460A in *MTND1*, and T14484C in *MTND6*) (Gorman & Taylor, 2011; Schrier & Falk, 2011). MERRF is another neurological syndrome caused by mutations in mtDNA, most commonly associated with an A-to-G transition at nucleotide position (np) 8344 within a gene encoding mitochondrial transfer RNA for lysine (*MTTK*) (Fukuhara, Tokiguchi, Shirakawa, & Tsubaki, 1980; Gorman & Taylor, 2011). A common retinal abnormality observed in LHON and MERRF is optic atrophy (degeneration of the optic nerve) (Gorman & Taylor, 2011), suggesting that the optic nerve is particularly sensitive to mitochondrial mutations. Patients with KSS carry a large-scale deletion of mtDNA often affecting multiple mitochondrial genes, which result in pigmentary retinopathy as one of multiple symptoms (Gorman & Taylor, 2011). Histological analysis suggested that RPE cells are primarily affected by KSS prior to photoreceptor degeneration (McKechnie, King, & Lee, 1985). MELAS is a syndrome mostly affecting the nervous system and muscles (Majamaa et al., 1998). About 80% of patients with MELAS carry an A-to-G transition at np 3243 in the *MTTL* gene encoding the mitochondrial transfer RNA for leucine and others have T-to-C transitions at np 3271 in the same gene

(Tsukuda et al., 1997). RPE cell abnormality appears to be a relatively common feature in MELAS patients, despite variations in phenotypic severity (Gorman & Taylor, 2011; Jones, Mitchell, Wang, & Sue, 2004; Schrier & Falk, 2011; Sue et al., 1997). These findings on inherited mtDNA mutations strongly indicate that each tissue/cell type shows different sensitivity to mtDNA mutations. While all cells contain mitochondria, disease symptoms caused by mtDNA mutations vary greatly, depending on the cell type. It is important to note that abnormalities in the optic nerve and RPE cells are frequently observed in the retina of patients with inherited mtDNA mutations, indicating their high sensitivity. Considering that the optic nerve and RPE cells are major targets of age-related diseases in the retina, sensitivity to mtDNA mutations may be associated with the mechanisms through which accumulation of somatic mtDNA mutations with age leads to abnormalities in those tissues.

Oxidative stress and antioxidant defense

As mitochondria produce energy through respiration, the cellular levels of ROS increase as by-products. In aged tissues, increased ROS production is observed (Capel et al., 2005; Sawada & Carlson, 1987). Increased ROS can induce cellular damage at different levels, including oxidative and nitrative modifications to cellular proteins, lipid peroxidation, DNA damage, and ultimately neuronal death (Chakravarti & Chakravarti, 2007; Cooke, Evans, Dizdaroglu, & Lunec, 2003; Evans, Dizdaroglu, & Cooke, 2004; Filipcik, Cente, Ferencik, Hulin, & Novak, 2006; Nita & Grzybowski, 2016). As mentioned earlier, retinal cells have abundant mitochondria and high mitochondrial activity, suggesting high ROS production. High levels of ROS may, in turn, cause cellular damage and pathologies in the retina. In fact, accumulating evidence suggests that increased oxidative stress is associated with age-related retinal diseases (Nita & Grzybowski, 2016).

RPE cells are considered to be major targets for AMD pathogenesis (Fisher & Ferrington, 2018; Kinnunen, Petrovski, Moe, Berta, & Kaarniranta, 2012). Recent progress in the technology that allows the differentiation of induced pluripotent stem (iPS) cells into RPE cells has allowed researchers to generate RPE cells derived from AMD patients and to test RPE cell-specific changes in AMD (Fields, Cai, Gong, & Del Priore, 2016; Golestaneh et al., 2016). iPS-RPE cells from AMD patients showed increases in ROS production under stress (Golestaneh et al., 2016) and reduced antioxidant expression and ability to resist oxidative stress (Chang et al., 2014; Golestaneh et al., 2016; Yang et al., 2014). One study also observed lower expression levels of peroxisome proliferator-activated receptor gamma coactivator 1-α (PGC-1α), a regulator of mitochondrial biogenesis and function, and NAD-dependent deacetylase sirtuin1 (SIRT1), which is known to deacetylate and activate PGC-1α in AMD patient-derived iPS-RPE cells, suggesting that this pathway, causing impaired mitochondrial activity, contributes to AMD pathophysiology (Golestaneh et al., 2016). A mouse genetic study has shown that mitochondrial antioxidant defense in RPE cells is critical for age-dependent abnormalities of the retina (Mao et al., 2014). RPE-specific conditional knockout of the mitochondrial superoxide dismutase 2 (*Sod2*) gene, encoding the mitochondrial antioxidant enzyme, induced oxidative stress in RPE cells and resulted in RPE dysfunction, damage to the choroid, and death of photoreceptor cells (Mao et al., 2014).

Progressive loss of optic nerve axons and retinal ganglion cell death are central characteristics of vision loss in glaucoma. In general, increased IOP is a major factor for initiation of ganglion cell damage. Increased ROS has been observed in the optic nerve of retinal ischemia models induced by acute IOP elevation (Bonne, Muller, & Villain, 1998; Muller et al., 1997) and by chronic IOP elevation (Moreno et al., 2004). In the model with chronic experimental IOP elevation induced by hyaluronic acid injection, significant decreases in retinal antioxidants and increases in retinal lipid peroxidation (a sign of increased ROS) were observed (Moreno et al., 2004). However, not all cases of glaucoma can be explained by elevation of IOP. There are a significant number of cases in which glaucoma manifests without any changes in IOP (Tezel, 2006). Another contributing factor for glaucoma that could be considered is hypoxia (Tezel, 2006). It has been hypothesized that in the area of the optic nerve head, retinal ganglion cells undergo hypoxia due to compromised local blood flow. A deficit in oxygen would result in a dramatic increase in the production of ROS, causing oxidative stress. Increased ROS due to elevated IOP and/or hypoxia would cause retinal ganglion cell degeneration, at least partially through increasing caspase activity (Tezel, 2006). Molecular pathways through which increased ROS causes retinal ganglion cell degeneration have been extensively reviewed elsewhere (Tezel, 2006).

There are two main sources of oxidative stress. One is through an enzymatic mechanism in which a multi-subunit enzyme, consisting of membrane and cytosolic components, NADPH oxidase (NOX), catalyzes the one-electron reduction of oxygen to superoxide anion in a process that involves oxidation of cytosolic NADPH to $NADP^+$ (Lambeth, 2004; Panday, Sahoo, Osorio, & Batra, 2015). In diabetes, NOX activity is increased in multiple tissues, including pancreatic β-cells and retina (Kowluru & Kowluru, 2014; Kowluru et al., 2014), suggesting its involvement in the pathogenesis of glaucoma. Mitochondria are the other major source of oxidative stress as a consequence of respiration (Murphy, 2009). Electron transfer in the mitochondrial electron-transport chain generates a proton gradient across the mitochondrial intermembrane, which drives the synthesis of ATP as the cellular energy source. Under normal physiological conditions, electrons are transferred from complex I or II to complex III of the electron transport chain, and then transferred onto complex IV and onto molecular oxygen, the final electron acceptor (Kowluru & Mishra, 2015; Murphy, 2009). However, some electrons leak out from complex I or III and interact with molecular oxygen to generate superoxide anion (O_2^-). This system of ROS production also occurs in diabetes. Increased glucose-derived pyruvate and its oxidation in the TCA cycle would increase the flux of electron donors into the electron-transport chain, which may overwhelm the electron-transport chain and increase generation of ROS (Brownlee, 2005; Kowluru & Mishra, 2015).

Cells have mechanisms to reduce ROS through the antioxidant defense system. Mitochondrial abnormalities, such as increased oxidative stress, can be the signal to induce nuclear transcriptional factors that regulate genes involved in this defense system (Finley & Haigis, 2009). Nuclear factor erythroid 2-related factor 2 (NRF2), a member of a family of basic leucine transcription factors, has been identified as one of those transcription factors that mediate mitochondrial stress and is a master regulator of cellular redox homeostasis (Ryoo & Kwak, 2018). NRF2 binds to antioxidant response elements (AREs) in the promoter region of genes involved in cellular homeostasis, including redox regulation (Itoh et al., 1997; Jaiswal, 2004; Nguyen, Sherratt, & Pickett, 2003). The gradual reduction of NRF2 expression and activity with age suggests that a decline of NRF2 function and the resulting decrease in antioxidant capacity are critical for the process of aging and the pathogenesis of age-related diseases, such as neurodegeneration and cancer (Schmidlin, Dodson, Madhavan, & Zhang, 2019; Silva-Palacios, Ostolga-Chavarria, Zazueta, & Konigsberg, 2018; Zhang, Davies, & Forman, 2015). In the eye, involvement of NRF2 in aging and age-dependent diseases, including AMD, glaucoma, cataract, and diabetic retinopathy, has been widely recognized (Batliwala, Xavier, Liu, Wu, & Pang, 2017). Mutations in *NRF2* have been associated with a higher risk of AMD development (Sliwinski, Kolodziejska, Szaflik, Blasiak, & Szaflik, 2013). In mice, Sachdeva et al. showed that aging RPE cells are vulnerable to oxidative damage due to impaired NRF2 signaling (Sachdeva, Cano, & Handa, 2014). Furthermore, mice deficient for *Nrf2* displayed AMD-like retinal pathologies, including druse deposition, degeneration of RPE, Bruch's membrane, and choriocapillaris, increased autofluorescence and development of spontaneous choroidal neovascularization (Zhao et al., 2011). The role of NRF2 has been also studied in animal models for diabetic retinopathy and glaucoma (Himori et al., 2013; Wei et al., 2011; Xu et al., 2014; Zhong, Mishra, & Kowluru, 2013). In the retina of streptozotocin-induced diabetic rats, DNA-binding activity of NRF2 to the regulatory sequence of the antioxidant gene glutamylcysteine ligase (*GCLC*), is impaired, resulting in decreased expression of GCLC (Zhong et al., 2013). *Nrf2* deficiency in mice has been shown to increase retinal ganglion cell death by optic nerve injury (Himori et al., 2013). A similar protective function of NRF2 for retinal ganglion cells was also shown in an ischemia-reperfusion model (Xu et al., 2015). These findings suggest that NRF2 and the NRF2-regulated antioxidant pathway can be potential targets for therapeutic approaches to treat age-related ocular diseases (Batliwala et al., 2017). Natural and synthetic compounds have been tested for the ability to stimulate the NRF2 pathway (Batliwala et al., 2017; Bellezza, 2018). The synthetic triterpenoid RTA 408 was shown to protect human RPE cells against H_2O_2-induced cell injury through activation of NRF2 (Liu et al., 2016). A natural compound found in turmeric, curcumin, protects cultured retinal cells from H_2O_2 exposure through the NRF2-driven expression of hemeoxygenase 1 (HO-1) (Mandal et al., 2009). Carotenoids such as zeaxanthin and lutein protect photoreceptors against light damage (Yu, Yan, & Beight, 2018). Lutein has been shown to activate NRF2 in ARPE-19 cells (RPE cell line) (Frede, Ebert, Kipp, Schwerdtle, & Baldermann, 2017). In animal models of glaucoma (optic nerve injury and ischemia-reperfusion models), the synthetic triterpenoid 1-(2-cyano-3,12-dioxooleana-1,9(11)-dien-28-oyl)imidazole (CDDO-Im) was shown to promote neural survival through the NRF2 pathway (Himori et al., 2013; Xu et al., 2015).

Mitochondrial dynamics

Mitochondria are highly dynamic organelles with central importance for ATP production in most eukaryotic cells (Chan, 2006). They are observed to undergo fission and fusion events continuously (mitochondrial dynamics), leading to a diverse range of morphological changes, from fragmented mitochondria to fused mitochondrial networks. Mitochondrial fusion in mammals is mediated by dynamin-related GTPases, mitofusin 1 and 2 (MFN1 and MFN2), that are responsible for fusion of mitochondrial outer membranes

(Chen et al., 2003), and OPA1 that facilitates fusion of mitochondrial inner membranes (Olichon et al., 2006). Mitochondrial fission in mammals is mediated by dynamin-related protein 1 (DRP1 or DNM1L), also a large GTPase. DRP1 is a cytosolic protein that is recruited to the mitochondrial outer membrane by multiple factors (Loson, Song, Chen, & Chan, 2013) to constrict mitochondria, resulting in division of a mitochondrion into two separate organelles (Mears et al., 2011). An increasing number of studies have shown that fission and fusion of mitochondria are important for many biological functions. Mitochondrial fission is required for inheritance and partitioning of organelles during cell division (Ishihara et al., 2009; Taguchi, Ishihara, Jofuku, Oka, & Mihara, 2007), for the release of proapoptotic factors from the intermembrane space (Frank et al., 2001; Scorrano, 2005; Scorrano et al., 2002), for intracellular distribution by cytoskeleton-mediated transport (Ishihara et al., 2009; Verstreken et al., 2005), and for turnover of damaged organelles by mitophagy (Narendra, Tanaka, Suen, & Youle, 2008; Twig et al., 2008). Mitochondrial fusion also plays important roles in mitochondrial function. Fused mitochondrial networks are important for the dissipation of metabolic energy through transmission of membrane potential along mitochondrial filaments (Amchenkova, Bakeeva, Chentsov, Skulachev, & Zorov, 1988; Skulachev, 2001) and for the complementation of mtDNA gene products in heteroplasmic cells, to counteract the decline of respiratory functions in aging (Chan, 2006; Nakada et al., 2001; Ono, Isobe, Nakada, & Hayashi, 2001). Thus, mitochondrial dynamics are key to regulating mitochondrial function and quality.

Changes in mitochondrial dynamics have been associated with aging and altered life span in yeast (Bernhardt, Muller, Reichert, & Osiewacz, 2015), *Caenorhabditis elegans* (Yang, Chen, Lee, & Walter, 2011), and *Drosophila* (McQuibban, Lee, Zheng, Juusola, & Freeman, 2006). Alterations in mitochondrial fusion and fission proteins have been also associated with several age-dependent diseases in humans and mice (Sebastian, Palacin, & Zorzano, 2017), suggesting dysregulation of mitochondrial dynamics as a contributing factor in the pathogenesis of these diseases. Autosomal optic atrophy type 1 is caused by mutations in the *OPA1* gene that is involved in mitochondrial dynamics and is characterized by an insidious onset of visual impairment in early childhood, with moderate to severe loss of visual acuity, temporal optic disc pallor, color vision deficits, and centrocecal scotoma of variable density (Alexander et al., 2000; Delettre et al., 2000). These findings suggest that mitochondrial dynamics are crucial to the physiology of retinal cells. In addition, RPE cells in AMD patients show decreased numbers of mitochondria compared to those of nonaffected control donors (Feher et al., 2006), suggesting that the regulation of mitochondrial dynamics may influence AMD pathologies. A mouse model displayed accelerated aging phenotypes in the retina including abnormal ectopic synapses, increased inflammatory cells in the subretinal space, photoreceptor degeneration, and accumulation of lipofuscin-like autofluorescent materials in RPE cells (Lee et al., 2016). The gene responsible for these age-related disease phenotypes encodes transmembrane protein 135 (TMEM135), which is involved in mitochondrial fission. Enlarged mitochondria, decreased ATP production, increased ROS, and sensitivity to oxidative stress were observed in *Tmem135* mutant mice (Lee et al., 2016). These results indicated that mitochondrial dynamics may have critical roles in regulating normal aging, impairment of which causes age-related retinal disease phenotypes (Lee et al., 2016). Other observations from glaucoma research suggest the potential involvement of the regulation of mitochondrial dynamics in the pathogenesis of age-related retinal disease. In a glaucoma model, IOP elevation was found to induce mitochondrial fission (Ju et al., 2008). Moreover, injection of a DRP1 inhibitor increased retinal ganglion cell survival in acute ischemic mouse retina (Park et al., 2011). These findings suggest that retinal ganglion cell degeneration can be caused by altering mitochondrial dynamics and functions.

Mitochondrial apoptosis and mitochondrial quality control

As described in the auditory system section, the mitochondrial apoptotic pathway that is regulated by BCL2 family members has a significant role in aging and age-related diseases. The retina is no exception in terms of the significance of this apoptotic pathway. The role of BAX in retinal ganglion cell degeneration has been extensively studied using *Bax*-deficient mice (reviewed in Maes, Schlamp, & Nickells, 2017). *Bax*-deficiency in mouse glaucoma models suppressed degeneration of retinal ganglion cells (Maes et al., 2017). For example, DBA/2J mice are the most widely studied mouse model, with spontaneous development of increased IOP and retinal ganglion cell pathologies resembling human glaucoma (Libby, Anderson, et al., 2005). DBA/2J mice that are deficient for *Bax* ($Bax^{-/-}$) were protected from retinal ganglion cell death (Libby, Li, et al., 2005). In addition, Li et al. showed that retinal ganglion cell death is dependent on BAX in the optic nerve crush model (Li, Schlamp, Poulsen, & Nickells, 2000). The effect on retinal degeneration of BAX-inhibiting peptides (BIPs) that belong to the peptide group of cell-penetrating peptides (CPPs) has been shown in multiple studies. BIP protects retinal ganglion cells from degeneration in both the rat optic nerve transection model (Qin, Patil, & Sharma, 2004) and the hypoxic-ischemic injury model (Chen et al., 2007). BIP also protects other cell types in the retina. All-*trans*-retinal (atRAL)-induced death of the cultured human RPE cell line, ARPE-19, and photoreceptor cell death in *Abca4*$^{-/-}$ *Rdh8*$^{-/-}$ mice (known to exhibit light-induced retinal degeneration with delayed clearance of atRAL), were rescued by BIP treatments (Sawada et al., 2014). In summary, BAX activation is critical for some forms of age-dependent retinal disease phenotypes and can be a therapeutic target for these diseases.

Mitochondrial quality control is one of the key mechanisms for cells to avoid apoptosis/degeneration (Palikaras, Daskalaki, Markaki, & Tavernarakis, 2017). As an essential part of mitochondrial quality control, mitophagy is a mechanism of eliminating severely damaged mitochondria (Hyttinen, Viiri, Kaarniranta, & Blasiak, 2018; Palikaras et al., 2017). Mitophagy refers to degradation of mitochondria through autophagy, a proteolytic pathway for cells to eliminate large structures such as cytoplasmic materials and organelles. Age-related decline of mitophagy increases the accumulation of damaged mitochondria and leads to age-related diseases. The age-related neurodegenerative disorder, Parkinson disease, is associated with two genes, *PINK1* (phosphatase and tennis homolog-induced putative kinase 1) and *PRKN* [Parkin RING-IBR-RING (RBR) E3 ubiquitin ligase] (Kitada et al., 1998; Valente et al., 2004). Both PINK1 and PRKN are involved in the degradation of damaged mitochondria through mitophagy (Rakovic et al., 2013). More recently, it has been shown that heat shock protein 70 (HSP70) participates in PINK1-mediated mitophagy by regulating the stability of PINK1 (Zheng et al., 2018). Interestingly, Subrizi and colleagues showed that HSP70 protects ARPE-19 cells from oxidative stress, suggesting the potential of HSP70 delivery to RPE cells as a new therapeutic strategy against RPE degeneration (Subrizi et al., 2015). This RPE cell protection by HSP70 may occur at least partially through the mitophagy process.

On the other hand, excessive mitophagy could reduce the number of mitochondria, which results in impairment of total mitochondrial function (for example, total energy production). Therefore cells also possess the mitochondrial biogenesis pathway to compensate for the loss of mitochondria and to balance the number of healthy mitochondria (Palikaras et al., 2017). PGC-1α is a transcriptional coactivator of a number of genes involved in energy metabolism, mitochondrial biogenesis, and ROS control, and its function is associated with age-dependent pathologies in RPE cells (reviewed in Kaarniranta, Kajdanek, Morawiec, Pawlowska, & Blasiak, 2018). Although the role of PGC-1α related to mitochondrial biogenesis in age-related diseases such as AMD has to be elucidated further, several lines of evidence suggest that PGC-1α has essential roles in maintaining the health of retinal cells, including RPE cells (Kaarniranta et al., 2018). PGC-1α is highly expressed in the mouse retina, especially in photoreceptor cells, and the retina of mice deficient for PGC-1α becomes more sensitive to light damage (Egger et al., 2012). Human RPE cells differentiated from AMD patient-derived iPS cells displayed repression of PGC-1α compared with those from normal donors (Golestaneh et al., 2016). RPE cells from AMD patients also showed lower expression of SIRT1 that deacetylates and activates PGC-1α (Golestaneh et al., 2016). These findings suggest that the SIRT1/PGC-1α pathway dysfunction may contribute to AMD pathophysiology.

Conclusions

Our senses, including hearing and vision, play critical roles for survival. For example, the auditory system, which detects sound, and the visual system, which detects visible light, play essential roles in locating food, avoiding predators, and/or finding a mate in a variety of species. Aging can affect all of these sensory systems, but the auditory and visual systems are thought to be more vulnerable. The auditory and visual systems contain specialized sensory receptors: hair cells that detect sound waves in the cochleae (Hudspeth, 1997) and photoreceptor cells that detect visible light in the eyes (Sung & Chuang, 2010). Age-related changes in the structure and functions of these sensory receptors and cochlear and retinal ganglion neurons result in hearing and visual impairments, including AHL, the most common form of hearing impairment (Yamasoba et al., 2013), AMD, the most common cause of visual impairment (Pennington & DeAngelis, 2016), and glaucoma, caused by retinal ganglion cell degeneration (Almasieh et al., 2012).

A central role for mitochondrial dysfunction in hearing and vision impairments is supported by the numerous reports that a large number of genetic syndromes associated with both hearing and vision disorders are due to defects in mitochondria (Biousse & Newman, 2001; Chinnery et al., 2000; Fischel-Ghodsian, 2003; Gorman & Taylor, 2011; Hudson & Chinnery, 2006; Kenney et al., 2010; Kokotas et al., 2007; Kujoth et al., 2007; McKechnie et al., 1985; Someya & Prolla, 2010); these mitochondrial diseases includes MELAS, MERRF, KSS, and PEO. This observation strongly suggests that photoreceptors, hair cells,

TABLE 14.1 Mutations associated with mitochondrial dysfunction and hearing impairments in humans and in mouse models.

Gene or disease	Function	References
KSS	Multiple mitochondrial genes are deleted	Chinnery et al. (2000), Kokotas et al. (2007)
MERRF	Mitochondrial transfer RNA lysine	Chinnery et al. (2000), Kokotas et al. (2007)
MELAS	Mitochondrial transfer RNA leucine	Chinnery et al. (2000), Kokotas et al. (2007)
POLG1 (PEO)	Mitochondrial DNA repair	Hudson and Chinnery (2006), Kujoth et al. (2005)
OPA1	Mitochondrial fusion	Hudson et al. (2008), Liguori et al. (2008)
CISD2 (Wolfram syndrome 2)	Calcium homeostasis, autophagy	Amr et al. (2007), Chen et al. (2009)
CS	TCA cycle	Johnson et al. (2012)
Idh2	TCA cycle	White et al. (2018)
Sirt3	Antioxidant defense	Someya et al. (2010)
Gpx1	Antioxidant defense	Ohlemiller et al. (2000)
Cat	Antioxidant defense	Someya et al. (2009)
Bak	Mitochondrial apoptosis	Someya et al. (2009)
Bcl11b	Mitochondrial apoptosis	Okumura et al. (2011)
PTEN	Mitochondrial apoptosis	Sha et al. (2010)
UCP2	Mitochondrial apoptosis	Sugiura et al. (2010)

KSS, Kearns–Sayre syndrome; *MELAS*, mitochondrial encephalomyopathy, lactic acidosis, and stroke-like episodes; *MERRF*, myoclonic epilepsy with ragged-red fibers; *OPA1*, optic atrophy 1; *PEO*, progressive external ophthalmoplegia.

TABLE 14.2 Mutations associated with mitochondrial dysfunction and retinal abnormalities in humans and in mouse models.

Gene or disease	Function	Retinal diseases	References
POLG (PEO)	Mitochondrial DNA repair	PEO	Copeland et al. (2003), Hudson and Chinnery (2006)
MTND4 (LHON)	mitochondrial respiratory Complex I	Optic atrophy	Gorman and Taylor (2011), Schrier and Falk (2011)
MTND1 (LHON)	mitochondrial respiratory Complex I	Optic atrophy	Gorman and Taylor (2011), Schrier and Falk (2011)
MTND6 (LHON)	mitochondrial respiratory Complex I	Optic atrophy	Gorman and Taylor (2011), Schrier and Falk (2011)
MTTK (MERRF)	mitochondrial transfer RNA lysine	Optic atrophy	Fukuhara et al. (1980)), Gorman and Taylor (2011)
KSS	Multiple mitochondrial genes are deleted	Pigmentary retinopathy, Photoreceptor degeneration	Gorman and Taylor (2011), McKechnie et al. (1985)
MTTL (MELAS)	Mitochondrial transfer RNA leucine	RPE cell abnormality	Gorman and Taylor (2011), Tsukuda et al. (1997)
Sod2	Mitochondrial antioxidant enzyme	RPE dysfunction and photoreceptor degeneration	Mao et al. (2014)
Nrf2	Transcription factor	AMD-like retinal pathologies	Zhao et al. (2011)
OPA1	Mitochondrial fusion	Optic atrophy	Alexander et al. (2000), Delettre et al. (2000)
Tmem135	Mitochondrial fission	Accelerated retinal aging phenotypes	Lee et al. (2016)
Bax	Mitochondrial apoptosis	Rescue of RGC degeneration	Li, Schlamp et al. (2000), Libby, Li, et al. (2005)
Ppargc1a (PGC-1a)	Mitochondrial energy metabolism	Increased retinal damage and apoptosis by light stimulation	Egger et al. (2012)

LHON, Leber hereditary optic neuropathy; *MELAS*, mitochondrial encephalomyopathy, lactic acidosis, and stroke-like episodes; *MERRF*, myoclonic epilepsy with ragged-red fibers; *OPA1*, optic atrophy 1; *PEO*, progressive external ophthalmoplegia.

and cochlear and retinal ganglion neurons in the cochleae and eyes are exquisitely sensitive to disturbances in energy metabolism and that mitochondrial decay associated with aging may selectively impact the cochleae and eyes over the course of the lifetime. In summary, this chapter reviewed what has been learned about the roles of mitochondrial dysfunction in age-related hearing and visual impairments. Tables 14.1 and 14.2 summarize the genes and pathways that have been uncovered. These findings have significantly advanced the field of age-related hearing and visual impairments and contributed to the understanding of the molecular mechanisms underlying aging of the auditory and visual systems in laboratory animals and humans.

Abbreviations

ABR	auditory brainstem response
AHL	age-related hearing loss
AKT	protein kinase B
AMD	age-related macular degeneration
AREs	antioxidant response elements
atRAL	all-trans-retinal
BCL11B	B-cell leukemia/lymphoma 11B gene
CAT	catalase
Complex I	NADH dehydrogenase
Complex III	ubiquinone-cytochrome c reductase
CR	calorie restriction
CS	citrate synthase
DRP1	dynamin-related protein 1
GCLC	glutamylcysteine ligase
GPX1	glutathione peroxidase 1
GSH	reduced glutathione
GSR	glutathione reductase
GSSG: GSTM1	glutathione S-transferase mu 1
GSTP1	glutathione S-transferase pi 1, oxidized glutathione
GSTT1	glutathione S-transferase theta 1
H_2O_2	hydrogen peroxide
HO-1	hemeoxygenase 1
HSP70	heat shock protein 70
IDH2	isocitrate dehydrogenase 2
IHC	inner hair cells
IOP	intraocular pressure
iPS cells	induced pluripotent stem cells
KSS	Kearns–Sayre syndrome
LHON	Leber hereditary optic neuropathy
MCAT	catalase targeted to the mitochondria
MELAS	mitochondrial encephalomyopathy, lactic acidosis, and stroke-like episodes
MERRF	myoclonic epilepsy with ragged red fibers
MIDD	maternally inherited diabetes and deafness
MFN1	mitofusin 1
MFN2	mitofusin 2
mtDNA	mitochondrial DNA
NCAT	catalase targeted to the nucleus
NOX	NADPH oxidase
NIHL	noise-induced hearing loss
NRF2	nuclear factor erythroid 2-related factor 2
$\bullet O_2^-$	superoxide
•OH	hydroxyl radical
OHC	outer hair cells
OPA1	optic atrophy 1
PCAT	catalase targeted to the peroxisome
PEO	progressive external ophthalmoplegia
PGC-1α	peroxisome proliferator-activated receptor gamma coactivator 1-alpha
PINK1	phosphatase and tennis homolog-induced putative kinase 1
PIP3	phosphatidylinositol 3,4,5-trisphosphate
POLG	DNA polymerase gamma
PRKN	Parkin RING-IBR-RING [RBR] E3 ubiquitin ligase
PTEN	phosphatase and tensin homolog
ROS	reactive oxygen species
RPE	retinal pigment epithelium
SGN	spiral ganglion neuron
SIRT1	sirtuin1
SIRT3	sirtuin3
SOD1	superoxide dismutase 1
SOD2	superoxide dismutase 2
SV	stria vascularis
TCA	tricarboxylic acid
TMEM135	transmembrane protein 135
TNF	tumor necrosis factors
TXNRD2	thioredoxin reductase 2
UCP2	uncoupling protein 2
WT	wild-type

References

Ahn, B. H., Kim, H. S., Song, S., Lee, I. H., Liu, J., Vassilopoulos, A., ... Finkel, T. (2008). A role for the mitochondrial deacetylase Sirt3 in regulating energy homeostasis. *Proceedings of the National Academy of Sciences of the United States of America*, *105*(38), 14447–14452. Available from https://doi.org/10.1073/pnas.0803790105.

Ahn, J. H., Kang, H. H., Kim, T. Y., Shin, J. E., & Chung, J. W. (2008). Lipoic acid rescues DBA mice from early-onset age-related hearing impairment. *Neuroreport*, *19*(13), 1265–1269. Available from https://doi.org/10.1097/WNR.0b013e328308b33800001756-200808270-00004, [pii].

Alder, V. A., Ben-Nun, J., & Cringle, S. J. (1990). PO2 profiles and oxygen consumption in cat retina with an occluded retinal circulation. *Investigative Ophthalmology & Visual Science*, *31*(6), 1029–1034. Available from https://www.ncbi.nlm.nih.gov/pubmed/2354908.

Alexander, C., Votruba, M., Pesch, U. E., Thiselton, D. L., Mayer, S., Moore, A., ... Wissinger, B. (2000). OPA1, encoding a dynamin-related GTPase, is mutated in autosomal dominant optic atrophy linked to chromosome 3q28. *Nature Genetics*, *26*(2), 211–215. Available from https://doi.org/10.1038/79944.

Almasieh, M., Wilson, A. M., Morquette, B., Cueva Vargas, J. L., & Di Polo, A. (2012). The molecular basis of retinal ganglion cell death in glaucoma. *Progress in Retinal and Eye Research*, *31*(2), 152–181. Available from https://doi.org/10.1016/j.preteyeres.2011.11.002.

Amchenkova, A. A., Bakeeva, L. E., Chentsov, Y. S., Skulachev, V. P., & Zorov, D. B. (1988). Coupling membranes as energy-transmitting cables. I. Filamentous mitochondria in fibroblasts and mitochondrial clusters in cardiomyocytes. *The Journal of Cell Biology*, *107*(2), 481–495. Available from https://doi.org/10.1083/jcb.107.2.481.

Amr, S., Heisey, C., Zhang, M., Xia, X. J., Shows, K. H., Ajlouni, K., ... Shiang, R. (2007). A homozygous mutation in a novel zinc-finger protein, ERIS, is responsible for Wolfram syndrome 2. *American Journal of Human Genetics*, *81*(4), 673–683. Available from https://doi.org/10.1086/520961.

Anderson, M. E. (1998). Glutathione: An overview of biosynthesis and modulation. *Chemico-Biological Interactions, 111–112*, 1–14. Available from http://www.ncbi.nlm.nih.gov/pubmed/9679538.

Angelaki, D. E., & Cullen, K. E. (2008). Vestibular system: The many facets of a multimodal sense. *Annual Review of Neuroscience, 31*, 125–150. Available from https://doi.org/10.1146/annurev.neuro.31.060407.125555.

Bai, U., Seidman, M. D., Hinojosa, R., & Quirk, W. S. (1997). Mitochondrial DNA deletions associated with aging and possibly presbycusis: A human archival temporal bone study. *American Journal of Otology, 18*(4), 449–453. Available from https://www.ncbi.nlm.nih.gov/pubmed/9233484.

Balaban, R. S., Nemoto, S., & Finkel, T. (2005). Mitochondria, oxidants, and aging. *Cell, 120*(4), 483–495. Available from https://doi.org/10.1016/j.cell.2005.02.001.

Banaclocha, M. M. (2001). Therapeutic potential of N-acetylcysteine in age-related mitochondrial neurodegenerative diseases. *Medical Hypotheses, 56*(4), 472–477. Available from https://doi.org/10.1054/mehy.2000.1194.

Batliwala, S., Xavier, C., Liu, Y., Wu, H., & Pang, I. H. (2017). Involvement of Nrf2 in ocular diseases. *Oxidative Medicine and Cellular Longevity, 2017*, 1703810. Available from https://doi.org/10.1155/2017/1703810.

Beckman, J. A., & Creager, M. A. (2016). Vascular complications of diabetes. *Circulation Research, 118*(11), 1771–1785. Available from https://doi.org/10.1161/CIRCRESAHA.115.306884.

Beckman, K. B., & Ames, B. N. (1998). The free radical theory of aging matures. *Physiological Reviews, 78*(2), 547–581. http://www.ncbi.nlm.nih.gov/pubmed/9562038.

Bellezza, I. (2018). Oxidative stress in age-related macular degeneration: Nrf2 as therapeutic target. *Frontiers in Pharmacology, 9*, 1280. Available from https://doi.org/10.3389/fphar.2018.01280.

Bender, A., Krishnan, K. J., Morris, C. M., Taylor, G. A., Reeve, A. K., Perry, R. H., ... Turnbull, D. M. (2006). High levels of mitochondrial DNA deletions in substantia nigra neurons in aging and Parkinson disease. *Nature Genetics, 38*(5), 515–517. Available from https://doi.org/10.1038/ng1769.

Bernhardt, D., Muller, M., Reichert, A. S., & Osiewacz, H. D. (2015). Simultaneous impairment of mitochondrial fission and fusion reduces mitophagy and shortens replicative lifespan. *Science Reports, 5*, 7885. Available from https://doi.org/10.1038/srep07885.

Biousse, V., & Newman, N. J. (2001). Neuro-ophthalmology of mitochondrial diseases. *Seminars in Neurology, 21*(3), 275–291. Available from https://doi.org/10.1055/s-2001-17945.

Birch, D. G., & Anderson, J. L. (1992). Standardized full-field electroretinography. Normal values and their variation with age. *Archives of Ophthalmology, 110*(11), 1571–1576. Available from https://doi.org/10.1001/archopht.1992.01080230071024.

Bonne, C., Muller, A., & Villain, M. (1998). Free radicals in retinal ischemia. *General Pharmacology, 30*(3), 275–280. Available from https://doi.org/10.1016/s0306-3623(97)00357-1.

Bonnel, S., Mohand-Said, S., & Sahel, J. A. (2003). The aging of the retina. *Experimental Gerontology, 38*(8), 825–831. Available from https://doi.org/10.1016/s0531-5565(03)00093-7.

Bristow, E. A., Griffiths, P. G., Andrews, R. M., Johnson, M. A., & Turnbull, D. M. (2002). The distribution of mitochondrial activity in relation to optic nerve structure. *Archives of Ophthalmology, 120*(6), 791–796. Available from https://doi.org/10.1001/archopht.120.6.791.

Brownlee, M. (2005). The pathobiology of diabetic complications: A unifying mechanism. *Diabetes, 54*(6), 1615–1625. Available from https://doi.org/10.2337/diabetes.54.6.1615.

Buono, R. J., & Sheffield, J. B. (1991). Changes in distribution of mitochondria in the developing chick retina. *Experimental Eye Research, 53*(2), 187–198. Available from https://doi.org/10.1016/0014-4835(91)90073-n.

Capel, F., Rimbert, V., Lioger, D., Diot, A., Rousset, P., Mirand, P. P., ... Mosoni, L. (2005). Due to reverse electron transfer, mitochondrial H_2O_2 release increases with age in human vastus lateralis muscle although oxidative capacity is preserved. *Mechanisms of Ageing and Development, 126*(4), 505–511. Available from https://doi.org/10.1016/j.mad.2004.11.001.

Chakravarti, B., & Chakravarti, D. N. (2007). Oxidative modification of proteins: Age-related changes. *Gerontology, 53*(3), 128–139. Available from https://doi.org/10.1159/000097865.

Chan, D. C. (2006). Mitochondria: Dynamic organelles in disease, aging, and development. *Cell, 125*(7), 1241–1252. Available from https://doi.org/10.1016/j.cell.2006.06.010.

Chang, Y. C., Chang, W. C., Hung, K. H., Yang, D. M., Cheng, Y. H., Liao, Y. W., ... Chen, S. J. (2014). The generation of induced pluripotent stem cells for macular degeneration as a drug screening platform: Identification of curcumin as a protective agent for retinal pigment epithelial cells against oxidative stress. *Frontiers in Aging Neuroscience, 6*, 191. Available from https://doi.org/10.3389/fnagi.2014.00191.

Chau, B. N., Chen, T. T., Wan, Y. Y., DeGregori, J., & Wang, J. Y. (2004). Tumor necrosis factor alpha-induced apoptosis requires p73 and c-ABL activation downstream of RB degradation. *Molecular and Cellular Biology, 24*(10), 4438–4447. Available from https://doi.org/10.1128/mcb.24.10.4438-4447.2004.

Chen, E., Soderberg, P. G., & Lindstrom, B. (1989). Activity distribution of cytochrome oxidase in the rat retina. A quantitative histochemical study. *Acta Ophthalmologica, 67*(6), 645–651. Available from https://doi.org/10.1111/j.1755-3768.1989.tb04396.x.

Chen, G., & Goeddel, D. V. (2002). TNF-R1 signaling: A beautiful pathway. *Science, 296*(5573), 1634–1635. Available from https://doi.org/10.1126/science.1071924.

Chen, H., Detmer, S. A., Ewald, A. J., Griffin, E. E., Fraser, S. E., & Chan, D. C. (2003). Mitofusins Mfn1 and Mfn2 coordinately regulate mitochondrial fusion and are essential for embryonic development. *The Journal of Cell Biology, 160*(2), 189–200. Available from https://doi.org/10.1083/jcb.200211046.

Chen, T., He, J., Shen, L., Fang, H., Nie, H., Jin, T., ... Bai, Y. (2011). The mitochondrial DNA 4,977-bp deletion and its implication in copy number alteration in colorectal cancer. *BMC Medical Genetics, 12*, 8. Available from https://doi.org/10.1186/1471-2350-12-8.

Chen, Y. F., Kao, C. H., Chen, Y. T., Wang, C. H., Wu, C. Y., Tsai, C. Y., ... Tsai, T. F. (2009). Cisd2 deficiency drives premature aging and causes mitochondria-mediated defects in mice. *Genes & Development, 23*(10), 1183–1194. Available from https://doi.org/10.1101/gad.1779509.

Chen, Y. N., Yamada, H., Mao, W., Matsuyama, S., Aihara, M., & Araie, M. (2007). Hypoxia-induced retinal ganglion cell death and the neuroprotective effects of beta-adrenergic antagonists. *Brain Research, 1148*, 28–37. Available from https://doi.org/10.1016/j.brainres.2007.02.027.

Chinnery, P. F., Elliott, C., Green, G. R., Rees, A., Coulthard, A., Turnbull, D. M., & Griffiths, T. D. (2000). The spectrum of hearing loss due to mitochondrial DNA defects. *Brain, 123*(Pt 1), 82–92. Available from https://doi.org/10.1093/brain/123.1.82.

Cooke, M. S., Evans, M. D., Dizdaroglu, M., & Lunec, J. (2003). Oxidative DNA damage: Mechanisms, mutation, and disease. *The FASEB Journal, 17*(10), 1195–1214. Available from https://doi.org/10.1096/fj.02-0752rev.

Copeland, W. C., & Longley, M. J. (2014). Mitochondrial genome maintenance in health and disease. *DNA Repair, 19*, 190–198. Available from https://doi.org/10.1016/j.dnarep.2014.03.010.

Copeland, W. C., Ponamarev, M. V., Nguyen, D., Kunkel, T. A., & Longley, M. J. (2003). Mutations in DNA polymerase gamma cause error prone DNA synthesis in human mitochondrial disorders. *Acta Biochimica Polonica, 50*(1), 155–167, 035001155.

Curcio, C. A., & Drucker, D. N. (1993). Retinal ganglion cells in Alzheimer's disease and aging. *Annals of Neurology*, *33*(3), 248–257. Available from https://doi.org/10.1002/ana.410330305.

Curcio, C. A., Medeiros, N. E., & Millican, C. L. (1996). Photoreceptor loss in age-related macular degeneration. *Investigative Ophthalmology & Visual Science*, *37*(7), 1236–1249. Available from https://www.ncbi.nlm.nih.gov/pubmed/8641827.

Curcio, C. A., Millican, C. L., Allen, K. A., & Kalina, R. E. (1993). Aging of the human photoreceptor mosaic: Evidence for selective vulnerability of rods in central retina. *Investigative Ophthalmology & Visual Science*, *34*(12), 3278–3296. Available from https://www.ncbi.nlm.nih.gov/pubmed/8225863.

Curcio, C. A., Millican, C. L., Bailey, T., & Kruth, H. S. (2001). Accumulation of cholesterol with age in human Bruch's membrane. *Investigative Ophthalmology & Visual Science*, *42*(1), 265–274. Available from https://www.ncbi.nlm.nih.gov/pubmed/11133878.

Datta, S., Cano, M., Ebrahimi, K., Wang, L., & Handa, J. T. (2017). The impact of oxidative stress and inflammation on RPE degeneration in non-neovascular AMD. *Progress in Retinal and Eye Research*, *60*, 201–218. Available from https://doi.org/10.1016/j.preteyeres.2017.03.002.

Delettre, C., Lenaers, G., Griffoin, J. M., Gigarel, N., Lorenzo, C., Belenguer, P., ... Hamel, C. P. (2000). Nuclear gene OPA1, encoding a mitochondrial dynamin-related protein, is mutated in dominant optic atrophy. *Nature Genetics*, *26*(2), 207–210. Available from https://doi.org/10.1038/79936.

Deschauer, M., Muller, T., Wieser, T., Schulte-Mattler, W., Kornhuber, M., & Zierz, S. (2001). Hearing impairment is common in various phenotypes of the mitochondrial DNA A3243G mutation. *Archives of Neurology*, *58*(11), 1885–1888. Available from https://doi.org/10.1001/archneur.58.11.1885.

Eells, J. T. (2019). Mitochondrial dysfunction in the aging retina. *Biology*, *8*(2). Available from https://doi.org/10.3390/biology8020031.

Egger, A., Samardzija, M., Sothilingam, V., Tanimoto, N., Lange, C., Salatino, S., ... Handschin, C. (2012). PGC-1alpha determines light damage susceptibility of the murine retina. *PLoS One*, *7*(2), e31272. Available from https://doi.org/10.1371/journal.pone.0031272.

Evans, M. D., Dizdaroglu, M., & Cooke, M. S. (2004). Oxidative DNA damage and disease: Induction, repair and significance. *Mutation Research*, *567*(1), 1–61. Available from https://doi.org/10.1016/j.mrrev.2003.11.001.

Feeney-Burns, L., Hilderbrand, E. S., & Eldridge, S. (1984). Aging human RPE: Morphometric analysis of macular, equatorial, and peripheral cells. *Investigative Ophthalmology & Visual Science*, *25*(2), 195–200, Retrieved from. Available from https://www.ncbi.nlm.nih.gov/pubmed/6698741.

Feher, J., Kovacs, I., Artico, M., Cavallotti, C., Papale, A., & Balacco Gabrieli, C. (2006). Mitochondrial alterations of retinal pigment epithelium in age-related macular degeneration. *Neurobiology of Aging*, *27*(7), 983–993. Available from https://doi.org/10.1016/j.neurobiolaging.2005.05.012.

Ferrington, D. A., Ebeling, M. C., Kapphahn, R. J., Terluk, M. R., Fisher, C. R., Polanco, J. R., ... Montezuma, S. R. (2017). Altered bioenergetics and enhanced resistance to oxidative stress in human retinal pigment epithelial cells from donors with age-related macular degeneration. *Redox Biology*, *13*, 255–265. Available from https://doi.org/10.1016/j.redox.2017.05.015.

Fields, M., Cai, H., Gong, J., & Del Priore, L. (2016). Potential of induced pluripotent stem cells (iPSCs) for treating age-related macular degeneration (AMD). *Cells*, *5*(4), 44. Available from https://doi.org/10.3390/cells5040044.

Filipcik, P., Cente, M., Ferencik, M., Hulin, I., & Novak, M. (2006). The role of oxidative stress in the pathogenesis of Alzheimer's disease. *Bratislavske Lekarske Listy*, *107*(9-10), 384–394, Retrieved from. Available from https://www.ncbi.nlm.nih.gov/pubmed/17262991.

Filosto, M., Mancuso, M., Nishigaki, Y., Pancrudo, J., Harati, Y., Gooch, C., ... DiMauro, S. (2003). Clinical and genetic heterogeneity in progressive external ophthalmoplegia due to mutations in polymerase gamma. *Archives of Neurology*, *60*(9), 1279–1284. Available from https://doi.org/10.1001/archneur.60.9.1279.

Finkel, T., Deng, C. X., & Mostoslavsky, R. (2009). Recent progress in the biology and physiology of sirtuins. *Nature*, *460*(7255), 587–591. Available from https://doi.org/10.1038/nature08197.

Finkel, T., & Holbrook, N. J. (2000). Oxidants, oxidative stress and the biology of ageing. *Nature*, *408*(6809), 239–247. Available from https://doi.org/10.1038/35041687.

Finley, L. W., & Haigis, M. C. (2009). The coordination of nuclear and mitochondrial communication during aging and calorie restriction. *Ageing Research Reviews*, *8*(3), 173–188. Available from https://doi.org/10.1016/j.arr.2009.03.003.

Fischel-Ghodsian, N. (2003). Mitochondrial deafness. *Ear and Hearing*, *24*(4), 303–313. Available from https://doi.org/10.1097/01.AUD.0000079802.82344.B5.

Fischel-Ghodsian, N., Bykhovskaya, Y., Taylor, K., Kahen, T., Cantor, R., Ehrenman, K., ... Keithley, E. (1997). Temporal bone analysis of patients with presbycusis reveals high frequency of mitochondrial mutations. *Hearing Research*, *110*(1–2), 147–154. Available from https://doi.org/10.1016/s0378-5955(97)00077-4.

Fisher, C. R., & Ferrington, D. A. (2018). Perspective on AMD pathobiology: A bioenergetic crisis in the RPE. *Investigative Ophthalmology & Visual Science*, *59*(4), AMD41–AMD47. Available from https://doi.org/10.1167/iovs.18-24289.

Fontana, L., Partridge, L., & Longo, V. D. (2010). Extending healthy life span--from yeast to humans. *Science*, *328*(5976), 321–326. Available from https://doi.org/10.1126/science.1172539.

Frank, S., Gaume, B., Bergmann-Leitner, E. S., Leitner, W. W., Robert, E. G., Catez, F., ... Youle, R. J. (2001). The role of dynamin-related protein 1, a mediator of mitochondrial fission, in apoptosis. *Developmental Cell*, *1*(4), 515–525. Available from https://doi.org/10.1016/s1534-5807(01)00055-7.

Frede, K., Ebert, F., Kipp, A. P., Schwerdtle, T., & Baldermann, S. (2017). Lutein activates the transcription factor Nrf2 in human retinal pigment epithelial cells. *Journal of Agricultural and Food Chemistry*, *65*(29), 5944–5952. Available from https://doi.org/10.1021/acs.jafc.7b01929.

Fukuhara, N., Tokiguchi, S., Shirakawa, K., & Tsubaki, T. (1980). Myoclonus epilepsy associated with ragged-red fibres (mitochondrial abnormalities): Disease entity or a syndrome? Light-and electron-microscopic studies of two cases and review of literature. *Journal of the Neurological Sciences*, *47*(1), 117–133. Available from https://doi.org/10.1016/0022-510x(80)90031-3.

Gao, H., & Hollyfield, J. G. (1992). Aging of the human retina. Differential loss of neurons and retinal pigment epithelial cells. *Investigative Ophthalmology & Visual Science*, *33*(1), 1–17. Available from https://www.ncbi.nlm.nih.gov/pubmed/1730530.

Georgescu, M. M. (2010). PTEN tumor suppressor network in PI3K-Akt pathway control. *Genes Cancer*, *1*(12), 1170–1177. Available from https://doi.org/10.1177/1947601911407325.

Gerth, C., Garcia, S. M., Ma, L., Keltner, J. L., & Werner, J. S. (2002). Multifocal electroretinogram: Age-related changes for different luminance levels. *Graefe's Archive for Clinical and Experimental Ophthalmology = Albrecht von Graefes Archiv fur Klinische und Experimentelle Ophthalmologie*, *240*(3), 202–208. Available from https://doi.org/10.1007/s00417-002-0442-6.

Golestaneh, N., Chu, Y., Cheng, S. K., Cao, H., Poliakov, E., & Berinstein, D. M. (2016). Repressed SIRT1/PGC-1alpha pathway and mitochondrial disintegration in iPSC-derived RPE disease model of age-related macular degeneration. *Journal of Translational*

Medicine, 14(1), 344. Available from https://doi.org/10.1186/s12967-016-1101-8.

Gorman, G. S., & Taylor, R. W. (2011). Mitochondrial DNA abnormalities in ophthalmological disease. *Saudi Journal of Ophthalmology, 25*(4), 395–404. Available from https://doi.org/10.1016/j.sjopt.2011.02.002.

Grabarczyk, P., Przybylski, G. K., Depke, M., Volker, U., Bahr, J., Assmus, K., ... Schmidt, C. A. (2007). Inhibition of BCL11B expression leads to apoptosis of malignant but not normal mature T cells. *Oncogene, 26*(26), 3797–3810. Available from https://doi.org/10.1038/sj.onc.1210152.

Gratton, M. A., & Schulte, B. A. (1995). Alterations in microvasculature are associated with atrophy of the stria vascularis in quiet-aged gerbils. *Hearing Research, 82*(1), 44–52. Available from https://doi.org/10.1016/0378-5955(94)00161-i.

Guastini, L., Mora, R., Dellepiane, M., Santomauro, V., Giorgio, M., & Salami, A. (2011). Water-soluble coenzyme Q10 formulation in presbycusis: Long-term effects. *Acta Oto-Laryngologica, 131*(5), 512–517. Available from https://doi.org/10.3109/00016489.2010.539261.

Hageman, G. S., Luthert, P. J., Victor Chong, N. H., Johnson, L. V., Anderson, D. H., & Mullins, R. F. (2001). An integrated hypothesis that considers drusen as biomarkers of immune-mediated processes at the RPE-Bruch's membrane interface in aging and age-related macular degeneration. *Progress in Retinal and Eye Research, 20*(6), 705–732. Available from https://doi.org/10.1016/s1350-9462(01)00010-6.

Halliwell, B., & Gutteridge, J. M. C. (2007). *Free radicals in biology and medicine* (4th ed.). Oxford; New York: Oxford University Press.

Haramizu, S., Nagasawa, A., Ota, N., Hase, T., Tokimitsu, I., & Murase, T. (2009). Different contribution of muscle and liver lipid metabolism to endurance capacity and obesity susceptibility of mice. *Journal of Applied Physiology, 106*(3), 871–879. Available from https://doi.org/10.1152/japplphysiol.90804.2008.

Hart, A. M., Terenghi, G., Kellerth, J. O., & Wiberg, M. (2004). Sensory neuroprotection, mitochondrial preservation, and therapeutic potential of N-acetyl-cysteine after nerve injury. *Neuroscience, 125*(1), 91–101. Available from https://doi.org/10.1016/j.neuroscience.2003.12.040S0306452203009539, [pii].

Himori, N., Yamamoto, K., Maruyama, K., Ryu, M., Taguchi, K., Yamamoto, M., & Nakazawa, T. (2013). Critical role of Nrf2 in oxidative stress-induced retinal ganglion cell death. *Journal of Neurochemistry, 127*(5), 669–680. Available from https://doi.org/10.1111/jnc.12325.

Hohn, A., & Grune, T. (2013). Lipofuscin: Formation, effects and role of macroautophagy. *Redox Biology, 1*, 140–144. Available from https://doi.org/10.1016/j.redox.2013.01.006.

Hudson, G., Amati-Bonneau, P., Blakely, E. L., Stewart, J. D., He, L., Schaefer, A. M., ... Taylor, R. W. (2008). Mutation of OPA1 causes dominant optic atrophy with external ophthalmoplegia, ataxia, deafness and multiple mitochondrial DNA deletions: A novel disorder of mtDNA maintenance. *Brain, 131*(Pt 2), 329–337. Available from https://doi.org/10.1093/brain/awm272.

Hudson, G., & Chinnery, P. F. (2006). Mitochondrial DNA polymerase-gamma and human disease. *Human Molecular Genetics, 15*(Spec No 2), R244–R252. Available from https://doi.org/10.1093/hmg/ddl233.

Hudspeth, A. J. (1997). How hearing happens. *Neuron, 19*(5), 947–950. Available from https://doi.org/10.1016/s0896-6273(00)80385-2.

Hyttinen, J. M. T., Viiri, J., Kaarniranta, K., & Blasiak, J. (2018). Mitochondrial quality control in AMD: Does mitophagy play a pivotal role? *Cellular and Molecular Life Sciences: CMLS, 75*(16), 2991–3008. Available from https://doi.org/10.1007/s00018-018-2843-7.

Ishihara, N., Nomura, M., Jofuku, A., Kato, H., Suzuki, S. O., Masuda, K., ... Mihara, K. (2009). Mitochondrial fission factor Drp1 is essential for embryonic development and synapse formation in mice. *Nature Cell Biology, 11*(8), 958–966. Available from https://doi.org/10.1038/ncb1907.

Itoh, K., Chiba, T., Takahashi, S., Ishii, T., Igarashi, K., Katoh, Y., ... Nabeshima, Y. (1997). An Nrf2/small Maf heterodimer mediates the induction of phase II detoxifying enzyme genes through antioxidant response elements. *Biochemical and Biophysical Research Communications, 236*(2), 313–322. Available from https://doi.org/10.1006/bbrc.1997.6943.

Jackson, G. R., & Owsley, C. (2000). Scotopic sensitivity during adulthood. *Vision Research, 40*(18), 2467–2473. Available from https://doi.org/10.1016/s0042-6989(00)00108-5.

Jackson, G. R., Owsley, C., & Curcio, C. A. (2002). Photoreceptor degeneration and dysfunction in aging and age-related maculopathy. *Ageing Research Reviews, 1*(3), 381–396. Available from https://doi.org/10.1016/s1568-1637(02)00007-7.

Jaiswal, A. K. (2004). Nrf2 signaling in coordinated activation of antioxidant gene expression. *Free Radical Biology & Medicine, 36*(10), 1199–1207. Available from https://doi.org/10.1016/j.freeradbiomed.2004.02.074.

Jiang, H., Talaska, A. E., Schacht, J., & Sha, S. H. (2007). Oxidative imbalance in the aging inner ear. *Neurobiology of Aging, 28*(10), 1605–1612. Available from https://doi.org/10.1016/j.neurobiolaging.2006.06.025.

Johnson, K. R., Gagnon, L. H., Longo-Guess, C., & Kane, K. L. (2012). Association of a citrate synthase missense mutation with age-related hearing loss in A/J mice. *Neurobiology of Aging, 33*(8), 1720–1729. Available from https://doi.org/10.1016/j.neurobiolaging.2011.05.009.

Jones, M., Mitchell, P., Wang, J. J., & Sue, C. (2004). MELAS A3243G mitochondrial DNA mutation and age related maculopathy. *American Journal of Ophthalmology, 138*(6), 1051–1053. Available from https://doi.org/10.1016/j.ajo.2004.06.026.

Ju, W. K., Kim, K. Y., Lindsey, J. D., Angert, M., Duong-Polk, K. X., Scott, R. T., ... Weinreb, R. N. (2008). Intraocular pressure elevation induces mitochondrial fission and triggers OPA1 release in glaucomatous optic nerve. *Investigative Ophthalmology & Visual Science, 49*(11), 4903–4911. Available from https://doi.org/10.1167/iovs.07-1661.

Kaarniranta, K., Kajdanek, J., Morawiec, J., Pawlowska, E., & Blasiak, J. (2018). PGC-1alpha protects RPE cells of the aging retina against oxidative stress-induced degeneration through the regulation of senescence and mitochondrial quality control. The significance for AMD pathogenesis. *International Journal of Molecular Sciences, 19*(8), 2317. Available from https://doi.org/10.3390/ijms19082317.

Kaguni, L. S. (2004). DNA polymerase gamma, the mitochondrial replicase. *Annual Review of Biochemistry, 73*, 293–320. Available from https://doi.org/10.1146/annurev.biochem.72.121801.161455.

Kauppila, T. E. S., Kauppila, J. H. K., & Larsson, N. G. (2017). Mammalian mitochondria and aging: An update. *Cell Metabolism, 25*(1), 57–71. Available from https://doi.org/10.1016/j.cmet.2016.09.017.

Kenney, M. C., Atilano, S. R., Boyer, D., Chwa, M., Chak, G., Chinichian, S., ... Udar, N. S. (2010). Characterization of retinal and blood mitochondrial DNA from age-related macular degeneration patients. *Investigative Ophthalmology & Visual Science, 51*(8), 4289–4297. Available from https://doi.org/10.1167/iovs.09-4778.

Kim, M. J., Haroon, S., Chen, G. D., Ding, D., Wanagat, J., Liu, L., ... Someya, S. (2019). Increased burden of mitochondrial DNA deletions and point mutations in early-onset age-related hearing loss in mitochondrial mutator mice. *Experimental Gerontology, 125*, 110675. Available from https://doi.org/10.1016/j.exger.2019.110675.

Kinnunen, K., Petrovski, G., Moe, M. C., Berta, A., & Kaarniranta, K. (2012). Molecular mechanisms of retinal pigment epithelium damage and development of age-related macular degeneration.

Acta Ophthalmologica, *90*(4), 299–309. Available from https://doi.org/10.1111/j.1755-3768.2011.02179.x.

Kitada, T., Asakawa, S., Hattori, N., Matsumine, H., Yamamura, Y., Minoshima, S., ... Shimizu, N. (1998). Mutations in the parkin gene cause autosomal recessive juvenile parkinsonism. *Nature*, *392*(6676), 605–608. Available from https://doi.org/10.1038/33416.

Kitahara, T., Li, H. S., & Balaban, C. D. (2004). Localization of the mitochondrial uncoupling protein family in the rat inner ear. *Hearing Research*, *196*(1–2), 39–48. Available from https://doi.org/10.1016/j.heares.2004.02.002.

Klapcinska, B., Derejczyk, J., Wieczorowska-Tobis, K., Sobczak, A., Sadowska-Krepa, E., & Danch, A. (2000). Antioxidant defense in centenarians (a preliminary study). *Acta Biochimica Polonica*, *47*(2), 281–292. Available from https://www.ncbi.nlm.nih.gov/pubmed/11051193.

Klein, B. E., Klein, R., & Linton, K. L. (1992). Intraocular pressure in an American community. The Beaver Dam Eye Study. *Investigative Ophthalmology & Visual Science*, *33*(7), 2224–2228. Available from https://www.ncbi.nlm.nih.gov/pubmed/1607232.

Klein, R., & Klein, B. E. (2013). The prevalence of age-related eye diseases and visual impairment in aging: Current estimates. *Investigative Ophthalmology & Visual Science*, *54*(14), ORSF5–ORSF13. Available from https://doi.org/10.1167/iovs.13-12789.

Klein, R., Klein, B. E., Moss, S. E., & Cruickshanks, K. J. (1994). The Wisconsin Epidemiologic Study of diabetic retinopathy. XIV. Ten-year incidence and progression of diabetic retinopathy. *Archives of Ophthalmology*, *112*(9), 1217–1228. Available from https://doi.org/10.1001/archopht.1994.01090210105023.

Klein, R., Klein, B. E., Moss, S. E., Davis, M. D., & DeMets, D. L. (1984). The Wisconsin epidemiologic study of diabetic retinopathy. III. Prevalence and risk of diabetic retinopathy when age at diagnosis is 30 or more years. *Archives of Ophthalmology*, *102*(4), 527–532. Available from https://doi.org/10.1001/archopht.1984.01040030405011.

Kokotas, H., Petersen, M. B., & Willems, P. J. (2007). Mitochondrial deafness. *Clinical Genetics*, *71*(5), 379–391. Available from https://doi.org/10.1111/j.1399-0004.2007.00800.x.

Kong, G. Y., Van Bergen, N. J., Trounce, I. A., & Crowston, J. G. (2009). Mitochondrial dysfunction and glaucoma. *Journal of Glaucoma*, *18*(2), 93–100. Available from https://doi.org/10.1097/IJG.0b013e318181284f.

Kong, Y., Trabucco, S. E., & Zhang, H. (2014). Oxidative stress, mitochondrial dysfunction and the mitochondria theory of aging. *Interdisciplinary Topics in Gerontology*, *39*, 86–107. Available from https://doi.org/10.1159/000358901.

Kowluru, A., & Kowluru, R. A. (2014). Phagocyte-like NADPH oxidase [Nox2] in cellular dysfunction in models of glucolipotoxicity and diabetes. *Biochemical Pharmacology*, *88*(3), 275–283. Available from https://doi.org/10.1016/j.bcp.2014.01.017.

Kowluru, R. A., & Abbas, S. N. (2003). Diabetes-induced mitochondrial dysfunction in the retina. *Investigative Ophthalmology & Visual Science*, *44*(12), 5327–5334. Available from https://doi.org/10.1167/iovs.03-0353.

Kowluru, R. A., & Chan, P. S. (2007). Oxidative stress and diabetic retinopathy. *Experimental Diabetes Research*, *2007*, 43603. Available from https://doi.org/10.1155/2007/43603.

Kowluru, R. A., Kowluru, A., Veluthakal, R., Mohammad, G., Syed, I., Santos, J. M., & Mishra, M. (2014). TIAM1-RAC1 signalling axis-mediated activation of NADPH oxidase-2 initiates mitochondrial damage in the development of diabetic retinopathy. *Diabetologia*, *57*(5), 1047–1056. Available from https://doi.org/10.1007/s00125-014-3194-z.

Kowluru, R. A., & Mishra, M. (2015). Oxidative stress, mitochondrial damage and diabetic retinopathy. *Biochimica et Biophysica Acta*, *1852*(11), 2474–2483. Available from https://doi.org/10.1016/j.bbadis.2015.08.001.

Kraytsberg, Y., Kudryavtseva, E., McKee, A. C., Geula, C., Kowall, N. W., & Khrapko, K. (2006). Mitochondrial DNA deletions are abundant and cause functional impairment in aged human substantia nigra neurons. *Nature Genetics*, *38*(5), 518–520. Available from https://doi.org/10.1038/ng1778.

Krishnan, K. J., Reeve, A. K., Samuels, D. C., Chinnery, P. F., Blackwood, J. K., Taylor, R. W., ... Turnbull, D. M. (2008). What causes mitochondrial DNA deletions in human cells? *Nature Genetics*, *40*(3), 275–279. Available from https://doi.org/10.1038/ng.f.94.

Kujoth, G. C., Bradshaw, P. C., Haroon, S., & Prolla, T. A. (2007). The role of mitochondrial DNA mutations in mammalian aging. *PLoS Genetics*, *3*(2), e24. Available from https://doi.org/10.1371/journal.pgen.0030024.

Kujoth, G. C., Hiona, A., Pugh, T. D., Someya, S., Panzer, K., Wohlgemuth, S. E., ... Prolla, T. A. (2005). Mitochondrial DNA mutations, oxidative stress, and apoptosis in mammalian aging. *Science*, *309*(5733), 481–484. Available from https://doi.org/10.1126/science.1112125.

Kurtenbach, A., & Weiss, M. (2002). Effect of aging on multifocal oscillatory potentials. *Journal of the Optical Society of America A Optics, Image Science, and Vision*, *19*(1), 190–196. Available from https://doi.org/10.1364/josaa.19.000190.

Laloi-Michelin, M., Meas, T., Ambonville, C., Bellanne-Chantelot, C., Beaufils, S., Massin, P., ... Mitochondrial Diabetes French Study, G. (2009). The clinical variability of maternally inherited diabetes and deafness is associated with the degree of heteroplasmy in blood leukocytes. *The Journal of Clinical Endocrinology and Metabolism*, *94*(8), 3025–3030. Available from https://doi.org/10.1210/jc.2008-2680.

Lambeth, J. D. (2004). NOX enzymes and the biology of reactive oxygen. *Nature Reviews Immunology*, *4*(3), 181–189. Available from https://doi.org/10.1038/nri1312.

Lee, H. C., & Wei, Y. H. (2012). Mitochondria and aging. *Advances in Experimental Medicine and Biology*, *942*, 311–327. Available from https://doi.org/10.1007/978-94-007-2869-1_14.

Lee, W. H., Higuchi, H., Ikeda, S., Macke, E. L., Takimoto, T., Pattnaik, B. R., ... Ikeda, A. (2016). Mouse Tmem135 mutation reveals a mechanism involving mitochondrial dynamics that leads to age-dependent retinal pathologies. *Elife*, *5*, e19264. Available from https://doi.org/10.7554/eLife.19264.

Li, Y., Schlamp, C. L., Poulsen, K. P., & Nickells, R. W. (2000). Bax-dependent and independent pathways of retinal ganglion cell death induced by different damaging stimuli. *Experimental Eye Research*, *71*(2), 209–213. Available from https://doi.org/10.1006/exer.2000.0873.

Libby, R. T., Anderson, M. G., Pang, I. H., Robinson, Z. H., Savinova, O. V., Cosma, I. M., ... John, S. W. (2005). Inherited glaucoma in DBA/2J mice: Pertinent disease features for studying the neurodegeneration. *Visual Neuroscience*, *22*(5), 637–648. Available from https://doi.org/10.1017/S0952523805225130.

Libby, R. T., Li, Y., Savinova, O. V., Barter, J., Smith, R. S., Nickells, R. W., & John, S. W. (2005). Susceptibility to neurodegeneration in a glaucoma is modified by Bax gene dosage. *PLoS Genetics*, *1*(1), 17–26. Available from https://doi.org/10.1371/journal.pgen.0010004.

Liberman, M. C., & Kujawa, S. G. (2017). Cochlear synaptopathy in acquired sensorineural hearing loss: Manifestations and mechanisms. *Hearing Research*, *349*, 138–147. Available from https://doi.org/10.1016/j.heares.2017.01.003.

Liebmann, J. M., & Lee, J. K. (2017). Current therapeutic options and treatments in development for the management of primary open-angle glaucoma. *The American Journal of Managed Care*, *23*(15

Suppl.), S279–S292. Available from https://www.ncbi.nlm.nih.gov/pubmed/29164845.

Lightfoot, J. T., Turner, M. J., Debate, K. A., & Kleeberger, S. R. (2001). Interstrain variation in murine aerobic capacity. *Medicine and Science in Sports and Exercise*, *33*(12), 2053–2057. Available from https://doi.org/10.1097/00005768-200112000-00012.

Liguori, M., La Russa, A., Manna, I., Andreoli, V., Caracciolo, M., Spadafora, P., ... Quattrone, A. (2008). A phenotypic variation of dominant optic atrophy and deafness (ADOAD) due to a novel OPA1 mutation. *Journal of Neurology*, *255*(1), 127–129. Available from https://doi.org/10.1007/s00415-008-0571-x.

Lim, M. C., Lee, S. Y., Cheng, B. C., Wong, D. W., Ong, S. G., Ang, C. L., & Yeo, I. Y. (2008). Diabetic retinopathy in diabetics referred to a tertiary centre from a nationwide screening programme. *Annals of the Academy of Medicine, Singapore*, *37*(9), 753–759. Available from https://www.ncbi.nlm.nih.gov/pubmed/18989491.

Lin, C. Y., Wu, J. L., Shih, T. S., Tsai, P. J., Sun, Y. M., Ma, M. C., & Guo, Y. L. (2010). N-Acetyl-cysteine against noise-induced temporary threshold shift in male workers. *Hearing Research*, *269*(1–2), 42–47. Available from https://doi.org/10.1016/j.heares.2010.07.005, S0378-5955(10)00332-1 [pii].

Lin, F. R., Metter, E. J., O'Brien, R. J., Resnick, S. M., Zonderman, A. B., & Ferrucci, L. (2011). Hearing loss and incident dementia. *Archives of Neurology*, *68*(2), 214–220. Available from https://doi.org/10.1001/archneurol.2010.362.

Lin, F. R., Niparko, J. K., & Ferrucci, L. (2011). Hearing loss prevalence in the United States. *Archives of Internal Medicine*, *171*(20), 1851–1852. Available from https://doi.org/10.1001/archinternmed.2011.506.

Lin, J. B., Tsubota, K., & Apte, R. S. (2016). A glimpse at the aging eye. *NPJ Aging and Mechanisms of Disease*, *2*, 16003. Available from https://doi.org/10.1038/npjamd.2016.3.

Lindsten, T., Ross, A. J., King, A., Zong, W. X., Rathmell, J. C., Shiels, H. A., ... Thompson, C. B. (2000). The combined functions of proapoptotic Bcl-2 family members bak and bax are essential for normal development of multiple tissues. *Molecular Cell*, *6*(6), 1389–1399. Available from https://doi.org/10.1016/s1097-2765(00)00136-2.

Liu, X., Ward, K., Xavier, C., Jann, J., Clark, A. F., Pang, I. H., & Wu, H. (2016). The novel triterpenoid RTA 408 protects human retinal pigment epithelial cells against H2O2-induced cell injury via NF-E2-related factor 2 (Nrf2) activation. *Redox Biology*, *8*, 98–109. Available from https://doi.org/10.1016/j.redox.2015.12.005.

Liu, X. Z., & Yan, D. (2007). Ageing and hearing loss. *The Journal of Pathology*, *211*(2), 188–197. Available from https://doi.org/10.1002/path.2102.

Loson, O. C., Song, Z., Chen, H., & Chan, D. C. (2013). Fis1, Mff, MiD49, and MiD51 mediate Drp1 recruitment in mitochondrial fission. *Molecular Biology of the Cell*, *24*(5), 659–667. Available from https://doi.org/10.1091/mbc.E12-10-0721.

Maes, M. E., Schlamp, C. L., & Nickells, R. W. (2017). BAX to basics: How the BCL2 gene family controls the death of retinal ganglion cells. *Progress in Retinal and Eye Research*, *57*, 1–25. Available from https://doi.org/10.1016/j.preteyeres.2017.01.002.

Majamaa, K., Moilanen, J. S., Uimonen, S., Remes, A. M., Salmela, P. I., Karppa, M., ... Hassinen, I. E. (1998). Epidemiology of A3243G, the mutation for mitochondrial encephalomyopathy, lactic acidosis, and strokelike episodes: Prevalence of the mutation in an adult population. *American Journal of Human Genetics*, *63*(2), 447–454. Available from https://doi.org/10.1086/301959.

Mancuso, M., Filosto, M., Bellan, M., Liguori, R., Montagna, P., Baruzzi, A., ... Carelli, V. (2004). POLG mutations causing ophthalmoplegia, sensorimotor polyneuropathy, ataxia, and deafness. *Neurology*, *62*(2), 316–318. Available from https://doi.org/10.1212/wnl.62.2.316.

Mandal, M. N., Patlolla, J. M., Zheng, L., Agbaga, M. P., Tran, J. T., Wicker, L., ... Anderson, R. E. (2009). Curcumin protects retinal cells from light-and oxidant stress-induced cell death. *Free Radical Biology & Medicine*, *46*(5), 672–679. Available from https://doi.org/10.1016/j.freeradbiomed.2008.12.006.

Mao, H., Seo, S. J., Biswal, M. R., Li, H., Conners, M., Nandyala, A., ... Lewin, A. S. (2014). Mitochondrial oxidative stress in the retinal pigment epithelium leads to localized retinal degeneration. *Investigative Ophthalmology & Visual Science*, *55*(7), 4613–4627. Available from https://doi.org/10.1167/iovs.14-14633.

Mari, M., Morales, A., Colell, A., Garcia-Ruiz, C., & Fernandez-Checa, J. C. (2009). Mitochondrial glutathione, a key survival antioxidant. *Antioxidants & Redox Signaling*, *11*(11), 2685–2700. Available from https://doi.org/10.1089/ARS.2009.2695.

Mattiasson, G., Shamloo, M., Gido, G., Mathi, K., Tomasevic, G., Yi, S., ... Wieloch, T. (2003). Uncoupling protein-2 prevents neuronal death and diminishes brain dysfunction after stroke and brain trauma. *Nature Medicine*, *9*(8), 1062–1068. Available from https://doi.org/10.1038/nm903.

Mattson, M. P. (2000). Apoptosis in neurodegenerative disorders. *Nature Reviews. Molecular Cell Biology*, *1*(2), 120–129. Available from https://doi.org/10.1038/35040009.

McFadden, S. L., Ding, D., Reaume, A. G., Flood, D. G., & Salvi, R. J. (1999). Age-related cochlear hair cell loss is enhanced in mice lacking copper/zinc superoxide dismutase. *Neurobiology of Aging*, *20*(1), 1–8. Available from https://doi.org/10.1016/s0197-4580(99)00018-4.

McKechnie, N. M., King, M., & Lee, W. R. (1985). Retinal pathology in the Kearns-Sayre syndrome. *The British Journal of Ophthalmology*, *69*(1), 63–75. Available from https://doi.org/10.1136/bjo.69.1.63.

McLaren, N. C., & Moroi, S. E. (2003). Clinical implications of pharmacogenetics for glaucoma therapeutics. *The Pharmacogenomics Journal*, *3*(4), 197–201. Available from https://doi.org/10.1038/sj.tpj.6500181.

McQuibban, G. A., Lee, J. R., Zheng, L., Juusola, M., & Freeman, M. (2006). Normal mitochondrial dynamics requires rhomboid-7 and affects Drosophila lifespan and neuronal function. *Current Biology: CB*, *16*(10), 982–989. Available from https://doi.org/10.1016/j.cub.2006.03.062.

Mears, J. A., Lackner, L. L., Fang, S., Ingerman, E., Nunnari, J., & Hinshaw, J. E. (2011). Conformational changes in Dnm1 support a contractile mechanism for mitochondrial fission. *Nature Structural & Molecular Biology*, *18*(1), 20–26. Available from https://doi.org/10.1038/nsmb.1949.

Mitchell, P., Liew, G., Gopinath, B., & Wong, T. Y. (2018). Age-related macular degeneration. *Lancet*, *392*(10153), 1147–1159. Available from https://doi.org/10.1016/S0140-6736(18)31550-2.

MITOMAP. (2019). A human mitochondrial genome database. <https://www.mitomap.org/MITOMAP>.

Moreno, M. C., Campanelli, J., Sande, P., Sanez, D. A., Keller Sarmiento, M. I., & Rosenstein, R. E. (2004). Retinal oxidative stress induced by high intraocular pressure. *Free Radical Biology & Medicine*, *37*(6), 803–812. Available from https://doi.org/10.1016/j.freeradbiomed.2004.06.001.

Muller, A., Pietri, S., Villain, M., Frejaville, C., Bonne, C., & Culcas, M. (1997). Free radicals in rabbit retina under ocular hyperpressure and functional consequences. *Experimental Eye Research*, *64*(4), 637–643. Available from https://doi.org/10.1006/exer.1996.0277.

Murphy, M. P. (2009). How mitochondria produce reactive oxygen species. *The Biochemical Journal*, *417*(1), 1–13. Available from https://doi.org/10.1042/BJ20081386.

Myers, M. P., Pass, I., Batty, I. H., Van der Kaay, J., Stolarov, J. P., Hemmings, B. A., ... Tonks, N. K. (1998). The lipid phosphatase activity of PTEN is critical for its tumor supressor function.

Proceedings of the National Academy of Sciences of the United States of America, *95*(23), 13513–13518. Available from https://doi.org/10.1073/pnas.95.23.13513.

Nakada, K., Inoue, K., Ono, T., Isobe, K., Ogura, A., Goto, Y. I., ... Hayashi, J. I. (2001). Inter-mitochondrial complementation: Mitochondria-specific system preventing mice from expression of disease phenotypes by mutant mtDNA. *Nature Medicine*, *7*(8), 934–940. Available from https://doi.org/10.1038/90976.

Namperumalsamy, P., Kim, R., Vignesh, T. P., Nithya, N., Royes, J., Gijo, T., ... Vijayakumar, V. (2009). Prevalence and risk factors for diabetic retinopathy: A population-based assessment from Theni District, south India. *The British Journal of Ophthalmology*, *93*(4), 429–434. Available from https://doi.org/10.1136/bjo.2008.147934.

Narendra, D., Tanaka, A., Suen, D. F., & Youle, R. J. (2008). Parkin is recruited selectively to impaired mitochondria and promotes their autophagy. *The Journal of Cell Biology*, *183*(5), 795–803. Available from https://doi.org/10.1083/jcb.200809125.

Nguyen, T., Sherratt, P. J., & Pickett, C. B. (2003). Regulatory mechanisms controlling gene expression mediated by the antioxidant response element. *Annual Review of Pharmacology and Toxicology*, *43*, 233–260. Available from https://doi.org/10.1146/annurev.pharmtox.43.100901.140229.

Nita, M., & Grzybowski, A. (2016). The role of the reactive oxygen species and oxidative stress in the pathomechanism of the age-related ocular diseases and other pathologies of the anterior and posterior eye segments in adults. *Oxidative Medicine and Cellular Longevity*, *2016*, 3164734. Available from https://doi.org/10.1155/2016/3164734.

Niu, X., Trifunovic, A., Larsson, N. G., & Canlon, B. (2007). Somatic mtDNA mutations cause progressive hearing loss in the mouse. *Experimental Cell Research*, *313*(18), 3924–3934. Available from https://doi.org/10.1016/j.yexcr.2007.05.029.

Niven, J. E., & Laughlin, S. B. (2008). Energy limitation as a selective pressure on the evolution of sensory systems. *The Journal of Experimental Biology*, *211*(Pt 11), 1792–1804. Available from https://doi.org/10.1242/jeb.017574.

Ohinata, Y., Yamasoba, T., Schacht, J., & Miller, J. M. (2000). Glutathione limits noise-induced hearing loss. *Hearing Research*, *146*(1–2), 28–34.

Ohlemiller, K. K., McFadden, S. L., Ding, D. L., Lear, P. M., & Ho, Y. S. (2000). Targeted mutation of the gene for cellular glutathione peroxidase (Gpx1) increases noise-induced hearing loss in mice. *Journal of the Association for Research in Otolaryngology: JARO*, *1*(3), 243–254. Available from https://doi.org/10.1007/s101620010043.

Ohlemiller, K. K., Wright, J. S., & Dugan, L. L. (1999). Early elevation of cochlear reactive oxygen species following noise exposure. *Audiology & Neuro-otology*, *4*(5), 229–236. Available from https://doi.org/10.1159/000013846.

Okumura, H., Miyasaka, Y., Morita, Y., Nomura, T., Mishima, Y., Takahashi, S., & Kominami, R. (2011). Bcl11b heterozygosity leads to age-related hearing loss and degeneration of outer hair cells of the mouse cochlea. *Experimental Animals/Japanese Association for Laboratory Animal Science*, *60*(4), 355–361. Available from https://doi.org/10.1538/expanim.60.355.

Olichon, A., Guillou, E., Delettre, C., Landes, T., Arnaune-Pelloquin, L., Emorine, L. J., ... Belenguer, P. (2006). Mitochondrial dynamics and disease, OPA1. *Biochimica et Biophysica Acta*, *1763*(5–6), 500–509. Available from https://doi.org/10.1016/j.bbamcr.2006.04.003.

Ono, T., Isobe, K., Nakada, K., & Hayashi, J. I. (2001). Human cells are protected from mitochondrial dysfunction by complementation of DNA products in fused mitochondria. *Nature Genetics*, *28*(3), 272–275. Available from https://doi.org/10.1038/90116.

Palaniappan, A. R., & Dai, A. (2007). Mitochondrial ageing and the beneficial role of alpha-lipoic acid. *Neurochemical Research*, *32*(9), 1552–1558. Available from https://doi.org/10.1007/s11064-007-9355-4.

Palikaras, K., Daskalaki, I., Markaki, M., & Tavernarakis, N. (2017). Mitophagy and age-related pathologies: Development of new therapeutics by targeting mitochondrial turnover. *Pharmacology & Therapeutics*, *178*, 157–174. Available from https://doi.org/10.1016/j.pharmthera.2017.04.005.

Panday, A., Sahoo, M. K., Osorio, D., & Batra, S. (2015). NADPH oxidases: An overview from structure to innate immunity-associated pathologies. *Cellular & Molecular Immunology*, *12*(1), 5–23. Available from https://doi.org/10.1038/cmi.2014.89.

Park, S. W., Kim, K. Y., Lindsey, J. D., Dai, Y., Heo, H., Nguyen, D. H., ... Ju, W. K. (2011). A selective inhibitor of drp1, mdivi-1, increases retinal ganglion cell survival in acute ischemic mouse retina. *Investigative Ophthalmology & Visual Science*, *52*(5), 2837–2843. Available from https://doi.org/10.1167/iovs.09-5010.

Pennington, K. L., & DeAngelis, M. M. (2016). Epidemiology of age-related macular degeneration (AMD): Associations with cardiovascular disease phenotypes and lipid factors. *Eye and Vision*, *3*, 34. Available from https://doi.org/10.1186/s40662-016-0063-5.

Qin, Q., Patil, K., & Sharma, S. C. (2004). The role of Bax-inhibiting peptide in retinal ganglion cell apoptosis after optic nerve transection. *Neuroscience Letters*, *372*(1-2), 17–21. Available from https://doi.org/10.1016/j.neulet.2004.08.075.

Raimundo, N., Baysal, B. E., & Shadel, G. S. (2011). Revisiting the TCA cycle: Signaling to tumor formation. *Trends in Molecular Medicine*, *17*(11), 641–649. Available from https://doi.org/10.1016/j.molmed.2011.06.001.

Rakovic, A., Shurkewitsch, K., Seibler, P., Grunewald, A., Zanon, A., Hagenah, J., ... Klein, C. (2013). Phosphatase and tensin homolog (PTEN)-induced putative kinase 1 (PINK1)-dependent ubiquitination of endogenous Parkin attenuates mitophagy: Study in human primary fibroblasts and induced pluripotent stem cell-derived neurons. *The Journal of Biological Chemistry*, *288*(4), 2223–2237. Available from https://doi.org/10.1074/jbc.M112.391680.

Rebrin, I., & Sohal, R. S. (2008). Pro-oxidant shift in glutathione redox state during aging. *Advanced Drug Delivery Reviews*, *60*(13–14), 1545–1552. Available from https://doi.org/10.1016/j.addr.2008.06.001.

Reitman, Z. J., & Yan, H. (2010). Isocitrate dehydrogenase 1 and 2 mutations in cancer: Alterations at a crossroads of cellular metabolism. *Journal of the National Cancer Institute*, *102*(13), 932–941. Available from https://doi.org/10.1093/jnci/djq187.

Ryoo, I. G., & Kwak, M. K. (2018). Regulatory crosstalk between the oxidative stress-related transcription factor Nfe2l2/Nrf2 and mitochondria. *Toxicology and Applied Pharmacology*, *359*, 24–33. Available from https://doi.org/10.1016/j.taap.2018.09.014.

Sachdeva, M. M., Cano, M., & Handa, J. T. (2014). Nrf2 signaling is impaired in the aging RPE given an oxidative insult. *Experimental Eye Research*, *119*, 111–114. Available from https://doi.org/10.1016/j.exer.2013.10.024.

Sawada, M., & Carlson, J. C. (1987). Changes in superoxide radical and lipid peroxide formation in the brain, heart and liver during the lifetime of the rat. *Mechanisms of Ageing and Development*, *41*(1–2), 125–137. Available from https://doi.org/10.1016/0047-6374(87)90057-1.

Sawada, O., Perusek, L., Kohno, H., Howell, S. J., Maeda, A., Matsuyama, S., & Maeda, T. (2014). All-trans-retinal induces Bax activation via DNA damage to mediate retinal cell apoptosis. *Experimental Eye Research*, *123*, 27–36. Available from https://doi.org/10.1016/j.exer.2014.04.003.

Schmidlin, C. J., Dodson, M. B., Madhavan, L., & Zhang, D. D. (2019). Redox regulation by NRF2 in aging and disease. *Free Radical Biology & Medicine*, *134*, 702–707. Available from https://doi.org/10.1016/j.freeradbiomed.2019.01.016.

Schon, E. A., DiMauro, S., & Hirano, M. (2012). Human mitochondrial DNA: Roles of inherited and somatic mutations. *Nature Reviews. Genetics*, *13*(12), 878–890. Available from https://doi.org/10.1038/nrg3275.

Schon, E. A., & Przedborski, S. (2011). Mitochondria: The next (neurode)generation. *Neuron*, *70*(6), 1033–1053. Available from https://doi.org/10.1016/j.neuron.2011.06.003.

Schrier, S. A., & Falk, M. J. (2011). Mitochondrial disorders and the eye. *Current Opinion in Ophthalmology*, *22*(5), 325–331. Available from https://doi.org/10.1097/ICU.0b013e328349419d.

Schriner, S. E., Linford, N. J., Martin, G. M., Treuting, P., Ogburn, C. E., Emond, M., ... Rabinovitch, P. S. (2005). Extension of murine life span by overexpression of catalase targeted to mitochondria. *Science*, *308*(5730), 1909–1911. Available from https://doi.org/10.1126/science.1106653.

Scorrano, L. (2005). Proteins that fuse and fragment mitochondria in apoptosis: Con-fissing a deadly con-fusion? *Journal of Bioenergetics and Biomembranes*, *37*(3), 165–170. Available from https://doi.org/10.1007/s10863-005-6572-x.

Scorrano, L., Ashiya, M., Buttle, K., Weiler, S., Oakes, S. A., Mannella, C. A., & Korsmeyer, S. J. (2002). A distinct pathway remodels mitochondrial cristae and mobilizes cytochrome c during apoptosis. *Developmental Cell*, *2*(1), 55–67. Available from https://doi.org/10.1016/s1534-5807(01)00116-2.

Sebastian, D., Palacin, M., & Zorzano, A. (2017). Mitochondrial dynamics: Coupling mitochondrial fitness with healthy aging. *Trends in Molecular Medicine*, *23*(3), 201–215. Available from https://doi.org/10.1016/j.molmed.2017.01.003.

Seidman, M. D., Khan, M. J., Bai, U., Shirwany, N., & Quirk, W. S. (2000). Biologic activity of mitochondrial metabolites on aging and age-related hearing loss. *The American Journal of Otology*, *21* (2), 161–167. Available from http://www.ncbi.nlm.nih.gov/pubmed/10733178.

Sha, S. H., Chen, F. Q., & Schacht, J. (2010). PTEN attenuates PIP3/Akt signaling in the cochlea of the aging CBA/J mouse. *Hearing Research*, *264*(1–2), 86–92. Available from https://doi.org/10.1016/j.heares.2009.09.002.

Silva-Palacios, A., Ostolga-Chavarria, M., Zazueta, C., & Konigsberg, M. (2018). Nrf2: Molecular and epigenetic regulation during aging. *Ageing Research Reviews*, *47*, 31–40. Available from https://doi.org/10.1016/j.arr.2018.06.003.

Skulachev, V. P. (2001). Mitochondrial filaments and clusters as intracellular power-transmitting cables. *Trends in Biochemical Sciences*, *26*(1), 23–29. Available from https://doi.org/10.1016/s0968-0004(00)01735-7.

Sliwinski, T., Kolodziejska, U., Szaflik, J. P., Blasiak, J., & Szaflik, J. (2013). Association between the 25129A > C polymorphism of the nuclear respiratory factor 2 gene and age-related macular degeneration. *Klinika Oczna*, *115*(2), 96–102. Available from https://www.ncbi.nlm.nih.gov/pubmed/24059022.

Sohal, R. S., & Forster, M. J. (2007). Coenzyme Q, oxidative stress and aging. *Mitochondrion*, *7*(Suppl.), S103–111. Available from https://doi.org/10.1016/j.mito.2007.03.006.

Sohal, R. S., & Weindruch, R. (1996). Oxidative stress, caloric restriction, and aging. *Science*, *273*(5271), 59–63. Available from https://doi.org/10.1126/science.273.5271.59.

Someya, S., & Prolla, T. A. (2010). Mitochondrial oxidative damage and apoptosis in age-related hearing loss. *Mechanisms of Ageing and Development*, *131*(7–8), 480–486. Available from https://doi.org/10.1016/j.mad.2010.04.006.

Someya, S., Xu, J., Kondo, K., Ding, D., Salvi, R. J., Yamasoba, T., ... Prolla, T. A. (2009). Age-related hearing loss in C57BL/6J mice is mediated by Bak-dependent mitochondrial apoptosis. *Proceedings of the National Academy of Sciences of the United States of America*, *106*(46), 19432–19437. Available from https://doi.org/10.1073/pnas.0908786106, 0908786106 [pii].

Someya, S., Yamasoba, T., Kujoth, G. C., Pugh, T. D., Weindruch, R., Tanokura, M., & Prolla, T. A. (2008). The role of mtDNA mutations in the pathogenesis of age-related hearing loss in mice carrying a mutator DNA polymerase gamma. *Neurobiology of Aging*, *29*(7), 1080–1092. Available from https://doi.org/10.1016/j.neurobiolaging.2007.01.014.

Someya, S., Yu, W., Hallows, W. C., Xu, J., Vann, J. M., Leeuwenburgh, C., ... Prolla, T. A. (2010). Sirt3 mediates reduction of oxidative damage and prevention of age-related hearing loss under caloric restriction. *Cell*, *143*(5), 802–812. Available from https://doi.org/10.1016/j.cell.2010.10.002.

Stratton, I. M., Kohner, E. M., Aldington, S. J., Turner, R. C., Holman, R. R., Manley, S. E., & Matthews, D. R. (2001). UKPDS 50: Risk factors for incidence and progression of retinopathy in Type II diabetes over 6 years from diagnosis. *Diabetologia*, *44*(2), 156–163. Available from https://doi.org/10.1007/s001250051594.

Strauss, O. (2005). The retinal pigment epithelium in visual function. *Physiological Reviews*, *85*(3), 845–881. Available from https://doi.org/10.1152/physrev.00021.2004.

Subrizi, A., Toropainen, E., Ramsay, E., Airaksinen, A. J., Kaarniranta, K., & Urtti, A. (2015). Oxidative stress protection by exogenous delivery of rhHsp70 chaperone to the retinal pigment epithelium (RPE), a possible therapeutic strategy against RPE degeneration. *Pharmaceutical Research*, *32*(1), 211–221. Available from https://doi.org/10.1007/s11095-014-1456-6.

Sue, C. M., Mitchell, P., Crimmins, D. S., Moshegov, C., Byrne, E., & Morris, J. G. (1997). Pigmentary retinopathy associated with the mitochondrial DNA 3243 point mutation. *Neurology*, *49*(4), 1013–1017. Available from https://doi.org/10.1212/wnl.49.4.1013.

Sugiura, S., Uchida, Y., Nakashima, T., Ando, F., & Shimokata, H. (2010). The association between gene polymorphisms in uncoupling proteins and hearing impairment in Japanese elderly. *Acta Oto-Laryngologica*, *130*(4), 487–492. Available from https://doi.org/10.3109/00016480903283758.

Sun, J., Folk, D., Bradley, T. J., & Tower, J. (2002). Induced overexpression of mitochondrial Mn-superoxide dismutase extends the life span of adult Drosophila melanogaster. *Genetics*, *161*(2), 661–672. Available from https://www.ncbi.nlm.nih.gov/pubmed/12072463.

Sundaresan, N. R., Samant, S. A., Pillai, V. B., Rajamohan, S. B., & Gupta, M. P. (2008). SIRT3 is a stress-responsive deacetylase in cardiomyocytes that protects cells from stress-mediated cell death by deacetylation of Ku70. *Molecular and Cellular Biology*, *28*(20), 6384–6401. Available from https://doi.org/10.1128/MCB.00426-08.

Sung, C. H., & Chuang, J. Z. (2010). The cell biology of vision. *The Journal of Cell Biology*, *190*(6), 953–963. Available from https://doi.org/10.1083/jcb.201006020.

Taguchi, N., Ishihara, N., Jofuku, A., Oka, T., & Mihara, K. (2007). Mitotic phosphorylation of dynamin-related GTPase Drp1 participates in mitochondrial fission. *The Journal of Biological Chemistry*, *282*(15), 11521–11529. Available from https://doi.org/10.1074/jbc.M607279200.

Tan, C. S., Gay, E. M., & Ngo, W. K. (2010). Is age a risk factor for diabetic retinopathy? *The British Journal of Ophthalmology*, *94*(9), 1268. Available from https://doi.org/10.1136/bjo.2009.169326.

Taylor, R. W., & Turnbull, D. M. (2005). Mitochondrial DNA mutations in human disease. *Nature Reviews. Genetics*, *6*(5), 389–402. Available from https://doi.org/10.1038/nrg1606.

Terman, A., & Brunk, U. T. (1998). Lipofuscin: Mechanisms of formation and increase with age. *APMIS: Acta Pathologica, Microbiologica, et Immunologica Scandinavica*, *106*(2), 265–276. Available from https://doi.org/10.1111/j.1699-0463.1998.tb01346.x.

Tezel, G. (2006). Oxidative stress in glaucomatous neurodegeneration: Mechanisms and consequences. *Progress in Retinal and Eye Research*, *25*(5), 490–513. Available from https://doi.org/10.1016/j.preteyeres.2006.07.003.

Tham, Y. C., Li, X., Wong, T. Y., Quigley, H. A., Aung, T., & Cheng, C. Y. (2014). Global prevalence of glaucoma and projections of glaucoma burden through 2040: A systematic review and meta-analysis. *Ophthalmology*, *121*(11), 2081–2090. Available from https://doi.org/10.1016/j.ophtha.2014.05.013.

Tsukuda, K., Suzuki, Y., Kameoka, K., Osawa, N., Goto, Y., Katagiri, H., ... Oka, Y. (1997). Screening of patients with maternally transmitted diabetes for mitochondrial gene mutations in the tRNA [Leu(UUR)] region. *Diabetic Medicine: A Journal of the British Diabetic Association*, *14*(12), 1032–1037.

Twig, G., Elorza, A., Molina, A. J., Mohamed, H., Wikstrom, J. D., Walzer, G., ... Shirihai, O. S. (2008). Fission and selective fusion govern mitochondrial segregation and elimination by autophagy. *The EMBO Journal*, *27*(2), 433–446. Available from https://doi.org/10.1038/sj.emboj.7601963.

Vajaranant, T. S., Wu, S., Torres, M., & Varma, R. (2012). The changing face of primary open-angle glaucoma in the United States: Demographic and geographic changes from 2011 to 2050. *American Journal of Ophthalmology*, *154*(2), 303–314. Available from https://doi.org/10.1016/j.ajo.2012.02.024, e303.

Valente, E. M., Abou-Sleiman, P. M., Caputo, V., Muqit, M. M., Harvey, K., Gispert, S., ... Wood, N. W. (2004). Hereditary early-onset Parkinson's disease caused by mutations in PINK1. *Science*, *304*(5674), 1158–1160. Available from https://doi.org/10.1126/science.1096284.

Varma, R., Ying-Lai, M., Klein, R., Azen, S. P., & Los Angeles Latino Eye Study, G. (2004). Prevalence and risk indicators of visual impairment and blindness in Latinos: The Los Angeles Latino Eye Study. *Ophthalmology*, *111*(6), 1132–1140. Available from https://doi.org/10.1016/j.ophtha.2004.02.002.

Vermulst, M., Bielas, J. H., Kujoth, G. C., Ladiges, W. C., Rabinovitch, P. S., Prolla, T. A., & Loeb, L. A. (2007). Mitochondrial point mutations do not limit the natural lifespan of mice. *Nature Genetics*, *39*(4), 540–543. Available from https://doi.org/10.1038/ng1988.

Vermulst, M., Wanagat, J., Kujoth, G. C., Bielas, J. H., Rabinovitch, P. S., Prolla, T. A., & Loeb, L. A. (2008). DNA deletions and clonal mutations drive premature aging in mitochondrial mutator mice. *Nature Genetics*, *40*(4), 392–394. Available from https://doi.org/10.1038/ng.95.

Verstreken, P., Ly, C. V., Venken, K. J., Koh, T. W., Zhou, Y., & Bellen, H. J. (2005). Synaptic mitochondria are critical for mobilization of reserve pool vesicles at Drosophila neuromuscular junctions. *Neuron*, *47*(3), 365–378. Available from https://doi.org/10.1016/j.neuron.2005.06.018.

Wallace, D. C. (2005). A mitochondrial paradigm of metabolic and degenerative diseases, aging, and cancer: A dawn for evolutionary medicine. *Annual Review of Genetics*, *39*, 359–407. Available from https://doi.org/10.1146/annurev.genet.39.110304.095751.

Wassell, J., Davies, S., Bardsley, W., & Boulton, M. (1999). The photoreactivity of the retinal age pigment lipofuscin. *The Journal of Biological Chemistry*, *274*(34), 23828–23832. Available from https://doi.org/10.1074/jbc.274.34.23828.

Wei, Y., Gong, J., Yoshida, T., Eberhart, C. G., Xu, Z., Kombairaju, P., ... Duh, E. J. (2011). Nrf2 has a protective role against neuronal and capillary degeneration in retinal ischemia-reperfusion injury. *Free Radical Biology & Medicine*, *51*(1), 216–224. Available from https://doi.org/10.1016/j.freeradbiomed.2011.04.026.

Weindruch, R., & Sohal, R. S. (1997). Seminars in medicine of the Beth Israel Deaconess Medical Center. Caloric intake and aging. *The New England Journal of Medicine*, *337*(14), 986–994. Available from https://doi.org/10.1056/NEJM199710023371407.

Weinreb, R. N., Aung, T., & Medeiros, F. A. (2014). The pathophysiology and treatment of glaucoma: A review. *JAMA: The Journal of the American Medical Association*, *311*(18), 1901–1911. Available from https://doi.org/10.1001/jama.2014.3192.

Weng, L., Brown, J., & Eng, C. (2001). PTEN induces apoptosis and cell cycle arrest through phosphoinositol-3-kinase/Akt-dependent and -independent pathways. *Human Molecular Genetics*, *10* (3), 237–242. Available from https://doi.org/10.1093/hmg/10.3.237.

White, K., Kim, M. J., Han, C., Park, H. J., Ding, D., Boyd, K., ... Someya, S. (2018). Loss of IDH2 accelerates age-related hearing loss in male mice. *Science Reports*, *8*(1), 5039. Available from https://doi.org/10.1038/s41598-018-23436-w.

WHO. (2019). Prevention of blindness and deafness. <https://www.who.int/pbd/deafness/estimates/en/>.

Wiktor, A., Sarna, M., Wnuk, D., & Sarna, T. (2018). Lipofuscin-mediated photodynamic stress induces adverse changes in nanomechanical properties of retinal pigment epithelium cells. *Science Reports*, *8*(1), 17929. Available from https://doi.org/10.1038/s41598-018-36322-2.

Wong, T. Y., Cheung, C. M., Larsen, M., Sharma, S., & Simo, R. (2016). Diabetic retinopathy. *Nature Reviews Disease Primers*, *2*, 16012. Available from https://doi.org/10.1038/nrdp.2016.12.

Wu, H. P., Hsu, C. J., Cheng, T. J., & Guo, Y. L. (2010). N-acetylcysteine attenuates noise-induced permanent hearing loss in diabetic rats. *Hearing Research*, *267*(1-2), 71–77. Available from https://doi.org/10.1016/j.heares.2010.03.082.

Xing, G., Chen, Z., & Cao, X. (2007). Mitochondrial rRNA and tRNA and hearing function. *Cell Research*, *17*(3), 227–239. Available from https://doi.org/10.1038/sj.cr.7310124.

Xu, Z., Cho, H., Hartsock, M. J., Mitchell, K. L., Gong, J., Wu, L., ... Duh, E. J. (2015). Neuroprotective role of Nrf2 for retinal ganglion cells in ischemia-reperfusion. *Journal of Neurochemistry*, *133*(2), 233–241. Available from https://doi.org/10.1111/jnc.13064.

Xu, Z., Wei, Y., Gong, J., Cho, H., Park, J. K., Sung, E. R., ... Duh, E. J. (2014). NRF2 plays a protective role in diabetic retinopathy in mice. *Diabetologia*, *57*(1), 204–213. Available from https://doi.org/10.1007/s00125-013-3093-8.

Yamasoba, T., Lin, F. R., Someya, S., Kashio, A., Sakamoto, T., & Kondo, K. (2013). Current concepts in age-related hearing loss: Epidemiology and mechanistic pathways. *Hearing Research*, *303*, 30–38. Available from https://doi.org/10.1016/j.heares.2013.01.021.

Yang, C. C., Chen, D., Lee, S. S., & Walter, L. (2011). The dynamin-related protein DRP-1 and the insulin signaling pathway cooperate to modulate *Caenorhabditis elegans* longevity. *Aging Cell*, *10*(4), 724–728. Available from https://doi.org/10.1111/j.1474-9726.2011.00711.x.

Yang, J., Li, Y., Chan, L., Tsai, Y. T., Wu, W. H., Nguyen, H. V., ... Tsang, S. H. (2014). Validation of genome-wide association study (GWAS)-identified disease risk alleles with patient-specific stem cell lines. *Human Molecular Genetics*, *23*(13), 3445–3455. Available from https://doi.org/10.1093/hmg/ddu053.

Youle, R. J., & Strasser, A. (2008). The BCL-2 protein family: Opposing activities that mediate cell death. *Nature Reviews:. Molecular Cell Biology*, *9*(1), 47–59. Available from https://doi.org/10.1038/nrm2308.

Yu, M., Yan, W., & Beight, C. (2018). Lutein and zeaxanthin isomers protect against light-induced retinopathy via decreasing oxidative and endoplasmic reticulum stress in BALB/cJ mice. *Nutrients*, *10* (7), 842. Available from https://doi.org/10.3390/nu10070842.

Zhang, H., Davies, K. J. A., & Forman, H. J. (2015). Oxidative stress response and Nrf2 signaling in aging. *Free Radical Biology & Medicine*, *88*(Pt B), 314–336. Available from https://doi.org/10.1016/j.freeradbiomed.2015.05.036.

Zhang, X., Saaddine, J. B., Chou, C. F., Cotch, M. F., Cheng, Y. J., Geiss, L. S., ... Klein, R. (2010). Prevalence of diabetic retinopathy in the United States, 2005–2008. *JAMA: The Journal of the American Medical Association*, *304*(6), 649–656. Available from https://doi.org/10.1001/jama.2010.1111.

Zhao, Z., Chen, Y., Wang, J., Sternberg, P., Freeman, M. L., Grossniklaus, H. E., & Cai, J. (2011). Age-related retinopathy in NRF2-deficient mice. *PLoS One*, *6*(4), e19456. Available from https://doi.org/10.1371/journal.pone.0019456.

Zheng, Q., Huang, C., Guo, J., Tan, J., Wang, C., Tang, B., & Zhang, H. (2018). Hsp70 participates in PINK1-mediated mitophagy by regulating the stability of PINK1. *Neuroscience Letters*, *662*, 264–270. Available from https://doi.org/10.1016/j.neulet.2017.10.051.

Zhong, Q., Mishra, M., & Kowluru, R. A. (2013). Transcription factor Nrf2-mediated antioxidant defense system in the development of diabetic retinopathy. *Investigative Ophthalmology & Visual Science*, *54*(6), 3941–3948. Available from https://doi.org/10.1167/iovs.13-11598.

Zhu, Y., Hoell, P., Ahlemeyer, B., & Krieglstein, J. (2006). PTEN: A crucial mediator of mitochondria-dependent apoptosis. *Apoptosis: An International Journal on Programmed Cell Death*, *11*(2), 197–207. Available from https://doi.org/10.1007/s10495-006-3714-5.

Ziegler, D. V., Wiley, C. D., & Velarde, M. C. (2015). Mitochondrial effectors of cellular senescence: Beyond the free radical theory of aging. *Aging Cell*, *14*(1), 1–7. Available from https://doi.org/10.1111/acel.12287.

CHAPTER

15

Cardiac aging

Ying Ann Chiao[1], Dao-Fu Dai[2], Robert J. Wessells[3] and Peter S. Rabinovitch[4]

[1]Aging and Metabolism Research Program, Oklahoma Medical Research Foundation, Oklahoma City, OK, United States [2]Department of Pathology, Carver College of Medicine, University of Iowa, Iowa City, IA, United States [3]Department of Physiology, Wayne State University School of Medicine, Detroit, MI, United States [4]Department of Laboratory Medicine and Pathology, University of Washington, Seattle, WA, United States

OUTLINE

Introduction

Aging is a dominant risk factor for cardiovascular disease, the leading cause of death in the United States (Benjamin et al., 2019). In addition to the increased prevalence of cardiovascular disease, aging is accompanied by progressive structural changes and functional declines of the heart (Lakatta & Levy, 2003b). Due to the increasing age of the population, cardiac aging and age-related cardiovascular diseases are becoming more clinically important. Therefore it is critical to understand the molecular mechanisms of cardiac aging and aging-induced predisposition to cardiovascular diseases, and to develop interventions to target these mechanisms. This chapter provides an overview of the cardiac aging process in humans, and murine and other animal models. The chapter also discusses the molecular mechanisms involved in the pathogenesis of cardiac aging, and recent advances in the development of interventions for cardiac aging.

Handbook of the Biology of Aging.
DOI: https://doi.org/10.1016/B978-0-12-815962-0.00015-9

Cardiac aging in humans

Changes in left ventricular structure and function

The data from the Framingham Heart Study (FHS) and the Baltimore Longitudinal Study on Aging (BLSA) in apparently healthy adult men and women demonstrate left ventricular (LV) hypertrophy with aging, as evidenced by increased LV wall thickness by echocardiography. The diastolic function declines with age, evidenced by reduced LV filling in early diastole measured as decreased peak E wave by Doppler echocardiography and increased dependency on atrial contraction (increased A wave) to maintain LV filling. There are several potential causes of diastolic dysfunction in the elderly. Reduced rates of calcium reuptake by myocardial sarcoplasmic reticulum (SR) ATPase (SERCA2a) delay ventricular relaxation, thus compromising LV filling during early diastole. Decreased elasticity/compliance due to LV hypertrophy and increased ventricular fibrosis further impaired LV filling leading to diastolic dysfunction. Increased atrial contraction during late diastole leads to increased atrial pressure and inadvertently causes atrial hypertrophy, which also increases the risk of atrial fibrillation. The prevalence of atrial fibrillation also increased with aging (Lakatta, 2003; Lakatta & Levy, 2003a, 2003b). Doppler measurement of the E/A ratio, the ratio between early (E) and late (A) diastolic LV filling, declines with age, according to FHS and BLSA. This decline in the E/A ratio suggests that a greater portion of LV filling results from late diastolic filling from atrial contraction as opposed to early diastolic filling by active LV relaxation, which is interpreted clinically as an indication of diastolic dysfunction. Diastolic dysfunction is highly prevalent in the elderly, especially women (Bursi et al., 2006), and it adversely affects exercise capacity in the geriatric population. It also predisposes to the development of diastolic heart failure, which is defined as heart failure with preserved ejection fraction (HFpEF). HFpEF accounts for more than half of the heart failure cases in patients older than 75 years old, independent of the presence of clinical hypertension, suggesting that aging per se is a risk factor for HFpEF.

Decreased exercise capacity is another hallmark of cardiac aging. In response to increased metabolic demand during exercise, youthful hearts augment cardiac output by increasing both heart rate (chronotropic response) and stroke volume (inotropic response) mediated through activation of the sympathetic nervous system (adrenergic system). The increase in stroke volume is primarily mediated by increased myocardial contractility and decreased peripheral vascular resistance. During intense exercise in young subjects, stroke volume increases proportionally with exercise intensity up to approximately 40%–50% of maximal capacity, after which the additional augmentation of cardiac output is driven by an increase in heart rate (Vella & Robergs, 2005). Reduced maximal heart rate is a major contributor to exercise intolerance in the elderly population. Two mechanisms may contribute to this chronotropic incompetence: degeneration of the cardiac conduction system with age and impaired autonomic regulation of the cardiovascular system. Interestingly, the decline in maximal heart rate correlates with decreased exercise capacity and is an independent predictor of cardiovascular mortality and morbidity (Bhatheja et al., 2005; Lauer et al., 1999). This impairment of cardiac reserve to exercise stress in the elderly is not obvious at rest, as shown by relatively preserved resting heart rate when in the supine position (Fleg et al., 1995) and the relatively preserved cardiac ejection fraction at rest. Exercise capacity significantly declines with age, as does cardiovascular reserve after prolonged exercise (Correia et al., 2002). This is attributed to a modest decrease in ejection fraction (inotropic) and a prominent decline in maximal heart rate (chronotropic incompetence) at peak exercise.

A decline in the postsynaptic β-adrenergic signaling efficiency has been reported with aging (Gerstenblith et al., 1977). This is supported by the observation that cardiovascular responses to β-adrenergic agonist infusions at rest decrease with age. Another study indicated that a diminished efficacy of postsynaptic β-adrenergic receptor signaling in aging changes the exercise hemodynamic profile of younger persons to make it resemble that of older individuals.

Apparent deficits in sympathetic modulation of cardiac and arterial functions with aging occur in the presence of exaggerated neurotransmitter levels. Plasma levels of norepinephrine and epinephrine during any perturbation from the supine basal state increase to a greater extent in older compared with younger healthy humans. The age-associated increase in plasma levels of norepinephrine results from an increased spillover into the circulation and, to a lesser extent, reduced plasma clearance. The degree of norepinephrine spillover into the circulation differs among body organs; increased spillover is prevalent within the aging heart. Deficient norepinephrine reuptake at nerve endings is a primary mechanism for increased spillover during acute graded exercise. During prolonged exercise, however, diminished neurotransmitter reuptake might also be associated with depletion and reduced release and spillover.

Rhythm changes

Heart rate variability (HRV), which is defined by beat-to-beat variation of heart rate at rest, has been shown to decline with age. A decline in HRV indicates cardiac autonomic dysregulation, which is commonly found in the elderly. Decreased HRV is well documented to associate with increased risk of cardiovascular mortality and morbidity (Tsuji et al., 1996). In addition to decreased HRV, there is an increase in the prevalence and the complexity of both supraventricular and ventricular premature or ectopic beats in apparently healthy older individuals, as detected by ECG during resting, ambulatory, or exercise performance. The clinical significance of these ectopic beats is unclear. Although these ectopic beats are generally benign and not associated with heart diseases, they do indicate aging changes in cardiac automaticity and the conduction system, as these ectopic beats are not obvious in younger persons (Fleg & Kennedy, 1992).

Aging has been associated with increased prevalence of atrial fibrillation. The American Heart Association's Heart Disease and Stroke Statistics showed that the elderly population accounts for more than 70% of patients with atrial fibrillation. Atrial fibrillation imposes a significant risk of intracardiac thromboembolism and subsequent stroke (Rosamond et al., 2007). The potential mechanisms of increased atrial fibrillation in aging include impaired diastolic filling in the aged heart, which subsequently increases end-diastolic volume and left atrial pressure. Increased left atrial pressure induces left atrial dilation and fibrosis. In addition, age-dependent myxomatous changes of the mitral valve may cause mitral regurgitation, which also results in left atrial enlargement. Left atrial enlargement predispose to atrial fibrillation, which further induces left atrial remodeling and dilation, contributing to a vicious cycle.

Valvular changes

Age-dependent valvular changes mainly include myxomatous degeneration and collagen deposition, called valvular sclerosis. Aortic valve sclerosis is present in 30%–80% of elderly individuals (Karavidas et al., 2010; Nassimiha et al., 2001; Stewart et al., 1997), which includes calcification of aortic valve leaflets and the aortic annulus that increases with age (Freeman & Otto, 2005; Otto et al., 1999). The risk factors of age-dependent aortic sclerosis include chronic hypertension, LV hypertrophy, hyperlipidemia, smoking, end-stage renal disease, and congenital bicuspid aortic valves (Olsen et al., 2005). Of note, elderly patients with aortic sclerosis have a concurrent increase in cardiovascular events and mortality (Aronow et al., 1999; Otto et al., 1999), as aortic sclerosis often develops in parallel with progression of atherosclerosis in other vessels (Otto, 2004). At earlier stages, aortic valve sclerosis does not obstruct blood flow, but it can progress into clinical aortic stenosis when severe thickening, stiffening, and calcification of the leaflet results in obstruction of the aortic valve. Increased leaflet calcification and decreased leaflet mobility may serve as early signs of this progression.

In the elderly, fibrosis and valvular calcification are the most common factors contributing to the development of aortic stenosis, which occurs when the aortic valve opening narrows due to the stiffening and calcification of the aortic valve leaflets (Olsen et al., 2005). This narrowing prevents blood from being pumped effectively, creating a pressure gradient between the aorta and the left ventricle. To compensate for this hindrance, the walls of the left ventricle thicken with prominent myocardial hypertrophy to maintain sufficient systolic function. Later in progression, the increased wall stress causes the left ventricle to dilate, leading to deterioration of systolic function. The prevalence of aortic valve stenosis increases in an age-dependent manner, with an estimated prevalence of 2% for severe stenosis, 5% for moderate stenosis, and 9% for mild stenosis in elderly patients. In addition, aortic regurgitation, also related to the calcification of the aortic cusps and annulus, also increases with age, and is present in 13%–16% of the elderly population (Nassimiha et al., 2001). The presence of aortic regurgitation indicates that the aortic valve does not close properly; the flow of blood back to the left ventricle from the aorta during diastole results in the left ventricle working harder and increasing in size (volume overload).

Mitral annular calcification (MAC), commonly found together with mitral valve sclerosis (Faggiano et al., 2003; Jeon et al., 2001), is a degenerative process involving the fibrous annulus of the mitral valve, and is associated with aging. Frequently found during an echocardiographic examination, MAC results when calcium deposits accumulate along and beneath the mitral valve annulus (Fulkerson et al., 1979). Individuals with hypertension, end-stage renal disease, and aortic stenosis, as well as mitral valve prolapse, are commonly found to have MAC. Individuals with MAC have increased risks of mitral stenosis and regurgitation, heart failure, atrial fibrillation, conduction system diseases, stroke, coronary and vascular diseases, mortality, and adverse cardiovascular events (Karavidas et al., 2010).

Mitral valve regurgitation is another abnormality common in the aging population. This condition occurs when the mitral valve fails to seal firmly, allowing the backward flow of blood during systole, and subsequently resulting in an insufficient delivery of blood to the rest of the body. Physical symptoms include shortness of breath and fatigue. Two of the leading causes of mitral valve regurgitation are age-dependent myxomatous degeneration of the valve and ischemic heart disease (Jebara et al., 1992). Chronic mitral valve regurgitation is one of the most common indications for valve surgery in elderly patients (Akins et al., 1997).

These ventricular and valvular changes in cardiac aging result in a severe compromise in the cardiac functional reserve capacity, as well as a lowered threshold for symptoms and signs of heart failure (Correia et al., 2002). Together with increased exposure to other cardiovascular risk factors, this makes the aged heart much more susceptible to stresses and disease-related challenges, thus contributing to increased heart failure and cardiovascular mortality in the elderly.

Cardiac aging in the mouse model

Rodent models, in particular the mouse model, are widely used in cardiac aging studies. The phenotypes of cardiac aging in mice closely recapitulate the phenotypes of human cardiac aging (Dai et al., 2009; Lakatta & Levy, 2003a, 2003b). Longitudinal echocardiography performed on a mouse longevity cohort by Dai et al. has showed that there is a 47% decline in the ratio of early to late diastolic mitral annular velocity (Ea/Aa) from 4–6 = month-old mice to 24-month-old C57BL6 mice, and that the frequency of diastolic dysfunction, defined as Ea/Aa <1 (Khouri et al., 2004), was increased to 55% in mice over 24 months of age (Dai et al., 2009). In addition, they observed a modest reduction of systolic function from young adult to the oldest group. They also demonstrated an increase (worsening) with age of the myocardial performance index (MPI), calculated as the ratio of the sum of isovolumetric contraction and relaxation time to LV ejection time (Dai et al., 2009). The increase in MPI indicates that a greater fraction of the cardiac cycle is spent to cope with the pressure changes between ventricle contraction and relaxation during isovolumetric phases, and has been shown to reflect both LV systolic and/or diastolic dysfunction (Tei et al., 1997). For geometric changes, aging is associated with significant increases in LV mass index (LVMI) and left atrial dimension, which indicate age-related LV and LA hypertrophy (Dai et al., 2009).

At a histopathological level, old murine hearts display interstitial fibrosis, increased variation in myocyte fiber size, hypercellularity, cytoplasmic vacuolization and hyalinization, collapse of sarcomeres, mineralization, arteriosclerosis and arteriolosclerosis, and these changes were collectively designated as age-associated cardiomyopathy (Treuting et al., 2008). Aged hearts also showed cardiomyocyte hypertrophy (increased myocardial fiber size), increased cardiomyocyte apoptosis (Yan et al., 2007), and increased deposition of collagen and amyloid (Mohammed et al., 2014; Westermark et al., 1979).

The relatively short lifespan of mice allows aging studies to be conducted in a manageable time frame in this model. The availability of genetically modified mice and tools to generate new mouse models (both by transgenesis and by viral transduction) are also advantageous for studying the molecular mechanisms of cardiac aging. In addition, while sharing similar cardiac aging phenotypes with humans, laboratory mice do not develop elevated blood pressure or adverse blood glucose and lipid profiles with age (Dai et al., 2009; Zheng et al., 2003). This allows the intrinsic cardiac changes of aging to be investigated without the confounding effects of hypertension and diabetes. Currently, the majority of our knowledge of the molecular basis of cardiac aging comes from studies using mouse and other animal models. The findings from these studies should be interpreted cautiously, taking into account the different species, strains, and definitions of age groups used by different studies.

Other models of cardiac aging

Large animal models of cardiac aging

Eichelberg and Seine studied the life expectancy and causes of death in over 9000 dogs of various breeds and showed that cardiovascular diseases are the second most frequent cause of death in dogs (Eichelberg & Seine, 1996). Aging is associated with lowered contractile function, prolonged contraction, and increased systolic and diastolic stiffness in dogs (Templeton et al., 1976, 1979). Pathologies of aged canine hearts include cardiac hypertrophy, increased myocardial stiffness, prolonged action potential, and reduced responsiveness to β-adrenergic stimulation. These age-related pathologies are associated with progressive loss of cardiac reserve and adaptability, and increased susceptibility to heart diseases in the aged dogs (Guglielmini, 2003). Common age-related cardiovascular diseases in dogs include chronic degenerative valvular disease, dilated cardiomyopathy, amyloidosis, lipofuscinosis, and sick sinus syndrome, with chronic degenerative valvular disease being the most prevalent, reaching about 75% for dogs over 16 years old (Van Vleet, 2001). Canine models have also been used to study multiple cardiovascular diseases including cardiac arrhythmia, ischemia reperfusion injury, and genetic cardiomyopathy, such as in Duchene muscular dystrophy (Hamlin, 2007; Jugdutt, 2002; Koh et al., 1995).

Recently, an NIH-supported long-term longitudinal study of aging in companion dogs, named the "Dog Aging Project," was launched (Kaeberlein et al., 2016). As a part of the project, a short-term veterinary clinical trial of rapamycin, an intervention previously showed to rescue cardiac aging in mice (Chiao et al., 2016; Dai et al., 2014; Flynn et al., 2013), was conducted on 24 companion dogs. The trial observed a trend of improved diastolic and systolic function in middle-aged companion dogs after 10 weeks of rapamycin treatment (Urfer et al., 2017). Future trials with larger sample sizes will be underway to validate these cardiac benefits.

Nonhuman primates are phylogenetically closer to humans compared to other animal models and are useful for the study of complex physiology. As in humans, cardiovascular disease is the leading cause of death in captive great apes, and the prevalence of cardiovascular disease increases with age in apes (Lowenstine et al., 2016). The longitudinal study of aging in rhesus monkeys (*Macaca mulatta*) conducted by the National Institute of Aging demonstrated that under normal diets rhesus monkeys develop age-dependent aortic and mitral valve degenerative calcifications, loss of myocardial fibers with compensatory hypertrophy of the remaining cardiomyocytes, lipofuscin accumulation, variable degree of myocarditis, multifocal interstitial fibrosis, and an increased frequency of myocardial infarction and congestive heart failure (Lane et al., 1999, 2002; Mattison et al., 2003; Roth et al., 2004).

The common marmoset (*Callithrix jacchus*) is another nonhuman primate model. Tardif et al. observed myocardial fibrosis at necropsy in 60% of marmosets over 10 years of age (Tardif et al., 2011). Tardif and colleagues later showed that cardiomyopathy is a common cause of death for marmosets over 6 years old but not for younger marmosets, suggesting an increased prevalence of cardiac failure in aged marmosets (Ross et al., 2012).

Due to their longer lifespan, cardiac aging studies in large animal models generally require longer time frames. However, studies in larger animals are valuable as a bridge for translating preclinical studies in rodent models to human trials.

Drosophila: An invertebrate model of cardiac aging

A major obstacle to generating data on normal aging from human subjects or vertebrate models is the length of time necessary to complete longitudinal observations. The insect model system of *Drosophila* is uniquely well-suited for studies across ages. It has a simple, tubular working heart with extensive developmental and functional homology to mammalian hearts (Blice-Baum et al., 2019; Sujkowski & Wessells, 2015), but has a lifespan of only a few weeks, and is easy to house in large numbers, facilitating longitudinal observations of age-related functional decline.

Wild-type flies experience a gradual decline in heart rate and tolerance for pacing stress across ages (Kaushik et al., 2015; Wessells et al., 2004), and an increase in spontaneous arrhythmias (Ocorr et al., 2007). These changes are, in part, regulated by signaling pathways associated with longevity, and can be modulated by tissue-autonomous genetic modulation. For example, cardiac-specific knockdowns of insulin receptor (Wessells et al., 2004) or TOR kinase (Luong et al., 2006) activity can dramatically slow age-related decline in cardiac function, and overexpression of downstream modulators such as *dFoxo* (Blice-Baum et al., 2017; Wessells et al., 2004), *d4eBP* (Wessells et al., 2009) or *dSesn* (Lee et al., 2010) can mimic these protective effects.

These signaling changes during aging are manifested at the level of the cardiomyocytes by distinct changes to the transcriptome and proteome (Cammarato et al., 2011), including expression of genes encoding contractile proteins (Cannon et al., 2017). Structural changes also occur, as reflected in the increase in passive stiffness of the heart with age (Kaushik et al., 2012, 2015).

In recent years, the fly model has been increasingly used to study changes in extracellular matrix composition and organization during aging, and the role of these changes in modulating function. For example, accumulation of type IV collagens around the aging heart has been associated with age-related cardiomyopathy, and reduction of this accumulation by genetic manipulation is strongly cardioprotective (Hartley et al., 2016; Sessions et al., 2017).

Decreased turnover of normal and/or damaged proteins and other molecules is thought to be a conserved component of cardiac aging, and accumulation of misfolded proteins has been observed in the aged *Drosophila* heart (Blice-Baum et al., 2017). This accumulation correlated with a reduction in gene expression of genes encoding components of the proteasome, and could be rescued by upregulation of *dFoxo* in the myocardium (Blice-Baum et al., 2017). While the role of decreasing autophagy in functional aging has been well-established in *Drosophila* skeletal muscle (Demontis & Perrimon, 2010), its role in cardiomyocytes is less clear. Evidence of reduced autophagy or broadly reduced expression of genes encoding autophagy-related proteins has not been reported in fly cardiac muscle. Turnover of mitochondria by components of the autophagosome, known as mitophagy, may be a critical factor during normal aging, however, as accumulation of oxidized mitochondria in the aging heart has been observed (Laker et al., 2014), and knockdowns of several components of the autophagosome have been shown to reduce cardiac function prematurely (Xu et al., 2019). Conversely, interventions that increase mitophagy, such as exercise, ameliorate age-related cardiac functional decline (Laker et al., 2017; Sujkowski et al., 2017).

The *Drosophila* model is especially amenable for the study of environmental factors that impact progression of cardiac aging, such as diet and exercise. These factors have long been understood to play an important role in the rate of aging, as well as the etiology of various cardiac pathologies. The short lifespan and ability to generate large sample sizes make the fly model a useful system for examining long-term effects of diet and exercise on cardiac function (Sujkowski & Wessells, 2018).

Flies can be raised on a simple diet consisting of yeast plus a sugar source. This facilitates creation of diet matrices where the percentage of protein and carbohydrates can be varied, and the effects on cardiac performance measured in large cohorts. Such experiments have shown that heart performance in aged flies is best on diets where protein and carbohydrates are relatively balanced, and among balanced diets, lower-calorie diets seem to be cardioprotective (Bazzell et al., 2013).

In contrast, chronic exposure to high-fat diets promotes accumulation of lipid droplets in the fly myocardium and leads to changes to myofilament structure (Birse et al., 2010). These changes, in turn, lead to decreased heart period, reduced fractional shortening, and increased arrhythmias (Birse et al., 2010; Diop et al., 2015). These changes can be rescued by reduction of TOR kinase activity (Birse et al., 2010), or by cardiac overexpression of *d4eBP* or the PGC1-α homolog *spargel* (Diop et al., 2015). Likewise, a high-sugar diet promoted hyperglycemia and accumulation of triglycerides in the fly myocardium (Musselman et al., 2011; Na et al., 2013). The effects of these dietary modulations qualitatively resemble changes seen during normal aging, suggesting that dietary modulation has substantial potential for cardioprotection during the aging process. Further supporting this idea, flies on time-restricted feeding, a form of dietary restriction, are strongly protected against both normal aging and the effects of a high-fat diet (Gill et al., 2015).

The cardioprotective effects of chronic exercise have also been examined in the fly model system. Wild-type flies that perform a ramped daily exercise program for 3 weeks have better fractional shortening and resistance to cardiac pacing stress than unexercised siblings (Piazza et al., 2009). Exercise also reduces cardiac lipid accumulation (Sujkowski et al., 2012), increases mitochondrial turnover (Laker et al., 2014), and protects mitochondrial function (Sujkowski et al., 2019) in aging flies.

The fly system has allowed identification of specific genetic factors that mediate the cardioprotective effects of exercise, including adrenergic receptors (Sujkowski et al., 2017) and the TOR-modulating protein Sestrin (Kim et al., 2020).

Recently, additional *Drosophila* models of exercise training have been generated (Mendez et al., 2016) that will facilitate further examination of genetic factors that modulate the effects of chronic exercise on cardiac aging.

The *Drosophila* model has historically been important in understanding developmental pathways and cardiac specification and has recently been increasingly utilized as a system for examining phenotypes related to wild-type aging and age-related disease. In the years to come, the fly system will likely continue to play an important role in understanding the interplay between genetics, environment, and age-related functional changes.

Molecular mechanisms of cardiac aging

In the last two decades, researchers have revealed several key molecular mechanisms of cardiac aging and have begun to target these mechanisms to develop cardiovascular healthspan interventions (summarized in Fig. 15.1).

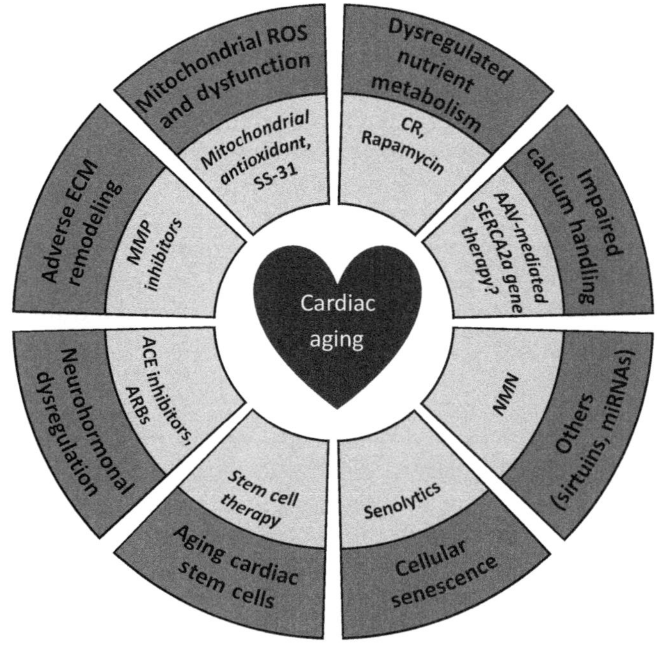

FIGURE 15.1 A schematic summary of the molecular mechanisms (*outer ring*) of cardiac aging and potential interventions (*inner ring*) targeting these mechanisms. The interventions in italics have been demonstrated to target the corresponding mechanisms in models of cardiovascular disease but not in aging models.

Mitochondrial dysfunction and mitochondrial ROS

The heart has a high energy demand and is rich in mitochondria, and is therefore especially susceptible to mitochondrial dysfunction and oxidative stress. In aged cardiomyocytes, enlarged mitochondria with swelling, loss of cristae, and even destruction of inner membranes have been observed (Terman & Brunk, 2004). Other studies have shown a reduced inner membrane (cristae) area or exceptionally wide cristae in mitochondria in aged rat, mouse, or fly hearts (Brandt et al., 2017; El'darov et al., 2015). However, while Hoppel and colleagues have shown reduced content of interfibrillar mitochondria (IFM) in aged rat hearts (there are two populations of mitochondria in the myocardium, the IFM and the sarcolemma mitochondria, SSM), they have reported no differences in mitochondrial ultrastructure in the two types of mitochondria in aged rat hearts (Fannin et al., 1999; Lesnefsky et al., 2016; Riva et al., 2006).

In addition to changes in mitochondrial structure, aging is associated with increased cardiac mitochondrial ROS production (Judge et al., 2005). The abnormal mitochondrial ROS production and impaired detoxification lead to mitochondrial dysfunction in aged hearts (see reviews in Mammucari & Rizzuto, 2010; Terzioglu & Larsson, 2007; Trifunovic & Larsson, 2008).

Direct evidence of the role of mitochondrial ROS in cardiac aging came from mice overexpressing catalase targeted to the mitochondria (mCAT). Mice expressing mCAT have an 18% extension of mean and maximal lifespan, while mice overexpressing wild-type human catalase (naturally delivered to peroxisomes, pCAT) or catalase with a nuclear targeting signal (nCAT) had 0%–5% extension of mean lifespan, with no effect on maximal lifespan (Schriner et al., 2005). As shown in Fig. 15.2, for each of the examined echocardiographic parameters, the mCAT mice exhibit attenuation of cardiac aging phenotypes (Figs. 15.2A–F), particularly for LV hypertrophy (Fig. 15.2A), diastolic dysfunction (Figs. 15.2C, D) and impairment of myocardial performance (Fig. 15.2F). The protective effect of mCAT expression is also evidenced in improved cardiac histopathology seen in older mice (Dai et al., 2009) or at the end of life (Treuting et al., 2008), including significant amelioration of age-dependent cardiomyocyte hypertrophy, interstitial fibrosis, and mitochondrial ultrastructural damage. These cardiac benefits are

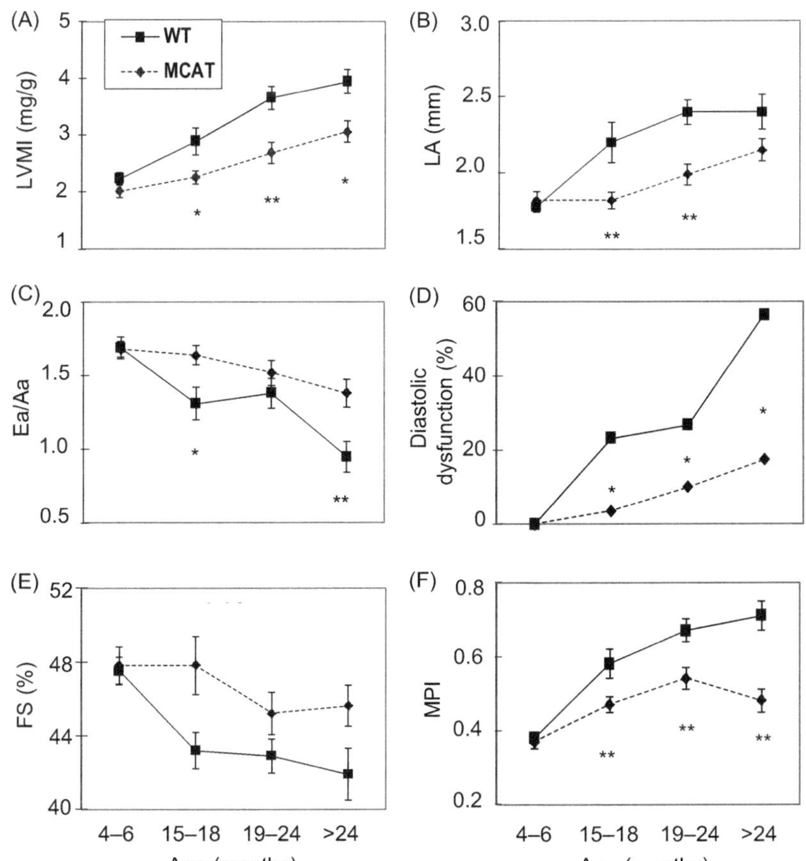

FIGURE 15.2 Echocardiographic parameters of cardiac aging in WT and mCAT C57Bl/6 mice in the four age ranges shown (~20 mice per group/genotype, $N = 170$ total). Left ventricular mass index (A), left atrial dimension (B), ratio of early to late diastolic mitral annular velocity (Ea/Aa) (C), the proportion of mice with diastolic dysfunction, defined as Ea/Aa < 1 (D), fractional shorting (FS) of the left ventricle (E) and the myocardial performance index (F) were measured. $^{*}P < .05$, $^{**}P < .01$. The linear trend between genotypes is significant, $P < .01$ in all cases, except fractional shortening, $P < .05$. *Modified from Dai, D. F., Santana, L. F., Vermulst, M., Tomazela, D. M., Emond, M. J., MacCoss, M. J., ... Rabinovitch, P. S. (2009). Overexpression of catalase targeted to mitochondria attenuates murine cardiac aging.* Circulation, *119(21), 2789–2797.*

accompanied by significantly reduced mitochondrial protein oxidative damage and mitochondrial DNA mutation and deletion frequencies in the hearts of mCAT mice (Dai et al., 2009; Schriner et al., 2005).

On the other hand, mice with homozygous mutation of mitochondrial polymerase γ (Polg$^{m/m}$) have significantly increased mtDNA mutations and deletions with age (Kujoth et al., 2005; Trifunovic et al., 2004). Polg$^{m/m}$ mice exhibit shortened lifespan and several progeroid phenotypes, and at middle age (13–14 months) they display cardiac aging phenotypes that are even more severe than those of wild-type mice of 24–30 months age. At middle age, Polg$^{m/m}$ mice develop cardiac hypertrophy and reduced systolic and diastolic function, suggesting that mtDNA mutations and deletions accelerate cardiac aging (Dai et al., 2010; Trifunovic et al., 2004). Interestingly, mCAT overexpression partially rescues the mitochondrial damage and cardiomyopathy in middle-age Polg$^{m/m}$ mice (Dai et al., 2010). Another study also showed that endurance exercise can prevent cardiac aging phenotypes in Polg$^{m/m}$ mice (Safdar et al., 2011), in part by augmentation of mitochondrial biogenesis to compensate for the defective mitochondria in these mice. These studies demonstrated the interaction between mtDNA damage, mitochondrial ROS, and mitochondrial dysfunction in cardiac aging.

Nutrient signaling in cardiac aging

Mechanistic target of rapamycin (mTOR) is an important modulator of aging and age-related disease that integrates nutrient and hormonal cues to regulate growth and longevity (Kennedy et al., 2007). Under conditions of abundant nutrients, mTOR is activated and this favors cell growth and cell division; under limited nutrients, mTOR activity is suppressed and this leads to reduced growth, enhanced resistance to stress, and increased lifespan. mTOR forms two distinct complexes, TORC1 and TORC2, to regulate two diverse sets of cellular processes. TORC1 regulates the protein homeostasis branch of the mTOR pathway and is sensitive to pharmacologic inhibition by rapamycin. Two well-studied downstream targets of TORC1, p70 ribosomal S6 kinase 1 (p70S6K1) and eukaryotic translation initiation factor 4E (eIF4E), are activated in pressure overload-induced cardiac hypertrophy (Volkers et al., 2013). Studies in *Drosophila* and mouse models have demonstrated that increased mTOR signaling accelerates, and reduced mTOR signaling attenuates, cardiac aging. Using the *Drosophila* model, Luong et al. showed that inhibition of the mTOR pathway attenuates age-related cardiac dysfunction (Luong et al., 2006). They later showed that overexpression of eIF4E binding protein (4EBP), which inhibits eIF4E, attenuates age-related cardiac dysfunction to a similar extent as overexpression of the TOR antagonist tuberous sclerosis complex (TSC), and that overexpression of eIF4E leads to a more rapid decline of myocardial function with age (Wessells et al., 2009). These findings implicate a major role for mTOR/eIF4E signaling in cardiac aging. In the mouse model, Meikle et al. demonstrated that mice with cardiac-specific deletion of TSC1, a model of increased mTOR signaling, develop dilated cardiomyopathy and have a median lifespan of 6 months (Meikle et al., 2005). Additional evidence for the role of mTOR signaling in cardiac aging came from studies that inhibition of mTOR signaling by caloric restriction (CR) or rapamycin (see Section Interventions for cardiac aging), both of which have been shown to be protective against cardiac aging in mice (Chiao et al., 2016; Dai et al., 2014; Flynn et al., 2013).

Neurohormonal regulation of cardiac aging (renin–angiotensin–aldosterone system, IGF)

Renin–angiotensin–aldosterone system

The renin–angiotensin–aldosterone system (RAAS) is a key endocrine system regulating hypertension, fluid and electrolyte balance, and systemic vascular resistance. The RAAS has been shown to play an important role in stress-induced cardiac hypertrophy, cardiac aging, and various cardiovascular diseases. The direct effect of angiotensin II (Ang) on the heart includes inducing cardiomyocyte hypertrophy and apoptosis, increasing cardiac fibrosis, and impairing cardiomyocyte relaxation (Domenighetti et al., 2005). These changes recapitulate the phenotypes found in cardiac aging. Indeed, we and others have reported that cardiac tissue Ang levels are significantly increased in the aged rodent heart (Dai et al., 2009; Groban et al., 2006), which may be secondary to increased tissue levels of angiotensin II converting enzyme (ACE) (Lakatta, 2003). The role of RAAS in cardiac aging is furthered reinforced by studies which showed that long-term inhibition of Ang either by ACE inhibitors, angiotensin receptor blockers, or genetic disruption of angiotensin receptor type I reduce age-dependent cardiac pathology and prolong rodent survival (Basso et al., 2007; Benigni et al., 2009).

Adrenergic signaling

Chronic activation of β-adrenergic signaling is deleterious to the heart. Stimulation of β-adrenergic signaling increases cardiac metabolic demand due to increased heart rate (chronotropic effect), contractility (inotropic effect), afterload (blood pressure), and wall stress. Several clinical trials have shown the survival benefit of inhibition of β-adrenergic signaling by β-blockers in patients with heart failure. Adrenergic signaling involves adrenergic receptors, which are G-protein-coupled receptors coupled with various adenylate cyclases. Adenylate cyclase type 5 (AC5) is a key enzyme downstream of β-adrenergic signaling in the heart. In mice, disruption of AC5 mitigated chronic pressure overload-induced cardiac hypertrophy, apoptosis, and failure induced by either chronic catecholamine stimulation or by transverse aortic banding (Okumura et al., 2003, 2007). AC-5 knockout mice have also been shown to extend lifespan and to attenuate LV hypertrophy, systolic dysfunction, apoptosis, and fibrosis (Yan et al., 2007) in aged mice. The aging-protective mechanisms may involve upregulation of the Raf-1/pMEK/pERK pathway, which confers protection against various stresses (Yan et al., 2007).

Insulin/IGF-1 signaling

Insulin/IGF-1 signaling is one of the best characterized determinants of lifespan regulation in vertebrate and invertebrate animal models. Deficiency in insulin/IGF-1 signaling has been shown to increase lifespans in worms (Apfeld and Kenyon, 1998; Dorman et al., 1995), flies (Tatar et al., 2001), and mammals. The latter was demonstrated by lifespan extension in Ames and Snell dwarf mice and growth hormone receptor and IGF-1 receptor knockout (GHRKO) mice, all having deficiency in growth hormone or IGF-1 signaling (Liang et al., 2003), as well as mice with overexpression of the Klotho insulin/insulin-like growth factor signaling inhibitor (Kurosu et al., 2005). Deficiency in insulin/IGF-1 signaling improved cardiac performance at advanced age in *Drosophila* and attenuated age-associated cardiomyocyte dysfunction in mice (Li et al., 2008; Wessells et al., 2004).

In contrast, it has been shown in humans that an age-dependent decline in serum IGF-1 level (Corpas et al., 1993) correlates with an *increased* risk of heart failure among elderly patients without prior history of heart disease (Vasan et al., 2003) (see a review see Khan et al., 2002). Interventions that increase IGF-1 signaling, including growth hormone therapy, may actually be beneficial in some patients with heart failure (Broglio et al., 1999). These beneficial effects of IGF-1 on cardiovascular disease may be related, at least in part, to mitochondrial protective mechanisms. This is supported by the observation that in vitro treatment of endothelial cells and cardiomyocytes with IGF-1 decreases mitochondrial superoxide production (Csiszar et al., 2008). Concomitant with this, low plasma levels of GH and IGF-1 in Ames dwarf mice are associated with increased mitochondrial oxidative stress both in the vasculature and the heart (Csiszar et al., 2008), mimicking the aging phenotype. Interestingly, mitochondrial oxidative stress in the heart of Ames dwarf mice appears to be associated with impaired contractile function (Ren and Brown-Borg, 2002). Recent studies show that treatment of aged rats with IGF-1 confers mitochondrial protection, including an attenuation of mitochondrial ROS production in the liver (Puche et al., 2008). The available data suggest that treatments that increase circulating IGF-1 levels exert cardiovascular protective effects in aging (Groban et al., 2006; Lopez-Lopez et al., 2007; Rivera et al., 2005). Thus, further studies are warranted to determine the role of mitochondrial mechanisms in the beneficial effects of GH replacement and/or IGF-1 treatment in the aged heart and vasculature.

Impaired calcium homeostasis

Active relaxation of the cardiomyocyte is a major determinant of diastolic function (Kass et al., 2004; Zile & Brutsaert, 2002). Ca^{2+} acts as a regulator and a messenger of excitation–contraction (EC) coupling. During depolarization (action potential), trace amounts of Ca^{2+} enter the cardiomyocyte through the L-type calcium channel (LTCC) and the increase in cytoplasmic Ca^{2+} concentration triggers a larger scale Ca^{2+} release from SR Ca^{2+} storage via the ryanodine receptor. Cytoplasmic Ca^{2+} binds to and activates troponin C within the myofilaments and induces myocyte contraction. During myocardial relaxation, Ca^{2+} dissociates from the actin–myosin complex, followed by reuptake into the SR through phospholamban-regulated sarco/endoplasmic reticulum Ca^{2+}-ATPase (SERCA) or extruded outside the cardiomyocyte through the Na^+/Ca^{2+} exchanger (NCX). It has been shown that aged murine hearts have reduced SERCA2 expression (Dai et al., 2009) and activity (Janczewski and Lakatta, 2010), with a compensatory increase in the levels of the NCX (Koban et al., 1998). Aged hearts also exhibit a compensatory increase in the L-type Ca^{2+} currents (Josephson et al., 2002) and prolonged action potential duration to preserve SR loading and to maintain intracellular Ca^{2+}-transients and contractions in old cardiomyocytes (Janczewski et al., 2002). Posttranslational modifications of SERCA2, including age-related oxidation and nitration, have also been demonstrated (Knyushko et al., 2005; Sharov et al., 2006), although their roles in cardiac aging require further investigation.

Adverse extracellular matrix remodeling

Extracellular matrix (ECM) is a complex network of proteins located outside the cells and provides structural and biological support to the surrounding cells. ECM provides structural support to the surrounding cells in the heart, and its composition is a major determinant of myocardial stiffness, which is a major regulator of diastolic function (Ouzounian et al., 2008). ECM remodeling is a dynamic process and ECM composition is tightly regulated by the balance of synthesis and degradation of ECM by matrix metalloproteinases (MMPs) and other proteases. Dysregulation of ECM synthesis and degradation have both been implicated in cardiac aging and pathology.

Transforming growth factor-β (TGF-β) is a family of profibrotic cytokines that induce ECM expression and suppress matrix degradation (Bujak and Frangogiannis, 2007). TGF-β1 heterozygous mice have reduced myocardial fibrosis and stiffness and increased LV compliance at 24 months of age (Brooks and Conrad, 2000). The expression of connective tissue growth factor (CTGF), a downstream mediator of TGF-β, increases with aging (Wang et al., 2010). Mice overexpressing cardiomyocyte-specific CTGF exhibit age-related cardiac dysfunction as early as 7 months of age (Panek et al., 2009). In a senescence-accelerated mouse model, increased TGF-β and CTGF expression was associated with myocardial fibrosis and diastolic dysfunction at 6 months of age (Reed et al., 2011). Bradshaw et al. showed that expression of a matricellular protein, secreted protein acidic and rich in cysteine (SPARC), increased with age and that deletion of SPARC reduced fibrillar collagen content and reduced LV diastolic stiffness (Bradshaw et al., 2010). This evidence suggests a link between aging, ECM synthesis, and diastolic dysfunction.

MMPs are a family of 25 zinc-dependent enzymes that regulate ECM protein degradation. MMP proteolytic activity is inhibited specifically in tissues by the tissue inhibitors of matrix metalloproteinase (TIMPs), a family composed of TIMP-1, -2, -3, and -4 (Tayebjee et al., 2005). Bonnema et al. showed that levels of MMP-2, MMP-7, TIMP-1, TIMP-2, and TIMP-4 increase but levels of MMP-9 decrease in human plasma (Bonnema et al., 2007). Lindsey and colleagues showed that the levels of MMPs-3, -9, -14 and TIMP-4 increase from middle age to old age in CB6F1 mice (Lindsey et al., 2005). Chiao et al. showed that MMP-9 levels increase in the LV and plasma of aged C57Bl6 mice (Chiao et al., 2011). Aged MMP-9 null mice have reduced collagen deposition and preserved diastolic function, and these attenuated cardiac aging phenotypes are accompanied by reduced expression of profibrotic proteins and a compensatory increase in MMP-8 levels (Chiao et al., 2012). Spinale et al. showed that mice overexpressing cardiac-specific MT1-MMP have increased collagen deposition and LV dysfunction in middle age, suggesting accelerated cardiac aging (Spinale et al., 2009). Despite the roles of MMPs in ECM degradation, the two studies showed that deletion of an MMP leads to reduced fibrosis while overexpression of another MMP promotes fibrosis. These findings suggest that ECM degradation is under complex regulation by MMPs during cardiac aging.

Cellular senescence of cardiac cells

Senescent cells are characterized by irreversible growth arrest, senescence-associated secretory phenotypes (SASPs) and resistance to apoptosis, and they have been showed to accumulate with age in tissues and organs that display age-related dysfunction or pathology (Campisi, 2013; van Deursen, 2014). Several studies have demonstrated accumulation of senescent cells or increased levels of cellular senescence markers in aged hearts (Anderson et al., 2019; Baker et al., 2016; Jazbutyte et al., 2013; Wang et al., 2015). Senescence of cardiac fibroblasts likely regulates cardiac fibrosis in aging and contributes to cardiac remodeling in myocardial infarction and atrial fibrillation (Jazbutyte et al., 2013; Wang et al., 2015; Xie et al., 2017; Zhu et al., 2013). Cardiac progenitor cells (CPCs) isolated from human subjects showed age-related increases in senescence-associated markers, including $p16^{INK4A}$ and SA-β-gal, and senescent CPCs displayed impaired proliferation and regenerative capacity (Lewis-McDougall et al., 2019). Although there are still debates on whether postmitotic cells like cardiomyocytes undergo cellular senescence, markers of cellular senescence have been detected in cardiomyocytes isolated from aged mice and rats (Anderson et al., 2019; Maejima et al., 2008). Baker et al. have demonstrated that clearance of senescent cells can extend murine lifespan and delay or attenuate age-related deterioration of several organs including hearts (Baker et al., 2016; 2011). They showed that while systolic function and LV hypertrophy were not altered, clearance of senescent cells reduced cardiomyocyte hypertrophy in 18-month-old mice (Baker et al., 2016). Removal of senescent cells also enhanced resistance to death from arrhythmogenesis caused by a toxic dose of isoproterenol and protection against isoproterenol-induced LV hypertrophy (Baker et al., 2016). In another study, removal of senescent cells in 27-month-old mice reduced cardiomyocyte hypertrophy and cardiac fibrosis (Anderson et al., 2019). While the role of cellular senescence in age-related cardiac dysfunction still requires further investigation, these findings have highlighted the therapeutic potentials of senolytics in cardiac aging.

Aging of resident stem cells and cardiomyocytes

Since the initial report in 2003 (Beltrami et al., 2003), the existence and the role of adult cardiac stem cells in cardiac physiology and disease have been controversial. Although an independent group reported that adult c-kit-positive "cardiac stem cells" are necessary and sufficient for functional cardiac regeneration and repair (Ellison et al., 2013), it is not reproducible yet. Several other investigators have applied various lineage-tracing methods and failed to demonstrate the significance of c-kit$^+$ cells in cardiac regeneration. Van Berlo et al. reported that c-kit$^+$ cells only minimally contribute to cardiomyocyte regeneration during development, aging, or in response to injury (van Berlo et al., 2014). By targeting the c-kit locus with multiple reporter genes in mice, Sultana et al. reported that c-kit cells rarely coexpress markers of the cardiac progenitor Nkx2.5, or the myocardial marker cardiac troponin T (cTnT). Instead, c-kit$^+$ cells predominantly coexpress a cardiac endothelial cell marker in developing and adult hearts (Sultana et al., 2015). Regardless of their role during normal physiology, these c-kit-positive cells are obviously insufficient to prevent the progression of cardiovascular disease with age or to spontaneously regenerate following acute ischemic events in human. Stem cell antigen 1 (Sca-1), a surface marker of murine hematopoietic stem cells, is another marker reported to indicate resident cardiac progenitor cells (Oh et al., 2003). A previous study reported a significant contribution of Sca-1$^+$ cells to adult cardiomyocytes (Uchida et al., 2013), however, a more recent fate-mapping study using an endogenous Sca-1 reporter demonstrated that endogenous Sca-1$^+$ cells predominantly mark endothelial cells (CD31$^+$) and only minimally mark cardiac troponin-positive cardiomyocytes ($<0.01\%$) under physiological and after myocardial infarction (Neidig et al., 2018).

According to a recent consensus statement on cardiomyocyte regeneration, most scientists agree that de novo cardiomyocyte turnover is approximately 1% per year in the adult mammalian heart (Eschenhagen et al., 2017). Bergmann et al. used data from environmental ^{14}C of human tissues (caused by increases in atmospheric ^{14}C following the atomic bomb testing in the mid-20th century) and mathematical modeling to calculate the rate of human cardiac DNA turnover. They estimated that cardiomyocyte turnover is approximately 1% per year at the age of 25, and the turnover rate decrease to 0.45% per year at the age of 75 in adult human hearts (Bergmann et al., 2009), indicating an aging effect on spontaneous cardiomyocyte renewal in human hearts. They further reported that cardiomyocyte turnover is highest in early childhood, until approximately 1 month of age, when the final number of cardiomyocytes was reached ($3.2 \times 10^9 \pm 0.75 \times 10^9$ cells). The number of cardiomyocytes remained constant over the lifetime (Bergmann et al., 2015). These observations are consistent with the findings in rodent hearts. The rodent heart continues to grow by means of cardiomyocyte proliferation only during the early postnatal period, up to a postnatal window of 7 days in rodents (Porrello et al., 2011). During the first postnatal week, myocardial injury induces a regenerative response, resulting in replacement of lost cardiomyocytes by new ones (Porrello et al., 2011). Fate-mapping studies suggest that this type of myocardial regeneration is mediated primarily by cardiomyocyte proliferation (Porrello et al., 2011; Senyo et al., 2013), which is limited by ROS and oxidative DNA damage. Postnatal hypoxemia, ROS scavenging by mitochondrial catalase, or inhibition of DNA damage response all extend the postnatal proliferative window of cardiomyocytes (Puente et al., 2014). Interestingly, they recently reported that gradual exposure to severe systemic hypoxemia inhibited oxidative metabolism, decreased ROS and oxidative DNA damage, leading to reactivation of cardiomyocyte regeneration from previously existing cardiomyocytes in adult mouse hearts (Nakada et al., 2017). It is still unclear whether this cardiomyocyte regenerative potential can be induced in large animals or humans. It is noteworthy that although there is evidence that cardiomyocytes have the potential to renew throughout life, this happens at a very low rate and the dominant mechanism of postnatal heart growth in mammals is through the mechanism of hypertrophy (increase in cardiomyocyte size, not proliferation).

Even though c-kit$^+$ or Sca-1$^+$ cells may not regenerate cardiomyocytes during physiological or pathological conditions, they are indeed resident stem cells within the hearts that give rise to endothelial cells. Intrinsic aging of these resident stem cells is supported by several experiments in rodents, which reveal that c-kit$^+$ stem cells in the older hearts had a higher rate of apoptosis and shorter telomeres (Anversa et al., 2005). In a rodent model of diabetic cardiomyopathy, c-kit$^+$ cells display telomere shortening, increased expression of senescence markers p53 and p16^{INK4a}, and increased apoptosis. All of the above changes in diabetic cardiomyopathy were attenuated by the ablation of the *p66^{Shc}* gene (Rota et al., 2006), suggesting a central role of mitochondrial ROS in aging of resident c-kit$^+$ stem cells. Studies using cardiosphere-derived cells, another type of resident cardiac stem cell, also demonstrate a significant age-dependent decline in the number and function of such stem cells derived from aged mouse atrial explants (Hsiao et al., 2014).

Intrinsic aging of stem cells has been widely documented in various organ systems (Geiger et al., 2014; Liang & Ghaffari, 2014; Oh et al., 2014). From the point of view of regenerative therapeutics, studies using other types of stem cells have also shown that the age of the stem cell donor is crucial. A recent study showed that mesenchymal stem cells from old donors were more susceptible to ROS-induced damage when transplanted into rats with a myocardial infarct (Liu et al., 2014). Antiaging interventions, such as Sirtuin 1 (SIRT1) overexpression, have been shown to ameliorate aged mesenchymal stem cell senescent phenotypes and improve their regenerative efficacy in experimental myocardial infarcts (Liu et al., 2014). Finally, an extrinsic hostile microenvironment and persistent adverse stimuli associated with advanced age may also impair regenerative capacity in aged heart and warrants future investigation.

Other potential mechanisms

Nicotinamide adenine dinucleotide (NAD^+) is both a coenzyme for electron transfer reactions and a substrate for NAD^+-consuming enzymes, including sirtuins (Belenky et al., 2007). SIRT1 is a NAD^+-dependent protein deacetylase that regulates aging and lifespan in model organisms. The Sadoshima lab showed that expression of SIRT1 increases in aged heart, and that low to moderate levels (2.5–7.5-fold) of SIRT1 overexpression can attenuate cardiac aging responses, while high-level (12.5-fold) overexpression resulted in cardiac hypertrophy and myocardial fibrosis at young age and death within a year (Alcendor et al., 2007). In a recent study, mice deficient in the mitochondrial deacetylase SIRT3 displayed accelerated cardiac aging changes, including early age onset of hypertrophy, associated with fibrosis, age-dependent increased mitochondrial swelling due to increased mPTP opening, and increased mortality after transverse aortic constriction (Hafner et al., 2010). NAD levels decline in multiple organs with age (Yoshino et al., 2011). Administration of NAD^+ precursor nicotinamide mononucleotide (NMN) can elevate intracellular NAD^+ levels in the heart (Lee et al., 2016). In a different study, NMN treatment promoted angiogenesis and exercise capacity in old mice via SIRT1 activation (Das et al., 2018). The therapeutic effect of NAD supplementation for cardiac aging remains to be elucidated.

Previous studies suggest that microRNAs (miRNAs) are important regulators of aging and cardiovascular diseases (Quiat and Olson, 2013; Smith-Vikos & Slack, 2012), and recently miRNAs have been implicated in cardiac aging. In aged cardiomyocyte cultures, van Almen et al. showed that miR-18a and miR-19b regulate expression of CTGF, TSP-1, and collagen, suggesting that these miRNAs mediate age-related ECM remodeling in the hearts (van Almen et al., 2011). Jazbutyte et al. showed that miR-22 expression increases with age in hearts and that miR-22 regulates cardiac fibroblast senescence (Jazbutyte et al., 2013). Boon et al. detected an age-related increase in miR-34a expression in mouse hearts and showed that in vivo silencing of miR-34a for 1 week rescued the increase in cardiomyocyte cell death in aged mice (Boon et al., 2013). In addition, aged miR-34a knockout mice exhibited improved contractile function and reduced cardiac hypertrophy compared to age-matched wild-type littermates, further suggesting a role of miR-34a in cardiac aging (Boon et al., 2013).

Interventions for cardiac aging

Calorie restriction

Calorie restriction (CR) is the most-studied longevity intervention (Colman et al., 2009; Cruzen & Colman, 2009; Fowler et al., 2010; Kastman et al., 2010; McKiernan et al., 2011). CR protects against a variety of age-related pathologies, including cardiovascular disease (Cruzen and Colman, 2009 ; Niemann et al., 2010; Shinmura et al., 2011a). Taffet et al. showed that CR protects against age-related impairment of diastolic function in mice (Taffet et al., 1997). Kemi et al. showed that moderate dietary restriction (35% calorie reduction) can attenuate age-related cardiomyopathy in male Sprague-Dawley rats (Kemi et al., 2000). In humans, volunteers undertaking CR for a mean of 6.5 years had reduced blood pressure and systemic oxidative stress, and improved diastolic function (Meyer et al., 2006). Similar improvements in diastolic function have been demonstrated in individuals on 1-year CR (Riordan et al., 2008). In mice, our laboratory recently showed that CR for 10 weeks can reverse age-related cardiac hypertrophy and diastolic dysfunction, in part by proteomic and metabolomic remodeling to a more youthful state (Dai et al., 2014).

Inhibition of mTOR signaling, normalization of mitochondrial biogenesis ((Lopez-Lluch et al., 2006), attenuation of mitochondrial ROS production, and the subsequent ROS-induced signaling (Csiszar et al., 2009; Nisoli et al., 2005; Shinmura et al., 2011b; Ungvari et al., 2008) and increased SIRT1 signaling (Lopez-Lluch et al., 2006; Lopez-Lluch et al., 2008) have all been implicated in the beneficial effects of CR. All of these areas have strong potential for the study of pharmacologic interventions to enhance healthspan.

Rapamycin

Due to the challenges of CR in humans, developing CR mimetics that mimic the beneficial effects of CR by targeting cellular metabolic and stress response pathways has been of great interest to the aging research community. The National Institute on Aging Intervention Testing Program has demonstrated that long-term rapamycin treatment initiated at 9 or 18 months of age extends lifespan in mice (Harrison et al., 2009; Miller et al., 2011). The lifespan benefits have been confirmed by subsequent study and other healthspan benefits have been demonstrated (Wilkinson et al., 2012).

With increasing evidence supporting the role of mTOR in aging, the effects of rapamycin on cardiac aging have drawn the attention of the aging research community. Neff et al. showed that rapamycin treatment for 1 year attenuated the increased LV dimensions in aged hearts but failed to show any effect on systolic function in male mice (Neff et al., 2013). Flynn et al. showed that rapamycin treatment for 12 weeks can attenuate age-related cardiac hypertrophy and marginally improve systolic function in aged female mice (Flynn et al., 2013). Our laboratory demonstrated that short-term rapamycin (10 weeks) recapitulated the effect of CR and substantially improved both diastolic function and LV hypertrophy in old mice (Dai et al., 2014). Recently, we showed that late-life rapamycin treatment exerts differential temporal regulation on mTOR downstream targets and transiently induces autophagy and mitochondrial biogenesis to promote mitochondrial turnover and rejuvenate cardiac energy metabolism (Chiao et al., 2016). The rapamycin-induced protection is accompanied by rejuvenated cardiac proteome and substrate metabolism (Chiao et al., 2016; Dai et al., 2014).

Further investigations of the mechanisms and optimum timing of rapamycin treatment will be required to evaluate the translational potential of rapamycin as a pharmacological intervention for cardiac aging. The efficacy of other pharmacological inhibitors of mTOR signaling, for example, rapalogs and RTB101, on age-related cardiac dysfunction also warrants further investigation.

Mitochondrial interventions

MitoQ and SkQ

The demonstration that mitochondrial localization of catalase expression is critical for protection from cardiac aging underscored the importance of mitochondrial specificity in antioxidant intervention (Dai et al., 2011). To take advantage of the negative potential gradient across the inner mitochondrial membrane (IMM), triphenylalkylphosphonium ions (TPP^+) have been conjugated to coenzyme Q (MitoQ) and plastoquinone (SkQ1) (Skulachev et al., 2009; Smith et al., 2012) to deliver these redox-active compounds into the mitochondrial matrix. MitoQ and SkQ1 offer protection in models of ischemia–reperfusion and cardiac hypertrophy (Adlam et al., 2005; Bakeeva et al., 2008; Dikalova et al., 2010; Graham et al., 2009), but their effects on cardiac aging remain to be demonstrated. A limitation of TPP^+-conjugated antioxidants is their dependence on mitochondrial potential, as that is often compromised in pathological conditions. Studies have shown that MitoQ and SkQ inhibit respiration and disrupt mitochondrial potential at concentrations above 25 μM, limiting their uptake (Antonenko et al., 2008; Kelso et al., 2001). Moreover, the pro-oxidant activity of MitoQ must be carefully evaluated when considering this intervention (Murphy & Smith, 2007; O'Malley et al., 2006; Scatena et al., 2007).

SS-31 peptide (Elamipretide)

SS-31 peptide (D-Arg-2′,6′-dimethyltyrosine-Lys-Phe-NH2) is a tetrapeptide that preferentially concentrates in the IMM over 1000-fold compared with the cytosolic concentration (Bakeeva et al., 2008; Doughan & Dikalov, 2007; Zhao et al., 2004). SS-31 prevents pressure overload-induced cardiac hypertrophy as well as failure in a highly parallel manner to mCAT overexpression (Dai et al., 2011; 2012, 2013). Originally thought to exert its beneficial effect by the free radical scavenging activity of dimethyl tyrosine in mitochondria (Graham et al., 2009), recent studies have revealed that SS-31 selectively binds to cardiolipin on the IMM (Birk et al., 2013; 2014; Szeto, 2013). This binding explains its targeting, as the mitochondrial inner membrane is unique in its cardiolipin content. SS-31 has been shown to alter the interaction of cardiolipin with cytochrome c and favors cytochrome c electron carrier function while inhibiting its peroxidase activity (Birk et al., 2014; Szeto, 2013). In aged murine skeletal SS-31 treatment acutely increases the efficiency and magnitude of ATP production, inhibits ROS generation, and improves exercise endurance (Siegel et al., 2013). In other studies it has been shown to prevent cardiolipin peroxidation and loss of cristae membranes (Birk et al., 2014). The protective effects of SS-31 on cardiac aging have not been reported, but unpublished studies from our laboratory have demonstrated that treatment with SS-31 for 8 weeks can reverse age-related diastolic dysfunction in old mice (Chiao et al., 2020), supporting the therapeutic potential of SS-31 in cardiac aging.

Inhibition of renin–angiotensin–aldosterone signaling

As described earlier, age-related changes in the RAAS system regulate cardiac aging phenotypes. Studies have demonstrated that long-term inhibition of Ang signaling by ACE inhibitors, angiotensin receptor blockers, or genetic disruption of angiotensin receptor type I can extend lifespan and delay age-dependent cardiac pathology in rodent models (Basso et al., 2005, 2007; Benigni et al., 2009).

Long-term inhibition of Ang II signaling by the angiotensin converting enzyme inhibitor enalapril or by the angiotensin receptor type I blocker losartan extends the lifespan of male Wistar rats and alleviates age-related cardiovascular pathologies (Basso et al., 2007). Another study also showed that life-long treatment with enalapril can attenuate cardiac hypertrophy and interstitial fibrosis in hearts of 24-month-old mice without significant changes in blood pressure (Inserra et al., 1995). Stein et al. showed that 10-month treatment with losartan, beginning at 12 months of age, reduces myocardial fibrosis and fibrosis-related arrhythmias in aged mice (Stein et al., 2010). Groban et al. showed that while both low-dose enalapril and losartan can reduce cardiac oxidative stress, only enalapril was able to improve diastolic dysfunction (Groban et al., 2012).

Senolytics

The positive outcomes of genetic clearance of senescent cells on lifespan and healthspan spark interest from the geroscience community on the therapeutic potentials of senolytics, a class of drugs that selectively kill senescent cells (Baker et al., 2016; 2011; Zhu et al., 2015). The relevance of senescence to cardiac aging has been described earlier. Intermittent treatment with Navitoclax reduced cardiomyocyte senescence and attenuated cardiomyocyte hypertrophy and cardiac fibrosis in 23-month-old mice (Anderson et al., 2019). Old mice that received pretreatment with Navitoclax also showed improved survival and mild attenuation of systolic dysfunction 4 weeks after myocardial infarction (Walaszczyk et al., 2019). These findings support the therapeutic potential of senolytics on cardiac aging and age-related cardiovascular diseases.

Stem cell therapies

The potential of stem cell therapy for cardiovascular diseases and aging has attracted increasing attention. Two methods for stem cell therapy are (1) direct delivery of cardiac stem/progenitor cells to the heart; and (2) administration of agents to enhance the function of endogenous cardiac stem/progenitor cells (Ballard and Edelberg, 2007). Stromal cell-derived factor (SDF-1), platelet-derived growth factor (PDGF), vascular endothelial growth factor (VEGF), and IGF-1 have been shown to enhance the function of endogenous cardiac stem/progenitor cells (Ballard and Edelberg, 2007). The effects of these factors on cardiac function in aging models require further investigation. Mohsin et al. overexpressed Pim-1 kinase in cardiac progenitor cells isolated from a 68-year-old heart failure patient, and showed increased survival, proliferation, differentiation, and persistence of the cardiac progenitor cells after transplantation into an immunocompromised mouse model of myocardial infarction. Transplantation of Pim-1-expressing progenitor cells significantly improved myocardial healing and function of the infarcted heart at 8 weeks (Mohsin et al., 2012). This ex vivo rejuvenation of human cardiac progenitor cells by Pim-1 overexpression is highly encouraging to the development of stem cell therapy for cardiac aging (Mohsin et al., 2013). However, the therapeutic effects of direct stem cell delivery are limited by the proliferation, engraftment, survival, and persistence of the transplanted cells. Recent success in cardiac regeneration in nonhuman primates following ischemic injury, however, appear highly promising and suggest that there may be a substantial future potential for cardiac stem cell therapy (Chien et al., 2019).

Therapies targeting miRNAs

As discussed earlier, recent studies demonstrated the regulatory roles of miRNAs in cardiac aging and these findings implicate the therapeutic potential of gene therapy targeting the age-related changes in miRNAs as cardiac aging intervention. As one miRNA can have multiple targets, gene therapy targeting miRNA may, however, cause undesirable side effects. An alternative approach is to identify the miRNA targets that mediate cardiac aging responses and to target these specific target genes for therapies.

Effects of therapies derived from geroscience on the spectrum of cardiovascular diseases

Multiple molecular mechanisms of cardiac aging have been shown to promote pathogenesis of age-related cardiovascular diseases. Calcium handling is impaired in models of heart failure due to reduced SERCA2 activity (Hobai & O'Rourke, 2001; O'Rourke et al., 1999). Genetic manipulation that increased SERCA2 activity (Miyamoto et al., 2000) and gene therapy delivering SERCA2 or disruption of phospholamban have been showed to attenuate heart failure (Hoshijima et al., 2002; Minamisawa et al., 1999; see the review in Pleger et al., 2013). Also, studies have shown that mitochondrial DNA mutations and deletions increase in experimental models of heart failure (Marin-Garcia et al., 2001), and mutations of genes encoding mitochondrial enzymes and mitochondrial DNA deletions in mitochondrial diseases are associated with idiopathic cardiomyopathies (DiMauro and Schon, 2003). Dysregulation of the mTOR pathway has been implicated in models of cardiac hypertrophy and heart failure (Sciarretta et al., 2018). Rapamycin attenuates the development of pressure overload-induced cardiac hypertrophy (Shioi et al., 2003) and can even cause regression of established pressure overload-induced LV hypertrophy, improve LV systolic function, and induce regression of LV fibrosis in mouse models of hypertrophy and heart failure (Gao et al., 2006; McMullen et al., 2004). The age-related changes in these pathogenic mechanisms likely contribute to the increased susceptibility of aged hearts to cardiovascular diseases. As previous studies of experimental models of cardiovascular diseases mostly focused on young animals, future studies of these models in aged animals will provide new insights into the interaction of cardiac aging and cardiovascular disease pathogenesis. This knowledge will be immensely valuable for developing or improving therapies for aging-associated cardiovascular diseases.

Concluding remarks and future perspectives

With the elderly population continuing to increase, there will be a huge economic burden of age-related cardiovascular disabilities. Therefore it is urgent to understand the molecular mechanisms of cardiac aging and their interactions with cardiovascular disease pathogenesis, and identify novel pathways that could be targeted for interventions aiming at retardation or attenuation of these age-associated alterations. The recent studies on the roles of different hallmarks of aging in cardiac aging have advanced our understanding of this process and have shed light on potential therapeutic strategies. Further understanding of the mechanism of cardiac aging will facilitate future translation of therapeutics to treat age-related cardiovascular disease and to improve cardiac function in the elderly population.

References

Adlam, V. J., Harrison, J. C., Porteous, C. M., James, A. M., Smith, R. A., Murphy, M. P., & Sammut, I. A. (2005). Targeting an antioxidant to mitochondria decreases cardiac ischemia-reperfusion injury. *The FASEB Journal*, *19*(9), 1088–1095.

Akins, C. W., Daggett, W. M., Vlahakes, G. J., Hilgenberg, A. D., Torchiana, D. F., Madsen, J. C., & Buckley, M. J. (1997). Cardiac operations in patients 80 years old and older. *The Annals of Thoracic Surgery*, *64*(3), 606–614, discussion 614-605.

Alcendor, R. R., Gao, S., Zhai, P., Zablocki, D., Holle, E., Yu, X., … Sadoshima, J. (2007). Sirt1 regulates aging and resistance to oxidative stress in the heart. *Circulation Research*, *100*(10), 1512–1521.

Anderson, R., Lagnado, A., Maggiorani, D., Walaszczyk, A., Dookun, E., Chapman, J., … Victorelli, S. (2019). Length-independent telomere damage drives post-mitotic cardiomyocyte senescence. *The EMBO Journal*, *38*, 5.

Antonenko, Y. N., Avetisyan, A. V., Bakeeva, L. E., Chernyak, B. V., Chertkov, V. A., Domnina, L. V., … Muntyan, M. S. (2008). Mitochondria-targeted plastoquinone derivatives as tools to interrupt execution of the aging program. 1. Cationic plastoquinone derivatives: Synthesis and in vitro studies. *Biochemistry*, *73*(12), 1273–1287.

Anversa, P., Rota, M., Urbanek, K., Hosoda, T., Sonnenblick, E. H., Leri, A., … Bolli, R. (2005). Myocardial aging—a stem cell problem. *Basic Research in Cardiology*, *100*(6), 482–493.

Apfeld, J., & Kenyon, C. (1998). Cell nonautonomy of *C. elegans* daf-2 function in the regulation of diapause and life span. *Cell*, *95*(2), 199–210.

Aronow, W. S., Ahn, C., Shirani, J., & Kronzon, I. (1999). Comparison of frequency of new coronary events in older subjects with and without valvular aortic sclerosis. *The American Journal of Cardiology*, *83*(4), 599–600, A598.

Bakeeva, L. E., Barskov, I. V., Egorov, M. V., Isaev, N. K., Kapelko, V. I., Kazachenko, A. V., … Saprunova, V. B. (2008). Mitochondria-targeted plastoquinone derivatives as tools to interrupt execution of the aging program. 2. Treatment of some ROS- and age-related diseases (heart arrhythmia, heart infarctions, kidney ischemia, and stroke). *Biochemistry Biokhimiia*, *73*(12), 1288–1299.

Baker, D. J., Childs, B. G., Durik, M., Wijers, M. E., Sieben, C. J., Zhong, J., … van Deursen, J. M. (2016). Naturally occurring p16(Ink4a)-positive cells shorten healthy lifespan. *Nature*, *530*(7589), 184–189.

Baker, D. J., Wijshake, T., Tchkonia, T., LeBrasseur, N. K., Childs, B. G., van de Sluis, B., … van Deursen, J. M. (2011). Clearance of p16Ink4a-positive senescent cells delays ageing-associated disorders. *Nature*, *479*(7372), 232–236.

Ballard, V. L., & Edelberg, J. M. (2007). Stem cells and the regeneration of the aging cardiovascular system. *Circulation Research*, *100* (8), 1116–1127.

Basso, N., Cini, R., Pietrelli, A., Ferder, L., Terragno, N. A., & Inserra, F. (2007). Protective effect of long-term angiotensin II inhibition. *American Journal of Physiology. Heart and Circulatory Physiology*, *293* (3), H1351–H1358.

Basso, N., Paglia, N., Stella, I., de Cavanagh, E. M., Ferder, L., del Rosario Lores Arnaiz, M., & Inserra, F. (2005). Protective effect of the inhibition of the renin-angiotensin system on aging. *Regulatory Peptides*, *128*(3), 247–252.

Bazzell, B., Ginzberg, S., Healy, L., & Wessells, R. J. (2013). Dietary composition regulates *Drosophila* mobility and cardiac physiology. *The Journal of Experimental Biology*, *216*(Pt 5), 859–868.

Belenky, P., Bogan, K. L., & Brenner, C. (2007). NAD^+ metabolism in health and disease. *Trends in Biochemical Sciences*, *32*(1), 12–19.

Beltrami, A. P., Barlucchi, L., Torella, D., Baker, M., Limana, F., Chimenti, S., … Nadal-Ginard, B. (2003). Adult cardiac stem cells are multipotent and support myocardial regeneration. *Cell*, *114*(6), 763–776.

Benigni, A., Corna, D., Zoja, C., Sonzogni, A., Latini, R., Salio, M., … Remuzzi, G. (2009). Disruption of the Ang II type 1 receptor promotes longevity in mice. *The Journal of Clinical Investigation*, *119*(3), 524–530.

Benjamin, E. J., Muntner, P., Alonso, A., Bittencourt, M. S., Callaway, C. W., Carson, A. P., … Elkind, M. S. V. (2019). Heart disease and stroke statistics-2019 update: A report from the American Heart Association. *Circulation*, *139*(10), e56–e528.

Bergmann, O., Bhardwaj, R. D., Bernard, S., Zdunek, S., Barnabe-Heider, F., Walsh, S., … Frisen, J. (2009). Evidence for cardiomyocyte renewal in humans. *Science*, *324*(5923), 98–102.

Bergmann, O., Zdunek, S., Felker, A., Salehpour, M., Alkass, K., Bernard, S., … Jashari, R. (2015). Dynamics of cell generation and turnover in the human heart. *Cell*, *161*(7), 1566–1575.

Bhatheja, R., Francis, G. S., Pothier, C. E., & Lauer, M. S. (2005). Heart rate response during dipyridamole stress as a predictor of mortality in patients with normal myocardial perfusion and normal electrocardiograms. *The American Journal of Cardiology*, *95*(10), 1159–1164.

Birk, A. V., Chao, W. M., Bracken, W. C., Warren, J. D., & Szeto, H. H. (2014). Targeting mitochondrial cardiolipin and the cytochrome c/cardiolipin complex to promote electron transport and optimize mitochondrial ATP synthesis. *British Journal of Pharmacology*, *171*(8), 2017–2028.

Birk, A. V., Liu, S., Soong, Y., Mills, W., Singh, P., Warren, J. D., ... Szeto, H. H. (2013). The mitochondrial-targeted compound SS-31 re-energizes ischemic mitochondria by interacting with cardiolipin. *Journal of the American Society of Nephrology: JASN, 24*(8), 1250–1261.

Birse, R. T., Choi, J., Reardon, K., Rodriguez, J., Graham, S., Diop, S., ... Oldham, S. (2010). High-fat-diet-induced obesity and heart dysfunction are regulated by the TOR pathway in *Drosophila*. *Cell Metabolism, 12*(5), 533–544.

Blice-Baum, A. C., Guida, M. C., Hartley, P. S., Adams, P. D., Bodmer, R., & Cammarato, A. (2019). As time flies by: Investigating cardiac aging in the short-lived *Drosophila* model. *Biochimica et Biophysica Acta Molecular Basis of Disease, 1856*(7), 1831–1844.

Blice-Baum, A. C., Zambon, A. C., Kaushik, G., Viswanathan, M. C., Engler, A. J., Bodmer, R., & Cammarato, A. (2017). Modest overexpression of FOXO maintains cardiac proteostasis and ameliorates age-associated functional decline. *Aging Cell, 16*(1), 93–103.

Bonnema, D. D., Webb, C. S., Pennington, W. R., Stroud, R. E., Leonardi, A. E., Clark, L. L., ... Zile, M. R. (2007). Effects of age on plasma matrix metalloproteinases (MMPs) and tissue inhibitor of metalloproteinases (TIMPs). *Journal of Cardiac Failure, 13*(7), 530–540.

Boon, R. A., Iekushi, K., Lechner, S., Seeger, T., Fischer, A., Heydt, S., ... Girmatsion, Z. (2013). MicroRNA-34a regulates cardiac ageing and function. *Nature, 495*(7439), 107–110.

Bradshaw, A. D., Baicu, C. F., Rentz, T. J., Van Laer, A. O., Bonnema, D. D., & Zile, M. R. (2010). Age-dependent alterations in fibrillar collagen content and myocardial diastolic function: Role of SPARC in post-synthetic procollagen processing. *American Journal of Physiology. Heart and Circulatory Physiology, 298*(2), H614–H622.

Brandt, T., Mourier, A., Tain, L. S., Partridge, L., Larsson, N. G., & Kuhlbrandt, W. (2017). Changes of mitochondrial ultrastructure and function during ageing in mice and *Drosophila*. *Elife, 6*, e24662.

Broglio, F., Fubini, A., Morello, M., Arvat, E., Aimaretti, G., Gianotti, L., ... Ghigo, E. (1999). Activity of GH/IGF-I axis in patients with dilated cardiomyopathy. *Clinical Endocrinology, 50*(4), 417–430.

Brooks, W. W., & Conrad, C. H. (2000). Myocardial fibrosis in transforming growth factor beta(1)heterozygous mice. *Journal of Molecular and Cellular Cardiology, 32*(2), 187–195.

Bujak, M., & Frangogiannis, N. G. (2007). The role of TGF-eta signaling in myocardial infarction and cardiac remodeling. *Cardiovascular Research, 74*(2), 184–195.

Bursi, F., Weston, S. A., Redfield, M. M., Jacobsen, S. J., Pakhomov, S., Nkomo, V. T., ... Roger, V. L. (2006). Systolic and diastolic heart failure in the community. *JAMA: The Journal of the American Medical Association, 296*(18), 2209–2216.

Cammarato, A., Ahrens, C. H., Alayari, N. N., Qeli, E., Rucker, J., Reedy, M. C., ... Bernstein, S. I. (2011). A mighty small heart: The cardiac proteome of adult *Drosophila* melanogaster. *PLoS One, 6*(4), e18497.

Campisi, J. (2013). Aging, cellular senescence, and cancer. *Annual Review of Physiology, 75*, 685–705.

Cannon, L., Zambon, A. C., Cammarato, A., Zhang, Z., Vogler, G., Munoz, M., ... Bodmer, R. (2017). Expression patterns of cardiac aging in *Drosophila*. *Aging Cell, 16*(1), 82–92.

Chiao, Y. A., Dai, Q., Zhang, J., Lin, J., Lopez, E. F., Ahuja, S. S., ... Jin, Y. F. (2011). Multi-analyte profiling reveals matrix metalloproteinase-9 and monocyte chemotactic protein-1 as plasma biomarkers of cardiac aging. *Circulation-Cardiovascular Genetics, 4*(4), 455–462.

Chiao, Y. A., Kolwicz, S. C., Basisty, N., Gagnidze, A., Zhang, J., Gu, H., ... Rabinovitch, P. S. (2016). Rapamycin transiently induces mitochondrial remodeling to reprogram energy metabolism in old hearts. *Aging, 8*(2), 314–327.

Chiao, Y. A., Ramirez, T. A., Zamilpa, R., Okoronkwo, S. M., Dai, Q., Zhang, J., ... Lindsey, M. L. (2012). Matrix metalloproteinase-9 deletion attenuates myocardial fibrosis and diastolic dysfunction in ageing mice. *Cardiovascular Research, 96*(3), 444–455.

Chiao, Y.A., Zhang, H., Sweetwyne, M., Whitson, J., Ting, Y.S., Basisty, N., Pino, L.K., Quarles, E., Nguyen, N.H., Campbell, M.D., et al. (2020). Late-life restoration of mitochondrial function reverses cardiac dysfunction in old mice. *Elife 9.*: e55513.

Chien, K. R., Frisen, J., Fritsche-Danielson, R., Melton, D. A., Murry, C. E., & Weissman, I. L. (2019). Regenerating the field of cardiovascular cell therapy. *Nature Biotechnology, 37*(3), 232–237.

Colman, R. J., Anderson, R. M., Johnson, S. C., Kastman, E. K., Kosmatka, K. J., Beasley, T. M., ... Weindruch, R. (2009). Caloric restriction delays disease onset and mortality in rhesus monkeys. *Science, 325*(5937), 201–204.

Corpas, E., Harman, S. M., & Blackman, M. R. (1993). Human growth hormone and human aging. *Endocrine Reviews, 14*(1), 20–39.

Correia, L. C., Lakatta, E. G., O'Connor, F. C., Becker, L. C., Clulow, J., Townsend, S., ... Fleg, J. L. (2002). Attenuated cardiovascular reserve during prolonged submaximal cycle exercise in healthy older subjects. *Journal of the American College of Cardiology, 40*(7), 1290–1297.

Cruzen, C., & Colman, R. J. (2009). Effects of caloric restriction on cardiovascular aging in non-human primates and humans. *Clinics in Geriatric Medicine, 25*(4), 733–743, ix-x.

Csiszar, A., Labinskyy, N., Jimenez, R., Pinto, J. T., Ballabh, P., Losonczy, G., ... Ungvari, Z. (2009). Anti-oxidative and anti-inflammatory vasoprotective effects of caloric restriction in aging: Role of circulating factors and SIRT1. *Mechanisms of Ageing and Development, 130*(8), 518–527.

Csiszar, A., Labinskyy, N., Perez, V., Recchia, F. A., Podlutsky, A., Mukhopadhyay, P., ... Ungvari, Z. (2008). Endothelial function and vascular oxidative stress in long-lived GH/IGF-deficient Ames dwarf mice. *American Journal of Physiology. Heart and Circulatory Physiology, 295*(5), H1882–H1894.

Dai, D. F., Chen, T., Szeto, H., Nieves-Cintron, M., Kutyavin, V., Santana, L. F., & Rabinovitch, P. S. (2011). Mitochondrial targeted antioxidant peptide ameliorates hypertensive cardiomyopathy. *Journal of the American College of Cardiology, 58*(1), 73–82.

Dai, D. F., Chen, T., Wanagat, J., Laflamme, M., Marcinek, D. J., Emond, M. J., ... Rabinovitch, P. S. (2010). Age-dependent cardiomyopathy in mitochondrial mutator mice is attenuated by overexpression of catalase targeted to mitochondria. *Aging Cell, 9*(4), 536–544.

Dai, D. F., Hsieh, E. J., Chen, T., Menendez, L. G., Basisty, N. B., Tsai, L., ... Rabinovitch, P. S. (2013). Global proteomics and pathway analysis of pressure-overload-induced heart failure and its attenuation by mitochondrial-targeted peptides. *Circulation-Heart Failure, 6*(5), 1067–1076.

Dai, D. F., Hsieh, E. J., Liu, Y., Chen, T., Beyer, R. P., Chin, M. T., ... Rabinovitch, P. S. (2012). Mitochondrial proteome remodelling in pressure overload-induced heart failure: The role of mitochondrial oxidative stress. *Cardiovascular Research, 93*(1), 79–88.

Dai, D. F., Karunadharma, P. P., Chiao, Y. A., Basisty, N., Crispin, D., Hsieh, E. J., ... Rabinovitch, P. S. (2014). Altered proteome turnover and remodeling by short-term caloric restriction or rapamycin rejuvenate the aging heart. *Aging Cell, 13*(3), 529–539.

Dai, D. F., Santana, L. F., Vermulst, M., Tomazela, D. M., Emond, M. J., MacCoss, M. J., ... Rabinovitch, P. S. (2009). Overexpression of catalase targeted to mitochondria attenuates murine cardiac aging. *Circulation, 119*(21), 2789–2797.

Das, A., Huang, G. X., Bonkowski, M. S., Longchamp, A., Li, C., Schultz, M. B., ... Hung, T. T. (2018). Impairment of an endothelial NAD($^+$)-H2S signaling network is a reversible cause of vascular. *Aging Cell, 173*(1), 74–89, e20.

Demontis, F., & Perrimon, N. (2010). FOXO/4E-BP signaling in *Drosophila* muscles regulates organism-wide proteostasis during aging. *Cell, 143*(5), 813–825.

Dikalova, A. E., Bikineyeva, A. T., Budzyn, K., Nazarewicz, R. R., McCann, L., Lewis, W., ... Dikalov, S. I. (2010). Therapeutic targeting of mitochondrial superoxide in hypertension. *Circulation Research, 107*(1), 106–116.

DiMauro, S., & Schon, E. A. (2003). Mitochondrial respiratory-chain diseases. *The New England Journal of Medicine*, *348*(26), 2656–2668.

Diop, S. B., Bisharat-Kernizan, J., Birse, R. T., Oldham, S., Ocorr, K., & Bodmer, R. (2015). PGC-1/spargel counteracts high-fat-diet-induced obesity and cardiac lipotoxicity downstream of TOR and brummer ATGL lipase. *Cell Reports*, *10*(9), 1572–1584.

Domenighetti, A. A., Wang, Q., Egger, M., Richards, S. M., Pedrazzini, T., & Delbridge, L. M. (2005). Angiotensin II-mediated phenotypic cardiomyocyte remodeling leads to age-dependent cardiac dysfunction and failure. *Hypertension*, *46*(2), 426–432.

Dorman, J. B., Albinder, B., Shroyer, T., & Kenyon, C. (1995). The age-1 and daf-2 genes function in a common pathway to control the lifespan of Caenorhabditis elegans. *Genetics*, *141*(4), 1399–1406.

Doughan, A. K., & Dikalov, S. I. (2007). Mitochondrial redox cycling of mitoquinone leads to superoxide production and cellular apoptosis. *Antioxidants & Redox Signaling*, *9*(11), 1825–1836.

Eichelberg, H., & Seine, R. (1996). [Life expectancy and cause of death in dogs. I. The situation in mixed breeds and various dog breeds]. *Berliner und Munchener Tierarztliche Wochenschrift*, *109*(8), 292–303.

El'darov, Ch. M., Vays, V. B., Vangeli, I. M., Kolosova, N. G., & Bakeeva, L. E. (2015). Morphometric examination of mitochondrial ultrastructure in aging cardiomyocytes. *Biochemistry. Biokhimiia*, *80*(5), 604–609.

Ellison, G. M., Vicinanza, C., Smith, A. J., Aquila, I., Leone, A., Waring, C. D., ... Condorelli, G. (2013). Adult c-kit(pos) cardiac stem cells are necessary and sufficient for functional cardiac regeneration and repair. *Cell*, *154*(4), 827–842.

Eschenhagen, T., Bolli, R., Braun, T., Field, L. J., Fleischmann, B. K., Frisen, J., ... Molkentin, J. D. (2017). Cardiomyocyte regeneration: A consensus statement. *Circulation*, *136*(7), 680–686.

Faggiano, P., Antonini-Canterin, F., Erlicher, A., Romeo, C., Cervesato, E., Pavan, D., ... Nicolosi, G. L. (2003). Progression of aortic valve sclerosis to aortic stenosis. *The American Journal of Cardiology*, *91*(1), 99–101.

Fannin, S. W., Lesnefsky, E. J., Slabe, T. J., Hassan, M. O., & Hoppel, C. L. (1999). Aging selectively decreases oxidative capacity in rat heart interfibrillar mitochondria. *Archives of Biochemistry and Biophysics*, *372*(2), 399–407.

Fleg, J. L., & Kennedy, H. L. (1992). Long-term prognostic significance of ambulatory electrocardiographic findings in apparently healthy subjects greater than or equal to 60 years of age. *The American Journal of Cardiology*, *70*(7), 748–751.

Fleg, J. L., O'Connor, F., Gerstenblith, G., Becker, L. C., Clulow, J., Schulman, S. P., & Lakatta, E. G. (1995). Impact of age on the cardiovascular response to dynamic upright exercise in healthy men and women. *Journal of Applied Physiology*, *78*(3), 890–900.

Flynn, J. M., O'Leary, M. N., Zambataro, C. A., Academia, E. C., Presley, M. P., Garrett, B. J., ... Kennedy, B. K. (2013). Late-life rapamycin treatment reverses age-related heart dysfunction. *Aging Cell*, *12*(5), 851–862.

Fowler, C. G., Chiasson, K. B., Leslie, T. H., Thomas, D., Beasley, T. M., Kemnitz, J. W., & Weindruch, R. (2010). Auditory function in rhesus monkeys: Effects of aging and caloric restriction in the Wisconsin monkeys five years later. *Hearing Research*, *261*(1–2), 75–81.

Freeman, R. V., & Otto, C. M. (2005). Spectrum of calcific aortic valve disease: Pathogenesis, disease progression, and treatment strategies. *Circulation*, *111*(24), 3316–3326.

Fulkerson, P. K., Beaver, B. M., Auseon, J. C., & Graber, H. L. (1979). Calcification of the mitral annulus: Etiology, clinical associations, complications and therapy. *The American Journal of Medicine*, *66*(6), 967–977.

Gao, X. M., Wong, G., Wang, B., Kiriazis, H., Moore, X. L., Su, Y. D., ... Du, X. J. (2006). Inhibition of mTOR reduces chronic pressure-overload cardiac hypertrophy and fibrosis. *Journal of Hypertension*, *24*(8), 1663–1670.

Geiger, H., Denkinger, M., & Schirmbeck, R. (2014). Hematopoietic stem cell aging. *Current Opinion in Immunology*, *29*, 86–92.

Gerstenblith, G., Frederiksen, J., Yin, F. C., Fortuin, N. J., Lakatta, E. G., & Weisfeldt, M. L. (1977). Echocardiographic assessment of a normal adult aging population. *Circulation*, *56*(2), 273–278.

Gill, S., Le, H. D., Melkani, G. C., & Panda, S. (2015). Time-restricted feeding attenuates age-related cardiac decline in *Drosophila*. *Science*, *347*(6227), 1265–1269.

Graham, D., Huynh, N. N., Hamilton, C. A., Beattie, E., Smith, R. A., Cocheme, H. M., ... Dominiczak, A. F. (2009). Mitochondria-targeted antioxidant MitoQ10 improves endothelial function and attenuates cardiac hypertrophy. *Hypertension*, *54*(2), 322–328.

Groban, L., Lindsey, S., Wang, H., Lin, M. S., Kassik, K. A., Machado, F. S., & Carter, C. S. (2012). Differential effects of late-life initiation of low-dose enalapril and losartan on diastolic function in senescent Fischer 344 x Brown Norway male rats. *Age*, *34*(4), 831–843.

Groban, L., Pailes, N. A., Bennett, C. D., Carter, C. S., Chappell, M. C., Kitzman, D. W., & Sonntag, W. E. (2006). Growth hormone replacement attenuates diastolic dysfunction and cardiac angiotensin II expression in senescent rats. *The Journals of Gerontology—Series A, Biological Sciences and Medical Sciences*, *61*(1), 28–35.

Guglielmini, C. (2003). Cardiovascular diseases in the ageing dog: Diagnostic and therapeutic problems. *Veterinary Research Communications*, *27*(Suppl. 1)), 555–560.

Hafner, A. V., Dai, J., Gomes, A. P., Xiao, C. Y., Palmeira, C. M., Rosenzweig, A., & Sinclair, D. A. (2010). Regulation of the mPTP by SIRT3-mediated deacetylation of CypD at lysine 166 suppresses age-related cardiac hypertrophy. *Aging*, *2*(12), 914–923.

Hamlin, R. L. (2007). Animal models of ventricular arrhythmias. *Pharmacology & Therapeutics*, *113*(2), 276–295.

Harrison, D. E., Strong, R., Sharp, Z. D., Nelson, J. F., Astle, C. M., Flurkey, K., ... Fernandez, E. (2009). Rapamycin fed late in life extends lifespan in genetically heterogeneous mice. *Nature*, *460*(7253), 392–395.

Hartley, P. S., Motamedchaboki, K., Bodmer, R., & Ocorr, K. (2016). SPARC-dependent cardiomyopathy in *Drosophila*. *Circulation Cardiovascular Genetics*, *9*(2), 119–129.

Hobai, I. A., & O'Rourke, B. (2001). Decreased sarcoplasmic reticulum calcium content is responsible for defective excitation-contraction coupling in canine heart failure. *Circulation*, *103*(11), 1577–1584.

Hoshijima, M., Ikeda, Y., Iwanaga, Y., Minamisawa, S., Date, M. O., Gu, Y., ... Chien, K. R. (2002). Chronic suppression of heart-failure progression by a pseudophosphorylated mutant of phospholamban via in vivo cardiac rAAV gene delivery. *Nature Medicine*, *8*(8), 864–871.

Hsiao, L. C., Perbellini, F., Gomes, R. S., Tan, J. J., Vieira, S., Faggian, G., ... Carr, C. A. (2014). Murine cardiosphere-derived cells are impaired by age but not by cardiac dystrophic dysfunction. *Stem Cells and Development*, *23*(9), 1027–1036.

Inserra, F., Romano, L., Ercole, L., de Cavanagh, E. M., & Ferder, L. (1995). Cardiovascular changes by long-term inhibition of the renin-angiotensin system in aging. *Hypertension*, *25*(3), 437–442.

Janczewski, A. M., & Lakatta, E. G. (2010). Modulation of sarcoplasmic reticulum $Ca^{(2+)}$ cycling in systolic and diastolic heart failure associated with aging. *Heart Failure Reviews*, *15*(5), 431–445.

Janczewski, A. M., Spurgeon, H. A., & Lakatta, E. G. (2002). Action potential prolongation in cardiac myocytes of old rats is an adaptation to sustain youthful intracellular Ca^{2+} regulation. *Journal of Molecular and Cellular Cardiology*, *34*(6), 641–648.

Jazbutyte, V., Fiedler, J., Kneitz, S., Galuppo, P., Just, A., Holzmann, A., ... Thum, T. (2013). MicroRNA-22 increases senescence and activates cardiac fibroblasts in the aging heart. *Age*, *35*(3), 747–762.

Jebara, V. A., Dervanian, P., Acar, C., Grare, P., Mihaileanu, S., Chauvaud, S., ... Carpentier, A. (1992). Mitral valve repair using carpentier techniques in patients more than 70 years old. Early and late results. *Circulation*, *86*(5 Suppl.), II53–II59.

Jeon, D. S., Atar, S., Brasch, A. V., Luo, H., Mirocha, J., Naqvi, T. Z., ... Siegel, R. J. (2001). Association of mitral annulus calcification, aortic valve sclerosis and aortic root calcification with abnormal myocardial perfusion single photon emission tomography in subjects age < or = 65 years old. *Journal of the American College of Cardiology*, *38*(7), 1988–1993.

Josephson, I. R., Guia, A., Stern, M. D., & Lakatta, E. G. (2002). Alterations in properties of L-type Ca channels in aging rat heart. *Journal of Molecular and Cellular Cardiology*, *34*(3), 297–308.

Judge, S., Jang, Y. M., Smith, A., Hagen, T., & Leeuwenburgh, C. (2005). Age-associated increases in oxidative stress and antioxidant enzyme activities in cardiac interfibrillar mitochondria: Implications for the mitochondrial theory of aging. *The FASEB Journal*, *19*(3), 419–421.

Jugdutt, B. I. (2002). The dog model of left ventricular remodeling after myocardial infarction. *Journal of Cardiac Failure*, *8*(6 Suppl.), S472–S475.

Kaeberlein, M., Creevy, K. E., & Promislow, D. E. (2016). The dog aging project: Translational geroscience in companion animals. *Mammalian Genome: Official Journal of the International Mammalian Genome Society*, *27*(7–8), 279–288.

Karavidas, A., Lazaros, G., Tsiachris, D., & Pyrgakis, V. (2010). Aging and the cardiovascular system. *Hellenic Journal of Cardiology: HJC = Hellenike Kardiologike Epitheorese*, *51*(5), 421–427.

Kass, D. A., Bronzwaer, J. G., & Paulus, W. J. (2004). What mechanisms underlie diastolic dysfunction in heart failure? *Circulation Research*, *94*(12), 1533–1542.

Kastman, E. K., Willette, A. A., Coe, C. L., Bendlin, B. B., Kosmatka, K. J., McLaren, D. G., ... Colman, R. J. (2010). A calorie-restricted diet decreases brain iron accumulation and preserves motor performance in old rhesus monkeys. *The Journal of Neuroscience*, *30* (23), 7940–7947.

Kaushik, G., Spenlehauer, A., Sessions, A. O., Trujillo, A. S., Fuhrmann, A., Fu, Z., ... Bodmer, R. (2015). Vinculin network-mediated cytoskeletal remodeling regulates contractile function in the aging heart. *Science Translational Medicine*, *7*(292), 292ra299.

Kaushik, G., Zambon, A. C., Fuhrmann, A., Bernstein, S. I., Bodmer, R., Engler, A. J., & Cammarato, A. (2012). Measuring passive myocardial stiffness in *Drosophila* melanogaster to investigate diastolic dysfunction. *Journal of Cellular and Molecular Medicine*, *16*(8), 1656–1662.

Kelso, G. F., Porteous, C. M., Coulter, C. V., Hughes, G., Porteous, W. K., Ledgerwood, E. C., ... Murphy, M. P. (2001). Selective targeting of a redox-active ubiquinone to mitochondria within cells: Antioxidant and antiapoptotic properties. *The Journal of Biological Chemistry*, *276*(7), 4588–4596.

Kemi, M., Keenan, K. P., McCoy, C., Hoe, C. M., Soper, K. A., Ballam, G. C., & van Zwieten, M. J. (2000). The relative protective effects of moderate dietary restriction versus dietary modification on spontaneous cardiomyopathy in male Sprague-Dawley rats. *Toxicologic Pathology*, *28*(2), 285–296.

Kennedy, B. K., Steffen, K. K., & Kaeberlein, M. (2007). Ruminations on dietary restriction and aging. *Cellular and Molecular Life Sciences: CMLS*, *64*(11), 1323–1328.

Khan, A. S., Sane, D. C., Wannenburg, T., & Sonntag, W. E. (2002). Growth hormone, insulin-like growth factor-1 and the aging cardiovascular system. *Cardiovascular Research*, *54*(1), 25–35.

Khouri, S. J., Maly, G. T., Suh, D. D., & Walsh, T. E. (2004). A practical approach to the echocardiographic evaluation of diastolic function. *Journal of the American Society of Echocardiography: Official Publication of the American Society of Echocardiography*, *17*(3), 290–297.

Kim, M., Sujkowski, A., Namkoong, S., Gu, B., Cobb, T., Kim, B., ... Lee JH. (2020). Sestrins are evolutionarily conserved mediators of exercise benefits. *Nature Communications 11, 190.*

Knyushko, T. V., Sharov, V. S., Williams, T. D., Schoneich, C., & Bigelow, D. J. (2005). 3-Nitrotyrosine modification of SERCA2a in the aging heart: A distinct signature of the cellular redox environment. *Biochemistry*, *44*(39), 13071–13081.

Koban, M. U., Moorman, A. F., Holtz, J., Yacoub, M. H., & Boheler, K. R. (1998). Expressional analysis of the cardiac Na-Ca exchanger in rat development and senescence. *Cardiovascular Research*, *37*(2), 405–423.

Koh, G. Y., Soonpaa, M. H., Klug, M. G., Pride, H. P., Cooper, B. J., Zipes, D. P., & Field, L. J. (1995). Stable fetal cardiomyocyte grafts in the hearts of dystrophic mice and dogs. *The Journal of Clinical Investigation*, *96*(4), 2034–2042.

Kujoth, G. C., Hiona, A., Pugh, T. D., Someya, S., Panzer, K., Wohlgemuth, S. E., ... Sedivy, J. M. (2005). Mitochondrial DNA mutations, oxidative stress, and apoptosis in mammalian aging. *Science*, *309*(5733), 481–484.

Kurosu, H., Yamamoto, M., Clark, J. D., Pastor, J. V., Nandi, A., Gurnani, P., ... Herz, J. (2005). Suppression of aging in mice by the hormone Klotho. *Science*, *309*(5742), 1829–1833.

Lakatta, E. G. (2003). Arterial and cardiac aging: Major shareholders in cardiovascular disease enterprises: Part III: Cellular and molecular clues to heart and arterial aging. *Circulation*, *107*(3), 490–497.

Lakatta, E. G., & Levy, D. (2003a). Arterial and cardiac aging: Major shareholders in cardiovascular disease enterprises: Part I: Aging arteries: A "set up" for vascular disease. *Circulation*, *107*(1), 139–146.

Lakatta, E. G., & Levy, D. (2003b). Arterial and cardiac aging: Major shareholders in cardiovascular disease enterprises: Part II: The aging heart in health: Links to heart disease. *Circulation*, *107*(2), 346–354.

Laker, R. C., Drake, J. C., Wilson, R. J., Lira, V. A., Lewellen, B. M., Ryall, K. A., ... Yan, Z. (2017). Ampk phosphorylation of Ulk1 is required for targeting of mitochondria to lysosomes in exercise-induced mitophagy. *Nature Communications*, *8*(1), 548.

Laker, R. C., Xu, P., Ryall, K. A., Sujkowski, A., Kenwood, B. M., Chain, K. H., ... Saucerman, J. J. (2014). A novel MitoTimer reporter gene for mitochondrial content, structure, stress, and damage in vivo. *The Journal of Biological Chemistry*, *289*(17), 12005–12015.

Lane, M. A., Ingram, D. K., & Roth, G. S. (1999). Calorie restriction in nonhuman primates: Effects on diabetes and cardiovascular disease risk. *Toxicological Sciences: An Official Journal of the Society of Toxicology*, *52*(2 Suppl.), 41–48.

Lane, M. A., Mattison, J., Ingram, D. K., & Roth, G. S. (2002). Caloric restriction and aging in primates: Relevance to humans and possible CR mimetics. *Microscopy Research and Technique*, *59*(4), 335–338.

Lauer, M. S., Francis, G. S., Okin, P. M., Pashkow, F. J., Snader, C. E., & Marwick, T. H. (1999). Impaired chronotropic response to exercise stress testing as a predictor of mortality. *JAMA: The Journal of the American Medical Association*, *281*(6), 524–529.

Lee, C. F., Chavez, J. D., Garcia-Menendez, L., Choi, Y., Roe, N. D., Chiao, Y. A., ... Tian, R. (2016). Normalization of NAD^+ redox balance as a therapy for heart failure. *Circulation*, *134*(12), 883–894.

Lee, J. H., Budanov, A. V., Park, E. J., Birse, R., Kim, T. E., Perkins, G. A., ... Karin, M. (2010). Sestrin as a feedback inhibitor of TOR that prevents age-related pathologies. *Science*, *327*(5970), 1223–1228.

Lesnefsky, E. J., Chen, Q., & Hoppel, C. L. (2016). Mitochondrial metabolism in aging heart. *Circulation Research*, *118*(10), 1593–1611.

Lewis-McDougall, F. C., Ruchaya, P. J., Domenjo-Vila, E., Shin Teoh, T., Prata, L., Cottle, B. J., ... Ellison-Hughes, G. M. (2019). Aged-senescent cells contribute to impaired heart regeneration. *Aging Cell*, *18*(3), e12931.

Li, L., Guo, Y., Zhai, H., Yin, Y., Zhang, J., Chen, H., ... Xia, Y. (2014). Aging increases the susceptivity of MSCs to reactive oxygen species and impairs their therapeutic potency for myocardial infarction. *PLoS One, 9*(11), e111850.

Li, Q., Ceylan-Isik, A. F., Li, J., & Ren, J. (2008). Deficiency of insulin-like growth factor 1 reduces sensitivity to aging-associated cardiomyocyte dysfunction. *Rejuvenation Research, 11*(4), 725–733.

Liang, H., Masoro, E. J., Nelson, J. F., Strong, R., McMahan, C. A., & Richardson, A. (2003). Genetic mouse models of extended lifespan. *Experimental Gerontology, 38*(11–12), 1353–1364.

Liang, R., & Ghaffari, S. (2014). Stem cells, redox signaling, and stem cell aging. *Antioxidants & Redox Signaling, 20*(12), 1902–1916.

Lindsey, M. L., Goshorn, D. K., Squires, C. E., Escobar, G. P., Hendrick, J. W., Mingoia, J. T., ... Spinale, F. G. (2005). Age-dependent changes in myocardial matrix metalloproteinase/tissue inhibitor of metalloproteinase profiles and fibroblast function. *Cardiovascular Research, 66*(2), 410–419.

Liu, X., Chen, H., Zhu, W., Hu, X., Jiang, Z., Xu, Y., ... Zhang, N. (2014). Transplantation of SIRT1-engineered aged mesenchymal stem cells improves cardiac function in a rat myocardial infarction model. *The Journal of Heart and Lung Transplantation: The Official Publication of the International Society for Heart Transplantation, 33*(10), 1083–1092.

Lopez-Lluch, G., Hunt, N., Jones, B., Zhu, M., Jamieson, H., Hilmer, S., ... de Cabo, R. (2006). Calorie restriction induces mitochondrial biogenesis and bioenergetic efficiency. *Proceedings of the National Academy of Sciences of United States of America, 103*(6), 1768–1773.

Lopez-Lluch, G., Irusta, P. M., Navas, P., & de Cabo, R. (2008). Mitochondrial biogenesis and healthy aging. *Experimental Gerontology, 43*(9), 813–819.

Lopez-Lopez, C., Dietrich, M. O., Metzger, F., Loetscher, H., & Torres-Aleman, I. (2007). Disturbed cross talk between insulin-like growth factor I and AMP-activated protein kinase as a possible cause of vascular dysfunction in the amyloid precursor protein/presenilin 2 mouse model of Alzheimer's disease. *The Journal of Neuroscience, 27*(4), 824–831.

Lowenstine, L. J., McManamon, R., & Terio, K. A. (2016). Comparative pathology of aging great apes: Bonobos, chimpanzees, gorillas, and orangutans. *Veterinary Pathology, 53*(2), 250–276.

Luong, N., Davies, C. R., Wessells, R. J., Graham, S. M., King, M. T., Veech, R., ... Oldham, S. M. (2006). Activated FOXO-mediated insulin resistance is blocked by reduction of TOR activity. *Cell Metabolism, 4*(2), 133–142.

Maejima, Y., Adachi, S., Ito, H., Hirao, K., & Isobe, M. (2008). Induction of premature senescence in cardiomyocytes by doxorubicin as a novel mechanism of myocardial damage. *Aging Cell, 7* (2), 125–136.

Mammucari, C., & Rizzuto, R. (2010). Signaling pathways in mitochondrial dysfunction and aging. *Mechanisms of Ageing and Development, 131*(7–8), 536–543.

Marin-Garcia, J., Goldenthal, M. J., & Moe, G. W. (2001). Abnormal cardiac and skeletal muscle mitochondrial function in pacing-induced cardiac failure. *Cardiovascular Research, 52*(1), 103–110.

Mattison, J. A., Lane, M. A., Roth, G. S., & Ingram, D. K. (2003). Calorie restriction in rhesus monkeys. *Experimental Gerontology, 38* (1-2), 35–46.

McKiernan, S. H., Colman, R. J., Lopez, M., Beasley, T. M., Aiken, J. M., Anderson, R. M., & Weindruch, R. (2011). Caloric restriction delays aging-induced cellular phenotypes in rhesus monkey skeletal muscle. *Experimental Gerontology, 46*(1), 23–29.

McMullen, J. R., Sherwood, M. C., Tarnavski, O., Zhang, L., Dorfman, A. L., Shioi, T., & Izumo, S. (2004). Inhibition of mTOR signaling with rapamycin regresses established cardiac hypertrophy induced by pressure overload. *Circulation, 109*(24), 3050–3055.

Meikle, L., McMullen, J. R., Sherwood, M. C., Lader, A. S., Walker, V., Chan, J. A., & Kwiatkowski, D. J. (2005). A mouse model of cardiac rhabdomyoma generated by loss of Tsc1 in ventricular myocytes. *Human Molecular Genetics, 14*(3), 429–435.

Mendez, S., Watanabe, L., Hill, R., Owens, M., Moraczewski, J., Rowe, G. C., ... Reed, L. K. (2016). The tread wheel: A novel apparatus to measure genetic variation in response to gently induced exercise for *Drosophila*. *PLoS One, 11*(10), e0164706.

Meyer, T. E., Kovacs, S. J., Ehsani, A. A., Klein, S., Holloszy, J. O., & Fontana, L. (2006). Long-term caloric restriction ameliorates the decline in diastolic function in humans. *Journal of the American College of Cardiology, 47*(2), 398–402.

Miller, R. A., Harrison, D. E., Astle, C. M., Baur, J. A., Boyd, A. R., de Cabo, R., ... Sharp, Z. D. (2011). Rapamycin, but not resveratrol or simvastatin, extends life span of genetically heterogeneous mice. *The Journals of Gerontology—Series A, Biological Sciences and Medical Sciences, 66*(2), 191–201.

Minamisawa, S., Hoshijima, M., Chu, G., Ward, C. A., Frank, K., Gu, Y., ... Chien, K. R. (1999). Chronic phospholamban-sarcoplasmic reticulum calcium ATPase interaction is the critical calcium cycling defect in dilated cardiomyopathy. *Cell, 99*(3), 313–322.

Miyamoto, M. I., del Monte, F., Schmidt, U., DiSalvo, T. S., Kang, Z. B., Matsui, T., ... Hajjar, R. J. (2000). Adenoviral gene transfer of SERCA2a improves left-ventricular function in aortic-banded rats in transition to heart failure. *Proceedings of the National Academy of Sciences of United States of America, 97*(2), 793–798.

Mohammed, S. F., Mirzoyev, S. A., Edwards, W. D., Dogan, A., Grogan, D. R., Dunlay, S. M., ... Redfield, M. M. (2014). Left ventricular amyloid deposition in patients with heart failure and preserved ejection fraction. *JACC Heart Failure, 2*(2), 113–122.

Mohsin, S., Khan, M., Nguyen, J., Alkatib, M., Siddiqi, S., Hariharan, N., ... Sussman, M. A. (2013). Rejuvenation of human cardiac progenitor cells with Pim-1 kinase. *Circulation Research, 113*(10), 1169–1179.

Mohsin, S., Khan, M., Toko, H., Bailey, B., Cottage, C. T., Wallach, K., ... Quijada, P. (2012). Human cardiac progenitor cells engineered with Pim-I kinase enhance myocardial repair. *Journal of the American College of Cardiology, 60*(14), 1278–1287.

Murphy, M. P., & Smith, R. A. (2007). Targeting antioxidants to mitochondria by conjugation to lipophilic cations. *Annual Review of Pharmacology and Toxicology, 47*, 629–656.

Musselman, L. P., Fink, J. L., Narzinski, K., Ramachandran, P. V., Hathiramani, S. S., Cagan, R. L., & Baranski, T. J. (2011). A high-sugar diet produces obesity and insulin resistance in wild-type *Drosophila*. *Disease Models & Mechanisms, 4*(6), 842–849.

Na, J., Musselman, L. P., Pendse, J., Baranski, T. J., Bodmer, R., Ocorr, K., & Cagan, R. A. (2013). *Drosophila* model of high sugar diet-induced cardiomyopathy. *PLoS Genetics, 9*(1), e1003175.

Nakada, Y., Canseco, D. C., Thet, S., Abdisalaam, S., Asaithamby, A., Santos, C. X., ... Hu, Z. (2017). Hypoxia induces heart regeneration in adult mice. *Nature, 541*(7636), 222–227.

Nassimiha, D., Aronow, W. S., Ahn, C., & Goldman, M. E. (2001). Association of coronary risk factors with progression of valvular aortic stenosis in older persons. *The American Journal of Cardiology, 87*(11), 1313–1314.

Neff, F., Flores-Dominguez, D., Ryan, D. P., Horsch, M., Schroder, S., Adler, T., ... Holtmeier, R. (2013). Rapamycin extends murine lifespan but has limited effects on aging. *The Journal of Clinical Investigation, 123*(8), 3272–3291.

Neidig, L. E., Weinberger, F., Palpant, N. J., Mignone, J., Martinson, A. M., Sorensen, D. W., ... van Berlo, J. H. (2018). Evidence for minimal cardiogenic potential of stem cell antigen 1-positive cells in the adult mouse heart. *Circulation, 138*(25), 2960–2962.

Niemann, B., Chen, Y., Issa, H., Silber, R. E., & Rohrbach, S. (2010). Caloric restriction delays cardiac ageing in rats: Role of mitochondria. *Cardiovascular Research, 88*(2), 267–276.

Nisoli, E., Tonello, C., Cardile, A., Cozzi, V., Bracale, R., Tedesco, L., ... Carruba, M. O. (2005). Calorie restriction promotes mitochondrial biogenesis by inducing the expression of eNOS. *Science, 310*(5746), 314–317.

Ocorr, K., Akasaka, T., & Bodmer, R. (2007). Age-related cardiac disease model of *Drosophila*. *Mechanisms of Ageing and Development, 128*(1), 112–116.

Oh, H., Bradfute, S. B., Gallardo, T. D., Nakamura, T., Gaussin, V., Mishina, Y., ... Schneider, M. D. (2003). Cardiac progenitor cells from adult myocardium: Homing, differentiation, and fusion after infarction. *Proceedings of the National Academy of Sciences of United States of America, 100*(21), 12313–12318.

Oh, J., Lee, Y. D., & Wagers, A. J. (2014). Stem cell aging: Mechanisms, regulators and therapeutic opportunities. *Nature Medicine, 20*(8), 870–880.

Okumura, S., Takagi, G., Kawabe, J., Yang, G., Lee, M. C., Hong, C., ... Ishikawa, Y. (2003). Disruption of type 5 adenylyl cyclase gene preserves cardiac function against pressure overload. *Proceedings of the National Academy of Sciences of United States of America, 100*(17), 9986–9990.

Okumura, S., Vatner, D. E., Kurotani, R., Bai, Y., Gao, S., Yuan, Z., ... Ishikawa, Y. (2007). Disruption of type 5 adenylyl cyclase enhances desensitization of cyclic adenosine monophosphate signal and increases Akt signal with chronic catecholamine stress. *Circulation, 116*(16), 1776–1783.

Olsen, M. H., Wachtell, K., Bella, J. N., Gerdts, E., Palmieri, V., Nieminen, M. S., ... Devereux, R. B. (2005). Aortic valve sclerosis relates to cardiovascular events in patients with hypertension (a LIFE substudy). *The American Journal of Cardiology, 95*(1), 132–136.

O'Malley, Y., Fink, B. D., Ross, N. C., Prisinzano, T. E., & Sivitz, W. I. (2006). Reactive oxygen and targeted antioxidant administration in endothelial cell mitochondria. *The Journal of Biological Chemistry, 281*(52), 39766–39775.

O'Rourke, B., Kass, D. A., Tomaselli, G. F., Kaab, S., Tunin, R., & Marban, E. (1999). Mechanisms of altered excitation-contraction coupling in canine tachycardia-induced heart failure, I: Experimental studies. *Circulation Research, 84*(5), 562–570.

Otto, C. M. (2004). Why is aortic sclerosis associated with adverse clinical outcomes? *Journal of the American College of Cardiology, 43*(2), 176–178.

Otto, C. M., Lind, B. K., Kitzman, D. W., Gersh, B. J., & Siscovick, D. S. (1999). Association of aortic-valve sclerosis with cardiovascular mortality and morbidity in the elderly. *The New England Journal of Medicine, 341*(3), 142–147.

Ouzounian, M., Lee, D. S., & Liu, P. P. (2008). Diastolic heart failure: Mechanisms and controversies. *Nature Clinical Practice. Cardiovascular Medicine, 5*(7), 375–386.

Panek, A. N., Posch, M. G., Alenina, N., Ghadge, S. K., Erdmann, B., Popova, E., ... Ozcelik, C. (2009). Connective tissue growth factor overexpression in cardiomyocytes promotes cardiac hypertrophy and protection against pressure overload. *PLoS One, 4*(8), e6743.

Piazza, N., Gosangi, B., Devilla, S., Arking, R., & Wessells, R. (2009). Exercise-training in young *Drosophila* melanogaster reduces age-related decline in mobility and cardiac performance. *PLoS One, 4*(6), e5886.

Pleger, S. T., Brinks, H., Ritterhoff, J., Raake, P., Koch, W. J., Katus, H. A., & Most, P. (2013). Heart failure gene therapy: The path to clinical practice. *Circulation Research, 113*(6), 792–809.

Porrello, E. R., Mahmoud, A. I., Simpson, E., Hill, J. A., Richardson, J. A., Olson, E. N., & Sadek, H. A. (2011). Transient regenerative potential of the neonatal mouse heart. *Science, 331*(6020), 1078–1080.

Puche, J. E., Garcia-Fernandez, M., Muntane, J., Rioja, J., Gonzalez-Baron, S., & Castilla Cortazar, I. (2008). Low doses of insulin-like growth factor-I induce mitochondrial protection in aging rats. *Endocrinology, 149*(5), 2620–2627.

Puente, B. N., Kimura, W., Muralidhar, S. A., Moon, J., Amatruda, J. F., Phelps, K. L., ... Mori, E. (2014). The oxygen-rich postnatal environment induces cardiomyocyte cell-cycle arrest through DNA damage response. *Cell, 157*(3), 565–579.

Quiat, D., & Olson, E. N. (2013). MicroRNAs in cardiovascular disease: From pathogenesis to prevention and treatment. *The Journal of Clinical Investigation, 123*(1), 11–18.

Reed, A. L., Tanaka, A., Sorescu, D., Liu, H., Jeong, E. M., Sturdy, M., ... Sutliff, R. L. (2011). Diastolic dysfunction is associated with cardiac fibrosis in the senescence-accelerated mouse. *American Journal of Physiology. Heart and Circulatory Physiology, 301*(3), H824–H831.

Ren, J., & Brown-Borg, H. M. (2002). Impaired cardiac excitation-contraction coupling in ventricular myocytes from Ames dwarf mice with IGF-I deficiency. *Growth Hormone & IGF Research: Official Journal of the Growth Hormone Research Society and the International IGF Research Society, 12*(2), 99–105.

Riordan, M. M., Weiss, E. P., Meyer, T. E., Ehsani, A. A., Racette, S. B., Villareal, D. T., ... Kovacs, S. J. (2008). The effects of caloric restriction- and exercise-induced weight loss on left ventricular diastolic function. *American Journal of Physiology. Heart and Circulatory Physiology, 294*(3), H1174–H1182.

Riva, A., Tandler, B., Lesnefsky, E. J., Conti, G., Loffredo, F., Vazquez, E., & Hoppel, C. L. (2006). Structure of cristae in cardiac mitochondria of aged rat. *Mechanisms of Ageing and Development, 127*(12), 917–921.

Rivera, E. J., Goldin, A., Fulmer, N., Tavares, R., Wands, J. R., & de la Monte, S. M. (2005). Insulin and insulin-like growth factor expression and function deteriorate with progression of Alzheimer's disease: Link to brain reductions in acetylcholine. *Journal of Alzheimer's Disease: JAD, 8*(3), 247–268.

Rosamond, W. F. K., Friday, G., et al. (2007). Heart disease and stroke statistics—2007 update: A report from the American Heart Association statistics committee and stroke statistics subcommittee. *Circulation, 115*, e69–e171.

Ross, C. N., Davis, K., Dobek, G., & Tardif, S. D. (2012). Aging phenotypes of common marmosets (*Callithrix jacchus*). *Journal of Aging Research, 2012*, 567143.

Rota, M., LeCapitaine, N., Hosoda, T., Boni, A., De Angelis, A., Padin-Iruegas, M. E., ... Pelicci, P. G. (2006). Diabetes promotes cardiac stem cell aging and heart failure, which are prevented by deletion of the p66shc gene. *Circulation Research, 99*(1), 42–52.

Roth, G. S., Mattison, J. A., Ottinger, M. A., Chachich, M. E., Lane, M. A., & Ingram, D. K. (2004). Aging in rhesus monkeys: Relevance to human health interventions. *Science, 305*(5689), 1423–1426.

Safdar, A., Bourgeois, J. M., Ogborn, D. I., Little, J. P., Hettinga, B. P., Akhtar, M., ... Tarnopolsky, M. A. (2011). Endurance exercise rescues progeroid aging and induces systemic mitochondrial rejuvenation in mtDNA mutator mice. *Proceedings of the National Academy of Sciences of United States of America, 108*(10), 4135–4140.

Scatena, R., Bottoni, P., Botta, G., Martorana, G. E., & Giardina, B. (2007). The role of mitochondria in pharmacotoxicology: A reevaluation of an old, newly emerging topic. *American Journal of Physiology. Cell Physiology, 293*(1), C12–C21.

Schriner, S. E., Linford, N. J., Martin, G. M., Treuting, P., Ogburn, C. E., Emond, M., ... Rabinovitch, P. S. (2005). Extension of murine life span by overexpression of catalase targeted to mitochondria. *Science, 308*(5730), 1909–1911.

Sciarretta, S., Forte, M., Frati, G., & Sadoshima, J. (2018). New insights into the role of mTOR signaling in the cardiovascular system. *Circulation Research, 122*(3), 489–505.

Senyo, S. E., Steinhauser, M. L., Pizzimenti, C. L., Yang, V. K., Cai, L., Wang, M., ... Lee, R. T. (2013). Mammalian heart renewal by pre-existing cardiomyocytes. *Nature*, *493*(7432), 433–436.

Sessions, A. O., Kaushik, G., Parker, S., Raedschelders, K., Bodmer, R., Van Eyk, J. E., & Engler, A. J. (2017). Extracellular matrix downregulation in the *Drosophila* heart preserves contractile function and improves lifespan. *Matrix Biology: Journal of the International Society for Matrix Biology*, *62*, 15–27.

Sharov, V. S., Dremina, E. S., Galeva, N. A., Williams, T. D., & Schoneich, C. (2006). Quantitative mapping of oxidation-sensitive cysteine residues in SERCA in vivo and in vitro by HPLC-electrospray-tandem MS: Selective protein oxidation during biological aging. *The Biochemical Journal*, *394*(Pt 3), 605–615.

Shinmura, K., Tamaki, K., Sano, M., Murata, M., Yamakawa, H., Ishida, H., & Fukuda, K. (2011a). Impact of long-term caloric restriction on cardiac senescence: Caloric restriction ameliorates cardiac diastolic dysfunction associated with aging. *Journal of Molecular and Cellular Cardiology*, *50*(1), 117–127.

Shinmura, K., Tamaki, K., Sano, M., Nakashima-Kamimura, N., Wolf, A. M., Amo, T., ... Takeshi, A. (2011b). Caloric restriction primes mitochondria for ischemic stress by deacetylating specific mitochondrial proteins of the electron transport chain. *Circulation Research*, *109*(4), 396–406.

Shioi, T., McMullen, J. R., Tarnavski, O., Converso, K., Sherwood, M. C., Manning, W. J., & Izumo, S. (2003). Rapamycin attenuates load-induced cardiac hypertrophy in mice. *Circulation*, *107*(12), 1664–1670.

Siegel, M. P., Kruse, S. E., Percival, J. M., Goh, J., White, C. C., Hopkins, H. C., ... Marcinek, D. J. (2013). Mitochondrial-targeted peptide rapidly improves mitochondrial energetics and skeletal muscle performance in aged mice. *Aging Cell*, *12*(5), 763–771.

Skulachev, V. P., Anisimov, V. N., Antonenko, Y. N., Bakeeva, L. E., Chernyak, B. V., Erichev, V. P., ... Lichinitser, M. R. (2009). An attempt to prevent senescence: A mitochondrial approach. *Biochimica et Biophysica Acta*, *1787*(5), 437–461.

Smith, R. A., Hartley, R. C., Cocheme, H. M., & Murphy, M. P. (2012). Mitochondrial pharmacology. *Trends in Pharmacological Sciences*, *33*(6), 341–352.

Smith-Vikos, T., & Slack, F. J. (2012). MicroRNAs and their roles in aging. *Journal of Cell Science*, *125*(Pt 1), 7–17.

Spinale, F. G., Escobar, G. P., Mukherjee, R., Zavadzkas, J. A., Saunders, S. M., Jeffords, L. B., ... Stroud, R. E. (2009). Cardiac-restricted overexpression of membrane type-1 matrix metalloproteinase in mice: Effects on myocardial remodeling with aging. *Circulation Heart Failure*, *2*(4), 351–360.

Stein, M., Boulaksil, M., Jansen, J. A., Herold, E., Noorman, M., Joles, J. A., ... van Rijen, H. V. (2010). Reduction of fibrosis-related arrhythmias by chronic renin-angiotensin-aldosterone system inhibitors in an aged mouse model. *American Journal of Physiology. Heart and Circulatory Physiology*, *299*(2), H310–H321.

Stewart, B. F., Siscovick, D., Lind, B. K., Gardin, J. M., Gottdiener, J. S., Smith, V. E., ... Otto, C. M. (1997). Clinical factors associated with calcific aortic valve disease. Cardiovascular Health Study. *Journal of the American College of Cardiology*, *29*(3), 630–634.

Sujkowski, A., Ramesh, D., Brockmann, A., & Wessells, R. (2017). Octopamine drives endurance exercise adaptations in *Drosophila*. *Cell Reports*, *21*(7), 1809–1823.

Sujkowski, A., Saunders, S., Tinkerhess, M., Piazza, N., Jennens, J., Healy, L., ... Wessells, R. (2012). dFatp regulates nutrient distribution and long-term physiology in *Drosophila*. *Aging Cell*, *11*(6), 921–932.

Sujkowski, A., Spierer, A. N., Rajagopalan, T., Bazzell, B., Safdar, M., Imsirovic, D., ... Wessells, R. (2019). Mito-nuclear interactions modify *Drosophila* exercise performance. *Mitochondrion*, *47*, 188–205.

Sujkowski, A., & Wessells, R. (2015). Drosphila models of cardiac aging and disease. In A. Vaiserman, A. Moskalev, & J. Pasyukova (Eds.), *Life extension: Lessons from Drosophila* (pp. 127–150). Switzerland: Springer.

Sujkowski, A., & Wessells, R. (2018). Using *Drosophila* to understand biochemical and behavioral responses to exercise. *Exercise and Sport Sciences Reviews*, *46*(2), 112–120.

Sultana, N., Zhang, L., Yan, J., Chen, J., Cai, W., Razzaque, S., ... Zhou, B. (2015). Resident c-kit($^+$) cells in the heart are not cardiac stem cells. *Nature Communication*, *6*, 8701.

Szeto, H. H. (2013). First-in-class cardiolipin therapeutic to restore mitochondrial bioenergetics. *British Journal of Pharmacology*, *171*(8), 2029–2050.

Taffet, G. E., Pham, T. T., & Hartley, C. J. (1997). The age-associated alterations in late diastolic function in mice are improved by caloric restriction. *The Journals of Gerontology—Series A, Biological Sciences and Medical Sciences*, *52*(6), B285–B290.

Tardif, S. D., Mansfield, K. G., Ratnam, R., Ross, C. N., & Ziegler, T. E. (2011). The marmoset as a model of aging and age-related diseases. *ILAR Journal/National Research Council, Institute of Laboratory Animal Resources*, *52*(1), 54–65.

Tatar, M., Kopelman, A., Epstein, D., Tu, M. P., Yin, C. M., & Garofalo, R. S. (2001). A mutant *Drosophila* insulin receptor homolog that extends life-span and impairs neuroendocrine function. *Science*, *292*(5514), 107–110.

Tayebjee, M. H., Lip, G. Y. H., Blann, A. D., & MacFadyen, R. J. (2005). Effects of age, gender, ethnicity, diurnal variation and exercise on circulating levels of matrix metalloproteinases (MMP)-2 and -9, and their inhibitors, tissue inhibitors of matrix metalloproteinases (TIMP)-1 and -2. *Thrombosis Research*, *115*(3), 205–210.

Tei, C., Nishimura, R. A., Seward, J. B., & Tajik, A. J. (1997). Noninvasive Doppler-derived myocardial performance index: Correlation with simultaneous measurements of cardiac catheterization measurements. *Journal of the American Society of Echocardiography: Official Publication of the American Society of Echocardiography*, *10*(2), 169–178.

Templeton, G. H., Platt, M. R., Willerson, J. T., & Weisfeldt, M. L. (1979). Influence of aging on left ventricular hemodynamics and stiffness in beagles. *Circulation Research*, *44*(2), 189–194.

Templeton, G. H., Willerson, J. T., Platt, M. R., & Weisfeldt, M. (1976). Contraction duration and diastolic stiffness in aged canine left ventricle. *Recent Advances in Studies on Cardiac Structure and Metabolism*, *11*, 169–173.

Terman, A., & Brunk, U. T. (2004). Myocyte aging and mitochondrial turnover. *Experimental Gerontology*, *39*(5), 701–705.

Terzioglu, M., & Larsson, N. G. (2007). Mitochondrial dysfunction in mammalian ageing. *Novartis Foundation Symposium*, *287*, 197–208, discussion 208-113.

Treuting, P. M., Linford, N. J., Knoblaugh, S. E., Emond, M. J., Morton, J. F., Martin, G. M., ... Ladiges, W. C. (2008). Reduction of age-associated pathology in old mice by overexpression of catalase in mitochondria. *The Journals of Gerontology. Series A, Biological Sciences and Medical Sciences*, *63*(8), 813–822.

Trifunovic, A., & Larsson, N. G. (2008). Mitochondrial dysfunction as a cause of ageing. *Journal of Internal Medicine*, *263*(2), 167–178.

Trifunovic, A., Wredenberg, A., Falkenberg, M., Spelbrink, J. N., Rovio, A. T., Bruder, C. E., ... Larsson, N. G. (2004). Premature ageing in mice expressing defective mitochondrial DNA polymerase. *Nature*, *429*(6990), 417–423.

Tsuji, H., Larson, M. G., Venditti, F. J., Jr., Manders, E. S., Evans, J. C., Feldman, C. L., & Levy, D. (1996). Impact of reduced heart rate variability on risk for cardiac events. The Framingham Heart Study. *Circulation*, *94*(11), 2850–2855.

Uchida, S., De Gaspari, P., Kostin, S., Jenniches, K., Kilic, A., Izumiya, Y., ... Braun, T. (2013). Sca1-derived cells are a source of myocardial renewal in the murine adult heart. *Stem Cell Reports*, *1*(5), 397–410.

Ungvari, Z., Parrado-Fernandez, C., Csiszar, A., & de Cabo, R. (2008). Mechanisms underlying caloric restriction and lifespan regulation: Implications for vascular aging. *Circulation Research*, *102*(5), 519–528.

Urfer, S. R., Kaeberlein, T. L., Mailheau, S., Bergman, P. J., Creevy, K. E., Promislow, D. E. L., & Kaeberlein, M. (2017). A randomized controlled trial to establish effects of short-term rapamycin treatment in 24 middle-aged companion dogs. *Geroscience*, *39*(2), 117–127.

van Almen, G. C., Verhesen, W., van Leeuwen, R. E., van de Vrie, M., Eurlings, C., Schellings, M. W., ... Schroen, B. (2011). MicroRNA-18 and microRNA-19 regulate CTGF and TSP-1 expression in age-related heart failure. *Aging Cell*, *10*(5), 769–779.

van Berlo, J. H., Kanisicak, O., Maillet, M., Vagnozzi, R. J., Karch, J., Lin, S. C., ... Molkentin, J. D. (2014). c-kit^{+} cells minimally contribute cardiomyocytes to the heart. *Nature*, *509*(7500), 337–341.

van Deursen, J. M. (2014). The role of senescent cells in ageing. *Nature*, *509*(7501), 439–446.

Van Vleet, J. F. (2001). Age-related non-neoplastic lesions of the heart. In U. Mohr, et al. (Eds.), *Pathobiology of the Aging Dog.* (vol. 2, pp. 101–107). Ames: Iowa State University Press.

Vasan, R. S., Sullivan, L. M., D'Agostino, R. B., Roubenoff, R., Harris, T., Sawyer, D. B., ... Wilson, P. W. (2003). Serum insulin-like growth factor I and risk for heart failure in elderly individuals without a previous myocardial infarction: The Framingham Heart Study. *Annals of Internal Medicine*, *139*(8), 642–648.

Vella, C. A., & Robergs, R. A. (2005). A review of the stroke volume response to upright exercise in healthy subjects. *British Journal of Sports Medicine*, *39*(4), 190–195.

Volkers, M., Toko, H., Doroudgar, S., Din, S., Quijada, P., Joyo, A. Y., ... Sussman, M. A. (2013). Pathological hypertrophy amelioration by PRAS40-mediated inhibition of mTORC1. *Proceedings of the National Academy of Sciences of United States of America*, *110*(31), 12661–12666.

Walaszczyk, A., Dookun, E., Redgrave, R., Tual-Chalot, S., Victorelli, S., Spyridopoulos, I., ... Richardson, G. D. (2019). Pharmacological clearance of senescent cells improves survival and recovery in aged mice following acute myocardial infarction. *Aging Cell*, *18*(3), e12945.

Wang, M., Zhang, J., Walker, S. J., Dworakowski, R., Lakatta, E. G., & Shah, A. M. (2010). Involvement of NADPH oxidase in age-associated cardiac remodeling. *Journal of Molecular and Cellular Cardiology*, *48*(4), 765–772.

Wang, X., Guo, Z., Ding, Z., Khaidakov, M., Lin, J., Xu, Z., ... Mehta, J. L. (2015). Endothelin-1 upregulation mediates aging-related cardiac fibrosis. *Journal of Molecular and Cellular Cardiology*, *80*, 101–109.

Wessells, R., Fitzgerald, E., Piazza, N., Ocorr, K., Morley, S., Davies, C., ... Bodmer, R. (2009). d4eBP acts downstream of both dTOR and dFoxo to modulate cardiac functional aging in *Drosophila*. *Aging Cell*, *8*(5), 542–552.

Wessells, R. J., Fitzgerald, E., Cypser, J. R., Tatar, M., & Bodmer, R. (2004). Insulin regulation of heart function in aging fruit flies. *Nature Genetics*, *36*(12), 1275–1281.

Westermark, P., Johansson, B., & Natvig, J. B. (1979). Senile cardiac amyloidosis: Evidence of two different amyloid substances in the ageing heart. *Scandinavian Journal of Immunology*, *10*(4), 303–308.

Wilkinson, J. E., Burmeister, L., Brooks, S. V., Chan, C. C., Friedline, S., Harrison, D. E., ... Miller, R. A. (2012). Rapamycin slows aging in mice. *Aging Cell*, *11*(4), 675–682.

Xie, J., Chen, Y., Hu, C., Pan, Q., Wang, B., Li, X., ... Xu, B. (2017). Premature senescence of cardiac fibroblasts and atrial fibrosis in patients with atrial fibrillation. *Oncotarget*, *8*(35), 57981–57990.

Xu, P., Damschroder, D., Zhang, M., Ryall, K. A., Adler, P. N., Saucerman, J. J., ... Yan, Z. (2019). Atg2, Atg9 and Atg18 in mitochondrial integrity, cardiac function and healthspan in *Drosophila*. *Journal of Molecular and Cellular Cardiology*, *127*, 116–124.

Yan, L., Vatner, D. E., O'Connor, J. P., Ivessa, A., Ge, H., Chen, W., ... Vatner, S. F. (2007). Type 5 adenylyl cyclase disruption increases longevity and protects against stress. *Cell*, *130*(2), 247–258.

Yoshino, J., Mills, K. F., Yoon, M. J., & Imai, S. (2011). Nicotinamide mononucleotide, a key NAD($^{+}$) intermediate, treats the pathophysiology of diet- and age-induced diabetes in mice. *Cell Metabolism*, *14*(4), 528–536.

Zhao, K., Zhao, G. M., Wu, D., Soong, Y., Birk, A. V., Schiller, P. W., & Szeto, H. H. (2004). Cell-permeable peptide antioxidants targeted to inner mitochondrial membrane inhibit mitochondrial swelling, oxidative cell death, and reperfusion injury. *The Journal of Biological Chemistry*, *279*(33), 34682–34690.

Zheng, F., Plati, A. R., Potier, M., Schulman, Y., Berho, M., Banerjee, A., ... Striker, G. E. (2003). Resistance to glomerulosclerosis in B6 mice disappears after menopause. *The American Journal of Pathology*, *162*(4), 1339–1348.

Zhu, F., Li, Y., Zhang, J., Piao, C., Liu, T., Li, H. H., & Du, J. (2013). Senescent cardiac fibroblast is critical for cardiac fibrosis after myocardial infarction. *PLoS One*, *8*(9), e74535.

Zhu, Y., Tchkonia, T., Pirtskhalava, T., Gower, A. C., Ding, H., Giorgadze, N., ... Miller, J. D. (2015). The Achilles' heel of senescent cells: From transcriptome to senolytic drugs. *Aging Cell*, *14*(4), 644–658.

Zile, M. R., & Brutsaert, D. L. (2002). New concepts in diastolic dysfunction and diastolic heart failure: Part II: Causal mechanisms and treatment. *Circulation*, *105*(12), 1503–1508.

CHAPTER

16

The aging immune system: Dysregulation, compensatory mechanisms, and prospects for intervention

Ludmila Müller[1] *and Graham Pawelec*[2,3]

[1]Department Lifespan Psychology, Max Planck Institute for Human Development, Berlin, Germany [2]Department of Immunology, University of Tübingen, Tübingen, Germany [3]Health Sciences North Research Institute, Sudbury, Ontario, Canada

OUTLINE

Introduction

Many of our immune defense mechanisms can be traced back to unicellular organisms, whereas others have developed after the emergence of multicellularity to protect the whole organism (Danilova, 2013). It is commonly assumed that protection of the host against pathogens must obviously be the prime purpose of immunity. In animals, even in the multicellular form of slime molds, certain cells are specialized to defend the whole (Chen, Zhuchenko, & Kuspa, 2007). Perhaps reflecting their evolutionary origin, such cells are equipped with a relatively limited range of pathogen-recognition receptors and mostly act by engulfing and destroying the target or by secreting toxic substances with a limited range of action. Such "innate" immunity is present in both invertebrates and vertebrates (Muller, Fulop, & Pawelec, 2013)

"Adaptive" immunity appears evolutionarily to be more recent, originating around 500 million years ago with the appearance of two phenotypically similar but genetically totally different systems of clonotypic immunity in jawed and jawless vertebrates (Boehm et al., 2018;

Handbook of the Biology of Aging.
DOI: https://doi.org/10.1016/B978-0-12-815962-0.00016-0

Cooper & Herrin, 2010). This suggests some essential function of this form of immune system in vertebrates which is not required in invertebrates. What this might be remains a subject of speculation. A unique feature of an adaptive immune system in both cases is the somatic development of a pool of clonally diverse circulating cells with a myriad of unique antigen-recognition receptors (Bonilla & Oettgen, 2010; Muller et al., 2013). When ligated under appropriate conditions, these and other coreceptors trigger cell activation and clonal expansion to amplify the originally small number of responding cells and their differentiation from naïve cells (unexposed to the target) to effector cells (for target destruction) and memory cells (for rapid responses to later rechallenge by the same invader; hence the term "adaptive" immunity). This system relies on the generation of an extremely diverse antigen receptor repertoire by genetic recombination of a relatively small number of variants, which multiplies antigen recognition capacity to an almost infinite number of potential targets, and can therefore mediate highly specific responses (Litman, Rast, & Fugmann, 2010).

Such an adaptive immune system based on random recombination of antigen receptor elements may easily generate cells recognizing the organism's own tissues. Hence, very tight control over the generation of cells of the adaptive immune system must be maintained to avoid destructive autoimmunity. In aging organisms both the generation of naïve cells and their education to prevent autoimmunity may be compromised, as well as other aspects of immune functionality. The latter include effects seen at the hematopoietic stem cell level and in the development of immune cells in the bone marrow affecting both adaptive and innate immunity. In addition, functional decline has been observed in the processing of thymic-derived lymphocytes altered in the process of thymic involution, the processing and recognition of antigens, clonal expansion of lymphocytes, and the production of effector molecules, all of which are affected by age.

This chapter provides an overview of age-associated effects on immunity primarily in humans, as summarized in Figs. 16.1 and 16.2, focusing on the thymic-derived T cell, which appears to be more affected by aging than most other immune cell types. We consider the clinical implications of such "immunosenescence" and possible approaches to amelioration.

Innate and adaptive immunity

Multiple types of innate immune cells distinguish invaders by means of their "pattern recognition receptors," which recognize highly conserved molecules on microorganisms that are not present on mammalian cells (Janeway & Medzhitov, 2002). The innate response is rapid, because it does not require clonal expansion and can directly trigger phagocytosis, cell cytotoxicity, activation of the complement system, release of acute-phase proteins, and the production of inflammatory mediators, such as cytokines and chemokines. Sequential recruitment and activation of granulocytic cells begin the process, accompanied by inflammation and activation of adaptive responses (Solana et al., 2012).

The antigen-specific adaptive immune response mediated by both cells and antibodies is slower and may take days to become effective due to the requirement for clonal expansion of the originally small numbers of cells with specific receptors (Bonilla & Oettgen, 2010; Fülop et al., 2017; Müller, Di Benedetto, & Pawelec, 2019). These include B cells, which modify their receptors during clonal expansion to select for the secretion of soluble factors (antibodies) with improving high specificity for three-dimensional antigen structures (affinity maturation). Also included are T cells, which do not alter antigen receptor characteristics and retain the same properties, recognizing short peptides presented on self-major histocompatibility molecules, throughout their clonal expansion (Fig. 16.2). T cells regulate B-cell responses, and other immune cell responses, as well as exerting direct cytotoxic effects themselves against, for example, virus-infected cells. Almost all B-cell responses are dependent on "help" from T cells, which are themselves only activated by specialized "antigen-presenting cells" (APCs) able to process and present the short peptide fragments from pathogens or other targets bound to self-MHC molecules. The APCs are usually components of the innate immune system, especially dendritic cells (DCs). Broadly speaking, this trio of cell types are those playing key roles in adaptive immune protection against pathogens, and in the development and maintenance of peripheral immune tolerance to self. This is a finely balanced, highly regulated networked system of great complexity and with the potential to fail with disastrous consequences if it becomes dysregulated. Indeed, age-associated dysregulation thereof might be expected not only to compromise antipathogen defense, but also to wreak havoc on organismal functioning, as reflected by the "immunologic theory of aging" first proposed many years ago by Roy Walford (Walford, 1964).

Age and immunity

Differences between young and old individuals reflect changes to parameters of the immune system commonly designated "immunosenescence" (Pawelec, 2018a). However, this term should principally be reserved for age-associated differences in immune parameters that have been shown to have detrimental effects on the organism. Currently, such "biomarkers" of immune aging are, however, rare or absent and the

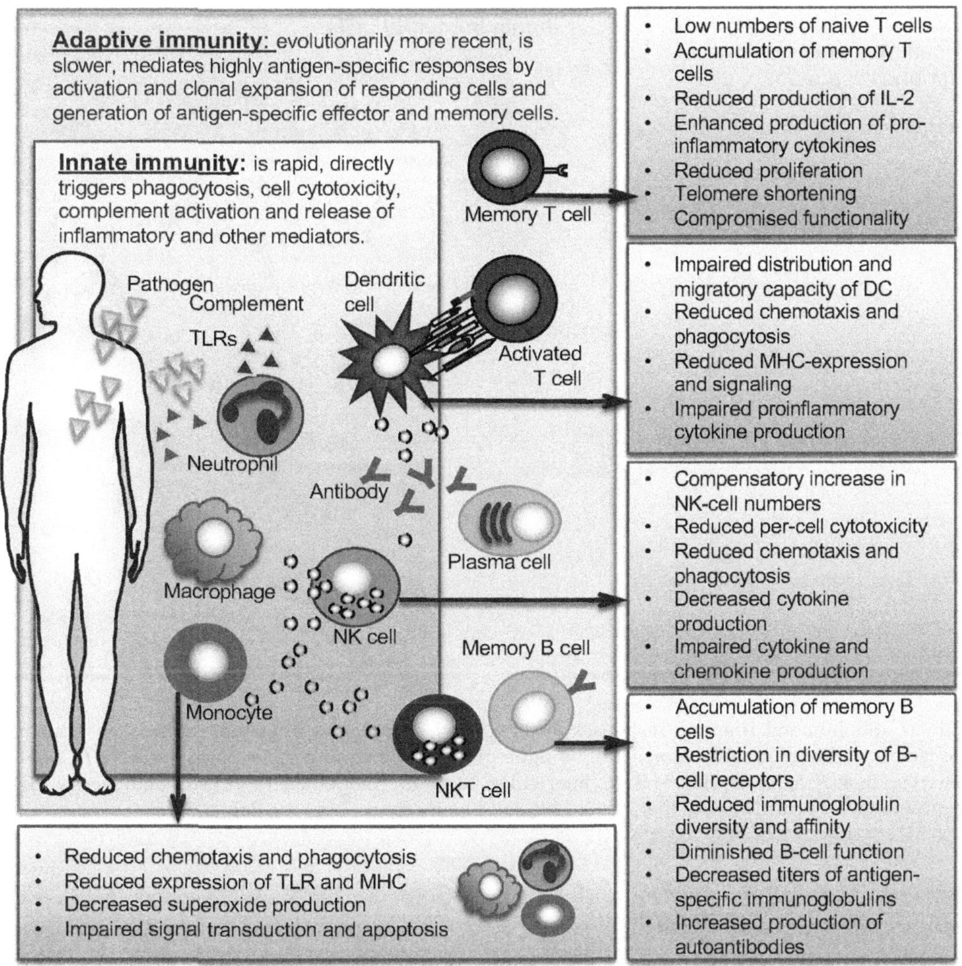

FIGURE 16.1 Impact of aging on innate and adaptive immunity. *MHC*, major histocompatibility complex; *NK cell*, natural killer cell; *NKT*, natural killer T cell; *TLR*, toll-like receptor.

term tends to be used indiscriminately to describe the different state of immune parameters in old versus young individuals. Thus, the term should be used with caution, especially because in adaptive immunity, adaptation with age is not obviously detrimental—indeed, aging per se does not necessarily lead to an unavoidable decline in all immune functions. Instead, a complex process of remodeling occurs depending on the individual's exposures and other factors.

It is often difficult to distinguish between beneficial adaptations and detrimental dysregulation (Pawelec, 2012). Immune remodeling will be dictated by both genetic and intrinsic events, such as the physiological process of thymic involution. Remodeling is also influenced by environmental factors, primarily mediated by encounters with foreign antigens over the lifetime mostly borne by pathogens but also likely to include cancer antigens and other products of mutated or modified self-components which increase with age. With the passage of chronological time, accumulating challenges and insults may synergize and result in dramatic alterations to immune functions, which could cause multiple detrimental outcomes.

Thus, immunosenescence is reflected not only in diminished immunity with decreased response against pathogens, but also in increased inappropriate immune activation, leading to autoimmunity and a subsequent increase in the incidence of autoimmune diseases as well as possibly other detrimental manifestations, as alluded to above. Definitively demonstrating which age-associated differences in immune parameters correlate with detrimental physiological outcomes, let alone cause them, is a great challenge even in mice, and much more so in humans.

The situation is further complicated by the great deal of heterogeneity in the biological consequences of the aging process, especially in outbred long-lived animals like humans. There are multiple biological, psychological, and other stressors experienced over the life course, acting in the presence of different individual genetic

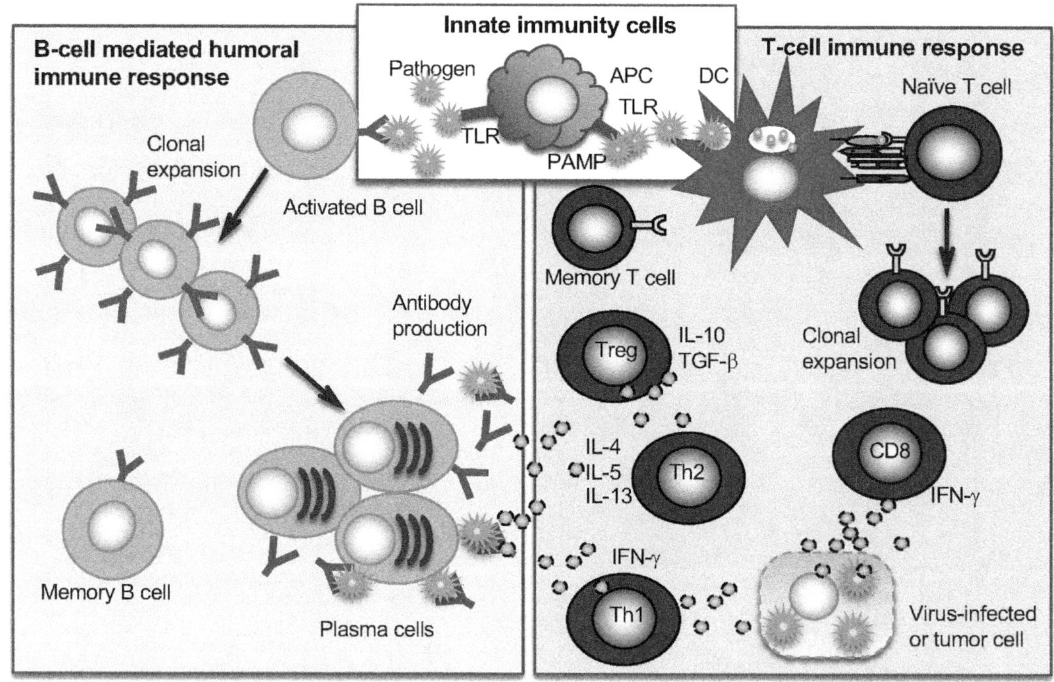

FIGURE 16.2 Cellular and humoral immune responses are mediated by a highly regulated networked system consisting of different immune cells, soluble factors and immune mediators. *APC*, antigen presenting cell; *CD8*, CD8-positive T cell; *DC*, dendritic cell; *IFN-γ*, interferon gamma; *IL-4*, interleukin 4: *IL-5*, interleukin 5; *IL-10*, interleukin 10; *IL-13*, interleukin 13; *PAMP*, pathogen-associated molecular pattern; *TGF-β*, transforming growth factor beta; *Th2*, T helper 2 cell; *TLR*, toll-like receptor; *Treg*, regulatory T cell.

backgrounds. Part of the latter reflects differences between males and females (Caruso, Accardi, Virruso, & Candore, 2013). Sex hormones broadly affect immune cells quantitatively and qualitatively, and influence cytokine as well as antibody production and the activity of natural killer (NK) cells, monocytes, and granulocytes (Gubbels Bupp, Potluri, Fink, & Klein, 2018; Oertelt-Prigione, 2012; Puchta et al., 2016). Aged females were shown to demonstrate more proinflammatory tendencies compared with older men (Di Benedetto, Gaetjen, & Muller, 2019; Kovats, 2015; Ostan et al., 2016), for example. Moreover, some aspects of male and female immunity might be different because certain genes involved in immunity are mapped to the X chromosome (Caruso et al., 2013). Surveys showing a slower rate of decline in certain immunological parameters in women than men have been suggested to contribute to the explanation of why women live longer (Garcia Verdecia et al., 2013; Hirokawa et al., 2013). Hence, we must bear in mind that different genetic backgrounds, including sex and race, interacting with environmental exposures and stresses, are likely to have a strong influence on immunosenescence. Caution must be exerted when generalizing from one reported study to the next. In any event, one must avoid falling into the trap of extrapolating from WEIRD (Western, educated, industrialized, rich, democratic) populations, which may not be representative (Alam, Goldeck, Larbi, & Pawelec, 2013). Even between closely related populations, unexpected differences in immune parameters associated with mortality at advanced age have been noted (Adriaensen et al., 2017), possibly reflecting differences between different birth cohorts (Pawelec, 2019).

In the following sections, we will briefly review current knowledge on the effects of aging on the different components of the mammalian, primarily human, immune system, beginning with the generation of all hematopoietic cells in the bone marrow, and then focusing particularly on adaptive immunity, where the greatest age differences are seen.

Effect of age on hematopoiesis

Should stem cell aging occur, it would be expected to be a central factor driving the senescence of the immune system, a biological system characterized by a very high cellular turnover. Age-related impairment in the function of hematopoietic stem cells (HSCs), which give rise to both myeloid and lymphoid lineages, might be responsible for much of the difference observed in both the innate and adaptive immune systems (Geiger, de Haan, & Florian, 2013; Weinberger & Grubeck-Loebenstein, 2012). More recent studies have revealed that HSCs from old mice transplanted to young mice reconstitute an immune system less able to respond to vaccination than young controls, and have identified pharmacological treatments to restore functionality of old HSCs (Leins et al., 2018). Downstream effects of the osteopontin-Wnt signaling pathway converging on this target, cdc42, may also be influential in genetic determination of human longevity (Pawelec, 2018d).

Whatever the reason, there is a lower output of cells of the lymphoid lineage in the elderly, whereas myeloid output tends to be increased relative to HSCs from younger individuals (Fig. 16.3). Interestingly, a high proportion of aged HSCs produce practically exclusively platelets (Konieczny & Arranz, 2018), the gene-expression program of which may apparently contribute to the suppression of the lymphoid lineage (Grover et al., 2016). The age-associated skewing of hematopoietic cell lineage production seems to involve a disorder of early HSC differentiation steps (Beerman, Maloney, Weissmann, & Rossi, 2010b; Konieczny & Arranz, 2018). The senescent HSC phenotype appears to be either stem cell-intrinsic, affecting all output, and/or caused by some systemic factors in old animals impinging on stem cell functionality (Conboy, Conboy, & Rando, 2013). It is therefore likely that a complex network of primarily HSC-intrinsic, but also extrinsic, factors is responsible for the age-related skewing of HSC differentiation (Florian et al., 2018; Geiger et al., 2013). Based on differential expression of CD150, two functionally distinct HSC subsets have been identified (Beerman et al., 2010a; Morita, Ema, & Nakauchi, 2010). One of these subsets seems to be more predisposed to lymphoid differentiation ($CD150^{lo}$), whereas the other is more predisposed to myeloid differentiation ($CD150^{hi}$). This may represent a mechanism for the predominance of myeloid-biased HSCs in aged individuals with the parallel age-associated loss of lymphoid-biased HSCs (Mandal & Rossi, 2012).

The criteria used to identify HSC aging include: (1) increased numbers of cells with an HSC-associated cell surface phenotype; (2) a reduced capacity of each HSC for self-renewal; (3) lineage skewing; (4) enhanced mobilization of HSCs from the bone marrow into the blood; and (5) reduced homing of HSCs back to the bone marrow (Fig. 16.3). Moreover, those HSCs that do home back to the bone marrow demonstrated a distinct niche selectivity (Geiger et al., 2013). The aging HSC niche also manifests multiple changes, including decreased numbers of osteoblasts leading to reduced osteogenesis, an increase in adipocytes due to the skewed differentiation of aged mesenchymal stem cells, and finally, the composition of the extracellular

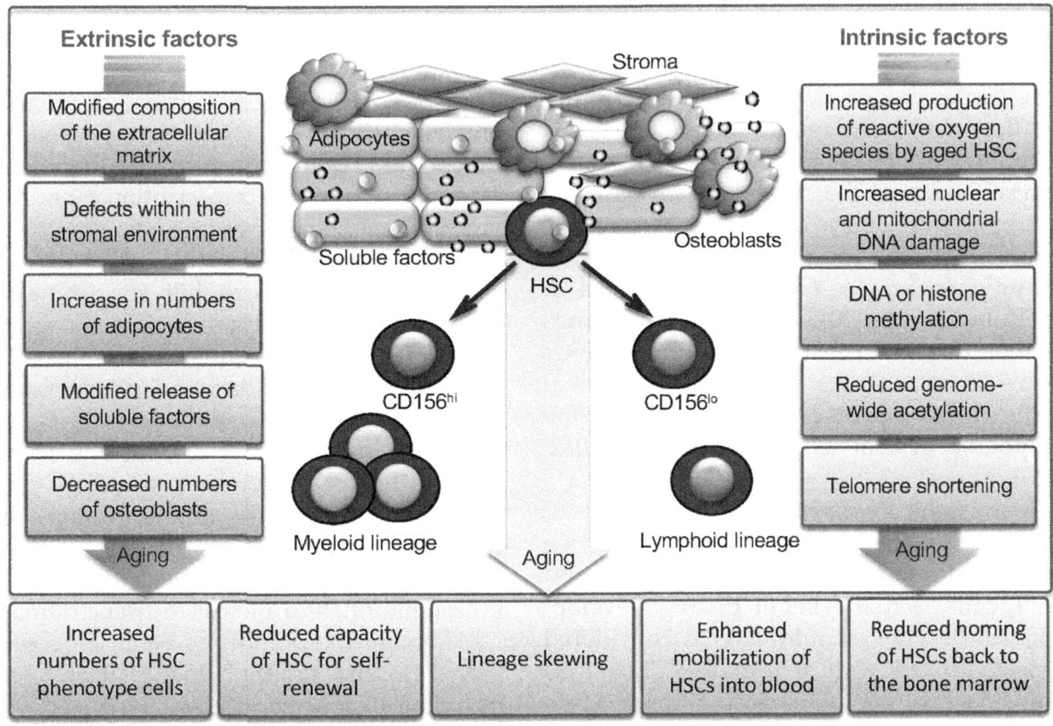

FIGURE 16.3 Effect of age on hematopoietic stem cells and their environment. *CD*, cluster of differentiation; *DNA*, deoxyribonucleic acid; *HSC*, hematopoietic stem cell.

matrix is modified, leading to a reduced self-renewal capacity (Yahata et al., 2011).

The microenvironment plays an important role in the development and activation of different immune cell types through the release of soluble factors and cell–cell interactions (Pangrazzi et al., 2017). Therefore defects within the stromal environment might greatly impact the function and development of immune cells and make an important contribution to immunosenescence (Su, Aw, & Palmer, 2013). In addition, stem cell-intrinsic mechanisms contribute markedly to HSC aging (Fig. 16.3) in that aged HSCs tend to show higher production of reactive oxygen species (ROS) with resultant damage to mitochondrial DNA (Ito et al., 2006). There is also increased nuclear DNA damage and telomere shortening in aged HSCs. Consequently, the resulting DNA damage signaling might cause senescence in the cell biological sense, apoptosis, or differentiation (Norddahl et al., 2011; Rube et al., 2011). Telomere shortening has been related to impairment in stem cell function in general; it may activate cell-intrinsic checkpoints as well as alter the stem cell microenvironment (Konieczny & Arranz, 2018; Tumpel & Rudolph, 2012).

The importance of epigenetic regulation in maintaining gene expression and determination of cell fate is increasingly accepted, and it will come as no surprise that this is also important in aging HSCs (Tollervey & Lunyak, 2012). Changes to the chromatin occur, reflected as DNA or histone methylation, or histone acetylation, resulting in inaccurate replication of epigenetic marks also measurable as reduced genome-wide acetylation. Because HSCs are long-lived cells, they may be at greater risk of accumulating DNA damage and mutations in critical genes. Therefore DNA damage checkpoint activity in HSCs seems to be essential for the preservation of their self-renewal capacity and for the suppression of malignant transformation, contributing to cancer protection and possibly promoting tissue aging (Moehrle & Geiger, 2016; Naka & Hirao, 2011; Rube et al., 2011). These checkpoints appear to be evolutionarily conserved in human HSCs and could serve as a target for preventing functional decline and the development of malignancies in the aging hematopoietic system (Mandal & Rossi, 2012; Wang et al., 2012).

However, it must be borne in mind that many of the differences in the hematopoietic system between old and young, including myeloid cell output bias and fewer B and T cells, actually begin during development (Snoeck, 2013). They should not be considered only as a process linked with senescence. They may reflect mechanisms evolved to maximize fitness early in life, but are detrimental later in life. This would be an example of the commonly assumed phenomenon of antagonistic pleiotropy hypothesized for many systems. A similar argument applies to the normal physiological process of thymic involution, which may or may not contribute to immunosenescence in later life, depending on the individual circumstances.

Effect of age on innate immunity

Although it is generally accepted that some aspects of innate immunity seem to be relatively well preserved (or even enhanced) during aging, accumulating evidence in the last decade supports the notion that immunosenescence not only affects adaptive immunity but also innate immunity. This will be briefly discussed here, before turning to adaptive immunity.

Innate immunity is mediated by a complex network of different cell types, which consist of such polymorphonuclear leukocytes as granulocytes, and mononuclear cells such as monocytes, macrophages, NK cells, and DCs. Activation of the innate immune system results in pathogen phagocytosis and the production of immune mediators (such as reactive oxygen and nitrogen species, defensins, and complement), as well as proinflammatory cytokines and chemokines. The latter are able to mediate the initial host response to pathogens and stimulate recruitment of cells of the adaptive immune system (Shaw et al., 2011). Recent evidence implies that, in contrast to previous assumptions, cells of the innate immune system may retain a memory of former stimulations, termed "trained immunity," altering the response upon new stimuli and becoming able to respond to a larger number of microbes upon secondary challenges (Crisan, Netea, & Joosten, 2016). This functional readjustment of immunological memory involves changes in transcriptional programs by reprogramming epigenetic marks (Jasiulionis, 2018).

Components of the innate immune response, including such cells as neutrophils, macrophages, and monocytes, are the first line of defense against infections. They are rapidly recruited to the site of infection, acting via pathogen killing and phagocytosis. Phagocytic cells initiate inflammation, and recruit NK cells and DCs that initiate adaptive immune responses (Solana et al., 2012). Phagocytosis is therefore considered to be an integral part of the innate immune system, but its reduced functional activity in the elderly leads to accumulation of unphagocytosed debris, chronic sterile inflammation, and increased tissue damage (Li, 2013).

Although the number of neutrophils seems to be preserved in the healthy elderly, their several functions, including phagocytosis, chemotaxis, and intracellular killing via free radical production, are reduced (Solana et al., 2012). The phagocytic and killing capacities of macrophages are also reduced with age, due to lower production of ROS (such as NO_2 and H_2O_2) and decreased levels of cytokines such as TNF and IL-1 (Fernandez-Morera, Calvanese, Rodriguez-Rodero, Menendez-Torre, & Fraga, 2010; Gonzalo, 2010; Goronzy & Weyand, 2012; Solana et al., 2012). Chemotaxis is compromised in neutrophils, monocytes/macrophages, and DCs, and their microbicidal function and phagocytic capacity may be lower as well.

Mechanistically, age-associated alterations in toll-like receptor (TLR) signaling that could be responsible for some of this dysfunctionality have been determined (Weinberger & Grubeck-Loebenstein, 2012). Studies have demonstrated that the expression pattern and function of certain TLRs on immune cells seem to be different in the elderly. Whereas the expression of some TLRs on monocytes, such as TLR1 and TLR2, increased with age, the function of other TLRs, such as TLR4 and TLR8, is decreased (Alvarez-Rodriguez et al., 2012). Positive associations of loss of function variants of TLR-1 and -6 with healthy aging have been reported, differing in men and women, and potentially mediated via less inflammaging in the elderly (Hamann et al., 2015). Extension of studies on TLR functions in older adults is very important in the context of improving vaccine adjuvants (Del Giudice, Rappuoli, & Didierlaurent, 2018)

Due to the skewed production of more myeloid cells and fewer lymphoid cells described above, the elderly tend to maintain peripheral levels of the former, as well as similar or even increased numbers of NK cells, but their functionality may be decreased on a *per cell* basis (see next section). These findings illustrate another general concept which should be considered when contemplating differences in measured parameters between young and old individuals: that is, the observed differences may be compensatory, with an accumulation of a certain cell type occurring because of the overriding necessity for maintaining the function that those cells mediate.

Natural killer cells

NK cells are cytotoxic lymphocytes involved in early defense. They recognize virus-infected as well as virally or nonvirally modified tumor cells in an MHC-unrestricted manner (Manser & Uhrberg, 2016). Studies of NK cell numbers and functionality in the very elderly illustrate the above concept regarding compensation with age; elderly people tend to possess higher percentages and absolute numbers of NK cells than the young. There is a different distribution of NK cell subsets, with expansion of the mature $CD56^{dim}$ NK-cell subset. This may partly explain the decreased proliferative response to cytokines in the elderly as well as the altered expression of some NK-cell receptors. It might also indirectly modulate adaptive immunity via altered cytokine production affecting the activation of DCs and monocytes, and promoting inflammation (Solana et al., 2012). Although overall NK-cell numbers tend to increase with age, their function on a *per cell* basis decreases, partly due to redistribution of the different maturation stages in the periphery. NK cells from the elderly produce lower levels of cytokines and chemokines (such as RANTES, MIP1α, and IL-8). Therefore, the overall increase in the number of NK cells could represent a compensatory mechanism in order to maintain an essential level of functionality with advanced age.

Dendritic cells

DCs are potent "professional" APC, constituting an important bridge between the innate and adaptive immune systems, without which adaptive immune responses cannot take place. The quality and intensity of adaptive immune responses as well as the type of T-cell response initiated, are dictated by the manner in which DCs present antigen and costimulate T cells. They are responsible for recognizing pathogen-associated molecules, for taking up, processing, and presenting these pathogen antigens in the form of short peptides bound to self-MHC molecules to T cells. DCs play crucial roles in the immune response, regulating the balance between immunity and tolerance (Agrawal, Agrawal, & Gupta, 2017; Clements, Maijo, Ivory, Nicoletti, & Carding, 2017).

The functions of DCs have been demonstrated to be marginally altered with age. DCs promote or inhibit immunity, crucial regulatory functions depending on a combination of factors including subset distribution, activation, and maturation. Therefore, age-related alterations in these DC features can result either in aberrant immune suppression or in hyperactivation of the immune response with impaired proinflammatory cytokine production, and may represent one of the factors contributing to morbidity and mortality in aged populations (Wong, Magnusson, & Ho, 2010; Zacca et al., 2015). Although no major effects on the numbers and phenotype of DC subsets have been reported in aged individuals, their ability to phagocytose antigens is impaired with age. Subtle differences in cytokine secretion (with increased basal levels of proinflammatory cytokines) may result in decreased incremental responses to foreign antigens as well as increased responses to self-antigens. This has been suggested to contribute to the loss of tolerance and increased autoimmunity commonly seen in the elderly (Agrawal & Gupta, 2011; Agrawal et al., 2017).

DCs appear to be affected by aging in terms of their distribution and migratory capacity, in their antigen-processing ability as well as in costimulatory signal expression and cytokine production. Although antigen presentation by DCs seems to be only subtly disturbed in the elderly, fewer DCs are found in peripheral blood and follicles, resulting in reduced efficiency of this line of defense against infections as well as in tumors (Adema, 2009; Della Bella et al., 2007; Derhovanessian, Solana, Larbi, & Pawelec, 2008; Gonzalo, 2010).

Effect of age on adaptive immunity

One of the hallmarks of immune aging is the low numbers and percentages of naive T cells in peripheral blood and lymphoid organs, with a reciprocal accumulation of antigen-experienced T cells. As we age, the T-cell compartment, especially the CD8 MHC class I-restricted subset, is skewed more toward highly differentiated effector T cells with a more limited T-cell antigen receptor repertoire. Analysis of chromatin accessibility in PBMCs identified memory $CD8^+$ T cells as the subpopulation with the most profound chromatin remodeling with aging (Ucar et al., 2017). A shift toward a more differentiated state of chromatin openness was also observed in naïve and central memory cells from older individuals (Moskowitz et al., 2017).

The highly differentiated effector T cells produce less of the T-cell growth factor interleukin 2 (IL-2) required for clonal expansion, and more proinflammatory cytokines (such as IFN-γ and TNF) contributing to the low-grade proinflammatory state commonly observed in the elderly. The fact that these T cells share some defining characteristics with "replicatively senescent" fibroblasts (short telomeres, production of proinflammatory cytokines, inability to proliferate) has led to the proposal that they must be "senescent" themselves. However, in the context of the dynamic processes involved in immune differentiation from naïve to effector to memory cells, such cells in the immune system are more likely to be late-stage differentiated cells still retaining required functionality. This does not exclude that some of them may indeed be senescent, as defined by compromised functionality within the context of their defined purpose (Ouyang et al., 2003) but it is a mistake to refer to whole populations of such cells, defined by the downregulation of certain surface receptors (e.g., CD28) and the upregulation of certain others (e.g., CD57) as definitively senescent. Indeed, in humans, many of these late-differentiated $CD8^+$ T cells are specific for the products of a single herpesvirus, with which almost all older people are infected (usually asymptomatically) and which may be essential for maintaining the virus in a latent state (Jackson et al., 2017; Pawelec, 2017).

Antibody responses to infectious agents and to vaccines tend to be decreased in most elderly people, partly explained by deficient T-cell help, but also resulting from intrinsic defects in B cells. Considering the role of epigenetic mechanisms in B-cell differentiation and function, age-related epigenetic alterations could be responsible for the decline of humoral immunity in elderly individuals (Jasiulionis, 2018). Accumulations of memory B cells resulting from exposures over the lifetime, as well as defects in class-switch recombination (to generate mature antibody responses) have been demonstrated in mice, but are more controversial in humans. However, the diversity of B-cell receptors is restricted in the very elderly, due to accumulation of memory cells, and this may be associated with frailty (Weinberger & Grubeck-Loebenstein, 2012). Similar types of data on subset differentiation, repertoire diversity, and function are available in greater profusion for T cells than B cells and are discussed in detail in the next section.

Impact of thymic involution and thymectomy on T cells

Not only is the output of immune cells from the bone marrow skewed with age, as we saw above, resulting in fewer B cells and more myeloid cells, but T-cell progenitors released to the periphery require additional processing in the thymus in order to generate functional mature naïve T cells. This process is also markedly affected by age. It may be that the progenitors themselves are compromised by age, but the impact of thymic involution is certainly of major importance for decreased naïve cell generation (Fig. 16.4), and here it is the thymic epithelial cells that are mostly responsible (Thapa & Farber, 2019). The thymus is most active before puberty, but thereafter undergoes the progressive anatomical changes referred to as "involution." What triggers thymic involution remains unclear, but given the hormonal changes at puberty, and the fact that castration or androgen blockade can lead to at least partial thymic regeneration in both humans and mice (Heng et al., 2005; Sutherland et al., 2005) it would seem that sex hormone production plays a major role. This process can hardly be considered as "aging" because it is a normal part of developmental programming, which begins even before puberty.

Conservation of this process in animals other than mammals (including birds, fish, and amphibians) suggests some important evolutionarily conserved reason for this event (Shanley, Aw, Manley, & Palmer, 2009). In young animals, it is of paramount importance to

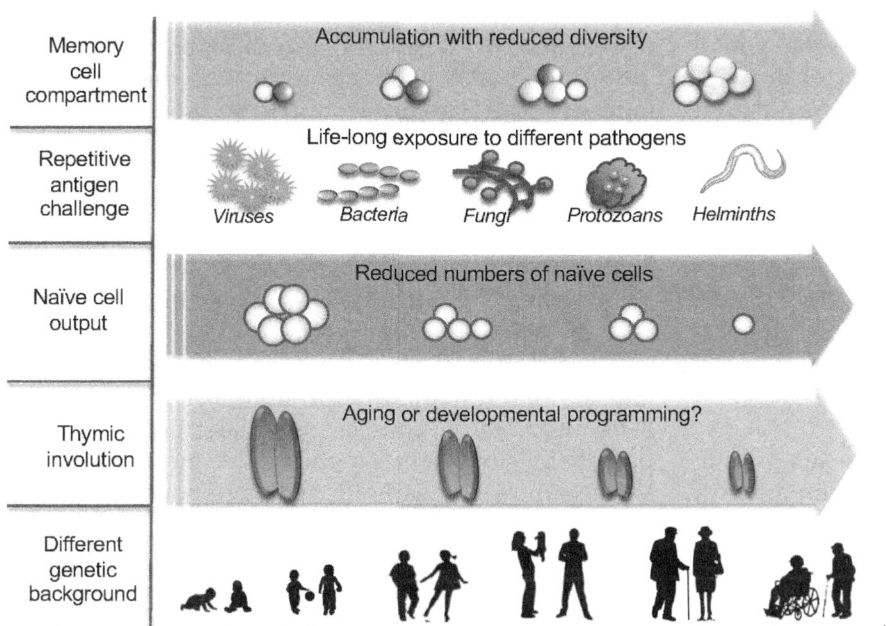

FIGURE 16.4 Impact of thymic involution and life-long exposure to antigens on immunosenescence.

generate a diverse naïve T-cell receptor (TCR) repertoire rapidly, in order to maximize the probability of recognizing the plethora of pathogens to which the individual is being constantly exposed after birth, and when protection via antibodies in the mother's milk ceases at weaning. Older animals that have survived these exposures will have generated memory T cells specific for the universe of antigens from pathogens that are most common in their environment. Their antigen-specific TCR repertoire will thus be already be "constricted," that is, focused on these "relevant" antigens. The requirement for generating more naïve T cells specific for antigens to which the individual may never be exposed is then a subsidiary concern vis-á-vis maintaining memory for those pathogens that are certainly present (Pawelec, 2017).

This explanation for thymic involution relies on a cost–benefit hypothesis proposing that the continued generation of new naïve T cells is a wasteful and dangerous process. It is resource-intensive because many of the progenitor T cells die in the thymus during the selection process. Because T-cell progenitors must randomly rearrange their TCR genes in the thymus, they have to be positively selected for weak binding to self-MHC molecules and for the lack of binding to autoantigens. The majority of T cells fails to fulfill these criteria and are eliminated. This makes the process expensive, and also dangerous in the case of imperfect negative selection. Even in young animals, negative selection certainly is imperfect because self-reactive cells are common in the periphery (Legoux et al., 2015). They are kept in check by regulatory T cells (Treg), which are also generated in the thymus, and act to suppress damaging peripheral autoreactivity. Thymic naïve Treg induction is also deceased with age (Valmori, Merlo, Souleimanian, Hesdorffer, & Ayyoub, 2005), so the "safest" and most conservative option is greatly to reduce thymic productivity at an early age. The peripheral TCR repertoire is maintained for many decades in humans (Naylor et al., 2005; Qi et al., 2014), certainly exceeding the usual life span of early humans, most of whom would not have lived long enough to become "immunosenescent." Thus, clonal diversity may be maintained for a longer period than evolutionarily "necessary" without the requirement for maintaining thymic function, but with a repertoire "crash" occurring after the age of 70 or so, according to some models (Johnson, Yates, Goronzy, & Antia, 2012).

The consensus scenario is that the thymus becomes markedly less active after puberty, and slowly decreased output thereafter until, by 50–60 years in humans, central generation of new naïve T cells essentially ceases (Fig. 16.4). The individual relies on memory to previously encountered pathogens, and residual naïve cells in case of exposure to new pathogens. According to these ideas, older people should have few naïve cells, more memory cells, and be more susceptible to novel infections. The latter is certainly the case, and the first two predictions also apply to $CD8^+$ T cells, if not necessarily to $CD4^+$ T cells. This raises the question as to whether the elderly do indeed suffer any negative consequences from having only low levels of naive T cells. This has been the assumption, and is likely to be true for exposure to emerging pathogens, such as SARS and West Nile virus, which are more serious in the elderly than in the young (Jean, Honarmand, Louie, & Glaser, 2007; Kilpatrick et al., 2008; Peiris et al., 2003). Recent evidence (based on the mathematical modeling of immune system decline) showed that thymic involution increased disease incidence and also provided mechanistical insight into carcinogenesis, suggesting that rather the decline in the T-cell output than mutation accumulations represents a major risk factor (Palmer, Albergante, Blackburn, & Newman, 2018). This argues for attempting to rejuvenate the thymus to increase the output of naïve cells and result in some restoration of TCR repertoire diversity.

The growing understanding of mechanisms driving thymic involution may result in the development of safe and effective strategies to rejuvenate the thymus for use in clinical settings (Ventevogel & Sempowski, 2013). As commonly the case when considering interventions, monitoring their success requires validated methodology, in this instance the ability to identify recent thymic emigrants (RTEs). However, this is a challenge that can only be imperfectly met at the present time. The frequency of TCR excision circles (TRECs) is commonly used to identify T cells that have undergone little peripheral proliferation. TRECs are DNA fragments formed when the TCR genes are rearranged, and are diluted out at each cell division (Douek et al., 1998). Determining TREC numbers before and after interventions would be an informative measure of thymic reactivation, but would be affected by cell division in the periphery. Naïve cells can be found within populations expressing clusters of surface molecules such as CD45RA (protein tyrosine phosphatase, receptor type, C isoform RA), CCR7 (C-C chemokine receptor type 7), CD31 (platelet endothelial cell adhesion molecule 1, PECAM-1), and PTK7 (tyrosine-protein kinase-like 7). There is no single definitive marker available, but more recent studies have confirmed the utility of combinations of these markers (Cunningham, Helm, & Fink, 2018).

The best method for identifying naïve T cells which have recently exited the thymus would be to determine TREC levels in $CD45RA^+CCR7^+CD31^+PTK7^+$ cells with long telomeres. However, there are almost as many favored phenotypes as investigators, so this area is ripe for harmonization and standardization. Nonetheless, combining PTK7 and CD31 marker analysis in peripheral cells may be informative for monitoring RTEs. T cells with this phenotype have high numbers of TRECs, are present at lower frequencies in older subjects (Haines et al., 2009), and are greatly decreased after thymectomy (Sauce et al., 2009). It is important to distinguish RTEs from persistent naïve T cells because the latter may be more functionally compromised as a result of chronological aging processes and/or peripheral homeostatic proliferation. Indeed, there is some evidence that the very small numbers of RTEs still produced by adults "compete" preferentially with mature naive cells for "space" in the periphery, but these technically challenging experiments relying on deuterium labeling were only performed in people up to 25 years of age (Vrisekoop et al., 2008).

In humans, neonatal thymectomy for cardiac surgery offers the opportunity to ask how serious a dramatic reduction of thymic output early in life might be (Stosio, Ruszkowski, Mikosik-Roczynska, Haponiuk, & Witkowski, 2017). Thus far, there are few data on this potentially important question affecting many people thymectomized in early life, who are now adults. They are not yet old enough, but longitudinal studies have so far shown little impact on infections and health at 1 and 3 years post-thymectomy (Mancebo et al., 2008). However, this is probably too short a time and exposures have not yet multiplied. This is a crucial question requiring long-term follow-up, which might also illuminate some aspects of immunosenescence in the elderly. Nevertheless, the alterations in the pools of immune cells and the disturbances in cytokine expression indicate that thymectomy has a significant influence on the immune function of peripheral T and B cells. Further clinical consequences of these changes have to be investigated (Roosen, Oosterlinck, & Meyns, 2015). The fact that some patients show signs of thymic regrowth 5–10 years after thymectomy is particularly encouraging and indicates that the thymus may have previously unsuspected regenerative capability (Stosio et al., 2017; van den Broek et al., 2016).

We have insufficient data to be sure that fewer naïve T cells commonly present in elderly people are indeed detrimental. However, some data are consistent with a lower proportion of naïve T cells, less repertoire diversity, and a delayed response to vaccination against tick-borne encephalitis 20 years after infant thymectomy compared to age-matched nonthymectomized controls (Prelog et al., 2008). A seminal study of vaccination against yellow fever using attenuated live virus has shown that elderly people exposed to this new infection do indeed respond more poorly to neoantigens, related to a dearth of $CD4^+$ naïve helper T cells, and poorer APC function (Schulz et al., 2015). On the other hand, in two series of longitudinal studies of the (not thymectomized) very elderly (over 85) in two different European populations (Derhovanessian et al., 2013; Larbi et al., 2008), low numbers of naïve $CD8^+$ T cells were not associated with excess mortality in a 2–9-year follow-up (see below). This is probably because these very old people were not traveling much and were not often exposed to pathogens that they had not previously encountered, again illustrating the general principle that studies on this subject must always be interpreted in the context of the populations investigated, and their particular circumstances.

Immune cell function

As we have seen above, bone marrow output of immune cells is different in the elderly relative to the young, with more myeloid cells and fewer B cells, fewer T cell progenitors, and low or absent thymic output of naïve T cells. As a cumulative result of exposures throughout life, the representation of different subsets and differentiation stages in peripheral lymphocytes can change dramatically, especially in the $CD8^+$ subset. In addition, the functional integrity of the individual cells may be altered, that is, cells in the same subset in young and old individuals behave differently, as described in the following section. A number of mechanisms have been proposed as possible causes for these alterations (such as the increased chronic inflammation, oxidative stress, metabolic disruption), but strong data establishing a cause and effect relationship are still missing (Goronzy & Weyand, 2013; Swain & Blomberg, 2013).

T-cell function

T cells from aged individuals demonstrate reduced functional ability compared to cells from young individuals. The crucial initial step in T-cell activation—the reorganization of the plasma membrane ("immune synapse" formation)—has been shown to be impaired in T cells from older people, reducing recruitment of signaling molecules following activation (Fülop et al., 2017; Larbi et al., 2011). Even T cells with a naïve phenotype from elderly people exhibit defective differentiation into effector cells after antigen stimulation, along with other functional impairments, such as reduced cytokine production and considerably restricted TCR repertoire (Ferrando-Martinez et al., 2011; Weiskopf, Weinberger, & Grubeck-Loebenstein, 2009). However, as noted above, the identification of what is truly a naïve T-cell subset may be problematic, as there is a great deal of plasticity in this respect (Fülop et al., 2017; Larbi & Fulop, 2014). Current evidence also suggests that epigenetic age-related alterations (such as methylation of cytokine gene promoters) can also contribute to disturbed immune T-cell function (Calvanese, Lara, Kahn, & Fraga, 2009; Gonzalo, 2010; Jasiulionis, 2018). Altered DNA methylation profiles have been demonstrated to be responsible for age-associated changes in immune-specific genes (such as surface molecules or the cytokines IL-2 and IFN-γ) as well as in genes involved in general cell homeostasis pathways (Fernandez-Morera et al., 2010; Jasiulionis, 2018; Shanley et al., 2009).

Interestingly, the bone marrow may act not only as the source of hematopoiesis, but represents also a long-term survival environment for antigen-experienced T cells and for the precursors of plasma B cells. Age-related impairments in the BM niche itself may lead to the accumulation of highly differentiated, senescent-like T cells in the BM. These in turn cause inflammation in the BM, inducing oxidative stress and senescence, and promote age-related changes in T- and B-cell phenotypes (Naismith & Pangrazzi, 2019).

B-cell function

As alluded to above, aged bone marrow produces less pro-B lymphocytes, and these have a reduced ability to differentiate into pre-B lymphocytes. Immunoglobulin diversity and affinity are reduced in the elderly, because of impaired somatic hypermutation inside the germinal center, which is the main site of B-cell proliferation and maturation (Ma, Wang, Mao, & Hao, 2019; Müller & Pawelec, 2014; Müller et al., 2019). B-cell function may be diminished as a consequence of a lack of optimal help from aged T cells, but intrinsic changes in B cells also occur and have a significant impact on antibody production (Frasca, Diaz, Romero, Landin, & Blomberg, 2011). As with T cells, there are also shifts in the proportions of naïve and antigen-experienced peripheral B-cell subpopulations. Antibody responses to both T-cell-dependent and T-cell-independent antigens are generally diminished with age, with wide variation between individuals, possibly reflecting underlying genetic, epigenetic, and environmental differences (Scholz, Diaz, Riley, Cancro, & Frasca, 2013; Swain & Blomberg, 2013).

Recently, a subset of B cells, so-called age-associated B cells (ABCs), has been discovered. This subset was initially isolated from elderly individuals and was found to be closely related to immunosenescence (Ma et al., 2019). The ABCs and ABC-like cells appear in the course of certain autoimmune diseases and following some viral infections (Jenks et al., 2018; Knox et al., 2017; Wang et al., 2018), indicating the possible feasibility of ABC-targeted therapeutic approaches (Ma et al., 2019).

Age-related deteriorations in adaptive immunity may also reflect changes in the repertoire and function of T and B cells, as well as changes in the repertoire of antigen receptor genes expressed by these cells, which are available to respond to antigenic challenges. Using high-throughput DNA sequencing, initial studies have now demonstrated age-associated alterations in the receptor expression within lymphocyte subsets, as well as in receptor repertoires responding to vaccination

(Boyd, Liu, Wang, Martin, & Dunn-Walters, 2013). In aged individuals the generation of long-term immunoglobulin-producing B lymphocytes is impaired, leading to decreased numbers of functional antibody-secreting B cells and consequentially, decreased titers of antigen-specific immunoglobulins. This results in a humoral immune response of shorter duration and generally decreased specificity in elderly people (Ademokun, Wu, & Dunn-Walters, 2010; Colonna-Romano et al., 2008). Aging-dependent alterations in epigenetic status might also be responsible for altered B-cell function in elderly population, due to modifications in the differentiation pathways and in regulation and rearrangement of BCR-Ig genes (Colonna-Romano et al., 2008; Shanley et al., 2009).

Clinical consequences of immunosenescence

Human aging is associated with many geriatric problems, the most important ones relating to infections, sarcopenia, and cancer. In general, high morbidity and mortality rates associated with disease in the elderly are common across a spectrum of pathogens. Probably associated with immunosenescence, the elderly population is more prone to infections, especially to influenza, pneumococci, respiratory syncytial virus, and streptococci, but also to opportunistic infections and re-emergent chronic viral infections, such as herpes zoster (John & Canaday, 2017). The most obvious example for this association can be the age-specific mortality rates in elderly related to influenza and pneumonia (Fulop, Larbi, & Pawelec, 2013b). These circumstantial data imply that both immune memory and the repertoire for response to new challenges are compromised in the elderly.

Aging is characterized by a chronic state of low-grade inflammation, at least some mediators of which are likely to be produced by cells of the immune system. Such "inflammaging" (Franceschi et al., 2000) may also have detrimental consequences for the development of other pathophysiological mechanisms underlying sarcopenia, risk of cardiovascular events, and tumorigenesis (Fulop, Larbi, Kotb, & Pawelec, 2013a), a further reflection of the "immunologic theory of aging" (Walford, 1964). High levels of such interleukins as IL-6, IL-1, TNF, and C-reactive protein are associated in elderly individuals with increased risk of morbidity and mortality, a manifestation of the "inflammaging" which appears to be a universal age-associated phenomenon, associated with frailty, morbidity, and mortality in aged individuals (Jenny, 2012).

As a result of the aging process some individuals accumulate dysfunctional characteristics in various systems and organs that are jointly defined as the frailty syndrome. Frailty is characterized by a state of low tolerance or predisposition to functional decline with high potential impact on morbidity and mortality (Rutenberg, Mitnitski, Farrell, & Rockwood, 2018). In other words, frailty may be considered as a kind of overall exhaustion of the reserves of the organism, as an individual's inability to combat biological or psychological stress without loss of function decreases (Freitas, Carvalho, Moura, & Sposito, 2012). In particular, cohort studies have indicated TNF and IL-6 levels as markers of frailty (Michaud et al., 2013). The study findings with 866 participants support an association between markers of inflammaging, such as systemic IL-6 levels, and low muscle mass and strength, which characterized sarcopenia—an important cause of morbidity and mortality in older adults (Kilgour et al., 2013).

Infection with the persistent herpesvirus HHV5 (cytomegalovirus, CMV), possibly increasing basal levels of inflammatory mediators, has also been shown to be associated with frailty and reduced survival in several studies, such as within a cohort of 511 individuals aged 65 years and older, followed up for 18 years. CMV infection has also been associated with a near doubling of cardiovascular deaths in healthy older individuals (Savva et al., 2013). Intriguingly, other herpes viruses do not have these effects. For example, in a prospective cohort study CMV infection was associated with frailty, while high IgG titers to HSV and EBV were associated with diabetes mellitus (Haeseker et al., 2013). There may, however be some similarities between the effects on immune status, inflammation, and frailty caused by chronic HIV infection (Tavenier, Margolick, & Leng, 2019).

Dysregulation of the inflammatory pathway may also affect the central nervous system and be involved in such neurodegenerative disorders as Parkinson disease, multiple sclerosis, major depression, and Alzheimer's disease (Chou & Effros, 2013; Daulatzai, 2016; Fulop et al., 2018; King et al., 2018; Le Page et al., 2018; Michaud et al., 2013; Muthuramalingam, Menon, Rajkumar, & Negi, 2016; Pedemonte et al., 2006) as well as playing a key role in bone loss (Redlich & Smolen, 2012). A correlational study implicated late-stage differentiated T cells in influencing the development and progression of Alzheimer's disease, as telomere length in T lymphocytes of these patients has been negatively associated with worsening disease status (Chou & Effros, 2013). Accumulations of late-stage differentiated T cells with short telomeres have also been recorded in patients with bipolar disorders, suggesting that immune alterations similar to those seen in chronological aging are also seen in many disease states without an obvious immunological etiology (Rizzo et al., 2013). Thus, the age-dependent dysregulation of immunity is associated with systemic effects on many organ systems, contributing to pathology and having important clinical consequences for elderly populations.

Effect of age on vaccination

Vaccination is the most efficient strategy for preventing infectious disease. The increased age-related susceptibility to infection of elderly individuals makes them a particularly important target population for vaccination. It is here that the clearest clinical consequences of immunosenescence are manifest, as most vaccines are thought to be less efficient in the elderly (Pawelec & McElhaney, 2018; Weinberger & Grubeck-Loebenstein, 2012; Weston, Friedland, Wu, & Howe, 2012). Nonetheless, vaccination might also help compensate for diminished immunity, and enhance protective immunity in the elderly. Even simple strategies such as higher antigen dose in the vaccine, as well as improved adjuvants and alternative routes of immunization may yield promising results (Andrew et al., 2019; Derhovanessian & Pawelec, 2012). More precisely defined conditions and mechanisms of such strategies will facilitate the design of better vaccines for the elderly to overcome immunosenescence and to ameliorate its consequences (Oviedo-Orta, Li, & Rappuoli, 2013; Pawelec & McElhaney, 2018). One key strategy could be to plan vaccinations for future protection of the elderly already in middle age (before the onset of immune senescence), while they still have sufficient naïve T-cell function for effective immune response and are able to produce memory T cells. This strategy is not applicable to influenza vaccines, which require new formulations each year. Possible strategies for improving results of vaccination might also be aimed at enhancing hematopoiesis and, consequently, T- and B-cell development. In this way, newly developed lymphocytes would be present at the time of vaccine administration resulting in maximizing of the naïve T- and B-cell responses (Ventevogel & Sempowski, 2013).

Prediction of vaccine responsiveness is especially difficult in older individuals, who exhibit high levels of variation compared to the young in many components of immunity and are often less responsive to vaccination (Shen-Orr & Furman, 2013). It is necessary to consider that for assessment of vaccine efficacy, in addition to antibody titers also the cellular immune response may be important, especially in the elderly population (McElhaney et al., 2006; Prelog, 2013).

A T-cell-inducing vaccine may therefore improve vaccination outcome in the elderly (Antrobus et al., 2012). However, in many instances, we are not sure just how well or how poorly vaccines really do work in the elderly. While vaccination against herpes zoster and pertussis has been demonstrated to improve quality of life, vaccination against pneumococcal disease and hepatitis appears to prevent mortality. It has been calculated that only vaccination against herpes zoster and pneumococcal disease seem to be cost-effective (Eilers et al., 2013). Four vaccines are now recommended in many countries for people over 60 years of age and for protecting them from influenza, pneumococcal, tetanus, and pertussis infections and for prevention of herpes zoster reactivation. Of the four vaccines, only the antitetanus/pertussis vaccine shows a satisfactory protective antibody response in the elderly (Weston et al., 2012), whereas vaccines against influenza and pneumococcal diseases may alleviate disease burden to some degree, often not inducing protective immunity in most elderly people (Andrew et al., 2019; Van Buynder & Booy, 2018). Until recently, vaccination against varicella zoster virus was able to only partly prevent reactivation of herpes zoster or slightly diminish the intensity of post-herpetic neuralgia (Goronzy & Weyand, 2013; Oxman et al., 2005), whereas a newly developed vaccine seems to be more effective (McElhaney, Verschoor, & Pawelec, 2019). Herpes zoster is a disease caused by reactivation of latent varicella zoster virus and affecting mainly the elderly population (John & Canaday, 2017). Like some other herpes viruses, the genome of this virus is latently present in the sensory ganglia and can be reactivated upon age-related decline of cell-mediated immunity (Levin, 2012). To achieve protection, the previously used vaccine for the elderly applied approximately 14 times higher doses, which were needed to compensate for the decrease in cell-mediated immunity observed in the aged (Amanna, 2012; Gagliardi, Gomes Silva, Torloni, & Soares, 2012). In contrast, the recently developed new vaccines and adjuvants are now more effective so that unprecedented success rates of over 90% have been achieved (McElhaney et al., 2019).

Additional approaches to promote Th1 responses and to enhance immune functions in the elderly should be undertaken, such as conjugation to carrier proteins, concomitant vaccination with other vaccines, and change in antigen concentrations or dose intervals (McElhaney et al., 2019; Oviedo-Orta et al., 2013). Strategies aimed at optimization of vaccines in targeting the aged population have to consider also their socio-psychological conditions. For example, it is known that depressed individuals have diminished cell-mediated responses to zoster vaccine, and treatment with antidepressant medication was associated with normalization of these responses (Irwin et al., 2013). Finally, a further obstacle complicating this issue is that some vaccines themselves may represent a substantial safety risk in elderly individuals. Despite the fact that any healthcare intervention (especially prophylactic ones) must ideally counterbalance risk and advantage of such intervention, safety and immunogenicity are often only poorly characterized in older populations (Amanna, 2012).

Thus, new vaccination strategies should not only include rational antigen and adjuvant designs, improved delivery systems and immunization schedules, and alternative expression technologies, but also consider specific psychological and medical conditions of the elderly population (Li, Rappuoli, & Xu, 2013; Thomas-Crusells, McElhaney, & Aguado, 2012). They should reverse memory T-cell exhaustion, as well as decrease inhibitory effects of regulatory T cells and myeloid-derived suppressor cells (MDSCs) to be effective in older adults (McElhaney et al., 2019).

Immune senescence and all-cause mortality

Age-related changes in the immune system and low-grade inflammation were shown to contribute to the risk of cardiovascular diseases, which play a major role in the morbidity and mortality of old individuals (Fülop, Witkowski, Olivieri, & Larbi, 2018). Age-associated dysregulation of inflammatory pathways also appears to affect the central nervous system and be involved in the pathophysiological mechanisms of neurodegenerative disorders, such as Alzheimer's disease (King et al., 2018; Michaud et al., 2013).

As aging is associated with dysregulated immunity, which appears to contribute to poor vaccination responsiveness and to increased morbidity and all-cause mortality, it would be of great clinical as well academic interest to identify immune correlates that could be causally related with response to vaccination and mortality (Pawelec, 2018b, c). This represents an enormous challenge, as it will be essential to integrate assessments of immune parameters into on-going and planned longitudinal studies, in order to identify the mechanisms and biomarkers of immunosenescence and possible intervention to restore appropriate immunity (Derhovanessian, Larbi, & Pawelec, 2009). Many longitudinal studies are currently in progress, but unfortunately only few include immunology or any type of biomarker analysis.

One series of studies of the very elderly identified a simple cluster of immunological parameters predictive of 2-, 4-, and 6-year mortality at follow-up (Wikby et al., 2005). The main factors were related to an accumulation of late-differentiated CD8 cells (resulting in an inverted CD4:CD8 ratio), high levels of $CD8^+$ $CD28^-$ T cells as well as poor proliferative responses. Additionally, lower numbers of B cells and infection with CMV were also part of this so-called "immune risk profile," or IRP as defined in these Swedish OCTO/NONA studies (Pawelec, Derhovanessian, Larbi, Strindhall, & Wikby, 2009). Although in these studies there were no data on the outcome of vaccination, many independent reports have suggested that some of these IRP biomarkers, such as age-related accumulation of $CD8^+$ $CD28^-$ T cells as well as CMV seropositivity correlate with poor responsiveness to vaccination (Derhovanessian & Pawelec, 2012). The decreased vaccine response in the elderly represents an important clinical consequence of such clonally expanded pools of CMV-specific $CD8^+$ T cells with the concomitant decrease in the naïve T-cell diversity. This means that individuals having a large number of $CD8^+$ $CD28^-$ T cells produce much less protective antibody following influenza vaccination than those having less of these cells (McElhaney et al., 2012), possibly due to decline in the cytolytic capacity of $CD8^+$ T cells responsible for clearing influenza virus from infected cells. Therefore an ideal vaccine should be capable of eliciting of long-lasting T- and B-cell-mediated immunity (Pawelec & McElhaney, 2018).

A negative influence of the population of $CD4^+$ $CD28^-$ T cells has been demonstrated in multiple age-related diseases. These effector memory T cells with cytotoxic activity have been recently described as having pathogenic potential (Lee & Lee, 2016). They were found to infiltrate target tissues of patients with multiple sclerosis, rheumatoid arthritis, myopathies, acute coronary syndromes, stable angina, atherosclerosis, coronary artery disease, inflammatory bowel disease, and viral hepatic infections (Broadley, Pera, Morrow, Davies, & Kern, 2017; Broux, Markovic-Plese, Stinissen, & Hellings, 2012).

Repetitive antigen challenge by persistent viruses and tumor antigens is among the prominent causal agents of T-cell senescence or exhaustion, often associated with CMV infection (Chou & Effros, 2013; Lang, Govind, & Aspinall, 2013; Pawelec et al., 2009). The most dramatic functional changes are those which occur only in CMV-infected people (although this is the majority or all of the elderly in most studies). These affect the $CD4^+$ subset to some degree, but are most noticeable in the $CD8^+$ T-cell subset, where nearly 25% of the total CD8 T cells may be specific for a single CMV-immunodominant epitope (Larbi & Fulop, 2014; Pawelec et al., 2009). Most of these CMV-specific $CD8^+$ T cells have decreased CD28, CD27, CCR7, CD62L, increased CD57, and reexpressed CD45RA surface markers (Wills et al., 2011).

This process of the CMV-driven expansion of CD8-positive T cells, accompanied by the loss of the costimulatory CD28 molecule, is considered to be a key predictor of immune incompetence in elderly individuals (Fulop et al., 2013b). The age-dependent accumulation of exhausted memory T cells, releasing the proinflammatory cytokines TNF and IFN-γ, together with components and factors of the innate immune system, is thought to contribute to the low-grade inflammation commonly observed in elderly people (Jenny, 2012). Accumulation of such large numbers of late-stage CD8 T cells characterizing the cluster of immune parameters defining the IRP in the OCTO/NONA studies was not seen in people uninfected with CMV, even when they were infected with other persistent herpes viruses (Derhovanessian et al., 2011).

Certain other exposures may drive similar phenotypes, as possibly in the case of HIV infection where the proportion of senescent CD8 T cells increases progressively with age and often consists of oligoclonal populations (Deeks, Verdin, & McCune, 2012). What is of particular interest is that even in individuals with HIV disease, a major fraction of late-differentiated, senescent, memory CD8 T cells are nonetheless specific for CMV antigens (Dock & Effros, 2011; Effros, 2016). The notion therefore emerges that these biomarkers of "senescence" are actually indicators of an adaptive response required to keep persistent infections in check. Some direct evidence for this idea comes from a study in which it was demonstrated that very elderly people with the highest levels of late-stage differentiated CMV-specific proinflammatory CD8 + T cells showed superior survival over 8 years' follow-up (Derhovanessian et al., 2013). This suggests that in the terminal phases of life, the highest priority of the immune system is to control chronic infections, regardless of the collateral damage this may cause.

Interventions to restore appropriate immunity

Having identified differences between young and old populations and zeroed in on those thought to be deleterious in the elderly, there is always the temptation to intervene to return the "senescent" system to the superior "young" one. Here, one must be completely convinced that any radical interventions would indeed be beneficial. Certain "common sense" interventions resulting in a general improvement of health also include, unsurprisingly, an immune improvement. Thus, interventions of this type for restoration of appropriate immunity in elderly populations, such as improved vaccination regimens, nutritional interventions, and physical activity programs, are likely to be safe, appear to be cost-effective, and should be rapidly implementable (Di Benedetto, Müller, Wenger, Duzel, & Pawelec, 2017; Fülop et al., 2017; Müller & Pawelec, 2014).

For more specific and likely more effective interventions in humans it is necessary to know which surrogate biomarker to monitor as an indication of success—mortality as a very robust and unequivocal clinical outcome is nonetheless not very informative (Müller & Pawelec, 2014). In the case of attempts to rejuvenate the thymus, for example by engineering re-expression of FOXN-1 (Zook et al., 2011), the output of naïve T cells is the obvious marker. Approaches could include androgen blockade as mentioned above, and also modulating inflammation to reduce systemic levels of factors like IL-6, Oncostatin M, TNF, and others which are implicated in thymic atrophy (Ferrando-Martinez et al., 2009; Sempowski et al., 2000). As thymic architecture is modified toward fatty tissue infiltration in old thymi, agents blocking adipogenesis could also be employed (Ross et al., 2000). All these systemic interventions would affect multiple different organ systems, in some cases perhaps also beneficially, but experimentation in the elderly would need to be pursued with extreme caution.

As the age-related clonal expansions of CD8 memory cells may crowd the available "immunological space" to reduce potential response to new strains of pathogens, to vaccines and to tumors (although perhaps not as much as previously believed—at least in mice; Vezys et al., 2009), preventing this might be beneficial. However, these cells may be there for a reason and their removal could do more harm than good. For example, one study found that the presence of such "senescent" cells was associated with improved survival in the very elderly (Derhovanessian et al., 2013). A less dangerous approach that could prevent repeated antigenic stimulation and the accumulation of exhausted memory T cells might be the childhood vaccination against persistent viruses, especially CMV. One could consider long-term antiviral treatment for those already infected, which might prevent reactivation of persistent CMV infection; but it is unlikely that antiviral therapy would completely eradicate persistent viruses and it is possible that long-term treatment with antiviral agents would be associated with serious side effects (Brunner, Herndler-Brandstetter, Weinberger, & Grubeck-Loebenstein, 2011). On the other hand, new approaches to the elimination of persistent viruses like CMV may be available sooner than we think (Weekes et al., 2013).

A great deal of interest is currently centered on the delicate balance of microbial species in the gut required to maintain healthy metabolism and adequate immune function, especially in elderly people. Disturbance in these homeostatic conditions can have negative consequences for the elderly, leading to increased infections and elevated inflammation, and thus, contributing to cancer, diabetes, and frailty (Nagpal et al., 2018). Therefore restoration of the microbial balance in the elderly by applying prebiotic and probiotic strategies may be an important intervention for improving and maintaining health and appropriate immunity (Mishra et al., 2019; Molina-Tijeras, Galvez, & Rodriguez-Cabezas, 2019). "Nutritional therapy" may be considered as the consumption of a probiotic drink (containing strains of *Lactobacillus casei*), which, after just 4 weeks, has demonstrated positive effects on immune function with improved NK-cell activity and restored antiinflammatory cytokine profile in a healthy nonimmunocompromised older group of individuals (Dong, Rowland, Thomas, & Yaqoob, 2013). Although changes in diversity of microbial flora may be an indicator of aging, there are few in-depth studies that have investigated the composition of the gut microbiota and its role in health and disease onset in the elderly (Claesson et al., 2012). Future progress requires the combinations of gut microbiology, gastroenterology, and epidemiology, together with developments in the rapid analysis of metabolites and microbial markers as well as detection of molecular signals and their pathways, in order to develop successful interventions (Duncan and Flint, 2013).

Physical activity and moderate exercise as possible intervention strategies for restoring appropriate immunity have also been investigated over the years. Unfortunately data gathered from the literature concerning the effects of physical activity on immunosenescence are still limited and conflicting, with existing reports either advocating benefits or asserting a lack of evidence. These conflicting data may reflect the heterogeneity of study protocols (with different types, intensities, and program lengths of exercises), as well the immunological parameters assayed (Walsh et al., 2011). Nevertheless, exercise as part of a healthy lifestyle clearly provides long-term benefits with regard to cardiovascular, cognitive, psychosocial, and other aspects of the elderly (de Araujo, Silva, Fernandes, & Benard, 2013; Di Benedetto et al., 2017). The positive impact of exercise on the immune system has been reported by several investigators (Gleeson et al., 2011; Hong, 2011; Irwin, Pike, Cole, & Oxman, 2003; Pedersen, 2011; Simpson et al., 2012; Spielmann et al., 2011; Walsh et al., 2011). Acute moderate aerobic exercise immediately prior to influenza vaccination has been reported to increase the efficacy of the vaccine in older women (Ranadive et al., 2014). The results of a randomized, controlled trial demonstrated that Tai Chi alone induces an increase in VZV-specific cell-mediated immunity. This increase was comparable in magnitude with that induced by varicella vaccine, and the two were additive (Irwin, Olmstead, & Oxman, 2007).

Interventions likely to be more effective are also likely to be more dangerous. For example, it was reported that the elderly may possess more immature myeloid cells (with the ability to suppress immune responses). These are similar to the MDSCs shown to be crucial in tumor immunity (Weide et al., 2013). Thus, the age-related elevation in the frequency of MDSCs may be associated with increased incidence of these diseases in elderly people (Gorgun et al., 2013; Verschoor et al., 2013). The temptation here would be to deplete these cells to restore function, but this could also result in autoimmunity. Similarly, alterations of the TH17:Treg ratios together with age-associated changes in cytokine secretion may lead to an imbalance between the proinflammatory and antiinflammatory immune responses, contributing to a higher susceptibility to develop inflammatory diseases with increasing age (Schmitt, Rink, & Uciechowski, 2013). Addressing imbalance in interactive systems as complex as the immune system will be problematic.

However, an entirely new approach, also derived from the enormous body of work in clinical cancer research may hold out hope that manipulating systems with so much inertia as the immune system can yield impressive results. This approach employs immunomodulatory monoclonal antibodies to specifically interfere with cell–cell interactions, as reported outside of the cancer context in the case of HIV (Macatangay & Rinaldo, 2009). Therefore further studies are required in well-characterized diverse human populations, with adequate follow-up: to better understand the nature of aging defects and to allow some mechanistic insights into causative processes in human aging; to prove the relevance of the biomarkers studied and, finally, to develop safe interventions for restoration or at least maintenance of appropriate immunity into extreme old age.

Perspectives

It is an intriguing question as to why invertebrates are able to protect themselves perfectly well using exclusively innate immunity, whereas all vertebrates invest considerable resources in maintaining additional complex and potentially dangerous adaptive immune systems. One possible explanation for this dichotomy might be that vertebrates have co-opted a much larger range of gut bacteria than invertebrates (in order to increase the efficacy of nutrient processing) (Lee & Mazmanian, 2010). Because innate immunity recognizes both useful and pathogenic microorganisms equally well, it can be hypothesized that adaptive immunity arose to help to distinguish between the two, by regulating innate immunity and targeting specific antigens derived from pathogens but not from symbionts (Muller et al., 2013; Pawelec, 2018a). Thus, symbiotic microorganisms may have influenced both the features of adaptive immune system evolution as well as their functions more profoundly than pathogens, in order to protect not only the host from pathogens but also the beneficial microbiota from destruction by innate immunity (Fagundes, Souza, Nicoli, & Teixeira, 2011; Lazar et al., 2018; Lazar et al., 2019; Lee & Mazmanian, 2010; Macpherson, Geuking, & McCoy, 2012). This hypothesis assigns a greater deal of importance to immune–microbiome interactions than hitherto studied in the context of aging and immunosenescence.

There are currently many other intriguing questions concerning the impact of age versus exposures on immune function and dysfunction, starting with stem cell aging in the bone marrow and how to remedy deficits thereof, and whether there are any "intrinsic" aging processes affecting all tissues in aging individuals. Some simple preliminary data comparing T-cell immune measures with fibroblasts suggest the unsurprising conclusion that organismal aging processes have little obvious impact on T-cell immunosenescence (Waaijer et al., 2019). A most important lesson at the moment is that the very marked heterogeneity between human populations and individuals within populations emphasizes the need for personalized assessments of immune function in order to identify where and how interventions might be targeted. Recent technological developments, driven particularly by efforts focused on cancer rather than aging [e.g., identifying cancer neoantigens and targeting them with individually tailored vaccines or adoptive immunotherapy, as illustrated in the case studies of Hilf et al. (2019)], encourage the belief that the application of targeted genetic manipulations could eventually be applied in less desperate circumstances. Thus, one could envisage individual T- and B-cell receptor repertoire analyses undertaken to identify holes in the repertoire for neoantigens likely to be important for that specific individual. Receptors recognizing these neoantigens could then be cloned into functional immune cells and reintroduced into the person by techniques being developed for cancer therapy. This approach would directly reconstitute a necessary repertoire for antigen recognition, bypassing HSC problems, thymic involution or deficient B-cell development in the BM, problems with antigen presentation due to defective APC or LN architecture, and immune effector cell differentiation blockade, and thus solve one of the main problems of immunosenescence. At the moment, this is practically beyond our reach, but is technically feasible and currently being actively applied in cancer. As in the latter, many other hurdles would remain to be overcome, but major advances in this respect seem likely in the near future.

References

Adema, G. J. (2009). Dendritic cells from bench to bedside and back. *Immunology Letters, 122*, 128–130.

Ademokun, A., Wu, Y. C., & Dunn-Walters, D. (2010). The ageing B cell population: Composition and function. *Biogerontology, 11*, 125–137.

Adriaensen, W., Pawelec, G., Vaes, B., Hamprecht, K., Derhovanessian, E., van Pottelbergh, G., ... Mathei, C. (2017). CD4:8 ratio above 5 is associated with all-cause mortality in CMV-seronegative very old women: Results from the BELFRAIL study. *The Journals of Gerontology. Series A Biological Sciences and Medical Sciences, 72*, 1155–1162.

Agrawal, A., Agrawal, S., & Gupta, S. (2017). Role of dendritic cells in inflammation and loss of tolerance in the elderly. *Frontiers in Immunology, 8*, 896.

Agrawal, A., & Gupta, S. (2011). Impact of aging on dendritic cell functions in humans. *Ageing Research Reviews, 10*, 336–345.

Alam, I., Goldeck, D., Larbi, A., & Pawelec, G. (2013). Aging affects the proportions of T and B cells in a group of elderly men in a developing country--a pilot study from Pakistan. *Age, 35*, 1521–1530.

Alvarez-Rodriguez, L., Lopez-Hoyos, M., Garcia-Unzueta, M., Amado, J. A., Cacho, P. M., & Martinez-Taboada, V. M. (2012). Age and low levels of circulating vitamin D are associated with impaired innate immune function. *Journal of Leukocyte Biology, 91*, 829–838.

Amanna, I. J. (2012). Balancing the efficacy and safety of vaccines in the elderly. *Open Longevity Science, 6*, 64–72.

Andrew, M. K., Bowles, S. K., Pawelec, G., Haynes, L., Kuchel, G. A., McNeil, S. A., & McElhaney, J. E. (2019). Influenza vaccination in older adults: Recent innovations and practical applications. *Drugs & Aging, 36*, 29–37.

Antrobus, R. D., Lillie, P. J., Berthoud, T. K., Spencer, A. J., McLaren, J. E., Ladell, K., ... Gilbert, S. C. (2012). A T cell-inducing influenza vaccine for the elderly: Safety and immunogenicity of MVA-NP + M1 in adults aged over 50 years. *PLoS One, 7*, e48322.

Beerman, I., Bhattacharya, D., Zandi, S., Sigvardsson, M., Weissman, I. L., Bryder, D., & Rossi, D. J. (2010a). Functionally distinct hematopoietic stem cells modulate hematopoietic lineage potential during aging by a mechanism of clonal expansion. *Proceedings of the National Academy of Sciences of the United States of America, 107*, 5465–5470.

Beerman, I., Maloney, W. J., Weissmann, I. L., & Rossi, D. J. (2010b). Stem cells and the aging hematopoietic system. *Current Opinion in Immunology, 22*, 500–506.

Boehm, T., Hirano, M., Holland, S. J., Das, S., Schorpp, M., & Cooper, M. D. (2018). Evolution of alternative adaptive immune systems in vertebrates. *Annual Review of Immunology, 36*, 19–42.

Bonilla, F. A., & Oettgen, H. C. (2010). Adaptive immunity. *The Journal of Allergy and Clinical Immunology, 125*, S33–S40.

Boyd, S. D., Liu, Y., Wang, C., Martin, V., & Dunn-Walters, D. K. (2013). Human lymphocyte repertoires in ageing. *Current Opinion in Immunology, 25*, 511–515.

Broadley, I., Pera, A., Morrow, G., Davies, K. A., & Kern, F. (2017). Expansions of cytotoxic CD4(+)CD28(-) T cells drive excess cardiovascular mortality in rheumatoid arthritis and other chronic inflammatory conditions and are triggered by CMV infection. *Frontiers in Immunology, 8*, 195.

Broux, B., Markovic-Plese, S., Stinissen, P., & Hellings, N. (2012). Pathogenic features of CD4 + CD28- T cells in immune disorders. *Trends in Molecular Medicine, 18*, 446–453.

Brunner, S., Herndler-Brandstetter, D., Weinberger, B., & Grubeck-Loebenstein, B. (2011). Persistent viral infections and immune aging. *Ageing Research Reviews, 10*, 362–369.

Calvanese, V., Lara, E., Kahn, A., & Fraga, M. F. (2009). The role of epigenetics in aging and age-related diseases. *Ageing Research Reviews, 8*, 268–276.

Caruso, C., Accardi, G., Virruso, C., & Candore, G. (2013). Sex, gender and immunosenescence: A key to understand the different lifespan between men and women? *Immunity & Ageing, 10*, 20.

Chen, G., Zhuchenko, O., & Kuspa, A. (2007). Immune-like phagocyte activity in the social amoeba. *Science, 317*, 678–681.

Chou, J. P., & Effros, R. B. (2013). T cell replicative senescence in human aging. *Current Pharmaceutical Design, 19*, 1680–1698.

Claesson, M. J., Jeffery, I. B., Conde, S., Power, S. E., O'Connor, E. M., Cusack, S., ... O'Toole, P. W. (2012). Gut microbiota composition correlates with diet and health in the elderly. *Nature, 488*, 178–184.

Clements, S. J., Maijo, M., Ivory, K., Nicoletti, C., & Carding, S. R. (2017). Age-associated decline in dendritic cell function and the impact of mediterranean diet intervention in elderly subjects. *Frontiers in Nutrition*, *4*, 65.

Colonna-Romano, G., Bulati, M., Aquino, A., Vitello, S., Lio, D., Candore, G., & Caruso, C. (2008). B cell immunosenescence in the elderly and in centenarians. *Rejuvenation Research*, *11*, 433–439.

Conboy, M. J., Conboy, I. M., & Rando, T. A. (2013). Heterochronic parabiosis: Historical perspective and methodological considerations for studies of aging and longevity. *Aging Cell*, *12*, 525–530.

Cooper, M. D., & Herrin, B. R. (2010). How did our complex immune system evolve? *Nature Reviews Immunology*, *10*, 2–3.

Crisan, T. O., Netea, M. G., & Joosten, L. A. (2016). Innate immune memory: Implications for host responses to damage-associated molecular patterns. *European Journal of Immunology*, *46*, 817–828.

Cunningham, C. A., Helm, E. Y., & Fink, P. J. (2018). Reinterpreting recent thymic emigrant function: Defective or adaptive? *Current Opinion in Immunology*, *51*, 1–6.

Danilova, N. (2013). *Evolution of the human immune system*. Chichester: John Wiley & Sons Ltd.

Daulatzai, M. A. (2016). Fundamental role of pan-inflammation and oxidative-nitrosative pathways in neuropathogenesis of Alzheimer's disease. *American Journal of Neurodegenerative Disease*, *5*, 1–28.

de Araujo, A. L., Silva, L. C., Fernandes, J. R., & Benard, G. (2013). Preventing or reversing immunosenescence: Can exercise be an immunotherapy? *Immunotherapy*, *5*, 879–893.

Deeks, S. G., Verdin, E., & McCune, J. M. (2012). Immunosenescence and HIV. *Current Opinion in Immunology*, *24*, 501–506.

Del Giudice, G., Rappuoli, R., & Didierlaurent, A. M. (2018). Correlates of adjuvanticity: A review on adjuvants in licensed vaccines. *Seminars in Immunology*, *39*, 14–21.

Della Bella, S., Bierti, L., Presicce, P., Arienti, R., Valenti, M., Saresella, M., ... Villa, M. L. (2007). Peripheral blood dendritic cells and monocytes are differently regulated in the elderly. *Clinical Immunology*, *122*, 220–228.

Derhovanessian, E., Larbi, A., & Pawelec, G. (2009). Biomarkers of human immunosenescence: Impact of Cytomegalovirus infection. *Current Opinion in Immunology*, *21*, 440–445.

Derhovanessian, E., Maier, A. B., Hahnel, K., Beck, R., de Craen, A. J., Slagboom, E. P., ... Pawelec, G. (2011). Infection with cytomegalovirus but not herpes simplex virus induces the accumulation of late-differentiated CD4+ and CD8+ T-cells in humans. *The Journal of General Virology*, *92*, 2746–2756.

Derhovanessian, E., Maier, A. B., Hahnel, K., Zelba, H., de Craen, A. J., Roelofs, H., ... Pawelec, G. (2013). Lower proportion of naive peripheral CD8+ T cells and an unopposed pro-inflammatory response to human Cytomegalovirus proteins in vitro are associated with longer survival in very elderly people. *Age*, *35*, 1387–1399.

Derhovanessian, E., & Pawelec, G. (2012). Vaccination in the elderly. *Microbial Biotechnology*, *5*, 226–232.

Derhovanessian, E., Solana, R., Larbi, A., & Pawelec, G. (2008). Immunity, ageing and cancer. *Immunity & Ageing*, *5*, 11.

Di Benedetto, S., Gaetjen, M., & Muller, L. (2019). The modulatory effect of gender and cytomegalovirus-seropositivity on circulating inflammatory factors and cognitive performance in elderly individuals. *International Journal of Molecular Sciences*, *20*(4), 990.

Di Benedetto, S., Müller, L., Wenger, E., Duzel, S., & Pawelec, G. (2017). Contribution of neuroinflammation and immunity to brain aging and the mitigating effects of physical and cognitive interventions. *Neuroscience and Biobehavioral Reviews*, *75*, 114–128.

Dock, J. N., & Effros, R. B. (2011). Role of CD8 T cell replicative senescence in human aging and in HIV-mediated immunosenescence. *Aging and Disease*, *2*, 382–397.

Dong, H., Rowland, I., Thomas, L. V., & Yaqoob, P. (2013). Immunomodulatory effects of a probiotic drink containing *Lactobacillus casei* Shirota in healthy older volunteers. *European Journal of Nutrition*, *52*, 1853–1863.

Douek, D. C., McFarland, R. D., Keiser, P. H., Gage, E. A., Massey, J. M., Haynes, B. F., ... Koup, R. A. (1998). Changes in thymic function with age and during the treatment of HIV infection. *Nature*, *396*, 690–695.

Duncan, S. H., & Flint, H. J. (2013). Probiotics and prebiotics and health in ageing populations. *Maturitas*, *75*, 44–50.

Effros, R. B. (2016). The silent war of CMV in aging and HIV infection. *Mechanisms of Ageing and Development*, *158*, 46–52.

Eilers, R., Krabbe, P. F., van Essen, T. G., Suijkerbuijk, A., van Lier, A., & de Melker, H. E. (2013). Assessment of vaccine candidates for persons aged 50 and older: A review. *BMC Geriatrics*, *13*, 32.

Fagundes, C. T., Souza, D. G., Nicoli, J. R., & Teixeira, M. M. (2011). Control of host inflammatory responsiveness by indigenous microbiota reveals an adaptive component of the innate immune system. *Microbes and Infection*, *13*, 1121–1132.

Fernandez-Morera, J. L., Calvanese, V., Rodriguez-Rodero, S., Menendez-Torre, E., & Fraga, M. F. (2010). Epigenetic regulation of the immune system in health and disease. *Tissue Antigens*, *76*, 431–439.

Ferrando-Martinez, S., Franco, J. M., Hernandez, A., Ordonez, A., Gutierrez, E., Abad, A., & Leal, M. (2009). Thymopoiesis in elderly human is associated with systemic inflammatory status. *Age*, *31*, 87–97.

Ferrando-Martinez, S., Ruiz-Mateos, E., Hernandez, A., Gutierrez, E., Rodriguez-Mendez Mdel, M., Ordonez, A., & Leal, M. (2011). Age-related deregulation of naive T cell homeostasis in elderly humans. *Age*, *33*, 197–207.

Florian, M. C., Klose, M., Sacma, M., Jablanovic, J., Knudson, L., Nattamai, K. J., ... Geiger, H. (2018). Aging alters the epigenetic asymmetry of HSC division. *PLoS Biology*, *16*, e2003389.

Franceschi, C., Bonafe, M., Valensin, S., Olivieri, F., De Luca, M., Ottaviani, E., & De Benedictis, G. (2000). Inflamm-aging. An evolutionary perspective on immunosenescence. *Annals of the New York Academy of Sciences*, *908*, 244–254.

Frasca, D., Diaz, A., Romero, M., Landin, A. M., & Blomberg, B. B. (2011). Age effects on B cells and humoral immunity in humans. *Ageing Research Reviews*, *10*, 330–335.

Freitas, W. M., Carvalho, L. S., Moura, F. A., & Sposito, A. C. (2012). Atherosclerotic disease in octogenarians: A challenge for science and clinical practice. *Atherosclerosis*, *225*, 281–289.

Fülop, T., Larbi, A., Dupuis, G., Le Page, A., Frost, E. H., Cohen, A. A., ... Franceschi, C. (2017). Immunosenescence and inflamm-aging as two sides of the same coin: Friends or foes? *Frontiers in Immunology*, *8*, 1960.

Fulop, T., Larbi, A., Kotb, R., & Pawelec, G. (2013a). Immunology of aging and cancer development. *Interdisciplinary Topics in Gerontology*, *38*, 38–48.

Fulop, T., Larbi, A., & Pawelec, G. (2013b). Human T cell aging and the impact of persistent viral infections. *Frontiers in Immunology*, *4*, 271.

Fulop, T., Witkowski, J. M., Bourgade, K., Khalil, A., Zerif, E., Larbi, A., ... Frost, E. H. (2018). Can an infection hypothesis explain the beta amyloid hypothesis of Alzheimer's disease? *Frontiers in Aging Neuroscience*, *10*, 224.

Fülop, T., Witkowski, J. M., Olivieri, F., & Larbi, A. (2018). The integration of inflammaging in age-related diseases. *Seminars in Immunology*, *40*, 17–35.

Gagliardi, A. M., Gomes Silva, B. N., Torloni, M. R., & Soares, B. G. (2012). Vaccines for preventing herpes zoster in older adults. *The Cochrane Database of Systematic Reviews*, *10*, CD008858.

Garcia Verdecia, B., Saavedra Hernandez, D., Lorenzo-Luaces, P., de Jesus Badia Alvarez, T., Leonard Rupale, I., Mazorra Herrera, Z., ... Lage Davila, A. (2013). Immunosenescence and gender: A study in healthy Cubans. *Immunity & Ageing*, *10*, 16.

Geiger, H., de Haan, G., & Florian, M. C. (2013). The ageing haematopoietic stem cell compartment. *Nature Reviews Immunology, 13*, 376–389.

Gleeson, M., Bishop, N. C., Stensel, D. J., Lindley, M. R., Mastana, S. S., & Nimmo, M. A. (2011). The anti-inflammatory effects of exercise: Mechanisms and implications for the prevention and treatment of disease. *Nature Reviews Immunology, 11*, 607–615.

Gonzalo, S. (2010). Epigenetic alterations in aging. *Journal of Applied Physiology, 109*, 586–597.

Gorgun, G. T., Whitehill, G., Anderson, J. L., Hideshima, T., Maguire, C., Laubach, J., ... Anderson, K. C. (2013). Tumor-promoting immune-suppressive myeloid-derived suppressor cells in the multiple myeloma microenvironment in humans. *Blood, 121*, 2975–2987.

Goronzy, J. J., & Weyand, C. M. (2012). Immune aging and autoimmunity. *Cellular and Molecular Life Sciences: CMLS, 69*, 1615–1623.

Goronzy, J. J., & Weyand, C. M. (2013). Understanding immunosenescence to improve responses to vaccines. *Nature Immunology, 14*, 428–436.

Grover, A., Sanjuan-Pla, A., Thongjuea, S., Carrelha, J., Giustacchini, A., Gambardella, A., ... Nerlov, C. (2016). Single-cell RNA sequencing reveals molecular and functional platelet bias of aged haematopoietic stem cells. *Nature Communications, 7*, 11075.

Gubbels Bupp, M. R., Potluri, T., Fink, A. L., & Klein, S. L. (2018). The confluence of sex hormones and aging on immunity. *Frontiers in Immunology, 9*, 1269.

Haeseker, M. B., Pijpers, E., Dukers-Muijrers, N. H., Nelemans, P., Hoebe, C. J., Bruggeman, C. A., ... Goossens, V. J. (2013). Association of cytomegalovirus and other pathogens with frailty and diabetes mellitus, but not with cardiovascular disease and mortality in psycho-geriatric patients; a prospective cohort study. *Immunity & Ageing, 10*, 30.

Haines, C. J., Giffon, T. D., Lu, L. S., Lu, X., Tessier-Lavigne, M., Ross, D. T., & Lewis, D. B. (2009). Human CD4+ T cell recent thymic emigrants are identified by protein tyrosine kinase 7 and have reduced immune function. *The Journal of Experimental Medicine, 206*, 275–285.

Hamann, L., Kupcinskas, J., Berrocal Almanza, L. C., Skieceviciene, J., Franke, A., Nothlings, U., & Schumann, R. R. (2015). Less functional variants of TLR-1/-6/-10 genes are associated with age. *Immunity & Ageing, 12*, 7.

Heng, T. S., Goldberg, G. L., Gray, D. H., Sutherland, J. S., Chidgey, A. P., & Boyd, R. L. (2005). Effects of castration on thymocyte development in two different models of thymic involution. *Journal of Immunology, 175*, 2982–2993.

Hilf, N., Kuttruff-Coqui, S., Frenzel, K., Bukur, V., Stevanovic, S., Gouttefangeas, C., ... Wick, W. (2019). Actively personalized vaccination trial for newly diagnosed glioblastoma. *Nature, 565*, 240–245.

Hirokawa, K., Utsuyama, M., Hayashi, Y., Kitagawa, M., Makinodan, T., & Fulop, T. (2013). Slower immune system aging in women versus men in the Japanese population. *Immunity & Ageing, 10*, 19.

Hong, S. (2011). Can we jog our way to a younger-looking immune system? *Brain, Behavior, and Immunity, 25*, 1519–1520.

Irwin, M. R., Levin, M. J., Laudenslager, M. L., Olmstead, R., Lucko, A., Lang, N., ... Oxman, M. N. (2013). Varicella zoster virus-specific immune responses to a herpes zoster vaccine in elderly recipients with major depression and the impact of antidepressant medications. *Clinical Infectious Diseases, 56*, 1085–1093.

Irwin, M. R., Olmstead, R., & Oxman, M. N. (2007). Augmenting immune responses to varicella zoster virus in older adults: A randomized, controlled trial of Tai Chi. *Journal of the American Geriatrics Society, 55*, 511–517.

Irwin, M. R., Pike, J. L., Cole, J. C., & Oxman, M. N. (2003). Effects of a behavioral intervention, Tai Chi Chih, on varicella-zoster virus specific immunity and health functioning in older adults. *Psychosomatic Medicine, 65*, 824–830.

Ito, K., Hirao, A., Arai, F., Takubo, K., Matsuoka, S., Miyamoto, K., ... Suda, T. (2006). Reactive oxygen species act through p38 MAPK to limit the lifespan of hematopoietic stem cells. *Nature Medicine, 12*, 446–451.

Jackson, S. E., Redeker, A., Arens, R., van Baarle, D., van den Berg, S. P. H., Benedict, C. A., ... Wills, M. R. (2017). CMV immune evasion and manipulation of the immune system with aging. *Geroscience, 39*, 273–291.

Janeway, C. A., Jr., & Medzhitov, R. (2002). Innate immune recognition. *Annual Review of Immunology, 20*, 197–216.

Jasiulionis, M. G. (2018). Abnormal epigenetic regulation of immune system during aging. *Frontiers in Immunology, 9*, 197.

Jean, C. M., Honarmand, S., Louie, J. K., & Glaser, C. A. (2007). Risk factors for West Nile virus neuroinvasive disease, California, 2005. *Emerging Infectious Diseases, 13*, 1918–1920.

Jenks, S. A., Cashman, K. S., Zumaquero, E., Marigorta, U. M., Patel, A. V., Wang, X., ... Sanz, I. (2018). Distinct effector B cells induced by unregulated toll-like receptor 7 contribute to pathogenic responses in systemic lupus erythematosus. *Immunity, 49*, 725–739, e726.

Jenny, N. S. (2012). Inflammation in aging: Cause, effect, or both? *Discovery Medicine, 13*, 451–460.

John, A. R., & Canaday, D. H. (2017). *Herpes zoster* in the older adult. *Infectious Disease Clinics of North America, 31*, 811–826.

Johnson, P. L., Yates, A. J., Goronzy, J. J., & Antia, R. (2012). Peripheral selection rather than thymic involution explains sudden contraction in naive CD4 T-cell diversity with age. *Proceedings of the National Academy of Sciences of the United States of America, 109*, 21432–21437.

Kilgour, A. H., Firth, C., Harrison, R., Moss, P., Bastin, M. E., Wardlaw, J. M., ... Starr, J. M. (2013). Seropositivity for CMV and IL-6 levels are associated with grip strength and muscle size in the elderly. *Immunity & Ageing, 10*, 33.

Kilpatrick, R. D., Rickabaugh, T., Hultin, L. E., Hultin, P., Hausner, M. A., Detels, R., ... Jamieson, B. D. (2008). Homeostasis of the naive CD4+ T cell compartment during aging. *Journal of Immunology, 180*, 1499–1507.

King, E., O'Brien, J. T., Donaghy, P., Morris, C., Barnett, N., Olsen, K., ... Thomas, A. J. (2018). Peripheral inflammation in prodromal Alzheimer's and Lewy body dementias. *Journal of Neurology, Neurosurgery, and Psychiatry, 89*, 339–345.

Knox, J. J., Buggert, M., Kardava, L., Seaton, K. E., Eller, M. A., Canaday, D. H., ... Betts, M. R. (2017). T-bet+ B cells are induced gp140 response. *JCI Insight, 2*(8), e92943.

Konieczny, J., & Arranz, L. (2018). Updates on old and weary haematopoiesis. *International Journal of Molecular Sciences, 19*(9), 2567.

Kovats, S. (2015). Estrogen receptors regulate innate immune cells and signaling pathways. *Cellular Immunology, 294*, 63–69.

Lang, P. O., Govind, S., & Aspinall, R. (2013). Reversing T cell immunosenescence: Why, who, and how. *Age, 35*, 609–620.

Larbi, A., Franceschi, C., Mazzatti, D., Solana, R., Wikby, A., & Pawelec, G. (2008). Aging of the immune system as a prognostic factor for human longevity. *Physiology, 23*, 64–74.

Larbi, A., & Fulop, T. (2014). From "truly Naive" to "exhausted senescent" T cells: When markers predict functionality. *Cytometry. Part A: The Journal of the International Society for Analytical Cytology, 85*, 25–35.

Larbi, A., Pawelec, G., Wong, S. C., Goldeck, D., Tai, J. J., & Fulop, T. (2011). Impact of age on T cell signaling: A general defect or specific alterations? *Ageing Research Reviews, 10*, 370–378.

Lazar, V., Ditu, L. M., Pircalabioru, G. G., Gheorghe, I., Curutiu, C., Holban, A. M., ... Chifiriuc, M. C. (2018). Aspects of gut microbiota and immune system interactions in infectious diseases, immunopathology, and cancer. *Frontiers in Immunology, 9*, 1830.

Lazar, V., Ditu, L. M., Pircalabioru, G. G., Picu, A., Petcu, L., Cucu, N., & Chifiriuc, M. C. (2019). Gut microbiota, host organism, and diet trialogue in diabetes and obesity. *Frontiers in Nutrition, 6*, 21.

Le Page, A., Dupuis, G., Frost, E. H., Larbi, A., Pawelec, G., Witkowski, J. M., & Fulop, T. (2018). Role of the peripheral innate immune system in the development of Alzheimer's disease. *Experimental Gerontology, 107*, 59–66.

Lee, G. H., & Lee, W. W. (2016). Unusual CD4(+)CD28(-) T cells and their pathogenic role in chronic inflammatory disorders. *Immune Network, 16*, 322–329.

Lee, Y. K., & Mazmanian, S. K. (2010). Has the microbiota played a critical role in the evolution of the adaptive immune system? *Science, 330*, 1768–1773.

Legoux, F. P., Lim, J. B., Cauley, A. W., Dikiy, S., Ertelt, J., Mariani, T. J., ... Moon, J. J. (2015). CD4 + T cell tolerance to tissue-restricted self antigens is mediated by antigen-specific regulatory T cells rather than deletion. *Immunity, 43*, 896–908.

Leins, H., Mulaw, M., Eiwen, K., Sakk, V., Liang, Y., Denkinger, M., ... Schirmbeck, R. (2018). Aged murine hematopoietic stem cells drive aging-associated immune remodeling. *Blood, 132*, 565–576.

Levin, M. J. (2012). Immune senescence and vaccines to prevent herpes zoster in older persons. *Current Opinion in Immunology, 24*, 494–500.

Li, C. K., Rappuoli, R., & Xu, X. N. (2013). Correlates of protection against influenza infection in humans--on the path to a universal vaccine? *Current Opinion in Immunology, 25*, 470–476.

Li, W. (2013). Phagocyte dysfunction, tissue aging and degeneration. *Ageing Research Reviews, 12*, 1005–1012.

Litman, G. W., Rast, J. P., & Fugmann, S. D. (2010). The origins of vertebrate adaptive immunity. *Nature Reviews. Immunology, 10*, 543–553.

Ma, S., Wang, C., Mao, X., & Hao, Y. (2019). B cell dysfunction associated with aging and autoimmune diseases. *Frontiers in Immunology, 10*, 318.

Macatangay, B. J., & Rinaldo, C. R. (2009). PD-1 blockade: A promising immunotherapy for HIV? *Cellscience, 5*, 61–65.

Macpherson, A. J., Geuking, M. B., & McCoy, K. D. (2012). Innate and adaptive immunity in host-microbiota mutualism. *Frontiers in Bioscience, 4*, 685–698.

Mancebo, E., Clemente, J., Sanchez, J., Ruiz-Contreras, J., De Pablos, P., Cortezon, S., ... Allende, L. M. (2008). Longitudinal analysis of immune function in the first 3 years of life in thymectomized neonates during cardiac surgery. *Clinical and Experimental Immunology, 154*, 375–383.

Mandal, P. K., & Rossi, D. J. (2012). DNA-damage-induced differentiation in hematopoietic stem cells. *Cell, 148*, 847–848.

Manser, A. R., & Uhrberg, M. (2016). Age-related changes in natural killer cell repertoires: Impact on NK cell function and immune surveillance. *Cancer Immunology, Immunotherapy: CII, 65*, 417–426.

McElhaney, J. E., Verschoor, C., & Pawelec, G. (2019). Zoster vaccination in older adults: Efficacy and public health implications. *The Journals of Gerontology. Series A Biological Sciences and Medical Sciences, 74*(8), 1239–1243.

McElhaney, J. E., Xie, D., Hager, W. D., Barry, M. B., Wang, Y., Kleppinger, A., ... Bleackley, R. C. (2006). T cell responses are better correlates of vaccine protection in the elderly. *Journal of Immunology, 176*, 6333–6339.

McElhaney, J. E., Zhou, X., Talbot, H. K., Soethout, E., Bleackley, R. C., Granville, D. J., & Pawelec, G. (2012). The unmet need in the elderly: How immunosenescence, CMV infection, co-morbidities and frailty are a challenge for the development of more effective influenza vaccines. *Vaccine, 30*, 2060–2067.

Michaud, M., Balardy, L., Moulis, G., Gaudin, C., Peyrot, C., Vellas, B., ... Nourhashemi, F. (2013). Proinflammatory cytokines, aging, and age-related diseases. *Journal of the American Medical Directors Association, 14*, 877–882.

Mishra, S. P., Wang, S., Nagpal, R., Miller, B., Singh, R., Taraphder, S., & Yadav, H. (2019). Probiotics and prebiotics for the amelioration of Type 1 diabetes: Present and future perspectives. *Microorganisms, 7*, 67.

Moehrle, B. M., & Geiger, H. (2016). Aging of hematopoietic stem cells: DNA damage and mutations? *Experimental Hematology, 44*, 895–901.

Molina-Tijeras, J. A., Galvez, J., & Rodriguez-Cabezas, M. E. (2019). The immunomodulatory properties of extracellular vesicles derived from probiotics: A novel approach for the management of gastrointestinal diseases. *Nutrients, 11*, 1038.

Morita, Y., Ema, H., & Nakauchi, H. (2010). Heterogeneity and hierarchy within the most primitive hematopoietic stem cell compartment. *The Journal of Experimental Medicine, 207*, 1173–1182.

Moskowitz, D. M., Zhang, D. W., Hu, B., Le Saux, S., Yanes, R. E., Ye, Z., ... Goronzy, J. J. (2017). Epigenomics of human CD8 T cell differentiation and aging. *Science Immunology, 2*(8), eaag0192.

Müller, L., Di Benedetto, S., & Pawelec, G. (2019). The immune system and its dysregulation with aging. *Sub-cellular Biochemistry, 91*, 21–43.

Muller, L., Fulop, T., & Pawelec, G. (2013). Immunosenescence in vertebrates and invertebrates. *Immunity & Ageing, 10*, 12.

Müller, L., & Pawelec, G. (2014). Aging and immunity - impact of behavioral intervention. *Brain, Behavior, and Immunity* 39, 8–22.

Muthuramalingam, A., Menon, V., Rajkumar, R. P., & Negi, V. S. (2016). Is depression an inflammatory disease? Findings from a cross-sectional study at a tertiary care center. *Indian Journal of Psychological Medicine, 38*, 114–119.

Nagpal, R., Mainali, R., Ahmadi, S., Wang, S., Singh, R., Kavanagh, K., ... Yadav, H. (2018). Gut microbiome and aging: Physiological and mechanistic insights. *Nutrition Healthy Aging, 4*, 267–285.

Naismith, E., and Pangrazzi, L. (2019). Impact of oxidative stress, inflammation, and senescence on the maintenance of immunological memory in the bone marrow in old age. *Bioscience Reports*, 39, BSR20190371.

Naka, K., & Hirao, A. (2011). Maintenance of genomic integrity in hematopoietic stem cells. *International Journal of Hematology, 93*, 434–439.

Naylor, K., Li, G., Vallejo, A. N., Lee, W. W., Koetz, K., Bryl, E., ... Goronzy, J. J. (2005). The influence of age on T cell generation and TCR diversity. *Journal of Immunology, 174*, 7446–7452.

Norddahl, G. L., Pronk, C. J., Wahlestedt, M., Sten, G., Nygren, J. M., Ugale, A., ... Bryder, D. (2011). Accumulating mitochondrial DNA mutations drive premature hematopoietic aging phenotypes distinct from physiological stem cell aging. *Cell Stem Cell, 8*, 499–510.

Oertelt-Prigione, S. (2012). The influence of sex and gender on the immune response. *Autoimmunity Reviews, 11*, A479–A485.

Ostan, R., Monti, D., Gueresi, P., Bussolotto, M., Franceschi, C., & Baggio, G. (2016). Gender, aging and longevity in humans: An update of an intriguing/neglected scenario paving the way to a gender-specific medicine. *Clinical Science, 130*, 1711–1725.

Ouyang, Q., Wagner, W. M., Wikby, A., Walter, S., Aubert, G., Dodi, A. I., ... Pawelec, G. (2003). Large numbers of dysfunctional CD8 + T lymphocytes bearing receptors for a single dominant CMV epitope in the very old. *Journal of Clinical Immunology, 23*, 247–257.

Oviedo-Orta, E., Li, C. K., & Rappuoli, R. (2013). Perspectives on vaccine development for the elderly. *Current Opinion in Immunology, 25*, 529–534.

Oxman, M. N., Levin, M. J., Johnson, G. R., Schmader, K. E., Straus, S. E., Gelb, L. D., ... Shingles Prevention Study, G. (2005). A vaccine to prevent herpes zoster and postherpetic neuralgia in older adults. *The New England Journal of Medicine, 352*, 2271–2284.

Palmer, S., Albergante, L., Blackburn, C. C., & Newman, T. J. (2018). Thymic involution and rising disease incidence with age. *Proceedings of the National Academy of Sciences of the United States of America*, *115*, 1883–1888.

Pangrazzi, L., Meryk, A., Naismith, E., Koziel, R., Lair, J., Krismer, M., ... Grubeck-Loebenstein, B. (2017). "Inflamm-aging" influences immune cell survival factors in human bone marrow. *European Journal of Immunology*, *47*, 481–492.

Pawelec, G. (2012). Hallmarks of human "immunosenescence": Adaptation or dysregulation? *Immunity & Ageing*, *9*, 15.

Pawelec, G. (2017). Does the human immune system ever really become "senescent"? F1000Res, 6:F1000 Faculty Rev-1323.

Pawelec, G. (2018a). Age and immunity: What is "immunosenescence"? *Experimental Gerontology*, *105*, 4–9.

Pawelec, G. (2018b). Immune correlates of clinical outcome in melanoma. *Immunology*, *153*, 415–422.

Pawelec, G. (2018c). Immune signatures predicting responses to immunomodulatory antibody therapy. *Current Opinion in Immunology*, *51*, 91–96.

Pawelec, G. P. (2018d). CASIN the joint: Immune aging at the stem cell level. *Blood*, *132*, 553–554.

Pawelec, G. (2019). Immune signatures associated with mortality differ in elderly populations from different birth cohorts and countries even within northern Europe. *Mechanisms of Ageing and Development*, *177*, 182–185.

Pawelec, G., Derhovanessian, E., Larbi, A., Strindhall, J., & Wikby, A. (2009). Cytomegalovirus and human immunosenescence. *Reviews in Medical Virology*, *19*, 47–56.

Pawelec, G., & McElhaney, J. (2018). Vaccines for improved cellular immunity to influenza. *EBioMedicine*, *30*, 12–13.

Pedemonte, E., Mancardi, G., Giunti, D., Corcione, A., Benvenuto, F., Pistoia, V., & Uccelli, A. (2006). Mechanisms of the adaptive immune response inside the central nervous system during inflammatory and autoimmune diseases. *Pharmacology & Therapeutics*, *111*, 555–566.

Pedersen, B. K. (2011). Exercise-induced myokines and their role in chronic diseases. *Brain, Behavior, and Immunity*, *25*, 811–816.

Peiris, J. S., Chu, C. M., Cheng, V. C., Chan, K. S., Hung, I. F., Poon, L. L., ... Group, H. U. S. S. (2003). Clinical progression and viral load in a community outbreak of coronavirus-associated SARS pneumonia: A prospective study. *Lancet*, *361*, 1767–1772.

Prelog, M. (2013). Differential approaches for vaccination from childhood to old age. *Gerontology*, *59*, 230–239.

Prelog, M., Wilk, C., Keller, M., Karall, T., Orth, D., Geiger, R., ... Wuerzner, R. (2008). Diminished response to tick-borne encephalitis vaccination in thymectomized children. *Vaccine*, *26*, 595–600.

Puchta, A., Naidoo, A., Verschoor, C. P., Loukov, D., Thevaranjan, N., Mandur, T. S., ... Bowdish, D. M. (2016). TNF drives monocyte dysfunction with age and results in impaired anti-pneumococcal immunity. *PLoS Pathogens*, *12*, e1005368.

Qi, Q., Liu, Y., Cheng, Y., Glanville, J., Zhang, D., Lee, J. Y., ... Goronzy, J. J. (2014). Diversity and clonal selection in the human T-cell repertoire. *Proceedings of the National Academy of Sciences of the United States of America*, *111*, 13139–13144.

Ranadive, S. M., Cook, M., Kappus, R. M., Yan, H., Lane, A. D., Woods, J. A., ... Fernhall, B. (2014). Effect of acute aerobic exercise on vaccine efficacy in older adults. *Medicine and Science in Sports and Exercise*, *46*, 455–461.

Redlich, K., & Smolen, J. S. (2012). Inflammatory bone loss: Pathogenesis and therapeutic intervention. *Nature Reviews. Drug Discovery*, *11*, 234–250.

Rizzo, L. B., Do Prado, C. H., Grassi-Oliveira, R., Wieck, A., Correa, B. L., Teixeira, A. L., & Bauer, M. E. (2013). Immunosenescence is associated with human cytomegalovirus and shortened telomeres in type I bipolar disorder. *Bipolar Disorders*, *15*, 832–838.

Roosen, J., Oosterlinck, W., & Meyns, B. (2015). Routine thymectomy in congenital cardiac surgery changes adaptive immunity without clinical relevance. *Interactive Cardiovascular and Thoracic Surgery*, *20*, 101–106.

Ross, S. E., Hemati, N., Longo, K. A., Bennett, C. N., Lucas, P. C., Erickson, R. L., & MacDougald, O. A. (2000). Inhibition of adipogenesis by WNT signaling. *Science*, *289*, 950–953.

Rube, C. E., Fricke, A., Widmann, T. A., Furst, T., Madry, H., Pfreundschuh, M., & Rube, C. (2011). Accumulation of DNA damage in hematopoietic stem and progenitor cells during human aging. *PLoS One*, *6*, e17487.

Rutenberg, A. D., Mitnitski, A. B., Farrell, S. G., & Rockwood, K. (2018). Unifying aging and frailty through complex dynamical networks. *Experimental Gerontology*, *107*, 126–129.

Sauce, D., Larsen, M., Fastenackels, S., Duperrier, A., Keller, M., Grubeck-Loebenstein, B., ... Appay, V. (2009). Evidence of premature immune aging in patients thymectomized during early childhood. *The Journal of Clinical Investigation*, *119*, 3070–3078.

Savva, G. M., Pachnio, A., Kaul, B., Morgan, K., Huppert, F. A., Brayne, C., ... Ageing, S. (2013). Cytomegalovirus infection is associated with increased mortality in the older population. *Aging Cell*, *12*, 381–387.

Schmitt, V., Rink, L., & Uciechowski, P. (2013). The Th17/Treg balance is disturbed during aging. *Experimental Gerontology*, *48*, 1379–1386.

Scholz, J. L., Diaz, A., Riley, R. L., Cancro, M. P., & Frasca, D. (2013). A comparative review of aging and B cell function in mice and humans. *Current Opinion in Immunology*, *25*, 504–510.

Schulz, A. R., Malzer, J. N., Domingo, C., Jurchott, K., Grutzkau, A., Babel, N., ... Thiel, A. (2015). Low thymic activity and dendritic cell numbers are associated with the immune response to primary viral infection in elderly humans. *Journal of Immunology*, *195*, 4699–4711.

Sempowski, G. D., Hale, L. P., Sundy, J. S., Massey, J. M., Koup, R. A., Douek, D. C., ... Haynes, B. F. (2000). Leukemia inhibitory factor, oncostatin M, IL-6, and stem cell factor mRNA expression in human thymus increases with age and is associated with thymic atrophy. *Journal of Immunology*, *164*, 2180–2187.

Shanley, D. P., Aw, D., Manley, N. R., & Palmer, D. B. (2009). An evolutionary perspective on the mechanisms of immunosenescence. *Trends in Immunology*, *30*, 374–381.

Shaw, A. C., Panda, A., Joshi, S. R., Qian, F., Allore, H. G., & Montgomery, R. R. (2011). Dysregulation of human Toll-like receptor function in aging. *Ageing Research Reviews*, *10*, 346–353.

Shen-Orr, S. S., & Furman, D. (2013). Variability in the immune system: Of vaccine responses and immune states. *Current Opinion in Immunology*, *25*, 542–547.

Simpson, R. J., Lowder, T. W., Spielmann, G., Bigley, A. B., LaVoy, E. C., & Kunz, H. (2012). Exercise and the aging immune system. *Ageing Research Reviews*, *11*, 404–420.

Snoeck, H. W. (2013). Aging of the hematopoietic system. *Current Opinion in Hematology*, *20*, 355–361.

Solana, R., Tarazona, R., Gayoso, I., Lesur, O., Dupuis, G., & Fulop, T. (2012). Innate immunosenescence: Effect of aging on cells and receptors of the innate immune system in humans. *Seminars in Immunology*, *24*, 331–341.

Spielmann, G., McFarlin, B. K., O'Connor, D. P., Smith, P. J., Pircher, H., & Simpson, R. J. (2011). Aerobic fitness is associated with lower proportions of senescent blood T-cells in man. *Brain, Behavior, and Immunity*, *25*, 1521–1529.

Stosio, M., Ruszkowski, J., Mikosik-Roczynska, A., Haponiuk, I., & Witkowski, J. M. (2017). The significance of neonatal thymectomy for shaping the immune system in children with congenital heart defects. *Kardiochir Torakochirurgia Polska*, *14*, 258–262.

Su, D. M., Aw, D., & Palmer, D. B. (2013). Immunosenescence: A product of the environment? *Current Opinion in Immunology*, *25*, 498–503.

Sutherland, J. S., Goldberg, G. L., Hammett, M. V., Uldrich, A. P., Berzins, S. P., Heng, T. S., ... Boyd, R. L. (2005). Activation of thymic regeneration in mice and humans following androgen blockade. *Journal of Immunology, 175*, 2741–2753.

Swain, S. L., & Blomberg, B. B. (2013). Immune senescence: New insights into defects but continued mystery of root causes. *Current Opinion in Immunology, 25*, 495–497.

Tavenier, J., Margolick, J. B., & Leng, S. X. (2019). T-cell immunity against cytomegalovirus in HIV infection and aging: Relationships with inflammation, immune activation, and frailty. *Medical Microbiology and Immunology, 208*, 289–294.

Thapa, P., & Farber, D. L. (2019). The role of the thymus in the immune response. *Thoracic Surgery Clinics, 29*, 123–131.

Thomas-Crusells, J., McElhaney, J. E., & Aguado, M. T. (2012). Report of the ad-hoc consultation on aging and immunization for a future WHO research agenda on life-course immunization. *Vaccine, 30*, 6007–6012.

Tollervey, J. R., & Lunyak, V. V. (2012). Epigenetics: Judge, jury and executioner of stem cell fate. *Epigenetics, 7*, 823–840.

Tumpel, S., & Rudolph, K. L. (2012). The role of telomere shortening in somatic stem cells and tissue aging: Lessons from telomerase model systems. *Annals of the New York Academy of Sciences, 1266*, 28–39.

Ucar, D., Marquez, E. J., Chung, C. H., Marches, R., Rossi, R. J., Uyar, A., ... Banchereau, J. (2017). The chromatin accessibility signature of human immune aging stems from CD8(+) T cells. *The Journal of Experimental Medicine, 214*, 3123–3144.

Valmori, D., Merlo, A., Souleimanian, N. E., Hesdorffer, C. S., & Ayyoub, M. (2005). A peripheral circulating compartment of natural naive CD4 Tregs. *The Journal of Clinical Investigation, 115*, 1953–1962.

Van Buynder, P., & Booy, R. (2018). Pneumococcal vaccination in older persons: Where are we today? *Pneumonia, 10*, 1.

van den Broek, T., Delemarre, E. M., Janssen, W. J., Nievelstein, R. A., Broen, J. C., Tesselaar, K., ... van Wijk, F. (2016). Neonatal thymectomy reveals differentiation and plasticity within human naive T cells. *The Journal of Clinical Investigation, 126*, 1126–1136.

Ventevogel, M. S., & Sempowski, G. D. (2013). Thymic rejuvenation and aging. *Current Opinion in Immunology, 25*, 516–522.

Verschoor, C. P., Johnstone, J., Millar, J., Dorrington, M. G., Habibagahi, M., Lelic, A., ... Bowdish, D. M. (2013). Blood CD33(+)HLA-DR(-) myeloid-derived suppressor cells are increased with age and a history of cancer. *Journal of Leukocyte Biology, 93*, 633–637.

Vezys, V., Yates, A., Casey, K. A., Lanier, G., Ahmed, R., Antia, R., & Masopust, D. (2009). Memory CD8 T-cell compartment grows in size with immunological experience. *Nature, 457*, 196–199.

Vrisekoop, N., den Braber, I., de Boer, A. B., Ruiter, A. F., Ackermans, M. T., van der Crabben, S. N., ... Tesselaar, K. (2008). Sparse production but preferential incorporation of recently produced naive T cells in the human peripheral pool. *Proceedings of the National Academy of Sciences of the United States of America, 105*, 6115–6120.

Waaijer, M. E. C., Goldeck, D., Gunn, D. A., van Heemst, D., Westendorp, R. G. J., Pawelec, G., & Maier, A. B. (2019). Are skin senescence and immunosenescence linked within individuals? *Aging Cell, 18*, e12956.

Walford, R. L. (1964). The immunologic theory of aging. *The Gerontologist, 4*, 195–197.

Walsh, N. P., Gleeson, M., Shephard, R. J., Gleeson, M., Woods, J. A., Bishop, N. C., ... Simon, P. (2011). Position statement. Part one: Immune function and exercise. *Exercise Immunology Review, 17*, 6–63.

Wang, J., Sun, Q., Morita, Y., Jiang, H., Gross, A., Lechel, A., ... Rudolph, K. L. (2012). A differentiation checkpoint limits hematopoietic stem cell self-renewal in response to DNA damage. *Cell, 148*, 1001–1014.

Wang, S., Wang, J., Kumar, V., Karnell, J. L., Naiman, B., Gross, P. S., ... Ettinger, R. (2018). IL-21 drives expansion and plasma cell differentiation of autoreactive CD11c(hi)T-bet(+) B cells in SLE. *Nature Communications, 9*, 1758.

Weekes, M. P., Tan, S. Y., Poole, E., Talbot, S., Antrobus, R., Smith, D. L., ... Lehner, P. J. (2013). Latency-associated degradation of the MRP1 drug transporter during latent human cytomegalovirus infection. *Science, 340*, 199–202.

Weide, B., Martens, A., Zelba, H., Stutz, C., Derhovanessian, E., Di Giacomo, A. M., ... Pawelec, G. (2013). Myeloid-derived suppressor cells predict survival of advanced melanoma patients: Comparison with regulatory T cells and NY-ESO-1- or Melan-A-specific T cells. *Clinical Cancer Research, 20*, 1601–1609.

Weinberger, B., & Grubeck-Loebenstein, B. (2012). Vaccines for the elderly. *Clinical Microbiology and Infection, 18*(Suppl. 5), 100–108.

Weiskopf, D., Weinberger, B., & Grubeck-Loebenstein, B. (2009). The aging of the immune system. *Transplant International, 22*, 1041–1050.

Weston, W. M., Friedland, L. R., Wu, X., & Howe, B. (2012). Vaccination of adults 65 years of age and older with tetanus toxoid, reduced diphtheria toxoid and acellular pertussis vaccine (Boostrix®): Results of two randomized trials. *Vaccine, 30*, 1721–1728.

Wikby, A., Ferguson, F., Forsey, R., Thompson, J., Strindhall, J., Lofgren, S., ... Johansson, B. (2005). An immune risk phenotype, cognitive impairment, and survival in very late life: Impact of allostatic load in Swedish octogenarian and nonagenarian humans. *The Journals of Gerontology. Series A Biological Sciences and Medical Sciences, 60*, 556–565.

Wills, M., Akbar, A., Beswick, M., Bosch, J. A., Caruso, C., Colonna-Romano, G., ... Pawelec, G. (2011). Report from the second cytomegalovirus and immunosenescence workshop. *Immunity & Ageing, 8*, 10.

Wong, C. P., Magnusson, K. R., & Ho, E. (2010). Aging is associated with altered dendritic cells subset distribution and impaired proinflammatory cytokine production. *Experimental Gerontology, 45*, 163–169.

Yahata, T., Takanashi, T., Muguruma, Y., Ibrahim, A. A., Matsuzawa, H., Uno, T., ... Ando, K. (2011). Accumulation of oxidative DNA damage restricts the self-renewal capacity of human hematopoietic stem cells. *Blood, 118*, 2941–2950.

Zacca, E. R., Crespo, M. I., Acland, R. P., Roselli, E., Nunez, N. G., Maccioni, M., ... Moron, G. (2015). Aging impairs the ability of conventional dendritic cells to cross-prime CD8 + T cells upon stimulation with a TLR7 ligand. *PLoS One, 10*, e0140672.

Zook, E. C., Krishack, P. A., Zhang, S., Zeleznik-Le, N. J., Firulli, A. B., Witte, P. L., & Le, P. T. (2011). Overexpression of Foxn1 attenuates age-associated thymic involution and prevents the expansion of peripheral CD4 memory T cells. *Blood, 118*, 5723–5731.

CHAPTER

17

Microbiome changes in aging

Kelly R. Reveles[1,2], *Eric H. Young*[1,2], *Amina R.A.L. Zeidan*[1,2] *and Qunfeng Dong*[3]

[1]College of Pharmacy, The University of Texas at Austin, Austin, TX, United States [2]Pharmacotherapy Education & Research Center, UT Health San Antonio, San Antonio, TX, United States [3]Department of Public Health Sciences, Loyola University Chicago, Chicago, IL, United States

OUTLINE

Introduction

Due to the dynamic increase in human population needs in the near future, aging researchers no longer simply focus on extending human lifespan; they are interested in increasing quality of life, extending the length of time people are living independently, and the overall healthspan for the aging population. It is commonly believed that human healthspan is determined by complex interactions between genetic, epigenetic, and environmental (primarily diet and lifestyle) factors; however, over the past two decades, scientists have uncovered another potential important contributor to human health—the microbiome. Research in this area has increased exponentially over the last two decades (Fig. 17.1). Studies have predominantly focused on the relationship between the microbiome and health, but more recent studies are also evaluating the role of the microbiome in promoting lifespan. This chapter will provide an overview of the

Handbook of the Biology of Aging.
DOI: https://doi.org/10.1016/B978-0-12-815962-0.00017-2

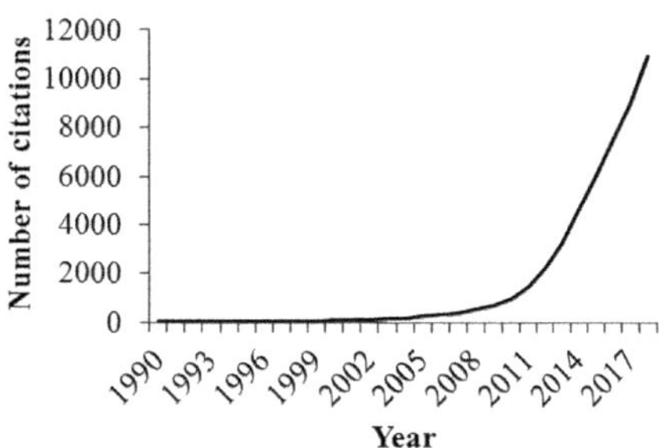

FIGURE 17.1 Number of PubMed citations by year using the keyword search term "microbiome."

human microbiome and describe how its composition and function change with age. We will also discuss the relationship between the microbiome and aging-related conditions, as well as microbiome-targeted therapies as potential interventions to improve health and the aging process.

Overview of the structure and function of the microbiome

The microbiome refers to the collection of microbial genomes within an ecosystem, whereas microbiota represent the individual microbes (bacteria, archaea, fungi, and viruses) that inhabit the site. While definitions differ, these two terms are often used interchangeably. The human microbiome is very large in scope, despite it often being overlooked in human disease or treatment (hence the name the "forgotten organ": O'Hara & Shanahan, 2006). Prior studies estimated that the human body contains approximately 10 times more microbial cells than human cells (i.e., roughly 100 trillion microbial cells and 10 trillion human cells) (Ley, Peterson, & Gordon, 2006). In more recent years, this ratio has been assumed to be closer to 1:1 to account for the number of human red blood cells (Sender, Fuchs, & Milo, 2016). The human intestinal tract alone is inhabited by over 1000 species of known bacteria (Gill et al., 2006). Importantly, it has been estimated that gut microbes encode greater than 150 times more genes that the human genome, meaning that microbes are involved in numerous functions within the human body (Ley, Peterson, et al., 2006).

Studying the microbiome

Our understanding of the microbiome has increased dramatically over the last two decades due to a shift from culture-dependent techniques to culture-independent technologies, which allow for identification of microbes that could not be previously identified using culture alone. Methods to study the microbiome include those that describe the community structure (i.e., what microbes are present and in what abundances) or the function of the ecosystem.

There are two main approaches for identifying which organisms are present in a sample: 16S rRNA gene sequencing and whole-genome sequencing. The 16S gene that encodes for the 16S rRNA molecule is unique only to bacteria and archaea; human cells do not possess this gene. 16S rRNA gene sequencing uses next-generation sequencing to identify the community composition. This approach is cost-effective, semiquantitative, and permits resolution to the genus level (and in some cases species); however, limitations exist. For example, short reads make accurate classification difficult. Additionally, this approach does not provide information beyond community structure (e.g., functionality). On the other hand, whole-genome sequencing provides a comprehensive phylogenetic view of community composition including eukaryotic organisms and viruses, identification of variants and polymorphisms, and function information. Other "omic" approaches can also be used to catalog microbiome function, including metatranscriptomics, metaproteomics, and metabolomics. These approaches characterize gene expression, protein expression, and metabolic productivity, respectively (Lynch & Pedersen, 2016).

Two popular measurements for studying the microbiome composition include alpha and beta diversity. Alpha diversity refers to the within-sample diversity, often indicated by the total number of bacterial taxa present in a sample (i.e., sample richness) and the extent to which bacterial taxa are evenly distributed (i.e., evenness). Alternatively, beta diversity indicates how similar or dissimilar microbial communities are between a pair of samples (Knights, Costello, & Knight, 2011). This is often documented as the average proportional abundance of bacterial taxa between groups or summarized microbial structure using principal coordinates analysis. A significant change in the structure or function of the microbiome is often referred to as dysbiosis.

Structure of the human microbiome

Many multidisciplinary projects over the last several years have contributed greatly to our understanding of the human microbiome. Notable collaborative projects have included the MetaHit project and the Human Microbiome Project, which aimed to characterize the microbiome of volunteers from multiple countries. These studies have found that the human microbiome is very diverse and differs substantially by body site. The predominant bacterial phyla include Firmicutes, Bacteroidetes, Actinobacteria, Proteobacteria, and

Verrucomicrobia. In fact, Firmicutes and Bacteroidetes constitute over 90% of the human gut microbiota (Backhed, Ley, Sonnenburg, Peterson, & Gordon, 2005; Eckburg et al., 2005). The primary determinant of community composition is anatomical location. The average bacterial composition of the gastrointestinal tract differs significantly compared to the oral cavity, for example. Even within a continuous body site, like the skin or gastrointestinal tract, the human microbiome is significantly impacted by biogeography and individuality (Oh et al., 2014). In humans, the microbiota composition is normally conserved at high taxonomic levels, but variation progressively increases at lower levels (Huse et al., 2008; Ley, Lozupone, Hamady, Knight, & Gordon, 2008).

In humans, the gastrointestinal tract is the largest microbial ecosystem, encompassing trillions of microbial cells (Ley, Peterson, et al., 2006). It is also the most well-studied, and hence, will be the primary focus of the chapter. Bacteria dominate the gut microbiota, while fungi and archaea makeup only about 1% of the species present. Some of the more highly conserved bacterial genera in high abundance in the gut include *Bacteroides, Faecalibacterium, Blautia, Prevotella, Clostridium, Ruminococcus,* and *Bifidobacterium* (Qin et al., 2010; Van den Abbeele et al., 2013).

Functions of the human microbiome

Despite extensive differences in microbiome composition between individuals, especially at lower taxonomic levels, the core functions of the microbiome are largely conserved (Turnbaugh et al., 2006). The functions of the microbiome are numerous and fall into three major categories: (1) protection, (2) biosynthesis, and (3) metabolism.

First, human microbiota play an important role in the maturation and education of the immune response (Fulde & Hornef, 2014). The immune system maintains a symbiotic relationship between the host and the commensal microbiota, as well as providing protective responses to pathogens and maintaining regulatory pathways involved in antigen tolerance (Belkaid & Hand, 2014; Kamada, Chen, Inohara, & Nunez, 2013). In addition, the microbiota are also known to play a protective role in acute inflammatory responses to injury, at least in part through toll-like receptor (TLR) activation, to promote tissue repair and survival (Buford, 2017).

Next, the microbiota influence the biosynthesis of several important molecules in humans. The microbiota are known to synthesize certain vitamins, including vitamin K and some B group vitamins (e.g., biotin, folates, riboflavin, thiamine) (Hill, 1997). They also play a role in regulating intestinal endocrine functions and production of steroid hormones (Neuman, Debelius, Knight, & Koren, 2015). Commensal microbiota produce neuroactive molecules, including serotonin, kynurenine, melatonin, γ-aminobutyric acid (GABA), catecholamines, histamine, and acetylcholine (Barrett, Ross, O'Toole, Fitzgerald, & Stanton, 2012; Lyte, 2011; Yano et al., 2015). The microbiota also produce other important compounds for energy biosynthesis, such as short-chain fatty acids (SCFAs) (Canfora, Jocken, & Blaak, 2015).

Lastly, the microbiota play a role in metabolism, including metabolism of bile salts and modification of specific drugs and elimination of exogenous toxins (Devlin & Fischbach, 2015; Haiser et al., 2013). End-products of bacterial metabolism include the SCFAs acetate, butyrate, and propionate, which are important for energy metabolism and signaling in the gut, as well as immune homeostasis (Biagi et al., 2010; Claesson et al., 2012; Cuervo, Salazar, Ruas-Madiedo, Gueimonde, & Gonzalez, 2013; Rampelli et al., 2013; Smith et al., 2013). Given the wide range of functions of the human microbiota (Fig. 17.2), it is not surprising that significant changes in the ecosystem composition could influence human health.

The healthy microbiome

The interindividual differences in microbiome composition, even among relatively homogeneous populations, has made it somewhat difficult to define what constitutes a "healthy" microbiome. Hallmarks of a healthy microbiome have been previously described to include: (1) a functional core that provides for all microbial housekeeping functions and production of compounds necessary to support human life, (2) microbial diversity, which should, in turn, lead to

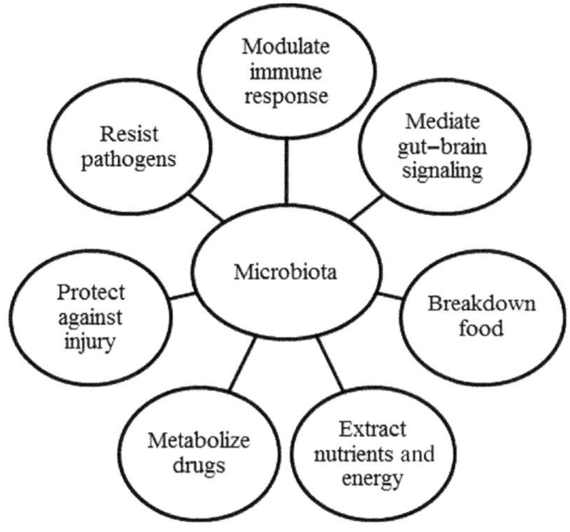

FIGURE 17.2 Core functions of the human gut microbiome.

greater functional capacity, and (3) resistance, resilience, and stability of the microbiome (e.g., ability of the microbiome to recover following perturbation) (Lloyd-Price, Abu-Ali, & Huttenhower, 2016). Certain bacterial taxa have also been generally associated with health, including those involved in SCFA production (e.g., *Faecalibacterium*, *Eubacterium*, and *Roseburia*), while others have been negatively associated with health, such as lipopolysaccharide producers and organisms with pathogenic potential (pathobionts; e.g., Enterobacteriaceae) (Zechner, 2017). A more detailed description of how certain bacterial taxa differ by chronological and biological age and by aging-related conditions is provided in the following sections.

Relationship between the microbiome and aging

The microbiome's role in regulation of longevity and aging

Gut microbiota are critical in promoting healthspan via energy and nutrient extraction, host immune system modulation, and protection against pathogens. Importantly, the gut microbiota also play an important role in regulating longevity and aging by altering cellular senescence, preventing immunosenescence, and by mediating host metabolism (Candela et al., 2009; Ikeda, Yasui, Hoshino, Arikawa, & Nishikawa, 2007; Komura, Yasui, Miyamoto, & Nishikawa, 2010; Weksler, Pawelec, & Franceschi, 2009). Certain bacterial species have been previously associated with improved lifespan in animals. In studies of *C. elegans*, dietary supplementation with *Bifidobacterium* and *Lactobacillus*, the most common forms of nonpathogenic microbiota, resulted in a significantly extended lifespan and increased protection against pathogenic bacteria when compared to nonsupplemented animals. (Ikeda et al., 2007; Komura et al., 2010). In *Drosophila*, the presence of bacteria during the first week of adult life enhanced longevity by 30%–35% (Brummel, Ching, Seroude, Simon, & Benzer, 2004). Conversely, studies in *Drosophila* have shown that dysbiosis favoring expansion of Gammaproteobacteria has been tightly linked to intestinal barrier dysfunction, systemic inflammation, and shorter lifespan (Clark et al., 2015). Mouse models have shown that age-related changes in the gut microbiome include an altered abundance of microbial taxa that are associated with dysbiosis (Kain et al., 2019).

Significantly, there is emerging evidence that the gut microbiota produce compounds that alter cellular senescence, an underlying regulator of the aging rate in mammals (Kibe et al., 2014). For example, the gut microbiota produce polyamines, which have been previously associated with longevity in bacteria, yeast, worms, flies, and mice (Eisenberg et al., 2009). Upregulation of intestinal polyamines with probiotic supplementation can suppress inflammation, improve longevity, and protect from age-induced memory impairment (Kibe et al., 2014).

Next, selected strains mediate the inflammatory response. Supplementation with *Bifidobacterium* has been previously correlated with reduced serum IL-10 and TNF-α concentrations (Ouwehand et al., 2008). Other studies have also demonstrated a reduction with age in antiinflammatory Firmicutes, such as *Clostridium* cluster XIVa and *F. parausnitzii*, which might contribute to an overall proinflammatory profile among older adults (Frank et al., 2007). Similarly, germ-free mice have been shown to be protected from inflammation; however, cohousing with old mice resulted in increased levels of proinflammatory cytokines, macrophage dysfunction, and permeability of the intestine (Thevaranjan et al., 2017). A proinflammatory state is a hallmark of human aging; thus, it is hypothesized that alteration in the gut microbiota would have a strong impact on aging and age-related diseases.

Several other key aging pathways, including nutrient sensing and metabolism, are regulated in part by the host microbiome. For example, in *Drosophila*, microbiota modulate host development and metabolic homeostasis via the target of rapamycin (TOR) pathway (Storelli et al., 2011). Studies have shown that alteration of microbial folate and methionine metabolism with metformin increases lifespan (Cabreiro et al., 2013). Animal models have also helped to identify specific dynamics of aging pathways within the gut microbiome, including bacterial polysaccharides and worm mitochondrial dynamics, and extensions in *Drosophila* lifespan through colanic acid supplementation (Han et al., 2017).

Given the importance of the host microbiome with regard to the development and programming of metabolic pathways and the immune response, it is critical to understand how the microbiome changes with age and how those changes may affect healthspan and lifespan.

Microbiome changes with chronological aging

The complexity and diversity of the microbiome develop throughout the human lifespan, with notable changes occurring in infancy, in the elderly, and even in centenarians (those aged 100–105 years) (Table 17.1). This section describes changes in the gut microbiome structure and function throughout chronological aging.

TABLE 17.1 Summary of studies comparing gut microbiome changes by chronological age*.

Age group	N	Population	Microbiota	Age effect in microbiota	Study
Newborns	98	Swedish	Firmicutes	↑	Backhed et al. (2015)
			Bacteroidetes	↑	
			Actinobacteria	↑	
			Proteobacteria	↑	
12–48 months (changes in 12–48 months vs adults aged 21–60)	28	American	Actinobacteria	↑	Cheng et al. (2016)
			Bacilli	↑	
			Bacteroidetes	=	
			Clostridium	Varies by cluster	
			Proteobacteria	=	
7–12 years (changes in 7–12 years vs adults)	37	American	Bacteroidetes	↓	Hollister et al. (2015)
			Bifidobacterium	↑	
			Dialister	↑	
			Faecalibacterium	↑	
			Roseburia	↑	
			Ruminococcus	↑	
19–35 years	28	British	Bacterioides	↓	Woodmansey et al. (2004)
67–75 years			Bifidobacteria (# and diversity)	↓	
73–101 years					
21–34 years	15	British	*Bacteroides* species	↑	Hopkins and Macfarlane (2002)
67–88 years			Bifidobacteria species	↓	
67–73 years CDAD*					
28–46 years	170	Irish	Firmicutes	↓	Claesson et al. (2011)
64–102 years			*Clostridium* cluster IV	↑	
			Ruminococcaceae	↑	
38–102 years	9	Italian	Proteobacteria *Eubacterium* *Bifidobacterium* *Faecalibacterium*	↑ ↓ ↓ ↓	Rampelli et al. (2013)
79–84 years (compared to prior young adult studies)	6	Japanese	*Clostridium* rRNA subcluster XIVa Bifidobacteria # *Ruminococcus obeum*	↓ ↓ ↓	Hayashi et al. (2003)
0–104 years:	367	Japanese			Odamaki et al. (2016)
In 0–1 year			Firmicutes	↑	
In 21–69 years			*Blautia*	↑	
In 21–69 years			*Bifidobacterium*	↑	
In 21–69 years			*Bacteroides*	↑	
In 100 + years			Actinobacteria	↑	
25–104 years	84	Italian	*Clostridium* cluster XIVa Bacilli Proteobacteria	↓ ↑ ↑	Biagi et al. (2010)

**Clostridium difficile-associated diarrhea.*

While a fetus may be exposed to commensal microbiota in utero, the rapid development of the microbiome begins following birth (Aagaard et al., 2014; DiGiulio et al., 2010). Notable differences in the infant microbiome have been seen based on method of delivery, infant feeding practices, and other factors. A Swedish study found that newborns (0–28 days old) delivered through vaginal birth have gut microbiota compositions similar in taxonomy to their mother's gut and vaginal microbiome (Backhed et al., 2015). Newborns born via cesarean section have a delayed fecal colonization rate, with rates of *Bifidobacterium*-like and *Lactobacillus*-like bacterial taxa not mirroring vaginally delivered infants until approximately 1 month and 10 days (Gronlund, Lehtonen, Eerola, & Kero, 1999). Cesarean-delivered infants also had significantly lower abundance of *Bacteroides fragilis* compared to vaginally delivered infants (Gronlund et al., 1999).

The second stage of gut microbiome development is generally shaped by the type of feeding used. The microbiota of infants vary greatly between breastfed and bottle-fed infants, with colonization rates differing even in instances when modified cow's milk was used to closely resemble human milk (Balmer & Wharton, 1989; Orrhage & Nord, 1999). Fecal samples collected from breastfed infants have shown a gut microbiome environment dominated by *Bifidobacterium* and *Staphylococcus* species, whereas bottle-fed infants had higher colonization rates of *Enterococcus* and *Clostridium* species (Orrhage & Nord, 1999). One study found a significant decrease in the relative abundance of Actinobacteria in infants prior to and after weaning (Odamaki et al., 2016).

In infancy, gut microbiome diversity expands rapidly, as does the functional capacity of the microbiome. Gut microbiome studies in newborns, infants, and toddlers have shown that a maturation process occurs during this short time period, creating a microbiome that begins to resemble the structure of adult gut microbiome environments by the age of 2–5 years (Bergstrom et al., 2014; Cheng et al., 2016; Odamaki et al., 2016; Yatsunenko et al., 2012). This childhood microbiome development includes increased stability in bacterial makeup of the environment, with *Bacteroidetes* becoming more established (Cheng et al., 2016). The growth occurring during these age periods is critical, as this development creates a microbial environment that, aside from specific intervention, does not deviate much until one reaches older age (generally 60–70 years), when the microbiome environment may begin to "revert" back, or become more taxonomically similar to infancy (Odamaki et al., 2016).

While microbiome diversity has generally reached its peak in preadolescence, the bacterial composition and functional genes are distinct from adults. The preadolescent microbiome is enriched with *Bifidobacterium*, *Faecalibacterium*, and *Lachnospiraceae* and functional pathways for folate and vitamin B_{12} biosynthesis (Hollister et al., 2015). After this time, the adult microbiome is dominated by Firmicutes and Bacteroidetes, as described in the chapter introduction, and tends to remain stable (unless major microbiome-disrupting exposures are experienced) until older age.

Sizable changes in the gut microbiome occur as adults reach the elderly stage, wherein the microbiota can become less diverse and lose compositional stability; these events are seen to coexist in cases of the conditions mentioned previously, while declines are seen in immunocompetence, for example, aging-related health deterioration (Claesson et al., 2011, 2012). Due to the lack of detailed nutrition and lifestyle information on human subjects over long periods of time, it is difficult to determine why exactly these changes occur with age; however, it is believed that nutrients can evoke the differences seen between adult-enriched and infant/elderly-enriched gut microbiome compositions (Odamaki et al., 2016). Other notable changes in the elderly population include a decrease in the relative abundance of the core microbiota (e.g., Firmicutes, Actinobacteria) and an expansion of Proteobacteria (Fig. 17.3). The large

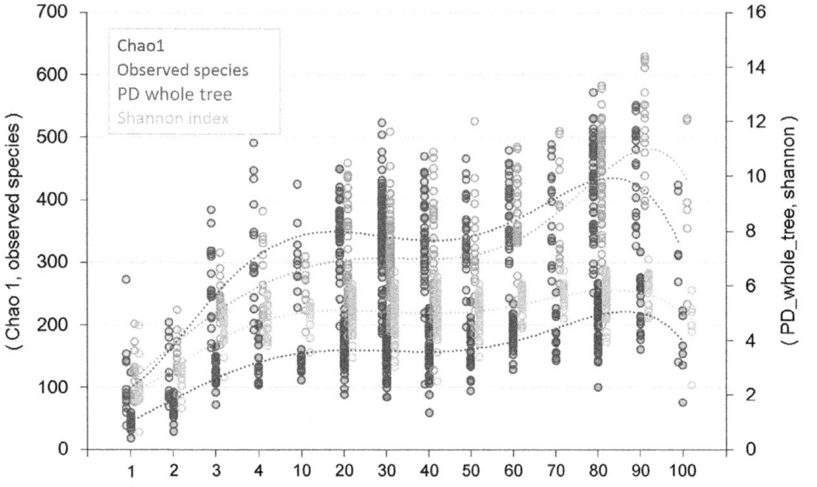

FIGURE 17.3 Age-related change in alpha-diversities of gut microbiota. *Source: Odamaki, T., Kato, K., Sugahara, H., Hashikura, N., Takahashi, S., Xiao, J. Z., … Osawa, R. (2016). Age-related changes in gut microbiota composition from newborn to centenarian: A cross-sectional study.* BMC Microbiology, *16, 90.*

proportional reduction in Actinobacteria with age is important as this phylum contains the health- and longevity-promoting Bifidobacteriaceae family. The enrichment for Proteobacteria with age is also concerning as this phylum contains many opportunistic bacteria, *H. pylori* and other *Enterobacteriaceae*, that can overtake commensal bacteria.

Another large deviation in microbial taxa occurs in centenarians (age 100 years and older). Centenarians generally have reached their aging milestones by avoiding many health-related problems. Centenarians have lower rates of morbidity and chronic illness and an extended healthspan when compared to octogenarians (aged 80–89 years) and nonagenarians (aged 90–99 years) in the same cohort (Franceschi & Bonafe, 2003; Kheirbek et al., 2017). Centenarians, therefore, are considered a type of "supercontrol" group, or an extreme phenotype, as they are at the extreme end of the health trait distribution.

Studies comparing young adults, the elderly, and centenarians have highlighted the changes that occur in the diversity and composition of the gut microbiome environment and how these do not show a linear change with age (Biagi et al., 2010; Odamaki et al., 2016). The microbiome changes seen in the elderly are even more pronounced in centenarians. Microbiome diversity tends to decrease in the extreme age group and the core microbiome composition declines (Fig. 17.4). Proteobacteria and Bacteroidetes are especially enriched in this population. In 2010, the Human Intestinal Tract Chip (HITChip) was used to characterize fecal microbiota in young adults, the elderly, and centenarians. This study found that the differences between centenarians and all other groups was greater than the differences between young adults and the elderly (Biagi et al., 2010). It seems that although those reaching "extreme aging" in years 100 and beyond are losing some of the core lifetime components of the gut microbiome, they acquire a wealth of new microbial taxa and components, which includes allochthonous microorganisms and pathobionts that can only colonize under seemingly abnormal conditions (Santoro et al., 2018).

Microbiome changes with biological aging

While chronological age is an important factor in predicting microbiome structure and the development of health-related conditions, it does not necessarily indicate a person's overall health status. Thus, biological aging has been proposed as an alternative marker of health with aging. Importantly, biological aging has been discovered as a more relevant indicator of a changing microbiome structure versus chronological age, as biological aging correlates with growth of coabundant organisms that differ from simple chronologically aged environments (Maffei et al., 2017). Biological aging has been associated with changes in alpha diversity, beta diversity, genus level subcommunities, and age-associated subcommunities with distinct biochemical functions (Maffei et al., 2017).

Frailty is a particularly important factor in regard to gut microbiome changes throughout the aging process. The literature refers to frailty as a measurable clinical syndrome that is not a function solely of chronological aging, but as a syndrome that is seen with increased biological aging and is associated with sizable deviations in the diversity of the gut microbiome (Maffei et al., 2017). This is an important determination, as frailty outperforms other factors, such as chronological aging, as a better estimate for a subject's risk of experiencing adverse health outcomes, including loss of normal microbiota function and gut dysbiosis (Langille et al., 2014; Maffei et al., 2017). The gut microbiome modulates a number of conditions, including changes in innate immunity, cognitive function, and sarcopenia—all related elements of frailty (O'Toole & Jeffery, 2015). A loss in diversity of the gut microbiome

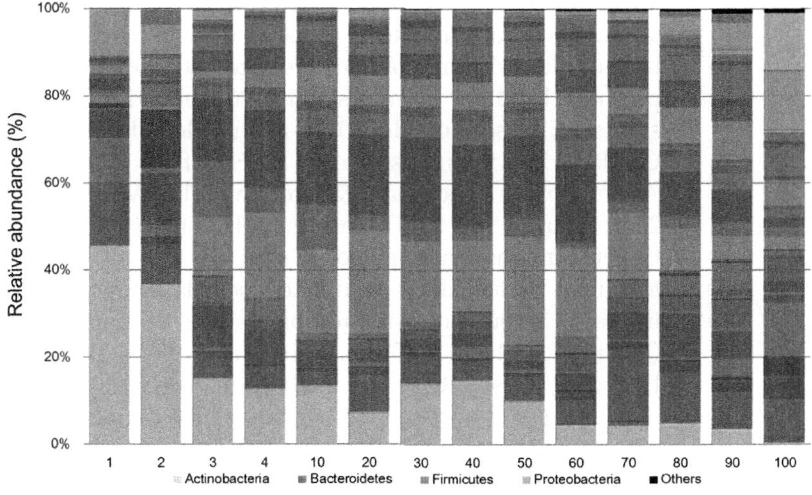

FIGURE 17.4 Age-related changes in gut microbiota composition. *Source: Odamaki, T., Kato, K., Sugahara, H., Hashikura, N., Takahashi, S., Xiao, J. Z., ... Osawa, R. (2016). Age-related changes in gut microbiota composition from newborn to centenarian: A cross-sectional study. BMC Microbiology,* 16, 90.

bacterial environment can affect the aging process; those who have a loss of microbial diversity in the distinct core groups of bacterial taxa experience reduced cognitive performance and increased frailty (O'Toole & Jeffery, 2015). A prior study noted a significant negative relationship between a 34-item frailty index score and alpha diversity, after correcting for sex, body mass index, antibiotic use, family membership, read count, and chronological age (Maffei et al., 2017). This same study found that the effect of biological age on beta diversity was similar in direction to chronological age, but was greater in magnitude, suggesting that the relative phylogenetic similarity in microbial communities is primarily based on biological aging, not chronological aging (Maffei et al., 2017).

Overall, understanding the development and stability of the gut microbiome is a complex endeavor that begins at birth and continues for the remainder of life. The gut microbiome will experience at least three major remodels during the lifespan, occurring at infancy, during older age, and during extremes of age. Importantly, no two individuals will experience the same changes over time. As described in the next section, the changes over time may be, in part, driving the development of many aging-related conditions.

Factors that affect the microbiome in aging

While human studies have noted a significant decline in microbiome diversity and altered composition with age that might contribute to disease, several other environmental and behavioral changes occur with age that may confound the relationships seen in many human microbiome studies. Importantly, there appears to be a bidirectional relationship between dysbiosis and aging mediated by external factors and inflammation (Fig. 17.5).

First, medication exposure can significantly alter the gut microbiome. Antibiotics in particular can severely disrupt the microbiota and it may never fully recover to its original composition (Dethlefsen, Huse, Sogin, & Relman, 2008; Dethlefsen & Relman, 2011). Newborns who are exposed to antibiotic drug therapies during this crucial period of microbiome development can experience long-term changes in bacterial diversity, such as decreased abundance of beneficial Bifidobacteriaceae and increased abundance of pathogenic Enterobacteriaceae (Gibson, Crofts, & Dantas, 2015). Use of broad-spectrum antibiotics, which target many species of aerobic and anaerobic bacteria, has increased among older adults in recent years, which could contribute to alterations in the microbiome seen in this group (Lee et al., 2014).

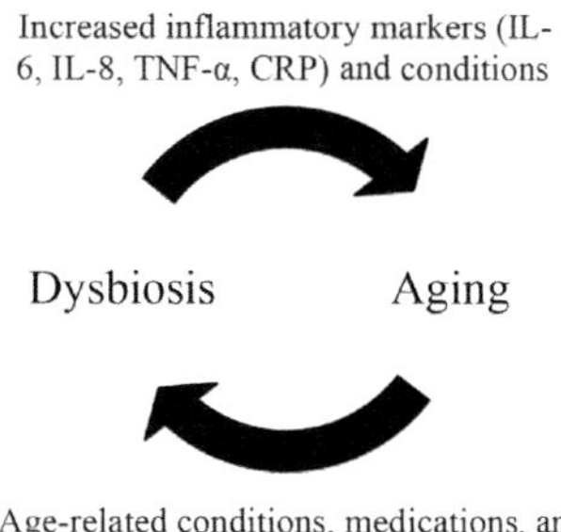

FIGURE 17.5 Bidirectional relationship between dysbiosis and aging.

Nonantibiotic medications can also affect the microbiome. In fact, a recent study screened more than 1000 marketed medications against 40 ubiquitous gut bacterial strains. This study found that 78% of antibiotics and 24% of the drugs with human targets inhibited the growth of at least one commensal bacterial strain (Maier et al., 2018). Some classes of medications that particularly affected the microbiome included gastric acid suppressants and chemotherapeutic agents. Gastric acid suppressants alter the microenvironment of the gastrointestinal tract by increasing the pH. They also target bacterial proton pumps, which could serve as a second potential mechanism for altering the gut microbiome. Prior studies have demonstrated a significant shift in microbiome composition following PPI use (Freedberg et al., 2015; Reveles, Ryan, Chan, Cosimi, & Haynes, 2018; Seto, Jeraldo, Orenstein, Chia, & DiBaise, 2014). Next, the molecular targets of certain chemotherapeutic classes, such as antimetabolites, are conserved in bacteria and may contribute to antibacterial activity (Bodet, Jorgensen, & Drutz, 1985). Other medications that are commonly used in older adults that have been studied as potential microbiome-mediating agents include neuropsychiatric medications (e.g., selective serotonin reuptake inhibitors, antipsychotics), opioids, metformin, and probiotics (Kristensen et al., 2016; Maier et al., 2018; Wang et al., 2018; Wu et al., 2017).

Dietary changes may also occur with age, which may ultimately alter the microbiome. The ability of diet to affect the gut microbiome has been noted in multiple studies, with arguments in favor of dietary interventions to modulate the microbiome, reduce inflammation, and promote healthier aging as adults develop into elderly stages (Claesson et al., 2011; Faith, McNulty, Rey, & Gordon, 2011; Guigoz, Dore, & Schiffrin, 2008; Kau, Ahern, Griffin, Goodman, & Gordon, 2011; Walker et al., 2011; Wu et al., 2011). Interestingly, studies in older, frailer humans have noted a shift in the microbiome toward a Bacteroidetes-dominated population, as well as increased populations of pathobionts among centenarians (Jeffery, Lynch, & O'Toole, 2016; Rampelli et al., 2013). However, these alterations in the gut microbial community are also associated with changes to low-fiber diets with less macronutrient diversity typical of nursing-home meals. Therefore it is unclear whether the microbial shift is a result of dietary change, or whether the microbial shift is associated with the aging process and leads to inflammation and disease.

Environmental changes (e.g., nursing home residence) with aging can also impact microbiome composition (Blaser & Falkow, 2009). Studies of patients in nursing homes have demonstrated significant differences in microbiome composition compared to community-dwelling older adults (Claesson et al., 2012). Even within a nursing home environment, the microbiome is affected by the residents' age, frailty, nutritional status, and physical location in relation to other residents (Haran, Bucci, Dutta, Ward, & McCormick, 2018). Other studies have found that within 18 months of being admitted to a long-term care facility, patients' microbiome composition gradually changes, gaining long-stay-type microbiota while simultaneously losing community-associated genera, with these subjects becoming more susceptible to large microbiota changes (Jeffery et al., 2016).

Limitations to studying age-related changes in the human microbiome

Despite the major knowledge gain from recent microbiome studies, there are a number of challenges that remain, including the variability between individuals in regard to lifestyle factors, study reproducibility, and studies which include limited cohorts that generally lack diversity in phenotypes and other microbiological factors (Lyte, 2011). Due to the severe complexity of the gut microbiome, few studies have been able to demonstrate causal relationships in humans in regard to mechanistic pathways and how they impact microbiome dysbiosis, clinical disease states, and aging or longevity.

Given the difficulty in following humans prospectively over years to decades of life, most human microbiome aging studies have been cross-sectional in design. This makes it especially difficult to control for environmental and behavioral confounders. For example, diet is one of the critical factors in shaping the microbiome composition; however, it is incredibly difficult to measure and control for in study designs. As with all clinical studies, the randomized controlled trial design is optimal for studying the effects of an intervention on the microbiome, as the design facilitates roughly equal distribution of known and unknown confounders across groups. The use of animal models to evaluate the interactions between the microbiome and phenotypic aging offers the advantage of being able to control current environment, diet, and exposure to medications. Furthermore, animal models allow for longitudinal study of age-related changes given the shorter lifespan. However, it must be noted that the findings from these studies might not be generalizable to human studies.

Contribution of the microbiome to aging-related conditions

The complexity of the human immune system and the gut microbiome aids in promoting healthy aging and, in turn, prevention of disease. The immune system comprises innate and adaptive properties which, over time, recognize the commensal bacteria residing in our bodies, forming a symbiotic alliance. This relationship further provides the human body with optimal protection from foreign bodies and pathogens. Although this relationship ultimately protects the host from most harm, disruption of the normal gut microbiota can predispose to many diseases. In an aging population, changes to the gut microbiota can increase the risk for a number of chronic conditions, such as cardiovascular disease (CVD), metabolic disease, and neurological disorders like Alzheimer disease (AD) and Parkinson disease (PD). In this section, we summarize aging-related conditions associated with the human microbiome (Table 17.2).

Metabolic diseases (obesity and diabetes)

Obesity is a global health concern. In the United States alone, obesity affects over 90 million adults and the prevalence is increasing, particularly in the elderly population (Centers for Disease Control & Prevention,

TABLE 17.2 Overview of disease states associated with gut microbiome dysbiosis.

Disease category	Disease state
Metabolic diseases	Obesity
	Type II diabetes mellitus
Cardiovascular diseases	Coronary artery disease
	Cerebrovascular disease
	Congestive heart failure
	Hypertension
Neurological diseases	Alzheimer's disease
	Parkinson's disease
	Depression
Infectious diseases	*Clostridioides difficile* infection
	Spontaneous bacterial peritonitis
	Other enteric infections
Oncological diseases	Gastrointestinal cancers
	Hepatocellular carcinomas
Other diseases	Inflammatory bowel disease
	Irritable bowel syndrome

2018). The percentage of obese individuals 60 years of age and older increased from 23.6% (9.9 million people) in 1990 to 37.4% (20.9 million people) in 2010 (Mathus-Vliegen, 2012).

While several factors can predispose an individual to obesity (e.g., diet, genetics, metabolism), the gut microbiome is increasingly recognized for its role in metabolism and contribution to metabolic diseases. Prior animal studies found an association between metabolism and energy storage following the introduction of a microbiome to the host. For example, one study noted an increase in total body fat and suppression of fasting-induced adipocyte factor (FIAF) after germ-free mice were colonized with microbiota from conventional mice. FIAF is crucial in the regulation of hepatic triglycerides, and when this factor is suppressed, hepatic triglyceride production increases, promoting higher caloric storage and ultimately weight gain (Backhed et al., 2004). A second study also noted significantly lower gross fecal energy, an increase in acetate and butyrate production, and increased total body fat when germ-free mice were colonized with the gut microbiota of obese mice (Turnbaugh et al., 2006).

Despite these known metabolic effects, the bacterial makeup of individuals with obesity has yielded varying results, although all involve two major phyla: Firmicutes and Bacteroidetes. Animal models have found an increase in the Firmicutes to Bacteroidetes ratio (an indicator of energy harvest) in obese mice compared to their lean counterparts (Turnbaugh et al., 2006). This was also seen in another mouse study, whereby obesity was associated with significantly greater abundance of Firmicutes and a 50% reduction in Bacteroidetes (Ley et al., 2005). This association was also observed in one human study, where obese patients initially had significantly more Firmicutes and less Bacteroidetes compared to lean patients. Once patients began dieting, the Firmicutes to Bacteroidetes ratio began to decrease (Ley, Turnbaugh, Klein, & Gordon, 2006). Although these studies point to an association between obesity and an increased Firmicutes to Bacteroidetes ratio, one study involving human subjects had contradictory findings. This study found that overweight subjects had greater concentrations of *Bacteroides* (from the phylum Bacteroidetes) compared to lean subjects, although not compared to obese subjects. However, both overweight and obese patients had lower numbers of *Ruminococcus flavefaciens* subgroup, and obese individuals had lower concentrations of *Clostridium leptum*; both of these bacteria belong to the Firmicutes phylum. Overall, the Firmicutes to Bacteroidetes ratio significantly decreased in both overweight and obese patients (Schwiertz et al., 2010). Lastly, several studies also found no association or significant differences between any bacterial phyla (Bondia-Pons et al., 2014; Duncan et al., 2008).

TABLE 17.3 Alterations of the gut microbiota in obese patients.

References	Microbiota in obese patients
Turnbaugh et al. (2006), Ley et al. (2005), Ley et al. (2006)	Increased Firmicutes to Bacteroidetes ratio
Schwiertz et al. (2010)	Decreased Firmicutes to Bacteroidetes ratio
Duncan et al. (2008), Bondia-Pons et al. (2014)	No differences between phyla

The results from these studies are summarized in Table 17.3.

Like obesity, T2DM is a chronic condition that is a major worldwide problem. Obesity and T2DM diabetes are tightly linked, as more than 80% of patients with T2D are classified as overweight (Wen & Duffy, 2017). According to the American Diabetes Association, 30.3 million Americans suffered from diabetes in 2015, with 25.2% of those 65 years of age and older (American Diabetes Association, 2016). Although many of these patients most likely were suffering from this disease in middle age or earlier, mortality and risk of hospitalization begin to increase in the elderly population (Brown, Mangione, Saliba, Sarkisian, & California Healthcare Foundation/American Geriatrics Society Panel on Improving Care for Elders with, 2003). T2DM is commonly characterized by a state of insulin resistance, leading to a dysregulation of blood glucose. The gut microbiota are proposed to play a role in T2DM pathogenesis via several mechanisms, including host signaling through bacterial lipopolysaccharides, metabolisms of dietary fibers to SCFAs, and bacterial modulation of bile acids (Allin, Nielsen, & Pedersen, 2015).

Several studies have evaluated the association between T2DM and the gut microbiome. Compositional differences have included lower relative abundance of Firmicutes and higher abundance of Bacteroidetes and Proteobacteria in the T2DM group compared to controls (Larsen et al., 2010). Another study found enrichment for *Lactobacillus* species and a reduction in *Clostridium* species in T2DM. Researchers also noted a positive correlation between *Lactobacillus* and fasting glucose and glycated hemoglobin (HbA1c), while there was a negative correlation between *Clostridium* and the same biomarkers (Karlsson et al., 2013). The Firmicutes to Bacteroidetes ratio has also been significantly associated with plasma glucose levels (Larsen et al., 2010).

Cardiovascular diseases

Like obesity, CVD is a global health concern. According to the World Health Organization, 17.9 million people die every year from CVD, which accounts for 31% of deaths worldwide, with heart attacks and stroke being the most common causes (World Health Organization, 2019). By 2030, it is estimated that 40% of all deaths in patients over the age of 65 years will be CVD-related (North & Sinclair, 2012). Genetic and environmental factors strongly influence an individual's predisposition to CVD, as it is commonly associated with diet and exercise. Aging is an additional important risk factor for developing CVD, as it has a profound effect on the heart's physiology. As cells in the cardiovascular system age, they predispose elderly patients to an increased risk of several diseases, including atherosclerosis, hypertension, myocardial infarction, and cerebrovascular disease (i.e., stroke) (North & Sinclair, 2012). Although an aging cardiovascular system is the primary predictor for developing CVD, the influence of the gut microbiome and choline metabolism is another area of interest in CVD pathogenesis.

In CVD, it has been shown that risk for disease is potentiated through the metabolism of phospholipid components, namely choline. Some metabolites of interest include trimethylamine *N*-oxide (TMAO), an oxidized product of trimethylamine, and betaine, which forms gaseous trimethylamine through intestinal microbe metabolism. Formation of TMAO from the gut bacteria is thought to increase CVD risk, particularly in the promotion of atherosclerosis. One mouse study found that the gut microbiota provide an obligate role in generating TMAO through the metabolism of dietary phosphatidylcholine. Mice with CVD had elevated levels of these phospholipid metabolites (Wang et al., 2011). In a separate mouse study, atherosclerosis-prone mice fed a choline-enriched diet without antibiotics developed atherosclerosis faster than those with a suppressed microbiota. Those that were fed a choline diet also had raised levels of TMAO and peritoneal macrophage cholesterol content (Wang et al., 2011).

Although evidence is limited in human models, it has been shown that dysbiosis of the gut microbiome plays a role in a number of CVDs, namely hypertension and heart failure. Yang et al. compared the gut microbiome in spontaneously hypertensive rats with that of wild-type rats and saw a significantly increased relative abundance of Firmicutes and a significant decrease in Bacteroidetes (Yang et al., 2015). Another study found that Bacteroidetes was significantly less abundant and Proteobacteria were more abundant in older patients with heart failure compared to younger patients with heart failure (Kamo et al., 2017).

Neurological diseases (Alzheimer disease, Parkinson disease, depression)

The gut microbiome can influence the development of several neurological diseases via the gut–brain axis, a two-way communication highway between the enteric and central nervous systems. Commensal gut bacteria play a large role in the expression of certain neurotransmitters and hormones, which can therefore influence neurodegenerative and psychiatric disorders, namely AD, PD, and depression (Foster & McVey Neufeld, 2013; Sampson et al., 2016; Sochocka et al., 2019). Neurotransmitters such as GABA, serotonin, melatonin, histamine, and acetylcholine are also produced by bacteria, which can further influence the enteric nervous system by modulating gut motility, ion channel opening and closing, and pain perception (Iyer, Aravind, Coon, Klein, & Koonin, 2004; Kunze et al., 2009). Lack of bacterial growth and colonization can lead to alteration of neurotransmitters, resulting in myriad gut sensory–motor functions, such as delayed gastric emptying and prolonged transit time through the gastrointestinal tract (Abrams & Bishop, 1967; Barbara et al., 2005; Clarke et al., 2013; Diaz Heijtz et al., 2011; Iwai, Ishihara, Yamanaka, & Ito, 1973; Stilling, Dinan, & Cryan, 2014).

AD is a neurodegenerative disease that is the leading cause of dementia in the United States. Almost six million Americans are affected by the disease, most of whom are over the age of 65 years (Alzheimer's Association, 2018). AD is characterized by a decline in cognition and memory, later resulting in poor independent functioning and eventually death. The disease was thought to be solely a result of amyloid-beta (Aβ) deposition, leading to immune responses in the brain via microglia activation and ultimately leading to a prolonged inflammatory response, brain injury, and neuronal death (Heneka et al., 2015; Sochocka, Diniz, & Leszek, 2017). However, growing evidence points to Aβ not being the only contributing factor in the development of AD. It has been proposed that changes in the gut microbiota can increase intestine epithelial barrier permeability as well as blood–brain barrier dysfunction that may lead to increased production of Aβ. As a result of Aβ overproduction, the brain then becomes saturated with Aβ, further leading to chronic inflammatory responses (Sochocka et al., 2019).

In addition to inflammatory reactions occurring in the brain, there is also a correlation between gut microbiome composition and AD. Animal models of mice susceptible to Aβ production have noted changes in several bacterial phyla, including a decrease in Firmicutes and an increase in Bacteroidetes as they aged (Harach et al., 2017). In a human study, patients with AD had lower alpha diversity and a decreased relative abundance of Firmicutes and Actinobacteria, as well as an increased relative abundance of Bacteroidetes compared to healthy controls, a result consistent with other chronic diseases affected by dysbiosis in addition to the mouse study mentioned earlier (Vogt et al., 2017).

Like AD, PD is a neurodegenerative disease, although it differs in its pathogenesis. PD commonly occurs in patients over the age of 60 years (Van Den Eeden et al., 2003). The disease originates from decreased dopamine production from the basal ganglia, resulting in disruptions in neural connections to the thalamus and motor cortex (Olanow & Tatton, 1999). Although PD mainly affects the motor system, a major nonmotor symptom is idiopathic constipation, a condition that is caused by neurodegenerative changes in the enteric nervous system (Cersosimo & Benarroch, 2012; Pfeiffer, 2011). It is believed that the accumulation of α-synuclein (αSyn), a neuronal protein, is commonly linked to several neurodegenerative diseases including PD. Accumulation of αSyn begins in the gut and ends in the brain through the vagus nerve (Del Tredici & Braak, 2008). One mouse study demonstrated that the gut microbiome plays a role in promoting motor dysfunction and neuro-inflammation through modulating microglia activation, which is also mediated by αSyn (Sampson et al., 2016). Current evidence regarding the influence of the gut microbiome in PD remains limited. One study in China found changes in the relative abundance of some phyla in patients with PD, although their findings were not statistically significant (Lin et al., 2018).

Lastly, depression is a neuropsychiatric disorder that can be affected by the gut microbiome. Although depression often affects younger individuals, late-life depression can lead to decreased cognitive and social functioning, as well as greater self-neglect, all of which can lead to increased mortality (Blazer, 2003). The etiology of depression is an amalgam of several imbalances, ranging from a deficiency or dysregulation of neuroendocrines to neuroimmune and neurotransmitter systems (Berton & Nestler, 2006; Dowlati et al., 2010; Stetler & Miller, 2011). It is known from a growing body of evidence in animal studies that the gut microbiota has an effect on the hypothalamic–pituitary–adrenal axis and altered tryptophan metabolism, the precursor to serotonin, supporting a potential role for the gut microbiota in depression pathogenesis (Clarke et al., 2013; Sudo et al., 2004).

Studies have shown that the gut microbiome is altered in depressed individuals. In one study, the gut microbiome composition of depressed mice was similar to that of healthy mice at the phylum level, but Prevotellaceae were decreased and Thermoanaerobacteriaceae were increased at the family level. When the gut microbiota of depressed mice were transferred to microbiota-depleted rats, the recipients exhibited a change in behavior, including anhedonia and anxiety-like behaviors. Rats that received donor feces also showed a decrease in alpha diversity, a significant decrease in the abundance of bacteria at the family level, including Bifidobacteriaceae and Coriobacteriaceae, and a significant increase of Propionibacteriaceae (Kelly et al., 2016). This study helps support a causal relationship between the gut microbiome and depression.

Infectious diseases

The gut microbiota and immune system play a key role in preventing pathogens from colonizing the gastrointestinal tract. However, dysbiosis in the gut can cause an imbalance between commensal and pathogenic bacteria, predisposing an individual to several infections. Elderly patients are more prone to developing infections due to physiological changes associated with aging, like immunosenescence (Yoshikawa, 1981). Certain medications, like antibiotics and proton pump inhibitors, can also predispose elderly patients to greater risk of infection. Prolonged exposure to these agents can allow several opportunistic pathogens to grow and proliferate due to an altered microbiome, potentially causing a number of infections, like *Clostroidioides difficile* infection, spontaneous bacterial peritonitis, respiratory infections (e.g., pneumonia), and other enteric infections (e.g., *Salmonella*) (Borriello, 1990; Lachar & Bajaj, 2016; Santos, 2014; de Steenhuijsen Piters et al., 2016).

One of the most well-known organisms known to proliferate as a result of dysbiosis is *Clostridioides difficile*, a Gram-positive spore-forming anaerobe. The CDC (Centers for Disease Control and Prevention) estimates that there are nearly half a million infections per year, predominately in older adults (Kelly & LaMont, 2008; Lessa et al., 2015). In 2013, the CDC reported that one of every three CDIs occur in patients 65 years of age or older, and more than 90% of deaths occur those over the age of 65 years (Antibiotic Resistance Threats in the United States, 2013).

Several studies involving animal models show how the gut microbiome changes post-CDI. Overall, these studies noted an overall decrease in bacterial diversity, a decrease in relative abundance of Firmicutes and Bacteroidetes and an increase in *Proteobacteria* (Chen et al., 2008; Peterfreund et al., 2012; Reeves et al., 2011). When comparing the fecal composition of human patients with CDI and those with antibiotic-associated diarrhea, CDI patients showed a decrease in relative abundance of Firmicutes, while both groups exhibited a decrease in butyrate-producing bacteria (Antharam et al., 2013).

Oncological diseases

The gut microbiome plays a role in the regulation of immune cells in the intestinal system, specifically through the production of antigens. For example, when the body is in a proinflammatory state, the gut microbiome responds by releasing antigens that lead to the downregulation of proinflammatory factors. The process is similar to when the body is under antiinflammatory conditions (Rea et al., 2018). Dysbiosis of intestinal microbiota can lead to the dysregulation of inflammatory responses, which can promote carcinogenesis in several organs (Grivennikov, Greten, & Karin, 2010). For example, in patients with chronic liver disease, hepatocellular carcinoma can be caused by the translocation of intestinal microbiota through the portal vein, thereby triggering inflammatory responses via TLR (Dapito et al., 2012; Round et al., 2011; Seki et al., 2007).

While organisms typically found in the gut microbiome do not normally promote carcinogenesis, there are some pathogens that are capable of doing so by prolonged injury to the epithelium along with inflammation. Infection by *H. pylori* can lead to gastritis, gastric ulcers, and potentially gastric cancer (Fox & Wang, 2007). *Salmonella enterica* subsp. *enterica* serovar Typhi and *Salmonella enterica* subsp. *enterica* serovar Paratyphi infections are also carcinogenic infectious agents, promoting gallbladder cancer as well as mucosa-associated lymphoid tissue (MALT) lymphomas (Caygill, Hill, Braddick, & Sharp, 1994; Welton, Marr, & Friedman, 1979). These infectious agents promote carcinogenesis through adaptive immune responses against these pathogens. Other infectious agents can also promote carcinogenesis, particularly due to prolonged exposure and no antibiotic treatment. These agents include *Borrelia burgdorferi*, which can result in skin MALT lymphomas, *Campylobacter jejuni*, which can cause immunoproliferative small intestinal disease, and *Chlamydia psittaci*, which can lead to ocular adnexal lymphomas (Ferreri et al., 2012; Lecuit et al., 2004; Senff et al., 2008). In addition to infectious pathogens contributing to carcinogenesis, changes in the normal gut microbiome have been seen in several types of cancers and their tissues, including breast, esophageal, colorectal, and prostate cancer (Goodman & Gardner, 2018). In particular, *Fusobacterium* species were seen to be increased in the microbiota of patients with colorectal cancer (Bullman et al., 2017; Kostic et al., 2013).

Other inflammatory diseases

There are also several diseases that are not directly correlated with aging, but that are associated with gut microbiota dysbiosis. Two main types of inflammatory diseases commonly associated with a change in the gut microbiome include inflammatory bowel disease (IBD), which consists of Crohn's disease and ulcerative colitis, and irritable bowel syndrome (IBS), which consists of IBS with constipation (IBS-C), IBS with diarrhea (IBS-D), or IBS with a mixture of both (IBS-M) (Pozuelo et al., 2015). Although the mechanisms behind both inflammatory diseases are not exactly clear, what is understood is that these diseases can affect the makeup of the gut microbiome. In IBD, there is reduced diversity within the microbiome, and other studies showed that SCFA-producing (SCFA) bacteria, like *Faecalibacterium prausnitzii* and *Akkermansia municiphila*, were among those with decreased abundance in patients with IBD (Ott et al., 2004; Rajilic-Stojanovic, Shanahan, Guarner, & de Vos, 2013; Sokol et al., 2008). Individuals with IBS have decreased diversity compared to healthy individuals, particularly those with IBS-D. In this subset, at the family and order levels, Ruminococcaceae, Erysipelotrichaceae, Methanobacteriaceae, and Clostridiales were all significantly reduced (Pozuelo et al., 2015).

Microbiome-targeted therapies as potential antiaging interventions

Biological aging is a complicated process involving many potential causes and cascading failures. While we continue to investigate the cellular processes linked to aging, interventional studies are focused on identifying pharmaceutical agents and other interventional strategies that target human health (Lopez-Otin, Blasco, Partridge, Serrano, & Kroemer, 2013). Due to the connection between dysbiosis and disease, microbiome-targeted interventions hold promise for altering and sustaining microbiota homeostasis in aging individuals. As described in more detail in this section, microbiome-targeted therapies potentially mediate the healthspan by improving gut function and immune homeostasis, while decreasing oxidative stress and inflammation. A summary of the most commonly studied microbiome-targeted interventions is provided in Fig. 17.6

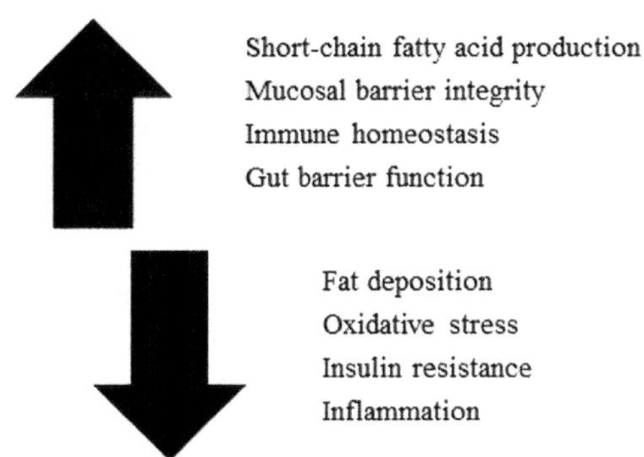

FIGURE 17.6 Mechanisms by which microbiome-targeted therapies might improve the healthspan.

Diet

While it is well known that diet influences the health of an individual, in recent years it has become clear that this association may be mediated in part through interactions with the host gut microbiome. In adults, diets rich in vegetable and fruit consumption, as well as low consumption of meat, have been associated with high gut microbial diversity, favoring a greater abundance of *Prevotella* compared to *Bacteroides* (Claesson et al., 2012; De Filippo et al., 2010; Wu et al., 2011). High-fat diets have been previously associated with a change in the microbiome, favoring a higher Firmicutes:Bacteroidetes ratio and a change in metabolic function (Murphy et al., 2010; Turnbaugh et al., 2006). Because of these associations, major changes in diet should have an impact on human healthspan and possibly lifespan. Caloric restriction and consumption of prebiotics are two potential approaches to altering the microbiome through diet.

Caloric restriction refers to limiting average daily caloric intake, below what is typical, without malnutrition (i.e., deprivation of essential nutrients). Caloric restriction is one of the few interventions that has been shown to effectively lengthen lifespan in multiple animal models, and more recently, to attenuate biomarkers of biologic aging in humans (Belsky, Huffman, Pieper, Shalev, & Kraus, 2017; Colman et al., 2014; Weindruch & Sohal, 1997). The etiology of these associations remains unclear, but evidence suggests the beneficial effects of caloric restriction may be mediated by changes in the gut microbiota.

Numerous studies have demonstrated the ability for caloric restriction to induce dramatic changes in the host gut microbiome. In mice, life-long caloric restriction has been associated with a significant change in the composition of the gut microbiome, specifically enriching for bacterial taxa associated with longevity (e.g., *Lactobacillus*) and reducing the relative abundance of bacterial taxa negatively associated with longevity (e.g., *Lactococcus*, Bacteroidales, Peptostreptococcaceae) (Zhang et al., 2013). This study also identified bacterial taxa specifically associated with lifespan in mice fed low-fat diets. Most of the taxa positively correlated with lifespan belonged to the phyla Firmicutes. Of note, eight bacterial taxa in the *Lactobacillus* genus were strongly correlated with lifespan. Additionally, within 30 days, the composition of the gut microbiota in mice fed a high-fat diet showed an increased abundance of Firmicutes, Proteobacteria, and Verrucomicrobia, and a decrease of Bacteroidetes (Tomas et al., 2016).

Notably, recent studies have documented important physiological changes associated with microbial composition in calorie-restricted animals. Caloric restriction in mice results in a microbial community dominated by *Lactobacillus* species within 2 weeks of the intervention (Pan et al., 2018). This change is associated with a decrease in both circulating microbial antigens and systemic inflammatory markers (e.g., tumor necrosis factor). The bacterial strain *Lactobacillus murinus* was specifically promoted in calorie-restricted mice and contributed to a reduction in age-related inflammation. A second study also noted enrichment in the gut microbiome for *Lactobacillus* in calorie-restricted mice compared to those on an ad libitum diet (Fraumene et al., 2018).

Few studies have assessed the impact of caloric restriction on the gut microbiome in humans. One study evaluated calorie restriction plus physical activity on gut microbial composition in humans and found that patients who lost little weight (<2 kg) did not see a significant shift in microbiome structure; however, patients who had higher weight loss (>4 kg) had a significantly increased abundance of certain bacterial taxa, including *Bacteroides fragilis* and *Lactobacillus* (Santacruz et al., 2009). In a second study, 1-year calorie restriction induced significant shifts in gut microbiota composition, reducing the Firmicutes:Bacteroidetes ratio and enhancing the growth of beneficial microorganisms such as *Bacteroides*, *Roseburia*, *Faecalibacterium*, and *Clostridium* XIVa (Ruiz et al., 2017).

While caloric restriction holds a great deal of promise, it remains controversial due to a lack of well-designed human clinical studies. Additionally, malnutrition is a concern if the diet is not carefully planned to include a sufficient quantity of essential nutrients. Further evidence of safety and effectiveness are needed before recommending a calorie restriction or fasting diet, particularly in older adults, unless under the supervision of an experienced registered dietician.

A second way to specifically target the microbiome through the diet is with prebiotics. A prebiotic has been previously defined as "a substrate that is selectively utilized by host microorganisms conferring a health benefit" (Gibson et al., 2017). Many compounds have been reported as prebiotics, including oligofructose, inulin, lactulose, and galactooligosaccharides (Roberfroid et al., 2010). More recently, resistant starch, pectin, and other fiber components have been proposed as prebiotics and have been the focus of many microbiome studies (Bird, Conlon, Christophersen, & Topping, 2010; Rastall & Gibson, 2015).

Fibers such as resistant starch and inulin comprise one of the more widely studied types of prebiotics known to have significant effects on the host microbiome in animal models of aging and more recently in human studies. Fiber has the potential for improving healthspan in the elderly by: (1) increasing SCFA production via alterations of the host microbiome;

(2) improving gut barrier function, and (3) and regulating glucose and lipid metabolism (Keenan, Marco, Ingram, & Martin, 2015). Diets that are mostly plant-based contain resistant starch and oligosaccharides and have been shown to positively affect microbiome diversity and the abundance of certain bacterial taxa (Flint, 2012; Flint, Scott, Duncan, Louis, & Forano, 2012).

Studies evaluating the impact of prebiotics on longevity have been conducted in animal models, as well as in human observational cohort studies. A study by Rozan et al. found that lifelong supplementation with an oligofructose-enriched inulin in rats improved biomarkers of aging and survival (Rozan et al., 2008). Interestingly, a second mouse model noted significantly reduced brain inflammation with aging in mice fed a high-fiber diet of 5% inulin (Matt et al., 2018). These changes were associated with a significant shift in gut microbiome composition, as well as increases in SCFA production. Human studies have also demonstrated relatively compelling data for fiber and health and mortality. A meta-analysis of seven prospective cohort studies found a 23% reduced risk of mortality associated with high dietary fiber intake (Kim & Je, 2014). Given the potential health benefits and limited gastrointestinal side effects (primarily flatulence and bloating), dietary fiber can be recommended as part of a healthy diet in older adults.

Probiotics

Given the specific microbiome changes with aging and the functional changes that are seen concomitantly, it is plausible that supplementation of microbiota could restore those organisms lost with age and potentially impact human health. According to the World Health Organization, probiotics are defined as "live microorganisms which, when administered in adequate amounts, confer a health benefit on the host" (Hill et al., 2014). The microorganisms most commonly used in human nutrition are certain strains belonging to the genera *Lactobacillus*, *Bifidobacterium*, and *Enterococcus*, as well as the yeast *Saccharomyces boulardii*. Probiotic effects on health are proposed to be mediated through: (1) interference with pathogenic organisms via competition for resources, (2) improvement of gastrointestinal epithelial barrier function, (3) immunomodulation, and (4) production of neurotransmitters (Sanchez et al., 2017). Much of the research on probiotics and health has focused on gastrointestinal conditions. Probiotic use has been associated with a benefit in infectious diarrhea, CDI, IBD, and *Helicobacter pylori* infection (Ritchie & Romanuk, 2012). Probiotics have also been studied in nongastrointestinal conditions, including AD, depression, metabolic diseases (including type II diabetes), CVD, allergic diseases, and respiratory tract infections (Huang, Wang, & Hu, 2016; Nimgampalle & Kuna, 2017; Samah, Ramasamy, Lim, & Neoh, 2016; Sun & Buys, 2015; Wang, Anvari, & Anagnostou, 2019; Wang et al., 2016). However, it is important to note that there is insufficient evidence at this time to support recommendations for probiotic use for nongastrointestinal conditions.

While studies have documented mixed results regarding alteration of the gut microbiome composition with probiotic use, many of the health effects have been attributed to the metabolic changes occurring due to improving gut ecosystem homeostasis (Kristensen et al., 2016). For example, a recent study noted that supplementation with *L. acidophilus* DDS-1 in young and aging mice resulted in an increased relative abundance of certain beneficial bacteria, such as *Akkermansia muciniphila* and *Lactobacillus* spp., and reduced the relative levels of opportunistic bacteria such as *Proteobacteria* spp. (Vemuri et al., 2018). Importantly, 10 metabolic pathways involved in amino acid, protein, carbohydrate, and butanoate metabolism were altered. Animal studies in *Drosophila* have found that probiotic and synbiotic formulations reduced markers of physiological stress, oxidative stress, and inflammation, thereby targeting several important aging mechanisms (Westfall, Lomis, & Prakash, 2018). In a recent human randomized controlled trial of older adult women, synbiotic use resulted in significant shifts in the gut microbiome composition, promotion of innate immunity through increased natural killer cells, decreased proinflammatory cytokine IL-6, and reduced total and LDL cholesterol levels (Costabile et al., 2017).

Importantly, work in animal models, primarily the nematode *Caenorhabditis elegans*, suggests that probiotics may also play a role in longevity. These lifespan-altering effects were associated with innate immune response signaling, improved resistance to oxidative stress, and modulated serotonin signaling (Grompone et al., 2012; Kwon, Lee, & Lim, 2016; Nakagawa et al., 2016; Park et al., 2015). There is also evidence that probiotic treatment can promote longevity in mice, possibly through suppression of chronic low-grade inflammatory processes in the colon (Matsumoto, Kurihara, Kibe, Ashida, & Benno, 2011). Interestingly, scientists are pursuing the possibility of genetically engineered probiotics to promote longevity. Recently, researchers screened 3983 *E. coli* mutants and discovered that 29 bacterial genes, when deleted, increase longevity in *C. elegans* (Han et al., 2017). Additionally, rationally designed combinations of microbes (e.g., Ecobiotics) are being developed to target known microbial deficiencies.

While the evidence is promising, there are important limitations with probiotic use. First, although most clinical studies have found probiotics to be relatively safe, potential adverse effects exists. Some of the risks associated with probiotic use include gastrointestinal side effects, unfavorable metabolic profile, excessive immune stimulation, and systemic infections in susceptible individuals, as well as horizontal transfer of genes (Doron & Snydman, 2015). Next, the optimal strain(s) to use, as well as doses and dosage forms are not well understood. Given the heterogeneity in host genome and microbiome, there might also be factors that influence the effectiveness of probiotics, but little is known in this area. Therefore further studies are required to optimize probiotic use, further quantify the risk to benefit ratio, as well as identify the potential mechanisms involved in the health benefits and adverse effects of probiotics.

Fecal microbiota transplantation

One important limitation of probiotics is that oftentimes supplementation is only with one or few bacterial species; however, dysbiosis may involve a deficiency in several important bacterial taxa. Fecal microbiota transplantation (FMT) addresses this limitation. FMT refers to the administration of healthy donor feces into the intestinal tract of a recipient in order to directly change the recipient's microbial composition and confer a health benefit (Bakken et al., 2011; Smits, Bouter, de Vos, Borody, & Nieuwdorp, 2013). The process involves first selecting a donor without a family history of autoimmune, metabolic, and malignant diseases and screening for any potential pathogens. The feces are then prepared by mixing with water or normal saline, followed by a filtration step to remove any particulate matter. The mixture can be administered through a nasogastric tube, nasojejunal tube, esophagogastroduodenoscopy, colonoscopy, or retention enema (Gupta, Allen-Vercoe, & Petrof, 2016). In 2012, the United States Food and Drug Administration classified FMT as a new investigational therapeutic option in the treatment of particular diseases (Vyas, Aekka, & Vyas, 2015).

Prior mouse models of FMT offer the best evidence of a causal relationship between microbial communities and disease to date. Evaluations of gnotobiotic mouse recipients of human donor fecal material as well as mouse-derived fecal material show weight gain and metabolic dysfunction, if they received material from an obese donor (Boulange, Neves, Chilloux, Nicholson, & Dumas, 2016; Lai et al., 2018; Ussar et al., 2015). Interestingly, transplanting fecal microbiota from lean individuals to obese individuals results in weight loss and rescued metabolic function (Di Luccia et al., 2015; Marotz & Zarrinpar, 2016). Further transplanting fecal microbiota from animals that were fed high-fat diets resulted in increased anxiety, increased stereotypical behavior, and decreased cognition associated with increased neuro-inflammation in mice that did not yet show changes in obesity or metabolic function (Bruce-Keller et al., 2015). These findings in mice suggest that shifts in microbial communities and the by-products that they provide are associated with inflammation and many disease states. They also suggest that FMT may be an interesting interventional treatment to stabilize microbial communities and restore homeostasis.

FMT has been used successfully for decades to treat CDI in geriatric people (Kassam, Lee, Yuan, & Hunt, 2013). In a recent meta-analysis of 37 studies, FMT was more effective than vancomycin (RR: 0.23; 95%CI: 0.07–0.80) in resolving recurrent and refractory CDI, regardless of the route of administration (Quraishi et al., 2017). Importantly, FMT promotes recovery of the microbiome post-CDI (Shahinas et al., 2012). Prior studies have noted primarily expected, short-term adverse events (e.g., bloating, diarrhea, constipation, flatulence) following FMT. Serious adverse events (e.g., infections, aspiration, ileus) are uncommon. Given these findings, further testing of FMT as a potential bacteriotherapy option for healthy aging are needed. FMT may allow for the restoration of a stable, highly diverse gut community typically found in young adults consisting of Firmicutes and Bacteroidetes, but with high diversity in other phyla of bacteria (Wilson, Vatanen, Cutfield, & O'Sullivan, 2019).

FMT's potential effectiveness and safety has been demonstrated in the prevention and treatment of other gastrointestinal and nongastrointestinal conditions including those commonly associated with aging (e.g., atherosclerosis, metabolic syndrome, type II diabetes, neurodegenerative disease) (Choi & Cho, 2016; Konturek et al., 2015). First, FMT has been studied in humans with metabolic syndrome, beneficially affecting the microbiota composition in recipients, as manifested by increased amounts of butyrate-producing bacteria and improved insulin sensitivity 6 weeks post-FMT (Vrieze et al., 2012). Interestingly, not all lean donors exerted similar effects on insulin sensitivity. Several donors demonstrated very significant effects, while others showed no effect. The authors suggest that "super-fecal donor" effects are most likely to occur through high amounts of butyrate-producing bacteria in the donors' fecal matter (Udayappan, Hartstra, Dallinga-Thie, & Nieuwdorp, 2014). Next, FMT has been extensively studied as a treatment for ulcerative colitis. In a recent meta-analysis of 18 studies ($n = 446$), FMT was found to be more effective (OR: 2.73) in treating ulcerative colitis compared to traditional therapies (Cao et al., 2018). Lastly, FMT is being studied as a potential intervention for neuropsychiatric and neurodegenerative diseases, such as Parkinson disorder, multiple sclerosis, and depression; however, supportive effectiveness data are currently lacking and are primarily limited to case reports or animal models (Kurokawa et al., 2018; Makkawi, Camara-Lemarroy, & Metz, 2018; Sun et al., 2018).

Conclusions

As lifespan improves and the proportion of people who reach older age increases exponentially, there is an urgent need to lessen the burden of aging-related conditions and improve quality of life with aging. Elderly adults experience a loss in microbiome diversity and core bacterial phyla, which predisposes to aging-related conditions, including metabolic, cardiovascular, and inflammatory diseases and infections. Given these changes in the microbiome with age and the established association between microbiome perturbations and disease, microbiome-targeted therapies may be important for extending human healthspan and lifespan. The field of microbiome research, as it relates to the aging process, has great opportunities for continued investigation. Robust research studies utilizing translational animal and human models are needed to further define the role of the microbiome on cellular aging and immunosenescence, to validate the causal relationship between the microbiome and aging-related conditions, and to determine the extent to which microbiome-targeted therapies can improve overall human health.

References

Aagaard, K., Ma, J., Antony, K. M., Ganu, R., Petrosino, J., & Versalovic, J. (2014). The placenta harbors a unique microbiome. *Science Translational Medicine*, *6*(237), 237–265.

Abrams, G. D., & Bishop, J. E. (1967). Effect of the normal microbial flora on gastrointestinal motility. *Proceedings of the Society for Experimental Biology and Medicine*, *126*(1), 301–304.

Allin, K. H., Nielsen, T., & Pedersen, O. (2015). Mechanisms in endocrinology: Gut microbiota in patients with type 2 diabetes mellitus. *European Journal of Endocrinology*, *172*(4), R167–R177.

Alzheimer's Association. (2018). 2018 Alzheimer's disease facts and figures. <https://www.alz.org/media/homeoffice/facts%20and%20figures/facts-and-figures.pdf> Accessed 07.06.20.

American Diabetes Association. (2016). Diabetes overview. <https://www.diabetes.org/diabetes> Accessed 07.06.20.

Antharam, V. C., Li, E. C., Ishmael, A., Sharma, A., Mai, V., Rand, K. H., & Wang, G. P. (2013). Intestinal dysbiosis and depletion of butyrogenic bacteria in *Clostridium* difficile infection and nosocomial diarrhea. *Journal of Clinical Microbiology*, *51*(9), 2884–2892.

Antibiotic Resistance Threats in the United States. (2013). <https://www.cdc.gov/drugresistance/pdf/ar-threats-2013-508.pdf> Accessed 07.06.20.

Backhed, F., Ding, H., Wang, T., Hooper, L. V., Koh, G. Y., Nagy, A., ... Gordon, J. I. (2004). The gut microbiota as an environmental factor that regulates fat storage. *Proceedings of the National Academy of Sciences of the United State of America*, *101*(44), 15718–15723.

Backhed, F., Ley, R. E., Sonnenburg, J. L., Peterson, D. A., & Gordon, J. I. (2005). Host-bacterial mutualism in the human intestine. *Science*, *307*(5717), 1915–1920.

Backhed, F., Roswall, J., Peng, Y., Feng, Q., Jia, H., Kovatcheva-Datchary, P., ... Wang, J. (2015). Dynamics and stabilization of the human gut microbiome during the first year of life. *Cell Host & Microbe*, *17*(6), 852.

Bakken, J. S., Borody, T., Brandt, L. J., Brill, J. V., Demarco, D. C., Franzos, M. A., ... Fecal Microbiota Transplantation, W. (2011). Treating *Clostridium difficile* infection with fecal microbiota transplantation. *Clinical Gastroenterology and Hepatology*, *9*(12), 1044–1049.

Balmer, S. E., & Wharton, B. A. (1989). Diet and faecal flora in the newborn: Breast milk and infant formula. *Archives of Disease in Childhood*, *64*(12), 1672–1677.

Barbara, G., Stanghellini, V., Brandi, G., Cremon, C., Di Nardo, G., De Giorgio, R., & Corinaldesi, R. (2005). Interactions between commensal bacteria and gut sensorimotor function in health and disease. *American Joural of Gastroenterology*, *100*(11), 2560–2568.

Barrett, E., Ross, R. P., O'Toole, P. W., Fitzgerald, G. F., & Stanton, C. (2012). gamma-Aminobutyric acid production by culturable bacteria from the human intestine. *Journal of Applied Microbiology*, *113*(2), 411–417.

Belkaid, Y., & Hand, T. W. (2014). Role of the microbiota in immunity and inflammation. *Cell*, *157*(1), 121–141.

Belsky, D. W., Huffman, K. M., Pieper, C. F., Shalev, I., & Kraus, W. E. (2017). Change in the rate of biological aging in response to caloric restriction: Calerie biobank analysis. *Journal of Gerontology—Series A Biological Sciences and Medical Sciences*, *73*(1), 4–10.

Bergstrom, A., Skov, T. H., Bahl, M. I., Roager, H. M., Christensen, L. B., Ejlerskov, K. T., ... Licht, T. R. (2014). Establishment of intestinal microbiota during early life: A longitudinal, explorative study of a large cohort of Danish infants. *Applied and Environmental Microbiology*, *80*(9), 2889–2900.

Berton, O., & Nestler, E. J. (2006). New approaches to antidepressant drug discovery: Beyond monoamines. *Nature Reviews Neuroscience*, *7*(2), 137–151.

Biagi, E., Nylund, L., Candela, M., Ostan, R., Bucci, L., Pini, E., ... De Vos, W. (2010). Through ageing, and beyond: Gut microbiota and inflammatory status in seniors and centenarians. *PLoS One*, *5*(5), e10667. Available from https://doi.org/10.1371/journal.pone.0010667.

Bird, A. R., Conlon, M. A., Christophersen, C. T., & Topping, D. L. (2010). Resistant starch, large bowel fermentation and a broader perspective of prebiotics and probiotics. *Beneficial Microbes*, *1*(4), 423–431.

Blaser, M. J., & Falkow, S. (2009). What are the consequences of the disappearing human microbiota? *Nature Reviews Microbiology*, *7*(12), 887–894.

Blazer, D. G. (2003). Depression in late life: Review and commentary. *Journal of Gerontology—Series A Biological Sciences and Medical Sciences*, *58*(3), 249–265.

Bodet, C. A., 3rd, Jorgensen, J. H., & Drutz, D. J. (1985). Antibacterial activities of antineoplastic agents. *Antimicrobial Agents and Chemotherapy*, *28*(3), 437–439.

Bondia-Pons, I., Maukonen, J., Mattila, I., Rissanen, A., Saarela, M., Kaprio, J., ... Oresic, M. (2014). Metabolome and fecal microbiota in monozygotic twin pairs discordant for weight: A big mac challenge. *FASEB Journal*, *28*(9), 4169–4179.

Borriello, S. P. (1990). The influence of the normal flora on *Clostridium difficile* colonisation of the gut. *Annals of Medicine*, *22*(1), 61–67.

Boulange, C. L., Neves, A. L., Chilloux, J., Nicholson, J. K., & Dumas, M. E. (2016). Impact of the gut microbiota on inflammation, obesity, and metabolic disease. *Genome Medicine*, *8*(1), 42.

Brown, A. F., Mangione, C. M., Saliba, D., Sarkisian, C. A., & California Healthcare Foundation/American Geriatrics Society Panel on Improving Care for Elders with, D. (2003). Guidelines for improving the care of the older person with diabetes mellitus.

Journal of the American Geriatrics Society, *51*(5 Suppl. Guidelines), S265–S280.

Bruce-Keller, A. J., Salbaum, J. M., Luo, M., Blanchard, E. t, Taylor, C. M., Welsh, D. A., & Berthoud, H. R. (2015). Obese-type gut microbiota induce neurobehavioral changes in the absence of obesity. *Biological Psychiatry*, *77*(7), 607–615.

Brummel, T., Ching, A., Seroude, L., Simon, A. F., & Benzer, S. (2004). *Drosophila* lifespan enhancement by exogenous bacteria. *Proceedings of the National Academy of Sciences of United States of America*, *101*(35), 12974–12979.

Buford, T. W. (2017). (Dis)Trust your gut: The gut microbiome in age-related inflammation, health, and disease. *Microbiome*, *5*(1), 80.

Bullman, S., Pedamallu, C. S., Sicinska, E., Clancy, T. E., Zhang, X., Cai, D., ... Meyerson, M. (2017). Analysis of fusobacterium persistence and antibiotic response in colorectal cancer. *Science*, *358* (6369), 1443–1448.

Cabreiro, F., Au, C., Leung, K. Y., Vergara-Irigaray, N., Cocheme, H. M., Noori, T., ... Gems, D. (2013). Metformin retards aging in *C. elegans* by altering microbial folate and methionine metabolism. *Cell*, *153*(1), 228–239.

Candela, M., Biagi, E., Centanni, M., Turroni, S., Vici, M., Musiani, F., ... Brigidi, P. (2009). Bifidobacterial enolase, a cell surface receptor for human plasminogen involved in the interaction with the host. *Microbiology*, *155*(Pt. 10), 3294–3303.

Canfora, E. E., Jocken, J. W., & Blaak, E. E. (2015). Short-chain fatty acids in control of body weight and insulin sensitivity. *Nature Reviews Endocrinology*, *11*(10), 577–591.

Cao, Y., Zhang, B., Wu, Y., Wang, Q., Wang, J., & Shen, F. (2018). The value of fecal microbiota transplantation in the treatment of ulcerative colitis patients: A systematic review and meta-analysis. *Gastroenterology Research and Practice*, *2018*, 5480961.

Caygill, C. P., Hill, M. J., Braddick, M., & Sharp, J. C. (1994). Cancer mortality in chronic typhoid and paratyphoid carriers. *Lancet*, *343* (8889), 83–84.

Centers for Disease Control and Prevention. (2018). Adult obesity facts. <https://www.cdc.gov/obesity/data/adult.html> Accessed 07.06.20.

Cersosimo, M. G., & Benarroch, E. E. (2012). Pathological correlates of gastrointestinal dysfunction in Parkinson's disease. *Neurobiology of Disease*, *46*(3), 559–564.

Chen, X., Katchar, K., Goldsmith, J. D., Nanthakumar, N., Cheknis, A., Gerding, D. N., & Kelly, C. P. (2008). A mouse model of *Clostridium difficile*-associated disease. *Gastroenterology*, *135*(6), 1984–1992.

Cheng, J., Ringel-Kulka, T., Heikamp-de Jong, I., Ringel, Y., Carroll, I., de Vos, W. M., ... Satokari, R. (2016). Discordant temporal development of bacterial phyla and the emergence of core in the fecal microbiota of young children. *ISME Journal*, *10*(4), 1002–1014.

Choi, H. H., & Cho, Y. S. (2016). Fecal microbiota transplantation: Current applications, effectiveness, and future perspectives. *Clinical Endoscopy*, *49*(3), 257–265.

Claesson, M. J., Cusack, S., O'Sullivan, O., Greene-Diniz, R., de Weerd, H., Flannery, E., ... O'Toole, P. W. (2011). Composition, variability, and temporal stability of the intestinal microbiota of the elderly. *Proceedings of the National Academy of Sciences of United States of America*, *108*(Suppl. 1)), 4586–4591.

Claesson, M. J., Jeffery, I. B., Conde, S., Power, S. E., O'Connor, E. M., Cusack, S., ... O'Toole, P. W. (2012). Gut microbiota composition correlates with diet and health in the elderly. *Nature*, *488* (7410), 178–184.

Clark, R. I., Salazar, A., Yamada, R., Fitz-Gibbon, S., Morselli, M., Alcaraz, J., ... Walker, D. W. (2015). Distinct shifts in microbiota composition during *Drosophila* aging impair intestinal function and drive mortality. *Cell Reports*, *12*(10), 1656–1667. Available from https://doi.org/10.1016/j.celrep.2015.08.004.

Clarke, G., Grenham, S., Scully, P., Fitzgerald, P., Moloney, R. D., Shanahan, F., ... Cryan, J. F. (2013). The microbiome-gut-brain axis during early life regulates the hippocampal serotonergic system in a sex-dependent manner. *Molecular Psychiatry*, *18*(6), 666–673.

Colman, R. J., Beasley, T. M., Kemnitz, J. W., Johnson, S. C., Weindruch, R., & Anderson, R. M. (2014). Caloric restriction reduces age-related and all-cause mortality in rhesus monkeys. *Nature Communications*, *5*, 3557.

Costabile, A., Bergillos-Meca, T., Rasinkangas, P., Korpela, K., de Vos, W. M., & Gibson, G. R. (2017). Effects of soluble corn fiber alone or in synbiotic combination with *Lactobacillus rhamnosus* GG and the pilus-deficient derivative GG-PB12 on fecal microbiota, metabolism, and markers of immune function: A randomized, double-blind, placebo-controlled, crossover study in healthy elderly (Saimes study). *Frontiers in Immunology*, *8*, 1443.

Cuervo, A., Salazar, N., Ruas-Madiedo, P., Gueimonde, M., & Gonzalez, S. (2013). Fiber from a regular diet is directly associated with fecal short-chain fatty acid concentrations in the elderly. *Nutrition Research Reviews*, *33*(10), 811–816.

Dapito, D. H., Mencin, A., Gwak, G. Y., Pradere, J. P., Jang, M. K., Mederacke, I., ... Schwabe, R. F. (2012). Promotion of hepatocellular carcinoma by the intestinal microbiota and TLR4. *Cancer Cell*, *21*(4), 504–516.

De Filippo, C., Cavalieri, D., Di Paola, M., Ramazzotti, M., Poullet, J. B., Massart, S., ... Lionetti, P. (2010). Impact of diet in shaping gut microbiota revealed by a comparative study in children from Europe and rural Africa. *Proceedings of the National Academy of Sciences of United States of America*, *107*(33), 14691–14696.

de Steenhuijsen Piters, W. A., Huijskens, E. G., Wyllie, A. L., Biesbroek, G., van den Bergh, M. R., Veenhoven, R. H., ... Bogaert, D. (2016). Dysbiosis of upper respiratory tract microbiota in elderly pneumonia patients. *ISME Journal*, *10*(1), 97–108.

Del Tredici, K., & Braak, H. (2008). A not entirely benign procedure: Progression of Parkinson's disease. *Acta Neuropathologica*, *115*(4), 379–384.

Dethlefsen, L., Huse, S., Sogin, M. L., & Relman, D. A. (2008). The pervasive effects of an antibiotic on the human gut microbiota, as revealed by deep 16S rRNA sequencing. *PLoS Biology*, *6*(11), e280.

Dethlefsen, L., & Relman, D. A. (2011). Incomplete recovery and individualized responses of the human distal gut microbiota to repeated antibiotic perturbation. *Proceedings of the National Academy of Sciences of United States of America*, *108*(Suppl. 1), 4554–4561.

Devlin, A. S., & Fischbach, M. A. (2015). A biosynthetic pathway for a prominent class of microbiota-derived bile acids. *Nature Chemical Biology*, *11*(9), 685–690.

Di Luccia, B., Crescenzo, R., Mazzoli, A., Cigliano, L., Venditti, P., Walser, J. C., ... Iossa, S. (2015). Rescue of fructose-induced metabolic syndrome by antibiotics or faecal transplantation in a rat model of obesity. *PLoS One*, *10*(8), e0134893.

Diaz Heijtz, R., Wang, S., Anuar, F., Qian, Y., Bjorkholm, B., Samuelsson, A., ... Pettersson, S. (2011). Normal gut microbiota modulates brain development and behavior. *Proceedings of the National Academy of Sciences of United States of America*, *108*(7), 3047–3052.

DiGiulio, D. B., Romero, R., Kusanovic, J. P., Gomez, R., Kim, C. J., Seok, K. S., ... Relman, D. A. (2010). Prevalence and diversity of microbes in the amniotic fluid, the fetal inflammatory response, and pregnancy outcome in women with preterm pre-labor rupture of membranes. *American Journal of Reproductive Immunology*, *64*(1), 38–57.

Doron, S., & Snydman, D. R. (2015). Risk and safety of probiotics. *Clinical Infectious Diseases*, *60*(Suppl. 2), S129–S134.

Dowlati, Y., Herrmann, N., Swardfager, W., Liu, H., Sham, L., Reim, E. K., & Lanctot, K. L. (2010). A meta-analysis of cytokines in major depression. *Biology Psychiatry, 67*(5), 446–457.

Duncan, S. H., Lobley, G. E., Holtrop, G., Ince, J., Johnstone, A. M., Louis, P., & Flint, H. J. (2008). Human colonic microbiota associated with diet, obesity and weight loss. *International Journal of Obesity, 32*(11), 1720–1724.

Eckburg, P. B., Bik, E. M., Bernstein, C. N., Purdom, E., Dethlefsen, L., Sargent, M., ... Relman, D. A. (2005). Diversity of the human intestinal microbial flora. *Science, 308*(5728), 1635–1638.

Eisenberg, T., Knauer, H., Schauer, A., Buttner, S., Ruckenstuhl, C., Carmona-Gutierrez, D., ... Madeo, F. (2009). Induction of autophagy by spermidine promotes longevity. *Nature Cell Biology, 11* (11), 1305–1314.

Faith, J. J., McNulty, N. P., Rey, F. E., & Gordon, J. I. (2011). Predicting a human gut microbiota's response to diet in gnotobiotic mice. *Science, 333*(6038), 101–104.

Ferreri, A. J., Govi, S., Pasini, E., Mappa, S., Bertoni, F., Zaja, F., ... Dolcetti, R. (2012). Chlamydophila psittaci eradication with doxycycline as first-line targeted therapy for ocular adnexae lymphoma: Final results of an international phase II trial. *Journal of Clinical Oncology, 30*(24), 2988–2994.

Flint, H. J. (2012). The impact of nutrition on the human microbiome. *Nutrition Review, 70*(Suppl. 1), S10–S13.

Flint, H. J., Scott, K. P., Duncan, S. H., Louis, P., & Forano, E. (2012). Microbial degradation of complex carbohydrates in the gut. *Gut Microbes, 3*(4), 289–306.

Foster, J. A., & McVey Neufeld, K. A. (2013). Gut-brain axis: How the microbiome influences anxiety and depression. *Trends Neuroscience, 36*(5), 305–312.

Fox, J. G., & Wang, T. C. (2007). Inflammation, atrophy, and gastric cancer. *Journal of Clinical Investigation, 117*(1), 60–69.

Franceschi, C., & Bonafe, M. (2003). Centenarians as a model for healthy aging. *Biochemical Society Transactions, 31*(2), 457–461.

Frank, D. N., St Amand, A. L., Feldman, R. A., Boedeker, E. C., Harpaz, N., & Pace, N. R. (2007). Molecular-phylogenetic characterization of microbial community imbalances in human inflammatory bowel diseases. *Proceedings of the National Academy of Sciences of United States of America, 104*(34), 13780–13785.

Fraumene, C., Manghina, V., Cadoni, E., Marongiu, F., Abbondio, M., Serra, M., ... Uzzau, S. (2018). Caloric restriction promotes rapid expansion and long-lasting increase of *Lactobacillus* in the rat fecal microbiota. *Gut Microbes, 9*(2), 104–114.

Freedberg, D. E., Toussaint, N. C., Chen, S. P., Ratner, A. J., Whittier, S., Wang, T. C., ... Abrams, J. A. (2015). Proton pump inhibitors alter specific taxa in the human gastrointestinal microbiome: A crossover trial. *Gastroenterology, 149*(4), 883–885, e889.

Fulde, M., & Hornef, M. W. (2014). Maturation of the enteric mucosal innate immune system during the postnatal period. *Immunological Review, 260*(1), 21–34.

Gibson, M. K., Crofts, T. S., & Dantas, G. (2015). Antibiotics and the developing infant gut microbiota and resistome. *Current Opinion Microbiology, 27*, 51–56.

Gibson, G. R., Hutkins, R., Sanders, M. E., Prescott, S. L., Reimer, R. A., Salminen, S. J., ... Reid, G. (2017). Expert consensus document: The International Scientific Association for Probiotics and Prebiotics (ISAPP) consensus statement on the definition and scope of prebiotics. *Nature Reviews Gastroenterol & Hepatology, 14* (8), 491–502.

Gill, S. R., Pop, M., Deboy, R. T., Eckburg, P. B., Turnbaugh, P. J., Samuel, B. S., ... Nelson, K. E. (2006). Metagenomic analysis of the human distal gut microbiome. *Science, 312*(5778), 1355–1359.

Goodman, B., & Gardner, H. (2018). The microbiome and cancer. *Journal of Pathology, 244*(5), 667–676.

Grivennikov, S. I., Greten, F. R., & Karin, M. (2010). Immunity, inflammation, and cancer. *Cell, 140*(6), 883–899.

Grompone, G., Martorell, P., Llopis, S., Gonzalez, N., Genoves, S., Mulet, A. P., ... Ramon, D. (2012). Anti-inflammatory *Lactobacillus rhamnosus* CNCM I-3690 strain protects against oxidative stress and increases lifespan in *Caenorhabditis elegans*. *PLoS One, 7*(12), e52493.

Gronlund, M. M., Lehtonen, O. P., Eerola, E., & Kero, P. (1999). Fecal microflora in healthy infants born by different methods of delivery: Permanent changes in intestinal flora after cesarean delivery. *Journal of Pediatric Gastroenterology and Nutrition, 28*(1), 19–25.

Guigoz, Y., Dore, J., & Schiffrin, E. J. (2008). The inflammatory status of old age can be nurtured from the intestinal environment. *Current Opinion in Clinical Nutrition and Metabolic Care, 11*(1), 13–20.

Gupta, S., Allen-Vercoe, E., & Petrof, E. O. (2016). Fecal microbiota transplantation: In perspective. *Therapeutic Advances in Gastroenterology, 9*(2), 229–239. Available from https://doi.org/10.1177/1756283X15607414.

Haiser, H. J., Gootenberg, D. B., Chatman, K., Sirasani, G., Balskus, E. P., & Turnbaugh, P. J. (2013). Predicting and manipulating cardiac drug inactivation by the human gut bacterium *Eggerthella Lenta*. *Science, 341*(6143), 295–298.

Han, B., Sivaramakrishnan, P., Lin, C. J., Neve, I. A. A., He, J., Tay, L. W. R., ... Wang, M. C. (2017). Microbial genetic composition tunes host longevity. *Cell, 169*(7), 1249–1262.

Harach, T., Marungruang, N., Duthilleul, N., Cheatham, V., Mc Coy, K. D., Frisoni, G., ... Bolmont, T. (2017). Reduction of abeta amyloid pathology in APPPS1 transgenic mice in the absence of gut microbiota. *Scientific Reports, 7*, 41802.

Haran, J. P., Bucci, V., Dutta, P., Ward, D., & McCormick, B. (2018). The nursing home elder microbiome stability and associations with age, frailty, nutrition and physical location. *Journal of Medical Microbiology, 67*(1), 40–51.

Hayashi, H., Sakamoto, M., Kitahara, M., & Benno, Y. (2003). Molecular analysis of fecal microbiota in elderly individuals using 16S rDNA library and T-RFLP. *Microbiology* and *Immunology, 47*(8), 557–570.

Heneka, M. T., Carson, M. J., El Khoury, J., Landreth, G. E., Brosseron, F., Feinstein, D. L., ... Kummer, M. P. (2015). Neuroinflammation in Alzheimer's disease. *Lancet Neurology, 14*(4), 388–405.

Hill, M. J. (1997). Intestinal flora and endogenous vitamin synthesis. *European Journal of Cancer Prevention, 6*(Suppl. 1)), S43–S45.

Hill, C., Guarner, F., Reid, G., Gibson, G. R., Merenstein, D. J., Pot, B., ... Sanders, M. E. (2014). Expert consensus document. The International Scientific Association for Probiotics and Prebiotics consensus statement on the scope and appropriate use of the term probiotic. *Nature Reviews Gastroenterology & Hepatology, 11*(8), 506–514.

Hollister, E. B., Riehle, K., Luna, R. A., Weidler, E. M., Rubio-Gonzales, M., Mistretta, T. A., ... Versalovic, J. (2015). Structure and function of the healthy pre-adolescent pediatric gut microbiome. *Microbiome, 3*, 36.

Hopkins, M. J. & Macfarlane, G. T. (2002). Changes in predominant bacterial populations in human faeces with age and with Clostridium difficile infection. *Journal of Medical Microbiology 51* (5), 448–454.

Huang, R., Wang, K., & Hu, J. (2016). Effect of probiotics on depression: A systematic review and meta-analysis of randomized controlled trials. *Nutrients, 8*(8), 483.

Huse, S. M., Dethlefsen, L., Huber, J. A., Mark Welch, D., Relman, D. A., & Sogin, M. L. (2008). Exploring microbial diversity and taxonomy using SSU rRNA hypervariable tag sequencing. *PLoS Genetics, 4*(11), e1000255. Available from https://doi.org/10.1371/journal.pgen.1000255.

Ikeda, T., Yasui, C., Hoshino, K., Arikawa, K., & Nishikawa, Y. (2007). Influence of lactic acid bacteria on longevity of caenorhabditis elegans and host defense against salmonella enterica serovar enteritidis. *Applied and Environmental Microbiology, 73*(20), 6404–6409.

Iwai, H., Ishihara, Y., Yamanaka, J., & Ito, T. (1973). Effects of bacterial flora on cecal size and transit rate of intestinal contents in mice. *Japanese Journal of Experimental Medicine, 43*(4), 297–305.

Iyer, L. M., Aravind, L., Coon, S. L., Klein, D. C., & Koonin, E. V. (2004). Evolution of cell-cell signaling in animals: Did late horizontal gene transfer from bacteria have a role? *Trends Genetics, 20* (7), 292–299.

Jeffery, I. B., Lynch, D. B., & O'Toole, P. W. (2016). Composition and temporal stability of the gut microbiota in older persons. *ISME Journal, 10*(1), 170–182.

Kain, V., Van Der Pol, W., Mariappan, N., Ahmad, A., Eipers, P., Gibson, D. L., ... Halade, G. V. (2019). Obesogenic diet in aging mice disrupts gut microbe composition and alters neutrophil:lymphocyte ratio, leading to inflamed milieu in acute heart failure. *FASEB Journal, 33*, 6456–6469.

Kamada, N., Chen, G. Y., Inohara, N., & Nunez, G. (2013). Control of pathogens and pathobionts by the gut microbiota. *Nature Immunology, 14*(7), 685–690.

Kamo, T., Akazawa, H., Suda, W., Saga-Kamo, A., Shimizu, Y., Yagi, H., ... Komuro, I. (2017). Dysbiosis and compositional alterations with aging in the gut microbiota of patients with heart failure. *PLoS One, 12*(3), e0174099.

Karlsson, F. H., Tremaroli, V., Nookaew, I., Bergstrom, G., Behre, C. J., Fagerberg, B., ... Backhed, F. (2013). Gut metagenome in European women with normal, impaired and diabetic glucose control. *Nature, 498*(7452), 99–103.

Kassam, Z., Lee, C. H., Yuan, Y., & Hunt, R. H. (2013). Fecal microbiota transplantation for *Clostridium* difficile infection: Systematic review and meta-analysis. *American Journal of Gastroenterology, 108*(4), 500–508.

Kau, A. L., Ahern, P. P., Griffin, N. W., Goodman, A. L., & Gordon, J. I. (2011). Human nutrition, the gut microbiome and the immune system. *Nature, 474*(7351), 327–336.

Keenan, M. J., Marco, M. L., Ingram, D. K., & Martin, R. J. (2015). Improving healthspan via changes in gut microbiota and fermentation. *Age, 37*(5), 98.

Kelly, J. R., Borre, Y. ,C. ,O. B., Patterson, E., El Aidy, S., Deane, J., ... Dinan, T. G. (2016). Transferring the blues: Depression-associated gut microbiota induces neurobehavioural changes in the rat. *Journal of Psychiatric Research, 82*, 109–118.

Kelly, C. P., & LaMont, J. T. (2008). *Clostridium difficile*—more difficult than ever. *New England Journal of Medicine, 359*(18), 1932–1940.

Kheirbek, R. E., Fokar, A., Shara, N., Bell-Wilson, L. K., Moore, H. J., Olsen, E., ... Llorente, M. D. (2017). Characteristics and incidence of chronic illness in community-dwelling predominantly male U. S. veteran centenarians. *Journal of the American Geriatric Society, 65* (9), 2100–2106.

Kibe, R., Kurihara, S., Sakai, Y., Suzuki, H., Ooga, T., Sawaki, E., ... Matsumoto, M. (2014). Upregulation of colonic luminal polyamines produced by intestinal microbiota delays senescence in mice. *Scientific Reports, 4*, 4548.

Kim, Y., & Je, Y. (2014). Dietary fiber intake and total mortality: A meta-analysis of prospective cohort studies. *American Journal of Epidemiology, 180*(6), 565–573.

Knights, D., Costello, E. K., & Knight, R. (2011). Supervised classification of human microbiota. *FEMS Microbiology Review, 35*(2), 343–359.

Komura, T., Yasui, C., Miyamoto, H., & Nishikawa, Y. (2010). Caenorhabditis elegans as an alternative model host for legionella pneumophila, and protective effects of Bifidobacterium infantis. *Applied and Environmental Microbiology, 76*(12), 4105–4108.

Konturek, P. C., Haziri, D., Brzozowski, T., Hess, T., Heyman, S., Kwiecien, S., ... Koziel, J. (2015). Emerging role of fecal microbiota therapy in the treatment of gastrointestinal and extra-gastrointestinal diseases. *Journal of Physiology and Pharmacology, 66*(4), 483–491.

Kostic, A. D., Chun, E., Robertson, L., Glickman, J. N., Gallini, C. A., Michaud, M., ... Garrett, W. S. (2013). Fusobacterium nucleatum potentiates intestinal tumorigenesis and modulates the tumor-immune microenvironment. *Cell Host & Microbe, 14*(2), 207–215.

Kristensen, N. B., Bryrup, T., Allin, K. H., Nielsen, T., Hansen, T. H., & Pedersen, O. (2016). Alterations in fecal microbiota composition by probiotic supplementation in healthy adults: A systematic review of randomized controlled trials. *Genome Medicine, 8*(1), 52.

Kunze, W. A., Mao, Y. K., Wang, B., Huizinga, J. D., Ma, X., Forsythe, P., & Bienenstock, J. (2009). *Lactobacillus reuteri* enhances excitability of colonic AH neurons by inhibiting calcium-dependent potassium channel opening. *Journal of Cellular and Molecular Medicine, 13*(8B), 2261–2270.

Kurokawa, S., Kishimoto, T., Mizuno, S., Masaoka, T., Naganuma, M., Liang, K. C., ... Mimura, M. (2018). The effect of fecal microbiota transplantation on psychiatric symptoms among patients with irritable bowel syndrome, functional diarrhea and functional constipation: An open-label observational study. *Journal of Affective Disorders, 235*, 506–512.

Kwon, G., Lee, J., & Lim, Y. H. (2016). Dairy propionibacterium extends the mean lifespan of *Caenorhabditis elegans* via activation of the innate immune system. *Scientific Reports, 6*, 31713.

Lachar, J., & Bajaj, J. S. (2016). Changes in the microbiome in cirrhosis and relationship to complications: Hepatic encephalopathy, spontaneous bacterial peritonitis, and sepsis. *Seminars in Liver Disease, 36*(4), 327–330.

Lai, Z. L., Tseng, C. H., Ho, H. J., Cheung, C. K. Y., Lin, J. Y., Chen, Y. J., ... Wu, C. Y. (2018). Fecal microbiota transplantation confers beneficial metabolic effects of diet and exercise on diet-induced obese mice. *Scientific Reports, 8*(1), 15625.

Langille, M. G., Meehan, C. J., Koenig, J. E., Dhanani, A. S., Rose, R. A., Howlett, S. E., & Beiko, R. G. (2014). Microbial shifts in the aging mouse gut. *Microbiome, 2*(1), 50.

Larsen, N., Vogensen, F. K., van den Berg, F. W., Nielsen, D. S., Andreasen, A. S., Pedersen, B. K., ... Jakobsen, M. (2010). Gut microbiota in human adults with type 2 diabetes differs from non-diabetic adults. *PLoS One, 5*(2), e9085.

Lecuit, M., Abachin, E., Martin, A., Poyart, C., Pochart, P., Suarez, F., ... Lortholary, O. (2004). Immunoproliferative small intestinal disease associated with *Campylobacter jejuni*. *New England Journal of Medicine, 350*(3), 239–248.

Lee, G. C., Reveles, K. R., Attridge, R. T., Lawson, K. A., Mansi, I. A., Lewis, J. S., 2nd, & Frei, C. R. (2014). Outpatient antibiotic prescribing in the United States: 2000 to 2010. *BMC Medicine, 12*, 96.

Lessa, F. C., Mu, Y., Bamberg, W. M., Beldavs, Z. G., Dumyati, G. K., Dunn, J. R., ... McDonald, L. C. (2015). Burden of *Clostridium difficile* infection in the United States. *New England Journal of Medicine, 372*(9), 825–834.

Ley, R. E., Backhed, F., Turnbaugh, P., Lozupone, C. A., Knight, R. D., & Gordon, J. I. (2005). Obesity alters gut microbial ecology. *Proceedings of the National Academy of Sciences of United States of America, 102*(31), 11070–11075.

Ley, R. E., Lozupone, C. A., Hamady, M., Knight, R., & Gordon, J. I. (2008). Worlds within worlds: Evolution of the vertebrate gut microbiota. *Nature Review Microbiology, 6*(10), 776–788.

Ley, R. E., Peterson, D. A., & Gordon, J. I. (2006). Ecological and evolutionary forces shaping microbial diversity in the human intestine. *Cell, 124*(4), 837–848.

Ley, R. E., Turnbaugh, P. J., Klein, S., & Gordon, J. I. (2006). Microbial ecology: Human gut microbes associated with obesity. *Nature, 444*(7122), 1022–1023.

Lin, A., Zheng, W., He, Y., Tang, W., Wei, X., He, R., ... Xie, H. (2018). Gut microbiota in patients with Parkinson's disease in southern China. *Parkinsonism Related Disorders, 53*, 82–88.

Lloyd-Price, J., Abu-Ali, G., & Huttenhower, C. (2016). The healthy human microbiome. *Genome Medicine, 8*(1), 51.

Lopez-Otin, C., Blasco, M. A., Partridge, L., Serrano, M., & Kroemer, G. (2013). The hallmarks of aging. *Cell, 153*(6), 1194–1217.

Lynch, S. V., & Pedersen, O. (2016). The human intestinal microbiome in health and disease. *New England Journal of Medicine, 375* (24), 2369–2379.

Lyte, M. (2011). Probiotics function mechanistically as delivery vehicles for neuroactive compounds: Microbial endocrinology in the design and use of probiotics. *Bioessays, 33*(8), 574–581.

Maffei, V. J., Kim, S., Blanchard, E. t, Luo, M., Jazwinski, S. M., Taylor, C. M., & Welsh, D. A. (2017). Biological aging and the human gut microbiota. *Journal of Gerontology—Series A Biology Sciences and Medical Sciences, 72*(11), 1474–1482.

Maier, L., Pruteanu, M., Kuhn, M., Zeller, G., Telzerow, A., Anderson, E. E., ... Typas, A. (2018). Extensive impact of non-antibiotic drugs on human gut bacteria. *Nature, 555*(7698), 623–628.

Makkawi, S., Camara-Lemarroy, C., & Metz, L. (2018). Fecal microbiota transplantation associated with 10 years of stability in a patient with SPMS. *Neurology—Neuroimmunology and Neuroinflammation, 5* (4), e459.

Marotz, C. A., & Zarrinpar, A. (2016). Treating obesity and metabolic syndrome with fecal microbiota transplantation. *Yale Journal of Biology & Medicine, 89*(3), 383–388.

Mathus-Vliegen, E. M. (2012). Obesity and the elderly. *Journal of Clinical Gastroenterology, 46*(7), 533–544.

Matsumoto, M., Kurihara, S., Kibe, R., Ashida, H., & Benno, Y. (2011). Longevity in mice is promoted by probiotic-induced suppression of colonic senescence dependent on upregulation of gut bacterial polyamine production. *PLoS One, 6*(8), e23652.

Matt, S. M., Allen, J. M., Lawson, M. A., Mailing, L. J., Woods, J. A., & Johnson, R. W. (2018). Butyrate and dietary soluble fiber improve neuroinflammation associated with aging in mice. *Frontiers in Immunology, 9*, 1832.

Murphy, E. F., Cotter, P. D., Healy, S., Marques, T. M., O'Sullivan, O., Fouhy, F., ... Shanahan, F. (2010). Composition and energy harvesting capacity of the gut microbiota: Relationship to diet, obesity and time in mouse models. *Gut, 59*(12), 1635–1642.

Nakagawa, H., Shiozaki, T., Kobatake, E., Hosoya, T., Moriya, T., Sakai, F., ... Miyazaki, T. (2016). Effects and mechanisms of prolongevity induced by *Lactobacillus gasseri* SBT2055 in *Caenorhabditis elegans*. *Aging Cell, 15*(2), 227–236.

Neuman, H., Debelius, J. W., Knight, R., & Koren, O. (2015). Microbial endocrinology: The interplay between the microbiota and the endocrine system. *FEMS Microbiology Review, 39*(4), 509–521.

Nimgampalle, M., & Kuna, Y. (2017). Anti-Alzheimer properties of probiotic, *Lactobacillus plantarum* MTCC 1325 in Alzheimer's disease induced albino rats. *Journal of Clinical and Diagnostic Research, 11*(8), KC01–KC05.

North, B. J., & Sinclair, D. A. (2012). The intersection between aging and cardiovascular disease. *Circulation Research, 110*(8), 1097–1108. Available from https://doi.org/10.1161/CIRCRESAHA.111.246876.

O'Hara, A. M., & Shanahan, F. (2006). The gut flora as a forgotten organ. *EMBO Reports, 7*(7), 688–693.

O'Toole, P. W., & Jeffery, I. B. (2015). Gut microbiota and aging. *Science, 350*(6265), 1214–1215.

Odamaki, T., Kato, K., Sugahara, H., Hashikura, N., Takahashi, S., Xiao, J. Z., ... Osawa, R. (2016). Age-related changes in gut microbiota composition from newborn to centenarian: A cross-sectional study. *BMC Microbiology, 16*, 90.

Oh, J., Byrd, A. L., Deming, C., Conlan, S., Program, N. C. S., Kong, H. H., & Segre, J. A. (2014). Biogeography and individuality shape function in the human skin metagenome. *Nature, 514*(7520), 59–64.

Olanow, C. W., & Tatton, W. G. (1999). Etiology and pathogenesis of Parkinson's disease. *Annual Review of Neuroscience, 22*, 123–144.

Orrhage, K., & Nord, C. E. (1999). Factors controlling the bacterial colonization of the intestine in breastfed infants. *Acta Paediatrica, 88*(430), 47–57.

Ott, S. J., Musfeldt, M., Wenderoth, D. F., Hampe, J., Brant, O., Folsch, U. R., ... Schreiber, S. (2004). Reduction in diversity of the colonic mucosa associated bacterial microflora in patients with active inflammatory bowel disease. *Gut, 53*(5), 685–693.

Ouwehand, A. C., Bergsma, N., Parhiala, R., Lahtinen, S., Gueimonde, M., Finne-Soveri, H., ... Salminen, S. (2008). Bifidobacterium microbiota and parameters of immune function in elderly subjects. *FEMS Immunology and Medical Microbiology, 53* (1), 18–25.

Pan, F., Zhang, L., Li, M., Hu, Y., Zeng, B., Yuan, H., ... Zhang, C. (2018). Predominant gut *Lactobacillus murinus* strain mediates anti-inflammaging effects in calorie-restricted mice. *Microbiome, 6*(1), 54.

Park, M. R., Oh, S., Son, S. J., Park, D. J., Oh, S., Kim, S. H., ... Kim, Y. (2015). *Bacillus licheniformis* isolated from traditional Korean food resources enhances the longevity of *Caenorhabditis elegans* through serotonin signaling. *Journal of Agricultural and Food Chemistry, 63*(47), 10227–10233.

Peterfreund, G. L., Vandivier, L. E., Sinha, R., Marozsan, A. J., Olson, W. C., Zhu, J., & Bushman, F. D. (2012). Succession in the gut microbiome following antibiotic and antibody therapies for *Clostridium difficile*. *PLoS One, 7*(10), e46966.

Pfeiffer, R. F. (2011). Gastrointestinal dysfunction in Parkinson's disease. *Parkinsonism Related Disorders, 17*(1), 10–15. Available from https://doi.org/10.1016/j.parkreldis.2010.08.003.

Pozuelo, M., Panda, S., Santiago, A., Mendez, S., Accarino, A., Santos, J., ... Manichanh, C. (2015). Reduction of butyrate- and methane-producing microorganisms in patients with irritable bowel syndrome. *Scientific Reports, 5*, 12693.

Qin, J., Li, R., Raes, J., Arumugam, M., Burgdorf, K. S., Manichanh, C., ... Wang, J. (2010). A human gut microbial gene catalogue established by metagenomic sequencing. *Nature, 464*(7285), 59–65.

Quraishi, M. N., Widlak, M., Bhala, N., Moore, D., Price, M., Sharma, N., & Iqbal, T. H. (2017). Systematic review with meta-analysis: The efficacy of faecal microbiota transplantation for the treatment of recurrent and refractory *Clostridium difficile* infection. *Alimentary Pharmacology Therapeutics, 46*(5), 479–493.

Rajilic-Stojanovic, M., Shanahan, F., Guarner, F., & de Vos, W. M. (2013). Phylogenetic analysis of dysbiosis in ulcerative colitis during remission. *Inflammatory Bowel Diseases, 19*(3), 481–488.

Rampelli, S., Candela, M., Turroni, S., Biagi, E., Collino, S., Franceschi, C., ... Brigidi, P. (2013). Functional metagenomic profiling of intestinal microbiome in extreme ageing. *Aging, 5*(12), 902–912.

Rastall, R. A., & Gibson, G. R. (2015). Recent developments in prebiotics to selectively impact beneficial microbes and promote intestinal health. *Current Opinion in Biotechnology, 32*, 42–46.

Rea, D., Coppola, G., Palma, G., Barbieri, A., Luciano, A., Del Prete, P., ... Arra, C. (2018). Microbiota effects on cancer: From risks to therapies. *Oncotarget, 9*(25), 17915–17927.

Reeves, A. E., Theriot, C. M., Bergin, I. L., Huffnagle, G. B., Schloss, P. D., & Young, V. B. (2011). The interplay between microbiome dynamics and pathogen dynamics in a murine model of *Clostridium difficile* infection. *Gut Microbes*, *2*(3), 145–158.

Reveles, K. R., Ryan, C. N., Chan, L., Cosimi, R. A., & Haynes, W. L. (2018). Proton pump inhibitor use associated with changes in gut microbiota composition. *Gut*, *67*(7), 1369–1370.

Ritchie, M. L., & Romanuk, T. N. (2012). A meta-analysis of probiotic efficacy for gastrointestinal diseases. *PLoS One*, *7*(4), e34938. Available from https://doi.org/10.1371/journal.pone.0034938.

Roberfroid, M., Gibson, G. R., Hoyles, L., McCartney, A. L., Rastall, R., Rowland, I., ... Meheust, A. (2010). Prebiotic effects: Metabolic and health benefits. *British Journal of Nutrition*, *104*(Suppl. 2), S1–S63. Available from https://doi.org/10.1017/S0007114510003363.

Round, J. L., Lee, S. M., Li, J., Tran, G., Jabri, B., Chatila, T. A., & Mazmanian, S. K. (2011). The Toll-like receptor 2 pathway establishes colonization by a commensal of the human microbiota. *Science*, *332*(6032), 974–977.

Rozan, P., Nejdi, A., Hidalgo, S., Bisson, J. F., Desor, D., & Messaoudi, M. (2008). Effects of lifelong intervention with an oligofructose-enriched inulin in rats on general health and lifespan. *British Journal of Nutrition*, *100*(6), 1192–1199.

Ruiz, A., Cerdo, T., Jauregui, R., Pieper, D. H., Marcos, A., Clemente, A., ... Suarez, A. (2017). One-year calorie restriction impacts gut microbial composition but not its metabolic performance in obese adolescents. *Environmental Microbiology*, *19*(4), 1536–1551.

Samah, S., Ramasamy, K., Lim, S. M., & Neoh, C. F. (2016). Probiotics for the management of type 2 diabetes mellitus: A systematic review and meta-analysis. *Diabetes Research and Clinical Practice*, *118*, 172–182. Available from https://doi.org/10.1016/j.diabres.2016.06.014.

Sampson, T. R., Debelius, J. W., Thron, T., Janssen, S., Shastri, G. G., Ilhan, Z. E., ... Mazmanian, S. K. (2016). Gut microbiota regulate motor deficits and neuroinflammation in a model of Parkinson's disease. *Cell*, *167*(6), 1469–1480, e1412.

Sanchez, B., Delgado, S., Blanco-Miguez, A., Lourenco, A., Gueimonde, M., & Margolles, A. (2017). Probiotics, gut microbiota, and their influence on host health and disease. *Molecular Nutrition and Food Research*, *61*(1).

Santacruz, A., Marcos, A., Warnberg, J., Marti, A., Martin-Matillas, M., Campoy, C., ... Group, E. S. (2009). Interplay between weight loss and gut microbiota composition in overweight adolescents. *Obesity*, *17*(10), 1906–1915.

Santoro, A., Ostan, R., Candela, M., Biagi, E., Brigidi, P., Capri, M., & Franceschi, C. (2018). Gut microbiota changes in the extreme decades of human life: A focus on centenarians. *Cellular and Molecular Life Sci*, *75*(1), 129–148.

Santos, R. L. (2014). Pathobiology of salmonella, intestinal microbiota, and the host innate immune response. *Frontiers in Immunology*, *5*, 252.

Schwiertz, A., Taras, D., Schafer, K., Beijer, S., Bos, N. A., Donus, C., & Hardt, P. D. (2010). Microbiota and SCFA in lean and overweight healthy subjects. *Obesity*, *18*(1), 190–195.

Seki, E., De Minicis, S., Osterreicher, C. H., Kluwe, J., Osawa, Y., Brenner, D. A., & Schwabe, R. F. (2007). TLR4 enhances TGF-beta signaling and hepatic fibrosis. *Nature Medicine*, *13*(11), 1324–1332.

Sender, R., Fuchs, S., & Milo, R. (2016). Are we really vastly outnumbered? Revisiting the ratio of bacterial to host cells in humans. *Cell*, *164*(3), 337–340.

Senff, N. J., Noordijk, E. M., Kim, Y. H., Bagot, M., Berti, E., Cerroni, L., ... International Society for Cutaneous, Lymphoma. (2008). European organization for research and treatment of cancer and international society for cutaneous lymphoma consensus recommendations for the management of cutaneous B-cell lymphomas. *Blood*, *112*(5), 1600–1609.

Seto, C. T., Jeraldo, P., Orenstein, R., Chia, N., & DiBaise, J. K. (2014). Prolonged use of a proton pump inhibitor reduces microbial diversity: Implications for *Clostridium difficile* susceptibility. *Microbiome*, *2*, 42.

Shahinas, D., Silverman, M., Sittler, T., Chiu, C., Kim, P., Allen-Vercoe, E., ... Pillai, D. R. (2012). Toward an understanding of changes in diversity associated with fecal microbiome transplantation based on 16S rRNA gene deep sequencing. *MBio*, *3*(5), e00338-12.

Smith, P. M., Howitt, M. R., Panikov, N., Michaud, M., Gallini, C. A., Bohlooly, Y. M., ... Garrett, W. S. (2013). The microbial metabolites, short-chain fatty acids, regulate colonic treg cell homeostasis. *Science*, *341*(6145), 569–573.

Smits, L. P., Bouter, K. E., de Vos, W. M., Borody, T. J., & Nieuwdorp, M. (2013). Therapeutic potential of fecal microbiota transplantation. *Gastroenterology*, *145*(5), 946–953.

Sochocka, M., Diniz, B. S., & Leszek, J. (2017). Inflammatory response in the CNS: Friend or foe? *Molecular Neurobiology*, *54*(10), 8071–8089.

Sochocka, M., Donskow-Lysoniewska, K., Diniz, B. S., Kurpas, D., Brzozowska, E., & Leszek, J. (2019). The gut microbiome alterations and inflammation-driven pathogenesis of Alzheimer's disease-a critical review. *Molecular Neurobiology*, *56*(3), 1841–1851.

Sokol, H., Pigneur, B., Watterlot, L., Lakhdari, O., Bermudez-Humaran, L. G., Gratadoux, J. J., ... Langella, P. (2008). *Faecalibacterium prausnitzii* is an anti-inflammatory commensal bacterium identified by gut microbiota analysis of Crohn disease patients. *Proceedings of the National Academy of Sciences of United States of America*, *105*(43), 16731–16736.

Stetler, C., & Miller, G. E. (2011). Depression and hypothalamic-pituitary-adrenal activation: A quantitative summary of four decades of research. *Psychosomatic Medicine*, *73*(2), 114–126.

Stilling, R. M., Dinan, T. G., & Cryan, J. F. (2014). Microbial genes, brain & behaviour—epigenetic regulation of the gut-brain axis. *Genes Brain and Behavior*, *13*(1), 69–86.

Storelli, G., Defaye, A., Erkosar, B., Hols, P., Royet, J., & Leulier, F. (2011). *Lactobacillus plantarum* promotes *Drosophila* systemic growth by modulating hormonal signals through TOR-dependent nutrient sensing. *Cell Metabolism*, *14*(3), 403–414.

Sudo, N., Chida, Y., Aiba, Y., Sonoda, J., Oyama, N., Yu, X. N., ... Koga, Y. (2004). Postnatal microbial colonization programs the hypothalamic-pituitary-adrenal system for stress response in mice. *Journal of Physiology*, *558*(Pt 1), 263–275.

Sun, J., & Buys, N. (2015). Effects of probiotics consumption on lowering lipids and CVD risk factors: A systematic review and meta-analysis of randomized controlled trials. *Annals of Medicine*, *47*(6), 430–440.

Sun, M. F., Zhu, Y. L., Zhou, Z. L., Jia, X. B., Xu, Y. D., Yang, Q., ... Shen, Y. Q. (2018). Neuroprotective effects of fecal microbiota transplantation on MPTP-induced Parkinson's disease mice: Gut microbiota, glial reaction and TLR4/TNF-alpha signaling pathway. *Brain Behavior, and Immunity*, *70*, 48–60.

Thevaranjan, N., Puchta, A., Schulz, C., Naidoo, A., Szamosi, J. C., Verschoor, C. P., ... Bowdish, D. M. E. (2017). Age-associated microbial dysbiosis promotes intestinal permeability, systemic inflammation, and macrophage dysfunction. *Cell Host & Microbe*, *21*(4), 455–466, e454.

Tomas, J., Mulet, C., Saffarian, A., Cavin, J. B., Ducroc, R., Regnault, B., ... Pedron, T. (2016). High-fat diet modifies the PPAR-gamma pathway leading to disruption of microbial and physiological ecosystem in murine small intestine. *Proceedings of the National Academy of Sciences of United States of America*, *113*(40), E5934–E5943.

Turnbaugh, P. J., Ley, R. E., Mahowald, M. A., Magrini, V., Mardis, E. R., & Gordon, J. I. (2006). An obesity-associated gut microbiome with increased capacity for energy harvest. *Nature*, *444*(7122), 1027–1031.

Udayappan, S. D., Hartstra, A. V., Dallinga-Thie, G. M., & Nieuwdorp, M. (2014). Intestinal microbiota and faecal transplantation as treatment modality for insulin resistance and type 2 diabetes mellitus. *Clinical and Experimental Immunology, 177*(1), 24–29.

Ussar, S., Griffin, N. W., Bezy, O., Fujisaka, S., Vienberg, S., Softic, S., ... Kahn, C. R. (2015). Interactions between gut microbiota, host genetics and diet modulate the predisposition to obesity and metabolic syndrome. *Cell Metabolism, 22*(3), 516–530.

Van den Abbeele, P., Belzer, C., Goossens, M., Kleerebezem, M., De Vos, W. M., Thas, O., ... de Wiele, T. (2013). Butyrate-producing *Clostridium cluster* XIVa species specifically colonize mucins in an in vitro gut model. *ISME Journal, 7*(5), 949–961.

Van Den Eeden, S. K., Tanner, C. M., Bernstein, A. L., Fross, R. D., Leimpeter, A., Bloch, D. A., & Nelson, L. M. (2003). Incidence of Parkinson's disease: Variation by age, gender, and race/ethnicity. *American Journal Epidemiology, 157*(11), 1015–1022.

Vemuri, R., Shinde, T., Shastri, M. D., Perera, A. P., Tristram, S., Martoni, C. J., ... Eri, R. (2018). A human origin strain *Lactobacillus acidophilus* DDS-1 exhibits superior in vitro probiotic efficacy in comparison to plant or dairy origin probiotics. *International Journal of Medical Sciences, 15*(9), 840–848.

Vogt, N. M., Kerby, R. L., Dill-McFarland, K. A., Harding, S. J., Merluzzi, A. P., Johnson, S. C., ... Rey, F. E. (2017). Gut microbiome alterations in Alzheimer's disease. *Scientific Reports, 7*(1), 13537.

Vrieze, A., Van Nood, E., Holleman, F., Salojarvi, J., Kootte, R. S., Bartelsman, J. F., ... Nieuwdorp, M. (2012). Transfer of intestinal microbiota from lean donors increases insulin sensitivity in individuals with metabolic syndrome. *Gastroenterology, 143*(4), 913–916, e917.

Vyas, D., Aekka, A., & Vyas, A. (2015). Fecal transplant policy and legislation. *World Journal of Gastroenterology, 21*(1), 6–11.

Walker, A. W., Ince, J., Duncan, S. H., Webster, L. M., Holtrop, G., Ze, X., ... Flint, H. J. (2011). Dominant and diet-responsive groups of bacteria within the human colonic microbiota. *ISME Journal, 5*(2), 220–230.

Wang, F., Meng, J., Zhang, L., Johnson, T., Chen, C., & Roy, S. (2018). Morphine induces changes in the gut microbiome and metabolome in a morphine dependence model. *Scientific Reports, 8*(1), 3596.

Wang, H. T., Anvari, S., & Anagnostou, K. (2019). The role of probiotics in preventing allergic disease. *Children, 6*(2), 24.

Wang, Y., Li, X., Ge, T., Xiao, Y., Liao, Y., Cui, Y., ... Zhang, T. (2016). Probiotics for prevention and treatment of respiratory tract infections in children: A systematic review and meta-analysis of randomized controlled trials. *Medicine, 95*(31), e4509.

Wang, Z., Klipfell, E., Bennett, B. J., Koeth, R., Levison, B. S., Dugar, B., ... Hazen, S. L. (2011). Gut flora metabolism of phosphatidylcholine promotes cardiovascular disease. *Nature, 472*(7341), 57–63.

Weindruch, R., & Sohal, R. S. (1997). Seminars in medicine of the Beth Israel deaconess medical center. Caloric intake and aging. *New England Journal of Medicine, 337*(14), 986–994.

Weksler, M. E., Pawelec, G., & Franceschi, C. (2009). Immune therapy for age-related diseases. *Trends Immunology, 30*(7), 344–350.

Welton, J. C., Marr, J. S., & Friedman, S. M. (1979). Association between hepatobiliary cancer and typhoid carrier state. *Lancet, 1*(8120), 791–794.

Wen, L., & Duffy, A. (2017). Factors influencing the gut microbiota, inflammation, and type 2 diabetes. *Journal of Nutrition, 147*(7), 1468S–1475S.

Westfall, S., Lomis, N., & Prakash, S. (2018). Longevity extension in *Drosophila* through gut-brain communication. *Scientific Reports, 8*(1), 8362.

Wilson, B. C., Vatanen, T., Cutfield, W. S., & O'Sullivan, J. M. (2019). The super-donor phenomenon in fecal microbiota transplantation. *Frontiers in Cellular and Infection Microbiology, 9*, 2.

Woodmansey, E.J., Mcmurdo, M.E., Macfarlane, G.T., & Macfarlane, S. (2004). Comparison of compositions and metabolic activities of fecal microbiotas in young adults and in antibiotic-treated and non-antibiotic-treated elderly subjects. *Applied and Environmental Microbiology 70*(10), 6113–6122.

World Health Organization. (2019). Cardiovascular disease. <https://www.who.int/health-topics/cardiovascular-diseases/#tab=tab_1> Accessed 07.06.20.

Wu, G. D., Chen, J., Hoffmann, C., Bittinger, K., Chen, Y. Y., Keilbaugh, S. A., ... Lewis, J. D. (2011). Linking long-term dietary patterns with gut microbial enterotypes. *Science, 334*(6052), 105–108.

Wu, H., Esteve, E., Tremaroli, V., Khan, M. T., Caesar, R., Manneras-Holm, L., ... Backhed, F. (2017). Metformin alters the gut microbiome of individuals with treatment-naive type 2 diabetes, contributing to the therapeutic effects of the drug. *Nature Medicine, 23*(7), 850–858.

Yang, T., Santisteban, M. M., Rodriguez, V., Li, E., Ahmari, N., Carvajal, J. M., ... Mohamadzadeh, M. (2015). Gut dysbiosis is linked to hypertension. *Hypertension, 65*(6), 1331–1340.

Yano, J. M., Yu, K., Donaldson, G. P., Shastri, G. G., Ann, P., Ma, L., ... Hsiao, E. Y. (2015). Indigenous bacteria from the gut microbiota regulate host serotonin biosynthesis. *Cell, 161*(2), 264–276.

Yatsunenko, T., Rey, F. E., Manary, M. J., Trehan, I., Dominguez-Bello, M. G., Contreras, M., ... Gordon, J. I. (2012). Human gut microbiome viewed across age and geography. *Nature, 486*(7402), 222–227.

Yoshikawa, T. T. (1981). Important infections in elderly persons. *Western Journal of Medicine, 135*(6), 441–445.

Zechner, E. L. (2017). Inflammatory disease caused by intestinal pathobionts. *Current Opinion in Microbiology, 35*, 64–69.

Zhang, C., Li, S., Yang, L., Huang, P., Li, W., Wang, S., ... Zhao, L. (2013). Structural modulation of gut microbiota in life-long calorie-restricted mice. *Nature Communication, 4*, 2163.

CHAPTER

18

Lipidomics of aging

Juan Pablo Palavicini[1,2] *and Xianlin Han*[1,2]

[1]Barshop Institute for Longevity and Aging Studies, University of Texas Health Science Center at San Antonio, San Antonio, Texas, United States [2]Division of Diabetes, Department of Medicine, University of Texas Health Science Center at San Antonio, San Antonio, Texas, United States

OUTLINE

Introduction

Lipids are broadly defined as a group of organic compounds in living organisms that are soluble in nonpolar solvents, including hydrophobic and amphiphilic biomolecules. Lipids are one of the main constituents of cells, and as such, they play essential biological functions that include:

1. Structural components of cellular membranes that provide hydrophobic barriers to separate cellular compartments, serving as an optimal matrix that facilitates transmembrane protein function; examples of membrane lipids include phospholipids, glycolipids, and cholesterol. The amphiphilic nature of some lipids allows them to form structures such as vesicles, liposomes, or membranes in an aqueous environment.
2. Major energy storage depots (triacylglycerols) and fuel sources (fatty acids) that can either generate energy through fatty acid oxidation, or synthesize new lipids from smaller constituent molecules.
3. Signaling molecules and cellular messengers, such as sphingosine-1-phosphate, diacylglycerol, and phosphatidylinositol phosphates, prostaglandins, oxysterols, among others.
4. Essential nutrients stored in the liver and fatty tissues, such as "fat-soluble" vitamins or isoprene-based lipids.
5. Transporters, like acylcarnitines that transport fatty acids in and out of mitochondria or polyprenols and their phosphorylated derivatives that transport oligosaccharides across membranes.
6. Enzyme activators like cardiolipins that activate enzymes involved in oxidative phosphorylation.
7. Lipids also form the basis of steroid hormones; among other important biological roles.

Classification of lipids

The consortium of lipid metabolites and pathways strategy (lipid MAPS) has classified lipids into eight categories: fatty acyls, glycerolipids, glycerophospholipids, sphingolipids, sterol lipids, prenol lipids, saccharolipids, and polyketides (Fahy et al., 2005). *Fatty acyls* represent a diverse group of molecules synthesized by chain elongation of an acetyl coenzyme A (acetyl-CoA) primer with malonyl-CoA (or methylmalonyl-CoA) groups that may contain a cyclic

Handbook of the Biology of Aging.
DOI: https://doi.org/10.1016/B978-0-12-815962-0.00018-4

functionality and/or are substituted with heteroatoms. Fatty acyls are characterized by a repeating series of methylene groups and are structurally the simplest lipids. This category includes various classes of fatty acids, eicosanoids, docosanoids, fatty alcohols, fatty aldehydes, fatty esters, fatty amides, fatty nitriles, fatty ethers, and hydrocarbons (Fahy et al., 2005). Fatty acyls in general and fatty acids in particular are the basic building blocks of more complex lipids such as glycerolipids and glycerophospholipids.

Glycerolipids encompass lipid species that can only be hydrolyzed into glycerol, a sugar group, fatty acid(s), and/or alkyl variants. Glycerolipids include the species of monoacylglycerols, diacylglycerols, triacylglycerols, and glycolipids (Fahy et al., 2005). *Glycerophospholipids* are defined by the presence of at least one phosphate (or phosphonate) group esterified to one of the glycerol hydroxyl groups. Glycerophospholipids are ubiquitous in nature, are key/abundant components of cellular membranes, and are also involved in metabolism and signaling. The complexity of glycerophospholipids is illustrated with the presence of different classes, subclasses, and individual molecular species with different acyl chain structures (Fig. 18.1). Lipids in this category contain three components: at least one glycerol molecule, at least one phosphate or phosphodiester is linked to a hydroxyl group of glycerol at the *sn*-3 position, and one or two aliphatic chains are connected to the *sn*-1, *sn*-2, or both hydroxyl groups of glycerol. There are multiple types of moieties that can be esterified with the phosphate that define over 10 different classes, and over 30 types of naturally occurring fatty acyl chains containing different numbers of carbon atoms (i.e., chain length), different degrees of unsaturation, and different locations of these double bonds (Fig. 18.1). In addition, there are three different subclasses defined by different types of linkages or bonds between the fatty acyl chain and the hydroxyl group of glycerol at the *sn*-1 position (Fahy et al., 2005).

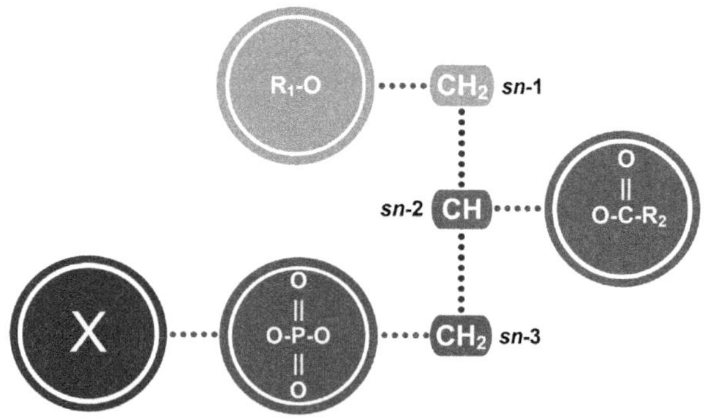

X (Headgroup)	Glycerophospholipid Class
Choline	Choline glycerophospholipid (PC)
Ethanolamine	Ethanolamine glycerophospholipid (PE)
Glycerol	Phosphatidylglycerol
Myoinositol	Phosphatidylinositol (PI)
Serine	Serine glycerophospholipid (PS)
Hydrogen oxide	Phosphatidic acid (PA)

FIGURE 18.1 **Classification of glycerophospholipids.** Glycerophospholipids are classified based on their polar headgroups, composed of specific moieties (X) connected to a phosphate group that is linked to the *sn*-3 position hydroxyl group of glycerol. R_1 and R_2 represent aliphatic chains connected to the *sn*-1 and *sn*-2 position hydroxyl groups of glycerol. Several glycerophospholipid classes are listed as examples.

Sphingolipids represent another major category of complex lipids that are also ubiquitous constituents of membranes in eukaryotes and encompass a large and complex family of molecules that contain a sphingoid base as its core structure (Bartke & Hannun, 2009; Fahy et al., 2005; Hannun & Obeid, 2018; Kolter, 2011). These sphingoid bases are synthesized from serine and a long-chain fatty acyl CoA to yield sphinganine and then dihydroceramide, which is converted into ceramide, phosphosphingolipids, glycosphingolipids, and other species (Fig. 18.2). Different polar moieties, which are linked to the hydroxyl group of the sphingoid base at position C1, represent individual sphingolipid classes. *Sterol lipids* are a group of compounds that carry a core signature of four fused rings and are subdivided into cholesterol and its derivatives, steroids, secosteroids, and bile acids and its derivatives, among others (Fahy et al., 2005). Cholesterol and its derivatives, the most widely studied sterol lipids in mammalian systems, constitute an important component of membrane lipids, along with glycerophospholipids and sphingomyelin.

Prenol lipids are synthesized from five carbon precursors (isopentenyl diphosphate and dimethylallyl diphosphate), produced mainly via the mevalonic pathway (Kuzuyama & Seto, 2003). *Saccharolipids* are lipids where fatty acids are linked directly to a sugar backbone, instead of a glycerol backbone as is the case in glycerolipids and glycerophospholipids. The best known saccharolipids are the acylated glucosamine precursors of the lipid A component of the lipopolysaccharides in Gram-negative bacteria (Raetz & Whitfield, 2002). *Polyketides* are a diverse group of secondary metabolites that contain either alternating carbonyl and methylene groups, or are derived from precursors that contain such alternating groups, and are found in bacteria, fungi, and plants that display diverse chemical structures and biological functions (Han, Choi, & Kim, 2018; Hu et al., 2010). For example, many mycotoxins produced by fungi are polyketides (Huffman, Gerber, & Du, 2010).

Lipids in biology of aging

Overview

Biological aging is a universal process characterized by a progressive deterioration of tissue structure and

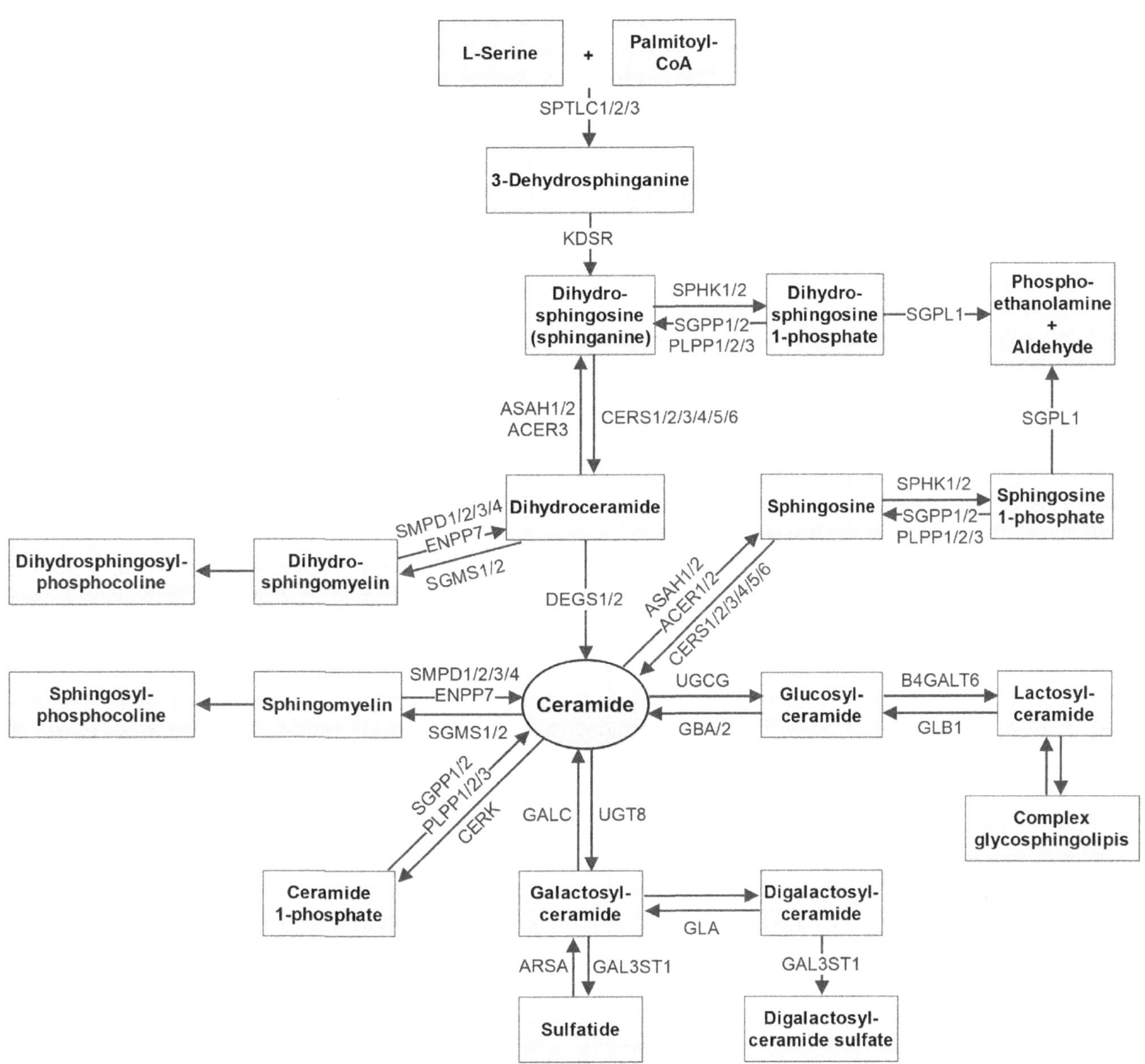

FIGURE 18.2 **Simplified schematic of mammalian sphingolipid metabolism.** Simplified sphingolipid-related network, adapted from the Kyoto Encyclopedia of Genes and Genomes (KEGG) human sphingolipid metabolism pathway map. The pathways indicate the origins of sphingoid base core structures and their derivatives. Human genes that code for the enzymes catalyzing specific reactions are depicted in red. Other sphingoid bases can be biosynthesized from other fatty acyl CoA (besides palmitoyl-CoA) or other amino acids (besides serine).

function (Balcombe & Sinclair, 2001), associated with increased risk for developing multiple chronic conditions such as cancer, cardiovascular, metabolic, neurodegenerative, and autoimmune diseases, eventually leading to death (Burkle et al., 2007; Campisi et al., 2009; Stepanova et al., 2015). As aging is the single greatest risk factor for the development of disease, understanding the biological mechanisms that modulate aging is critical for the development of health-maximizing interventions. In fact, understanding the molecular basis of aging is a major challenge that the biomedical science field currently faces.

Aging is regarded as a complex, multifactorial process where diverse deleterious changes accumulate in cells and tissues with advancing age that are responsible for the increased risk of disease and death and are driven by both intrinsic (genetic) and extrinsic (environmental) factors (reviewed in Balcombe & Sinclair, 2001; Harman, 2003;Kirkwood & Austad, 2000; Medvedev, 1990; Tosato et al., 2007). A wide number of hypotheses/models of aging have been proposed over the past decades (reviewed in Hayflick, 1998; Holliday, 2006; Tosato et al., 2007), including the oxidative stress or free radical theory (Harman, 2003), the mitochondrial theory (Barja, 2014;

Cadenas & Davies, 2000; Harman, 1972), the inflammation hypothesis (Chung et al., 2001), the immunological theory (inflammaging) (Franceschi et al., 2000; Fulop et al., 2014), the network theory (Kowald & Kirkwood, 1994; Kriete et al., 2006; Promislow, 2004), among others. These different theories of aging should not be considered mutually exclusive, but rather complementary (reviewed in Sergiev, Dontsova, & Berezkin, 2015; Weinert & Timiras, 2003).

This chapter focuses on the role of lipids in aging, we aim to summarize the multiple lines of research on animal models and humans that have revealed a strong association between specific lipid profiles and longevity (reviewed in Bustos & Partridge, 2017; de Diego, Peleg, & Fuchs, 2019; Schroeder & Brunet, 2015). The continuous use of lipidomics in concert with genetic and mechanistic studies is likely to reveal major roles for lipids in the regulation of longevity and increase our understanding of the complex relationship between lipids and aging.

Ceramide

Ceramide structure/function

Ceramides are at the center of sphingolipid metabolism, as all sphingolipids are generated from ceramide or its analogs and broken down into ceramide (Fig. 18.2; Han, 2016; Hannun & Obeid, 2011; Hannun & Obeid, 2018). The ceramide family encompasses numerous lipid molecular species that vary on the chain length, unsaturation, and hydroxylation of both the sphingoid base (the most abundant being sphingosine) and fatty acyl moieties (Hannun & Obeid, 2011). For example, it is estimated that there are more than 200 mammalian ceramide species that act as either the substrate or product of more than 28 different enzymes (Hannun & Obeid, 2011).

As a sum of all of its species, ceramide is present in cells at levels that usually range from 0.1% to 1.0% of total phospholipids (Hannun & Obeid, 2011). Because ceramide is highly hydrophobic, it tends to reside in the membranes where it is generated unless it is transported (Hanada, 2010; Sonnino et al., 2006). Ceramides are key players of lipid cell signaling and regulation of cell death and survival that play essential roles as both intermediates in the biosynthesis of more complex sphingolipids and as signaling molecules that participate in a plethora of biological processes. For example, ceramides have been associated with signal transduction, ER stress, autophagy, protein and lipid transport, exosome secretion, inflammation, mitochondria function, telomerase activity, cell migration, apoptosis, cell growth, differentiation, cellular senescence, among other cellular pathways (Bartke & Hannun, 2009; Wang & Bieberich, 2018). The physiological relevance of sphingolipid and ceramide metabolism is witnessed by the deleterious consequences of genetic alterations in ceramide metabolism that lead to inherited human monogenic disorders due to ceramide synthesis and degradation defects (reviewed in Dubot et al., 2019; Dunn, Tifft, & Proia, 2019). Alterations of ceramide metabolism are also associated with the development of multiple pathobiological disorders, such as cancer, asthma, diabetes, metabolic syndrome, inflammation, liver disease, cardiovascular diseases, neurodegenerative diseases, and infectious diseases (reviewed in Arana et al., 2010; Cowart, 2009; Han, 2005; Hannun & Obeid, 2011; Nikolova-Karakashian, 2018; Plotegher et al., 2019; Wang & Bieberich, 2018; Yuyama, Mitsutake, & Igarashi, 2014).

Ceramides, longevity, and aging

The first longevity gene discovered in the yeast *Saccharomyces cerevisiae* was the longevity-assurance gene (*LAG1*) (Dubot et al., 2019), which was subsequently shown to encode a ceramide synthase (Guillas et al., 2001; Schorling et al., 2001). Deletion of LAG1 results in an increase of mean and maximum life span in ascomycetes (*S. cerevisiae* and *Neurospora crassa*) (Brunson et al., 2016; D'Mello, 1994). Similarly, simultaneous deletion or reduced expression of two of the three ceramide synthase genes in *Caenorhabditis elegans* also results in life span extension (Mosbech et al., 2013; Tedesco et al., 2008). Furthermore, pharmacological inhibition of ceramide synthesis using myriocin, a potent inhibitor of serine palmitoyltransferase (SPT), the enzyme that catalyzes the first step in ceramide biosynthesis, also leads to an increase of life span in yeast and *C. elegans* (Cutler et al., 2014; Huang, Liu, & Dickson, 2012). Similarly, downregulation of yeast SPT subunits (LCB1 and LCB2) or worm SPT by siRNA also increases life span (Cutler et al., 2014; Huang et al., 2012). In mammals (mice), SPT inhibition improves cardiac function and cardiovascular disease (Hojjati et al., 2005; Walls et al., 2018), reverses insulin resistance and hepatic steatosis (Campana et al., 2018; Kurek et al., 2015; Schmitz-Peiffer, 2010), and reduces Alzheimer's disease (AD) pathology (Geekiyanage, Upadhye, & Chan, 2013).

Ceramide levels and its metabolism follow a development-aging continuum with increases in basal ceramide levels being a general feature of the aging process in yeast, worms, and mammals (Cutler & Mattson, 2001; Cutler et al., 2014; Nikolova-Karakashian, Karakashian, & Rutkute, 2008; Sacket et al., 2009; Yi et al., 2016). It has been shown that ceramides accumulate systemically during aging, with multiple model organisms (from yeast to nonhuman primates) and humans studies consistently reporting an age-related increase in ceramide levels (including specific ceramide species and/or total content) at a systemic level (blood, heart, skeletal muscle, liver, kidney, brain, and other organs/tissues) ((Babenko et al., 2016; Couttas et al., 2018; Giusto, Roque, & Ilincheta, 1992; Khayrullin et al., 2019; Kobayashi et al., 2013; Lightle,

Oakley, & Nikolova-Karakashian, 2000; Mielke et al., 2015; Monette et al., 2011; Noh et al., 2019; Sogaard et al., 2019) and unpublished data from our laboratory).

Ceramides and telomeres

Telomerase is a ribonucleoprotein complex, composed of a reverse transcriptase (TERT) and an RNA partner (TR), that is responsible for maintaining chromosome ends via addition of tandem repeat sequences to the ends of chromosomes (Blackburn, Greider, & Szostak, 2006; Weinrich et al., 1997). Telomerase expression, usually absent in somatic cells, allows malignant cells to maintain telomerase length insuring infinite replicative capacity. Similarly, increased telomerase activity confers protection against apoptosis. It has been reported that stimulation of endogenous ceramide production, exogenous short-chain ceramides, and ceramide analogs inhibit telomerase activity in multiple cell lines (Kraveka et al., 2003; Ogretmen et al., 2001a, 2001b; Rossi et al., 2005; Wooten & Ogretmen, 2005; Wooten-Blanks et al., 2007).

Ceramides and senescence

Cellular senescence refers to the limited capacity of cells to undergo population doublings; as a result, cells have a finite life span beyond which they can no longer proliferate. This finite life span correlates with the age of an organism and with its life expectancy; such that the older the age or the shorter the life span, the lower ability cells have to undergo population doubling (Hayflick, 1965). Ceramide has been proposed as a mediator of cellular senescence, with endogenous ceramide levels and ceramide metabolism markedly elevated in senescent cells, as well as in senescence-accelerated mice (Khayrullin et al., 2019; Rodriguez-Calvo et al., 2007; Venable et al., 1995). Addition of ceramide to young cells inhibits phospholipase D and recapitulates all the standard measures of cellular senescence (Venable & Obeid, 1999; Venable et al., 1995; Webb, Arnholt, & Venable, 2010). Inhibition of de novo ceramide biosynthesis reversed stromal cell and neuronal senescence (Granzotto et al., 2019; Kim et al., 2019).

Ceramides and inflammaging

Aging is associated with activation of the innate immune system leading to a condition known as "inflammaging" (Franceschi et al., 2007). The "danger theory" argues that certain molecules released from senescent or dying cells might constitute signals that trigger an immune response, often termed damage-associated molecular patterns (DAMPs) (Dela Cruz & Kang, 2018; Seong & Matzinger, 2004; Zhang et al., 2010). It has been proposed that ceramides act as DAMPs and contribute to age-related inflammation (inflammaging). In fact, it has been shown that ceramides act as TLR4 agonists and to induce caspase-1 cleavage and IL-1β secretion (Fischer et al., 2007; Franceschi & Campisi, 2014; Plociennikowska et al., 2015; Scheiblich et al., 2017; Vandanmagsar et al., 2011). Growing evidence points to ceramides playing a critical role in NACHT, LRR, and PYD domains containing protein 3 (NLRP3) inflammasome activation (Scheiblich et al., 2017).

In summary, ceramides are notable candidates for modulating aging not only because their concentration increases with age, but also because they been mechanistically linked to longevity, cellular senescence, telomerase activity, autophagy, oxidative stress, mitochondrial dysfunction, and insulin resistance. Furthermore, ceramide metabolism is disrupted in age-related diseases including diabetes, cardiovascular and immune dysfunction, cancer, and neurodegeneration (Arana et al., 2010; Babenko et al., 2016; Campana et al., 2018; Chien et al., 2009; Cutler & Mattson, 2001; Geekiyanage et al., 2013; Han, 2005; Khayrullin et al., 2019; Kraveka et al., 2003; Kurek et al., 2015; Lightle et al., 2000; Monette et al., 2011; Mosbech et al., 2013; Schmitz-Peiffer, 2010; Venable et al., 1995; Yuyama et al., 2014).

Cardiolipin

Cardiolipin structure/function

Cardiolipin is a complex polyglycerophospholipid that is found exclusively in bacterial and mitochondrial membranes, and is intricately involved in mitochondrial functionality (Hoch, 1992; Paradies et al., 2014b; Ren, Phoon, & Schlame, 2014). In eukaryotes, cardiolipin is dominantly found in the inner mitochondrial membrane, the site of its de novo synthesis. Cardiolipin constitutes about 20% of the total mitochondrial lipid composition and plays an essential role in mitochondrial structure, function, and bioenergetics (Ikon & Ryan, 2017; Schlame & Greenberg, 2017). Specifically, it has been shown that cardiolipin is critical for: (1) maintaining inner membrane fluidity, contact point curvature, and osmotic stability (Robinson, 1993; Schlame, Rua, & Greenberg, 2000); (2) decreasing the energy required to create folds or cristae in the inner mitochondrial membrane (Acehan et al., 2007; Ikon & Ryan, 2017; Khalifat et al., 2011; Nichols-Smith, Teh, & Kuhl, 2004; Shibata et al., 1994); and (3) stabilizing the respiratory complexes into supercomplexes to facilitate optimal electron transfer among the redox partners (Bazan et al., 2013; Jussupow, Di Luca, & Kaila, 2019; Kiebish et al., 2012; Mileykovskaya & Dowhan, 2009; Pfeiffer et al., 2003; Zhang, Mileykovskaya, & Dowhan, 2002).

Cardiolipin is a unique class of phospholipids that contains four fatty acyl chains and two negative charges (De Bruijn, 1966). The fatty acyl composition of cardiolipin varies markedly between tissues, developmental stages, sex, and disease states (reviewed in Acaz-Fonseca et al., 2019; Chicco & Sparagna, 2007; Claypool & Koehler, 2012; He & Han, 2014; Paradies et al., 2019; Semba et al., 2019; Shen et al., 2015; Shi, 2010). In peripheral high-energy-demanding tissues, like cardiac/skeletal muscle and the liver, the majority of cardiolipin is composed of linoleic acid (18:2) (Han et al., 2005; Schlame et al., 1999; Schlame et al., 2005). On the other hand, in the brain and the nervous system in general, the fatty acyl composition is highly diverse (Acaz-Fonseca et al., 2019; Cheng et al., 2008; Pointer & Klegeris, 2017). It has been found that reduced cardiolipin content, increased cardiolipin peroxidation, and altered cardiolipin remodeling are associated with aging and several cardiac, metabolic, and neurodegenerative diseases. Specifically, disruption of cardiolipin metabolism plays a causal role in the etiology of Barth syndrome, an inherited cardiomyopathy (reviewed in Gaspard & McMaster, 2015; Schlame & Ren, 2006); and has been implicated with cardiovascular disease, obesity, diabetes, liver disease, Parkinson disease, and traumatic brain injury (reviewed in Chicco & Sparagna, 2007; Claypool & Koehler, 2012; He & Han, 2014; Paradies et al., 2014a; Pointer & Klegeris, 2017; Shen et al., 2015).

Cardiolipin depletion and peroxidation in aging

Historically, mitochondrial dysfunction has been considered a central regulator of the aging process (Harman, 1972; Jang et al., 2018; Kauppila, Kauppila, & Larsson, 2017). Mitochondrial cardiolipin molecules are likely and early targets of oxygen-free radical attack, because: (1) they are almost exclusively located in the inner mitochondrial membrane near to the site of reactive oxygen species (ROS) production; and (2) have a high content of polyunsaturated fatty acids (PUFAs) and the sensitivity to oxidation increases exponentially as a function of the number of double bonds per fatty acid molecule (Bielski, Arudi, & Sutherland, 1983; Boveris & Chance, 1973; Turrens, Alexandre, & Lehninger, 1985). Consequently, mitochondrial oxidative stress leads to cardiolipin peroxidation and hence, impairment of mitochondrial function (Paradies et al., 2009). Depletion and oxidation of cardiolipin with aging negatively affect the biochemical function of mitochondrial membranes, leading to cellular dysfunction and eventually cell death (unpublished data from our laboratory).

During aging and age-related diseases, heart cardiolipin homeostasis is markedly impaired, with major (18:2-containing) cardiolipin species being dramatically reduced, and oxidized cardiolipin species being significantly increased (He & Han, 2014; Lee et al., 2006; Petrosillo et al., 2009; Sparagna et al., 2007). Similarly, declines in cardiolipin content with age have also been reported in liver and brain tissue (Paradies et al., 2011; Ruggiero et al., 1992; Sen et al., 2007; Vorbeck et al., 1982). It has been found that the loss of cardiolipin is associated with changes in cardiolipin synthase activity, alterations in cardiolipin remodeling, cardiolipin peroxidation, and decreased mitochondrial membrane fluidity (Nicolson & Ash, 2014).

A loss of cardiolipin destabilizes respiratory supercomplexes and further increases electron leak and ROS production (Maranzana et al., 2013). Cardiolipin peroxidation can also change the lipid packing in cristae membranes and increase proton leak (Ikon & Ryan, 2017; Koshkin & Greenberg, 2002), and provides substrates for production of oxylins (Kiebish et al., 2012). The high concentration of PUFAs in cardiolipin not only makes them a prime target for reactions with oxidizing agents, but also enables them to participate in long free-radical chain reactions, generating hydroperoxides as well as endoperoxides (Paradies et al., 2014a). Upon fragmentation, these compounds produce a broad range of reactive intermediates including malondialdehyde (MDA) and 4-hydroxy-trans-2-nonenal (HNE) (Ayala et al., 2014; Gawel et al., 2004; Musatov et al., 2002; Paradies et al., 2014a).

In addition to its important role in maintaining cell viability, cardiolipin can also play an important role in cell death, especially when it is oxidized. The oxidation of cardiolipin produces kinks in the acyl chain that disturb cardiolipin microdomains in the internal mitochondrial membrane and cause a loss of curvature (Szeto, 2014). Cardiolipin peroxidation also disrupts supercomplexes and causes cytochrome *c* to be detached from the internal mitochondrial membrane (Hanske et al., 2012; Kagan et al., 2005). All of these result in inhibition of mitochondrial respiration and set the stage for apoptosis (Gonzalvez & Gottlieb, 2007; Schug & Gottlieb, 2009).

Cardiolipin and inflammaging

Moreover, mitochondria have emerged as a major driver of the inflammaging phenotype (Krysko et al., 2011; Zhou et al., 2011). It has been proposed that cardiolipin is a putative mitochondrial DAMP (Dela Cruz & Kang, 2018; Jang et al., 2018). In fact, there is evidence that cardiolipin can directly bind and activate the NLRP3 inflammasome (Iyer et al., 2013).

Sulfatide

Sulfatide structure/function

Sulfatide is a class of sulfated galactocerebroside with a sphingoid backbone and different lengths of acyl chains which can be hydroxylated (Takahashi & Suzuki, 2012). Sulfatides, which are commonly found at the surface of most eukaryotic cells, are not just structural components of the plasma membrane, but also participate in a wide range of cellular processes including axon–myelin interactions, neural plasticity, protein trafficking, cell adhesion/aggregation, and immune responses among others (Xiao, Finkielstein, & Capelluto, 2013).

Sulfatides play an important role in the physiology of the peripheral and central nervous systems (PNS and CNS, respectively). Sulfatides are predominantly synthesized by oligodendrocytes in the CNS and Schwann cells in the PNS, and are highly enriched in the myelin sheath (Takahashi & Suzuki, 2012), where they make up about a third of total galactolipids (Wang et al., 2014). In the CNS, sulfatides are negative regulators for the differentiation and survival of oligodendrocytes during myelination. Studies on a sulfatide-deficient mouse model, cerebroside sulfotransferase (*Cst*) KO mice, have implicated sulfatide with myelin maintenance and axonal conduction (Honke et al., 2002; Ishibashi et al., 2002; Palavicini et al., 2016; Wang et al., 2014). Abnormal sulfatide degradation leads to a rare lysosomal storage neurological disease known as metachromatic leukodystrophy, characterized by progressive central and peripheral demyelination (Kohlschutter, 2013).

Importantly, during the last two decades, it has been consistently demonstrated that brain sulfatide content is specifically and dramatically reduced at the earliest clinically recognizable stages of AD in human subjects (Cheng et al., 2013; Gonzalez de San Roman et al., 2017; Han, 2005, 2007; Han et al., 2002, 2003a), as well as in multiple AD mouse models (Cheng et al., 2010; Hong et al., 2016). Abnormal sulfatide concentrations have also been associated with white matter lesions and vascular dementia (Fredman et al., 1992; Jonsson et al., 2010).

Besides its well-established biological role in the nervous system, sulfatide is a multifunctional molecule that also plays important roles in insulin secretion, the immune system, hemostasis/thrombosis, cancer, and virus infections (reviewed in Takahashi & Suzuki, 2012; Xiao et al., 2013).

Loss of brain sulfatide in aging

Previous studies have revealed that murine brain sulfatide levels are exponentially increased during the first 2 months of age and continue to increase at a slower rate during adulthood, eventually plateauing during middle age (~12 months) (Cheng et al., 2010; Han et al., 2003b; Palavicini et al., 2016; Wang et al., 2014). Interestingly, in old age (after >50% mortality or 28 months) this pattern reverses, and brain sulfatide levels decline (Han et al., 2003b; Torello et al., 1986). This decline in brain sulfatide content may be associated with a loss of myelin in old age, as other myelin-enriched lipids (e.g., galactocerebroide and ethanolamine plasmalogen) also decline in aged mice (Torello et al., 1986). Consistently, a progressive brain sulfatide deficiency in old age has also been reported in humans (Couttas et al., 2018; Svennerholm et al., 1994; Svennerholm, Bostrom, & Jungbjer, 1997).

It has been demonstrated that sulfatide metabolism in the brain is modulated by apolipoprotein E (apoE), a protein involved in lipid metabolism that has been strongly linked to aging and AD (Cheng et al., 2010; Han et al., 2003b). Interestingly, the *APOE4* gene variant, which is associated with neurodegeneration, decreased life span, and has the strongest and most reproducible genetic association with AD, accelerates this age- and AD-induced sulfatide deficiency (Han et al., 2003b). Conversely, the *APOE2* gene variant, which has the strongest and most consistent genetic association with human longevity, confers neuroprotection and protects against AD, prevents against sulfatide deficiency (unpublished data from our laboratory). Sulfatide-depleted animals display neuroinflammation and reduced life spans, further supporting a role for brain sulfatide in aging (unpublished data from our laboratory).

Plasmalogen

Plasmalogen structure/function

Plasmalogens are a class of naturally occurring membrane glycerophospholipids with unique properties, primarily present as phosphatidylethanolamine (PE) and phosphatidylcholine (PC) species (Braverman & Moser, 2012; Gross, 1984; Paul, Lancaster, & Meikle, 2019). They contain a vinyl-ether linked aliphatic chain (i.e., an alkenyl chain) at the *sn*-1 position of the glycerol backbone and an ester-linked fatty acid at the *sn*-2 position. The alkenyl chain at the *sn*-1 position consists primarily of 16:0, 18:0, or 18:1 alkyl chain, while the *sn*-2 position is predominantly occupied by a polyunsaturated fatty acyl chain, typically arachidonic acid (AA, 20:4) or docosahexaenoic acid (DHA, 22:6).

Plasmalogens are abundant in mammalian biological membranes, constituting ~8%–10% of total phospholipid mass in most human organs, with even higher abundances in specific organs like heart and brain (Lohner, 1996; Nagan & Zoeller, 2001). Plasmalogens are particularly enriched in plasma membrane lipid rafts (Pike et al., 2002) and are also significant components of different subcellular membranes including the nucleus, endoplasmic reticulum (ER), post-Golgi network, and mitochondria (Honsho et al., 2008). In most tissues, the levels of PE plasmalogens predominate over PC plasmalogens (Messias et al., 2018). In fact, PE plasmalogens comprise ≥50% of total PE in multiple human tissues, including nervous, cardiovascular, immune (neutrophils and eosinophils), and reproductive (spermatozoa) systems (Brouwers et al., 1998; Evans, Weaver, & Clegg, 1980; Farooqui & Horrocks, 2001; Lee, 1998; Lohner, 1996). High plasmalogen PE levels are also found in kidney, lung, spleen, skeletal muscle, and other organs (Nagan & Zoeller, 2001; Paul et al., 2019). On the other hand, mammalian cardiac tissue is unique in its high proportion of plasmalogen PC content (35%–40%) (Diagne et al., 1984; Panganamala et al., 1971).

Plasmalogens are critical for human health and have established roles in neuronal development, the immune response, and as endogenous antioxidants (Paul et al., 2019). First, as important components of cell membranes, plasmalogens influence membrane dynamics by forming more condensed and thicker lipid bilayers (Han & Gross, 1990; Rog & Koivuniemi, 2016). Second, plasmalogens may act as potent antioxidants given the enhanced electron density of the vinyl ether bond at the *sn*-1 position in the hydrophilic domain of the membrane that makes them susceptible to ROS attack. In vitro studies have supported the role of plasmalogens as antioxidants (reviewed in Engelmann, 2004; Murphy, 2001). For example, plasmalogen-deficient cells showed increased susceptibility to chemically induced hypoxia and ROS, while restoration of plasmalogen species promoted survival from chemical hypoxia (Zoeller et al., 1999). However, the idea of plasmalogens acting as "scavengers" for ROS still needs to be verified in vivo, and has been challenged by the finding that plasmalogen oxidation products (e.g., α-hydroxy aldehydes and plasmalogen epoxides), which could themselves be harmful for the organism, accumulate in chronic diseases (Felde & Spiteller, 1995; Lessig & Fuchs, 2009; Loidl-Stahlhofen et al., 1995). Third, since plasmalogens typically contain a PUFA in the *sn*-2 position, they act as reservoirs of these biologically active lipid mediators (Nagan & Zoeller, 2001). Upon plasmalogen cleavage via action of phospholipase A2, the released lysoplasmalogens and PUFAs (particularly AA and DHA) act as lipid mediators for cell signaling (Nagan & Zoeller, 2001; Paul et al., 2019). Fourth, it has been proposed that plasmalogens regulate cholesterol metabolism (synthesis, transport, efflux, and oxidation) by different mechanisms in multiple contexts (Honsho, Abe, & Fujiki, 2015; Maeba & Ueta, 2003; Mandel et al., 1998; Munn et al., 2003), consistent with the fact that both plasmalogens and cholesterol are enriched in lipid rafts (Pike et al., 2002). Fifth, it has been shown that plasmalogens regulate macrophage phagocytic activity (Rubio et al., 2018).

The importance of plasmalogens in physiological processes is emphasized by a series of human genetic peroxisomal disorders exhibiting plasmalogen deficiency (reviewed in Dorninger, Forss-Petter, & Berger, 2017; Schrakamp et al., 1988). In addition, plasmalogens have been implicated in a variety of disease states, including AD (Ginsberg et al., 1995; Grimm et al., 2011; Han, Holtzman, & McKeel, 2001; Igarashi et al., 2011) and metabolic diseases (Graessler et al., 2009; Meikle et al., 2013; Puri et al., 2009).

Plasmalogen loss in aging

Animal studies have shown that plasmalogen content in the brain evolves with development, first increasing during myelination stages and later decreasing in old and senescence-accelerated mice (Andre et al., 2005, 2006a, 2006b). Consistently, it has been reported that the human brain undergoes "lipid raft aging" during nonpathological aging, with major alterations in plasmalogen, DHA, and AA content (Diaz et al., 2018). In addition, plasmalogens in human serum markedly decrease with aging and correlate with high-density lipoprotein (Maeba et al., 2007).

Lipotoxicity in aging

Lipotoxicity can be defined as a metabolic syndrome that results from the accumulation of lipid species (including free fatty acids, diacylglycerols, triacylglycerols, and ceramides) in nonadipose tissue as a result of an imbalance between de novo lipogenesis and fatty acid oxidation, leading to cellular dysfunction, apoptosis, and eventual loss of function/death (Lelliott & Vidal-Puig, 2004). The safest place to store lipids is in white adipose tissue, where highly lipotoxic free fatty acids are neutralized in the form of "inert" triacylglycerols. However, free fatty acids and other lipids get "overspilled" to nonadipose tissues when (1) basal fat cell lipolysis is increased, (2) adipocyte storage capacity becomes impaired or saturated, and/or (3) de novo lipogenesis is increased. Adipocyte lipolysis is increased during overnutrition and is closely associated with insulin resistance (reviewed in Morigny et al., 2016). Increased hepatic de novo lipogenesis can also lead to lipotoxicity, as is the case in fatty liver disease (reviewed in Sanders & Griffin, 2016; Softic, Cohen, & Kahn, 2016). Finally, a substantial reduction in fat storage capacity secondary to poor differentiation and expandability of adipose tissue in response to nutritional demands for storage can also lead to lipotoxicity, as in lipodystrophy, a pathological state that is apparently opposite to obesity (Slawik & Vidal-Puig, 2006). When lipid excess is redirected toward peripheral organs (such as the skeletal muscle, heart, liver, and pancreas), the initial response is to facilitate the storage of the free fatty acid surplus in the form of triacylglycerols, however the buffer capacity of this ectopic fat accumulation quickly becomes saturated. Under these conditions, there is an excess of lipotoxic lipid species, including free fatty acids, that induce organ-specific toxic responses leading to apoptosis (Slawik & Vidal-Puig, 2006).

Insulin resistance and insulin insufficiency, characteristic of metabolic disorders (e.g., obesity, diabetes, and lipodystrophy), are associated with increases in net fatty acid flux toward peripheral organs, eventually resulting in lipotoxicity (Lelliott & Vidal-Puig, 2004). This is because insulin has two specific effects on white adipose tissue: (1) it stimulates uptake of glucose and fatty acids into white adipose tissue and (2) it suppresses lipolysis, where triacylglycerols are hydrolyzed into glycerol and free fatty acids; both of which are impaired under insulin-resistant/deficient conditions (Holm et al., 2000; Molina et al., 1989).

Aging is associated with insulin resistance (reviewed in Carrascosa et al., 2011; Ryan, 2000) as well as a "physiological" increase in triacylglycerols and free fatty acids in nonadipose tissues, even in lean individuals (Murthy et al., 1986; Pararasa, Bailey, & Griffiths, 2015; Slawik & Vidal-Puig, 2006). This might be a causal link between age and the substantially increased prevalence of chronic diseases like type II diabetes, fatty liver, and cardiovascular disease. Aging may exacerbate the effects of lipotoxicity directly by increasing food intake, decreasing energy expenditure, and/or impairing adipose tissue expandability (Horber et al., 1997; Slawik & Vidal-Puig, 2006). For example, the physiological reduction of lean body mass associated with aging may facilitate changes in energy balance, promoting fat deposition versus fatty acid oxidation. Increased fat availability may increase the pool of lipids redirected to form reactive lipid species (diglycerides, free fatty acids, and/or ceramides) in ectopic locations. Conversely, the beneficial effects of calorie restriction and exercise (through their effects on energy balance) may occur as a result of reduced lipid toxicity in nonadipose tissues (Slawik & Vidal-Puig, 2006).

The age-related loss of muscle mass and function, referred to as sarcopenia, is a universal characteristic of the aging process (reviewed in Kamel, 2003; Larsson et al., 2019) that is often accompanied by an increase in whole-body adiposity (Visser et al., 1998). These changes in body composition have important clinical implications, given that loss of muscle and gain of fat mass are both significantly and independently associated with declining physical performance and increase the risk for disability, hospitalization, and mortality in older individuals (Goodpaster et al., 2001; Newman et al., 2003).

Acknowledgments

This work was partially supported by National Institute of Aging Grant RF1 AG061872. The authors thank the members of the Han lab, particularly Nancy Gonzalez and Ziying Xu, for reading this chapter and providing advice on how to improve its clarity and quality.

References

Acaz-Fonseca, E., et al. (2020). Sex differences and gonadal hormone regulation of brain cardiolipin, a key mitochondrial phospholipid. *Journal of Neuroendocrinology, 32*(1), e12774.

Acehan, D., et al. (2007). Comparison of lymphoblast mitochondria from normal subjects and patients with Barth syndrome using electron microscopic tomography. *Laboratory Investigation; a Journal of Technical Methods and Pathology, 87*(1), 40–48.

Andre, A., et al. (2005). Effects of aging and dietary n-3 fatty acids on rat brain phospholipids: Focus on plasmalogens. *Lipids, 40*(8), 799–806.

Andre, A., et al. (2006a). Cerebral plasmalogens and aldehydes in senescence-accelerated mice P8 and R1: A comparison between weaned, adult and aged mice. *Brain Research, 1085*(1), 28–32.

Andre, A., et al. (2006b). Plasmalogen metabolism-related enzymes in rat brain during aging: Influence of n-3 fatty acid intake. *Biochimie, 88*(1), 103–111.

Arana, L., et al. (2010). Ceramide and ceramide 1-phosphate in health and disease. *Lipids in Health and Disease, 9*, 15.

Ayala, A., Munoz, M. F., & Arguelles, S. (2014). Lipid peroxidation: Production, metabolism, and signaling mechanisms of malondialdehyde and 4-hydroxy-2-nonenal. *Oxidative Medicine and Cellular Longevity, 2014*, 360438.

Babenko, N. A., et al. (2016). Role of acid sphingomyelinase in the age-dependent dysregulation of sphingolipids turnover in the tissues of rats. *General Physiology and Biophysics, 35*(2), 195–205.

Balcombe, N. R., & Sinclair, A. (2001). Ageing: Definitions, mechanisms and the magnitude of the problem. *Best Practice & Research. Clinical Gastroenterology, 15*(6), 835–849.

Barja, G. (2014). The mitochondrial free radical theory of aging. *Progress in Molecular Biology and Translational Science, 127*, 1–27.

Bartke, N., & Hannun, Y. A. (2009). Bioactive sphingolipids: Metabolism and function. *Journal of Lipid Research, 50*, S91–S96.

Bazan, S., et al. (2013). Cardiolipin-dependent reconstitution of respiratory supercomplexes from purified Saccharomyces cerevisiae complexes III and IV. *The Journal of Biological Chemistry, 288*(1), 401–411.

Bielski, B. H., Arudi, R. L., & Sutherland, M. W. (1983). A study of the reactivity of HO2/O2- with unsaturated fatty acids. *The Journal of Biological Chemistry, 258*(8), 4759–4761.

Blackburn, E. H., Greider, C. W., & Szostak, J. W. (2006). Telomeres and telomerase: The path from maize, Tetrahymena and yeast to human cancer and aging. *Nature Medicine, 12*(10), 1133–1138.

Boveris, A., & Chance, B. (1973). The mitochondrial generation of hydrogen peroxide. General properties and effect of hyperbaric oxygen. *Biochemical Journal, 134*(3), 707–716.

Braverman, N. E., & Moser, A. B. (2012). Functions of plasmalogen lipids in health and disease. *Biochimica et Biophysica Acta, 1822*(9), 1442–1452.

Brouwers, J. F., et al. (1998). Quantitative analysis of phosphatidylcholine molecular species using HPLC and light scattering detection. *Journal of Lipid Research, 39*(2), 344–353.

Brunson, J. K., et al. (2016). lac-1 and lag-1 with ras-1 affect aging and the biological clock in Neurospora crassa. *Ecology and Evolution, 6*(23), 8341–8351.

Burkle, A., et al. (2007). Pathophysiology of ageing, longevity and age related diseases. *Immunity & Ageing: I & A, 4*, 4.

Bustos, V., & Partridge, L. (2017). Good Ol' fat: Links between lipid signaling and longevity. *Trends in Biochemical Sciences, 42*(10), 812–823.

Cadenas, E., & Davies, K. J. (2000). Mitochondrial free radical generation, oxidative stress, and aging. *Free Radical Biology & Medicine, 29*(3-4), 222–230.

Campana, M., et al. (2018). Inhibition of central de novo ceramide synthesis restores insulin signaling in hypothalamus and enhances beta-cell function of obese Zucker rats. *Molecular Metabolism, 8*, 23–36.

Campisi, G., et al. (2009). Pathophysiology of age-related diseases. *Immunity & Ageing: I & A, 6*, 12.

Carrascosa, J. M., et al. (2011). Development of insulin resistance during aging: Involvement of central processes and role of adipokines. *Current Protein & Peptide Science, 12*(4), 305–315.

Cheng, H., et al. (2008). Shotgun lipidomics reveals the temporally dependent, highly diversified cardiolipin profile in the mammalian brain: Temporally coordinated postnatal diversification of cardiolipin molecular species with neuronal remodeling. *Biochemistry, 47*(21), 5869–5880.

Cheng, H., et al. (2010). Apolipoprotein E mediates sulfatide depletion in animal models of Alzheimer's disease. *Neurobiology of Aging, 31*(7), 1188–1196.

Cheng, H., et al. (2013). Specific changes of sulfatide levels in individuals with pre-clinical Alzheimer's disease: An early event in disease pathogenesis. *Journal of Neurochemistry, 127*(6), 733–738.

Chicco, A. J., & Sparagna, G. C. (2007). Role of cardiolipin alterations in mitochondrial dysfunction and disease. *American Journal of Physiology. Cell Physiology, 292*(1), C33–C44.

Chien, C. C., et al. (2009). Activation of telomerase and cyclooxygenase-2 in PDGF and FGF inhibition of C2-ceramide-induced apoptosis. *Journal of Cellular Physiology, 218*(2), 405–415.

Chung, H. Y., et al. (2001). The inflammation hypothesis of aging: Molecular modulation by calorie restriction. *Annals of the New York Academy of Sciences, 928*, 327–335.

Claypool, S. M., & Koehler, C. M. (2012). The complexity of cardiolipin in health and disease. *Trends in Biochemical Sciences, 37*(1), 32–41.

Couttas, T. A., et al. (2018). Age-dependent changes to sphingolipid balance in the human hippocampus are gender-specific and may sensitize to neurodegeneration. *Journal of Alzheimer's Disease: JAD, 63*(2), 503–514.

Cowart, L. A. (2009). Sphingolipids: Players in the pathology of metabolic disease. *Trends in Endocrinology and Metabolism: TEM, 20*(1), 34–42.

Cutler, R. G., et al. (2014). Sphingolipid metabolism regulates development and lifespan in Caenorhabditis elegans. *Mechanisms of Ageing and Development, 143-144*, 9–18.

Cutler, R. G., & Mattson, M. P. (2001). Sphingomyelin and ceramide as regulators of development and lifespan. *Mechanisms of Ageing and Development, 122*(9), 895–908.

De Bruijn, J. H. (1966). Chemical structure and serological activity of natural and synthetic cardiolipin and related compounds. *The British Journal of Venereal Diseases, 42*(2), 125–128.

Dela Cruz, C. S., & Kang, M. J. (2018). Mitochondrial dysfunction and damage associated molecular patterns (DAMPs) in chronic inflammatory diseases. *Mitochondrion, 41*, 37–44.

Diagne, A., et al. (1984). Studies on ether phospholipids. II. Comparative composition of various tissues from human, rat and guinea pig. *Biochimica et Biophysica Acta, 793*(2), 221–231.

Diaz, M., et al. (2018). "Lipid raft aging" in the human frontal cortex during nonpathological aging: Gender influences and potential implications in Alzheimer's disease. *Neurobiology of Aging, 67*, 42–52.

de Diego, I., Peleg, S., & Fuchs, B. (2019). The role of lipids in aging-related metabolic changes. *Chemistry and Physics of Lipids, 222*, 59–69.

Dorninger, F., Forss-Petter, S., & Berger, J. (2017). From peroxisomal disorders to common neurodegenerative diseases—the role of ether phospholipids in the nervous system. *FEBS Letters, 591*(18), 2761–2788.

D'Mello, N. P., et al. (1994). Cloning and characterization of LAG1, a longevity-assurance gene in yeast. *The Journal of Biological Chemistry, 269*(22), 15451–15459.

Dubot, P., et al. (2019). Inherited monogenic defects of ceramide metabolism: Molecular bases and diagnoses. *Clinica Chimica Acta; International Journal of Clinical Chemistry, 495*, 457–466.

Dunn, T. M., Tifft, C. J., & Proia, R. L. (2019). A perilous path: The inborn errors of sphingolipid metabolism. *Journal of Lipid Research, 60*(3), 475–483.

Engelmann, B. (2004). Plasmalogens: Targets for oxidants and major lipophilic antioxidants. *Biochemical Society Transactions, 32*(Pt 1), 147–150.

Evans, R. W., Weaver, D. E., & Clegg, E. D. (1980). Diacyl, alkenyl, and alkyl ether phospholipids in ejaculated, in utero-, and in vitro-incubated porcine spermatozoa. *Journal of Lipid Research, 21*(2), 223–228.

Fahy, E., et al. (2005). A comprehensive classification system for lipids. *Journal of Lipid Research, 46*(5), 839–861.

Farooqui, A. A., & Horrocks, L. A. (2001). Plasmalogens: Workhorse lipids of membranes in normal and injured neurons and glia. *The Neuroscientist: A Review Journal Bringing Neurobiology, Neurology and Psychiatry, 7*(3), 232–245.

Felde, R., & Spiteller, G. (1995). Plasmalogen oxidation in human serum lipoproteins. *Chemistry and Physics of Lipids, 76*(2), 259–267.

Fischer, H., et al. (2007). Ceramide as a TLR4 agonist; a putative signalling intermediate between sphingolipid receptors for microbial ligands and TLR4. *Cellular Microbiology, 9*(5), 1239–1251.

Franceschi, C., et al. (2000). Inflammaging. An evolutionary perspective on immunosenescence. *Annals of the New York Academy of Sciences, 908*, 244–254.

Franceschi, C., et al. (2007). Inflammaging and anti-inflammaging: A systemic perspective on aging and longevity emerged from studies in humans. *Mechanisms of Ageing and Development, 128*(1), 92–105.

Franceschi, C., & Campisi, J. (2014). Chronic inflammation (inflammaging) and its potential contribution to age-associated diseases. *The Journals of Gerontology. Series A, Biological Sciences and Medical Sciences, 69*(Suppl 1), S4–S9.

Fredman, P., et al. (1992). Sulfatide as a biochemical marker in cerebrospinal fluid of patients with vascular dementia. *Acta Neurologica Scandinavica, 85*(2), 103–106.

Fulop, T., et al. (2014). On the immunological theory of aging. *Interdisciplinary Topics in Gerontology, 39*, 163–176.

Gaspard, G. J., & McMaster, C. R. (2015). Cardiolipin metabolism and its causal role in the etiology of the inherited cardiomyopathy Barth syndrome. *Chemistry and Physics of Lipids, 193*, 1–10.

Gawel, S., et al. (2004). [Malondialdehyde (MDA) as a lipid peroxidation marker]. *Wiadomosci Lekarskie (Warsaw, Poland: 1960), 57*(9-10), 453–455.

Geekiyanage, H., Upadhye, A., & Chan, C. (2013). Inhibition of serine palmitoyltransferase reduces Abeta and tau hyperphosphorylation in a murine model: A safe therapeutic strategy for Alzheimer's disease. *Neurobiology of Aging, 34*(8), 2037–2051.

Ginsberg, L., et al. (1995). Disease and anatomic specificity of ethanolamine plasmalogen deficiency in Alzheimer's disease brain. *Brain Research, 698*(1-2), 223–226.

Giusto, N. M., Roque, M. E., & Ilincheta de Boschero, M. G. (1992). Effects of aging on the content, composition and synthesis of sphingomyelin in the central nervous system. *Lipids, 27*(11), 835–839.

Gonzalez de San Roman, E., et al. (2017). Imaging mass spectrometry (IMS) of cortical lipids from preclinical to severe stages of Alzheimer's disease. *Biochim Biophys Acta Biomembr, 1859*(9 Pt B), 1604–1614.

Gonzalvez, F., & Gottlieb, E. (2007). Cardiolipin: Setting the beat of apoptosis. *Apoptosis: An International Journal on Programmed Cell Death, 12*(5), 877–885.

Goodpaster, B. H., et al. (2001). Attenuation of skeletal muscle and strength in the elderly: The Health ABC Study. *Journal of Applied Physiology (1985)*, *90*(6), 2157–2165.

Graessler, J., et al. (2009). Top-down lipidomics reveals ether lipid deficiency in blood plasma of hypertensive patients. *PLoS One*, *4* (7), e6261.

Granzotto, A., et al. (2019). Inhibition of de novo ceramide biosynthesis affects aging phenotype in an in vitro model of neuronal senescence. *Aging (Albany NY)*, *11*, 6336–6357.

Grimm, M. O., et al. (2011). Plasmalogen synthesis is regulated via alkyl-dihydroxyacetonephosphate-synthase by amyloid precursor protein processing and is affected in Alzheimer's disease. *Journal of Neurochemistry*, *116*(5), 916–925.

Gross, R. W. (1984). High plasmalogen and arachidonic acid content of canine myocardial sarcolemma: A fast atom bombardment mass spectroscopic and gas chromatography-mass spectroscopic characterization. *Biochemistry*, *23*(1), 158–165.

Guillas, I., et al. (2001). C26-CoA-dependent ceramide synthesis of Saccharomyces cerevisiae is operated by Lag1p and Lac1p. *The EMBO Journal*, *20*(11), 2655–2665.

Han, J. W., Choi, G. J., & Kim, B. S. (2018). Antimicrobial aromatic polyketides: A review of their antimicrobial properties and potential use in plant disease control. *World Journal of Microbiology and Biotechnology*, *34*(11), 163.

Han, X. (2005). Lipid alterations in the earliest clinically recognizable stage of Alzheimer's disease: Implication of the role of lipids in the pathogenesis of Alzheimer's disease. *Current Alzheimer Research*, *2*(1), 65–77.

Han, X. (2007). Potential mechanisms contributing to sulfatide depletion at the earliest clinically recognizable stage of Alzheimer's disease: A tale of shotgun lipidomics. *Journal of Neurochemistry*, *103* (Suppl 1), 171–179.

Han, X. (2016). Lipidomics for studying metabolism. *Nature Reviews Endocrinology*, *12*(11), 668–679.

Han, X., et al. (2002). Substantial sulfatide deficiency and ceramide elevation in very early Alzheimer's disease: Potential role in disease pathogenesis. *Journal of Neurochemistry*, *82*(4), 809–818.

Han, X., et al. (2003a). Cerebrospinal fluid sulfatide is decreased in subjects with incipient dementia. *Annals of Neurology*, *54*(1), 115–119.

Han, X., et al. (2003b). Novel role for apolipoprotein E in the central nervous system. Modulation of sulfatide content. *Journal of Biological Chemistry*, *278*(10), 8043–8051.

Han, X., et al. (2005). Shotgun lipidomics identifies cardiolipin depletion in diabetic myocardium linking altered substrate utilization with mitochondrial dysfunction. *Biochemistry*, *44*(50), 16684–16694.

Han, X., & Gross, R. W. (1990). Plasmenylcholine and phosphatidylcholine membrane bilayers possess distinct conformational motifs. *Biochemistry*, *29*(20), 4992–4996.

Han, X., Holtzman, D. M., & McKeel, D. W., Jr. (2001). Plasmalogen deficiency in early Alzheimer's disease subjects and in animal models: Molecular characterization using electrospray ionization mass spectrometry. *Journal of Neurochemistry*, *77*(4), 1168–1180.

Hanada, K. (2010). Intracellular trafficking of ceramide by ceramide transfer protein. *Proceeding of the Japan Academy Series B-Physical and Biological Sciences*, *86*(4), 426–437.

Hannun, Y. A., & Obeid, L. M. (2011). Many ceramides. *The Journal of Biological Chemistry*, *286*(32), 27855–27862.

Hannun, Y. A., & Obeid, L. M. (2018). Sphingolipids and their metabolism in physiology and disease. *Nature Reviews. Molecular Cell Biology*, *19*(3), 175–191.

Hanske, J., et al. (2012). Conformational properties of cardiolipin-bound cytochrome c. *Proceedings of the National Academy of Science of the United States of America*, *109*(1), 125–130.

Harman, D. (1972). The biologic clock: The mitochondria? *Journal of the American Geriatrics Society*, *20*(4), 145–147.

Harman, D. (2003). The free radical theory of aging. *Antioxidants & Redox Signaling*, *5*(5), 557–561.

Hayflick, L. (1965). The limited in vitro lifetime of human diploid cell strains. *Experimental Cell Research*, *37*, 614–636.

Hayflick, L. (1998). How and why we age. *Experimental Gerontology*, *33*(7-8), 639–653.

He, Q., & Han, X. (2014). Cardiolipin remodeling in diabetic heart. *Chemistry and Physics of Lipids*, *179*, 75–81.

Hoch, F. L. (1992). Cardiolipins and biomembrane function. *Biochimica et Biophysica Acta*, *1113*(1), 71–133.

Hojjati, M. R., et al. (2005). Effect of myriocin on plasma sphingolipid metabolism and atherosclerosis in apoE-deficient mice. *The Journal of Biological Chemistry*, *280*(11), 10284–10289.

Holliday, R. (2006). Aging is no longer an unsolved problem in biology. *Annals of the New York Academy of Sciences*, *1067*, 1–9.

Holm, C., et al. (2000). Molecular mechanisms regulating hormone-sensitive lipase and lipolysis. *Annual Review of Nutrition*, *20*, 365–393.

Hong, J. H., et al. (2016). Global changes of phospholipids identified by MALDI imaging mass spectrometry in a mouse model of Alzheimer's disease. *Journal of Lipid Research*, *57*(1), 36–45.

Honke, K., et al. (2002). Paranodal junction formation and spermatogenesis require sulfoglycolipids. *Proceedings of the National Academy of Science of the United States of America*, *99*(7), 4227–4232.

Honsho, M., et al. (2008). Isolation and characterization of mutant animal cell line defective in alkyl-dihydroxyacetonephosphate synthase: Localization and transport of plasmalogens to post-Golgi compartments. *Biochimica et Biophysica Acta*, *1783*(10), 1857–1865.

Honsho, M., Abe, Y., & Fujiki, Y. (2015). Dysregulation of plasmalogen homeostasis impairs cholesterol biosynthesis. *The Journal of Biological Chemistry*, *290*(48), 28822–28833.

Horber, F. F., et al. (1997). Effect of sex and age on bone mass, body composition and fuel metabolism in humans. *Nutrition (Burbank, Los Angeles County, Calif.)*, *13*(6), 524–534.

Hu, W., et al. (2010). Polyketides from marine dinoflagellates of the genus Prorocentrum, biosynthetic origin and bioactivity of their okadaic acid analogues. *Mini Reviews in Medicinal Chemistry*, *10* (1), 51–61.

Huang, X., Liu, J., & Dickson, R. C. (2012). Down-regulating sphingolipid synthesis increases yeast lifespan. *PLoS Genetics*, *8*(2), e1002493.

Huffman, J., Gerber, R., & Du, L. (2010). Recent advancements in the biosynthetic mechanisms for polyketide-derived mycotoxins. *Biopolymers*, *93*(9), 764–776.

Igarashi, M., et al. (2011). Disturbed choline plasmalogen and phospholipid fatty acid concentrations in Alzheimer's disease prefrontal cortex. *Journal of Alzheimer's Disease: JAD*, *24*(3), 507–517.

Ikon, N., & Ryan, R. O. (2017). Cardiolipin and mitochondrial cristae organization. *Biochimica Biophysica Acta Biomembranes*, *1859*(6), 1156–1163.

Ishibashi, T., et al. (2002). A myelin galactolipid, sulfatide, is essential for maintenance of ion channels on myelinated axon but not essential for initial cluster formation. *The Journal of Neuroscience*, *22*(15), 6507–6514.

Iyer, S. S., et al. (2013). Mitochondrial cardiolipin is required for Nlrp3 inflammasome activation. *Immunity*, *39*(2), 311–323.

Jang, J. Y., et al. (2018). The role of mitochondria in aging. *The Journal of Clinical Investigation*, *128*(9), 3662–3670.

Jonsson, M., et al. (2010). Cerebrospinal fluid biomarkers of white matter lesions - cross-sectional results from the LADIS study. *European Journal of Neurology: The Official Journal of the European Federation of Neurological Societies*, *17*(3), 377–382.

Jussupow, A., Di Luca, A., & Kaila, V. R. I. (2019). How cardiolipin modulates the dynamics of respiratory complex I. *Science Advances*, *5*(3), eaav1850.

Kagan, V. E., et al. (2005). Cytochrome c acts as a cardiolipin oxygenase required for release of proapoptotic factors. *Nature Chemical Biology*, *1*(4), 223–232.

Kamel, H. K. (2003). Sarcopenia and aging. *Nutrition Reviews*, *61*(5 Pt 1), 157–167.

Kauppila, T. E. S., Kauppila, J. H. K., & Larsson, N. G. (2017). Mammalian mitochondria and aging: An update. *Cell Metabolism*, *25*(1), 57–71.

Khalifat, N., et al. (2011). Lipid packing variations induced by pH in cardiolipin-containing bilayers: The driving force for the cristae-like shape instability. *Biochimica et Biophysica Acta*, *1808*(11), 2724–2733.

Khayrullin, A., et al. (2019). Very long-chain C24:1 ceramide is increased in serum extracellular vesicles with aging and can induce senescence in bone-derived mesenchymal stem cells. *Cells*, *8*(1), 37.

Kiebish, M. A., et al. (2012). Myocardial regulation of lipidomic flux by cardiolipin synthase: Setting the beat for bioenergetic efficiency. *The Journal of Biological Chemistry*, *287*(30), 25086–25097.

Kim, M. K., et al. (2019). Links between accelerated replicative cellular senescence and down-regulation of SPHK1 transcription. *BMB Reports*, *52*(3), 220–225.

Kirkwood, T. B., & Austad, S. N. (2000). Why do we age? *Nature*, *408* (6809), 233–238.

Kobayashi, K., et al. (2013). Increase in secretory sphingomyelinase activity and specific ceramides in the aorta of apolipoprotein E knockout mice during aging. *Biological & Pharmaceutical Bulletin*, *36*(7), 1192–1196.

Kohlschutter, A. (2013). Lysosomal leukodystrophies: Krabbe disease and metachromatic leukodystrophy. *Handbook of Clinical Neurology*, *113*, 1611–1618.

Kolter, T. (2011). A view on sphingolipids and disease. *Chemistry and Physics of Lipids*, *164*(6), 590–606.

Koshkin, V., & Greenberg, M. L. (2002). Cardiolipin prevents rate-dependent uncoupling and provides osmotic stability in yeast mitochondria. *The Biochemical Journal*, *364*(Pt 1), 317–322.

Kowald, A., & Kirkwood, T. B. (1994). Towards a network theory of ageing: A model combining the free radical theory and the protein error theory. *Journal of Theoretical Biology*, *168*(1), 75–94.

Kraveka, J. M., et al. (2003). Involvement of endogenous ceramide in the inhibition of telomerase activity and induction of morphologic differentiation in response to all-trans-retinoic acid in human neuroblastoma cells. *Archives of Biochemistry and Biophysics*, *419*(2), 110–119.

Kriete, A., et al. (2006). Systems approaches to the networks of aging. *Ageing Research Reviews*, *5*(4), 434–448.

Krysko, D. V., et al. (2011). Emerging role of damage-associated molecular patterns derived from mitochondria in inflammation. *Trends in Immunology*, *32*(4), 157–164.

Kurek, K., et al. (2015). Inhibition of ceramide de novo synthesis ameliorates diet induced skeletal muscles insulin resistance. *Journal of Diabetes Research*, *2015*, 154762.

Kuzuyama, T., & Seto, H. (2003). Diversity of the biosynthesis of the isoprene units. *Natural Product Reports*, *20*(2), 171–183.

Larsson, L., et al. (2019). Sarcopenia: Aging-related loss of muscle mass and function. *Physiological Reviews*, *99*(1), 427–511.

Lee, H. J., et al. (2006). Selective remodeling of cardiolipin fatty acids in the aged rat heart. *Lipids in Health and Disease*, *5*, 2.

Lee, T. C. (1998). Biosynthesis and possible biological functions of plasmalogens. *Biochimica et Biophysica Acta*, *1394*(2-3), 129–145.

Lelliott, C., & Vidal-Puig, A. J. (2004). Lipotoxicity, an imbalance between lipogenesis de novo and fatty acid oxidation. *International Journal of Obesity and Related Metabolic Disorders: Journal of the International Association for the Study of Obesity*, *28*(Suppl 4), S22–S28.

Lessig, J., & Fuchs, B. (2009). Plasmalogens in biological systems: Their role in oxidative processes in biological membranes, their contribution to pathological processes and aging and plasmalogen analysis. *Current Medicinal Chemistry*, *16*(16), 2021–2041.

Lightle, S. A., Oakley, J. I., & Nikolova-Karakashian, M. N. (2000). Activation of sphingolipid turnover and chronic generation of ceramide and sphingosine in liver during aging. *Mechanisms of Ageing and Development*, *120*(1-3), 111–125.

Lohner, K. (1996). Is the high propensity of ethanolamine plasmalogens to form non-lamellar lipid structures manifested in the properties of biomembranes? *Chemistry and Physics of Lipids*, *81*(2), 167–184.

Loidl-Stahlhofen, A., et al. (1995). Epoxidation of plasmalogens: Source for long-chain alpha-hydroxyaldehydes in subcellular fractions of bovine liver. *The Biochemical Journal*, *309*(Pt 3), 807–812.

Maeba, R., et al. (2007). Plasmalogens in human serum positively correlate with high- density lipoprotein and decrease with aging. *Journal of Atherosclerosis and Thrombosis*, *14*(1), 12–18.

Maeba, R., & Ueta, N. (2003). Ethanolamine plasmalogens prevent the oxidation of cholesterol by reducing the oxidizability of cholesterol in phospholipid bilayers. *Journal of Lipid Research*, *44*(1), 164–171.

Mandel, H., et al. (1998). Plasmalogen phospholipids are involved in HDL-mediated cholesterol efflux: Insights from investigations with plasmalogen-deficient cells. *Biochemical and Biophysical Research Communications*, *250*(2), 369–373.

Maranzana, E., et al. (2013). Mitochondrial respiratory supercomplex association limits production of reactive oxygen species from complex I. *Antioxidants & Redox Signaling*, *19*(13), 1469–1480.

Medvedev, Z. A. (1990). An attempt at a rational classification of theories of ageing. *Biological Reviews of the Cambridge Philosophical Society*, *65*(3), 375–398.

Meikle, P. J., et al. (2013). Plasma lipid profiling shows similar associations with prediabetes and type 2 diabetes. *PLoS One*, *8*(9), e74341.

Messias, M. C. F., et al. (2018). Plasmalogen lipids: Functional mechanism and their involvement in gastrointestinal cancer. *Lipids in Health and Disease*, *17*(1), 41.

Mielke, M. M., et al. (2015). Demographic and clinical variables affecting mid- to late-life trajectories of plasma ceramide and dihydroceramide species. *Aging Cell*, *14*(6), 1014–1023.

Mileykovskaya, E., & Dowhan, W. (2009). Cardiolipin membrane domains in prokaryotes and eukaryotes. *Biochimica et Biophysica Acta*, *1788*(10), 2084–2091.

Molina, J. M., et al. (1989). Decreased activation rate of insulin-stimulated glucose transport in adipocytes from obese subjects. *Diabetes*, *38*(8), 991–995.

Monette, J. S., et al. (2011). R)-alpha-Lipoic acid treatment restores ceramide balance in aging rat cardiac mitochondria. *Pharmacological Research: The Official Journal of the Italian Pharmacological Society*, *63* (1), 23–29.

Morigny, P., et al. (2016). Adipocyte lipolysis and insulin resistance. *Biochimie*, *125*, 259–266.

Mosbech, M. B., et al. (2013). Functional loss of two ceramide synthases elicits autophagy-dependent lifespan extension in C. elegans. *PLoS One*, *8*(7), e70087.

Munn, N. J., et al. (2003). Deficiency in ethanolamine plasmalogen leads to altered cholesterol transport. *Journal of Lipid Research*, *44* (1), 182–192.

Murphy, R. C. (2001). Free-radical-induced oxidation of arachidonoyl plasmalogen phospholipids: Antioxidant mechanism and precursor pathway for bioactive eicosanoids. *Chemical Research in Toxicology*, *14*(5), 463–472.

Murthy, V. K., et al. (1986). Increased fatty acid uptake, a factor in increased hepatic triacylglycerol synthesis in aging rats. *Mechanisms of Ageing and Development*, *37*(1), 49–54.

Musatov, A., et al. (2002). Identification of bovine heart cytochrome c oxidase subunits modified by the lipid peroxidation product 4-hydroxy-2-nonenal. *Biochemistry*, *41*(25), 8212–8220.

Nagan, N., & Zoeller, R. A. (2001). Plasmalogens: Biosynthesis and functions. *Progress in Lipid Research, 40*(3), 199–229.

Newman, A. B., et al. (2003). Strength and muscle quality in a well-functioning cohort of older adults: The health, aging and body composition study. *Journal of the American Geriatrics Society, 51*(3), 323–330.

Nichols-Smith, S., Teh, S. Y., & Kuhl, T. L. (2004). Thermodynamic and mechanical properties of model mitochondrial membranes. *Biochimica et Biophysica Acta, 1663*(1-2), 82–88.

Nicolson, G. L., & Ash, M. E. (2014). Lipid replacement therapy: A natural medicine approach to replacing damaged lipids in cellular membranes and organelles and restoring function. *Biochimica et Biophysica Acta, 1838*(6), 1657–1679.

Nikolova-Karakashian, M. (2018). Alcoholic and non-alcoholic fatty liver disease: Focus on ceramide. *Advances in Biological Regulation, 70*, 40–50.

Nikolova-Karakashian, M., Karakashian, A., & Rutkute, K. (2008). Role of neutral sphingomyelinases in aging and inflammation. *Sub-cellular Biochemistry, 49*, 469–486.

Noh, S. A., et al. (2019). Alterations in lipid profile of the aging kidney identified by MALDI imaging mass spectrometry. *Journal of Proteome Research, 18*(7), 2803–2812.

Ogretmen, B., et al. (2001a). Molecular mechanisms of ceramide-mediated telomerase inhibition in the A549 human lung adenocarcinoma cell line. *The Journal of Biological Chemistry, 276*(35), 32506–32514.

Ogretmen, B., et al. (2001b). Role of ceramide in mediating the inhibition of telomerase activity in A549 human lung adenocarcinoma cells. *The Journal of Biological Chemistry, 276*(27), 24901–24910.

Palavicini, J. P., et al. (2016). Novel molecular insights into the critical role of sulfatide in myelin maintenance/function. *Journal of Neurochemistry, 139*(1), 40–54.

Panganamala, R. V., et al. (1971). Positions of double bonds in the monounsaturated alk-1-enyl groups from the plasmalogens of human heart and brain. *Chemistry and Physics of Lipids, 6*(2), 97–102.

Paradies, G., et al. (2009). Role of cardiolipin peroxidation and Ca^{2+} in mitochondrial dysfunction and disease. *Cell Calcium, 45*(6), 643–650.

Paradies, G., et al. (2011). Mitochondrial dysfunction in brain aging: Role of oxidative stress and cardiolipin. *Neurochemistry International, 58*(4), 447–457.

Paradies, G., et al. (2014a). Oxidative stress, cardiolipin and mitochondrial dysfunction in nonalcoholic fatty liver disease. *World Journal of Gastroenterology: WJG, 20*(39), 14205–14218.

Paradies, G., et al. (2014b). Functional role of cardiolipin in mitochondrial bioenergetics. *Biochimica et Biophysica Acta, 1837*(4), 408–417.

Paradies, G., et al. (2019). Role of cardiolipin in mitochondrial function and dynamics in health and disease: Molecular and pharmacological aspects. *Cells, 8*(7), 728.

Pararasa, C., Bailey, C. J., & Griffiths, H. R. (2015). Ageing, adipose tissue, fatty acids and inflammation. *Biogerontology, 16*(2), 235–248.

Paul, S., Lancaster, G. I., & Meikle, P. J. (2019). Plasmalogens: A potential therapeutic target for neurodegenerative and cardiometabolic disease. *Progress in Lipid Research, 74*, 186–195.

Petrosillo, G., et al. (2009). Mitochondrial complex I dysfunction in rat heart with aging: Critical role of reactive oxygen species and cardiolipin. *Free Radical Biology & Medicine, 46*(1), 88–94.

Pfeiffer, K., et al. (2003). Cardiolipin stabilizes respiratory chain supercomplexes. *The Journal of Biological Chemistry, 278*(52), 52873–52880.

Pike, L. J., et al. (2002). Lipid rafts are enriched in arachidonic acid and plasmenylethanolamine and their composition is independent of caveolin-1 expression: A quantitative electrospray ionization/mass spectrometric analysis. *Biochemistry, 41*(6), 2075–2088.

Plociennikowska, A., et al. (2015). Co-operation of TLR4 and raft proteins in LPS-induced pro-inflammatory signaling. *Cellular and Molecular Life Sciences: CMLS, 72*(3), 557–581.

Plotegher, N., et al. (2019). Ceramides in Parkinson's disease: From recent evidence to new hypotheses. *Frontiers in Neuroscience, 13*, 330.

Pointer, C. B., & Klegeris, A. (2017). Cardiolipin in central nervous system physiology and pathology. *Cellular and Molecular Neurobiology, 37*(7), 1161–1172.

Promislow, D. E. (2004). Protein networks, pleiotropy and the evolution of senescence. *Proceedings of the Royal Society. Biological Sciences, 271*(1545), 1225–1234.

Puri, P., et al. (2009). The plasma lipidomic signature of nonalcoholic steatohepatitis. *Hepatology (Baltimore, MD), 50*(6), 1827–1838.

Raetz, C. R., & Whitfield, C. (2002). Lipopolysaccharide endotoxins. *Annual Review of Biochemistry, 71*, 635–700.

Ren, M., Phoon, C. K., & Schlame, M. (2014). Metabolism and function of mitochondrial cardiolipin. *Progress in Lipid Research, 55*, 1–16.

Robinson, N. C. (1993). Functional binding of cardiolipin to cytochrome c oxidase. *Journal of Bioenergetics and Biomembranes, 25*(2), 153–163.

Rodriguez-Calvo, R., et al. (2007). Peroxisome proliferator-activated receptor alpha down-regulation is associated with enhanced ceramide levels in age-associated cardiac hypertrophy. *The Journals of Gerontology. Series A, Biological Sciences and Medical Sciences, 62*(12), 1326–1336.

Rog, T., & Koivuniemi, A. (2016). The biophysical properties of ethanolamine plasmalogens revealed by atomistic molecular dynamics simulations. *Biochimica et Biophysica Acta, 1858*(1), 97–103.

Rossi, M. J., et al. (2005). Inhibition of growth and telomerase activity by novel cationic ceramide analogs with high solubility in human head and neck squamous cell carcinoma cells. *Otolaryngology—Head and Neck Surgery: Official Journal of American Academy of Otolaryngology-Head and Neck Surgery, 132*(1), 55–62.

Rubio, J. M., et al. (2018). Regulation of phagocytosis in macrophages by membrane ethanolamine plasmalogens. *Frontiers in Immunology, 9*, 1723.

Ruggiero, F. M., et al. (1992). Lipid composition in synaptic and non-synaptic mitochondria from rat brains and effect of aging. *Journal of Neurochemistry, 59*(2), 487–491.

Ryan, A. S. (2000). Insulin resistance with aging: Effects of diet and exercise. *Sports Medicine (Auckland, N.Z.), 30*(5), 327–346.

Sacket, S. J., et al. (2009). Increase in sphingolipid catabolic enzyme activity during aging. *Acta Pharmacologica Sinica, 30*(10), 1454–1461.

Sanders, F. W., & Griffin, J. L. (2016). De novo lipogenesis in the liver in health and disease: More than just a shunting yard for glucose. *Biological Reviews of the Cambridge Philosophical Society, 91*(2), 452–468.

Scheiblich, H., et al. (2017). Activation of the NLRP3 inflammasome in microglia: The role of ceramide. *Journal of Neurochemistry, 143*(5), 534–550.

Schlame, M., et al. (1999). Microanalysis of cardiolipin in small biopsies including skeletal muscle from patients with mitochondrial disease. *Journal of Lipid Research, 40*(9), 1585–1592.

Schlame, M., et al. (2005). Molecular symmetry in mitochondrial cardiolipins. *Chemistry and Physics of Lipids, 138*(1-2), 38–49.

Schlame, M., & Greenberg, M. L. (2017). Biosynthesis, remodeling and turnover of mitochondrial cardiolipin. *Biochim Biophys Acta Mol Cell Biol Lipids, 1862*(1), 3–7.

Schlame, M., & Ren, M. (2006). Barth syndrome, a human disorder of cardiolipin metabolism. *FEBS Letters, 580*(23), 5450–5455.

Schlame, M., Rua, D., & Greenberg, M. L. (2000). The biosynthesis and functional role of cardiolipin. *Progress in Lipid Research, 39*(3), 257–288.

Schmitz-Peiffer, C. (2010). Targeting ceramide synthesis to reverse insulin resistance. *Diabetes, 59*(10), 2351–2353.

Schorling, S., et al. (2001). Lag1p and Lac1p are essential for the Acyl-CoA-dependent ceramide synthase reaction in Saccharomyces cerevisae. *Molecular Biology of the Cell*, *12*(11), 3417–3427.

Schrakamp, G., et al. (1988). Plasmalogen biosynthesis in peroxisomal disorders: Fatty alcohol versus alkylglycerol precursors. *Journal of Lipid Research*, *29*(3), 325–334.

Schroeder, E. A., & Brunet, A. (2015). Lipid profiles and signals for long life. *Trends in Endocrinology and Metabolism: TEM*, *26*(11), 589–592.

Schug, Z. T., & Gottlieb, E. (2009). Cardiolipin acts as a mitochondrial signalling platform to launch apoptosis. *Biochimica et Biophysica Acta*, *1788*(10), 2022–2031.

Semba, R. D., et al. (2019). Tetra-linoleoyl cardiolipin depletion plays a major role in the pathogenesis of sarcopenia. *Medical Hypotheses*, *127*, 142–149.

Sen, T., et al. (2007). Depolarization and cardiolipin depletion in aged rat brain mitochondria: Relationship with oxidative stress and electron transport chain activity. *Neurochemistry International*, *50*(5), 719–725.

Seong, S. Y., & Matzinger, P. (2004). Hydrophobicity: An ancient damage-associated molecular pattern that initiates innate immune responses. *Nature Reviews. Immunology*, *4*(6), 469–478.

Sergiev, P. V., Dontsova, O. A., & Berezkin, G. V. (2015). Theories of aging: An ever-evolving field. *Acta Naturae*, *7*(1), 9–18.

Shen, Z., et al. (2015). The role of cardiolipin in cardiovascular health. *Biomed Research International*, *2015*, 891707.

Shi, Y. (2010). Emerging roles of cardiolipin remodeling in mitochondrial dysfunction associated with diabetes, obesity, and cardiovascular diseases. *Journal of Biomedical Research*, *24*(1), 6–15.

Shibata, A., et al. (1994). Significant stabilization of the phosphatidylcholine bilayer structure by incorporation of small amounts of cardiolipin. *Biochimica et Biophysica Acta*, *1192*(1), 71–78.

Slawik, M., & Vidal-Puig, A. J. (2006). Lipotoxicity, overnutrition and energy metabolism in aging. *Ageing Research Reviews*, *5*(2), 144–164.

Softic, S., Cohen, D. E., & Kahn, C. R. (2016). Role of dietary fructose and hepatic de novo lipogenesis in fatty liver disease. *Digestive Diseases and Sciences*, *61*(5), 1282–1293.

Sogaard, D., et al. (2019). Muscle-saturated bioactive lipids are increased with aging and influenced by high-intensity interval training. *International Journal of Molecular Sciences*, *20*(5), 1240.

Sonnino, S., et al. (2006). Dynamic and structural properties of sphingolipids as driving forces for the formation of membrane domains. *Chemical Reviews*, *106*(6), 2111–2125.

Sparagna, G. C., et al. (2007). Loss of cardiac tetralinoleoyl cardiolipin in human and experimental heart failure. *Journal of Lipid Research*, *48*(7), 1559–1570.

Stepanova, M., et al. (2015). Age-independent rise of inflammatory scores may contribute to accelerated aging in multi-morbidity. *Oncotarget*, *6*(3), 1414–1421.

Svennerholm, L., et al. (1994). Membrane lipids of adult human brain: Lipid composition of frontal and temporal lobe in subjects of age 20 to 100 years. *Journal of Neurochemistry*, *63*(5), 1802–1811.

Svennerholm, L., Bostrom, K., & Jungbjer, B. (1997). Changes in weight and compositions of major membrane components of human brain during the span of adult human life of Swedes. *Acta Neuropathologica*, *94*(4), 345–352.

Szeto, H. H. (2014). First-in-class cardiolipin-protective compound as a therapeutic agent to restore mitochondrial bioenergetics. *British Journal of Pharmacology*, *171*(8), 2029–2050.

Takahashi, T., & Suzuki, T. (2012). Role of sulfatide in normal and pathological cells and tissues. *Journal of Lipid Research*, *53*(8), 1437–1450.

Tedesco, P., et al. (2008). Genetic analysis of hyl-1, the C. elegans homolog of LAG1/LASS1. *Age (Dordr)*, *30*(1), 43–52.

Torello, L. B., et al. (1986). A comparative-evolutionary study of lipids in the aging brain of mice. *Neurobiology of Aging*, *7*(5), 337–346.

Tosato, M., et al. (2007). The aging process and potential interventions to extend life expectancy. *Clinical Interventions in Aging*, *2*(3), 401–412.

Turrens, J. F., Alexandre, A., & Lehninger, A. L. (1985). Ubisemiquinone is the electron donor for superoxide formation by complex III of heart mitochondria. *Archives of Biochemistry and Biophysics*, *237*(2), 408–414.

Vandanmagsar, B., et al. (2011). The NLRP3 inflammasome instigates obesity-induced inflammation and insulin resistance. *Nature Medicine*, *17*(2), 179–188.

Venable, M. E., et al. (1995). Role of ceramide in cellular senescence. *The Journal of Biological Chemistry*, *270*(51), 30701–30708.

Venable, M. E., & Obeid, L. M. (1999). Phospholipase D in cellular senescence. *Biochimica et Biophysica Acta*, *1439*(2), 291–298.

Visser, M., et al. (1998). High body fatness, but not low fat-free mass, predicts disability in older men and women: The cardiovascular health study. *The American Journal of Clinical Nutrition*, *68*(3), 584–590.

Vorbeck, M. L., et al. (1982). Aging-dependent modification of lipid composition and lipid structural order parameter of hepatic mitochondria. *Archives of Biochemistry and Biophysics*, *217*(1), 351–361.

Walls, S. M., et al. (2018). Ceramide-protein interactions modulate ceramide-associated lipotoxic cardiomyopathy. *Cell Reports*, *22*(10), 2702–2715.

Wang, C., et al. (2014). Alterations in mouse brain lipidome after disruption of CST gene: A lipidomics study. *Molecular Neurobiology*, *50*(1), 88–96.

Wang, G., & Bieberich, E. (2018). Sphingolipids in neurodegeneration (with focus on ceramide and S1P). *Advances in Biological Regulation*, *70*, 51–64.

Webb, L. M., Arnholt, A. T., & Venable, M. E. (2010). Phospholipase D modulation by ceramide in senescence. *Molecular and Cellular Biochemistry*, *337*(1-2), 153–158.

Weinert, B. T., & Timiras, P. S. (2003). Invited review: Theories of aging. *Journal of Applied Phisiology (1985)*, *95*(4), 1706–1716.

Weinrich, S. L., et al. (1997). Reconstitution of human telomerase with the template RNA component hTR and the catalytic protein subunit hTRT. *Nature Genetics*, *17*(4), 498–502.

Wooten, L. G., & Ogretmen, B. (2005). Sp1/Sp3-dependent regulation of human telomerase reverse transcriptase promoter activity by the bioactive sphingolipid ceramide. *The Journal of Biological Chemistry*, *280*(32), 28867–28876.

Wooten-Blanks, L. G., et al. (2007). Mechanisms of ceramide-mediated repression of the human telomerase reverse transcriptase promoter via deacetylation of Sp3 by histone deacetylase 1. *The FASEB Journal*, *21*(12), 3386–3397.

Xiao, S., Finkielstein, C. V., & Capelluto, D. G. (2013). The enigmatic role of sulfatides: New insights into cellular functions and mechanisms of protein recognition. *Advances in Experimental Medicine and Biology*, *991*, 27–40.

Yi, J. K., et al. (2016). Aging-related elevation of sphingoid bases shortens yeast chronological life span by compromising mitochondrial function. *Oncotarget*, *7*(16), 21124–21144.

Yuyama, K., Mitsutake, S., & Igarashi, Y. (2014). Pathological roles of ceramide and its metabolites in metabolic syndrome and Alzheimer's disease. *Biochimica et Biophysica Acta*, *1841*(5), 793–798.

Zhang, M., Mileykovskaya, E., & Dowhan, W. (2002). Gluing the respiratory chain together. Cardiolipin is required for supercomplex formation in the inner mitochondrial membrane. *The Journal of Biological Chemistry*, *277*(46), 43553–43556.

Zhang, Q., et al. (2010). Circulating mitochondrial DAMPs cause inflammatory responses to injury. *Nature*, *464*(7285), 104–107.

Zhou, R., et al. (2011). A role for mitochondria in NLRP3 inflammasome activation. *Nature*, *469*(7329), 221–225.

Zoeller, R. A., et al. (1999). Plasmalogens as endogenous antioxidants: Somatic cell mutants reveal the importance of the vinyl ether. *The Biochemical Journal*, *338*(Pt 3), 769–776.

CHAPTER

19

Trends in morbidity, healthy life expectancy, and the compression of morbidity

Eileen M. Crimmins[1], Yuan S. Zhang[2], Jung Ki Kim[1] and Morgan E. Levine[3]

[1]Davis School of Gerontology, University of Southern California, Los Angeles, CA, United States [2]Carolina Population Center, The University of North Carolina at Chapel Hill, Chapel Hill, NC, United States [3]Department of Pathology, Yale University School of Medicine, New Haven, CT, United States

OUTLINE

Introduction

Fries introduced the compression of morbidity hypothesis in 1980 when he suggested that as the age of death was postponed, the age of onset of chronic illness would be postponed more than death so that the period of morbidity and disability at the end of life would be shorter (Fries, 1980). Seldom has a more intriguing hypothesis been introduced to the scientific community. If compression of morbidity occurred, then many of the problems that we associate with an aging population would be solved. If people led longer healthy lives, they could be productive and independent longer and would require fewer resources to deal with deteriorating health and loss of independence. This hypothesis has generated a significant body of empirical research with results that are often not consistent with each other but which provide little support for the hypothesis as a general description of what has been happening in countries with long-lived populations who have had significant increases in life expectancy in recent decades. Our charge in this chapter is to assess the evidence for compression of morbidity and to clarify how likely this is in the future.

Dimensions of morbidity

Interest in the length of healthy life as well as the total length of life began after infectious diseases were largely replaced by chronic diseases as the most important causes of morbidity and death. When death was caused by infection it generally occurred quickly and from clearly proximate causes; in contrast, death and morbidity due to chronic conditions generally arise over decades and are caused by distal conditions. The complexity of the process of morbidity from chronic conditions is one reason why the empirical evidence on the compression of morbidity reported in the literature is somewhat confusing.

Handbook of the Biology of Aging.
DOI: https://doi.org/10.1016/B978-0-12-815962-0.00019-6

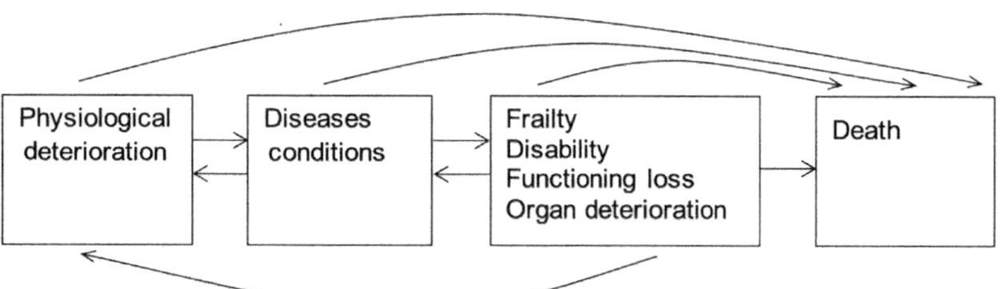

FIGURE 19.1 Dimensions and health changes in the morbidity process. *Adapted from Crimmins, E. M., Vasunilashorn, S., & Kim, J.K. (2010). Biodemography: New approaches to understanding trends and differences in population health and mortality. Demography, 47S, S41–S64.*

We provide a heuristic model of the process of morbidity change for populations in Fig. 19.1. The multiple dimensions of morbidity are shown in the boxes. In the population physiological dysregulation precedes the onset of clinical disease, and disease precedes the onset of disability and death (Crimmins, Vasunilashorn, & Kim, 2010). The clinical recognition of morbidity related to chronic diseases may come decades after the process of physiological change begins. Some of the confusion about whether there is empirical evidence of a compression of morbidity comes from the fact that most research uses one but not all of the dimensions of morbidity to assess health, for example, disability, disease, etc. The various dimensions of health do not have to change in the same way at the same time (Crimmins, 1996; Spiers, Jagger, & Clarke, 1996). There can be expansion of one aspect of morbidity and compression in another aspect.

The model in Fig. 19.1 portrays the average population-level change; no single individual needs to go through all the stages of this process. Not all morbidity results in disability or functioning loss; for instance, many people live close to the end of their lives with cancer and maintain functioning. Another reason change in morbidity and mortality may be asynchronous is that a significant amount of morbidity is not very closely related to mortality. For instance, osteoarthritis is a major cause of disability and functioning loss but is not an important cause of mortality.

The arrows in Fig. 19.1 represent the potential changes between health states at an individual level. Changes in the rates of these processes underlie changes in population health, death rates, and life expectancy. Morbidity can change because the initial onset of physiological deterioration is delayed or because the progression through any of the states is delayed. It can also change because mortality changes from any health state; mortality tends to eliminate the sickest people from the population. Therefore a reduction in mortality can leave the population with more sick people. Note that there is also recovery or movement toward better health from some of the morbid states. It is this complex set of changes that combines to produce changes in both life expectancy and population health.

The length of life cycles and population health

The second reason why the literature presenting evidence relevant to the compression of morbidity is confusing is that some people address the question by looking at trends in life cycle indicators, for example, the average length of life, the average length of healthy life, the average length of disabled life, and others examine indicators of population health, for example, the percentage of the population with disability or disease. Preston (1982) has clarified theoretically that mortality change is likely to have substantial effects on individual life cycles but small effects on the characteristics of the population, for example, health status. Demographers have also argued that the relationship of mortality change to change in the health of the population depends on where change occurs in the process of morbidity change (Crimmins, Hayward, & Saito, 1994). If mortality declines because its cause is eliminated or delayed, then the length of life, the length of healthy life, and population health are likely to improve. On the other hand, if mortality declines because someone was treated for a condition that would have caused death but now they stay alive with the condition, the length of life will increase, the length of healthy life may increase or decrease, and population health may improve or deteriorate. For example, if a treatment is found for a cause like scarlet fever, there would likely be improvements in both the length of healthy life and population health; but if the treatment is for heart disease, the length of life with heart disease is increased and the prevalence of heart disease in the population can increase. On the other hand, if the initial physiological dysregulation leading to disease and disability is delayed, the length of life should be increased and the likelihood of disease reduced.

Trends in population prevalence of physiological dysregulation, diseases and conditions, functioning loss and disability, and life expectancy

Before we briefly discuss evidence on changes in the last couple of decades in each dimension of

TABLE 19.1 Life expectancy at ages 65 and 85, United States, 2017–1970.

	Age	2017	2010	2000	1990	1980	1970
Total	65	19.4	19.1	17.8	17.3	16.5	15.0
	85	6.6	6.5	6.2	6.2	6.0	5.3
Male	65	18.0	17.7	16.1	15.1	14.2	13.0
	85	5.9	5.8	5.5	5.3	5.1	4.7
Female	65	20.6	20.3	19.1	19.0	18.4	16.8
	85	6.8	6.9	6.6	6.7	6.4	5.6

Source: 1970–2000, Arias, E. (2012). United States life tables, 2008. National Vital Statistics Report, 61*(3); Arias (2010); Arias, E. (2014). United States life tables, 2010.* National Vital Statistics Reports, 63*(7), Arias (2017); Arias, E., & Xu, J. (2019). United States life tables, 2017.* National Vital Statistics Reports, 68*(3).*

morbidity for persons at older ages, we begin with changes in life expectancy at the older ages. In the United States rapid mortality decline among the older population began in the late 1960s. Between 1970 and 2017 life expectancy at age 65 increased from 15 years to 19.4 years; at age 85, the increase was from 5.3 to 6.6 years (Table 19.1). Life expectancy at the older ages has continued to increase for both men and women but the trend was more consistent for men. We should note that in the United States life expectancy at birth stopped increasing in 2010, and decreased in the 3 years after 2014 (Woolf & Shoomaker, 2019)

In our discussion of trends in morbidity we will both cite recent literature and illustrate trends with data from the last 15 or 20 years for older persons from the Health and Retirement Study (HRS) and the National Health and Nutrition Examination Survey (NHANES), large nationally representative studies of the US population.

Most of the research on compression of morbidity for older persons has focused on trends in disability and this body of literature leads to different conclusions depending on what measures of disability are used, the severity of the disability examined, the age group examined, and the period of time studied (Cambois, Clavel, Romieu, & Robine, 2008). In general, there have been decreases in the proportion of people with less severe disability and mixed trends for those with more severe disability, that is, periods of increase, stability, and improvement (Freedman et al., 2012; Seeman, Merkin, Crimmins, & Karlamangla, 2010).

In the HRS study we show the trend from 1998 to 2016 in the proportion of persons with less severe functioning problems represented by ability to perform Nagi functions at ages 75 and above (Fig. 19.2A). These problems include walking a distance, stooping, lifting weights, climbing stairs, sitting for an extended period, picking up small objects, and reaching over the head. For those 75–84 the pattern is one of stability; for those above 85 there is slight upward movement in the last decade. When we look at trends in inability to perform numbers of activities of daily living (ADLs) and instrumental ADLs (IADLs), indicators of more serious disability, there is some decrease up through 2012 and then some increase in the last 4 years.

The literature on trends in diseases has been consistent in showing an increase over time in the proportion of people with major diseases such as heart disease, cancer, stroke, and diabetes at least up to the year 2000 (Crimmins & Beltran-Sanchez, 2011; Cutler, Richardson, Keeler, & Steiger, 1997; Martin, Freedman, Schoeni, & Andreski, 2009). Among Americans, there has also been an increase in the average number of diseases per person (Crimmins, Garcia, & Kim, 2010). If we examine trends in the prevalence of heart disease, cancer, and stroke after 1999 using the nationally representative NHANES data we see very small changes in the prevalence of these conditions; stability characterizes this period (Fig. 19.3).

In a further investigation of changes in disease across two cohorts followed longitudinally we examine survival without specific disease and age at onset of diseases in two birth cohorts of the HRS. Cohort survival provides a somewhat fuller picture of the process of health change over time than the trends in repeated cross-sections shown above. If disease onset is being eliminated or delayed to later ages, we should find the later cohort has longer survival without disease. The earlier birth cohort was born in 1931–1941 and was first interviewed in 1992 when they were 51–61. At this time they reported the presence of disease and the age at onset of existing conditions (some information on age of onset of existing conditions is collected in a later wave) and from then on they reported every 2 years on the onset of diseases. We use data for 8 years of the survey. The second cohort was born in 1943–1953 and was interviewed at age 51–61 in 2004 and then followed for 8 years until 2012.

When we look at the data for these two cohorts born 12 years apart, we see there is less survival without cancer (Fig. 19.4A), no change in survival

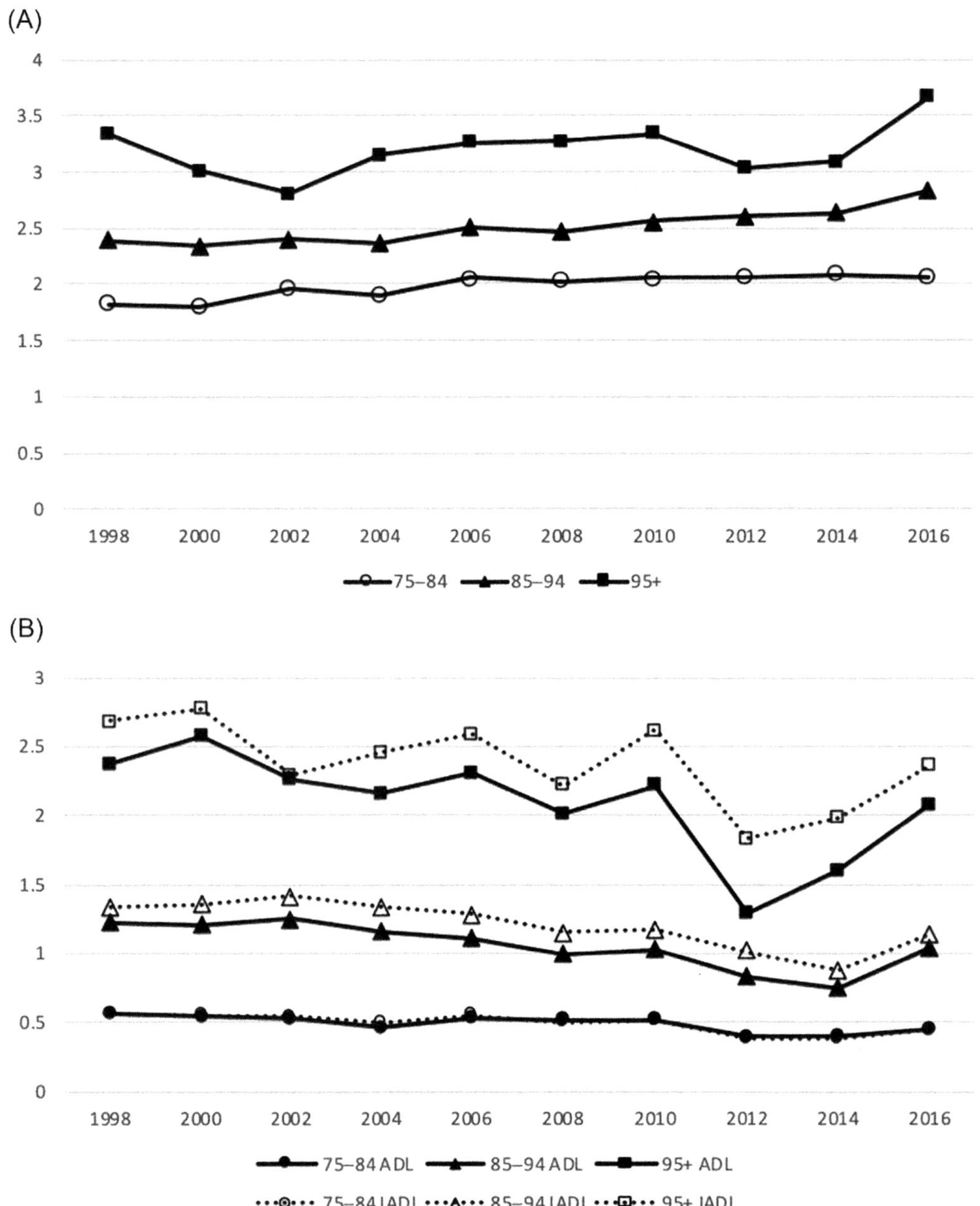

FIGURE 19.2 Trends in functioning and disability: HRS 1998–2012, age 75+. (A) Number of Nagi problems. (B) Number of ADL and IADL difficulties. *Source: HRS data.*

with heart disease (Fig. 19.4B), a small decrease in survival without stroke (Fig. 19.4C), and an increase in survival without having had a myocardial infarction (Fig. 19.4D). So only having the likelihood of having a heart attack appears to be reduced or delayed in a fashion that might be considered improving health or compression of morbidity (Crimmins, Zhang, Kim, & Levine, 2019). The other changes would be indicators of deteriorating health or increased morbidity.

We can also look at age of onset of these conditions in the two cohorts. There is no clear change in the likelihood of getting heart disease, cancer, or stroke at a specific age; however, the likelihood of having a heart attack at each age appears substantially reduced in the later cohort (Fig. 19.5).

Next we examine change over time in the prevalence of the precursors of disease and mortality by looking at the average number of risk factors measured as clinically high. We find a marked decrease over two decades in the average number of eight risk factors (i.e., total cholesterol, HDL, LDL, triglycerides, obesity, HbA1C, systolic blood pressure, diastolic blood pressure), at clinical high risk

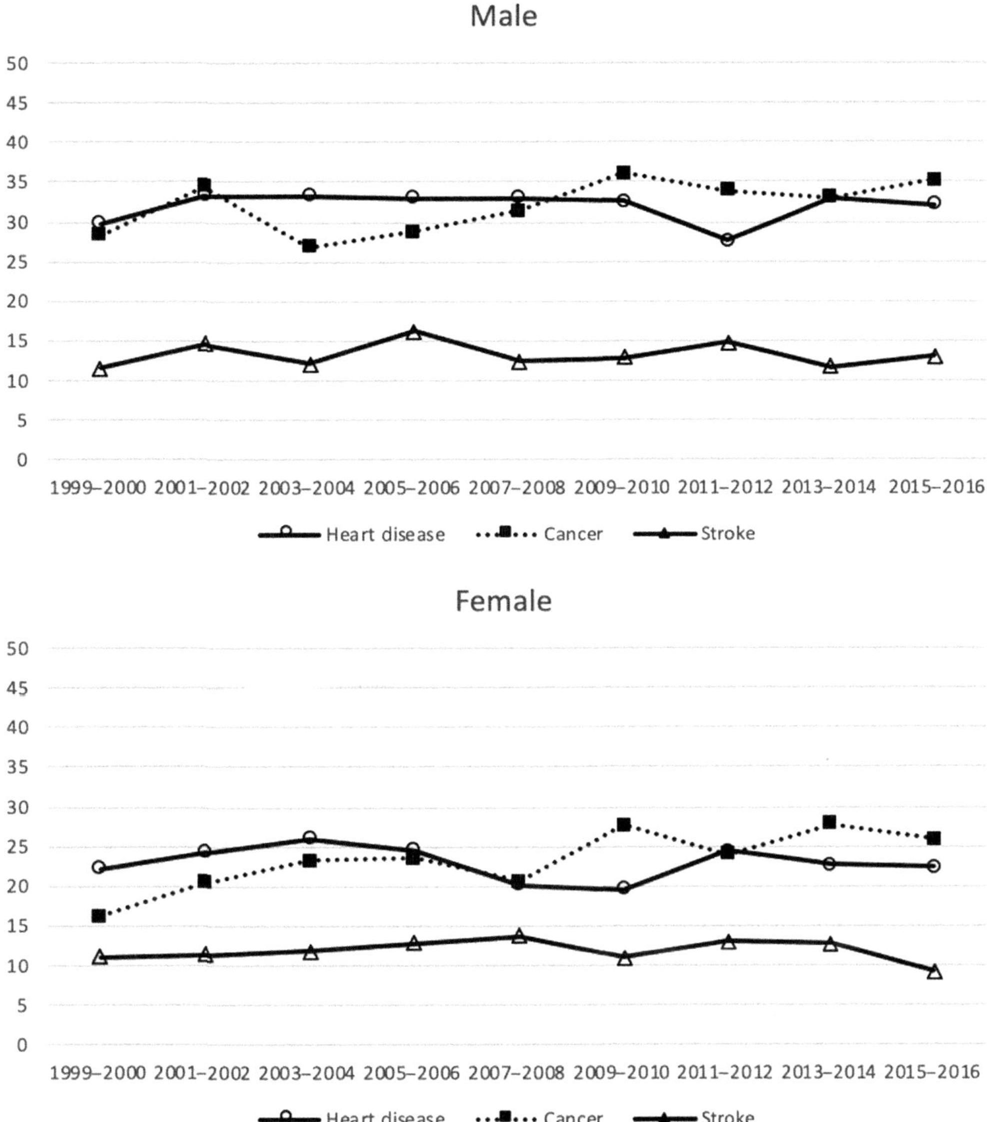

FIGURE 19.3 Chronic disease trend, percent with heart disease, cancer, stroke: 1999–2000 to 2015–2016, age 75 + . *Source: NHANES sample.*

above age 40 for males and above age 50 for females between 1990 and 2010 (Fig. 19.6). The recent drop in risk is largely due to an increase in the control of both high blood pressure and high cholesterol through the increased use of medication rather than a decrease in the percentage of people ever presenting with high blood pressure or high cholesterol (Kim, Ailshire, & Crimmins, 2019).

Length of life and length of healthy life

Evidence for the compression of morbidity most appropriately comes from estimates of the length of life and the length of healthy life which combine mortality and morbidity measures. The original statement of the compression of morbidity hypothesis regarded the length of life as fixed and posited that as this length of life was approached, the length of healthy life would increase and the length of morbid life would decrease (Fries, 1980). Most research, including more recent work by Fries, recognizes that life expectancy continues to increase and views compression of morbidity when the average age of the morbidity onset is put off more than the increase in life expectancy, so that the length of life and the proportion of life with morbidity are shortened (Fries, 1984; Nusselder, 2003).

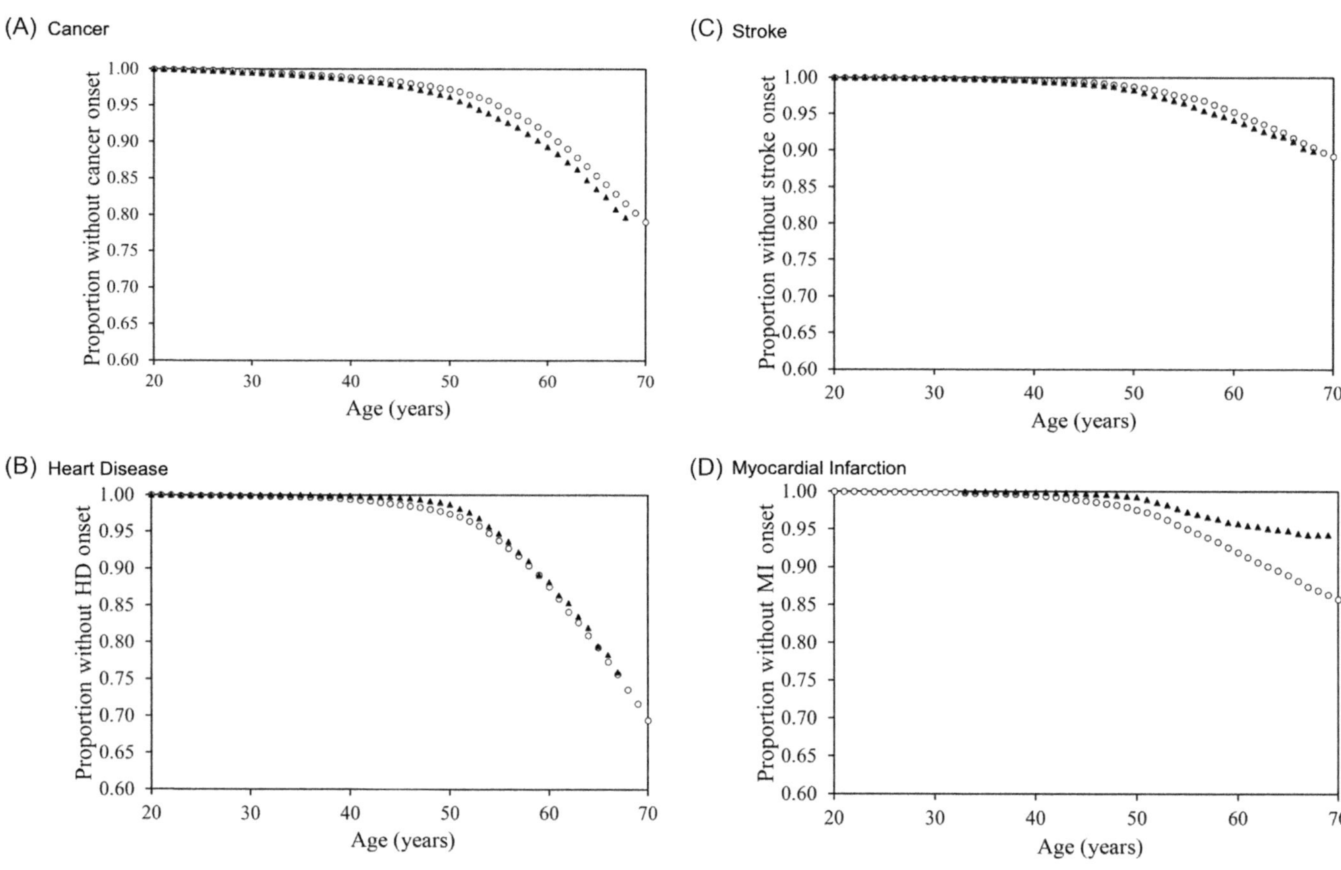

FIGURE 19.4 Survival without disease onset for two cohorts, 1931–1941 and 1943–1953: (A) cancer; (B) heart disease; (C) stroke; (D) myocardial infarction.

As indicated above, the age at which life expectancy is examined and the methods and data used to make estimates of healthy life all affect the conclusions about the compression of morbidity. Most analyses of life with and without severe disability among older persons in the United States have found increases in the length of disability-free life, at least in recent decades (Crimmins, Kim, & Saito, 2016; Crimmins, Saito, & Ingegneri, 1997). As an example, we can examine results from two recent longitudinal cohorts of Americans 70 years of age and over followed for multiple interviews (Crimmins, Hayward, Hagedorn, Saito, & Brouard, 2009). These two cohorts were each collected by the National Center for Health Statistics and are called the Longitudinal Study of Aging I and the Longitudinal Study of Aging II (LSOA I and II). The estimated increase in life expectancy at age 70 for the second cohort was 0.6 years, which was the same as the increase in disability-free life expectancy or life expectancy without ADL or IADL difficulty. There was no change in the length of disabled life expectancy. The proportion of disability-free life remained roughly constant at age 70 (80%–81%). The use of the two longitudinal cohorts allowed us to examine the processes that were changing and there was a significant decline in disability onset, a significant increase in recovery from disability, and a significant decline in mortality among the most disabled across the cohorts. Similar results have been found in another dataset (Cai & Lubitz, 2007). However, an examination of life expectancy with and without loss of mobility functioning, which is less severe disability than that considered above, found increases in life with functioning problems that were much greater than increases in life expectancy (Crimmins & Beltran-Sanchez, 2011). On the other hand, some examinations of even more mild disability have reported increases in the absolute length of nondisabled life and reductions in disabled life that result in some compression of the proportion of disabled life at least for some periods and some subgroups of the population (Crimmins & Saito, 2001); however, for adults up to age 85, it appears that any recent compression of morbidity may be limited to white males (Solé-Auró, Beltrán-Sánchez, & Crimmins, 2014).

Investigations using the presence of disease to examine compression of morbidity generally report an increase in the length of life with many diseases; a decrease in life without a number of diseases and a growing proportion

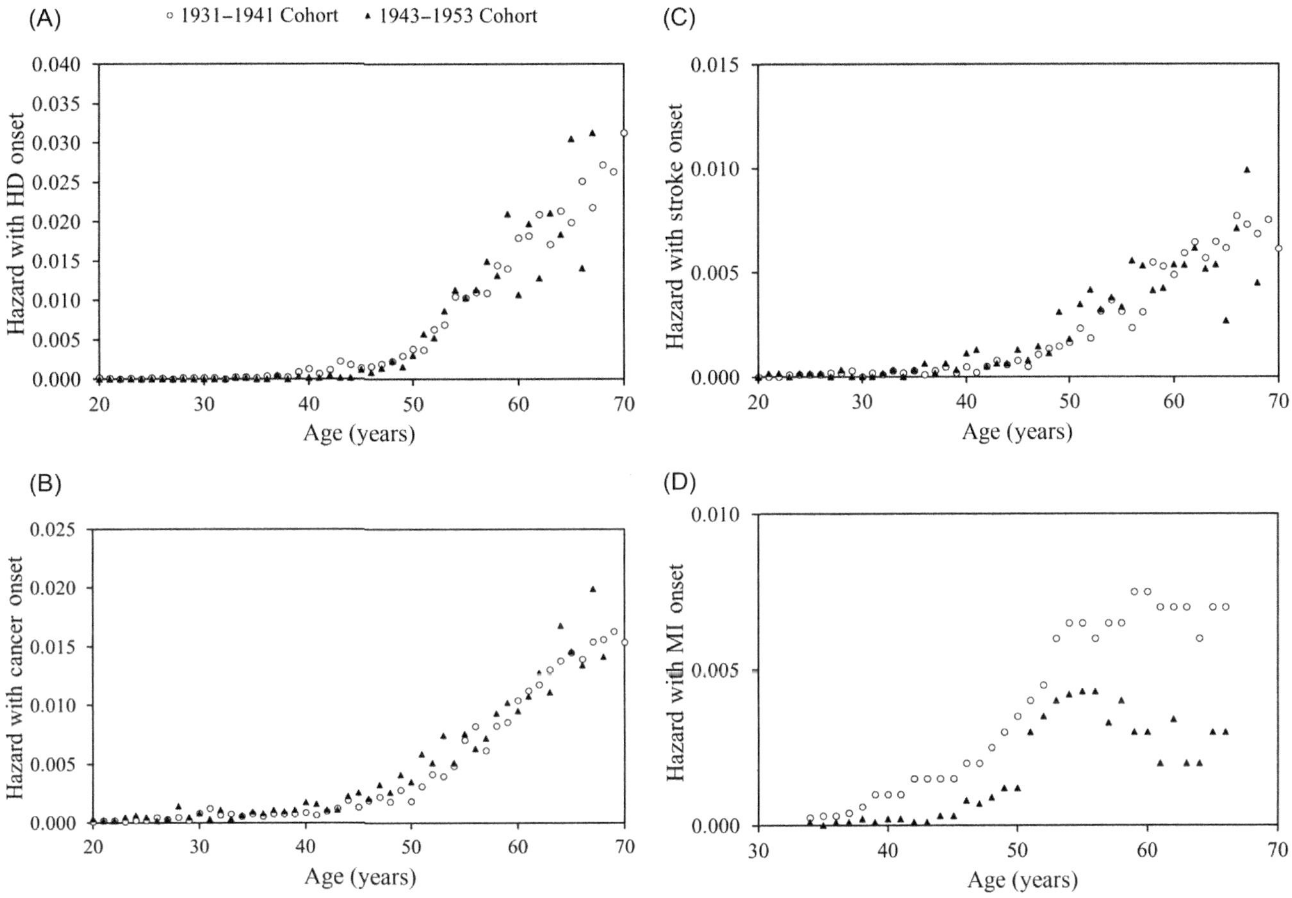

FIGURE 19.5 Hazard of disease onset for two cohorts, 1931–1941 and 1943–1953: (A) heart disease; (B) cancer; (C) stroke; (D) myocardial infarction.

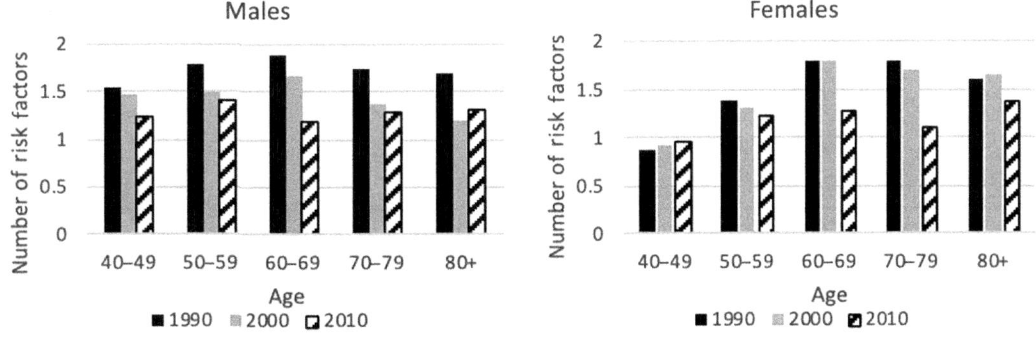

FIGURE 19.6 Mean number of high-risk cardiovascular risk factors for 1990, 2000, and 2010 by age (males and females). Note: Cardiovascular risk indicators include systolic and diastolic blood pressure, body mass index, total cholesterol, high-density lipoprotein cholesterol, low-density lipoprotein cholesterol, triglycerides, and glycated hemoglobin (HbA1c). Risk levels for each indicator are defined as measured levels above (or below) the clinical cutoff. Data are from NHANES. *Source: Kim, J.K., Ailshire, J., & Crimmins, E.M. (2019). Twenty year trends in cardiovascular risk among men and women in the United States. Aging Clinical and Experimental Research, 31(1), 135–143.*

of life spent with disease although the pattern varies by condition. We can use the data from the HRS study we reported earlier to make estimates of life expectancy with and without a number of diseases for the US population at age 65 in 1998 and 2014 (Table 19.2). Length of life with and without disease as well as the proportion of remaining life spent with each disease are shown for hypertension, diabetes, stroke, heart disease, cancer, and arthritis. The length of life with each disease, except cancer, increases over the period. The length of life without disease increases for cancer, heart disease, and stroke. These combine so that the proportion of life spent with

TABLE 19.2 Expected length of life lived with and without disease at age 65, HRS 1998–2014.

	1998			2014		
	Years of life lived without disease	**Years of life lived with disease**	**% of disease-free years**	**Years of life lived without disease**	**Years of life lived with disease**	**% of disease-free years**
Male						
Hypertension	8.8	7.3	54.7	5.5	13.0	29.7
Diabetes	13.6	2.5	84.5	13.2	5.3	71.4
Cancer	13.6	2.6	84.0	17.4	1.1	94.1
Heart disease	10.8	5.4	66.7	11.2	7.3	60.5
Stroke	14.4	1.8	88.9	16.1	2.4	87.0
Arthritis	7.7	8.4	47.8	6.7	11.8	36.2
Female						
Hypertension	9.3	10.0	48.2	6.5	14.6	30.8
Diabetes	16.8	2.5	87.0	16.0	5.1	75.8
Cancer	16.9	2.4	87.6	19.5	1.6	92.4
Heart disease	14.3	5.0	74.1	14.6	6.5	69.2
Stroke	17.3	2.0	89.6	18.8	2.3	89.1
Arthritis	7.5	11.7	39.1	4.8	16.3	22.7

Calculations from the HRS and US life tables.

TABLE 19.3 US global burden of disease estimates of life expectancy and health-adjusted life expectancy (HALE) at birth and at age 65 in the years 1990 and 2017.

	1990	**2017**	**Increase**
At birth			
LE	75.6	78.6	3.0
HALE	64.7	66.6	1.9
At age 65			
LE	17.4	19.4	2.0
HALE	12.8	13.9	1.1

GBD 2017 DALYs and HALE Collaborators. (2018). Global, regional, and national disability-adjusted life-years (DALYs) for 359 diseases and injuries and healthy life expectancy (HALE) for 195 countries and territories, 1990–2017: A systematic analysis for the global burden of disease study 2017. Lancet, 392(10159), 1859–1922.

disease is increased for each cause except cancer, where it increased, and stroke where it stayed constant.

The Global Burden of Disease group has tried to be very comprehensive in including most diseases and disability in its estimates of health-adjusted life expectancy (HALE). They consider almost 300 diseases and associated disability in their estimates that span all ages and are done for most countries. They have recently estimated life expectancy and health-adjusted life expectancy at birth for the United States (GBD 2017 DALYs & HALE Collaborators, 2018) (Table 19.3). Their conclusion is that between 1990 and 2017 both life expectancy and health-adjusted life expectancy increased, but that the increases in life expectancy of 3 years at birth and 2 years at age 65 were greater than the increases in health-adjusted life expectancy of 1.9 years and 1.1 years. This would be regarded as an expansion of morbidity rather than a compression.

Conclusions

There is little evidence that the decline in mortality and increase in life expectancy at older ages is primarily due to a delay in the morbidity process. However, it is clear that we have made strides in preventing some of the progression of morbidity process. For instance, while the proportion of the population ever having high blood pressure or high cholesterol has not been reduced, the proportion who are measured high has been reduced markedly in recent years due to pharmaceutical interventions. The fact that there is evidence of an increase in the age of myocardial infarction and a reduction in the age-specific likelihood of having a heart attack may be, at least partially, a result of the increased control of high blood pressure and high cholesterol.

There clearly has been an increase in the proportion of older persons with disease in the population as well as an increase in the length of life with disease. One of the major factors in causing an increase in disease prevalence in the population is that diseases have become less progressive and less lethal. We have saved people from death and disability related to disease (Crimmins & Saito, 2001; Freedman et al., 2012). Reductions in mortality leading to increases in life expectancy have occurred in good part because those with disability and diseases live longer. One estimate is that about half the decline in cardiovascular deaths between 1980 and 2000 was due to disease management and treatment (Ford et al., 2007). Strides in reducing cancer death rates have also been important in recent decades. Our progress in increasing life expectancy by saving people from death from cardiovascular disease and cancer means that more people with disease live longer years in the population with disease. The reduction in the disability level associated with disease is one of the reasons that the increases in disability-free life expectancy have better been able to keep up with increases in life expectancy at least during some periods. Even though recent mortality reductions have been concentrated among those with severe disability, the proportion of life with this type of disability and the prevalence of disability in the population have generally not increased.

Progress in controlling morbidity and reducing mortality at older ages is likely to continue for the foreseeable future. At present this progress is much more likely to lead to a further delay in the progression of the morbidity process rather than delaying the onset of the process. Delay of the onset of the process is much more dependent on delaying the process of "aging" rather than the process of disease. Delays in the process are likely to result in increases in the length of healthy life but it is hard to say how they will affect the length of life with disease, functioning loss, and disability. It is hard to envision an end of life that does not involve a process of deterioration even if it occurs at a later age than now. It would be a mistake to assume that any increase in the length of life with morbidity is a failure. The aim of much of our medical care has been to allow those with diseases and disability to live longer; and this appears to be at the root of the trends in population health and healthy life expectancy.

References

Arias, E. (2012). United States life tables, 2008. *National Vital Statistics Report*, *61*(3).

Arias, E. (2014). United states life tables, 2010. *National Vital Statistics Reports*, *63*(7).

Arias, E., & Xu, J. (2019). United states life tables, 2017. *National Vital Statistics Reports*, *68*(3).

Cai, L., & Lubitz, J. (2007). Was there compression of disability for older Americans from 1992 to 2003? *Demography*, *4*, 479–495.

Cambois, E., Clavel, A., Romieu, I., & Robine, J. M. (2008). Trends in disability-free life expectancy at age 65 in France: Consistent and diverging patterns according to the underlying disability measure. *European Journal of Ageing*, *5*, 287–298.

Crimmins, E. M. (1996). Mixed trends in population health among older adults. *Journal of Gerontology: Social Sciences*, *51B*, 223–225.

Crimmins, E. M., & Beltran-Sanchez, H. (2011). Trends in mortality and morbidity: Is there a compression of morbidity? *Journal of Gerontology: Social Sciences*, *66*, 75–86.

Crimmins, E. M., & Saito, Y. (2001). Trends in healthy life expectancy in the United States, 1970–1990: Gender, racial, and educational differences. *Social Science and Medicine*, *52*, 1629–1641.

Crimmins, E. M., Garcia, K., & Kim, J. K. (2010). Are international differences in health similar to international differences in life-expectancy? In E. M. Crimmins, S. H. Preston, & B. Cohen (Eds.), *International differences in mortality at older ages: Dimensions and sources* (pp. 68–101). Washington, DC: National Research Council, National Academy of Sciences.

Crimmins, E. M., Kim, J. K. ., & Saito, Y. (2016). Change in cognitively healthy and cognitively impaired life expectancy in the United States: 2000–2010. *Social Science and Medicine: Population Health*, *2*, 793–797.

Crimmins, E. M., Zhang, Y., Kim, J. K., & Levin, M. (2019). Changing disease prevalence, incidence, and mortality among older cohorts: The health and retirement study. *Journals of Gerontology A: Biological and Medical Sciences*, *74*(Suppl. 1), S21–S26.

Crimmins, E. M., Hayward, M. D., Hagedorn, A., Saito, Y., & Brouard, N. (2009). Change in disability-free life expectancy for Americans 70 years old and over. *Demography*, *40*, 627–646.

Crimmins, E. M., Hayward, M. D., & Saito, Y. (1994). Changing mortality and morbidity rates and the health status and life expectancy of the older U.S. population. *Demography*, *31*, 159–175.

Crimmins, E. M., Saito, Y., & Ingegneri, D. (1997). Trends in disability-free life expectancy in the United States, 1970–1990. *Population and Development Review*, *23*(3), 555–572.

Crimmins, E. M., Vasunilashorn, S., & Kim, J. K. (2010). Biodemography: New approaches to understanding trends and differences in population health and mortality. *Demography*, *47S*, S41–S64.

Cutler, D. M., Richardson, E., Keeler, T. E., & Staiger, D. (1997). *Measuring the health of the U.S. population. Brookings papers on economic activity. Microeconomics* (pp. 217–282). Washington, DC: Brookings Institute.

Ford, E. S., Ajani, U. A., Croft, J. B., Critchley, J. A., Labarthe, D. R., Kottke, T. E., ... Capewell, S. (2007). Explaining the decrease in U.S. deaths from coronary disease, 1980–2000. *New England Journal of Medicine*, *356*, 2388–2398.

Freedman, V. A., Spillman, B. C., Andreski, P. M., Cornman, J. C., Crimmins, E. M., Kramarow, E., ... Waidmann, T. A. (2012).

Trends in late-life activity limitations in the United States: An update from five national surveys. *Demography*, *50*, 661–671.

Fries, J. F. (1980). Aging, natural death, and the compression of morbidity. *New England Journal of Medicine*, *303*, 1369–1370.

Fries, J. F. (1984). The compression of morbidity: Miscellaneous comments about a theme. *The Gerontologist*, *24*, 354–359.

GBD 2017 DALYs and HALE Collaborators. (2018). Global, regional, and national disability-adjusted life-years (DALYs) for 359 diseases and injuries and healthy life expectancy (HALE) for 195 countries and territories, 1990–2017: A systematic analysis for the global burden of disease study 2017. *Lancet*, *392*(10159), 1859–1922.

Kim, J. K., Ailshire, J., & Crimmins, E. M. (2019). Twenty year trends in cardiovascular risk among men and women in the United States. *Aging Clinical and Experimental Research*, *31*(1), 135–143.

Martin, L., Freedman, V. A., Schoeni, R., & Andreski, P. (2009). Health and functioning of the baby boom approaching 60. *Journal of Gerontology: Social Sciences*, *64*, 369–377.

Nusselder, W. J. (2003). Compression of morbidity. In J. M. Robine, C. Jagger, C. D. Mathers, E. M. Crimmins, & R. M. Suzman (Eds.), *Determining health expectancies* (2, pp. 35–58). West Sussex: John Wiley & Sons, Ltd.

Preston, S. H. (1982). Relations between individual life cycles and population characteristics. *American Sociological Review*, *47*, 253–264.

Seeman, T., Merkin, S. S., Crimmins, E., & Karlamangla, A. (2010). Disability trends among older Americans: National health and nutrition examination surveys, 1988–1994 and 1999–2004. *American Journal of Public Health*, *100*, 100–107.

Solé-Auró, A., Beltrán-Sánchez, H., & Crimmins, E. M. (2014). Are differences in disability-free life expectancy by gender, race and education widening at older ages? *Population and Policy Review*, *34*, 1–18.

Spiers, N., Jagger, C., & Clarke, M. (1996). Physical function and perceived health. *Journal of Gerontology B: Psychological Sciences and Social Sciences*, *51B*(S), S226–233.

Woolf, S., & Schoomaker, H. (2019). Life expectancy and mortality rates in the United States, 1959–2017. *JAMA*, *322*(20), 1996–2016.

Further reading

Crimmins, E. M. (2018). *Trends in mortality, disease, and physiological status in the older population. National academies of sciences, engineering, and medicine*. Future directions in the demography of aging (pp. 1–28). Washington, DC: National Academies Press, Chapter 1.

Crimmins, E. M., & Zhang, Y. (2019). Aging populations, mortality and life expectancy. *Annual Review of Sociology*, *45*, 69–89.

Crimmins, E. M., Zhang, Y., & Saito, Y. (2016). Trends over four decades in disability-free life expectancy in the United States. *American Journal of Public Health*, *106*, 1287–1293.

Langa, K. M., Larson, E. B., Crimmins, E. M., Faul, J. D., Levine, D. A., Kabeto, M. U., & Weir, D. R. (2017). A comparison of the prevalence of dementia in the United States in 2000 and 2012. *Journal of the American Medical Association: Internal Medicine*, *177* (1), 51–58.

US Burden of Disease Collaborators. (2013). The state of US health, 1990 2010: Burden of diseases, injuries, and risk factors. *JAMA*, *310*, 591–608.

Author Index

Note: Page numbers followed by "*f*" and "*t*" refer to figures and tables, respectively.

B

C

D

E

F

G

H

L

M

N

O

P

T

X

Y

Z

Subject Index

Note: Page numbers followed by "*f*" and "*t*" refer to figures and tables, respectively.

Q

R

S

9780128159620